ASME HANDBOOK

Engineering Tables

ASME HANDBOOK

Engineering Tables

Edited by JESSE HUCKERT
Professor of Mechanical Engineering
The Ohio State University

Sponsored by
the Metals Engineering Handbook Board
of The American Society of Mechanical Engineers

FIRST EDITION

McGRAW-HILL BOOK COMPANY, INC.

NEW YORK TORONTO LONDON

1956

ASME Handbook
ENGINEERING TABLES

Library of Congress Catalog Card Number: 52-5326

ISBN 07-0 01516-3

10 1112 HIMB 7654

FOREWORD

The ASME Handbook of Engineering Tables is one of the many services of the Society. It came into being as a result of a survey in 1941 by the Metals Engineering Division which revealed the need of the mechanical engineer and designer for a ready reference to the properties and characteristics of metals. A preliminary study was instituted, and as a result, in August, 1945, the Executive Committee of the Council authorized the publication of a Metals Engineering Handbook and appointed a Handbook Board. This Board was set up as a continuing body, the members to be selected upon the recommendation of the Metals Engineering Division with the approval of the standing Committee on Professional Divisions and the concurrence of the Publications Committee. The personnel of the Board has changed from time to time.

The Society is grateful to the members of the Board and to the many others who have made valuable contributions to the text of the Handbook.

The Society is particularly happy to recognize the active cooperation of the American Society for Metals, which has permanent representation on the Board through a nominee of its own choice.

FREDERICK S. BLACKALL, JR., *President,* 1953
THE AMERICAN SOCIETY OF MECHANICAL ENGINEERS

This Handbook has been prepared to fill the urgent need for a reference manual related to the design engineer's point of view. A wealth of information of direct interest to him has been compiled from many authentic sources and is presented in a form which, it is hoped, will prove most useful to the experienced engineer and the embryo designer alike.

This volume comprises 15 sections dealing with *engineering tables* to supplement the designer's knowledge of standards for shape, dimension, gears, and the like. A volume has already been published dealing with the *design* function. Another deals with the *processes* by which metals are converted to finished product. Another tabulates the *properties of metals* about which a design engineer needs information.

The Advisory Committee, consisting of Messrs. H. B. Lewis, O. J. Horger, C. L. Tutt, Jr., and J. F. Young, has reviewed under the Board's direction matters relating to content, quality, format, and courses of action on the Handbook, and has reported findings and recommendations to the Board. The work of this committee will continue as it will have the responsibility of recommending to the Board necessary revisions to keep this Handbook abreast of the ever-changing need for current design data.

Contributions have been made by members of The American Society of Mechanical Engineers and other societies. Industrial organizations have been most generous in furnishing data and in permitting the use of material already in print. In each case, proper recognition is given to these sources.

LEWIS K. SILLCOX, *Chairman*
ASME METALS ENGINEERING HANDBOOK BOARD

Members of the Metals Engineering Handbook Board of The American Society of Mechanical Engineers

---

Lewis K. Sillcox, *Chairman,* Honorary Vice Chairman of the Board. The New York Air Brake Company, New York, N.Y.

Paul E. Needham, *Executive Vice Chairman,* 16 Kenmare Road, Larchmont, N.Y.

Davitt S. Bell, President, Edgewater Steel Company, Pittsburgh, Pa.

Frank B. Bell, Chairman of the Board, Edgewater Steel Company, Pittsburgh, Pa. (*Deceased May* 6, 1949.)

Oscar J. Horger, Chief Engineer, Railway Division, The Timken Roller Bearing Company, Canton, Ohio.

Herbert B. Lewis, Marketing Service Engineer, Lukens Steel Company, Coatesville, Pa. (*Deceased August* 1, 1952.)

Taylor Lyman, Editor, *Metals Handbook,* American Society for Metals, Cleveland, Ohio

Erik Oberg, Editor Emeritus. "Machinery," Industrial Press, New York, N.Y. (*Deceased October* 22, 1951.)

Charles L. Tutt, Jr., Administrative Chairman of Fifth-Year and Thesis Programs, General Motors Institute, Flint, Michigan

Thomas H. Wickenden, Consultant, The International Nickel Company, Inc., New York, N.Y.

James F. Young, Consultant–Energy Conversion, Engineering Services Division, General Electric Company, New York, N.Y.

Jean M. Meyer, *Secretary,* Research Administrator. The American Society of Mechanical Engineers, New York, N.Y.

PREFACE

Five years ago when the compilation of these tables was begun, I was unaware that so few of the data regularly used by designers could be condensed into a single volume. The idea at first was to assemble, into one ready reference, most of the standard tables that apply to everyday mechanical design.

When the hopelessness of the idea became clear, the aim was scaled down to embody a collection of tables, up-to-date and generally recognized as standards, which are often wanted by engineers and technicians, but which are not commonly found in handbooks. Thus a table of involute functions is included; tables of squares and cubes and trigonometric functions, being readily found elsewhere, are not. Methods for finding the loads on ball bearings are included; the loads that ball bearings can safely carry are not. Data upon bars and tubes suitable for reworking into gears, levers, shafts, or screws are included; the sizes of rolled sections as used in structures are not.

Tables applicable to the design of specific parts are grouped together. Frequently they are cross-referenced to facilitate the use of them in design calculations. They aim to supply details and to supplement the design information in the companion volume on "Metals Engineering—Design." Nor do they treat properties of materials—that being the sphere of the companion volume on "Metals Properties." The Index by Sections is a cross section of the matter in this volume.

In general a single source is given for a table—a primary one, such as the American Standards Association. Identical data may be available from more than one source, which accounts for the listing of sometimes two or even three sources. Occasionally none is given if the table was prepared from more than two sources. By stating the source I trust I pay the debt of using the data. The source is indicative of the quality of the data and serves as a reference for additional information. Rarely is any source republished here in full.

Section 12 was compiled and edited by Dr. Arthur M. Wahl, the authority on springs. Section 10 was compiled and edited by Mr. Edward Fitzgerald.

Besides the national bodies, more than fifty companies made engineering data and company standards available to me. I acknowledge with appreciation the wholesome cooperation of them. I regret that but a fraction of their contributions is published here.

I conclude by naming four persons to whom I owe much: Mrs. Maxine Fitzgerald, who typed the manuscript and read and reread proof; Mrs. Gene M. Weeks and Mr. Frank Philippbar, who handled layout and printing details; and Miss Jean M. Meyer, Secretary to the Board.

JESSE HUCKERT

The Indexes and Ways to Use Them

Table Numbers: Most items, including exposition and diagrams, are indexed by hyphenated table numbers, as 3-24. The 3 before the hyphen designates the section as 3; the 24 singles out the particular table within the section. Pages have been similarly designated, the number before the hyphen indicating the section and the number following indicating the sequence of pages numbered consecutively within the section.

Index by Sections: Data pertinent to the design of a machine element are grouped together. To illustrate, everything in Section 5 pertains to the design of bevel gears. A good part of Section 1 applies to the design of shafts, beginning with available bar stock and ending with shoulder heights for bearings.

Sequent Index by Tables: An index by tables subdivides each section, showing the gist of tabular matter, table and page numbers within the section. The use first of the Index by Sections and then a sequent index is a quick way to find specific tabular data. For example: Suppose a designer wants to verify the beam strength of a pair of spur gears. The Index by Sections defines the Section as 3. A glance at the sequent index of Section 3 points to Tables 3-33 to 3-39.

The Table Titles: The table titles are no longer than usual. They qualify the tabular matter, for there is no text for that purpose. Read them to acquire information. In cases where the titles alone seem inadequate, titles are often amplified by subtitles or footnotes. The footnotes, therefore, may further qualify the data. Read them too.

Bibliography: A bibliography has been provided in the back of the book showing full title, source, and date of material used. Brief references are shown beneath each table title to facilitate identification.

CONTENTS

INDEX TO SECTION 1

Bar Stock and Shafting
Formulas for Stress and Strain

Conversion Factors
Properties of Sections and Cylinders

TABLE 1–1

American Standard Sizes and Tolerances of Finished Transmission and Machinery Shafting

ASA B–17.1 1943

Stock Diameters, inches		Tolerance[1] on Diameter (-)	Weight lbs. per lineal ft.	Stock Diameters, inches		Tolerance[1] on Diameter (-)	Weight lbs. per lineal ft.
Transmission Shafting	Machinery Shafting			Transmission Shafting	Machinery Shafting		
Stock	1/2	0.002	0.667	2 7/16	2 7/16	0.004	15.96
Lengths of	9/16	0.002	0.844		2 1/2	0.004	16.68
16, 20 and	5/8	0.002	1.04		2 5/8	0.004	18.36
24 feet	11/16	0.002	1.26		2 3/4	0.004	19.16
	3/4	0.002	1.50		2 7/8	0.004	22.08
	13/16	0.002	1.76	2 15/16	3	0.004	24.00
	7/8	0.002	2.04		3 1/8	0.004	26.04
15/16	15/16	0.002	2.34		3 1/4	0.004	28.20
	1	0.002	2.64		3 3/8	0.004	30.36
	1 1/16	0.003	3.00	3 7/16	3 1/2	0.004	32.64
	1 1/8	0.003	3.36		3 5/8	0.004	35.04
1 3/16	1 3/16	0.003	3.72		3 3/4	0.004	37.56
	1 1/4	0.003	4.20		3 7/8	0.004	40.08
	1 5/16	0.003	4.56	3 15/16	4	0.004	42.72
	1 3/8	0.003	5.04		4 1/4	0.005	48.12
1 7/16	1 7/16	0.003	5.52	4 7/16	4 1/2	0.005	54.00
	1 1/2	0.003	6.00		4 3/4	0.005	60.12
	1 9/16	0.003	6.48	4 15/16	5	0.005	66.72
	1 5/8	0.003	7.08		5 1/4	0.005	77.56
1 11/16	1 11/16	0.003	7.56	5 7/16	5 1/2	0.005	80.64
	1 3/4	0.003	8.16		5 3/4	0.005	88.20
	1 13/16	0.003	8.76	5 15/16	6	0.005	96.00
	1 7/8	0.003	9.36		6 1/4	0.006	104.2
1 15/16	1 15/16	0.003	9.96	6 1/2	6 1/2	0.006	112.7
	2	0.003	10.68		6 3/4	0.006	121.6
	2 1/16	0.004	11.28	7	7	0.006	130.8
	2 1/8	0.004	12.00		7 1/4	0.006	140.4
2 3/16	2 3/16	0.004	12.72	7 1/2	7 1/2	0.006	150.0
	2 1/4	0.004	13.56		7 3/4	0.006	160.0
	2 5/16	0.004	14.28	8	8	0.006	171.6
	2 3/8	0.004	15.00				

[1]NOTE: These tolerances are *negative* and represent the maximum allowable variation *below* the exact nominal size. For example the maximum diameter of the 1½ inch shaft is 1.500 inches and its minimum allowable diameter is 1.497 inches.

TABLE 1-2

Tolerances for Cold-Finished Carbon-Steel Bars and Shafting

Steel Products Manual AISI, Sec. 9, Sept, 1952
ASTM Designation: A108-52T

Size, in inches	Maximum of Carbon Range 0.28% or less	Maximum of Carbon Range Over 0.28% to 0.55% incl.	*All Carbons Stress Relieved	Maximum of Carbon Range Over 0.55% or All Carbons Heat Treated
All tolerances are in inches and are minus.				
Rounds -- Cold Drawn or Turned and Polished				
To 1 incl.	0.002	0.003	0.004	0.006
Over 1 to 2 incl.	0.003	0.004	0.006	0.008
Over 2 to 4 incl.	0.004	0.005	0.008	0.010
Over 4 to 6 incl.	0.005	0.006	0.010	0.012
Over 6 to 7-3/4 incl.	0.006	0.008	0.012	0.016
Hexagons -- Cold Drawn				
To 5/16 incl.	0.002	0.003	0.004	0.006
Over 5/16 to 1 incl.	0.003	0.004	0.006	0.008
Over 1 to 2-1/2 incl.	0.004	0.005	0.008	0.010
Over 2-1/2 to 3-1/8 incl.	0.005	0.006	0.010	0.012
Squares -- Cold Drawn				
To 5/16 incl.	0.003	0.004	0.006	0.008
Over 5/16 to 1 incl.	0.004	0.005	0.008	0.010
Over 1 to 2-1/2 incl.	0.005	0.006	0.010	0.012
Over 2-1/2 to 4 incl.	0.006	0.008	0.012	0.016
†Flats -- Cold Finished				
Width, in inches				
To 3/4 incl.	0.003	0.004	0.006	0.008
Over 3/4 to 1-1/2 incl.	0.004	0.005	0.008	0.010
Over 1-1/2 to 3 incl.	0.005	0.006	0.010	0.012
Over 3 to 4 incl.	0.006	0.008	0.011	0.016
Over 4 to 6 incl.	0.008	0.010	0.012	0.020
Over 6	0.013	--	--	--

†The tolerances for flats apply to thickness as well as width. ASTM Designation A108-52T gives somewhat different tolerances than these for flats.

*ASTM Designation: A311-52T, Stress-Relief-Annealed Cold-Drawn Carbon-Steel Bars, provides somewhat different tolerances than these taken from the Steel Products Manual.

TABLE 1-3

Tolerances for Cold-Finished Alloy-Steel Bars

Steel Products Manual AISI, Sec. 29, May, 1949
ASTM Designation: A331-50T

Size, in inches	Maximum of Carbon Range 0.28 per cent or less	Maximum of Carbon Range Over 0.28 to 0.55 per cent incl. or All Carbons Annealed	Maximum of Carbon Range Over 0.55 per cent or All Carbons Heat Treated or Stress Relieved
All tolerances are in inches and are minus.			
Rounds -- Cold Drawn or Turned and Polished			
To 1 incl.	0.003	0.005	0.007
Over 1 to 2 incl.	0.004	0.006	0.009
Over 2 to 4 incl.	0.005	0.007	0.011
Over 4 to 6 incl.	0.006	0.008	0.013
Over 6 to 7-3/4	0.007	0.010	0.017
Hexagons -- Cold Drawn			
To 5/16 incl.	0.003	0.005	0.007
Over 5/16 to 1 incl.	0.004	0.006	0.009
Over 1 to 2-1/2 incl.	0.005	0.007	0.011
Over 2-1/2 to 3-1/8 incl.	0.006	0.008	0.013
Squares -- Cold Drawn			
To 5/16 incl.	0.004	0.006	0.009
Over 5/16 to 1 incl.	0.005	0.007	0.011
Over 1 to 2-1/2 incl.	0.006	0.008	0.013
Over 2-1/2 to 4	0.007	0.010	0.017
*Flats -- Cold Finished			
Width, in inches			
To 3/4 incl.	0.004	0.006	0.009
Over 3/4 to 1-1/2 incl.	0.005	0.007	0.011
Over 1-1/2 to 3 incl.	0.006	0.008	0.013
Over 3 to 4 incl.	0.007	0.010	0.017
Over 4 to 6 incl.	0.009	0.012	0.021
Over 6	0.014	--	--

*The tolerances for flats apply to thickness as well as width.

TABLE 1-4

Permissible Variations in Sizes of Turned, Ground and Polished Rounds and Round Sections Ground and Polished from Cold-Drawn Rounds

ASTM Designation: A108-52T; A331-50T
Steel Products Manual AISI, Sections 9 August 1952 and 29 May, 1949

Specified Diameter, In.	Permissible Variation in Specified Size, In. plus	minus
Under 2-1/2	0.000	0.002
2-1/2 and over	0.000	0.003

See Tables 1-2 and 1-3

TABLE 1-5

Simplified Practice Recommended Sizes of Hot-Rolled Round, Carbon-Steel Bars for all Purposes

Bulletin R-222-46, Supt. of Documents, Gov't Printing Office
Steel Products Manual AISI, Sec. 8, Aug., 1952

Nominal Sizes, inclusive inches	Fractional Increments inches
1/4 to 29/32	Advancing by sixty-fourths
29/32 to 2-1/16	Advancing by thirty-seconds
2-1/16 to 4-1/8	Advancing by sixteenths
4-1/8 to 6-1/4	Advancing by eighths
6-1/4 to 8-1/4	Advancing by fourths

TABLE 1-6

Sizes and Shapes of Cold-Finished Carbon-Steel and Alloy-Steel Bar Commonly Available

Steel Products Manual AISI, Sections 9 and 29
ASTM Designations: A 108–52T; A331-50T

Shape	Size Range	Size Steps
Rounds	up to 7-3/4, inclusive	to 1 in., inclusive, by 64ths
Squares	up to 4, inclusive	over 1 to 2 in., inclusive, by 32nds
Hexagons	up to 3-1/8 inches	over 2 in., inclusive, by 16ths
Flats	*1/4 inch and over in specified thickness, and up to 12 inches in specified width.	

*Minimum of 1/8 inch in ASTM Designation A108–52T

TABLE 1-7

Intermediate Simplified Practice Sizes of Hot-Rolled, Round, Carbon-Steel Bars

Bulletin R–222–46, Supt. of Documents. Gov't. Printing Office
Steel Products Manual AISI, Sec. 8, Aug, 1952

For Bolts and Rivets Decimal Sizes, Inches		For Heat-Treated Studs Decimal Sizes, Inches
0.365	1.047	0.507
0.445	1.110	0.632
0.490	1.172	0.758
0.615	1.235	0.883
0.680	1.297	1.009
0.740	1.360	1.135
0.865	1.422	1.261
0.912	1.485	1.387
0.990		1.514

TABLE 1-8

Simplified Practice Recommended Sizes of Hot-Rolled Steel Squares and Round-Cornered Squares

Bulletin R-222-46, Supt. of Documents, Gov't. Printing Office
Steel Products Manual AISI, Sec. 8 Aug, 1952

Squares

Nominal Sizes, inclusive inches	Fractional Increments inches
1/4 to 1-5/16	Advancing by thirty-seconds
1-5/16 to 4-1/4	Advancing by sixteenths
4-1/4 to 5-1/2	Advancing by fourths

Round-Cornered Squares*

Nominal Sizes inclusive inches	Fractional Increments inches	Nominal Corner-Radii inches
3/8 to 1/2	Advancing by thirty-seconds	1/16
17/32 to 13/16	Advancing by thirty-seconds	3/32
27/32 to 1-15/32	Advancing by thirty-seconds	1/8
1-1/2 to 1-15/16	Advancing by sixteenths	1/4
2 to 2-7/16	Advancing by sixteenths	5/16
2-1/2 to 2-7/8	Advancing by sixteenths	3/8
3 to 3-3/8	Advancing by eighths	7/16
3-1/2 to 3-7/8	Advancing by eighths	1/2
4 to 4-1/4	Advancing by fourths	5/8
4-1/2 to 5-1/2	Advancing by fourths	3/4

* Sizes are face to face. Round-cornered squares shall be rolled to dimensions, not to weights per linear foot.

TABLE 1–9

Simplified Practice Recommended Sizes of Hot-Rolled, Hexagon Steel Bars

Bulletin R-222-46, Supt. of Documents, Gov't. Printing Office
Steel Products Manual AISI, Sec. 8, Aug, 1952

Nominal Sizes, inclusive inches	Fractional Increments inches
1/4 to 2-1/16	Advancing by thirty-seconds
2-1/16 to 4-1/16	Advancing by sixteenths

TABLE 1–10

Permissible Variations in the Sizes of Hot-Rolled, Carbon Steel and Alloy Steel Rounds, Squares and Round-Cornered Squares

ASTM Designation: A107-52aT
Steel Products Manual AISI,
Sections 8 Aug, 1952 and 10 May, 1949

Specified Sizes inches	Variations from Size Over	Variations from Size Under	Out-of-Round or Out-of-Square Section
To 5/16 incl.	0.005	0.005	0.008
Over 5/16 to 7/16 incl.	0.006	0.006	0.009
Over 7/16 to 5/8 incl.	0.007	0.007	0.010
Over 5/8 to 7/8 incl.	0.008	0.008	0.012
Over 7/8 to 1 incl.	0.009	0.009	0.013
Over 1 to 1-1/8 incl.	0.010	0.010	0.015
Over 1-1/8 to 1-1/4 incl.	0.011	0.011	0.016
Over 1-1/4 to 1-3/8 incl.	0.012	0.012	0.018
Over 1-3/8 to 1-1/2 incl.	0.014	0.014	0.021
Over 1-1/2 to 2 incl.	1/64	1/64	0.023
Over 2 to 2-1/2 incl.	1/32	0	0.023
Over 2-1/2 to 3-1/2 incl.	3/64	0	0.035
Over 3-1/2 to 4-1/2 incl.	1/16	0	0.046
Over 4-1/2 to 5-1/2 incl.	5/64	0	0.058
Over 5-1/2 to 6-1/2 incl.	1/8	0	0.070
Over 6-1/2 to 8-1/4 incl.	5/32	0	0.085

NOTE: Out-of-round is the difference between the maximum and minimum diameters of the bar, measured at the same cross section. Out-of-square section is the difference in the two dimensions at the same cross section of a square bar between opposite faces.

TABLE 1–11

Permissible Variations in the Sizes of Hot-Rolled Carbon Steel and Alloy Steel Hexagons

Steel Products Manual AISI, Sections 8 Aug, 1952 and 10 May, 1949

Specified Sizes between Opposite Sides, inches	Variation from Size Over	Variation from Size Under	Out of Hexagon Section
To 1/2 incl.	0.007	0.007	0.011
Over 1/2 to 1 incl.	0.010	0.010	0.015
Over 1 to 1-1/2 incl.	0.021	0.013	0.025
Over 1-1/2 to 2 incl.	1/32	1/64	1/32
Over 2 to 2-1/2 incl.	3/64	1/64	3/64
Over 2-1/2 to 3-1/2 incl.	1/16	1/64	1/16

NOTE: Out-of-hexagon section is the greatest difference between any two dimensions at the same cross section between opposite faces.

TABLE 1–12

Permissible Variations in the Sizes of Hot-Rolled Carbon Steel and Alloy Steel Flats

Steel Products Manual AISI, Sections 8 Aug, 1952 and 10 May, 1949

Specified Widths	Variations from Thickness, for Thickness Given, Over and Under					Variations from Width	
	Under 1/4	1/4 to 1/2, Incl.	Over 1/2 to 1, Incl.	Over 1 to 2, Incl.	Over 2	Over	Under
To 1 incl.	0.007	0.008	0.010	--	--	1/64	1/64
Over 1 to 2 incl.	0.007	0.012	0.015	1/32	--	1/32	1/32
Over 2 to 4 incl.	0.008	0.015	0.020	1/32	3/64	1/16	1/32
Over 4 to 6 incl.	0.009	0.015	0.020	1/32	1/16	3/32	1/16

NOTE: All measurements in inches.

TABLE 1–13

Allowances for Machining Hot-Rolled *Alloy-Steel Rounds

Steel Products Manual AISI, Section 10 May 1949

Specified Size	Minimum Stock Allowance on Surface	on Diameter
Up to 5/8 incl.	0.016	0.032
Over 5/8 to 7/8 incl.	0.021	0.042
Over 7/8 to 1 incl.	0.023	0.046
Over 1 to 1-1/8 incl.	0.025	0.050
Over 1-1/8 to 1-1/4 incl.	0.028	0.056
Over 1-1/4 to 1-3/8 incl.	0.030	0.060
Over 1-3/8 to 1-1/2 incl.	0.033	0.066
Over 1-1/2 to 2 incl.	0.042	0.084
Over 2 to 2-1/2 incl.	0.052	0.104
Over 2-1/2 to 3-1/2 incl.	0.072	0.144
Over 3-1/2 to 4-1/2 incl.	0.090	0.180
Over 4-1/2 to 5-1/2 incl.	0.110	0.220
Over 5-1/2 to 6-1/2 incl.	0.125	0.250
Over 6-1/2 to 8 incl.	0.155	0.310

NOTE: All measurements in inches.

*Section 8 of the Steel Products Manual recommends a machining allowance of 1/8 inch for turning hot-rolled carbon-steel bars of 1½- to 3-inch diameters inclusive, and 1/4 for hot-rolled bars over 3 inches in diameter.

TABLE 1–14

Permissible Variations in Hot-Rolled Bars of Tool Steel Rounds, Squares, Octagons, Hexagons

Steel Products Manual AISI, Sec. 25, April 1949

Specified Sizes, In.	Variations From Size, In. Under	Over	Out of Section, Max.
To 1/2 incl.	0.005	0.012	0.010
Over 1/2 to 1 incl.	0.010	0.016	0.013
Over 1 to 1-1/2 incl.	0.012	0.020	0.018
Over 1-1/2 to 2 incl.	0.015	0.025	0.021
Over 2 to 2-1/2 incl.	0.020	0.030	0.025
Over 2-1/2 to 3 incl.	0.020	0.040	0.030
Over 3 to 4 incl.	0.025	0.050	0.035
Over 4 to 5 incl.	0.025	0.060	0.035

TABLE 1–15

Permissible Variations in Hot-Rolled Flat Bars of Tool Steel

Steel Products Manual AISI, Sec. 25, April 1949

Specification Width	Width Variation From Size		Variation From Thickness According to Thickness							
			To 1/4		Over 1/4 to 1/2		Over 1/2 to 1		Over 1 to 2	
	Under	Over	Under	Over	Under	Over	Under	Over	Under	Over
To 1 incl.	1/64	1/32	0.006	0.010	0.008	0.012	0.010	0.016	0.020	0.020
Over 1 to 2 incl.	1/32	3/64	0.006	0.014	0.008	0.016	0.010	0.020	0.020	0.024
Over 2 to 3 incl.	1/32	3/64	0.006	0.018	0.008	0.020	0.010	0.024	0.020	0.027
Over 3 to 4 incl.	3/64	1/16	0.008	0.020	0.010	0.022	0.013	0.024	0.024	0.030
Over 4 to 5 incl.	3/64	1/16	0.010	0.020	0.012	0.024	0.015	0.030	0.027	0.035
Over 5 to 6 incl.	1/16	3/32	0.012	0.020	0.014	0.030	0.018	0.030	0.030	0.035

All dimensions are in inches.

TABLE 1–16

Permissible Variations in Hammered Bars of Tool Steel Rounds, Squares, Octagons, Hexagons

Steel Products Manual AISI, Sec. 25, April 1949

Specified Sizes, In.	Variations From Size, In. Under	Over
Over 1 to 2 incl.	0.030	0.060
Over 2 to 3 incl.	0.030	0.080
Over 3 to 5 incl.	0.060	0.125
Over 5 to 7 incl.	0.125	0.187
Over 7	0.187	0.312

TABLE 1–17

Permissible Variations in Hammered Flat Bars of Tool Steel

Steel Products Manual AISI, Sec. 25, April 1949

Specification Width	Width Variation From Size		Variations From Thickness, for Thickness Given									
			To 1		Over 1 to 3		Over 3 to 5		Over 5 to 7		Over 7	
	Under	Over	Under	Over	Under	Over	Under	Over	Under	Over	Under	Over
Over 1 to 3 incl.	0.031	0.078	0.016	0.031	0.031	0.078	--	--	--	--	--	--
Over 3 to 5 incl.	0.062	0.125	0.031	0.062	0.047	0.094	0.062	0.125	--	--	--	--
Over 5 to 7 incl.	0.125	0.187	0.047	0.094	0.062	0.125	0.078	0.156	0.125	0.187	--	--
Over 7	0.187	0.312	0.062	0.125	0.078	0.156	0.094	0.187	0.156	0.219	0.187	0.312

All dimensions are in inches.

TABLE 1–18

Diameter Tolerances on Free-Cutting, Brass Rod and Bar, for Use in Screw Machines, and on Naval Brass Rods

ASTM Designations: B16–52; B21–52

Diameter or Distance Between Parallel Surfaces In.	[a]Tolerance, Plus and Minus, In. Rounds	Hexagons Octagons	[b]Piston Finish
Up to 0.150, incl.	0.0013	0.0025	
Over 0.150 to 0.500, incl.	0.0015	0.003	
Over 0.500 to 1.00, incl.	0.002	0.004	0.0013
Over 1.00 to 2.00, incl.	0.0025	0.005	0.0015
Over 2.00	0.15 per cent[c]	0.30 per cent[c]	0.10 per cent[d]

[a] When tolerances are specified as all plus or all minus, double the values given.
[b] Applies to ASTM Designation B21 only.
[c] Expressed to the nearest 0.001 in.
[d] Expressed to the nearest multiple of 0.0005 in.

TABLE 1–19

Thickness Tolerances for Naval Brass Rods and Rectangular and Square Bars of Free-Cutting, Brass for Screw Machines

ASTM Designations: B16–52; B21–52

Thickness, In.	*Tolerance, Plus and Minus, on Width: 1/2 and Under	Over 1/2 to 1-1/4 Incl	Over 1-1/4 to 2.00 Incl	Over 2.00 to 4.00 Incl	Over 4.00 to 8.00 Incl	Over 8.00 to 12.00 Incl
Over 0.188 to 0.500 incl.	0.0035	0.004	0.0045	0.0045	0.006	0.008
Over 0.500 to 1.00 incl.		0.0045	0.005	0.005	0.007	0.009
Over 1.00 to 2.00 incl.		0.005	0.005	0.006	0.008	
Over 2.00 to 4.00 incl.				0.30 per cent, expressed to nearest 0.001		

*When tolerances are specified as all plus or all minus double the values given. Not all ASTM Designations use the thickness tolerances given here for naval brass.

TABLE 1-20

Width Tolerances for Rectangular Bars

ASTM Designations: B249-52T

Width, In.	*Tolerance, Plus and Minus, according to ASTM Designation	
	B16, B21 B98 (Alloy B), B133, B140	B98 (Alloys A, C, D), B138, B139, B150, B151, and B196
Over 0.188 to 0.500 incl.	0.0035	0.005
Over 0.500 to 1.25 incl.	0.005	0.007
Over 1.25 to 2.00 incl.	0.008	0.010
Over 2.00 to 4.00 incl.	0.012	0.015
Over 4.00 to 12.00 incl.	0.30 per cent, expressed to nearest 0.001 in.	0.50 per cent expressed to nearest 0.001 in.

*When tolerances are specified as all plus or all minus, double the values given.

TABLE 1-21

Straightness Tolerances for Non-Ferrous Rod, Bar, and Shapes

ASTM Designations: B249-52T

Specific ASTM designations to which this tentative specification applies are B16-52, B21-52, B98-52, B133-52T, B138-52, B139-52, B140-52, B150-52, B151-52, and B196-52.

Applicable to Any Longitudinal Surface or Edge	Maximum Curvature (Depth of Arc) In.	Portion of Total Length in Which Depth of Arc Is Measured In.
Drawn rods	1/2	120
Drawn bars and shapes	1/2	72

TABLE 1-22

Straightness Tolerances for Shafting

ASTM Designation: B249-52T

Specific ASTM designations to which this tentative specification applies are

B21-52, B138-52, B139-52 and B150-52

Length of Shaft, Feet	Maximum Permissible Departure From Straightness of Either Center or End Portions	Minimum Diameter Applicable for Length Indicated, In.
Up to 6, incl.	0.005	1/2
Up to 7, incl.	0.007	1/2
Up to 8, incl.	0.009	1/2
Up to 9, incl.	0.012	1/2
Up to 10, incl.	0.014	1/2
Up to 11, incl.	0.017	1/2
Up to 12, incl.	0.020	1/2
Up to 14, incl.	0.028	5/8
Up to 16, incl.	0.036	3/4
Up to 18, incl.	0.045	1
Up to 20, incl.	0.055	1-1/4
Up to 22, incl.	0.068	1-1/2
Up to 24, incl.	0.078	1-3/4
Up to 26, incl.	0.094	2

TABLE 1-23

Section Moduli and Moments of Inertia of Round Shafting

Dodge Catalog D 55

Shaft Size	Section Modulus		Moment of Inertia	
	Bending	Torsion	Bending	Torsion
1/16	.000024	.000048	.000001	.000002
1/8	.000192	.000383	.000012	.000024
3/16	.000647	.001294	.000061	.000121
1/4	.001534	.003068	.000192	.000383
5/16	.002996	.005992	.000468	.000936
3/8	.005177	.010354	.000971	.001941
7/16	.008221	.016442	.001798	.003597
1/2	.0123	.0245	.0031	.0061
9/16	.0175	.0349	.0049	.0098
5/8	.0240	.0479	.0075	.0150
11/16	.0319	.0638	.0110	.0219
3/4	.0414	.0828	.0155	.0311
13/16	.0527	.1053	.0214	.0428
7/8	.0658	.1315	.0288	.0575
15/16	.0809	.1618	.0379	.0758
1	.0982	.1963	.0491	.0982
1-1/16	.1178	.2355	.0626	.1251
1-1/8	.1398	.2796	.0786	.1573
1-3/16	.1644	.3288	.0976	.1952
1-1/4	.1917	.3835	.1198	.2397
1-5/16	.2220	.4439	.1457	.2913
1-3/8	.2552	.5104	.1755	.3509
1-7/16	.2916	.5832	.2096	.4192
1-1/2	.3313	.6627	.2485	.4970
1-9/16	.3745	.7490	.2926	.5852
1-5/8	.4213	.8425	.3423	.6846
1-11/16	.4718	.9435	.3981	.7961
1-3/4	.5262	1.052	.4604	.9208
1-13/16	.5846	1.169	.5298	1.060
1-7/8	.6471	1.294	.6067	1.213
1-15/16	.7140	1.428	.6917	1.384
2	.7854	1.571	.7854	1.571
2-1/16	.8614	1.723	.8883	1.777
2-1/8	.9421	1.884	1.001	2.002
2-3/16	1.028	2.055	1.124	2.248
2-1/4	1.118	2.237	1.258	2.516
2-5/16	1.214	2.428	1.404	2.808
2-3/8	1.315	2.630	1.562	3.124

continued on next page

TABLE 1-23, continued

Shaft Size	Section Modulus Bending	Section Modulus Torsion	Moment of Inertia Bending	Moment of Inertia Torsion
2-7/16	1.422	2.844	1.733	3.466
2-1/2	1.534	3.068	1.918	3.835
2-9/16	1.652	3.304	2.117	4.233
2-5/8	1.776	3.552	2.331	4.661
2-11/16	1.906	3.811	2.561	5.122
2-3/4	2.042	4.084	2.807	5.615
2-13/16	2.184	4.368	3.071	6.143
2-7/8	2.333	4.666	3.354	6.707
2-15/16	2.489	4.977	3.655	7.310
3	2.651	5.301	3.976	7.952
3-1/16	2.820	5.640	4.318	8.636
3-1/8	2.996	5.992	4.681	9.363
3-3/16	3.179	6.359	5.067	10.13
3-1/4	3.370	6.740	5.477	10.95
3-5/16	3.568	7.137	5.910	11.82
3-3/8	3.774	7.548	6.369	12.74
3-7/16	3.988	7.976	6.854	13.71
3-1/2	4.209	8.419	7.366	14.73
3-9/16	4.439	8.878	7.907	15.81
3-5/8	4.677	9.353	8.476	16.95
3-11/16	4.923	9.845	9.076	18.15
3-3/4	5.177	10.35	9.707	19.41
3-13/16	5.440	10.88	10.37	20.74
3-7/8	5.712	11.42	11.07	22.14
3-15/16	5.993	11.99	11.80	23.60
4	6.283	12.57	12.57	25.13
4-1/16	6.582	13.16	13.37	26.74
4-1/8	6.891	13.78	14.21	28.42
4-3/16	7.209	14.42	15.09	30.19
4-1/4	7.536	15.07	16.01	32.03
4-5/16	7.874	15.75	16.98	33.96
4-3/8	8.221	16.44	17.98	35.97
4-7/16	8.579	17.16	19.03	38.07
4-1/2	8.946	17.89	20.13	40.26
4-9/16	9.324	18.65	21.27	42.54
4-5/8	9.713	19.43	22.46	44.92
4-11/16	10.11	20.22	23.70	47.40
4-3/4	10.52	21.04	24.99	49.98
4-13/16	10.94	21.88	26.33	52.66
4-7/8	11.37	22.75	27.72	55.45
4-15/16	11.82	23.63	29.17	58.35
5	12.27	24.54	30.68	61.36
5-1/16	12.74	25.48	32.24	64.49

continued on next page

TABLE 1-23, continued

Shaft Size	Section Modulus Bending	Section Modulus Torsion	Moment of Inertia Bending	Moment of Inertia Torsion
5-1/8	13.22	26.43	33.86	67.73
5-3/16	13.70	27.41	35.55	71.09
5-1/4	14.21	28.41	37.29	74.58
5-5/16	14.72	29.44	39.10	78.20
5-3/8	15.25	30.49	40.97	81.94
5-7/16	15.78	31.57	42.91	85.82
5-1/2	16.33	32.67	44.92	89.84
5-9/16	16.90	33.79	46.99	93.99
5-5/8	17.47	34.95	49.14	98.29
5-11/16	18.06	36.12	51.36	102.7
5-3/4	18.66	37.33	53.66	107.3
5-13/16	19.28	38.56	56.03	112.1
5-7/8	19.91	39.82	58.48	117.0
5-15/16	20.55	41.10	61.01	122.0
6	21.21	42.41	63.62	127.2
6-1/16	21.88	43.75	66.31	132.6
6-1/8	22.56	45.12	69.09	138.2
6-3/16	23.26	46.51	71.95	143.9
6-1/4	23.97	47.94	74.90	149.8
6-5/16	24.69	49.39	77.94	155.9
6-3/8	25.44	50.87	81.08	162.2
6-7/16	26.19	52.38	84.30	168.6
6-1/2	26.96	53.92	87.62	175.2
6-5/8	28.55	57.09	94.56	189.1
6-3/4	30.19	60.39	101.9	203.8
6-7/8	31.90	63.80	109.7	219.3
7	33.67	67.35	117.9	235.7
7-1/8	35.51	71.02	126.5	253.0
7-1/4	37.41	74.82	135.6	271.2
7-3/8	39.38	78.76	145.2	290.4
7-1/2	41.42	82.84	155.3	310.6
7-5/8	43.52	87.05	165.9	331.9
7-3/4	45.70	91.40	177.1	354.2
7-7/8	47.95	95.89	188.8	377.6
8	50.27	100.5	201.1	402.1
8-1/8	52.66	105.3	213.9	427.9
8-1/4	55.13	110.3	227.4	454.8
8-3/8	57.67	115.3	241.5	483.0
8-1/2	60.29	120.6	256.2	512.5
8-5/8	62.99	126.0	271.6	543.3
8-3/4	65.77	131.6	287.7	575.5
8-7/8	68.63	137.3	304.5	609.1
9	71.57	143.1	322.1	644.1
9-1/8	74.59	149.2	340.3	680.7
9-1/4	77.70	155.4	359.4	718.7

continued on next page

TABLE 1-23, continued

Shaft Size	Section Modulus Bending	Section Modulus Torsion	Moment of Inertia Bending	Moment of Inertia Torsion
9-3/8	80.89	161.8	379.2	758.4
9-1/2	84.17	168.3	399.8	799.6
9-5/8	87.54	175.1	421.3	842.6
9-3/4	90.99	182.0	443.6	887.2
9-7/8	94.54	189.1	466.8	933.6
10	98.17	196.3	490.9	981.7
10-1/4	105.72	211.4	541.8	1084
10-1/2	113.65	227.3	596.7	1193
10-3/4	121.96	243.9	655.5	1311
11	130.67	261.3	718.7	1437
11-1/4	139.78	279.6	786.3	1573
11-1/2	149.31	298.6	858.5	1717
11-3/4	159.26	318.5	935.7	1871
12	169.65	339.3	1018	2036
12-1/4	180.47	360.9	1105	2211
12-1/2	191.75	383.5	1198	2397
12-3/4	203.48	407.0	1297	2594
13	215.69	431.4	1402	2804
13-1/4	228.37	456.7	1513	3026
13-1/2	241.5	483.1	1630	3261
13-3/4	255.2	510.4	1755	3509
14	269.4	538.8	1886	3771
14-1/4	284.1	568.2	2024	4048
14-1/2	299.3	598.6	2170	4340
14-3/4	315.0	630.1	2324	4647
15	331.3	662.7	2485	4970
15-1/4	348.2	696.4	2655	5310
15-1/2	365.6	731.2	2833	5667
15-3/4	383.6	767.1	3021	6041
16	402.1	804.2	3217	6434
16-1/4	421.3	842.5	3422	6846
16-1/2	441.0	882.0	3638	7277
16-3/4	461.4	922.7	3864	7728
17	482.3	964.7	4100	8200
17-1/4	503.9	1008	4346	8693
17-1/2	526.2	1052	4604	9208
17-3/4	549.1	1098	4873	9745
18	572.6	1145	5153	10306
18-1/4	596.7	1193	5445	10891
18-1/2	621.6	1243	5750	11500
18-3/4	647.1	1294	6067	12134
19	673.4	1347	6397	12794
19-1/4	700.3	1401	6741	13481
19-1/2	728.0	1456	7098	14195

continued on next page

TABLE 1-23, continued

Shaft Size	Section Modulus		Moment of Inertia	
	Bending	Torsion	Bending	Torsion
19-3/4	756.3	1513	7469	14937
20	785.4	1571	7854	15708
20-1/4	815.2	1630	8254	16508
20-1/2	845.8	1692	8669	17339
20-3/4	877.1	1754	9100	18200
21	909.2	1818	9547	19093
21-1/4	942.1	1884	10009	20019
21-1/2	975.7	1951	10489	20978
21-3/4	1010	2020	10985	21970
22	1045	2091	11499	22998
22-1/4	1081	2163	12031	24061
22-1/2	1118	2237	12581	25161
22-3/4	1156	2312	13149	26298
23	1194	2389	13737	27473
23-1/4	1234	2468	14344	28687
23-1/2	1274	2548	14971	29941
23-3/4	1315	2630	15618	31236
24	1357	2714	16286	32572
24-1/4	1400	2800	16975	33951
24-1/2	1444	2888	17686	35372
24-3/4	1488	2977	18419	36838
25	1534	3068	19175	38350
25-1/4	1580	3161	19954	39907
25-1/2	1628	3256	20755	41511
25-3/4	1676	3352	21581	43163
26	1726	3451	22432	44864
26-1/4	1776	3552	23307	46614
26-1/2	1827	3654	24208	48415
26-3/4	1879	3758	25134	50268
27	1932	3865	26087	52174
27-1/2	2042	4083	28074	56148
28	2155	4310	30172	60344
28-1/2	2273	4545	32385	64771
29	2394	4789	34719	69437
29-1/2	2520	5041	37176	74351
30	2651	5301	39761	79522
30-1/2	2785	5571	42479	84957
31	2925	5849	45333	90666
31-1/2	3069	6137	48329	96659
32	3217	6434	51472	102944
32-1/2	3370	6740	54765	109530
33	3528	7056	58214	116428
34	3859	7717	65597	131194
35	4209	8418	73662	147324

TABLE 1–24

Properties of a Plane Area

Roark "Formulas for Stress and Strain"
McGraw-Hill

Form of section	Area A	Distance from centroid to extremities of section y_1, y_2	Moments of inertia I_1 and I_2 about principal central axes 1 and 2	Radii of gyration, r_1 and r_2, about principal central axes
1. Square 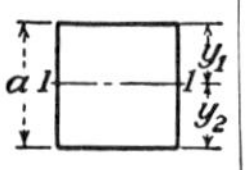	$A = a^2$	$y_1 = y_2 = \frac{1}{2}a$	$I_1 = I_2 = I_3 = \frac{1}{12}a^4$	$r_1 = r_2 = r_3 = 0.289a$
2. Rectangle 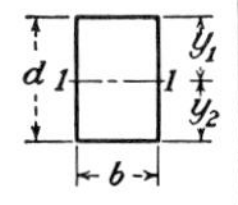	$A = bd$	$y_1 = y_2 = \frac{1}{2}d$	$I_1 = \frac{1}{12}bd^3$	$r_1 = 0.289d$
3. Triangle 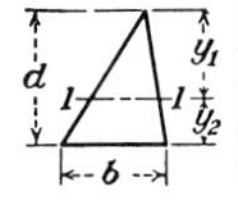	$A = \frac{1}{2}bd$	$y_1 = \frac{2}{3}d$ $y_2 = \frac{1}{3}d$	$I_1 = \frac{1}{36}bd^3$	$r_1 = 0.2358d$
4. Trapezoid 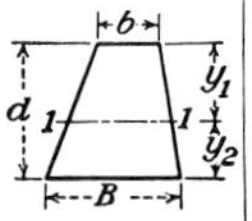	$A = \frac{1}{2}(B + b)d$	$y_1 = d\dfrac{2B + b}{3(B + b)}$ $y_2 = d\dfrac{B + 2b}{3(B + b)}$	$I_1 = \dfrac{d^3(B^2 + 4Bb + b^2)}{36(B + b)}$	$r_1 = \dfrac{d}{6(B + b)}\sqrt{2(B^2 + 4Bb + b^2)}$
5. Regular polygon with n sides 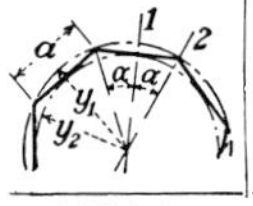	$A = \frac{1}{4}na^2 \cot \alpha$	$y_1 = \dfrac{a}{2 \sin \alpha}$ $y_2 = \dfrac{a}{2 \tan \alpha}$	$I_1 = \dfrac{A(6y_1^2 - a^2)}{24}$ $I_2 = \dfrac{A(12y_2^2 + a^2)}{48}$	$r_1 = \sqrt{\dfrac{6y_1^2 - a^2}{24}}$ $r_2 = \sqrt{\dfrac{12y_2^2 + a^2}{48}}$
6. Solid circle 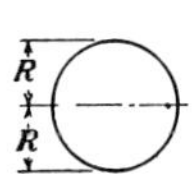	$A = \pi R^2$	$y_1 = y_2 = R$	$I = \frac{1}{4}\pi R^4$	$r = \frac{1}{2}r$
7. Hollow circle 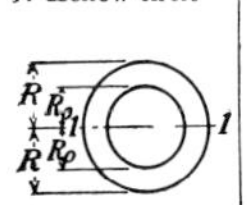	$A = \pi(R^2 - R_0^2)$	$y_1 = y_2 = R$	$I = \frac{1}{4}\pi(R^4 - R_0^4)$	$r = \sqrt{\frac{1}{4}(R^2 + R_0^2)}$
8. Solid semicircle 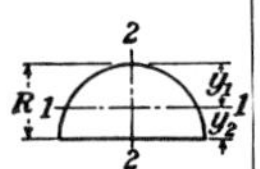	$A = \frac{1}{2}\pi R^2$	$y_1 = 0.5756R$ $y_2 = 0.4244R$	$I_1 = 0.1098R^4$ $I_2 = \frac{1}{8}\pi R^4$	$r_1 = 0.2643R$ $r_2 = \frac{1}{2}r$
9. Circular sector 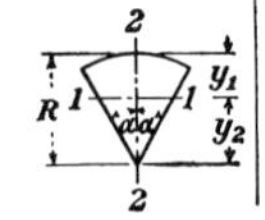	$A = \alpha R^2$	$y_1 = R\left(1 - \dfrac{2 \sin \alpha}{3\alpha}\right)$ $y_2 = 2R\dfrac{\sin \alpha}{3\alpha}$	$I_1 = \dfrac{1}{4}R^4\left[\alpha + \sin \alpha \cos \alpha - \dfrac{16 \sin^2 \alpha}{9\alpha}\right]$ $I_2 = \frac{1}{4}R^4[\alpha - \sin \alpha \cos \alpha]$	$r_1 = \dfrac{1}{2}R\sqrt{1 + \dfrac{\sin \alpha \cos \alpha}{\alpha} - \dfrac{16 \sin^2 \alpha}{9\alpha^2}}$ $r_2 = \dfrac{1}{2}R\sqrt{1 - \dfrac{\sin \alpha \cos \alpha}{\alpha}}$

continued on next page

TABLE 1–24, continued

Form of section	Area A	Distance from centroid to extremities of section y_1, y_2	Moments of inertia I_1 and I_2 about principal central axes 1 and 2	Radii of gyration, r_1 and r_2, about principal central axes
10. Circular segment 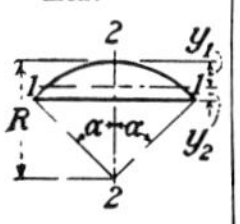	$A = \frac{1}{2}R^2(2\alpha - \sin 2\alpha)$	$y_1 = R\left(1 - \frac{4 \sin^3 \alpha}{6\alpha - 3 \sin 2\alpha}\right)$ $y_2 = R\left(\frac{4 \sin^3 \alpha}{6\alpha - 3 \sin 2\alpha} - \cos \alpha\right)$	$I_1 = R^4\left[\frac{1}{8}(2\alpha - \sin 2\alpha)\left(1 + \frac{2 \sin^3 \alpha \cos \alpha}{\alpha - \sin \alpha \cos \alpha}\right) - \frac{8}{9}\frac{\sin^6 \alpha}{2\alpha - \sin 2\alpha}\right]$ $I_2 = R^4\left[\frac{1}{8}(2\alpha - \sin 2\alpha) - \frac{1}{12}\frac{(2\alpha - \sin 2\alpha) \sin^3 \alpha \cos \alpha}{\alpha - \sin \alpha \cos \alpha}\right]$	$r_1 = \frac{1}{2}R\sqrt{1 + \frac{2 \sin^3 \alpha \cos \alpha}{\alpha - \sin \alpha \cos \alpha} - \frac{64}{9}\frac{\sin^6 \alpha}{(2\alpha - \sin 2\alpha)^2}}$ $r_2 = \frac{1}{2}R\sqrt{1 - \frac{2 \sin^3 \alpha \cos \alpha}{3(\alpha - \sin \alpha \cos \alpha)}}$
11. Very thin annulus 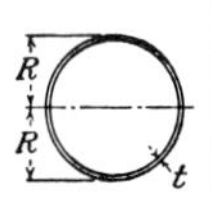	$A = 2\pi Rt$	$y_1 = y_2 = R$	$I = \pi R^3 t$	$r = 0.707R$
12. Sector of thin annulus 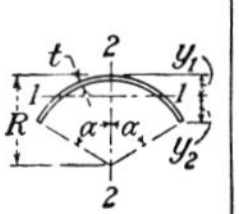	$A = 2\alpha Rt$	$y_1 = R\left(1 - \frac{\sin \alpha}{\alpha}\right)$ $y_2 = R\left(\frac{\sin \alpha}{\alpha} - \cos \alpha\right)$	$I_1 = R^3 t\left(\alpha + \sin \alpha \cos \alpha - \frac{2 \sin^2 \alpha}{\alpha}\right)$ $I_2 = R^3 t(\alpha - \sin \alpha \cos \alpha)$	$r_1 = R\sqrt{\frac{\alpha + \sin \alpha \cos \alpha - 2 \sin^2 \alpha/\alpha}{2\alpha}}$ $r_2 = R\sqrt{\frac{\alpha - \sin \alpha \cos \alpha}{2\alpha}}$
13. Solid ellipse 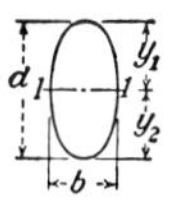	$A = \frac{1}{4}\pi bd$	$y_1 = y_2 = \frac{1}{2}d$	$I_1 = \frac{1}{64}\pi bd^3$	$r_1 = \frac{1}{4}d$
14. Hollow ellipse 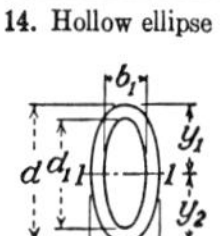	$A = \frac{1}{4}\pi(bd - b_1 d_1)$	$y_1 = y_2 = \frac{1}{2}d$	$I_1 = \frac{1}{64}\pi(bd^3 - b_1 d_1^3)$	$r_1 = \frac{1}{4}\sqrt{\frac{bd^3 - b_1 d_1^3}{bd - b_1 d_1}}$

TABLE 1–25

Shear, Moment, and Deflection Formulas for Beams

Roark "Formulas for Stress and Strain"
McGraw-Hill

Notation: W = load (lb); w = unit load (lb per linear in.). M is positive when clockwise; V is positive when upward; y is positive when upward. Constraining moments, applied couples, loads, and reactions are positive when acting as shown. All forces are in pounds, all moments in inch-pounds; all deflections and dimensions in inches. θ is in radians and tan $\theta = \theta$.

Statically Determinate Cases

Loading, support, and reference number	Reactions R_1 and R_2, vertical shear V	Bending moment M and maximum bending moment	Deflection y, maximum deflection, and end slope θ
1. Cantilever, end load 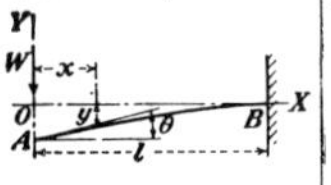	$R_2 = +W$ $V = -W$	$M = -Wx$ Max $M = -Wl$ at B	$y = -\frac{1}{6}\frac{W}{EI}(x^3 - 3l^2x + 2l^3)$ Max $y = -\frac{1}{3}\frac{Wl^3}{EI}$ at A $\theta = +\frac{1}{2}\frac{Wl^2}{EI}$ at A
2. Cantilever, intermediate load 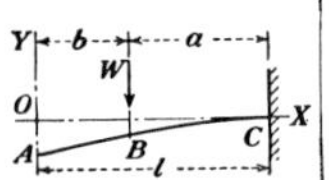	$R_2 = +W$ $(A$ to $B)$ $V = 0$ $(B$ to $C)$ $V = -W$	$(A$ to $B)$ $M = 0$ $(B$ to $C)$ $M = -W(x - b)$ Max $M = -Wa$ at C	$(A$ to $B)$ $y = -\frac{1}{6}\frac{W}{EI}(-a^3 + 3a^2l - 3a^2x)$ $(B$ to $C)$ $y = -\frac{1}{6}\frac{W}{EI}[(x - b)^3 - 3a^2(x - b) + 2a^3]$ Max $y = -\frac{1}{6}\frac{W}{EI}(3a^2l - a^3)$ $\theta = +\frac{1}{2}\frac{Wa^2}{EI}$ $(A$ to $B)$
3. Cantilever, uniform load 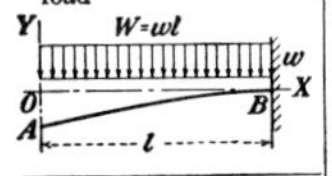	$R_2 = +W$ $V = -\frac{W}{l}x$	$M = -\frac{1}{2}\frac{W}{l}x^2$ Max $M = -\frac{1}{2}Wl$ at B	$y = -\frac{1}{24}\frac{W}{EIl}(x^4 - 4l^3x + 3l^4)$ Max $y = -\frac{1}{8}\frac{Wl^3}{EI}$ $\theta = +\frac{1}{6}\frac{Wl^2}{EI}$ at A
4. Cantilever, partial uniform load 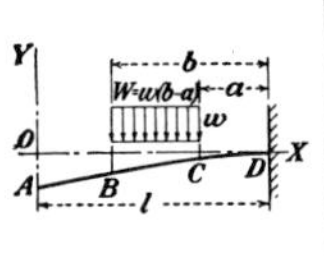	$R_2 = +W$ $(A$ to $B)$ $V = 0$ $(B$ to $C)$ $V = -\frac{W}{b - a}(x - l + b)$ $(C$ to $D)$ $V = -W$	$(A$ to $B)$ $M = 0$ $(B$ to $C)$ $M = -\frac{1}{2}\frac{W}{b - a}(x - l + b)^2$ $(C$ to $D)$ $M = -\frac{1}{2}W(2x - 2l + a + b)$ Max $M = -\frac{1}{2}W(a + b)$ at D	$(A$ to $B)$ $y = -\frac{1}{24}\frac{W}{EI}[4(a^2 + ab + b^2)(l - x) - a^3 - ab^2 - a^2b - b^3]$ $(B$ to $C)$ $y = -\frac{1}{24}\frac{W}{EI}\left[6(a + b)(l - x)^2 - 4(l - x)^3 + \frac{(l - x - a)^4}{b - a}\right]$ $(C$ to $D)$ $y = -\frac{1}{12}\frac{W}{EI}[3(a + b)(l - x)^2 - 2(l - x)^3]$ Max $y = -\frac{1}{24}\frac{W}{EI}[4(a^2 + ab + b^2)l - a^3 - ab^2 - a^2b - b^3]$ at A $\theta = +\frac{1}{6}\frac{W}{EI}[a^2 + ab + b^2]$ $(A$ to $B)$
5. Cantilever, triangular load 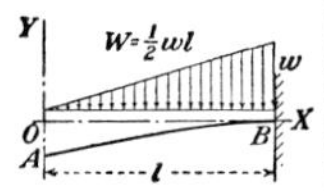	$R_2 = +W$ $V = -\frac{W}{l^2}x^2$	$M = -\frac{1}{3}\frac{W}{l^2}x^3$ Max $M = -\frac{1}{3}Wl$ at B	$y = -\frac{1}{60}\frac{W}{EIl^2}(x^5 - 5l^4x + 4l^5)$ Max $y = -\frac{1}{15}\frac{Wl^3}{EI}$ at A $\theta = +\frac{1}{12}\frac{Wl^2}{EI}$ at A
6. Cantilever, partial triangular load 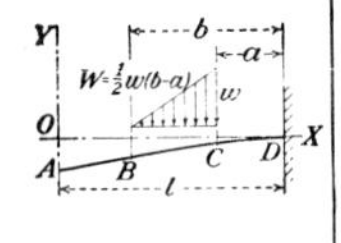	$R_2 = +W$ $(A$ to $B)$ $V = 0$ $(B$ to $C)$ $V = -\frac{W(x - l + b)^2}{(b - a)^2}$ $(C$ to $D)$ $V = -W$	$(A$ to $B)$ $M = 0$ $(B$ to $C)$ $M = -\frac{1}{3}\frac{W(x - l + b)^3}{(b - a)^2}$ $(C$ to $D)$ $M = -\frac{1}{3}W(3x - 3l + b + 2a)$ Max $M = -\frac{1}{3}W(b + 2a)$ at D	$(A$ to $B)$ $y = -\frac{1}{60}\frac{W}{EI}[(5b^2 + 10ba + 15a^2)(l - x) - 4a^3 - 2ab^2 - 3a^2b - b^3]$ $(B$ to $C)$ $y = -\frac{1}{60}\frac{W}{EI}\left[(20a + 10b)(l - x)^2 - 10(l - x)^3 + 5\frac{(l - x - a)^4}{b - a} - \frac{(l - x - a)^5}{(b - a)^2}\right]$ $(C$ to $D)$ $y = -\frac{1}{6}\frac{W}{EI}[(2a + b)(l - x)^2 - (l - x)^3]$ Max $y = -\frac{1}{60}\frac{W}{EI}[(5b^2 + 10ba + 15a^2)l - 4a^3 - 2ab^2 - 3a^2b - b^3]$ at A $\theta = +\frac{1}{12}\frac{W}{EI}(3a^2 + 2ab + b^2)$ $(A$ to $B)$
7. Cantilever, triangular load 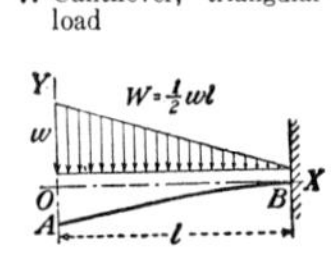	$R_2 = +W$ $V = -W\left(\frac{2lx - x^2}{l^2}\right)$	$M = -\frac{1}{3}\frac{W}{l^2}(3lx^2 - x^3)$ Max $M = -\frac{2}{3}Wl$ at B	$y = -\frac{1}{60}\frac{W}{EIl^2}(-x^5 - 15l^4x + 5lx^4 + 11l^5)$ Max $y = -\frac{11}{60}\frac{Wl^3}{EI}$ at A $\theta = +\frac{1}{4}\frac{W}{EI}l^2$ at A

continued on next page

TABLE 1–25, continued

Loading, support, and reference number	Reactions R_1 and R_2, vertical shear V	Bending moment M and maximum bending moment	Deflection y, maximum deflection, and end slope θ
8. Cantilever, partial triangular load 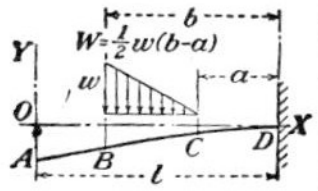	$R_2 = +W$ $(A \text{ to } B)\ V = 0$ $(B \text{ to } C)$ $V = -W\left[1 - \frac{(l-a-x)^2}{(b-a)^2}\right]$ $(C \text{ to } D)\ V = -W$	$(A \text{ to } B)\ M = 0$ $(B \text{ to } C)\ M = -\frac{1}{3}W\left[\frac{3(x-l+b)^2}{b-a} - \frac{(x-l+b)^3}{(b-a)^2}\right]$ $(C \text{ to } D)\ M = -\frac{1}{3}W(-3l+3x+2b+a)$ $\text{Max } M = -\frac{1}{3}W(2b+a) \text{ at } D$	$(A \text{ to } B)\ y = -\frac{1}{60}\frac{W}{EI}[(5a^2+10ab+15b^2)(l-x) - a^3 - 2a^2b - 3ab^2 - 4b^3]$ $(B \text{ to } C)\ y = -\frac{1}{60}\frac{W}{EI}\left[\frac{(l-x-a)^5}{(b-a)^2} - 10(l-x)^3 + (10a+20b)(l-x)^2\right]$ $(C \text{ to } D)\ y = -\frac{1}{6}\frac{W}{EI}[(a+2b)(l-x)^2 - (l-x)^3]$ $\text{Max } y = -\frac{1}{60}\frac{W}{EI}[(5a^2+10ab+15b^2)l - a^3 - 2a^2b - 3ab^2 - 4b^3] \text{ at } A$ $\theta = +\frac{1}{12}\frac{W}{EI}(a^2+2ab+3b^2)\ (A \text{ to } B)$
9. Cantilever, end couple 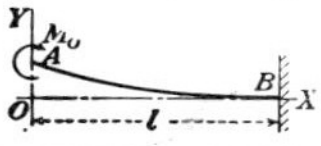	$R_2 = 0$ $V = 0$	$M = M_0$ $\text{Max } M = M_0 (A \text{ to } B)$	$y = \frac{1}{2}\frac{M_0}{EI}(l^2 - 2lx + x^2)$ $\text{Max } y = +\frac{1}{2}\frac{M_0 l^2}{EI} \text{ at } A$ $\theta = -\frac{M_0 l}{EI} \text{ at } A$
10. Cantilever, intermediate couple 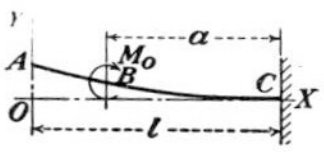	$R_2 = 0$ $V = 0$	$(A \text{ to } B)\ M = 0$ $(B \text{ to } C)\ M = M_0$ $\text{Max } M = M_0\ (B \text{ to } C)$	$(A \text{ to } B)\ y = \frac{M_0 a}{EI}\left(l - \frac{1}{2}a - x\right)$ $(B \text{ to } C)\ y = \frac{1}{2}\frac{M_0}{EI}[(x-l+a)^2 - 2a(x-l+a) + a^2]$ $\text{Max } y = \frac{M_0 a}{EI}\left(l - \frac{1}{2}a\right) \text{ at } A$ $\theta = -\frac{M_0 a}{EI}\ (A \text{ to } B)$
11. End supports, center load 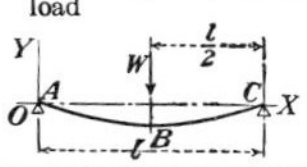	$R_1 = +\frac{1}{2}W \quad R_2 = +\frac{1}{2}W$ $(A \text{ to } B)\ V = +\frac{1}{2}W$ $(B \text{ to } C)\ V = -\frac{1}{2}W$	$(A \text{ to } B)\ M = +\frac{1}{2}Wx$ $(B \text{ to } C)\ M = +\frac{1}{2}W(l-x)$ $\text{Max } M = +\frac{1}{4}Wl \text{ at } B$	$(A \text{ to } B)\ y = -\frac{1}{48}\frac{W}{EI}(3l^2x - 4x^3)$ $\text{Max } y = -\frac{1}{48}\frac{Wl^3}{EI} \text{ at } B$ $\theta = -\frac{1}{16}\frac{Wl^2}{EI} \text{ at } A, \quad \theta = +\frac{1}{16}\frac{Wl^2}{EI} \text{ at } C$
12. End supports, intermediate load 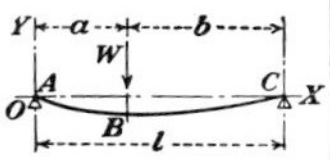	$R_1 = +W\frac{b}{l} \quad R_2 = +W\frac{a}{l}$ $(A \text{ to } B)\ V = +W\frac{b}{l}$ $(B \text{ to } C)\ V = -W\frac{a}{l}$	$(A \text{ to } B)\ M = +W\frac{b}{l}x$ $(B \text{ to } C)\ M = +W\frac{a}{l}(l-x)$ $\text{Max } M = +W\frac{ab}{l} \text{ at } B$	$(A \text{ to } B)\ y = -\frac{Wbx}{6EIl}[2l(l-x) - b^2 - (l-x)^2]$ $(B \text{ to } C)\ y = -\frac{Wa(l-x)}{6EIl}[2lb - b^2 - (l-x)^2]$ $\text{Max } y = -\frac{Wab}{27EIl}(a+2b)\sqrt{3a(a+2b)} \text{ at } x = \sqrt{\frac{1}{3}a(a+2b)} \text{ when } a > b$ $\theta = -\frac{1}{6}\frac{W}{EI}\left(bl - \frac{b^3}{l}\right) \text{ at } A; \quad \theta = +\frac{1}{6}\frac{W}{EI}\left(2bl + \frac{b^3}{l} - 3b^2\right) \text{ at } C$
13. End supports, uniform load 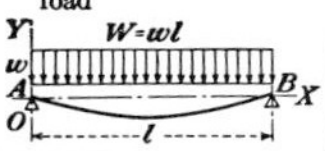	$R_1 = +\frac{1}{2}W \quad R_2 = +\frac{1}{2}W$ $V = \frac{1}{2}W\left(1 - \frac{2x}{l}\right)$	$M = \frac{1}{2}W\left(x - \frac{x^2}{l}\right)$ $\text{Max } M = +\frac{1}{8}Wl \text{ at } x = \frac{1}{2}l$	$y = -\frac{1}{24}\frac{Wx}{EIl}(l^3 - 2lx^2 + x^3)$ $\text{Max } y = -\frac{5}{384}\frac{Wl^3}{EI} \text{ at } x = \frac{1}{2}l$ $\theta = -\frac{1}{24}\frac{Wl^2}{EI} \text{ at } A \quad \theta = +\frac{1}{24}\frac{Wl^2}{EI} \text{ at } B$
14. End supports, partial uniform load 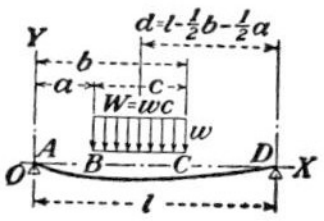	$R_1 = W\frac{d}{l}$ $R_2 = \frac{W}{l}\left(a + \frac{1}{2}c\right)$ $(A \text{ to } B)\ V = R_1$ $(B \text{ to } C)\ V = R_1 - W\frac{x-a}{c}$ $(C \text{ to } D)\ V = R_1 - W$	$(A \text{ to } B)\ M = R_1 x$ $(B \text{ to } C)\ M = R_1 x - W\frac{(x-a)^2}{2c}$ $(C \text{ to } D)\ M = R_1 x - W(x - \frac{1}{2}a - \frac{1}{2}b)$ $\text{Max } M = W\frac{d}{l}\left(a + \frac{cd}{2l}\right) \text{ at } x = a + \frac{cd}{l}$	$(A \text{ to } B)\ y = \frac{1}{48EI}\left\{8R_1(x^3 - l^2x) + Wx\left[\frac{8d^3}{l} - \frac{2bc^2}{l} + \frac{c^3}{l} + 2c^2\right]\right\}$ $(B \text{ to } C)\ y = \frac{1}{48EI}\left\{8R_1(x^3 - l^2x) + Wx\left[\frac{8d^3}{l} - \frac{2bc^2}{l} + \frac{c^3}{l} + 2c^2\right] - 2W\frac{(x-a)^4}{c}\right\}$ $(C \text{ to } D)\ y = \frac{1}{48EI}\left\{8R_1(x^3 - l^2x) + Wx\left[\frac{8d^3}{l} - \frac{2bc^2}{l} + \frac{c^3}{l}\right] - 8W(x - \frac{1}{2}a - \frac{1}{2}b)^3 + W(2bc^2 - c^3)\right\}$ $\theta = \frac{1}{48EI}\left[-8R_1l^2 + W\left(\frac{8d^3}{l} - \frac{2bc^2}{l} + \frac{c^3}{l} + 2c^2\right)\right] \text{ at } A;$ $\theta = \frac{1}{48EI}\left[16R_1l^2 - W\left(24d^2 - \frac{8d^3}{l} + \frac{2bc^2}{l} - \frac{c^3}{l}\right)\right] \text{ at } B$

continued on next page

Loading, support, and reference number	Reactions R_1 and R_2, vertical shear V	Bending moment M and maximum bending moment	Deflection y, maximum deflection, and end slope θ
15. End supports, triangular load 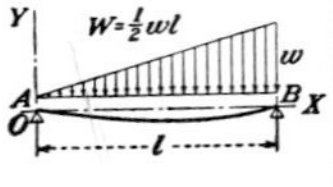	$R_1 = \frac{1}{3}W$ $R_2 = \frac{2}{3}W$ $V = W\left(\frac{1}{3} - \frac{x^2}{l^2}\right)$	$M = \frac{1}{3}W\left(x - \frac{x^3}{l^2}\right)$ Max $M = 0.128Wl$ at $x = l\left(\frac{\sqrt{3}}{3}\right) = 0.5774l$	$y = -\frac{1}{180}\frac{Wx}{EIl^2}(3x^4 - 10l^2x^2 + 7l^4)$ Max $y = -0.01304\frac{Wl^3}{EI}$ at $x = 0.519l$ $\theta = -\frac{7}{180}\frac{Wl^2}{EI}$ at A; $\theta = +\frac{8}{180}\frac{Wl^2}{EI}$ at B.
16. End supports, partial triangular load 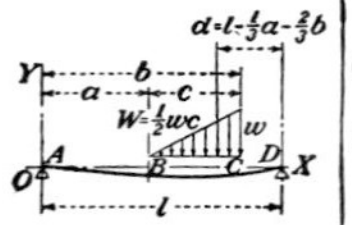	$R_1 = W\frac{d}{l}$ $R_2 = W\frac{l-d}{l}$ (A to B) $V = +R_1$ (B to C) $V = R_1 - \left(\frac{x-a}{c}\right)^2 W$ (C to D) $V = R_1 - W$	(A to B) $M = R_1x$ (B to C) $M = R_1x - W\frac{(x-a)^3}{3c^2}$ (C to D) $M = R_1x - \frac{1}{3}W(3x - a - 2b)$ Max. $M = W\frac{d}{l}\left(a + \frac{2}{3}c\sqrt{\frac{d}{l}}\right)$ at $x = a + c\sqrt{\frac{d}{l}}$	(A to B) $y = \frac{1}{6EI}\left\{R_1(x^3 - l^2x) + Wx\left[\frac{d^3}{l} + \frac{1}{6}c^2\left(1 - \frac{b}{l}\right) + \frac{17}{270}\frac{c^3}{l}\right]\right\}$ (B to C) $y = \frac{1}{6EI}\left[R_1(x^3 - l^2x) - \frac{1}{10}W\frac{(x-a)^5}{c^2} + Wx\left(\frac{d^3}{l} + \frac{1}{6}c^2 - \frac{1}{6}c^2\frac{b}{l} + \frac{17}{270}\frac{c^3}{l}\right)\right]$ (C to D) $y = \frac{1}{6EI}\left\{R_1(x^3 - l^2x) - W\left[d^3 - d^3\frac{x}{l} - \frac{1}{6}bc^2\left(1 - \frac{x}{l}\right) + \frac{17}{270}c^3\left(1 - \frac{x}{l}\right)\right]\right\}$ $\theta = \frac{1}{6EI}\left[-R_1l^2 + W\left(\frac{d^3}{l} + \frac{1}{6}c^2 + \frac{17}{270}\frac{c^3}{l} - \frac{1}{6}\frac{c^2b}{l}\right)\right]$ at A $\theta = \frac{1}{6EI}\left[2R_1l^2 + W\left(\frac{d^3}{l} + \frac{17}{270}\frac{c^3}{l} - \frac{1}{6}\frac{c^2b}{l} - 3d^2\right)\right]$ at D
17. End supports, triangular load 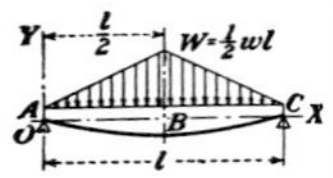	$R = \frac{1}{2}W$ $R_2 = \frac{1}{2}W$ (A to B) $V = \frac{1}{2}W\left(1 - 4\frac{x^2}{l^2}\right)$ (B to C) $V = -\frac{1}{2}W\left(1 - 4\frac{(l-x)^2}{l^2}\right)$	(A to B) $M = \frac{1}{6}W\left(3x - 4\frac{x^3}{l^2}\right)$ (B to C) $M = \frac{1}{6}W\left[3(l-x) - 4\frac{(l-x)^3}{l^2}\right]$ Max $M = \frac{1}{6}Wl$ at B	(A to B) $y = \frac{1}{6}\frac{Wx}{EIl^2}\left(\frac{1}{2}l^2x^2 - \frac{1}{5}x^4 - \frac{5}{16}l^4\right)$ Max $y = -\frac{1}{60}\frac{Wl^3}{EI}$ at B $\theta = -\frac{5}{96}\frac{Wl^2}{EI}$ at A; $\theta = +\frac{5}{96}\frac{Wl^2}{EI}$ at C
18. End supports, triangular load 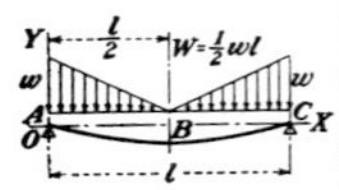	$R_1 = \frac{1}{2}W$ $R_2 = \frac{1}{2}W$ (A to B) $V = \frac{1}{2}W\left(\frac{l-2x}{l}\right)^2$ (B to C) $V = -\frac{1}{2}W\left(\frac{2x-l}{l}\right)^2$	(A to B) $M = \frac{1}{2}W\left(x - 2\frac{x^2}{l} + \frac{4}{3}\frac{x^3}{l^2}\right)$ (B to C) $M = \frac{1}{2}W\left[(l-x) - 2\frac{(l-x)^2}{l} + \frac{4}{3}\frac{(l-x)^3}{l^2}\right]$ Max $M = \frac{1}{12}Wl$ at B	(A to B) $y = \frac{1}{12}\frac{W}{EI}\left(x^3 - \frac{x^4}{l} + \frac{2}{5}\frac{x^5}{l^2} - \frac{3}{8}l^2x\right)$ Max $y = -\frac{3}{320}\frac{Wl^3}{EI}$ at B $\theta = -\frac{1}{32}\frac{Wl^2}{EI}$ at A; $\theta = +\frac{1}{32}\frac{Wl^2}{EI}$ at B
19. End supports, end couple 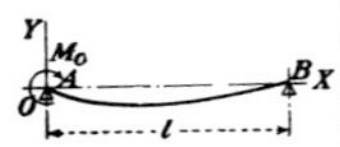	$R_1 = -\frac{M_0}{l}$ $R_2 = +\frac{M_0}{l}$ $V = R_1$	$M = M_0 + R_1x$ Max $M = M_0$ at A	$y = \frac{1}{6}\frac{M_0}{EI}\left(3x^2 - \frac{x^3}{l} - 2lx\right)$ Max $y = -0.0642\frac{M_0l^2}{EI}$ at $x = 0.422l$ $\theta = -\frac{1}{3}\frac{M_0l}{EI}$ at A $\theta = +\frac{1}{6}\frac{M_0l}{EI}$ at B
20. End supports, intermediate couple 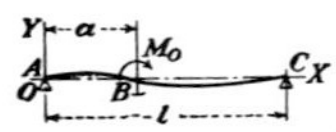	$R_1 = -\frac{M_0}{l}$ $R_2 = +\frac{M_0}{l}$ (A to C) $V = R_1$	(A to B) $M = R_1x$ (B to C) $M = R_1x + M_0$ Max $-M = R_1a$ just left of B Max $+M = R_1a + M_0$ just right of B	(A to B) $y = \frac{1}{6}\frac{M_0}{EI}\left[\left(6a - 3\frac{a^2}{l} - 2l\right)x - \frac{x^3}{l}\right]$ (B to C) $y = \frac{1}{6}\frac{M_0}{EI}\left[3a^2 + 3x^2 - \frac{x^3}{l} - \left(2l + 3\frac{a^2}{l}\right)x\right]$ $\theta = -\frac{1}{6}\frac{M_0}{EI}\left(2l - 6a + 3\frac{a^2}{l}\right)$ at A; $\theta = +\frac{1}{6}\frac{M_0}{EI}\left(l - 3\frac{a^2}{l}\right)$ at C $\theta = \frac{M_0}{EI}\left(a - \frac{a^2}{l} - \frac{1}{3}l\right)$ at B

TABLE 1-26

Formulas for Torsional Deformation and Stress

Roark "Formulas for Stress and Strain" McGraw-Hill

General formulas: $\theta = \frac{TL}{KG}$, $s = \frac{T}{Q}$, where θ = angle of twist (rad); T = twisting moment (in.-lb); L = length (in.); s = unit shear stress (lb. per sq. in.); G = modulus of rigidity (lb. per sq. in.); K (in.^{4}) and Q (in.^{3}) are functions of the cross section.

Form and dimensions of cross sections, other quantities involved, and case number	Formula for K in $\theta = \frac{TL}{KG}$	Formula for shear stress
1. Solid circular section 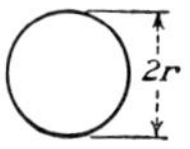	$K = \frac{1}{2}\pi r^4$	Max $s = \frac{2T}{\pi r^3}$ at boundary
2. Solid elliptical section 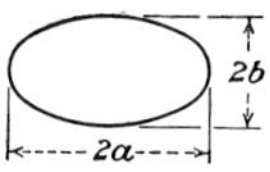	$K = \frac{\pi a^3 b^3}{a^2 + b^2}$	Max $s = \frac{2T}{\pi a b^2}$ at ends of minor axis
3. Solid square section 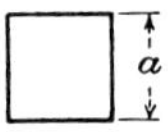	$K = 0.1406a^4$	Max $s = \frac{T}{0.208a^3}$ at mid-point of each side
4. Solid rectangular section 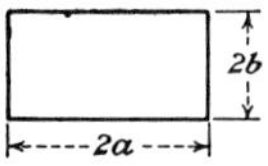	$K = ab^3\left[\frac{16}{3} - 3.36\frac{b}{a}\left(1 - \frac{b^4}{12a^4}\right)\right]$	Max $s = \frac{T(3a + 1.8b)}{8a^2b^2}$ at mid-point of each longer side
5. Solid triangular section (equilateral) 	$K = \frac{a^4\sqrt{3}}{80}$	Max $s = \frac{20T}{a^3}$ at mid-point of each side
6. **Hollow** concentric circular section 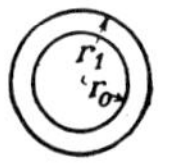	$K = \frac{1}{2}\pi(r_1^4 - r_0^4)$	Max $s = \frac{2Tr_1}{\pi(r_1^4 - r_0^4)}$ at outer boundary
7. Hollow elliptical section, outer and inner boundaries similar ellipses $q = \frac{a_0}{a} = \frac{b_0}{b}$ 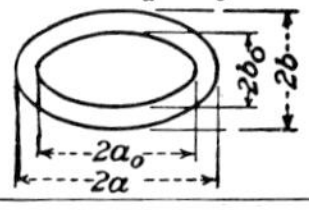	$K = \frac{\pi a^3 b^3}{a^2 + b^2}(1 - q^4)$	Max $s = \frac{2T}{\pi a b^2(1 - q^4)}$ at ends of minor axis on outer surface
8. **Hollow**, thin-walled elliptical section of uniform thickness. U = length of median boundary, shown dotted $U = \pi(a + b - t)\left[1 + 0.27\frac{(a-b)^2}{(a+b)^2}\right]$, approx. 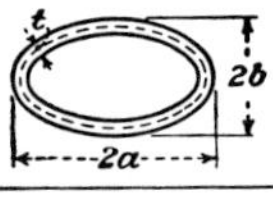	$K = \frac{4\pi^2 t[(a - \frac{1}{2}t)^2(b - \frac{1}{2}t)^2]}{U}$	Average $s = \frac{T}{2\pi t(a - \frac{1}{2}t)(b - \frac{1}{2}t)}$ (stress nearly uniform if t is small)

continued on next page

TABLE 1–26, continued

Form and dimensions of cross sections, other quantities involved, and case number	Formula for K in $\theta = \frac{TL}{KG}$	Formula for shear stress
9. Any thin tube of uniform thickness. U = length of median boundary, A = mean of areas enclosed by outer and inner boundaries, or (approx.) area within median boundary 	$K = \frac{4A^2t}{U}$	Average $s = \frac{T}{2tA}$ (stress nearly uniform if t is small)
10. Any thin tube. U and A as for Case 9; t = thickness at any point 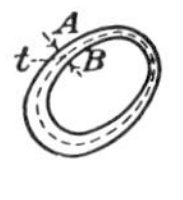	$K = \frac{4A^2}{\int \frac{dU}{t}}$	Average s on any thickness $AB = \frac{T}{2tA}$ (Max s where t is a minimum)
11. Hollow rectangle 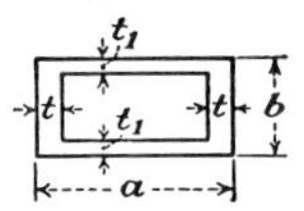	$K = \frac{2tt_1(a-t)^2(b-t_1)^2}{at + bt_1 - t^2 - t_1^2}$	Average $s = \frac{T}{2t(a-t)(b-t_1)}$ near mid-length of short sides Average $s = \frac{T}{2t_1(a-t)(b-t_1)}$ near mid-length of long sides (There will be higher stresses at inner corners unless fillets of fairly large radius are provided)
12. Thin circular open tube of uniform thickness. r = mean radius 	$K = \frac{2}{3}\pi r t^3$	Max $s = \frac{T(6\pi r + 1.8t)}{4\pi^2 r^2 t^2}$, along both edges remote from ends (this assumes t small compared with mean radius; otherwise use formulas given for Cases 14 to 20)
13. Any thin open tube of uniform thickness. U = length of median line, shown dotted 	$K = \frac{1}{3}Ut^3$	Max $s = \frac{T(3U + 1.8t)}{U^2t^2}$, along both edges remote from ends (this assumes t small compared with least radius of curvature of median line; otherwise use formulas given for Cases 14 to 20)
14. Any elongated section with axis of symmetry OX. U = length, A = area of section, I_x = moment of inertia about axis of symmetry. 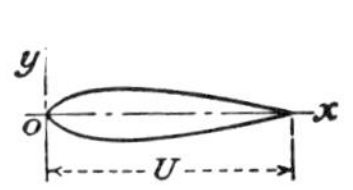	$K = \frac{4I_x}{\left(1 + 16\frac{I_x}{AU^2}\right)}$	For all solid sections of irregular form (Cases 14 to 20, inclusive) the max shear stress occurs at or very near one of the points where the largest inscribed circle touches the boundary,* and of these, at the one where the curvature of the boundary is algebraically least. (Convexity represents positive, concavity negative, curvature of the boundary.) At a point where the curvature is positive (boundary of section straight or convex) this max stress is given approximately by: $s = G\frac{\theta}{L}c$ or $s = \frac{T}{K}c$ where $c = \frac{D}{1 + \frac{\pi^2 D^4}{16A^2}}\left[1 + 0.15\left(\frac{\pi^2 D^4}{16A^2} - \frac{D}{2r}\right)\right]$, where D = diameter of largest inscribed circle r = radius of curvature of boundary at the point (positive for this case) A = area of the section At a point where the curvature is negative (boundary of section concave, or reentrant) this max stress is given approximately by $s = G\frac{\theta}{L}c$ or $s = \frac{T}{K}c$ where $c = \frac{D}{1 + \frac{\pi^2 D^4}{16A^2}}\left[1 + \left\{0.118 \log_e\left(1 - \frac{D}{2r}\right) - 0.238\frac{D}{2r}\right\} \tanh \frac{2\phi}{\pi}\right]$ where D, A, and r have same meaning as before and ϕ = angle through which a tangent to the boundary rotates in turning or traveling around the reentrant portion, measured in radians. (Here r is *negative.*) The above formulas should also be used for Cases 12 and 13 when t is relatively large compared with radius of median line * Unless at some other point on boundary there is **a sharp reentrant angle,** causing high local stress.
15. Any elongated section or thin open tube. dU = elementary length along median line, t = thickness normal to median line, A = area of section 	$K = \frac{\frac{1}{3}F}{\left(1 + \frac{4}{3}\frac{F}{AU^2}\right)}$ where $F = \int_0^U t^3\, dU$	
16. Any solid, fairly compact section without reentrant angles. J = polar moment of inertia about centroidal axis; A = area of section 	$K = \frac{A^4}{40J}$	

continued on next page

Form and dimensions of cross sections, other quantities involved, and case number	Formula for K in $\theta = \dfrac{TL}{KG}$	Formula for shear stress
17. I section, flange thickness uniform. r = fillet radius, D = diameter largest inscribed circle, $t = b$ if $b < d$, $t = d$ if $d < b$, $t_1 = b$ if $b > d$, $t_1 = d$ if $d > b$ 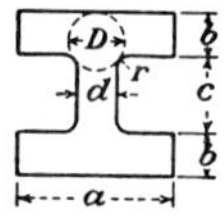	$K = 2K_1 + K_2 + 2\alpha D^4$ where $K_1 = ab^3\left[\frac{1}{3} - 0.21\frac{b}{a}\left(1 - \frac{b^4}{12a^4}\right)\right]$ $K_2 = \frac{1}{3}cd^3$ $\alpha = \frac{t}{t_1}\left(0.15 + 0.1\frac{r}{b}\right)$	
18. T section, flange thickness uniform; r, D, t and t_1 as for Case 17 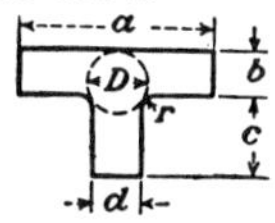	$K = K_1 + K_2 + \alpha D^4$ where $K_1 = ab^3\left[\frac{1}{3} - 0.21\frac{b}{a}\left(1 - \frac{b^4}{12a^4}\right)\right]$ $K_2 = cd^3\left[\frac{1}{3} - 0.105\frac{d}{c}\left(1 - \frac{d^4}{192c^4}\right)\right]$ $\alpha = \frac{t}{t_1}\left(0.15 + 0.10\frac{r}{b}\right)$	
19. L section, r and D as for Cases 17 and 18, $b \gtreqless d$ 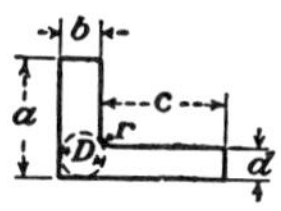	$K = K_1 + K_2 + \alpha D^4$ where $K_1 = ab^3\left[\frac{1}{3} - 0.21\frac{b}{a}\left(1 - \frac{b^4}{12a^4}\right)\right]$ $K_2 = cd^3\left[\frac{1}{3} - 0.105\frac{d}{c}\left(1 - \frac{d^4}{192c^4}\right)\right]$ $\alpha = \frac{b}{d}\left(0.07 + 0.076\frac{r}{b}\right)$	
20. U section or Z section 	K = sum of K's of constituent L sections, computed as for Case 19	
21. Eccentric hollow circular section 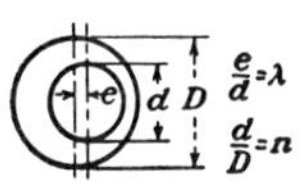 $\frac{e}{d} = \lambda$, $\frac{d}{D} = n$	$K = \pi(D^4 - d^4)/32Q$ where $Q = 1 + \left[\frac{16n^2}{(1-n^2)(1-n^4)}\right]\lambda^2 + \left[\frac{384n^4}{(1-n^2)^2(1-n^4)^4}\right]\lambda^4$	Max $S = 16TDF/\pi(D^4 - d^4)$ where $F = 1 + \left[\frac{4n^2}{1-n^2}\right]\lambda + \left[\frac{32n^2}{(1-n^2)(1-n^4)}\right]\lambda^2 + \left[\frac{48n^2(1+2n^2+3n^4+2n^6)}{(1-n^2)(1-n^4)(1-n^6)}\right]\lambda^3 + \left[\frac{64n^2(2+12n^2+19n^4+28n^6+18n^8+14n^{10}+3n^{12})}{(1-n^2)(1-n^4)(1-n^6)(1-n^8)}\right]\lambda^4$ (Ref. 10) Reference: Wilson, T.S.; "The Eccentric Circular Tube," Aircraft Eng., Vol. 14, No. 157, March, 1942

TABLE 1-27

Formulas for Stress and Strain Caused by Pressure On or Between Elastic Bodies

Roark "Formulas for Stress and Strain" McGraw-Hill

Notation: s_c = unit compressive stress; s_s = unit shear stress; s_t = unit tensile stress; a = radius of circular contact area for cases 1, 2, and 3; b = width of rectangular contact area for cases 4, 5, and 6; c = major semiaxis and d = minor semiaxis of elliptical contact area for cases 7 and 8; y = combined deformation of both bodies at each contact, along axis of load; ν = Poisson's ratio; E = modulus of elasticity. Subscripts 1 and 2 refer to bodies 1 and 2, respectively. All dimensions in inches, all forces in pounds.

Conditions and Case No.	Formulas for dimensions of contact area and for a maximum stress
1. Sphere on a flat plate. P = total load 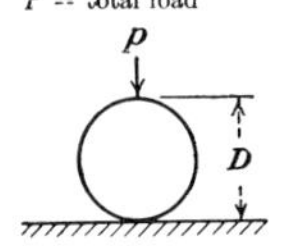	$a = 0.721\sqrt[3]{PD\left[\frac{1-\nu_1^2}{E_1}+\frac{1-\nu_2^2}{E_2}\right]}$ Max $s_c = 0.918\sqrt[3]{\frac{P}{D^2\left[\frac{1-\nu_1^2}{E_1}+\frac{1-\nu_2^2}{E_2}\right]^2}}$ If $E_1 = E_2 = E$ and $\nu_1 = \nu_2 = 0.3$, $a = 0.881\sqrt[3]{\frac{PD}{E}}$, Max $s_c = 0.616\sqrt[3]{\frac{PE^2}{D^2}}$, Max $s_t = 0.133$ (Max s_c), $y = 1.55\sqrt[3]{\frac{P^2}{E^2D}}$ Max $s_s = \frac{1}{3}$(Max s_c), at depth $\frac{1}{4}D$ below surface of plate (approximate values, from Refs. 3 and 6) Ref.: Timoshenko, S.: "Theory of Elasticity" Engineering Societies Monograph, McGraw-Hill, 1934
2. Sphere on a sphere. P = total load 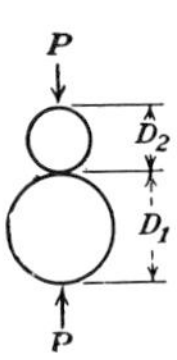	$a = 0.721\sqrt[3]{P\left(\frac{D_1D_2}{D_1+D_2}\right)\left[\frac{1-\nu_1^2}{E_1}+\frac{1-\nu_2^2}{E_2}\right]}$ Max $s_c = 0.918\sqrt[3]{P\frac{\left(\frac{D_1+D_2}{D_1D_2}\right)^2}{\left[\frac{1-\nu_1^2}{E_1}+\frac{1-\nu_2^2}{E_2}\right]^2}}$, $y = 1.04\sqrt[3]{\frac{P^2(D_1+D_2)}{D_1D_2}\left(\frac{1-\nu_1^2}{E_1}+\frac{1-\nu_2^2}{E_2}\right)^2}$ If $E_1 = E_2 = E$ and $\nu_1 = \nu_2 = 0.3$, $a = 0.881\sqrt[3]{\frac{P}{E}\frac{D_1D_2}{D_1+D_2}}$, Max $s_c = 0.616\sqrt[3]{PE^2\left(\frac{D_1+D_2}{D_1D_2}\right)^2}$, Max $s_s = \frac{1}{3}$ (Max s_c), Max $s_t = 0.133$ (Max s_c), $y = 1.55\sqrt[3]{\frac{P^2}{E^2}\frac{(D_1+D_2)}{D_1D_2}}$ Ref.: Timoshenko, S.: "Theory of Elasticity" Engineering Societies Monograph, McGraw-Hill, 1934
3. Sphere in spherical socket. P = total load 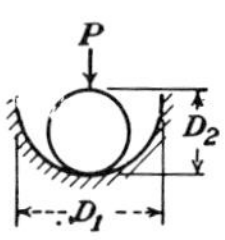	$a = 0.721\sqrt[3]{P\frac{D_1D_2}{D_1-D_2}\left[\frac{1-\nu_1^2}{E_1}+\frac{1-\nu_2^2}{E_2}\right]}$ Max $s_c = 0.918\sqrt[3]{P\frac{\left(\frac{D_1-D_2}{D_1D_2}\right)^2}{\left[\frac{1-\nu_1^2}{E_1}+\frac{1-\nu_2^2}{E_2}\right]^2}}$ If $E_1 = E_2 = E$ and $\nu_1 = \nu_2 = 0.3$, $a = 0.881\sqrt[3]{\frac{P}{E}\frac{D_1D_2}{D_1-D_2}}$, Max $s_c = 0.616\sqrt[3]{PE^2\left(\frac{D_1-D_2}{D_1D_2}\right)^2}$, Max $s_s = \frac{1}{3}$ (Max s_c), Max $s_t = 0.133$ (Max s_c), $y = 1.55\sqrt[3]{\frac{P^2}{E^2}\frac{(D_1-D_2)}{D_1D_2}}$
4. Cylinder on flat plate. p = load per linear in. 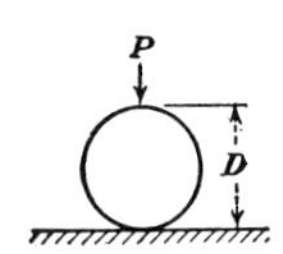	$b = 1.6\sqrt{pD\left[\frac{1-\nu_1^2}{E_1}+\frac{1-\nu_2^2}{E_2}\right]}$ Max $s_c = 0.798\sqrt{\frac{p}{D\left[\frac{1-\nu_1^2}{E_1}+\frac{1-\nu_2^2}{E_2}\right]}}$ Total compression of cylinder between two plates is: $\Delta D = 4p\left(\frac{1-\nu^2}{\pi E}\right)\left(\frac{1}{3}+\log_e\frac{2D}{b}\right)$ If $E_1 = E_2 = E$ and $\nu_1 = \nu_2 = 0.3$, $b = 2.15\sqrt{\frac{pD}{E}}$, Max $s_c = 0.591\sqrt{\frac{pE}{D}}$ For $E = 30{,}000{,}000$, $\nu_1 = \nu_2 = 0.25$, $b = 0.0004\sqrt{pD}$, Max $s_c = 3190\sqrt{\frac{p}{D}}$, Max $s_s = 958\sqrt{\frac{p}{D}}$ at depth $0.393b$ below surface of plane (Approximate formula from Thomas, H. R., and V. A. Hoersch: Stresses Due to the Pressure of one Elastic Solid Upon Another. Eng. Exp. Sta.. Univ. Ill. Bull. 212, 1930) Ref. : Föppl, A.: "Technische Mechanik" 4th ed., Vol. 5, p 350
5. Cylinder on cylinder. Axes parallel. p = load per linear in. 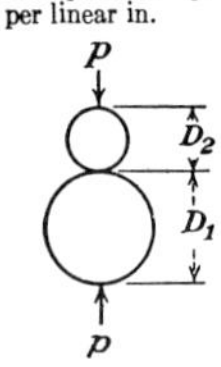	$b = 1.6\sqrt{p\frac{D_1D_2}{D_1+D_2}\left[\frac{1-\nu_1^2}{E_1}+\frac{1-\nu_2^2}{E_2}\right]}$ Max $s_c = 0.798\sqrt{\frac{p\frac{D_1+D_2}{D_1D_2}}{\left[\frac{1-\nu_1^2}{E_1}+\frac{1-\nu_2^2}{E_2}\right]}}$ If $E_1 = E_2 = E$ and $\nu_1 = \nu_2 = 0.3$, $b = 2.15\sqrt{\frac{p}{E}\frac{D_1D_2}{D_1+D_2}}$, Max $s_c = 0.591\sqrt{pE\frac{D_1+D_2}{D_1D_2}}$, $y = \frac{2(1-\nu^2)}{E}\frac{p}{\pi}\left(\frac{2}{3}+\log_e\frac{2D_1}{b}+\log_e\frac{2D_2}{b}\right)$
6. Cylinder in circular groove. p = load per linear in. 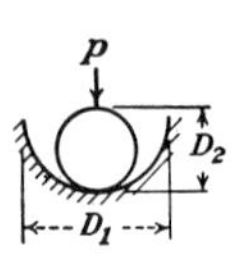	$b = 1.6\sqrt{p\frac{D_1D_2}{D_1-D_2}\left[\frac{1-\nu_1^2}{E_1}+\frac{1-\nu_2^2}{E_2}\right]}$ Max $s_c = 0.798\sqrt{\frac{p\frac{D_1-D_2}{D_1D_2}}{\left[\frac{1-\nu_1^2}{E_1}+\frac{1-\nu_2^2}{E_2}\right]}}$ If $E_1 = E_2 = E$ and $\nu_1 = \nu_2 = 0.3$, $b = 2.15\sqrt{\frac{p}{E}\frac{D_1D_2}{D_1-D_2}}$, Max $s_c = 0.591\sqrt{pE\frac{D_1-D_2}{D_1D_2}}$

continued on next page

Table 1–27, continued

Conditions and Case No.	Formulas for dimensions of contact area and for a maximum stress

7. Cylinder on cylinder. Axes at right angles. P = total load

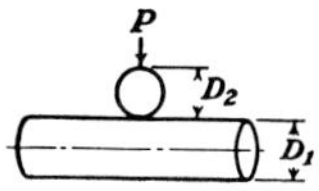

$$c = \alpha \sqrt[3]{P \frac{D_1 D_2}{D_1 + D_2}\left[\frac{1-\nu_1^2}{E_1} + \frac{1-\nu_2^2}{E_2}\right]},\quad d = \beta c,\quad \text{Max } s_c = \frac{1.5P}{\pi c d} \qquad y = \lambda \sqrt[3]{\frac{P^2}{\left(\frac{E_1}{1-\nu_1^2} + \frac{E_2}{1-\nu_2^2}\right)^2} \frac{(D_1 + D_2)}{D_1 D_2}}$$

where α and β and λ depend on ratio $\frac{D_1}{D_2}$ and have values as follows:

$\frac{D_1}{D_2}$	1	$1\frac{1}{2}$	2	3	4	6	10
α	0.908	1.045	1.158	1.350	1.505	1.767	2.175
β	1	0.765	0.632	0.482	0.400	0.308	0.221
λ	2.080	2.060	2.025	1.950	1.875	1.770	1.613

If $E_1 = E_2 = 30{,}000{,}000$, $\nu_1 = \nu_2 = 0.25$, $c = 0.00397\alpha \sqrt[3]{P \frac{D_1 D_2}{D_1 + D_2}}$

For these values of E and ν and for values of $\frac{D_1}{D_2}$ between 1 and 8, $\text{Max } s_c = \frac{11{,}750}{\left(\frac{R_1}{R_2}\right)^{0.271}} \sqrt[3]{\frac{P}{R_2^2}}$, where $R_1 = \frac{1}{2}D_1$, $R_2 = \frac{1}{2}D_2$

(Approximate formula from Thomas, H. R., and V. A. Hoersch: Stresses Due to the Pressure of one Elastic Solid Upon Another, Eng. Exp. Sta., Univ. Ill. Bull. 212, 1930)

8. General case of two bodies in contact. P = total pressure

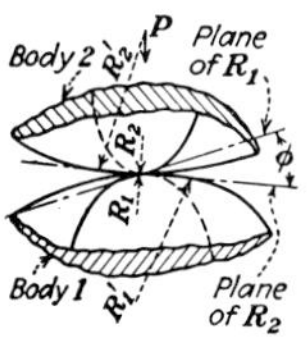

At point of contact minimum and maximum radii of curvature are R_1 and R_1' for Body 1, R_2 and R_2' for Body 2. Then $\frac{1}{R_1}$ and $\frac{1}{R_1'}$ are *principal curvatures* of Body 1, and $\frac{1}{R_2}$ and $\frac{1}{R_2'}$ of Body 2, and in each body the principal curvatures are mutually perpendicular. The plane containing curvature $\frac{1}{R_1}$ in Body 1 makes with the plane containing curvature $\frac{1}{R_2}$ in Body 2 the angle ϕ. Then:

$$\text{Max } s_c = \frac{1.5P}{\pi c d},\quad c = \alpha \sqrt[3]{\frac{P\delta}{K}},\quad d = \beta \sqrt[3]{\frac{P\delta}{K}},\ \text{and } y = \lambda \sqrt[3]{\frac{P^2}{K^2\delta}},\ \text{where } \delta = \frac{4}{\frac{1}{R_1} + \frac{1}{R_2} + \frac{1}{R_1'} + \frac{1}{R_2'}} \text{ and } K = \frac{8}{3} \frac{E_1 E_2}{E_2(1-\nu_1^2) + E_1(1-\nu_2^2)}$$

α and β are given by the following table, where $\theta = \text{arc cos } \frac{1}{4}\delta \sqrt{\left(\frac{1}{R_1} - \frac{1}{R_1'}\right)^2 + \left(\frac{1}{R_2} - \frac{1}{R_2'}\right)^2 + 2\left(\frac{1}{R_1} - \frac{1}{R_1'}\right)\left(\frac{1}{R_2} - \frac{1}{R_2'}\right) \cos 2\phi}$

θ	0°	10°	20°	30°	35°	40°	45°	50°	55°	60°	65°	70°	75°	80°	85°	90°
α	∞	6.612	3.778	2.731	2.397	2.136	1.926	1.754	1.611	1.486	1.378	1.284	1.202	1.128	1.061	1.00
β	0	0.319	0.408	0.493	0.530	0.567	0.604	0.641	0.678	0.717	0.759	0.802	0.846	0.893	0.944	1.00
λ	—	0.851	1.220	1.453	1.550	1.637	1.709	1.772	1.828	1.875	1.912	1.944	1.967	1.985	1.996	2.00

Values taken from Tech. Paper, Bureau of Standards, No. 201, 1921

9. Rigid knife-edge across edge of semi-infinite plate. Load p lb. per linear in.

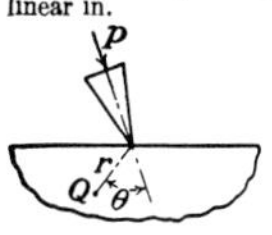

At any point Q, $s_c = \frac{2p \cos \theta}{\pi r}$

Ref.: Timoshenko, S.: "Theory of Elasticity" Engineering Societies Monograph, McGraw-Hill, 1934

10. Rigid block of width $2b$ across edge of semi-infinite plate. Load p lb. per linear in.

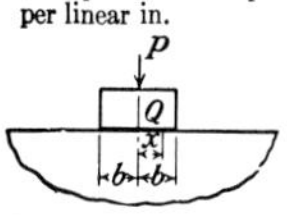

At any point Q on surface of contact, $s_c = \frac{p}{\pi \sqrt{b^2 - x^2}}$

Ref.: Timoshenko, S.: "Theory of Elasticity" Engineering Societies Monograph, McGraw-Hill, 1934

11. Uniform pressure p lb. per sq. in. over length L across edge of semi-infinite plate

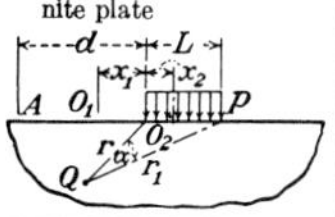

At any point O_1 outside loaded area, $y = \frac{2p}{\pi E}\left[(L + x_1)\log_e \frac{d}{L + x_1} - x_1 \log_e \frac{d}{x_1}\right] + pL\left(\frac{1-\nu}{\pi E}\right)$

At any point O_2 inside loaded area, $y = \frac{2p}{\pi E}\left[(L - x_2)\log_e \frac{d}{L - x_2} + x_2 \log_e \frac{d}{x_2}\right] + pL\left(\frac{1-\nu}{\pi E}\right)$

Where y = deflection relative to a remote point A distant d from edge of loaded area

At any point Q, $S_c = 0.318\, p(\alpha + \sin \alpha)$

$S_s = 0.318\, p \sin \alpha$

Ref.: Timoshenko, S.: "Theory of Elasticity" Engineering Societies Monograph, McGraw-Hill, 1934

12. Rigid cylindrical die of radius R on surface of semi-infinite body, total load P lb.

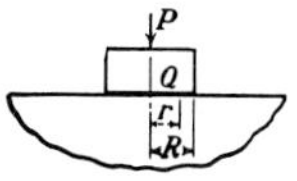

$y = \frac{P(1-\nu^2)}{2RE}$

At any point Q on surface of contact $s_c = \frac{P}{2\pi R \sqrt{R^2 - r^2}}$

Max $s_c = \infty$ at edge

Min $s_c = \frac{P}{2\pi R^2}$ at center

Ref.: Timoshenko, S.: "Theory of Elasticity" Engineering Societies Monograph, McGraw-Hill, 1934

continued on next page

Conditions and Case No.	Formulas for dimensions of contact area and for a maximum stress
13. Uniform pressure p lb. per sq. in. over circular area of radius R on surface of semi-infinite body 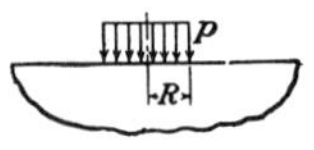	Max $y = \frac{2pR(1 - \nu^2)}{E}$ at center y at edge $= \frac{4pR(1 - \nu^2)}{\pi E}$ Max $s_s = 0.33\ p$ at point $0.638R$ below center of loaded area
14. Uniform pressure p lb. per sq. in. over square area of sides $2b$ on surface of semi-infinite body 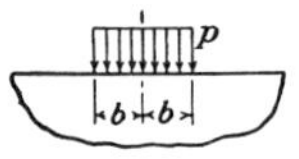	Max $y = \frac{2.24pb(1 - \nu^2)}{E}$ at center $y = \frac{1.12pb(1 - \nu^2)}{E}$ at corners Average $y = \frac{1.90pb(1 - \nu^2)}{E}$ Ref.: Timoshenko, S.: "Theory of Elasticity" Engineering Societies Monograph, McGraw-Hill, 1934

TABLE 1-28

Formulas for Computing Pound-Feet-Square Magnitudes of Certain Solids

Product Engineering, March, 1948

Part	Radius of Gyration R, Feet	Weight W, lb. w = Weight per Cu In. of Material, lb.	WR^2 lb-ft^2
CIRCULAR CYLINDER 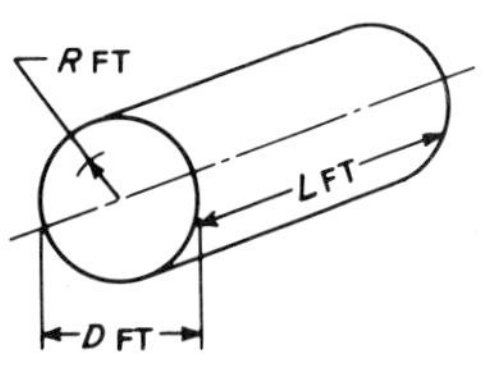	$0.354\ D$	$1.360\ w\,L\,D^2$	$170.4\ w\,L\,D^4$
HOLLOW CIRCULAR CYLINDER 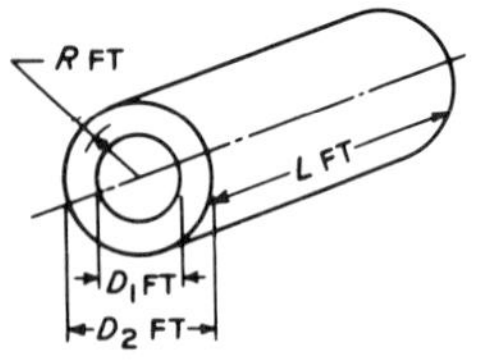	$0.354\sqrt{D_2^2 + D_1^2}$	$1{,}360\ wL(D_2^2 - D_1^2)$	$170.4\,wL(D_2^4 - D_1^4)$

continued on next page

Part	Radius of Gyration R, Feet	Weight W, lb. w Weight per Cu In. of Material, lb.	WR^2 lb-ft^2
ELLIPTICAL CYLINDER 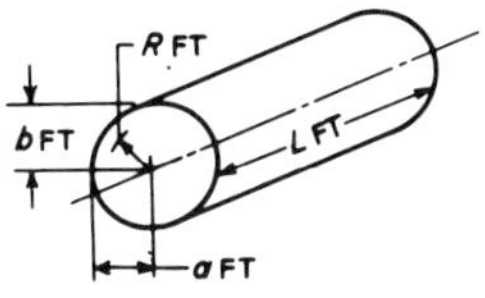	$\sqrt{\frac{a^2+b^2}{2}}$	$5{,}426\, ab\, wL$	$5{,}426\, ab\, wl \left(\frac{a^2+b^2}{2}\right)$
CIRCULAR CONE 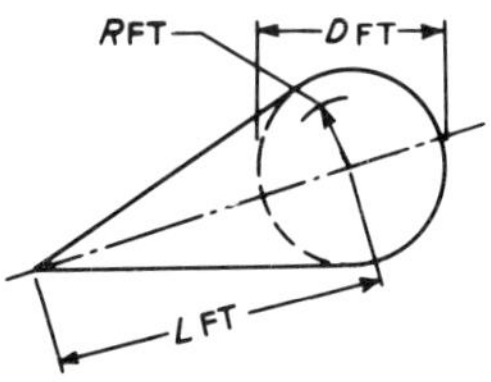	$\frac{D}{3.66}$	$452\, wD^2 L$	$33.7\, wD^4 L$
FRUSTUM OF CONE 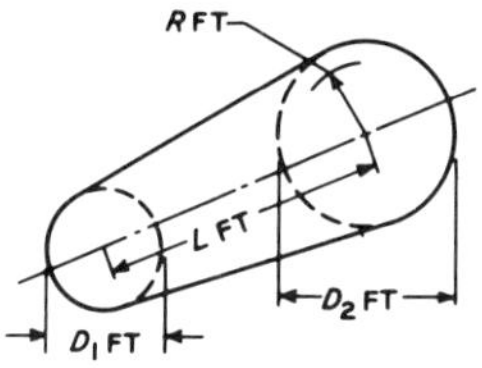	$0.275\sqrt{\frac{D_2^5 - D_1^5}{D_2^3 - D_1^3}}$	$[453\,(D_2^2 + D_1^2) + 511\sqrt{D_2^2\, D_1^2}]\, wL$	$\left(\frac{D_2^5 - D_1^5}{D_2^3 - D_1^3}\right)\left[34(D_2^2 + D_1^2) + 38.3\sqrt{D_2^2\, D_1^2}\right] wL$
TORUS 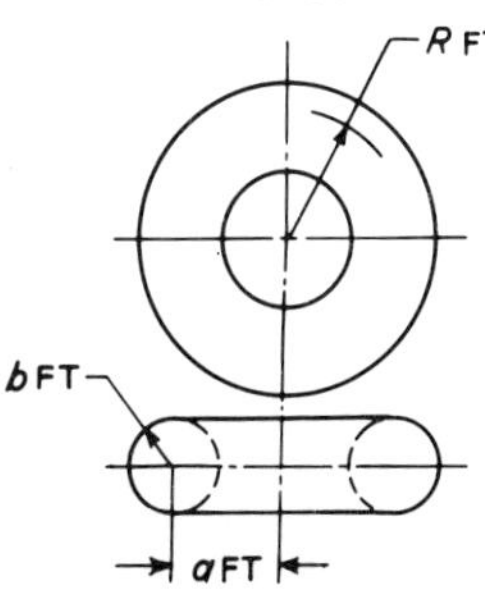	$0.5\sqrt{4a^2 + 3b^2}$	$33{,}938\, wab^2$	$8{,}485\, wab^2(4a^2 + 3b^2)$

TABLE 1-29

Work and (Mechanical*) Energy

Quantity	Definition of Quantity	Units	
Work	A scalar quantity - the scalar product of a vector force and a vector displacement	foot-pounds	Product of the magnitude of a force and the distance moved in the direction of the force.
Potential Energy	Capacity of a body for doing work as the consequence of position	Wh, foot pounds or $\frac{M}{g}h$	W = weight in pounds h = height above datum plane, feet M = mass g = force of gravity
Kinetic Energy	Capacity for doing work by reason of motion of a body	$1/2\, M\, V^2$, foot-pounds	$M = W/g$ V = speed in feet per second
Kinetic Energy of Rotation	Work done by torque about a fixed axis.	$1/2\, I\, \omega^2$, foot-pounds $\frac{WR^2 N^2}{5872}$ ft lb	I = mass moment of inertia of body, pound-feet-square ω = angular speed, radians per second W = weight or part, pounds R = Radius of gyration of part, feet N = revolutions per min. g = 32.2 fps^2
Strain Energy	Capacity for doing work because of elastic properties of a body	$1/2\, Pe$, inch-pounds $1/2\, (S^2/E)\, al$, inch-pounds **For rod of uniform section in tension**	P = axial tensile load, pounds e = elongation, inches S = axial tensile stress, psi a = cross sectional area, sq. in. l = length of bar
Power	Time rate of doing work	1 Horsepower = 550 ft lb per second = 33000 ft lb per minute	
Speed ratio of a mechanism	Ratio of distance (or angle) moved by load to distance (or angle) moved by effort.	ω_G / ω_P	ω_G= angular speed of **driven** ω_P= angular speed of driver
Mechanical Advantage	Ratio of magnitude of load to magnitude of effort.		

* Many other forms of energy, as atomic, chemical, electrical, heat, light are excluded by this restriction to mechanical forms.

TABLE 1-30

Amplitude Ratio Versus Frequency Ratio

Den Hartog, "Mechanical Vibrations" Third Edition. McGraw-Hill Book Co.

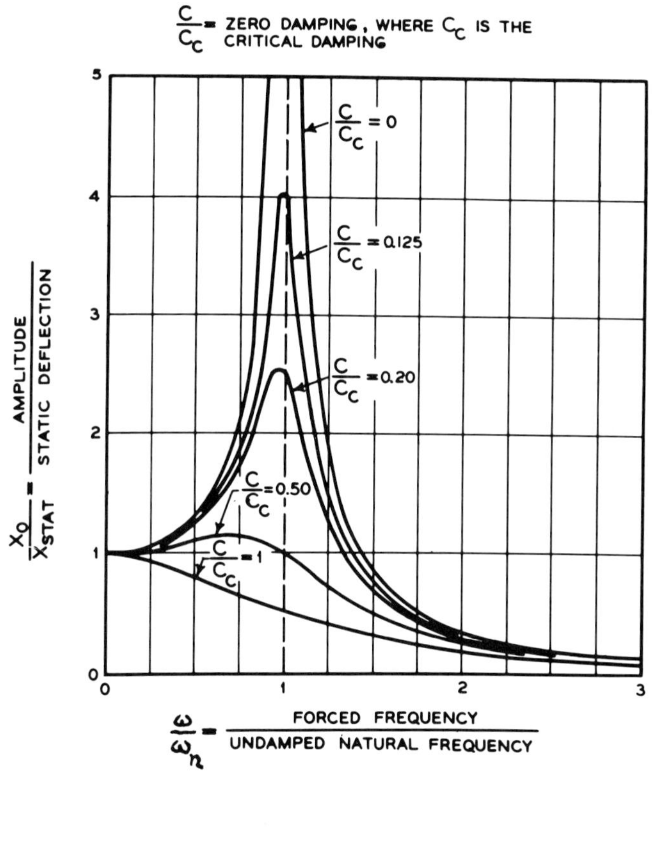

TABLE 1–31

Relationship Between Static Deflection and Natural Frequency

Bulletin No. 103 Lord Mfg. Co.

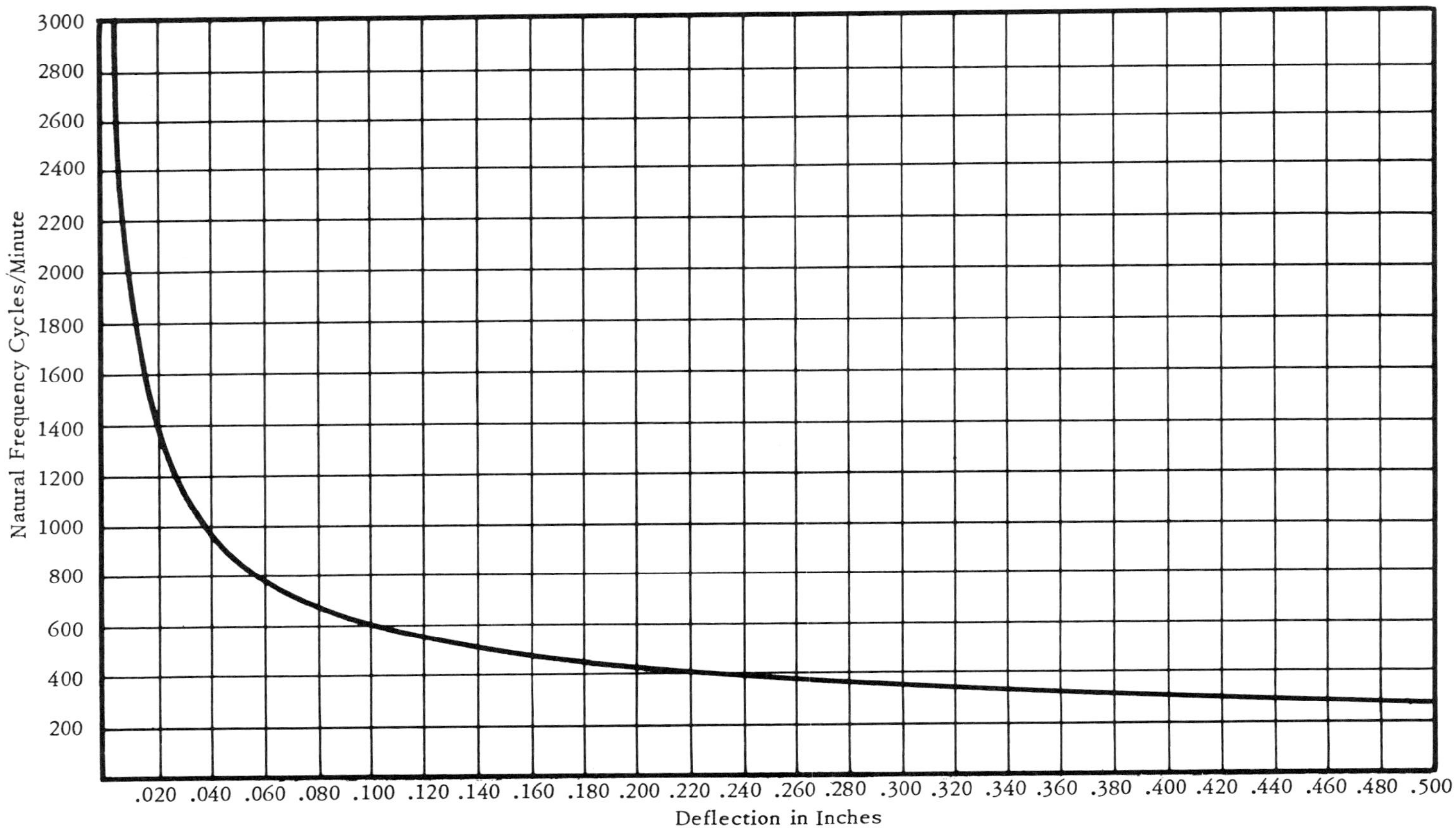

Formula: $f_n = 9.55\sqrt{\frac{g}{d}}$

where f_n = the natural frequency of the mounting system in cycles per minute

g = acceleration of gravity, 386.4 inches per second per second

d = deflection in inches

TABLE 1–32

Vibration Formulas

I. Linear Spring Constants (pounds per inch deflection)

"Mechanical Vibrations," Third Edition, Den Hartog, McGraw-Hill Book Co.

Coil dia. D; wire dia. d; n turns $\quad k = \dfrac{Gd^4}{8nD^3}$ (1)

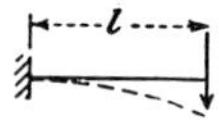
Cantilever $\quad k = \dfrac{3EI}{l^3}$ (2)

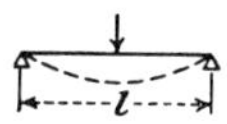
Beam on two supports; centrally loaded $\quad k = \dfrac{48EI}{l^3}$ (3)

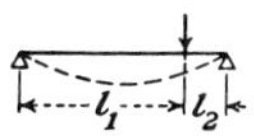
Beam on two supports; load off center $\quad k = \dfrac{3EIl}{l_1^2 l_2^2}$ (4)

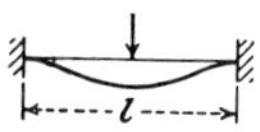
Clamped-clamped beam; centrally loaded $\quad k = \dfrac{192EI}{l^3}$ (5)

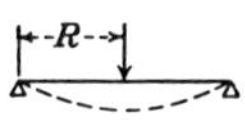
Circular plate, thickness t; centrally loaded; circumferential edge simply supported $\quad k = \dfrac{16\pi D}{R^2}\,\dfrac{1+\mu}{3+\mu}$ (6)

in which the plate constant is

$$D = \frac{Et^3}{12(1-\mu^2)} \qquad (6a)$$

μ = Poisson's ratio ≈ 0.3

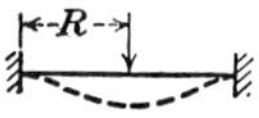
Circular plate; circumferential edge clamped $\quad k = \dfrac{16\pi D}{R^2}$ (7)

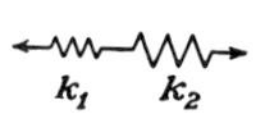
Two springs in series $\quad k = \dfrac{1}{1/k_1 + 1/k_2}$ (8)

II. Rotational Spring Constants (inch-pounds torque per radian rotation)

"Mechanical Vibrations," Third Edition, Den Hartog, McGraw-Hill Book Co.

(Inch-pounds torque per radian rotation)

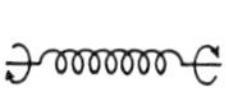
Twist of coil spring; wire dia. d; coil dia. D; n turns $\quad k = \dfrac{Ed^4}{64nD}$ (9)

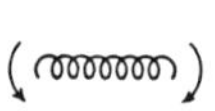
Bending of coil spring $\quad k = \dfrac{Ed^4}{32nD} \cdot \dfrac{1}{1 + E/2G}$ (10)

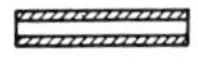
Twist of hollow circular shaft, outer dia. D, inner dia. d, length l $\quad k = \dfrac{GI_p}{l} = \dfrac{\pi}{32}\,\dfrac{G(D^4 - d^4)}{l}$ (11)

For steel $\quad k = 1.18 \times 10^6 \times \dfrac{D^4 - d^4}{l}$

continued on next page

III. Natural Frequencies of Simple Systems

"Mechanical Vibrations," Third Edition, Den Hartog, McGraw-Hill Book Co.

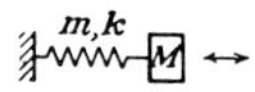

End mass M; spring mass m, spring stiffness k

$$\omega_n = \sqrt{k/(M + m/3)} \quad (12)$$

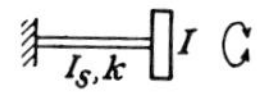

End inertia I; shaft inertia I_s, shaft stiffness k

$$\omega_n = \sqrt{k/(I + I_s/3)} \quad (13)$$

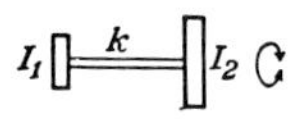

Two disks on a shaft

$$\omega_n = \sqrt{\frac{k(I_1 + I_2)}{I_1 I_2}} \quad (14)$$

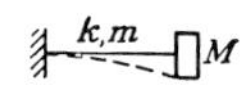

Cantilever; end mass M; beam mass m, stiffness by formula (2)

$$\omega_n = \sqrt{\frac{k}{M + 0.23m}} \quad (15)$$

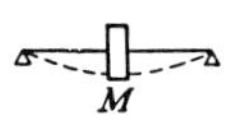

Simply supported beam; central mass M; beam mass m; stiffness by formula (3)

$$\omega_n = \sqrt{\frac{k}{M + 0.5m}} \quad (16)$$

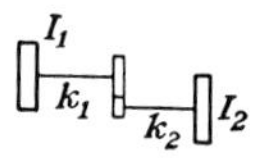

Massless gears, speed of I_2 n times as large as speed of I_1

$$\omega_n = \sqrt{\frac{1}{\frac{1}{k_1} + \frac{1}{n^2 k_2}} \times \frac{I_1 + n^2 I_2}{I_1 \cdot n^2 I_2}} \quad (17)$$

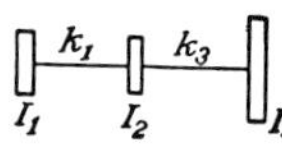

$$\omega_n^2 = \frac{1}{2}\left(\frac{k_1}{I_1} + \frac{k_3}{I_3} + \frac{k_1 + k_3}{I_2}\right) \pm \frac{1}{2}\sqrt{\left(\frac{k_1}{I_1} + \frac{k_3}{I_3} + \frac{k_1 + k_3}{I_2}\right)^2 - 4\frac{k_1 k_3}{I_1 I_2 I_3}(I_1 + I_2 + I_3)} \quad (18)$$

IV. Uniform Beams (longitudinal and torsional vibration)

"Mechanical Vibrations," Third Edition, Den Hartog, McGraw-Hill Book Co.

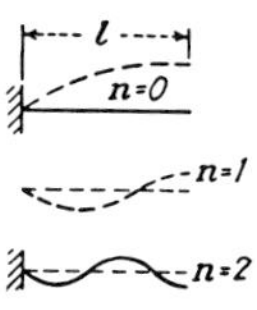

Longitudinal vibration of cantilever: A = cross section, E = modulus of elasticity. μ_1 = mass per unit length, n = 0,1,2,3 = number of nodes

$$\omega_n = \left(n + \frac{1}{2}\right)\pi\sqrt{\frac{AE}{\mu_1 l^2}} \quad (19)$$

For steel and l in inches this becomes

$$f = \frac{\omega_n}{2\pi} = (1 + 2n)\frac{51{,}000}{l}$$

cycles per second (19*a*)

For air at atm. pressure, l in inches:

Organ pipe open at one end, closed at the other

$$f = \frac{\omega_n}{2\pi} = (1 + 2n)\frac{3{,}300}{l}$$

cycles per second (19*b*)

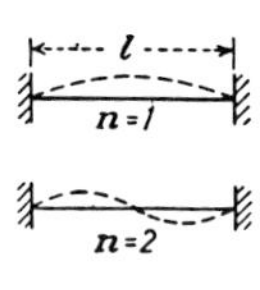

Longitudinal vibration of beam clamped at both ends; n = number of half waves along length

$$\omega_n = n\pi\sqrt{\frac{AE}{\mu_1 l^2}} \quad (20)$$

For steel, l in inches:

$$f = \frac{\omega_n}{2\pi} = \frac{102{,}000}{l}$$

cycles per second (20*a*)

Organ pipe closed at both ends (air)

$$f = \frac{\omega_n}{2\pi} = \frac{6{,}600}{l}$$

cycles per second (20*b*)

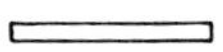

Torsional vibration of beams

Same as (19) and (20); replace tensional stiffness AE by torsional stiffness GI_p; replace μ_1 by the moment of inertia per unit length $i_1 = I_{bar}/l$.

continued on next page

V. Uniform Beams (transverse or bending vibrations)

"Mechanical Vibrations," Third Edition, Den Hartog, McGraw-Hill Book Co.

The same general formula holds for all the following cases,

$$\omega_n = a_n\sqrt{\frac{EI}{\mu_1 l^4}} \tag{21}$$

where EI is the bending stiffness of the section, l is the length of the beam, μ_1 is the mass per unit length $= W/gl$, and a_n is a numerical constant, different for each case and listed below

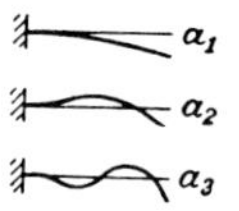

Cantilever or "clamped-free" beam

$a_1 = 3.52$
$a_2 = 22.4$
$a_3 = 61.7$
$a_4 = 121.0$
$a_5 = 200.0$

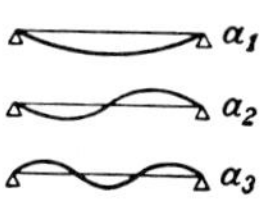

Simply supported or "hinged-hinged" beam

$a_1 = \pi^2 = 9.87$
$a_2 = 4\pi^2 = 39.5$
$a_3 = 9\pi^2 = 88.9$
$a_4 = 16\pi^2 = 158.$
$a_5 = 25\pi^2 = 247.$

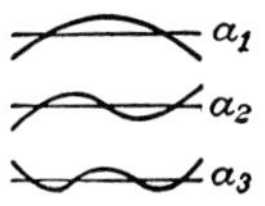

"Free-free" beam or floating ship

$a_1 = 22.4$
$a_2 = 61.7$
$a_3 = 121.0$
$a_4 = 200.0$
$a_5 = 298.2$

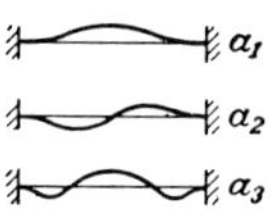

"Clamped-clamped" beam has same frequencies as "free-free"

$a_1 = 22.4$
$a_2 = 61.7$
$a_3 = 121.0$
$a_4 = 200.0$
$a_5 = 298.2$

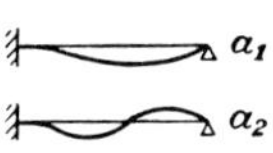

"Clamped-hinged" beam may be considered as half a "clamped-clamped" beam for even a-numbers

$a_1 = 15.4$
$a_2 = 50.0$
$a_3 = 104.$
$a_4 = 178.$
$a_5 = 272.$

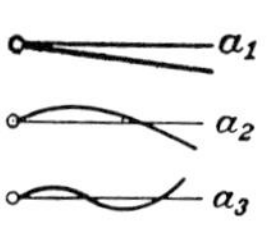

"Hinged-free" beam or wing of autogyro may be considered as half a "free-free" beam for even a-numbers

$a_1 = 0$
$a_2 = 15.4$
$a_3 = 50.0$
$a_4 = 104.$
$a_5 = 178.$

TABLE 1-33

Internal Friction or Damping in Engineering Materials

Trans ASME, Vol 68, 1946, Robertson and Yorgiadis

Material: Description	Approx[a] Rockwell Hardness	Yield Strength 1000 psi: Direct	Yield Strength 1000 psi: Shear	*Experimental Data: $\Delta W/\sigma^3$ 10^{-12}	Stresses 1000 psi	$\Delta W/\tau_m^3$ 10^{-12}	Stresses 1000 psi	K
Lucite, methyl — methacrylate resin.		6[b]		390	0.4-1.8			
Bakelite, grade X laminated-phenolic (paper base).		6[c]		65	0.3-3			
Plywood (1/48 in. birch, resin-bonded under 1000 psi); specific gravity, 1.05.		5[d]		16	0.5-5			
Magnesium alloy *M* (1.5 per cent *Mn*) extruded tubing . .	F 17	9 C[e] (21T)		1.0	1.8-4.5	9.2	1.0-4.0	0.48
Magnesium alloy J-1 (6.5 per cent Al, 1 per cent Zn, 0.2 per cent Mn) extruded tubing	B 25 (F 77)	12 C[e] (21 T)		0.55	2.2-9.0	3.9	1.4-5.0	0.51
Monel metal (67 per cent Ni, 30 per cent Cu, 1.4 per cent Fe, 1 per cent Mn) seamless tubing:								
3/4 in. diam	B 89	68[f]	48[b]	0.05	8.0-34	0.27	3.0-30	0.57
1/2 in. diam	B 93	74[f]	54[b]	0.03	10 -60	0.19	9.0-18	0.54
SAE 1025 steel, seam-welded tubing, "as welded"	B 64	64[b]	22[b] (17[g]	0.05	9 -34	0.4	4.0-20	0.51
SAE X4130 steel, seam-welded tubing, "normalized"	B 96	80[b]	52[b]	0.043	9 -45	0.2	5.0-37	0.60
Steel tubing — Rowett (0.17 per cent C, 0.24 per cent Mn):								
Annealed.			12.5			0.29	4.5-10	
Hard drawn			31			0.027	up to 11	

[a]As measured by authors. [b]Approximate yield point or yield strength estimated from producer's data book. [c]Estimated from 12,500 psi tensile strength given by producer's data book. [d]Compression test by authors. [e]From producer's data book, *C* is in compression, *T* in tension. [f]From producer's book, proportional limits are 48 and 54. [g]Rough Torsion test by authors.

Note: All data in inch-pound-second units.

*ΔW is the damping capacity in inch-pounds per cubic inch per cycle; σ is the maximum direct stress in psi; τ_m is the maximum shear stress. All test data were obtained with completely reversed stress cycles.

TABLE 1–34

Linear Expansion of Steel Shafting

Dodge Catalog D 55

Length Feet	Temperature Increase - Degrees Fahr. 20°	40°	60°	80°	100°
1	.0016	.0032	.0048	.0063	.0079
2	.0032	.0063	.0095	.0127	.0158
3	.0048	.0095	.0143	.0190	.0238
4	.0063	.0127	.0190	.0253	.0317
5	.0079	.0158	.0238	.0317	.0396
6	.0095	.0190	.0285	.0380	.0475
7	.0111	.0222	.0333	.0444	.0554
8	.0127	.0253	.0380	.0507	.0634
9	.0143	.0285	.0428	.0570	.0713
10	.0158	.0317	.0475	.0634	.0792
12	.0190	.0380	.0570	.0760	.0950
14	.0222	.0444	.0665	.0887	.1109
16	.025	.051	.076	.101	.127
18	.029	.057	.086	.114	.143
20	.032	.063	.095	.127	.158
25	.040	.079	.119	.158	.198
30	.048	.095	.143	.190	.238
35	.055	.111	.166	.222	.277
40	.063	.127	.190	.253	.317
45	.071	.143	.214	.285	.356
50	.079	.158	.238	.317	.396
55	.087	.174	.261	.348	.436
60	.095	.190	.285	.380	.475
65	.103	.206	.309	.412	.515
70	.111	.222	.333	.444	.554
75	.119	.238	.356	.475	.594
80	.127	.253	.380	.507	.634
85	.135	.269	.404	.539	.673
90	.143	.285	.428	.570	.713
95	.150	.301	.451	.602	.752
100	.158	.317	.475	.634	.792
110	.174	.348	.523	.697	.871
120	.190	.380	.570	.760	.950
130	.206	.412	.618	.824	1.030
140	.222	.444	.665	.887	1.109
150	.238	.475	.713	.950	1.188

All dimensions are in inches except first column.

TABLE 1–35

Shafting Size Chart for Combined Torsion and Bending Moments

General Catalog 900 Link-Belt Company

6,000 psi shear stress. Multiply value from chart by shear factor from Table 1–36 to get shaft size corresponding to another allowable shear stress.

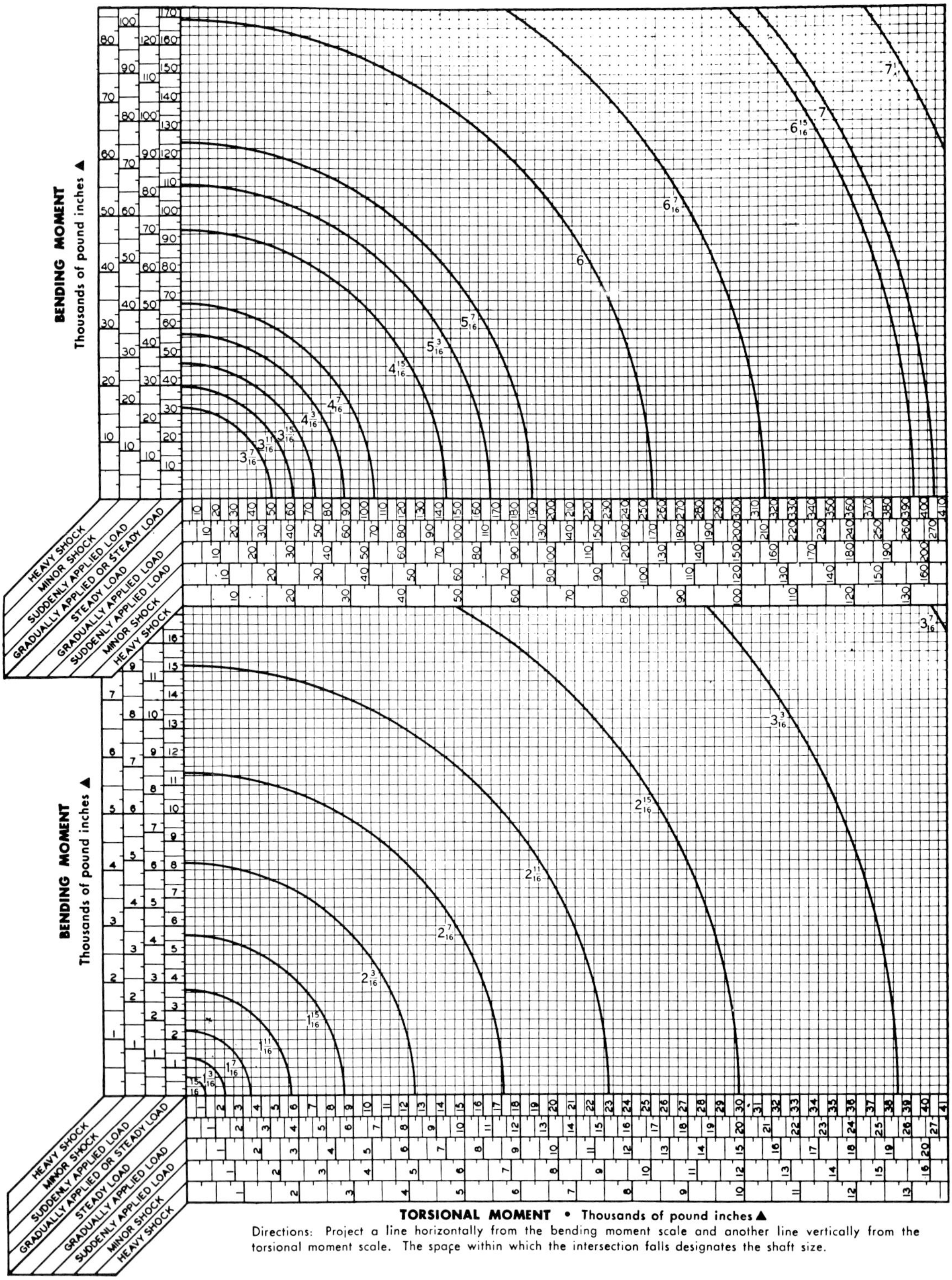

Directions: Project a line horizontally from the bending moment scale and another line vertically from the torsional moment scale. The space within which the intersection falls designates the shaft size.

TABLE 1-36

Shear Factors When Allowable Design Stress Differs from 6000 psi
General Catalog 900, Link Belt Company

Multiply shaft size obtained from Tables 1–35, –37, –38 or –39 by shear factor to get shaft size at desired shear stress.

Shear Stress	Shear Factor	Shear Stress	Shear Factor
500	2.2894	7000	.9499
1000	1.8171	8000	.9086
1500	1.5874	9000	.8736
2000	1.4422	10000	.8434
2500	1.3389	11000	.8171
3000	1.2599	12000	.7937
3500	1.1968	13000	.7728
4000	1.1447	14000	.7539
4500	1.1006	15000	.7368
5000	1.0626	16000	.7211
5500	1.0294	17000	.7067
6000	1.0000	18000	.6934

TABLE 1-37

Approximate Horsepower Transmitted by Steel Shafting Under Combined Torsion and Heavy Bending
General Catalog 900 – Link Belt Company

Shaft Diam Inches	Horsepower of Shafting* Revolutions per minute														
	25	50	75	100	125	150	175	200	225	250	275	300	350	400	450
$^{15}/_{16}$	.15	.3	.45	.6	.7	.9	1.0	1.2	1.3	1.5	1.6	1.8	2.1	2.4	2.7
$1^{3}/_{16}$	.3	.6	.9	1.2	1.5	1.8	2.1	2.5	2.8	3.1	3.4	3.7	4.3	5.0	5.6
$1^{7}/_{16}$	.55	1.1	1.7	2.2	2.7	3.3	3.8	4.4	4.9	5.5	6.1	6.6	7.7	8.8	9.9
$1^{11}/_{16}$	.9	1.8	2.7	3.5	4.4	5.3	6.2	7.1	8.0	8.9	9.8	10.7	12.5	14.3	16.1
$1^{15}/_{16}$	1.3	2.7	3.9	5.4	6.7	8.1	9.5	10.8	12.2	13.5	14.9	16.3	19.0	21.7	24.4
$2^{3}/_{16}$	1.9	3.9	5.7	7.8	9.7	11.7	13.7	15.6	17.6	19.5	21.5	23.4	27.4	31.3	35.2
$2^{7}/_{16}$	2.7	5.4	8.1	10.8	13.5	16.2	18.9	21.6	24.3	27.0	29.7	32.4	37.9	43.3	48.7
$2^{11}/_{16}$	3.6	7.3	10.8	14.5	18.1	21.7	25.4	29.0	32.6	36.2	39.9	43.5	50.8	58.0	65.3
$2^{15}/_{16}$	4.7	9.5	14.1	18.9	23.6	28.4	33.1	37.9	42.6	47.3	52.1	56.8	66.3	75.8	85.3
$3^{7}/_{16}$	7.6	15.2	22.8	30.3	37.9	45.5	53.1	60.7	68.3	75.9	83.5	91.1	106.0	121.0	136.0
$3^{15}/_{16}$	11.4	22.8	34.2	45.6	57.0	68.4	79.9	91.3	102.0	114.0	125.0	136.0	159.0	182.0	205.0
$4^{7}/_{16}$	16.3	32.7	48.9	65.3	81.6	98.0	114.0	130.0	147.0	163.0	179.0	196.0	228.0	261.0	294.0
$4^{15}/_{16}$	22.5	45.0	67.5	90.0	112.0	135.0	157.0	180.0	202.0	225.0	247.0	270.0	315.0	360.0	405.0
$5^{7}/_{16}$	30.0	60.0	90.0	120.0	150.0	180.0	210.0	240.0	270.0	300.0	330.0	360.0	420.0	480.0	541.0
$5^{15}/_{16}$	39.0	78.0	117.0	156.0	195.0	234.0	273.0	313.0	352.0	391.0	430.0	469.0	547.0	626.0	704.0
$6^{1}/_{2}$	51.2	103.0	153.6	205.0	256.0	308.0	359.0	410.0	462.0	513.0	564.0	616.0	718.0	821.0	924.0
7	64.0	128.0	192.0	256.0	320.0	384.0	448.0	513.0	577.0	641.0	705.0	769.0	897.0	1026.0	1154.0
$7^{1}/_{2}$	79.0	158.0	237.0	315.0	394.0	473.0	552.0	631.0	709.0	788.0	867.0	946.0	1104.0	1262.0	1419.0
8	96.0	191.0	288.0	382.0	478.0	574.0	670.0	765.0	861.0	957.0	1053.0	1148.0	1340.0	1531.0	1723.0
$8^{1}/_{2}$	115.0	230.0	345.0	459.0	574.0	688.0	803.0	918.0	1033.0	1148.0	1263.0	1377.0	1607.0	1837.0	2066.0
9	136.0	273.0	408.0	545.0	681.0	817.0	954.0	1090.0	1226.0	1363.0	1499.0	1635.0	1908.0	2181.0	2453.0
$9^{1}/_{2}$	160.2	321.0	481.0	641.0	801.0	961.0	1122.0	1282.0	1442.0	1603.0	1763.0	1923.0	2244.0	2565.0	2885.0
10	187.0	374.0	561.0	747.0	934.0	1121.0	1308.0	1495.0	1682.0	1869.0	2056.0	2243.0	2617.0	2991.0	3365.0

* Based on uniform loads and 6000 psi shear stress in shafts with keyseats. Multiply value from table by shear factor from Table 1–36 to get shaft size corresponding to another allowable shear stress.

TABLE 1-38

Approximate Horsepower Transmitted by Steel Shafting Under Combined Torsion and Moderate Bending

General Catalog 900 – Link Belt Company

Shaft Diam Inches	Horsepower of Shafting* Revolutions per minute 25	50	75	100	125	150	175	200	225	250	275	300	350	400	450	500
$\frac{15}{16}$	.2	.5	.7	1.0	1.2	1.5	1.7	2.0	2.3	2.5	2.8	3.0	3.5	4.1	4.6	5.1
$1\frac{3}{16}$	.5	1.0	1.5	2.0	2.6	3.1	3.6	4.1	4.7	5.2	5.7	6.2	7.3	8.3	9.4	10.4
$1\frac{7}{16}$	.9	1.9	2.7	3.7	4.6	5.5	6.4	7.4	8.3	9.2	10.1	11.1	12.9	14.8	16.6	18.5
$1\frac{11}{16}$	1.5	2.9	4.5	5.9	7.4	8.9	10.4	11.9	13.4	14.9	16.4	17.9	20.9	23.9	26.9	29.9
$1\frac{15}{16}$	2.3	4.5	6.9	9.0	11.3	13.6	15.8	18.1	20.4	22.6	24.9	27.2	31.7	36.2	40.8	45.3
$2\frac{3}{16}$	3.3	6.5	9.9	13.0	16.3	19.5	22.8	26.1	29.3	32.6	35.8	39.1	45.6	52.2	58.7	65.2
$2\frac{7}{16}$	4.5	9.0	13.5	18.0	22.5	27.0	31.6	36.1	40.6	45.1	49.6	54.1	63.2	72.2	81.2	90.2
$2\frac{11}{16}$	6.1	12.1	18.3	24.2	30.2	36.3	42.3	48.4	54.4	60.5	66.5	72.6	84.7	96.8	108.0	121.0
$2\frac{15}{16}$	7.9	15.8	23.7	31.6	39.5	47.4	55.3	63.2	71.1	79.0	86.9	94.8	110.0	126.0	142.0	158.0
$3\frac{7}{16}$	12.7	25.3	38.1	50.6	63.3	75.9	88.6	101.0	113.0	126.0	139.0	151.0	177.0	202.0	227.0	253.0
$3\frac{15}{16}$	19.0	38.1	57.0	76.1	94.1	114.0	133.0	152.0	171.0	190.0	209.0	228.0	266.0	304.0	342.0	380.0
$4\frac{7}{16}$	27.0	54.0	81.0	108.0	136.0	163.0	190.0	217.0	245.0	272.0	299.0	326.0	381.0	435.0	490.0	544.0
$4\frac{15}{16}$	37.5	75.0	112.5	150.0	187.0	225.0	262.0	300.0	337.0	375.0	412.0	450.0	525.0	600.0	675.0	750.0
$5\frac{7}{16}$	50.0	100.0	150.0	200.0	250.0	330.0	350.0	400.0	451.0	501.0	551.0	601.0	701.0	801.0	902.0	1002.0
$5\frac{15}{16}$	65.2	131.0	195.6	261.0	326.0	391.0	456.0	522.0	587.0	652.0	717.0	783.0	913.0	1044.0	1174.0	1305.0

* Based on uniform loads and 6000 psi shear stress in shafts with keyseats. Multiply value from table by shear factor from Table 1-36 to get shaft size corresponding to another allowable shear stress.

TABLE 1-39

Approximate Horsepower Transmitted by Steel Shafting Under Uniform Torsion and Without Bending

General Catalog 900 – Link Belt Company

Shaft Diam Inches	Horsepower of Shafting* Revolutions per Minute 25	50	75	100	125	150	175	200	225	250	275	300	350	400	450	500
$\frac{15}{16}$	.4	.8	1.2	1.5	1.9	2.3	2.6	3.0	3.4	3.8	4.2	4.6	5.3	6.1	6.9	7.7
$1\frac{3}{16}$	.8	1.6	2.4	3.1	3.9	4.6	5.4	6.2	7.0	7.8	8.6	9.3	10.9	12.5	14.0	15.6
$1\frac{7}{16}$	1.4	2.8	4.2	5.5	6.9	8.3	9.7	11.1	12.4	13.8	15.2	16.6	19.4	22.2	24.9	27.7
$1\frac{11}{16}$	2.2	4.5	6.6	8.9	11.2	13.4	15.7	17.9	20.2	22.4	24.7	26.9	31.4	35.9	40.4	44.9
$1\frac{15}{16}$	3.4	6.8	10.2	13.5	16.9	20.3	23.7	27.1	30.5	33.9	37.3	40.7	47.5	54.3	61.1	67.9
$2\frac{3}{16}$	4.9	9.8	14.7	19.5	24.4	29.3	34.2	39.1	44.0	48.9	53.8	58.6	68.4	78.2	88.0	97.8
$2\frac{7}{16}$	6.8	13.5	18.4	27.0	33.8	40.6	47.3	54.1	60.9	67.6	74.4	81.2	94.7	108.0	121.0	135.0
$2\frac{11}{16}$	9.1	18.1	27.3	36.2	45.3	54.4	63.4	72.5	81.6	90.7	99.7	108.0	126.0	145.0	163.0	181.0
$2\frac{15}{16}$	11.8	23.7	35.4	47.3	59.2	71.0	82.9	94.7	106.0	118.0	130.0	142.0	165.0	189.0	213.0	236.0
$3\frac{7}{16}$	18.9	37.9	56.7	75.9	94.9	113.0	132.0	151.0	170.0	189.0	208.0	227.0	265.0	303.0	341.0	379.0
$3\frac{15}{16}$	28.5	57.0	85.5	114.0	142.0	171.0	199.0	228.0	256.0	285.0	313.0	342.0	399.0	456.0	513.0	570.0

* Based on uniform loads and 6000 psi shear stress in shafts with keyseats. Multiply value from table by shear factor from Table 1-36 to get shaft size corresponding to another allowable shear stress.

TABLE 1–40

Minimum Shaft and Maximum Housing Shoulder Diameters For Metric Annular Ball Bearings Except Type BM (Magneto) AFBMA Standards, Section 7 March, 1951

This table represents the minimum shaft shoulder diameters which will properly locate the bearings on the shafts, and the maximum housing shoulder diameters which will properly locate the bearings in the housings, when single row bearings are mounted for use under plain radial loads or light thrustloads. Other types of loads and bearings may require higher shoulders, resulting in larger diameter shaft shoulders or smaller diameter housing shoulders, to carry their thrust loads. Maximum shaft shoulder heights are dependent upon individual manufacturer's design of bearings used in the application.

Bore	Minimum Shaft Diameter				Maximum Housing Shoulder Diameter			
mm	10 Series	02 Series	03 Series	04 Series	10 Series	02 Series	03 Series	04 Series
4			.22				.55	
5			.27				.67	
6		.30				.67		
7		.34				.79		
8	.38				.79			
9		.45				.83		
10	.47	.50	.50		.95	.98	1.18	
12	.55	.58	.63		1.02	1.06	1.22	
15	.67	.69	.75		1.18	1.18	1.42	
17	.75	.77	.83	.95	1.30	1.34	1.61	2.17
20	.89	.94	.94	1.06	1.46	1.61	1.77	2.56
25	1.08	1.14	1.14	1.34	1.65	1.81	2.17	2.80
30	1.34	1.34	1.34	1.54	1.93	2.21	2.56	3.19
35	1.53	1.53	1.69	1.73	2.21	2.56	2.80	3.58
40	1.73	1.73	1.93	1.97	2.44	2.87	3.19	3.94
45	1.94	1.94	2.13	2.17	2.72	3.07	3.58	4.33
50	2.13	2.13	2.36	2.44	2.91	3.27	3.94	4.65
55	2.33	2.47	2.56	2.64	3.27	3.58	4.33	5.04
60	2.53	2.67	2.84	2.84	3.47	3.98	4.65	5.43
65	2.72	2.86	3.03	3.03	3.66	4.37	5.04	5.83
70	2.91	3.06	3.23	3.31	4.06	4.57	5.43	6.54
75	3.11	3.25	3.43	3.50	4.25	4.76	5.83	6.93
80	3.31	3.55	3.62	3.70	4.65	5.12	6.22	7.32
85	3.50	3.75	3.90	4.06	4.84	5.51	6.54	7.56
90	3.84	3.94	4.09	4.25	5.16	5.91	6.93	8.15
95	4.05	4.21	4.29		5.35	6.22	7.32	
100	4.23	4.41	4.49		5.55	6.61	7.91	
105	4.53	4.61	4.69		5.91	7.01	8.31	
110	4.72	4.80	4.88		6.30	7.40	8.90	
120	5.12	5.20	5.28		6.69	7.99	9.69	
130	5.51	5.67	5.83		7.48	8.50	10.32	
140	5.91	6.06	6.22		7.87	9.29	11.10	
150	6.38	6.46	6.61		8.39	10.08	11.89	
160	6.77	6.85	7.01		8.98	10.87	12.68	
170	7.17	7.40	7.40		9.76	11.50	13.47	
180	7.56	7.80	7.80		10.55	11.89	14.25	
190	7.95	8.19	8.35		10.95	12.68	14.88	
200	8.35	8.58	8.74		11.73	13.47	15.67	
220	9.21	9.37	9.53		12.84	15.04	17.24	
240	10.00	10.16	10.32		13.62	16.61	18.82	
260	10.95	11.10	11.34		15.04	18.03	20.16	
280	11.73	11.89	12.13		15.83	18.82	21.73	
300	12.52	12.68			17.40	20.39		
320	13.31	13.47			18.19	21.97		

TABLE 1–41

Minimum Shaft and Maximum Housing Shoulder Diameters For Type BM (Magneto), Ball Bearings
AFBMA Standards, Section 7 March, 1951

Bearing Size Bore, mm	Shaft Shoulder Diameter Minimum, In.	Housing Shoulder Diameter Maximum, In.
5	1/4	37/64
6	5/16	7/8
7	23/64	7/8
8	25/64	7/8
9	7/16	1-1/32
10	15/32	1-1/32
11	33/64	1-3/16
12	9/16	1-3/16
13	19/32	1-7/64
14	5/8	1-9/32
15	11/16	1-9/32
16	49/64	1-23/64
17	51/64	1-19/32
19	57/64	1-7/16
20	59/64	1-23/32
25	1-1/8	1-29/32

TABLE 1–42

Minimum Shaft and Maximum Housing Shoulder Diameters for Type BIC (formerly Inch Type S) Ball Bearings
AFBMA Standards, Section 7 March, 1951

Bearing Size, In. Bore	O.D.	Width	Shaft Shoulder Diameter Minimum, In.	Housing Shoulder Diameter Maximum, In.
1/8	3/8	5/32	3/16	5/16
3/16	1/2	5/32	1/4	7/16
1/4	5/8	.196	5/16	9/16
1/8	1/2	11/64	3/16	7/16
1/4	3/4	7/32	5/16	11/16
3/8	7/8	7/32	7/16	13/16
1/2	1-1/8	1/4	9/16	1-1/16
5/8	1-3/8	9/32	3/4	1-1/4
3/4	1-5/8	5/16	7/8	1-1/2
7/8	1-7/8	3/8	1	1-3/4
1	2	3/8	1-1/8	1-7/8
1-1/8	2-1/8	3/8	1-1/4	2
1-1/4	2-1/4	3/8	1-3/8	2-1/8
1-3/8	2-1/2	7/16	1-1/2	2-3/8
1-1/2	2-5/8	7/16	1-5/8	2-1/2

TABLE 1–43

Minimum Shaft and Maximum Housing Shoulder Diameters for Cylindrical Roller Bearings

AFBMA Standards, Section 7 – March, 1951

Bore	Shaft Shoulder *Diameter Inches		Housing Shoulder *Diameter Inches		Bore	Shaft Shoulder *Diameter Inches		Housing Shoulder *Diameter Inches	
	Min	Max	Max	Min		Min	Max	Max	Min
02 Dimension Series			(Table 2–15)			02 Dimension Series			
10	.55	.60	1.08	1.02	95	4.29	4.46	6.12	6.06
12	.62	.66	1.15	1.08	100	4.56	4.71	6.48	6.35
15	.74	.75	1.26	1.20	105	4.77	4.96	6.84	6.66
17	.84	.86	1.44	1.41	110	5.01	5.21	7.20	7.07
20	1.02	1.04	1.67	1.59	120	5.47	5.52	7.74	7.67
25	1.22	1.23	1.86	1.85	130	5.86	6.06	8.28	8.07
30	1.42	1.48	2.24	2.18	140	6.35	6.62	9.00	8.75
35	1.66	1.71	2.61	2.50	150	6.84	7.06	9.73	9.44
40	1.87	1.96	2.88	2.83	160	7.29	7.62	10.48	10.13
45	2.08	2.15	3.06	3.04	170	7.74	8.09	11.18	10.82
50	2.28	2.31	3.24	3.20	180	8.16	8.51	11.51	11.25
55	2.52	2.60	3.60	3.55	190	8.63	9.01	12.23	11.88
60	2.73	2.85	3.96	3.88	200	9.12	9.43	12.97	12.55
65	3.00	3.02	4.32	4.20	220	10.02	10.46	14.41	13.95
70	3.19	3.23	4.50	4.41	240	10.96	11.46	15.87	15.32
75	3.37	3.44	4.68	4.62	260	11.97	12.54	17.29	16.71
80	3.60	3.73	5.05	4.94	280	12.66	13.20	18.00	17.38
85	3.85	3.96	5.40	5.29	300	13.71	14.32	19.45	19.00
90	4.05	4.11	5.76	5.64	320	14.63	15.30	20.95	20.14
03 Dimension Series			(Table 2–15)			03 Dimension Series			
10	.59	.65	1.24	1.12	80	3.79	3.95	6.02	6.00
12	.66	.72	1.31	1.27	85	4.04	4.22	6.38	6.25
15	.80	.82	1.49	1.41	90	4.25	4.44	6.73	6.63
17	.90	.97	1.67	1.58	95	4.54	4.73	7.09	7.00
20	1.02	1.10	1.82	1.77	100	4.83	4.97	7.62	7.50
25	1.25	1.31	2.20	2.13	105	5.05	5.24	7.97	7.75
30	1.50	1.51	2.55	2.50	110	5.37	5.48	8.50	8.25
35	1.73	1.77	2.83	2.76	120	5.82	6.01	9.21	8.91
40	1.94	1.99	3.19	3.13	130	6.32	6.48	9.92	9.66
45	2.20	2.24	3.54	3.50	140	6.74	6.96	10.63	10.37
50	2.42	2.52	3.90	3.86	150	7.11	7.43	11.34	11.08
55	2.65	2.74	4.25	4.13	160	7.65	7.99	12.05	11.75
60	2.88	2.99	4.61	4.46	170	8.07	8.26	12.76	12.38
65	3.11	3.21	4.96	4.88	180	8.61	9.12	13.46	12.93
70	3.32	3.46	5.31	5.25	190	9.02	9.53	14.17	13.70
75	3.57	3.74	5.67	5.63	200	9.56	10.12	14.88	14.38

* Shoulder diameters outside these limits may be used where machine assembly consideration permits, or upon advice of the particular bearing manufacturer whose bearings are to be used. The minimum diameter of shaft shoulder specified here will satisfy the maximum corner contour on the inner races of the bearings of any manufacturer; the maximum diameter of shaft shoulder will clear the diameter under the rollers on the bearings of any manufacturer. Likewise, the maximum diameter of housing shoulder will satisfy the maximum corner contour on the outer races of the bearings of any manufacturer; and the minimum diameter of housing shoulder will clear the diameter over the rollers on the bearings of any manufacturer.

TABLE 1-44

Minimum Shaft and Maximum Housing Shoulder Diameters For Type TS Tapered Roller Bearings

AFBMA Standards, Section 7 March, 1951

Bearing Number	Shoulder Diameter		Bearing Number	Shoulder Diameter	
	Cone Minimum	Cup Maximum		Cone Minimum	Cup Maximum
A2037-A2126	19/32	1	02872-02820	1-7/16	2-5/16
A2047-A2126	21/32	1	2690-2631	1-5/8	2-1/4
A4050-A4138	23/32	1-3/32	17118-17244	1-7/16	2-1/16
A4059-A4138	25/32	1-3/32	15117-15250	1-7/16	2-5/32
A6062-A6157	7/8	1-9/32	14117A-14276	1-11/16	2-5/16
17580-17520	7/8	1-3/8	17119-17244	1-7/16	2-1/16
05062-05185	15/16	1-9/16	2558-2523	1-9/16	2-3/8
09062-09196	7/8	1-9/16	3191-3120	1-11/16	2-5/16
A6067-A6157	7/8	1-9/32	15118-15250	1-5/8	2-5/32
A6075-A6157	31/32	1-9/32	14116-14276	1-7/16	2-5/16
05075-05185	1	1-9/16	1674-1620	1-9/16	2-7/32
09067-09195	1	1-5/8	08125-08231	1-7/16	2-1/32
09078-09196	1	1-9/16	15123-15245	1-11/16	2-3/32
1775-1729	1-1/16	1-7/8	02475-02420	1-11/16	2-1/4
05079-05185	1-1/32	1-9/16	14125A-14276	1-3/4	2-5/16
07079-07196	1-1/8	1-23/32	2582-2523	1-11/16	2-3/8
12580-12520	1-1/16	1-19/32	3193-3120	1-3/4	2-5/16
3660-3620	1-3/16	2	02875-02820	1-3/4	2-5/16
07087-07204	1-1/8	1-3/4	2875-2820	1-3/4	2-3/8
1380-1329	1-1/8	1-3/4	3476-3420	1-5/8	2-1/2
1755-1729	1-1/8	1-7/8	346-332	1-9/16	2-13/16
1280-1220	1-1/8	1-7/8	26126-26283	1-9/16	2-3/8
1779-1729	1-1/8	1-7/8	14130-14276	1-3/4	2-5/16
3659-3620	1-5/16	2	2585-2523	1-3/4	2-3/8
2685-2631	1-1/4	2-1/4	3196-3120	1-13/16	2-5/16
26093-26283	1-3/8	2-3/8	2876-2820	1-13/16	2-3/8
07098-07204	1-1/4	1-3/4	2785-2720	1-13/16	2-1/2
07100-07204	1-7/32	1-3/4	31590-31520	1-5/8	2-3/8
15578-15520	1-1/4	1-29/32	14137A-14276	1-11/16	2-5/16
1986-1932	1-1/4	2	02877-02820	1-7/8	2-5/16
15100-15250	1-1/2	2-5/32	2878-2820	1-5/8	2-3/8
2687-2631	1-5/16	2-1/4	25878-25820	1-13/16	2-13/32
26100-26283	1-3/8	2-3/8	2786-2720	2	2-1/2
3189-3120	1-5/16	2-5/16	36137-36300	1-11/16	2-1/2
15580-15520	1-1/2	1-29/32	31593-31520	1-7/8	2-3/8
1985-1932	1-5/16	2	3478-3420	1-7/8	2-1/2
15112-15250	1-9/16	2-5/32	335-332	1-11/16	2-13/16
2689-2631	1-3/8	2-1/4	3379-3320	1-7/8	2-11/16
02474-02420	1-3/8	2-1/4	417-414	1-11/16	3
2578-2523	1-1/2	2-3/8	449-432	1-3/4	3-1/8
26112-26283	1-7/16	2-3/8	19138-19283	1-11/16	2-3/8
3198-3120	1-7/16	2-5/16	339-332	1-11/16	2-13/16

Table 2-17 gives dimensions of bearings; Table 2-25 describes Type TS. All dimensions are in inches.

continued on next page

TABLE 1-44, continued

Bearing Number	Shoulder Diameter Cone Minimum	Shoulder Diameter Cup Maximum	Bearing Number	Shoulder Diameter Cone Minimum	Shoulder Diameter Cup Maximum
26883-26820	1-11/16	2-5/8	355-354A	2-1/8	3
2794-2720	1-15/16	2-1/2	3578-3525	2-1/4	2-13/16
19143-19283	1-3/4	2-3/8	49175-49368	2-1/4	3-1/16
16143-16282	1-7/8	2-7/16	46176-46368	2-5/16	3
3878-3820	1-13/16	2-3/4	438-432	2-1/4	3-1/8
13889-13836	1-3/4	2-5/16	527-522	2-5/16	3-3/8
13685-13621	1-15/16	2-5/16	460-453X	2-3/8	3-9/16
19150-19283	1-3/4	2-3/8	59175-59412	2-3/8	3-1/2
16150-16284	1-15/16	2-7/16	535-532X	2-3/8	3-5/8
2788-2720	1-15/16	2-1/2	65385-65321	2-3/8	3-7/8
3490-3420	2	2-1/2	615-612	2-7/16	4
28151-28315	2	2-5/8	25584-25520	2-1/16	2-7/8
337-332	1-3/4	2-13/16	3776-3720	2-5/16	3-1/8
3381-3320	2	2-11/16	358-354A	2-1/16	3
3876-3820	2-1/16	2-3/4	376-372A	2-1/8	3-5/16
418-414	2	3	18690-18620	2-3/16	2-3/4
49150-49368	2-1/16	3-1/16	13181-13318	2-1/16	2-3/4
444-432	2-1/16	3-1/8	359S-354A	2-3/16	3
542-532X	2-3/16	3-5/8	2984-2924	2-5/16	2-15/16
26881-26820	2-1/16	2-5/8	436-432	2-5/16	3-1/8
3382-3320	2	2-11/16	369A-362A	2-3/8	3-5/32
422-414	2-1/16	3	3778-3720	2-5/8	3-1/8
11157-11300	1-7/8	2-9/16	3779-3720	2-3/8	3-1/8
28158-28315	1-7/8	2-5/8	49580-49520	2-7/16	3-5/16
344-332	2-1/16	2-13/16	528-522	2-7/16	3-3/8
420-414	2-1/16	3	5358-5335	2-1/4	3-3/8
24780-24720	2-1/16	2-21/32	463-453X	2-9/16	3-9/16
11162-11315	1-15/16	2-5/8	536-532X	2-7/16	3-5/8
342-332	2-1/8	2-13/16	617-612	2-9/16	4
26882-26820	2-1/8	2-5/8	3781-3720	2-7/16	3-1/8
3877-3820	2-1/8	2-3/4	5395-5335	2-1/2	3-3/8
3577-3525	2-1/8	2-13/16	366-362A	2-5/16	3-5/32
419-414	2-1/16	3	465-453X	2-7/16	3-9/16
4388-4335	2-1/8	2-15/16	18200-18337	2-5/16	2-15/16
49162-49368	2-3/16	3-1/16	368-362A	2-5/16	3-5/32
46162-46368	2	3	368A-362A	2-7/16	3-5/32
447-432	2-1/8	3-1/8	3780-3720	2-1/2	3-1/8
526-522	2-1/4	3-3/8	375-372A	2-3/8	3-5/16
12168-12303	2	2-5/8	49585-49520	2-9/16	3-5/16
25578-25520	2-1/16	2-7/8	529X-522	2-1/2	3-3/8
3579-3525	2-3/16	2-13/16	455-453X	2-3/8	3-9/16
342S-332	2-1/8	2-13/16	59200-59412	2-9/16	3-1/2
26884-26820	2-1/8	2-5/8	4580-4535	2-9/16	3-7/16
25577-25520	2-3/16	2-7/8	537-532X	2-9/16	3-5/8
12175-12303	2-1/16	2-5/8	3975-3920	2-11/16	3-3/4
13175-13318	1-15/16	2-3/4	619-612	2-5/8	4
35175-35326	2-3/16	2-31/32	555-552A	2-5/8	4-1/8
25580-25520	2-1/4	2-7/8	6279-6220	2-11/16	4-1/8

Table 2-17 gives dimensions of bearings; Table 2-25 describes Type TS. All dimensions are in inches.

continued on next page

Bearing Number	Shoulder Diameter Cone Minimum	Shoulder Diameter Cup Maximum	Bearing Number	Shoulder Diameter Cone Minimum	Shoulder Diameter Cup Maximum
3767-3720	2-7/16	3-1/8	399A-394A	3-1/16	3-15/16
33890-33821	2-3/8	3-1/4	480-472A	3-1/4	4-1/16
377-372A	2-7/16	3-5/16	560S-552A	3-1/4	4-1/8
540-532X	2-5/8	3-5/8	570-563	3-1/4	4-5/16
389A-382	2-3/8	3-1/2	33275-33462	3-5/16	3-15/16
456-453X	2-11/16	3-9/16	482-472A	3-1/4	4-1/16
4595-4535	2-11/16	3-7/16	566-563	3-5/16	4-5/16
539-532X	2-5/8	3-5/8	643-632	3-5/16	4-1/2
621-612	2-3/4	4	6454-6420	3-9/16	4-15/16
557S-552A	2-13/16	4-1/8	655-652	3-1/2	5-1/8
6280-6220	2-3/4	4-1/8	835-832	3-5/8	5-3/4
636-632	2-7/8	4-1/2	34275-34478	3-3/16	4-1/4
385-382	2-9/16	3-1/2	33281-33462	3-5/16	3-15/16
466S-453X	2-5/8	3-9/16	567A-563	3-3/8	4-5/16
389-382	2-9/16	3-1/2	645-632	3-5/8	4-1/2
28682-28622	2-3/4	3-7/16	567-563	3-7/16	4-5/16
387-382	2-5/8	3-1/2	6460-6420	3-9/16	4-15/16
387A-382	2-11/16	3-1/2	744-742	3-9/16	5-1/8
462-453X	2-11/16	3-9/16	657-652	3-9/16	5-1/8
390-394A	2-3/4	3-15/16	568-563	3-1/4	4-5/16
3979-3920	2-7/8	3-3/4	34300-34478	3-3/8	4-1/4
623-612	2-13/16	4	42687-42620	3-9/16	4-3/8
555S-552A	2-7/8	4-1/8	47680-47620	3-3/8	4-9/16
65225-65500	2-7/8	4-1/16	495A-493	3-5/8	4-11/16
6375-6320	3	4-7/16	575-572	3-5/8	4-13/16
29582-29520	2-3/4	3-11/16	6461-6420	3-5/8	4-15/16
397-394A	2-11/16	3-15/16	748S-742	3-11/16	5-1/8
28985-28920	2-7/8	3-7/16	659-652	3-11/16	5-1/8
3980-3920	2-15/16	3-3/4	6576-6535	3-13/16	5-7/16
5582-5535	2-13/16	4-1/16	843-832	4	5-3/4
6376-6320	3-1/16	4-7/16	34306-34478	3-9/16	4-1/4
39250-39412	2-7/8	3-11/16	47681-47620	3-3/4	4-9/16
395-394A	3	3-15/16	496-493	3-3/4	4-11/16
3982-3920	3-1/16	3-3/4	581-572	3-3/4	4-13/16
33251-33462	2-7/8	3-15/16	740-742	3-15/16	5-1/8
477-472A	2-7/8	4-1/16	838-832	3-11/16	5-3/4
5584-5535	3-1/8	4-1/16	47686-47620	3-13/16	4-9/16
559-552A	3-1/16	4-1/8	495-493	3-13/16	4-11/16
565-563	3-1/8	4-5/16	580-572	3-13/16	4-13/16
639-632	3-3/16	4-1/2	663-652	3-7/8	5-1/8
6379-6320	3-3/16	4-7/16	6559-6535	4	5-7/16
395S-394A	3-1/8	3-15/16	842-832	3-15/16	5-3/4
3984-3920	3-1/8	3-3/4	27690-27620	3-13/16	4-7/16
33262-33462	3-3/16	3-15/16	498-493	3-7/8	4-11/16
479-472A	3-1/16	4-1/16	749-742	3-15/16	5-1/8
560-552A	3-3/16	4-1/8	497-493	3-7/8	4-11/16
6386-6320	3-5/16	4-7/16	596-592A	4	5-3/16
641-632	3-1/4	4-1/2	665-652	3-15/16	5-1/8

Table 2-17 gives dimensions of bearings; Table 2-25 describes Type TS. All dimensions are in inches.

continued on next page

TABLE 1-44, continued

Bearing Number	Shoulder Diameter Cone Minimum	Shoulder Diameter Cup Maximum	Bearing Number	Shoulder Diameter Cone Minimum	Shoulder Diameter Cup Maximum
42350-42584	4-1/16	5-1/8	EE153044-153100	6-1/8	8-9/16
593-592A	4-1/8	5-3/16	67388-67320	5-11/16	7-1/16
759-752	4-1/8	5-9/16	74500-74850	5-7/8	7-9/16
6580-6535	4-3/16	5-7/16	95500-95925	6	8-1/16
855-854	4-5/8	6-9/16	EE116050-116097	5-3/4	8-3/8
69354-69630	4-1/16	5-1/2	EE153050-153100	6-3/8	8-9/16
47890-47820	4-3/16	5	EE540502-541162	6-7/8	9-3/4
42362-42584	4-3/16	5-1/8	EE455051-455116	6-1/4	9-3/4
598-592A	4-3/16	5-3/16	EE580500-581200	6-3/8	10-3/8
77362-77675	4-5/16	5-7/8	EE750502-751200	6-1/4	10-3/8
42368-42584	4-1/4	5-1/8	48506-48750	5-11/16	6-9/16
42375-42584	4-1/4	5-1/8	799-792	5-3/4	7-3/16
594-592A	4-5/16	5-3/16	797-792	5-13/16	7-3/16
683-672	4-7/16	5-11/16	67390-67320	5-7/8	7-1/16
77375-77675	4-3/8	5-7/8	74525-74850	6	7-9/16
864-854	4-7/8	6-9/16	95525-95925	6-1/2	8-1/16
52387-52618	4-1/2	5-7/16	48393-48320	5-15/16	6-13/16
685-672	4-1/2	5-11/16	74537-74850	6-1/8	7-9/16
779-772	4-9/16	6-3/16	EE580537-581200	7-1/4	10-3/8
52400-52618	4-9/16	5-7/16	74550-74850	6-3/16	7-9/16
687-672	4-5/8	5-11/16	73551-73875	6-3/16	7-7/8
780-772	4-11/16	6-3/16	898-892	6-1/4	7-15/16
861-854	5	6-9/16	82550-82950	6-3/8	8-3/8
782-772	4-3/4	6-3/16	99550-99100	6-3/4	8-13/16
56418-56650	4-13/16	5-3/4	EE540550-541162	7	9-3/4
37425-37625	4-13/16	5-1/2	EE750558-751200	6-3/8	10-3/8
56425-56650	4-13/16	5-3/4	EE450551-451212	7-1/8	10-3/8
71425-71750	5	6-5/8	73562-73875	6-1/4	7-7/8
936-932	5-3/8	7-1/4	82562-82950	6-7/16	8-3/8
64433-64700	5	6-1/8	82576-82950	6-1/2	8-3/8
71437-71750	5-1/16	6-5/8	81574-81962	6-9/16	8-3/4
64450-64700	5-1/8	6-1/8	99575-99100	6-7/8	8-13/16
71450-71750	5-3/16	6-5/8	EE107057-107105	6-7/8	9-1/16
938-932	5-1/2	7-1/4	EE217056-217112	6-7/8	9-5/8
68462-68712	5-3/16	6-1/4	EE750576-751200	6-5/8	10-3/8
795-792	5-1/2	7-3/16	EE450577-451212	7-1/4	10-3/8

Table 2-17 gives dimensions of bearings; Table 2-25 describes Type TS. All dimensions are in inches.

continued on next page

TABLE 1–44, continued

Bearing Number	Shoulder Diameter Cone Minimum	Shoulder Diameter Cup Maximum	Bearing Number	Shoulder Diameter Cone Minimum	Shoulder Diameter Cup Maximum
82587–82950	6–5/8	8–3/8	93750–93125	8–5/8	11–1/8
99587–99100	6–15/16	8–13/16	EE210753–211300	8–3/4	11–11/16
EE560590–561275	7–7/8	10–7/8	EE420751–421437	9	12–7/8
81599–81962	6–3/4	8–3/4	93787–93125	8–7/8	11–1/8
99600–99100	7–1/16	8–13/16	EE132083–132125	8–7/8	11–7/16
EE107060–107105	7–1/8	9–1/16	93800–93125	9	11–1/8
EE217060–217112	6–3/4	9–5/8	EE122080–122125	8–9/16	11–7/16
EE450601–451212	7–1/2	10–3/8	EE420801–421437	9–1/8	12–7/8
EE560600–561275	8–1/4	10–7/8	EE710806–711600	9–5/8	14–1/4
EE560629–561275	7–7/8	10–7/8	EE132084–132125	8–15/16	11–7/16
EE590638–591350	7–7/8	11–1/2	93825–93125	9–1/8	11–1/8
86650–86100	7–5/16	9–1/16	543085–543114	9–3/16	10–9/16
94649–94113	7–3/4	10–1/16	EE130851–131400	9–3/4	12–7/8
EE219065–219122	7–3/4	10–1/2	543086–543114	9–1/4	10–9/16
EE590650–591350	8–3/4	11–1/2	544090–544118	9–5/8	10–15/16
EE618065–618136	8–1/8	12	88900–88128	10	11–3/4
EE780655–781400	7–3/4	12–3/8	96900–96140	10–1/4	12–3/8
EE108065–108142	8–1/2	12–5/8	EE130902–131400	10–1/8	12–7/8
86669–86100	7–7/16	9–1/16	EE430900–431575	10–5/8	14
EE590675–591350	8–1/8	11–1/2	EE710906–711600	10–3/8	14–1/4
EE780676–781400	8	12–3/8	EE700091–700167	10–1/2	14–3/4
67787–67720	7–9/16	8–7/8	88925–88128	10–3/16	11–3/4
94687–94113	8	10–1/16	EE8575–8520	10–3/16	11–7/8
EE219068–219122	8	10–1/2	96925–96140	10–1/2	12–3/8
EE780688–781400	8	12–3/8	EE125094–125145	10–5/8	13–5/16
67790–67720	7–11/16	8–7/8	EE127095–127140	10–1/2	12–3/4
EE91702–91112	8	10–1/8	EE170950–171436	10–5/8	13–3/16
94700–94113	8–1/8	10–1/16	EE125095–125145	10–5/8	13–5/16
EE280702–281200	8–1/8	10–7/8	EE923095–923175	10–7/8	15–5/8
EE470078–470132	8–1/4	11–1/2	EE295950–295193	11–1/4	17–1/4
EE780705–781400	8–1/8	12–3/8	M249749–M249710	10–13/16	13–1/16
EE420701–421437	9–1/8	12–7/8	EE251001–251575	10–7/8	14–1/2
87737–87111	8–1/8	10–1/8	EE722110–722185	12–5/8	16–3/4
222075–222126	8–1/2	11–1/8	EE128111–128160	12–1/8	14–3/4
67885–67820	8–1/4	9–1/2	EE224115–224204	13	18–1/8
87750–87111	8–1/4	10–1/8	EE724119–724195	14–1/8	17–3/8
			EE201250–201800	13–3/4	16–7/16

Table 2-17 gives dimensions of bearings; Table 2-25 describes Type TS. All dimensions are in inches.

TABLE 1-45

Minimum Shaft and Maximum Housing Shoulder Diameters For Types TSS and TST Tapered Roller Bearings

AFBMA Standards, Section 7 March, 1951

Bearing Number	Shoulder Diameter Cone Minimum	Shoulder Diameter Cup Maximum	Bearing Number	Shoulder Diameter Cone Minimum	Shoulder Diameter Cup Maximum
11590-11520	15/16	1-9/32	9385-9321	4-1/4	5-5/8
21075-21212	1-3/16	1-9/16	98335-98788	4-3/8	6-3/4
23100-23256	1-7/16	2	98350-98788	4-1/2	6-3/4
41100-41286	1-9/16	2-5/16	90381-90744	4-13/16	6-1/4
41125-41286	1-13/16	2-5/16	98400-98788	4-7/8	6-3/4
43112-43312	1-5/8	2-9/16	EE215040-215096	5-1/2	8
43125-43312	1-3/4	2-9/16	97450-97900	5-3/8	7-3/4
44150-44348	2-1/16	2-7/8	EE514045-514110	6-1/8	9
44158-44348	2-3/16	2-7/8	97493-97900	5-3/4	7-3/4
44162-44348	2-3/16	2-7/8	97500-97900	5-13/16	7-3/4
53162-53375	2-3/16	3-1/8	EE516050-516120	6-5/8	9-3/4
53177-53375	2-7/16	3-1/8	97503-97900	5-7/8	7-3/4
55200-55443	2-11/16	3-1/2	EE516057-516120	7-1/4	9-3/4
72200-72487	2-7/8	3-7/8	EE518061-518135	7-3/4	11
72212-72487	2-15/16	3-7/8	EE117065-117146	8	12-1/4
78215-78551	3-1/8	4-1/2	EE117063-117148	8-1/8	12-1/2
78225-78551	3-3/16	4-1/2	EE607070-607140	8-3/8	12
66585-66520	3	4	EE350701-351687	9	14
9180-9121	3-7/16	5-1/16	EE350750-351687	9-1/2	14
78250-78551	3-1/4	4-1/2	EE380080-380190	10-1/4	15-1/2
9185-9120	3-5/8	5-1/16	EE390090-390200	11-1/8	16-3/8
9285-9220	3-15/16	5-5/16	EE390095-390200	11-5/8	16-3/8
9380-9321	4	5-5/8	EE620100-620220	12-3/8	18-1/2
98316-98788	4-1/4	6-3/4	EE991251-992550	15-3/8	20-7/8

Table 2-18 gives dimensions of bearings; Table 2-25 describes types. All dimensions are in inches.

continued on next page

TABLE 1–45, continued

Bearing Number	Shoulder Diameter Cup	Bearing Number	Shoulder Diameter Cup	Bearing Number	Shoulder Diameter Cup
15575T–15520	1–29/32	3490T–3420	2–1/2	65212T–65500	4–1/16
07098T–07204	1–3/4	3381T–3320	2–11/16	467T–453X	3–9/16
1757T–1729	1–7/8	415T–414	3	468T–453X	3–9/16
2688T–2631	2–1/4	26879T–26820	2–5/8	388T–382	3–1/2
26112T–26283	2–3/8	11156T–11315	2–5/8	559T–552A	4–1/8
26117T–26283	2–3/8	3382T–3320	2–11/16	560T–552A	4–1/8
2583T–2523	2–3/8	3575T–3525	2–13/16	399T–394A	3–15/16
14123T–14276	2–5/16	422T–414	3	485T–472A	4–1/16
2580T–2523	2–3/8	3384T–3320	2–11/16	488T–472A	4–1/16
14582T–14525	2–9/32	419T–414	3	755T–752	5–9/16
2581T–2523	2–3/8	439T–432	3–1/8	5784T–5735	4–9/16
14132T–14276	2–5/16	3578T–3525	2–13/16	758T–752	5–9/16
2790T–2720	2–1/2	435T–432	3–1/8	764T–752	5–9/16
25877T–25820	2–13/32	349T–332	2–13/16	52398T–52618	5–7/16
2787T–2720	2–1/2	25584T–25520	2–7/8	780T–772	6–3/16
3379T–3320	2–11/16	359T–354A	3	74551T–74850	7–9/16
2791T–2720	2–1/2	463T–453X	3–9/16	74563T–74850	7–9/16
25879T–25820	2–13/32	536T–532X	3–5/8	EE217061T–217114	9–3/4
3380T–3320	2–11/16	537T–532X	3–5/8	99619T–99100	8–13/16
2798T–2720	2–1/2	65199T–65500	4–1/16	94719T–94113	10–1/16
26876T–26820	2–5/8	377T–372A	3–5/16	EE470077T–470133	11–1/2
19150T–19283	2–3/8	28677T–28622	3–7/16	EE122083T–122125	11–7/16
2882T–2820	2–3/8	368T–362A	3–5/32	EE420826T–421450	12–7/8
2788T–2720	2–1/2	539T–532X	3–5/8	88919T–88128	11–3/4

Table 2-18 gives dimensions of bearings; Table 2-25 describes types. All dimensions are in inches.

TABLE 1–46

Degrees into Radians

Deg	Radians	Deg	Radians	Deg	Radians	Deg	Radians	Deg	Radians	Deg	Radians
1	0.01745	31	0.54105	61	1.06465	91	1.58825	121	2.11185	151	2.63545
2	.03491	32	.55851	62	1.08210	92	1.60570	122	2.12930	152	2.65290
3	.05236	33	.57596	63	1.09956	93	1.62316	123	2.14676	153	2.67035
4	.06981	34	.59341	64	1.11701	94	1.64061	124	2.16421	154	2.68781
5	.08727	35	.61087	65	1.13446	95	1.65806	125	2.18166	155	2.70526
6	.10472	36	.62832	66	1.15192	96	1.67552	126	2.19912	156	2.72271
7	.12217	37	.64577	67	1.16937	97	1.69297	127	2.21657	157	2.74017
8	.13963	38	.66323	68	1.18682	98	1.71042	128	2.23402	158	2.75762
9	.15708	39	.68068	69	1.20428	99	1.72788	129	2.25148	159	2.77507
10	.17453	40	.69813	70	1.22173	100	1.74533	130	2.26893	160	2.79253
11	.19199	41	.71559	71	1.23918	101	1.76278	131	2.28638	161	2.80998
12	.20944	42	.73304	72	1.25664	102	1.78024	132	2.30384	162	2.82743
13	.22689	43	.75049	73	1.27409	103	1.79769	133	2.32129	163	2.84489
14	.24435	44	.76795	74	1.29154	104	1.81514	134	2.33874	164	2.86234
15	.26180	45	.78540	75	1.30900	105	1.83260	135	2.35620	165	2.87979
16	.27925	46	.80285	76	1.32645	106	1.85005	136	2.37365	166	2.89725
17	.29671	47	.82031	77	1.34390	107	1.86750	137	2.39110	167	2.91470
18	.31416	48	.83776	78	1.36136	108	1.88496	138	2.40856	168	2.93215
19	.33161	49	.85521	79	1.37881	109	1.90241	139	2.42601	169	2.94961
20	.34907	50	.87267	80	1.39626	110	1.91986	140	2.44346	170	2.96706
21	.36652	51	.89012	81	1.41372	111	1.93732	141	2.46092	171	2.98451
22	.38397	52	.90757	82	1.43117	112	1.95477	142	2.47837	172	3.00197
23	.40143	53	.92502	83	1.44862	113	1.97222	143	2.49582	173	3.01942
24	.41888	54	.94248	84	1.46608	114	1.98968	144	2.51328	174	3.03687
25	.43633	55	.95993	85	1.48353	115	2.00713	145	2.53073	175	3.05433
26	.45379	56	.97738	86	1.50098	116	2.02458	146	2.54818	176	3.07178
27	.47124	57	.99484	87	1.51844	117	2.04204	147	2.56564	177	3.08923
28	.48869	58	1.01299	88	1.53589	118	2.05949	148	2.58309	178	3.10669
29	.50615	59	1.02974	89	1.55334	119	2.07694	149	2.60054	179	3.12414
30	.52360	60	1.04720	90	1.57080	120	2.09440	150	2.61800	180	3.14159

TABLE 1–47

Decimals of a Degree into Radians

Decimal	.00	.01	.02	.03	.04	.05	.06	.07	.08	.09
0.0	0.00000	0.00017	0.00035	0.00052	0.00070	0.00087	0.00105	0.00122	0.00140	0.00157
.1	.00174	.00192	.00209	.00227	.00244	.00262	.00279	.00297	.00314	.00332
.2	.00349	.00367	.00384	.00401	.00419	.00436	.00454	.00471	.00489	.00506
.3	.00524	.00541	.00558	.00576	.00593	.00611	.00628	.00646	.00663	.00681
.4	.00698	.00716	.00733	.00750	.00768	.00785	.00803	.00820	.00838	.00855
.5	.00873	.00890	.00908	.00925	.00942	.00960	.00977	.00995	.01012	.01030
.6	.01047	.01065	.01082	.01100	.01117	.01134	.01152	.01169	.01187	.01204
.7	.01222	.01239	.01257	.01274	.01292	.01309	.01326	.01344	.01361	.01379
.8	.01396	.01414	.01431	.01449	.01466	.01484	.01501	.01518	.01536	.01553
.9	.01571	.01588	.01606	.01623	.01641	.01658	.01676	.01693	.01710	.01728

TABLE 1–48

Minutes into Radians

Min	Radians	Min	Radians	Min	Radians	Min	Radians
1	0.00029	16	0.00465	31	0.00902	46	0.01338
2	0.00058	17	.00495	32	.00931	47	.01367
3	.00087	18	.00524	33	.00960	48	.01396
4	.00116	19	.00553	34	.00989	49	.01425
5	.00145	20	.00582	35	.01018	50	.01454
6	.00175	21	.00611	36	.01047	51	.01484
7	.00204	22	.00640	37	.01076	52	.01513
8	.00233	23	.00669	38	.01105	53	.01542
9	.00262	24	.00698	39	.01134	54	.01571
10	.00291	25	.00727	40	.01164	55	.01600
11	.00320	26	.00756	41	.01193	56	.01629
12	.00349	27	.00785	42	.01222	57	.01658
13	.00378	28	.00814	43	.01251	58	.01687
14	.00407	29	.00844	44	.01280	59	.01716
15	.00436	30	.00873	45	.01309	60	.01745

TABLE 1–49

Decimals of a Degree into Minutes and Seconds

Deci-	.00		.01		.02		.03		.04		.05		.06		.07		.08		.09	
mal	Min	Sec	Min	Sec	Min	Sec	Min	Sec	Min	Sec	Min	Sec	Min	Sec	Min	Sec	Min	Sec	Min	Sec
0.0	0	0	0	36	1	12	1	48	2	24	3	0	3	36	4	12	4	48	5	24
.1	6	0	6	36	7	12	7	48	8	24	9	0	9	36	10	12	10	48	11	24
.2	12	0	12	36	13	12	13	48	14	24	15	0	15	36	16	12	16	48	17	24
.3	18	0	18	36	19	12	19	48	20	24	21	0	21	36	22	12	22	48	23	24
.4	24	0	24	36	25	12	25	48	26	24	27	0	27	36	28	12	28	48	29	24
.5	30	0	30	36	31	12	31	48	32	24	33	0	33	36	34	12	34	48	35	24
.6	36	0	36	36	37	12	37	48	38	24	39	0	39	36	40	12	40	48	41	24
.7	42	0	42	36	43	12	43	48	44	24	45	0	45	36	46	12	46	48	47	24
.8	48	0	48	36	49	12	49	48	50	24	51	0	51	36	52	12	52	48	53	24
.9	54	0	54	36	55	12	55	48	56	24	57	0	57	36	58	12	58	48	59	24

TABLE 1–50

Minutes into Decimals of a Degree

	0″	10″	20″	30″	40″	50″	
0	.00000	.00278	.00556	.00833	.01111	.01389	0
1	.01667	.01944	.02222	.02500	.02778	.03056	1
2	.03333	.03611	.03889	.04167	.04444	.04722	2
3	.05000	.05278	.05556	.05833	.06111	.06389	3
4	.06667	.06944	.07222	.07500	.07778	.08056	4
5	.08333	.08611	.08889	.09167	.09444	.09722	5
6	.10000	.10278	.10556	.10833	.11111	.11389	6
7	.11667	.11944	.12222	.12500	.12778	.13056	7
8	.13333	.13611	.13889	.14167	.14444	.14722	8
9	.15000	.15278	.15556	.15833	.16111	.16389	9
10	.16667	.16944	.17222	.17500	.17778	.18056	10
11	.18333	.18611	.18889	.19167	.19444	.19722	11
12	.20000	.20278	.20556	.20833	.21111	.21389	12
13	.21667	.21944	.22222	.22500	.22778	.23056	13
14	.23333	.23611	.23889	.24167	.24444	.24722	14
15	.25000	.25278	.25556	.25833	.26111	.26389	15
16	.26667	.26944	.27222	.27500	.27778	.28056	16
17	.28333	.28611	.28889	.29167	.29444	.29722	17
18	.30000	.30278	.30556	.30833	.31111	.31389	18
19	.31667	.31944	.32222	.32500	.32778	.33056	19
20	.33333	.33611	.33889	.34167	.34444	.34722	20
21	.35000	.35278	.35556	.35833	.36111	.36389	21
22	.36667	.36944	.37222	.37500	.37778	.38056	22
23	.38333	.38611	.38889	.39167	.39444	.39722	23
24	.40000	.40278	.40556	.40833	.41111	.41389	24
25	.41667	.41944	.42222	.42500	.42778	.43056	25
26	.43333	.43611	.43889	.44167	.44444	.44722	26
27	.45000	.45278	.45556	.45833	.46111	.46389	27
28	.46667	.46944	.47222	.47500	.47778	.48056	28
29	.48333	.48611	.48889	.49167	.49444	.49722	29
30	.50000	.50278	.50556	.50833	.51111	.51389	30
31	.51667	.51944	.52222	.52500	.52778	.53056	31
32	.53333	.53611	.53889	.54167	.54444	.54722	32
33	.55000	.55278	.55556	.55833	.56111	.56389	33
34	.56667	.56944	.57222	.57500	.57778	.58056	34
35	.58333	.58611	.58889	.59167	.59444	.59722	35
36	.60000	.60278	.60556	.60833	.61111	.61389	36
37	.61667	.61944	.62222	.62500	.62778	.63056	37
38	.63333	.63611	.63889	.64167	.64444	.64722	38
39	.65000	.65278	.65556	.65833	.66111	.66389	39
40	.66667	.66944	.67222	.67500	.67778	.68056	40
41	.68333	.68611	.68889	.69167	.69444	.69722	41
42	.70000	.70278	.70556	.70833	.71111	.71389	42
43	.71667	.71944	.72222	.72500	.72778	.73056	43
44	.73333	.73611	.73889	.74167	.74444	.74722	44
45	.75000	.75278	.75556	.75833	.76111	.76389	45
46	.76667	.76944	.77222	.77500	.77778	.78056	46

continued on next page

TABLE 1–50 (continued)

	0″	10″	20″	30″	40″	50″	
47	.78333	.78611	.78889	.79167	.79444	.79722	47
48	.80000	.80278	.80556	.80833	.81111	.81389	48
49	.81667	.81944	.82222	.82500	.82778	.83056	49
50	.83333	.83611	.83889	.84167	.84444	.84722	50
51	.85000	.85278	.85556	.85833	.86111	.86389	51
52	.86667	.86944	.87222	.87500	.87778	.88056	52
53	.88333	.88611	.88889	.89167	.89444	.89722	53
54	.90000	.90278	.90556	.90833	.91111	.91389	54
55	.91667	.91944	.92222	.92500	.92778	.93056	55
56	.93333	.93611	.93889	.94167	.94444	.94722	56
57	.95000	.95278	.95556	.95833	.96111	.96389	57
58	.96667	.96944	.97222	.97500	.97778	.98056	58
59	.98333	.98611	.98889	.99167	.99444	.99722	59

TABLE 1–51

Radians into Degrees

Radians	.00	.01	.02	.03	.04	.05	.06	.07	.08	.09
	Deg	Deg	Deg	Deg	Deg	Deg	Deg	Deg	Deg	Deg
0.0	0.0000	0.5730	1.1459	1.7189	2.2918	2.8648	3.4377	4.0107	4.5837	5.1566
.1	5.7296	6.3025	6.8755	7.4485	8.0214	8.5944	9.1673	9.7403	10.3132	10.8862
.2	11.4591	12.0321	12.6051	13.1780	13.7510	14.3239	14.8969	15.4699	16.0428	16.6158
.3	17.1887	17.7617	18.3346	18.9076	19.4806	20.0535	20.6265	21.1994	21.7724	22.3454
.4	22.9183	23.4913	24.0642	24.6372	25.2101	25.7831	26.3561	26.9290	27.5020	28.0749
.5	28.6479	29.2208	29.7938	30.3668	30.9397	31.1527	32.0856	32.6586	33.2316	33.8045
.6	34.3775	34.9504	35.5234	36.0963	36.6693	37.2423	37.8152	38.3882	38.9611	39.5341
.7	40.1070	40.6800	41.2530	41.8259	42.3989	42.9718	43.5448	44.1178	44.6907	45.2637
.8	45.8366	46.4096	46.9825	47.5555	48.1285	48.7014	49.2744	49.8473	50.4203	50.9932
.9	51.5662	52.1392	52.7121	53.2851	53.8580	54.4310	55.0039	55.5769	56.1499	56.7228

1 radian = 57.29578 deg 2 radians = 114.59156 deg 3 radians = 171.88734 deg

TABLE 1–52
Linear Measure in the Metric System

1 micron (μ)	=	0.000001 meter 0.001 millimeter
10 millimeters (mm)	=	1 centimeter
10 centimeters (cm)	=	1 decimeter
10 decimeters (dm)	=	1 meter
10 meters (m)	=	1 dekameter
10 dekameters	=	1 hectometer
10 hectometers	=	1 kilometer
10 kilometers	=	1 myriameter
10 myriameters	=	1 megameter

TABLE 1–53
Selected Equivalents in Metric and English Standards of Measure

1 inch = 2.5400 millimeters
1 foot = 304.800 millimeters
1 foot = 0.3048 meter
1 yard = 0.914402 meter
1 mile = 1.609347 kilometers
1 square inch = 6.45163 square centimeters
1 square foot = 929.034 square centimeters
1 square yard = 0.836131 square meters

1 millimeter = 0.03937 inch
1 millimeter = 0.003281 foot
1 meter = 3.2808 feet
1 meter = 1.0936 yard
1 kilometer = 0.621370 mile
1 square centimeter = 0.155000 square inch
1 square centimeter = 0.001076 square foot
1 square meter = 1.19599 square yard
1 square meter = 10.7639 square feet

TABLE 1–54

Millimeters into Inches

Milli-meters	Inches	Milli-meters	Inches	Milli-meters	Inches	Milli-meters	Inches
1	.03937	26	1.02362	51	2.00787	76	2.99212
2	.07874	27	1.06299	52	2.04724	77	3.03149
3	.11811	28	1.10236	53	2.08661	78	3.07086
4	.15748	29	1.14173	54	2.12598	79	3.11023
5	.19685	30	1.18110	55	2.16535	80	3.14960
6	.23622	31	1.22047	56	2.20472	81	3.18897
7	.27559	32	1.25984	57	2.24409	82	3.22834
8	.31496	33	1.29921	58	2.28346	83	3.26771
9	.35433	34	1.33858	59	2.32283	84	3.30708
10	.39370	35	1.37795	60	2.36220	85	3.34645
11	.43307	36	1.41732	61	2.40157	86	3.38582
12	.47244	37	1.45669	62	2.44094	87	3.42519
13	.51181	38	1.49606	63	2.48031	88	3.46456
14	.55118	39	1.53543	64	2.51968	89	3.50393
15	.59055	40	1.57480	65	2.55905	90	3.54330
16	.62992	41	1.61417	66	2.59842	91	3.58267
17	.66929	42	1.65354	67	2.63779	92	3.62204
18	.70866	43	1.69291	68	2.67716	93	3.66141
19	.74803	44	1.73228	69	2.71653	94	3.70078
20	.78740	45	1.77165	70	2.75590	95	3.74015
21	.82677	46	1.81102	71	2.79527	96	3.77952
22	.86614	47	1.85039	72	2.83464	97	3.81889
23	.90551	48	1.88976	73	2.87401	98	3.85826
24	.94488	49	1.92913	74	2.91338	99	3.89763
25	.98425	50	1.96850	75	2.95275	100	3.93700

TABLE 1-55

Multiples and Roots of π and Base e of Natural Logarithms

π = 3.141592654 $\quad$ $1/\pi$ = 0.3183098862

π^2 = 9.869604401 $\quad$ $1/\pi^2$ = 0.1013211836

π^3 = 31.006276680

$\sqrt{\pi}$ = 1,772453831 $\quad$ $1/\sqrt{\pi}$ = 0.5641895835

$\sqrt[3]{\pi}$ = 1.464591888 $\quad$ $\sqrt[3]{1/\pi}$ = 0.6827840632

$\log_e \pi$ = 1.144729886 $\quad$ $\log_{10} \pi$ = 0.4971498727

e = 2.718281828 $\quad$ $1/e$ = 0.3678794412

$1/M = \log_e 10$ = 2.302585093 $\quad$ $M = \log_{10} e$ = 0.4342944819

1 radian = 57.29577951 degrees

1 radian = 3437.746771 minutes

$\sqrt{2}$ = 1.414213562

$\sqrt{3}$ = 1.7320550808

INDEX TO
SECTION 2

TABLE 2–1

Millimeter Bore-Sizes of Ball Bearings, Universally Available

Bore		Bore		Bore		Bore	
mm	Inch	mm	Inch	mm	Inch	mm	Inch
*10	.3937	45	1.7717	95	3.7402	170	6.6929
*12	.4724	50	1.9685	100	3.9370	180	7.0866
15	.5906	55	2.1654	105	4.1339	190	7.4803
17	.6693	60	2.3622	110	4.3307	200	7.8740
		65	2.5591			220	8.6614
20	.7874	70	2.7559	120	4.7244	240	9.4488
25	.9843	75	2.9528	130	5.1181	260	10.2362
30	1.1811	80	3.1496	140	5.5118	280	11.0236
35	1.3780	85	3.3465	150	5.9055	300	11.8110
40	1.5748	90	3.5433	160	6.2992	320	12.5984

*Except for the 10 and 12 mm sizes, cylindrical roller bearings are regularly available in same bore sizes.

TABLE 2–2

Inch Bore-Sizes of Ball Bearings, Commonly Available*

Bore		Bore	
Fraction	Decimal	Fraction	Decimal
1/8	.1250	3/4	.7500
3/16	.1875	7/8	.8750
1/4	.2500	1	1.0000
3/8	.3750	1-1/8	1.1250
1/2	.5000	1-1/4	1.2500
5/8	.6250	1-3/8	1.3750
		1-1/2	1.5000

*See Table 2-4 for other miniature ball bearings having inch dimensions.

TABLE 2-3

Boundary Dimensions, Type BM (Magneto) Ball Bearings, Separable†

AFBMA Standards, Section 2 – Nov 1952

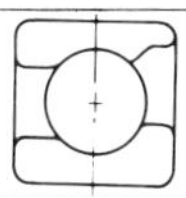

Bearing Number	Bore		Outer Diameter		Individual Ring Width		Fillet Radius	
	Nominal		Nominal		Nominal		r*	
	mm	inch	mm	inch	mm	inch	mm	inch
5BM02	5	.1969	16	.6299	5	.1969	.2	.008
6BM002	6	.2362	24	.9449	7	.2756	.3	.012
7BM002	7	.2756	24	.9449	7	.2756	.3	.012
8BM016	8	.3150	24	.9449	7	.2756	.3	.012
9BM001	9	.3543	28	1.1024	8	.3150	.3	.012
10BM001	10	.3937	28	1.1024	8	.3150	.3	.012
11BM002	11	.4331	32	1.2598	7	.2756	.4	.016
12BM002	12	.4724	32	1.2598	7	.2756	.4	.016
13BM001	13	.5118	30	1.1811	7	.2756	.3	.012
14BM001	14	.5512	35	1.3780	8	.3150	.5	.020
15BM001	15	.5906	35	1.3780	8	.3150	.5	.020
16BM001	16	.6299	38	1.4961	10	.3937	1.0	.040
17BM003	17	.6693	44	1.7323	11**	.4331	1.0	.040

†Separable means that bearing may be disassembled and the rings fitted separately to the housing and shaft.
*The corner radius or chamfer on bearings must clear the maximum fillet radius given in the table. This specification does not control bearing corner contours.
**Outer Ring Width — 10 mm.

TABLE 2-4

Standardized Dimensions for Miniature Ball Bearings* – Inch Dimensions

Bore No.	Bore Inch	O.D. Inch	Width Inch	Bore No.	Bore Inch	O.D. Inch	Width Inch
01	0.0250	1/10	1/32	04	0.0550	3/16	5/64
02	.0400	1/8	3/64	05	.0781	1/4	3/32
03	.0469	5/32	1/16	06	.0937	5/16	7/64

*The AFBMA defines a miniature ball bearing as one having an outside diameter less than 9 mm and 3/8 in.

TABLE 2–5

AFBMA* Coding of Antifriction Bearings

Antifriction bearings, particularly the conventional varieties of ball and roller bearings, can be positively identified and fully described by the AFBMA number-letter code. Thus 50BC02JO33B identifies and describes a single-row ball bearing. The 50BC02 is that portion of the code number pertinent to identification. It is called the **basic number.** The remainder of the code is called the **supplementary number.** Since seals, shields, snap ring modifications and lubrication, as well as cage construction, clearances and tolerances, are the matter within the supplementary number, the prudent designer may find the context of Section 5, AFBMA Standards, well worth careful study, for space here is not available for adequate explanation of it.

The *B* of the basic number identifies the bearing as a ball bearing. The *C* stands for Conrad or non-filling-slot type. The 50 defines the bore as 50 mm; the 02 is a width-OD dimension series, indicating indirectly by the zero that the width of the ball bearing is 20 mm and by the 2 that the OD is 90 mm. Other dimension series are illustrated by Tables 2-8 to 2-16 inclusive. These dimension series are replacing the older forms of designating series, as for example, light, medium, extra light, and consequently they are positioned conspicuously in the headings of many of the tables to follow on ball and roller bearings.

Besides ball bearings the AFBMA identification code aims toward the standardization of all kinds of antifriction bearings. Such bearings are divided into five main groups with certain letters to distinguish each type as follows:

Group	Kind of Bearing	First letter of type code	Remarks and Examples
I	Annular ball bearings	*B* Airframe bearings are exceptions	Table 2-23
	Cylindrical roller bearings	*R*	Table 2-24
	Self-aligning roller bearings	*S* Airframe bearings are exceptions	
II	Inch dimensioned tapered roller bearings		Table 2-25, Note that the letter *T* is reserved by AFBMA for "thrust" rather than "taper".
III	Journal roller bearings – inch dimensions	*J*	
	Journal roller bearings – metric dimension	*M*	
	Needle roller bearings	*N*	Table 2-26
	Cam roller bearings	*C*	Table 2-22
	Loose rollers	*L*	
IV	Thrust bearings – ball and roller	*T*	
	Clutch release bearings	*T*	
	Tapered roller thrust bearings	*T*	
V	Unground bearings	*U*	

*The Anti-Friction Bearing Manufacturers Association, Inc, 60 East 42nd St, New York 17, N.Y.

TABLE 2–6

Dimension Series 10 (Extra Light) Ball Bearings

AFBMA Standards, Section 2 – Nov 1952

Ball Bearing Code Numbers					Ball Bearing Dimensions							
Type BA	Type BN	Type BT	Type BC	Type BH	Bore *B* Nominal mm	Bore *B* Nominal Inch	Outside Diam *D* Nominal mm	Outside Diam *D* Nominal Inch	Individual Ring Width *W* Nominal mm	Individual Ring Width *W* Nominal Inch	Fillet Radius r^* mm	Fillet Radius r^* Inch
10BA10	10BN10	10BT10	10BC10	10BH10	10	.3937	26	1.0236	8	.3150	.3	.012
12BA10	12BN10	12BT10	12BC10	12BH10	12	.4724	28	1.1024	8	.3150	.3	.012
15BA10	15BN10	15BT10	15BC10	15BH10	15	.5906	32	1.2598	9	.3543	.3	.012
17BA10	17BN10	17BT10	17BC10	17BH10	17	.6693	35	1.3780	10	.3937	.3	.012
20BA10	20BN10	20BT10	20BC10	20BH10	20	.7874	42	1.6535	12	.4724	.6	.025
25BA10	25BN10	25BT10	25BC10	25BH10	25	.9843	47	1.8504	12	.4724	.6	.025
30BA10	30BN10	30BT10	30BC10	30BH10	30	1.1811	55	2.1654	13	.5118	1.0	.04
35BA10	35BN10	35BT10	35BC10	35BH10	35	1.3780	62	2.4409	14	.5512	1.0	.04
40BA10	40BN10	40BT10	40BC10	40BH10	40	1.5748	68	2.6772	15	.5906	1.0	.04
45BA10	45BN10	45BT10	45BC10	45BH10	45	1.7717	75	2.9528	16	.6299	1.0	.04
50BA10	50BN10	50BT10	50BC10	50BH10	50	1.9685	80	3.1496	16	.6299	1.0	.04
55BA10	55BN10	55BT10	55BC10	55BH10	55	2.1654	90	3.5433	18	.7087	1.0	.04
60BA10	60BN10	60BT10	60BC10	60BH10	60	2.3622	95	3.7402	18	.7087	1.0	.04
65BA10	65BN10	65BT10	65BC10	65BH10	65	2.5591	100	3.9370	18	.7087	1.0	.04
70BA10	70BN10	70BT10	70BC10	70BH10	70	2.7559	110	4.3307	20	.7874	1.0	.04
75BA10	75BN10	75BT10	75BC10	75BH10	75	2.9528	115	4.5276	20	.7874	1.0	.04
80BA10	80BN10	80BT10	80BC10	80BH10	80	3.1496	125	4.9213	22	.8661	1.0	.04
85BA10	85BN10	85BT10	85BC10	85BH10	85	3.3465	130	5.1181	22	.8661	1.0	.04
90BA10	90BN10	90BT10	90BC10	90BH10	90	3.5433	140	5.5118	24	.9449	1.5	.06
95BA10	95BN10	95BT10	95BC10	95BH10	95	3.7402	145	5.7087	24	.9449	1.5	.06
100BA10	100BN10	100BT10	100BC10	100BH10	100	3.9370	150	5.9055	24	.9449	1.5	.06
105BA10	105BN10	105BT10	105BC10	105BH10	105	4.1339	160	6.2992	26	1.0236	2.0	.08
110BA10	110BN10	110BT10	110BC10	110BH10	110	4.3307	170	6.6929	28	1.1024	2.0	.08
120BA10	120BN10	120BT10	120BC10	120BH10	120	4.7244	180	7.0866	28	1.1024	2.0	.08
130BA10	130BN10	130BT10	130BC10	130BH10	130	5.1181	200	7.8740	33	1.2992	2.0	.08
140BA10	140BN10	140BT10	140BC10	140BH10	140	5.5118	210	8.2677	33	1.2992	2.0	.08
150BA10	150BN10	150BT10	150BC10	150BH10	150	5.9055	225	8.8583	35	1.3780	2.0	.08
160BA10	160BN10	160BT10	160BC10	160BH10	160	6.2992	240	9.4488	38	1.4961	2.0	.08
170BA10	170BN10	170BT10	170BC10	170BH10	170	6.6929	260	10.2362	42	1.6535	2.0	.08
180BA10	180BN10	180BT10	180BC10	180BH10	180	7.0866	280	11.0236	46	1.8110	2.0	.08
190BA10	190BN10	190BT10	190BC10	190BH10	190	7.4803	290	11.4173	46	1.8110	2.0	.08
200BA10	200BN10	200BT10	200BC10	200BH10	200	7.8740	310	12.2047	51	2.0079	2.0	.08
220BA10	220BN10	220BT10	220BC10	220BH10	220	8.6614	340	13.3858	56	2.2047	2.5	.10
240BA10	240BN10	240BT10	240BC10	240BH10	240	9.4488	360	14.1732	56	2.2047	2.5	.10
260BA10	260BN10	260BT10	260BC10	260BH10	260	10.2362	400	15.7480	65	2.5591	3.0	.12
280BA10	280BN10	280BT10	280BC10	280BH10	280	11.0236	420	16.5354	65	2.5591	3.0	.12
300BA10	300BN10	300BT10	300BC10	300BH10	300	11.8110	460	18.1102	74	2.9134	3.0	.12
320BA10	320BN10	320BT10	320BC10	320BH10	320	12.5984	480	18.8976	74	2.9134	3.0	.12

*The corner radius or chamfer on bearings must clear the maximum fillet radius given in the table. This specification does not control bearing corner contours.

TABLE 2–7

Dimension Series 02 (Light) Ball Bearings

AFBMA Standards, Section 2 – Nov 1952

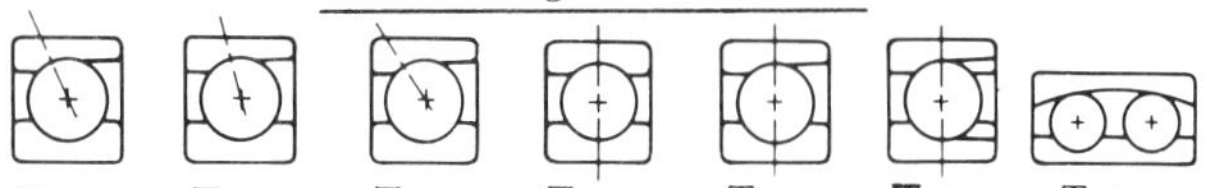

Ball Bearing Code Numbers							Ball Bearing Dimensions							
							Bore *B*		Outside Diam *D*		Individual Ring Width *W*		Fillet Radius *r**	
Type BA	Type BN	Type BT	Type BC	Type BH	Type BL	Type BS	Nominal mm	Nominal Inch	Nominal mm	Nominal Inch	Nominal mm	Nominal Inch	mm	Inch
			4BC03				4	.1575	16	.6299	5	.1969	.3	.012
			5BC03			5BS03	5	.1969	19	.7480	6	.2362	.3	.012
			6BC02			6BS02	6	.2362	19	.7480	6	.2362	.3	.012
			7BC02			7BS02	7	.2756	22	.8661	7	.2756	.3	.012
			8BC10			8BS10	8	.3150	22	.8661	7	.2756	.3	.012
			9BC02			9BS02	9	.3543	26	1.0236	8	.3150	.6	.025
10BA02	10BN02	10BT02	10BC02	10BH02	10BL02	10BS02	10	.3937	30	1.1811	9	.3543	.6	.025
12BA02	12BN02	12BT02	12BC02	12BH02	12BL02	10BS02	12	.4724	32	1.2598	10	.3937	.6	.025
15BA02	15BN02	15BT02	15BC02	15BH02	15BL02	15BS02	15	.5906	35	1.3780	11	.4331	.6	.025
17BA02	17BN02	17BT02	17BC02	17BH02	17BL02	17BS02	17	.6693	40	1.5748	12	.4724	.6	.025
20BA02	20BN02	20BT02	20BC02	20BH02	20BL02	20BS02	20	.7874	47	1.8504	14	.5512	1.0	.04
25BA02	25BN02	25BT02	25BC02	25BH02	25BL02	25BS02	25	.9843	52	2.0472	15	.5906	1.0	.04
30BA02	30BN02	30BT02	30BC02	30BH02	30BL02	30BS02	30	1.1811	62	2.4409	16	.6299	1.0	.04
35BA02	35BN02	35BT02	35BC02	35BH02	35BL02	35BS02	35	1.3780	72	2.8346	17	.6693	1.0	.04
40BA02	40BN02	40BT02	40BC02	40BH02	40BL02	40BS02	40	1.5748	80	3.1496	18	.7087	1.0	.04
45BA02	45BN02	45BT02	45BC02	45BH02	45BL02	45BS02	45	1.7717	85	3.3465	19	.7480	1.0	.04
50BA02	50BN02	50BT02	50BC02	50BH02	50BL02	50BS02	50	1.9685	90	3.5435	20	.7874	1.0	.04
55BA02	55BN02	55BT02	55BC02	55BH02	55BL02	55BS02	55	2.1654	100	3.9370	21	.8268	1.5	.06
60BA02	60BN02	60BT02	60BC02	60BH02	60BL02	60BS02	60	2.3622	110	4.3307	22	.8661	1.5	.06
65BA02	65BN02	65BT02	65BC02	65BH02	65BL02	65BS02	65	2.5591	120	4.7244	23	.9055	1.5	.06
70BA02	70BN02	70BT02	70BC02	70BH02	70BL02	70BS02	70	2.7559	125	4.9213	24	.9449	1.5	.06
75BA02	75BN02	75BT02	75BC02	75BH02	75BL02	75BS02	75	2.9528	130	5.1181	25	.9843	1.5	.06
80BA02	80BN02	80BT02	80BC02	80BH02	80BL02	80BS02	80	3.1496	140	5.5118	26	1.0236	2.0	.08
85BA02	85BN02	85BT02	85BC02	85BH02	85BL02	85BS02	85	3.3465	150	5.9055	28	1.1024	2.0	.08
90BA02	90BN02	90BT02	90BC02	90BH02	90BL02	90BS02	90	3.5433	160	6.2992	30	1.1811	2.0	.08
95BA02	95BN02	95BT02	95BC02	95BH02	95BL02	95BS02	95	3.7402	170	6.6929	32	1.2598	2.0	.08
100BA02	100BN02	100BT02	100BC02	100BH02	100BL02	100BS02	100	3.9370	180	7.0866	34	1.3386	2.0	.08
105BA02	105BN02	105BT02	105BC02	105BH02	105BL02	105BS02	105	4.1339	190	7.4803	36	1.4173	2.0	.08
110BA02	110BN02	110BT02	110BC02	110BH02	110BL02	110BS02	110	4.3307	200	7.8740	38	1.4961	2.0	.08
120BA02	120BN02	120BT02	120BC02	120BH02	120BL02		120	4.7244	215	8.4646	40	1.5748	2.0	.08
130BA02	130BN02	130BT02	130BC02	130BH02	130BL02		130	5.1181	230	9.0551	40	1.5748	2.5	.10
140BA02	140BN02	140BT02	140BC02	140BH02	140BL02		140	5.5118	250	9.8425	42	1.6535	2.5	.10
150BA02	150BN02	150BT02	150BC02	150BH02	150BL02		150	5.9055	270	10.6299	45	1.7717	2.5	.10
160BA02	160BN02	160BT02	160BC02	160BH02	160BL02		160	6.2992	290	11.4173	48	1.8898	2.5	.10
170BA02	170BN02	170BT02	170BC02	170BH02	170BL02		170	6.6929	310	12.2047	52	2.0472	3.0	.12
180BA02	180BN02	180BT02	180BC02	180BH02	180BL02		180	7.0866	320	12.5984	52	2.0472	3.0	.12
190BA02	190BN02	190BT02	190BC02	190BH02	190BL02		190	7.4803	340	13.3858	55	2.1654	3.0	.12
200BA02	200BN02	200BT02	200BC02	200BH02	200BL02		200	7.8740	360	14.1732	58	2.2835	3.0	.12
220BA02	220BN02	220BT02	220BC02	220BH02	220BL02		220	8.6614	400	15.7480	65	2.5591	3.0	.12
240BA02	240BN02	240BT02	240BC02	240BH02	240BL02		240	9.4488	440	17.3228	72	2.8346	3.0	.12
260BA02	260BN02	260BT02	260BC02	260BH02	260BL02		260	10.2362	480	18.8976	80	3.1496	4.0	.16
280BA02	280BN02	280BT02	280BC02	280BH02	280BL02		280	11.0236	500	19.6850	80	3.1496	4.0	.16
300BA02	300BN02	300BT02	300BC02	300BH02	300BL02		300	11.8110	540	21.2598	85	3.3465	4.0	.16
320BA02	320BN02	320BT02	320BC02	320BH02	320BL02		320	12.5984	580	22.8346	92	3.6220	4.0	.16

*The corner radius or chamfer on bearings must clear the maximum fillet radius given in the table. This specification does not control bearing corner contours.

TABLE 2-8

Dimension Series 03 (Medium) Ball Bearings

AFBMA Standards, Section 2—Nov 1952

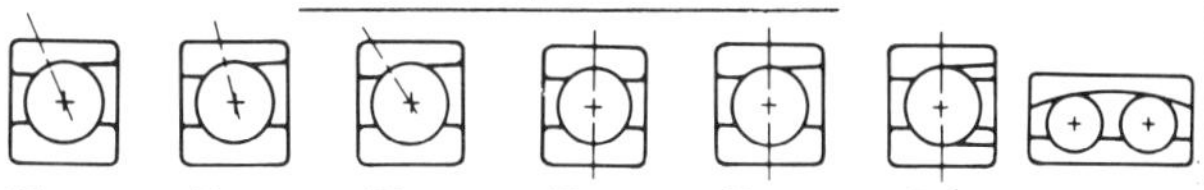

Ball Bearing Code Numbers							Ball Bearing Dimensions							
							Bore B		Outside Diam D		Individual Ring Width W		Fillet Radius r*	
Type BA	Type BN	Type BT	Type BC	Type BH	Type BL	Type BS	Nominal		Nominal		Nominal			
							mm	Inch	mm	Inch	mm	Inch	mm	Inch
10BA03	10BN03	10BT03	10BC03	10BH03	10BL03	10BS03	10	.3937	35	1.3780	11	.4331	.6	.025
12BA03	12BN03	12BT03	12BC03	12BH03	12BL03	12BS03	12	.4724	37	1.4567	12	.4724	1.0	.04
15BA03	15BN03	15BT03	15BC03	15BH03	15BL03	15BS03	15	.5906	42	1.6535	13	.5118	1.0	.04
17BA03	17BN03	17BT03	17BC03	17BH03	17BL03	17BS03	17	.6693	47	1.8504	14	.5512	1.0	.04
20BA03	20BN03	20BT03	20BC03	20BH03	20BL03	20BS03	20	.7874	52	2.0472	15	.5906	1.0	.04
25BA03	25BN03	25BT03	25BC03	25BH03	25BL03	25BS03	25	.9843	62	2.4409	17	.6693	1.0	.04
30BA03	30BN03	30BT03	30BC03	30BH03	30BL03	30BS03	30	1.1811	72	2.8346	19	.7480	1.0	.04
35BA03	35BN03	35BT03	35BC03	35BH03	35BL03	35BS03	35	1.3780	80	3.1496	21	.8268	1.5	.06
40BA03	40BN03	40BT03	40BC03	40BH03	40BL03	40BS03	40	1.5748	90	3.5433	23	.9055	1.5	.06
45BA03	45BN03	45BT03	45BC03	45BH03	45BL03	45BS03	45	1.7717	100	3.9370	25	.9843	1.5	.06
50BA03	50BN03	50BT03	50BC03	50BH03	50BL03	50BS03	50	1.9685	110	4.3307	27	1.0630	2.0	.08
55BA03	55BN03	55BT03	55BC03	55BH03	55BL03	55BS03	55	2.1654	120	4.7244	29	1.1417	2.0	.08
60BA03	60BN03	60BT03	60BC03	60BH03	60BL03	60BS03	60	2.3622	130	5.1181	31	1.2205	2.0	.08
65BA03	65BN03	65BT03	65BC03	65BH03	65BL03	65BS03	65	2.5591	140	5.5118	33	1.2992	2.0	.08
70BA03	70BN03	70BT03	70BC03	70BH03	70BL03	70BS03	70	2.7559	150	5.9055	35	1.3780	2.0	.08
75BA03	75BN03	75BT03	75BC03	75BH03	75BL03	75BS03	75	2.9528	160	6.2992	37	1.4567	2.0	.08
80BA03	80BN03	80BT03	80BC03	80BH03	80BL03	80BS03	80	3.1496	170	6.6929	39	1.5354	2.0	.08
85BA03	85BN03	85BT03	85BC03	85BH03	85BL03	85BS03	85	3.3465	180	7.0866	41	1.6142	2.5	.10
90BA03	90BN03	90BT03	90BC03	90BH03	90BL03	90BS03	90	3.5433	190	7.4803	43	1.6929	2.5	.10
95BA03	95BN03	95BT03	95BC03	95BH03	95BL03	95BS03	95	3.7402	200	7.8740	45	1.7717	2.5	.10
100BA03	100BN03	100BT03	100BC03	100BH03	100BL03	100BS03	100	3.9370	215	8.4646	47	1.8504	2.5	.10
105BA03	105BN03	105BT03	105BC03	105BH03	105BL03	105BS03	105	4.1339	225	8.8583	49	1.9291	2.5	.10
110BA03	110BN03	110BT03	110BC03	110BH03	110BL03	110BS03	110	4.3307	240	9.4488	50	1.9685	2.5	.10
120BA03	120BN03	120BT03	120BC03	120BH03	120BL03		120	4.7244	260	10.2362	55	2.1654	2.5	.10
130BA03	130BN03	130BT03	130BC03	130BH03	130BL03		130	5.1181	280	11.0236	58	2.2835	3.0	.12
140BA03	140BN03	140BT03	140BC03	140BH03	140BL03		140	5.5118	300	11.8110	62	2.4409	3.0	.12
150BA03	150BN03	150BT03	150BC03	150BH03	150BL03		150	5.9055	320	12.5984	65	2.5591	3.0	.12
160BA03	160BN03	160BT03	160BC03	160BH03	160BL03		160	6.2992	340	13.3858	68	2.6772	3.0	.12
170BA03	170BN03	170BT03	170BC03	170BH03	170BL03		170	6.6929	360	14.1732	72	2.8346	3.0	.12
180BA03	180BN03	180BT03	180BC03	180BH03	180BL03		180	7.0866	380	14.9606	75	2.9528	3.0	.12
190BA03	190BN03	190BT03	190BC03	190BH03	190BL03		190	7.4803	400	15.7480	78	3.0709	4.0	.16
200BA03	200BN03	200BT03	200BC03	200BH03	200BL03		200	7.8740	420	16.5354	80	3.1496	4.0	.16
220BA03	220BN03	220BT03	220BC03	220BH03	220BL03		220	8.6614	460	18.1102	88	3.4646	4.0	.16
240BA03	240BN03	240BT03	240BC03	240BH03	240BL03		240	9.4488	500	19.6850	95	3.7402	4.0	.16
260BA03	260BN03	260BT03	260BC03	260BH03	260BL03		260	10.2362	540	21.2598	102	4.0157	5.0	.20
280BA03	280BN03	280BT03	280BC03	280BH03	280BL03		280	11.0236	580	22.8346	108	4.2520	5.0	.20

*The corner radius or chamfer on bearings must clear the maximum fillet radius given in the table. This specification does not control bearing corner contours.

TABLE 2-9

Dimension Series 04 (Heavy) Ball Bearings

AFBMA Standards, Section 2 – Nov 1952

Ball Bearing Code Numbers							Ball Bearing Dimensions							
	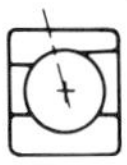	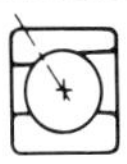	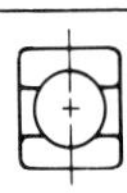	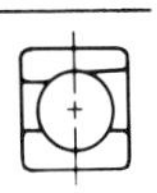	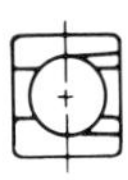	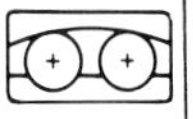	Bore B		Outside Diam D		Individual Ring Width W		Fillet Radius r*	
Type BA	Type BN	Type BT	Type BC	Type BH	Type BL	Type BS	Nominal mm	Nominal Inch	Nominal mm	Nominal Inch	Nominal mm	Nominal Inch	mm	Inch
17BA04	17BN04	17BT04	17BC04	17BH04	17BL04	17BS04	17	.6693	62	2.4409	17	.6693	1.0	.04
20BA04	20BN04	20BT04	20BC04	20BH04	20BL04	20BS04	20	.7874	72	2.8346	19	.7480	1.0	.04
25BA04	25BN04	25BT04	25BC04	25BH04	25BL04	25BS04	25	.9843	80	3.1496	21	.8268	1.5	.06
30BA04	30BN04	30BT04	30BC04	30BH04	30BL04	30BS04	30	1.1811	90	3.5433	23	.9055	1.5	.06
35BA04	35BN04	35BT04	35BC04	35BH04	35BL04	35BS04	35	1.3780	100	3.9370	25	.9843	1.5	.06
40BA04	40BN04	40BT04	40BC04	40BH04	40BL04	40BS04	40	1.5748	110	4.3307	27	1.0630	2.0	.08
45BA04	45BN04	45BT04	45BC04	45BH04	45BL04	45BS04	45	1.7717	120	4.7244	29	1.1417	2.0	.08
50BA04	50BN04	50BT04	50BC04	50BH04	50BL04	50BS04	50	1.9685	130	5.1181	31	1.2205	2.0	.08
55BA04	55BN04	55BT04	55BC04	55BH04	55BL04	55BS04	55	2.1654	140	5.5118	33	1.2992	2.0	.08
60BA04	60BN04	60BT04	60BC04	60BH04	60BL04	60BS04	60	2.3622	150	5.9055	35	1.3780	2.0	.08
65BA04	65BN04	65BT04	65BC04	65BH04	65BL04	65BS04	65	2.5591	160	6.2992	37	1.4567	2.0	.08
70BA04	70BN04	70BT04	70BC04	70BH04	70BL04	70BS04	70	2.7559	180	7.0866	42	1.6535	2.5	.10
75BA04	75BN04	75BT04	75BC04	75BH04	75BL04	75BS04	75	2.9528	190	7.4803	45	1.7717	2.5	.10
80BA04	80BN04	80BT04	80BC04	80BH04	80BL04	80BS04	80	3.1496	200	7.8740	48	1.8898	2.5	.10
85BA04	85BN04	85BT04	85BC04	85BH04	85BL04	85BS04	85	3.3465	210	8.2677	52	2.0472	3.0	.12
90BA04	90BN04	90BT04	90BC04	90BH04	90BL04	90BS04	90	3.5433	225	8.8583	54	2.1260	3.0	.12
95BA04	95BN04	95BT04	95BC04	95BH04	95BL04		95	3.7402	240	9.4488	55	2.1654	3.0	.12
100BA04	100BN04	100BT04	100BC04	100BH04	100BL04		100	3.9370	250	9.8425	58	2.2835	3.0	.12
105BA04	105BN04	105BT04	105BC04	105BH04	105BL04		105	4.1339	260	10.2362	60	2.3622	3.0	.12
110BA04	110BN04	110BT04	110BC04	110BH04	110BL04		110	4.3307	280	11.0236	65	2.5591	3.0	.12
120BA04	120BN04	120BT04	120BC04	120BH04	120BL04		120	4.7244	310	12.2047	72	2.3846	4.0	.16
130BA04	130BN04	130BT04	130BC04	130BH04	130BL04		130	5.1181	340	13.3858	78	3.0709	4.0	.16
140BA04	140BN04	140BT04	140BC04	140BH04	140BL04		140	5.5118	360	14.1732	82	3.2283	4.0	.16
150BA04	150BN04	150BT04	150BC04	150BH04	150BL04		150	5.9055	380	14.9606	85	3.3465	4.0	.16

*The corner radius or chamfer on bearings must clear the maximum fillet radius given in the table. This specification does not control bearing corner contours.

TABLE 2–10

Dimension Series 19, Type BC, Ball Bearings, Single Row, Deep Groove

AFBMA Standards, Section 2–Nov 1952

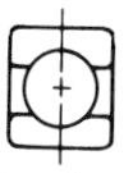 Bearing Number	Bore mm	Bore Inch	Outside Diameter mm	Outside Diameter Inch	Width mm	Width Inch	Fillet Radius r* Inch
15BC19	15	.5906	28	1.1024	7	.2756	.012
17BC19	17	.6693	30	1.1811	7	.2756	.012
20BC19	20	.7874	37	1.4567	9	.3543	.012
25BC19	25	.9843	42	1.6535	9	.3543	.012
30BC19	30	1.1811	47	1.8504	9	.3543	.012
35BC19	35	1.3780	55	2.1654	10	.3937	.024
40BC19	40	1.5748	62	2.4409	12	.4724	.024
45BC19	45	1.7717	68	2.6772	12	.4724	.024
50BC19	50	1.9685	72	2.8346	12	.4724	.024
55BC19	55	2.1654	80	3.1496	13	.5118	.039
60BC19	60	2.3622	85	3.3465	13	.5118	.039
65BC19	65	2.5591	90	3.5433	13	.5118	.039
70BC19	70	2.7559	100	3.9370	16	.6299	.039
75BC19	75	2.9528	105	4.1339	16	.6299	.039
80BC19	80	3.1496	110	4.3307	16	.6299	.039
85BC19	85	3.3465	120	4.7244	18	.7087	.039
90BC19	90	3.5433	125	4.9213	18	.7087	.039
95BC19	95	3.7402	130	5.1181	18	.7087	.039
100BC19	100	3.9370	140	5.5118	20	.7874	.039
105BC19	105	4.1339	145	5.7087	20	.7874	.039
110BC19	110	4.3307	150	5.9055	20	.7874	.039
120BC19	120	4.7244	165	6.4961	22	.8661	.039
130BC19	130	5.1181	180	7.0866	24	.9449	.059
140BC19	140	5.5118	190	7.4803	24	.9449	.059
150BC19	150	5.9055	210	8.2677	28	1.1024	.079
160BC19	160	6.2992	220	8.6614	28	1.1024	.079
170BC19	170	6.6929	230	9.0551	28	1.1024	.079
180BC19	180	7.0866	250	9.8425	33	1.2992	.079
190BC19	190	7.4803	260	10.2362	33	1.2992	.079
200BC19	200	7.8740	280	11.0236	38	1.4961	.079
220BC19	220	8.6614	300	11.8110	38	1.4961	.079
240BC19	240	9.4488	320	12.5984	38	1.4961	.079
260BC19	260	10.2362	360	14.1732	46	1.8110	.079
280BC19	280	11.0236	380	14.9606	46	1.8110	.079
300BC19	300	11.8110	420	16.5354	56	2.2047	.098

*The corner radius or chamfer on bearings must clear the maximum fillet radius given in the table. This specification does not control bearing corner contours.

TABLE 2-11

Dimension Series 22 (Light) and 23 (Medium), Type BS, Ball Bearings, Double Row, Radial, Self-Aligning

AFBMA Standards, Section 2—Nov 1952

Dimension Series 22 (Light)

Bore Nominal mm	Bore Nominal Inch	Bearing Number	Outer Diameter Nominal mm	Outer Diameter Nominal Inch	Individual Ring Width Nominal mm	Individual Ring Width Nominal Inch	Fillet Radius r* Nominal mm	Fillet Radius r* Nominal Inch
10	.3937	10BS22	30	1.1811	14	.5512	.6	.025
12	.4724	12BS22	32	1.2598	14	.5512	.6	.025
15	.5906	15BS22	35	1.3780	14	.5512	.6	.025
17	.6693	17BS22	40	1.5748	16	.6299	1.0	.04
20	.7874	20BS22	47	1.8504	18	.7087	1.0	.04
25	.9843	25BS22	52	2.0472	18	.7087	1.0	.04
30	1.1811	30BS22	62	2.4409	20	.7874	1.0	.04
35	1.3780	35BS22	72	2.8346	23	.9055	1.0	.04
40	1.5748	40BS22	80	3.1496	23	.9055	1.0	.04
45	1.7717	45BS22	85	3.3465	23	.9055	1.0	.04
50	1.9685	50BS22	90	3.5433	23	.9055	1.0	.04
55	2.1654	55BS22	100	3.9370	25	.9843	1.5	.06
60	2.3622	60BS22	110	4.3307	28	1.1024	1.5	.06
65	2.5591	65BS22	120	4.7244	31	1.2205	1.5	.06
70	2.7559	70BS22	125	4.9213	31	1.2205	1.5	.06
75	2.9528	75BS22	130	5.1181	31	1.2205	1.5	.06
80	3.1496	80BS22	140	5.5118	33	1.2992	2.0	.08
85	3.3465	85BS22	150	5.9055	36	1.4173	2.0	.08
90	3.5433	90BS22	160	6.2992	40	1.5748	2.0	.08
95	3.7402	95BS22	170	6.6929	43	1.6929	2.0	.08
100	3.9370	100BS22	180	7.0866	46	1.8110	2.0	.08
105	4.1339	105BS22	190	7.4803	50	1.9685	2.0	.08
110	4.3307	110BS22	200	7.8740	53	2.0866	2.0	.08
120	4.7244	120BS22	215	8.4646	58	2.2835	2.0	.08
130	5.1181	130BS22	230	9.0551	64	2.5197	2.5	.10
140	5.5118	140BS22	250	9.8425	68	2.6772	2.5	.10
150	5.9055	150BS22	270	10.6299	73	2.8740	2.5	.10
160	6.2992	160BS22	290	11.4173	80	3.1496	2.5	.10
170	6.6929	170BS22	310	12.2047	86	3.3858	3.0	.12
180	7.0866	180BS22	320	12.5984	86	3.3858	3.0	.12
190	7.4803	190BS22	340	13.3858	92	3.6220	3.0	.12
200	7.8740	200BS22	360	14.1732	98	3.8583	3.0	.12
220	8.6614	220BS22	400	15.7480	108	4.2520	3.0	.12
240	9.4488	240BS22	440	17.3228	120	4.7244	3.0	.12
260	10.2362	260BS22	480	18.8976	130	5.1181	4.0	.16
280	11.0236	280BS22	500	19.6850	130	5.1181	4.0	.16
300	11.8110	300BS22	540	21.2598	140	5.5118	4.0	.16
320	12.5984	320BS22	580	22.8346	150	5.9055	4.0	.16

Dimension Series 23 (Medium)

Bearing Number	Outer Diameter Nominal mm	Outer Diameter Nominal Inch	Individual Ring Width Nominal mm	Individual Ring Width Nominal Inch	Fillet Radius r* Nominal mm	Fillet Radius r* Nominal Inch
10BS23	35	1.3780	17	.6693	.6	.025
12BS23	37	1.4567	17	.6693	1.0	.04
15BS23	42	1.6535	17	.6693	1.0	.04
17BS23	47	1.8504	19	.7480	1.0	.04
20BS23	52	2.0472	21	.8268	1.0	.04
25BS23	62	2.4409	24	.9449	1.0	.04
30BS23	72	2.8346	27	1.0630	1.0	.04
35BS23	80	3.1496	31	1.2205	1.5	.06
40BS23	90	3.5433	33	1.2992	1.5	.06
45BS23	100	3.9370	36	1.4173	1.5	.06
50BS23	110	4.3307	40	1.5748	2.0	.08
55BS23	120	4.7244	43	1.6929	2.0	.08
60BS23	130	5.1181	46	1.8110	2.0	.08
65BS23	140	5.5118	48	1.8898	2.0	.08
70BS23	150	5.9055	51	2.0079	2.0	.08
75BS23	160	6.2992	55	2.1654	2.0	.08
80BS23	170	6.6929	58	2.2835	2.0	.08
85BS23	180	7.0866	60	2.3622	2.5	.10
90BS23	190	7.4803	64	2.5197	2.5	.10
95BS23	200	7.8740	67	2.6378	2.5	.10
100BS23	215	8.4646	73	2.8740	2.5	.10
105BS23	225	8.8583	77	3.0315	2.5	.10
110BS23	240	9.4488	80	3.1496	2.5	.10
120BS23	260	10.2362	86	3.3858	2.5	.10
130BS23	280	11.0236	93	3.6614	3.0	.12
140BS23	300	11.8110	102	4.0157	3.0	.12
150BS23	320	12.5984	108	4.2520	3.0	.12
160BS23	340	13.3858	114	4.4882	3.0	.12
170BS23	360	14.1732	120	4.7244	3.0	.12
180BS23	380	14.9606	126	4.9606	3.0	.12
190BS23	400	15.7480	132	5.1968	4.0	.16
200BS23	420	16.5354	138	5.4331	4.0	.16
220BS23	460	18.1102	145	5.7087	4.0	.16
240BS23	500	19.6850	155	6.1024	4.0	.16
260BS23	540	21.2598	165	6.4961	5.0	.20
280BS23	580	22.8346	175	6.8898	5.0	.20

*The corner radius or chamfer on bearings must clear the maximum fillet radius given in the table. This specification does not control bearing corner contours.

TABLE 2–12

Dimension Series 32 (Light) and 92 (Extended Light) Ball Bearings

AFBMA Standard, Section 2 – Nov 1952

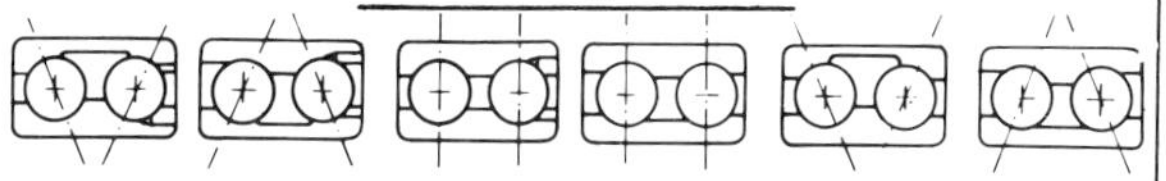

Ball Bearing Numbers						Ball Bearing Dimensions							
						Bore		Outside Diameter		Ring Width		Fillet Radius r*	
Type BD	Type BE	Type BF	Type BK	Type BJ	Type BG	mm	Inch	mm	Inch	mm	Inch	mm	Inch
Dimension Series 32 (Light)													
10BD32	10BE32	10BF32	10BK32	10BJ32	10BG32	10	.3937	30	1.1811	14.3	9/16	.6	.025
12BD32	12BE32	12BF32	12BK32	12BJ32	12BG32	12	.4724	32	1.2598	15.9	5/8	.6	.025
15BD32	15BE32	15BF32	15BK32	15BJ32	15BG32	15	.5906	35	1.3780	15.9	5/8	.6	.025
17BD32	17BE32	17BF32	17BK32	17BJ32	17BG32	17	.6693	40	1.5748	17.5	11/16	.6	.025
20BD32	20BE32	20BF32	20BK32	20BJ32	20BG32	20	.7874	47	1.8504	20.6	13/16	1.0	.04
25BD32	25BE32	25BF32	25BK32	25BJ32	25BG32	25	.9843	52	2.0472	20.6	13/16	1.0	.04
30BD32	30BE32	30BF32	30BK32	30BJ32	30BG32	30	1.1811	62	2.4409	23.8	15/16	1.0	.04
35BD32	35BE32	35BF32	35BK32	35BJ32	35BG32	35	1.3780	72	2.8346	27.0	1- 1/16	1.0	.04
40BD32	40BE32	40BF32	40BK32	40BJ32	40BG32	40	1.5748	80	3.1496	30.2	1- 3/16	1.0	.04
45BD32	45BE32	45BF32	45BK32	45BJ32	45BG32	45	1.7717	85	3.3465	30.2	1- 3/16	1.0	.04
50BD32	50BE32	50BF32	50BK32	50BJ32	50BG32	50	1.9685	90	3.5433	30.2	1- 3/16	1.0	.04
55BD32	55BE32	55BF32	55BK32	55BJ32	55BG32	55	2.1654	100	3.9370	33.3	1- 5/16	1.5	.06
60BD32	60BE32	60BF32	60BK32	60BJ32	60BG32	60	2.3622	110	4.3307	36.5	1- 7/16	1.5	.06
65BD32	65BE32	65BF32	65BK32	65BJ32	65BG32	65	2.5591	120	4.7244	38.1	1- 1/2	1.5	.06
70BD32	70BE32	70BF32	70BK32	70BJ32	70BG32	70	2.7559	125	4.9213	39.7	1- 9/16	1.5	.06
75BD32	75BE32	75BF32	75BK32	75BJ32	75BG32	75	2.9528	130	5.1181	41.3	1- 5/8	1.5	.06
80BD32	80BE32	80BF32	80BK32	80BJ32	80BG32	80	3.1496	140	5.5118	44.4	1- 3/4	2.0	.08
85BD32	85BE32	85BF32	85BK32	85BJ32	85BG32	85	3.3465	150	5.9055	49.2	1-15/16	2.0	.08
90BD32	90BE32	90BF32	90BK32	90BJ32	90BG32	90	3.5433	160	6.2992	52.4	2- 1/16	2.0	.08
95BD32	95BE32	95BF32	95BK32	95BJ32	95BG32	95	3.7402	170	6.6929	55.6	2- 3/16	2.0	.08
100BD32	100BE32	100BF32	100BK32	100BJ32	100BG32	100	3.9370	180	7.0866	60.3	2- 3/8	2.0	.08
105BD32	105BE32	105BF32	105BK32	105BJ32	105BG32	105	4.1339	190	7.4803	65.1	2- 9/16	2.0	.08
110BD32	110BE32	110BF32	110BK32	110BJ32	110BG32	110	4.3307	200	7.8740	69.8	2- 3/4	2.0	.08
Dimension Series 92 (Extended Light)													
120BD92	120BE92	120BF92	120BK92	120BJ92	120BG92	120	4.7244	215	8.4646	76.2	3	2.0	.08
130BD92	130BE92	130BF92	130BK92	130BJ92	130BG92	130	5.1181	230	9.0551	79.4	3- 1/8	2.5	.10
140BD92	140BE92	140BF92	140BK92	140BJ92	140BG92	140	5.5118	250	9.8425	82.6	3- 1/4	2.5	.10
150BD92	150BE92	150BF92	150BK92	150BJ92	150BG92	150	5.9055	270	10.6299	88.9	3- 1/2	2.5	.10
160BD92	160BE92	160BF92	160BK92	160BJ92	160BG92	160	6.2992	290	11.4173	98.4	3- 7/8	2.5	.10
170BD92	170BE92	170BF92	170BK92	170BJ92	170BG92	170	6.6929	310	12.2047	104.8	4- 1/8	3.0	.12
180BD92	180BE92	180BF92	180BK92	180BJ92	180BG92	180	7.0866	320	12.5984	108.0	4- 1/4	3.0	.12
190BD92	190BE92	190BF92	190BK92	190BJ92	190BG92	190	7.4803	340	13.3858	114.3	4- 1/2	3.0	.12
200BD92	200BE92	200BF92	200BK92	200BJ92	200BG92	200	7.8740	360	14.1732	120.7	4- 3/4	3.0	.12
220BD92	220BE92	220BF92	220BK92	220BJ92	220BG92	220	8.6614	400	15.7480	133.4	5- 1/4	3.0	.12
240BD92	240BE92	240BF92	240BK92	240BJ92	240BG92	240	9.4488	440	17.3228	146.0	5- 3/4	3.0	.12
260BD92	260BE92	260BF92	260BK92	260BJ92	260BG92	260	10.2362	480	18.8976	158.8	6- 1/4	4.0	.16
280BD92	280BE92	280BF92	280BK92	280BJ92	280BG92	280	11.0236	500	19.6850	165.1	6- 1/2	4.0	.16
300BD92	300BE92	300BF92	300BK92	300BJ92	300BG92	300	11.8110	540	21.2598	177.8	7	4.0	.16
320BD92	320BE92	320BF92	320BK92	320BJ92	320BG92	320	12.5984	580	22.8346	190.5	7- 1/2	4.0	.16

*The corner radius or chamfer on bearings must clear the maximum fillet radius given in the table. This specification does not control bearing corner contours.

TABLE 2–13

Dimension Series 33 (Medium) and 93 (Extended Medium) Ball Bearings

AFBMA Standards, Section 2–Nov 1952

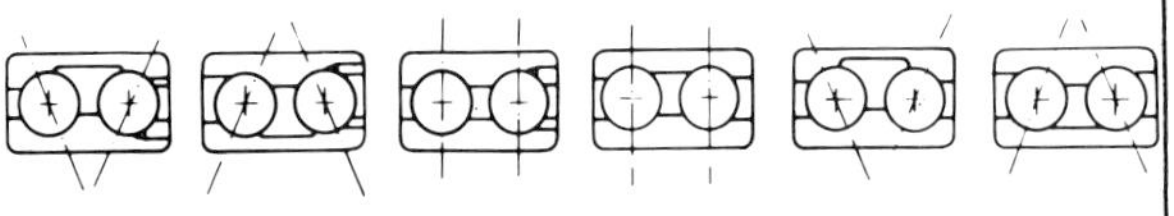

Ball Bearing Numbers						Ball Bearing Dimensions							
						Bore		Outside Diameter		Ring Width		Fillet Radius r*	
Type BD	Type BE	Type BF	Type BK	Type BJ	Type BG	mm	Inch	mm	Inch	mm	Inch	mm	Inch
Dimension Series 33 (Medium)													
10BD33	10BE33	10BF33	10BK33	10BJ33	10BG33	10	.3937	35	1.3780	19.0	3/4	.6	.025
12BD33	12BE33	12BF33	12BK33	12BJ33	12BG33	12	.4724	37	1.4567	19.0	3/4	1.0	.04
15BD33	15BE33	15BF33	15BK33	15BJ33	15BG33	15	.5906	42	1.6535	19.0	3/4	1.0	.04
17BD33	17BE33	17BF33	17BK33	17BJ33	17BG33	17	.6693	47	1.8504	22.2	7/8	1.0	.04
20BD33	20BE33	20BF33	20BK33	20BJ33	20BG33	20	.7874	52	2.0472	22.2	7/8	1.0	.04
25BD33	25BE33	25BF33	25BK33	25BJ33	25BG33	25	.9843	62	2.4409	25.4	1	1.0	.04
30BD33	30BE33	30BF33	30BK33	30BJ33	30BG33	30	1.1811	72	2.8346	30.2	1- 3/16	1.0	.04
35BD33	35BE33	35BF33	35BK33	35BJ33	35BG33	35	1.3780	80	3.1496	34.9	1- 3/8	1.5	.06
40BD33	40BE33	40BF33	40BK33	40BJ33	40BG33	40	1.5748	90	3.5433	36.5	1- 7/16	1.5	.06
45BD33	45BE33	45BF33	45BK33	45BJ33	45BG33	45	1.7717	100	3.9370	39.7	1- 9/16	1.5	.06
50BD33	50BE33	50BF33	50BK33	50BJ33	50BG33	50	1.9685	110	4.3307	44.4	1- 3/4	2.0	.08
55BD33	55BE33	55BF33	55BK33	55BJ33	55BG33	55	2.1654	120	4.7244	49.2	1-15/16	2.0	.08
60BD33	60BE33	60BF33	60BK33	60BJ33	60BG33	60	2.3622	130	5.1181	54.0	2- 1/8	2.0	.08
65BD33	65BE33	65BF33	65BK33	65BJ33	65BG33	65	2.5591	140	5.5118	58.7	2- 5/16	2.0	.08
70BD33	70BE33	70BF33	70BK33	70BJ33	70BG33	70	2.7559	150	5.9055	63.5	2- 1/2	2.0	.08
75BD33	75BE33	75BF33	75BK33	75BJ33	75BG33	75	2.9528	160	6.2992	68.3	2-11/16	2.0	.08
80BD33	80BE33	80BF33	80BK33	80BJ33	80BG33	80	3.1496	170	6.6929	68.3	2-11/16	2.0	.08
85BD33	85BE33	85BF33	85BK33	85BJ33	85BG33	85	3.3465	180	7.0866	73.0	2- 7/8	2.5	.10
90BD33	90BE33	90BF33	90BK33	90BJ33	90BG33	90	3.5433	190	7.4803	73.0	2- 7/8	2.5	.10
95BD33	95BE33	95BF33	95BK33	95BJ33	95BG33	95	3.7402	200	7.8740	77.8	3- 1/16	2.5	.10
100BD33	100BE33	100BF33	100BK33	100BJ33	100BG33	100	3.9370	215	8.4646	82.6	3- 1/4	2.5	.10
105BD33	105BE33	105BF33	105BK33	105BJ33	105BG33	105	4.1339	225	8.8583	87.3	3- 7/16	2.5	.10
110BD33	110BE33	110BF33	110BK33	110BJ33	110BG33	110	4.3307	240	9.4488	92.1	3- 5/8	2.5	.10
Dimension Series 93 (Extended Medium)													
120BD93	120BE93	120BF93	120BK93	120BJ93	120BG93	120	4.7244	260	10.2362	104.8	4- 1/8	2.5	.10
130BD93	130BE93	130BF93	130BK93	130BJ93	130BG93	130	5.1181	280	11.0236	111.1	4- 3/8	3.0	.12
140BD93	140BE93	140BF93	140BK93	140BJ93	140BG93	140	5.5118	300	11.8110	114.3	4- 1/2	3.0	.12
150BD93	150BE93	150BF93	150BK93	150BJ93	150BG93	150	5.9055	320	12.5984	123.9	4- 7/8	3.0	.12
160BD93	160BE93	160BF93	160BK93	160BJ93	160BG93	160	6.2992	340	13.3858	133.4	5- 1/4	3.0	.12
170BD93	170BE93	170BF93	170BK93	170BJ93	170BG93	170	6.6929	360	14.1732	139.7	5- 1/2	3.0	.12
180BD93	180BE93	180BF93	180BK93	180BJ93	180BG93	180	7.0866	380	14.9606	146.0	5- 3/4	3.0	.12
190BD93	190BE93	190BF93	190BK93	190BJ93	190BG93	190	7.4803	400	15.7480	152.4	6	4.0	.16
200BD93	200BE93	200BF93	200BK93	200BJ93	200BG93	200	7.8740	420	16.5354	165.1	6- 1/2	4.0	.16

*The corner radius or chamfer on bearings must clear the maximum fillet radius given in the table. This specification does not control bearing corner contours.

TABLE 2–14

Dimension Series 10, Types RN and RU, Cylindrical Roller Bearings

AFBMA Standards, Section 2 – Nov 1952

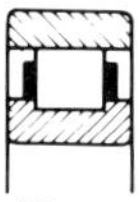
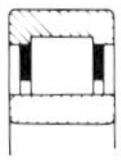

Type RN	Type RU	Bore (Metric Dimensions) mm	Bore Inch	Outside Diam mm	Outside Diam Inch	Ring Width mm	Ring Width Inch	Fillet Radius r* mm	Fillet Radius r* Inch
17RN10	17RU10	17	.6693	35	1.3780	10	.3937	.3	.012
20RN10	20RU10	20	.7874	42	1.6535	12	.4724	.6	.025
25RN10	25RU10	25	.9843	47	1.8504	12	.4724	.6	.025
30RN10	30RU10	30	1.1811	55	2.1654	13	.5118	1.0	.04
35RN10	35RU10	35	1.3780	62	2.4409	14	.5512	1.0	.04
40RN10	40RU10	40	1.5748	68	2.6772	15	.5906	1.0	.04
45RN10	45RU10	45	1.7717	75	2.9528	16	.6299	1.0	.04
50RN10	50RU10	50	1.9685	80	3.1496	16	.6299	1.0	.04
55RN10	55RU10	55	2.1654	90	3.5433	18	.7087	1.0	.04
60RN10	60RU10	60	2.3622	95	3.7402	18	.7087	1.0	.04
65RN10	65RU10	65	2.5591	100	3.9370	18	.7087	1.0	.04
70RN10	70RU10	70	2.7559	110	4.3307	20	.7874	1.0	.04
75RN10	75RU10	75	2.9528	115	4.5276	20	.7874	1.0	.04
80RN10	80RU10	80	3.1496	125	4.9213	22	.8661	1.0	.04
85RN10	85RU10	85	3.3465	130	5.1181	22	.8661	1.0	.04
90RN10	90RU10	90	3.5433	140	5.5118	24	.9449	1.5	.06
95RN10	95RU10	95	3.7402	145	5.7087	24	.9449	1.5	.06
100RN10	100RU10	100	3.9370	150	5.9055	24	.9449	1.5	.06
105RN10	105RU10	105	4.1339	160	6.2992	26	1.0236	2.0	.08
110RN10	110RU10	110	4.3307	170	6.6929	28	1.1024	2.0	.08
120RN10	120RU10	120	4.7244	180	7.0866	28	1.1024	2.0	.08
130RN10	130RU10	130	5.1181	200	7.8740	33	1.2992	2.0	.08
140RN10	140RU10	140	5.5118	210	8.2677	33	1.2992	2.0	.08
150RN10	150RU10	150	5.9055	225	8.8583	35	1.3780	2.0	.08
160RN10	160RU10	160	6.2992	240	9.4488	38	1.4961	2.0	.08
170RN10	170RU10	170	6.6929	260	10.2362	42	1.6535	2.0	.08
180RN10	180RU10	180	7.0866	280	11.0236	46	1.8110	2.0	.08
190RN10	190RU10	190	7.4803	290	11.4173	46	1.8110	2.0	.08
200RN10	200RU10	200	7.8740	310	12.2047	51	2.0079	2.0	.08
220RN10	220RU10	220	8.6614	340	13.3858	56	2.2047	2.5	.10
240RN10	240RU10	240	9.4488	360	14.1732	56	2.2047	2.5	.10
260RN10	260RU10	260	10.2362	400	15.7480	65	2.5591	3.0	.12
280RN10	280RU10	280	11.0236	420	16.5354	65	2.5591	3.0	.12
300RN10	300RU10	300	11.8110	460	18.1102	74	2.9134	3.0	.12
320RN10	320RU10	320	12.5984	480	18.8976	74	2.9134	3.0	.12

*The corner radius or chamfer on bearings must clear the maximum fillet radius given in the table. This specification does not control bearing corner contours.

TABLE 2–15

Dimension Series 02 (Light) and 03 (Medium) Cylindrical Roller Bearings

AFBMA Standards, Section 2 – Nov 1952

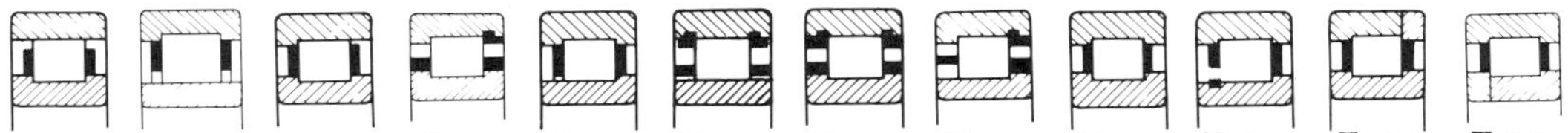

Type RN	Type RU	Type RF	Type RY	Type RJ	Type RM	Type RK	Type RS	Type RC	Type RG	Type RP	Type RT	Bore mm	Bore Inch	Outside Diam mm	Outside Diam Inch	Ring Width mm	Ring Width Inch	Fillet Radius r^* mm	Fillet Radius r^* Inch
Dimension Series 02 (Light)																			
15RN02	......	......	15RY02	......	15RM02	......	15RS02	15RC02	15RG02	15RP02	15RT02	15	.5906	35	1.3780	11	.4331	.6	.025
17RN02	17RU02	17RF02	17RY02	17RJ02	17RM02	17RK02	17RS02	17RC02	17RG02	17RP02	17RT02	17	.6693	40	1.5748	12	.4724	.6	.025
20RN02	20RU02	20RF02	20RY02	20RJ02	20RM02	20RK02	20RS02	20RC02	20RG02	20RP02	20RT02	20	.7874	47	1.8504	14	.5512	1.0	.04
25RN02	25RU02	25RF02	25RY02	25RJ02	25RM02	25RK02	25RS02	25RC02	25RG02	25RP02	25RT02	25	.9843	52	2.0472	15	.5906	1.0	.04
30RN02	30RU02	30RF02	30RY02	30RJ02	30RM02	30RK02	30RS02	30RC02	30RG02	30RP02	30RT02	30	1.1811	62	2.4409	16	.6299	1.0	.04
35RN02	35RU02	35RF02	35RY02	35RJ02	35RM02	35RK02	35RS02	35RC02	35RG02	35RP02	35RT02	35	1.3780	72	2.8346	17	.6693	1.0	.04
40RN02	40RU02	40RF02	40RY02	40RJ02	40RM02	40RK02	40RS02	40RC02	40RG02	40RP02	40RT02	40	1.5748	80	3.1496	18	.7087	1.0	.04
45RN02	45RU02	45RF02	45RY02	45RJ02	45RM02	45RK02	45RS02	45RC02	45RG02	45RP02	45RT02	45	1.7717	85	3.3465	19	.7480	1.0	.04
50RN02	50RU02	50RF02	50RY02	50RJ02	50RM02	50RK02	50RS02	50RC02	50RG02	50RP02	50RT02	50	1.9685	90	3.5433	20	.7874	1.0	.04
55RN02	55RU02	55RF02	55RY02	55RJ02	55RM02	55RK02	55RS02	55RC02	55RG02	55RP02	55RT02	55	2.1654	100	3.9370	21	.8268	1.5	.06
60RN02	60RU02	60RF02	60RY02	60RJ02	60RM02	60RK02	60RS02	60RC02	60RG02	60RP02	60RT02	60	2.3622	110	4.3307	22	.8661	1.5	.06
65RN02	65RU02	65RF02	65RY02	65RJ02	65RM02	65RK02	65RS02	65RC02	65RG02	65RP02	65RT02	65	2.5591	120	4.7244	23	.9055	1.5	.06
70RN02	70RU02	70RF02	70RY02	70RJ02	70RM02	70RK02	70RS02	70RC02	70RG02	70RP02	70RT02	70	2.7559	125	4.9213	24	.9449	1.5	.06
75RN02	75RU02	75RF02	75RY02	75RJ02	75RM02	75RK02	75RS02	75RC02	75RG02	75RP02	75RT02	75	2.9528	130	5.1181	25	.9843	1.5	.06
80RN02	80RU02	80RF02	80RY02	80RJ02	80RM02	80RK02	80RS02	80RC02	80RG02	80RP02	80RT02	80	3.1496	140	5.5118	26	1.0236	2.0	.08
85RN02	85RU02	85RF02	85RY02	85RJ02	85RM02	85RK02	85RS02	85RC02	85RG02	85RP02	85RT02	85	3.3465	150	5.9055	28	1.1024	2.0	.08
90RN02	90RU02	90RF02	90RY02	90RJ02	90RM02	90RK02	90RS02	90RC02	90RG02	90RP02	90RT02	90	3.5433	160	6.2992	30	1.1811	2.0	.08
95RN02	95RU02	95RF02	95RY02	95RJ02	95RM02	95RK02	95RS02	95RC02	95RG02	95RP02	95RT02	95	3.7402	170	6.6929	32	1.2598	2.0	.08
100RN02	100RU02	100RF02	100RY02	100RJ02	100RM02	100RK02	100RS02	100RC02	100RG02	100RP02	100RT02	100	3.9370	180	7.0866	34	1.3386	2.0	.08
105RN02	105RU02	105RF02	105RY02	105RJ02	105RM02	105RK02	105RS02	105RC02	105RG02	105RP02	105RT02	105	4.1339	190	7.4803	36	1.4173	2.0	.08
110RN02	110RU02	110RF02	110RY02	110RJ02	110RM02	110RK02	110RS02	110RC02	110RG02	110RP02	110RT02	110	4.3307	200	7.8740	38	1.4961	2.0	.08
120RN02	120RU02	120RF02	120RY02	120RJ02	120RM02	120RK02	120RS02	120RC02	120RG02	120RP02	120RT02	120	4.7244	215	8.4646	40	1.5748	2.0	.08
130RN02	130RU02	130RF02	130RY02	130RJ02	130RM02	130RK02	130RS02	130RC02	130RG02	130RP02	130RT02	130	5.1181	230	9.0551	40	1.5748	2.5	.10
140RN02	140RU02	140RF02	140RY02	140RJ02	140RM02	140RK02	140RS02	140RC02	140RG02	140RP02	140RT02	140	5.5118	250	9.8425	42	1.6535	2.5	.10
150RN02	150RU02	150RF02	150RY02	150RJ02	150RM02	150RK02	150RS02	150RC02	150RG02	150RP02	150RT02	150	5.9055	270	10.6299	45	1.7717	2.5	.10
160RN02	160RU02	160RF02	....	160RJ02	....	....	....	160RC02	160RG02	160RP02	160RT02	160	6.2992	290	11.4173	48	1.8898	2.5	.10
170RN02	170RU02	170RF02	....	170RJ02	....	....	....	170RC02	170RG02	170RP02	170RT02	170	6.6929	310	12.2047	52	2.0472	3.0	.12
180RN02	180RU02	180RF02	....	180RJ02	....	....	....	180RC02	180RG02	180RP02	180RT02	180	7.0866	320	12.5984	52	2.0472	3.0	.12
190RN02	190RU02	190RF02	....	190RJ02	....	....	....	190RC02	190RG02	190RP02	190RT02	190	7.4803	340	13.3858	55	2.1654	3.0	.12
200RN02	200RU02	200RF02	....	200RJ02	....	....	....	200RC02	200RG02	200RP02	200RT02	200	7.8740	360	14.1732	58	2.2835	3.0	.12

*The corner radius or chamfer on bearings must clear the maximum fillet radius given in the table. This specification does not control bearing corner contours.

continued on next page

TABLE 2–15, continued

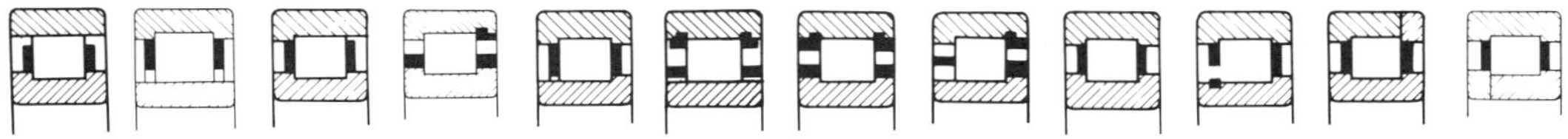

Type RN	Type RU	Type RF	Type RY	Type RJ	Type RM	Type RK	Type RS	Type RC	Type RG	Type RP	Type RT	Bore mm	Bore Inch	Outside Diam mm	Outside Diam Inch	Ring Width mm	Ring Width Inch	Fillet Radius r* mm	Fillet Radius r* Inch
220RN02	220RU02	220RF02	....	220RJ02	....	....	....	220RC02	220RG02	220RP02	220RT02	220	8.6614	400	15.7480	65	2.5591	3.0	.12
240RN02	240RU02	240RF02	....	240RJ02	....	....	....	240RC02	240RG02	240RP02	240RT02	240	9.4488	440	17.3228	72	2.8346	3.0	.12
260RN02	260RU02	260RF02	....	260RJ02	....	....	....	260RC02	260RG02	260RP02	260RT02	260	10.2362	480	18.8976	80	3.1496	4.0	.16
280RN02	280RU02	280RF02	....	280RJ02	....	....	....	280RC02	280RG02	280RP02	280RT02	280	11.0236	500	19.6850	80	3.1496	4.0	.16
300RN02	300RU02	300RF02	....	300RJ02	....	....	....	300RC02	300RG02	300RP02	300RT02	300	11.8110	540	21.2598	85	3.3465	4.0	.16
320RN02	320RU02	320RF02	....	320RJ02	....	....	....	320RC02	320RG02	320RP02	320RT02	320	12.5984	580	22.8346	92	3.6220	4.0	.16
Dimension Series 03 (Medium)																			
15RN03	15RU03	15RF03	15RY03	15RJ03	15RM03	15RK03	15RS03	15RC03	15RG03	15RP03	15RT03	15	.5906	42	1.6535	13	.5118	1.0	.04
17RN03	17RU03	17RF03	17RY03	17RJ03	17RM03	17RK03	17RS03	17RC03	17RG03	17RP03	17RT03	17	.6693	47	1.8504	14	.5512	1.0	.04
20RN03	20RU03	20RF03	20RY03	20RJ03	20RM03	20RK03	20RS03	20RC03	20RG03	20RP03	20RT03	20	.7874	52	2.0472	15	.5906	1.0	.04
25RN03	25RU03	25RF03	25RY03	25RJ03	25RM03	25RK03	25RS03	25RC03	25RG03	25RP03	25RT03	25	.9843	62	2.4409	17	.6693	1.0	.04
30RN03	30RU03	30RF03	30RY03	30RJ03	30RM03	30RK03	30RS03	30RC03	30RG03	30RP03	30RT03	30	1.1811	72	2.8346	19	.7480	1.0	.04
35RN03	35RU03	35RF03	35RY03	35RJ03	35RM03	35RK03	35RS03	35RC03	35RG03	35RP03	35RT03	35	1.3780	80	3.1496	21	.8268	1.5	.06
40RN03	40RU03	40RF03	40RY03	40RJ03	40RM03	40RK03	40RS03	40RC03	40RG03	40RP03	40RT03	40	1.5748	90	3.5433	23	.9055	1.5	.06
45RN03	45RU03	45RF03	45RY03	45RJ03	45RM03	45RK03	45RS03	45RC03	45RG03	45RP03	45RT03	45	1.7717	100	3.9370	25	.9843	1.5	.06
50RN03	50RU03	50RF03	50RY03	50RJ03	50RM03	50RK03	50RS03	50RC03	50RG03	50RP03	50RT03	50	1.9685	110	4.3307	27	1.0630	2.0	.08
55RN03	55RU03	55RF03	55RY03	55RJ03	55RM03	55RK03	55RS03	55RC03	55RG03	55RP03	55RT03	55	2.1654	120	4.7244	29	1.1417	2.0	.08
60RN03	60RU03	60RF03	60RY03	60RJ03	60RM03	60RK03	60RS03	60RC03	60RG03	60RP03	60RT03	60	2.3622	130	5.1181	31	1.2205	2.0	.08
65RN03	65RU03	65RF03	65RY03	65RJ03	65RM03	65RK03	65RS03	65RC03	65RG03	65RP03	65RT03	65	2.5591	140	5.5118	33	1.2992	2.0	.08
70RN03	70RU03	70RF03	70RY03	70RJ03	70RM03	70RK03	70RS03	70RC03	70RG03	70RP03	70RT03	70	2.7559	150	5.9055	35	1.3780	2.0	.08
75RN03	75RU03	75RF03	75RY03	75RJ03	75RM03	75RK03	75RS03	75RC03	75RG03	75RP03	75RT03	75	2.9528	160	6.2992	37	1.4567	2.0	.08
80RN03	80RU03	80RF03	80RY03	80RJ03	80RM03	80RK03	80RS03	80RC03	80RG03	80RP03	80RT03	80	3.1496	170	6.6929	39	1.5354	2.0	.08
85RN03	85RU03	85RF03	85RY03	85RJ03	85RM03	85RK03	85RS03	85RC03	85RG03	85RP03	85RT03	85	3.3465	180	7.0866	41	1.6142	2.5	.10
90RN03	90RU03	90RF03	90RY03	90RJ03	90RM03	90RK03	90RS03	90RC03	90RG03	90RP03	90RT03	90	3.5433	190	7.4803	43	1.6929	2.5	.10
95RN03	95RU03	95RF03	95RY03	95RJ03	95RM03	95RK03	95RS03	95RC03	95RG03	95RP03	95RT03	95	3.7402	200	7.8740	45	1.7717	2.5	.10
100RN03	100RU03	100RF03	100RY03	100RJ03	100RM03	100RK03	100RS03	100RC03	100RG03	100RP03	100RT03	100	3.9370	215	8.4646	47	1.8504	2.5	.10
105RN03	105RU03	105RF03	105RY03	105RJ03	105RM03	105RK03	105RS03	105RC03	105RG03	105RP03	105RT03	105	4.1339	225	8.8583	49	1.9291	2.5	.10
110RN03	110RU03	110RF03	110RY03	110RJ03	110RM03	110RK03	110RS03	110RC03	110RG03	110RP03	110RT03	110	4.3307	240	9.4488	50	1.9685	2.5	.10
120RN03	120RU03	120RF03	120RY03	120RJ03	120RM03	120RK03	120RS03	120RC03	120RG03	120RP03	120RT03	120	4.7244	260	10.2362	55	2.1654	2.5	.10
130RN03	130RU03	130RF03	130RY03	130RJ03	130RM03	130RK03	130RS03	130RC03	130RG03	130RP03	130RT03	130	5.1181	280	11.0236	58	2.2835	3.0	.12
140RN03	140RU03	140RF03	140RY03	140RJ03	140RM03	140RK03	140RS03	140RC03	140RG03	140RP03	140RT03	140	5.5118	300	11.8110	62	2.4409	3.0	.12
150RN03	150RU03	150RF03	150RY03	150RJ03	150RM03	150RK03	150RS03	150RC03	150RG03	150RP03	150RT03	150	5.9055	320	12.5984	65	2.5591	3.0	.12
160RN03	160RU03	160RF03	....	160RJ03	....	160RK03	....	160RC03	160RG03	160RP03	160RT03	160	6.2992	340	13.3858	68	2.6772	3.0	.12
170RN03	170RU03	170RF03	....	170RJ03	....	170RK03	....	170RC03	170RG03	170RP03	170RT03	170	6.6929	360	14.1732	72	2.8346	3.0	.12
180RN03	180RU03	180RF03	....	180RJ03	....	180RK03	....	180RC03	180RG03	180RP03	180RT03	180	7.0866	380	14.9606	75	2.9528	3.0	.12
190RN03	190RU03	190RF03	....	190RJ03	....	190RK03	....	190RC03	190RG03	190RP03	190RT03	190	7.4803	400	15.7480	78	3.0709	4.0	.16
200RN03	200RU03	200RF03	....	200RJ03	....	200RK03	....	200RC03	200RG03	200RP03	200RT03	200	7.8740	420	16.5354	80	3.1496	4.0	.16

*The corner radius or chamfer on bearings must clear the maximum fillet radius given in the table. This specification does not control bearing corner contours.

TABLE 2-16

Dimension Series 32, 92, 33 and 93 Cylindrical Roller Bearings

AFBMA Standards, Section 2—Nov 1952

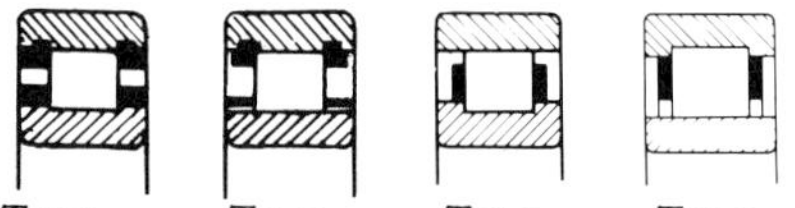

Type RK	Type RM	Type RN	Type RU	Bore mm	Bore Inch	Outside Diam mm	Outside Diam Inch	Width mm	Width Inch	Fillet Radius r* mm	Fillet Radius r* Inch
Dimension Series 32											
17RK32	17RM32	17RN32	17RU32	17	.6693	40	1.5748	17.5	11/16	1.0	.04
20RK32	20RM32	20RN32	20RU32	20	.7874	47	1.8504	20.6	13/16	1.0	.04
25RK32	25RM32	25RN32	25RU32	25	.9843	52	2.0472	20.6	13/16	1.0	.04
30RK32	30RM32	30RN32	30RU32	30	1.1811	62	2.4409	23.8	15/16	1.0	.04
35RK32	35RM32	35RN32	35RU32	35	1.3780	72	2.8346	27.0	1- 1/16	1.0	.04
40RK32	40RM32	40RN32	40RU32	40	1.5748	80	3.1496	30.2	1- 3/16	1.0	.04
45RK32	45RM32	45RN32	45RU32	45	1.7717	85	3.3465	30.2	1- 3/16	1.0	.04
50RK32	50RM32	50RN32	50RU32	50	1.9685	90	3.5433	30.2	1- 3/16	1.0	.04
55RK32	55RM32	55RN32	55RU32	55	2.1654	100	3.9370	33.3	1- 5/16	1.5	.06
60RK32	60RM32	60RN32	60RU32	60	2.3622	110	4.3307	36.5	1-7/16	1.5	.06
65RK32	65RM32	65RN32	65RU32	65	2.5591	120	4.7244	38.1	1 - 1/2	1.5	.06
70RK32	70RM32	70RN32	70RU32	70	2.7559	125	4.9213	39.7	1- 9/16	1.5	.06
75RK32	75RM32	75RN32	75RU32	75	2.9528	130	5.1181	41.3	1 - 5/8	1.5	.06
80RK32	80RM32	80RN32	80RU32	80	3.1496	140	5.5118	44.4	1 - 3/4	2.0	.08
85RK32	85RM32	85RN32	85RU32	85	3.3465	150	5.9055	49.2	1-15/16	2.0	.08
90RK32	90RM32	90RN32	90RU32	90	3.5433	160	6.2992	52.4	2- 1/16	2.0	.08
95RK32	95RM32	95RN32	95RU32	95	3.7402	170	6.6929	55.6	2- 3/16	2.0	.08
100RK32	100RM32	100RN32	100RU32	100	3.9370	180	7.0866	60.3	2 - 3/8	2.0	.08
105RK32	105RM32	105RN32	105RU32	105	4.1339	190	7.4803	65.1	2- 9/16	2.0	.08
110RK32	110RM32	110RN32	110RU32	110	4.3307	200	7.8740	69.8	2 - 3/4	2.0	.08
Dimension Series 92											
120RK92	120RM92	120RN92	120RU92	120	4.7244	215	8.4646	76.2	3	2.0	.08
130RK92	130RM92	130RN92	130RU92	130	5.1181	230	9.0551	79.4	3 - 1/8	2.5	.10
140RK92	140RM92	140RN92	140RU92	140	5.5118	250	9.8425	82.6	3 - 1/4	2.5	.10
150RK92	150RM92	150RN92	150RU92	150	5.9055	270	10.6299	88.9	3 - 1/2	2.5	.10
160RK92	160RM92	160RN92	160RU92	160	6.2992	290	11.4173	98.4	3 - 7/8	2.5	.10
170RK92	170RM92	170RN92	170RU92	170	6.6929	310	12.2047	104.8	4 - 1/8	3.0	.12
180RK92	180RM92	180RN92	180RU92	180	7.0866	320	12.5984	108.0	4 - 1/4	3.0	.12
190RK92	190RM92	190RN92	190RU92	190	7.4803	340	13.3858	114.3	4 - 1/2	3.0	.12
200RK92	200RM92	200RN92	200RU92	200	7.8740	360	14.1732	120.7	4 - 3/4	3.0	.12
220RK92	220RM92	220RN92	220RU92	220	8.6614	400	15.7480	133.4	5 - 1/4	3.0	.12
240RK92	240RM92	240RN92	240RU92	240	9.4488	440	17.3228	146.1	5 - 3/4	3.0	.12
260RK92	260RM92	260RN92	260RU92	260	10.2362	480	18.8976	158.8	6 - 1/4	4.0	.16
280RK92	280RM92	280RN92	280RU92	280	11.0236	500	19.6850	165.1	6 - 1/2	4.0	.16
300RK92	300RM92	300RN92	300RU92	300	11.8110	540	21.2598	177.8	7	4.0	.16
320RK92	320RM92	320RN92	320RU92	320	12.5984	580	22.8346	190.5	7 - 1/2	4.0	.16

*The corner radius or chamfer on bearings must clear the maximum fillet radius given in the table. This specification does not control bearing corner contours.

continued on next page

TABLE 2–16, continued

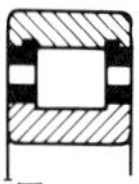 Type RK	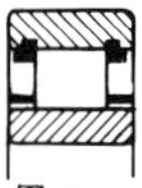 Type RM	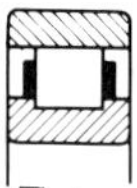 Type RN	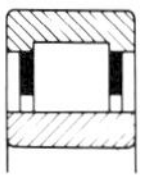 Type RU	Bore mm	Bore Inch	Outside Diam mm	Outside Diam Inch	Width mm	Width Inch	Fillet Radius r* mm	Fillet Radius r* Inch
Dimension Series 33											
17RK33	17RM33	17RN33	17RU33	17	.6693	47	1.8504	22.2	7/8	1.0	.04
20RK33	20RM33	20RN33	20RU33	20	.7874	52	2.0472	22.2	7/8	1.0	.04
25RK33	25RM33	25RN33	25RU33	25	.9843	62	2.4409	25.4	1	1.0	.04
30RK33	30RM33	30RN33	30RU33	30	1.1811	72	2.8346	30.2	1- 3/16	1.0	.04
35RK33	35RM33	35RN33	35RU33	35	1.3780	80	3.1496	34.9	1- 3/8	1.5	.06
40RK33	40RM33	40RN33	40RU33	40	1.5748	90	3.5433	36.5	1- 7/16	1.5	.06
45RK33	45RM33	45RN33	45RU33	45	1.7717	100	3.9370	39.7	1- 9/16	1.5	.06
50RK33	50RM33	50RN33	50RU33	50	1.9685	110	4.3307	44.4	1 - 3/4	2.0	.08
55RK33	55RM33	55RN33	55RU33	55	2.1654	120	4.7244	49.2	1-15/16	2.0	.08
60RK33	60RM33	60RN33	60RU33	60	2.3622	130	5.1181	54.0	2 - 1/8	2.0	.08
65RK33	65RM33	65RN33	65RU33	65	2.5591	140	5.5118	58.7	2- 5/16	2.0	.08
70RK33	70RM33	70RN33	70RU33	70	2.7559	150	5.9055	63.5	2 - 1/2	2.0	.08
75RK33	75RM33	75RN33	75RU33	75	2.9528	160	6.2992	68.3	2-11/16	2.0	.08
80RK33	80RM33	80RN33	80RU33	80	3.1496	170	6.6929	68.3	2-11/16	2.0	.08
85RK33	85RM33	85RN33	85RU33	85	3.3465	180	7.0866	73.0	2 - 7/8	2.5	.10
90RK33	90RM33	90RN33	90RU33	90	3.5433	190	7.4803	73.0	2 - 7/8	2.5	.10
95RK33	95RM33	95RN33	95RU33	95	3.7402	200	7.8740	77.8	3- 1/16	2.5	.10
100RK33	100RM33	100RN33	100RU33	100	3.9370	215	8.4646	82.6	3 - 1/4	2.5	.10
105RK33	105RM33	105RN33	105RU33	105	4.1339	225	8.8583	87.3	3- 7/16	2.5	.10
110RK33	110RM33	110RN33	110RU33	110	4.3307	240	9.4488	92.1	3 - 5/8	2.5	.10
Dimension Series 93											
120RK93	120RM93	120RN93	120RU93	120	4.7244	260	10.2362	104.8	4 - 1/8	2.5	.10
130RK93	130RM93	130RN93	130RU93	130	5.1181	280	11.0236	111.1	4 - 3/8	3.0	.12
140RK93	140RM93	140RN93	140RU93	140	5.5118	300	11.8110	114.3	4 - 1/2	3.0	.12
150RK93	150RM93	150RN93	150RU93	150	5.9055	320	12.5984	123.8	4 - 7/8	3.0	.12
160RK93	160RM93	160RN93	160RU93	160	6.2992	340	13.3858	133.4	5 - 1/4	3.0	.12
170RK93	170RM93	170RN93	170RU93	170	6.6929	360	14.1732	139.7	5 - 1/2	3.0	.12
180RK93	180RM93	180RN93	180RU93	180	7.0866	380	14.9606	146.1	5 - 3/4	3.0	.12
190RK93	190RM93	190RN93	190RU93	190	7.4803	400	15.7480	152.4	6	4.0	.16
200RK93	200RM93	200RN93	200RU93	200	7.8740	420	16.5354	165.1	6 - 1/2	4.0	.16

*The corner radius or chamfer on bearings must clear the maximum fillet radius given in the table. This specification does not control bearing corner contours.

TABLE 2-17

Tapered Roller Bearings, Type TS

AFBMA Standards, Section 2 — Nov 1952 and Section 5 – April 1953

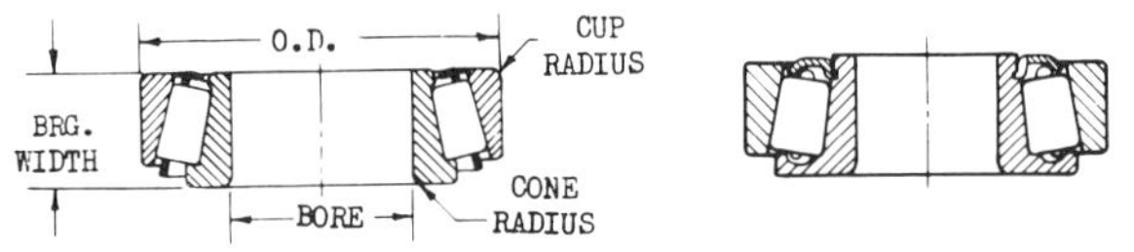

Bearing Number	Bore	O D	Brg. Width	Radius Cone	Radius Cup
A2037-A2126	.3750	1.2595	.3940	3/64	3/64
A2047-A2126	.4720	1.2595	.3940	1/32	3/64
A4050-A4138	.5000	1.3775	.4330	3/64	3/64
A4059-A4138	.5900	1.3775	.4330	1/32	3/64
A6062-A6157	.6250	1.5745	.4730	3/64	3/64
17580-17520	.6250	1.6875	.6563	1/16	1/16
05062-05185	.6250	1.8504	.5662	1/16	3/64
09062-09196	.6250	1.9380	.9063	1/32	1/16
A6067-A6157	.6690	1.5745	.4730	1/32	3/64
A6075-A6157	.7500	1.5745	.4730	.040	3/64
05075-05185	.7500	1.8504	.5662	3/64	3/64
09067-09195	.7500	1.9380	.7100	3/64	3/64
09078-09196	.7500	1.9380	.9063	3/64	1/16
1775-1729	.7500	2.2400	.7625	1/16	3/64
05079-05185	.7870	1.8504	.5662	1/16	3/64
07079-07196	.7874	1.9687	.5313	1/16	.040
12580-12520	.8125	1.9380	.7813	1/16	1/16
3660-3620	.8125	2.4375	1.1250	3/32	1/8
07087-07204	.8750	2.0470	.5910	3/64	3/64
1380-1329	.8750	2.1250	.7625	1/16	1/16
1755-1729	.8750	2.2400	.7625	3/64	3/64
1280-1220	.8750	2.2500	.8750	1/32	1/16
1779-1729	.9375	2.2400	.7625	1/32	3/64
3659-3620	.9375	2.4375	1.1250	3/32	1/8
2685-2631	.9375	2.6150	.9375	1/32	3/64
26093-26283	.9375	2.8345	.7480	3/32	1/16
07098-07204	.9835	2.0470	.5910	1/16	3/64
07100-07204	1.0000	2.0470	.5910	.040	3/64
15578-15520	1.0000	2.2500	.6875	3/64	1/16
1986-1932	1.0000	2.3125	.7500	3/64	3/64
15100-15250	1.0000	2.5000	.8125	9/64	3/64
2687-2631	1.0000	2.6150	.9375	3/64	3/64
26100-26283	1.0000	2.8345	.7480	1/16	1/16
3189-3120	1.0000	2.8592	1.1875	1/32	1/8
15580-15520	1.0625	2.2500	.6875	9/64	1/16
1985-1932	1.1250	2.3125	.7500	1/32	3/64
15112-15250	1.1250	2.5000	.8125	9/64	9/64
2689-2631	1.1250	2.6150	.9375	3/64	3/64
02474-02420	1.1250	2.6875	.8750	1/32	1/16
2578-2523	1.1250	2.7500	.9375	3/32	3/64
26112-26283	1.1250	2.8345	.7480	1/16	1/16
3198-3120	1.1250	2.8593	1.1875	3/64	1/8
02872-02820	1.1250	2.8750	.8750	1/32	1/8
2690-2631	1.1562	2.6150	.9375	9/64	3/64
17118-17244	1.1805	2.4410	.6300	1/16	1/16
15117-15250	1.1805	2.5000	.8125	3/64	3/64
14117A-14276	1.1810	2.7170	.7813	9/64	3/64
17119-17244	1.1875	2.4410	.6300	1/16	1/16
2558-2523	1.1875	2.7500	.9375	3/32	3/64
3191-3120	1.1875	2.8593	1.1875	9/64	1/8
15118-15250	1.1895	2.5000	.8125	9/64	3/64
14116-14276	1.1900	2.7170	.7813	1/32	3/64
1674-1620	1.2450	2.6250	.8125	1/16	1/16
08125-08231	1.2500	2.3125	.5781	.040	.040
15123-15245	1.2500	2.4410	.7150	9/64*	3/64
02475-02420	1.2500	2.6875	.8750	9/64	1/16
14125A-14276	1.2500	2.7170	.7813	9/64	3/64
2582-2523	1.2500	2.7500	.9375	9/64	3/64
3193-3120	1.2500	2.8593	1.1875	9/64	1/8
02875-02820	1.2500	2.8750	.8750	9/64	1/8
2875-2820	1.2500	2.8750	.8750	9/64	1/8
3476-3420	1.2500	3.1250	1.1563	3/64	1/8
346-332	1.2500	3.1496	.8268	1/32	3/64
26126-26283	1.2600	2.8345	.7480	1/16	1/16
14130-14276	1.3125	2.7170	.7813	9/64	3/64
2585-2523	1.3125	2.7500	.9375	9/64	3/64
3196-3120	1.3125	2.8593	1.1875	9/64	1/8
2876-2820	1.3125	2.8750	.8750	9/64	1/8
2785-2720	1.3125	3.0000	.9375	9/64	1/8
31590-31520	1.3125	3.0000	1.1563	.025	1/8
14137A-14276	1.3750	2.7170	.7813	1/16	3/64
02877-02820	1.3750	2.8750	.8750	9/64	1/8
2878-2820	1.3750	2.8750	.8750	1/32	1/8
25878-25820	1.3750	2.8750	.9375	9/64	3/32
2786-2720	1.3750	3.0000	.9375	13/64	1/8
36137-36300	1.3750	3.0000	1.1563	1/16	1/8
31593-31520	1.3750	3.0000	1.1563	9/64	1/8
3478-3420	1.3750	3.1250	1.1563	9/64	1/8
335-332	1.3750	3.1496	.8268	1/32	3/64
3379-3320	1.3750	3.1562	1.1563	9/64	1/8
417-414	1.3750	3.4843	1.0625	1/32	1/16
449-432	1.3750	3.7500	1.0938	1/32	3/32
19138-19283	1.3770	2.8345	.6700	1/16	1/16
339-332	1.3779	3.1496	.8268	1/32	3/64
26883-26820	1.3779	3.1562	1.0000	1/32	1/8
2794-2720	1.4365	3.0000	.9375	9/64	1/8
19143-19283	1.4375	2.8345	.6700	1/16	1/16
16143-16282	1.4375	2.8345	.7480	9/64	1/16
3878-3820	1.4375	3.3750	1.1875	1/32	1/8
13889-13836	1.5000	2.5625	.5000	1/16	1/32
13685-13621	1.5000	2.7170	.7500	9/64	3/32
19150-19283	1.5000	2.8345	.6700	.060	1/16
16150-16284	1.5000	2.8440	.8125	9/64	3/64
2788-2720	1.5000	3.0000	.9375	9/64	1/8
3490-3420	1.5000	3.1250	1.1563	9/64	1/8
28151-28315	1.5000	3.1495	.8270	9/64	1/16
337-332	1.5000	3.1495	.8268	1/32	3/64
3381-3320	1.5000	3.1562	1.1563	9/64	1/8
3876-3820	1.5000	3.3750	1.1875	9/64	1/8
418-414	1.5000	3.4843	1.0625	9/64	1/16
49150-49368	1.5000	3.6875	1.2500	9/64	1/8
444-432	1.5000	3.7500	1.0938	9/64	3/32
542-532X	1.5000	4.2500	1.4375	9/64	1/8
26881-26820	1.5625	3.1562	1.0000	9/64	1/8
3382-3320	1.5625	3.1562	1.1563	9/64	1/8
422-414	1.5625	3.4843	1.0625	9/64	1/16
11157-11300	1.5740	3.0000	.7090	1/16	1/16
28158-28315	1.5748	3.1495	.8270	1/16	1/16
344-332	1.5748	3.1496	.8268	9/64	3/64
420-414	1.5748	3.4843	1.0625	9/64	1/16
24780-24720	1.6250	3.0000	.8750	9/64	1/32
11162-11315	1.6250	3.1495	.7090	1/16	1/16
342-332	1.6250	3.1496	.8268	9/64	3/64
26882-26820	1.6250	3.1562	1.0000	9/64	1/8
3877-3820	1.6250	3.3750	1.1875	9/64	1/8
3577-3525	1.6250	3.4375	1.1875	9/64	1/8
419-414	1.6250	3.4843	1.0625	9/64	1/16
4388-4335	1.6250	3.5625	1.5625	9/64	1/8
49162-49368	1.6250	3.6875	1.2500	9/64	1/8
46162-46368	1.6250	3.6875	1.2500	1/32	1/8

*Compound Radius.

continued on next page

TABLE 2–17, continued

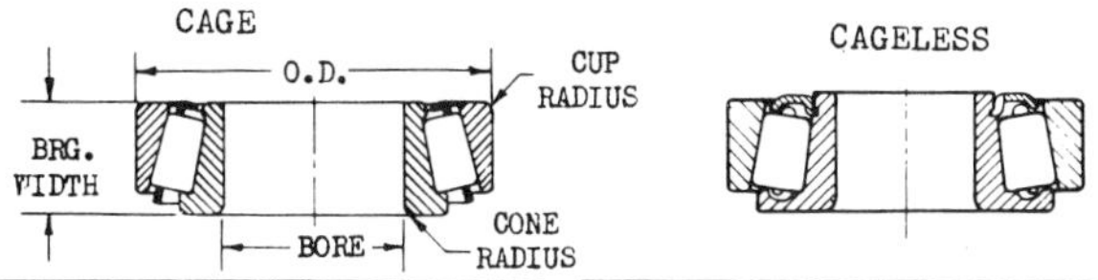

Bearing Number	Bore	O D	Brg. Width	Radius Cone	Radius Cup
447–432	1.6250	3.7500	1.0938	9/64	3/32
526–522	1.6250	4.0000	1.3750	9/64	1/8
12168–12303	1.6875	3.0312	.6875	.060	.060
25578–25520	1.6875	3.2650	.9375	.090	1/32
3579–3525	1.6875	3.4375	1.1875	9/64	1/8
342S–332	1.6880	3.1496	.8268	9/64	3/64
26884–26820	1.6880	3.1562	1.0000	9/64	1/8
25577–25520	1.6880	3.2650	.9375	9/64	1/32
12175–12303	1.7500	3.0312	.6875	.060	.060
13175–13318	1.7500	3.1875	.7500	0	1/16
35175–35326	1.7500	3.2650	.8750	9/64	1/32
25580–25520	1.7500	3.2650	.9375	9/64	1/32
355–354A	1.7500	3.3464	.8125	3/32	3/64
3578–3525	1.7500	3.4375	1.1875	9/64	1/8
49175–49368	1.7500	3.6875	1.2500	9/64	1/8
46176–46368	1.7500	3.6875	1.2500	9/64	1/8
438–432	1.7500	3.7500	1.0938	9/64	3/32
527–522	1.7500	4.0000	1.3750	9/64	1/8
460–453X	1.7500	4.1250	1.1875	9/64	1/8
59175–59412	1.7500	4.1250	1.4375	9/64	1/8
535–532X	1.7500	4.2500	1.4375	9/64	1/8
65385–65321	1.7500	4.5000	1.7500	9/64	1/32
615–612	1.7500	4.7500	1.6250	9/64	1/8
25584–25520	1.7710	3.2650	.9375	1/16	1/32
3776–3720	1.7710	3.6718	1.1875	9/64	1/8
358–354A	1.7716	3.3464	.8125	1/16	3/64
376–372A	1.7716	3.8125	.8750	1/32	1/16
18690–18620	1.8125	3.1250	.6875	7/64	1/16
13181–13318	1.8125	3.1875	.7500	1/32	1/16
359S–354A	1.8125	3.3464	.8125	3/32	3/64
2984–2924	1.8125	3.3464	1.0000	9/64	3/64
436–432	1.8125	3.7500	1.0938	9/64	3/32
369A–362A	1.8750	3.5000	.8125	9/64	3/64
3778––3720	1.8750	3.6718	1.1875	1/4	1/8
3779–3720	1.8750	3.6718	1.1875	9/64	1/8
49580–49520	1.8750	4.0000	1.2500	9/64	1/8
528–522	1.8750	4.0000	1.3750	9/64	1/8
5358–5335	1.8750	4.0625	1.7188	3/64	1/8
463–453X	1.8750	4.1250	1.1875	3/16	1/8
536–532X	1.8750	4.2500	1.4375	9/64	1/8
617–612	1.8750	4.7500	1.6250	9/64	1/8
3781–3720	1.9375	3.6718	1.1875	9/64	1/8
5395–5335	1.9375	4.0625	1.7188	9/64	1/8
366–362A	1.9685	3.5000	.8125	3/32	3/64
465–453X	1.9685	4.1250	1.1875	3/32	1/8
18200–18337	2.0000	3.3750	.7500	.060	.060
368–362A	2.0000	3.5000	.8125	1/16	3/64
368A–362A	2.0000	3.5000	.8125	9/64	3/64
3780–3720	2.0000	3.6718	1.1875	9/64	1/8
375–372A	2.0000	3.8125	.8750	3/32	1/16
49585–49520	2.0000	4.0000	1.2500	9/64	1/8
529X–522	2.0000	4.0000	1.3750	9/64	1/8
455–453X	2.0000	4.1250	1.1875	1/32	1/8
59200–59412	2.0000	4.1250	1.4375	9/64	1/8
4580–4535	2.0000	4.1250	1.5625	9/64	1/8
537–532X	2.0000	4.2500	1.4375	9/64	1/8
3975–3920	2.0000	4.4375	1.1875	9/64	1/8
619–612	2.0000	4.7500	1.6250	9/64	1/8
555–552A	2.0000	4.8750	1.5000	3/32	1/8
6279–6220	2.0000	5.0000	2.0000	9/64	1/8
3767–3720	2.0625	3.6718	1.1875	3/32	1/8
33890–33821	2.0625	3.7500	1.0938	1/16	3/32
377–372A	2.0625	3.8125	.8750	3/32	1/16
540–532X	2.0625	4.2500	1.4375	9/64	1/8
389A–382	2.1250	3.8750	.8268	1/32	1/32
456–453X	2.1250	4.1250	1.1875	9/64	1/8
4595–4535	2.1250	4.1250	1.5625	9/64	1/8
539–532X	2.1250	4.2500	1.4375	9/64	1/8
621–612	2.1250	4.7500	1.6250	9/64	1/8
557S–552A	2.1250	4.8750	1.5000	9/64	1/8
6280–6220	2.1250	5.0000	2.0000	9/64	1/8
636–632	2.1250	5.3750	1.6250	9/64	1/8
385–382	2.1653	3.8750	.8268	3/32	1/32
466S–453X	2.1875	4.1250	1.1875	3/32	1/8
389–382	2.1880	3.8750	.8268	3/32	1/32
28682–28622	2.2500	3.8437	.9688	9/64	1/32
387–382	2.2500	3.8750	.8268	3/32	1/32
387A–382	2.2500	3.8750	.8268	9/64	1/32
462–453X	2.2500	4.1250	1.1875	3/32	1/8
390–394A	2.2500	4.3307	.8661	3/32	3/64
3979–3920	2.2500	4.4375	1.1875	9/64	1/8
623–612	2.2500	4.7500	1.6250	9/64	1/8
555S–552A	2.2500	4.8750	1.5000	9/64	1/8
65225–65500	2.2500	5.0000	1.7500	9/64	1/8
6375–6320	2.2500	5.3447	2.1250	11/64	1/8
29582–29520	2.3622	4.2500	1.0000	1/32	1/8
397–394A	2.3622	4.3307	.8661	1/32	3/64
28985–28920	2.3750	4.0000	1.0000	9/64	1/8
3980–3920	2.3750	4.4375	1.1875	9/64	1/8
5582–5535	2.3750	4.8125	1.7188	1/32	1/8
6376–6320	2.3750	5.3447	2.1250	9/64	1/8
39250–39412	2.5000	4.1250	.8438	.080	.080
395–394A	2.5000	4.3307	.8661	9/64	3/64
3982–3920	2.5000	4.4375	1.1875	9/64	1/8
33251–33462	2.5000	4.6250	1.1875	1/32	1/8
477–472A	2.5000	4.7244	1.1418	1/32	1/8
5584–5535	2.5000	4.8125	1.7188	9/64	1/8
559–552A	2.5000	4.8750	1.5000	9/64	1/8
565–563	2.5000	5.0000	1.4375	9/64	1/8
639–632	2.5000	5.3750	1.6250	9/64	1/8
6379–6320	2.5625	5.3447	2.1250	9/64	1/8
395S–394A	2.6250	4.3307	.8661	9/64	3/64
3984–3920	2.6250	4.4375	1.1875	9/64	1/8
33262–33462	2.6250	4.6250	1.1875	9/64	1/8
479–472A	2.6250	4.7244	1.1418	3/32	1/8
560–552A	2.6250	4.8750	1.5000	9/64	1/8
6386–6320	2.6250	5.3447	2.1250	11/64	1/8
641–632	2.6250	5.3750	1.6250	9/64	1/8
399A–394A	2.6875	4.3307	.8661	3/32	3/64
480–472A	2.6875	4.7244	1.1418	9/64	1/8
560S–552A	2.6875	4.8750	1.5000	9/64	1/8
570–563	2.6875	5.0000	1.4375	9/64	1/8
33275–33462	2.7500	4.6250	1.1875	9/64	1/8
482–472A	2.7500	4.7244	1.1418	9/64	1/8
566–563	2.7500	5.0000	1.4375	9/64	1/8
643–632	2.7500	5.3750	1.6250	9/64	1/8
6454–6420	2.7500	5.8750	2.1250	13/64	1/8
655–652	2.7500	6.0000	1.6250	9/64	1/8
835–832	2.7500	6.6250	2.1250	9/64	1/8
34275–34478	2.7559	4.7812	.9688	.080	.080
33281–33462	2.8125	4.6250	1.1875	9/64	1/8
567A–563	2.8125	5.0000	1.4375	9/64	1/8
645–632	2.8125	5.3750	1.6250	1/4	1/8
567–563	2.8750	5.0000	1.4375	9/64	1/8
6460–6420	2.8750	5.8750	2.1250	9/64	1/8
744–742	2.8750	5.9090	1.7500	9/64	1/8
657–652	2.8750	6.0000	1.6250	9/64	1/8
568–563	2.9062	5.0000	1.4375	1/32	1/8
34300–34478	3.0000	4.7812	.9688	.080	.080
42687–42620	3.0000	5.0000	1.1875	9/64	1/8
47680–47620	3.0000	5.2500	1.3125	1/32	1/8
495A–493	3.0000	5.3750	1.1875	9/64	1/8
575–572	3.0000	5.5115	1.4375	9/64	1/8
6461–6420	3.0000	5.8750	2.1250	9/64	1/8
748S–742	3.0000	5.9090	1.7500	9/64	1/8

continued on next page

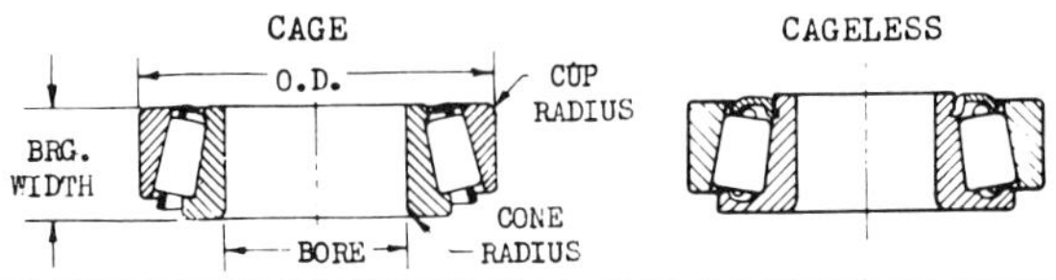

Bearing Number	Bore	O D	Brg. Width	Radius Cone	Radius Cup
659–652	3.0000	6.0000	1.6250	9/64	1/8
6576–6535	3.0000	6.3750	2.1250	9/64	1/8
843–832	3.0000	6.6250	2.1250	1/4	1/8
34306–34478	3.0625	4.7812	.9688	9/64	.080
47681–47620	3.1875	5.2500	1.3125	9/64	1/8
496–493	3.1875	5.3750	1.1875	9/64	1/8
581–572	3.1875	5.5115	1.4375	9/64	1/8
740–742	3.1875	5.9090	1.7500	13/64	1/8
838–832	3.1875	6.6250	2.1250	1/32	1/8
47686–47620	3.2500	5.2500	1.3125	9/64	1/8
495–493	3.2500	5.3750	1.1875	9/64	1/8
580–572	3.2500	5.5115	1.4375	9/64	1/8
663–652	3.2500	6.0000	1.6250	9/64	1/8
6559–6535	3.2500	6.3750	2.1250	9/64	1/8
842–832	3.2500	6.6250	2.1250	9/64	1/8
27690–27620	3.2813	4.9375	1.0000	9/64	1/16
498–493	3.3125	5.3750	1.1875	9/64	1/8
749–742	3.3475	5.9090	1.7500	9/64	1/8
497–493	3.3750	5.3750	1.1875	9/64	1/8
596–592A	3.3750	6.0000	1.5625	9/64	1/8
665–652	3.3750	6.0000	1.6250	9/64	1/8
42350–42584	3.5000	5.8437	1.1250	.120	.120
593–592A	3.5000	6.0000	1.5625	9/64	1/8
759–752	3.5000	6.3750	1.8750	9/64	1/8
6580–6535	3.5000	6.3750	2.1250	9/64	1/8
855–854	3.5000	7.5000	2.2500	5/16	1/8
69354–69630	3.5430	6.3030	1.1860	3/32	1/8
47890–47820	3.6250	5.7500	1.3125	9/64	1/8
42362–42584	3.6250	5.8437	1.1250	9/64	.120
598–592A	3.6250	6.0000	1.5625	9/64	1/8
77362–77675	3.6250	6.7500	1.8750	9/64	1/8
42368–42584	3.6875	5.8437	1.1250	.120	.120
42375–42584	3.7500	5.8437	1.1250	.120	.120
594–592A	3.7500	6.0000	1.5625	9/64	1/8
683–672	3.7500	6.6250	1.6250	9/64	1/8
77375–77675	3.7500	6.7500	1.8750	9/64	1/8
864–854	3.7500	7.5000	2.2500	5/16	1/8
52387–52618	3.8750	6.1875	1.4375	9/64	1/8
685–672	3.8750	6.6250	1.6250	9/64	1/8
779–772	3.8750	7.1250	1.8750	9/64	1/8
52400–52618	4.0000	6.1875	1.4375	9/64	1/8
687–672	4.0000	6.6250	1.6250	9/64	1/8
780–772	4.0000	7.1250	1.8750	9/64	1/8
861–854	4.0000	7.5000	2.2500	5/16	1/8
782–772	4.1250	7.1250	1.8750	9/64	1/8
56418–56650	4.1875	6.5000	1.4375	9/64	1/8
37425–37625	4.2500	6.2500	.9063	9/64	1/8
56425–56650	4.2500	6.5000	1.4375	9/64	1/8
71425–71750	4.2500	7.5000	1.8750	9/64	1/8
936–932	4.2500	8.3750	2.6250	5/16	1/8
64433–64700	4.3304	7.0000	1.6250	9/64	1/8
71437–71750	4.3750	7.5000	1.8750	9/64	1/8
64450–64700	4.5000	7.0000	1.6250	9/64	1/8
71450–71750	4.5000	7.5000	1.8750	9/64	1/8
938–932	4.5000	8.3750	2.6250	9/32	1/8
68462–68712	4.6250	7.1250	1.3750	9/64	1/8
795–792	4.7500	8.1250	1.8750	1/8	1/8
EE153044–153100	4.7500	10.0000	3.0625	3/8	1/4
67388–67320	5.0000	8.0000	1.8125	9/64	1/8
74500–74850	5.0000	8.5000	1.8750	9/64	1/8
95500–95925	5.0000	9.2500	2.5000	1/4	1/8
EE116050–116097	5.0000	9.7500	2.5000	1/8	3/16
EE153050–153100	5.0000	10.0000	3.0625	3/8	1/4
EE540502–541162	5.0000	11.6250	3.2500	17/32	1/4
EE455051–455116	5.0000	11.6250	3.3750	1/4	1/4
EE580500–581200	5.0000	12.0000	3.1250	1/4	1/8

Bearing Number	Bore	O D	Brg. Width	Radius Cone	Radius Cup
EE750502–751200	5.0000	12.0000	2.3750	1/4	1/4
48506–48750	5.0625	7.5000	1.3750	9/64	1/8
799–792	5.0625	8.1250	1.8750	1/8	1/8
797–792	5.1181	8.1250	1.8750	9/64	1/8
67390–67320	5.2500	8.0000	1.8125	9/64	1/8
74525–74850	5.2500	8.5000	1.8750	9/64	1/8
95525–95925	5.2500	9.2500	2.5000	3/8	1/8
48393–48320	5.3750	7.5000	1.5625	9/64	1/8
74537–74850	5.3750	8.5000	1.8750	9/64	1/8
EE580537–581200	5.3750	12.0000	3.1250	1/2	1/8
74550–74850	5.5000	8.5000	1.8750	9/64	1/8
73551–73875	5.5000	8.7500	1.3750	9/64	1/8
898–892	5.5000	9.0000	2.2500	9/64	1/8
82550–82950	5.5000	9.5000	2.2500	9/64	1/8
99550–99100	5.5000	10.0000	2.6250	9/32	1/8
EE540550–541162	5.5000	11.6250	3.2500	3/8	1/4
EE750558–751200	5.5000	12.0000	2.3750	1/8	1/4
EE450551–451212	5.5000	12.1250	3.5000	3/8	17/64
73562–73875	5.6250	8.7500	1.3750	9/64	1/8
82562–82950	5.6250	9.5000	2.2500	9/64	1/8
82576–82950	5.7500	9.5000	2.2500	9/64	1/8
81574–81962	5.7500	9.6250	1.8750	9/64	1/8
99575–99100	5.7500	10.0000	2.6250	9/32	1/8
EE107057–107105	5.7500	10.5625	2.9375	1/4	1/4
EE217056–217112	5.7500	11.2500	3.0000	1/4	1/4
EE750576–751200	5.7500	12.0000	2.3750	1/8	1/4
EE450577–451212	5.7500	12.1250	3.5000	3/8	17/64
82587–82950	5.8750	9.5000	2.2500	9/64	1/8
99587–99100	5.8750	10.0000	2.6250	9/32	1/8
EE560590–561275	5.9000	12.7500	3.0625	17/32	3/16
81599–81962	6.0000	9.6250	1.8750	9/64	1/8
99600–99100	6.0000	10.0000	2.6250	9/32	1/8
EE107060–107105	6.0000	10.5625	2.9375	1/4	1/4
EE217060–217112	6.0000	11.2500	3.0000	1/16	1/4
EE450601–451212	6.0000	12.1250	3.5000	3/8	17/64
EE560600–561275	6.0000	12.7500	3.0625	11/16	3/16
EE560629–561275	6.2960	12.7500	3.0625	13/32	3/16
EE590638–591350	6.3750	13.5000	3.1250	1/4	1/4
86650–86100	6.5000	10.0000	1.8125	3/16	1/8
94649–94113	6.5000	11.3750	2.5000	9/32	1/8
EE219065–219122	6.5000	12.2500	3.2500	1/4	1/4
EE590650–591350	6.5000	13.5000	3.1250	11/16	1/4
EE618065–618136	6.5000	13.6875	2.7500	3/8	1/4
EE780655–781400	6.5000	14.0000	2.4375	3/16	3/16
EE108065–108142	6.5000	14.2500	4.1875	17/32	1/8
86669–86100	6.6929	10.0000	1.8125	3/16	1/8
EE590675–591350	6.7500	13.5000	3.1250	1/4	1/4
EE780676–781400	6.7500	14.0000	2.4375	3/16	3/16
67787–67720	6.8750	9.7500	1.8750	9/64	1/8
94687–94113	6.8750	11.3750	2.5000	9/32	1/8
EE219068–219122	6.8750	12.2500	3.2500	1/4	1/4
EE780688–781400	6.8750	14.0000	2.4375	3/16	3/16
67790–67720	7.0000	9.7500	1.8750	9/64	1/8
EE91702–91112	7.0000	11.2500	2.5000	1/4	1/8
94700–94113	7.0000	11.3750	2.5000	9/32	1/8
EE280702–281200	7.0000	12.0000	2.6250	1/4	1/8
EE470078–470132	7.0000	13.2500	3.5625	1/4	1/4
EE780705–781400	7.0000	14.0000	2.4375	3/16	3/16
EE420701–421437	7.0000	14.3720	3.6250	1/2	1/8
87737–87111	7.3750	11.1250	2.0000	9/64	1/8
NA222075–222126	7.3750	12.5970	3.5000	7/32	3/16
67885–67820	7.5000	10.5000	1.8750	9/64	1/8
87750–87111	7.5000	11.1250	2.0000	9/64	1/8
93750–93125	7.5000	12.5000	2.5000	11/64	1/8
EE210753–211300	7.5000	13.0000	2.5000	9/32	1/8
EE420751–421437	7.5000	14.3720	3.6250	1/4	1/8

TABLE 2-18

Tapered Roller Bearings, Types TSS and TST

AFBMA Standards, Section 2—Nov 1952 and Section 5 – April 1953

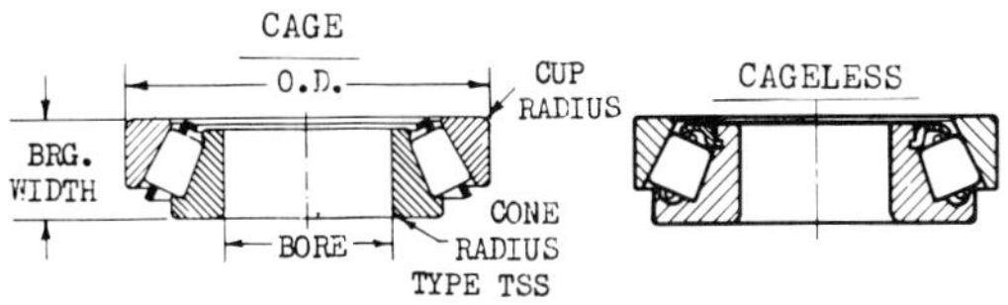

Bearing Number	Bore	O.D.	Brg. Width	Cone Radius	Cup Radius
11590-11520	.6250	1.6875	.5625	1/16	1/16
21075-21212	.7500	2.1250	.8750	1/16	3/32
23100-23256	1.0000	2.5625	.8750	1/16	1/16
41100-41286	1.0000	2.8593	.9688	3/32	1/16
41125-41286	1.1250	2.8593	.9688	3/16	1/16
43112-43312	1.1250	3.1250	1.0000	1/32	1/16
43125-43312	1.2500	3.1250	1.0000	1/16	1/16
44150-44348	1.5000	3.4843	1.0000	3/32	1/16
44158-44348	1.5625	3.4843	1.0000	9/64	1/16
44162-44348	1.6250	3.4843	1.0000	3/32	1/16
53162-53375	1.6250	3.7500	1.2188	1/16	1/32
53177-53375	1.7500	3.7500	1.2188	9/64	1/32
55200-55443	2.0000	4.4375	1.1875	9/64	1/8
72200-72487	2.0000	4.8750	1.4375	9/64	1/8
72212-72487	2.1250	4.8750	1.4375	9/64	1/8
78215-78551	2.1250	5.5130	1.4375	9/64	3/32
78225-78551	2.2500	5.5130	1.4375	9/64	3/32
66585-66520	2.3622	4.8125	1.3125	9/64	1/8
9180-9121	2.4375	6.0000	1.8750	9/64	1/8
78250-78551	2.5000	5.5130	1.4375	3/32	3/32
9185-9121	2.6875	6.0000	1.8750	9/64	1/8
9285-9220	3.0000	6.3750	1.9375	9/64	1/8
9380-9321	3.0000	6.7500	1.9375	9/64	1/8
98316-98788	3.1496	7.8740	2.0772	9/64	1/8
9385-9321	3.3125	6.7500	1.9375	9/64	1/8
98335-98788	3.3465	7.8740	2.0772	9/64	1/8
98350-98788	3.5000	7.8740	2.0772	9/64	1/8
90381-90744	3.8125	7.4375	2.0000	9/64	1/8
98400-98788	4.0000	7.8740	2.0772	9/64	1/8
EE215040-215096	4.0000	9.8750	3.0000	1/4	1/8
97450-97900	4.5000	9.0000	2.1250	9/64	1/8
EE514045-514110	4.5000	11.0000	3.2500	1/4	1/4
97493-97900	4.9330	9.0000	2.1250	9/64	1/8
97500-97900	5.0000	9.0000	2.1250	9/64	1/8
EE516050-516120	5.0000	12.0000	3.5000	1/4	1/4
97503-97900	5.0312	9.0000	2.1250	9/64	1/8
EE516057-516120	5.7500	12.0000	3.5000	1/4	1/4
EE518061-518135	6.1250	13.5000	3.3750	1/4	1/4
EE117065-117146	6.2992	14.7638	3.4375	1/4	1/4
EE117063-117148	6.3750	14.7500	3.4375	1/4	1/16
EE607070-607140	7.0000	14.0000	3.1250	1/4	1/4
EE350701-351637	7.0000	16.8750	4.1875	1/4	1/4
EE350750-351687	7.5000	16.8750	4.1875	1/4	1/4
EE380080-380190	8.0000	19.0000	4.6250	1/4	1/4
EE390090-390200	9.0000	20.0000	4.6250	1/4	1/4
EE390095-390200	9.5000	20.0000	4.6250	1/4	1/4
EE620100-620220	10.0000	22.0000	4.8750	5/16	5/16
EE991251-992550	12.5000	25.5000	5.5000	17/32	17/32

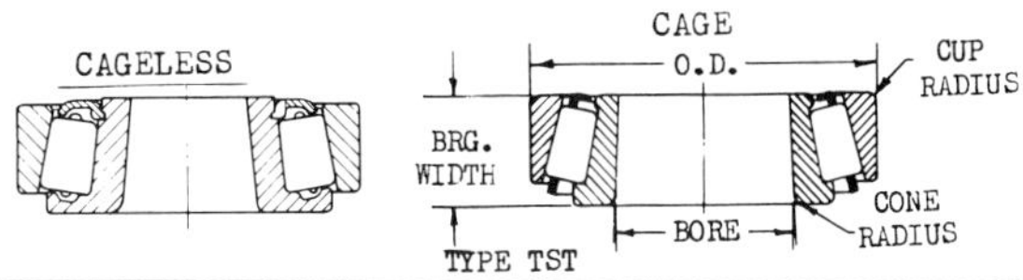

Bearing Number	Bore	Taper per Foot	O D	Brg. Width	Cone Radius	Cup Radius
15575T-15520	.9375	1.00	2.2500	.6875	1/16	1/16
07098T-07204	.9835	1.00	2.0470	.5910	1/16	3/64
1757T-1729	1.0000	1.00	2.2400	.7625	1/32	3/64
2688T-2631	1.0625	1.00	2.6150	.9375	1/16	3/64
26112T-26283	1.1250	1.00	2.8345	.7480	1/16	1/16
26117T-26283	1.1770	1.00	2.8345	.7480	1/16	1/16
2583T-2523	1.1875	1.00	2.7500	.9375	1/16	3/64
14123T-14276	1.2500	1.00	2.7170	.7813	1/16	3/64
2580T-2523	1.2500	1.00	2.7500	.9375	1/16	3/64
14582T-14525	1.3125	1.00	2.6875	.8125	1/32	3/32
2581T-2523	1.3125	1.00	2.7500	.9375	1/32	3/64
14132T-14276	1.3125	1.00	2.7170	.7813	1/16	3/64
2790T-2720	1.3125	1.00	3.0000	.9375	1/32	1/8
25877T-25820	1.3750	1.00	2.8750	.9375	1/16	3/32
2787T-2720	1.3750	1.00	3.0000	.9375	1/16	1/8
3379T-3320	1.3750	1.00	3.1562	1.1563	1/32	1/8
2791T-2720	1.4062	1.00	3.0000	.9375	9/64	1/8
25879T-25820	1.4375	1.00	2.8750	.9375	1/16	3/32
3380T-3320	1.4375	1.00	3.1562	1.1563	1/16	1/8
2798T-2720	1.4855	1.00	3.0000	.9375	1/16	1/8
26876T-26820	1.4855	1.00	3.1562	1.0000	1/16	1/8
19150T-19283	1.5000	1.00	2.8345	.6700	1/16	1/16
2882T-2820	1.5000	1.00	2.8750	.8750	1/32	1/8
2788T-2720	1.5000	1.00	3.0000	.9375	1/16	1/8
3490T-3420	1.5000	1.00	3.1250	1.1563	9/64	1/8
3381T-3320	1.5000	1.00	3.1562	1.1563	9/64	1/8
415T-414	1.5000	1.00	3.4843	1.0625	1/32	1/32
26879T-26820	1.5495	1.00	3.1562	1.0000	1/16	1/8
11156T-11315	1.5625	1.00	3.1495	.7090	1/16	1/16
3382T-3320	1.5625	1.00	3.1562	1.1563	9/64	1/8
3575T-3525	1.5625	1.00	3.4375	1.1875	1/16	1/8
422T-414	1.5625	1.00	3.4843	1.0625	3/32	1/32
3384T-3320	1.6250	1.00	3.1562	1.1563	1/16	1/8
419T-414	1.6250	1.00	3.4843	1.0625	9/64	1/32
439T-432	1.6250	1.00	3.7500	1.0938	1/32	3/32
3578T-3525	1.7500	1.00	3.4375	1.1875	9/64	1/8
435T-432	1.7500	1.00	3.7500	1.0938	1/32	3/32
349T-332	1.7688	9°0'	3.1496	.8268	1/32	3/64
25584T-25520	1.7708	1.00	3.2650	.9375	1/16	1/32
359T-354A	1.8125	1.00	3.3464	.8125	1/32	3/64
463T-453X	1.8438	1.00	4.1250	1.1875	1/16	1/8
536T-532X	1.8750	1.00	4.2500	1.4375	9/64	1/8
537T-532X	2.0000	1.00	4.2500	1.4375	9/64	1/8
65199T-65500	2.0000	2.00	5.0000	1.7500	9/64	1/8
377T-372A	2.0625	1.00	3.8125	.8750	3/32	1/16
28677T-28622	2.0625	1.00	3.8437	.9688	9/64	1/32
368T-362A	2.0852	9°0'	3.5000	.8125	1/32	3/64
539T-532X	2.1250	1.00	4.2500	1.4375	9/64	1/8
65212T-65500	2.1250	2.00	5.0000	1.7500	9/64	1/8
467T-453X	2.2500	1.00	4.1250	1.1875	9/64	1/8
468T-453X	2.3125	1.00	4.1250	1.1875	9/64	1/8
388T-382	2.3340	9°0'	3.8750	.8268	1/32	1/32
559T-552A	2.5000	1.00	4.8750	1.5000	1/32	1/8
560T-552A	2.6250	1.50	4.8750	1.5000	9/64	1/8
399T-394A	2.6463	9°0'	4.3307	.8661	1/32	3/64
485T-472A	2.8447	1.00	4.7244	1.1418	9/64	1/8
488T-472A	2.8773	9°0'	4.7244	1.1418	1/32	1/8
755T-752	3.0000	1.50	6.3750	1.8750	9/64	1/8
5784T-5735	3.1875	1.00	5.3438	1.7500	9/64	1/8
758T-752	3.3750	1.00	6.3750	1.8750	9/64	1/8

TABLE 2-19
Tapered Roller Bearings, Type TDI, Double Cone, Single Cups
AFBMA Standards, Section 2 — Nov 1952 and Section 5 – April 1953

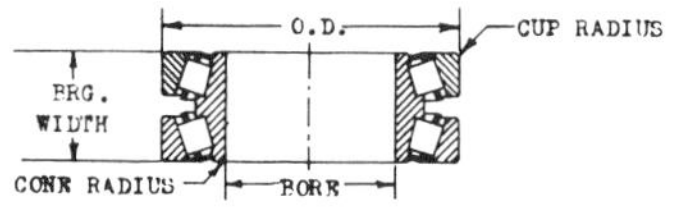

Bearing Number	Bore	O D	Brg. Width	Cone Radius	Cup Radius
17116D-17244	1.1875	2.4410	1.3306	1/32	1/16
14126D-14276	1.2500	2.7170	1.5625	1/16	3/64
14134D-14276	1.3125	2.7170	1.5625	1/16	3/64
19145D-19283	1.4375	2.8345	1.5392	1/32	1/16
19152D-19283	1.5000	2.8345	1.5392	1/32	1/16
13169D-13318	1.6875	3.1875	1.3750	1/32	1/16
358D-354A	1.6875	3.3464	1.9790	1/16	3/64
13176D-13318	1.7500	3.1875	1.3750	0	1/16
13182D-13318	1.8125	3.1875	1.3750	1/32	1/16
376DE-372A	1.8750	3.8125	2.0940	1/32	1/16
378DE-372A	1.9375	3.8125	2.0940	1/32	1/16
375D-372A	2.0000	3.8125	2.0940	1/32	1/16
78216D-78551	2.1650	5.5130	2.6020	3/32	3/32
399D-394A	2.4375	4.3307	2.1870	1/32	3/64
78251D-78551	2.5000	5.5130	2.6020	3/32	3/32
765D-752	3.0000	6.3750	4.0000	9/64	1/8
496D-493	3.1875	5.3750	2.3750	1/16	1/8
581D-572	3.1875	5.5115	3.1875	1/16	1/8
498D-493	3.3125	5.3750	3.0000	1/32	1/8
767D-752	3.5000	6.3750	4.0000	1/16	1/8
865D-854	3.5000	7.5000	4.6250	3/8	1/8
42362D-42584	3.6250	5.8437	2.2500	1/16	.120
867D-854	3.7500	7.5000	4.6250	1/4	1/8
52388D-52618	3.8750	6.1875	3.1563	1/16	1/8
779D-772	3.8750	7.1250	4.0000	1/16	1/8
52400D-52618	4.0000	6.1875	3.1563	1/16	1/8
868D-854	4.0000	7.5000	4.6250	1/16	1/8
945D-932	4.0000	8.3750	5.6250	13/32	1/8
782D-772	4.1250	7.1250	4.0000	1/16	1/8
71426D-71750	4.2500	7.5000	3.8750	1/16	1/8
946D-932	4.2500	8.3750	5.6250	1/8	1/8
95426D-95925	4.2500	9.2500	5.5000	33/64	1/8
71450D-71750	4.5000	7.5000	3.8750	1/16	1/8
938D-932	4.5000	8.3750	5.6250	1/8	1/8
95451D-95925	4.5000	9.2500	5.5000	33/64	1/8
EE116048D-116098	4.6250	9.8750	6.0000	5/16	3/16
95474D-95925	4.7500	9.2500	5.5000	1/4	1/8
EE153047D-153100	4.7500	10.0000	6.3750	1/2	1/4
EE153048D-153100	4.8750	10.0000	6.3750	7/16	1/4
97500D-97900	5.0000	9.0000	6.3125	1/16	1/8
95499D-95925	5.0000	9.2500	5.5000	13/64	1/8
EE153053D-153100	5.0000	10.0000	6.3750	1/8	1/4
EE540501D-541162	5.0000	11.6250	5.8750	33/64	1/4
EE455050D-455116	5.0000	11.6250	6.5000	17/32	1/4
74512D-74850	5.1250	8.5000	4.0000	1/16	1/8
73512D-73875	5.1250	8.7500	2.3850	9/64	1/8

Bearing Number	Bore	O D	Brg. Width	Cone Radius	Cup Radius
67390D-67320	5.2500	8.0000	3.6250	1/16	1/8
EE455052D-455116	5.2500	11.6250	6.5000	3/8	1/4
EE450531D-451250	5.3110	12.5000	6.3750	17/32	17/64
EE455053D-455116	5.3750	11.6250	6.5000	17/32	1/4
73550D-73875	5.5000	8.7500	2.3850	9/64	1/8
EE455048D-455116	5.5000	11.6250	6.5000	1/8	1/4
EE450550D-451250	5.5000	12.5000	6.3750	17/32	17/64
EE92558D-92988	5.5620	9.8750	4.3750	1/16	3/16
EE517056D-517117	5.6250	11.7500	4.2500	1/8	1/8
EE517057D-517117	5.6875	11.7500	4.2500	1/8	1/8
81576D-81962	5.7500	9.6250	3.4375	1/16	1/8
EE450575D-451250	5.7500	12.5000	6.3750	17/32	17/64
99587D-99100	5.8750	10.0000	4.7500	1/16	1/8
81601D-81962	6.0000	9.6250	3.4375	1/16	1/8
99603D-99100	6.0000	10.0000	6.2500	1/16	1/8
EE517061D-517117	6.0000	11.7500	4.2500	1/8	1/8
EE450600D-451250	6.0000	12.5000	6.3750	3/8	17/64
EE217063D-217114	6.2500	11.4375	4.9375	1/8	1/4
82680D-82620	7.0000	11.0000	4.4375	1/16	1/8
EE91700D-91112	7.0000	11.2500	4.1875	1/16	1/8
94704D-94113	7.0000	11.3750	6.2500	1/16	1/8
EE280700D-281200	7.0000	12.0000	4.3086	1/8	1/8
EE210700D-211300	7.0000	13.0000	4.3750	1/16	1/8
EE222074D-222126	7.3750	12.5970	6.6250	1/8	3/16
93751D-93125	7.5000	12.5000	5.2500	1/4	1/8
EE210750D-211300	7.5000	13.0000	4.3750	1/8	1/8
EE420750D-421450	7.5000	14.5000	6.2500	1/8	1/8
EE132078D-132125	7.8750	12.5000	3.7500	1/8	1/8
93580D-93520	8.0000	12.5000	4.1563	1/16	1/8
93801D-93125	8.0000	12.5000	5.2500	1/4	1/8
EE132081D-132125	8.0000	12.5000	3.7500	1/8	1/8
EE420800D-421450	8.0000	14.5000	6.2500	1/8	1/8
9975D-9920	8.5000	13.0000	8.0000	1/8	1/8
EE130850D-131400	8.5000	14.0000	4.7500	1/16	1/16
96851D-96140	8.5000	14.0000	5.0000	1/4	1/8
EE130888D-131400	8.8750	14.0000	6.5000	5/16	1/16
EE130903D-131400	9.0000	14.0000	6.5000	5/16	1/16
EE529091D-529157	9.0000	15.7500	5.5000	1/8	1/8
EE430901D-431575	9.0000	15.7500	6.2500	1/8	1/8
EE700090D-700167	9.0000	16.7500	7.0000	9/64	1/4
EE8575D-8520	9.2500	12.8750	3.6875	1/16	1/8
EE127094D-127140	9.4970	14.0000	4.2500	1/16	1/8
8880D-8820	9.5000	13.5000	3.6250	1/16	1/8
EE170951D-171450	9.5000	14.5000	3.6500	1/16	1/8
EE821096D-821165	9.5000	16.5000	7.0000	1/8	1/4

TABLE 2-20

Needle Roller Bearings, Types NAA and NBA

AFBMA Standards, Section 2 — Nov 1952

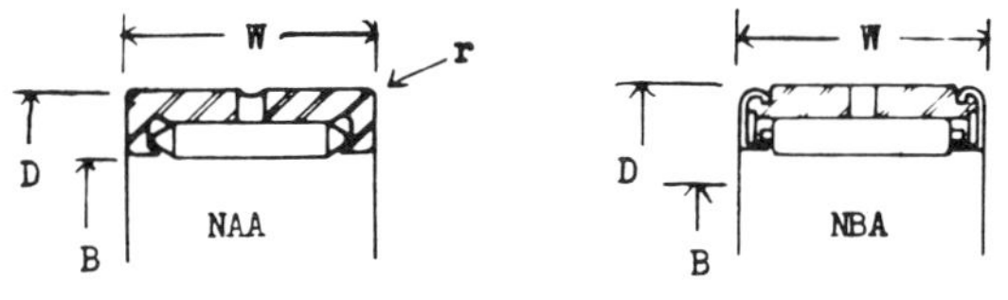

Bearing Numbers		*B* Bore	*D* Outside Diameter	*W* Width	*r* Fillet* Radius
6NAA1213	6NBA1213	.3750	.8125	.750	.025
8NAA1216	8NBA1216	.5000	1.0000	.750	.025
10NAA1218	10NBA1218	.6250	1.1250	.750	.025
12NAA1220	12NBA1220	.7500	1.2500	.750	.040
14NAA1222	14NBA1222	.8750	1.3750	.750	.040
16NAA1624	16NBA1624	1.0000	1.5000	1.000	.040
18NAA1626	18NBA1626	1.1250	1.6250	1.000	.040
20NAA1628	20NBA1628	1.2500	1.7500	1.000	.040
22NAA1630	22NBA1630	1.3750	1.8750	1.000	.040
24NAA2033	24NBA2033	1.5000	2.0625	1.250	.060
26NAA2035	26NBA2035	1.6250	2.1875	1.250	.060
28NAA2037	28NBA2037	1.7500	2.3125	1.250	.060
30NAA2039	30NBA2039	1.8750	2.4375	1.250	.060
32NAA2041	32NBA2041	2.0000	2.5625	1.250	.060
36NAA2448	36NBA2448	2.2500	3.0000	1.500	.060
40NAA2452	40NBA2452	2.5000	3.2500	1.500	.080
44NAA2456	44NBA2456	2.7500	3.5000	1.500	.080
48NAA2460	48NBA2460	3.0000	3.7500	1.500	.080
52NAA3268	52NBA3268	3.2500	4.2500	2.000	.080
56NAA3272	56NBA3272	3.5000	4.5000	2.000	.080
60NAA3276	60NBA3276	3.7500	4.7500	2.000	.100
64NAA3280	64NBA3280	4.0000	5.0000	2.000	.100
68NAA3284	68NBA3284	4.2500	5.2500	2.000	.100
72NAA4096	72NBA4096	4.5000	6.0000	2.500	.100
80NAA40104	80NBA40104	5.0000	6.5000	2.500	.100
88NAA40112	88NBA40112	5.5000	7.0000	2.500	.100
96NAA40120	96NBA40120	6.0000	7.5000	2.500	.120
104NAA40128	104NBA40128	6.5000	8.0000	2.500	.120
116NAA48146	116NBA48146	7.2500	9.1250	3.000	.120
124NAA48154	124NBA48154	7.7500	9.6250	3.000	.120
132NAA48162	132NBA48162	8.2500	10.1250	3.000	.120
140NAA48170	140NBA48170	8.7500	10.6250	3.000	.160
148NAA48178	148NBA48178	9.2500	11.1250	3.000	.160

All dimensions are given in inches

*The corner radius or chamfer on bearings must clear the maximum fillet radius given in the table.

This specification does not control bearing corner contours.

All rings to have oil holes centrally located.

TABLE 2-21

Needle Roller Bearings, Types NAB and NBB

AFBMA Standards, Section 2 – Nov 1952

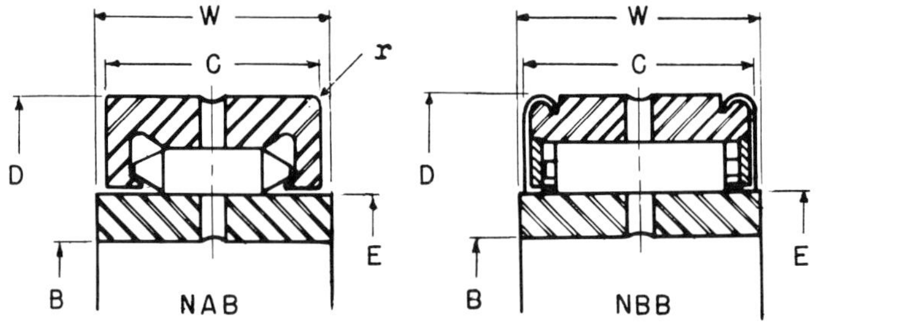

Bearing Numbers		B Bore	D Outside Diameter	E Inner Ring O. D.	C Outer Ring Width	W Over-all Width	r Fillet* Radius
6NAB1218	6NBB1218	.3750	1.1250	.625	.750	.760	.025
8NAB1220	8NBB1220	.5000	1.2500	.750	.750	.760	.040
10NAB1222	10NBB1222	.6250	1.3750	.875	.750	.760	.040
12NAB1624	12NBB1624	.7500	1.5000	1.000	1.000	1.010	.040
14NAB1626	14NBB1626	.8750	1.6250	1.125	1.000	1.010	.040
16NAB1628	16NBB1628	1.0000	1.7500	1.250	1.000	1.010	.040
18NAB1630	18NBB1630	1.1250	1.8750	1.375	1.000	1.010	.040
20NAB2033	20NBB2033	1.2500	2.0625	1.500	1.250	1.260	.060
22NAB2035	22NBB2035	1.3750	2.1875	1.625	1.250	1.260	.060
24NAB2037	24NBB2037	1.5000	2.3125	1.750	1.250	1.260	.060
26NAB2041	26NBB2041	1.6250	2.5625	2.000	1.250	1.260	.060
28NAB2448	28NBB2448	1.7500	3.0000	2.250	1.500	1.510	.060
32NAB2452	32NBB2452	2.0000	3.2500	2.500	1.500	1.510	.080
36NAB2456	36NBB2456	2.2500	3.5000	2.750	1.500	1.510	.080
40NAB2460	40NBB2460	2.5000	3.7500	3.000	1.500	1.510	.080
44NAB3268	44NBB3268	2.7500	4.2500	3.250	2.000	2.010	.080
48NAB3272	48NBB3272	3.0000	4.5000	3.500	2.000	2.010	.080
52NAB3280	52NBB3280	3.2500	5.0000	4.000	2.000	2.010	.100
56NAB3284	56NBB3284	3.5000	5.2500	4.250	2.000	2.010	.100
60NAB4096	60NBB4096	3.7500	6.0000	4.500	2.500	2.515	.100
64NAB40104	64NBB40104	4.0000	6.5000	5.000	2.500	2.515	.100
72NAB40112	72NBB40112	4.5000	7.0000	5.500	2.500	2.515	.100
80NAB40120	80NBB40120	5.0000	7.5000	6.000	2.500	2.515	.120
88NAB40127	88NBB40127	5.5000	8.0000	6.500	2.500	2.515	.120
96NAB48146	96NBB48146	6.0000	9.1250	7.250	3.000	3.015	.120
104NAB48154	104NBB48154	6.5000	9.6250	7.750	3.000	3.015	.120
112NAB48162	112NBB48162	7.0000	10.1250	8.250	3.000	3.015	.120
120NAB48170	120NBB48170	7.5000	10.6250	8.750	3.000	3.015	.160
128NAB48178	128NBB48178	8.0000	11.1250	9.250	3.000	3.015	.160

All dimensions are given in inches.

*The corner radius or chamfer on bearings must clear the maximum fillet radius given in the table. This specification does not control bearing corner contours.

All rings to have oil holes centrally located.

TABLE 2-22

AFBMA Standard Cam Followers

AFBMA Standards, Section 2 – Nov 1952

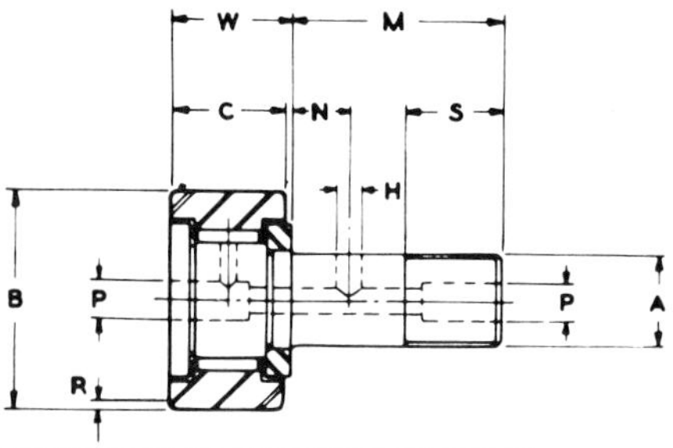

Bearing No.	A Stud Diam +.001 −.000	B Roller O D +.000 −.001	W Overall Width	C Roller Width +.000 −.005	M Stud Length	S Thread Length	R Rise	N	H	P	Bore Diameter for Stud Max	Bore Diameter for Stud Min
001CTA	.1900	.5000	3/8	.344	1/2	1/4	.01	NONE	NONE	1/8*	.1905	.1900
007CTA	.2160	.5625	13/32	.375	5/8	11/32	.015	NONE	NONE	1/8*	.2165	.2160
011CTA	.2500	.6250	7/16	.406	5/8	11/32	.02	NONE	NONE	1/8*	.2505	.2500
017CTA	.3125	.6875	15/32	.437	13/16	3/8	.02	NONE	NONE	1/8*	.3130	.3125
023CTA	.3750	.7500	17/32	.500	1	1/2	.03	1/4	3/32	3/16	.3755	.3750
029CTA	.3750	.8750	17/32	.500	1	1/2	.03	1/4	3/32	3/16	.3755	.3750
033CTA	.4375	1.0000	21/32	.625	1- 1/8	1/2	.05	5/16	1/8	3/16	.4380	.4375
037CTA	.4375	1.1250	21/32	.625	1- 1/8	1/2	.05	5/16	1/8	3/16	.4380	.4375
045CTA	.5000	1.2500	25/32	.750	1- 1/4	5/8	.07	5/16	1/8	3/16	.5005	.5000
047CTA	.5000	1.3750	25/32	.750	1- 1/4	5/8	.07	5/16	1/8	3/16	.5005	.5000
053CTA	.6250	1.5000	29/32	.875	1- 9/16	13/16	.09	3/8	5/32	3/16	.6255	.6250
059CTA	.6250	1.6250	29/32	.875	1- 9/16	13/16	.09	3/8	5/32	3/16	.6255	.6250
067CTA	.7500	1.7500	1- 1/32	1.000	1- 7/8	15/16	.10	15/32	5/32	3/16	.7505	.7500
071CTA	.7500	1.8750	1- 1/32	1.000	1- 7/8	15/16	.10	15/32	5/32	3/16	.7505	.7500
077CTA	.8750	2.0000	1- 9/32	1.250	2- 3/16	1- 1/16	.12	9/16	3/16	3/16	.8755	.8750
081CTA	.8750	2.2500	1- 9/32	1.250	2- 3/16	1- 1/16	.12	9/16	3/16	3/16	.8755	.8750
087CTA	1.0000	2.5000	1-17/32	1.500	2- 1/2	1- 3/16	.15	21/32	3/16	3/16	1.0005	1.0000
095CTA	1.0000	2.7500	1-17/32	1.500	2- 1/2	1- 3/16	.15	21/32	3/16	3/16	1.0005	1.0000
105CTA	1.2500	3.0000	1-25/32	1.750	3- 1/8	1- 1/2	.18	13/16	3/16	1/4	1.2505	1.2500
113CTA	1.2500	3.2500	1-25/32	1.750	3- 1/8	1- 1/2	.18	13/16	3/16	1/4	1.2505	1.2500
119CTA	1.3750	3.5000	2- 1/32	2.000	3- 7/16	1-11/16	.21	7/8	3/16	1/4	1.3755	1.3750
127CTA	1.5000	4.0000	2- 9/32	2.250	3- 3/4	1-13/16	.25	31/32	3/16	1/4	1.5005	1.5000

All dimensions are given in inches

*Grease fitting hole in head end only.

TABLE 2–23
Types and Characteristics of Ball Bearings

In code a ball bearing can be completely designated by a number and letter combination as 50BC02JPXEOM10. The basic number is 50BC02. The 50 indicates that the bore is 50 mm, the B identifies the bearing as a ball bearing, the C denotes the type as Conrad, and the 02 defines the width as 20 mm and outside diameter as 90 mm in accord with a dimensional series code. The JPXEOM10 define modifications, as cages, seals and shields, internal fit and tolerances and special requirements. See the Standards of the AFBMA for details.

Ball Bearing Type	Cross Section	Description
BA	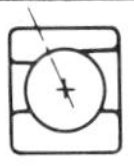	Single row, angular contact, self-contained, contact angle 22° to 32°, inclusive. Metric
BC		Single row, radial, non-filling slot assembly, (Conrad). Metric
BD	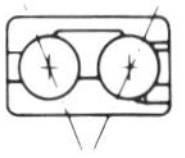	Double row, filling slot assembly, angular contact, vertex of contact angle inside bearing. Metric
BE	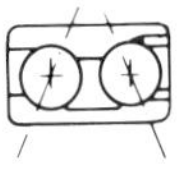	Double row, filling slot assembly, angular contact, vertex of contact angle outside bearing. Metric
BF	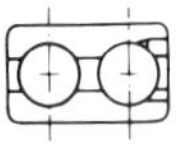	Double row, filling slot assembly, radial contact. Metric
BG		Double row, non-filling slot assembly, angular contact, vertex of contact angle outside bearing. Metric
BH	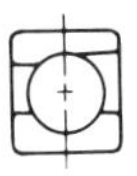	Single row, self-contained, radial contact. Metric
BIC		Inch dimensions. Single row, radial, non-filling slot assembly (same bearing as type BC in metric dimensions)
BJ		Double row, non-filling slot assembly, angular contact, vertex of contact angle inside bearing. Metric
BK	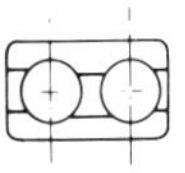	Double row, non-filling slot assembly, radial contact. Metric
BL	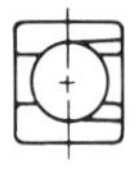	Single row, radial, filling slot assembly. Metric
BM	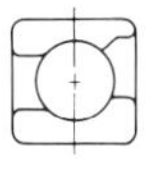	Single row, separable assembly. Metric
BN	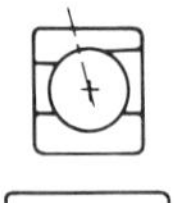	Single row, angular contact, self-contained, contact angle less than 22°. Metric
BS	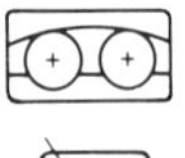	Double row, radial, self-aligning, raceway of outer ring spherical. Metric
BT	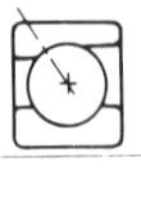	Single row, angular contact, self-contained, contact angle larger than 32° but less than 45°. Metric

TABLE 2–24
Types and Characteristics of Cylindrical Roller Bearings

Like ball bearings, cylindrical roller bearings can be completely designated by a number and letter symbol as 25RN02J112. The basic number is 25RN02. The R indicates that the bearing has cylindrical rollers; the 25 is the bore in millimeters; and the 02 is a dimensional series code that defines the width and OD, Tables 2–14 to 2–16. See the Standards of the AFBMA for details.

Roller Bearing Type	Cross Section	Description
RC	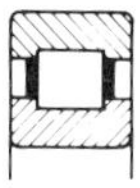	Single row, double flanged inner and outer rings, non-separable, two direction locating. Metric
RF RIF	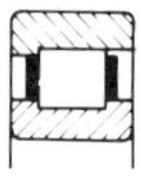	Single row, double flanged inner ring, single flanged outer ring, outer ring separable, one direction locating. Metric and inch dimensions.
RG	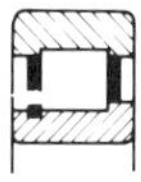	Single row, single flanged inner ring, double flanged outer ring, rollers retained by retainment ring recessed in inner ring, non-separable, one direction locating. Metric
RJ	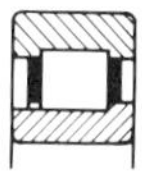	Single row, double flanged outer ring, single flanged inner ring, inner ring separable, one direction locating. Metric
RK RIK	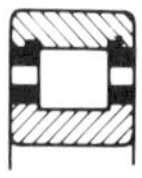	Single row, double flanged inner ring, rollers retained by retainment rings recessed in outer ring, non-separable, non-locating. Metric and inch dimensions.
RM	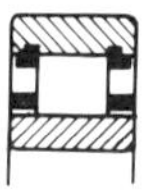	Single row, straight inner ring, rollers retained by cage end rings or retainment rings recessed in outer ring, inner ring separable, non-locating. Metric
RN RIN	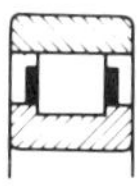	Single row, double flanged inner ring, straight outer ring, outer ring separable, non-locating. Metric and inch dimensions.
RP RIP	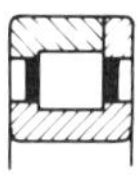	Single row, double flanged inner ring, double flanged outer ring, with one flange separable, outer ring separable, two direction locating. Metric and inch dimensions.
RS	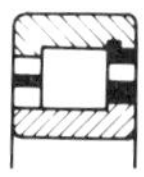	Single row, single flanged inner and outer rings, roller retained by flange and single retainment ring recessed in outer ring, inner ring separable, one direction locating. Metric
RT	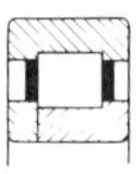	Single row, double flanged inner ring with one flange separable, double flanged outer ring, inner ring separable, two direction locating. Metric
RU RIU	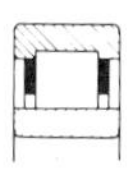	Single row, straight inner ring, double flanged outer ring, inner ring separable, non-locating. Metric and inch dimensions.
RY	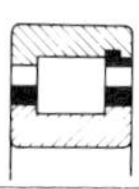	Single row, double flanged inner ring, single flanged outer ring, rollers retained by flange and single retainment ring recessed in outer ring, non-separable, one direction locating. Metric

TABLE 2–25

Types and Characteristics of Tapered Roller Bearings

So long as producers of tapered roller bearings — inch dimensions — continue to use the same parts numbers, the specification of the parts numbers corresponding to the cone and the cup quite definitely describes a bearing for design purposes. In the bearing numbers of Tables 2–17 to 2–19, that portion of the bearing number to the left of the hyphen identifies the part number of the cone, that on the right of the hyphen identifies the cup. The letter T is associated with tapered roller bearings, although the letter *T* connotes "Thrust" rather than "Taper" in AFBMA type symbols, Table 2-5. See the Timken Engineering Journal, the Timken Roller Bearing Company, for complete information about tapered roller bearings.

Type of Tapered Roller Bearing	Cross Section	Description
TDI	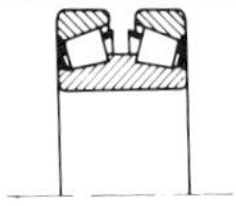	Double row, single cups (outer races), double cone (inner races). Widely used where load capacity of double-row bearing is required, particularly as anchor bearing. Simpler to mount than two (standard) single-row bearings.
TDO	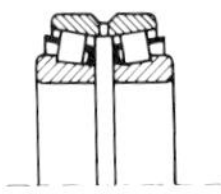	Double row, double cup, two single cones, adjustable. Often mounted so as to float in housing when capacity of two-row bearing is required.
TDOS	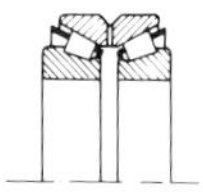	Double row, double cup, two single cones, steep angle, adjustable. Used where thrust load predominates.
TNA	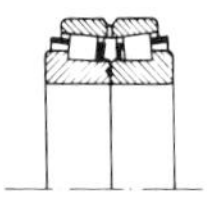	Double row, double cup, two single cones, non-adjustable, but otherwise similar to type TDO.
TNAS	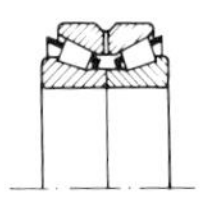	Double row, double cup, two single cones, steep angle, non-adjustable.
TS	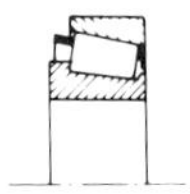	Single row, straight bore, recognized as the standard type of tapered roller bearing, and the most widely used. High radial load and moderate thrust.
TSF	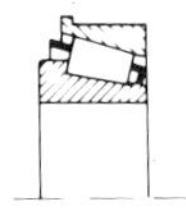	Single row, straight bore, flanged cup. Housing details and machining often can be planned to suit the flanged cup to advantage over the straight cylindrical cup.
TSS	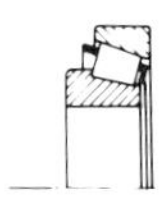	Single row, steep angle bearing used where thrust loads are equal to or exceed radial loads.
TST	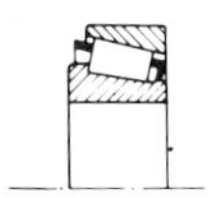	Single row, tapered bore. Caution is sensed regarding the free use of this bearing by these statements, which are quoted from the Timken Journal, "Most applications in which this type of bearing is used are of a special type of design requiring special handling. All applications in which this type bearing is required should be referred to the Engineering Department of the Timken Roller Bearing Company."

TABLE 2–26

Types and Characteristics of Needle Roller Bearings

Needle bearings, excepting those for airframes, have undergone less standardization among manufacturers than other antifriction bearings. The designer who makes his selection carelessly from a manufacturer's catalog therefore runs the risk of specifying products that may be procurable from only a single source. Those producers whose products conform to the Standards of the AFBMA are interchangeable dimensionwise. Most manufacturers' catalogs indicate clearly which products do or do not conform to AFBMA Standards.

In code 18 NAB1630 is a typical basic number for a needle bearing — inch dimensions. The 18 indicates that the bore is 18/16 = 1.1250 inches. The N denotes the type of bearing as needle; the A pertains to the type of roller; and the B indicates the bearing has an inner race. The 16 indicates the width of the bearing in units of 1/16 inch so that 16/16 = one inch. The 30 defines the OD as 30/16 = 1.875 inches. This number-letter symbol is used in an identical manner when the dimensions are metric.

The bearings described in this table and in Tables 2–20 and 2–21 have case-hardened shells that also retain the rollers, with or without inner races. Thus these bearings come as units. Another class of needle bearing is those having loose rollers. Journal roller bearings are separated as a class by AFBMA. Possibly this signifies that the standardization of needle bearings has still a long way to go.

Needle Bearing *Type	Cross Section	Description
NAA	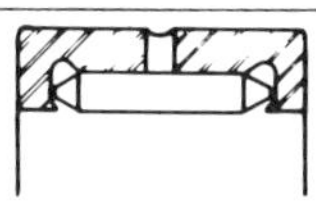	Single row, hardened outer race or shell retains rollers, inch dimensions. Since there is no inner race, load capacity of bearing depends upon shaft hardness — see Table 2–27.
NAB	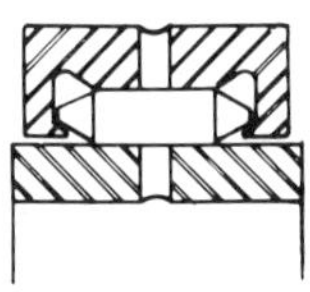	Single row, outer race retains rollers, separable inner race, inch dimensions.
NBA	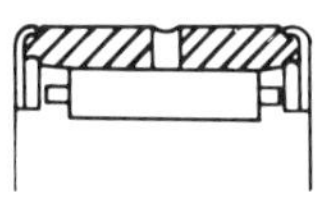	Single row, roller cage locked to outer race, inch dimensions. Since there is no inner race, load capacity of bearing depends upon shaft hardness — see Table 2–27.
NBB	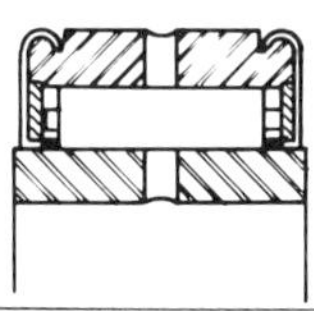	Single row, roller cage locked to outer race, separable inner race, inch dimensions.

*These are four of the most common types of established line needle bearings. See AFBMA Standards, Section 5—April 1953, for many others.

TABLE 2-27

Shaft-Hardness Load-Chart for Drawn-Cup Needle Bearings

Practice of the Torrington Co

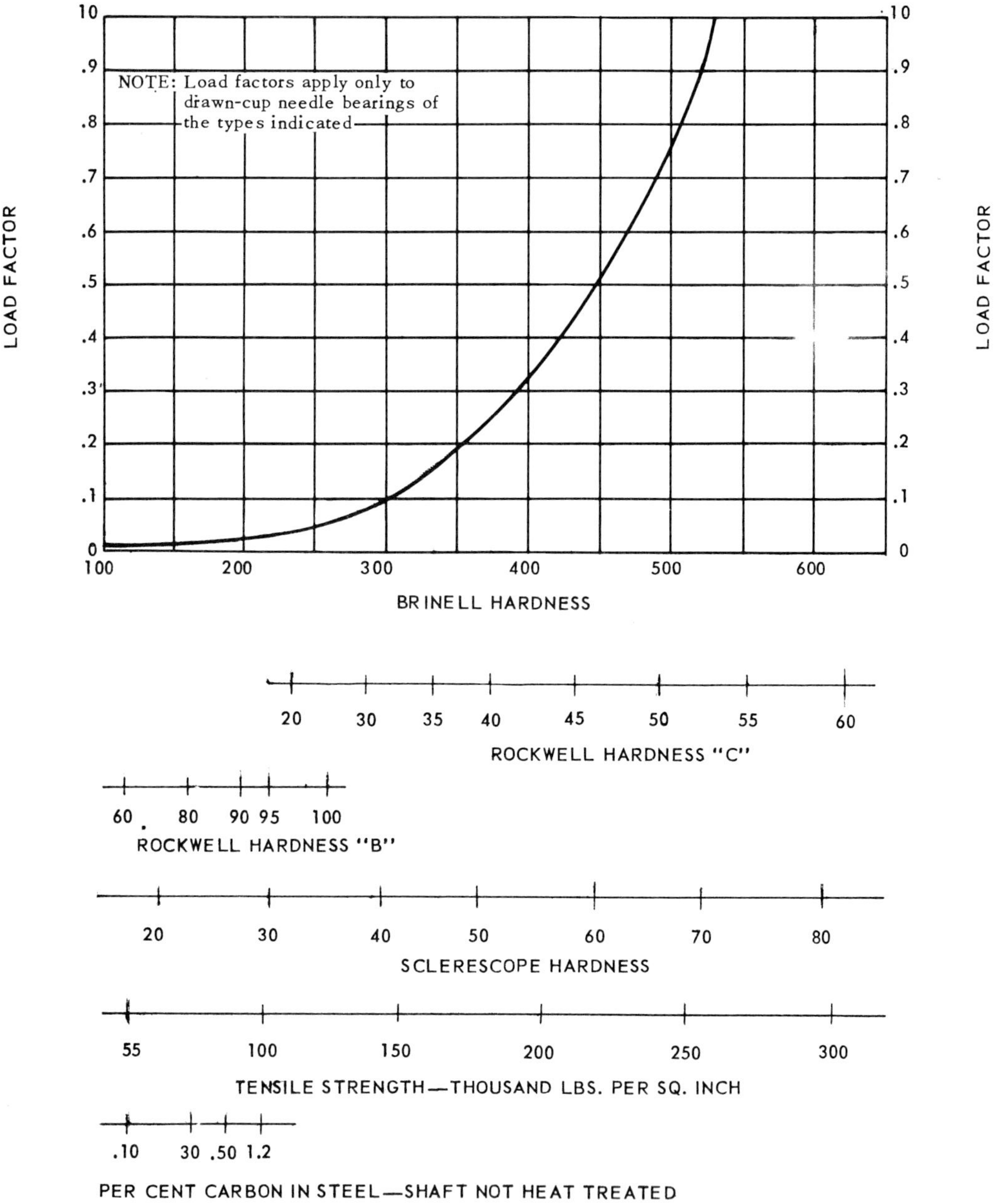

TABLE 2–28

Methods of Showing Anti-Friction Bearing on Drawings

See Section 7 for Fits and Tolerances and Section 1 for Shoulder Heights.

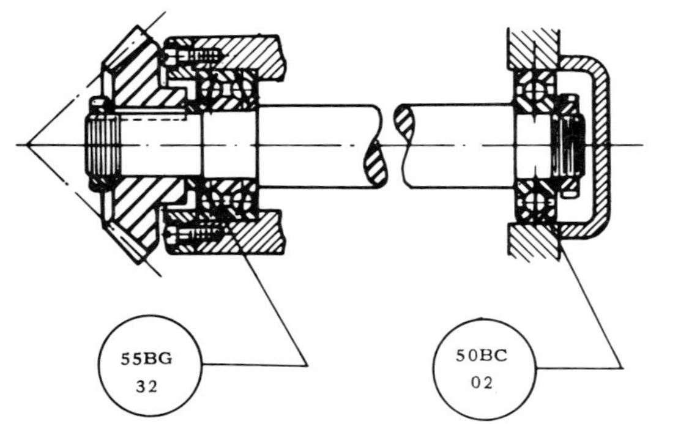

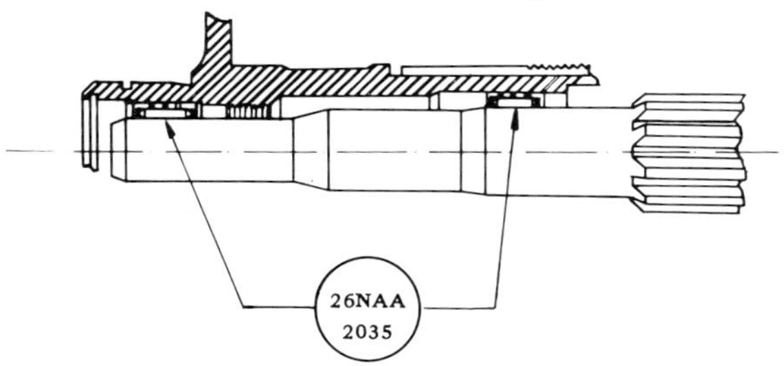

Since antifriction bearings are units already assembled, the design detailer generally is interested mainly in the boundary dimensions of the units, the fits with adjacent parts and adequate descriptions of the units for shop and procurement purposes.

Normally, antifriction bearings are drawn only on assembly drawings, and then in the manner illustrated by the figures. Bearing numbers, as indicated, quite adequately identify them. On parts lists and bills of material, the bearing number may well be supplemented by descriptive details as to rows of balls or rollers, type, alignment requirements, fit specifications, and so on. As more and more manufacturers and users of antifriction bearings bring identifications into agreement with AFBMA standards, the easier the job of specification and procurement of a desired unit will be.

Notes on drawings are recommended in cases where assembly is out-of-the-ordinary. For instance, the heating of the inner ring of a bearing to expand it before assembly onto a shaft. Caution needs to be taken to avoid heating the ring above 250°F, as overheating reduces the hardness. A note, therefore, should advise the assembler to heat the bearing in clean oil or in a controlled furnace to a temperature of between 200° to 250° to facilitate assembly.

Bearing manufacturers have boundary drawings of standard bearings in full, half, quarter and sometimes other sizes. These drawings can be had upon request. To use them the detailer merely places them beneath any transparent drawing and traces the outlines. Besides the saving in time and the use of conventional design detail, these "to-size" drawings clearly indicate shoulder heights and areas that must be kept clear for running purposes.

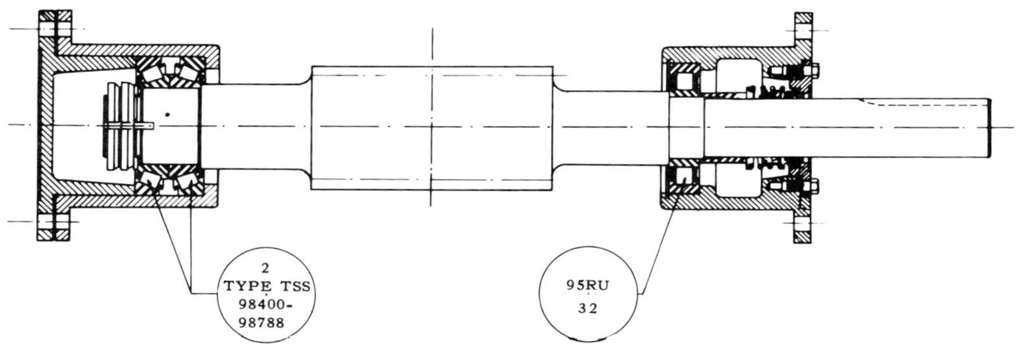

TABLE 2–29

Representative Sizes and Dimensions of 1/16-In. Wall* Split-Type Bushings
General Catalog The Cleveland Graphite Bronze Co

Materials—Steel-backed Bronzes and Babbitts or all Bronze.

Nominal Shaft Diameter	For Press Fit in Housing Hole Size ±0.0005	Inside Diam in Mean Housing Hole† ±0.001	Minimum Length ±0.010	Maximum Length ±0.010
0.250	0.375	0.252	1/4	3/4
.3125	.4375	.3145	1/4	3/4
.375	.500	.377	1/4	3/4
.4375	.5625	.4395	1/4	7/8
0.500	0.625	0.502	1/4	1
.5625	.6875	.5645	5/16	1-1/8
.625	.750	.627	5/16	1-1/4
.6875	.8125	.6895	3/8	1-3/8
0.750	0.875	0.752	3/8	1-1/2
.8125	.9375	.815	7/16	1-5/8
.875	1.000	.8775	7/16	1-3/4
.9375	1.0625	.940	1/2	1-7/8
1.000	1.125	1.0025	1/2	2
1.125	1.250	1.1275	9/16	2-1/4
1.250	1.375	1.2525	5/8	2-1/2
1.375	1.500	1.3775	11/16	2-3/4
1.500	1.625	1.5025	3/4	3
1.625	1.750	1.6275	13/16	3
1.750	1.875	1.7525	7/8	3
1.875	2.000	1.878	15/16	3
2.000	2.125	2.003	1	3
2.125	2.250	2.128	1-1/16	3
2.250	2.375	2.253	1-1/8	3
2.375	2.500	2.378	1-3/16	3
2.500	2.625	2.503	1-1/4	3
2.625	2.750	2.628	1-5/16	3
2.750	2.875	2.753	1-3/8	3
2.875	3.000	2.878	1-7/16	3
3.000	3.125	3.003	1-1/2	3

All dimensions in inches.

*Also available in 1/32 wall for same bore sizes up to 2 in. and in 3/32 wall in bore sizes from 1/2 to 4 in., incl. 1/8" wall, larger diameters and lengths outside range specified, can also be supplied.

† In designs where running clearances, alignment or concentricity demand closer tolerances, bushings are procurable with stock on inside diameter for boring, reaming, broaching or burnishing to size after assembly.

TABLE 2-30

Representative Sizes and Dimensions of Straight Shell Bearings

General Catalog The Cleveland Graphite Bronze Co

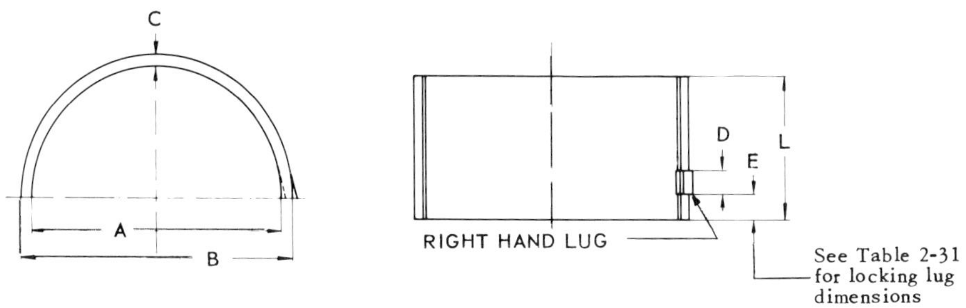

A Nominal	Shaft Diameter Actual		B Bore of Housing or Rod		C Wall Thick- ness	L Length Minimum ±0.005	 Maximum ±0.005	D Lug Width +0.005 −0.000
3/4	0.7485	0.749	0.8745	0.875	0.0625	5/8	3/4	0.175
7/8	.8735	.874	.9995	1.000	.0625	5/8	7/8	0.175
1	.9985	.999	1.1245	1.125	.0625	5/8	1	0.175
1-1/8	1.1235	1.124	1.2495	1.250	.0625	5/8	1-1/8	0.175
1-1/4	1.2485	1.249	1.3745	1.375	.0625	5/8	1-1/4	0.175
1-3/8	1.3735	1.374	1.4995	1.500	.0625	5/8	1-5/16	0.175
1-1/2	1.498	1.499	1.6245	1.625	0.0625	5/8	1-3/8	0.175
1-1/2			1.651	1.652	.075			
1-5/8	1.623	1.624	1.7495	1.750	.0625	5/8	1-7/16	0.175
1-5/8			1.776	1.777	.075			
1-3/4	1.748	1.749	1.8745	1.875	.0625	5/8	1-1/2	0.175
1-3/4			1.901	1.902	.075			
1-7/8	1.873	1.874	1.9995	2.000	0.0625	5/8	1-9/16	0.175
1-7/8			2.026	2.027	.075			
2	1.998	1.999	2.1245	2.125	.0625	5/8	1-5/8	0.175
2			2.151	2.152	.075			
2-1/8	2.123	2.124	2.2495	2.250	.0625	5/8	1-11/16	0.175
2-1/8			2.276	2.277	.075			
2-1/4	2.248	2.249	2.3745	2.375	0.0625	11/16	1-13/16	0.175
2-1/4			2.401	2.402	.075			
2-3/8	2.373	2.374	2.4995	2.500	.0625	11/16	1-7/8	0.175
2-3/8			2.526	2.527	.075			
2-1/2	2.498	2.499	2.6505	2.651	.075	3/4	2	0.175
2-1/2			2.691	2.692	.095			
2-5/8	2.623	2.624	2.7755	2.776	0.075	13/16	2-1/8	0.238
2-5/8			2.816	2.817	.095			
2-3/4	2.748	2.749	2.9005	2.901	.075	13/16	2-3/16	0.238
2-3/4			2.941	2.942	.095			
2-7/8	2.873	2.874	3.0255	3.026	.075	7/8	2-5/16	0.238
2-7/8			3.066	3.067	.095			

continued on next page

TABLE 2-30, continued

A			B		C	L		D
Shaft Diameter			Bore of Housing or Rod		Wall Thick-ness	Length		Lug Width +0.005 −0.000
Nominal	Actual					Minimum ±0.005	Maximum ±0.005	
3	2.998	2.999	3.1505	3.151	0.075	7/8	2-3/8	0.238
3			3.191	3.192	.095			
3-1/8	3.123	3.124	3.2755	3.276	.075	15/16	2-1/2	.238
3-1/4	3.248	3.249	3.4415	3.4425	.095	1	2-5/8	.238
3-1/4			3.502	3.503	.125			
3-3/8	3.373	3.374	3.5665	3.5675	0.095	1	2-11/16	0.238
3-1/2	3.498	3.499	3.6915	3.6925	.095	1-1/16	2-13/16	.238
3-1/2			3.752	3.753	.125			
3-5/8	3.623	3.624	3.8165	3.8175	.095	1-1/16	2-7/8	.363
3-3/4	3.748	3.749	3.942	3.943	0.095	1-1/8	3	0.363
3-3/4			4.002	4.003	.125			
4	3.998	3.999	4.192	4.193	.095	1-3/16	3-3/16	.363
4			4.252	4.253	.125			
4-1/4	4.248	4.249	4.502	4.503	.125	1-1/4	3-3/8	.363
4-1/4			4.562	4.563	.155			
4-1/2	4.498	4.499	4.752	4.753	0.125	1-3/8	3-5/8	0.363
4-1/2			4.812	4.813	.155			
4-3/4	4.748	4.749	5.002	5.003	.125	1-7/16	3-13/16	.363
4-3/4			5.062	5.063	.155			
5	4.998	4.999	5.252	5.253	.125	1-1/2	4	.363
5			5.312	5.313	.155			

TABLE 2-31

Locking Lug Dimensions for Shell and Flanged Bearings

General Catalog   The Cleveland Graphite Bronze Co

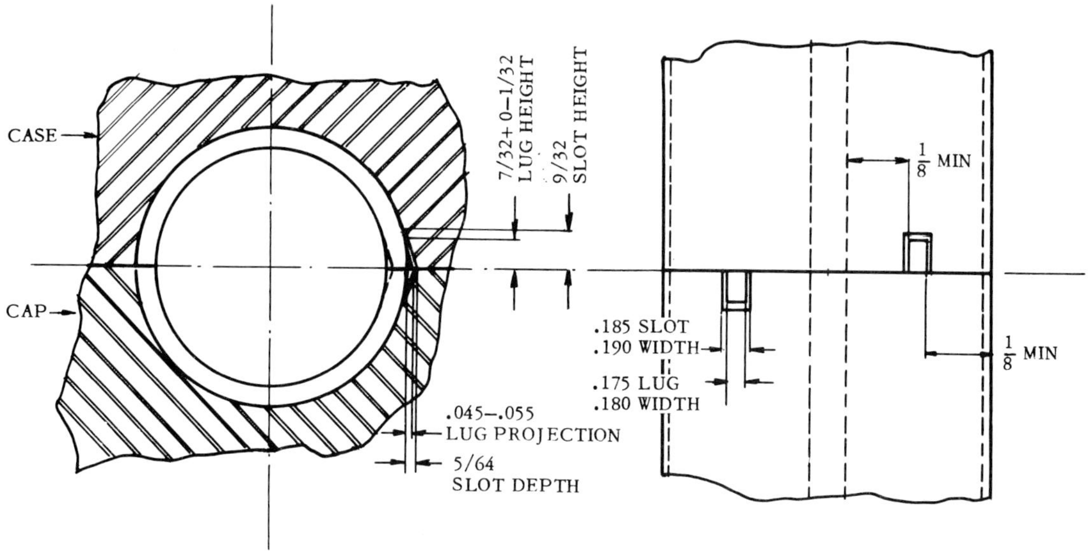

For Bearings With Following Shaft Diameter	Lug Width*		Slot Width		Lug Projection		Slot Depth	Lug Height +0-1/32	Slot Height
Small Diam and Extremely Short Bearings	.113	.118	.123	.128	.031	.041	1/16	5/32	7/32
3/4 Diam to 1-1/2 Diam Inclusive	.175	.180	.185	.190	.031	.041	1/16	5/32	7/32
1-9/16 Diam to 2-1/2 Diam Inclusive (Bearings with 1/16 wall)	.175	.180	.185	.190	.031	.041	1/16	7/32	9/32
1-9/16 Diam to 2-1/2 Diam Inclusive (Bearings with wall over 1/16)	.175	.180	.185	.190	.045	.055	5/64	7/32	9/32
2-9/16 Diam to 3-1/2 Diam Inclusive	.238	.243	.248	.253	.045	.055	5/64	7/32	9/32
3-9/16 Diam and over	.363	.368	.373	.378	.055	.065	3/32	11/32	13/32
Large Diam and Extra Long Bearings	.488	.493	.498	.503	.055	.065	3/32	11/32	13/32

All dimensions in inches.

*For bearings used in aluminum housing or for extremely long bearings, a wider lug than specified for that particular diameter should be used. For very short bearings, a narrower lug than specified for that particular diameter may be used.

TABLE 2-32

Representative Sizes and General* Dimensions of Flanged Thinwall Bearings

General Catalog   The Cleveland Graphite Bronze Co

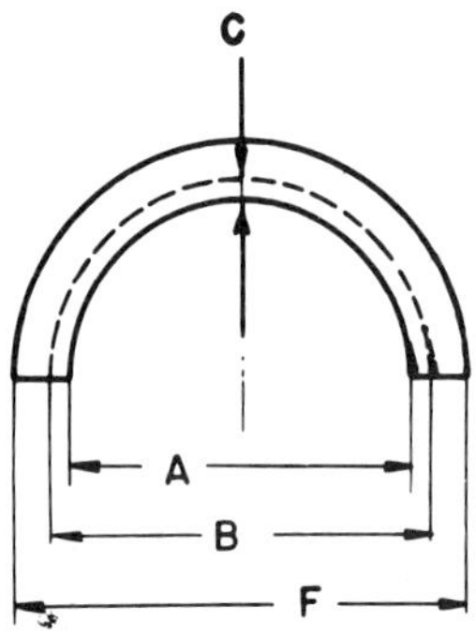

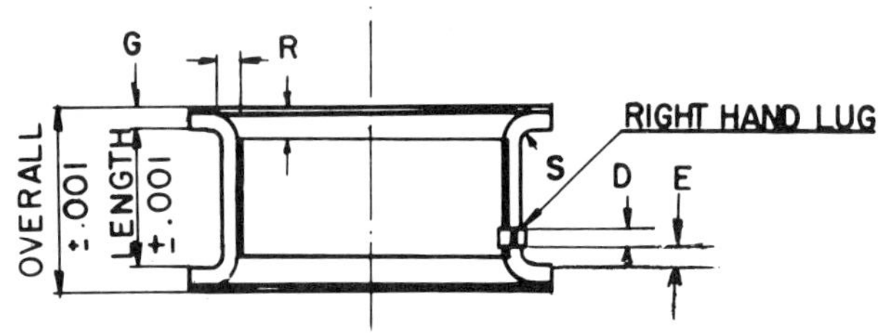

SEE TABLE 2-31

A			B		C	F	G	R	S	D
Shaft Diameter					Wall Thick-ness		Flange Thick-ness		Outer Diameter	Lug Width
Nominal	Actual		Housing Bore		Nomi-nal	Flange Diameter	±.001	Under-cut	Radius Maximum	+.005 −.000
1-5/8	1.623	1.624	1.776	1.777	.075	2-5/16	.082	1/8	1/16	.175
1-3/4	1.748	1.749	1.901	1.902	.075	2-15/32	.082	1/8	1/16	.175
1-7/8	1.873	1.874	2.026	2.027	.075	2-5/8	.082	1/8	1/16	.175
2	1.998	1.999	2.151	2.152	.075	2-25/32	.082	1/8	1/16	.175
2-1/8	2.123	2.124	2.276	2.277	.075	2-15/16	.082	1/8	1/16	.175
2-1/4	2.248	2.249	2.401	2.402	.075	3-3/32	.082	1/8	1/16	.175
2-3/8	2.373	2.374	2.526	2.527	.075	3-1/4	.082	1/8	1/16	.175
2-1/2	2.498	2.499	2.691	2.692	.095	3-7/16	.102	9/64	1/16	.175
2-5/8	2.623	2.624	2.816	2.817	.095	3-9/16	.102	9/64	1/16	.238
2-3/4	2.748	2.749	2.941	2.942	.095	3-3/4	.102	5/32	1/16	.238
2-7/8	2.873	2.874	3.066	3.067	.095	3-7/8	.102	5/32	1/16	.238
3	2.998	2.999	3.191	3.192	.095	4-1/32	.102	11/64	5/64	.238
3-1/8	3.123	3.124	3.316	3.317	.095	4-3/16	.102	11/64	5/64	.238
3-1/4	3.248	3.249	3.502	3.503	.125	4-3/8	.132	13/64	5/64	.238
3-3/8	3.373	3.374	3.627	3.628	.125	4-1/2	.132	13/64	5/64	.238
3-1/2	3.498	3.499	3.752	3.753	.125	4-21/32	.132	13/64	5/64	.238
3-3/4	3.748	3.749	4.002	4.003	.125	5	.132	13/64	5/64	.363
4	3.998	3.999	4.252	4.253	.125	5-5/16	.132	13/64	5/64	.363
4-1/4	4.248	4.249	4.562	4.563	.155	5-11/16	.162	15/64	5/64	.363

All dimensions are in inches.

*See Table 2–34 for illustration of dimensioning details.

TABLE 2–33

Sizes and General* Proportions of Heavy Wall Bearings

General Catalog The Cleveland Graphite Bronze Co

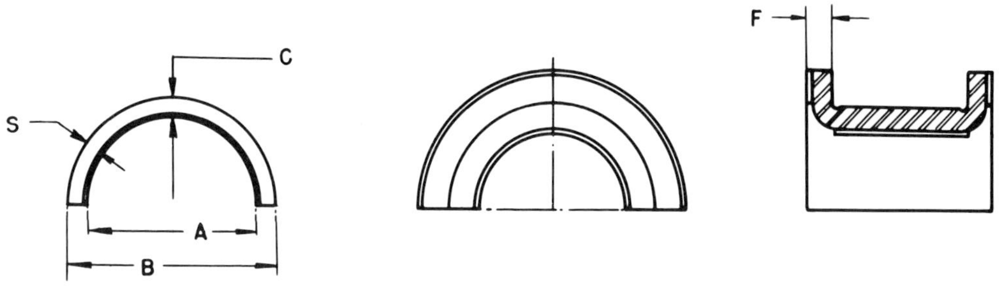

STRAIGHT BEARING FLANGED BEARING

A Shaft Diameter Nominal	Actual		B Housing Bore		C Wall Thickness Nominal	S Steel Thickness	F Flange Thickness Nominal
3	2.999	3.000	3.312	3.313	5/32	1/8	1/8
3-1/8	3.124	3.125	3.437	3.438	5/32	1/8	1/8
3-1/4	3.249	3.250	3.562	3.563	5/32	1/8	1/8
3-3/8	3.374	3.375	3.687	3.688	5/32	1/8	1/8
3-1/2	3.499	3.500	3.812	3.813	5/32	1/8	1/8
3-5/8	3.624	3.625	3.937	3.938	5/32	1/8	1/8
3-3/4	3.749	3.750	4.062	4.063	5/32	1/8	1/8
3-7/8	3.874	3.875	4.187	4.188	5/32	1/8	1/8
4	3.999	4.000	4.437	4.438	7/32	3/16	5/32
4-1/4	4.249	4.250	4.687	4.688	7/32	3/16	5/32
4-1/2	4.499	4.500	4.937	4.938	7/32	3/16	5/32
4-3/4	4.749	4.750	5.187	5.188	7/32	3/16	5/32
5	4.999	5.000	5.437	5.438	7/32	3/16	5/32
5-1/4	5.249	5.250	5.687	5.688	7/32	3/16	5/32
5-1/2	5.499	5.500	5.937	5.938	7/32	3/16	5/32
5-3/4	5.749	5.750	6.187	6.188	7/32	3/16	5/32
6	5.999	6.000	6.562	6.563	9/32	1/4	7/32
6-1/4	6.249	6.250	6.812	6.813	9/32	1/4	7/32
6-1/2	6.499	6.500	7.062	7.063	9/32	1/4	7/32
6-3/4	6.749	6.750	7.312	7.313	9/32	1/4	7/32
7	6.999	7.000	7.562	7.563	9/32	1/4	7/32
7-1/4	7.249	7.250	7.812	7.813	9/32	1/4	7/32
7-1/2	7.499	7.500	8.062	8.063	9/32	1/4	7/32
7-3/4	7.749	7.750	8.312	8.313	9/32	1/4	7/32
8	7.999	8.000	8.749	8.750	3/8	11/32	9/32
8-1/2	8.499	8.500	9.249	9.250	3/8	11/32	9/32
9	8.999	9.000	9.749	9.750	3/8	11/32	9/32
9-1/2	9.499	9.500	10.249	10.250	3/8	11/32	9/32
10	9.998	10.000	10.936	10.938	15/32	7/16	11/32
10-1/2	10.498	10.500	11.436	11.438	15/32	7/16	11/32

All dimensions are in inches.

***See Table 2–35 for illustration of dimensioning details.**

continued on next page

TABLE 2–33, continued

A Shaft Diameter Nominal	A Shaft Diameter Actual		B Housing Bore		C Wall Thickness Nominal	S Steel Thickness	F Flange Thickness Nominal
11	10.998	11.000	11.936	11.938	15/32	7/16	11/32
11-1/2	11.498	11.500	12.436	12.438	15/32	7/16	11/32
12	11.998	12.000	12.936	12.938	15/32	7/16	11/32
13	12.998	13.000	14.123	14.125	9/16	17/32	13/32
14	13.998	14.000	15.123	15.125	9/16	17/32	13/32
15	14.998	15.000	16.248	16.250	5/8	19/32	15/32
16	15.998	16.000	17.248	17.250	5/8	19/32	15/32

TABLE 2–34

Typical Dimensions on Drawing of a Flanged Thinwall Bearing. Cf Table 2–32

General Catalog The Cleveland Graphite Bronze Co.

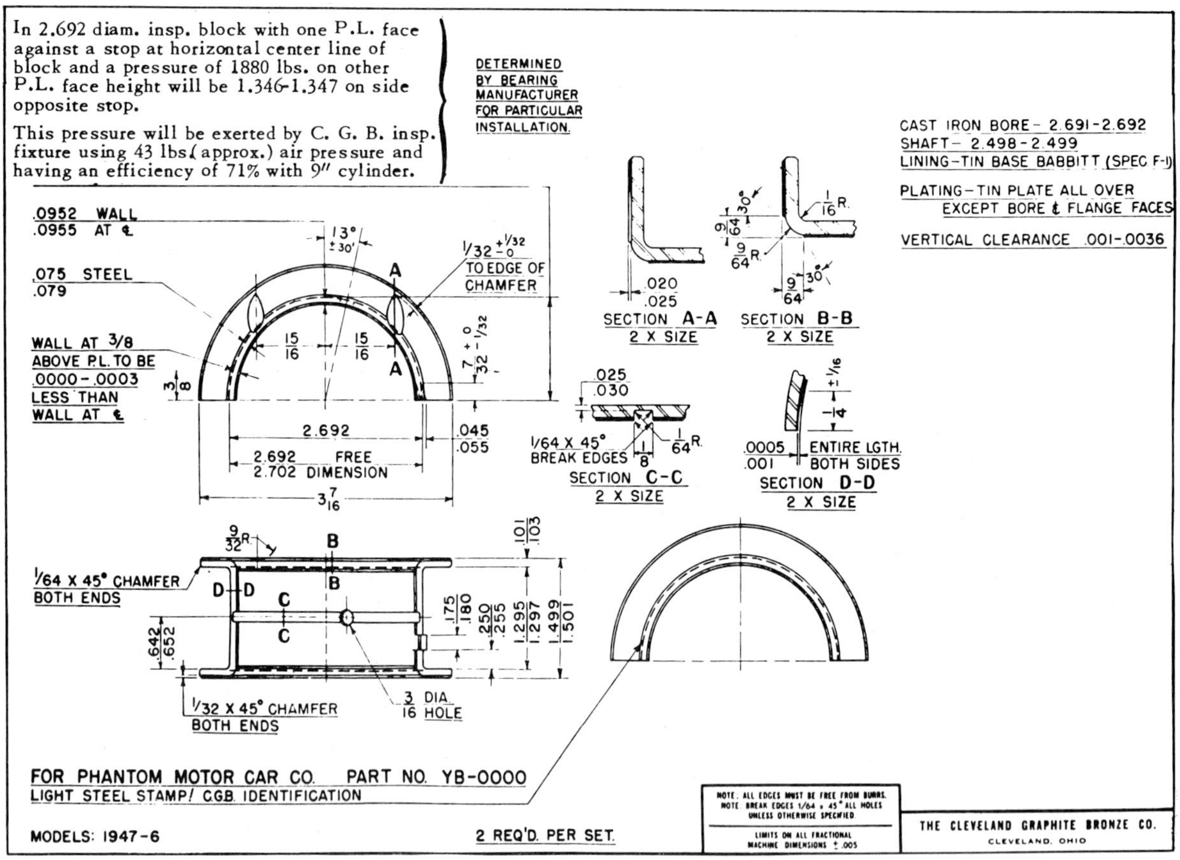

TABLE 2–35

Typical Dimensions on Drawing of a Flanged Heavy Wall Bearing. Cf. Table 2–33

General Catalog   The Cleveland Graphite Bronze Co.

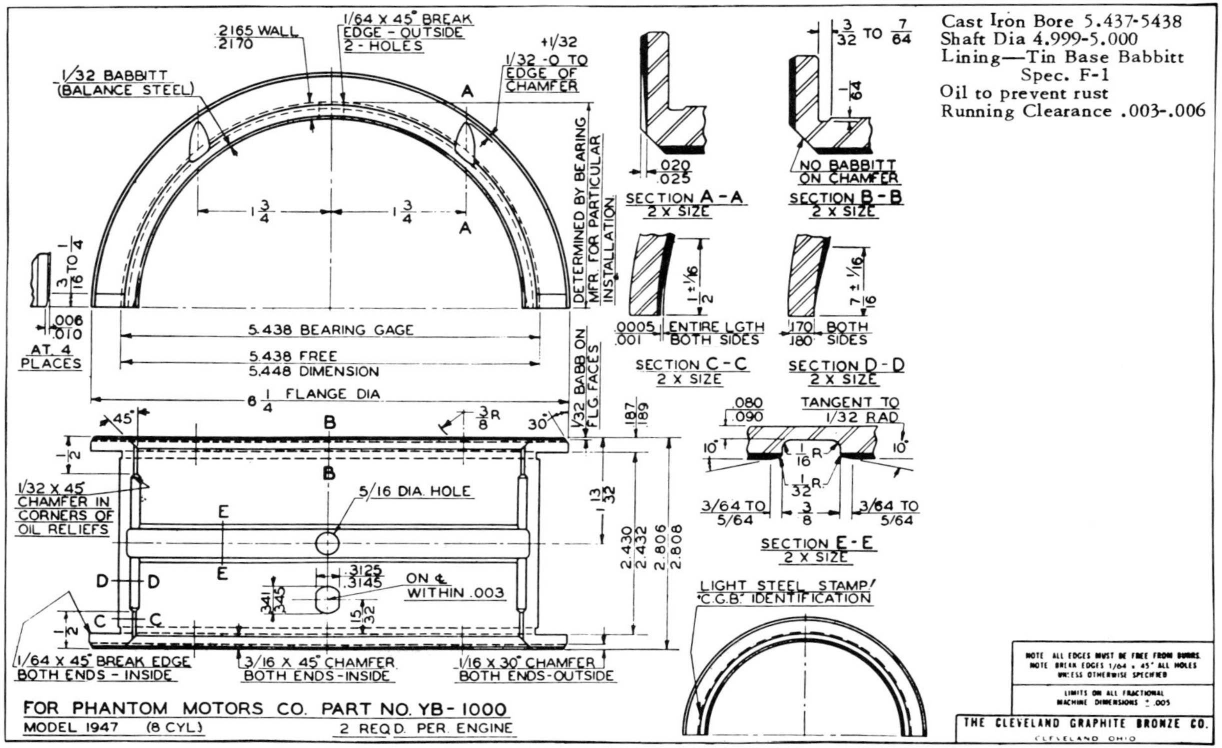

TABLE 2-36

Recommended Running Clearances for Normal Bearing Applications General Catalog The Cleveland Graphite Bronze Co.

Material	Minimum Clearance per Inch of Shaft Diameter
White metals	0.0005
Copper — lead	0.00075
Aluminum	0.0010

TABLE 2-37

Sleeve Bearing Clearances in Industrial Applications

Bearing Design Data Johnson Bronze

Class of Bearing	Running Clearance, Thousandths of an inch, for shaft dia under ½	1	2	3½	5½
Precision Spindle Practice—Hardened & ground spindle lapped into the bronze bushing. Below 500 ft./Min. & 500 psi	.00025 to .00075	.00075 to 0015.	.0015 to .0025	.0025 to .0035	.0035 to .005
Precision spindle practice—Hardened & ground spindle lapped into bro..ze bushing. Above 500 ft./Min. & 500 psi	.0005 to .001	.001 to .002	.002 to .003	.003 to .0045	.0045 to .0065
Electric Motor & Generator Practice—Ground Journal in broached or reamed bronze bushing or reamed Babbitt bushing	.0005 to .0015	.001 to .002	.0015 to .0035	.002 to .004	.003 to .006
General Machine Practice (Continuous rotating motion)—Turned steel or cold rolled steel Journals in bored & reamed bronze or poured & reamed babbitt bushings	.002 to .004	.0025 to .0045	.003 to .005	.004 to .007	.005 to .008
General Machine Practice (Oscillating Motion)—Journal & Bearing material as above	.0025 to .0045	.0025 to .0045	.003 to .005	.004 to .007	.005 to .008
Rough Machine Practice—Turned steel or cold-rolled steel Journals in Poured babbitt bearings	.003 to .006	.005 to .009	.008 to .012	.011 to .016	.014 to .020

NOTE: Cylindrical fits, allowances and tolerances are treated more fully in Section 7.

TABLE 2-38

Running Fits

Practice of Caterpillar Tractor Co

Nominal Size	*A 0.001 to .003 loose Hole	*A Shaft	†B 0.003 to .007 loose Hole	†B Shaft	‡C 1/64 loose Hole	‡C Shaft
1/8	0.126	0.124	0.126	0.120	0.126	7/64
	.125	.123	.123	.119	.121	
5/32	.1572	.1552	.1572	.1512	.1572	9/64
	.1562	.1542	.1542	.1502	.1522	
3/16	.1885	.1865	.1885	.1825	.1885	11/64
	.1875	.1855	.1855	.1815	.1835	
1/4	.251	.249	.251	.245	.251	15/64
	.250	.248	.248	.244	.246	
5/16	0.3135	0.3115	0.3135	0.3075	0.3135	19/64
	.3125	.3105	.3105	.3065	.3085	
3/8	.376	.374	.376	.370	.376	23/64
	.375	.373	.373	.369	.371	
7/16	.4385	.4365	.4385	.4325	.4385	27/64
	.4375	.4355	.4355	.4315	.4335	
1/2	.501	.499	.501	.495	.501	31/64
	.500	.498	.498	.494	.496	
9/16	0.5635	0.5615	0.5635	0.5575	0.5635	35/64
	.5625	.5605	.5605	.5565	.5585	
5/8	.626	.624	.626	.620	.626	39/64
	.625	.623	.623	.619	.621	
11/16	.6885	.6865	.6885	.6825	.6885	43/64
	.6875	.6855	.6855	.6815	.6835	
3/4	.751	.749	.751	.745	.751	47/64
	.750	.748	.748	.744	.746	
13/16	0.8135	0.8115	0.8135	0.8075	0.8135	51/64
	.8125	.8105	.8105	.8065	.8085	
7/8	.876	.874	.876	.870	.876	55/64
	.875	.873	.873	.869	.871	
15/16	.9385	.9365	.9385	.9325	.9385	59/64
	.9375	.9355	.9355	.9315	.9335	
1	1.001	.999	1.001	.995	1.001	63/64
	1.000	.998	.998	.994	.996	

All dimensions are in inches.
*Intended for high class bearing, well lubricated and properly aligned.
†Intended for use on intermediate running fit where very close tolerances are not necessary but clearance must be held reasonably close.
‡Intended for rough bearings where problems of alignment, lubrication or rusting may preclude the use of closer fits.
NOTE: Cylindrical fits, allowances and tolerances are treated more fully in Section 7.

continued on next page

Table 2-38, continued

Nominal Size	*A 0.0015 to .0035 loose Hole	Shaft	†B 0.005 to .009 loose Hole	Shaft	‡C 1/32 loose Hole	Shaft
1-1/16	1.0635	1.0610	1.0635	1.0555	1.0635	1-1/32
	1.0625	1.0600	1.0605	1.0545	1.0585	
1-1/8	1.126	1.1235	1.126	1.118	1.126	1-3/32
	1.125	1.1225	1.123	1.117	1.121	
1-3/16	1.1885	1.1860	1.1885	1.1805	1.1885	1-5/32
	1.1875	1.1850	1.1855	1.1795	1.1835	
1-1/4	1.251	1.2485	1.251	1.243	1.251	1-7/32
	1.250	1.2475	1.248	1.242	1.246	
1-5/16	1.3135	1.311	1.3135	1.3055	1.3135	1-9/32
	1.3125	1.310	1.3105	1.3045	1.3085	
1-3/8	1.376	1.3735	1.376	1.368	1.376	1-11/32
	1.375	1.3725	1.373	1.367	1.371	
1-7/16	1.4385	1.4360	1.4385	1.4305	1.4385	1-13/32
	1.4375	1.4350	1.4355	1.4295	1.4335	
1-1/2	1.501	1.4985	1.501	1.493	1.501	1-15/32
	1.500	1.4975	1.498	1.492	1.496	
1-9/16	1.5635	1.561	1.5635	1.5555	1.5635	1-17/32
	1.5625	1.560	1.5605	1.5545	1.5585	
1-5/8	1.626	1.6235	1.626	1.618	1.626	1-19/32
	1.625	1.6225	1.623	1.617	1.621	
1-11/16	1.6885	1.686	1.6885	1.6805	1.6885	1-21/32
	1.6875	1.685	1.6855	1.6795	1.6835	
1-3/4	1.751	1.7485	1.751	1.743	1.751	1-23/32
	1.750	1.7475	1.748	1.742	1.746	
1-13/16	1.8135	1.811	1.8135	1.8055	1.8135	1-25/32
	1.8125	1.810	1.8105	1.8045	1.8085	
1-7/8	1.876	1.8735	1.876	1.868	1.876	1-27/32
	1.875	1.8725	1.873	1.867	1.871	
1-15/16	1.9385	1.9360	1.9385	1.9305	1.9385	1-29/32
	1.9375	1.9350	1.9355	1.9295	1.9335	
2	2.0010	1.9985	2.0010	1.993	2.0010	1-31/32
	2.0000	1.9975	1.9980	1.992	1.9960	

All dimensions are in inches.
*Intended for high class bearing, well lubricated and properly aligned.
†Intended for use on intermediate running fit where very close tolerances are not necessary but clearance must be held reasonably close.
‡Intended for rough bearings where problems of alignment, lubrication or rusting may preclude the use of closer fits.
NOTE: Cylindrical fits, allowances and tolerances are treated more fully in Section 7.

TABLE 2-39

Journal and Bearing Diameters Manufacturer of Electrical Machinery
Trans. ASME, 1934 Vol. 56, p. 894

	Journal		Bearing					
			Horizontal		Vertical		Step	
Nominal Diam	Max Diam	Below Max Diam	Min Bore	Above Min Bore	Min Bore	Above Min Bore	Min Bore	Above Min Bore
3/8	0.375	0.0005	0.377	0.001	0.377	0.001	0.3755	0.0005
1/2	0.500	0.0005	0.502	0.001	0.502	0.001	0.5005	0.0005
5/8	0.625	0.0005	0.627	0.001	0.627	0.001	0.6255	0.0005
3/4	0.750	0.0005	0.752	0.001	0.752	0.001	0.7505	0.0005
7/8	0.875	0.0005	0.877	0.001	0.877	0.001	0.8755	0.0005
1	1.000	0.0005	1.002	0.001	1.002	0.001	1.0005	0.0005
1-1/8	1.125	0.0005	1.128	0.001	1.128	0.001	1.126	0.0005
1-1/4	1.250	0.0005	1.253	0.001	1.253	0.001	1.251	0.0005
1-1/2	1.500	0.0005	1.503	0.001	1.503	0.001	1.501	0.0005
1-3/4	1.750	0.0005	1.753	0.001	1.753	0.001	1.751	0.0005
2	2.000	0.0005	2.003	0.001	2.003	0.001	2.001	0.0005
2-1/4	2.250	0.0005	2.253	0.001	2.253	0.001	2.251	0.0005
2-1/2	2.500	0.0005	2.503	0.001	2.503	0.001	2.501	0.0005
2-3/4	2.750	0.0005	2.754	0.002	2.754	0.001	2.7515	0.0005
3	3.000	0.0005	3.004	0.002	3.004	0.001	3.0015	0.0005
3-1/4	3.250	0.0005	3.254	0.002	3.254	0.001	3.2515	0.0005
3-1/2	3.500	0.001	3.504	0.002	3.504	0.001	3.5015	0.0005
4	4.000	0.001	4.005	0.002	4.005	0.001	4.002	0.001
4-1/2	4.500	0.001	4.505	0.002	4.505	0.001	4.502	0.001
5	5.000	0.001	5.006	0.002	5.005	0.002	5.0025	0.001
5-1/2	5.500	0.001	5.507	0.002	5.505	0.002	5.503	0.001
6	6.000	0.001	6.009	0.002	6.006	0.002	6.003	0.001
7	7.000	0.001	7.011	0.002	7.006	0.002	7.0035	0.001
8	8.000	0.001	8.012	0.003	8.006	0.002	8.004	0.002
9	9.000	0.001	9.013	0.004	9.006	0.002	9.0045	0.002
10	10.000	0.0015	10.014	0.005	10.007	0.003	10.005	0.002
11	11.000	0.0015	11.015	0.005	11.007	0.003	11.0055	0.002
12	12.000	0.0015	12.016	0.005	12.007	0.003	12.006	0.002
13	13.000	0.0015	13.016	0.005	13.007	0.003	13.0065	0.002
14	14.000	0.0015	14.016	0.005	14.007	0.003	14.007	0.002
15	15.000	0.0015	15.016	0.005	15.007	0.003	15.0075	0.002
16	16.000	0.0015	16.016	0.005	16.007	0.003	16.008	0.002
17	17.000	0.0015	17.018	0.005	17.007	0.003	17.008	0.002
18	18.000	0.0015	18.018	0.005	18.007	0.003	18.008	0.002
19	19.000	0.0015	19.018	0.005	19.007	0.003	19.008	0.002
20	20.000	0.0015	20.018	0.005	20.007	0.003	20.008	0.002
21	21.000	0.002	21.018	0.005	21.007	0.003	21.008	0.002
22	22.000	0.002	22.020	0.008	22.007	0.003	22.008	0.002
23	23.000	0.002	23.020	0.008	23.007	0.003	23.008	0.002
24	24.000	0.002	24.020	0.008	24.007	0.003	24.008	0.002
25	25.000	0.003	25.020	0.008				
26	26.000	0.003	26.020	0.008				
28	28.000	0.003	28.022	0.008				
30	30.000	0.003	30.022	0.008				
32	32.000	0.003	32.024	0.010				
34	34.000	0.003	34.024	0.010				
36	36.000	0.003	36.024	0.010				

All dimensions are in inches.

TABLE 2-40

Principles of Mechanics Inherent to the Computation of Loads on Bearings

Straddle and overhung mountings of a multitude of machine elements with respect to two bearings — to say nothing about shafts supported by more than two bearings — provide an unending variety of conditions to be analyzed for bearing loads. The bearing loads caused by pulleys and sprockets are found in one way; those caused by cams and connecting rods in another. And gears alone provide many sets of conditions for study.

Certain sets of conditions naturally prevail more often than others. Formulas, derived to cover particular sets of conditions, expedite design calculations, especially when the same set of conditions occur over and over. On the other hand, when the conditions being studied are unlike those of the previous treatment, the designer has the task of matching the formulas to the conditions. Correct matching entails at least a fair understanding of just how the known or assumed forces and couples resolve into equivalent or resultant forces and couples at the bearings.

A good understanding of the few principles underlying, first the resolutions of the applied forces into components and then of the resolution of the components into equivalent forces and couples, empowers the designer to avoid matching altogether; instead, he can compute bearing loads directly by simple applications of principles of mechanics, and to suit either a peculiar or an ordinary set of conditions.

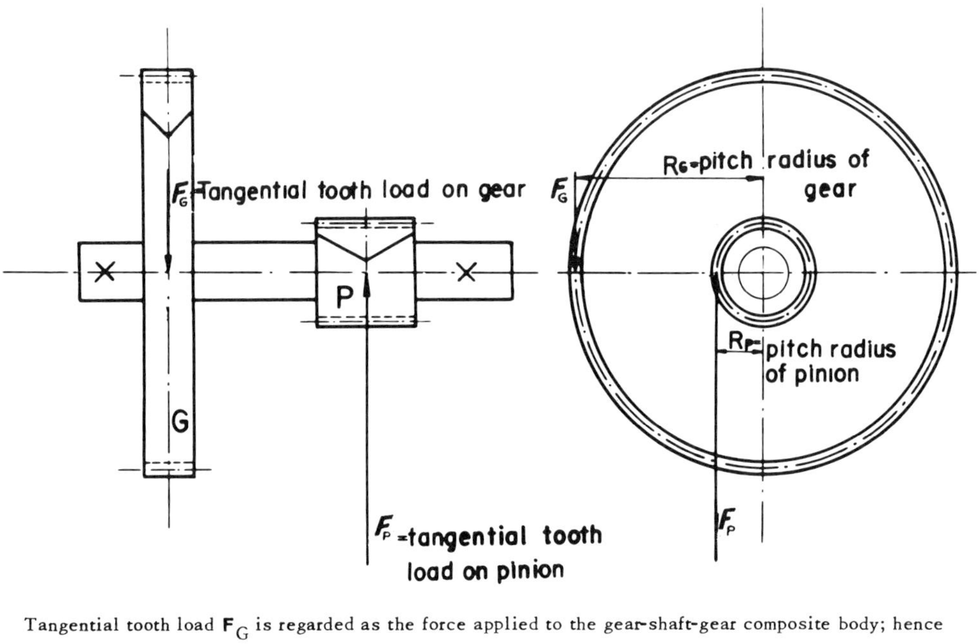

Tangential tooth load $\mathbf{F}_G$ is regarded as the force applied to the gear-shaft-gear composite body; hence $\mathbf{F}_P$ becomes the resisting force or resisting tooth load.

EQUILIBRIUM, or action equal reaction,is the first principle to recall. For the purposes of load analysis in machines, composite parts, rather than separate machine elements, become significant. Consider, for example, the intermediary shaft and two herringbone gears of a double-reduction speed reducer. Several pieces (the two gears, the shaft,keys, fastenings, et cetera) make up a composite body, for which the free-body diagram of mechanics can be imagined or constructed. For action to equal reaction there must exist a resisting tooth-load $\mathbf{F}_P^*$, else there could be no driving tooth-load $\mathbf{F}_G$. The equivalence of the resisting moment $F_P R_P$ and the driving moment $F_G R_G$ causes torque, i.e. a couple, within the shaft.

SUBSTITUTION OF A FORCE AND A COUPLE FOR ANOTHER FORCE is the second principle of mechanics to be remembered. Thus a force $\mathbf{F}_G$* applied at pitch radius R_G from the shaft centerline can be replaced by a couple $F_G R_G$ and an equal and parallel force $\mathbf{F}'_G$ perpendicular to the shaft centerline. Similarly force $\mathbf{F}_P$ can be resolved into a couple $F_P R_P$ and a force $\mathbf{F}'_P$ also perpendicular to the shaft through its centerline. Forces $\mathbf{F}'_G$ and $\mathbf{F}'_P$ act on the shaft like loads on a beam. Bearing reactions can be found by the methods of moments as applied to beams. In these tables bearing loads are distinguished from bearing reactions. Both have equal magnitudes but bearing reactions are directed toward the shaft to establish equilibrium of the forces acting upon the shaft; bearing loads, on the contrary, act toward the bearing.

*Bold-faced letters distinguish vectors from scalars.

continued on next page

The method of moments just mentioned is a good way to find bearing loads when the radial forces upon the shaft are parallel, or nearly so. But when gears are involved the radial forces are likely to be at angles to each other anywhere within 360 degrees. Under such conditions the method of moments can be applied by resolving the several applied loads into horizontal and vertical components, from which the horizontal and vertical components of load on each bearing can be found separately. The horizontal and vertical components at each bearing can then be combined into a resultant bearing load. A different order of resolving and combining the components – but yielding the same resultant loads at the bearings – is outlined on the opposite page under the headings: Application of Principles.

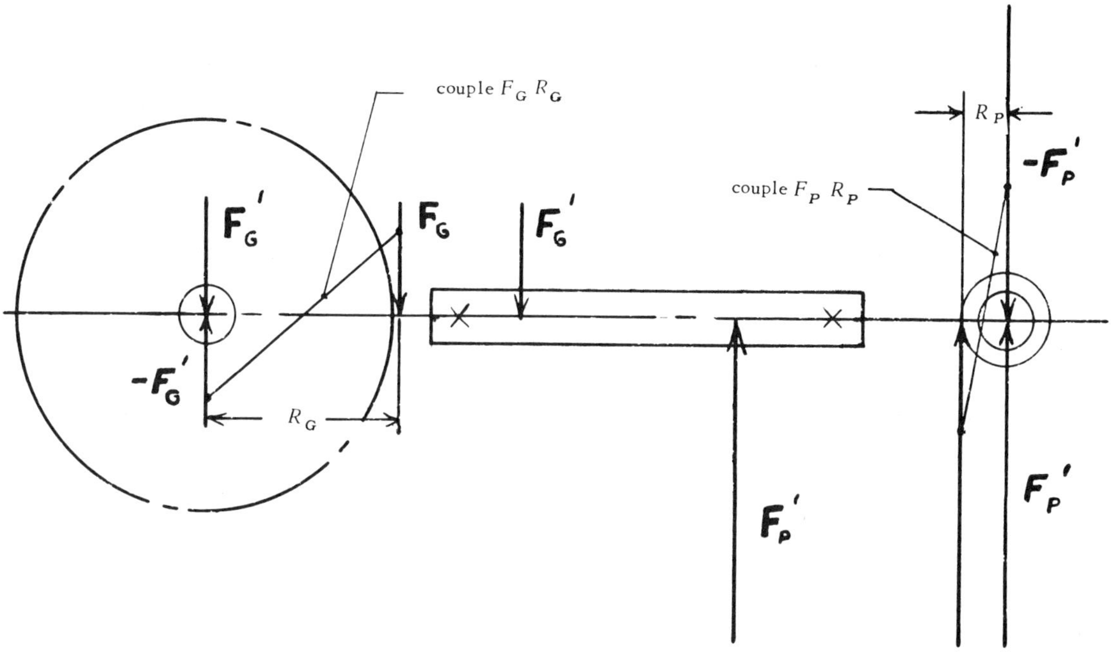

Tangential tooth loads $\mathbf{F}_G$ and $\mathbf{F}_P$ at pitch radii R_G and R_P of gear and pinion cause radial forces $\mathbf{F}'_G$ and $\mathbf{F}'_P$, respectively, upon the shaft. The forces $\mathbf{F}'_G$ and $\mathbf{F}'_P$ cause bearing loads; the couple $F_G R_G = F_P R_P$ within the shaft does not.

RESOLUTION OF A FORCE (LOAD) is the third principle to be recalled. Quite generally a load applied to a given composite body, for example the tangential load, $\mathbf{F}_G$ on a helical rack tooth, has a line of action that is oblique to a surface. For equilibrium, this requires at least one component at right angles to the known load, such that the vector sum of the right-angled components is a resultant force normal to the surface. On gearing in particular, the resolution of the applied load into components at the point of load application is one of the chief problems in the computation of bearing loads.

In the example of the helical rack the tooth profile is the surface to which the tangential force **F** is oblique. Assume that the rack is constrained to move in the customary direction, i.e., in the direction of applied force **F**. Constrained to move only in the direction of **F** means that the pitch plane must be arrested against displacements either perpendicular to itself, i.e., in the direction of **S** or widthwise in the direction **T**. The directions of **S** and **T** are mutually perpendicular to each other and to **F**. Moreover the direction **S** is such as to be perpendicular to an axis of rotation if the rack were regarded as a helical gear and had a shaft about which to rotate. Likewise the direction **T** is taken as if it were parallel to an axis of rotation.

Helix angle fixes the obliquity of **T** with respect to the tooth surface; pressure angle fixes the obliquity of **S** with respect to the tooth surface. Force **T** is known as thrust; **S** is quite commonly called separating force. Both cause components of bearing loads. Both depend upon **F** and can be determined when the helix and the pressure angles are known, Table 2-46. Using vectors the fundamental mathematical relationship is expressed by

$$\mathbf{N} = \mathbf{F} + \mathbf{T} + \mathbf{S}$$

Bold-faced letters distinguish vectors from scalars.

continued on next page

TABLE 2-40, continued

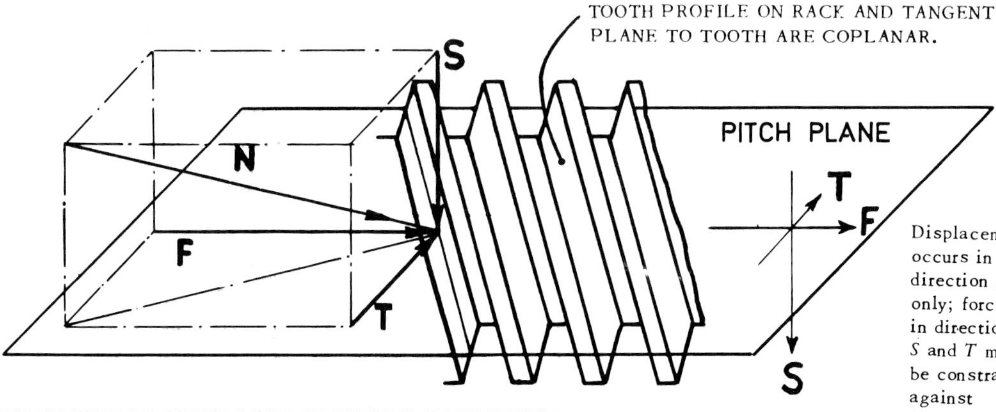

Displacement occurs in direction *F* only; forces in directions *S* and *T* must be constrained against displacement.

RESOLUTION OF TOOTH LOAD ON HELICAL RACK TOOTH

APPLICATION OF PRINCIPLES. Another way of finding the bearing loads caused by a composite body, as the speed reducer shaft with gears, is to find the components of load at each bearing as a consequence of each applied load, as *Q*, separately. The several components at each bearing can then be added geometrically to obtain a resultant bearing load. This is the procedure exemplified in many of the tables that follow, particularly those for gears. Four steps are involved:

First, the unknown legs of each normal tooth load, namely, **T** and **S**, are found from the known tangential tooth load, **F**.

Second, Each of the three legs, **F**, **T** and **S** is replaced by an equivalent force **F′**, **T′** or **S** having a line of action through the axis at the gear center plus a couple (Principle 2 above). Now **F′** and **S** are perpendicular to the axis, whereas **T′** is along it.

Third, Seldom is a bearing located along the axis at the gear center; hence each equivalent force, as **F′**, that is perpendicular to the axis will cause radial bearing-load components at both bearings. So will thrust couple *TR*, where *R* is the gear pitch radius. The several bearing-load components are found for each bearing.

Fourth, The components at each bearing from all causes are added geometrically to find a resultant bearing load.

Not all of these steps are carried through to completion in Tables 2–43 to 2–58 inclusive. There are a number of reasons for this: Tables become disproportionally complex from the great amount of duplication for straddle and overhung mountings. Each designer can carry forward as many or as few of the minor components as he wishes. Thrust couples and separating forces frequently can be omitted. For the sake of completeness, if for no other reason, these secondary quantities could not go unmentioned. Neither could they be omitted from the tables, for someone is sure to need them.

Manufacturers' literature — as Volume II, New Department Hand Book, Link-Belt General Catalog 900, Timken Engineering Journal, to mention three of many — are sources of information pertinent to the computation of bearing loads, both for particular types of bearings and for complex sets of conditions, as for example when hypoid gears are involved.

TABLE 2-41

Computation of Bearing Loads Due to Chain, Belt and Rope Drives

New Departure Handbook Vol. II

HP = Horsepower transmitted.

N = Revolutions per minute.

r = Effective radius of pulley or sprocket.

$$P = \frac{HP \times 63025}{N \times r} = \text{Effective pull on tight side of chain, belt or rope.}$$

L = Total load on shaft.

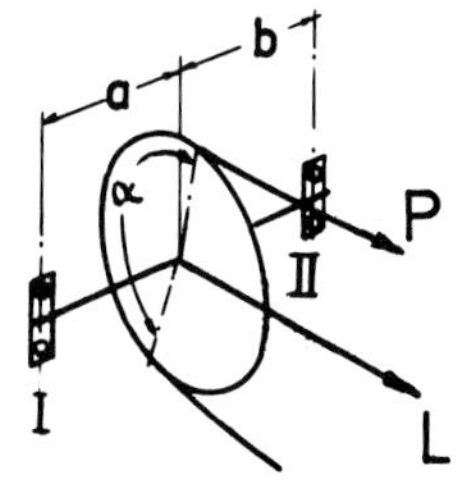

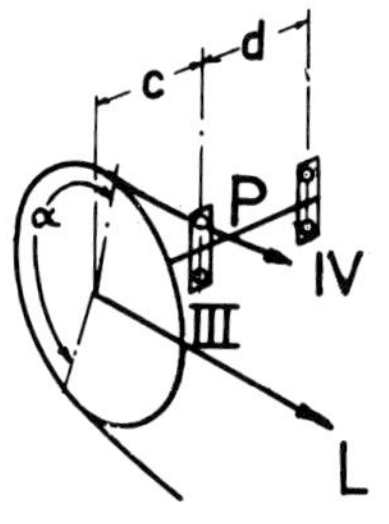

Individual bearing loads with center distances, as indicated are:

Brg. I	Brg. II	Brg. III	Brg. IV
$L\frac{b}{a+b}$	$L\frac{a}{a+b}$	$L\frac{c+d}{d}$	$L\frac{c}{d}$

Chain Drives

With chains running over properly shaped sprockets or sheaves, the total shaft load $L = P$.

Flat Belt Drives

For normal conditions where heavy belt tension is not required to transmit the power, $L = 2.5P$.

For severe conditions of belt tension, as in the case of small diameter pulleys, or where angle of wrap is considerably reduced, $L = 3P$.

V-Belt Drives

Friction in the pulley grooves with V-Belts permits proper transmission of power with comparatively low static tension and $L = 2P$.

Rope Drives

For average conditions with pulley grooves of 45° included angle, $L = 2P$.

When pulley grooves with included angle as high as 60° may be used and greater rope tension is required, $L = 3P$.

TABLE 2-42

Belt or Chain Pull Factors, F_x

The Timken Engineering Journal **The Timken Roller Bearing Co**

Type of Drive	F_x	Formula into which F_x is to be substituted
Chains, single	1.00	$BP = \frac{126{,}000 \times HP \times F_x}{D_M \times RPM}$
Chains, double	1.25	
"V" belts	1.50	where
Single ply belts	2.00	BP = The total belt or chain pull
Double ply belts	2.50	HP = Horsepower transmitted
Triple ply belts	3.00	D_M = Working outside diameter of pulleys and pitch diameters of sprockets

TABLE 2–43

Bearing-Load Components Caused by Spur Gears

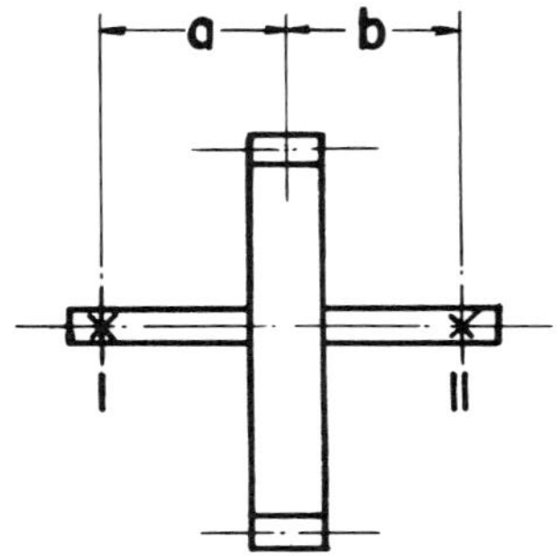

STRADDLE MOUNTED

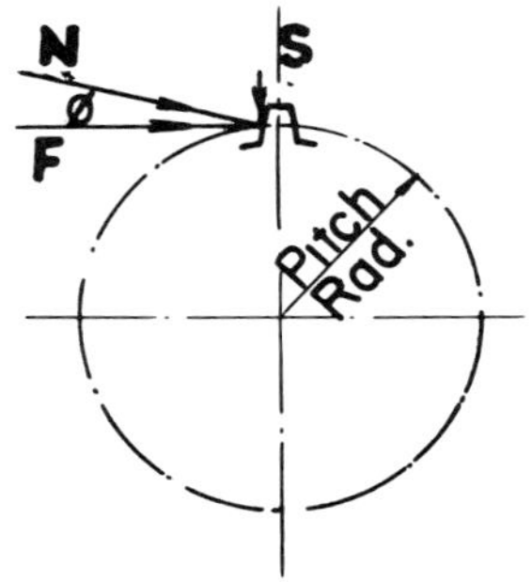

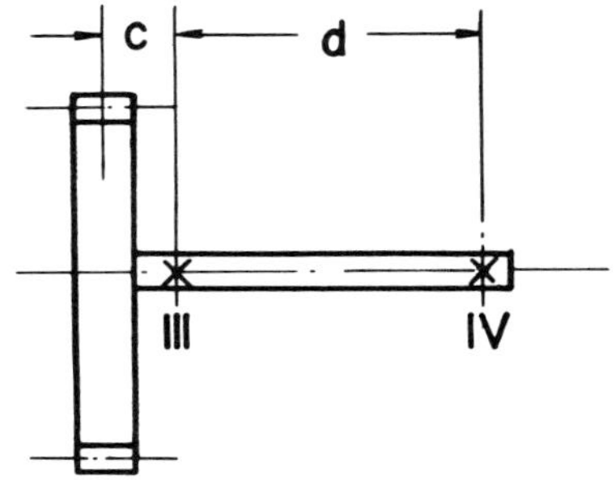

OVERHUNG MOUNTED

In terms of horsepower and speed, rpm, the tangential tooth load **F** at the pitch point on the tooth has a magnitude

$$F = \frac{63025 \times \text{horsepower}}{\text{Pitch radius (inches)} \times \text{rpm}}, \text{ pounds}$$

The separating force, **S**, mutually perpendicular to force **F** and acting along a line radial to the gear axis, has a magnitude $S = F \tan \phi$, pounds, where ϕ is the tooth pressure angle. Vector force **N**, which is normal to the tooth, i.e., along the line of action, is the sum (geometrical) of vector forces **F** and **S**; namely,

$$\mathbf{N} = \mathbf{F} + \mathbf{S}$$

Note: In accord with the conventional connotation of vectors, the **F** denotes the vector while the F represents only the magnitude of vector **F**.

(continued on next page)

TABLE 2–43, continued

Tooth-Load Vector	Bearing-Load Components*							
	Bearing I		Bearing II		Bearing III		Bearing IV	
	Magnitude	Direction‡	Magnitude	Direction‡	Magnitude	Direction‡	Magnitude	Direction‡
F	$F_I = \frac{F\,b}{a+b}$	Parallel to and same in sense as **F**	$F_{II} = \frac{F\,a}{a+b}$	Parallel to and same in sense as **F**	$F_{III} = \frac{F(c+d)}{d}$	Parallel to and same in sense as **F**	$F_{IV} = \frac{F\,c}{d}$	Parallel to but opposite in sense to **F**
S†	$S_I = \frac{S\,b}{a+b}$	Parallel to and same in sense as **S**	$S_{II} = \frac{S\,a}{a+b}$	Parallel to and same in sense as **S**	$S_{III} = \frac{S(c+d)}{d}$	Parallel to and same in sense as **S**	$S_{IV} = \frac{S\,c}{d}$	Parallel to but opposite in sense to **S**
N†	$N_I = \sqrt{F_I^2 + S_I^2}$	Parallel to and same in sense as **N**	$N_{II} = \sqrt{F_{II}^2 + S_{II}^2}$	Parallel to and same in sense as **N**	$N_{III} = \sqrt{F_{III}^2 + S_{III}^2}$	Parallel to and same in sense as **N**	$N_{IV} = \sqrt{F_{IV}^2 + F_{IV}^2}$	Parallel to but opposite in sense to **N**

*In machinery where power is transmitted or transformed, the weights of gears and shafts usually cause small bearing loads in comparison with the loads developed under service conditions. Thus components of loads as a consequence of weight of parts are omitted here. There are machines, however, in which the weights of parts cause significant bearing loads that should not be omitted from design calculations.

†With a pressure angle of 20° the inclusion of the components caused by the separating force gives total load components barely 6.5 per cent greater than the components caused by the tangential force alone. Design calculations therefore are often simplified by considering the tangential force as the total load. Cf. numerical values of example in Table 2–44.

‡For the same system of forces the direction of a resultant has the opposite sense to that of the equilibrant. Here bearing load has the direction of the resultant, the opposite direction being regarded as the bearing reaction.

TABLE 2-44

Example to Illustrate Computation of Bearing Loads Caused by Singly Mounted Spur Gears

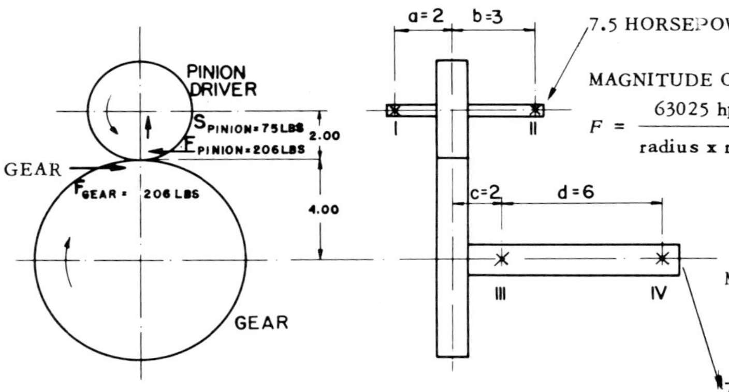

TANGENTIAL FORCES ON PINION AND GEAR ARE EQUAL AND OPPOSITE.

7.5 HORSEPOWER INPUT COUPLE AT 1150 RPM

MAGNITUDE OF TANGENTIAL FORCE,

$$F = \frac{63025 \text{ hp}}{\text{radius x rpm}} = 206 \text{ lbs}$$

MAGNITUDE OF SEPARATING FORCE,

$$S = F \tan 20^\circ = 75 \text{ lbs}$$

7.5 HORSEPOWER OUTPUT COUPLE AT 575 rpm

Component Loads on Bearings

Bearing I Magnitude	Direction*	Bearing II Magnitude	Direction*	Bearing III Magnitude	Direction*	Bearing IV Magnitude	Direction*
$F_I = \frac{F\,b}{a+b}$		$F_{II} = \frac{F\,a}{a+b}$		$F_{III} = \frac{F(c+d)}{d}$		$F_{IV} = \frac{F\,c}{d}$	
$= \frac{206 \times 3}{5}$	←	$= \frac{206 \times 2}{5}$	←	$= \frac{206 \times 8}{6}$	→	$= \frac{206 \times 2}{6}$	←
= 124 lbs		= 82 lbs		= 275 lbs		= 69 lbs	
$S_I = \frac{S\,b}{a+b}$	↑	$S_{II} = \frac{S\,a}{a+b}$	↑	$S_{III} = \frac{S(c+d)}{d}$	↓	$S_{IV} = \frac{S\,c}{d}$	↑
$= \frac{75 \times 3}{5} = 45$		$= \frac{75 \times 2}{5} = 30$		$= \frac{75 \times 8}{6} = 100$		$= \frac{75 \times 2}{6} = 25$	
N_I = 132		N_{II} = 87		N_{III} = 293		N_{IV} = 73	

*Note that bearing reactions, that is, forces upon shafts at bearings, are equal in magnitude but of opposite senses.

TABLE 2–45

Method for Finding Bearing Loads Caused by Two Spur Gears on Same Shaft

Inasmuch as the line of action of the tooth load on one gear may make any angle between zero and 360° with the line of action of the tooth load on the other, general formulas contain ± signs depending upon the quadrants in which the lines of action lie. But this is not all. Straddle and overhung mountings also reverse signs. Cf the directions in Table 2–44. Rather than list algebraic formulas expressing the total bearing load caused by both gears and by the mounting conditions, supplemented by other tables treating the choice of signs to suit conditions, the following procedure seems simpler and more reliable: First, find the bearing-load components caused by each gear singly. See Table 2–43. Second, add the components at each bearing geometrically. Further to simplify the calculations, the components caused by separating forces are neglected.

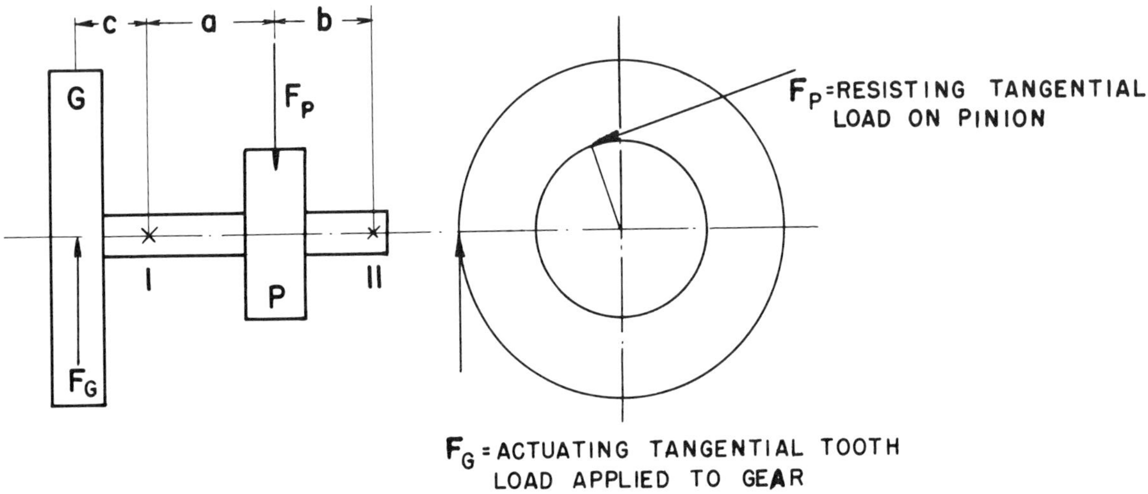

Gear and Tangential Tooth Load	Components of Load on Bearing, Cf Table 2–43			
	Bearing I		Bearing II	
	Magnitude	Direction	Magnitude	Direction
F_G, pounds	$F_{GI} = \frac{F_G(a+b+c)}{a+b}$		$F_{GII} = \frac{F_G c}{a+b}$	
F_P, pounds	$F_{PI} = \frac{F_P b}{a+b}$		$F_{PII} = \frac{F_P a}{a+b}$	
Resultant bearing loads Geometrical addition usually can be made graphically with sufficient accuracy, even when separating components are included.		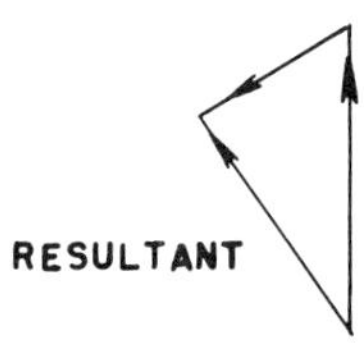		

Bearing reactions are equal and opposite the resultants.

TABLE 2–46

Resolution of Tooth Load on Helical Gear

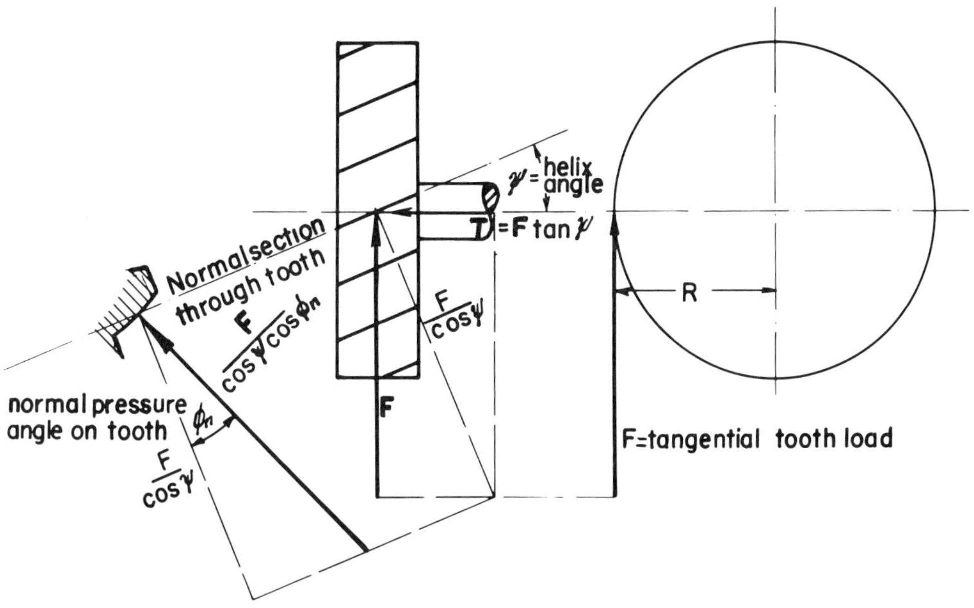

Quantity and Symbol	Formula for Computing* Magnitude of Force or Couple	Definition of Symbols and Comments
Tangential tooth load, F, pounds	$F = \dfrac{63{,}025 \text{ hp}}{R \times rpm}$	R = pitch radius, inches Force F applied to the tooth has as the equivalent an equal and parallel force through the shaft axis plus a couple FR. The couple, of course, is balanced by an equal and opposite twisting moment in the shaft. The force through the center of the shaft becomes the major components of bearing loads.
Axial thrust, T, pounds Use Table 2–47 to find direction of axial thrust.	$T = F \tan \Psi$	Ψ = pitch helix angle Axial thrust cannot be neglected in bearing design.

*Directions of axial thrusts can be found from Table 2–47; the directions of bearing components resulting from both the tangential force and the separating force are similar to those for spur gears, Table 2–43.

continued on next page

TABLE 2–46, continued

Quantity and Symbol	Formula for Computing* Magnitude of Force of Couple	Definition of Symbols and Comments
Radial separating force, S, pounds	$S = \dfrac{F \tan \phi_n}{\cos \psi}$	ϕ_n = normal pressure angle of tooth
	or	
	$S = F \tan \phi_t$	ϕ_t = transverse pressure angle of tooth
	See Table 4–7	Radial separating force is sometimes small enough in comparison with F to be neglected.
Thrust couple, TR, pound inches	$TR = F_c B$	F_c = magnitude, pounds, of radial load on bearings as consequence of thrust couple.
		B = distance between bearings, inches
		Bearing load components from thrust couple are mutually perpendicular to the gear axis and to the major radial loads caused by force **F**. Furthermore, they are parallel to separating force components, and like the latter are often neglected, especially if R is small and/or B is large.

*Directions of axial thrusts can be found from Table 2–47; the directions of bearing components resulting from both the tangential force and the separating force are similar to those for spur gears, Table 2–43.

TABLE 2–47

Thrust Diagrams for Helical Gears

Catalog No. 1000 D.O. James Manufacturing Co.

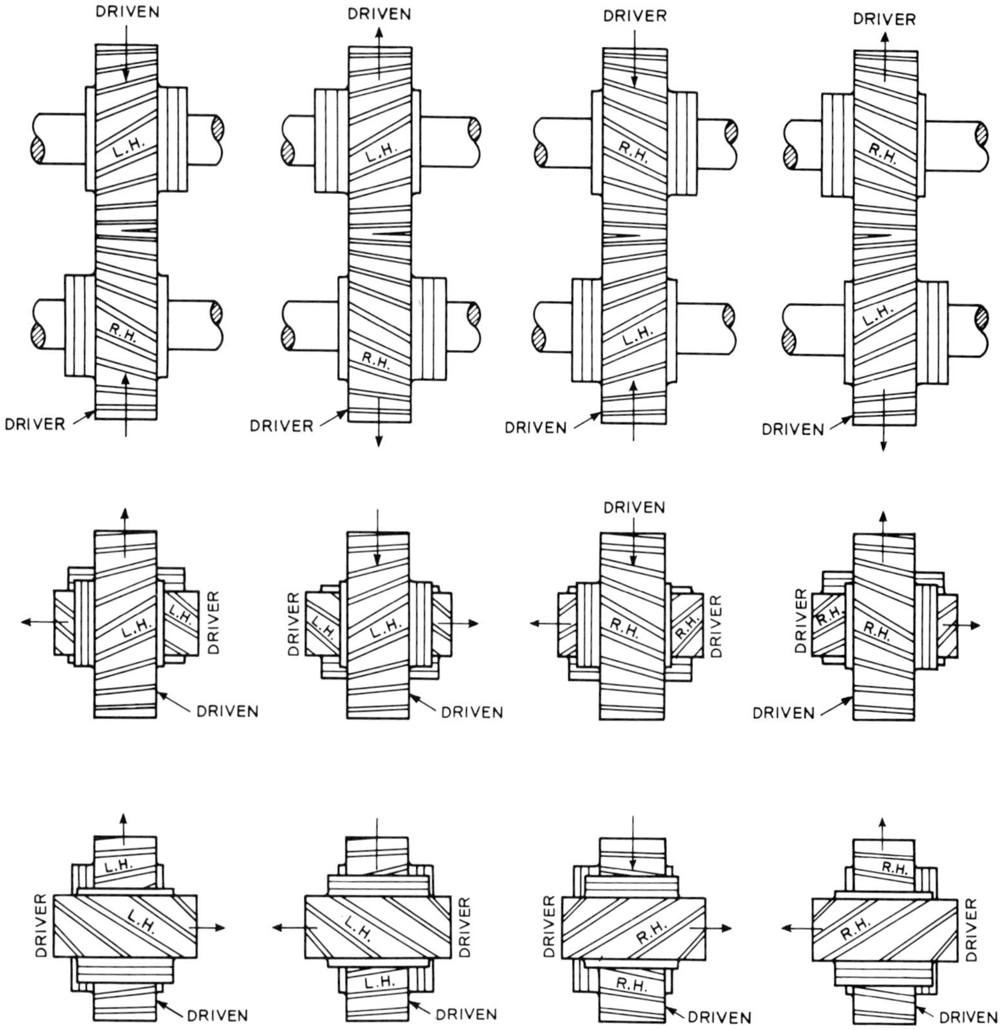

DIRECTION OF THRUST DEPENDS UPON DIRECTION OF ROTATION, RELATIVE POSITION OF DRIVER AND DRIVEN GEAR, AND DIRECTION OF HELIX

TABLE 2-48

Resolution of Tooth Load on Straight and Zerol Bevel Gear

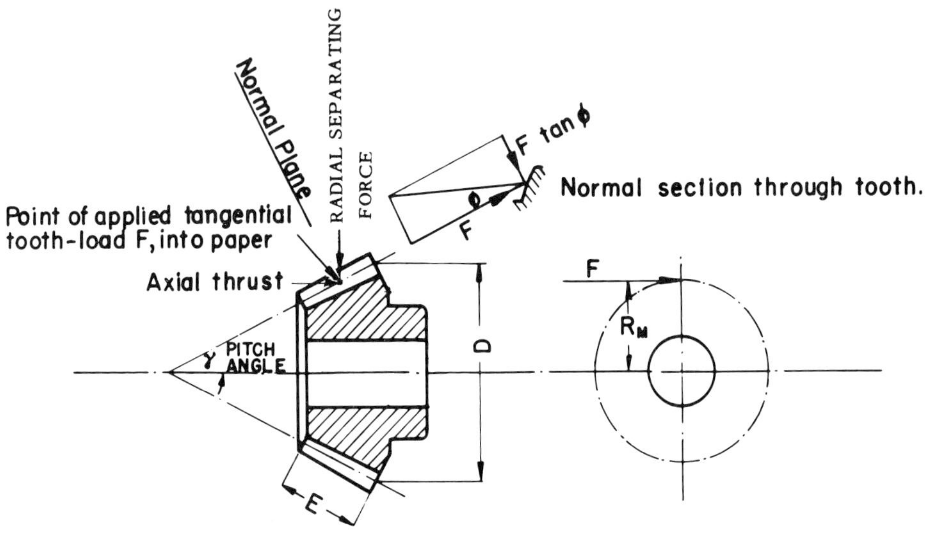

Quantity and Symbol	Formula for Computing Magnitude of Force or Couple	Definition of Symbols and Comments
Mean pitch radius, R_M, inches	$R_M = 1/2(D - E \sin \gamma)$	D = pitch diameter, inches E = face width, inches γ = pitch angle of bevel gear The mean pitch radius locates the center of pressure along the tooth axially as well as radially.
Tangential tooth load, F, pounds	$F = \dfrac{63{,}025 \text{ hp}}{R_M \times \text{rpm}}$	Force F applied to the tooth has as the equivalent an equal and parallel force through the shaft axis plus a couple FR_M. The couple, of course, is balanced by an equal and opposite twisting moment in the shaft. The force through the center of the shaft becomes the major component of bearing loads.
Normal plane separating force	$F \tan \phi$	ϕ = normal tooth pressure angle Since the normal plane makes an angle other than 90° with the axis, pressure angle combines with pitch angle to cause an axial as well as a radial component.
Axial thrust T, pounds	$T = F \tan \phi \sin \gamma$	Axial thrust on straight and zerol bevel gears always tends to push mating members out of mesh, known as the positive direction. This axial thrust is rarely negligible in bearing design.

continued on next page

TABLE 2–48, continued

Quantity and Symbol	Formula for Computing Magnitude of Force or Couple	Definition of Symbols and Comments
Radial separating force, S_r, pounds	$S_r = F \tan \phi \cos \gamma$	Frequently small enough in comparison with F to be neglected.
Thrust couple TR_M, pound inches	$TR_M = F_c R$	F_c = magnitude, pounds, of radial load on bearings as consequence of thrust couple B = distance between bearings, inches Bearing load components from thrust couple are mutually perpendicular to the gear axis and to the major radial loads caused by force F. Furthermore, they are parallel to radial separating force, S_r, and like the latter are often small enough to be neglected. See Table 2–40 for further comments about transforming the components of tooth load into bearing reactions.

TABLE 2–49

Comments on Resolution of Tooth-Loads on Spiral Bevel and Hypoid Gears

Spiral angles on spiral bevel and hypoid gears introduce additional tooth-load components when compared to straight bevel gears for the same reasons that the tangential tooth-load on a helical gear tooth resolves into more components than the tangential tooth load on a spur gear. In the case of the helical gear, the axial component is a function of the helix angle alone. But in the case of the spiral bevel or hypoid, spiral angle introduces both radial and axial components.

Furthermore, the direction of this axial component, i.e., away from or toward the cone center, depends upon the hand and magnitude of the spiral angle, the pitch angle, the direction of rotation and whether the gear is the driving or driven member. It follows that the resultant axial component, as a consequence of both pressure angle and spiral angle, may be such as to tend to displace the mating members toward the cone center, or more tightly into mesh. For straight and zerol bevel gears, on the other hand, these components always tend to force the pinion and gear out of mesh, which is regarded as the positive or + direction. See Tables 2-52 and 2-53 .

continued on next page

SELECTION OF HAND OF SPIRAL
Bevel and Hypoid Gear Design   Gleason Works

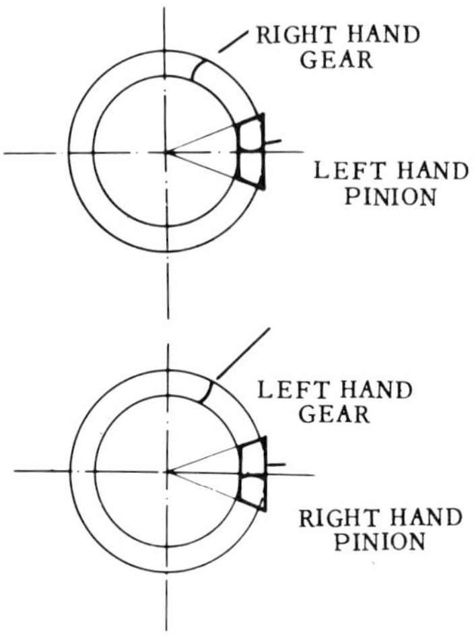

"The hand of spiral on spiral bevel and hypoid gears is denoted by the direction in which the teeth curve, that is, left-hand teeth incline away from the axis in the counter-clockwise direction when an observer looks at the face of the gear and right-hand teeth incline away from the axis in the clockwise direction. The hand of spiral of one member of a pair is always opposite to that of its mate. It is customary to use the hand of spiral of the pinion to identify the combination, that is, a left-hand combination is one with a left-hand spiral on the pinion and a right-hand spiral on the gear. The hand of spiral has no effect on the smoothness and quietness of operation or on the efficiency. Attention, however, is called to the difference in the effect of the thrust loads as stated in the following paragraph.

"A left-hand spiral pinion driving clockwise (viewed from the back) tends to move axially away from the cone center; while a right-hand pinion tends to move toward the center because of the oblique direction of the curved teeth. If there is excessive end play in the pinion shaft because of faulty assembly, the movement of a right-hand pinion driving clockwise will take up the back-lash under heavy load, and the teeth of the gear and pinion may wedge together, while a left-hand spiral pinion under the same conditions would back away and merely introduce additional backlash between the teeth, a condition which would not prevent the gears from functioning. When the ratio, pressure angle and spiral angle are such that it is possible, the hand of spiral should be selected to give an axial thrust that tends to move both the gear and the pinion out of mesh. Otherwise the hand of spiral should be selected to give an axial thrust that tends to move the pinion out of mesh. Often the mounting conditions will dictate the hand of spiral to be selected.

"In a reversible drive there is, of course, no choice unless the pair performs a heavier duty in one direction a greater part of the time.

"On hypoids, when the pinion is below center and to the right when facing the front of the gear, the pinion hand of spiral should always be left-hand. With the pinion above center and to the right, the pinion hand should always be right-hand." See the figure.

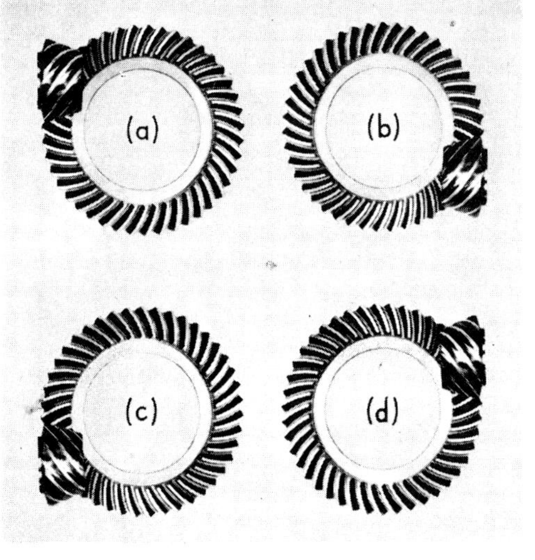

Hypoid gears and pinions. (a) and (b) are referred to as having an offset "below center," while those in (c) and (d) have an offset "above center." In determining the direction of offset, it is customary to look at the gear with the pinion at the right.

TABLE 2–50

An Enumeration of the Forces and Couples Caused by Tooth Load on Spiral Bevel and Hypoid Gears and Which Must be Supported by Bearings

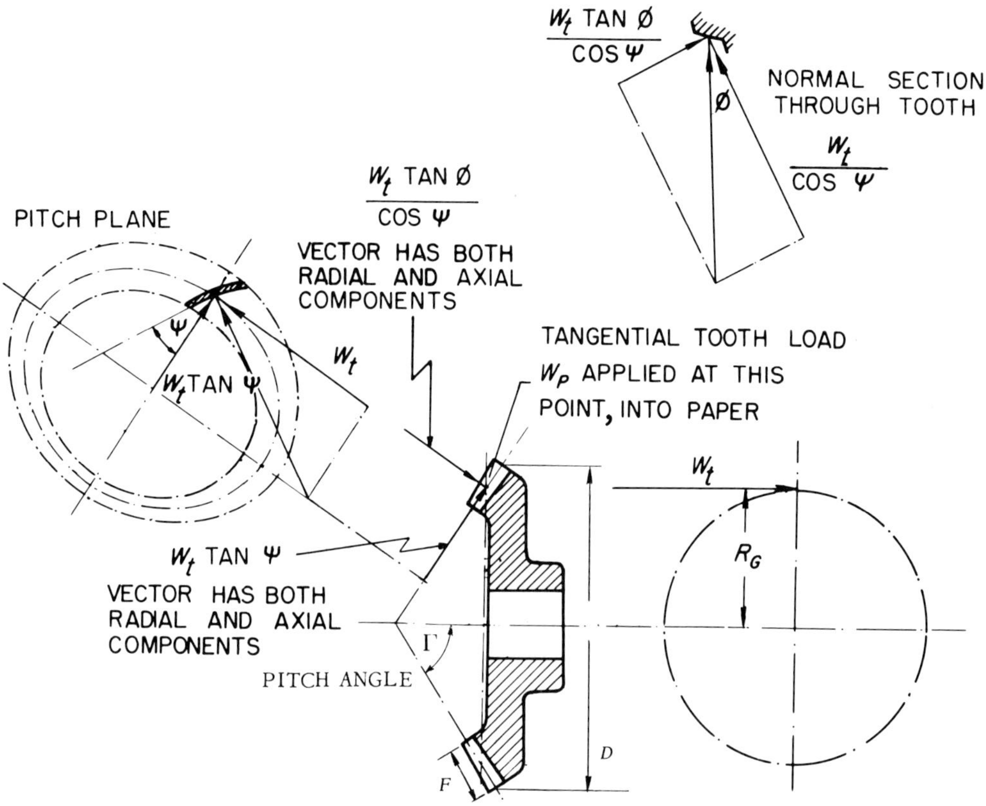

Item Number and Name of Quantity	Formula for Finding the Magnitude of Force or Couple	Definition of Symbols and Comments
1 Shaft Torque	$W_t R_G = \dfrac{63{,}025 \text{ hp}}{\text{rpm}}$ pound-inches Shaft torque caused by a bevel pinion is found by same formula merely by using dimensions and rpm of pinion.	W_t = tangential tooth load, pounds R_G = mean pitch radius of gear, inches $= 1/2\,(D - F \sin \Gamma)$, where D = pitch diameter gear, inches F = gear face width, inches Γ = gear pitch angle Torque is held in equilibrium by a resisting moment within shaft, and consequently does not appear in bearing loads directly.

continued on next page.

Item Number and Name of Quantity	Formula for Finding the Magnitude of Force or Couple	Definition of Symbols and Comments
2 Magnitude of Tangential Tooth Load, pounds	W_t, on gear W_{tP}, on pinion Table 2-51 gives the relationship of the tooth load on the pinion in terms of that on gear.	The biggest components of radial loads on bearings are produced by the tangential tooth load of magnitude W_t. Mutually perpendicular to the (vector) tangential tooth load are two other vectors, one in the pitch plane of magnitude $W_t \tan \Psi$ and the other, resembling a separating force, of magnitude $W_t \tan \phi / \cos \Psi$. The sum of these three mutually perpendicular components, of course, is the normal force on the tooth. Neither of the latter two vectors is either parallel or perpendicular to the axis; hence, each has both a radial and an axial component. Thus, the total axial thrust, item 3, is the sum (or difference) of two components. So is the total separating force, item 5.
3 Axial Thrust	W_x, pounds The formula and instructions on its use are given separately, Table 2–52, because of the several conditions involved.	Table 2–53 is an alternate method of finding axial thrust. Axial thrust on spiral bevel and hypoid gears differs from that on straight and zerol bevel gears, in that the direction of the thrust may be either toward or away from the cone center. The diagram in Table 2–52 indicates how the direction of thrust is related to hands of spiral and directions of rotation. Proper running of bevel gears requires adequate support against an axial displacement of either member, and consequently design calculations for thrust are always important.
4 Thrust Couple	$W_x R_G$ pound inches	The thrust couple causes radial components of bearing load that are perpendicular to those caused by the tangential tooth load and parallel to those caused by the separating force, item 5. If the distance between bearings is relatively large and/or the radius R_G is small, as for pinions, the components of bearing load as a consequence of thrust couple are small in comparison with other components. Accordingly they are sometimes neglected in design calculations.
5 Separating Force	W_r', pounds The formula and instructions on its use are given separately, Table 2–54.	Table 2–55 is an alternate method of finding the radial separating force.
6 Resultant Radial Loads on Bearings	The methods of Table 2–43 can be used to transform radial components of items 2 and 5 concentrated at the gear center on the axis to equivalent components at the bearings, and to suit mounting conditions.	The radial forces of items 2 and 5 each resolve into two equivalent loads, one on each bearing. When the thrust couple, item 4, is also taken into consideration, each bearing has three components of load. Two of these three components are parallel and can be added algebraically, after which the sum can be added geometrically to the third to obtain the total radial bearing load.

TABLE 2–51

Tangential Tooth Load on Spiral Bevel or Hypoid Pinion

Bevel and Hypoid Gear Design   Gleason Works

Magnitude To Be Found	Formula	In Which
Tangential tooth load, W_t on gear member, pounds	$W_t = \frac{126,050 \text{ horsepower}}{(D - F \sin \Gamma) \text{ rpm}}$	D = pitch diameter of gear, inches F = gear face width, inches Γ = { gear pitch angle (bevel gear); gear root angle (hypoid gear) }
Tangential tooth load, W_{tP} on mating bevel pinion pounds	$W_{tP} = W_t$	Figure in Table 2-50
Tangential tooth load, W_{tP} on mating hypoid pinion, pounds	$W_{tP} = W_t \frac{\cos \psi_P}{\cos \psi_G}$	ψ_P = pinion spiral angle ψ_G = gear spiral angle

TABLE 2–52

Formulas for Computing Axial Component of Tooth Load on Spiral Bevel and Hypoid Gears, or Take From Chart in Table 2–53

Bevel and Hypoid Gear Design — Gleason Works

Driving Member: Hand of Spiral	Driving Member: Rotation	Value of Axial Thrust
Right	Clockwise	Driving Member: $W_x = \frac{W_t}{\cos \Psi}(\tan \phi \sin \gamma - \sin \Psi \cos \gamma)$
Left	Counter-clockwise	Driven Member: $W_x = \frac{W_t}{\cos \Psi}(\tan \phi \sin \gamma + \sin \Psi \cos \gamma)$
Right	Counter-clockwise	Driving Member: $W_x = \frac{W_t}{\cos \Psi}(\tan \phi \sin \gamma + \sin \Psi \cos \gamma)$
Left	Clockwise	Driven Member: $W_x = \frac{W_t}{\cos \Psi}(\tan \phi \sin \gamma - \sin \Psi \cos \gamma)$

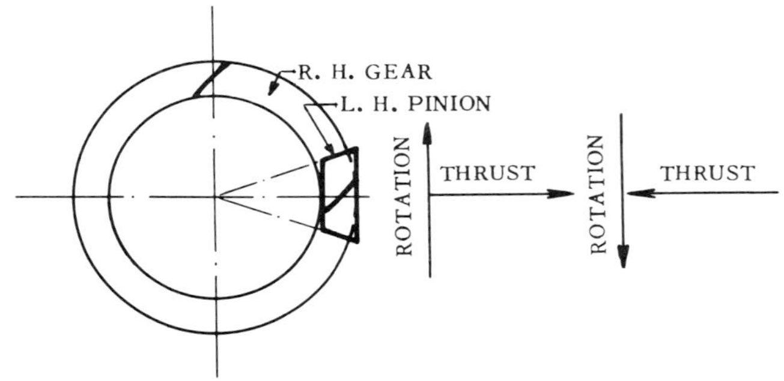

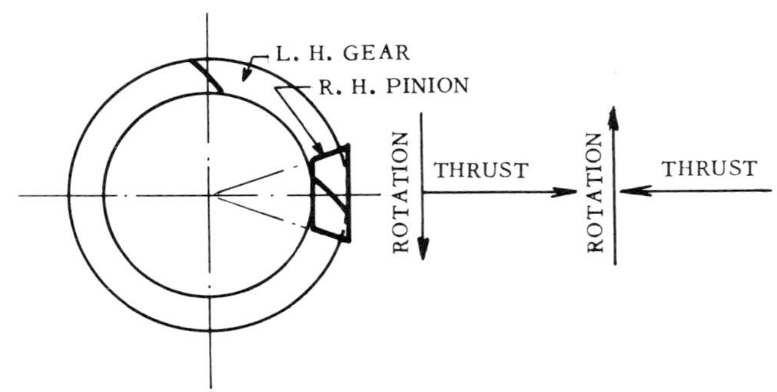

W_x = axial thrust load.

W_t = transmitted tooth load tangential to pitch circle.

ϕ = normal pressure angle. This is the pressure angle on the driving side of the tooth.

Ψ = spiral angle.

γ = pitch angle on bevel gears
= face angle of pinion and root angle of gear on hypoid gears.

A positive sign (+) indicates direction of thrust is away from cone center.

A negative sign (–) indicates direction of thrust is toward cone center.

When using the above formulas the tangential load, spiral angle, pitch angle, and pressure angle for the corresponding member must be used.

On hypoid gears the tangential load on the pinion is equal to the tangential load on the gear times the cosine of the pinion spiral angle divided by the cosine of the gear spiral angle.

TABLE 2–53

Chart for Finding Axial Component of Tooth Load on Spiral Bevel and Hypoid Gears, or Compute by Formulas in Table 2–52

Bevel and Hypoid Gear Design   Gleason Works

To get magnitude of axial thrust, W_x pounds, caused by a pinion or a gear, multiply the tangential tooth load, W_{tP} or W_t pounds, Table 2–51, by the per cent taken from the chart, that is, W_x = (reading from chart) W_t.

To get per cent reading from chart, lay a straight edge upon the chart to join a point on the pressure angle scale ϕ, at the right, and the point of intersection defined by the pitch angle and spiral angle curves corresponding to the pinion or gear being considered. The point at which the straight-edge cuts the per cent scale on the left is the per cent sought. Note that the intersection point of the pitch angle and spiral angle curves for a pinion lie on opposite sides of the horizontal centerline to the intersection point of those for the mating gear.

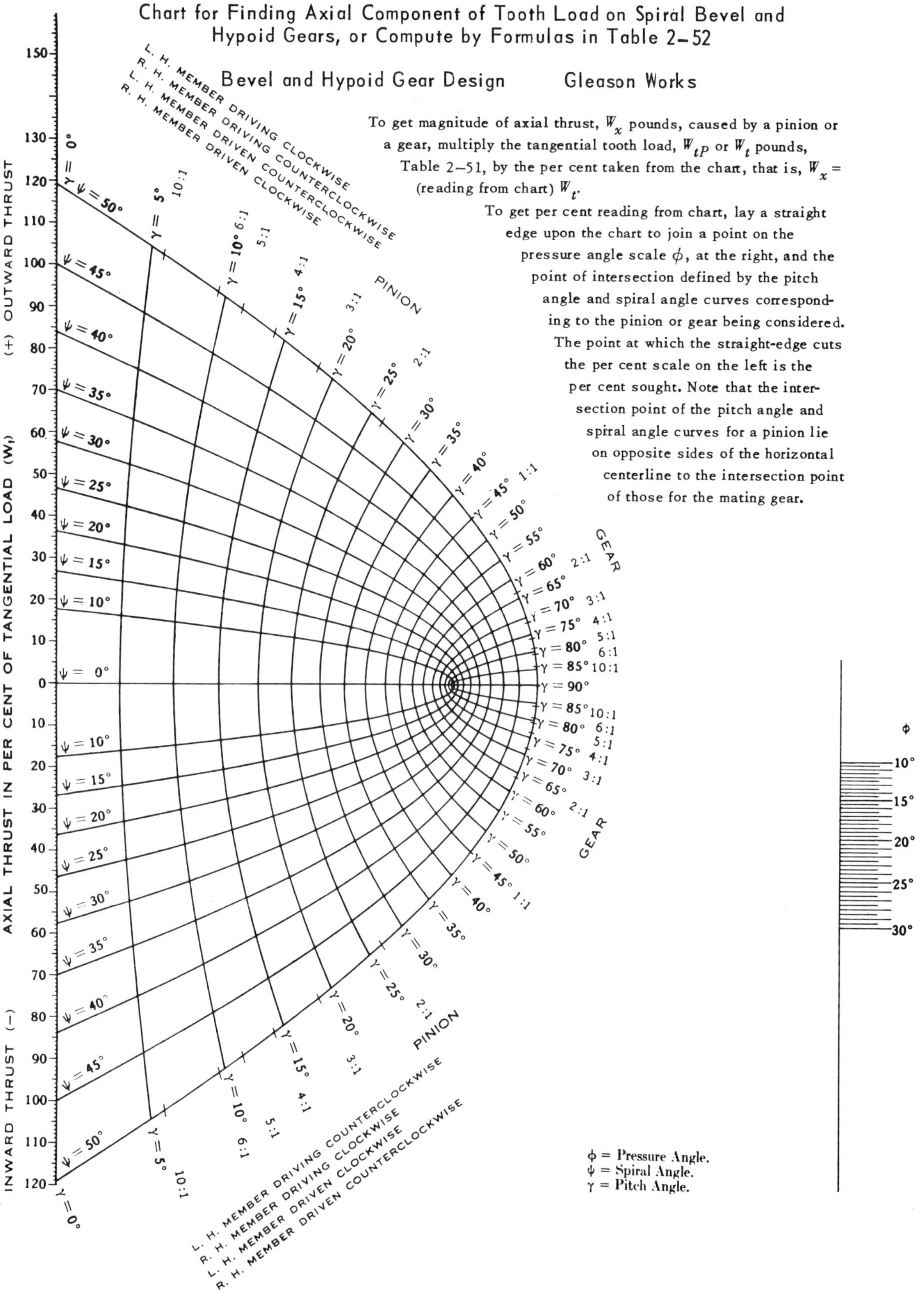

ϕ = Pressure Angle.
ψ = Spiral Angle.
γ = Pitch Angle.

TABLE 2–54

Formulas for Computing Radial Separating Force Caused by Tooth Load on Spiral Bevel or Hypoid Gears (or Take from Chart in Table 2–55)

Bevel and Hypoid Gear Design Gleason Works

Driving Member: Hand of Spiral	Driving Member: Rotation	Value of Separating Component of Radial Load
Right	Clockwise	Driving Member $W_r' = \dfrac{W_t}{\cos\Psi}(\tan\phi\cos\gamma + \sin\Psi\sin\gamma)$
Left	Counter-clockwise	Driven Member $W_r' = \dfrac{W_t}{\cos\Psi}(\tan\phi\cos\gamma - \sin\Psi\sin\gamma)$
Right	Counter-clockwise	Driving Member $W_r' = \dfrac{W_t}{\cos\Psi}(\tan\phi\cos\gamma - \sin\Psi\sin\gamma)$
Left	Clockwise	Driven Member $W_r' = \dfrac{W_t}{\cos\Psi}(\tan\phi\cos\gamma + \sin\Psi\sin\gamma)$

W_r' = separating component of the radial load.
W_t = transmitted tooth load tangential to pitch circle.
ϕ = normal pressure angle. This is the pressure angle on the driving side of the tooth.
Ψ = spiral angle.
γ = pitch angle on bevel gears.
= face angle of pinion and root angle of gear on hypoid gears.

A positive sign (+) indicates direction of force is away from the mating member (separating force).

A negative sign (–) indicates direction of force is toward the mating member (attracting force).

When using the above formulas the tangential load, spiral angle, pitch angle, and pressure angle for the corresponding member must be used.

On hypoid gears the tangential load on the pinion is equal to the tangential load on the gear times the cosine of the pinion spiral angle divided by the cosine of the gear spiral angle.

TABLE 2–55

Chart for Finding Radial Separating Force Caused by Tooth Load on Spiral Bevel or Hypoid Gears, or Compute from Formulas in Table 2–54

Bevel and Hypoid Gear Design Gleason Works

To get magnitude of radial separating force, W_r' pounds, multiply the tangential tooth load, W_t pounds, Table 2–51, by the per cent taken from the chart, i.e., $W_r' = (\text{reading from chart})\ W_t$.

To get per cent reading from chart, lay a straight edge upon the chart to join a point on the pressure angle scale ϕ, at the left, and the point of intersection defined by the pitch angle and spiral angle curves corresponding to the gear or pinion being considered. The point at which the straight edge cuts the per cent scale on the right is the per cent sought. Note that when the pitch angle and spiral angle curves for the pinion of a mating pair intersect below the horizontal centerline, those for the gear must intersect about it, and vice versa.

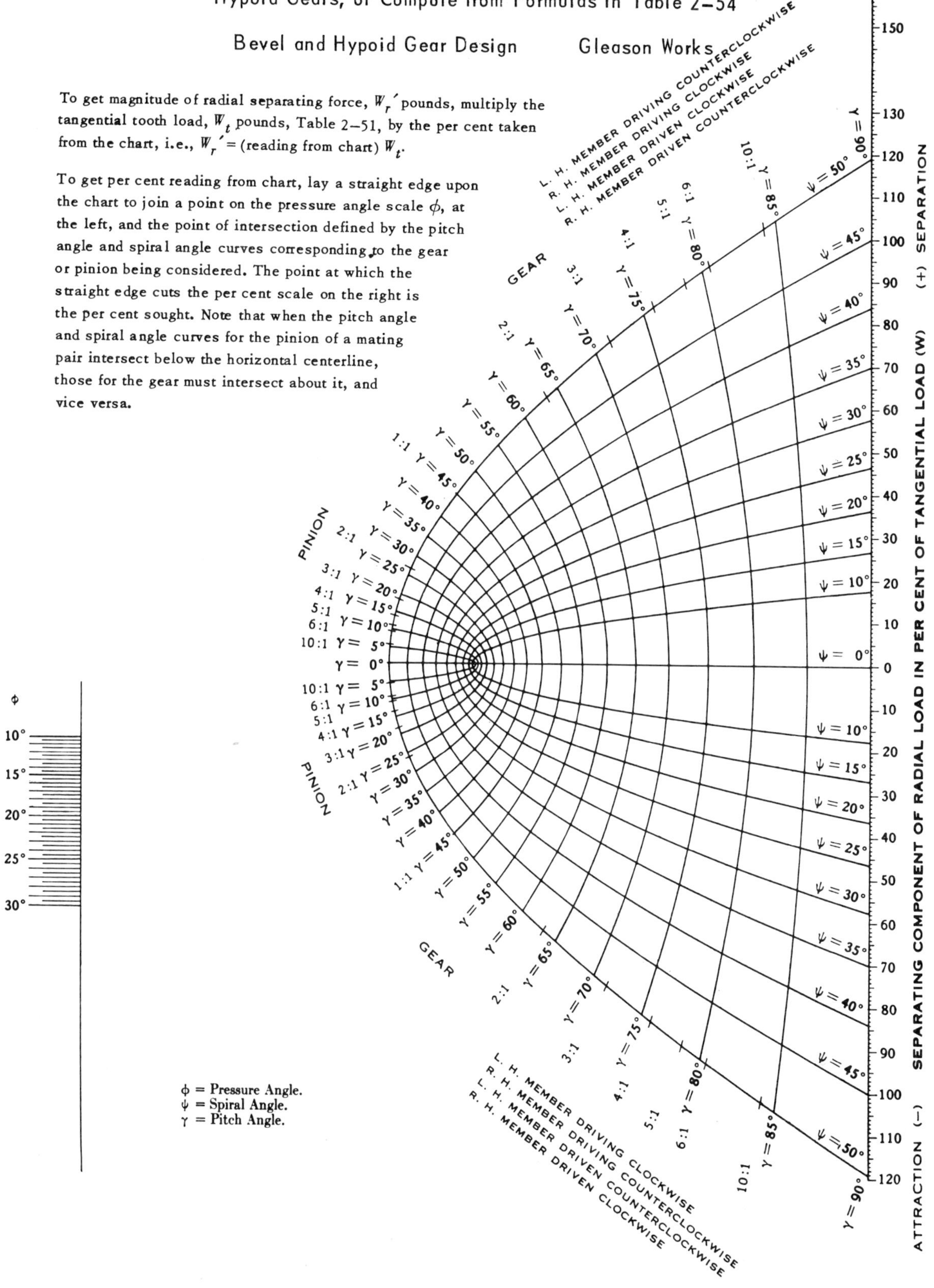

ϕ = Pressure Angle.
ψ = Spiral Angle.
γ = Pitch Angle.

TABLE 2-56

Resolution of Load on Worm Thread

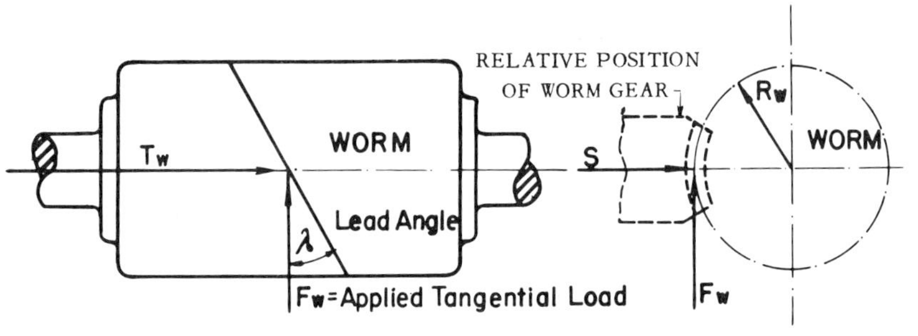

Quantity To be Found	Formula for Finding the Magnitude of Force or Couple	Definition of Symbols and Comments
Magnitude of tangential force F_W, pounds	$F_W = \dfrac{63{,}025 \text{ hp}}{R_W \; x \text{ rpm}}$	R_W = pitch radius of worm, inches. See Tables 6–11 and 6–18 for ways to establish pitch diameters of worms
Shaft Torque	$F_W R_W$, pound-inches	
Axial Thrust T_W, pounds	$T_W = \dfrac{F_W}{\tan \lambda}$	λ = lead angle, Tables 6–11 and 6–6. Neglecting friction losses, the axial thrust on the worm becomes the driving force on the worm gear. Axial thrust, generally concentrated at one bearing, is always a significant bearing load.
Thrust Couple pound-inches	$T_W R_W = \dfrac{F_W R_W}{\tan \lambda}$	Although T_W is comparatively large, R_w is correspondingly small for worms and the distance between bearings is generally large. Thus the radial components of bearing load, as the consequence of thrust couple, are usually negligible in comparison with other radial components.
Separating Force, S, pounds	$S = \dfrac{F_W \tan \phi_x}{\tan \lambda}$	ϕ_x = axial pressure angle. Relations for converting normal pressure angle and helix angle into axial pressure angle and lead angle are given in Table 4–7.
Efficiency		A few authors propose that friction losses be included in bearing-load calculations. And there is no denying that friction losses have some influence on bearing loads. The big question is where they ever modify bearing loads to a degree worth bothering with. Worm gearing, being the least efficient of the common types, particularly at small lead angles, Table 6–7, is undoubtedly the type to analyze first. Nevertheless, only the conventional methods of analysis are presented here because they seem to be more than adequate rather than inadequate, and because there is no abundance of evidence to show that the added calculations to include efficiencies might lead to savings in materials or sizes of bearings.

TABLE 2-57

Resolution of Tooth Load on Worm Gear

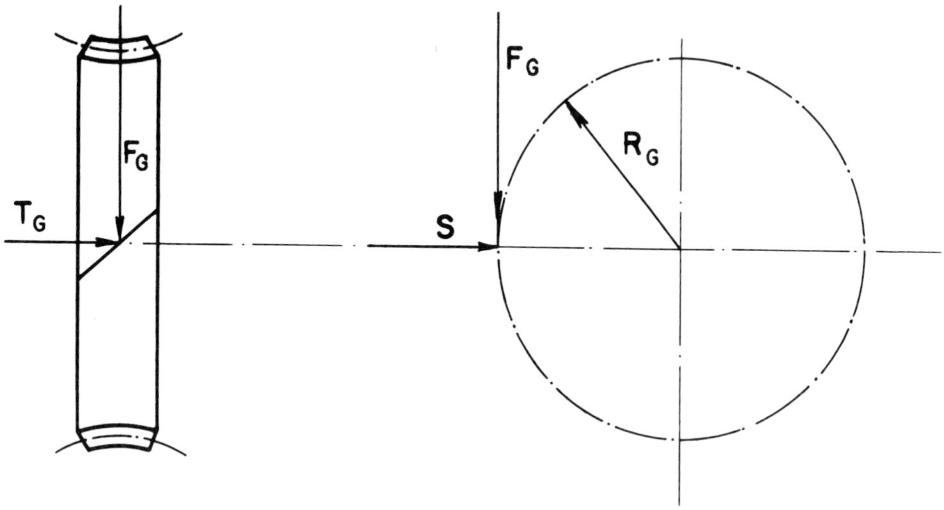

Quantity To be Found	Formula for Finding the Magnitude of Force or Couple	Definition of Symbols and Comments
Tangential Tooth Load, F_G, pounds	$F_G = \frac{63{,}025 \text{ hp}}{R_G \times \text{rpm}}$	R_G = pitch diameter of worm gear, inches T_W = axial thrust of worm, Table 2–56
	$F_G = T_W$, pounds	Vector directions of $\mathbf{F_G}$ and $\mathbf{T_W}$ are opposed.
Shaft Torque	$F_G R_G$, pound-inches	
Axial Thrust, T_G, pounds	$T_G = F_W$	F_W = tangential force on worm Vector directions of $\mathbf{T_G}$ and $\mathbf{F_W}$ are opposed.
Thrust Couple, pound-inches	$T_G R_G = F_c B$	F_c = magnitude of radial component on bearing, pounds B = distance between bearings, inches
Separating Force	S, pounds	Same in magnitude as separating force on worm but oppositely directed.

TABLE 2-58

Bearing Loads Caused by Planetary Spur Gearing

Volume II, New Departure Handbook

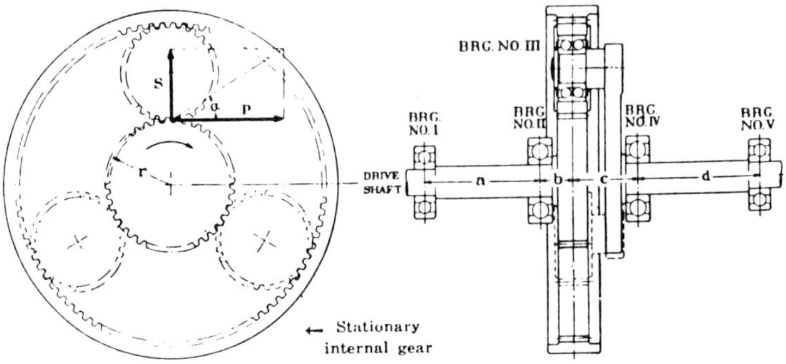

When the system employs two or more planet gears (usually three, as shown by dotted lines in diagram), the tangential and separating forces, due to the input torque, counterbalance each other insofar as they can produce appreciable loads on bearings I, II, IV and V. However, bearing III will be loaded because of the torque transmitted.

$Q = \dfrac{HP \times 63025}{N}$ = TORQUE INPUT, lbs inches, where HP = horsepower transmitted and N = rpm of driving gear.

$P = \dfrac{Q}{r}$ = TANGENTIAL FORCE, in pounds of driving sun gear, where r = Pitch radius of gear in inches.

$S = P \tan a$ = SEPARATING FORCE in pounds of the sun gear, where a = Tooth pressure angle.

For three planet gears, the load on each planet pin bearing, position III, due to driving torque and reaction, will (with equally distributed torque) be

$$\frac{2P}{3}$$

When torque is transmitted through a single planet gear, the following loads are produced:

BEARING LOADS

Due to	on Brg. I	on Brg. II	on Brg. III	on Brg. IV	on Brg. V
P	$P\frac{b}{a} = P_I$	$P\frac{a+b}{a} = P_{II}$	$2P$	$2P\frac{c+d}{d}$	$2P\frac{c}{d}$
S	$S\frac{b}{a} = S_I$	$S\frac{a+b}{a} = S_{II}$			
Total Load	$\sqrt{P_I^2 + S_I^2}$	$\sqrt{P_{II}^2 + S_{II}^2}$	$2P$	$2P\frac{c+d}{d}$	$2P\frac{c}{d}$

SPEED CHANGE

$$\text{Output rpm} = \frac{N}{2\left(1 + \dfrac{\text{Number of teeth in planet gear}}{\text{Number of teeth in sun gear}}\right).}$$

INDEX TO
SECTION 3

Spur Gears

TABLE 3–1

General Classification of Gears

Albert and Rogers, Kinematics of Machinery Wiley

Name of Gear Pair	Relation of Axes	Pitch Surfaces	Pitch Elements of Teeth	Teeth Described by Adjective
Spur gears	Parallel	Cylinders	Straight lines	Straight, meaning elements of teeth are parallel to axis
Parallel helical gears	Parallel	Cylinders	Helices, Table 4-5	Helical elements of teeth
Herringbone gears	Parallel	Cylinders	Helices, R and L hand	Double helical
Straight bevel gears	Intersecting	Cones	Straight lines	Straight, meaning elements of teeth converge to intersect axis
Coniflex bevel gears	Intersecting	Cones	Straight lines	Crowned straight
Spiral bevel gears	Intersecting	Cones	Curved lines	Spiral, meaning curved like a mathematical spiral
Zerol bevel gears	Intersecting	Cones	Curved lines	Spiral with zero mean helix angle
Crossed helical gears*	At an angle and not intersecting	Cylinders	Helices, Table 4-5	Helical
Worm and wormwheel	Usually at 90° and not intersecting	Cylinder for worm	Helices for worm	Helical for worm
Hypoid	Non-intersecting at any angle, but generally 90°	Hyperboloids of revolution	Straight or curved lines	Oblique, straight or curved

* A helical gear-pair is not clearly defined unless the angle between the shafts is specified. The adjective 'spur' in 'spur gears' defines gears that operate on parallel shafts. Likewise, 'bevel' in 'bevel gears' denotes gears that operate on intersecting shafts. Unlike 'spur' and 'bevel', the adjective 'helical' describes the kind of teeth on the mating pair, and consequently, the angular relationship of the shafts has to be expressly stated by other means. The adjective 'crossed', which modifies 'axes', is used in the 1950 AGMA Standard, 112.02, Gear Nomenclature, to denote helical-toothed gears for operation on non-parallel axes. Use of the names 'spiral gears' and 'screw gears' instead of crossed helical gears is discouraged, for neither of the two names is descriptive of either the kind of teeth or the axes relationship.

TABLE 3-2

Letter Symbols of Pitches in Gear Engineering and Formulas Expressing One Pitch in Terms of Two Others. Involute Gear Teeth

ASA B6.10-1950 AGMA 112.02

Symbol	*Pitch*
$P\ (P_d)$	Diametral pitch
P_{nd}	Normal diametral pitch
p	Circular pitch
p_t	Transverse circular pitch
p_b	Base pitch
p_n	Normal circular pitch
p_N	Normal base pitch
p_x	Axial pitch
p_X	Axial base pitch

From American Machinist of July 4 and 11, 1929

	p_t	p_b	p_n	p_N	p_x	p_X
p_t			$p_x = \frac{p_t\, p_n}{\sqrt{p_t^2 - p_n^2}}$	$p_X = \frac{p_t\, p_N}{\sqrt{p_t^2 - p_N^2}}$	$p_n = \frac{p_t\, p_x}{\sqrt{p_t^2 + p_x^2}}$	$p_N = \frac{p_t\, p_X}{\sqrt{p_t^2 + p_X^2}}$
p_b				$p_x = \frac{p_b\, p_N}{\sqrt{p_b^2 - p_N^2}}$	$p_N = \frac{p_b\, p_x}{\sqrt{p_b^2 + p_x^2}}$	
p_n	$p_x = \frac{p_t\, p_n}{\sqrt{p_t^2 - p_n^2}}$				$p_t = \frac{p_n\, p_x}{\sqrt{p_x^2 - p_n^2}}$	
p_N	$p_X = \frac{p_t\, p_N}{\sqrt{p_t^2 - p_N^2}}$	$p_x = \frac{p_b\, p_N}{\sqrt{p_b^2 - p_N^2}}$			$p_b = \frac{p_N\, p_x}{\sqrt{p_x^2 - p_N^2}}$	$p_t = \frac{p_N\, p_X}{\sqrt{p_X^2 - p_N^2}}$
p_x	$p_n = \frac{p_t\, p_x}{\sqrt{p_t^2 + p_x^2}}$	$p_N = \frac{p_b\, p_x}{\sqrt{p_b^2 + p_x^2}}$	$p_t = \frac{p_n\, p_x}{\sqrt{p_x^2 - p_n^2}}$	$p_b = \frac{p_N\, p_x}{\sqrt{p_x^2 - p_N^2}}$		
p_X	$p_N = \frac{p_t\, p_X}{\sqrt{p_t^2 + p_X^2}}$			$p_t = \frac{p_N\, p_X}{\sqrt{p_X^2 - p_N^2}}$		

Definitions of angular dimensions for helical involute gears are given in Tables 4-6 and 4-7.

TABLE 3-3

*Diametral Pitches in Common Use

For Spur and Helical Gears

			Gears of 20 diametral pitch and finer belong to the fine pitch series	
1	3	8	20	56
1¼	3½	9	24	64
1½	4	10	28	72
1¾	4½	12	32	80
2	5	14	36	96
2¼	6	16	40	
2½	7	18	48	128

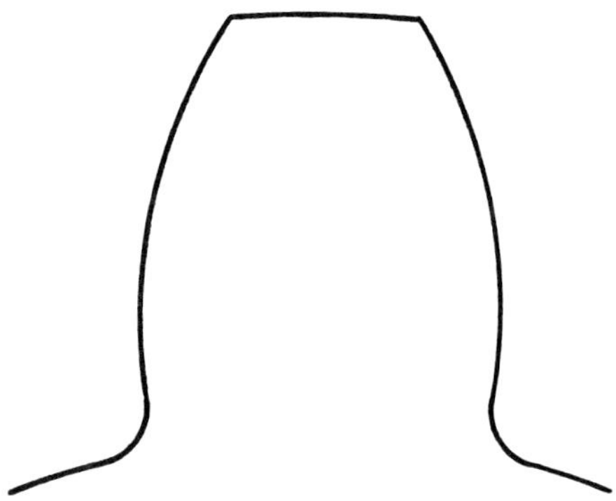

ACTUAL SIZE OF 1 DP 20° STUB TOOTH

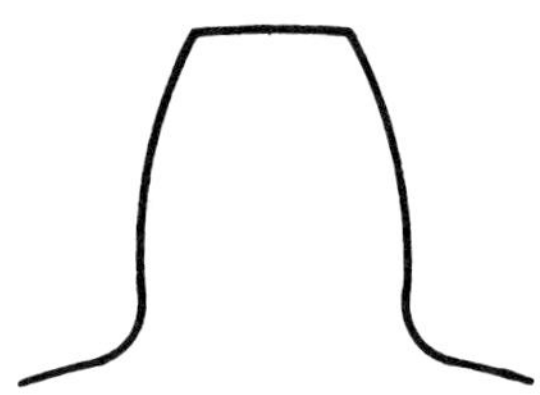

ACTUAL SIZE OF 2 DP 20° STUB TOOTH

* Usable diametral pitches for spur and helical gears are limited by the available cutters and hobs. Bevel gears are less restricted by diametral pitches since the same tools or cutter can be used just as well to cut a bevel gear of 5.125 diametral pitch as to cut one of 5 diametral pitch.

TABLE 3-4

American Standard Tooth Proportions for 20-Degree Involute Spur Gears

ASA B6.1 – 1932 and ASA B6.7 – 1950 AGMA 207.03 Fine-Pitch System

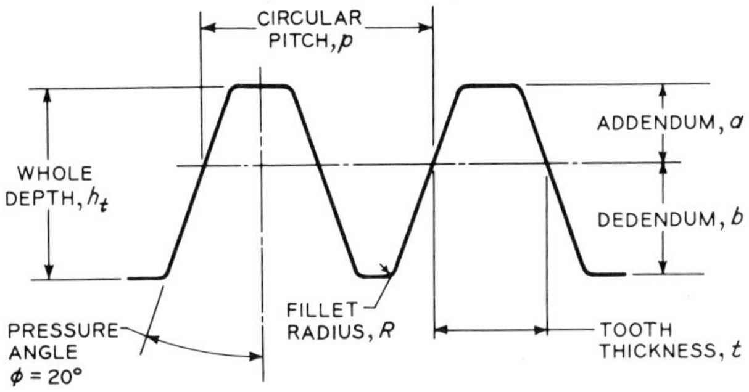

N_P = number teeth in pinion

N_G = number teeth in gear

Diametral pitch, P	P Coarser than 20		P = 20 and finer
Pressure angle, ϕ	20° Stub	20° Full Depth Teeth	
Addendum of pinion, a_P Addendum of gear, a_G	$\frac{0.800}{P}$	$\frac{1.000}{P}$	$\frac{1.000}{P}$
Dedendum of pinion, b_P Dedendum of gear, b_P	$\frac{1.000}{P}$	$\frac{1.157}{P}$	$\frac{1.200}{P} + 0.002$
Working depth, h_k	$\frac{1.600}{P}$	$\frac{2.000}{P}$	$\frac{2.000}{P}$
Whole depth, h_t	$\frac{1.800}{P}$	$\frac{2.157}{P}$	$\frac{2.200}{P} + 0.002$
Clearance, c	$\frac{0.200}{P}$	$\frac{0.157}{P}$	$\frac{0.200}{P} + 0.002$
Tooth circular thickness, t, no allowance for backlash	$\frac{1.5708}{P}$	$\frac{1.5708}{P}$	$\frac{1.5708}{P}$
Minimum fillet radius, R, at root of basic rack	$\frac{0.300}{P}$	$\frac{0.235}{P}$	Unspecified. Tool wear introduces variable amounts

TABLE 3–5

Basic Tooth Proportions for 20°, Stub-Tooth, Spur Gears
See Table 3–4 for Formulas.

Diametral Pitch, P	Circular Pitch, p	Tooth Thickness, t, on Pitch Circle, Zero Backlash	Addendum $a_P = a_G$	Dedendum $b_P = b_G$	Double Addendum, $2a$, and Working Depth b_k	Whole Depth, b_t
1	3.1416	1.5708	0.800	1.000	1.600	1.800
1-1/4	2.5133	1.2566	0.640	0.800	1.280	1.440
1-1/2	2.0944	1.0472	0.533	0.667	1.067	1.200
1-3/4	1.7952	0.8976	0.457	0.571	0.914	1.029
2	1.5708	0.7854	0.400	0.500	0.800	0.900
2-1/4	1.3963	0.6981	0.356	0.444	0.711	0.800
2-1/2	1.2566	0.6283	0.320	0.400	0.640	0.720
2-3/4	1.1424	0.5712	0.291	0.364	0.582	0.655
3	1.0472	0.5236	0.267	0.333	0.533	0.600
3-1/2	0.8976	0.4488	0.229	0.286	0.457	0.514
4	0.7854	0.3927	0.200	0.250	0.400	0.450
4-1/2	0.6980	0.3491	0.178	0.222	0.356	0.400
5	0.6283	0.3142	0.160	0.200	0.320	0.360
6	0.5236	0.2618	0.133	0.167	0.267	0.300
7	0.4488	0.2244	0.114	0.143	0.229	0.257
8	0.3927	0.1963	0.100	0.125	0.200	0.225
9	0.3491	0.1745	0.089	0.111	0.178	0.200
10	0.3142	0.1571	0.080	0.100	0.160	0.180
12	0.2618	0.1309	0.067	0.083	0.133	0.150
14	0.2244	0.1122	0.057	0.071	0.114	0.129
16	0.1963	0.0982	0.050	0.063	0.100	0.113
18	0.1745	0.0873	0.044	0.056	0.089	0.100
20	0.1571	0.0785	0.040	0.052*	0.080	0.092*
24	0.1309	0.0655	0.033	0.044	0.067	0.077
28	0.1122	0.0561	0.028	0.038	0.057	0.066
32	0.0982	0.0491	0.025	0.033	0.050	0.058
36	0.0873	0.0436	0.022	0.030	0.044	0.052
40	0.0785	0.0393	0.020	0.025	0.040	0.045
48	0.0655	0.0327	0.017	0.021	0.033	0.038
56	0.0561	9.0281	0.014	0.018	0.029	0.032
64	0.0491	0.0245	0.013	0.016	0.025	0.028
72	0.0436	0.0218	0.011	0.014	0.022	0.025
80	0.0393	0.0196	0.010	0.013	0.020	0.023
96	0.0327	0.0164	0.0083	0.010	0.017	0.019
112	0.0281	0.0140	0.0071	0.009	0.014	0.016
128	0.0245	0.0123	0.0062	0.008	0.012	0.014
144	0.0218	0.0109	0.0056	0.007	0.011	0.013

*Dedendums of 20 diametral pitch and finer contain 0.002 additional clearance.

TABLE 3–6

Basic Tooth Proportions for 20°, Full-Depth, Spur Gears
See Table 3–4 for Formulas.

Diametral Pitch, P	Circular Pitch, p	Tooth Thickness, t, on Pitch Circle, Zero Backlash	Addendum $a_P = a_G$	Dedendum $b_P = b_G$	Double Addendum, $2a$, and Working Depth, h_k	Whole Depth, h_t
1	3.1416	1.5708	1.000	1.157	2.000	2.157
1-1/4	2.5133	1.2566	0.800	0.926	1.600	1.726
1-1/2	2.0944	1.0472	0.667	0.771	1.333	1.438
1-3/4	1.7952	0.8976	0.571	0.661	1.143	1.233
2	1.5708	0.7854	0.500	0.579	1.000	1.079
2-1/4	1.3963	0.6981	0.444	0.514	0.889	0.959
2-1/2	1.2566	0.6283	0.400	0.463	0.800	0.863
2-3/4	1.1424	0.5712	0.364	0.421	0.727	0.784
3	1.0472	0.5236	0.333	0.386	0.667	0.719
3-1/2	0.8976	0.4488	0.286	0.331	0.571	0.616
4	0.7854	0.3927	0.250	0.289	0.500	0.539
4-1/2	0.6980	0.3491	0.222	0.257	0.444	0.479
5	0.6283	0.3142	0.200	0.231	0.400	0.431
6	0.5236	0.2618	0.167	0.193	0.333	0.360
7	0.4488	0.2244	0.143	0.165	0.286	0.308
8	0.3927	0.1963	0.125	0.145	0.250	0.270
9	0.3491	0.1745	0.111	0.129	0.222	0.240
10	0.3142	0.1571	0.100	0.116	0.200	0.216
12	0.2618	0.1309	0.083	0.096	0.167	0.180
14	0.2244	0.1122	0.071	0.083	0.143	0.154
16	0.1963	0.0982	0.063	0.072	0.125	0.135
18	0.1745	0.0873	0.056	0.064	0.111	0.120
20	0.1571	0.0785	0.050	0.062	0.100	0.112
24	0.1309	0.0655	0.042	0.052	0.083	0.094
28	0.1122	0.0561	0.036	0.045	0.071	0.081
32	0.0982	0.0491	0.031	0.040	0.063	0.071
36	0.0873	0.0436	0.028	0.035	0.056	0.063
40	0.0785	0.0393	0.025	0.032	0.050	0.057
48	0.0655	0.0327	0.021	0.027	0.042	0.048
56	0.0561	0.0281	0.018	0.023	0.036	0.041
64	0.0491	0.0245	0.016	0.021	0.031	0.036
72	0.0436	0.0218	0.014	0.019	0.028	0.033
80	0.0393	0.0196	0.013	0.017	0.025	0.030
96	0.0327	0.0164	0.010	0.015	0.021	0.025
112	0.0281	0.0140	0.009	0.013	0.018	0.022
128	0.0245	0.0123	0.008	0.011	0.016	0.019
144	0.0218	0.0109	0.007	0.010	0.014	0.017

TABLE 3–7

Tooth Proportions by Diametral Pitches in Fellows System for 20°, Stub-Tooth, Spur Gears

The Involute Curve and Involute Gearing The Fellows Gear Shaper Co.

Diametral Pitch,* P	Circular Pitch, p	Tooth Thickness, t, on Pitch Circle, Zero Backlash	Addendum $a = 1/P$*	Dedendum** $b_P = b_G$	Whole Depth, h_t	Double Depth, $2h_t$
3/4	1.0472	0.5236	0.250	0.313	0.563	1.1250
4/5	0.7854	0.3927	0.200	0.250	0.450	0.9000
5/7	0.6283	0.3142	0.143	0.179	0.322	0.6430
6/8	0.5236	0.2618	0.125	0.156	0.281	0.5626
7/9	0.4488	0.2244	0.111	0.139	0.250	0.5000
8/10	0.3927	0.1964	0.100	0.125	0.225	0.4500
9/11	0.3491	0.1745	0.091	0.114	0.205	0.4091
10/12	0.3142	0.1571	0.083	0.104	0.188	0.3750
11/14	0.2856	0.1428	0.071	0.089	0.161	0.3214
12/14	0.2618	0.1309	0.071	0.089	0.161	0.3214
13/16	0.2417	0.1208	0.063	0.078	0.141	0.2812
14/18	0.2244	0.1122	0.056	0.069	0.125	0.2500
16/21	0.1964	0.0982	0.048	0.059	0.107	0.2135
18/24	0.1745	0.0873	0.042	0.052	0.094	0.1873
20/26	0.1571	0.0785	0.038	0.048	0.087	0.1732
22/29	0.1428	0.0714	0.034	0.043	0.078	0.1557
24/32	0.1309	0.0654	0.031	0.039	0.071	0.1415
26/35	0.1208	0.0604	0.029	0.036	0.065	0.1297
28/37	0.1122	0.0561	0.027	0.034	0.061	0.1229
30/40	0.1047	0.0524	0.025	0.032	0.057	0.1140
32/42	0.0982	0.0491	0.024	0.031	0.054	0.1088
34/45	0.0924	0.0462	0.022	0.029	0.051	0.1018
36/48	0.0873	0.0436	0.021	0.027	0.048	0.0957
38/50	0.0827	0.0413	0.020	0.026	0.046	0.0920
40/54	0.0785	0.0393	0.019	0.024	0.043	0.0855

*In the Fellows system, the first numeral, as the 4 in four-five pitch, determines the pitch diameter, as for example, $d = N_P/4$; the second numeral, as the 5, determines addendum and dedendum, as $a_P = 1/5$ and $b_P = 1.25/5$.

**Dedendum $b = 1.250/P$ for $P \leqq 19$; $b = 1.200/P + 0.002$ for 20 diametral pitch and finer.

TABLE 3–8

One-Diametral-Pitch Dimensions for 20° Enlarged Spur Pinions

ASA B6.7–1950 AGMA 207.03

To get dimensions at another pitch, divide tabular values by diametral pitch.

Pinion			Short Addendum Gear at Standard Center Distance			Standard Gear at Enlarged Center Distance	
Number Teeth, N_P	Outside Diameter d_o	Circular Tooth Thickness at Standard Pitch Diameter	Decrease in Standard Outside Diameter	Circular Thickness at Standard Pitch Diameter	Recommended Minimum Number Teeth N_G	Increase Over Standard Center Distance	Contact Ratio Two Equal Pinions
10	12.8302	1.8730	0.8302	1.2686	33	0.4151	1.135
11	13.7132	1.8304	0.7132	1.3112	30	0.3566	1.186
12	14.5963	1.7878	0.5963	1.3538	27	0.2982	1.238
13	15.4793	1.7452	0.4793	1.3964	25	0.2397	1.290
14	16.3623	1.7027	0.3623	1.4389	23	0.1812	1.344
15	17.2453	1.6601	0.2453	1.4815	21	0.1227	1.398
16	18.1284	1.6175	0.1284	1.5241	19	0.0642	1.436
17	19.0114	1.5749	0.0114	1.5667	18	0.0057	1.511

All dimensions are given in inches

TABLE 3–9

Formulas for Calculating Dimensions of External, Equal Addendums, Straight Spur Gears

Straight (tooth) spur gears operate on parallel shafts. The distance between the shafts, the speed ratio, the angular speed of one member and the power to be transmitted are frequently the known requirements upon which to base the design of a pair. Pressure angle and standard or stub teeth can be chosen to suit hobs or cutters on hand. A diametral pitch is chosen to yield whole numbers of teeth on the specified center distance, teeth of adequate strength and durability, and to suit available machine tools, hobs and cutters.

To Find	From	Formula
Pinion pitch diameter, d	Center distance C, and ratio of angular speeds, ω_G/ω_P (External gear)	$d = \dfrac{2C\ \omega_G/\omega_P}{1 + \omega_G/\omega_P}$
Pinion pitch diameter, d	Number of teeth, N_P, and diametral pitch, P	$d = \dfrac{N_P}{P}$
Gear pitch diameter, D	Center distance, C, and pinion pitch diameter, d	$D = 2C - d$
Number teeth on pinion, N_P	Pitch diameter, d, and diametral pitch, P	$N_P = dP$
Number teeth on gear, N_G	Pitch diameter, D, and diametral pitch, P	$N_G = DP$
Center distance, C	Numbers of teeth, N_P, N_G, and diametral pitch, P (External gear)	$C = \dfrac{N_P + N_G}{2P}$
Addendum, $a_P = a_G$ or take from Tables 3–5, 3–6 or 3–7	Basic tooth proportions and diametral pitch, P	$a_P = \dfrac{\text{Constant, Table 3–4}}{P}$
Dedendum, $b_P = b_G$ Tables 3–5, 3–6 or 3–7	Basic tooth proportions and diametral pitch, P	$b_P = \dfrac{\text{Constant, Table 3–4}}{P}$
Clearance, c	Basic tooth proportions and diametral pitch, P	$c = \dfrac{\text{Constant, Table 3–4}}{P}$
Working depth, h_k	Addendums, $a_P = a_G$	$h_k = 2a_p = 2a_G$
Whole depth, h_t	Working depth, h_k, and clearance, c	$h_t = h_k + c$
Outside diameter of pinion, d_o	Pinion pitch diameter, d, and addendum, a_P	$d_o = d + 2a_P$
Outside diameter of gear, D_o	Gear pitch diameter, D, and addendum, a_G	$D_o = D + 2a_G$
Root diameters, d_R, D_R	Pitch diameters, d, D, and dedendums, b_p, b_G	$d_R = d - 2b_P$ $D_R = D - 2b_G$
Base diameter of pinion, d_B	Pinion pitch diameter, d, and pressure angle, ϕ (= 20°)	$d_B = d \cos \phi$
Base diameter of gear, D_B	Gear pitch diameter, D, and pressure angle, ϕ (= 20°)	$D_B = D \cos \phi$
Circular pitch, p	Diametral pitch, P	$p = 3.141{,}592{,}654/P$
Circular pitch, p	Numbers of teeth, N_P, N_G and pitch diameters, d, D	$p = \dfrac{\pi d}{N_P} = \dfrac{\pi D}{N_G}$
Tooth circular thickness, t, on pitch circle	Diametral pitch, P	$t = 1.5708/P$

continued on next page

Table 3–9; continued

To Find	From	Formula
Angle, θ, subtended by tooth circular thickness, t	Pitch circle circumference and number of teeth	$\theta = \frac{180}{N}$, degrees
Tooth chordal thickness, t_c. See Table 3–13	Pitch diameter, d, and angle θ	$t_c = d \sin \theta/2$
Chordal addendum, a_c. See Figure, Table 3–14	Addendum a_p, pitch diameter d, and angle θ	$a_c = a_p + \frac{d}{2}(1 - \cos \theta/2)$

TABLE 3–10

Formulas for Calculating Dimensions of an Internal, Straight Spur Gear

An internal gear always meshes with an external pinion. The pairing members have the same directions of angular rotations. Since the formulas of Table 3-9 do not cover fully the calculations for the internal gear, additional formulas are given here. To have satisfactory tooth action from standard tooth shapes, the difference between the number of teeth on the pinion and a mating internal gear should be at least eight teeth for 20° stub proportions and at least ten teeth for 20° full-depth proportions.

To Find	From	Formula
Pinion pitch diameter, d	Center distance C, and angular speed ratio, ω_P/ω_G	$d = \frac{2C}{\omega_P/\omega_G - 1}$
Gear pitch diameter, D	Number of teeth, N_G, and diametral pitch, P	$D = \frac{N_G}{P*}$
Center distance, C	Number of teeth, N_P, N_G, and diametral pitch, P	$C = \frac{N_G - N_P}{2P*}$
Addendum, a_G, or take from Tables 3–5, 3–6 or 3–7	Basic tooth proportions and diametral pitch, P	$a_G = \frac{\text{Constant, Table 3–4}}{P**}$
Dedendum, b_G Tables 3–5, 3–6, 3–7	Basic tooth proportions and diametral pitch, P	$b_G = \frac{\text{Constant,}^{\dagger} \text{ Table 3–4}}{P**}$
Internal diameter, D_i	Pitch diameter, D, and addendum, a_G	$D_i = D - 2a_G$
Whole depth, h_t	Addendum, a_G, and dedendum, b_G	$h_t = a_G + b_G$
Root diameter, D_R	Internal diameter, D_i and whole depth, h_t	$D_R = D_i + 2h_t$
Base diameter, D_b	Pitch diameter, D, and pressure angle, ϕ	$D_b = D \cos \phi$

* In Fellows system P is numerator of fraction.
** In Fellows system P is denominator of fraction.
† In Fellows system $b_G = 1.25/P$ for $P \leq 19$ and $b_G = 1.20/P + 0.002$ for 20 diametral pitch and finer.

TABLE 3–11

ASA Recommended Backlash between Assembled Gears (Spur Gears, Parallel and Crossed Helical Gears, Double-Helical or Herringbone Gears, Straight, Spiral, and Zerol Bevel Gears and Hypoid Gears) ASA B6.6 – 1946

Diametral Pitch	Backlash, Inches	Circular Pitch, Inches	Backlash, Inches
1	0.025 – 0.040	4	0.032 – 0.050
1 1/2	0.018 – 0.027	3	0.024 – 0.038
2	0.014 – 0.020	2	0.017 – 0.025
2 1/2	0.011 – 0.016	1 1/2	0.013 – 0.019
3	0.009 – 0.014	1	0.009 – 0.014
4	0.007 – 0.011	3/4	0.007 – 0.011
5	0.006 – 0.009	1/2	0.005 – 0.007
6	0.005 – 0.008	1/4	0.003 – 0.005
7	0.004 – 0.007	1/8	0.002 – 0.004
8 and 9	0.004 – 0.006		
10 to 13	0.003 – 0.005		
14 to 32	0.002 – 0.004		

TABLE 3–12

Specified Backlash and Center Distance Change Due to Backlash for Different Classes of Fine Pitch Gears ASA B6.11–1951 AGMA 236.03

Class	Diametral Pitch Range	Backlash in Mating Gears	*Approx Change in Center Distance for 20-Deg Pressure Angle
A	20 to 45	0.004 to 0.006	0.0055 to 0.0082
	46 to 70	0.003 to 0.005	0.0042 to 0.0068
	71 to 90	0.002 to 0.0035	0.0028 to 0.0046
B	20 to 60	0.002 to 0.004	0.0028 to 0.0055
	61 to 120	0.0015 to 0.003	0.002 to 0.0042
	121 and finer	0.001 to 0.002	0.0014 to 0.0028
C	20 to 60	0.001 to 0.002	0.0014 to 0.0028
	61 to 120	0.0007 to 0.0015	0.001 to 0.002
	121 and finer	0.0005 to 0.001	0.0008 to 0.0014
D	No measurable backlash at any pitch.		

*For helical gears of 20–deg normal pressure angle, divide these table values by the cosine of the helix angle to obtain the transverse backlash.

TABLE 3–13

Chordal Tooth Thickness, One Diametral Pitch

The Involute Curve and Involute Gearing – 1950 The Fellows Gear Shaper Co.

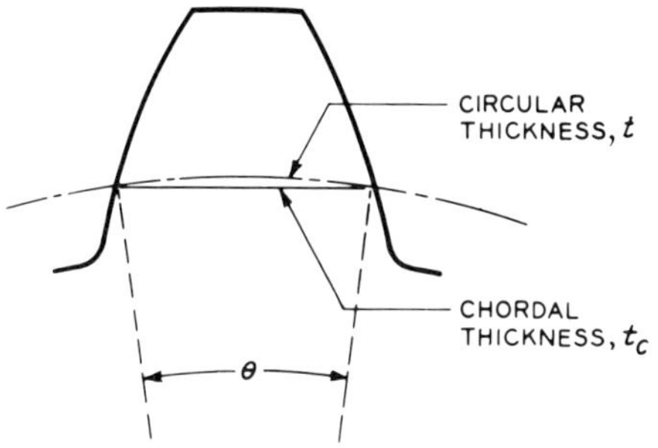

Divide value in table by diametral pitch to get chordal thickness at another diametral pitch.

Number Teeth N	Chordal Thickness t_c	Number Teeth N	Chordal Thickness t_c	Number Teeth N	Chordal Thickness t_c	Number Teeth N	Chordal Thickness t_c
8	1.56072	32	1.57016	56	1.57059	80	1.57068
9	1.56283	33	1.57020	57	1.57059	81	1.57068
10	1.56434	34	1.57024	58	1.57060	82	1.57069
11	1.56546	35	1.57027	59	1.57061	83	1.57069
12	1.56631	36	1.57030	60	1.57062	84	1.57069
13	1.56697	37	1.57032	61	1.57062	85	1.57069
14	1.56750	38	1.57035	62	1.57062	86	1.57070
15	1.56793	39	1.57037	63	1.57063	87	1.57070
16	1.56827	40	1.57039	64	1.57064	88	1.57070
17	1.56856	41	1.57041	65	1.57064	89	1.57070
18	1.56880	42	1.57043	66	1.57064	90	1.57070
19	1.56901	43	1.57045	67	1.57065	91	1.57071
20	1.56918	44	1.57046	68	1.57065	92	1.57071
21	1.56933	45	1.57048	69	1.57065	93	1.57071
22	1.56946	46	1.57049	70	1.57066	94	1.57071
23	1.56957	47	1.57050	71	1.57066	95	1.57072
24	1.56967	48	1.57051	72	1.57066	96	1.57072
25	1.56976	49	1.57052	73	1.57067	97	1.57072
26	1.56984	50	1.57053	74	1.57067	98	1.57072
27	1.56991	51	1.57053	75	1.57068	99	1.57072
28	1.56997	52	1.57055	76	1.57068	100	1.57073
29	1.57003	53	1.57056	77	1.57068	110	1.57074
30	1.57008	54	1.57057	78	1.57068	120	1.57075
31	1.57012	55	1.57058	79	1.57068	Rack	1.57080

TABLE 3–14

Chordal Addendum, One Diametral Pitch

The Involute Curve and Involute Gearing – 1950 The Fellows Gear Shaper Co.

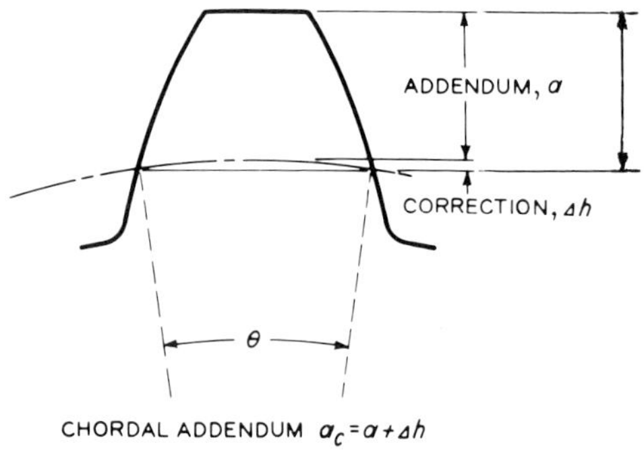

Divide value in table by diametral pitch to get chordal addendum at another diametral pitch, except for Fellows stub teeth. For the latter, divide the correction factor Δh by the numerator of the diametral pitch fraction and add the quotient to the addendum from Table 3–7.

Number Teeth, N	Correction Δh	Chordal Addendum, a_c Stub when $a = 0.8/P$	Full Depth $a = 1/P$
8	0.07686	0.87686	1.07686
9	0.06836	0.86836	1.06836
10	0.06158	0.86158	1.06158
11	0.05597	0.85597	1.05597
12	0.05133	0.85133	1.05133
13	0.04733	0.84733	1.04733
14	0.04401	0.84401	1.04401
15	0.04109	0.84109	1.04109
16	0.03852	0.83852	1.03852
17	0.03623	0.83623	1.03623
18	0.03425	0.83425	1.03425
19	0.03245	0.83245	1.03245
20	0.03083	0.83083	1.03083
21	0.02936	0.82936	1.02936
22	0.02803	0.82803	1.02803
23	0.02681	0.82681	1.02681
24	0.02569	0.82569	1.02569
25	0.02466	0.82466	1.02466
26	0.02372	0.82372	1.02372
27	0.02284	0.82284	1.02284
28	0.02202	0.82202	1.02202
29	0.02126	0.82126	1.02126
30	0.02056	0.82056	1.02056
31	0.01989	0.81989	1.01989
32	0.01927	0.81927	1.01927
33	0.01869	0.81869	1.01869
34	0.01814	0.81814	1.01814
35	0.01762	0.81762	1.01762
36	0.01713	0.81713	1.01713
37	0.01667	0.81667	1.01667
38	0.01623	0.81623	1.01623
39	0.01581	0.81581	1.01581
40	0.01541	0.81541	1.01541
41	0.01504	0.81504	1.01504
42	0.01468	0.81468	1.01468
43	0.01435	0.81435	1.01435
44	0.01401	0.81401	1.01401
45	0.01371	0.81371	1.01371
46	0.01318	0.81318	1.01318
47	0.01312	0.81312	1.01312
48	0.01285	0.81285	1.01285
49	0.01259	0.81259	1.01259
50	0.01234	0.81234	1.01234
51	0.01209	0.81209	1.01209
52	0.01186	0.81186	1.01186
53	0.01164	0.81164	1.01164
54	0.01142	0.81142	1.01142
55	0.01121	0.81121	1.01121

continued on next page

TABLE 3–14, continued

Number Teeth, N	Correction Δh	Chordal Addendum, a_c Stub when $a = 0.8/P$	Full Depth $a = 1/P$
56	0.01102	0.81102	1.01102
57	0.01082	0.81082	1.01082
58	0.01063	0.81063	1.01063
59	0.01045	0.81045	1.01045
60	0.01028	0.81028	1.01028
61	0.01011	0.81011	1.01011
62	0.00995	0.80995	1.00995
63	0.00979	0.80979	1.00979
64	0.00964	0.80964	1.00964
65	0.00949	0.80949	1.00949
66	0.00934	0.80934	1.00934
67	0.00921	0.80921	1.00921
68	0.00907	0.80907	1.00907
69	0.00884	0.80884	1.00884
70	0.00881	0.80881	1.00881
71	0.00869	0.80869	1.00869
72	0.00857	0.80857	1.00857
73	0.00845	0.80845	1.00845
74	0.00834	0.80834	1.00834
75	0.00822	0.80822	1.00822
76	0.00812	0.80812	1.00812
77	0.00801	0.80801	1.00801
78	0.00791	0.80791	1.00791
79	0.00781	0.80781	1.00781
80	0.00771	0.80771	1.00771
81	0.00762	0.80762	1.00762
82	0.00752	0.80752	1.00752
83	0.00743	0.80743	1.00743
84	0.00734	0.80734	1.00734
85	0.00725	0.80725	1.00725
86	0.00717	0.80717	1.00717
87	0.00713	0.80713	1.00713
88	0.00701	0.80701	1.00701
89	0.00693	0.80693	1.00693
90	0.00685	0.80685	1.00685
91	0.00678	0.80678	1.00678
92	0.00671	0.80671	1.00671
93	0.00664	0.80664	1.00664
94	0.00657	0.80657	1.00657
95	0.00649	0.80649	1.00649
96	0.00641	0.80641	1.00641
97	0.00634	0.80634	1.00634
98	0.00631	0.80631	1.00631
99	0.00623	0.80623	1.00623
100	0.00617	0.80617	1.00617
110	0.00561	0.80561	1.00561
120	0.00514	0.80514	1.00514
Rack	0.00000	0.80000	1.00000

TABLE 3–15

Change in Backlash, ΔB, Corresponding to a Change of 0.001 Inch in Center Distance, ΔC, at Various Pressure Angles on Spur Gears

The Involute Curve and Involute Gearing – 1950 The Fellows Gear Shaper Co.

Multiply value from table by number of thousands change in center distance to get total change in backlash.

Pressure Angle, ϕ, Deg	Change ΔB per $\Delta C = 0.001$ $\Delta B = 2(0.001) \tan \phi$
5	0.00017
10	0.00035
14-1/2	0.00052
15	0.00054
17-1/2	0.00063
20	0.00073
22-1/2	0.00083
25	0.00093
30	0.00115
35	0.00140

TABLE 3-16

Measuring Pin Diameters for Standard Involute Spur Gears AGMA Standard 236.03 for Fine-Pitch Gears ASA B6.11–1951

Diametral Pitch, P	Standard External $d_{pin} = \frac{1.7280}{P}$	Standard Internal $d_{pin} = \frac{1.6800}{P}$	Long Addendum Pinions $d_{pin} = \frac{1.9200}{P}$
1	1.7280	1.6800	1.9200
1-1/4	1.3824	1.3440	1.5360
1-1/2	1.1520	1.1200	1.2800
1-3/4	0.9874	0.9600	1.0971
2	0.8640	0.8400	0.9600
2-1/2	0.6912	0.6720	0.7680
3	0.5760	0.5600	0.6400
3-1/2	0.4937	0.4800	0.5486
4	0.4320	0.4200	0.4800
5	0.3456	0.3360	0.3840
6	0.2880	0.2800	0.3200
7	0.2469	0.2400	0.2743
8	0.2160	0.2100	0.2400
9	0.1920	0.1867	0.2133
10	0.1728	0.1680	0.1920
12	0.1440	0.1400	0.1600
16	0.1080	0.1050	0.1200
20	0.0864	0.0840	0.0960
24	0.0720	0.0700	0.0800
28	0.0617	0.0600	0.0686
32	0.0540	0.0525	0.0600
36	0.0480	0.0467	0.0533
40	0.0432	0.0420	0.0480
48	0.0360	0.0350	0.0400
60	0.0288	0.0280	0.0320
72	0.0240	0.0233	0.0267
96	0.0180	0.0175	0.0200
100	0.0173	0.0168	0.0192

TABLE 3–17

Dimensions for Measurements Over Pins
External Involute Spur Gears

ASA B6.11 – 1951 AGMA Standard 236.03

One* Diametral Pitch **1.728-Inch Pin Diameter** **20° Pressure Angle**

N = No. of Teeth M = Dimension Over Pins (Zero Backlash) k_m = Thickness Factor†

N	M*	k_m	N	M*	k_m	N	M*	k_m
10								
11								
12	For Standard Long-		46	48.4265	2.48	81	83.4262	2.58
13			47	49.4007	2.49	82	84.4418	2.58
14	Addendum Pinions		48	50.4279	2.49	83	85.4271	2.59
15			49	51.4031.	2.50	84	86.4423	2.59
	See Table 3–20		50	52.4292	2.50	85	87.4279	2.59
16								
17			51	53.4053	2.50	86	88.4428	2.59
			52	54.4304	2.51	87	89.4287	2.59
18	20.3840	2.23	53	55.4074	2.51	88	90.4433	2.59
19	21.3200	2.25	54	56.4315	2.52	89	91.4295	2.60
20	22.3900	2.26	55	57.4093	2.52	90	92.4437	3.60
21	23.3321	2.28	56	58.4325	2.52	91	93.4303	2.60
22	24.3952	2.29	57	59.4111	2.53	92	94.4441	2.60
23	25.3423	2.30	58	60.4335	2.53	93	95.4310	2.60
24	26.3997	2.32	59	61.4128	2.53	94	96.4445	2.60
25	27.3511	2.33	60	62.4344	2.53	95	97.4317	2.60
26	28.4036	2.34	61	63.4144	2.54	96	98.4449	2.61
27	29.3586	2.35	62	64.4352	2.54	97	99.4323	2.61
28	30.4071	2.36	63	65.4159	2.54	98	100.4453	2.61
29	31.3652	2.37	64	66.4361	2.55	99	101.4329	2.61
30	32.4102	2.38	65	67.4173	2.55	100	102.4456	2.61
31	33.3710	2.39	66	68.4369	2.55	101	103.4335	2.61
32	34.4130	2.40	67	69.4186	2.55	102	104.4460	2.61
33	35.3761	2.41	68	70.4376	2.56	103	105.4341	2.61
34	36.4155	2.41	69	71.4198	2.56	104	106.4463	2.62
35	37.3807	2.42	70	72.4383	2.56	105	107.4346	2.62
36	38.4178	2.43	71	73.4210	2.56	106	108.4466	2.62
37	39.3849	2.43	72	74.4390	2.57	107	109.4352	2.62
38	40.4198	2.44	73	75.4221	2.57	108	110.4469	2.62
39	41.3886	2.45	74	76.4396	2.57	109	111.4357	2.62
40	42.4217	2.45	75	77.4232	2.57	110	112.4472	2.62
41	43.3920	2.46	76	78.4402	2.57	111	113.4362	2.62
42	44.4234	2.46	77	79.4242	2.58	112	114.4475	2.62
43	45.3951	2.47	78	80.4408	2.58	113	115.4367	2.63
44	46.4250	2.47	79	81.4252	2.58	114	116.4478	2.63
45	47.3980	2.48	80	82.4413	2.58	115	117.4372	2.63

*To find no-backlash distance M at diametral pitch other than one, divide M value of table by diametral pitch. (Dimensions are in inches.)

†To correct dimension M for backlash, multiply thickness factor k_m from table by thousandth of an inch that the tooth is desired thinner than no-backlash thickness and subtract from M/P dimension. See Table 3–21 for sample calculation.

continued on next page

TABLE 3–17, continued

N = No. of Teeth M = Dimension Over Pins (Zero Backlash) k_m = Thickness Factor †

N	M*	k_m	N	M*	k_m	N	M*	k_m
116	118.4481	2.63	151	153.4435	2.65	186	188.4540	2.67
117	119.4376	2.63	152	154.4518	2.65	187	189.4474	2.67
118	120.4484	2.63	153	155.4438	2.65	188	190.4541	2.67
119	121.4380	2.63	154	156.4520	2.66	189	191.4476	2.67
120	122.4486	2.63	155	157.4440	2.66	190	192.4542	2.67
121	123.4384	2.63	156	158.4521	2.66	191	193.4478	2.67
122	124.4489	2.63	157	159.4443	2.66	192	194.4543	2.67
123	125.4388	2.63	158	160.4523	2.66	193	195.4480	2.67
124	126.4491	2.64	159	161.4445	2.66	194	196.4544	2.67
125	127.4392	2.64	160	162.4524	2.66	195	197.4482	2.67
126	128.4493	2.64	161	163.4448	2.66	196	198.4546	2.67
127	129.4396	2.64	162	164.4526	2.66	197	199.4483	2.67
128	130.4496	2.64	163	165.4450	2.66	198	200.4547	2.67
129	131.4400	2.64	164	166.4527	2.66	199	201.4485	2.68
130	132.4498	2.64	165	167.4453	2.66	200	202.4548	2.68
131	133.4404	2.64	166	168.4528	2.66	300	302.4579	2.70
132	134.4500	2.64	167	169.4455	2.66			
133	135.4408	2.64	168	170.4529	2.66	400	402.4596	2.71
134	136.4502	2.64	169	171.4457	2.66			
135	137.4411	2.64	170	172.4531	2.66	500	502.4606	2.72
136	138.4504	2.64	171	173.4459	2.66	201	203.4487	2.68
137	139.4414	2.64	172	174.4532	2.66	301	303.4538	2.70
138	140.4506	2.65	173	175.4461	2.66	401	403.4565	2.71
139	141.4418	2.65	174	176.4533	2.67	501	503.4581	2.72
140	142.4508	2.65	175	177.4463	2.67			
141	143.4421	2.65	176	178.4535	2.67			
142	144.4510	2.65	177	179.4465	2.67			
143	145.4424	2.65	178	180.4536	2.67			
144	146.4512	2.65	179	181.4467	2.67			
145	147.4427	2.65	180	182.4537	2.67			
146	148.4513	2.65	181	183.4469	2.67			
147	149.4430	2.65	182	184.4538	2.67			
148	150.4515	2.65	183	185.4471	2.67			
149	151.4433	2.65	184	186.4539	2.67			
150	152.4516	2.65	185	187.4473	2.67			

***To find no-backlash distance M at diametral pitch other than one, divide M value of table by diametral pitch. (Dimensions are in inches.)**

†To correct dimension M for backlash, multiply thickness factor k_m from table by thousandth of an inch that the tooth is desired thinner than no-backlash thickness and subtract from M/P dimension. See Table 3–21 for sample calculation.

TABLE 3–18

Dimensions for Measurements Between Pins
Internal Involute Spur Gears
AGMA Standard 236.03 1948
ASA B6.11 – 1951

One* Diametral Pitch **1.680-Inch Pin Diameter** **20° Pressure Angle**

N = No. of Teeth M = Dimension Between Pins (Zero Backlash) k_m ≐ Thickness Factor†

N	M*	k_m	N	M*	k_m	N	M*	k_m
30	27.6649	3.38						
31	28.6285	3.35	66	63.7011	2.97	101	98.6974	2.88
32	29.6699	3.32	67	64.6832	2.96	102	99.7096	2.88
33	30.6353	3.29	68	65.7018	2.96	103	100.6979	2.88
34	31.6739	3.27	69	66.6844	2.96	104	101.7099	2.88
35	32.6411	3.25	70	67.7025	2.95	105	102.6984	2.88
36	33.6773	3.23	71	68.6856	2.95	106	103.7102	2.88
37	34.6462	3.21	72	69.7032	2.95	107	104.6989	2.88
38	35.6804	3.20	73	70.6867	2.94	108	105.7105	2.87
39	36.6507	3.18	74	71.7038	2.94	109	106.6994	2.87
40	37.6831	3.16	75	72.6878	2.94	110	107.7107	2.87
41	38.6547	3.15	76	73.7044	2.94	111	108.6998	2.87
42	39.6855	3.14	77	74.6888	2.93	112	109.7110	2.87
43	40.6582	3.13	78	75.7049	2.93	113	110.7002	2.87
44	41.6875	3.11	79	76.6897	2.93	114	111.7112	2.87
45	42.6614	3.10	80	77.7054	2.92	115	112.7006	2.87
46	43.6893	3.09	81	78.6905	2.92	116	113.7114	2.86
47	44.6644	3.08	82	79.7059	2.92	117	114.7010	2.86
48	45.6910	3.08	83	80.6914	2.92	118	115.7117	2.86
49	46.6670	3.07	84	81.7064	2.91	119	116.7014	2.86
50	47.6926	3.06	85	82.6922	2.91	120	117.7119	2.86
51	48.6694	3.05	86	83.7068	2.91	121	118.7018	2.86
52	49.6940	3.04	87	84.6929	2.91	122	119.7121	2.86
53	50.6716	3.04	88	85.7072	2.91	123	120.7022	2.86
54	51.6953	3.03	89	86.6936	2.90	124	121.7123	2.86
55	52.6737	3.02	90	87.7076	2.90	125	122.7026	2.86
56	53.6965	3.02	91	88.6943	2.90	126	123.7125	2.85
57	54.6756	3.01	92	89.7080	2.90	127	124.7029	2.85
58	55.6975	3.01	93	90.6950	2.90	128	125.7127	2.85
59	56.6774	3.00	94	91.7084	2.89	129	126.7032	2.85
60	57.6985	3.00	95	92.6956	2.89	130	127.7129	2.85
61	58.6789	2.99	96	93.7087	2.89	131	128.7036	2.85
62	59.6994	2.99	97	94.6962	2.89	132	129.7130	2.85
63	60.6805	2.98	98	95.7090	2.89	133	130.7039	2.85
64	61.7003	2.98	99	96.6968	2.89	134	131.7132	2.85
65	62.6819	2.97	100	97.7093	2.88	135	132.7042	2.85

*To find no-backlash distance M at diametral pitch other than one, divide M value of table by diametral pitch. (Dimensions are in inches.)

†To correct dimension M for backlash, multiply thickness factor k_m from table by thousandth of an inch that the tooth is desired thinner than no-backlash thickness and add to M/P dimension. See Table 3-21 for sample calculation.

continued on next page

TABLE 3–18, continued

N = No. of Teeth M = Dimension Between Pins (Zero Backlash) k_m = Thickness Factor†

N	M*	k_m	N	M*	k_m	N	M*	k_m
136	133.7134	2.85	161	158.7074	2.83	186	183.7162	2.82
137	134.7045	2.84	162	159.7151	2.83	187	184.7097	2.82
138	135.7135	2.84	163	160.7076	2.83	188	185.7163	2.82
139	136.7047	2.84	164	161.7152	2.83	189	186.7098	2.82
140	137.7137	2.84	165	162.7078	2.83	190	187.7164	2.82
141	138.7050	2.84	166	163.7153	2.83	191	188.7100	2.82
142	139.7139	2.84	167	164.7080	2.83	192	189.7165	2.82
143	140.7053	2.84	168	165.7154	2.83	193	190.7101	2.82
144	141.7140	2.84	169	166.7082	2.83	194	191.7166	2.81
145	142.7055	2.84	170	167.7156	2.82	195	192.7103	2.81
146	143.7141	2.84	171	168.7084	2.82	196	193.7166	2.81
147	144.7058	2.84	172	169.7157	2.82	197	194.7104	2.81
148	145.7143	2.84	173	170.7086	2.82	198	195.7167	2.81
149	146.7061	2.84	174	171.7158	2.82	199	196.7106	2.81
150	147.7144	2.84	175	172.7087	2.82	200	197.7168	2.81
151	148.7063	2.84	176	173.7158	2.82	300	297.7192	2.79
152	149.7145	2.83	177	174.7089	2.82			
153	150.7065	2.83	178	175.7159	2.82	400	397.7203	2.78
154	151.7146	2.83	179	176.7090	2.82			
155	152.7068	2.83	180	177.7160	2.82	500	497.7210	2.77
156	153.7148	2.83	181	178.7092	2.82	201	198.7107	2.81
157	154.7070	2.83	182	179.7161	2.82	301	298.7151	2.79
158	155.7149	2.83	183	180.7094	2.82	401	398.7172	2.78
159	156.7072	2.83	184	181.7162	2.82	501	498.7185	2.77
160	157.7150	2.83	185	182.7095	2.82			

*To find no-backlash distance M at diametral pitch other than one, divide M value of table by diametral pitch. (Dimensions are in inches.)

†To correct dimension M for backlash, multiply thickness factor k_m from table by thousandth of an inch that the tooth is desired thinner than no-backlash thickness and add to M/P dimension. See Table 3–21 for sample calculation.

TABLE 3–19

Pin Measurement of Standard 20° Spur Rack

AGMA 236.03 ASA B6.11–1951

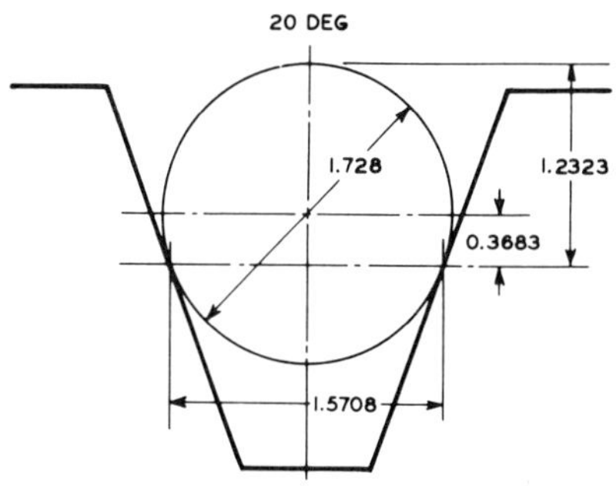

TABLE 3–20

Pin Measurements Over Standard Long-Addendum Pinions – External AGMA Standard 236.03 ASA B6.11–1951

One* Diametral Pitch		1.920-In. Pin Diameter		20° Pressure Angle
Number Teeth, N	Outside Diameter d_o	Circular Thickness t	Dimension Over Pins (Zero Backlash) M*	Thickness Factor** k_m
10	12.8302	1.8730	13.4408	1.62
11	13.7132	1.8304	14.2678	1.68
12	14.5963	1.7878	15.3428	1.73
13	15.4793	1.7452	16.1807	1.79
14	16.3623	1.7027	17.2233	1.84
15	17.2453	1.6601	18.0674	1.89
16	18.1284	1.6175	19.0851	1.95
17	19.0114	1.5749	19.9326	2.00

*To find no-backlash distance M at diametral pitch other than one, divide M value of table by diametral pitch.

**To correct dimension M for backlash, multiply thickness factor k_m from table by thousandth of an inch tooth is desired thinner than no-backlash thickness and subtract from M/P dimension. See Table 3–21 for sample calculation.

TABLE 3–21

Examples Illustrating Use of Tables to Find Pin Measurement Dimensions, Involute Spur Gears

Definition of Quantity and Symbol	Coarse Pitch Pair 30 Tooth External 80 Tooth Internal	Fine Pitch Pair 30 Tooth External 80 Tooth Internal
Diametral pitch, P	6	24
Pressure angle, ϕ	20°	20°
Backlash allowance, ΔB, each member of pair	0.004	0.002
M from Table 3-17 $N_P = 30$	32.4102	32.4102
k_m from Table 3-17 $N_P = 30$	2.38	2.38
M/P, theoretical or zero backlash	5.4017	1.3504
Correction, $\Delta M = \Delta B\, k_m$	0.0095	0.0048
Dimension over pins (subtract)	5.392 *	1.346
Pin diameters, Table 3–16	0.288	0.072
M from Table 3–18, $N_G = 80$	77.7054	77.7054
k_m from Table 3–18, $N_G = 80$	2.92	2.92
M/P, theoretical or zero backlash	12.9509	3.2377
Correction, $\Delta M = \Delta B\, k_m$	0.0117	0.0058
Dimension between pins (add)	12.963	3.244
Pin diameters, Table 3–16	0.280	0.070

* Procedure to provide thicker tooth than finished tooth in shaving and grinding is as follows: Suppose tooth is to be finished by shaving with 0.002 stock allowance. Then dimension over pins before shaving would be 5.397. After shaving, as calculated.

TABLE 3–22

Formulas for Calculating Dimensions for Measurements Over or Between Pins Involute Spur Gears

Quantity and Symbol	Formula	
Involute function, a	$\text{inv}\ a = \tan a - a$	
Diameter of pitch circle, D		
Diameter of base circle, D_b	$D_b = D \cos \phi$	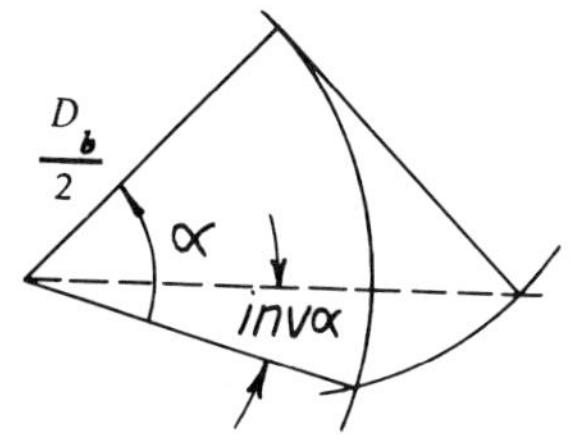
Pressure angle, ϕ		
$\text{inv}\ 20^\circ = 0.0149044,\ \cos 20^\circ = 0.9396926$		
Tooth circular thickness, t	$t = \dfrac{1.570796}{P}$	
Diametral pitch, P		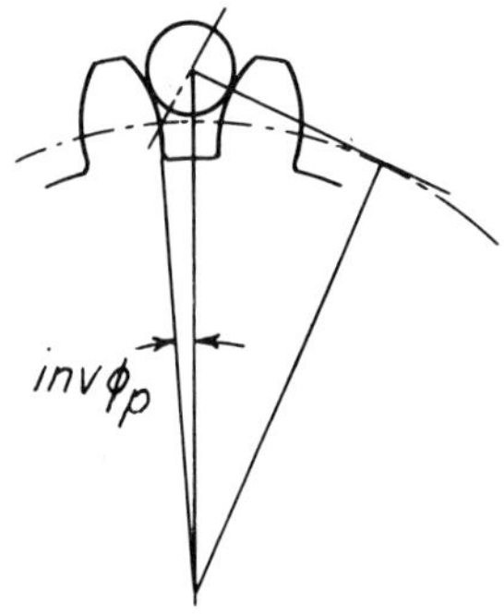
Number teeth, N		
Measuring pin diameter, d_p	Table 3–16	
EXTERNAL GEAR	(see Table 3–23 for numerical example)	
Involute function ϕ_P on circle of diameter D_P thru pin center	$\text{inv}\ \phi_P = t/D + \text{inv}\ \phi + d_p/D_b - \pi/N$	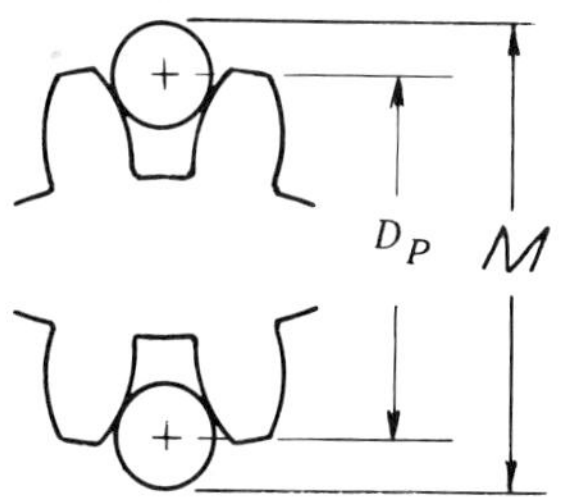
Dimension over pins M	$D_P = \dfrac{D_b}{\cos \phi_P}$	
External gear, N even	$M_{even} = D_P + d_p$	
External gear, N odd	$M_{odd} = D_P \cos \dfrac{90^\circ}{N} + d_p$	
INTERNAL GEAR	(See Table 3-23 for numerical example.)	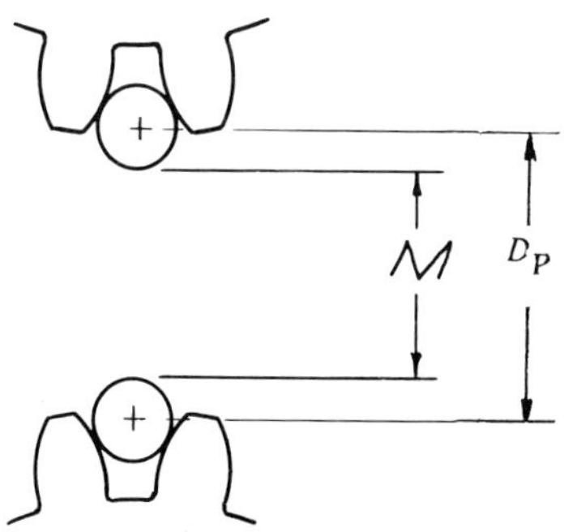
Involute function ϕ_P on circle of diameter D_P thru pin center	$\text{inv}\ \phi_P = \pi/N + \text{inv}\ \phi - t/D - d_p/D_b$	
Dimensions between pins, M	$D_P = \dfrac{D_b}{\cos \phi_P}$	
Internal gear, N even	$M_{even} = D_P - d_p$	
Internal gear, N odd	$M_{odd} = D_P \cos \dfrac{90^\circ}{N} - d_t$	

TABLE 3–23

Calculations to Illustrate Application of Formulas in Computing Dimensions for Pin Measurements External and Internal Involute Spur Gears

		Pinion		Gear
Number of teeth	N_P	30	N_G	87 internal
Diametral pitch, P		6		
Pressure angle, ϕ		20°		
Backlash allowance, pinion only		$B = 0.004$		zero
Circular thickness, t		$t - B = 0.2578$		$t = 0.2618$
Pitch diameter		$d = N_P/P = 5.0000$		$D = N_G/P = 14.5000$
Base Diameter		$d_b = d \cos\phi = 4.69846$		$D_b = D\cos\phi = 13.6255$

EXTERNAL for pinion, Table 3–22

$$\text{inv}\,\phi_P = \frac{t-B}{d} + \text{inv}\,\phi + \frac{d_p}{d_b} - \frac{\pi}{N_P}$$

d_p = pin diameter, Table 3–16

Since pinion has an even number of teeth, dimension M over pins, including backlash, is

$$M_{even} = D_P + d_p$$

Compare with answer in Table 3–21.

$\frac{t-B}{d} = \frac{0.2578}{5}$	=	0.051560
$\text{inv}\,\phi = \text{inv}\,20°$	=	0.014904
$\frac{d_p}{d_b} = \frac{0.2880}{4.69846}$	=	0.061297
sum		0.127761
π/N_P	=	0.104720
(subtract) inv ϕ_P	=	0.023041
ϕ_P (Table 3–24)	=	23.00°
$\cos\phi_P$	=	0.920510
$D_P = \frac{d_b}{\cos\phi_P} = \frac{4.69846}{0.92051}$	=	5.1042
d_p	=	0.2880
M_{even}		5.3922

INTERNAL gear, Table 3–22

$$\text{inv}\,\phi_P = \pi/N + \text{inv}\,\phi - t/D - d_p/D_b$$

$d_p = 0.2800$ Table 3–16

For odd number of teeth

$$M_{odd} = D_P \cos 90°/N - d_p$$

$$\cos\frac{90°}{N-87} = 0.999837$$

π/N_G	=	0.036110
inv ϕ	=	0.014904
sum	=	0.051014
$t/D = 0.2618/14.500$	=	0.018055
$d_p/D_b = 0.2800/13.6255$	=	0.020550
inv ϕ_P (difference)	=	0.012409
ϕ_P (Table 3–24)	=	18.85°
$\cos\phi_P$	=	0.946368
$D_P = D_b/\cos\phi_P$	=	14.3977
$D_P \cos 90°/N$	=	14.3954
d_p	=	0.280
M_{odd} (difference)	=	14.1154

TABLE 3–24

Involute Function inv $\phi = \tan \phi - \phi$

Buckingham Manual of Gear Design 1935 Industrial Press
Vogel Involutometry and Trigonometry 1945 Michigan Tool

$\phi°$	inv ϕ	$\phi°$	inv ϕ	$\phi°$	inv ϕ
0.50	0.0000002215	3.00	0.00004790	4.00	0.0001136
1.00	001772	3.20	05814	4.20	01316
1.50	00598	3.40	06975	4.40	01513
2.00	01418	3.60	08281	4.60	01729
2.50	02771	3.80	09742	4.80	01965
5.00	0.0002222	6.00	0.0003845	7.00	0.0006115
.10	02358	.10	04041	.10	06382
.20	02500	.20	04244	.20	06657
.30	02647	.30	04453	.30	06939
.40	02800	.40	04669	.40	07230
5.50	0.0002959	6.50	0.0004892	7.50	0.0007528
.60	03124	.60	05122	.60	07835
.70	03295	.70	05359	.70	08150
.80	03472	.80	05604	.80	08473
.90	03655	.90	05856	.90	08805

Decimal of Deg	8°	9°	10°	11°	12°	13°	Decimal of Deg
.00	0.0009145	0.0013048	0.0017941	0.0023941	0.0031170	0.0039754	.00
.05	9318	13268	18213	24272	31566	40221	.05
.10	9494	13491	18489	24606	31966	40692	.10
.15	9672	13616	18767	24944	32369	41166	.15
.20	0.0009852	0.0013944	0.0019048	0.0025285	0.0032775	0.0041644	.20
.25	10034	14174	19332	25628	33184	42126	.25
.30	10219	14407	19619	25975	33598	42612	.30
.35	10406	14642	19908	26325	34014	43102	.35
.40	0.0010595	0.0014880	0.0020201	0.0026678	0.0034434	0.0043595	.40
.45	10786	15120	20496	27035	34858	44092	.45
.50	10980	15363	20795	27394	35285	44593	.50
.55	11176	15609	21096	27757	35716	45098	.55
.60	0.0011375	0.0015857	0.0021400	0.0028123	0.0036150	0.0045607	.60
.65	11575	16108	21707	28493	36588	46120	.65
.70	11779	16362	22017	28865	37029	46636	.70
.75	11984	16618	22330	29241	37474	47157	.75
.80	0.0012192	0.0016877	0.0022646	0.0029620	0.0037923	0.0047681	.80
.85	12402	17139	22966	30003	38375	48210	.85
.90	12615	17403	23288	30389	38831	48742	.90
.95	12830	17671	23613	30778	39291	49279	.95

continued on next page

TABLE 3–24, continued

Decimal of Deg	14°	15°	16°	17°	18°	19°	Decimal of Deg
.00	0.0049819	0.0061498	0.0074927	0.0090247	0.010760	0.012715	.00
.05	50364	62127	75647	91065	10853	12819	.05
.10	50912	62760	76372	91889	10946	12923	.10
.15	51465	63397	77101	92717	11039	13028	.15
.20	0.0052022	0.0064039	0.0077835	0.0093551	0.011133	0.013134	.20
.25	52582	64686	78574	94390	11228	13240	.25
.30	53147	65337	79318	95234	11323	13346	.30
.35	53716	65992	80067	96083	11419	13454	.35
.40	0.0054290	0.0066652	0.0080820	0.0096937	0.011515	0.013562	.40
.45	54867	67316	81578	97797	11612	13670	.45
.50	55448	67985	82342	98662	11709	13779	.50
.55	56034	68658	83110	99532	11807	13889	.55
.60	0.0056624	0.0069336	0.0083883	0.010041	0.011906	0.013999	.60
.65	57218	70019	84661	10129	12005	14110	.65
.70	57817	70706	85444	10217	12105	14222	.70
.75	58420	71398	86232	10307	12205	14334	.75
.80	0.0059027	0.0072095	0.0087025	0.010396	0.012306	0.014447	.80
.85	59638	72796	87823	10486	12407	14560	.85
.90	60254	73501	88626	10577	12509	14674	.90
.95	60874	74212	89434	10669	12612	14789	.95

Decimal of Deg	20°	21°	22°	23°	24°	25°	Decimal of Deg
.00	0.014904	0.017345	0.020054	0.023049	0.026350	0.029975	.00
.05	15020	17474	20197	23207	26523	30166	.05
.10	15137	17603	20340	23365	26697	30357	.10
.15	15254	17734	20484	23524	26872	30549	.15
.20	0.015372	0.017865	0.020629	0.023684	0.027048	0.030741	.20
.25	15490	17996	20775	23845	27225	30935	.25
.30	15609	18129	20921	24006	27402	31130	.30
.35	15729	18262	21069	24169	27581	31325	.35
.40	0.015850	0.018395	0.021216	0.024332	0.027760	0.031521	.40
.45	15970	18530	21365	24495	27940	31718	.45
.50	16092	18665	21514	24660	28121	31917	.50
.55	16214	18800	21665	24825	28302	32116	.55
.60	0.016337	0.018937	0.021815	0.024992	0.028485	0.032315	.60
.65	16461	19074	21967	25159	28668	32516	.65
.70	16585	19212	22119	25326	28852	32718	.70
.75	16710	19350	22272	25495	29037	32920	.75
.80	0.016836	0.019490	0.022426	0.025664	0.029223	0.033124	.80
.85	16962	19630	22581	25834	29410	33328	.85
.90	17089	19770	22736	26005	29598	33534	.90
.95	17217	19912	22892	26177	29786	33740	.95

continued on next page

TABLE 3–24, continued

Decimal of Deg	26°	27°	28°	29°	30°	31°	Decimal of Deg
.00	0.033947	0.038287	0.043017	0.048164	0.053751	0.059809	.00
.05	34155	38514	43264	48432	54043	60124	.05
.10	34364	38742	43513	48702	54336	60441	.10
.05	34574	38971	43762	48973	54629	60759	.15
.20	0.034785	0.039201	0.044012	0.049245	0.054924	0.061079	.20
.25	34996	39432	44264	49518	55221	61400	.25
.30	35209	39664	44516	49792	55518	61721	.30
.35	35423	39897	44770	50068	55817	62045	.35
.40	0.035637	0.040131	0.045024	0.050344	0.056116	0.062369	.40
.45	35853	40366	45280	50622	56417	62695	.45
.50	36069	40602	45537	50901	56720	63022	.50
.55	36287	40838	45795	51181	57023	63350	.55
.60	0.036505	0.041076	0.046054	0.051462	0.057328	0.063680	.60
.65	36724	41316	46313	51744	57633	64011	.65
.70	36945	41556	46575	52027	57940	64343	.70
.75	37166	41797	46837	52312	58249	64677	.75
.80	0.037388	0.042039	0.047100	0.052597	0.058558	0.065012	.80
.85	37611	42282	47364	52884	58869	65348	.85
.90	37835	42526	47630	53172	59181	65685	.90
.95	38060	42771	47896	53461	59494	66024	.95

Decimal of Deg	32°	33°	34°	35°	36°	37°	Decimal of Deg
.00	0.066364	0.073449	0.081097	0.089342	0.098224	0.107782	.00
.05	66705	73818	81494	89771	98685	108279	.05
.10	67048	74188	81894	90201	99149	108777	.10
.15	67392	74559	82294	90633	99614	109277	.15
.20	0.067738	0.074932	0.082697	0.091066	0.100080	0.109779	.20
.25	68084	75307	83100	91502	100548	110282	.25
.30	68432	75683	83506	91938	101019	110788	.30
.35	68782	76060	83913	92377	101490	111295	.35
.40	0.069133	0.076439	0.084321	0.092816	0.101964	0.111805	.40
.45	69485	76819	84731	93258	102439	112316	.45
.50	69838	77200	85142	93701	102916	112828	.50
.55	70193	77583	85555	94146	103395	113343	.55
.60	0.070549	0.077968	0.085970	0.094592	0.103875	0.113860	.60
.65	70907	78354	86386	95041	104357	114378	.65
.70	71266	78741	86804	95490	104841	114899	.70
.75	71626	79130	87223	95942	105327	115421	.75
.80	0.071988	0.079520	0.087644	0.096395	0.105814	0.115945	.80
.85	72351	79912	88066	96850	106304	116471	.85
.90	72716	80306	88490	97306	106795	116999	.90
.95	73082	80700	88915	97764	107288	117529	.95

continued on next page

TABLE 3–24, continued

Decimal of Deg	38°	39°	40°	41°	42°	43°	Decimal of Deg
.00	0.118060	0.129106	0.140968	0.153702	0.167366	0.182024	.00
.05	118594	129679	141584	154362	168074	182784	.05
.10	119130	130254	142201	155025	168786	183546	.10
.15	119667	130832	142821	155691	169499	184312	.15
.20	0.120207	0.131411	0.143443	0.156358	0.170216	0.185080	.20
.25	120748	131992	144068	157028	170934	185851	.25
.30	121291	132576	144694	157700	171656	186625	.30
.35	121837	133162	145323	158375	172380	187401	.35
.40	0.122384	0.133750	0.145954	0.159052	0.173106	0.188180	.40
.45	122933	134339	146587	159732	173835	188962	.45
.50	123484	134931	147222	160414	174566	189746	.50
.55	124037	135525	147860	161098	175300	190534	.55
.60	0.124592	0.136122	0.148500	0.161785	0.176037	0.191324	.60
.65	125150	136720	149142	162474	176776	192116	.65
.70	125709	137320	149787	163165	177518	192912	.70
.75	126270	137923	150434	163859	178262	193710	.75
.80	0.126833	0.138528	0.151082	0.164556	0.179009	0.194511	.80
.85	127398	139134	151734	165254	179759	195315	.85
.90	127965	139743	152388	165956	180511	196122	.90
.95	128534	140355	153044	166660	181266	196932	.95

Decimal of Deg	44°	45°	46°	47°	48°	49°	Decimal of Deg
.00	0.197744	0.214602	0.232679	0.252064	0.272854	0.295157	.00
.05	198559	215476	233616	253069	273933	296314	.05
.10	199377	216353	234557	254078	275015	297475	.10
.15	200198	217234	235501	255090	276101	298640	.15
.20	0.201022	0.218117	0.236448	0.256106	0.277190	0.299809	.20
.25	201849	219004	237399	257126	278284	300983	.25
.30	202678	219893	238353	258149	279381	302160	.30
.35	203511	220786	239310	259176	280483	303342	.35
.40	0.204346	0.221682	0.240271	0.260206	0.281588	0.304527	.40
.45	205184	222580	241235	261240	282697	305717	.45
.50	206026	223483	242202	262277	283810	306912	.50
.55	206870	224388	243173	263318	284927	308110	.55
.60	0.207717	0.225296	0.244147	0.264363	0.286047	0.309313	.60
.65	208567	226208	245124	265412	287172	310520	.65
.70	209420	227123	246106	266464	288301	311731	.70
.75	210276	228041	247090	267519	289433	312946	.75
.80	0.211135	0.228962	0.248078	0.268579	0.290570	0.314166	.80
.85	211998	229886	249069	269642	291711	315390	.85
.90	212863	230714	250064	270709	292856	316619	.90
.95	213731	231745	251062	271780	294004	317852	.95

continued on next page

TABLE 3–24, continued

Decimal of Deg	50°	51°	52°	53°	54°	55°	Decimal of Deg
.00	0.319089	0.344779	0.372370	0.402020	0.433904	0.468217	.00
.05	320331	346112	373803	403560	435560	470000	.05
.10	321577	347450	375240	405105	437223	471790	.10
.15	322827	348793	376683	406656	438891	473586	.15
.20	0.324082	0.350141	0.378130	0.408212	0.440566	0.475390	.20
.25	325341	351493	379583	409775	442247	477200	.25
.30	326605	352850	381042	411342	443934	479016	.30
.35	327874	354212	382505	412916	445627	480840	.35
.40	0.329146	0.355579	0.383974	0.414495	0.447326	0.482670	.40
.45	330424	356951	385448	416080	449032	484507	.45
.50	331706	358328	386928	417671	450744	486351	.50
.55	332992	359709	388412	419268	452462	488202	.55
.60	0.334283	0.361096	0.389902	0.420870	0.454187	0.490060	.60
.65	335579	362488	391398	422479	455918	491925	.65
.70	336879	363884	392899	424093	457656	493797	.70
.75	338184	365286	394406	425713	459400	495676	.75
.80	0.339494	0.366693	0.395918	0.427340	0.461150	0.497562	.80
.85	340808	368105	397435	428972	462907	499455	.85
.90	342127	369521	398958	430610	464670	501355	.90
.95	343451	370943	400486	432254	466440	503262	.95

Decimal of Deg	56°	57°	58°	59°	60°	61°	Decimal of Deg
.00	0.505177	0.545027	0.588044	0.634535	0.684853	0.739397	.00
.05	507098	547100	590283	636957	687477	742243	.05
.10	509027	549182	592531	639389	690110	745101	.10
.15	510964	551271	594788	641830	692755	747970	.15
.20	0.512907	0.553368	0.597053	0.644281	0.695410	0.750852	.20
.25	514858	555473	599328	646741	698076	753745	.25
.30	516816	557586	601611	649212	700753	756651	.30
.35	518782	559708	603903	651692	703441	759568	.35
.40	0.520755	0.561837	0.606205	0.654182	0.706139	0.762498	.40
.45	522736	563975	608515	656682	708849	765439	.45
.50	524724	566121	610834	659192	711570	768393	.50
.55	526720	568276	613162	661712	714302	771360	.55
.60	0.528723	0.570438	0.615500	0.664242	0.717045	0.774338	.60
.65	530734	572609	617847	666783	719799	777330	.65
.70	532753	574789	620203	669333	722564	780333	.70
.75	534779	576976	622568	671894	725341	783350	.75
.80	0.536813	0.579173	0.624943	0.674465	0.728129	0.786379	.80
.85	538855	581378	627327	677046	730929	789420	.85
.90	540905	583591	629720	679638	733740	792475	.90
.95	542962	585813	632123	682240	736563	795542	.95

TABLE 3–25

Data* Ordinarily Put on Detail Drawing † of a Spur Gear

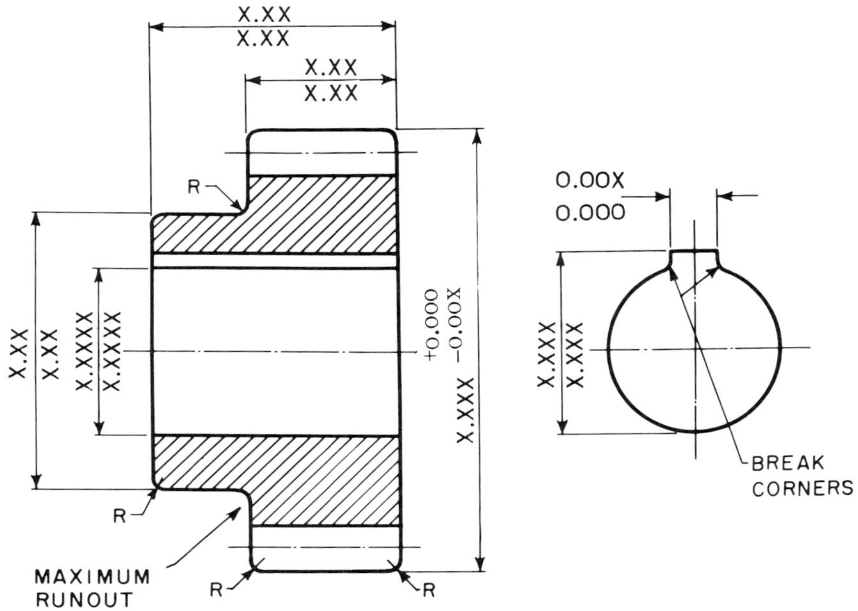

SPUR GEAR DATA	
TEETH (NUMBER)	XX
PITCH (DIAMETRAL)	X
PITCH DIAMETER	X.XXX
PRESSURE ANGLE	XX°XX'
ADDENDUM	0.XXX
‡ CHORDAL ADDENDUM	0.XXX
‡ CHORDAL THICKNESS	0.XXX
WHOLE DEPTH	X.XXX
BACKLASH ALLOWED	0.0XX
HOB OR CUTTER NO.	
MEASURING PIN DIAMETER	0.XXX
‡ DIAMETER OVER PINS	X.XXX

MATERIAL AND HEAT TREATMENT

MATERIAL
HEAT TREATMENT
DRAWING TEMPERATURE
HARDNESS

*Additional information is required on drawings of gears to be finished by shaving, grinding, lapping and burnishing and on drawings of precision and master gears, which are to be inspected for runout, pitch and profile errors.

† The gear in section is usually sufficient for simple gears. Good practice is to put on the drawing those dimensions, with tolerances, needed in making the blank. Specification of the material, heat treatment and the data on cutting the teeth are stated in tabular fashion and in notes. The part number of the mate, as well as that of the gear described, and mounting distance, are among the other information commonly found on gear drawings.

‡ Tooth thickness can be measured and checked more accurately by measuring pins and micrometers than by gear tooth calipers. Either method may be deleted to conform to the practice of the gear-cutting department. Sometimes a rough-cut thickness is desired in addition to the finished thickness. When size is determined by master gear, the class number can be specified instead of the thickness measurement.

TABLE 3-26

Stock Allowances for Various Methods of Finishing Profiles on Spur Gears

Method of Finish	Size Range by Diametral Pitch	Allowance on Circular Thickness	Remarks
Grinding	coarser than 4 4 to 12 12 to 64	0.012 to 0.006 0.008 to 0.003 0.004 to 0.002	Allow no more stock than enough to clean up cutting errors and distortions caused by heat treatment and hardening.
Shaving	4 5 to 7 8 to 20 20 to 40 40 to 64	0.005 0.004 0.003 0.002 0.001	Common practice is to shave gears to reduce errors left by cutting. Shaving is used also to make pitch line tooth thickness greater at mid-face than at edges. Such teeth are said to be crowned, eliptoid, or curve shaped.
Burnishing Lapping		nil nil	Gears are burnished primarily to get surface finish.
Heat treatment	Some gear steels, for example, SAE 2345, have been observed to grow with heat treatment. To make allowance for the effects of such growth on backlash and tooth thickness, determine the increase in pitch radius, Δ C, as the consequence of growth and find the corresponding change in tooth thickness, ΔB, from Table 3–15.		

TABLE 3-27

Tolerance on Outside Diameters of Cylindrical Gear Blanks Practice of Barber-Colman Co.

Range of Blank Diameters, Inches	Maximum Diameter	Tolerance from Theoretical OD: Precision Quality	Standard Quality	General Purpose Quality
Up to 1/2	theoretical	− 0.001	− 0.002	− 0.003
1/2 to 2	diameter	− 0.002	− 0.003	− 0.005
2 to 4	for all	− 0.003	− 0.004	− 0.007
4 to 8	qualities	− 0.005	− 0.007	− 0.010

TABLE 3–28

Tolerance on Outside Diameter for Fine-Pitch Spur Gears

ASA B6.7–1950 AGMA 207.03

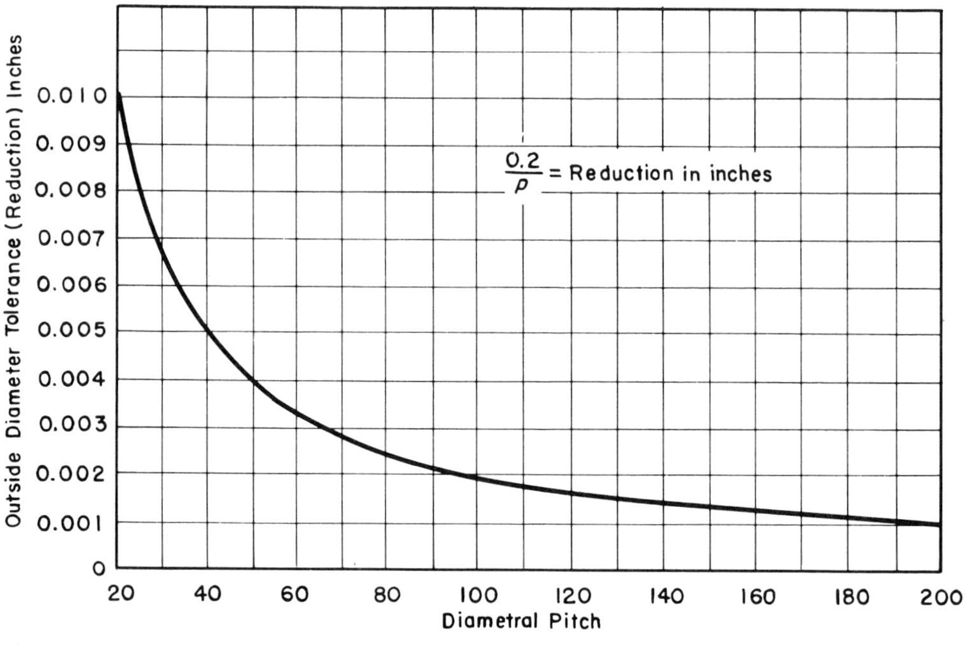

TABLE 3–29

Tolerances on Gear Blank Runout of Outside Diameter With Bore for Fine-Pitch Gears

AGMA 236.03 ASA B6.11–1951

Diametral Pitch	Class	Runout Indicator Reading
20 to 39	Commercial 1,2,3, and 4	0.003
40 to 79	Commercial 1,2,3, and 4	0.002
80 and finer	Commercial 1,2,3, and 4	0.001
20 to 39	Precision 1,2, and 3	0.002
40 to 79	Precision 1,2, and 3	0.0015
80 and finer	Precision 1,2, and 3	0.001

TABLE 3–30

Bore Sizes of Cylindrical Gear Blanks, with Tolerances, and Mating Shaft Diameters, with Fit Allowances, for 3 Qualities of Gears Practice* of Barber-Colman Co.

Nominal Diameter, Inches	Precision Quality Hole Diameter	Precision Quality Shaft Diameter	Standard Quality Hole Diameter	Standard Quality Shaft Diameter	General Purpose Quality Hole Diameter	General Purpose Quality Shaft Diameter
3/8	.3752/.3748	.3746/.3743	.3756/.3748	.3746/.3741	.3756/.3748	.3746/.3738
1/2	.5003/.4998	.4996/.4993	.5006/.4998	.4996/.4991	.5006/.4998	.4996/.4988
5/8	.6253/.6248	.6246/.6243	.6256/.6248	.6246/.6241	.6256/.6248	.6246/.6238
3/4	.7503/.7498	.7496/.7493	.7506/.7498	.7496/.7491	.7506/.7498	.7496/.7488
1	1.0004/.9998	.9995/.9991	1.0008/.9998	.9995/.9989	1.0008/.9998	.9995/.9987
1 1/4	1.2504/1.2498	1.2495/1.2491	1.2508/1.2498	1.2495/1.2489	1.2508/1.2498	1.2495/1.2485
1 1/2	1.5005/1.4998	1.4995/1.4990	1.5012/1.4998	1.4995/1.4988	1.5012/1.4998	1.4995/1.4985
2	2.0006/1.9998	1.9994/1.9989	2.0012/1.9998	1.9994/1.9987	2.0012/1.9998	1.9994/1.9983
2 1/2	2.5006/2.4998	2.4994/2.4989	2.5012/2.4998	2.4994/2.4987	2.5012/2.4998	2.4994/2.4983

*Minimum hole diameter is basic. This permits the use of reamers, arbors, and plug gages for more than a single quality, resulting in a lower investment in tools over shaft size as basic. The positive allowance is the same for all three qualities of a given size, both the tolerance on the hole and that on the shaft tending toward a looser fit.

TABLE 3–31

Tolerances* on Bore and Lateral Runout for Fine-Pitch Gears

ASA B6.11–1951 AGMA 236.03

Class	Tolerance** on Bore Diameter	Tolerance on Lateral Runout
Commercial 1	0.002	0.002 per inch of radius; max 0.004
Commercial 2	0.001	0.0015 per inch of radius; max 0.0025
Commercial 3, and 4	0.0007	0.001 per inch of radius; max 0.002
Precision 1	0.0005	0.0007 per inch of radius; max 0.0015
Precision 2, and 3	0.0002	0.0005 per inch of radius; max 0.001

*The AGMA Standard gives, in addition to the tolerances listed here, also tolerances on bore taper, concavity and convexity of mounting and registering surfaces, and parallelism.

**Tolerance is total deviation from nominal size. Bilateral or unilateral dimensioning may be used according to shop preference.

"Commercial fine-pitch gears can usually be produced on generally available gear-cutting machines and tools in good operating condition without resorting to special technique or subsequent refining operations.

"Precision gears will usually require the best available precision gear-production equipment in good operating condition. Subsequent refining operations such as shaving, grinding, lapping and sometimes special techniques are usually required".

RUNOUT. "Runout of a gear, as measured by methods other than by running two gears in intimate contact, is the total difference between high and low readings of a dial indicator suitably arranged to denote the off-center relation of the axis of the tooth profiles with respect to the gear journals or the axis about which the gear rotates. It is twice the eccentricity. It includes the effect of lateral runout or wobble.

"A gear cutting or grinding machine in proper operating condition should produce teeth without appreciable eccentricity in relation to the axis of rotation of the machine. When eccentricity occurs it is almost always caused by inaccuracies introduced when the gear is removed from the cutting or grinding machine and then mounted for inspection or operation. Runout can be held to close limits only by proper setup of workpiece on machine; this calls for good accurate fits in journals, bores and locating faces of blanks, shafts, arbors, and chucks".

TABLE 3–32

Tolerances* for Spur and Helical Gears ASA B6.6 1946

(All readings in ten-thousandths of an inch.)

Class	Diametral Pitch	Runout of Pitch Diameter, Total Indicator Reading								Pitch Error Measured on Pitch Circle in Plane of Rotation							
Pitch Diameter		¾	1½	3	6	12	25	50	100	¾	1½	3	6	12	25	50	100
Class 1	1	..	..	..	..	70	90	90	100	..	..	..	..	50	75	125	180
(Up to	2	..	..	..	60	60	80	80	90	..	..	..	40	50	70	100	150
80 fpm)	4	..	..	50	60	60	80	80	90	..	..	30	40	50	60	90	120
	8	..	30	50	60	60	80	80	..	..	15	20	25	30	35	35	...
	16	30	30	50	60	60	80	..	..	15	17	20	25	25	..	..	...
Class 2	1	..	..	..	..	30	35	40	45	..	..	..	..	25	35	45	60
(Up to	2	..	..	..	20	20	25	30	35	..	..	..	10	15	20	25	30
400 fpm)	4	..	..	20	20	20	25	30	35	..	..	5	6	7	8	9	10
	8	..	15	15	20	20	25	30	..	..	5	5½	6	6½	7	7	..
	16	15	15	15	20	20	25	30	..	3½	3½	4	4	4½	5	5½	..
Class 3	2	..	..	..	20	20	25	25	30	..	..	..	5	5	5½	6	6
(Up to	4	..	..	10	20	20	25	25	30	..	..	3½	3½	4	4	5	5
2000 fpm)	8	..	10	10	15	20	25	25	30	..	3½	3½	3½	4	4	5	5
	16	10	10	10	15	20	25	25	..	2½	2½	3	3	3½	3½	4	..
	32	10	10	10	15	20	25	..	..	2	2	2½	2½	3	3	..	..
Class 4	4	..	..	..	10	10	12	14	..	..	..	..	3½	4	4	4	..
(Over	8	..	..	10	10	10	12	14	16	..	..	2½	2½	3	3	3	3
2000 fpm)	16	..	10	10	10	10	12	14	16	..	2	2½	2½	2½	3	3	3
	32	10	10	10	10	10	12	..	..	2	2	2	2	2½	2½	..	..

*The ASA Standard contains tolerances on accumulated error, profile error, and lead error as well as the tolerances on runout and pitch error quoted here. Runout can be verified without special test equipment, especially if the design specifies proof surfaces against which the dial indicator is to be placed. Furthermore, runout is influenced greatly by the fits between bores and arbors and shafts, the squareness of clamping surfaces and the care with which the work piece is set in the machine and mounted afterward; the other errors, for the most part, depend upon the accuracy and rigidity of the machine tools in which the teeth are cut or finished.

TABLE 3-33

AGMA Standard Rating of 20-Deg Involute Spur Gears for Surface Durability and Beam Strength

The rating of a spur gear is the lower of the two ratings. NOTE: The first step in applying the formulas to rating determinations is to select materials in accordance with Tables 3–35 and 3–36. For more information on AGMA Specifications of Materials see AGMA 241.04–1954, *Gear Materials-Steel;* AGMA 242.02–1946, *Cast Iron Gear Blanks;* and AGMA 243.01–1954, *Cast Bronze Gear Blanks.*

AGMA Standard	Formula	In Which
210.01 1946 Surface Durability	EXTERNAL GEARS Horsepower rating, P $P = F_i K_r D_o C_f$ INTERNAL GEAR For internal gear, replace K_r with K_{ri} from Table 3–36. $D_o = \dfrac{d^2 (1 - \sqrt{V}/84) \text{ (rpm of pinion)}}{158{,}000}$	F_i = combined factor for face width and inbuilt factor (Table 3–34) K_r = combined factor for materials, tooth form and ratio (Table 3–35) D_o = combined factor for pinion diameter, speed factor and rpm d = pitch diameter of pinion, inches V = pitch line speed, fpm C_f = factor to correct for increased stress at start of single tooth contact (Table 3–37 or 3–38)
220.01 1946 Strength	Normal or continuous horsepower rating P, from 8 to 10 hours per day under uniform load, $P = \dfrac{0.5\, F_i\, S\, Y_k\, D_g}{P_d}$ AGMA Standard 220.01 recommends that Y_k be determined from a layout, and describes the method therefor. It also states that Y_k may be taken as 1/4 in preliminary or rough design calculations. Table 3–40 is a better approximation than is Y_k = 1/4; and not as reliable as the layout method recommended in the standard. $D_g = \dfrac{d (1 - \sqrt{V}/84) \text{ (rpm of pinion)}}{126{,}000}$	F_i = combined factor for face width and inbuilt factor (Table 3–34) S = allowable bending stress, psi (Table 3–39) P_d = diametral pitch Y_k = tooth form factor (Table 3–40). D_g = combined factor for pinion diameter, speed factor and rpm d = pitch diameter of pinion, inches V = pitch line speed, fpm

TABLE 3–34

Factor F_i for Use in Calculating Horsepower Ratings of Spur, Helical, and Herringbone Gears

AGMA 210.01, 211.01, 220.01

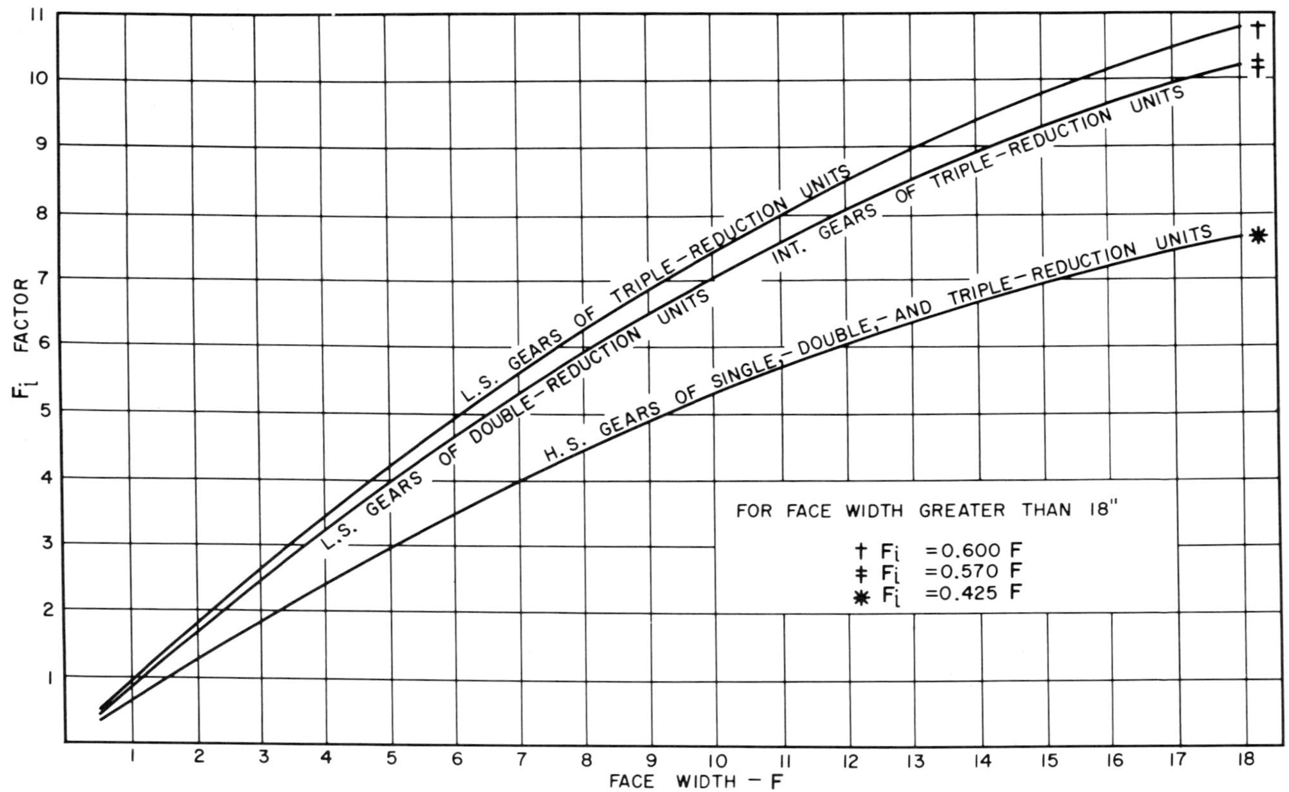

NOTATIONS ON CURVES APPLY TO ENCLOSED DRIVES. FOR OPEN GEARING USE LOWEST CURVE FOR ALL OVERHUNG PINIONS, FOR IMPERFECT LUBRICATION, AND FOR FIRST REDUCTIONS. USE CENTER AND TOP CURVES ONLY FOR SECOND AND THIRD REDUCTIONS RESPECTIVELY WHERE MOUNTING IS SUBSTANTIAL AND WHERE GEARS ARE ADEQUATELY LUBRICATED BY CLEAN OIL.

TABLE 3-35

Factor K_r for Use in Calculating Horsepower Ratings of External Spur, Helical and Herringbone Gears

AGMA 210.01, 211.01, 220.01

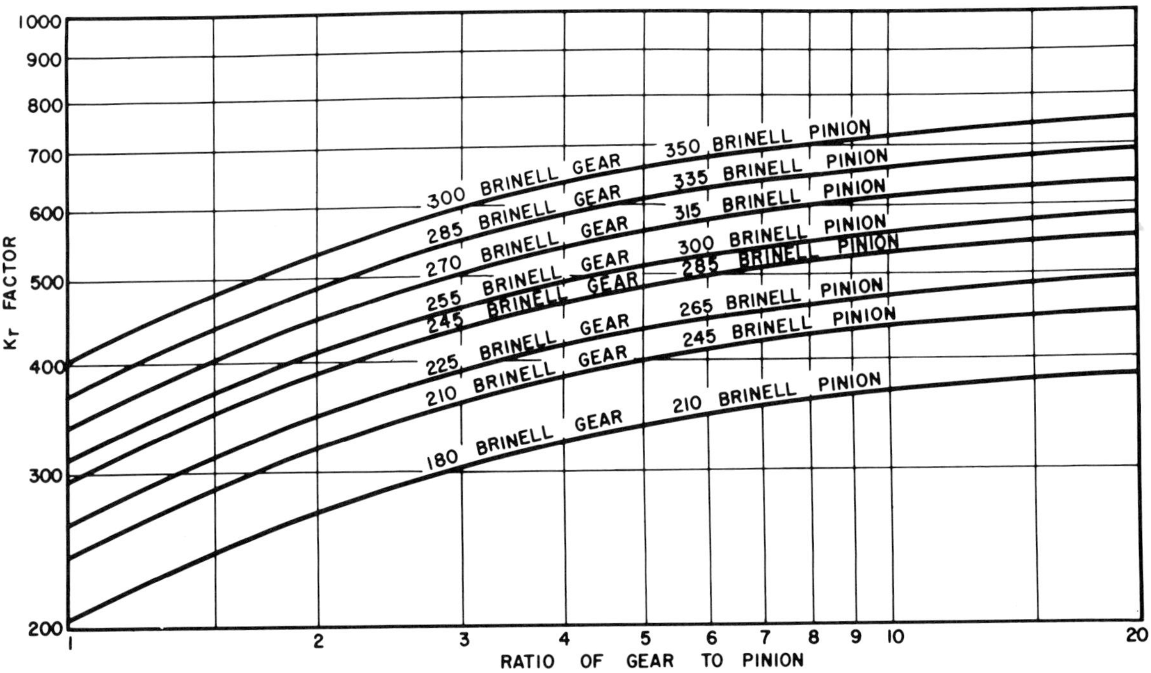

	CASE-HARDENED AND THROUGH-HARDENED STEEL						SURFACE HARDENED	CAST IRON	BRONZE
BRINELL GEAR	575	500	350	335	315	225	440	200	40,000 PSI
BRINELL PINION	575	500	450	380	360	450	440	210	180
K_r	1530 C_r 0.9	1350 C_r 0.9	1060 C_r 0.9	973 C_r 0.95	870 C_r 0.95	608 C_r 0.95	890 C_r 0.9	344 C_r 1.0	274 C_r 1.0

$$\text{GEAR RATIO} = \frac{\text{TEETH IN GEAR}}{\text{TEETH IN PINION}} \qquad C_r = \frac{\text{GEAR RATIO}}{\text{GEAR RATIO} + 1}$$

FOR MATERIAL SELECTIONS DIFFERING FROM THESE STANDARDIZED HARDNESS COMBINATIONS, AS A 245 BRINELL PINION 225 BRINELL GEAR, THE CURVE CORRESPONDING TO THE 210 BRINELL GEAR 245 BRINELL PINION WOULD BE USED.

FOR CASE-HARDENED AND THROUGH-HARDENED GEARS AND PINIONS, THE DEGREE OF DISTORTION FROM HEAT WILL VARY, BUT THE TECHNIQUE AND PROPORTIONS PERMIT EVALUATING CONTACT AT 0.95 WHEN ONE ELEMENT IS HARDENED AFTER CUTTING, AND 0.90 WHEN BOTH ELEMENTS ARE HARDENED AFTER CUTTING. USE A DISTORTION FACTOR OF UNITY FOR GROUND GEARS, WHICH ARE REGARDED AS CUT AFTER HARDENING.

TABLE 3–36

Factor K_{ri} For Use in Calculating Horsepower Ratings of Internal Spur and Helical Gears

AGMA 210.01, 211.01, 220.01

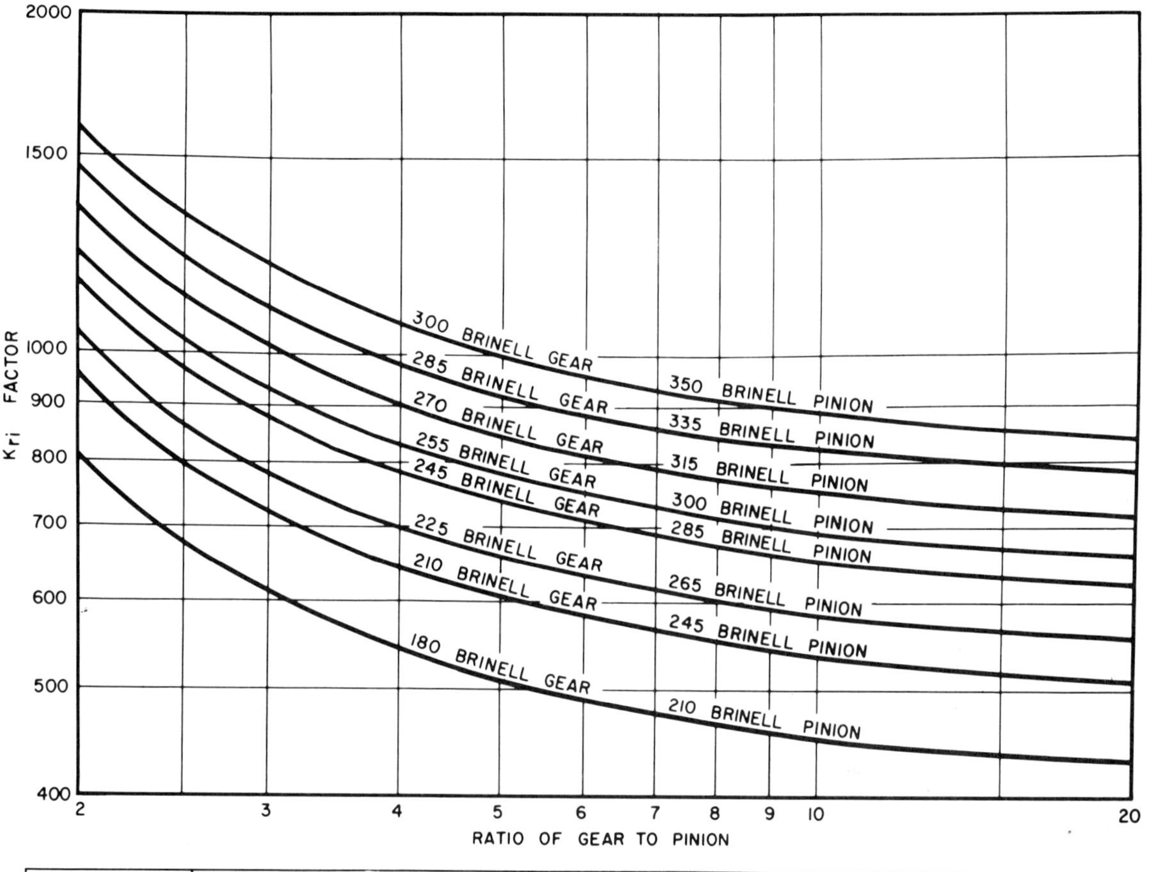

	CASE-HARDENED AND THROUGH-HARDENED STEEL						SURFACE HARDENED	CAST IRON	BRONZE
BRINELL GEAR	575	500	350	335	315	225	440	200	40,000 PSI
BRINELL PINION	575	500	450	380	360	450	440	210	180
Kri	1530 Cri 0.9	1350 Cri 0.9	1060 Cri 0.9	973 Cri 0.95	870 Cri 0.95	608 Cri 0.95	890 Cri 0.9	344 Cri 1.0	274 Cri 1.0

$$\text{GEAR RATIO} = \frac{\text{TEETH IN GEAR}}{\text{TEETH IN PINION}} \qquad C_{ri} = \frac{\text{GEAR RATIO}}{\text{GEAR RATIO} - 1}$$

FOR CASE-HARDENED AND THROUGH-HARDENED GEARS AND PINIONS, THE DEGREE OF DISTORTION FROM HEAT WILL VARY, BUT THE TECHNIQUE AND PROPORTIONS PERMIT EVALUATING CONTACT AT 0.95 WHEN ONE ELEMENT IS HARDENED AFTER CUTTING, AND 0.90 WHEN BOTH ELEMENTS ARE HARDENED AFTER CUTTING. USE A DISTORTION FACTOR OF UNITY FOR GROUND GEARS, WHICH ARE REGARDED AS CUT AFTER HARDENING.

FOR MATERIAL SELECTIONS DIFFERING FROM THESE STANDARDIZED HARDNESS COMBINATIONS, AS A 245 BRINELL PINION 225 BRINELL GEAR, THE CURVE CORRESPONDING TO THE 210 BRINELL GEAR 245 BRINELL PINION WOULD BE USED.

TABLE 3-37

Factor C_f For Use in Calculating Durability Rating of 20-Degree Stub Tooth Spur Gears

AGMA 210.01 1946

Teeth in Pinion	Standard Addendum Teeth Gear Ratio 1:1	1.5:1	2:1	3:1	5:1	10:1	Long Addendum Teeth Gear Ratio 3:1	10:1
12	0.659	0.579	0.535	0.505	0.469	0.437	0.810	0.734
13	0.724	0.637	0.598	0.557	0.525	0.493	0.841	0.784
14	0.770	0.685	0.642	0.603	0.573	0.543	0.869	0.821
15	0.806	0.720	0.685	0.641	0.610	0.580	0.889	0.850
16	0.832	0.750	0.718	0.673	0.645	0.616	0.904	0.872
17	0.853	0.777	0.743	0.700	0.673	0.646	0.918	0.891
18	0.872	0.798	0.765	0.724	0.698	0.670	0.930	0.907
19	0.888	0.815	0.783	0.745	0.720	0.693	0.937	0.920
20	0.900	0.830	0.800	0.762	0.740	0.713	0.944	0.930
21	0.912	0.843	0.815	0.779	0.755	0.732	0.950	0.939
22	0.923	0.856	0.827	0.793	0.770	0.747	0.957	0.945
23	0.931	0.867	0.838	0.804	0.783	0.760	0.962	0.951
24	0.938	0.876	0.848	0.816	0.794	0.773	0.967	0.958
25	0.944	0.884	0.857	0.827	0.805	0.785	0.971	0.963
27	0.953	0.900	0.872	0.843	0.823	0.804	0.979	0.972
29	0.960	0.910	0.884	0.859	0.840	0.820	0.984	0.980
31	0.966	0.920	0.893	0.870	0.851	0.833	0.989	0.985
33	0.971	0.927	0.902	0.881	0.863	0.846	0.994	0.992
35	0.974	0.933	0.909	0.890	0.872	0.856	0.998	0.996
38	0.979	0.941	0.917	0.900	0.884	0.870	1.003	1.002
41	0.982	0.948	0.924	0.909	0.893	0.880	1.006	1.006
44	0.985	0.954	0.931	0.916	0.902	0.889		
47	0.988	0.959	0.936	0.923	0.909	0.897		
50	0.990	0.963	0.941	0.928	0.915	0.904		
55	0.992	0.968	0.947	0.934	0.923	0.913		
60	0.994	0.972	0.952	0.941	0.930	0.921		
65	0.996	0.975	0.955	0.946	0.936	0.927		

Standard addendum teeth

$$\text{Addendum} = \frac{0.8}{P}$$

where P = diametral pitch

Long addendum teeth:

$$\text{Pinion addendum} = \frac{1.2}{P}$$

$$\text{Gear addendum} = \frac{0.4}{P}$$

TABLE 3–38

Factor C_f For Use in Calculating Durability Rating of 20-Deg Full-Depth Spur Gears

AGMA 210.01 1946

Teeth in Pinion	Standard Addendum Teeth Gear Ratio						Long Addendum Teeth Gear Ratio	
	1:1	1.5:1	2:1	3:1	5:1	10:1	3:1	10:1
12							0.990	0.990
13							1.007	1.010
14							1.020	1.028
15							1.033	1.040
16	0.930	0.860	0.830	0.800	0.770	0.750	1.042	1.051
17	0.940	0.874	0.848	0.820	0.790	0.770	1.049	1.060
18	0.950	0.889	0.863	0.838	0.810	0.790	1.054	1.066
19	0.957	0.900	0.875	0.852	0.828	0.810	1.058	1.072
20	0.963	0.910	0.888	0.864	0.842	0.823	1.062	1.076
21	0.969	0.920	0.898	0.877	0.853	0.839	1.064	1.079
22	0.973	0.928	0.906	0.887	0.865	0.849	1.066	1.081
23	0.977	0.934	0.913	0.895	0.874	0.859	1.066	1.083
24	0.980	0.940	0.920	0.902	0.883	0.868	1.068	1.085
25	0.983	0.945	0.925	0.908	0.892	0.877	1.069	1.086
27	0.986	0.952	0.933	0.918	0.904	0.890	1.070	1.088
29	0.989	0.958	0.940	0.927	0.913	0.900	1.070	1.089
31	0.991	0.963	0.946	0.933	0.922	0.909	1.069	1.090
33	0.993	0.967	0.952	0.940	0.930	0.917	1.068	1.090
35	0.994	0.970	0.957	0.945	0.935	0.924	1.067	1.090
38	0.996	0.974	0.963	0.952	0.943	0.933	1.065	1.090
41	0.997	0.978	0.967	0.957	0.949	0.940	1.063	1.089
44	0.998	0.981	0.971	0.962	0.954	0.946		
47	0.999	0.983	0.974	0.966	0.959	0.952		
50	0.999	0.985	0.977	0.969	0.963	0.956		
55	1.000	0.987	0.980	0.972	0.968	0.962		
60	1.000	0.989	0.982	0.975	0.972	0.966		
65	1.000	0.990	0.984	0.978	0.974	0.970		

Standard addendum teeth

Addendum $= \frac{1}{P}$

Where $P =$ diametral pitch

Long addendum teeth:

Pinion addendum $= \frac{1.5}{P}$

Gear addendum $= \frac{0.5}{P}$

TABLE 3–39

Allowable Bending Stress in Steels for Spur Gears

AGMA Standard 220.01 1946

Brinell Hardness No. of Steel	Allowable Stress, psi	Brinell Hardness No. of Steel	Allowable Stress, psi
160	40,000	315	70,000
210	50,000	335	77,000
245	60,000	360	83,000
270	65,000	440 and higher	90,000

TABLE 3–40

Form Factor Y_k for use in Calculating Strength Rating of 20-Deg Spur Gear Teeth

NOTE: Tip relief and root fillets, as well as numbers of teeth, pressure angle and addendums, influence the beam strength of the teeth on a gear pair. To include the proper weight on tooth form factor Y_k, of all these variables, the layout method of AGMA Standard 220.01 should be adhered to.

Number Teeth on Pinion	20° Stub, Table 3-5, With Hob Edge Radius = 0.304/P				20° Full-Depth, Table 3-6, With Hob Edge Radius = 0.24/P			
	1/1 Ratio	Gear of 50 Teeth	Gear of 100 Teeth	Gear of 150 Teeth	1/1 Ratio	Gear of 50 Teeth	Gear of 100 Teeth	Gear of 150 Teeth
17	0.344	0.367	0.374	0.388	0.330	0.351	0.363	0.367
18	0.351	0.371	0.380	0.393	0.336	0.356	0.368	0.372
19	0.358	0.375	0.385	0.398	0.341	0.361	0.372	0.376
20	0.364	0.378	0.390	0.402	0.346	0.366	0.376	0.380
21	0.369	0.382	0.394	0.406	0.351	0.370	0.380	0.384
22	0.374	0.386	0.398	0.410	0.356	0.374	0.384	0.388
23	0.379	0.389	0.402	0.414	0.361	0.378	0.388	0.392
24	0.383	0.392	0.406	0.417	0.365	0.382	0.391	0.396
25	0.387	0.395	0.410	0.420	0.369	0.386	0.394	0.400
26	0.391	0.398	0.413	0.423	0.374	0.390	0.398	0.404
27	0.395	0.401	0.416	0.426	0.378	0.394	0.401	0.408
28	0.398	0.404	0.419	0.429	0.382	0.398	0.404	0.411

continued on next page

Table 3–40, continued

Number Teeth on Pinion	20° Stub. Table 3–5, With Hob Edge Radius 0.304/*P*				20° Full-Depth, Table 3–6 With Hob Edge Radius 0.24/*P*			
	1/1 Ratio	Gear of 50 Teeth	Gear of 100 Teeth	Gear of 150 Teeth	1/1 Ratio	Gear of 50 Teeth	Gear of 100 Teeth	Gear of 150 Teeth
29	0.402	0.407	0.422	0.431	0.386	0.401	0.407	0.414
30	0.405	0.409	0.425	0.433	0.390	0.404	0.410	0.417
31	0.408	0.412	0.427	0.436	0.394	0.407	0.413	0.420
32	0.411	0.414	0.429	0.438	0.398	0.410	0.416	0.423
33	0.411	0.416	0.431	0.440	0.402	0.413	0.419	0.426
34	0.415	0.418	0.433	0.442	0.406	0.416	0.422	0.429
35	0.417	0.420	0.435	0.444	0.410	0.419	0.424	0.431
36	0.419	0.422	0.437	0.446	0.413	0.421	0.427	0.434
38	0.422	0.426	0.440	0.448	0.419	0.426	0.432	0.439
40	0.425	0.429	0.442	0.451	0.425	0.431	0.437	0.444
42	0.428	0.431	0.444	0.454	0.428	0.435	0.441	0.448
44	0.431	0.433	0.446	0.456	0.435	0.439	0.445	0.452
46	0.434	0.435	0.449	0.458	0.440	0.442	0.449	0.455
48	0.437	0.437	0.451	0.460	0.444	0.445	0.453	0.459
50	0.439	0.439	0.453	0.462	0.448	0.448	0.456	0.463
52	0.441		0.455	0.464	0.451		0.458	0.466
56	0.446		0.459	0.467	0.457		0.463	0.471
60	0.451		0.462	0.470	0.463		0.468	0.476
64	0.455		0.465	0.472	0.468		0.472	0.481
68	0.459		0.468	0.475	0.473		0.476	0.485
72	0.463		0.471	0.477	0.477		0.480	0.489
76	0.465		0.473	0.480	0.480		0.484	0.493
80	0.470		0.475	0.482	0.484		0.487	0.497
90	0.478		0.480	0.486	0.493		0.495	0.505
100	0.485		0.485	0.490	0.501		0.501	0.512
125	0.498			0.499	0.520			0.525
150	0.506			0.506	0.536			0.536

INDEX TO SECTION 4

Helical and Herringbone Gears

TABLE 4–1

Data* Ordinarily Put on Detail Drawing** of a Small Helical Gear

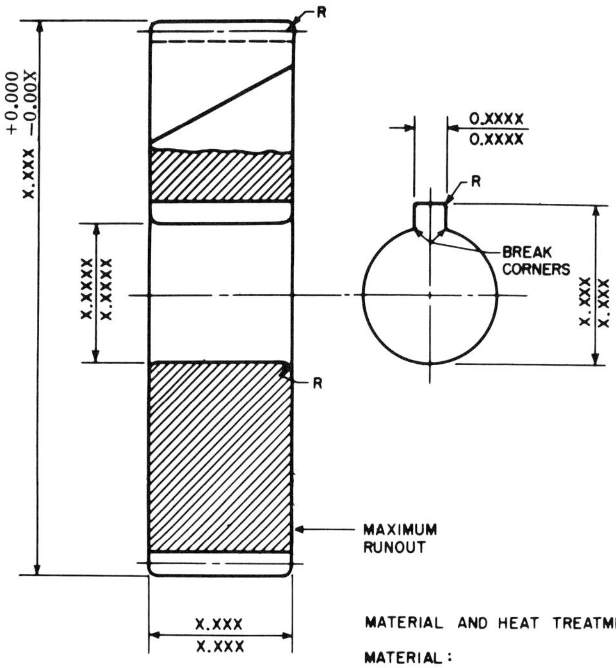

HELICAL GEAR DATA	
TEETH (NUMBER)	XX
NORMAL DIAMETRAL PITCH	X.XXX
PITCH DIAMETER	X.XXXX
NORMAL PRESSURE ANGLE	XX°XX'
HELIX ANGLE	XX°XX'XX"
LEAD	X.XXXX
HELIX HAND	RIGHT OR LEFT
ADDENDUM	0.XXX
WHOLE DEPTH	0.XXX
BACKLASH	0.0XX
HOB OR CUTTER NO.	
MEASURING PIN OR BALL DIAMETER	0.XXX
DIAMETER OVER PINS	X.XXX

MATERIAL AND HEAT TREATMENT

MATERIAL:
HEAT TREATMENT
DRAWING TEMPERATURE
HARDNESS

* METHODS OF FINISHING, GAGING, AND INSPECTION OF GEARS, AS WELL AS THE REQUIREMENTS ON QUALITY, VARY SO WIDELY THAT NO ATTEMPT IS MADE HERE TO INCLUDE ALL THE INFORMATION THE DESIGNER MUST PUT ON EVERY DRAWING. INSTEAD, THESE DATA ARE OFFERED AS A PATTERN, WHICH CAN BE ALTERED AND EXPANDED TO SUIT THE REQUIREMENTS OF DIFFERENT COMPANIES

** A SECTION THROUGH THE GEAR IS OFTEN ENOUGH. GOOD PRACTICE IS TO PUT ON THE DRAWING THOSE DIMENSIONS, WITH TOLERANCES, NEEDED IN MAKING THE BLANK. SPECIFICATION OF MATERIAL, HEAT TREATMENT, AND DATA ON CUTTING THE TEETH ARE STATED IN TABULAR FASHION AND IN NOTES

TABLE 4–2

Data Ordinarily Put on Detail Drawing of Fine Pitch Helical Gear

(See "Fine-Pitch Gear Drawing Data Required to Assure Quality," *Product Engineering*, July 1952)

ASA B6.11–1951 AGMA 236.03

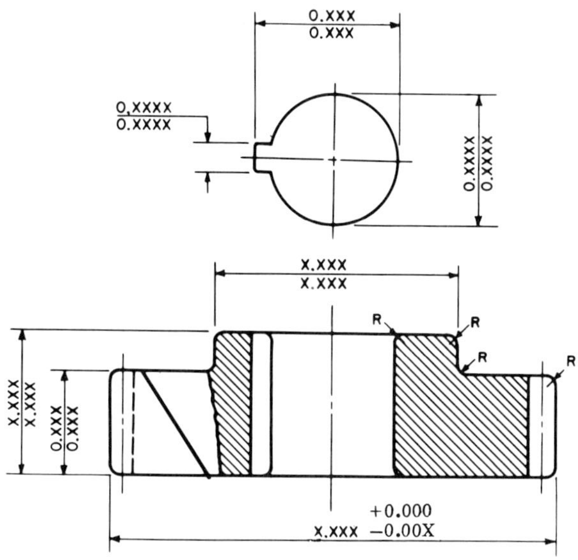

MATERIAL AND HEAT TREATMENT

FINE-PITCH HELICAL GEAR DATA	
TEETH (NUMBER)	XX
NORMAL DIAMETRAL PITCH	XX
TRANSVERSE PITCH	XX.XXX
NORMAL PRESSURE ANGLE	XX°XX'
HELIX ANGLE	XX°XX'
STANDARD PITCH DIAMETER	X.XXX
OPERATING PITCH DIAMETER	X.XXX
TESTING DIAMETER	X.XXX
CLASS (SEE TABLE 3-12)	
CENTER DISTANCE WITH MATING GEAR	X.XXX

INSPECTION DATA	
MAXIMUM ALLOWABLE PROFILE ERROR	0.00XX
MAXIMUM ALLOWABLE PITCH ERROR	0.00XX
MAXIMUM ALLOWABLE RUNOUT	0.00XX

Fine-pitch gears are used extensively in instruments and devices where precision is often paramount. The designer of high-quality gears can help himself not only by specifying clearly the requirements to be met but also by choosing blank proportions that contribute toward them. On the matter of blank accuracy ASA Standard B6.11–1951 has this to say:

"The accuracy to which gears can be produced is largely affected by the design of the gear blank. Unless the gear blank has the required accuracy and rigidity it penalizes the production of good gears. When a gear is designed with a hole, the hole should be large enough to support adequately the blank during the machining of the gear teeth and yet not so large as to cause distortion. Face widths should be wide enough,,in proportion to the outside diameters, to avoid springing, and to permit obtaining flatness in important surfaces. Very short bore lengths should be avoided wherever possible. It is possible, however, to machine relatively thin blanks in stacks provided the surfaces are flat and parallel with each other. When this technique is employed the whole stack of blanks is considered as a single blank rather than individual components. Quite frequently gear blanks must be designed with hubs. Attention should be given to wall sections of the hubs. Too thin a section will not permit proper clamping of the blank during the machining operation and may also affect proper mounting of the gear. Another fairly common practice is that of producing pinions, and on occasion, gears, integral with their shafts. This practice lends itself to increased accuracy if the proper precautions are taken. Among these precautions is that of avoiding deflection of the shaft by having the shaft length and shaft diameter well proportioned to the gear or pinion diameter.

"The following are the most common faults in gear-blank designs, which have proved troublesome:

(a) very small bores compared to the outside diameter
(b) very large bores compared to the outside diameter
(c) narrow face widths with large outside diameter
(d) short bore length
(e) thin hub walls
(f) gears integral with shafts that lack stiffness."

TABLE 4–3

Typical Detail Drawing of Medium-Sized Helical or Herringbone Pinion
Farrel-Birmingham Co.

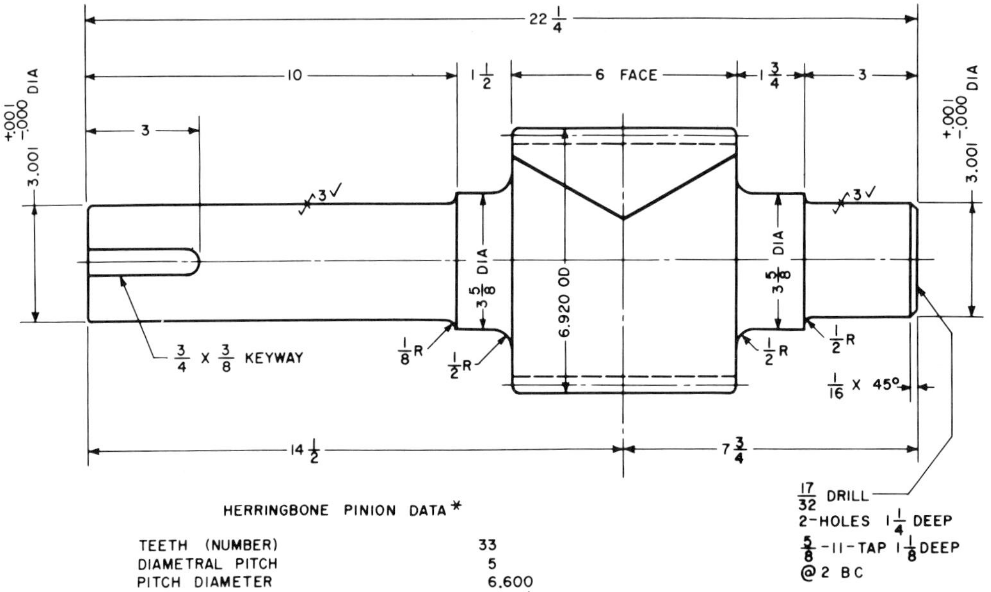

HERRINGBONE PINION DATA*

TEETH (NUMBER)	33
DIAMETRAL PITCH	5
PITCH DIAMETER	6.600
TRANSVERSE PRESSURE ANGLE	20°00'
HELIX ANGLE	30°00'
ADDENDUM	0.240
WHOLE DEPTH	0.320
BACKLASH	0.006
CUTTERS NO.	
PART NO. OF MATE	
NO. TEETH ON MATE	120
OPERATING CENTERS	

MATERIAL:
HARDNESS: 245/285 BRINELL

*Regular practice in gear drawings is to place data pertinent to the teeth in tabular form. The material and heat treatment are generally covered by notes. Blank dimensions, tolerances, finish, and other dimensions necessary to describe fully the length, size and shape of the part are placed upon the drawing proper.

TABLE 4–4

Detail Drawing of Large Helical or Herringbone Gear

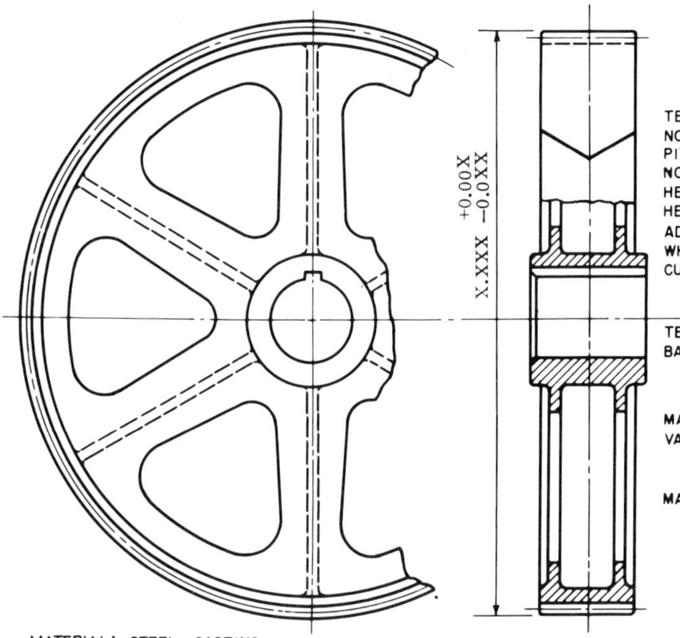

HERRINGBONE GEAR DATA

TEETH (NUMBER)	XXX
NORMAL DIAMETRAL PITCH	X.XXX
PITCH DIAMETER	XX.XXXX
NORMAL PRESSURE ANGLE	XX°XX'
HELIX ANGLE	XX°XX'
HELIX HANDS	RIGHT TO LEFT
ADDENDUM	0.XXX
WHOLE DEPTH	X.XXX
CUTTER OR HOB NO.	
TEETH ON MATE	XX
BACKLASH	0.0XX
MAXIMUM RUNOUT	0.00XX
VARIATION FROM TRUE INVOLUTE, OD TO PITCH DIAMETER	0.00XX–0.00XX
MAXIMUM SPACING ERROR	0.00XX

MATERIAL: STEEL CASTING
ANNEALED AT °F

To describe fully a large gear generally requires more than one view in the specification of the many details of construction. Although none is described here, welded constructions are becoming more common for large gear blanks. Welded construction has the advantage that the rim material can be a high-quality gear steel while the frame and hub need be steel of only structural quality. Tables 4–18 to 4–28, inclusive, can be used to find drawing dimensions and proportions when the gear blanks are castings.

TABLE 4-5

Hands of Helix on Helical and Herringbone Gears

ASA B6.10-1950 AGMA 112.03

Illustration	Item No. in Standard	Definition
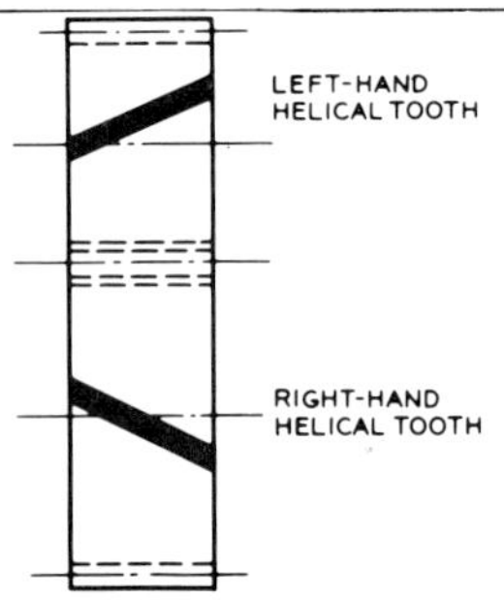 PARALLEL HELICAL GEARS	2.04	Parallel helical gears operate on parallel axes and the helices are of opposite hand.
	2.03	A helical gear is cylindrical in form and has helical teeth.
	2.06	Single-helical gears have teeth of only one hand on each gear.
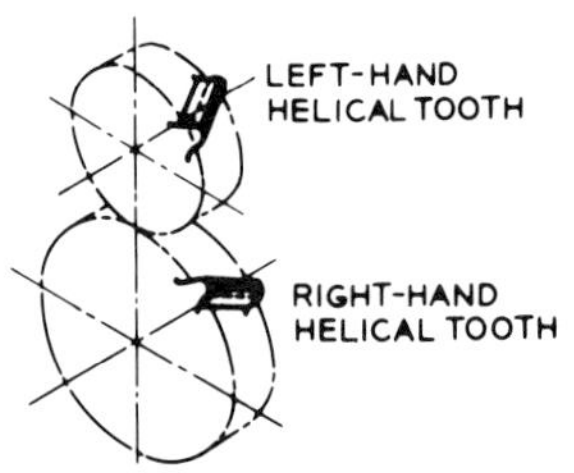 PARALLEL HELICAL GEARS - 2.04	2.25	A left-hand helical gear is one in which the teeth twist counterclockwise as they recede from an observer looking along the axis.
	2.24	A right-hand helical gear is one in which the teeth twist clockwise as they recede from an observer looking along the axis.
	2.25	Two external helical gears operating on parallel axes must be of opposite hand. An internal helical gear and its pinion must be of the same hand.
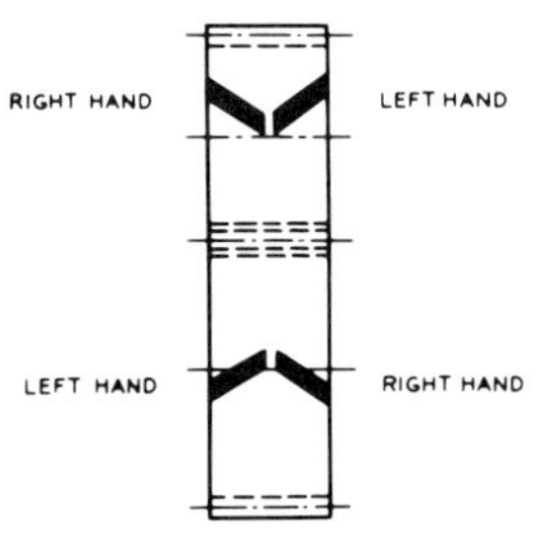	2.07	Double-Helical (Herringbone) Gears each have both right-hand and left-hand helical teeth, and operate on parallel axes.
	Comment	To mesh, the hands on mating (external) herringbone gears must be reversed.
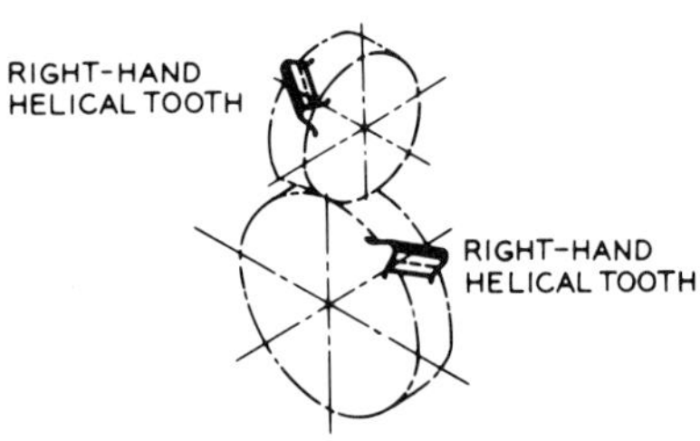 CROSSED HELICAL GEARS - 2.05	2.05	Crossed helical gears operate on crossed axes and may have teeth of the same or of opposite hand.
	Comment	Selection of the hands on crossed helical gears is influenced by the shaft angle, Σ. If Σ is large, or a right angle, as it frequently is, then the hands are made the same; the sum of the helix angles equals the shaft angle. If Σ is small, or the shafts are nearly parallel, opposite hands are chosen; the difference of the helix angles equals the shaft angle.

TABLE 4–6

Definitions of Angular Dimensions for Helical Involute Gears
ASA B6.10 – 1950 AGMA 112.03

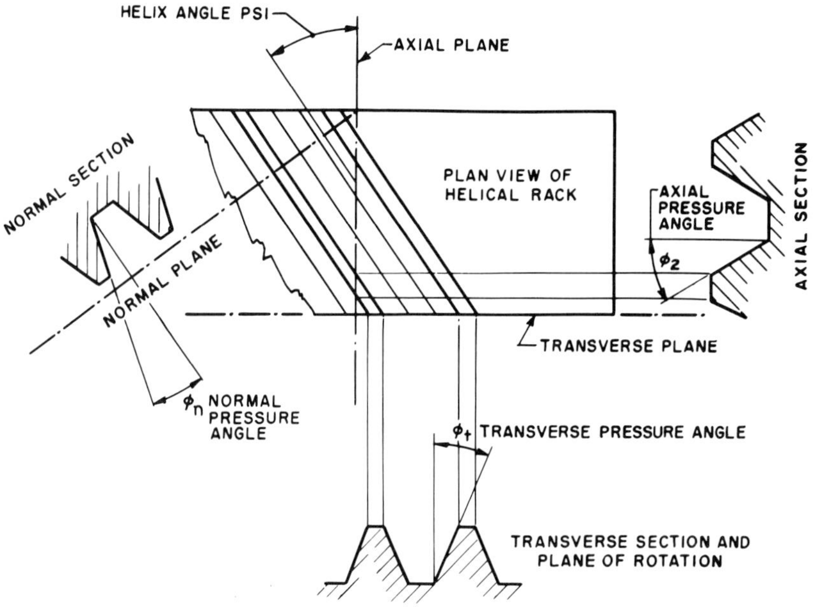

Symbol	Name	Definition
ϕ_t	Transverse pressure angle	The angle at the pitch point between the tooth profile and a radial line in a transverse plane
ϕ_n	Normal pressure angle	The angle at the pitch point between the tooth profile and a radial line in a normal plane
ϕ_x	Axial pressure angle	The angle at the pitch point between the tooth profile and a radial line in an axial plane
ψ_p	Pitch helix angle	The angle on the pitch cylinder between the helix and an element of the cylinder
ψ_b	Base helix angle	The angle on the base cylinder between the helix and an element of the cylinder
λ_p	Pitch lead angle (the complement of the pitch helix angle)	The angle on the pitch cylinder between the helix and a plane of rotation
λ_b	Base lead angle (the complement of the base helix angle	The angle on the base cylinder between the helix and a plane of rotation

Definitions of pitches and the relations of one pitch to another are given in Table 3–2.

TABLE 4–7

Formulas Expressing One Angular Dimension in Terms of Two Others — Helical Gears

Allan H. Candee Gear Geometry American Machinist, July 4 and 11, 1929

	ϕ_t Transverse pressure angle	ϕ_n Normal pressure angle	ϕ_x Axial pressure angle
ϕ_t		$\sin\phi_t = \dfrac{\sin\phi_n}{\cos\psi_b}$ $\tan\phi_t = \dfrac{\tan\phi_n}{\cos\psi_p}$	$\tan\phi_t = \tan\phi_x \tan\psi_p$
ϕ_n	$\sin\phi_n = \sin\phi_t \cos\psi_b$ $\tan\phi_n = \tan\phi_t \cos\psi_p$		$\tan\phi_n = \tan\phi_x \sin\psi_p$
ϕ_x	$\tan\phi_x = \dfrac{\tan\phi_t}{\tan\psi_p}$	$\tan\phi_x = \dfrac{\tan\phi_n}{\sin\psi_p}$	

	ψ_p Pitch helix angle	ψ_b Base helix angle	λ_b Base lead angle	λ_p Pitch lead angle	on Worms
ψ_p		$\sin\psi_p = \dfrac{\sin\psi_b}{\cos\phi_n}$ $\tan\psi_p = \dfrac{\tan\psi_b}{\cos\phi_t}$	$\cos\lambda_p = \dfrac{\cos\lambda_b}{\cos\phi_n}$ $\tan\lambda_p = \dfrac{\cos\phi_t}{\tan\lambda_b}$		λ_p
ψ_b	$\sin\psi_b = \sin\psi_p \cos\phi_n$ $\tan\psi_b = \tan\psi_p \cos\phi_t$			$\cos\lambda_b = \cos\lambda_p \cos\phi_n$ $\tan\lambda_b = \dfrac{\tan\lambda_p}{\cos\phi_t}$	λ_b

Symbols are defined in Table 4–6

TABLE 4–8

Standard Tooth Proportions for Involute Helical Gears

See Tables 3–3, 3–4, 3–5, 3–6, 3–7 and 3–8.

In theory, and quite generally in practice, the same generating tools (hobs or cutters) are used for helical as for spur gears, including the tools for generating fine-pitch pinions on the enlarged center distance system.

The envelope generated by a hob, or by a rack-shaped planing tool, is the normal basic rack. The helix angle is flexible and is obtained merely by tilting the hob in the machine tool. Tooth proportions and calculations are quite generally based on normal diametral pitch Table 4-10.

On the other hand, generation of a helical gear with a pinion-shaped cutter is analogous to mating two helical gears together; the lead on the generated gear is proportional to the lead on the cutter in the ratio of the numbers of teeth on gear and cutter. This relationship between cutter and lead accounts for the few available helix angles, which have been more or less standardized, on the machine tools for generating helical gears by this process. The two commonly available leads on Fellows machines are 25.904 and 41.270 inches Table 4-17. Design calculations can be based on diametral pitch just as for spur gears. See Table 4-11.

TABLE 4-10

Formulas for Calculating Dimensions of Parallel Helical Gears Using Normal Diametral Pitch (Hobbed or Equivalent Method of Generating)

To Find	From	Formula
Pinion pitch diameter, d	Number of teeth, N_P, the normal diametral pitch, P_n and the pitch helix angle, ψ	$d = \dfrac{N_P}{P_n \cos\psi}$
Gear pitch diameter, D	For parallel helical gears pitch **helices are equal but of opposite hands** on gear and pinion, Table 4–5	$D = \dfrac{N_G}{P_n \cos\psi}$
Center distance, C	**For external gears**	$C = \dfrac{d + D}{2}$
Addendum, $a_P = a_G$ or take from Tables 3-5 or 3-6	Normal diametral pitch, P_n and basic tooth proportions	$a_P = \dfrac{\text{constant, Table 3-4}}{P_n}$
Dedendum, $b_P = b_G$ Tables 3-5 or 3-6	Normal diametral pitch, P_n and basic tooth proportions	$b_P = \dfrac{\text{constant, Table 3-4}}{P_n}$
Clearance, c	Normal diametral pitch, P_n and basic tooth proportions	$c = \dfrac{\text{constant, Table 3-4}}{P_n}$
Working depth, h_k	Addendums, $a_P = a_G$	$h_k = 2a_P = 2a_G$
Whole depth, h_t	Working depth, h_k, and clearance, c	$h_t = h_k + c$
Outside diameter of pinion, d_o	Pinion pitch diameter, d and addendum, a_P	$d_o = d + 2a_P$
Outside diameter of gear, D_o	Gear pitch diameter, D and addendum, a_G	$D_o = D + 2a_G$
Root diameters, d_R, D_R	Pitch diameters, d, D, and dedendums, b_P, b_G	$d_R = d - 2b_P$ $D_R = D - 2b_G$
Transverse pressure angle, ϕ_t	Normal pressure angle, ϕ_n and pitch helix angle, ψ $\tan(\phi_n = 20^\circ) = 0.36397023$	$\tan\phi_t = \dfrac{\tan\phi_n}{\cos\psi}$
Base circle diameters, d_B, D_B	Pitch diameters, d, D, and transverse pressure angle, ϕ_t	$d_B = d \cos\phi_t$ $D_B = D \cos\phi_t$
Diametral pitch, P	Normal diametral pitch, P_n	$P = P_n \cos\psi$
Normal circular pitch, p_n	Normal diametral pitch, P_n $\pi = 3.1415927$	$p_n = \pi / P_n$

continued on next page

TABLE 4-9

Theoretical Outside Diameters on Sykes, Cut, Double-Helical Gears

Practice of Farrel-Birmingham Co , Inc.

Diametral Pitch P	For ratios of speed *reduction of 3 to 1 or less, and a minimum sum of numbers of teeth of 36, teeth have equal addendums. Add the following increments to the pitch diameters of both pinion and gear:	For ratios of speed *reduction in excess of 3 to 1, and a minimum sum of numbers of teeth of 55, add the following increments to the pitch diameters:	
		PINION	GEAR
1	1.600	2.400	0.800
$1\frac{1}{4}$	1.280	1.920	.640
$1\frac{1}{2}$	1.067	1.600	.533
$1\frac{3}{4}$	0.914	1.371	.457
2	0.800	1.200	.400
$2\frac{1}{2}$	0.640	0.960	0.320
3	.533	.800	.267
$3\frac{1}{2}$	.457	.686	.229
4	.400	.600	.200
5	.320	.480	.160
6	0.267	0.400	0.133
7	.229	.343	.114
8	.200	.300	.100
9	.178	.267	.089
10	.160	.240	.080

*Gears used to increase speed are regarded as exceptions to regular practice, requiring special tooth proportions. So are gears usedas pump rotors and gears driving rolls where a variation in center-distance is expected.

TABLE 4–10 (continued)

To Find	From	Formula
Circular pitch, p	Normal diametral pitch, P_n	$p = \dfrac{\pi}{P_n \cos \psi}$
Circular pitch, p	Numbers of teeth N_P, N_G and pitch diameters d, D	$p = \dfrac{\pi d}{N_P} = \dfrac{\pi D}{N_G}$
Normal tooth circular thickness, t_n	Normal diametral pitch, P_n $(a_P = a_G)$	$t_n = \dfrac{1.5707963}{P_n}$
Backlash: Allowance for backlash is made by diminishing t_n by desired amount. Backlash allowance should be included in circular thickness carried forward into calculations for measurements over pins.		
Base helix angle, ψ_b Table 4-7	**Normal and transverse pressure angles, ϕ_n, ϕ_t** $\sin(\phi_n = 20^\circ) = 0.34202014$	$\cos \psi_b = \dfrac{\sin \Phi_n}{\sin \Phi_t}$
Involute function, inv Φ_m, for measurement over pins or balls	$\text{inv}\,\Phi_m = \dfrac{t_n}{d \cos \psi} + \text{inv}\,\Phi_t + \dfrac{d_{pin}}{d_B \cos \psi_b} - \dfrac{\pi}{N_P}$	
Pinion and External Gear	d_{pin} is diameter of measuring pin or ball. See Table 3-24 for involute functions.	
Diameter, d_2 to centers of measuring pins	Base diameter, d_B and inv Φ_m	$d_2 = \dfrac{d_B}{\cos \Phi_m}$
Pinion diameter, d_m, for measurement over pins or balls	When N_P is even When N_P is odd	$d_m = d_2 + d_{pin}$ $d_m = d_2 \cos 90/N + d_{pin}$
Gear diameter, D_m, for measurement over pins or balls	Use three formulas, immediately above, substituting D for d, D_B for d_B, N_G for N_P, D_2 for d_2, and D_m for d_m.	
Lead (inches), l_P, l_G $\pi = 3.1415927$	Pitch diameters, d, D, and pitch helix angle, ψ	$l_P = \pi d \cot \psi$ $l_G = \pi D \cot \psi$
Lead	Numbers of teeth, N_P, N_G; Normal diametral pitch, P_n; and pitch helix, ψ	$l_P = \dfrac{\pi N_P}{P_n \sin \psi}$ $l_G = \dfrac{\pi N_G}{P_n \sin \psi}$

Since leads are directly proportional to numbers of teeth, the leads specified for the generation of mating helical gears should be precisely as the numbers of teeth.

continued on next page

TABLE 4–10, continued

EXAMPLE: (Taken from *Analytical Mechanics of Gears*, p. 155, Buckingham, McGraw-Hill)

	Pinion	Gear
Numbers of teeth	$N_P = 20$	$N_G = 55$
Normal diametral pitch, P_n	8	
$\cos \psi = \dfrac{N_P + N_G}{2\,C\,P_n}$	0.89286	
Pitch helix angle, ψ	26.765°	
Computed leads	17.44054	47.96149
Pinion lead rounded off to	17.440	
Then gear lead becomes		$\dfrac{55 \times 17.440}{20} = 47.960$

TABLE 4–11

Formulas for Calculating Dimensions of 20° Involute; Stub, Parallel Helical Gears Using (Transverse) Diametral Pitch (See Explanation, Table 4–8)

Manual of Gear Design, Section 3 Buckingham The Industrial Press

To Find	From	Formula
Pinion pitch diameter, d	Number of teeth, N_P, and diametral pitch, P	$d = \dfrac{N_P}{P}$
Gear pitch diameter, D	Number of teeth, N_G, and diametral pitch, P	$D = \dfrac{N_G}{P}$
Center distance, C	Pitch diameter, d and D	$C = \dfrac{d + D}{2}$
Outside diameters, d_o, D_o	Pitch diameters, d, D; stub tooth constant; and diametral pitch, P	$d_o = d + 1.600/P$ $D_o = D + 1.600/P$
Base circle diameter, d_B, D_B	Pitch diameters, d, D, and transverse pressure angle, $\phi = 20°$	$d_B = d \cos \phi$ $D_B = D \cos \phi$
Root diameters, d_R, D_R	Pitch diameters, d, D; stub tooth constant; and diametral pitch, P	$d_R = d - 2.000/P$ $D_R = D - 2.000/P$
Generation pressure angle, ϕ_g	Outside diameter of cutter, D_{OC}; base circle diameter of cutter, D_{BC}; and base circle diameter of pinion, d_B	$\tan \phi_g = \dfrac{\sqrt{D^2_{OC} - D^2_{BC}}}{D_{BC} + d_B}$
Center distance for generation, C_g	Base circle diameter of cutter, D_{BC}; pinion base circle diameter, d_B; and generation pressure angle, ϕ	$C_g = \dfrac{D_{BC} + d_B}{2 \cos \phi_g}$
Minimum root diameter to avoid undercutting, d_u	Outside diameter of cutter, D_{OC}, and generation center distance, C_g	$d_u = 2C_g - D_{OC}$

TABLE 4–12

One-Diametral-Pitch *Dimensions of 20° Stub, Involute Helical Pinions for Driving Internal Helical Gears

Manual of Gear Design, Section 3 Buckingham The Industrial Press

No. of Teeth N_P	Outside Radius r_o	Pitch Radius r	Root Radius (Shaped) R_{r1}	Base Radius r_b	$\sqrt{r_o^2 - r_b^2}$
16	9.050	8.00	7.237	7.51754	5.0387
17	9.550	8.50	7.737	7.98739	5.2348
18	10.050	9.00	8.238	8.45723	5.4293
19	10.550	9.50	8.738	8.92708	5.6223
20	11.050	10.00	9.239	9.39693	5.8137
21	11.550	10.50	9.739	9.86677	6.0041
22	12.050	11.00	10.239	10.33667	6.1932
23	12.550	11.50	10.739	10.80647	6.3814
24	13.050	12.00	11.239	11.27631	6.5686
25	13.550	12.50	11.740	11.74616	6.7550
26	14.050	13.00	12.240	12.21600	6.9405
27	14.550	13.50	12.740	12.68585	7.1254
28	15.050	14.00	13.240	13.15570	7.3095
29	15.550	14.50	13.740	13.62554	7.4931
30	16.050	15.00	14.241	14.09539	7.6760
31	16.550	15.50	14.741	14.56524	7.8585
32	17.050	16.00	15.241	15.03508	8.0404
33	17.550	16.50	15.741	15.50493	8.2219
34	18.050	17.00	16.241	15.97477	8.4029
35	18.550	17.50	16.741	16.44462	8.5835
36	19.050	18.00	17.242	16.91447	8.7637
37	19.550	18.50	17.742	17.38431	8.9436
38	20.050	19.00	18.242	17.85416	9.1231
39	20.550	19.50	18.742	18.32401	9.3023

No. of Teeth N_P	Outside Radius r_o	Pitch Radius r	Root Radius (Shaped) R_{r1}	Base Radius r_b	$\sqrt{r_o^2 - r_b^2}$
40	21.050	20.00	19.242	18.79385	9.4812
41	21.550	20.50	19.742	19.26370	9.6598
42	22.050	21.00	20.242	19.73355	9.8381
43	22.550	21.50	20.743	20.20339	10.0162
44	23.050	22.00	21.243	20.67324	10.1940
45	23.550	22.50	21.743	21.14308	10.3717
46	24.050	23.00	22.243	21.61293	10.5491
47	24.550	23.50	22.743	22.08278	10.7262
48	25.050	24.00	23.243	22.55262	10.9032
49	25.550	24.50	23.743	23.02247	11.0800
50	26.050	25.00	24.243	23.49232	11.2567
51	26.550	25.50	24.743	23.96216	11.4331
52	27.050	26.00	25.243	24.43201	11.6094
53	27.550	26.50	25.744	24.90185	11.7855
54	28.050	27.00	26.244	25.37170	11.9615
55	28.550	27.50	26.744	25.84155	12.1374
56	29.050	28.00	27.244	26.31139	12.3131
57	29.550	28.50	27.744	26.78124	12.4887
58	30.050	29.00	28.244	27.25109	12.6641
59	30.550	29.50	28.744	27.72093	12.8394
60	31.050	30.00	29.244	28.19078	13.0146
61	31.550	30.50	29.744	28.66062	13.1898
62	32.050	31.00	30.244	29.13047	13.3648
63	32.550	31.50	30.744	29.60032	13.5397

*To get the dimensions at another diametral pitch, divide the tabular values by the diametral pitch.

continued on next page

TABLE 4–12 (continued)

No. of Teeth N_p	Outside Radius r_o	Pitch Radius r	Root Radius (Shaped) R_{rl}	Base Radius r_b	$\sqrt{r_o^2 - r_b^2}$
64	33.050	32.00	31.244	30.07014	13.7145
65	33.550	32.50	31.744	30.54001	13.8892
66	34.050	33.00	32.244	31.00986	14.0638
67	34.550	33.50	32.744	31.47970	14.2383
68	35.050	34.00	33.244	31.94955	14.4127
69	35.550	34.50	33.744	32.41940	14.5871
70	36.050	35.00	34.244	32.88924	14.7614
71	36.550	35.50	34.744	33.35909	14.9356
72	37.050	36.00	35.244	33.82893	15.1097
73	37.550	36.50	35.745	34.29878	15.2838
74	38.050	37.00	36.245	34.76863	15.4578
75	38.550	37.50	36.745	35.23847	15.6317
76	39.050	38.00	37.245	35.70832	15.8056
77	39.550	38.50	37.745	36.17817	15.9794
78	40.050	39.00	38.245	36.64801	16.1531
79	40.550	39.50	38.745	37.11786	16.3268
80	41.050	40.00	39.245	37.58770	16.5005
81	41.550	40.50	39.745	38.05755	16.6741
82	42.050	41.00	40.245	38.52740	16.8476
83	42.550	41.50	40.745	38.99724	17.0210
84	43.050	42.00	41.245	39.46709	17.1944
85	43.550	42.50	41.746	39.93694	17.3678
86	44.050	43.00	42.246	40.40678	17.5412
87	44.550	43.50	42.746	40.87663	17.7145
88	45.050	44.00	43.246	41.34648	17.8877
89	45.550	44.50	43.746	41.81632	18.0609

No. of Teeth N_p	Outside Radius r_o	Pitch Radius r	Root Radius (Shaped) R_{rl}	Base Radius r_b	$\sqrt{r_o^2 - r_b^2}$
90	46.050	45.00	44.246	42.28617	18.2340
91	46.550	45.50	44.746	42.75601	18.4071
92	47.050	46.00	45.246	43.22586	18.5802
93	47.550	46.50	45.746	43.69571	18.7533
94	48.050	47.00	46.246	44.16555	18.9263
95	48.550	47.50	46.746	44.63540	19.0993
96	49.050	48.00	47.246	45.10525	19.2722
97	49.550	48.50	47.746	45.57509	19.4451
98	50.050	49.00	48.246	46.04494	19.6180
99	50.550	49.50	48.746	46.51478	19.7908
100	51.050	50.00	49.246	46.98463	19.9636
101	51.550	50.50	49.746	47.45448	20.1364
102	52.050	51.00	50.246	47.92432	20.3091
103	52.550	51.50	50.746	48.39417	20.4818
104	53.050	52.00	51.246	48.86402	20.6545
105	53.550	52.50	51.746	49.33386	20.8272
106	54.050	53.00	52.246	49.80371	20.9998
107	54.550	53.50	52.746	50.27356	21.1724
108	55.050	54.00	53.246	50.74340	21.3450
109	55.550	54.50	53.746	51.21325	21.5175
110	56.050	55.00	54.246	51.68309	21.6900
111	56.550	55.50	54.746	52.15294	21.8625
112	57.050	56.00	55.246	52.62279	22.0350
113	57.550	56.50	55.746	53.09263	22.2075
114	58.050	57.00	56.246	53.56248	22.3799
115	58.550	57.50	56.746	54.03233	22.5523

*To get the dimensions at another diametral pitch, divide the tabular values by the diametral pitch.

TABLE 4–13

One-Diametral-Pitch Dimensions* of 20° Stub, Involute INTERNAL Helical Gears

Manual of Gear Design, Section 3 Buckingham The Industrial Press

No. of Teeth N_i	Root Radius R_R	Pitch Radius R	Internal Radius R_i	Base Radius R_b	$\sqrt{R_i^2 - R_b^2}$	Largest Cutter N_c	Largest Pinion N_p
25	13.719	12.50	12.05	11.74616	2.6889	18	19
26	14.220	13.00	12.55	12.21600	2.8760	18	20
27	14.721	13.50	13.05	12.68585	3.0613	18	20
28	15.222	14.00	13.55	13.15570	3.2450	21	21
29	15.724	14.50	14.05	13.62554	3.4274	21	22
30	16.225	15.00	14.55	14.09539	3.6086	21	23
31	16.727	15.50	15.05	14.56524	3.7889	21	24
32	17.228	16.00	15.55	15.03508	3.9684	25	25
33	17.729	16.50	16.05	15.50493	4.1472	25	26
34	18.230	17.00	16.55	15.97477	4.3254	25	27
35	18.731	17.50	17.05	16.44462	4.5029	28	28
36	19.231	18.00	17.55	16.91447	4.6800	28	29
37	19.732	18.50	18.05	17.38431	4.8567	28	30
38	20.233	19.00	18.55	17.85416	5.0330	28	31
39	20.733	19.50	19.05	18.32401	5.2089	32	32
40	21.234	20.00	19.55	18.79385	5.3845	32	33
41	21.734	20.50	20.05	19.26370	5.5598	32	34
42	22.235	21.00	20.55	19.73355	5.7349	35	35
43	22.735	21.50	21.05	20.20339	5.9097	35	36
44	23.236	22.00	21.55	20.67324	6.0843	35	37
45	23.736	22.50	22.05	21.14308	6.2588	35	38
46	24.237	23.00	22.55	21.61293	6.4330	35	39
47	24.737	23.50	23.05	22.08278	6.6070	35	40
48	25.237	24.00	23.55	22.55262	6.7809	35	41
49	25.738	24.50	24.05	23.02247	6.9547	42	42
50	26.238	25.00	24.55	23.49232	7.1283	42	43
51	26.738	25.50	25.05	23.96216	7.3018	42	44
52	27.239	26.00	25.55	24.43201	7.4752	42	45
53	27.739	26.50	26.05	24.90185	7.6485	42	46
54	28.239	27.00	26.55	25.37170	7.8217	42	47
55	28.739	27.50	27.05	25.84155	7.9948	42	48
56	29.240	28.00	27.55	26.31139	8.1678	49	49
57	29.740	28.50	28.05	26.78124	8.3407	49	50
58	30.240	29.00	28.55	27.25109	8.5135	49	51
59	30.740	29.50	29.05	27.72093	8.6863	49	52
60	31.240	30.00	29.55	28.19078	8.8590	49	53
61	31.740	30.50	30.05	28.66062	9.0316	49	54
62	32.241	31.00	30.55	29.13047	9.2042	49	55
63	32.741	31.50	31.05	29.60032	9.3767	56	56
64	33.241	32.00	31.55	30.07016	9.5492	56	57
65	33.741	32.50	32.05	30.54001	9.7216	56	58
66	34.241	33.00	32.55	31.00986	9.8939	56	59
67	34.741	33.50	33.05	31.47970	10.0662	56	60
68	35.242	34.00	33.55	31.94955	10.2385	56	61
69	35.742	34.50	34.05	32.41940	10.4108	56	62
70	36.242	35.00	34.55	32.88924	10.5830	63	63
71	36.742	35.50	35.05	33.35909	10.7551	63	64
72	37.242	36.00	35.55	33.82893	10.9272	63	65

*To get the dimensions at another diametral pitch, divide the tabular values by the diametral pitch.

continued on next page

TABLE 4–13, continued

No. of Teeth N_i	Root Radius R_R	Pitch Radius R	Internal Radius R_i	Base Radius R_b	$\sqrt{R_i^2 - R_b^2}$	Largest Cutter N_c	Largest Pinion N_p
73	37.742	36.50	36.05	34.29878	11.0993	63	66
74	38.242	37.00	36.55	34.76863	11.2714	63	67
75	38.743	37.50	37.05	35.23847	11.4434	63	68
76	39.243	38.00	37.55	35.70832	11.6154	63	69
77	39.743	38.50	38.05	36.17817	11.7873	70	70
78	40.243	39.00	38.55	36.64801	11.9593	70	71
79	40.743	39.50	39.05	37.11786	12.1312	70	72
80	41.243	40.00	39.55	37.58770	12.3031	70	73
81	41.743	40.50	40.05	38.05755	12.4750	70	74
82	42.243	41.00	40.55	38.52740	12.6468	70	75
83	42.743	41.50	41.05	38.99724	12.8186	70	76
84	43.243	42.00	41.55	39.46709	12.9904	70	77
85	43.743	42.50	42.05	39.93694	13.1621	70	78
86	44.244	43.00	42.55	40.40678	13.3338	70	79
87	44.744	43.50	43.05	40.87663	13.5055	70	80
88	45.244	44.00	43.55	41.34648	13.6773	70	81
89	45.744	44.50	44.05	41.81632	13.8490	70	82
90	46.244	45.00	44.55	42.28617	14.0207	70	83
91	46.744	45.50	45.05	42.75601	14.1924	84	84
92	47.244	46.00	45.55	43.22586	14.3641	84	85
93	47.744	46.50	46.05	43.69571	14.5357	84	86
94	48.244	47.00	46.55	44.16555	14.7073	84	87
95	48.744	47.50	47.05	44.63540	14.8789	84	88
96	49.244	48.00	47.55	45.10525	15.0505	84	89
97	49.745	48.50	48.05	45.57509	15.2221	84	90
98	50.245	49.00	48.55	46.04494	15.3937	84	91
99	50.745	49.50	49.05	46.51478	15.5652	84	92
100	51.245	50.00	49.55	46.98463	15.7368	84	93
101	51.745	50.50	50.05	47.45448	15.9083	84	94
102	52.245	51.00	50.55	47.92432	16.0798	84	95
103	52.745	51.50	51.05	48.39417	16.2513	84	96
104	53.245	52.00	51.55	48.86402	16.4228	84	97
105	53.745	52.50	52.05	49.33386	16.5943	84	98
106	54.245	53.00	52.55	49.80371	16.7658	84	99
107	54.745	53.50	53.05	50.27356	16.9372	84	100
108	55.245	54.00	53.55	50.74340	17.1087	84	101
109	55.745	54.50	54.05	51.21325	17.2802	84	102
110	56.245	55.00	54.55	51.68309	17.4516	84	103
111	56.745	55.50	55.05	52.15294	17.6230	84	104
112	57.245	56.00	55.55	52.62279	17.7945	84	105
113	57.745	56.50	56.05	53.09263	17.9659	84	106
114	58.245	57.00	56.55	53.56248	18.1373	84	107
115	58.745	57.50	57.05	54.03233	18.3087	84	108
116	59.245	58.00	57.55	54.50217	18.4801	84	109
117	59.746	58.50	58.05	54.97202	18.6515	84	110
118	60.246	59.00	58.55	55.44186	18.8229	84	111
119	60.746	59.50	59.05	55.91171	18.9942	84	112
120	61.246	60.00	59.55	56.38156	19.1656	84	113
121	61.746	60.50	60.05	56.85140	19.3370	84	114
122	62.246	61.00	60.55	57.32125	19.5083	84	115
123	62.746	61.50	61.05	57.79110	19.6797	84	116
124	63.246	62.00	61.55	58.26094	19.8510	84	117
125	63.746	62.50	62.05	58.73079	20.0224	84	118

*To get the dimensions at another diametral pitch, divide the tabular values by the diametral pitch.

TABLE 4-14
Formulas for Calculating Dimensions of Crossed Helical Gears (Spiral Gears)

To Find	From	Formula
Shaft angle, Σ	Pitch helix angle of driver, ψ_D and of follower ψ_F	$\psi_D + \psi_F = \Sigma$ large shaft angle $\psi_D - \psi_F = \Sigma$ small shaft angle
Hands of helices	Both right or both left when shaft angle is sum of helix angles. The driver generally has the larger helix angle. The hands are opposite when the difference of the helix angles equals the shaft angle.	
Pitch diameter of driver, d	Number of teeth, N_D; normal diametral pitch, P_n; and helix angle, ψ_D	$d = \dfrac{N_D}{P_n \cos \psi_D}$
Pitch diameter of follower, D	Number of teeth, N_F; normal diametral pitch, P_n; and helix angle, ψ_F	$D = \dfrac{N_F}{P_n \cos \psi_F}$
Center distance, C	Pitch diameters, d, D	$C = \dfrac{d + D}{2}$

Any two external, involute helical gears of the same normal diametral pitch will mesh together, provided the shaft angle and center distance are varied to suit. If the quotient of the two foregoing formulas for pitch diameters is formed to eliminate P_n, the following fundamental relationship among numbers of teeth, helix angles and pitch diameters is obtained:

$$\frac{N_F}{N_D} = \frac{D \cos \psi_F}{d \cos \psi_D}$$

Any pair of crossed helical gears has to satisfy the last relationship. A pair of parallel helical gears also satisfy it as a special case, for $\psi_D = \psi_F$. This transcendental equality is readily solved so long as d and D constitute the quantities to be found. When, however, a pair of crossed helical gears are desired for shafts at a specified distance apart, finding suitable helix angles becomes a trial-and-error calculation. Graphical methods are helpful (see Section 3, *Manual of Gear Design,* by Buckingham or a textbook on kinematics of machinery) in finding an approximate solution, after which trial and error methods soon reduce the errors in angles to the desired degree of accuracy for design purposes.

To Find	From	Formula
Lead, inches, on driver, l_D	Pitch diameter of driver, d; pitch helix angle, ψ_D	$l_D = \dfrac{\pi d}{\tan \psi_D}$
Lead of follower, l_F	Number of teeth, N_F; normal diametral pitch, P_n; and pitch helix angle, ψ_F	$l_F = \dfrac{N_F P_n}{\sin \psi_F}$
Tooth proportions and other data	Same as for parallel helical gear	See Tables 4-10 and 4-11

TABLE 4–15

Definitions of Contact Ratio, Involute Helical Gears

ASA B6.10–1950 AGMA 112.03

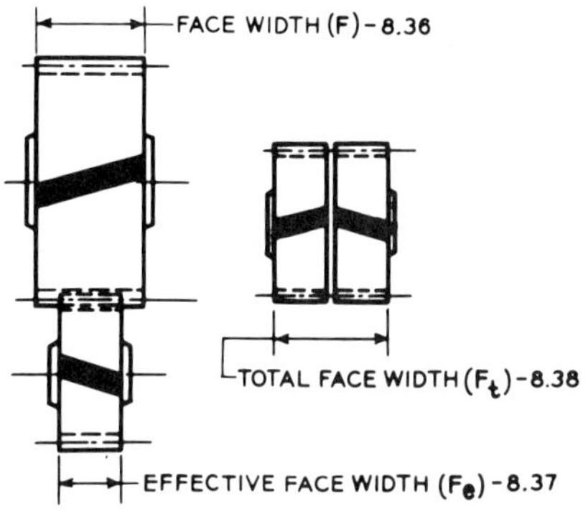

Illustration	Item No. in Standard	Symbol and Definition
(see figure above)	8.36	*(F)* Face Width is the length of the teeth in an axial plane.
	8.37	*(F_e)* Effective Face Width is the portion that may actually come into contact with mating teeth, as occasionally one member of a pair of gears may have a greater face width than the other.
	8.38	*(F_t)* Total Face Width is the actual dimension of a gear blank that exceeds the effective face width, or as in double-helical gears where the total face width includes any distance separating right-hand and left-hand helices.

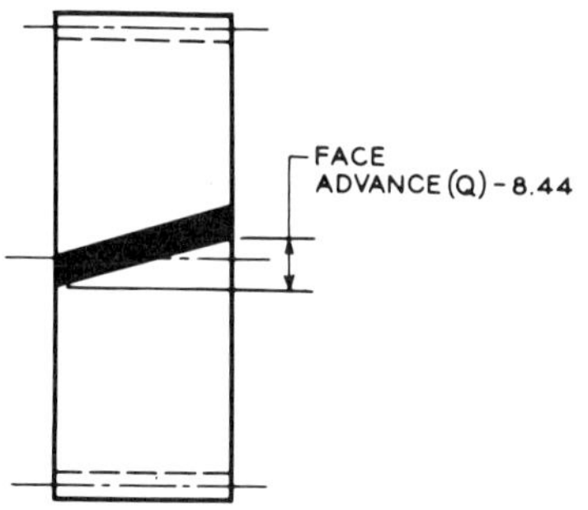

Illustration	Item No. in Standard	Symbol and Definition
(see figure above)	8.44	*(Q)* Face Advance is the distance on a pitch circle through which a helical or spiral tooth moves from the position at which contact begins at one end of the tooth curve to the position when contact ceases at the other end.
	11.16	*(m_c)* Contact Ratio is the ratio of the arc of action to the circular pitch, and sometimes is thought of as an average number of teeth in contact. For involute gears, the contact ratio is obtained most directly as the ratio of the length of action to the base pitch.
	11.17	*(m)* Transverse Contact Ratio is the contact ratio in the transverse plane.
	11.18	*(m_n)* Normal Contact Ratio is the contact ratio in the normal section.
	11.19	*(m_x)* Axial Contact Ratio is the ratio of face width to the axial pitch in helical teeth.
	11.20	*(m_F)* Face Contact Ratio is the ratio of the face advance to the circular pitch, usually having the same value as axial contact ratio.
	11.21	*(m_t)* Total Contact Ratio is the sum of the transverse contact ratio and axial contact ratio, which may be thought of as the average total number of teeth in contact in parallel helical gears or spiral bevel gears.

TABLE 4-16

Formulas for Contact Ratios, Involute Helical Gears

Manual of Gear Design, Section 3 Buckingham The Industrial Press

To Find	From	Formula
PARALLEL HELICAL GEARS (EXTERNAL OR INTERNAL)		
Face contact ratio, m_F	Face width, F; circular pitch, p; and pitch helix angle, ψ	$m_F = \dfrac{F \tan \psi}{p}$
PARALLEL HELICAL GEARS (EXTERNAL)		
Transverse contact ratio, m Item 11.17, Table 4–15	Pitch radii, r and R; base radii, r_b, R_b; outside radii r_o, R_o; center distance, C; circular pitch, p; and transverse pressure angle, ϕ	$m = \dfrac{\sqrt{r_o^2 - r_b^2} + \sqrt{R_o^2 - R_b^2} - C \sin \phi}{p \cos \phi}$
PARALLEL HELICAL GEARS (INTERNAL)		
	As above, except that R_i denotes the radius of the addendum circle of the internal gear	$m = \dfrac{\sqrt{r_o^2 - r_b^2} + C \sin \phi - \sqrt{R_i^2 - R_b^2}}{p \cos \phi}$
HELICAL PINION AND RACK		
Transverse contact ratio, m	Addendum of rack, a; and foregoing symbols for pinion	$m = \dfrac{a + \sin \phi \left[\sqrt{r_o^2 - r_b^2} - r \sin \phi\right]}{p \sin \phi \cos \phi}$
PARALLEL HELICAL GEARS (EXTERNAL OR INTERNAL)		
Total contact ratio, m_t	Face contact ratio, m_F, transverse contact ratio, m	$m_t = m_F + m$

HERRINGBONE GEARS

Apply the foregoing formulas to half the face width.

One reason why helical and herringbone gears run more quietly and smoothly than spur gears is the continuity of engagement of the teeth as the consequence of face advance. Since face contact ratio depends directly upon both face width and helix angle, either or both can be adjusted to ensure continuity of engagement, and provided the machine tools can be set to the required lead angles. Roughly the helix angle on a double helical or herringbone gear has to be about twice that on a single helical gear of the same face width to get equivalent contact ratios.

CROSSED HELICAL GEARS

The formulas are rather complex. The reader is referred to the Buckingham books, particularly Section 3, *Manual of Gear Design*.

TABLE 4–17

Minimum Width of Face for Continuous Helical Action Of Helical and Herringbone Gears with Helix Angles Of Approximately 15 and 23 Degrees

The Internal Gear The Fellows Gear Shaper Co

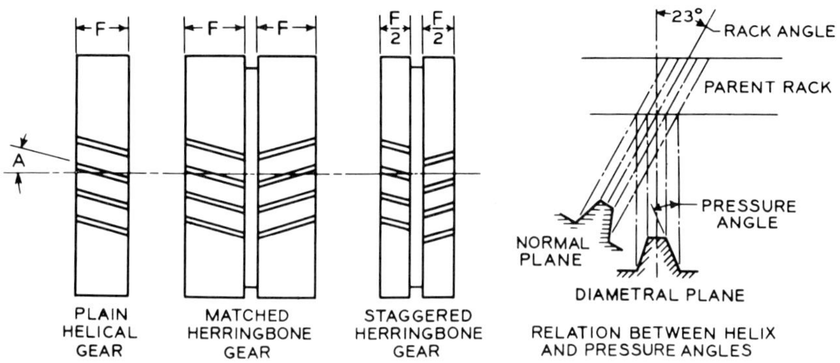

Diametral Pitch	Normal Pitch	Helix Angle "A"	Lead of Helix in Inches	Minimum Width of Face "F" in Inches
5/7	5.184	$15^\circ - 20'$	41.270	$2\frac{19}{64}$
5/7	5.456	$23^\circ - 35'$	25.904	$1\frac{7}{16}$
6/8	6.209	$14^\circ - 55'$	41.270	$1\frac{31}{32}$
6/8	6.518	23°	25.904	$1\frac{15}{64}$
7/9	7.254	$15^\circ - 12'$	41.270	$1\frac{21}{32}$
7/9	7.629	$23^\circ - 25'$	25.904	$1\frac{3}{64}$
8/10	8.279	$14^\circ - 55'$	41.270	$1\frac{31}{64}$
8/10	8.691	23°	25.904	$\frac{15}{16}$
9/11	9.324	$15^\circ - 9'$	41.270	$1\frac{19}{64}$
9/11	9.801	$23^\circ - 20'$	25.904	$\frac{13}{16}$
10/12	10.349	$14^\circ - 55'$	41.270	$1\frac{3}{16}$
10/12	10.863	23°	25.904	$\frac{3}{4}$
12/14	12.418	$14^\circ - 55'$	41.270	$\frac{63}{64}$
12/14	13.036	23°	25.904	$\frac{5}{8}$

TABLE 4-18

Proportions for Cut, Double-Helical Gears With 6 H-Type* Arms
Farrel-Birmingham Company

Dimension Defining Diagrams

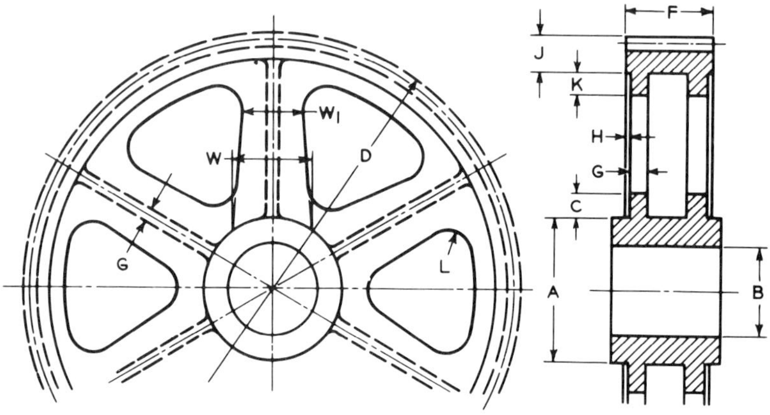

Formula for Computing Dimension

$A = 1.6\,B$ for steel casting
$\;\;= 1.8\,B$ for cast iron

$B = 6\sqrt[3]{\dfrac{\text{horsepower}}{\text{rpm}}}$ (average)

$\;\;= 5\sqrt[3]{\dfrac{\text{horsepower}}{\text{rpm}}}$ (minimum)

$\;\;= 7\sqrt[3]{\dfrac{\text{horsepower}}{\text{rpm}}}$ (maximum)

$C = 0.33\,W$
F^{*} = Face
G = Table 4-19
H = F/12
J = Table 4-19
$K = 0.33\,W$
$L = 0.33\,W$
$W = 0.33\sqrt{DF}$, Table 4-20
$W_1 = W - 0.8$ inch taper per foot (Table 7-22)

*Most gears of 6-inch face and over are H-type construction. See Table 4-21 for cross-arm and solid-web proportions.

TABLE 4-19

Rim and Web Thicknesses of Large Herringbone Gears in Terms of Diametral Pitch

Practice of Farrel-Birmingham Company

Diametral Pitch	Dimension in Table 4-18 *G*	*J*	Diametral Pitch	Dimension in Table 4-18 *G*	*J*
3/4		4-3/4	4	9/16	1
1	1-1/4	3-5/8	4-1/2	9/16	15/16
1-1/4	1	2-15/16	5	9/16	7/8
1-1/2	7/8	2-7/16	6	9/16	3/4
1-3/4	7/8	2-1/8	7	1/2	5/8
2	3/4	1-7/8	8		9/16
2-1/4	3/4	1-11/16	9		9/16
2-1/2	3/4	1-9/16	10		1/2
3	5/8	1-5/16	12		7/16
3-1/2	5/8	1-1/8	14		3/8
			16		3/8

TABLE 4–20

Arm Width *W* of Large Herringbone Gears in Terms of Diameter (In.) and Face (In.)

Practice of Farrel-Birmingham Co.

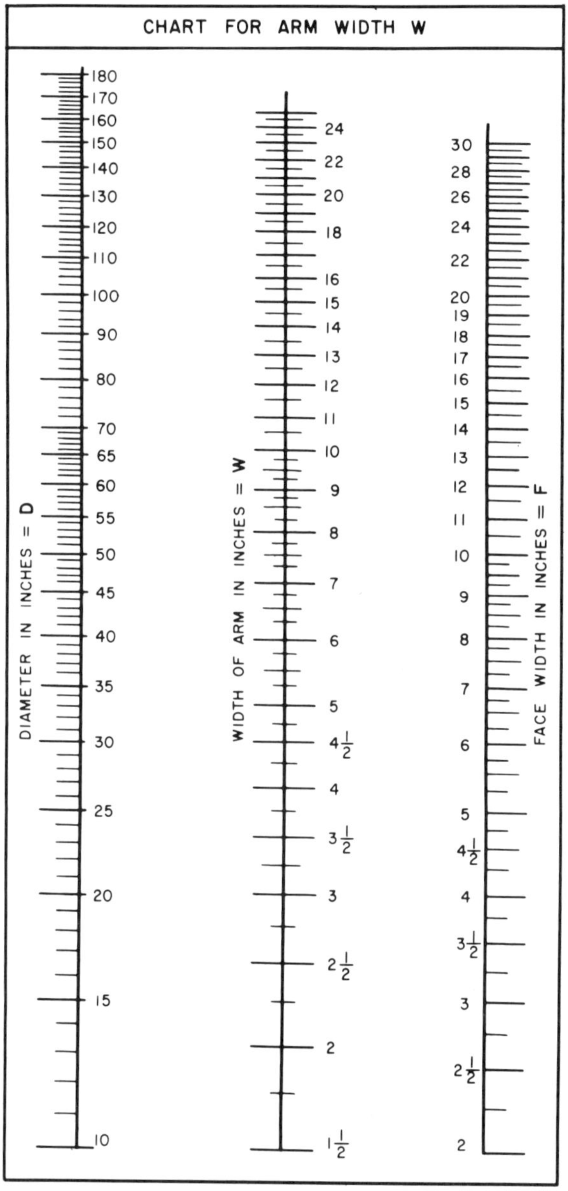

TABLE 4–21

Proportions of Cross-Arm and Solid-Web (Medium Size*) Herringbone Gears

Practice of Farrel-Birmingham Co.

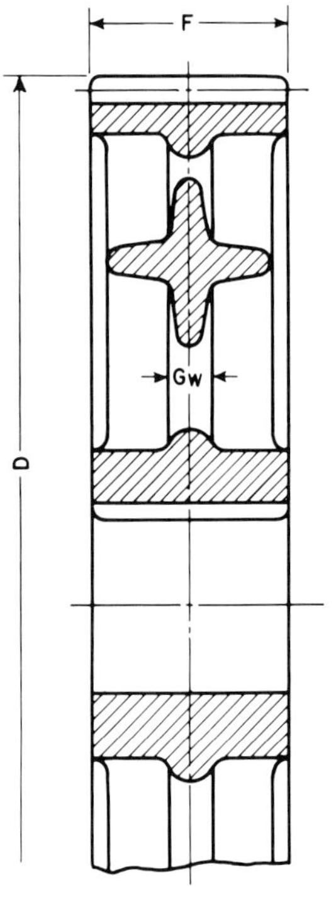

WIDTH OF ARM, W = 0.33 DF
(SEE DIAGRAM OF TABLE 4–18)
OR TAKE FROM TABLE 4–20

THICKNESS G_W OF CROSS ARM OR SOLID WEB IS NOT LESS THAN G OF TABLE 4–19 NOR GREATER THAN 1.5G.

* FOR FACE WIDTHS OF 6 INCHES AND OVER THE H–TYPE CONSTRUCTION IS PREFERRED.

TABLE 4–22

Rim Thickness and Set-Back on Cast-Steel Industrial Gear-Blanks of H-Arm Type Between 25 and 120 Inches in Diameter

Data from The Falk Corporation

EXAMPLE

GEAR 60" DIA $2\frac{1}{2}$" DP TRACE LINE A–A–A MIN RIM THICKNESS = 0.79"
GEAR 105" DIA $2\frac{1}{2}$" DP TRACE LINE B–B–B MIN RIM THICKNESS = 0.99"

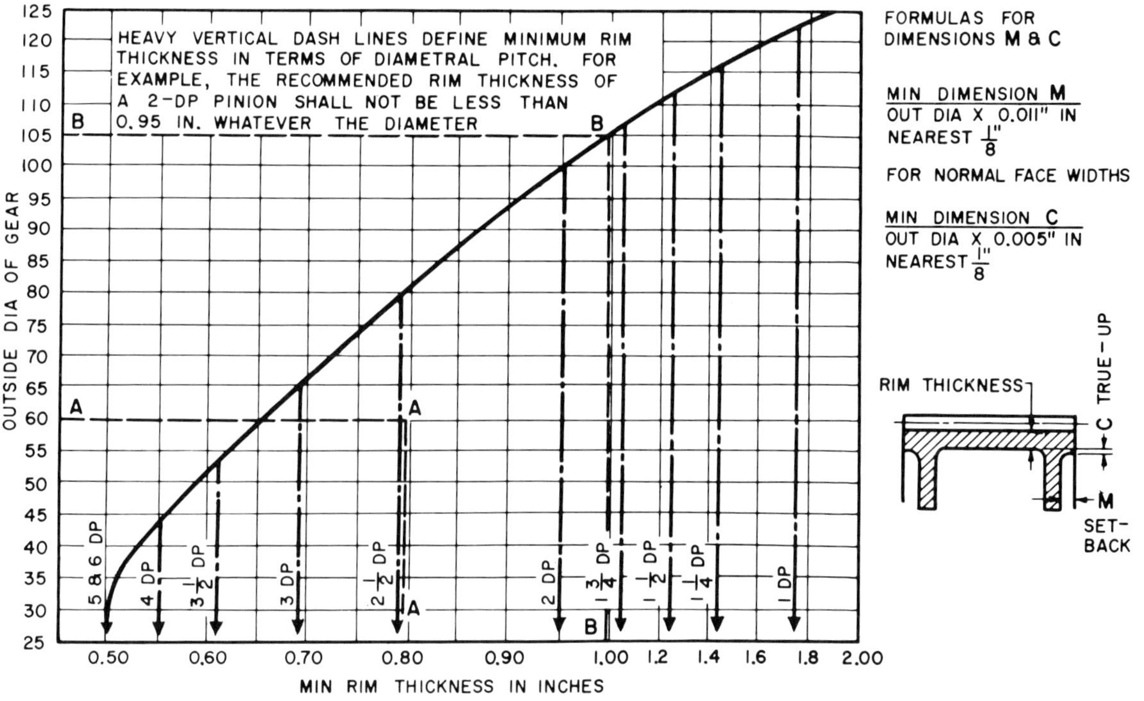

FORMULAS FOR DIMENSIONS M & C

MIN DIMENSION M
OUT DIA X 0.011" IN NEAREST $\frac{1}{8}$"

FOR NORMAL FACE WIDTHS

MIN DIMENSION C
OUT DIA X 0.005" IN NEAREST $\frac{1}{8}$"

TABLE 4–23

Number of H-Arms, Web Thickness, and Rim Flange Thickness on Cast-Steel Industrial Gear-Blanks Over 25 Inches in Diameter

Data from The Falk Corp.

VALUES SHOWN ARE FOR THE FOLLOWING:

MAXIMUM FACE WIDTHS

6-5 & 4 DP	10" FACE
$3\frac{1}{2}$ & 3 DP	12" FACE
$2\frac{1}{2}$ & 2 DP	15" FACE
$1\frac{3}{4}$ & $1\frac{1}{2}$ DP	18" FACE
$1\frac{1}{4}$ & 1 DP	21" FACE

WIDER FACE WIDTHS TO RECEIVE SPECIAL CONSIDERATION

STRESS RELIEF OPTIONAL $C = \frac{CP}{4} + \frac{1"}{2}$

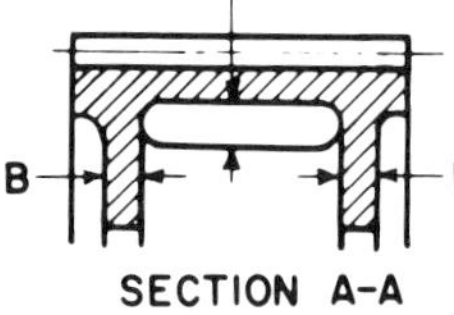

SECTION A-A

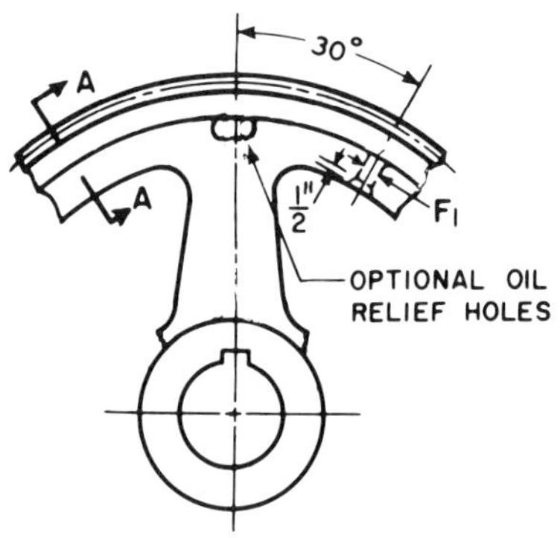

NUMBER OF ARMS

4 - ARMS UP TO 30" DIA
5 - ARMS 31" TO 45" DIA
6 - ARMS - GEARS ABOVE 45" DIA

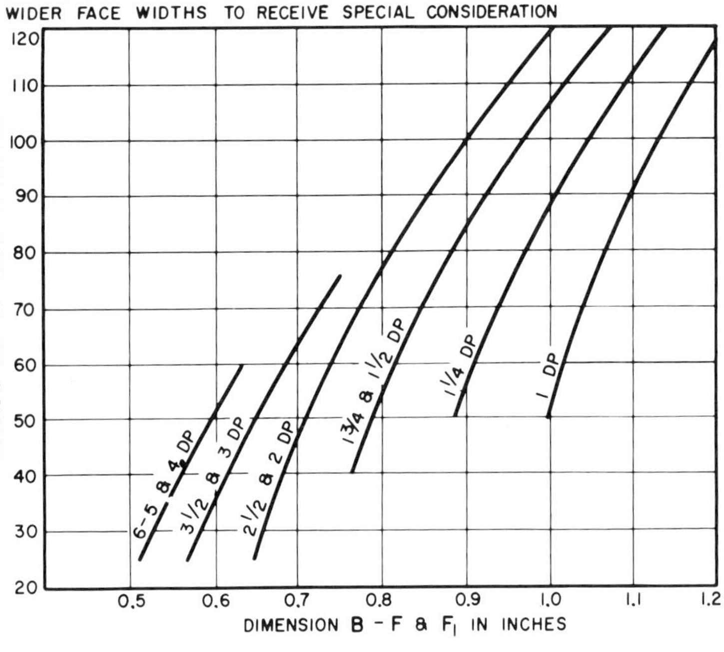

NOTE:-

READ VALUES TO NEAREST $\frac{1"}{16}$ FOR 6-DP TO 3-DP GEARS

READ VALUES TO NEAREST $\frac{1"}{8}$ FOR $2\frac{1}{2}$-DP TO 1-DP GEARS

TABLE 4–24

Flange Depths on Cast-Steel Industrial Gear-Blanks of Any Face Width And Over 25 Inches in Diameter

Data from The Falk Corp

READ VALUES TO NEAREST $\frac{1}{4}''$

EXAMPLE

100"–DIA GEAR, TRACE LINE A–A–A — MIN DIM. **A** = $4\frac{1}{2}''$

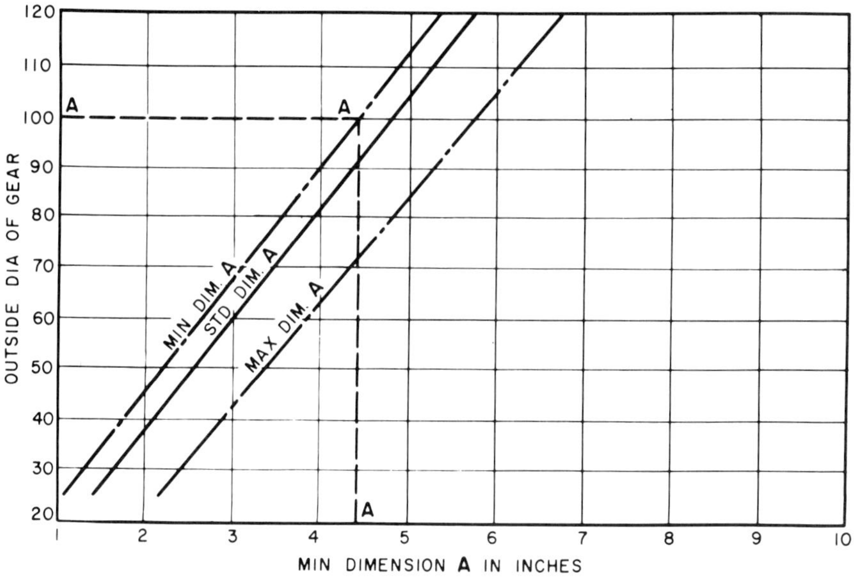

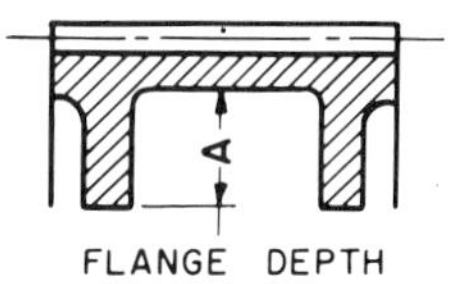

TABLE 4–25

Arm Proportions on Cast-Steel Industrial Gear-Blanks Having Ratios of Outside Diameter to Face Width Between 8 to 1 and 3 to 1

Data from The Falk Corp

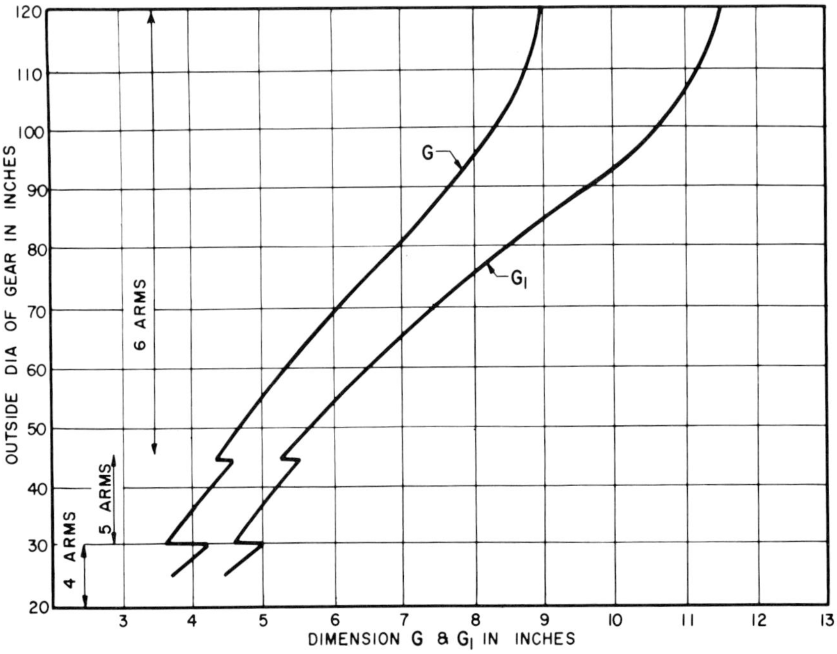

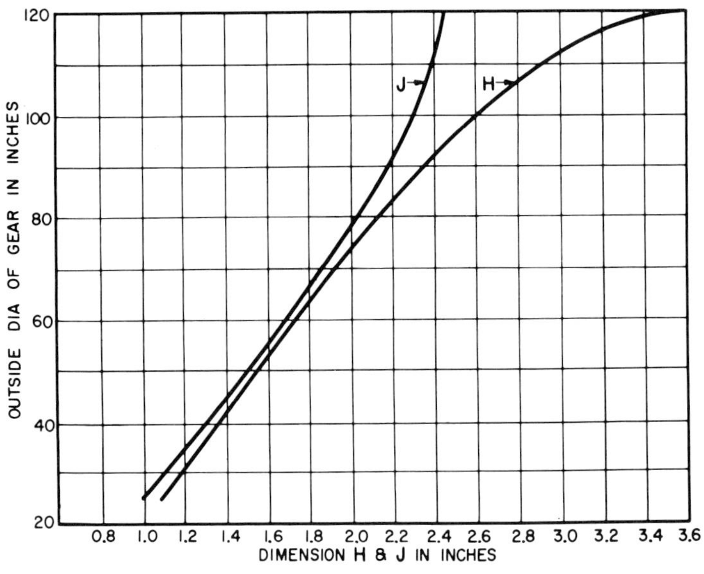

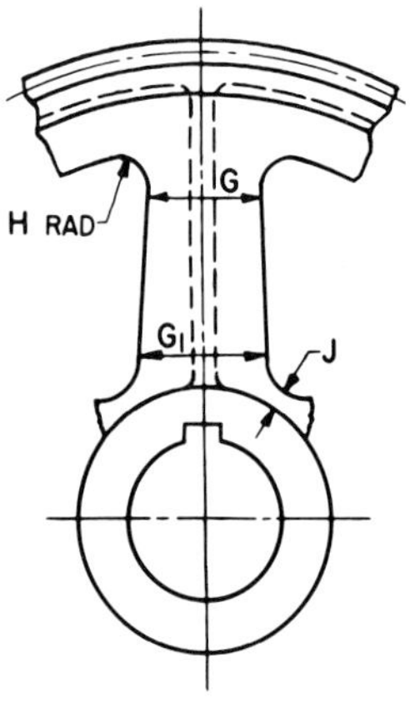

Inasmuch as arm proportions are influenced by pattern and foundry practices alternate designs are provided, Table 4–26. The proportions illustrated by this table are generally used for designs wherein the face width is changed by varying the widths of the rim and hub sections between arm faces.

TABLE 4–26

Alternate Arm Proportions on Cast-Steel Industrial Gear-Blanks
Data from The Falk Corp

See Table 4–25

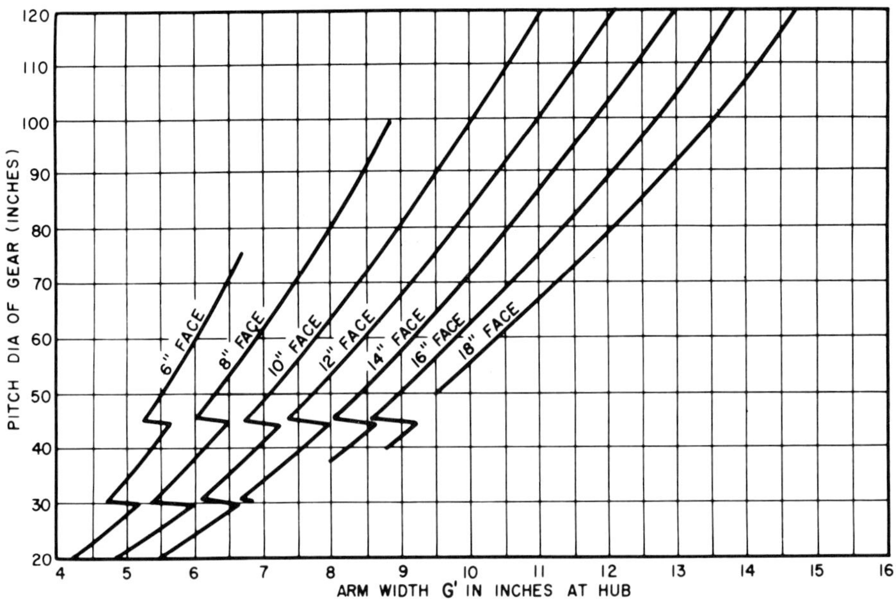

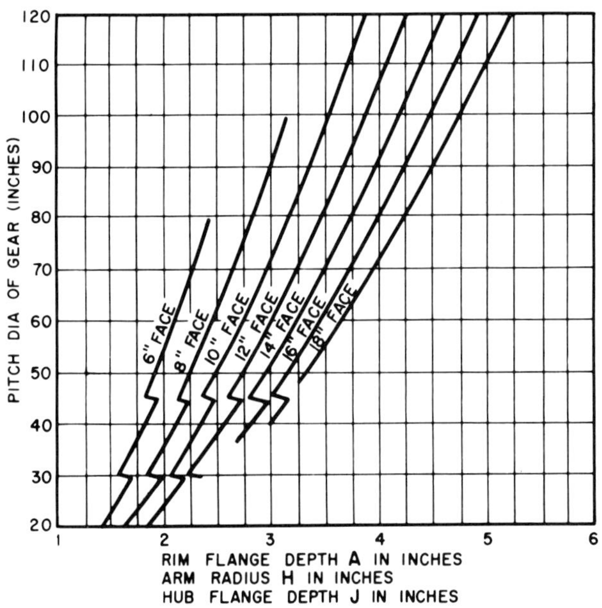

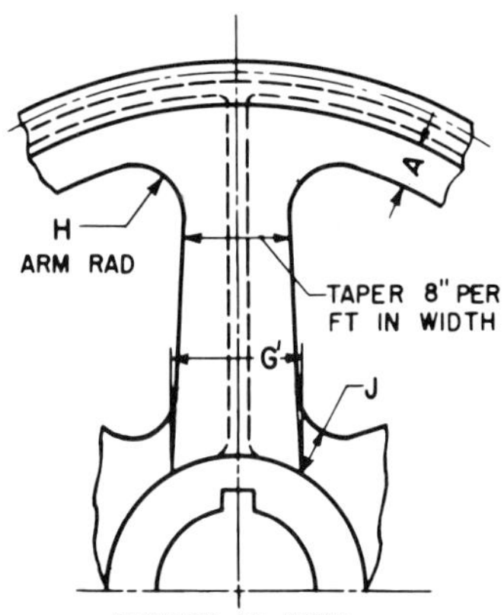

NUMBER OF ARMS

4 ARMS UP TO 30" DIA
5 ARMS 31" TO 45" DIA
6 ARMS 45" TO 120" DIA
OVER 120" DIA TO RECEIVE SPECIAL CONSIDERATION

The alternate proportions are generally used for single purpose, gear-blank design or "sweep pattern construction where modifications are simple.

TABLE 4–27

Minimum Hub Thickness of Cast-Steel Industrial Gear Blanks Over 25 Inches in Diameter

Data from the Falk Corporation

EXAMPLE

FOR 8" BORE 3-DP TRACE LINE A-A-A HUB THICKNESS = $2\frac{1}{2}$"

VALUES SHOWN ARE BASED ON BORES WITH STD AGMA KEYWAYS AND FOR THE FOLLOWING FACE WIDTHS

6-5	& 4 DP	10" FACE MAX
$3\frac{1}{2}$	& 3 DP	12" FACE MAX
$2\frac{1}{2}$	& 2 DP	15" FACE MAX
$1\frac{3}{4}$	& $1\frac{1}{2}$ DP	18" FACE MAX
$1\frac{1}{4}$	& 1 DP	21" FACE MAX

SEE TABLE 8-11 FOR AGMA STANDARD KEYWAYS

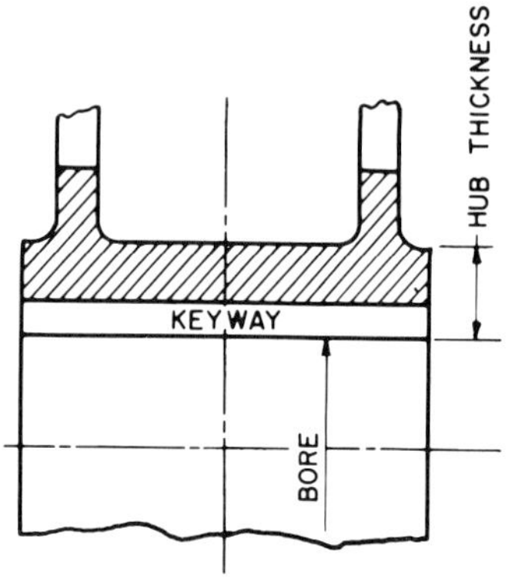

FACE WIDTHS IN EXCESS OF THESE MAXIMUMS REQUIRE SPECIAL CONSIDERATION. VALUES TO BE READ IN NEAREST $\frac{1}{8}$"

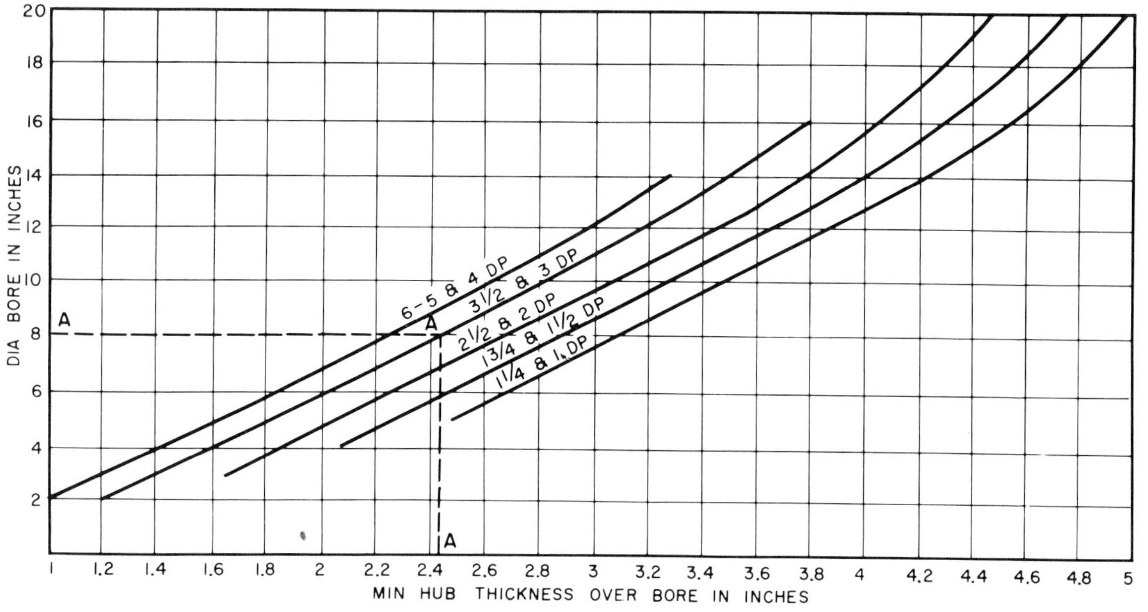

TABLE 4–28

Counterbore Dimensions Within Hubs of Cast-Steel Industrial Gear Blanks

Over 25 Inches in Diameter

Data from The Falk Corp

COUNTERBORES

1. UP TO 5" BORE
   NO COUNTERBORE UNLESS HUB LENGTH IS MORE THAN 2 TIMES THE BORE DIA
2. 5" TO 14" BORE
   WHEN HUB LENGTH EQUALS OR EXCEEDS $1\frac{1}{2}$ TIMES BORE DIA, COUNTERBORE IS TO BE PROVIDED
3. 14" BORE AND UP
   ALL HUBS TO BE COUNTERBORED IF POSSIBLE

DIMENSION A EQUALS APPROX $\frac{1}{3}$ OF HUB LENGTH
DIMENSION B EQUALS $\frac{1''}{32}$ FOR BORES UP TO 14" DIA
DIMENSION B EQUALS $\frac{1''}{16}$ FOR BORES OVER 14" DIA

EXAMPLES

1. 8"–BORE 12"–LONG HUB A = 4" B = $\frac{1''}{32}$
   COUNTERBORE EQUALS $8\frac{1''}{16}$ DIA
2. 15"–BORE 24"–LONG HUB A = 8" B = $\frac{1''}{16}$
   COUNTERBORE EQUALS $15\frac{1''}{8}$ DIA

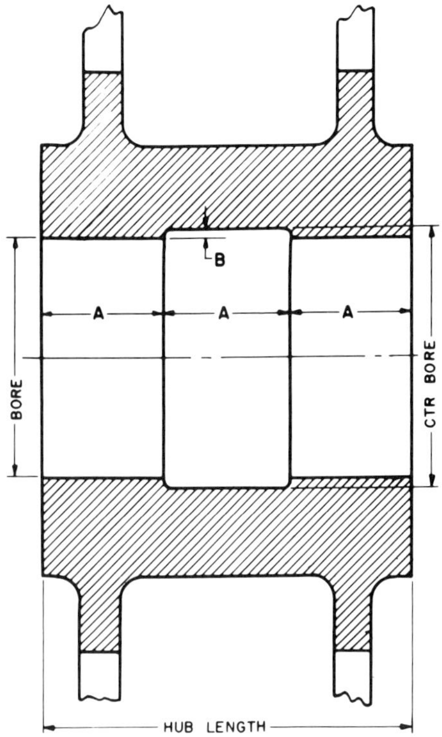

TABLE 4–29

Tolerances on Commercial Helical and Herringbone Gears for Pitch Line Speeds Less Than 2500 fpm

Data from The Falk Corporation

Diameter Range, Inches	Allowable Runout in Plane of Rotation Trued to Bore	Concentricity of Bore to Outside Diameter	Tolerance on Outside Diameter
2 to 10	0.002	0.002	+0.001, –0.005
10 to 30	0.002	0.002	+0.001, –0.010
30 to 60	0.003	0.003	+0.002, –0.015
60 to 90	0.003	0.003	+0.003, –0.020
90 to 120	0.004	0.004	+0.004, –0.025
120 to 150	0.004	0.004	+0.005, –0.030

Tables 3–27 to 3–32, inclusive, also give tolerances for spur and helical gears.

TABLE 4-30

AGMA Standard Ratings for Surface Durability and Strength of Helical and Herringbone Gears Operating at Linear Speeds Less Than 4000 Feet per Minute

AGMA Standard	Formula*	Definition of Terms
211.01 1944 Surface Durability	EXTERNAL Horsepower rating, P $P = F_i K_r{}^* D_s$ $D_s = \dfrac{d^2\left[\dfrac{78}{78+\sqrt{V}}\right](\text{rpm of pinion})}{126000}$ INTERNAL Substitute $K_{ri}{}^*$ for K_r, Table 3-36	F_i = combined factor for face width and inbuilt factor, Table 3-34 K_r = combined factor for materials, tooth form and ratio, Table 3-35 D_s = combined factor for pinion speed and pitch diameter d = pitch diameter of pinion V = pitch line speed, fpm
221.01 1948 Strength **not applicable to marine or high-speed gearing.** 1. Simple gear trains	Normal or continuous rating based on a service factor of one is $P = 0.5s\, Y_{hk}\, F\, Z\, M_h\, D_m$, which is half the peak horsepower rating. $D_m = \dfrac{d\left[\dfrac{78}{78+\sqrt{V}}\right](\text{rpm of pinion})}{395000}$	s = allowable bending stress: 250 times Brinell hardness for steel; 40 percent of ultimate tensile strength for bronze and alloy iron; 300 times core Brinell hardness for case-hardened gears. Y_{hk} = strength form factor, Table 4–31 F = effective face width in inches Z = length of line of action in inches, compute or obtain from layout M_h = angle factor, Table 4–32 D_m = combined factor for pinion speed and pitch diameter.
2. Multiple contact gear trains	In commercially manufactured planetary transmissions, the load division factor is taken as 0.9. If the load division is manually adjusted or automatically equalized, the load division factor is 0.95.	Planetary ratings are equal to 0.9 of the number of planet pinions multiplied by the rating computed for an equivalent simple set. The contact between the sun pinion and planets is critical. For double countershaft, simple gear trains use twice the rating of an equivalent single set multiplied by the load division factor.

* In the AGMA Standard this factor is the product of four components, two of which have been incorporated into Tables 3-34 and 3-35 on a basis of minimum values for good practice. An alternate and more exact method of calculation is provided in the Standard.

TABLE 4-31

Determination of Conservative Value of Tooth Form Factor Y_{hk} for Helical Gears

AGMA 221.01 – 1948

Layout	Measurement and Computations
Construct section of generated tooth profile in normal plane for one diametral pitch (or larger, as preferred)	Measure the thickness of the tooth t_W, at section where generated fillet joins the involute profile. Compute ratio r/t_W.
	Also measure distance X, and find $Y_h = \frac{2X}{3}$.

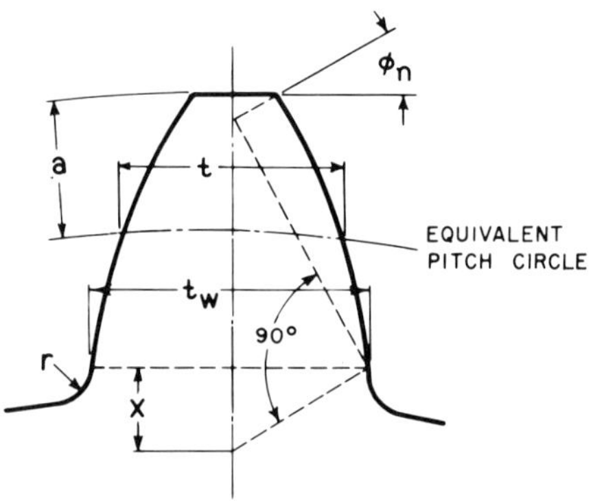

Form factor, Y_h, approaches the value $4X/3$ for steep angle helical gears; it approaches the value $2X/3$ for low helix angles. Accordingly the latter value is conservative.

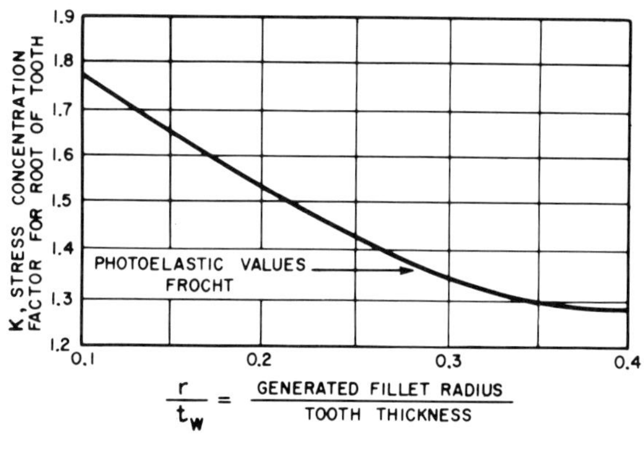

From curve for stress concentration factor, find K to correspond to computed r/t_W ratio

$$Y_{hk} = \frac{Y_h}{K}$$

For approximations, K may be taken as 1.33 for fillets cut with round nose hobs.

TABLE 4–32

Angle Factor M_h for Strength Calculations of Helical Gears

AGMA 221.01 1948

$$\text{Angle Factor } M_h = \frac{\text{cosine } \psi}{\text{cosine } \phi}$$

Helix Angle = ψ	Traverse Pressure Angle = ϕ 14½°	17°	20°	22½°	25°	Normal Pressure Angle = ϕ_n 14½°	17½°	20°	22°	25°
5°	1.030	1.041	1.061	1.079	1.099	1.031	1.046	1.064	1.072	1.100
7-1/2°	1.023	1.035	1.055	1.071	1.092	1.025	1.040	1.058	1.072	1.098
10°	1.018	1.029	1.049	1.068	1.087	1.018	1.034	1.050	1.063	1.090
12°	1.010	1.021	1.041	1.060	1.079	1.012	1.029	1.042	1.058	1.082
15°	.999	1.010	1.030	1.048	1.066	1.000	1.018	1.031	1.048	1.071
18°	.984	.995	1.012	1.030	1.049	.987	1.002	1.020	1.034	1.059
23°	.951	.962	.980	.997	1.014	.955	.972	.992	1.002	1.030
30°	.895	.906	.924	.939	.956	.905	.924	.940	.956	.983
35°	.846	.856	.872	.888	.904	.859	.877	.896	.914	.943
40°	.791	.801	.816	.831	.846	.809	.829	.849	.865	.898
45°	.731	.740	.754	.765	.781	.754	.775	.796	.815	.847

TABLE 4–33
AGMA Standard Helical and Herringbone Mill Gears

AGMA 321.03 1951

AGMA Standard	Formula	Definition of Terms
321.03 Standard 321.03 covers gears directly connected to grinding mills and kilns. It is applicable to semi-enclosed gearing but not enclosed speed reducers or gear motors.	*Surface durability* Horsepower rating is $P^* = F_i\ K_r\ D_s\ W_H$ $D_s = \dfrac{d^2\left[\dfrac{78}{78+\sqrt{V}}\right](\text{rpm of pinion})}{126000}$ *Beam strength rating* Same as AGMA Standard 221.01 in Table 4–30	F_i – Table 3–34 K_r – Table 3–35 d = pitch diameter of pinion V = pitch line speed, fpm W_H = factor depending upon speed ratio and hardnesses of materials, Table 4–34
	Service factor **equivalent** horsepower is the product of the specified or anticipated horsepower, not motor rating, and a suitable service factor. Surface durability and beam strength must be equal to or exceed this equivalent horsepower requirement.	Recommended service factors Driers 1.25 } use 1.0 if rpm Kilns 1.25 } is less than 5 Ball mills 1.5 Pebble mills 1.5 Tube mills 1.5 Rod mills 1.75

* Note that the horsepower rating for surface durability of 321.03 differs from that of Standard 211.01, Table 4–30 by the quantity W_H.

TABLE 4–34

Factor W_H for Computation of Surface Durability Rating of Helical and Herringbone Mill Gears

AGMA 321.03–1951

See Table 4–33

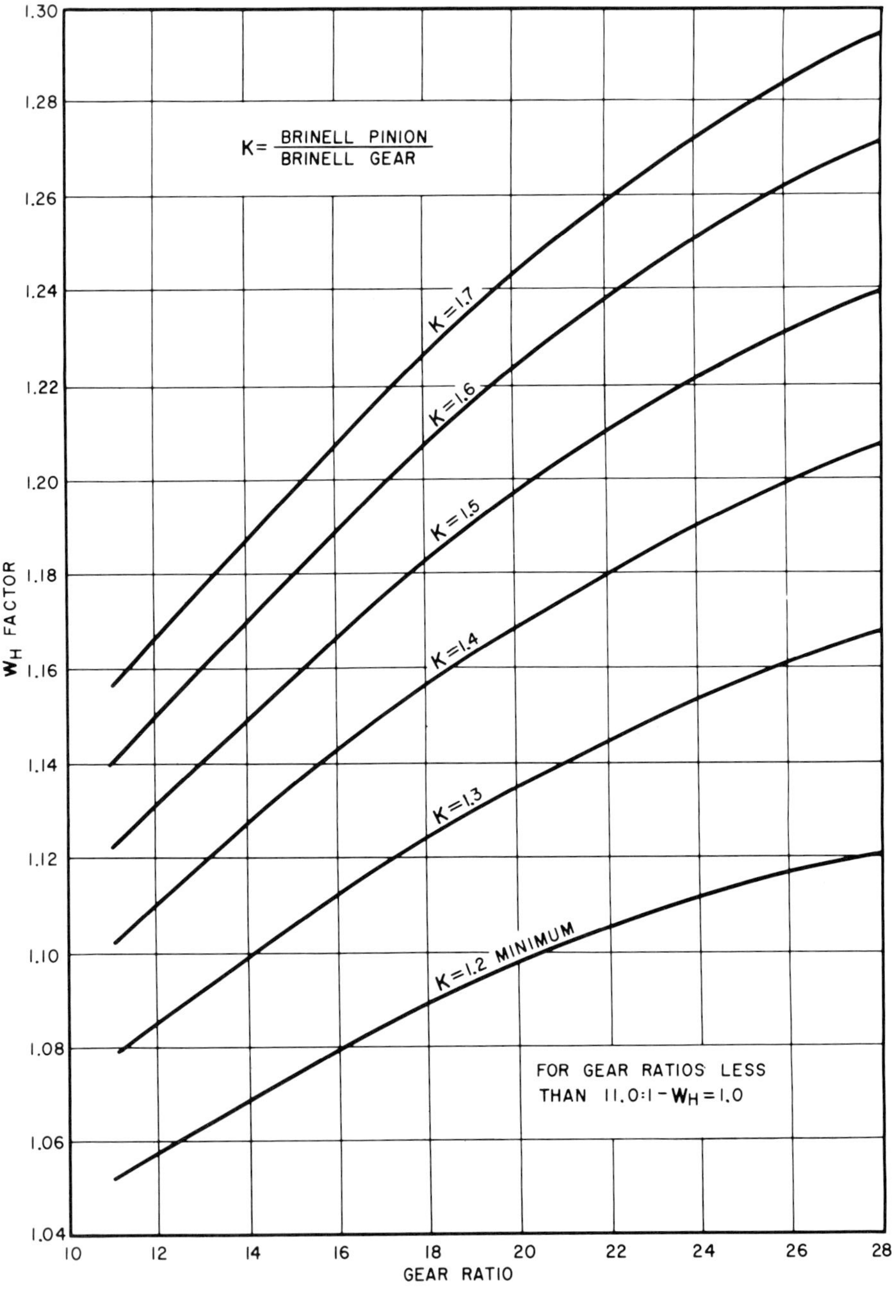

TABLE 4-35

Recommended Service Factors for Helical and Herringbone Speed Reducers†

AGMA 420.02 1951

Prime mover	Duration of service	Driven Machine Load Classifications		
		Uniform	Moderate Shock	Heavy Shock
	Occasional 1/2 hr per day	*0.50	**0.80	1.25
Electric	Intermittent 3 hr per day	**0.80	1.00	1.50
motor	8 to 10 hr per day	1.00	1.25	1.75
	24 hr per day	1.25	1.50	1.75
Multi-cylinder	Occasional 1/2 hr per day	**0.80	1.00	1.50
internal	Intermittent 3 hr per day	1.00	1.25	1.75
combustion	8 to 10 hr per day	1.25	1.50	2.00
engine	24 hr per day	1.50	1.75	2.00
Single-cylinder	Occasional 1/2 hr per day	1.00	1.25	1.75
internal	Intermittent 3 hr per day	1.25	1.50	2.00
combustion	8 to 10 hr per day	1.50	1.75	2.25
engine	24 hr per day	1.75	2.00	2.25

*Maximum momentary or starting load must not exceed 100 per cent of normal (100% overload).

**Maximum momentary or starting load must not exceed 160 per cent of normal (100% overload).

†Speed reducers as mechanical units are available in wide variety from several manufacturers. Units that comply with AGMA practices carry the AGMA monogram. Since the tables assembled here aim to cover machine design in general, rather than the design of particular units or machines, speed reducers as such are considered beyond the scope of this work. Furthermore, a designer of speed reducers will want the complete standard, not just portions of it as would be the case if it were condensed here. Table 4—35 is cited from AGMA 420.02 primarily for want of a more complete list of such service factors.

TABLE 4-36
AGMA Standard High-Speed Helical and Herringbone Gear Units

AGMA 421.02 1947

Pinion speeds of 3600 rpm and greater, or pitch line speeds of 4000 fpm and higher, or journal speeds of 1500 fpm and higher, exclusive of marine propulsion gearing.

Formula	Definition of Terms
Horsepower rating for surface durability $P = C^2 F M K$ (rpm of pinion) $M = \dfrac{N_G/N_P}{31500\,(N_G/N_P + 1)^3}$	C = center distance, inches F = face width of gears, inches N_G = number teeth on gear N_P = number teeth on pinion
$K = \dfrac{W(N_G/N_P + 1)}{d \quad N_G/N_P}$	W = tooth load, pounds per inch of face d = pitch diameter of pinion, inches
Among several other more or less general requirements is this: the maximum torsional stress, S_S, in any part of the shaft between journals shall not exceed $66.6K_t$ $S_S \leq 66.6\,K_t$	K_t = a constant, Table 4–37
The maximum S_S at the coupling, including due allowances for stress concentration and the effect of keyways, is to be based on nominal shaft diameter and shall not exceed 17 per cent of the yield strength of the material, that is, approximately $S_S \leq 0.509\,K_t\,B_n$	B_n = Brinell hardness number

TABLE 4-37

Torsional Stress Constants, K_t, for Design Calculations of High-Speed Helical and Herringbone Gear Units

AGMA 421.02 1947

Driven Machine	K_t Values Electric Motor	Steam Turbine	Internal Combustion Engine
Compressor – Centrifugal	85	85	63
Compressor – Rotary	63	63	55
Compressor – Refrigerant Air Conditioning Service	90		
Fan – Forced Draft	85	85	70
Fan – Induced Draft	75	75	63
Generator		110	63
Pump – Centrifugal circulating	95	95	75
Pump – Descaling (with Surge Tank)	55	55	
Pump – Pipe Line (Centrifugal)	75	75	63
Pump – Waterworks (Gen. Purpose)	75	75	63

TABLE 4-38

Single Helical Versus Double Helical or Herringbone Gears

There are some applications where single helical gears are definitely superior to double helical gears, and others where just the opposite is true. The comparisons that follow pertain, of course, to helical gears for operation on parallel shafts, commonly designated as parallel helical gears, since herringbone or double helical gears will not mesh and run together on other than parallel shafts.

(1) In machines where the gears are assembled onto shafts already in place by face-wise sliding along the teeth, there is no choice; single helical gears must be used. The same condition prevails in change-gear clusters and transmissions in which the gears are slid into mesh by axial displacements.

(2) In applications where both shafts sustain large axial thrusts, single helical gears are preferred to double gears because the pressure between the teeth can distribute itself evenly across the width of the single helical tooth but unequally on the double hands of the herringbone tooth.

(3) For the same reason herringbone gears are objectionable for applications wherein both shafts must be anchored against axial movement.

(4) In designs where only one shaft, or neither shaft, is subjected to significant axial thrust, double helical gears cause no additional thrust whereas single helical gears induce thrusts that have to be resisted by the bearings on both shafts.

INDEX TO SECTION 5

Straight Bevel Gears | Zerol Bevel Gears | Spiral Bevel Gears

TABLE 5–1

Data Ordinarily Put on Detail Drawing* of Straight Bevel Gear

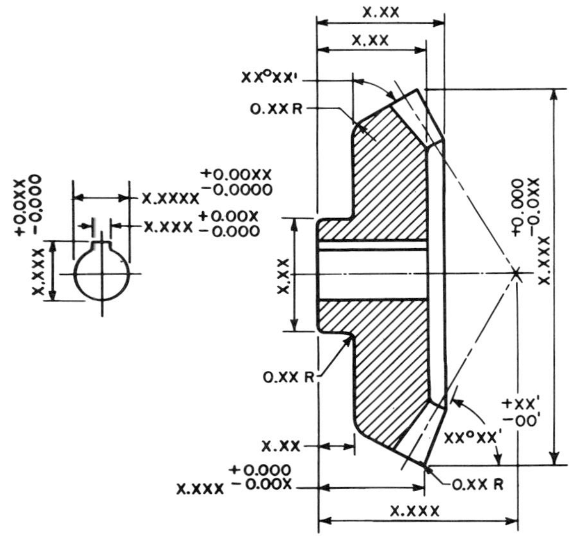

STRAIGHT BEVEL PINION DATA	
NUMBER TEETH	XX
DIAMETRAL PITCH	X.XXXX
PITCH DIAMETER	X.XXXX
SHAFT ANGLE	XX°XX'
PRESSURE ANGLE	XX°XX'
PITCH ANGLE	XX°XX'
ROOT ANGLE	XX°XX'
ADDENDUM ANGLE	X°XX'
ADDENDUM	0.XXX
CHORDAL ADDENDUM	0.XXX
CHORDAL THICKNESS	0.XXX
WHOLE DEPTH	0.XXX
CUTTING AND TEST DATA	
TOOTH ANGLE	XX°XX'
ROUGHING POINT WIDTH	0.XXX
BACKLASH IN ASSEMBLY	0.0XX TO 0.0XX
PART NUMBER OF MATE	
NUMBER TEETH ON MATE	XX —

MATERIAL & HEAT TREATMENT

MATERIAL
PRELIMINARY HEAT TREATMENT
DRAWING TEMPERATURE
HARDNESS

NOTES ON MACHINING

1. LIMITS ON FINISH DIMENSIONS ARE $\pm \frac{1}{64}$ UNLESS OTHERWISE SPECIFIED
2. BREAK ALL SHARP CORNERS

*A SECTION ON A DIAMETER IS OFTEN SUFFICIENT. GOOD PRACTICE IS TO PUT ON THE DRAWING THOSE DIMENSIONS, WITH TOLERANCES, NEEDED IN MAKING THE BLANK. SPECIFICATION OF THE MATERIAL, HEAT TREATMENT AND DATA FOR CUTTING THE TEETH ARE STATED IN TABULAR FASHION AND IN NOTES.

TABLE 5–2
Defining Tooth Proportions for Full-Depth, Long and Short Addendums, Straight and Zerol Bevel Gears

ASA B6.8 1950 AGMA 206.03 Gleason Systems

Working depth, h_k	$\dfrac{2.000}{P = \text{diametral pitch}}$
Whole depth, h_t	$\dfrac{2.188}{P} + 0.002$
Clearance, c	$\dfrac{0.188}{P} + 0.002$
Long addendum on pinion, a_P	$\dfrac{\text{Table 5–5, for ratio}}{P}$
Short addendum on gear, a_G	$\dfrac{\text{Table 5–5, for ratio}}{P}$
Ratio in Table 5–5, m	$\dfrac{N_G}{N_P} = \dfrac{\text{number teeth on gear}}{\text{number teeth on pinion}}$
Circular thicknesses of teeth, t_G, t_P	Compute for coarse-pitch gears, Tables 5–6, 5–9, 5–31 and 5–33. Table 5–10 for fine-pitch gears.

TABLE 5–3

Recommended Maximum Face Width Straight and Zerol Bevel Gears

	Face Width
Straight Bevel Gears	
Coarse Pitch	Not to exceed 1/3 cone distance nor greater than $\frac{10 \text{ inches}}{\text{diametral pitch}}$
20 Pitch & Finer	Not to exceed 3/10 cone distance nor greater than $\frac{8 \text{ inches}}{\text{diametral pitch}}$
Zerol Bevel Gears	Not to exceed 1/4 cone distance nor greater than $\frac{10 \text{ inches}}{\text{diametral pitch}}$

TABLE 5–4

Fewest Numbers of Teeth to Avoid Undercut, 90° Shaft Angle, 20° Pressure Angle, Straight Bevel Gears

Number Teeth on Pinion N_P	Minimum Number Teeth on Gear N_G	Ratio (Minimum) N_G/N_P
13	30	2.31
14	20	1.43
15	17	1.13
16	16	1.00

TABLE 5-5

Long and Short Addendums for One Diametral* Pitch According to Ratios 90° Shaft Angle Straight and Zerol Bevel Gears

ASA B6.8–1950 AGMA 206.03 Gleason Systems

Ratio $m = N_G/N_P$	Pinion Addendum a_P	Gear** Addendum a_G	Ratio $m = N_G/N_P$	Pinion Addendum a_P	Gear** Addendum a_G
1.00 to 1.00	1.000	1.000	1.42 to 1.45	1.240	0.760
1.00 to 1.02	1.010	0.990	1.45 to 1.48	1.250	0.750
1.02 to 1.03	1.020	0.980	1.48 to 1.52	1.260	0.740
1.03 to 1.04	1.030	0.970	1.52 to 1.56	1.270	0.730
1.04 to 1.05	1.040	0.960	1.56 to 1.60	1.280	0.720
1.05 to 1.06	1.050	0.950	1.60 to 1.65	1.290	0.710
1.06 to 1.08	1.060	0.940	1.65 to 1.70	1.300	0.700
1.08 to 1.09	1.070	0.930	1.70 to 1.76	1.310	0.690
1.09 to 1.11	1.080	0.920	1.76 to 1.82	1.320	0.680
1.11 to 1.12	1.090	0.910	1.82 to 1.89	1.330	0.670
1.12 to 1.14	1.100	0.900	1.89 to 1.97	1.340	0.660
1.14 to 1.15	1.110	0.890	1.97 to 2.06	1.350	0.650
1.15 to 1.17	1.120	0.880	2.06 to 2.16	1.360	0.640
1.17 to 1.19	1.130	0.870	2.16 to 2.27	1.370	0.630
1.19 to 1.21	1.140	0.860	2.27 to 2.41	1.380	0.620
1.21 to 1.23	1.150	0.850	2.41 to 2.58	1.390	0.610
1.23 to 1.25	1.160	0.840	2.58 to 2.78	1.400	0.600
1.25 to 1.27	1.170	0.830	2.78 to 3.05	1.410	0.590
1.27 to 1.29	1.180	0.820	3.05 to 3.41	1.420	0.580
1.29 to 1.31	1.190	0.810	3.41 to 3.94	1.430	0.570
1.31 to 1.33	1.200	0.800	3.94 to 4.82	1.440	0.560
1.33 to 1.36	1.210	0.790	4.82 to 6.81	1.450	0.550
1.36 to 1.39	1.220	0.780	6.81 to ∞	1.460	0.540
1.39 to 1.42	1.230	0.770			

**Use larger addendum on gear when ratio permits a choice.

*Divide value from table by diametral pitch to get addendum for another diametral pitch.

N_G = number of teeth on gear N_P = number of teeth on pinion

TABLE 5–6

Formulas for Calculating Dimensions of Straight Bevel Gears for Operation at 90° Shaft Angle

ASA B6.8–1950 AGMA 206.03 Gleason System

1 Number teeth on pinion	N_P, Table 5–4	5 Working depth	$h_k = \frac{2.000}{P}$
2 Number teeth on gear	N_G	6 Whole depth	$h_t = \frac{2.188}{P} + 0.002$
3 Diametral pitch	P	7 Clearance	$c = h_t - h_k$
4 Face width	F, Table 5–3	8 Pressure angle	$\phi = 20°$

	Pinion	Gear
9 Pitch diameter	$d = \frac{N_P}{P}$	$D = \frac{N_G}{P}$
10 Pitch angle	$\gamma = 90 - \Gamma$	$\Gamma = \tan^{-1} \frac{N_G}{N_P}$
11 Cone distance	$A_o = \frac{D}{2 \sin \Gamma}$	
12 Addendum	$a_P = \frac{\text{Table 5–5}}{P}$	$a_G = \frac{\text{Table 5–5}}{P}$
13 Dedendum	$b_P = \frac{2.188}{P} - a_P$	$b_G = \frac{2.188}{P} - a_G$
14 Dedendum angle	$\delta_P = \tan^{-1} \frac{b_P}{A_o}$	$\delta_G = \tan^{-1} \frac{b_G}{A_o}$
15 Face angle of blank	$\gamma_o = \gamma + \delta_G$	$\Gamma_o = \Gamma + \delta_P$
16 Root angle	$\gamma_R = \gamma - \delta_P$	$\Gamma_R = \Gamma - \delta_G$
17 Outside diameter	$d_o = d + 2a_P \cos \gamma$	$D_o = D + 2a_G \cos \Gamma$
18 Pitch apex to crown	$x_o = D/2 - a_P \sin \gamma$	$X_o = d/2 - a_G \sin \Gamma$
19 Circular thickness	* $t_P = \frac{3.1416}{P} - t_G$	$t_G = \frac{1.5708}{P} - (a_P - a_G) \tan \phi + \frac{K, \text{ Table 5–8}}{P}$
20 Backlash	B, Table 5–9	
21 Chordal thickness	$t_{CP} = t_P - \frac{t_P^3}{6d^2} - \frac{B}{2}$	$t_{CG} = t_G - \frac{t_G^3}{6D^2} - \frac{B}{2}$
22 Chordal addendum	$a_{CP} = a_P + \frac{t_P^2 \cos \gamma}{4d}$	$a_{CG} = a_G + \frac{t_G^2 \cos \Gamma}{4D}$
23 Tooth angle	$\frac{3438}{A_o} \left(\frac{t_P}{2} + b_P \tan \phi \right)$ minutes	$\frac{3438}{A_o} \left(\frac{t_G}{2} + b_G \tan \phi \right)$ minutes
24 Limit (max.) point width	$\frac{A_o - F}{A_o} (t_G - 2b_P \tan \phi) - 0.0015$	$\frac{A_o - F}{A_o} (t_P - 2b_G \tan \phi) - 0.0015$
25 Tool axial advance	0.002	0.002

* For fine pitch gears, 20 to 64 diametral pitch, circular thicknesses can be found by dividing tabular value from Table 5–10 by diametral pitch.

TABLE 5–7

Example to Illustrate Tabular Form for Calculating Dimensions of Straight Bevel Gears

Gleason 20° Straight Bevel Gear System—1951

See Table 5.6 for Formulas

1	Number teeth on pinion	$N_P = 16$	5	Working depth	$h_k = 0.400$
2	Number teeth on gear	$N_G = 49$	6	Whole depth	$h_t = 0.440$
3	Diametral pitch	$P = 5$	7	Clearance	$c = 0.040$
4	Face width	$F = 1.5$	8	Pressure angle	$\phi = 20°00'$

		Pinion		Gear
9	Pitch diameter	$d = 3.2000$		$D = 9.8000$
10	Pitch angle	$\gamma = 18°5'$		$\Gamma = 71°55'$
11	Cone distance		$A_o = 5.1546$	
12	Addendum	$a_P = 0.2840$		$a_G = 0.1160$
13	Dedendum	$b_P = 0.154$		$b_G = 0.322$
14	Dedendum angle	$\delta_P = 1°42'$		$\delta_G = 3°34'$
15	Face angle of blank	$\gamma_o = 21°39'$		$\Gamma_o = 73°37'$
16	Root angle	$\gamma_R = 16°23'$		$\Gamma_R = 68°21'$
17	Outside diameter	$d_o = 3.740$		$D_o = 9.872$
18	Pitch apex to crown	$x_o = 4.812$		$X_o = 1.490$
19	Circular thickness	$t_P = 0.3703$		$t_G = 0.2580$
20	Backlash		$B = 0.005$	
21	Chordal thickness	$t_{CP} = 0.367$		$t_{CG} = 0.255$
22	Chordal addendum	$a_{CP} = 0.294$		$a_{CG} = 0.117$
23	Tooth angle	$2°41'$		$2°44'$
24	Limit point width	0.102		0.095
25	Tool axial advance	0.002		0.002

TABLE 5–8

Gleason K Values for One Diametral Pitch for Circular Thickness Formula Straight and Zerol Bevel Gears

Ratio $\frac{N_G}{N_P}$	13 to 14	15 to 16	17 to 21	22 to 26	27 to 35	36 to 45
	Values of K in Inches					
1.000 to 1.020	--	+ 0.000	+ 0.000	+ 0.000	+ 0.000	+ 0.000
1.020 to 1.075	--	+ 0.020	+ 0.020	+ 0.015	+ 0.015	+ 0.010
1.075 to 1.140	--	+ 0.035	+ 0.035	+ 0.030	+ 0.025	+ 0.020
1.140 to 1.260	--	+ 0.055	+ 0.050	+ 0.045	+ 0.040	+ 0.030
1.260 to 1.855	+ 0.075	+ 0.070	+ 0.070	+ 0.060	+ 0.050	+ 0.040
1.855 to 2.250	+ 0.060	+ 0.060	+ 0.060	+ 0.050	+ 0.040	+ 0.030
2.250 to 2.645	+ 0.040	+ 0.045	+ 0.045	+ 0.035	+ 0.030	--
2.645 to 3.105	+ 0.020	+ 0.025	+ 0.025	+ 0.020	+ 0.015	--
3.105 to 3.650	+ 0.005	+ 0.010	+ 0.010	+ 0.005	+ 0.000	--
3.65 to 4.35	- 0.015	- 0.010	- 0.005	- 0.005	- 0.010	--
4.35 to 5.21	- 0.035	- 0.030	- 0.025	- 0.025	- 0.030	--
5.21 to 6.25	- 0.050	- 0.045	- 0.040	- 0.040	- 0.045	--
6.25 to 7.58	- 0.070	- 0.065	- 0.060	- 0.060	- 0.060	--
7.58 to 9.35	- 0.090	- 0.080	- 0.075	- 0.075	- 0.075	--
9.35 to 11.50	- 0.110	- 0.100	- 0.095	- 0.095	- 0.095	--

In case of choice, use smaller value
N_G = Number of teeth on gear
N_P = Number of teeth on pinion

TABLE 5–9

Backlash for Straight and Zerol Bevel Gears Gleason Systems

D.P.	Backlash
1.00 to 1.25	0.020 - 0.030
1.25 to 1.50	0.018 - 0.026
1.50 to 1.75	0.016 - 0.022
1.75 to 2.00	0.014 - 0.018
2.00 to 2.50	0.012 - 0.016
2.50 to 3.00	0.010 - 0.013
3.00 to 3.50	0.008 - 0.011
3.50 to 4.00	0.007 - 0.009
4 to 5	0.006 - 0.008
5 to 6	0.005 - 0.007
6 to 8	0.004 - 0.006
8 to 10	0.003 - 0.005
10 to 20	0.002 - 0.004
20 and finer	0.001 - 0.003

These backlashes are recommended between gears that are assembled ready to run. Normally design calculations are based on the smaller tolerance. Because of manufacturing tolerances and changes resulting from heat treatment, a backlash upon assembly is often smaller than that recommended in the table. Accordingly, the backlash used in design calculations should be increased enough to offset the changes during manufacture.

TABLE 5-10

Circular Thickness of Fine-Pitch, Straight Bevel Gears
ASA B6.8–1950 AGMA 206.03

One Diametral Pitch. To get circular thicknesses at another diametral pitch, divide tabular value by that diametral pitch.

Ratio $m = N_G/N_P$	Pinion Circular Thickness t_P	Gear Circular Thickness t_G	Ratio $m = N_G/N_P$	Pinion Circular Thickness t_P	Gear Circular Thickness t_G
1.00 to 1.00	1.5708	1.5708	1.42 to 1.45	1.7455	1.3961
1.00 to 1.02	1.5781	1.5635	1.45 to 1.48	1.7528	1.3888
1.02 to 1.03	1.5854	1.5562	1.48 to 1.52	1.7601	1.3815
1.03 to 1.04	1.5926	1.5490	1.52 to 1.56	1.7673	1.3743
1.04 to 1.05	1.5999	1.5417	1.56 to 1.60	1.7746	1.3670
1.05 to 1.06	1.6072	1.5344	1.60 to 1.65	1.7819	1.3597
1.06 to 1.08	1.6145	1.5271	1.65 to 1.70	1.7892	1.3524
1.08 to 1.09	1.6218	1.5198	1.70 to 1.76	1.7965	1.3451
1.09 to 1.11	1.6290	1.5126	1.76 to 1.82	1.8037	1.3379
1.11 to 1.12	1.6363	1.5053	1.82 to 1.89	1.8110	1.3306
1.12 to 1.14	1.6436	1.4980	1.89 to 1.97	1.8183	1.3233
1.14 to 1.15	1.6509	1.4907	1.97 to 2.06	1.8256	1.3160
1.15 to 1.17	1.6582	1.4834	2.06 to 2.16	1.8329	1.3087
1.17 to 1.19	1.6654	1.4762	2.16 to 2.27	1.8401	1.3015
1.19 to 1.21	1.6727	1.4689	2.27 to 2.41	1.8474	1.2942
1.21 to 1.23	1.6800	1.4616	2.41 to 2.58	1.8547	1.2869
1.23 to 1.25	1.6873	1.4543	2.58 to 2.78	1.8620	1.2796
1.25 to 1.27	1.6945	1.4471	2.78 to 3.05	1.8693	1.2723
1.27 to 1.29	1.7018	1.4398	3.05 to 3.41	1.8765	1.2651
1.29 to 1.31	1.7091	1.4325	3.41 to 3.94	1.8838	1.2578
1.31 to 1.33	1.7164	1.4252	3.94 to 4.82	1.8911	1.2505
1.33 to 1.36	1.7237	1.4179	4.82 to 6.81	1.8984	1.2432
1.36 to 1.39	1.7309	1.4107	6.81 to ∞	1.9057	1.2359
1.39 to 1.42	1.7382	1.4034			

N_G – Number of teeth on gear

N_P – Number of teeth on pinion

TABLE 5–11

Formulas for Calculating Dimensions of Angular, Straight Bevel Gears

ASA B6.8–1950 AGMA 206.03 Gleason System

1 Number teeth on pinion	N_P	5 Working depth	$h_k = \dfrac{2.000}{P}$
2 Number teeth on gear	N_G	6 Whole depth	$h_t = \dfrac{2.188}{P} + 0.002$
3 Diametral pitch	P	7 Shaft angle	Σ
4 Face width	F	8 Pressure angle	ϕ (see item 8 after item 18)

	Pinion	Gear
9 Pitch diameter	$d = \dfrac{N_P}{P}$	$D = \dfrac{N_G}{P}$
10 Pitch angle	(from item 10a or 10b)	(from item 10a or 10b) and less than 90°
10a When Σ is less than 90°	$\tan\gamma = \dfrac{\sin\Sigma}{N_G/N_P + \cos\Sigma}$	$\tan\Gamma = \dfrac{\sin\Sigma}{N_P/N_G + \cos\Sigma)}$
10b When Σ is greater than 90°	$\tan\gamma = \dfrac{\sin(180^\circ - \Sigma)}{N_G/N_P - \cos(180^\circ - \Sigma)}$	$\tan\Gamma = \dfrac{\sin(180^\circ - \Sigma)}{N_P/N_G - \cos(180^\circ - \Sigma)}$
10c Check calculation	$\Sigma = \gamma + \Gamma$	
11 Cone distance	$A_o = \dfrac{D}{2\sin\Gamma}$	
12 Addendum	$a_P = h_k - a_G$	$a_G = \dfrac{\text{From Table 5-4 for } m_{90}\text{, item 12a}}{P}$
12a ${}_{90}N_P$ = number teeth on equivalent 90° pinion; $m_{90} = \tan\Gamma_{90}$ = equivalent 90° ratio	${}_{90}N_P = \dfrac{N_P \sin\Gamma_{90}}{\cos\gamma}$	$m_{90} = \sqrt{\dfrac{N_G \cos\gamma}{N_P \cos\Gamma}} = \tan\Gamma_{90}$
13 Dedendum	$b_P = \dfrac{2.188}{P} - a_P$	$b_G = \dfrac{2.188}{P} - a_G$
14 Dedendum angle	$\delta_P = \tan^{-1}\dfrac{b_P}{A_o}$	$\delta_G = \tan^{-1}\dfrac{b_G}{A_o}$
15 Face angle of blank	$\gamma_o = \gamma + \delta_G$	$\Gamma_o = \Gamma + \delta_P$
16 Root angle	$\gamma_R = \gamma - \delta_P$	$\Gamma_R = \Gamma - \delta_G$
17 Outside diameter	$d_o = d + 2a_P\cos\gamma$	$D_o = D + 2a_G\cos\Gamma$
18 Pitch apex to crown	$x_o = A_o\cos\gamma - a_P\sin\gamma$	$X_o = A_o\cos\Gamma - a_G\sin\Gamma$

continued on next page

TABLE 5–11, continued

	Pinion	Gear
8 Pressure angle to avoid undercut ϕ		Enter Table 5–12 with δ_P and γ. Point of intersection must not be above curve of selected pressure angle.
19 Circular thickness	$t_P = \dfrac{3.1416}{P} - t_G$	$t_G = \dfrac{1.5708}{P} - (a_P - a_G)\tan\phi + \dfrac{K, \text{ Table 5–8, using } m_{90} \text{ and } {}_{90}N_{P'} \text{ item } 12a}{P}$
20 Backlash		B, Table 5–9
21 Chordal thickness	$t_{CP} = t_P - \dfrac{t_P^3}{6d^2} - \dfrac{B}{2}$	$t_{CG} = t_G - \dfrac{t_G^3}{6D^2} - \dfrac{B}{2}$
22 Chordal addendum	$a_{CP} = a_P + \dfrac{t_P^2 \cos\gamma}{4d}$	$a_{CG} = a_G + \dfrac{t_G^2 \cos\Gamma}{4D}$
23 Tooth angle	$\dfrac{3438}{A_o}\left(\dfrac{t_P}{2} + b_P \tan\phi\right)$ minutes	$\dfrac{3438}{A_o}\ \dfrac{t_G}{2} + b_G \tan\phi$ minutes
24 Limit point width	$\dfrac{A_o - F}{A_o}(t_G - 2b_P \tan\phi) - 0.0015$	$\dfrac{A_o - F}{A_o}(t_P - 2b_G \tan\phi) - 0.0015$
25 Tool axial advance	0.002	0.002

NOTE: Angular gears require special *ratio of roll gears* for generation on Gleason generators. The decimal ratio for the *NC/*75 ratio machines is found as follows: Decimal ratio of gears = $A_o\ P/37.5$.

TABLE 5–12

Relation Between the Dedendum Angle and Pitch Angle at Which Undercut Begins With Sharp-Cornered Tools on Generated Straight Bevel Gears

Gleason 20° Straight Bevel Gear System–1951 Gleason Works

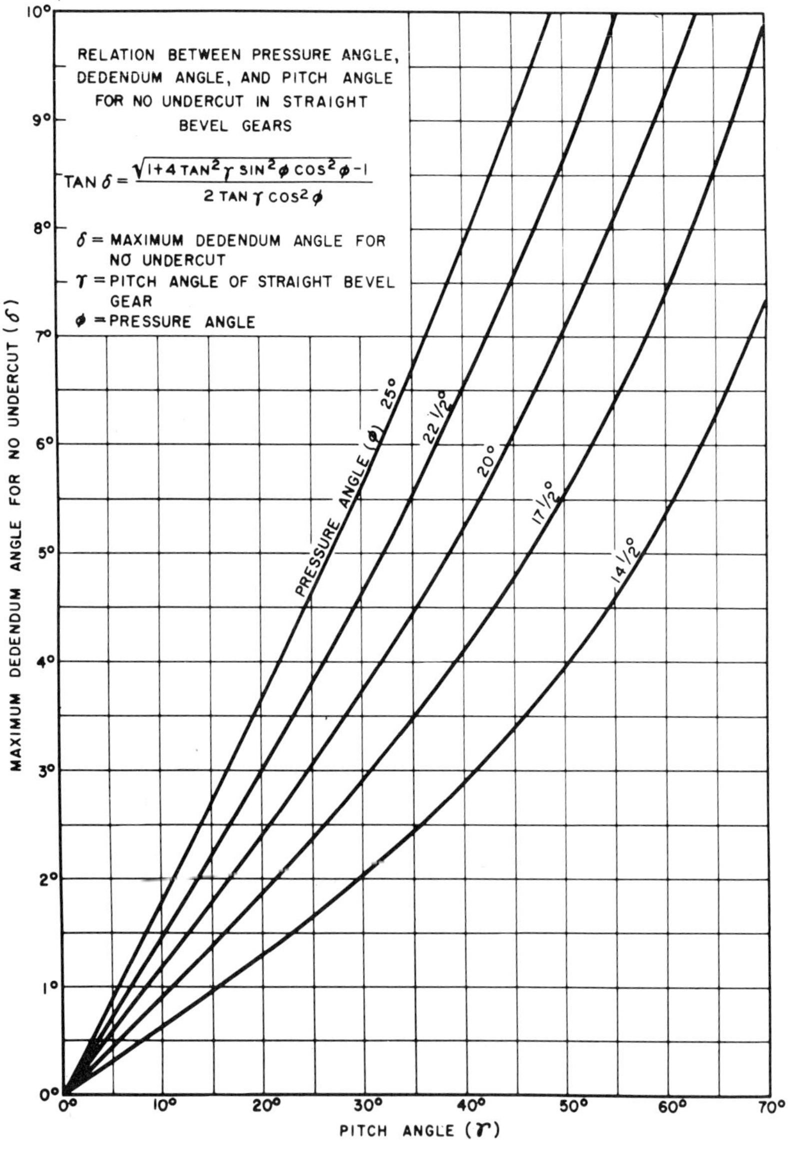

TABLE 5–13

Example to Illustrate Calculations of Dimensions of a Pair of Fine-Pitch, Angular, Straight, Bevel Gears

ASA B6.8–1950 AGMA 206.03

1	Number teeth on pinion	$N_P = 20$	5	Working depth	$h_k = 0.0666$
2	Number teeth on gear	$N_G = 30$	6	Whole depth	$h_t = 0.0749$
3	Diametral pitch	$P = 30$	7	Shaft angle	$\Sigma = 120°0'$
4	Face width	$F = 1/8$	8	Pressure angle	$\phi = 20°0'$

		Pinion	Gear
9	Pitch diameter	$d = 0.6667$	$D = 1.0000$
10	Pitch angle	$\gamma = 40°54'$	$\Gamma = 79°6'$
10b		$\tan\gamma = \dfrac{\sin 60°}{\frac{30}{20} - \cos 60°} = 0.866025$	$\tan\Gamma = \dfrac{0.866025}{0.166667} = 5.19589$
11	Cone distance	$A_o = 0.5092$	
12	Addendum	$a_P = 0.0463$	$a_G = 0.0203$
12a	Equivalent 90° ratio, m_{90}	*	$m_{90} = 2.45$
13	Dedendum	$b_P = 0.0266$	$b_G = 0.0526$
14	Dedendum angle	$\delta_P = 2°59'$	$\delta_G = 5°54'$
15	Face angle of blank	$\gamma_o = 46°48'$	$\Gamma_o = 82°5'$
16	Root angle	$\gamma_R = 37°55'$	$\Gamma_R = 73°12'$
17	Outside diameter	$d_o = 0.737$	$D_o = 1.008$
18	Pitch apex to crown	$x_o = 0.355$	$X_o = 0.076$
19	Circular thickness	$\dfrac{\text{Table 5–10 and } m_{90}\text{, item 12a}}{P}$ $t_P = 0.0618$	$t_G = 0.0429$
20	Backlash usually can be omitted in calculations for fine-pitch gears.		
21	Chordal thickness	$t_{CP} = 0.062$	$t_{CG} = 0.043$
22	Chordal addendum	$a_{CP} = 0.047$	$a_{CG} = 0.020$
23	Tooth angle	4°34'	
24	Limit point width	0.016	
25	Tool axial advance	0.002	

*Calculation of number of teeth on equivalent 90° pinion, ${}_{90}N_p$, can be omitted for fine-pitch gears, 20 to 64 diametral pitch, because teeth of fine-pitch combinations rarely need to be balanced in thickness to equalize strengths and consequently circular thicknesses, item 19, can be found from Table 5–10.

TABLE 5–14

Diagram Showing Important Bevel Gear Blank Dimensions

ASA B6.8–1950 AGMA 206.03

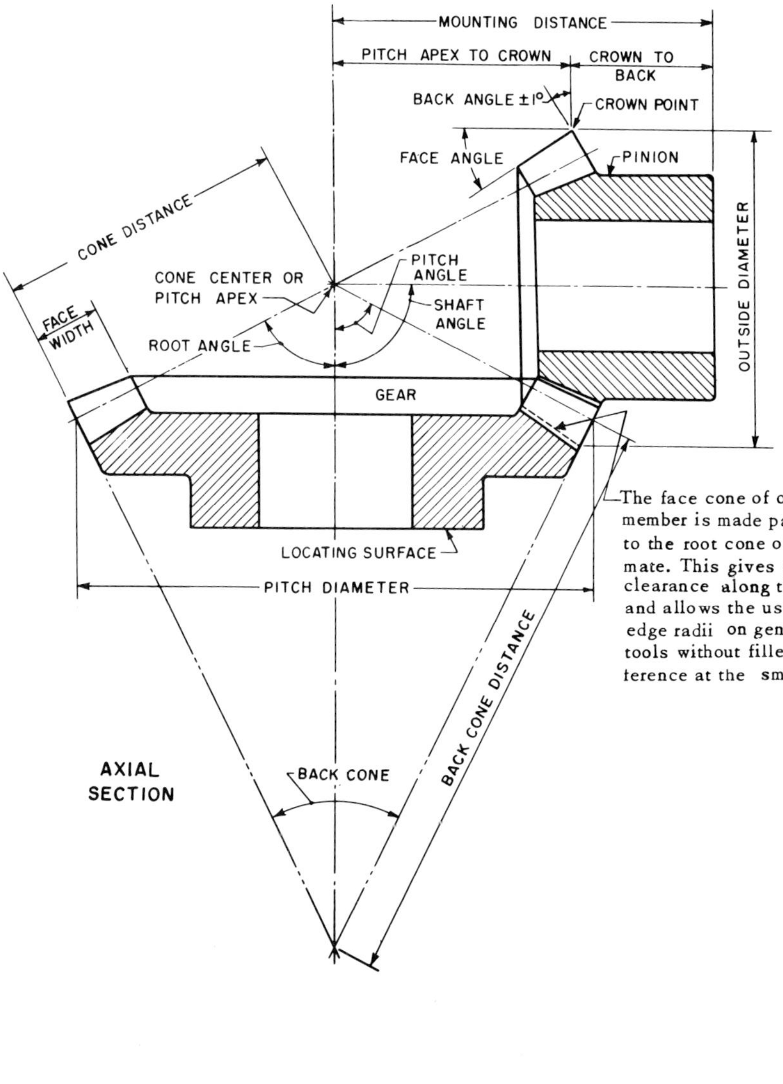

The face cone of one member is made parallel to the root cone of the mate. This gives constant clearance along the tooth and allows the use of large edge radii on generating tools without fillet interference at the small end.

TABLE 5–15

Gear Blank Tolerances on Bores and Shanks of Bevel Gears

Recommendation of Gleason Works

Nominal Size of Centering Bore or Shank	Super-Precision Gears		Precision Gears		General Quality Gears		Commercial Gears	
	Shank	Bore	Shank	Bore	Shank	Bore	Shank	Bore
up to 1 inch	wring fit in chuck	wring fit on arbor						
up to 4			+ 0.0000 – 0.0002	+ 0.0002 – 0.0000	+ 0.0000 – 0.0005	+ 0.0005 – 0.0000	+ 0.000 – 0.001	+ 0.001 – 0.000
4 to 10			+ 0.0000 – 0.0003	+ 0.0003 – 0.0000	+ 0.000 – 0.001	+ 0.001 – 0.000	+ 0.000 – 0.002	+ 0.002 – 0.000
10 to 20					+ 0.000 – 0.001	+ 0.001 – 0.000	+ 0.000 – 0.003	+ 0.003 – 0.000
over 20					+ 0.000 – 0.002	+ 0.002 – 0.000	+ 0.000 – 0.004	+ 0.004 – 0.000

Dimensions are in inches.

TABLE 5-16

Tolerances on OD and Crown-to-Back Dimensions of Bevel Gear Blank

Recommendations of Gleason Works

Diametral Pitch	Outside-Diameter Tolerance	Crown-to-Back* Tolerance
2.5 and coarser	+ 0.000 – 0.010	+ 0.000 – 0.004
2.5 to 24	+ 0.000 – 0.005	+ 0.000 – 0.002
24 and finer	+ 0.000 – 0.003	+ 0.000 – 0.002

*Tolerance may have to be increased for gears on which backing is ground to fit the mounting distance.

TABLE 5-17

Tolerances on Face and Back Angles of Bevel Gear Blanks
Recommendations of Gleason Works

Diametral Pitch	Face Angle Tolerance	Back Angle Tolerance
coarser than 18	*+ 8 minutes - 0	*± 15 minutes
18 to 35	+ 15 minutes - 0	± 30 minutes
35 and finer	+ 30 minutes - 0	± one degree

*A closer tolerance is usually necessary if face cone or back cone serves as a locating surface during manufacture.

Figure in Table 5-14

TABLE 5-18

Tolerances on OD and Crown to Back on Blanks for Fine-Pitch Bevel Gears

ASA B6.8-1950 AGMA 206.03

Diametral Pitch	Outside Diameter Tolerance
20 to 30	+ 0.000 - 0.005
31 to 40	+ 0.000 - 0.004
41 to 56	+ 0.000 - 0.003
57 to 94	+ 0.000 - 0.002
95 and finer	+ 0.000 - 0.001

Diametral Pitch	Crown-to-Back Tolerance
20 to 47	+ 0.000 - 0.002
47 and finer	+ 0.000 - 0.001

TABLE 5–19

Tolerances on Face and Back Angles of Blanks For Fine-Pitch Bevel Gears

ASA B6.8–1950 AGMA 206.03

Face Width, In.	Face Angle Tolerance	Back Angle Tolerance
1/2	+ 10 minutes - 0	
1/4	+ 20 minutes - 0	± one degree generally is satisfactory
3/16	+ 30 minutes - 0	
1/8	+ 40 minutes - 0	

TABLE 5–20

Runout and Other Blank Tolerances for Bevel Gears Gleason Works

PINION–SHANK TYPE

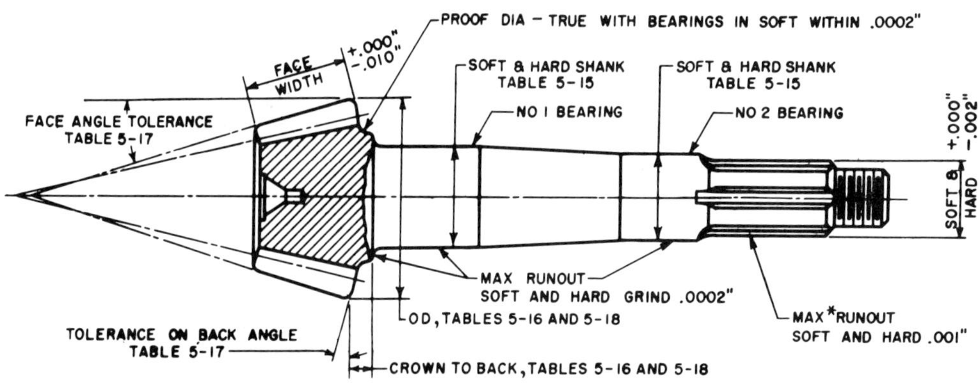

*These tolerances used only when this diameter is gripping surface for collet of cutting and testing chucks.

Clean centers after hardening and straighten to a maximum runout of .0005" on No. 1 bearing and .001" on No. 2 bearing. When bearing spacing exceeds 3", the maximum runout on No. 2 bearing is .002".

Proof diameter must run true within 0.0005" before hard grinding.

continued on next page

PINION-BORED TYPE

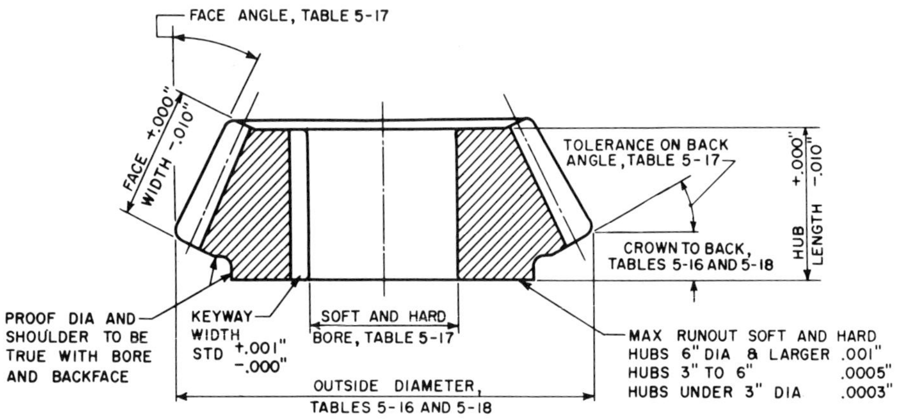

Back face of hub should not be ground in hard unless the distortion is excessive.

Diameter and shoulder proof must run true within .0002" before grinding.

GEARS-WEB TYPE-COUNTERBORED

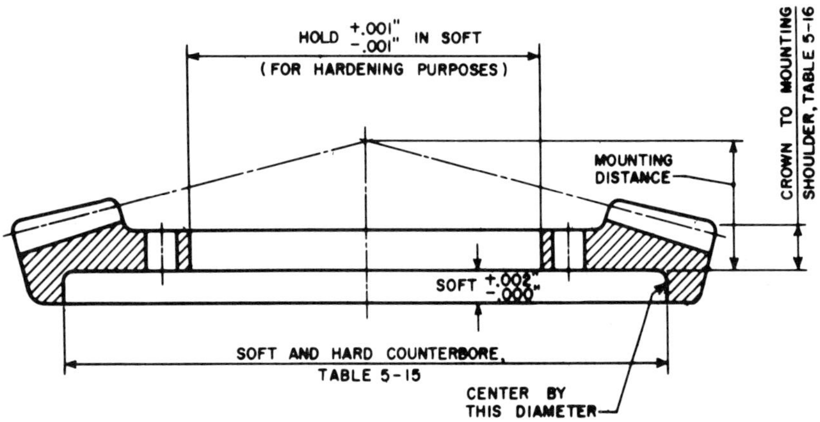

GEARS-HUB TYPE

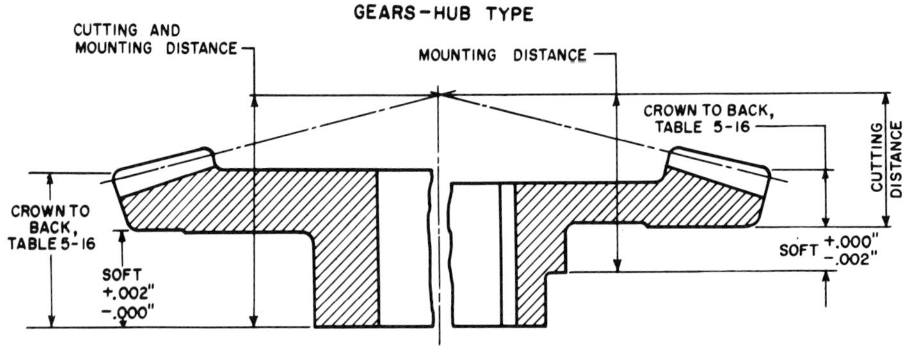

Dimension from shoulder to end of hub or mounting shoulder must be held to +.000", −.002" for cutting and hardening support.

Tolerance on other surfaces to be the same as shown on preceding page.

continued on next page

GEARS–WEBLESS TYPE

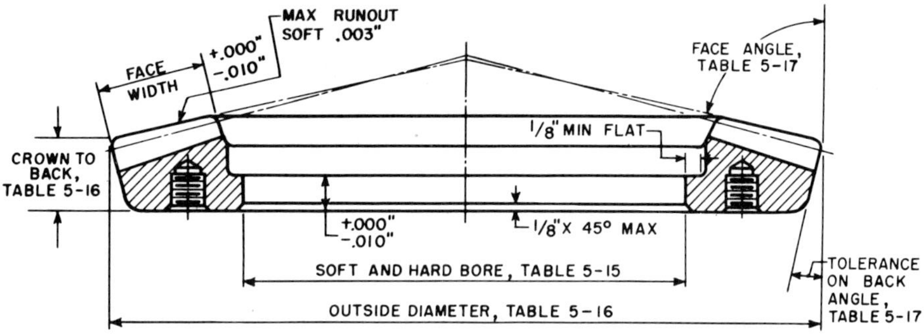

GEARS–WEB TYPE

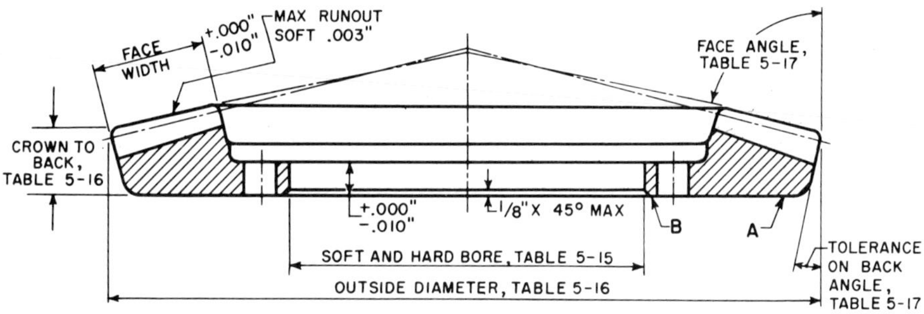

Maximum out-of-round in bore is .001″ in soft, .003″ after hardening but before grinding.

When placed on flat plate, back of gear should be flat within the following limits:

Webless Type

Soft—.001″ feeler must not go under anywhere.

Hard—.003″ feeler must not go under anywhere.

Web Type

Soft—.001″ feeler must not go under anywhere.

Hard—.003″ feeler must not go under at "A" and .005″ feeler must not go under at "B". If .004″ feeler goes under anywhere at "B," .002″ feeler must go under around the entire bore.

Back of gear should be hard ground only when teeth are to be hard ground and in special cases.

TABLE 5-21

Tolerance on Axes Intersection for Bevel Gears
Gleason Works

Gear Diameters, Inches	Tolerance on Intersection of Axes
up to 12	± 0.001
12 to 24	± 0.002
24 to 36	± 0.0025

TABLE 5-22

Coniflex Bevel Gears
Gleason Works

Gleason generators of recent design produce a localized tooth bearing on straight bevel gears. Such gears are known as "Coniflex" bevel gears. The localization of the tooth bearing permits a slight amount of adjustment of the gears in assembly and small deflections under loads without concentration of pressure on the ends of the teeth.

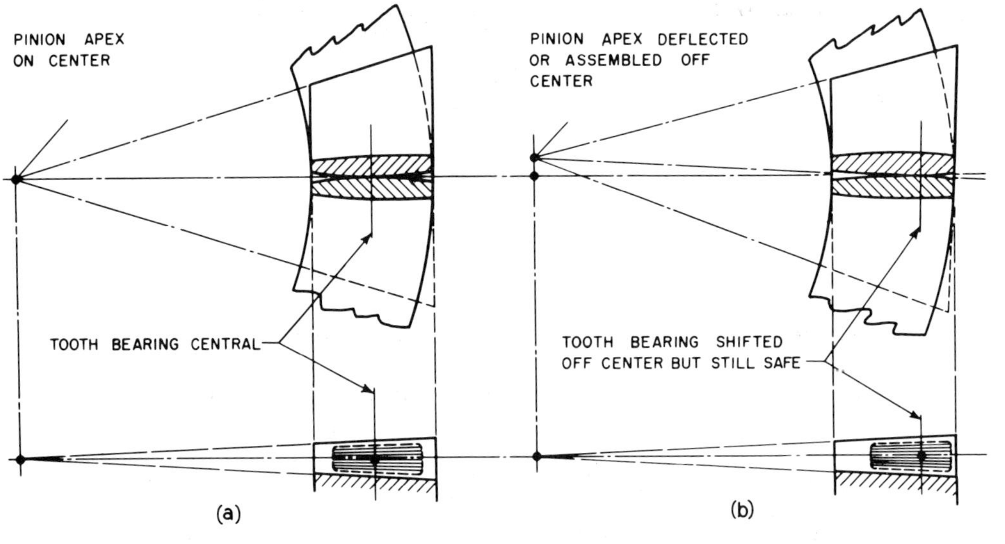

Advantage of Coniflex gears. The usual operating position of a straight bevel gear and pinion is shown at (a), while the position after a displacement is shown at (b). In the displaced position, the load still is not concentrated on the ends of the teeth, nor is the length of contact materially shortened. The gears will continue to run smoothly and quietly with a safe distribution of load.

TABLE 5–23

Commonly Applied Procedure of Gleason Works in Designing Straight Bevel Gears for Surface Durability and Fatigue Strength

Gear ratio, pinion speed in rpm, magnitude and character of loading are regarded as known data upon which to base design.

Step Number	Formula	In Which
First, compute design load	$P_2 = C_S P_1$ or $\frac{P_M}{2}$ whichever is greater	P_2 = design load in horsepower P_1 = normal operating load in horsepower P_M = momentary peak load in horsepower C_S = service factor, Table 5–24
	and $T_2 = \frac{63{,}025\,P_2}{\text{pinion rpm}}$	T_2 = design torque on pinion in inch-pounds
	Starting torque on pinion should not exceed $3T_2$.	
Second, find gear size based on surface durability	$P_{100} = \frac{100\,P_2}{C_M\,(\text{pinion rpm})}$	P_{100} = rated power per 100 rpm of pinion C_M = material factor for surface durability, Table 5–25
	Enter Table 5–26 with P_{100} and ratio to find an approximate pinion pitch diameter, d_{app} $V = 0.262\,d_{app}\,(\text{pinion rpm})$	NOTE: If pinion rpm and d_{app} gives a pitch line speed V greater than 1000 ft per min, increase d over that found from Table 5–26.
Third, find numbers of teeth and diametral pitch	Enter Table 5–28 with d_{app} and ratio to get an approximate number of teeth on pinion. Formula relationships are	
	$d = \frac{N_P}{P}$ $D = \frac{N_G}{P}$ $\text{Ratio} = \frac{N_G}{N_P}$	d = (exact) pitch diameter of pinion N_P = number teeth on pinion N_G = number teeth on gear P = diametral pitch
Fourth, compute fatigue strength	$P_3 = \frac{F\,Y_K\,V\,C_m}{6.3\,P}$	P_3 = maximum horsepower gears can safely transmit under normal operating conditions. P_3 should be greater than P_2. F = face width (not to exceed one-third the cone distance) Y_K = tooth form factor, Table 5–27 V = pitch line speed, fpm C_m = material factor for strength, Table 5–25 P = diametral pitch

TABLE 5–24

Service Factor,* C_S, For Use in Evaluating Loads on Straight Bevel Gears According to Gleason Procedure

Gleason 20° Straight Bevel Gear System – 1951 Gleason Works

Power Source	Character of Load on Driven Machine		
	Uniform	Moderate Shock	Heavy Shock
Uniform	1.00	1.25	1.75
Light Shock	1.10	1.35	1.80
Medium Shock	1.25	1.50	1.85

*This table is for speed-decreasing drives; for speed-increasing drives add 0.15 to these factors.

TABLE 5–25

Material Factors, C_m for Surface Durability and C_m for Strength, for Straight Bevel Gears

Gleason 20° Straight Bevel Gear System – 1951 Gleason Works

Gear	Brinell	Rockwell "C"	Pinion	Brinell	Rockwell "C"	C_M for Durability	C_m for Strength
Cast Iron	—	—	Cast Iron	—	—	0.30	0.10†
Cast Iron	—	—	Annealed Steel	160-200	—	0.30	0.10†
Cast Iron	—	—	Surface Hardened Steel	—	50*	0.40	0.10†
Cast Iron	—	—	Case Hardened Steel	—	55*	0.40	0.10†
Heat Treated Steel	210–245	—	Heat Treated Steel	245-280	—	0.35	0.50
Surface Hardened Steel	—	50*	Surface Hardened Steel	—	50*	1.00	0.50
Surface Hardened Steel	—	50*	Case Hardened Steel	—	55*	1.00	0.50
Case Hardened Steel	—	55*	Case Hardened Steel	—	55*	1.00	1.00

*Minimum values.
†Based on cast iron of good quality with tensile strength of at least 30,000 pounds per square inch.

TABLE 5–26

Load Capacity of Case-Hardened Straight Bevel Gears Operating at 90° Shaft Angle

Gleason 20° Straight Bevel Gear System – 1951 Gleason Works

CHART TO ACCOMPANY TABLES 5–23 AND 5–44

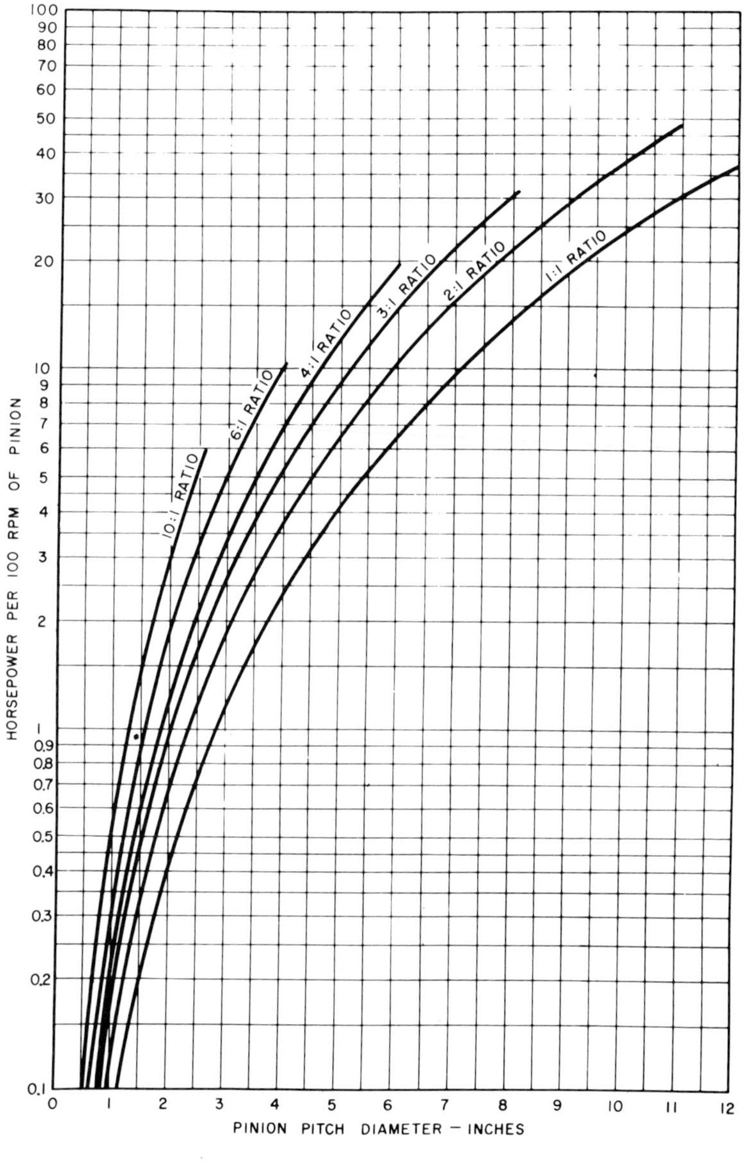

TABLE 5–27

Tooth Form Factors Y_K For the Gleason 20° Straight Bevel Gear System – 1951

Number of Teeth in Pinion	Ratios (90° Shaft Angle Only) 1.000 to 1.055	1.055 to 1.110	1.110 to 1.170	1.170 to 1.225	1.225 to 1.285	1.285 to 1.350	1.350 to 1.415	1.415 to 1.490	1.490 to 1.565	1.565 to 1.650	1.650 to 1.735	1.735 to 1.840	1.840 to 1.945	1.945 to 2.065	2.065 to 2.195
	Values of Y_K for Gleason Formula — 20° Pressure Angle Only														
13															
14								0.645	0.654	0.664	0.673	0.683	0.692	0.702	0.711
15			0.618	0.627	0.636	0.646	0.656	0.664	0.673	0.683	0.692	0.702	0.711	0.720	0.729
16	0.617	0.627	0.637	0.646	0.655	0.664	0.674	0.682	0.691	0.700	0.709	0.719	0.728	0.737	0.746
17	0.635	0.644	0.654	0.663	0.672	0.681	0.691	0.699	0.708	0.716	0.725	0.735	0.744	0.753	0.761
18	0.652	0.661	0.671	0.680	0.689	0.698	0.707	0.715	0.724	0.732	0.741	0.750	0.759	0.768	0.776
19	0.668	0.677	0.687	0.696	0.705	0.714	0.722	0.730	0.739	0.747	0.756	0.765	0.773	0.782	0.790
20	0.684	0.693	0.703	0.712	0.720	0.729	0.737	0.745	0.754	0.762	0.771	0.779	0.787	0.796	0.804
21	0.697	0.706	0.715	0.724	0.732	0.741	0.749	0.757	0.766	0.774	0.782	0.790	0.798	0.807	0.815
22	0.709	0.718	0.727	0.736	0.744	0.752	0.760	0.768	0.777	0.785	0.793	0.801	0.809	0.817	0.825
23	0.721	0.730	0.739	0.747	0.755	0.763	0.771	0.779	0.788	0.796	0.804	0.811	0.819	0.827	0.835
24	0.732	0.741	0.750	0.758	0.766	0.774	0.782	0.790	0.798	0.806	0.814	0.821	0.829	0.837	0.845
25	0.743	0.752	0.761	0.769	0.777	0.785	0.792	0.800	0.808	0.816	0.824	0.831	0.839	0.847	0.854
26	0.754	0.763	0.771	0.779	0.787	0.795	0.802	0.810	0.818	0.825	0.833	0.840	0.848	0.856	0.863
27	0.763	0.772	0.780	0.788	0.796	0.804	0.811	0.818	0.826	0.833	0.841	0.848	0.856	0.864	0.871
28	0.771	0.780	0.789	0.797	0.804	0.812	0.819	0.826	0.834	0.841	0.849	0.856	0.863	0.871	0.879
29	0.779	0.788	0.797	0.805	0.812	0.820	0.827	0.834	0.841	0.848	0.856	0.863	0.870	0.878	0.886
30	0.787	0.796	0.804	0.812	0.819	0.827	0.834	0.841	0.848	0.855	0.863	0.870	0.877	0.885	0.892
31-32	0.798	0.807	0.815	0.823	0.830	0.837	0.844	0.851	0.858	0.865	0.873	0.880	0.887	0.894	
33-34	0.811	0.820	0.828	0.835	0.842	0.850	0.857	0.864	0.870	0.877	0.884	0.891	0.898	0.905	
35-36	0.824	0.832	0.840	0.848	0.855	0.862	0.869	0.876	0.882	0.889	0.896	0.903	0.909	0.916	
37-38	0.836	0.844	0.852	0.859	0.866	0.874	0.881	0.887	0.893	0.900	0.907	0.914	0.920	0.927	
39-41	0.851	0.859	0.867	0.874	0.881	0.888	0.895	0.901	0.907	0.914	0.920	0.927	0.933	0.939	

The above table of tooth form factors is for 20-deg pressure angle throughout. The stress concentration factor for stress at the root fillet is based on a tool edge radius of $\frac{0.240\text{ in.}}{P}$. Table values of Y_K should be decreased by 11 per cent if the tool radius is reduced to $\frac{0.120\text{ in.}}{P}$.

continued on next page

TABLE 5–27, continued

Number of Teeth in Pinion	2.195 to 2.335	2.335 to 2.500	2.50 to 2.69	2.69 to 2.89	2.89 to 3.12	3.12 to 3.38	3.38 to 3.68	3.68 to 4.03	4.03 to 4.39	4.39 to 4.85	4.85 to 5.4	5.4 to 6.0	6.0 to 6.8	6.8 to 7.7	7.7 to 8.8	8.8 to 10.0
	Ratios (90° Shaft Angle Only)															
	Values of Y_K for Gleason Formula — 20° Pressure Angle Only															
13	0.701	0.711	0.721	0.730	0.740	0.750	0.760	0.770	0.779	0.789	0.799	0.809	0.819	0.829	0.840	0.850
14	0.721	0.730	0.740	0.749	0.759	0.769	0.779	0.789	0.798	0.808	0.818	0.828	0.838	0.848	0.859	0.869
15	0.739	0.748	0.758	0.767	0.777	0.786	0.796	0.806	0.815	0.825	0.835	0.845	0.855	0.865	0.876	0.886
16	0.755	0.764	0.774	0.783	0.793	0.802	0.811	0.821	0.831	0.840	0.850	0.861	0.871	0.881	0.891	0.901
17	0.770	0.779	0.789	0.798	0.808	0.817	0.826	0.836	0.845	0.854	0.864	0.875	0.885	0.895	0.906	0.916
18	0.785	0.794	0.803	0.812	0.822	0.831	0.840	0.850	0.859	0.868	0.878	0.889	0.899	0.909	0.920	0.930
19	0.799	0.808	0.817	0.826	0.835	0.844	0.853	0.863	0.872	0.881	0.891	0.902	0.912	0.922	0.933	0.944
20	0.812	0.821	0.830	0.839	0.848	0.857	0.866	0.876	0.885	0.894	0.904	0.915	0.925	0.935	0.946	0.957
21	0.823	0.832	0.840	0.850	0.858	0.867	0.876	0.886	0.895	0.904	0.914	0.925	0.935	0.945	0.956	0.967
22	0.833	0.842	0.850	0.860	0.868	0.877	0.886	0.895	0.904	0.913	0.923	0.934	0.945	0.955	0.966	0.977
23	0.843	0.852	0.860	0.870	0.878	0.886	0.895	0.904	0.913	0.922	0.932	0.943	0.954	0.964	0.976	0.986
24	0.852	0.861	0.869	0.879	0.887	0.895	0.904	0.913	0.922	0.931	0.941	0.952	0.963	0.973	0.985	0.996
25	0.862	0.870	0.878	0.888	0.896	0.904	0.913	0.922	0.931	0.940	0.950	0.961	0.971	0.982	0.994	1.005
26	0.871	0.879	0.887	0.897	0.904	0.913	0.921	0.930	0.939	0.948	0.958	0.969	0.979	0.990	1.002	1.013
27	0.879	0.887	0.895	0.904	0.911	0.920	0.928	0.937	0.946	0.955	0.965	0.976	0.986	0.997	1.009	1.020
28	0.886	0.895	0.902	0.911	0.918	0.927	0.935	0.944	0.953	0.962	0.972	0.983	0.993	1.004	1.016	1.027
29	0.893	0.902	0.909	0.918	0.925	0.934	0.942	0.951	0.960	0.969	0.978	0.989	0.999	1.011	1.023	1.034
30	0.900	0.908	0.915	0.924	0.931	0.940	0.948	0.957	0.966	0.975	0.984	0.995	1.005	1.017	1.029	1.040

The above table of tooth form factors is for 20-deg pressure angle throughout. The stress concentration factor for stress at the root fillet is based on a tool edge radius of $\frac{0.240 \text{ in.}}{P}$. Table values of Y_K should be decreased by 11 per cent if the tool radius is reduced to $\frac{0.120 \text{ in.}}{P}$

TABLE 5–28

Approximate Relationship Between Pinion Pitch Diameter and Numbers of Teeth for Selected Gear Ratios

Gleason 20° Straight Bevel Gear System – 1951 Gleason Works

CHART TO ACCOMPANY TABLE 5–23

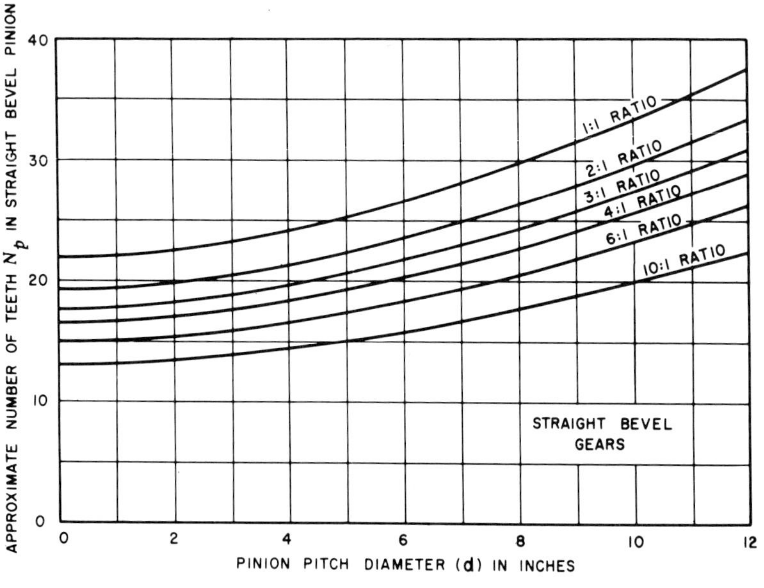

TABLE 5–29

Pressure Angle and Ratio on Zerol Bevel Gears at 90° Shaft Angle

Gleason Zerol Bevel Gear System–1954 Gleason Works

Number of Teeth on Pinion, N_P	Number of Teeth on Gear, N_G, and *20° Pressure angle	22½° Pressure angle	25° Pressure angle
17	17 or more		
16	20 or more	16 to 19, inclusive	
15	25 or more	15 to 24, inclusive	
14		14 or more	
13		15 or more	13 and 14

*The basic pressure angle is 20°, the larger pressure angles being used for the ratios as listed to avoid undercut. The face width of zerol bevel gears is limited to 25 percent of the cone distance or to $10/P$, whichever is the smaller.

TABLE 5–30

Data Ordinarily Put on Specification Drawing* of Zerol Bevel Gear

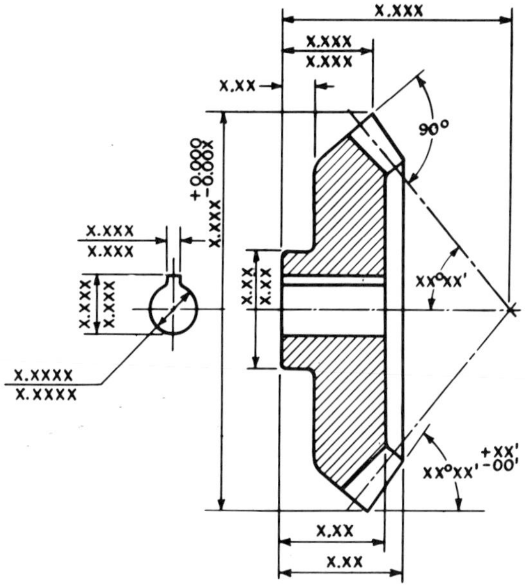

ZEROL BEVEL PINION DATA

NUMBER TEETH	XX
DIAMETRAL PITCH	X.XXXX
PITCH DIAMETER	X.XXX
SHAFT ANGLE	XX°XX'
PRESSURE ANGLE	XX°XX'
HAND OF SPIRAL	
ROOT ANGLE	XX°XX'
ADDENDUM	0.XXX
WHOLE DEPTH	0.XXX
CIRCULAR THICKNESS	0.XXXX

CUTTING AND TEST DATA

CUTTING SUMMARY NO.	
BACKLASH IN ASSEMBLY	0.0XX TO 0.0XX
PART NO. OF MATE	
NUMBER TEETH IN MATE	XX

MATERIAL & HEAT TREATMENT

SPECIFICATIONS
HEAT TREATMENT SPECIFICATION
DEPTH OF CASE
CASE HARDNESS
CORE HARDNESS

NOTES ON MACHINING

1. LIMITS ON FINISH DIMENSIONS ARE $\pm \frac{1}{64}$ UNLESS OTHERWISE SPECIFIED
2. BREAK ALL SHARP CORNERS

*A SECTION ON A DIAMETER IS OFTEN SUFFICIENT. GOOD PRACTICE IS TO PUT ON THE DRAWING PROPER THOSE DIMENSIONS, WITH TOLERANCES, NEEDED IN MAKING THE BLANK. SPECIFICATION OF THE MATERIAL, HEAT TREATMENT AND SOME DATA ON TOOTH PROPORTIONS ARE STATED IN TABULAR FASHION AND IN NOTES. DATA PERTINENT TO SETTINGS OF THE MACHINE TOOLS FOR CUTTING AND FINISHING THE TEETH ARE COMPILED ON A SUPPLEMENTARY SHEET, CALLED A CUTTING SUMMARY. VARIATIONS IN THE METHODS OF CUTTING AND GRINDING ZEROL BEVEL GEARS MAKE THE DUPLICATION AND MATING OF THEM IMPOSSIBLE WITHOUT A CUTTING SUMMARY. THIS IS ONE REASON WHY THE ORDINARY DIMENSIONAL DRAWINGS OF ZEROL AND SPIRAL BEVEL GEARS ARE COMMONLY REFERRED TO AS SPECIFICATION DRAWINGS RATHER THAN DETAIL DRAWINGS.

TABLE 5–31

Formulas for Calculating Dimensions of Zerol Bevel Gears* for Operation at 90° Shaft Angle

Gleason Zerol Bevel Gear System–1954 Gleason Works

1 Number of teeth on pinion	N_P, Table 5–29	5 Working depth	$h_k = \frac{2.000}{P}$
2 Number of teeth on gear	N_G	6 Whole depth	$h_t = \frac{2.188}{P} + 0.002$
3 Diametral pitch	P	7 Clearance	$c = h_t - h_k$
4 Face width	F, Table 5–3	8 Pressure angle ϕ, Table 5–29	

	Pinion	Gear
9 Pitch diameter	$d = N_P/P$	$D = N_G/P$
10 Pitch angle	$\gamma = \tan^{-1} \frac{N_P}{N_G}$	$\Gamma = 90^\circ - \gamma$
11 Cone distance	$A_o = \frac{D}{2 \sin \Gamma}$	
12 Addendum	$a_P = \frac{\text{Table 5–5}}{P}$	$a_G = \frac{\text{Table 5–5}}{P}$
13 Dedendum	$b_P = h_t - a_P$	$b_G = h_t - a_G$
14 Dedendum angle	$\delta_P = \tan^{-1} \frac{b_P}{A_o} + \Delta\delta$†, Table 5–33	$\delta_G = \tan^{-1} \frac{b_G}{A_o} + \Delta\delta$†
15 Face angle of blank	$\gamma_o = \gamma + \delta_G$	$\Gamma_o = \Gamma + \delta_P$
16 Root angle	$\gamma_R = \gamma - \delta_P$	$\Gamma_R = \Gamma - \delta_G$
17 Outside diameter	$d_o = d + 2\, a_P \cos \gamma$	$D_o = D + 2\, a_G \cos \Gamma$
18 Pitch apex to crown	$x_o = D/2 - a_P \sin \gamma$	$X_o = d/2 - a_G \sin \Gamma$
19 Circular thickness	$t_P = \frac{3.1416}{P} - t_G$	$t_G = \frac{1.5708}{P} - (a_P - a_G) \tan \phi + \frac{K\text{, Table 5–8}}{P}$

*Normally zerol bevel gears have the same blank proportions throughout and tooth action as straight **bevel gears** and may be assembled in the same mountings.

†The dedendum angle consists of two parts: the angle without regard for the Duplex taper plus the change in angle to give Duplex taper.

TABLE 5-32

Example to Illustrate Tabular Form of Calculating Dimensions of Zerol Bevel Gears

Gleason Zerol Bevel Gear System—1954 Gleason Works

See Table 5-31 for Formulas

1 Number teeth on pinion	$N_P = 16$	5 Working depth	$h_k = 0.200$
2 Number teeth on gear	$N_G = 49$	6 Whole depth	$h_t = 0.221$
3 *Diametral pitch	$P = 10$	7 Clearance	$C = 0.021$
4 Face width	$F = 0.625$	8 Pressure angle	$\phi = 20^\circ$

	Pinion		Gear
9 Pitch diameter	$d = 1.6000$		$D = 4.9000$
10 Pitch angle	$\gamma = 18^\circ 5'$		$\Gamma = 71^\circ 55'$
11 Cone distance		$A_o = 2.5773$	
12 Addendum	$a_P = 0.1420$		$a_G = 0.0580$
13 Dedendum	$b_P = 0.079$		$b_G = 0.163$
14 Dedendum angle	$\delta_P = 3^\circ 40'$		$\delta_G = 5^\circ 2'$
15 Face angle of blank	$\gamma_o = 23^\circ 37'$		$\Gamma_o = 75^\circ 35'$
16 Root angle	$\gamma_R = 14^\circ 25'$		$\Gamma_R = 66^\circ 23'$
17 Outside diameter	$d_o = 1.870$		$D_o = 4.936$
18 Cone center to crown	$x_o = 2.406$		$X_o = 0.745$
19 Circular thickness	$t_P = 0.1852$		$t_G = 0.1290$

*The Duplex Method for cutting Zerol bevel gears provides a rapid and economical method by which both the gear and the pinion are cut spread blade, i.e., both sides of a tooth space are finished simultaneously. In order to accomplish this, the root line of the gear blank is tilted to produce tooth bottoms of uniform width, while maintaining proper taper along the pitch line. For Duplex Zerols, the diametral pitch is limited to 10 and finer if the teeth are to be cut only. If the gear teeth are to be ground, a diametral pitch as coarse as 6 may be used.

TABLE 5-33

Dedendum Angle Increment, $\Delta\,\delta$, for Duplex Taper, Zerol Bevel Gears

Gleason Zerol Bevel Gear System—1954 Gleason Works

Pressure Angle	When Shaft Angle Is 90° Change in Dedendum Angle = $\Delta\,\delta$ (Minutes)	Where
20°	$\Delta\,\delta = \frac{6668}{N_C} - \frac{300\sqrt{d}}{N_C F} - \frac{14P}{N_C}$	N_C = number teeth in crown gear = $2PA_o$
22-1/2°	$\Delta\,\delta = \frac{4868}{N_C} - \frac{300\sqrt{d}}{N_C F} - \frac{14P}{N_C}$	F = face width d = pitch diameter of pinion
25°	$\Delta\,\delta = \frac{3412}{N_c} - \frac{300\sqrt{d}}{N_c F} - \frac{14P}{N_c}$	A_o = outside cone distance P = diametral pitch

Pressure Angle	When Shaft Angle is Less or Greater Than 90°, Change in Dedendum Angle = $\Delta\,\delta$ (Minutes)	Where
20°	$\Delta\,\delta = \frac{6668}{N_C} - \frac{300}{F}\sqrt{\frac{1}{N_C P\,(\tan\gamma + \tan\Gamma)}} - \frac{14P}{N_C}$	γ = pitch angle of pinion
22-1/2°	$\Delta\,\delta = \frac{4868}{N_C} - \frac{300}{F}\sqrt{\frac{1}{N_C P\,(\tan\gamma + \tan\Gamma)}} - \frac{14P}{N_C}$	Γ = pitch angle of gear
25°	$\Delta\,\delta = \frac{3412}{N_c} - \frac{300}{F}\sqrt{\frac{1}{N_c P\,(\tan\gamma + \tan\Gamma)}} - \frac{14P}{N_c}$	

TABLE 5-34

Formulas for Calculating Dimensions of Angular Zerol Bevel Gears

Gleason Bevel Gear System—1954 Gleason Works

1 Number of teeth on pinion	N_P, Table 5-29	5 Working depth	$h_k = \dfrac{2.000}{P}$
2 Number of teeth on gear	N_G	6 Whole depth	$h_t = \dfrac{2.188}{P} + 0.002$
3 Diametral pitch	P	7 Shaft angle	Σ
4 Face Width	F, Table 5-3	8 Pressure angle	ϕ (item 20)

	Pinion	Gear
Items 9 to 13, inclusive, as given in Table 5-11, can now be used to find the pitch angles, the equivalent 90-degree, bevel-gear ratio, and the dedendums.		
14 Dedendum angle	$\delta_P = \tan^{-1}\dfrac{b_P}{A_o} + \Delta\delta$, Table 5-33	$\delta_G = \tan^{-1}\dfrac{b_G}{A_o} + \Delta\delta$
15 Face angle of blank	$\gamma_o = \gamma + \delta_G$	$\Gamma_o = \Gamma + \delta_P$
16 Root angle	$\gamma_R = \gamma - \delta_P$	$\Gamma_R = \Gamma - \delta_G$
17 Outside diameter	$d_o = d + 2a_P \cos\gamma$	$D_o = D + 2a_G \cos\Gamma$
18 Pitch apex to crown	$x_o = A_o \cos\gamma - a_P \sin\gamma$	$X_o = A_o \cos\Gamma - a_G \sin\Gamma$
19 Circular thickness	$t_P = \dfrac{3.1416}{P} - t_G$	$t_G = \dfrac{1.5708}{P} - (a_P - a_G)\tan\phi + \dfrac{K\text{, Table 5-8, using } m_{90}\text{, item 12a}}{P}$
20 Minimum pressure angle, ϕ'	$\sin\phi' = \sqrt{\dfrac{1.15\, b_P}{A_o \tan\gamma}}$	Select the pressure angle $\phi = 20^\circ$ or $22\frac{1}{2}^\circ$ or 25° such that ϕ is not less than the minimum ϕ'

TABLE 5–35

Data Ordinarily Put on Specification Drawing* of Spiral Bevel Gear

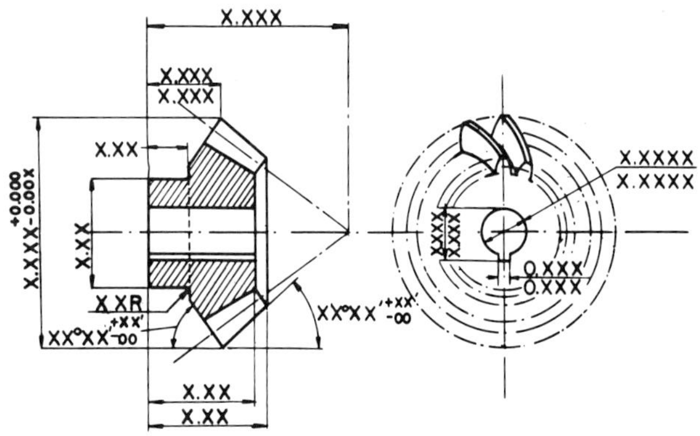

SPIRAL BEVEL GEAR DATA

NUMBER TEETH	XX
DIAMETRAL PITCH	X.XXXX
PITCH DIAMETER	X.XXXX
SHAFT ANGLE	XX° XX′
PRESSURE ANGLE	XX° XX′
SPIRAL ANGLE	XX° XX′
HAND OF SPIRAL	
ROOT ANGLE	XX° XX′
ADDENDUM	0.XXX
WHOLE DEPTH	0.XXX
CIRCULAR THICKNESS	0.XXXX
PART NO. OF MATE	
NUMBER TEETH IN MATE	XX

BEVEL TOOTH CUTTING & GRINDING DATA

DRIVER OR DRIVEN	
DIRECTION OF ROTATION	
SPEED IN RPM	
BACKLASH IN ASSEMBLY	0.0XX TO 0.0XX
SUMMARY NO.	

MATERIAL & HEAT TREATMENT

SPECIFICATIONS
PRELIMINARY HEAT TREATMENT SPEC
PRELIMINARY HARDNESS
HEAT TREATMENT SPEC
DEPTH OF CASE
CASE HARDNESS
CORE HARDNESS

NOTES ON MACHINING

1. LIMITS ON FINISH DIMENSION ARE $\pm \frac{1}{64}$ UNLESS OTHERWISE SPECIFIED
2. BREAK ALL SHARP CORNERS

Accordingly the cutting summary becomes as much a part of the detailed description of a spiral bevel gear as does the drawing for the blank. In fact a spiral bevel gear cannot be duplicated without a cutting summary.

*Good practice is to put on the drawing proper those dimensions, with tolerances, needed in making the blank. Specification of the material, heat treatment and some data on the tooth proportions are stated in tabular form and in notes. Data pertinent to the settings of the machine tools for cutting and finishing the teeth are compiled on a supplementary sheet, called a cutting summary.

TABLE 5–36

Defining Tooth Proportions for Spiral Bevel Gears Long and Short Addendums

Gleason †Spiral Bevel Gear System – 1952 Gleason Works

Working depth	h_k	$\dfrac{1.700}{P}$	P = Diametral pitch
Whole depth	h_t	$\dfrac{1.888}{P} + (0.005)$ *	
Clearance	c	$\dfrac{0.188}{P}$	
Short addendum on gear	a_G	$\dfrac{\text{Table 5–37, for ratio}}{P}$	
Limiting Ratios, Table 5–38		N_G = number teeth in gear N_P = number teeth in pinion	
Long addendum on pinion	a_P	$h_k - a_G$	
Circular thickness of tooth on gear	t_G	$\dfrac{1.5708}{P} - 1.22\ (a_P - a_G) \tan \phi - \dfrac{K,\ \text{Table 5–39}}{P}$	
Circular thickness of tooth on pinion	t_P	$\dfrac{3.1416}{P} - t_G$	
Pressure angle	ϕ	$\phi = 20^\circ$ is the basic pressure angle	

* It is common practice on gears of 10 diametral pitch and coarser to rough-cut 0.005 deeper than the calculated depth to avoid having finishing blades cut on ends.

† The data given in this table and the tables that follow pertain to spiral bevel gears for general industrial purposes, either with speed increasing or with speed decreasing, and which are more or less recognized as standard applications. The teeth are generated and of 12 diametral pitch and coarser. There are, on the contrary, many applications of spiral bevel gears that do not conform to these standard proportions, for one reason or another, and which may be regarded as Special Designs. The Gleason Works lists the following in the latter category: (1) automotive rear-axle drives; (2) formate gears; (3) gears and pinions of 12 diametral pitch and finer which are usually cut by one of the duplex spread-blade methods; (4) gear cut spread-blade and pinion cut single-side, with a spiral angle less than 20 degrees; (5) ratios having fewer teeth than those listed in Table 5–38.

TABLE 5–37

Gear Addendum for One Diametral Pitch Spiral Bevel Gears

Gleason

To obtain addendum select from table value corresponding to ratio given by the formula.*

$$\text{Ratio} = \frac{\text{Number of teeth in gear}}{\text{Number of teeth in pinion}}$$

Ratios		Add.	Ratios		Add.	Ratios		Add.	Ratios		Add.
From	To	Inch.	From	To	Inch	From	To	Inch	From	To	Inch
1.00	1.00	0.850	1.15	1.17	0.750	1.41	1.44	0.650	1.99	2.10	0.550
1.00	1.02	0.840	1.17	1.19	0.740	1.44	1.48	0.640	2.10	2.23	0.540
1.02	1.03	0.830	1.19	1.21	0.730	1.48	1.52	0.630	2.23	2.38	0.530
1.03	1.05	0.820	1.21	1.23	0.720	1.52	1.57	0.620	2.38	2.58	0.520
1.05	1.06	0.810	1.23	1.26	0.710	1.57	1.63	0.610	2.58	2.82	0.510
1.06	1.08	0.800	1.26	1.28	0.700	1.63	1.68	0.600	2.82	3.17	0.500
1.08	1.09	0.790	1.28	1.31	0.690	1.68	1.75	0.590	3.17	3.67	0.490
1.09	1.11	0.780	1.31	1.34	0.680	1.75	1.82	0.580	3.67	4.56	0.480
1.11	1.13	0.770	1.34	1.37	0.670	1.82	1.90	0.570	4.56	7.00	0.470
1.13	1.15	0.760	1.37	1.41	0.660	1.90	1.99	0.560	7.00	∞	0.460

*In case of choice, use the larger addendum.

TABLE 5–38

Fewest Numbers of Teeth to Avoid Undercut on Spiral Bevel Gears

Gleason Spiral Bevel Gear System–1952 Gleason Works

Number Teeth on Pinion N_P	Minimum Number Teeth on Gear N_G	Ratio $\frac{N_G}{N_P}$	Condition
17	17	1	These limiting numbers of teeth are based on a standard pressure angle of 20° and for a spiral angle of 35 degrees. If smaller spiral angles are used, undercut may occur and the contact ratio may be less.
16	18	1.12	
15	19	1.26	
14	20	1.43	
13	22	1.69	
12	26	2.16	

TABLE 5–39

Circular Thickness Factors for Spiral Bevel Gears with 20° Pressure Angle and 35° Spiral Angle

Gleason Spiral Bevel Gear System—1952 Gleason Works

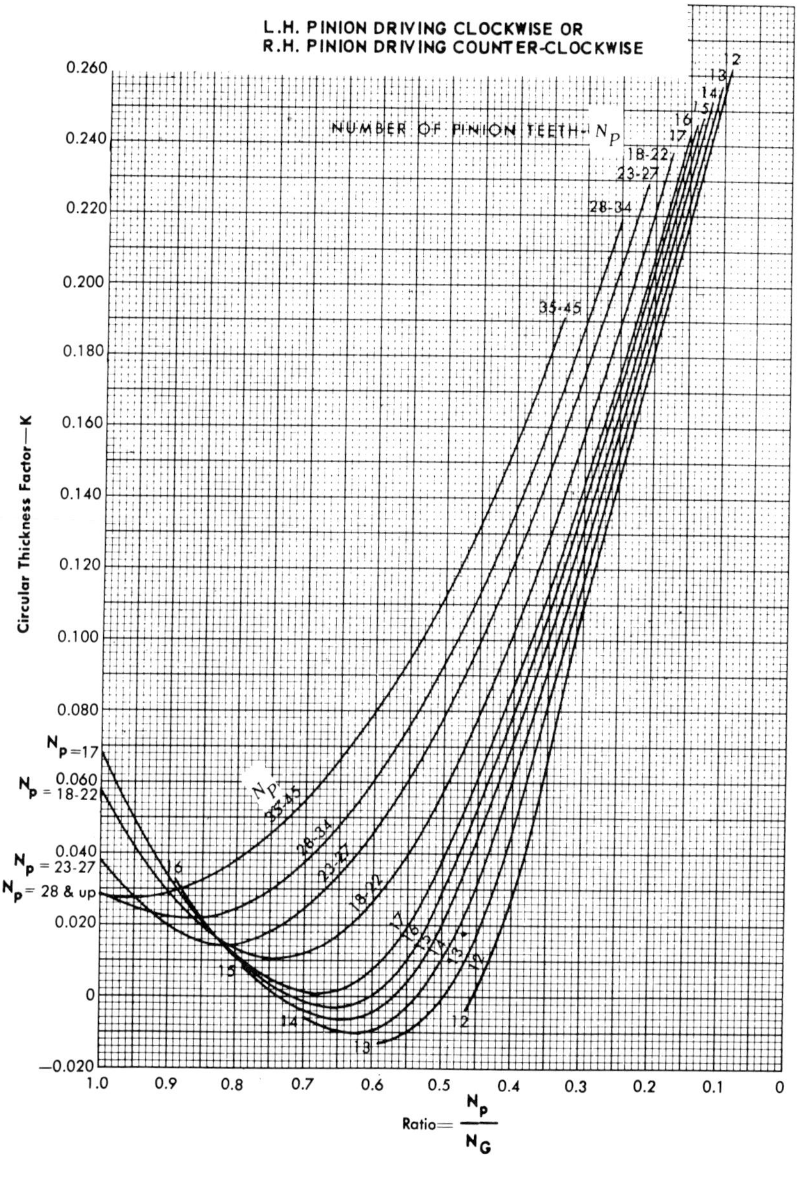

The tooth thicknesses are proportioned so that the stresses in the gear and pinion will be approximately equal with a left hand pinion driving clockwise or a right hand pinion driving counter clockwise. This will give a satisfactory balance of life for gears operating below the endurance limit. If the gears are to operate above the endurance limit, special proportions will be required. Also, on reversible drives which must be designed for optimum load capacity special proportions will be required. The method of determining the balance of strength for these special cases may be found in the Gleason publication "Strength of Bevel and Hypoid Gears".

TABLE 5-40

Formulas for Calculating Dimensions of Spiral Bevel Gears for Operation at 90° Shaft Angle

Gleason Spiral Bevel Gear System Gleason Works

1 Number teeth on pinion	N_P, Table 5–37		
2 Number teeth on gear	N_G	6 Whole depth	$h_t = \dfrac{1.888}{P}$
3 Diametral pitch	P		
4 *Face width ($0.3\ A_o$)	F	7 Clearance	$c = h_t - h_k$
5 Working depth	$h_k = \dfrac{1.700}{P}$	8 Pressure angle	$\phi = 20^\circ$

	Pinion	Gear
9 Pitch diameter	$d = N_P/P$	$D = N_G/P$
10 Pitch angle	$\gamma = \tan^{-1} \dfrac{N_P}{N_G}$	$\Gamma = \tan^{-1} \dfrac{N_G}{N_P}$
11 Cone distance	$A_o = \dfrac{D}{2 \sin \Gamma}$	
12 Addendum	$a_P = h_k - a_G$	$a_G = \dfrac{\text{Table 5–37}}{P}$
13 Dedendum	$b_P = h_t - a_P$	$b_G = h_t - a_G$
14 Dedendum angle	$\delta_P = \tan^{-1} \dfrac{b_P}{A_o}$	$\delta_G = \tan^{-1} \dfrac{b_G}{A_o}$
15 Root angle	$\gamma_R = \gamma - \delta_P$	$\Gamma_R = \Gamma - \delta_G$
16 †Face angle of blank	$\gamma = \gamma + \delta_G$	$\Gamma_o = \Gamma + \delta_P$
17 Outside diameter	$d_o = d + 2a_P \cos \gamma$	$D_o = D + 2a_G \cos \Gamma$
18 Pitch apex to crown	$x_o = D/2 - a_P \sin \gamma$	$X_o = d/2 - a_G \sin \Gamma$
19 Circular thickness	$t_P = \dfrac{3.1416}{P} - t_G$	$t_G = \dfrac{1.5708}{P} - 1.22\,(a_P - a_G) \tan \phi - \dfrac{K,\ \text{Table 5–39}}{P}$

*The recommended face width is 0.3 the cone distance or $10/P$, whichever is the smaller.

†The face cone element of a blank is made parallel to the root cone element of the mating gear. This gives constant clearance along the tooth and allows the use of larger edge radii on the cutters without fillet interference at the small end.

TABLE 5–41

Example to Illustrate Tabular Method of Calculating Dimensions of Spiral Bevel Gears

Gleason Spiral Bevel Gear System—1952 Gleason Works

1	Number teeth on pinion	$N_P = 14$	5	Working depth	$h_k = 0.425$
2	Number teeth on gear	$N_G = 43$	6	Whole depth	$h_t = 0.472$
3	Diametral pitch	$P = 4$	7	Clearance	$c = 0.047$
4	Face width	$F = 1.625$	8	Pressure angle	$\phi = 20^{\circ}$

		Pinion	Gear
9	Pitch diameter	$d = 3.5000$	$D = 10.7500$
10	Pitch angle	$\gamma = 18^{\circ}2'$	$\Gamma = 71^{\circ}58'$
11	Cone distance	$A_o = 5.6527$	
12	Addendum	$a_P = 0.300$	$a_G = 0.125$
13	Dedendum	$b_P = 0.172$	$b_G = 0.347$
14	Dedendum angle	$\delta_P = 1^{\circ}45'$	$\delta_G = 3^{\circ}31'$
15	Root angle	$\gamma_R = 16^{\circ}17'$	$\Gamma_R = 68^{\circ}27'$
16	Face angle of blank	$\gamma_o = 21^{\circ}33'$	$\Gamma_o = 73^{\circ}43'$
17	Outside diameter	$d_o = 4.071$	$D_o = 10.827$
18	Pitch apex to crown	$x_o = 5.282$	$X_o = 1.631$
19	Circular thickness	$t_P = 0.4974$	$t_G = 0.2880$

TABLE 5–42

Formulas for Calculating Dimensions of Angular, Spiral Bevel Gears

Gleason Spiral Bevel Gear System – 1952 Gleason Works

1 Number teeth on pinion	N_P	5 Working depth	$h_k = 1.700/P$
2 Number teeth on gear	N_G	6 Whole depth	$h_t = 1.888/P$
3 Diametral pitch	P	7 Shaft angle	Σ
4 Face width	F	8 Pressure angle	ϕ (see item 8 after item 18)

	Pinion	Gear
9 Pitch diameter	$d = N_P/P$	$D = N_G/P$
10 Pitch angle	γ (from item 10a or 10b)	Γ (item 10a or 10b and less than 90°)
10a When Σ is less than 90°	$\tan\gamma = \dfrac{\sin\Sigma}{N_G/N_P + \cos\Sigma}$	$\tan\Gamma = \dfrac{\sin\Sigma}{N_P/N_G + \cos\Sigma}$
or		
10b When Σ is greater than 90°	$\tan\gamma = \dfrac{\sin(180^\circ - \Sigma)}{N_G/N_P - \cos(180^\circ - \Sigma)}$	$\tan\Gamma = \dfrac{\sin(180^\circ - \Sigma)}{N_P/N_G - \cos(180^\circ - \Sigma)}$
10c Check calculation	$\Sigma = \gamma + \Gamma$	or $\sin\gamma/\sin\Gamma = N_P/N_G$
11 Cone distance	$A_o = \dfrac{D}{2\sin\Gamma}$	
12 Addendum	$a_P = h_k - a_G$	$a_G = \dfrac{\text{From Table 5–37 for } m_{90}\text{, item 12a}}{P}$
12a ${}_{90}N_P$ = number teeth on equivalent 90° pinion; $m_{90} = \tan\Gamma_{90}$ = equivalent 90° ratio	${}_{90}N_P = \dfrac{N_P\sin\Gamma_{90}}{\cos\gamma}$	$m_{90} = \sqrt{\dfrac{N_G\cos\gamma}{N_P\cos\Gamma}} = \tan\Gamma_{90}$
13 **Dedendum**	$b_P = h_t - a_P$	$b_G = h_t - a_G$
14 Dedendum angle	$\delta_P = \tan^{-1}\dfrac{b_P}{A_o}$	$\delta_G = \tan^{-1}\dfrac{b_G}{A_o}$
15 Face angle of blank	$\gamma_o = \gamma + \delta_G$	$\Gamma_o = \Gamma + \delta_P$
16 Root angle	$\gamma_R = \gamma - \delta_P$	$\Gamma_R = \Gamma - \delta_G$
17 Outside diameter	$d_o = d + 2a_P\cos\gamma$	$D_o = D + 2a_G\cos\Gamma$
18 Pitch apex to crown	$x_o = A_o\cos\gamma - a_P\sin\gamma$	$X_o = A_o\cos\Gamma - a_G\sin\Gamma$
8 Pressure angle ϕ to avoid undercut	Enter Table 5–43 with δ_P and γ. Point of intersection must not be above curve of chosen pressure angle.	
19 Circular thickness	$t_P = \dfrac{3.1416}{P} - t_G$	$t_G = \dfrac{1.5708}{P} - 1.22\,(a_P - a_G)\tan\phi - \dfrac{K\text{, Table 5–39, using } m_{90} \text{ and } {}_{90}N_P\text{, item 12a}}{P}$

TABLE 5–43

Relation Between Pressure Angle, Dedendum Angle, and Pitch Angle for No Undercut in Spiral Bevel Gears with 35-Deg Spiral Angle

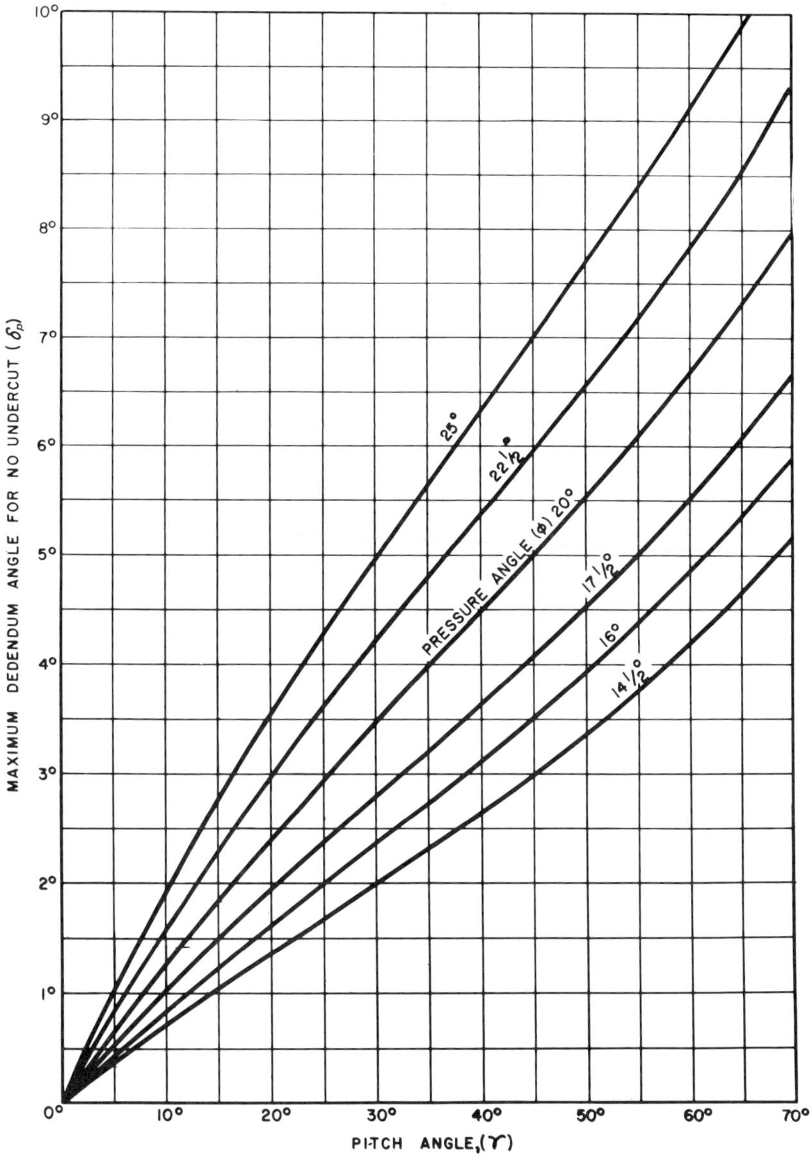

Relation between the dedendum angle and pitch angle at which undercut begins to occur in generating spiral gears of 35° spiral angle using sharp-cornered tools.

TABLE 5-44

Commonly Applied Procedure of Gleason Works in Designing Spiral Bevel Gears for Surface Durability*

Gleason Bevel and Hypoid Gear Design Gleason Works

Step Number	Formula and Instructions	Remarks and Definition of Symbols
(1) find service horsepower	(a) Service horsepower, P, is found by multiplying the actual horsepower by a service factor C_S, Table 5-24. (b) Starting torque should not exceed twice normal operating torque	
(2) find gear size	(a) Approximate pitch diameter of pinion is obtained by entering Table 5-26 with ratio and horsepower per 100 rpm of pinion. (b) Approximate number of teeth on pinion is obtained by entering Table 5-45 with ratio and approximate pinion pitch diameter. (c) Compute exact values from $d = N_P/P_d$; $D = N_G/P_d$ Ratio $= N_G/N_P$	N_P = Number teeth on pinion N_G = Number teeth on gear P_d = Diametral pitch d = Pitch diameter of pinion D = Pitch diameter of gear
(3) find face width	(a) Enter Table 5-46 with ratio and pinion pitch diameter to obtain approximate face width, F. (b) Round off approximate face width such that, $F \lesseqqgtr 0.3\, A_o$ and $F \lesseqqgtr 10/P_d$	A_o = Cone distance P_d = Diametral pitch
(4) verify or select spiral angle	(a) The spiral angle ψ should be such as to give a face contact ratio of at least 1.25. Enter Table 5-47 with the product of $F \times P_d$ and select a suitable spiral angle. (b) Hand of spiral is the opposite on mating members. See Table 2-49 for guide to selection.	F = Face width

*Bending stress calculations are omitted because surface durability generally limits maximum loading. For applications where loads are high and the repetitions of large stresses may cause fatigue failures, the fatigue life under bending may be determined as outlined in a pamphlet published by the Gleason Works entitled, "Stress Determination and Fatigue Life of Generated Bevel Gears".

continued on next page

Step Number	Formula and Instructions	Remarks and Definitions of Symbols
(5) compute horsepower on basis of surface durability	Maximum horsepower rating for surface durability is $P = F\,C_M\,C_C\,K_3$ $C_C = \sqrt{0.4\,(m_F + m_p)}$	F = Face width C_M = Material factor for surface durability, Table 5-25 C_C = Factor for contact ratio m_F = Face contact ratio Table 5–47, but not over 2.00 m_p = Profile contact ratio, Table 5-49 K_3 = Combined factor for pinion diameter, speed factor, rpm, and allowable load, Table 5–48.

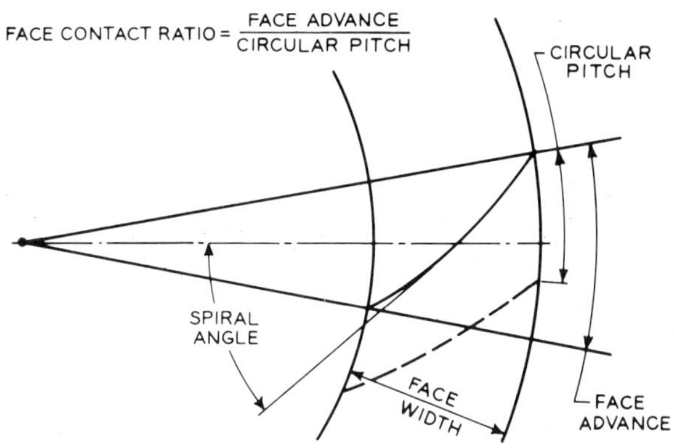

TABLE 5–45

Approximate Relationship Between Pinion Pitch Diameter and Numbers of Teeth by Ratios, Spiral Bevel Gears

Gleason Bevel and Hypoid Gear Design Gleason Works

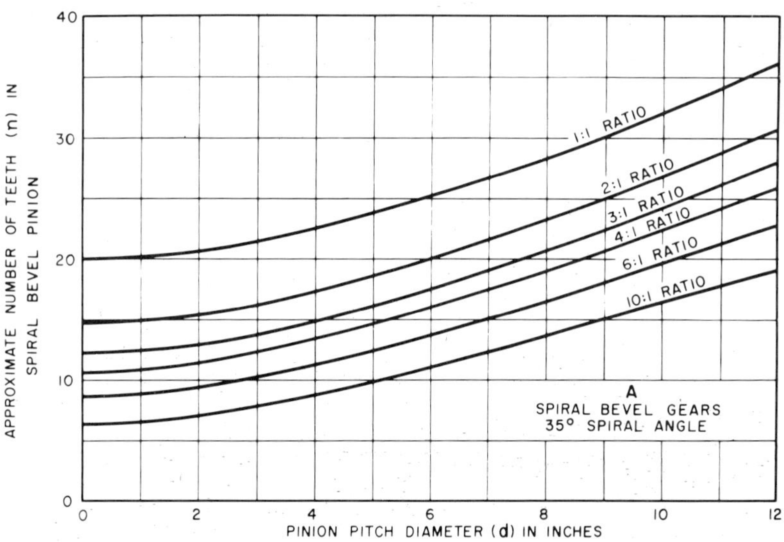

This chart, to accompany Table 5–44, is based on surface durability for general industrial drives and applies to case hardened gears and pinions. It is not intended to cover such applications as aircraft, marine and automotive gears where special processes may permit high loadings and where durability is determined by laboratory and field tests.

TABLE 5–46

Face Width as a Function of Ratio and Pinion Pitch Diameter on 90° Shaft Angle, Spiral Bevel Gears

Gleason Bevel and Hypoid Gear Design   Gleason Works

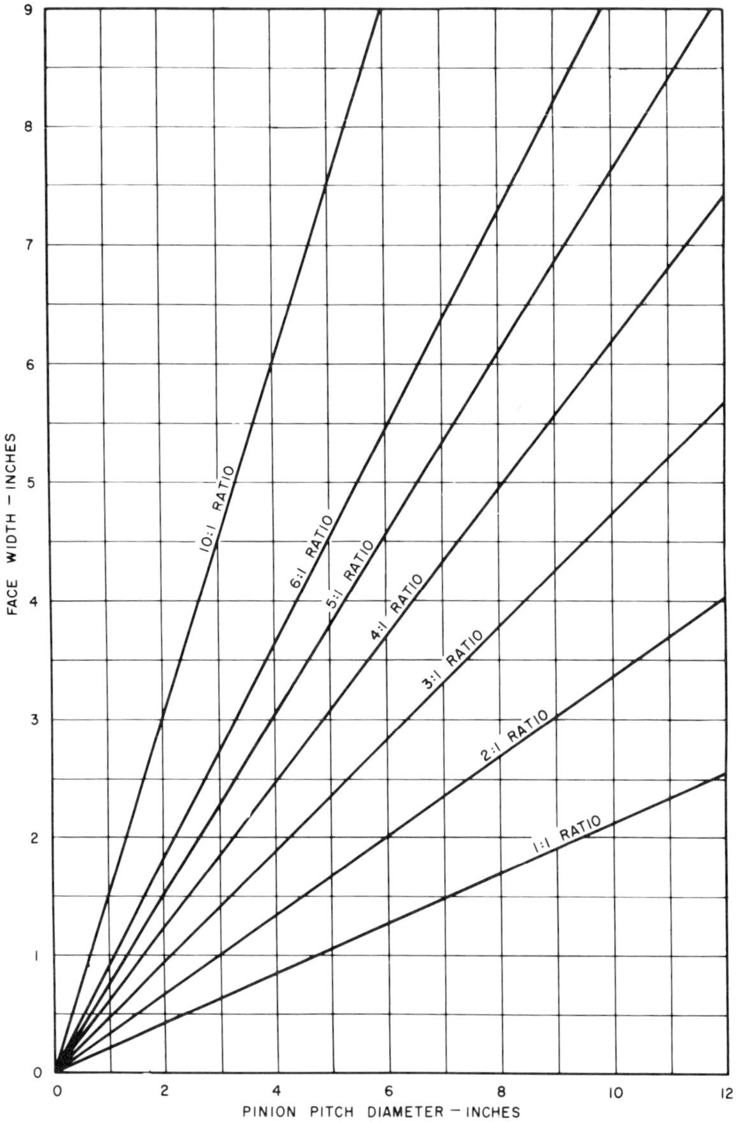

Chart to accompany Table 5–44

TABLE 5–47

Face Contact Ratio, Spiral Bevel Gears

Gleason Bevel and Hypoid Gear Design   Gleason Works

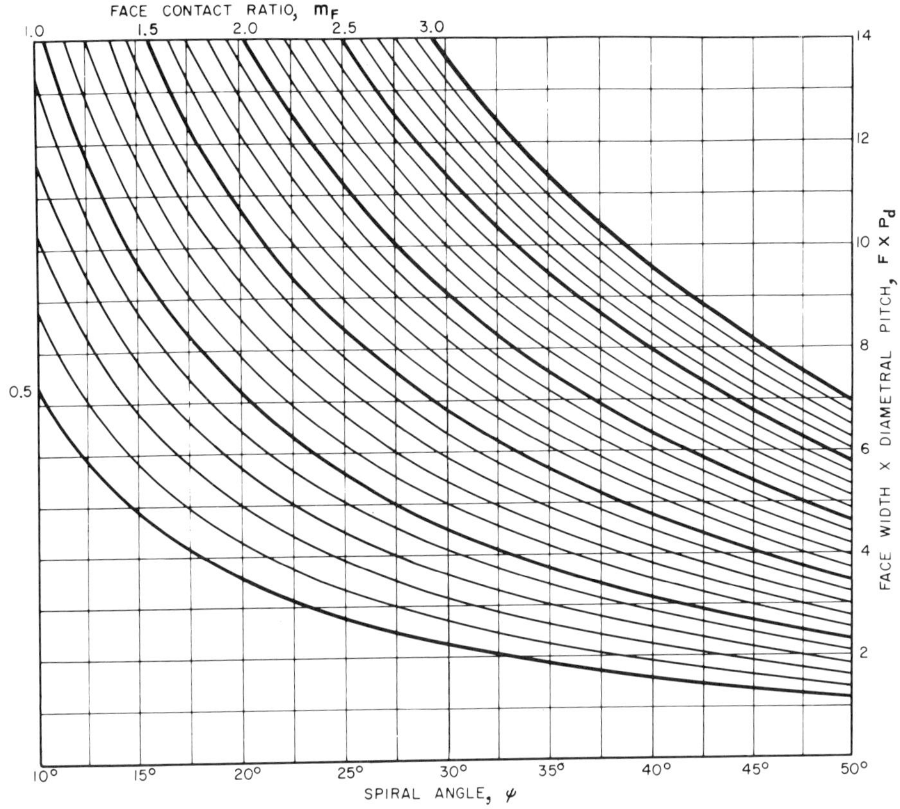

Chart to accompany Table 5–44

Diagram defining face contact ratio. The face advance should be at least 1.25 times the circular pitch.

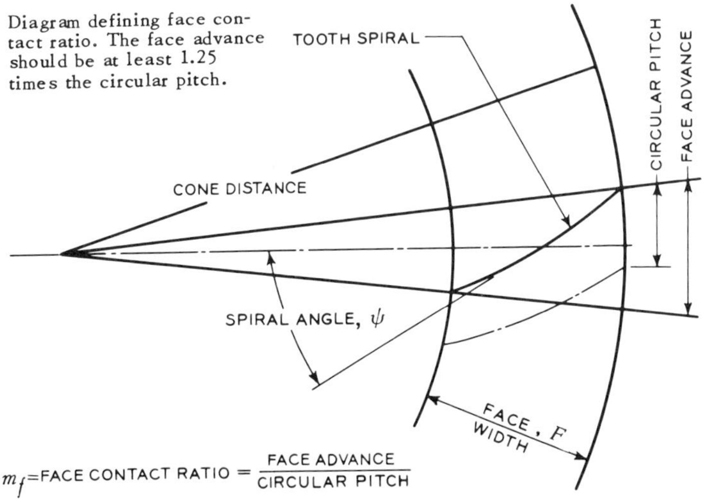

$$m_f = \text{FACE CONTACT RATIO} = \frac{\text{FACE ADVANCE}}{\text{CIRCULAR PITCH}}$$

TABLE 5–48

K_3 Factor for Pinion Diameter and Speed, Spiral Bevel Gears
Gleason Bevel and Hypoid Gear Design   Gleason Works

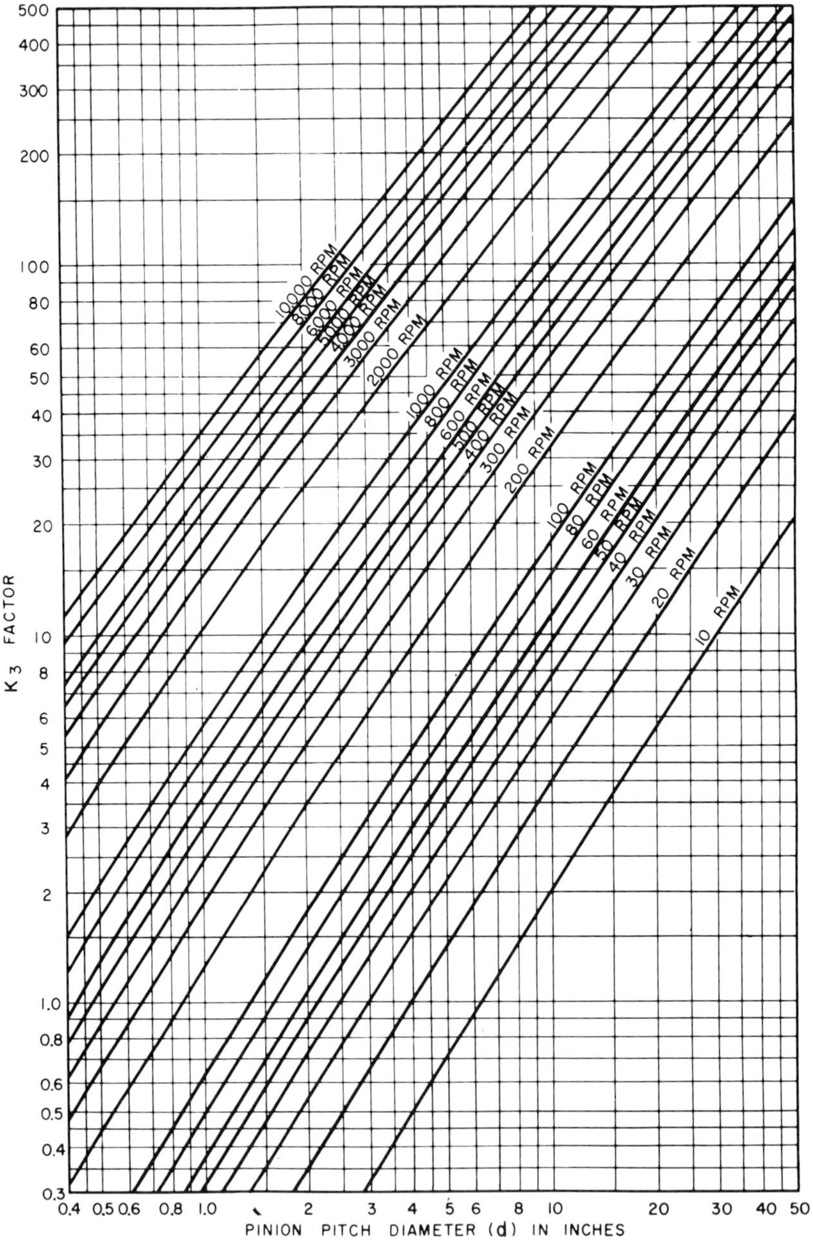

Chart to accompany Table 5–44

TABLE 5–49

Profile Contact Ratio, m_p, 20° Pressure Angle, Spiral Bevel Gears

Gleason Bevel and Hypoid Gear Design   Gleason Works

Number Teeth on Pinion	m_p	Number Teeth on Pinion	m_p
12	1.19	22	1.27
13	1.20	23 and 24	1.28
14	1.21	25 and 26	1.29
15	1.22	27,28 and 29	1.30
16	1.23	30 and 31	1.31
17	1.24	32,33,34 and 35	1.32
18 and 19	1.25	36 and 37	1.33
20 and 21	1.26	38,39 and 40	1.34

Tabular data for use with Table 5–44

TABLE 5–50

Maximum* Design Loads in Pounds Per Inch of Face Case-Hardened, Ground Teeth, †Spiral Bevel Gears for Aircraft

L. J. O'Brien, Aircraft Bevel Gears   SAE Journal, March 1945

Number Teeth on Pinion	Diametral Pitch					
	3	4	5	6	8	10
12 – 14	2450	2120	1900	1730	1500	1340
15 – 18	2740	2370	2120	1935	1675	1500
19 – 23	3085	2670	2390	2180	1890	1690
24 and more	3465	3000	2685	2450	2120	1900

NOTE: Mountings and lubrications must be in accordance with best practice in order to transmit the above loads satisfactorily.

† In the design of Zerol bevel gears for aircraft, the allowable loads are taken as one-half those for spiral bevel gears.

* Although the data listed here exceed the allowable loads ordinarily used in general industrial design, they are indicative of the extremes in loading when the conditions warrant it. Such data can be the base for judging quickly the approximate loads a given pair of gears might carry.

TABLE 5–51

Example to Illustrate Design of a Pair of Spiral Bevel Gears for Surface Durability

Gleason Bevel and Hypoid Gear Design Gleason Works

Item from Table 5–44 \ Specification of Example	A 65-horsepower, 1800-rpm electric motor is to drive a centrifugal pump continuously at 600 rpm by means of spiral bevel gears at 90° shaft angle. Starting torque is 4000 inch-pounds.
(1) Service horsepower, P	(a) Service factor, C_S, for uniform load, Table 5–24 is 1.00; hence, service horsepower = motor horsepower (b) Normal operating torque $= \dfrac{63{,}025\ P}{\text{rpm}} = 2276$ in.-lb which exceeds half the starting torque. Service horsepower P therefore is 65.
(2) Gear size	(a) Horsepower per 100 rpm of pinion $= \dfrac{100\ P}{\text{rpm}} = 3.61$ Ratio is 1800/600 = 3:1 From Table 5–26, approximate pinion pitch diameter $d_{\text{app}} = 3.43$ (b) Table 5–45 and $d_{\text{app}} = 3.43$ gives $N_{P\,(\text{app})}$ between 13 and 14, say 14. If exact 3:1 ratio must be maintained, then gear would have 42 teeth. For centrifugal pump drive, ratio need not be exactly 3:1. Moreover, an uneven ratio is preferable; hence, a 43/14 combination is chosen. (c) An approximate diametral pitch is 14/3.43 = 4.08, say $P_d = 4$ $d = N_P/P_d = 14/4 = 3.500$; $D = 43/4 = 10.750$ $N_G/N_P = 43/14 = 3.07$
(3) Face width, F	(a) From Table 5–46, ratio = 3.07 and d = 3.50, approximate face width is 1.65 (b) Say F = 1-5/8 = 1.625
(4) Verification of spiral angle	(a) Face width times diametral pitch = 1.625 × 4 = 6.5 Table 5–47 at 35° spiral angle, face contact ratio, m_F = 1.73, which is well in excess of 1.25 minimum.
(5) Horsepower rating for surface durability	C_M (Table 5–25) = 1.00 m_p (Table 5–49) = 1.21 $C_C = \sqrt{0.4\,(m_F + m_p)} = 1.085$ K_3 (Table 5–48) = 47 $P = RC_MC_CK_3 = 82.7$ horsepower, which is greater than the original 65 horsepower, but not unreasonably in excess thereof.

INDEX TO SECTION 6

Worm Gearing

TABLE 6-1

Data Ordinarily Put on Detail Drawing* of Cylindrical Worm

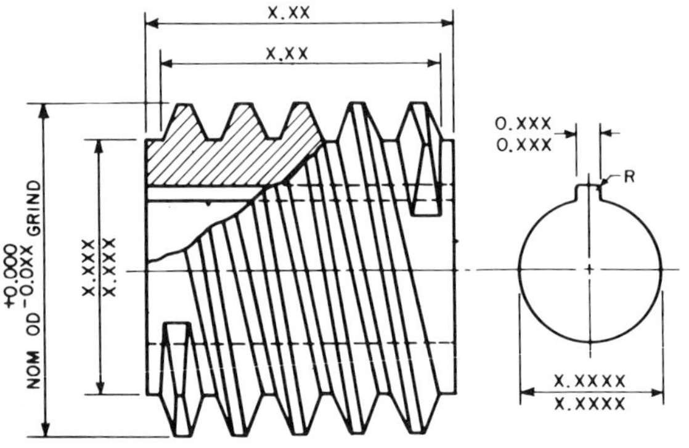

MATERIAL:
SAE 3115 STEEL
CASEHARDEN TO 52 ROCKWELL C

WORM DATA	
NUMBER OF THREADS	X
HAND	
AXIAL PITCH	X.XXXX
LEAD	X.XXXX
PRESSURE ANGLE (NORM)	XX° XX'
PITCH DIAMETER	X.XXX
ADDENDUM	0.XXX
WHOLE DEPTH	X.XXX
NORMAL TOOTH THICKNESS	0.XXX
MATING WORM GEAR	
DRAWING NO.	
NUMBER OF TEETH	XX
CENTER DISTANCE	XX.XXX

METHOD OF FINISHING WORM

IF INTERCHANGEABLE MANUFACTURE IS DESIRED, THE WORM MUST BE AN EXACT DUPLICATE OF THE HOB THAT IS USED ON THE MATING WORM GEAR. SEE TABLE 6-14 REGARDING TYPE OF NOTES FOR PRODUCTION PROCEDURES.

* A SHELL-TYPE WORM IS SHOWN HERE FOR SIMPLICITY. MORE OFTEN A WORM IS MADE INTEGRAL WITH THE SHAFT, IN WHICH CASE DATA AND INFORMATION TO DESCRIBE ADEQUATELY THE COMPOSITE UNIT ARE REQUIRED

TABLE 6-2

Data Ordinarily Put on Detail Drawing* of Worm Gear

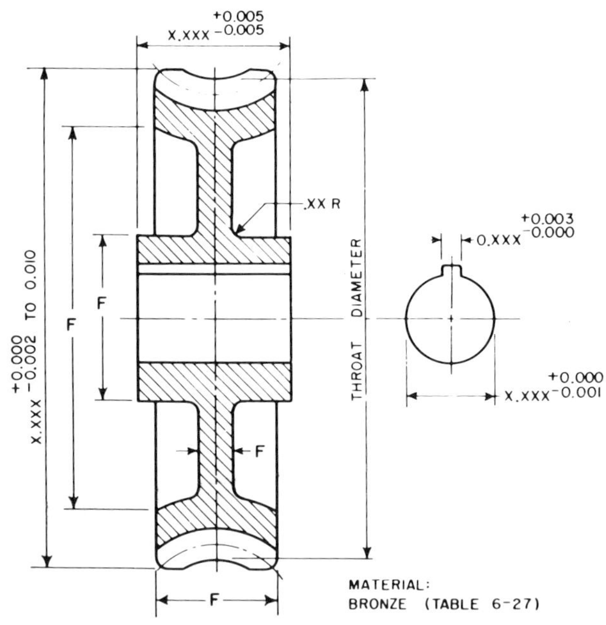

MATERIAL:
BRONZE (TABLE 6-27)

WORM GEAR DATA	
NUMBER OF TEETH	XX
HAND	
CIRCULAR PITCH	X.XXX
PRESSURE ANGLE	XX°XX'
HELIX ANGLE	XX°XX'
PITCH DIAMETER	XX.XXX
THROAT DIAMETER	XX.XXX
ADDENDUM	0.XXX
WHOLE DEPTH	X.XXX
CHORDAL THICKNESS	0.XXX
CUTTING INSTRUCTIONS	
HOB NUMBER	
MATING WORM DATA	
DRAWING NO.	
NUMBER OF THREADS	X
CENTER DISTANCE	XX.XXX
BACKLASH	0.0XX

DIMENSIONS F OFTEN ARE FRACTIONAL WITH GENERAL TOLERANCE OF ± 1/32 OR ± 1/64 OR ± 1/16

* A SECTION THROUGH A DIAMETER OF A SIMPLE WORM GEAR IS SOMETIMES SUFFICIENT, BUT GENERALLY OTHER VIEWS ARE NEEDED IF THE PART IS TO BE CLEARLY AND COMPLETELY DESCRIBED. COMMON PRACTICE IS TO PUT ON THE DRAWING PROPER THOSE DIMENSIONS, WITH TOLERANCES, NEEDED IN MAKING THE BLANK. SPECIFICATION OF THE MATERIAL, DATA FOR HOBBING THE TEETH, AND INSPECTION DATA ARE STATED IN TABULAR FASHION AND IN NOTES

TABLE 6-3

Axial Pitches in Common Use For Coarse*-Pitch Worm Gearing

Axial† Pitch, Inch Fraction	Axial† Pitch, Inch Decimal	Corresponding Module‡, Inch	Axial† Pitch, Inch Fraction	Axial† Pitch, Inch Decimal	Corresponding Module‡, Inch
1/4	0.2500	0.079577	7/8	0.8750	0.278521
5/16	0.3125	0.099472	1	1.0000	0.318310
3/8	0.3750	0.119366	1-1/4	1.2500	0.397887
1/2	0.5000	0.159155	1-1/2	1.5000	0.477465
5/8	0.6250	0.198944	1-3/4	1.7500	0.557042
3/4	0.7500	0.238732	2	2.0000	0.636620

*See Table 6—4 for fine-pitch worm gearing.

†Axial (linear) pitches of even fractions are commonly used for worm gearing to facilitate the selection of change gears in the machine tools for finishing the worm.

‡Module is the ratio of the pitch diameter of the worm gear divided by the number of teeth, and therefore the reciprocal of the diametral pitch. One inch module equals π inches axial pitch.

A technical paper, *Proposal for a Standard Design for General Industrial Coarse-Pitch Cylindrical Worm Gearing*, by F. G. East, recently published in the Trans. ASME, Feb. 1954, recommends axial pitches as follows: 3/16, 1/4, 5/16, 3/8, 1/2, 5/8, 3/4, 1 in., 1-1/4, 1-1/2, 1-3/4, 2 in., 2-1/4, 2-1/2, 2-3/4, 3 in.

TABLE 6–4

Standard Axial Pitches* for Fine-Pitch† Worm Gearing ASA B6.9–1950 AGMA 374.02

Symbol, ASA B6.10–1950	Axial Pitch, p_x, Inches			
p_x	0.030	0.050	0.080	0.130
	0.040	0.065	0.100	0.160

*"Axial pitch" is preferred to "linear pitch." The axial pitch of the worm is equal to the circular pitch of the worm gear in the central plane.

†Gears of 20 diametral pitch and finer are classed as fine pitch.

TABLE 6-5

Number of Threads or Starts on Cylindrical Worms

Number of Threads, Symbol, ASA B6.5-1949	Today's Practice*	Old Practice	Limitations proposed by Buckingham† to gain simplification and wide coverage	Proposed Standard by East Trans. ASME Feb. 1954
N_W	1, 2, 3, 4, 5,	1, 2, 3, 4	1, 3, 6 for majority of applications	1, 2, 3, 4
	6, 7, 8. 9. 10		12, 18, 24 for wide coverage	6 and 8

*During the last 10 or 20 years, worm-gear ratios have been extended by multiple-threaded worms to include reductions as low as 4 or 5 to 1. Prior thereto reductions ranging from 25 or 30 to 1 to as much as 100 or 200 to 1, and more, was regarded as the field of worm gearing. Spur and bevel gears are often limited to ratios of 8 or 10 to 1; hence there existed an in-between range, say from 8 to 1 to 25 to 1, that was not suitably covered. Worms of 5 to 10 threads, and more, now supply ratios over this intermediate range. Mr. East in a technical paper, *Proposal for a Standard Design for General Industrial Coarse-Pitch Cylindrical Worm Gearing*, discourages the use of 5 and 7 threads. He goes on to say that whenever a greater number of threads than 8 is required, odd and prime numbers should be avoided to facilitate the production of the worm without special indexing equipment.

† "Analytical Mechanics of Gears," McGraw-Hill Book Co.

TABLE 6-6

Standard Lead Angles on Fine-Pitch,* Cylindrical Worms

ASA B6.9-1950 AGMA 374.02

Symbol, ASA B6.10-1950	Pitch Lead Angle, λ, Deg.		
λ	0.5	4.0	14.0
or λ_p when the lead angle at the pitch	1.0	5.0	17.0
cylinder needs to be distinguished from	1.5	7.0	21.0
a lead angle at some other diameter.	2.0	9.0	25.0
	3.0	11.0	30.0

*Gears of 20 diametral pitch and finer are classed as fine pitch.

TABLE 6–7

Curves Showing How Theoretical Efficiency of Worm Gearing Varies With Lead Angle and Coefficients of Friction

Machine Design Drawing Room Problems, Albert, 4th Edition, John Wiley & Sons

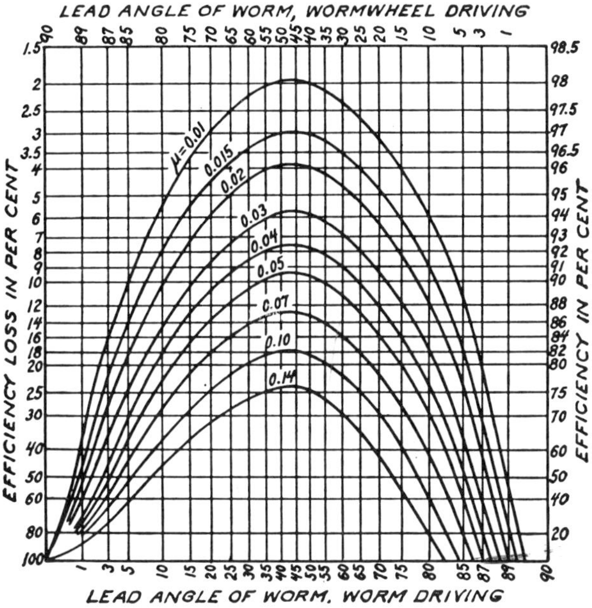

TABLE 6-8

Performance Efficiencies of Single-Reduction, Worm-Gear Speed Reducers for Power Transmission

Catalog 200 The Cleveland Worm and Gear Co.

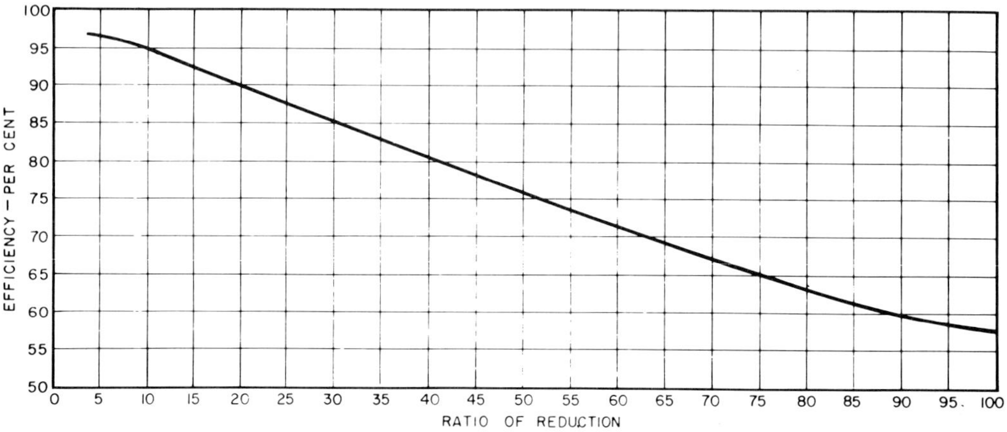

THIS EFFICIENCY CURVE, REPRESENTING OVER-ALL REDUCTION UNIT EFFICIENCY, HAS BEEN PLOTTED FROM DATA OBTAINED FROM A GREAT NUMBER OF ACTUAL TESTS. THE INDICATED VALUES ARE SUBSTANTIALLY CORRECT FOR THE OPERATION OF SINGLE-REDUCTION UNITS AT THE USUAL ELECTRIC MOTOR SPEEDS, AND AT THE FULL CLASS 1 RATED LOAD OF THE REDUCTION UNIT.

WORM-GEAR EFFICIENCY VARIES DIRECTLY WITH LEAD ANGLE, BEING GREATEST THEORETICALLY AT 45 DEG, TABLE 6-7. LARGE LEAD ANGLES, IN TURN, ARE COUPLED WITH MANY THREADS AND SMALL RATIOS.

TABLE 6-9

Worm Pressure Angle As a Function of Lead Angle to Avoid Undercutting on Worm Gear

Analytical Mechanics of Gears, Buckingham, McGraw-Hill Book Co.

Worm Pressure Angle* in Axial Section or Half Thread Angle	Maximum Lead Angle on Worm	Remarks
14-1/2 deg	up to 16 deg	A pressure angle that is much smaller than the lead angle results in excessive undercutting on the off-center sections of the worm gear on the leaving side.
20	25	
25	35	
30	45	

*The pressure angle in the axial section is to be distinguished from the normal pressure angle on the cutting tool or grinding wheel for producing a worm. Compare angular relationships in Table 4-7. Furthermore, the normal pressure angle actually produced on the worm may be slightly smaller than the normal pressure angle on the cutter or grinding wheel, depending upon worm diameter, lead angle, and diameter of cutter or grinding wheel.

TABLE 6–10

Worm-Thread Helix Angle in Degrees

Formulas for Gearing, 17th Edition, Brown and Sharpe Mfg. Co.

Lead of Worm, Inches	Pitch-line Diameter of Worm, Inches																				
	1	1⅛	1¼	1⅜	1½	1⅝	1¾	1⅞	2	2⅛	2¼	2⅜	2½	2⅝	2¾	2⅞	3	3⅛	3¼	3⅜	3½
¼	4½	4	3½	3¼	3	2¾	2½	2½	2¼	2¼	2	2	2	1¾	1¾	1½	1½	1½	1½	1½	1½
⅜	6¾	6	5½	5	4½	4¼	4	3¾	3½	3¼	3	3	2¾	2½	2½	2½	2¼	2¼	2¼	2	2
½	9	8	7¼	6¾	6	5½	5¼	5	4½	4¼	4¼	4	3¾	3½	3¼	3¼	3	3	2¾	2¾	2½
⅝	11¼	10	9	8¼	7½	7	6½	6	5¾	5¼	5	4¾	4½	4½	4¼	4	3¾	3¾	3½	3½	3¼
¾	13½	12	10¾	10	9	8½	7¾	7¼	6¾	6½	6	5¾	5½	5¼	5	4¾	4½	4½	4¼	4¼	4
⅞	15½	14	12½	11½	10½	9¾	9	8½	8	7½	7	6¾	6½	6	5¾	5½	5¼	5	5	4¾	4½
1	17¾	16	14¼	13	12	11¼	10¼	9¾	9	8½	8	7¾	7¼	7	6¾	6¼	6	5¾	5½	5½	5¼
1⅛	19¾	18	16	14¾	13½	12½	11½	11	10¼	9½	9	8½	8	7¾	7½	7¼	6¾	6½	6¼	6	6
1¼	21¾	19½	17¾	16¼	15	13¾	12¾	12	11¼	10¾	10	9½	9	8¾	8¼	8	7¾	7½	7	6¾	6½
1⅜	23¾	21¼	19¼	17¾	16¼	15	14	13	12½	11¾	11	10½	10	9½	9	8¾	8¼	8	7¾	7½	7
1½	25½	23	21	19¼	17½	16½	15¼	14¼	13½	12¾	12	11½	11	10¼	10	9½	9	8¾	8½	8	7¾
1⅝	27½	24¾	22½	20¾	19	17¾	16½	15½	14½	13¾	13	12¼	11¾	11¼	10¾	10¼	9¾	9½	9	8¾	8½
1¾	29¼	26½	24	22	20½	19	17¾	16½	15½	14¾	14	13¼	12½	12	11½	11	10½	10	9¾	9½	9
1⅞	30¾	28	25½	23½	21¾	20¼	19	17½	16¾	15¾	15	14¼	13½	13	12¼	11¾	11¼	11	10½	10	9¾
2	32½	29¾	27	25	23	21½	20	19	17¾	16¾	15¾	15	14¼	13¾	13¼	12½	12	11½	11	10¾	10½
2¼	35¾	32¾	30	27½	25½	24	22¼	21	19¾	18¾	17¾	17	16	15¼	14¾	14	13½	13	12½	12	11½
2½	38½	35½	32½	30	28	26¼	24½	23	21¾	20¾	19½	18½	17¾	17	16¼	15½	15	14¼	13¾	13½	12¾
2¾	41¼	38	35	32½	30¼	28½	26½	25	23¾	22½	21¼	20¼	19¼	18½	17¾	17	16¼	15¾	15	14¾	14
3	43¾	40½	37½	35	32½	30½	28¾	27	25½	24¼	23	22	21	20	19¼	18½	17¾	17	16½	16	15¼
3¼	46	42¾	39¾	37	34½	32½	30¾	29	27½	26	24¼	23½	22½	21½	20¾	19¾	19	18½	17¾	17	16½
3½	48¼	45	41¾	39¼	36¾	34½	32½	30¾	29¼	27¾	26¼	25¼	24	23	22	21¼	20½	19½	19	18¼	17¾
3¾	50¼	47	43¾	41	38¾	36½	34¼	32½	31	29½	28	26¾	25½	24½	23½	22½	21¾	21	20¼	19½	19
4	51¾	48½	45¾	43	40½	38¼	36	34¼	32¾	31	29½	28¼	27	26	25	24	23	22	21½	20¾	20
4¼	53½	50½	47½	44¾	42¼	39¾	37¾	36	34¼	32½	31	29¾	28½	27¼	26¼	25¼	24¼	23½	22½	22	21¼
4½	55	42	49	46½	43¾	41½	39½	37½	35¾	34	32½	31¼	30	28¾	27½	26½	25½	24½	23¾	23	22½
4¾	56½	53½	50½	48	45¼	43	40¾	39	37¼	35½	34	32½	31¼	30	29	27¾	26¾	25¾	25	24	23½
5	57¾	54¾	51¾	49¼	46¾	44½	42¼	40½	38¾	37	35¼	34	32½	31¼	30¼	29	28	27	26	25¼	24½
5¼	59¼	56¼	53¼	50¾	48	46	43¾	41¾	40	38¼	36½	35¼	34	32½	31½	30¼	29	28¼	27	26½	25½
5½	60¼	57¼	54¾	52	49½	47¼	45	43	41¼	39½	38	36½	35	33¾	32¾	31½	30¼	29¼	28¼	27½	26½
5¾	61¼	58½	55¾	53	50¾	48½	46¼	44¼	42½	40¾	39¼	37¾	36¼	35	33¾	32½	31½	30¼	29½	28½	27¾
6	62¼	59¾	57	54¼	52	49½	47½	45¾	43¾	42	40¼	39	37½	36¼	35	33¾	32½	31½	30½	29½	28¾

continued on next page

TABLE 6–10, continued

Lead of Worm, Inches	Pitch-line Diameter of Worm, Inches																			
	3⅝	3¾	3⅞	4	4⅛	4¼	4⅜	4½	4⅝	4¾	4⅞	5	5⅛	5¼	5⅜	5½	5⅝	5¾	5⅞	6
¼	1¼	1¼	1¼	1¼	1¼	1¼	1	1	1	1	1	1	1	1	1	1	1	1	1	1
⅜	2	2	2	1¾	1¾	1¾	1½	1½	1½	1½	1½	1½	1½	1½	1¼	1¼	1¼	1¼	1¼	1¼
½	2½	2½	2½	2¼	2¼	2¼	2	2	2	2	2	2	2	2	1¾	1¾	1¾	1¾	1¾	1½
⅝	3¼	3	3	2¾	2¾	2¾	2½	2½	2½	2½	2½	2¼	2¼	2¼	2¼	2¼	2¼	2	2	2
¾	3¾	3¾	3½	3¼	3¼	3¼	3¼	3	3	3	3	2¾	2¾	2¾	2¾	2½	2½	2¼	2¼	2¼
⅞	4½	4¼	4¼	4	4	4	3¾	3½	3½	3½	3¼	3¼	3¼	3	3	3	2¾	2¾	2¾	2¾
1	5	5	4¾	4½	4½	4½	4¼	4	4	3¾	3¾	3¾	3½	3½	3½	3¼	3¼	3¼	3¼	3
1⅛	5¾	5½	5½	5¼	5	5	4¾	4½	4½	4¼	4¼	4¼	4	4	3¾	3¾	3¾	3¾	3½	3½
1¼	6¼	6	6	5¾	5½	5½	5¼	5	5	4¾	4¾	4½	4½	4½	4¼	4¼	4¼	4	4	3¾
1⅜	7	6¾	6½	6¼	6	6	5¾	5½	5½	5¼	5¼	5	5	4¾	4¾	4½	4½	4½	4¼	4¼
1½	7½	7¼	7	6¾	6¾	6½	6¼	6	6	5¾	5½	5½	5¼	5¼	5	5	4¾	4¾	4¾	4½
1⅝	8¼	8	7¾	7½	7¼	7	6¾	6½	6½	6¼	6	6	5¾	5¾	5½	5½	5¼	5¼	5	5
1¾	8¾	8½	8¼	8	7¾	7½	7¼	7	7	6¾	6½	6½	6¼	6	6	5¾	5¾	5½	5½	5½
1⅞	9½	9	8¾	8½	8¼	8	7¾	7½	7½	7¼	7	7	6¾	6½	6½	6¼	6	6	5¾	5¾
2	10	9¾	9½	9	8¾	8½	8½	8¼	8	7¾	7½	7¼	7¼	7	6¾	6¾	6½	6½	6¼	6
2¼	11¼	11	10½	10¼	10	9½	9½	9	9	8¾	8½	8¼	8	8	7¾	7½	7½	7¼	7	7
2½	12½	12	11¾	11¼	11	10¾	10½	10	10	9½	9¼	9¼	9	8¾	8½	8¼	8¼	8	7¾	7½
2¾	13¾	13¼	12¾	12¼	12	11¾	11½	11	10¾	10½	10¼	10	9¾	9½	9¼	9¼	9	8¾	8½	8½
3	15	14½	14	13½	13	13	12¼	12	11½	11½	11	11	10¾	10½	10¼	10	9¾	9½	9½	9¼
3¼	16	15½	15	14½	14¼	14	13¼	13	12½	12¼	12	11¾	11½	11	11	10¾	10½	10¼	10¼	10
3½	17¼	16¾	16	15½	15½	15	14¼	14	13½	13¼	13	12½	12¼	12	11¾	11½	11¼	11	10¾	10½
3¾	18½	18	17¼	16½	16¼	15¾	15¼	15	14½	14¼	13¾	13½	13	13	12½	12¼	12	11¾	11½	11¼
4	19½	19	18½	17½	17¼	16¾	16¼	16	15½	15	14½	14¼	14	13¾	13¼	13	13	12½	12¼	12
4¼	20½	20	19½	18½	18½	17¾	17¼	16¾	16½	16	15½	15¼	15	14½	14	14	13¾	13½	13	12¾
4½	21½	21	20½	19¾	19½	18½	18¼	17½	17¼	17	16½	16	15¾	15¼	15	14¾	14½	14¼	13¾	13½
4¾	22¾	22	21½	20¾	20½	19½	19¼	18½	18¼	18	17¼	17	16½	16	16	15½	15¼	15	14½	14¼
5	23¾	23	22½	21¾	21¼	20½	20	19½	19	18½	18¼	18	17¼	17	16½	16¼	16	15½	15¼	15
5¼	24¾	24	23½	22½	22	21½	21	20½	20	19½	19	18½	18¼	17¾	17½	17	16¾	16¼	16	15¾
5½	25¾	25¼	24½	23½	23	22½	22	21½	21	20½	20	19½	19	18½	18¼	18	17½	17	16¾	16½
5¾	27	26¼	25½	24½	24	23½	23	22	21¾	21	20½	20¼	19¾	19½	19	18½	18¼	17¾	17½	17
6	27¾	27	26½	25½	25	24¼	23¾	23	22½	22	21½	21	20½	20	19½	19¼	19	18½	18	17¾

TABLE 6–11

Formulas for Calculating Dimensions of Worm

Data upon which to start the design of a worm-gear pair may be altogether different from those that formed the requirements of a prior design. What follows assumes that an approximate center distance and a reduction ratio are among the data from the beginning.

To Find	From	Formula
*Pitch diameter (or mean diameter) of worm, d	Center distance, C (See Table 6–21)	$d = \frac{C^{0.875} \text{(Table 6–25)}}{2.2}$ approximately
or		
Pitch diameter, d according to "Formulas in Gearing" B & S Mfg. Co.	Axial pitch, p_x (See Table 6–3)	For shell-type worms: $d = 2.4\ p_x + 1.1$ For integral worms: $d = 2.35\ p_x + 0.4$
Lead, l Figure in Table 6–13	Number of threads, N_W, and axial pitch, p_x	$l = N_W\ p_x$
Lead angle, λ Figure in Table 6–13	Lead, l, and pitch diameter, d	$\cot \lambda = \frac{\pi d}{l}$
Normal circular pitch, p_n	Axial pitch, p_x, and lead angle λ	$p_n = p_x \cos \lambda$
Addendum, a_W, "Formulas in Gearing," B & S Mfg. Co.	Axial pitch, p_x, or normal circular pitch, p_n	$a_W = 0.3183\ p_x$ for $\lambda \leq 18^\circ$ $a_W = 0.3183\ p_n$ for $\lambda > 18^\circ$
Whole depth, h_{tW}	Axial pitch, p_x, or normal circular pitch, p_n	$h_{tW} = 0.6866\ p_x$ for $\lambda \leq 18^\circ$ $h_{tW} = 0.6866\ p_n$ for $\lambda > 18^\circ$
Outside diameter, d_o	Pitch diameter, d, and addendum, a_W	$d_o = d + 2a_W$
Thickness of thread at pitch line, t_W	Axial pitch, p_x, lead angle λ, and backlash, B	$t_W = \frac{p_x \cos \lambda}{2} - B$
Minimum length of worm, F_W ("Formulas in Gearing" B & S Mfg. Co.)	Throat diameter of worm gear, D_t, and working depth equal to twice the addendum, or $2a$	$F_W = 2\sqrt{2a(D_t - 2a)}$
F_W	Circular pitch of gear, p, and factor from Table 6–15	$F_W = p$ (factor)
Pressure angle, ϕ	20° to 30° depending to a large degree on lead angle and to a lesser degree on face width of the gear.	See Table 6–9

*A technical paper, *Proposal for a Standard Design for General Industrial Coarse-Pitch Cylindrical Worm Gearing* by East, recently published in Trans. ASME, Feb. 1954, recommends that the worm nominal pitch diameter be made an integral number of half-axial pitches with multiples of full pitches being preferred. AGMA Standard 214.02-1954 recommends that the approximate pitch diameter of the worm $d = C^{0.875}/2.2$ be altered until it closely equals $d_R + 2b_G$, where d_R is a recommended root diameter of $C^{0.875}/3$ and b_G is the dedendum of the gear, the tooth proportions in the case of double enveloping worm being such that $b_G = 0.275\ p_n$. The standard gives detailed instructions on the verification of proportions by layout.

TABLE 6-12

Formulas for Calculating Dimensions of Worm Gear

To Find	From	Formula
Pitch diameter of worm gear, D	Center distance, C, and mean or pitch diameter of worm, d	$D = 2C - d$
D	Number of teeth, N_G, and circular pitch, p, where $p = p_x$, the axial pitch of worm	$D = \frac{pN_G}{\pi}$ Preferably N_G should not be less than 30 teeth to avoid undercutting.
Helix angle, ψ, of worm gear	Same as lead angle λ of worm	Table 6-11
Normal circular pitch, p_n	Circular pitch, p, and helix angle, ψ	$p_n = p \cos \psi$
Addendum, a_G	Circular pitch, p, or normal circular pitch, p_n	$A_G = 0.3183\, p$ for $\psi \leq 18^\circ$ $A_G = 0.3183\, p_n$ for $\psi > 18^\circ$
Whole depth, h_{tG}	Circular pitch, p, or normal circular pitch, p_n	$h_{tG} = 0.6866\, p$ for $\psi \leq 18^\circ$ $h_{tG} = 0.6866\, p_n$ for $\psi > 18^\circ$
Working depth, h_{kG}	Addendum, a_G	$h_{kG} = 2a_G$
Throat diameter, D_t See Table 6-13	Number of teeth, N_G, circular pitch, p, and addendum, a_G	$D_t = \frac{N_G\, p}{\pi} + 2a_G$
Radius of curvature of throat, R	Center distance, C, and throat diameter, D_t Figure in Table 6-13	$R = C - \frac{D_t}{2}$
Outside diameter of worm gear, D_o	Pitch diameter, D, and addendum, a_G Figure in Table 6-13	$D_o = D + 3a_G$
D_o	Pitch diameter, D, radius R and included angle α Figure in Table 6-13	$D_o = D + 2R\left(1 - \cos\frac{\alpha}{2}\right)$ $60^\circ < \alpha < 90^\circ$
Face width, F_G	Pitch diameter of worm, d, and outside diameter, d_o	$F_G < \frac{3d}{4}$ $F_G = \sqrt{d_o^{\,2} - d^2}$
F_G	Center distance, C Table 6-21	$F_G = \frac{C^{0.875}\text{(Table 6-25)}}{3}$

TABLE 6-13

Basic Dimensions of a Pair of Worm-Gears for Operation on Shafts at 90 Deg

(See Table 6–17 for Standard, Fine-Pitch Worm Gears)

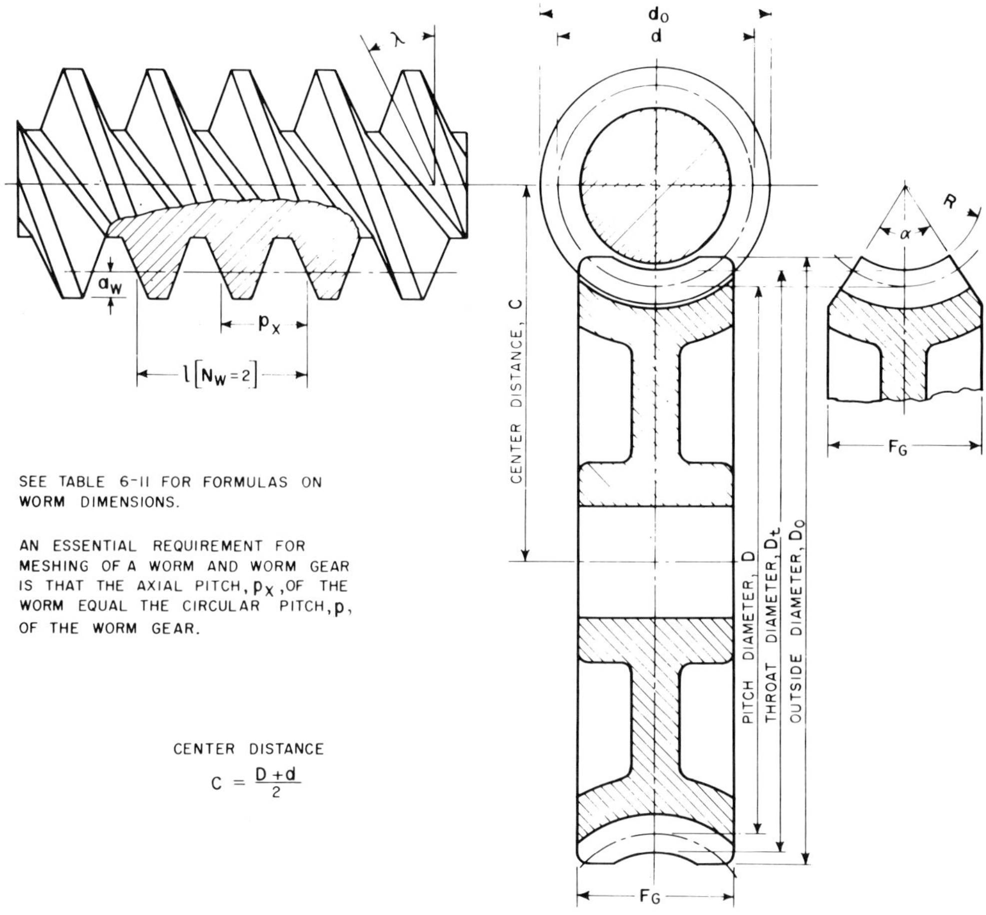

Since the efficiency of worm gearing varies so markedly with lead angle, Tables 6–7 and 6–8, worm gears of large lead angles are desirable when power is to be transmitted. The choice of lead angle, on the other hand, cannot be made on a basis of efficiency alone because the root diameter of the worm diminishes with lead angle, which not only reduces the relative strength of the worm but also results in a narrower face width, F_G, on the gear. Lead angles up to 45° are practical but usually they are compromises well under that maximum.

TABLE 6–14

Relative Sizes of the Worm and the Hob for the Gear

Conjugate teeth on a worm and the mating wheel are obtained by cutting the teeth in the wheel with a hob that is the exact counterpart of the worm. In effect, this means that the teeth on the hob and the threads on the worm must be finished in precisely the same manner. For instance, if the teeth on the hob are finished by grinding with a flat-sided wheel so as to envelope a true involute helicoid, then the worm threads should also be finished by grinding with a flat-sided wheel.

If the worm is to be milled with a double-angle cutter, then the hob should be ground with a wheel having the same included angle and diameter. This explains why a hob manufacturer asks for information pertaining to the method of making the worm before filling an order for worm-gear hobs. He has to have such data if he is to duplicate the worm in the hob.

Normally the hob is somewhat larger in diameter than the worm. It has to be to provide clearance and to allow for sharpening. How much larger the hob can be than the worm depends chiefly upon the lead angle. On this point, Buckingham, *Analytical Mechanics of Gears,* is quoted as follows:

> With a lead angle of 10 degrees or less and with a nominal pitch radius or effective radius of about 1.500 inch, the hob may be as much as 1/4 inch larger in diameter than the worm without affecting the form of the off-center sections appreciably. Any such difference is small enough to be soon wiped out by the plastic flow of the material of the worm gear in operation. As the lead angle increases, however, the amount of oversize of the hob must be reduced to maintain commensurate conditions. Thus with a lead angle of about 35 degrees, the hob should be held to within about 0.005 inch of the diameter of the worm. The ductility of the material of the worm gear also plays a part here. A more ductile material will permit more plastic deformation of the surface in the running-in period than will a harder material.

To preserve the counterpart relationship of worm and hob the design of a worm and gear are restricted further by requirements of manufacture. If the worm is integral with the shaft, the chances are that it is too small to have a shell-type hob as a counterpart. In short, the counterpart of an integral worm is generally an integral hob; that of a shell-type worm is a shell-type hob.

If quantities of a particular worm-gear pair are to be manufactured, tool cost may be relatively unimportant. But when the quantities wanted are limited or few, tool cost and the time delay in getting them may be of major importance; tools of more or less standard variety are desirable. In this respect the shell-type hob is probably the more versatile.

Worm gears are hobbed usually in one of two ways; either by feeding the hob radially into the gear blank, or by feeding it tangentially. The former is perhaps the faster. The latter quite generally roughs and finishes the teeth in one operation. It is the better suited to large numbers of threads.

List of Standards for Worm Gearing

American Standards Association, No. and Date	American Gear Mfgs. Assoc. No. and Date	Title of Standard
ASA B6.5–1949	AGMA 111.02–1949	Letter Symbols for Gear Engineering
	116.01–1950	Abbreviations for Gear Engineering
ASA B6.10–1950	AGMA 112.02–1950	Gear Nomenclature
	AGMA 213.02–1952	Surface Durability of Cylindrical-Worm Gearing
	AGMA 214.02–1954	Surface Durability of Double Enveloping Worm Gearing
	AGMA 243.01–1954	Cast Bronze Gear Blanks
	AGMA 250.01–1946	Lubrication of Closed and Open Gearing
ASA B6.9–1950	AGMA 374.02–1950	Design for Fine-Pitch Worm Gearing
	AGMA 440.02–1950	Cylindrical-Worm Gear Speed Reducers
	AGMA 441.02–1954	Double Enveloping-Worm Gear Speed Reducers

TABLE 6–15

Length of Worm Factor According to Number of Teeth on Gear
Catalog No. 1000 D. O. James Mfg. Co.

Number of Teeth Worm Gear	Factor for 1-Inch Circular Pitch	Number of Teeth Worm Gear	Factor for 1-Inch Circular Pitch	Number of Teeth Worm Gear	Factor for 1-Inch Circular Pitch	Number of Teeth Worm Gear	Factor for 1-Inch Circular Pitch
10	2.93	55	6.92	100	9.35	145	11.24
11	3.08	56	6.98	101	9.40	146	11.28
12	3.22	57	7.04	102	9.45	147	11.32
13	3.35	58	7.10	103	9.50	148	11.36
14	3.48	59	7.16	104	9.55	149	11.40
15	3.60			105	9.60		
16	3.72	60	7.22	106	9.65	150	11.44
17	3.84	61	7.28	107	9.70	151	11.48
18	3.95	62	7.34	108	9.75	152	11.52
19	4.06	63	7.40	109	9.80	153	11.56
		64	7.46			154	11.60
20	4.17	65	7.52	110	9.84	155	11.64
21	4.27	66	7.58	111	9.88	156	11.68
22	4.37	67	7.64	112	9.92	157	11.72
23	4.47	68	7.70	113	9.96	158	11.755
24	4.57	69	7.76	114	10.00	159	11.79
25	4.66			115	10.04		
26	4.75	70	7.82	116	10.08	160	11.825
27	4.84	71	7.88	117	10.12	161	11.86
28	4.93	72	7.94	118	10.16	162	11.895
29	5.02	73	8.00	119	10.20	163	11.93
		74	8.05			164	11.965
30	5.11	75	8.10	120	10.24	165	12.00
31	5.20	76	8.15	121	10.28	166	12.035
32	5.28	77	8.20	122	10.32	167	12.07
33	5.36	78	8.25	123	10.36	168	12.105
34	5.44	79	8.30	124	10.40	169	12.14
35	5.52			125	10.44		
36	5.60	80	8.35	126	10.48	170	12.175
37	5.68	81	8.40	127	10.52	171	12.21
38	5.76	82	8.45	128	10.56	172	12.245
39	5.83	83	8.50	129	10.60	173	12.28
		84	8.55			174	12.315
40	5.90	85	8.60	130	10.64	175	12.35
41	5.97	86	8.65	131	10.68	176	12.385
42	6.04	87	8.70	132	10.72	177	12.42
43	6.11	88	8.75	133	10.76	178	12.455
44	6.18	89	8.80	134	10.80	179	12.49
45	6.25			135	10.84		
46	6.32	90	8.85	136	10.88		
47	6.39	91	8.90	137	10.92		
48	6.46	92	8.95	138	10.96		
49	6.53	93	9.00	139	11.00		
		94	9.05				
50	6.60	95	9.10	140	11.04		
51	6.67	96	9.15	141	11.08		
52	6.74	97	9.20	142	11.12		
53	6.80	98	9.25	143	11.16		
54	6.86	99	9.30	144	11.20		

EXAMPLE: Multiply circular pitch by factor. 1½ in. C.P. and 70T in Gear = 7.82 in. x 1.5 = 11.73 or 11¾ in.

TABLE 6-16

Included Angles Recommended on Thread Milling Cutters

Formulas in Gearing   Brown and Sharpe Mfg. Co.

In Brown & Sharpe's practice it is usually desirable to cut worm threads in a thread miller using an angular milling cutter or formed cutter having straight sides, the included angle to be governed by the following table:

Worms having an angle of thread with axis of 78° or more require 29° included cutter.

Worms having an angle of thread with axis of 70°–78° require 40° included cutter.

Worms having angle of thread with axis of 65°–70° require 45° included cutter.

Worms having angle of thread with axis of 65° or less require 50° included cutter.

Worms mating with wheels of 24 teeth or less should be finished with cutter of 40°, or larger, included angle.

TABLE 6–17

Formulas for Calculating Dimensions of Standard Fine-Pitch Worm and Worm Gear, 90° Shaft Angle

ASA B6.9–1950   AGMA 374.02

Reduction ratio and distance between shafts are usually among the defining requirements in the design of a pair of worm gears. Next if a standard axial pitch or a pitch lead angle is selected, a few trial calculations and Table 6–18 soon narrow the possible combinations to a number where a specific combination can be chosen as suitable for final design.

1	Number threads on worm, N_W		
2	Number teeth on wheel, N_G		
3	Normal pressure angle on hob or grinding wheel, $\phi_n = 20°$		
4	Axial pitch of worm, p_x		
5	Circular pitch of gear, p		
6	Lead angle of worm, λ Table 6–18		
7	Helix angle of gear, $\psi = 90° - \lambda$		
		Worm	**Worm Gear or Wheel**
8	Lead l	$l = N_W p_x$ or Table 6–18	
9	Pitch diameter	$d = \dfrac{l}{\pi \tan \lambda}$ or Table 6–18	$D = \dfrac{N_G\, p}{\pi}$
10	Normal circular pitch	$p_n = p_x \cos \lambda$	$p_n = p \cos \psi$
11	Addendum	$a = 0.3183\, p_n$ or Table 6–19	
12	Whole depth	$h_t = 0.7003\, p_n + 0{,}002$ or Table 6–19	
13	Working depth	$h_k = 0.6366\, p_n$	
14	Clearance	$c = h_t - h_k$	
15	Tooth thickness (normal)	$t = 0.5\, p_n$	
16	Center distance	$C = 0.5\,(d + D)$	

continued on next page

TABLE 6-17, continued

		Worm	Worm Gear or Wheel
17	Outside diameter Note: Since fine-pitch gearing is primarily used to transmit motion, rather than power, sufficient contact can be had generally with cylindrical wheels and with worms that are shorter than normally recommended for power drives.	$d_o = d + 2a$	$D_o = 2C - d + 2a$ for cylindrical worm-gear blank. $D_o = 2C - 0.891d + 1.782a$ for throated, worm-gear blank. See Figure in Table 6–13.
18	Throat diameter of worm gear		$D_t = D + 2a$
19	Safe minimum length of threaded portion of worm	$F_W = \sqrt{D_o^2 - D^2}$	
20	Face width of worm gear		$F_G = 0.57735\ d_o$ for throated, worm-gear blank. See figure in Table 6–13. $F_{G_{min}} = 1.125\sqrt{(d_o + 2c)^2 - (d - 4a)^2}$ in the case of the cylindrical or non-throated type of gear blank

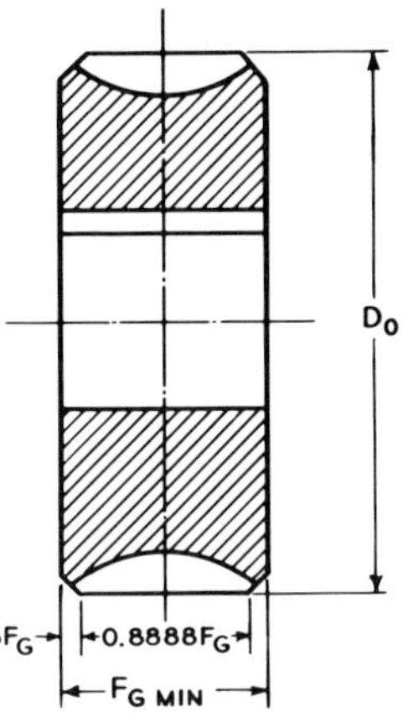

CYLINDRICAL,
WORM-GEAR BLANK

		Worm	Worm Gear or Wheel
21	Difference between normal pressure angles of thread and of cutter, minutes	$\Delta\phi = \frac{5400\ d \sin^3\lambda}{N_W(D_c \cos^2\lambda + d}$ where D_C is diameter of cutter or grinding wheel.	
22	Normal pressure angle on worm	$\phi_n - \Delta\phi$	

TABLE 6–18

Pitch Diameters of Fine Pitch Worms Corresponding to Each Standard Combination of Lead and Lead Angle

ASA B6.9–1950 AGMA 374.02

N_w	Lead (Inches)	LEAD ANGLES IN DEGREES, λ														
		0.5	1.0	1.5	2.0	3.0	4.0	5.0	7.0	9.0	11.0	14.0	17.0	21.0	25.0	30.0
1	0.030	1.0937	0.5472	0.3647	0.2735	......	......	......	......	......	......	......	......	......	......	......
1	0.040	1.4583	0.7297	0.4863	0.3646	0.2429	......	......	......	......	......	......	......	......	......	......
1	0.050	1.8228	0.9121	0.6079	0.4558	0.3037	0.2276	......	......	......						
2	0.060	2.1874	1.0945	0.7295	0.5469	0.3644	0.2731	......	......	......						
1	0.065	......	1.1857	0.7903	0.5925	0.3948	0.2959	0.2365	......	......						
1,2	0.080	......	1.4593	0.9726	0.7293	0.4859	0.3641	0.2911	......	......						
3	0.090	......	1.6417	1.0942	0.8204	0.5466	0.4097	0.3274	0.2333	......						
1,2	0.100	......	1.8242	1.2158	0.9116	0.6073	0.4552	0.3638	0.2592	......	......	......	......	......	......	......
3,4	0.120	......	2.1890	1.4590	1.0939	0.7288	0.5462	0.4366	0.3111	0.2412	......	......	......	......	......	......
1,2	0.130	......	......	1.5805	1.1851	0.7896	0.5917	0.4730	0.3370	0.2613	......	......	......	......	......	......
3,5	0.150	......	......	1.8237	1.3674	0.9110	0.6828	0.5457	0.3889	0.3015	0.2456	......	......	......	......	......
1,2,4	0.160	......	......	1.9453	1.4585	0.9718	0.7283	0.5821	0.4148	0.3216	0.2620	......	......	......	......	......
6	0.180	......	......	2.1884	1.6408	1.0932	0.8193	0.6549	0.4667	0.3618	0.2948	......	......	......	......	......
3	0.195	......	......	......	1.7776	1.1843	0.8876	0.7095	0.5055	0.3919	0.3193	0.2490	......	......	......	......
2,4,5	0.200	......	......	......	1.8232	1.2147	0.9104	0.7276	0.5185	0.4020	0.3275	0.2553	......	......	......	......
7	0.210	......	......	......	1.9143	1.2754	0.9559	0.7640	0.5444	0.4221	0.3439	0.2681	......	......	......	......
3,6,8	0.240	......	......	......	2.1878	1.4576	1.0924	0.8732	0.6222	0.4823	0.3930	0.3064	0.2499	......	......	......
5	0.250	......	......	......	......	1.5184	1.1380	0.9096	0.6481	0.5024	0.4094	0.3192	0.2603	......	......	......
2,4	0.260	......	......	......	......	1.5791	1.1835	0.9459	0.6741	0.5225	0.4258	0.3319	0.2707	......	......	......
9	0.270	......	......	......	......	1.6398	1.2290	0.9823	0.7000	0.5426	0.4421	0.3447	0.2811	......	......	......
7	0.280	......	......	......	......	1.7006	1.2745	1.0187	0.7259	0.5627	0.4585	0.3575	0.2915	......	......	......
3,6,10	0.300	......	......	......	......	1.8220	1.3656	1.0915	0.7778	0.6029	0.4913	0.3830	0.3123	0.2488	......	......
2,4,8	0.320	......	......	......	......	1.9435	1.4566	1.1642	0.8296	0.6431	0.5240	0.4085	0.3332	0.2654	......	......
5	0.325	......	......	......	......	1.9739	1.4794	1.1824	0.8426	0.6532	0.5322	0.4149	0.3384	0.2695	......	......
7	0.350	......	......	......	......	2.1257	1.5932	1.2734	0.9074	0.7034	0.5731	0.4468	0.3644	0.2902	......	......
9	0.360	......	......	......	......	......	1.6387	1.3098	0.9333	0.7235	0.5895	0.4596	0.3748	0.2985	0.2457	......
3,6	0.390	......	......	......	......	......	1.7752	1.4189	1.0111	0.7838	0.6387	0.4979	0.4060	0.3234	0.2662	......
4,5,8,10	0.400	......	......	......	......	......	1.8207	1.4553	1.0370	0.8039	0.6550	0.5107	0.4165	0.3317	0.2730	......
9	0.450	......	......	......	......	......	2.0483	1.6372	1.1666	0.9044	0.7369	0.5745	0.4685	0.3732	0.3072	0.2481
7	0.455	......	......	......	......	......	......	1.6554	1.1796	0.9144	0.7451	0.5809	0.4737	0.3773	0.3106	0.2509
3,6	0.480	......	......	......	......	......	......	1.7463	1.2444	0.9647	0.7860	0.6128	0.4998	0.3980	0.3277	0.2646
5,10	0.500	......	......	......	......	......	......	1.8191	1.2963	1.0049	0.8188	0.6383	0.5206	0.4146	0.3413	0.2757
4,8	0.520	......	......	......	......	......	......	1.8919	1.3481	1.0451	0.8515	0.6639	0.5414	0.4312	0.3550	0.2867
7	0.560	......	......	......	......	......	......	2.0374	1.4518	1.1255	0.9170	0.7149	0.5830	0.4644	0.3823	0.3087
9	0.585	......	......	......	......	......	......	......	1.5166	1.1757	0.9580	0.7469	0.6091	0.4851	0.3993	0.3225
6	0.600	......	......	......	......	......	......	......	1.5555	1.2059	0.9825	0.7660	0.6247	0.4975	0.4096	0.3308
4,8	0.640	......	......	......	......	......	......	......	1.6592	1.2863	1.0480	0.8171	0.6663	0.5307	0.4369	0.3529
5,10	0.650	......	......	......	......	......	......	......	1.6852	1.3064	1.0644	0.8298	0.6767	0.5390	0.4437	0.3584
7	0.700	......	......	......	......	......	......	......	1.8148	1.4068	1.1463	0.8937	0.7288	0.5805	0.4778	0.3859
9	0.720	......	......	......	......	......	......	......	1.8666	1.4470	1.1791	0.9192	0.7496	0.5971	0.4915	0.3970
6	0.780	......	......	......	......	......	......	......	2.0222	1.5676	1.2773	0.9958	0.8121	0.6468	0.5324	0.4300
5,8,10	0.800	......	......	......	......	......	......	......	......	1.6078	1.3101	1.0213	0.8329	0.6634	0.5461	0.4411
9	0.900	......	......	......	......	......	......	......	......	1.8088	1.4738	1.1490	0.9370	0.7463	0.6144	0.4962
7	0.910	......	......	......	......	......	......	......	......	1.8289	1.4902	1.1618	0.9474	0.7546	0.6212	0.5017
6	0.960	......	......	......	......	......	......	......	......	1.9294	1.5721	1.2256	0.9995	0.7961	0.6553	0.5293
10	1.000	......	......	......	......	......	......	......	......	2.0098	1.6376	1.2767	1.0412	0.8292	0.6826	0.5513
8	1.040	......	......	......	......	......	......	......	......	......	1.7031	1.3277	1.0828	0.8624	0.7099	0.5734
7	1.120	......	......	......	......	......	......	......	......	......	1.8341	1.4299	1.1661	0.9287	0.7645	0.6175
9	1.170	......	......	......	......	......	......	......	......	......	1.9160	1.4937	1.2181	0.9720	0.7987	0.6451
8	1.280	......	......	......	......	......	......	......	......	......	2.0961	1.6341	1.3327	1.0614	0.8738	0.7057
10	1.300	......	......	......	......	......	......	......	......	......	......	1.6597	1.3535	1.0780	0.8874	0.7167
9	1.440	......	......	......	......	......	......	......	......	......	......	1.8384	1.4993	1.1941	0.9830	0.7939
10	1.600	......	......	......	......	......	......	......	......	......	......	2.0427	1.6658	1.3268	1.0922	0.8821

This table gives the pitch diameter for each combination of lead and lead angle, together with the number of threads for a particular lead and diameter.

TABLE 6–19

Tooth Proportions of Fine-Pitch Worm Gearing for All Combinations of Standard Axial Pitches and Lead Angles

ASA B6.9–1950 AGMA 374.02

Standard Axial Pitch (Inches)	Tooth Propor-tions	LEAD ANGLE IN DEGREES														
		0.5	1	1.5	2	3	4	5	7	9	11	14	17	21	25	30
0.030	a	0.0095	0.0095	0.0095	0.0095	0.0095	0.0095	0.0095	0.0095	0.0094	0.0094	0.0093	0.0091	0.0089	......	......
	h_t	0.0229	0.0229	0.0229	0.0229	0.0229	0.0229	0.0229	0.0229	0.0227	0.0227	0.0225	0.0220	0.0216	......	......
	p_n	0.0300	0.0300	0.0300	0.0300	0.0300	0.0299	0.0299	0.0298	0.0296	0.0294	0.0291	0.0287	0.0280	......	......
0.040	a	0.0127	0.0127	0.0127	0.0127	0.0127	0.0127	0.0127	0.0126	0.0126	0.0125	0.0124	0.0122	0.0119	0.0115	......
	h_t	0.0299	0.0299	0.0299	0.0299	0.0299	0.0299	0.0299	0.0297	0.0297	0.0295	0.0293	0.0288	0.0282	0.0273	......
	p_n	0.0400	0.0400	0.0400	0.0400	0.0399	0.0399	0.0398	0.0397	0.0395	0.0393	0.0388	0.0383	0.0373	0.0363	......
0.050	a	0.0159	0.0159	0.0159	0.0159	0.0159	0.0159	0.0159	0.0158	0.0157	0.0156	0.0154	0.0152	0.0149	0.0144	0.0138
	h_t	0.0370	0.0370	0.0370	0.0370	0.0370	0.0370	0.0370	0.0368	0.0365	0.0363	0.0359	0.0354	0.0348	0.0337	0.0324
	p_n	0.0400	0.0500	0.0500	0.0500	0.0499	0.0499	0.0498	0.0496	0.0494	0.0491	0.0485	0.0478	0.0467	0.0453	0.0433
0.065	a	......	0.0207	0.0207	0.0207	0.0207	0.0206	0.0206	0.0205	0.0204	0.0203	0.0201	0.0198	0.0193	0.0188	0.0179
	h_t	......	0.0475	0.0475	0.0475	0.0475	0.0473	0.0473	0.0471	0.0469	0.0467	0.0462	0.0456	0.0445	0.0434	0.0414
	p_n	......	0.0650	0.0650	0.0650	0.0649	0.0648	0.0648	0.0645	0.0642	0.0638	0.0631	0.0622	0.0607	0.0589	0.0563
0.080	a	......	0.0255	0.0255	0.0254	0.0264	0.0254	0.0254	0.0253	0.0252	0.0250	0.0247	0.0244	0.0238	0.0231	0.0221
	h_t	......	0.0581	0.0581	0.0579	0.0579	0.0579	0.0579	0.0577	0.0574	0.0570	0.0563	0.0557	0.0544	0.0528	0.0506
	p_n	......	0.0800	0.0800	0.0800	0.0799	0.0798	0.0797	0.0794	0.0790	0.0785	0.0776	0.0765	0.0747	0.0725	0.0693
0.100	a	......	0.0318	0.0318	0.0318	0.0318	0.0318	0.0317	0.0316	0.0314	0.0312	0.0309	0.0304	0.0297	0.0288	0.0276
	h_t	......	0.0720	0.0720	0.0720	0.0720	0.0720	0.0717	0.0716	0.0711	0.0706	0.0700	0.0689	0.0673	0.0654	0.0627
	p_n	......	0.1000	0.1000	0.0999	0.0999	0.0998	0.0996	0.0993	0.0988	0.0982	0.0970	0.0956	0.0934	0.0906	0.0866
0.130	a	......	......	0.0414	0.0414	0.0413	0.0413	0.0412	0.0411	0.0409	0.0406	0.0402	0.0396	0.0386	0.0375	0.0358
	h_t	......	......	0.0931	0.0931	0.0929	0.0929	0.0926	0.0924	0.0920	0.0913	0.0904	0.0891	0.0869	0.0845	0.0808
	p_n	......	......	0.1300	0.1299	0.1298	0.1297	0.1295	0.1290	0.1284	0.1276	0.1261	0.1243	0.1214	0.1178	0.1126
0.160	a	......	......	0.0509	0.0509	0.0509	0.0508	0.0507	0.0506	0.0503	0.0500	0.0494	0.0487	0.0475	0.0462	0.0441
	h_t	......	......	0.1140	0.1140	0.1140	0.1138	0.1135	0.1133	0.1127	0.1120	0.1107	0.1091	0.1065	0.1036	0.0990
	p_n	......	......	0.1599	0.1599	0.1598	0.1596	0.1594	0.1588	0.1580	0.1571	0.1552	0.1530	0.1494	0.1450	0.1386

TABLE 6–20

Example to Illustrate Application of Formulas in Table 6–17 to Finding Dimensions of a Fine-Pitch Worm and Worm Gear

ASA B6.9–1950 AGMA 374.02

Suppose a 1-to-50-reduction worm-gear pair is to be designed to operate on an approximate center distance of 0.55 inch. The ratio is such as to suggest a single-thread worm, *i.e.*, $N_W = 1$. Table 6–18 is entered with $N_W = 1$ to find worm diameters d, and corresponding worm gear diameters D are computed roughly from item 9, Table 6–17. A few trials show that $l = 0.050$ approximates the desired center distance. Calculations for dimension now proceed as outlined in Table 6–17.

1	Number threads on worm	$N_W = 1$	4,5	Axial pitch	$p_x = p = 0.050$
2	Number of teeth on gear	$N_G = 50$	6	Lead angle	$\lambda = 3° 0'$
3	Normal pressure angle on 4-inch grinding wheel	20°	7	Helix angle	$\psi = 87° 0'$

		Worm				Worm Gear
8	Lead	$l = 0.050$				
9	Pitch diameter	$d = 0.3037$	Table 6–18			$D = 0.7958$
10	Normal circular pitch	$p_n = 0.0499$	or Table 6–19			$p_n = 0.0499$
11	Addendum			$a = 0.0159$	Table 6–19	
12	Whole depth			$h_t = 0.0370$	Table 6–19	
13	Working depth			$h_k = 0.0318$		
14	Clearance			$c = 0.0052$		
15	Tooth thickness			$t = 0.0250$		
16	Center distance			$C = 0.5498$		
17	Outside diameter	$d_o = 0.3355$				$D_o = 0.8277$
18	Omit for cylindrical gear					
19	Length of threaded portion of worm	$F_W = 0.2276$				
20	Face width of worm gear					$F_G = 0.2405$
21	Difference in normal pressure angles	$\Delta\phi = \text{nil}$				
22	Normal pressure angle on worm	20°				

TABLE 6-21

AGMA Standard Ratings for Surface Durability of Worm Gearing
AGMA 213.02 – 1952 AGMA 440.02 – 1952

AGMA Standard	Formula	Definition of Symbols and Limiting Conditions
213.02 Surface durability of cylindrical worm gearing for power transmission	Input horsepower rating for Class I service $= 0.0028\, C^{2.71}\, n_W\, S_v \left[\dfrac{\left(1 - \frac{1}{m}\right)^2}{m}\right]$ Providing Speed of sliding does not exceed 6000 feet per minute, $d = \dfrac{C^{0.875}}{2.2}$, **approximately** and $F_e = \sqrt{(d + b_k)^2 - d^2}$ Horsepower is proportionately reduced when the effective face width of gear is less than the maximum F_e allowed and overlapping tooth contact is retained.	(See Table 6–24 for other Classes) C = center distance, inches (Table 6–22 for values of $0.0028C^{2.71}$) n_W = worm speed in rpm, and over 100 S_v = speed-of-sliding factor (Table 6–23) $m = \text{ratio} = \dfrac{\text{No. teeth in gear}}{\text{No. threads on worm}}$ (Table 6–25 for values of $C^{0.875}$) d = mean worm diameter, *i.e.*, the diameter to the average working depth on the thread F_e = maximum effective face width of gear, inches, and not to exceed $0.75d$. b_k = working depth of tooth, inches.
440.02 Standard Practice for Cylindrical Worm-Gear Speed Reducers Standard limited to enclosed gears of bath lubrication types.	Besides the surface durability rating outlined above, AGMA 213.02 also recognizes a thermal horsepower rating as determined by AGMA 440.02. Thermal horsepower rating, which is independent of service factor classification, is $= \dfrac{K_T}{2.5 + m/2}$ where m is not less than 6. For worm speeds less than 100 rpm, rating is determined as output torque in inch-pounds. First: Compute input horsepower rating for surface durability using $n_W = 100$. Second: Convert input horsepower rating to input torque, inch pounds. Third: Get output torque by multiplying input torque by efficiency, E.	K_T = a thermal constant, designed to limit oil-bath rise in temperature to 100 F above ambient, and not over 200 F. The thermal rating formula is not applicable to worm speeds in excess of 2000 rpm. Take K_T from Table 6–26. $m = \text{ratio} = \dfrac{\text{No. teeth in gear}}{\text{No. threads on worm}}$ E = efficiency Efficiency of ratios 6 to 1 or less may be taken as 97 per cent; efficiency of high ratios as 100 per cent less half the ratio, or $E = 100 - (m/2)$

TABLE 6–22

Basic Pressure Factor, $0.0028C^{2.71}$
(See Table 6–21)

AGMA 213.01–1950

Center Distance C	$0.0028C^{2.71}$	Center Distance C	$0.0028C^{2.71}$	Center Distance C	$0.0028C^{2.71}$	Center Distance C	$0.0028C^{2.71}$
3.00	0.05498	10.00	1.4360	17.00	6.0489	24.00	15.400
3.25	.06830	10.25	1.5354	17.25	6.2930	24.25	15.838
3.50	.08348	10.50	1.6390	17.50	6.5432	24.50	16.285
3.75	.1006	10.75	1.7469	17.75	6.7996	24.75	16.740
4.00	.1199	11.00	1.8592	18.00	7.0623	25.00	17.202
4.25	.1413	11.25	1.9760	18.25	7.3313	25.25	17.672
4.50	.1650	11.50	2.0972	18.50	7.6066	25.50	18.150
4.75	.1910	11.75	2.2231	18.75	7.8884	25.75	18.637
5.00	.2195	12.00	2.3536	19.00	8.1767	26.00	19.131
5.25	.2505	12.25	2.4889	19.25	8.4716	26.25	19.634
5.50	.2841	12.50	2.6290	19.50	8.7731	26.50	20.144
5.75	.3205	12.75	2.7739	19.75	9.0812	26.75	20.663
6.00	.3597	13.00	2.9238	20.00	9.3961	27.00	21.191
6.25	.4018	13.25	3.0787	20.25	9.7178	27.25	21.727
6.50	.4468	13.50	3.2386	20.50	10.046	27.50	22.272
6.75	.4950	13.75	3.4038	20.75	10.382	27.75	22.824
7.00	.5462	14.00	3.5740	21.00	10.724	28.00	23.386
7.25	.6007	14.25	3.7497	21.25	11.074	28.50	24.535
7.50	.6585	14.50	3.9307	21.50	11.431	29.00	25.719
7.75	.7197	14.75	4.1170	21.75	11.794	29.50	26.939
8.00	.7844	15.00	4.3089	22.00	12.165	30.00	28.194
8.25	.8526	15.25	4.5063	22.25	12.544	30.50	29.485
8.50	.9244	15.50	4.7093	22.50	12.929	31.00	30.814
8.75	1.0000	15.75	4.9180	22.75	13.322	32.00	33.583
9.00	1.0793	16.00	5.1324	23.00	13.723	33.00	36.503
9.25	1.1625	16.25	5.3527	23.25	14.131	34.00	39.579
9.50	1.2497	16.50	5.5788	23.50	14.546	35.00	42.814
9.75	1.3408	16.75	5.8108	23.75	14.970	36.00	46.210

TABLE 6–23

Value of Speed-of-Sliding Factor, S_v
(See Table 6–21)

AGMA 213.02 – 1952

V is speed of sliding along mean helix in feet per minute. $V = \dfrac{\pi d n_w}{12 \cos \lambda}$ where d is worm pitch diameter, n_w is worm speed in rpm and λ is lead angle.

The values are based upon $d = \dfrac{C^{0.875}}{2.2}$ but for the purposes of calculating horsepower ratings, they are accurate enough over the entire range of diameter-center-distance ratios.

Speed of Sliding, V, fpm	$S_v = \dfrac{180}{180 + V^{0.85}}$	V	S_v	V	S_v	V	S_v
0	1.000	100	0.782	400	0.525	1600	0.254
5	0.979	110	.768	450	.500	1700	.244
10	.962	120	.755	500	.478	1800	.235
15	.947	130	.742	550	.458	1900	.227
20	.934	140	.730	600	.439	2000	.220
25	.921	150	.718	650	.423	2100	.213
30	.909	160	.707	700	.407	2200	.206
35	.898	170	.696	750	.394	2400	.194
40	.887	180	.685	800	.380	2600	.184
45	.877	190	.676	850	.369	2800	.175
50	.866	200	.666	900	.357	3000	.166
55	.857	220	.648	950	.347	3200	.159
60	.847	240	.631	1000	.337	3400	.152
65	.838	260	.615	1050	.328	3600	.146
70	.830	280	.600	1100	.319	3800	.140
75	.821	300	.585	1150	.311	4000	.135
80	.813	320	.572	1200	.303	4500	.125
85	.805	340	.559	1300	.289	5000	.114
90	.797	360	.547	1400	.276	5500	.107
95	.790	380	.536	1500	.264	6000	.100

TABLE 6–24

Service Factors to be Applied to Determination of Durability Rating of Worm Gearing for Power Transmission

(See Table 6–21)

AGMA 213.02–1952 Surface Durability of Cylindrical Worm Gearing

<table>
<tr><th>Class</th><th>Service Factor</th><th>Conditions of Operation</th></tr>
<tr><td>1</td><td>Unity
(I.e., durability rating as determined from methods of Standard 213.02 is true rating)</td><td>Normal 8- to 10-hr service, free from recurrent shocks, i.e., shock loads that recur at approximately even and frequent intervals.</td></tr>
<tr><td>2</td><td>Divide Class 1 determination by 1.2</td><td>8- to 10-hr service where recurrent shock loading is encountered, or 24-hr service without shock loading.</td></tr>
<tr><td>3</td><td>Divide Class 1 determination by 1.3</td><td>Twenty-four-hr service plus shock loads.</td></tr>
<tr><td>4</td><td>
<table>
<tr><th rowspan="2">Divide by</th><th colspan="3">When total minutes of operation are</th></tr>
<tr><th>Per hour, multiple cycles of</th><th>Per cycle, 1 cycle per hr</th><th>Per cycle, 1 cycle per 2 hr or more</th></tr>
<tr><td>0.6</td><td></td><td>5</td><td>10</td></tr>
<tr><td>.7</td><td>2</td><td>10</td><td>20</td></tr>
<tr><td>.8*</td><td>5</td><td>15</td><td>30</td></tr>
<tr><td>.9</td><td>10</td><td>20</td><td>40</td></tr>
</table>
</td><td>For intermittent service where worm speed is 100 rpm or more, divide the Class I durability rating by that factor in the column at the extreme left which most nearly corresponds to the minutes of operation and frequency stated in the other three columns.

Under Class 4 service the thermal horsepower rating of Table 6–21 does not apply.</td></tr>
<tr><td>5</td><td>Factors of Class 4 apply if service is intermittent and worm speed is 100 rpm or more.</td><td>When worm speed is less than 100 rpm, carry out the calculations with the output torque in inch-pounds as outlined in Table 6–21.</td></tr>
</table>

* As examples a service factor of 0.8 would apply: (a) More than one start and stop per hour with total time of operation not exceeding 5 minutes per hour; (b) one start and stop per hour with total time of operation not exceeding 15 minutes per cycle; (c) one start and stop in 2 hours or more with total time of operation not exceeding 30 minutes per cycle.

TABLE 6–25

Values* of Center Distance Quantity $C^{0.875}$

(See Tables 6–11 and 6–21)

Center Distance, C, Inches	$C^{0.875}$	Center Distance, C, Inches	$C^{0.875}$	Center Distance, C, Inches	$C^{0.875}$
1.000	1.000	3.000	2.615	5.000	4.089
.050	1.044	.050	2.653	.050	4.125
.100	1.088	.100	2.691	.100	4.161
.150	1.131	.150	2.729	.150	4.197
1.200	1.173	3.200	2.767	5.200	4.233
.250	1.216	.250	2.805	.250	4.269
.300	1.258	.300	2.843	.300	4.304
.350	1.300	.350	2.881	.350	4.339
1.400	1.342	3.400	2.918	5.400	4.374
.450	1.384	.450	2.956	.450	4.409
.500	1.426	.500	2.993	.500	4.444
.550	1.468	.550	3.030	.550	4.480
1.600	1.509	3.600	3.067	5.600	4.515
.650	1.550	.650	3.105	.650	4.550
.700	1.591	.700	3.142	.700	4.585
.750	1.632	.750	3.179	.750	4.620
1.800	1.673	3.800	3.216	5.800	4.656
.850	1.714	.850	3.253	.850	4.691
.900	1.754	.900	3.290	.900	4.726
.950	1.794	.950	3.327	.950	4.761
2.000	1.834	4.000	3.364	6.000	4.796
.050	1.874	.050	3.401	.050	4.831
.100	1.914	.100	3.438	.100	4.866
.150	1.954	.150	3.475	.150	4.901
2.200	1.993	4.200	3.512	6.200	4.936
.250	2.033	.250	3.549	.250	4.970
.300	2.073	.300	3.585	.300	5.005
.350	2.112	.350	3.621	.350	5.040
2.400	2.151	4.400	3.657	6.400	5.075
.450	2.190	.450	3.693	.450	5.110
.500	2.229	.500	3.729	.500	5.144
.550	2.268	.550	3.765	.550	5.179
2.600	2.307	4.600	3.801	6.600	5.214
.650	2.346	.650	3.837	.650	5.248
.700	2.385	.700	3.873	.700	5.283
.750	2.424	.750	3.909	.750	5.317
2.800	2.462	4.800	3.945	6.800	5.352
.850	2.501	.850	3.981	.850	5.386
.900	2.539	.900	4.017	.900	5.420
.950	2.577	.950	4.053	.950	5.455
3.000	2.615	5.000	4.089	7.000	5.489

*The tabular values are suitable for interpolation, if so desired. Ordinarily, dimensions are rounded off to suit fractional or decimal numbers and the nearest tabular value is adequate without interpolation.

continued on next page

Center Distance, C, Inches	$C^{0.875}$
7.000	5.489
.050	5.523
.100	5.557
.150	5.591
7.200	5.626
.250	5.660
.300	5.694
.350	5.728
7.400	5.762
.450	5.796
.500	5.830
.550	5.864
7.600	5.898
.650	5.932
.700	5.966
.750	6.000
7.800	6.033
.850	6.067
.900	6.101
.950	6.135
8.000	6.169
.050	6.203
.100	6.237
.150	6.270
8.200	6.304
.250	6.338
.300	6.372
.350	6.405
8.400	6.438
.450	6.471
.500	6.505
.550	6.538
8.600	6.572
.650	6.606
.700	6.639
.750	6.672
8.800	6.706
.850	6.739
.900	6.772
.950	6.806
9.000	6.839

Center Distance, C, Inches	$C^{0.875}$
9.000	6.839
.050	6.872
.100	6.905
.150	6.938
9.200	6.971
.250	7.005
.300	7.038
.350	7.071
9.400	7.104
.450	7.137
.500	7.170
.550	7.203
9.600	7.236
.650	7.269
.700	7.302
.750	7.335
9.800	7.368
.850	7.400
.900	7.433
.950	7.466
10.000	7.499
.100	7.565
.200	7.631
.300	7.696
.400	7.761
10.500	7.826
.600	7.891
.700	7.956
.800	8.021
.900	8.086
11.000	8.151
.100	8.216
.200	8.280
.300	8.345
.400	8.410
11.500	8.474
.600	8.539
.700	8.603
.800	8.668
.900	8.732
12.000	8.796

Center Distance, C, Inches	$C^{0.875}$
12.000	8.796
.100	8.860
.200	8.924
.300	8.988
.400	9.052
12.500	9.116
.600	9.180
.700	9.243
.800	9.307
.900	9.370
13.000	9.434
.100	9.497
.200	9.561
.300	9.624
.400	9.688
13.500	9.751
.600	9.814
.700	9.877
.800	9.940
.900	10.003
14.000	10.066
.100	10.129
.200	10.192
.300	10.254
.400	10.317
14.500	10.380
.600	10.443
.700	10.505
.800	10.568
.900	10.631
15.000	10.693
.100	10.755
.200	10.817
.300	10.880
.400	10.942
15.500	11.004
.600	11.066
.700	11.128
.800	11.190
.900	11.252
16.000	11.314

Center Distance, C, Inches	$C^{0.875}$
16.000	11.314
.100	11.376
.200	11.438
.300	11.500
.400	11.561
16.500	11.623
.600	11.685
.700	11.746
.800	11.807
.900	11.869
17.000	11.930
.100	11.991
.200	12.052
.300	12.114
.400	12.175
17.500	12.236
.600	12.297
.700	12.359
.800	12.420
.900	12.481
18.000	12.542
.100	12.603
.200	12.664
.300	12.725
.400	12.786
18.500	12.847
.600	12.907
.700	12.968
.800	13.029
.900	13.090
19.000	13.150
.100	13.210
.200	13.270
.300	13.331
.400	13.391
19.500	13.451
.600	13.512
.700	13.572
.800	13.632
.900	13.692
20.000	13.753

*The tabular values are suitable for interpolation, if so desired. Ordinarily, dimensions are rounded off to suit fractional or decimal numbers and the nearest tabular value is adequate without interpolation.

TABLE 6–26

Thermal Factor Chart, K_T

AGMA 440.02 – 1952 Standard Practice for Cylindrical-Worm Gear Speed Reducers

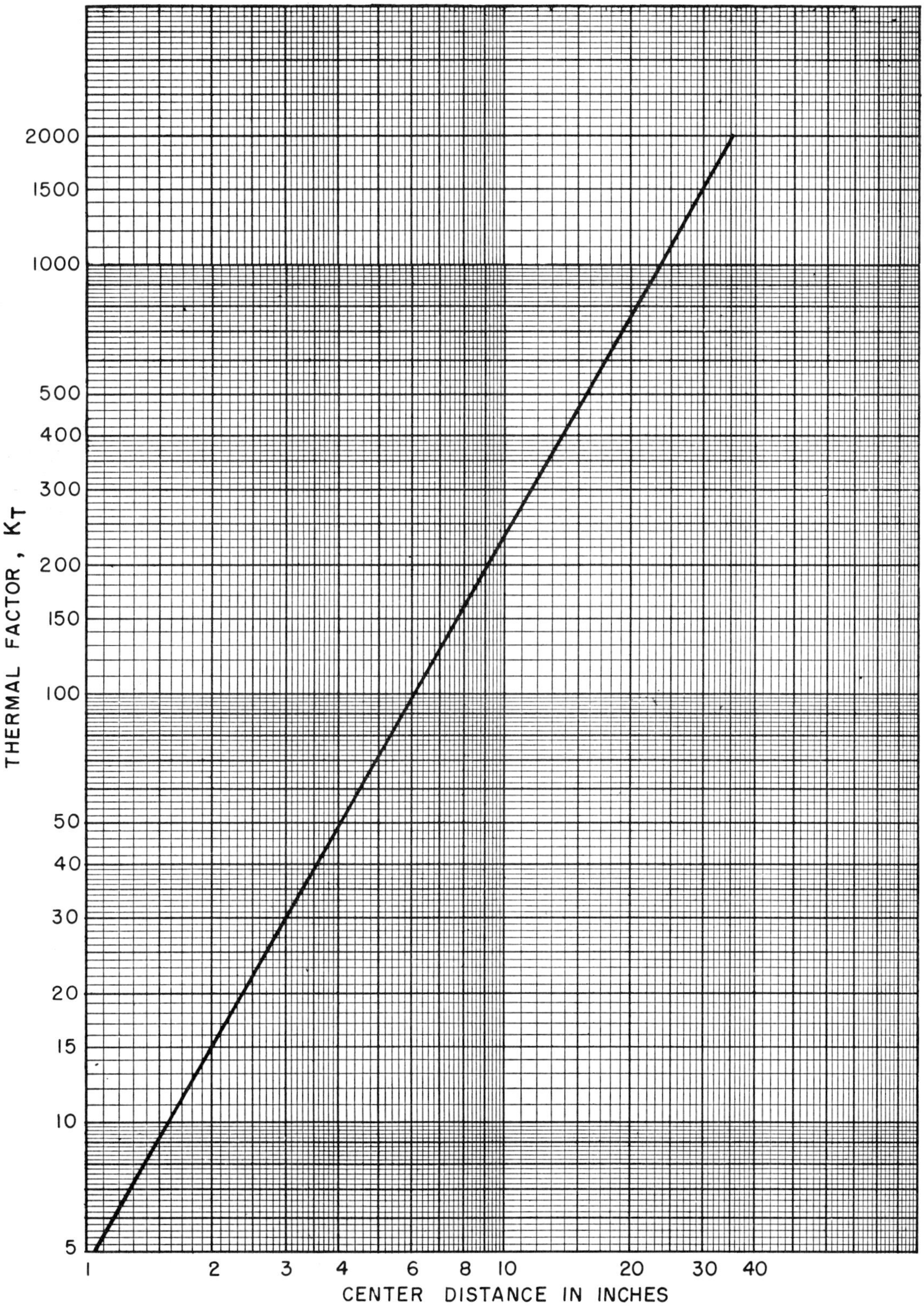

TABLE 6-27

AGMA Standard Bronzes for Cast, Worm-Gear Blanks
AGMA 243.01 – 1954

Class	Composition	Cu	Sn	Ni	Pb	Zn	Phos.	IMP	Min Br Hardness* on Toothed Portion
1**	Nickel Tin Bronze	Bal	9.75% 10.75%	1.25% 1.75%	--	--	0.030% Max	0.25% Max	500 Kg. 70
1c†	Chill Cast-Nickel	"	"	"	--	--	"	"	80
2**	Tin Bronze	"	10-12%	--	--	2% Max	"	"	70
2c†	Chill Cast Tin Br.	"	"	--	--	"	"	"	85
3‡	Leaded Bronze	"	9-11%	--	1-2	--	"	--	None required

* To be measured on side of rim.
** Widely used for general-purpose worm gears.
† Where production volume justifies the expense of making standard chill-rings, Class 1c followed by Class 2c is most commonly used.
‡ Used only for high-speed applications.

TABLE 6-28

Bore Tolerances on Shell-Type Cylindrical Worms
Practice of Boston Gear Works

Nominal Bore in inches		Tolerance on Basic Dimension	
over	inclusive	plus	minus
0	1-1/4	0.0002	– 0.0005
1-1/4	3	0.0002	– 0.0006
3	up	0.0005	– 0.0010

Keyway Tolerances on Shell-Type Cylindrical Worms

Dimension	Tolerance	
Width of keyway	+ 0.0015	– 0.0015
Depth	+ 0.005	– 0.0000

TABLE 6–29

Tolerance on Center Distance of Worm Gearing
Catalog 400 The Cleveland Worm and Gear Co.

Center Distance	Tolerance, All Plus
Under 4 inches	+ 0.003
4 inches and over	+ 0.005

TABLE 6–30

Some Dimensions, with Tolerances, on Worm Gearing
Practice of Caterpillar Tractor Co.

Worm		Worm Gear	
Pitch diameter	Decimal dimension (no tolerance)	Pitch diameter	Decimal dimension (no tolerance)
Outside diameter	High limit = nominal diameter, as calculated with a tolerance of ± .003	Outside diameter	High limit = nominal diameter, as calculated with a tolerance of ±.003
Axial pitch	Decimal dimension (no tolerance)	Center distance between worm	Decimal dimension Up to 8 in. ±.0005 Over 8 to 16 in. ±.0010
Lead	Decimal dimension (no tolerance)	Throat diameter	Decimal dimension
Pressure angle	Degrees, minutes, and seconds		
Chordal thickness at pitch diameter	Decimal dimension (no tolerance)	Pressure angle	Degrees, minutes, and seconds
Show chordal thickness and pressure angle in section view through teeth at right angle to helix angle.		Chordal thickness at pitch diameter	Decimal dimension (no tolerance)
		Corner radius	Fractional dimension to nearest 1/16
		Face width	Fractional dimension to nearest 1/16
		Show chordal thickness and pressure angle in sectional view through tooth at right angles to helix angle.	

TABLE 6–31

Proportions and Tolerances for Solid Bronze Worm-Gear Blanks

Catalog 400, First Edition – 1952 The Cleveland Worm and Gear Co.

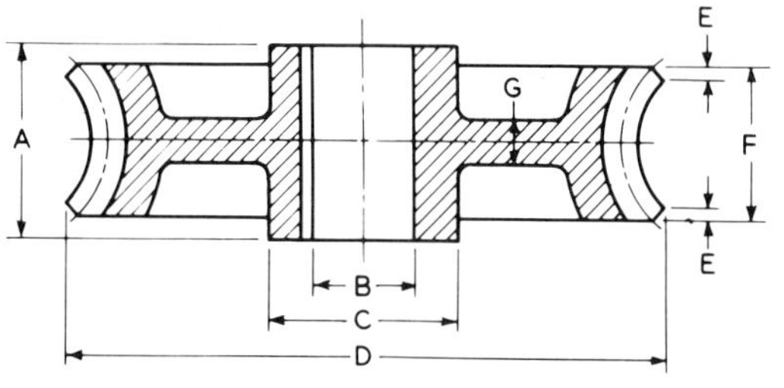

Center Distance Inches	Bore Diameter B	Hub Diameter C	Hub Length A	Face Width F	Web G	Outside Diameter D	Chamfer E	Keyway*
3.000	1.378 1.377	2-1/8	1.620 1.630	7/8	3/8	5-1/16	1/8 × 30°	5/16 × 5/32
3.500	1.500 1.499	2-3/8	2.245 2.255	1	1/2	6-1/16	1/8 × 30°	3/8 × 3/16
4.000 type AT	1.875 1.874	3	2.495 2.505	1-1/8	9/16	6-7/8	1/8 × 45°	1/2 × 1/4
4.000 type AH	2.164 2.163	3-1/4	2.370 2.380	1-1/8	9/16	6-7/8	1/8 × 45°	1/2 × 1/4
4.750	2.125 2.124	3-1/2	2.745 2.755	1-1/2	3/4	8-1/4	3/16 × 45°	1/2 × 1/4
5.500 type AT	2.875 2.874	4-5/8	2.995 3.005	1-3/4	7/8	9-3/8	3/16 × 45°	3/4 × 3/8
5.500 type AH	3.344 3.343	4-3/4	3.745 3.755	2	7/8	9-3/8	1/4 × 45°	1/2 × 1/4

*Tolerance on depth $^{+0.010}_{-0.000}$ on width $^{+0.003}_{-0.000}$

NOTE: On medium and large worm gears, flat rather than full-throated design is good practice to reduce the chance of tooth damage during handling.

TABLE 6–32

Proportions and Tolerances on Flanged, Bronze Rims and on Cast Iron Hubs of Composite Worm Wheels

Catalog 400, First Edition – 1952 The Cleveland Worm and Gear Co.

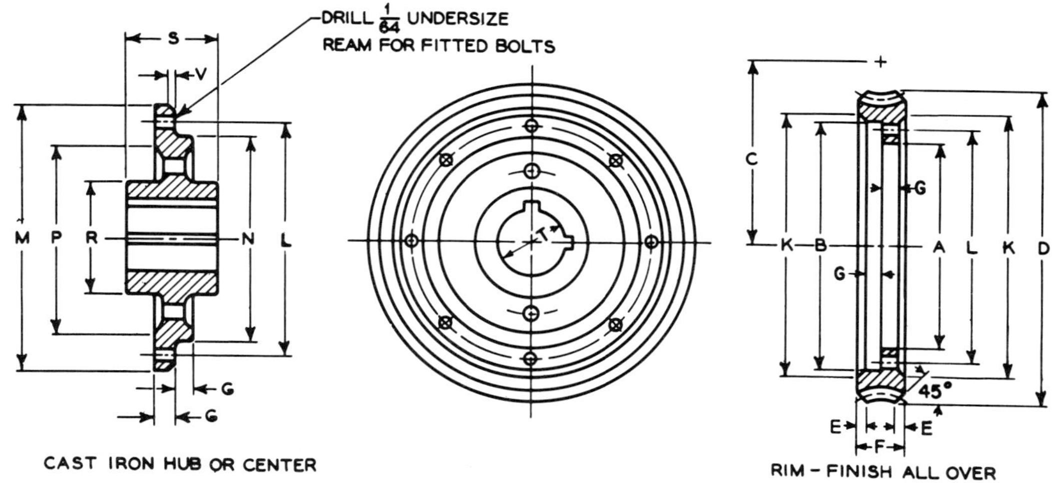

Center Distance Inches	A*	B†	D	E	F	G	K	L	M†	N*	P	R	S	T	V	Bolt Holes No	Bolt Holes Size	Keyway‡
	Dimension With Tolerance																	
6.8715	6.008 6.012	9.000 9.002	$11\frac{13}{16}$	$\frac{1}{4}$	2	$\frac{3}{4}$	9	$7\frac{1}{2}$	9.005 9.007	5.998 6.002		$4\frac{5}{8}$	2.995 3.005	2.874 2.875	$\frac{1}{16}$	6	$\frac{5}{8}$	$\frac{3}{4} \times \frac{3}{8}$
8.173	8.008 8.012	11.000 11.002	$14\frac{3}{16}$	$\frac{1}{4}$	$2\frac{1}{4}$	$\frac{3}{4}$	$11\frac{1}{4}$	$9\frac{1}{2}$	11.005 11.007	7.998 8.002		$5\frac{3}{4}$	4.745 4.755	3.249 3.250	$\frac{1}{16}$	6	$\frac{5}{8}$	$\frac{3}{4} \times \frac{3}{8}$

*Positive allowance provides clearance between diameters A and N.

†Shrink fit between diameters B and M takes into consideration the temperature expansion rates of bronze and cast iron so that looseness does not develop under normal operating temperatures of 160 to 180°F. Bronze rims having the dimensions specified will expand sufficiently in water at 200°F to drop into place over the hub diameter.

‡Tolerance on depth $^{+0.010}_{-0.000}$; on width $^{+0.003}_{-0.000}$

NOTE: On medium and large worm gears, flat rather than full-throated design is good practice to reduce the chance of tooth damage during handling.

Only one keyway is provided in hubs for center distances 6.875 and 8.173; all the others have two keyways 90° apart as illustrated.

continued on next page

TABLE 6–32, continued

Center Distance Inches	Dimension With Tolerance A*	B†	D	E	F	G	K	L	M†	N*	P	R	S	T	V	Bolt Holes No	Bolt Holes Size	Keyway‡
10.000	10.508 10.512	13.500 13.502	17⅜	¼	2½	¾	13¾	12	13.507 13.509	10.498 10.502	8½	6⅛	5.245 5.255	3.624 3.625	1/16	8	⅝	⅞ × 7/16
12.000	13.508 13.512	16.750 16.753	21	5/16	3	1⅛	17	15⅛	16.758 16.760	13.498 13.502	11½	7	5.745 5.755	4.124 4.125	⅛	8	¾	1 × ½
13.4365	15.258 15.262	19.000 19.003	25⅝	⅜	3¼	1¼	19½	17⅞	19.009 19.011	15.248 15.252	12¾	8¼	6.245 6.255	4.624 4.625	⅛	8	⅞	1¼ × ⅝
15.000	17.008 17.012	21.500 21.503	26½	⅜	3½	1¼	22	19¼	21.511 21.513	16.998 17.002	14	9¾	6.745 6.755	5.624 5.625	⅛	8	⅞	1½ × ¾
18.000	21.508 21.512	26.500 26.503	32⅛	½	4	1¼	27	24	26.513 26.515	21.498 21.502	18½	11½	6.745 6.755	6.374 6.375	⅛	12	1	1½ × ¾
19.518	24.508 24.512	29.500 29.503	35¼	½	4½	1¼	29½	27	29.514 29.516	24.498 24.502	21	11½	6.745 6.755	6.374 6.375	⅛	12	1	1½ × ¾
21.837	27.008 27.012	32.000 32.003	38⅝	⅝	5	1½	32½	29½	32.015 32.017	26.998 27.002	23	13	7.495 7.505	7.249 7.250	⅛	12	1	1½ × ¾

*Positive allowance provides clearance between diameters A and N.

†Shrink fit between diameters B and M takes into consideration the temperature expansion rates of bronze and cast iron so that looseness does not develop under normal operating temperatures of 160 to 180°F. Bronze rims having the dimensions specified will expand sufficiently in water at 200°F to drop into place over the hub diameter.

‡Tolerance on depth $^{+0.010}_{-0.000}$; on width $^{+0.003}_{-0.000}$

Only one keyway is provided in hubs for center distances 6.875 and 8.173; all the others have two keyways 90° apart as illustrated.

TABLE 6–33

Backlash of Worm Gears in Terms of Center Distance

Catalog 400, First Edition – 1952 The Cleveland Worm and Gear Co.

Center Distance, In.	Backlash in In. Minimum	Normal	Maximum
3, 3-1/2	0.003	0.005	0.008
4	0.003	0.005	0.010
4-3/4, 5-1/2	0.005	0.007	0.010
6-7/8	0.005	0.008	0.012
8-11/64	0.005	0.008	0.015
10	0.005	0.010	0.020
12	0.007	0.012	0.020
13-7/16	0.007	0.012	0.020
15 to 22	0.010	0.015	0.020

Beneath Table 6–14 there is a list of the current standards on worm gearing. ASA B6.9 – 1950 provides a standard of design for fine-pitch worm gears but there exists no comparable standard for coarse-pitch worm gears. The technical paper, *Proposal for a Standard Design for General Industrial Coarse-Pitch Cylindrical Worm Gearing,* by F.G. East published in the Trans. ASME, Feb – 1954, is a noteworthy contribution toward such a standard. Reference is made to Mr. East's proposal in footnotes to some of the foregoing Tables, but his paper was not available in time to incorporate the tables and charts from it into this publication.

INDEX TO SECTION 7

Cylindrical Fits

Allowances and Tolerances

Standard Tapers

TABLE 7–1

American Standard Preferred Basic Sizes

ASA B4.1–1947

	0.0100	5/16	0.3125	1 7/8	1.8750
	0.0125	3/8	0.3750	2	2.0000
1/64	0.01562	7/16	0.4375	2 1/8	2.1250
	0.0200	1/2	0.5000	2 1/4	2.2500
	0.0250	9/16	0.5625	2 3/8	2.3750
1/32	0.03125	5/8	0.6250	2 1/2	2.5000
	0.0400	11/16	0.6875	2 5/8	2.6250
	0.0500	3/4	0.7500	2 3/4	2.7500
1/16	0.0625	7/8	0.8750	2 7/8	2.8750
	0.0800	1	1.0000	3	3.0000
3/32	0.09375	1 1/8	1.1250	3 1/4	3.2500
	0.1000	1 1/4	1.2500	3 1/2	3.5000
1/8	0.1250	1 3/8	1.3750	3 3/4	3.7500
5/32	0.15625	1 1/2	1.5000	4	4.0000
3/16	0.1875	1 5/8	1.6250		
1/4	0.2500	1 3/4	1.7500		

All dimensions are given in inches.

TABLE 7–2

Recommended Tolerances and*Allowances in Specifying Fits

ASA B4.1–1947

0.0001	0.0006	0.0025	**0.0100**
0.00015	0.0008	0.0030	0.0120
0.0002	**0.0010**	0.0040	0.0150
0.00025	0.0012	**0.0050**	**0.0200**
0.0003	0.0015	0.0060	0.0250
0.0004	**0.0020**	0.0080	**0.0300**
0.0005			

All dimensions are given in inches.
* The values indicated in heavy type are the preferred values.

TABLE 7–3

American Standard Loose Fit, Class 1
Large Allowance, Interchangeable
ASA B4a–1925

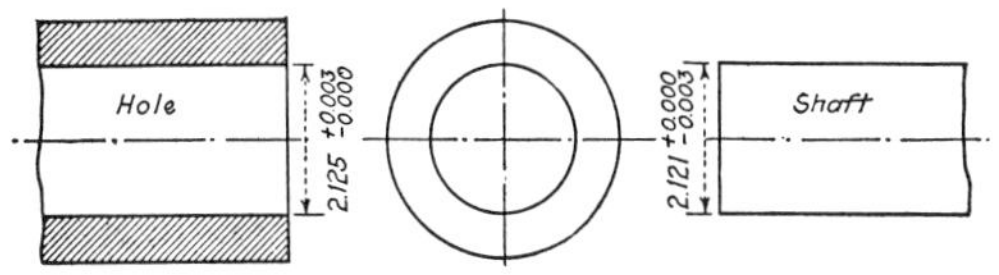

EXAMPLE OF LOOSE FIT (CLASS 1)

SUMMARY OF DATA

	Hole	Shaft	
Tightest Fit	2.125	2.121	0.004 Allowance
Loosest Fit	2.128	2.118	0.010 Allowance + Tolerances

FORMULAS

When d = mean size,

Hole Tolerance $= 0.0025\sqrt[3]{d}$

Shaft Tolerance $= 0.0025\sqrt[3]{d}$

Allowance $= 0.0025\sqrt[3]{d^2}$

This fit provides for considerable freedom and embraces certain fits where accuracy is not essential.

Size			Limits				Tightest Fit	Loosest Fit
From	Up to and Incl.	Mean	Hole or External Member		Shaft or Internal Member		Allowance	Allowance + Tolerances
			+		−	−	+*	+*
0	3/16	1/8	0.001	0.000	0.001	0.002	0.001	0.003
3/16	5/16	1/4	0.002	0.000	0.001	0.003	0.001	0.005
5/16	7/16	3/8	0.002	0.000	0.001	0.003	0.001	0.005
7/16	9/16	1/2	0.002	0.000	0.002	0.004	0.002	0.006
9/16	11/16	5/8	0.002	0.000	0.002	0.004	0.002	0.006
11/16	13/16	3/4	0.002	0.000	0.002	0.004	0.002	0.006
13/16	15/16	7/8	0.002	0.000	0.002	0.004	0.002	0.006
15/16	1 1/16	1	0.003	0.000	0.003	0.006	0.003	0.009
1 1/16	1 3/16	1 1/8	0.003	0.000	0.003	0.006	0.003	0.009
1 3/16	1 3/8	1 1/4	0.003	0.000	0.003	0.006	0.003	0.009
1 3/8	1 5/8	1 1/2	0.003	0.000	0.003	0.006	0.003	0.009
1 5/8	1 7/8	1 3/4	0.003	0.000	0.004	0.007	0.004	0.010
1 7/8	2 1/8	2	0.003	0.000	0.004	0.007	0.004	0.010
2 1/8	2 3/8	2 1/4	0.003	0.000	0.004	0.007	0.004	0.010
2 3/8	2 3/4	2 1/2	0.003	0.000	0.005	0.008	0.005	0.011
2 3/4	3 1/4	3	0.004	0.000	0.005	0.009	0.005	0.013
3 1/4	3 3/4	3 1/2	0.004	0.000	0.006	0.010	0.006	0.014
3 3/4	4 1/4	4	0.004	0.000	0.006	0.010	0.006	0.014
4 1/4	4 3/4	4 1/2	0.004	0.000	0.007	0.011	0.007	0.015
4 3/4	5 1/2	5	0.004	0.000	0.007	0.011	0.007	0.015
5 1/2	6 1/2	6	0.005	0.000	0.008	0.013	0.008	0.018
6 1/2	7 1/2	7	0.005	0.000	0.009	0.014	0.009	0.019
7 1/2	8 1/2	8	0.005	0.000	0.010	0.015	0.010	0.020

All dimensions in inches.
* NOTE: (+) denotes clearance or amount of looseness.

TABLE 7–4

American Standard Free Fit, Class 2
Liberal Allowance, Interchangeable
ASA B4a–1925

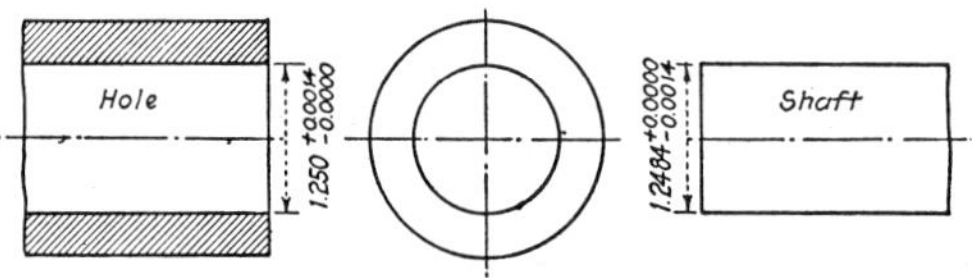

EXAMPLE OF FREE FIT (CLASS 2)

SUMMARY OF DATA

	Hole	Shaft	
Tightest Fit	1.2500	1.2484	0.0016 Allowance
Loosest Fit	1.2514	1.2470	0.0044 Allowance + Tolerances

FORMULAS

When d = mean size,

$$\text{Hole Tolerance} = 0.0013\sqrt[3]{d}$$

$$\text{Shaft Tolerance} = 0.0013\sqrt[3]{d}$$

$$\text{Allowance} = 0.0014\sqrt[3]{d^2}$$

For running fits with speeds of 600 r.p.m. or over, and journal pressures of 600 lb. per sq. in. or over.

Size			Limits				Tightest Fit	Loosest Fit
From	Up to and Incl.	Mean	Hole or External Member		Shaft or Internal Member		Allowance	Allowance + Tolerances
			+		−	−	+*	+*
0	3/16	1/8	0.0007	0.0000	0.0004	0.0011	0.0004	0.0018
3/16	5/16	1/4	0.0008	0.0000	0.0006	0.0014	0.0006	0.0022
5/16	7/16	3/8	0.0009	0.0000	0.0007	0.0016	0.0007	0.0025
7/16	9/16	1/2	0.0010	0.0000	0.0009	0.0019	0.0009	0.0029
9/16	11/16	5/8	0.0011	0.0000	0.0010	0.0021	0.0010	0.0032
11/16	13/16	3/4	0.0012	0.0000	0.0012	0.0024	0.0012	0.0036
13/16	15/16	7/8	0.0012	0.0000	0.0013	0.0025	0.0013	0.0037
15/16	1 1/16	1	0.0013	0.0000	0.0014	0.0027	0.0014	0.0040
1 1/16	1 3/16	1 1/8	0.0014	0.0000	0.0015	0.0029	0.0015	0.0043
1 3/16	1 3/8	1 1/4	0.0014	0.0000	0.0016	0.0030	0.0016	0.0044
1 3/8	1 5/8	1 1/2	0.0015	0.0000	0.0018	0.0033	0.0018	0.0048
1 5/8	1 7/8	1 3/4	0.0016	0.0000	0.0020	0.0036	0.0020	0.0052
1 7/8	2 1/8	2	0.0016	0.0000	0.0022	0.0038	0.0022	0.0054
2 1/8	2 3/8	2 1/4	0.0017	0.0000	0.0024	0.0041	0.0024	0.0058
2 3/8	2 3/4	2 1/2	0.0018	0.0000	0.0026	0.0044	0.0026	0.0062
2 3/4	3 1/4	3	0.0019	0.0000	0.0029	0.0048	0.0029	0.0067
3 1/4	3 3/4	3 1/2	0.0020	0.0000	0.0032	0.0052	0.0032	0.0072
3 3/4	4 1/4	4	0.0021	0.0000	0.0035	0.0056	0.0035	0.0077
4 1/4	4 3/4	4 1/2	0.0021	0.0000	0.0038	0.0059	0.0038	0.0080
4 3/4	5 1/2	5	0.0022	0.0000	0.0041	0.0063	0.0041	0.0085
5 1/2	6 1/2	6	0.0024	0.0000	0.0046	0.0070	0.0046	0.0094
6 1/2	7 1/2	7	0.0025	0.0000	0.0051	0.0076	0.0051	0.0101
7 1/2	8 1/2	8	0.0026	0.0000	0.0056	0.0082	0.0056	0.0108

All dimensions in inches.

* NOTE: (+) denotes clearance or amount of looseness.

TABLE 7–5

American Standard Medium Fit, Class 3
Medium Allowance, Interchangeable

ASA B4a–1925

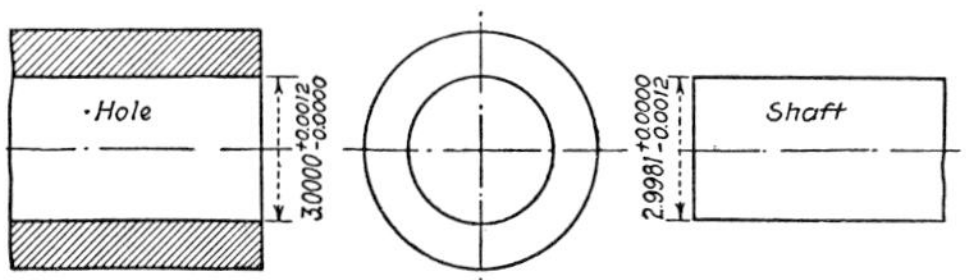

Example of Medium Fit (Class 3)

Summary of Data

	Hole	Shaft	
Tightest Fit	3.0000	2.9981	0.0019 Allowance
Loosest Fit	3.0012	2.9969	0.0043 Allowance + Tolerances

Formulas

When d = mean size,

Hole Tolerance $= 0.0008 \sqrt[3]{d}$

Shaft Tolerance $= 0.0008 \sqrt[3]{d}$

Allowance $= 0.0009 \sqrt[3]{d^2}$

For running fits under 600 r.p.m. and with journal pressures less than 600 lb. per sq. in.; also for sliding fits, and the more accurate machine-tool and automotive parts.

Size			Limits				Tightest Fit	Loosest Fit
From	Up to and Incl.	Mean	Hole or External Member		Shaft or Internal Member		Allowance	Allowance + Tolerances
			+		−	−	+*	+*
0	3/16	1/8	0.0004	0.0000	0.0002	0.0006	0.0002	0.0010
3/16	5/16	1/4	0.0005	0.0000	0.0004	0.0009	0.0004	0.0014
5/16	7/16	3/8	0.0006	0.0000	0.0005	0.0011	0.0005	0.0017
7/16	9/16	1/2	0.0006	0.0000	0.0006	0.0012	0.0006	0.0018
9/16	11/16	5/8	0.0007	0.0000	0.0007	0.0014	0.0007	0.0021
11/16	13/16	3/4	0.0007	0.0000	0.0007	0.0014	0.0007	0.0021
13/16	15/16	7/8	0.0008	0.0000	0.0008	0.0016	0.0008	0.0024
15/16	1 1/16	1	0.0008	0.0000	0.0009	0.0017	0.0009	0.0025
1 1/16	1 3/16	1 1/8	0.0008	0.0000	0.0010	0.0018	0.0010	0.0026
1 3/16	1 3/8	1 1/4	0.0009	0.0000	0.0010	0.0019	0.0010	0.0028
1 3/8	1 5/8	1 1/2	0.0009	0.0000	0.0012	0.0021	0.0012	0.0030
1 5/8	1 7/8	1 3/4	0.0010	0.0000	0.0013	0.0023	0.0013	0.0033
1 7/8	2 1/8	2	0.0010	0.0000	0.0014	0.0024	0.0014	0.0034
2 1/8	2 3/8	2 1/4	0.0010	0.0000	0.0015	0.0025	0.0015	0.0035
2 3/8	2 3/4	2 1/2	0.0011	0.0000	0.0017	0.0028	0.0017	0.0039
2 3/4	3 1/4	3	0.0012	0.0000	0.0019	0.0031	0.0019	0.0043
3 1/4	3 3/4	3 1/2	0.0012	0.0000	0.0021	0.0033	0.0021	0.0045
3 3/4	4 1/4	4	0.0013	0.0000	0.0023	0.0036	0.0023	0.0049
4 1/4	4 3/4	4 1/2	0.0013	0.0000	0.0025	0.0038	0.0025	0.0051
4 3/4	5 1/2	5	0.0014	0.0000	0.0026	0.0040	0.0026	0.0054
5 1/2	6 1/2	6	0.0015	0.0000	0.0030	0.0045	0.0030	0.0060
6 1/2	7 1/2	7	0.0015	0.0000	0.0033	0.0048	0.0033	0.0063
7 1/2	8 1/2	8	0.0016	0.0000	0.0036	0.0052	0.0036	0.0068

All dimensions in inches.

* Note: (+) denotes clearance or amount of looseness.

TABLE 7-6

American Standard Snug Fit, Class 4
Zero Allowance, Interchangeable
ASA B4a-1925

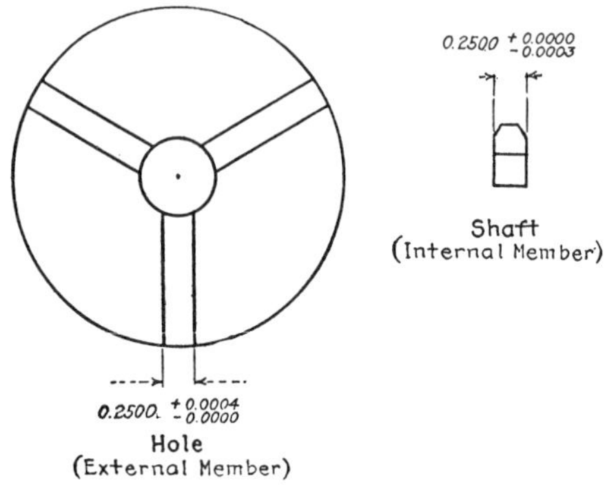

Drill Chuck Body and Jaws. Example of Snug Fit (Class 4)

Summary of Dimensions

	Hole	Shaft		
Tightest Fit	0.2500	0.2500	0.0000	Allowance
Loosest Fit	0.2504	0.2497	0.0007	Allowance + Tolerance

Formulas

When d = mean size,

Hole Tolerance = $0.0006\sqrt[3]{d}$

Shaft Tolerance = $0.0004\sqrt[3]{d}$

Allowance = 0.0000

This is the closest fit which can be assembled by hand and necessitates work of considerable precision. It should be used where no perceptible shake is permissible and where moving parts are not intended to move freely under load.

Size			Limits				Tightest Fit	Loosest Fit
From	Up to and Incl.	Mean	Hole or External Member		Shaft or Internal Member		Allowance	Allowance + Tolerances
			+			−		+*
0	3/16	1/8	0.0003	0.0000	0.0000	0.0002	0.0000	0.0005
3/16	5/16	1/4	0.0004	0.0000	0.0000	0.0003	0.0000	0.0007
5/16	7/16	3/8	0.0004	0.0000	0.0000	0.0003	0.0000	0.0007
7/16	9/16	1/2	0.0005	0.0000	0.0000	0.0003	0.0000	0.0008
9/16	11/16	5/8	0.0005	0.0000	0.0000	0.0003	0.0000	0.0008
11/16	13/16	3/4	0.0005	0.0000	0.0000	0.0004	0.0000	0.0009
13/16	15/16	7/8	0.0006	0.0000	0.0000	0.0004	0.0000	0.0010
15/16	1 1/16	1	0.0006	0.0000	0.0000	0.0004	0.0000	0.0010
1 1/16	1 3/16	1 1/8	0.0006	0.0000	0.0000	0.0004	0.0000	0.0010
1 3/16	1 3/8	1 1/4	0.0006	0.0000	0.0000	0.0004	0.0000	0.0010
1 3/8	1 5/8	1 1/2	0.0007	0.0000	0.0000	0.0005	0.0000	0.0012
1 5/8	1 7/8	1 3/4	0.0007	0.0000	0.0000	0.0005	0.0000	0.0012
1 7/8	2 1/8	2	0.0008	0.0000	0.0000	0.0005	0.0000	0.0013
2 1/8	2 3/8	2 1/4	0.0008	0.0000	0.0000	0.0005	0.0000	0.0013
2 3/8	2 3/4	2 1/2	0.0008	0.0000	0.0000	0.0005	0.0000	0.0013
2 3/4	3 1/4	3	0.0009	0.0000	0.0000	0.0006	0.0000	0.0015
3 1/4	3 3/4	3 1/2	0.0009	0.0000	0.0000	0.0006	0.0000	0.0015
3 3/4	4 1/4	4	0.0010	0.0000	0.0000	0.0006	0.0000	0.0016
4 1/4	4 3/4	4 1/2	0.0010	0.0000	0.0000	0.0007	0.0000	0.0017
4 3/4	5 1/2	5	0.0010	0.0000	0.0000	0.0007	0.0000	0.0017
5 1/2	6 1/2	6	0.0011	0.0000	0.0000	0.0007	0.0000	0.0018
6 1/2	7 1/2	7	0.0011	0.0000	0.0000	0.0008	0.0000	0.0019
7 1/2	8 1/2	8	0.0012	0.0000	0.0000	0.0008	0.0000	0.0020

All dimensions in inches.

* Note: (+) denotes clearance or amount of looseness.

TABLE 7–7

American Standard Wringing Fit, Class 5
Zero to Negative Allowance, Selective Assembly
ASA B4a–1925

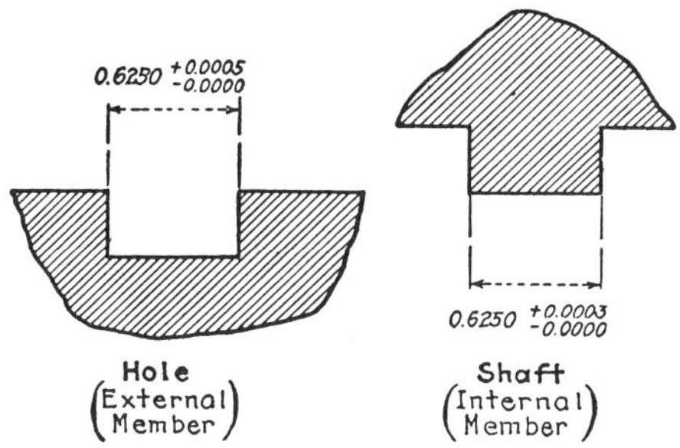

LOCATING KEYS, TONGUES OR DOWELS. EXAMPLE OF WRINGING FIT (CLASS 5)

SUMMARY OF DIMENSIONS

	Hole	Shaft	
Tightest Fit	0.6250	0.6253	−0.0003
Loosest Fit	0.6255	0.6250	+0.0005
Selected Fit	0.6250	0.6250	0.0000

FORMULAS

When d = mean size,

Hole Tolerance = $0.0006\sqrt[3]{d}$

Shaft Tolerance = $0.0004\sqrt[3]{d}$

Average interference of metal = 0.0000

The average interference of metal is the desired condition and must be obtained by selective assembly that is, by mating large shafts in large holes and small shafts in small holes.

This is also known as a "tunking fit" and it is practically metal-to-metal. Assembly is usually selective and not interchangeable.

Size			Limits				Tightest Fit	Loosest Fit	Selected Fit
From	Up to and Incl.	Mean	Hole or External Member		Shaft or Internal Member		Allowance	Allowance + Tolerances	Average Interference of metal
			+		+		−*	+*	
0	3/16	1/8	0.0003	0.0000	0.0002	0.0000	0.0002	0.0003	0.0000
3/16	5/16	1/4	0.0004	0.0000	0.0003	0.0000	0.0003	0.0004	0.0000
5/16	7/16	3/8	0.0004	0.0000	0.0003	0.0000	0.0003	0.0004	0.0000
7/16	9/16	1/2	0.0005	0.0000	0.0003	0.0000	0.0003	0.0005	0.0000
9/16	11/16	5/8	0.0005	0.0000	0.0003	0.0000	0.0003	0.0005	0.0000
11/16	13/16	3/4	0.0005	0.0000	0.0004	0.0000	0.0004	0.0005	0.0000
13/16	15/16	7/8	0.0006	0.0000	0.0004	0.0000	0.0004	0.0006	0.0000
15/16	1 1/16	1	0.0006	0.0000	0.0004	0.0000	0.0004	0.0006	0.0000
1 1/16	1 3/16	1 1/8	0.0006	0.0000	0.0004	0.0000	0.0004	0.0006	0.0000
1 3/16	1 3/8	1 1/4	0.0006	0.0000	0.0004	0.0000	0.0004	0.0006	0.0000
1 3/8	1 5/8	1 1/2	0.0007	0.0000	0.0005	0.0000	0.0005	0.0007	0.0000
1 5/8	1 7/8	1 3/4	0.0007	0.0000	0.0005	0.0000	0.0005	0.0007	0.0000
1 7/8	2 1/8	2	0.0008	0.0000	0.0005	0.0000	0.0005	0.0008	0.0000
2 1/8	2 3/8	2 1/4	0.0008	0.0000	0.0005	0.0000	0.0005	0.0008	0.0000
2 3/8	2 3/4	2 1/2	0.0008	0.0000	0.0005	0.0000	0.0005	0.0008	0.0000
2 3/4	3 1/4	3	0.0009	0.0000	0.0006	0.0000	0.0006	0.0009	0.0000
3 1/4	3 3/4	3 1/2	0.0009	0.0000	0.0006	0.0000	0.0006	0.0009	0.0000
3 3/4	4 1/4	4	0.0010	0.0000	0.0006	0.0000	0.0006	0.0010	0.0000
4 1/4	4 3/4	4 1/2	0.0010	0.0000	0.0007	0.0000	0.0007	0.0010	0.0000
4 3/4	5 1/2	5	0.0010	0.0000	0.0007	0.0000	0.0007	0.0010	0.0000
5 1/2	6 1/2	6	0.0011	0.0000	0.0007	0.0000	0.0007	0.0011	0.0000
6 1/2	7 1/2	7	0.0011	0.0000	0.0008	0.0000	0.0008	0.0011	0.0000
7 1/2	8 1/2	8	0.0012	0.0000	0.0008	0.0000	0.0008	0.0012	0.0000

All dimensions in inches.

* NOTE: (−) denotes interference of metal or negative allowance.

* NOTE: (+) denotes clearance or amount of looseness.

TABLE 7–8

American Standard Tight Fit, Class 6
Slight Negative Allowance, Selective Assembly
ASA B4a–1925

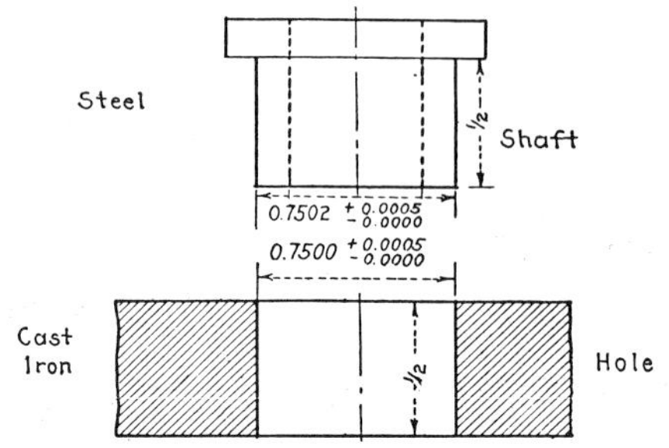

JIG BUSHING. EXAMPLE OF TIGHT FIT (CLASS 6)

SUMMARY OF DIMENSIONS

	Hole	Shaft	
Tightest Fit	0.7500	0.7507	−0.0007
Loosest Fit	0.7505	0.7502	+0.0003
†Selected Fit	0.7500	0.7502	−0.0002
††Selected Fit	0.7505	0.7507	−0.0002

Hole stress = 2608 lb./sq. in.

Force for pressing = 1/2 × 0.149 = 0.075 ton.

† Small shaft in small hole.

† Large shaft in large hole.

FORMULAS

When d = mean size,

Hole Tolerance = $0.0006\sqrt[3]{d}$

Shaft Tolerance = $0.0006\sqrt[3]{d}$

Average interference of metal = $0.00025d$.

The average interference of metal is the desired condition and must be obtained by selective assembly that is, by mating large shafts in large holes and small shafts in small holes.

Light pressure is required to assemble these fits and the parts are more or less permanently assembled, such as the fixed ends of studs for gears, pulleys, rocker arms, etc. These fits are used for drive fits in thin sections or extremely long fits in other sections, and also for shrink fits on very light sections. Used in automotive, ordnance, and general machine manufacturing.

Size			Limits				Tightest Fit	Loosest Fit	Selected Fit
From	Up to and Incl.	Mean	Hole or External Member		Shaft or Internal Member		Allowance	Allowance + Tolerances	Average Interference of metal
			+		+	+	−*	*	−*
0	3/16	1/8	0.0003	0.0000	0.0003	0.0000	0.0003	+0.0003	0.0000
3/16	5/16	1/4	0.0004	0.0000	0.0005	0.0001	0.0005	+0.0003	0.0001
5/16	7/16	3/8	0.0004	0.0000	0.0005	0.0001	0.0005	+0.0003	0.0001
7/16	9/16	1/2	0.0005	0.0000	0.0006	0.0001	0.0006	+0.0004	0.0001
9/16	11/16	5/8	0.0005	0.0000	0.0007	0.0002	0.0007	+0.0003	0.0002
11/16	13/16	3/4	0.0005	0.0000	0.0007	0.0002	0.0007	+0.0003	0.0002
13/16	15/16	7/8	0.0006	0.0000	0.0008	0.0002	0.0008	+0.0004	0.0002
15/16	1 1/16	1	0.0006	0.0000	0.0009	0.0003	0.0009	+0.0003	0.0003
1 1/16	1 3/16	1 1/8	0.0006	0.0000	0.0009	0.0003	0.0009	+0.0003	0.0003
1 3/16	1 3/8	1 1/4	0.0006	0.0000	0.0009	0.0003	0.0009	+0.0003	0.0003
1 3/8	1 5/8	1 1/2	0.0007	0.0000	0.0011	0.0004	0.0011	+0.0003	0.0004
1 5/8	1 7/8	1 3/4	0.0007	0.0000	0.0011	0.0004	0.0011	+0.0003	0.0004
1 7/8	2 1/8	2	0.0008	0.0000	0.0013	0.0005	0.0013	+0.0003	0.0005
2 1/8	2 3/8	2 1/4	0.0008	0.0000	0.0014	0.0006	0.0014	+0.0002	0.0006
2 3/8	2 3/4	2 1/2	0.0008	0.0000	0.0014	0.0006	0.0014	+0.0002	0.0006
2 3/4	3 1/4	3	0.0009	0.0000	0.0017	0.0008	0.0017	+0.0001	0.0008
3 1/4	3 3/4	3 1/2	0.0009	0.0000	0.0018	0.0009	0.0018	−0.0000	0.0009
3 3/4	4 1/4	4	0.0010	0.0000	0.0020	0.0010	0.0020	−0.0000	0.0010
4 1/4	4 3/4	4 1/2	0.0010	0.0000	0.0021	0.0011	0.0021	−0.0001	0.0011
4 3/4	5 1/2	5	0.0010	0.0000	0.0023	0.0013	0.0023	−0.0003	0.0013
5 1/2	6 1/2	6	0.0011	0.0000	0.0026	0.0015	0.0026	−0.0004	0.0015
6 1/2	7 1/2	7	0.0011	0.0000	0.0029	0.0018	0.0029	−0.0007	0.0018
7 1/2	8 1/2	8	0.0012	0.0000	0.0032	0.0020	0.0032	−0.0008	0.0020

All dimensions in inches.

* NOTE: (−) denotes interference of metal or negative allowance.

* NOTE: (+) denotes clearance or amount of looseness.

TABLE 7–9

Interference, Resultant Stresses, and Forces for Tight Fits, Class 6

ASA B4a–1925

Mean Size	Interference of Metal per Inch of Mean Size			Greatest Hole Stress, Steel Shaft in						Force for Pressing Steel Shaft into					
				Steel Hole			Cast-Iron Hole			Steel Hole			Cast-Iron Hole		
	Tightest Fit	Loosest Fit	Selected Fit	Tightest Fit	Loosest Fit	Selected Fit	Tightest Fit	Loosest Fit	Selected Fit	Tightest Fit	Loosest Fit	Selected Fit	Tightest Fit	Loosest Fit	Selected Fit
	Inches	Inches	Inches	Lb per sq. in.	Lb per sq. in.	Lb per sq. in.	Lb per sq. in.	Lb per sq. in.	Lb per sq. in.	Tons	Tons	Tons	Tons	Tons	Tons
1/8	0.00240	No values, due to smallest shaft member in largest hole member	0.00025	69600	No values, due to smallest shaft member in largest hole member	7250	25037	No values, due to smallest shaft member in largest hole member	2608	0.389	No values, due to smallest shaft member in largest hole member	0.000	0.224	No values, due to smallest shaft member in largest hole member	0.000
1/4	0.00200		0.00025	58000		7250	20864		2608	0.649		0.130	0.374		0.075
3/8	0.00133		0.00025	38700		7250	13909		2608	0.649		0.130	0.374		0.075
1/2	0.00120		0.00025	34800		7250	12518		2608	0.779		0.130	0.448		0.075
5/8	0.00112		0.00025	32480		7250	11684		2608	0.909		0.260	0.523		0.149
3/4	0.00093		0.00025	27100		7250	9737		2608	0.909		0.260	0.523		0.149
7/8	0.00091		0.00025	26500		7250	9538		2608	1.038		0.260	0.598		0.149
1	0.00090		0.00025	26100		7250	9389		2608	1.168		0.389	0.672		0.224
1 1/8	0.00080		0.00025	23200		7250	8346		2608	1.168		0.389	0.672		0.224
1 1/4	0.00072		0.00025	20880		7250	7511		2608	1.168		0.389	0.672		0.224
1 1/2	0.00073		0.00025	21300		7250	7650		2608	1.428		0.519	0.822		0.299
1 3/4	0.00063		0.00025	18240		7250	6557		2608	1.428		0.519	0.822		0.299
2	0.00065		0.00025	18850		7250	6781		2608	1.687		0.649	0.971		0.374
2 1/4	0.00062		0.00025	18040		7250	6491		2608	1.817		0.779	1.046		0.448
2 1/2	0.00056		0.00025	16200		7250	5842		2608	1.817		0.779	1.046		0.448
3	0.00057		0.00025	16433		7250	5911		2608	2.207		1.038	1.270		0.598
3 1/2	0.00051		0.00025	14910		7250	5365		2608	2.336		1.168	1.345		0.672
4	0.00050		0.00025	14500		7250	5216		2608	2.596		1.298	1.494		0.747
4 1/2	0.00047	0.00002	0.00025	13530	644	7250	4868	232	2608	2.726	0.130	1.428	1.569	0.075	0.822
5	0.00046	0.00006	0.00025	13340	1740	7250	4799	625	2608	2.985	0.389	1.687	1.719	0.224	0.971
6	0.00043	0.00007	0.00025	12570	1933	7250	4521	695	2608	3.375	0.519	1.947	1.942	0.299	1.121
7	0.00041	0.00010	0.00025	12010	2900	7250	4322	1043	2608	3.764	0.909	2.336	2.167	0.523	1.345
8	0.00040	0.00010	0.00025	11600	2900	7250	4173	1043	2608	4.154	1.038	2.596	2.391	0.598	1.494
				Stress = 29000000 A/d			Stress = 10432000 A/d			Force = 1298 A			Force = 747.2 A		

A = allowance, d = mean size.

Values for stress and force are true only when hub diameter equals twice the hole diameter.

The force values are for fit one inch long. For other lengths multiply by the length of fit in inches.

The values of greatest hole stress are for mean diameter sizes only. For other sizes in a step, the interference per inch of diameter and hence the greatest hole stress will vary from the values given in the table by less than 10 per cent. Where greater accuracy for the intermediate steps is required, the interference per inch of diameter should be obtained by use of the formula: Average interference of metal = 0.00025 d in which d is the diameter of the hole for which the size of the shaft is being computed.

TABLE 7–10
American Standard Medium Force Fit, Class 7 Negative Allowance, Selective Assembly
ASA B4a–1925

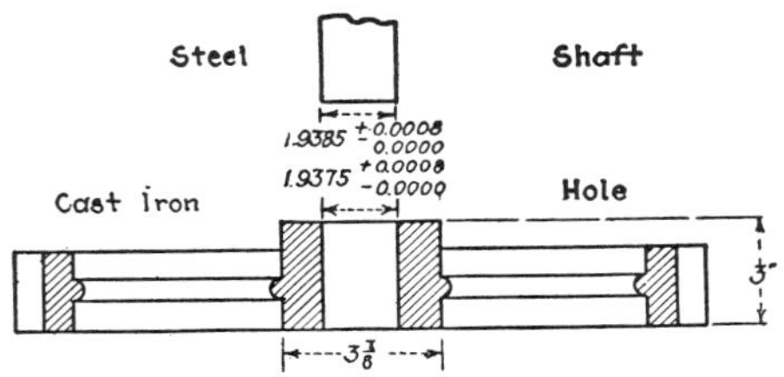

Cast-Iron Gear and Steel Shaft. Example of Medium Force Fit (Class 7)

Summary of Dimensions

	Hole	Shaft	
Tightest Fit	1.9375	1.9393	−0.0018
Loosest Fit	1.9383	1.9385	−0.0002
Selected Fit	1.9375	1.9385	−0.0010
Selected Fit	1.9383	1.9393	−0.0010

Hole stress = 5216 lb. per sq. in.

Force for pressing = 3 × 0.747 = 2.241 tons.

Formulas

When d = mean size,

Hole Tolerance = $0.0006 \sqrt[3]{d}$

Shaft Tolerance = $0.0006 \sqrt[3]{d}$

Average interference of metal = $0.0005d$.

The average interference of metal given is the desired condition and must be obtained by selective assembly, that is, by mating large shafts in large holes and small shafts in small holes.

For hole and shaft tolerances, the same formulas were used for sizes larger than 8 in., although there was no data available for these diameters.

Considerable pressure is required to assemble these fits and the parts are considered permanently assembled. These fits are used in fastening locomotive wheels, car wheels, armatures of dynamos and motors. and crank disks to their axles or shafts. They are also used for shrink fits on medium sections or long fits. These fits are the tightest which are recommended for cast-iron holes or external members as they stress cast iron to its elastic limit.

Size			Limit				Tightest Fit	Loosest Fit	Selected Fit
From	Up to and Incl.	Mean	Hole or External Member		Shaft or Internal Member		Allowance	Allowance + Tolerances	Average Interference of metal
			+		+	+	−*	*	−*
0	3/16	1/8	0.0003	0.0000	0.0004	0.0001	0.0004	+0.0002	0.0001
3/16	5/16	1/4	0.0004	0.0000	0.0005	0.0001	0.0005	+0.0003	0.0001
5/16	7/16	3/8	0.0004	0.0000	0.0006	0.0002	0.0006	+0.0002	0.0002
7/16	9/16	1/2	0.0005	0.0000	0.0008	0.0003	0.0008	+0.0002	0.0003
9/16	11/16	5/8	0.0005	0.0000	0.0008	0.0003	0.0008	+0.0002	0.0003
11/16	13/16	3/4	0.0005	0.0000	0.0009	0.0004	0.0009	+0.0001	0.0004
13/16	15/16	7/8	0.0006	0.0000	0.0010	0.0004	0.0010	+0.0002	0.0004
15/16	1 1/16	1	0.0006	0.0000	0.0011	0.0005	0.0011	+0.0001	0.0005
1 1/16	1 3/16	1 1/8	0.0006	0.0000	0.0012	0.0006	0.0012	0.0000	0.0006
1 3/16	1 3/8	1 1/4	0.0006	0.0000	0.0012	0.0006	0.0012	0.0000	0.0006
1 3/8	1 5/8	1 1/2	0.0007	0.0000	0.0015	0.0008	0.0015	−0.0001	0.0008
1 5/8	1 7/8	1 3/4	0.0007	0.0000	0.0016	0.0009	0.0016	−0.0002	0.0009
1 7/8	2 1/8	2	0.0008	0.0000	0.0018	0.0010	0.0018	−0.0002	0.0010
2 1/8	2 3/8	2 1/4	0.0008	0.0000	0.0019	0.0011	0.0019	−0.0003	0.0011
2 3/8	2 3/4	2 1/2	0.0008	0.0000	0.0021	0.0013	0.0021	−0.0005	0.0013
2 3/4	3 1/4	3	0.0009	0.0000	0.0024	0.0015	0.0024	−0.0006	0.0015
3 1/4	3 3/4	3 1/2	0.0009	0.0000	0.0027	0.0018	0.0027	−0.0009	0.0018
3 3/4	4 1/4	4	0.0010	0.0000	0.0030	0.0020	0.0030	−0.0010	0.0020
4 1/4	4 3/4	4 1/2	0.0010	0.0000	0.0033	0.0023	0.0033	−0.0013	0.0023
4 3/4	5 1/2	5	0.0010	0.0000	0.0035	0.0025	0.0035	−0.0015	0.0025
5 1/2	6 1/2	6	0.0011	0.0000	0.0041	0.0030	0.0041	−0.0019	0.0030
6 1/2	7 1/2	7	0.0011	0.0000	0.0046	0.0035	0.0046	−0.0024	0.0035
7 1/2	8 1/2	8	0.0012	0.0000	0.0052	0.0040	0.0052	−0.0028	0.0040
8 1/2	9 1/2	9	0.0012	0.0000	0.0057	0.0045	0.0057	−0.0033	0.0045
10 1/2	11 1/2	10	0.0013	0.0000	0.0063	0.0050	0.0063	−0.0037	0.0050
11 1/2	13	12	0.0014	0.0000	0.0074	0.0060	0.0074	0.0046	0.0060
13	15	14	0.0014	0.0000	0.0084	0.0070	0.0084	−0.0056	0.0070
15	17	16	0.0015	0.0000	0.0095	0.0080	0.0095	−0.0065	0.0080
17	19	18	0.0016	0.0000	0.0106	0.0090	0.0106	−0.0074	0.0090
19	22	20	0.0016	0.0000	0.0116	0.0100	0.0116	−0.0084	0.0100
22	26	24	0.0017	0.0000	0.0137	0.0120	0.0137	−0.0103	0.0120
26	30	28	0.0018	0.0000	0.0158	0.0140	0.0158	−0.0122	0.0140
30	34	32	0.0019	0.0000	0.0179	0.0160	0.0179	−0.0141	0.0160
34	38	36	0.0020	0.0000	0.0200	0.0180	0.0200	−0.0160	0.0180
38	44	40	0.0021	0.0000	0.0221	0.0200	0.0221	−0.0179	0.0200
44	52	48	0.0022	0.0000	0.0262	0.0240	0.0262	−0.0218	0.0240
52	60	56	0.0023	0.0000	0.0303	0.0280	0.0303	−0.0257	0.0280
60	68	64	0.0024	0.0000	0.0344	0.0320	0.0344	−0.0296	0.0320
68	76	72	0.0025	0.0000	0.0385	0.0360	0.0385	−0.0335	0.0360
76	88	80	0.0026	0.0000	0.0426	0.0400	0.0426	−0.0374	0.0400
88	104	96	0.0027	0.0000	0.0507	0.0480	0.0507	−0.0453	0.0480
104	120	112	0.0029	0.0000	0.0589	0.0560	0.0589	−0.0531	0.0560
120	136	128	0.0030	0.0000	0.0670	0.0640	0.0670	−0.0610	0.0640

All dimensions in inches.

* Note: (−) denotes interference of metal or negative allowance.

* Note: (+) denotes clearance or amount of looseness.

TABLE 7–11

Interference, Resultant Stresses, and Forces for Medium Force Fits, Class 7

ASA B4a-1925

Mean Size	Interference of Metal per Inch of Mean Size			Greatest Hole Stress, Steel Shaft in						Force for Pressing Steel Shaft into					
				Steel Hole			Cast Iron Hole			Steel Hole			Cast Iron Hole		
	Tightest Fit	Loosest Fit	Selected Fit	Tightest Fit	Loosest Fit	Selected Fit	Tightest Fit	Loosest Fit	Selected Fit	Tightest Fit	Loosest Fit	Selected Fit	Tightest Fit	Loosest Fit	Selected Fit
	Inch	Inch	Inch	Lb per sq. in.	Lb per sq. in.	Lb per sq. in.	Lb per sq. in.	Lb per sq. in.	Lb per sq. in.	Tons	Tons	Tons	Tons	Tons	Tons
1/8	0.00320	*	0.0005	92800	*	14500	33382	*	5216	0.519	*	0.130	0.299	*	0.075
1/4	0.00200	*	0.0005	58000	*	14500	20864	*	5216	0.649	*	0.130	0.374	*	0.075
3/8	0.00160	*	0.0005	46400	*	14500	16691	*	5216	0.779	*	0.260	0.448	*	0.149
1/2	0.00160	*	0.0005	46400	*	14500	16691	*	5216	1.038	*	0.389	0.598	*	0.224
5/8	0.00128	*	0.0005	37120	*	14500	13353	*	5216	1.038	*	0.389	0.598	*	0.224
3/4	0.00120	*	0.0005	34800	*	14500	12513	*	5216	1.168	*	0.519	0.672	*	0.299
7/8	0.00114	*	0.0005	33140	*	14500	11922	*	5216	1.298	*	0.519	0.747	*	0.299
1	0.00110	*	0.0005	31900	*	14500	11475	*	5216	1.428	*	0.649	0.822	*	0.374
1 1/8	0.00107	*	0.0005	30930	*	14500	11127	*	5216	1.558	*	0.779	0.897	*	0.448
1 1/4	0.00096	*	0.0005	27830	*	14500	10015	*	5216	1.558	*	0.779	0.897	*	0.448
1 1/2	0.00100	0.00007	0.0005	29000	1934	14500	10432	695	5216	1.947	0.130	1.038	1.121	0.075	0.598
1 3/4	0.00091	0.00011	0.0005	26510	3314	14500	9538	1192	5216	2.077	0.260	1.168	1.196	0.149	0.672
2	0.00090	0.00010	0.0005	26100	2900	14500	9389	1043	5216	2.336	0.260	1.298	1.345	0.149	0.747
2 1/4	0.00084	0.00013	0.0005	24490	3866	14500	8809	1391	5216	2.466	0.389	1.428	1.420	0.224	0.822
2 1/2	0.00084	0.00020	0.0005	24340	5800	14500	8763	2086	5216	2.726	0.649	1.687	1.569	0.373	0.971
3	0.00080	0.00020	0.0005	23200	5800	14500	8346	2086	5216	3.115	0.779	1.947	1.793	0.448	1.121
3 1/2	0.00077	0.00026	0.0005	22370	7457	14500	8048	2682	5216	3.505	1.168	2.336	2.017	0.672	1.345
4	0.00075	0.00025	0.0005	21750	7250	14500	7824	2608	5216	3.894	1.298	2.596	2.242	0.747	1.494
4 1/2	0.00073	0.00029	0.0005	21270	8377	14500	7652	3014	5216	4.283	1.687	2.985	2.466	0.971	1.719
5	0.00070	0.00030	0.0005	20300	8700	14500	7302	3130	5216	4.543	1.947	3.245	2.615	1.121	1.868
6	0.00068	0.00032	0.0005	19810	9183	14500	7129	3303	5216	5.322	2.466	3.894	3.064	1.420	2.242
7	0.00066	0.00034	0.0005	19060	9942	14500	6855	3577	5216	5.971	3.115	4.543	3.437	1.793	2.615
				Stress = 29000000 A/d			Stress = 10432000 A/d			Force = 1298 A			Force = 747.2 A		

continued on next page

TABLE 7–11, continued

Mean Size	Interference of Metal* per Inch of Mean Size			Greatest Hole Stress, Steel Shaft in						Force for Pressing Steel Shaft into					
				Steel Hole			Cast Iron Hole			Steel Hole			Cast Iron Hole		
	Tightest Fit	Loosest Fit	Selected Fit	Tightest Fit	Loosest Fit	Selected Fit	Tightest Fit	Loosest Fit	Selected Fit	Tightest Fit	Loosest Fit	Selected Fit	Tightest Fit	Loosest Fit	Selected Fit
	Inch	Inch	Inch	Lb per sq. in.	Lb per sq. in.	Lb per sq. in.	Lb per sq. in.	Lb per sq. in.	Lb per sq. in.	Tons	Tons	Tons	Tons	Tons	Tons
8	0.00065	0.00035	0.0005	18850	10150	14500	6781	3651	5216	6.750	3.634	5.192	• 3.885	2.092	2.989
9	0.00063	0.00037	0.0005	18370	10630	14500	6607	3875	5216	7.399	4.283	5.841	4.259	2.466	3.362
10	0.00063	0.00037	0.0005	18270	10730	14500	6572	3860	5216	8.177	4.803	6.490	4.707	2.765	3.736
12	0.00062	0.00038	0.0005	17890	11120	14500	6433	3999	5216	9.605	5.971	7.788	5.529	3.437	4.483
14	0.00060	0.00040	0.0005	17400	11600	14500	6259	4173	4216	10.903	7.269	9.068	6.276	4.184	5.230
16	0.00059	0.00041	0.0005	17220	11780	14500	6194	4238	5216	12.331	8.437	10.384	7.098	4.857	5.978
18	0.00059	0.00041	0.0005	17080	11920	14500	6143	4289	5216	13.759	9.605	11.682	7.920	5.529	6.725
20	0.00058	0.00042	0.0005	16820	12180	14500	6051	4381	5216	15.057	10.903	12.980	8.668	6.276	7.472
24	0.00057	0.00043	0.0005	16550	12450	14500	5955	4477	5216	17.783	13.369	15.576	10.236	7.696	8.966
28	0.00056	0.00044	0.0005	16360	12640	14500	5887	4173	5216	20.508	14.538	18.172	11.806	8.369	10.461
32	0.00056	0.00044	0.0005	16220	12780	14500	5835	4597	5216	23.234	18.302	20.768	13.375	10.536	11.955
36	0.00056	0.00044	0.0005	16110	12890	14500	5796	4636	5216	25.960	20.768	23.364	14.944	11.955	13.450
40	0.00055	0.00045	0.0005	16020	12980	14500	5764	4668	5216	28.686	23.234	25.960	16.513	13.375	14.944
48	0.00055	0.00045	0.0005	15830	13170	14500	5694	4738	5216	34.008	28.296	31.152	19.577	16.289	17.933
56	0.00054	0.00046	0.0005	15690	13310	14500	5644	4788	5216	39.329	33.359	36.344	22.640	19.203	20.922
64	0.00054	0.00046	0.0005	15570	13410	14500	5607	4825	5216	44.651	38.421	41.536	25.704	22.117	23.910
72	0.00053	0.00047	0.0005	15500	13490	14500	5578	4854	5216	49.973	43.483	46.728	28.767	25.031	26.899
80	0.00053	0.00047	0.0005	15440	13560	14500	5555	4877	5216	55.295	48.545	51.920	31.831	27.945	29.888
96	0.00053	0.00047	0.0005	15320	13680	14500	5509	4923	5216	65.809	58.799	62.304	37.883	33.848	35.866
112	0.00053	0.00047	0.0005	15250	13750	14500	5486	4946	5216	76.452	68.924	72.688	44.010	39.676	41.843
128	0.00052	0.00048	0.0005	15180	13820	14500	5461	4972	5216	86.966	79.178	83.072	50.062	45.579	47.821
				Stress = 29000000 A/d			Stress = 10432000 A/d			Force = 1298 A			Force = 747.2 A		

A = allowance, d = mean size.

Values for stress and force are true only when hub diameter equals twice the hole diameter.

The force values are for fit one inch long. For other lengths multiply by the length of fit in inches.

* **No values, due to smallest shaft member** being smaller than the largest hole member.

The values of Greatest Hole Stress **are for mean** diameter sizes only. For other sizes in a step, the interference per inch of diameter and hence the Greatest Hole Stress **will vary from** the values given in the table by less than 10 %. Where greater accuracy for the intermediate steps is required, the interference per inch of diameter should **be obtained by use of the formula:** Average interference of metal = 0.0005 d, in which d is the diameter of the hole for which the size of the shaft is being computed.

TABLE 7–12

American Standard Heavy Force and Shrink Fit, Class 8
Considerable Negative Allowance, Selective Assembly
ASA B4a-1925

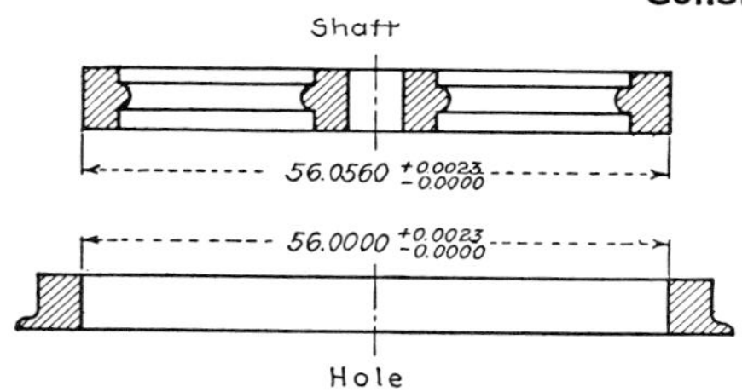

SHRINKING OF A LOCOMOTIVE TIRE ON ITS WHEEL.

EXAMPLE OF SHRINK FIT (CLASS 8)

SUMMARY OF DIMENSIONS

	Hole	Shaft	
Tightest Fit	56.0000	56.0583	−0.0583
Loosest Fit	56.0023	56.0560	−0.0537
† Selected Fit	56.0000	56.0560	−0.0560
†† Selected Fit	56.0023	56.0583	−0.0560

Hole stress = 29000 lb./sq. in.

† Small shaft in small hole.
†† Large shaft in large hole.

FORMULAS

When d = mean size,

Hole Tolerance = $0.0006\sqrt[3]{d}$

Shaft Tolerance = $0.0006\sqrt[3]{d}$

Average interference of metal = $0.001d$.

The average interference of metal given is the desired condition and must be obtained by selective assembly, that is, by mating large shafts in large holes and small shafts in small holes.

For hole and shaft tolerances, the same formulas were used for sizes larger than 8 in., although there was no data available for these diameters.

These fits are used for steel holes where the metal can be stressed to its elastic limit. These fits cause excessive stress for cast-iron holes. Shrink fits are used where heavy force fits are impractical, as on locomotive wheel tires, heavy crank disks of large engines, etc.

Size			Limits				Tightest Fit	Loosest Fit	Selected Fit
From	Up to and Incl.	Mean	Hole or External Member		Shaft or Internal Member		Allowance	Allowance + tolerances	Average interference of metal
			+		+	+	−*	*	−*
0	3/16	1/8	0.0003	0.0000	0.0004	0.0001	0.0004	+0.0002	0.0001
3/16	5/16	1/4	0.0004	0.0000	0.0007	0.0003	0.0007	+0.0001	0.0003
5/16	7/16	3/8	0.0004	0.0000	0.0008	0.0004	0.0008	−0.0000	0.0004
7/16	9/16	1/2	0.0005	0.0000	0.0010	0.0005	0.0010	0.0000	0.0005
9/16	11/16	5/8	0.0005	0.0000	0.0011	0.0006	0.0011	−0.0001	0.0006
11/16	13/16	3/4	0.0005	0.0000	0.0013	0.0008	0.0013	−0.0003	0.0008
13/16	15/16	7/8	0.0006	0.0000	0.0015	0.0009	0.0015	−0.0003	0.0009
15/16	1 1/16	1	0.0006	0.0000	0.0016	0.0010	0.0016	−0.0004	0.0010
1 1/16	1 3/16	1 1/8	0.0006	0.0000	0.0017	0.0011	0.0017	−0.0005	0.0011
1 3/16	1 3/8	1 1/4	0.0006	0.0000	0.0019	0.0013	0.0019	−0.0007	0.0013
1 3/8	1 5/8	1 1/2	0.0007	0.0000	0.0022	0.0015	0.0022	−0.0008	0.0015
1 5/8	1 7/8	1 3/4	0.0007	0.0000	0.0025	0.0018	0.0025	−0.0011	0.0018
1 7/8	2 1/8	2	0.0008	0.0000	0.0028	0.0020	0.0028	−0.0012	0.0020
2 1/8	2 3/8	2 1/4	0.0008	0.0000	0.0031	0.0023	0.0031	−0.0015	0.0023
2 3/8	2 3/4	2 1/2	0.0008	0.0000	0.0033	0.0025	0.0033	−0.0017	0.0025
2 3/4	3 1/4	3	0.0009	0.0000	0.0039	0.0030	0.0039	−0.0021	0.0030
3 1/4	3 3/4	3 1/2	0.0009	0.0000	0.0044	0.0035	0.0044	−0.0026	0.0035
3 3/4	4 1/4	4	0.0010	0.0000	0.0050	0.0040	0.0050	−0.0030	0.0040
4 1/4	4 3/4	4 1/2	0.0010	0.0000	0.0055	0.0045	0.0055	−0.0035	0.0045
4 3/4	5 1/2	5	0.0010	0.0000	0.0060	0.0050	0.0060	−0.0040	0.0050
5 1/2	6 1/2	6	0.0011	0.0000	0.0071	0.0060	0.0071	−0.0049	0.0060
6 1/2	7 1/2	7	0.0011	0.0000	0.0081	0.0070	0.0081	−0.0059	0.0070
7 1/2	8 1/2	8	0.0012	0.0000	0.0092	0.0080	0.0092	−0.0068	0.0080
8 1/2	9 1/2	9	0.0012	0.0000	0.0102	0.0090	0.0102	−0.0078	0.0090
10 1/2	11 1/2	10	0.0013	0.0000	0.0113	0.0100	0.0113	−0.0087	0.0100
11 1/2	13	12	0.0014	0.0000	0.0134	0.0120	0.0134	−0.0106	0.0120
13	15	14	0.0014	0.0000	0.0154	0.0140	0.0154	−0.0126	0.0140
15	17	16	0.0015	0.0000	0.0175	0.0160	0.0175	−0.0145	0.0160
17	19	18	0.0016	0.0000	0.0196	0.0180	0.0196	−0.0164	0.0180
19	22	20	0.0016	0.0000	0.0216	0.0200	0.0216	−0.0184	0.0200
22	26	24	0.0017	0.0000	0.0257	0.0240	0.0257	−0.0223	0.0240
26	30	28	0.0018	0.0000	0.0298	0.0280	0.0298	−0.0262	0.0280
30	34	32	0.0019	0.0000	0.0339	0.0320	0.0339	−0.0301	0.0320
34	38	36	0.0020	0.0000	0.0380	0.0360	0.0380	−0.0340	0.0360
38	44	40	0.0021	0.0000	0.0421	0.0400	0.0421	−0.0379	0.0400
44	52	48	0.0022	0.0000	0.0502	0.0480	0.0502	−0.0458	0.0480
52	60	56	0.0023	0.0000	0.0583	0.0560	0.0583	−0.0537	0.0560
60	68	64	0.0024	0.0000	0.0664	0.0640	0.0664	−0.0616	0.0640
68	76	72	0.0025	0.0000	0.0745	0.0720	0.0745	−0.0695	0.0720
76	88	80	0.0026	0.0000	0.0826	0.0800	0.0826	−0.0774	0.0800
88	104	96	0.0027	0.0000	0.0987	0.0960	0.0987	−0.0933	0.0960
104	120	112	0.0029	0.0000	0.1149	0.1120	0.1149	−0.1091	0.1120
120	136	128	0.0030	0.0000	0.1310	0.1280	0.1310	−0.1250	0.1280

All dimensions in inches.

* **NOTE: (−) denotes interference of metal or negative allowance.**

* **NOTE: (+) denotes clearance or amount of looseness.**

TABLE 7–13

Interference, Resultant Stresses and Forces for Heavy Force and Shrink Fits, Class 8

ASA B4a-1925

Mean Size	Interference of Metal per Inch of Mean Size			Greatest Hole Stress, Steel Shaft in Steel Hole			Force for Pressing, Steel Shaft into Steel Hole		
	Tightest Fit	Loosest Fit	Selected Fit	Tightest Fit	Loosest Fit	Selected Fit	Tightest Fit	Loosest Fit	Selected Fit
	Inch	Inch	Inch	Lb per sq. in.	Lb per sq. in.	Lb per sq. in.	Tons	Tons	Tons
1/8	0.00320	*	0.001	92800	*	29000	0.519	*	0.130
1/4	0.00280	*	0.001	81200	*	29000	0.909	*	0.389
3/8	0.00213	*	0.001	61870	*	29000	1.038	*	0.519
1/2	0.00200	*	0.001	58000	*	29000	1.298	*	0.649
5/8	0.00176	0.00016	0.001	51040	4640	29000	1.428	0.130	0.779
3/4	0.00173	0.00040	0.001	42270	11600	29000	1.687	0.389	1.038
7/8	0.00171	0.00034	0.001	49710	9943	29000	1.947	0.389	1.168
1	0.00160	0.00040	0.001	46400	11600	29000	2.077	0.519	1.298
1 1/8	0.00151	0.00044	0.001	43840	12890	29000	2.207	0.649	1.428
1 1/4	0.00152	0.00056	0.001	44080	16230	29000	2.466	0.909	1.687
1 1/2	0.00147	0.00053	0.001	42530	15470	29000	2.856	1.038	1.947
1 3/4	0.00143	0.00062	0.001	41430	18230	29000	3.245	1.428	2.336
2	0.00140	0.00060	0.001	40600	17400	29000	3.634	1.558	2.596
2 1/4	0.00138	0.00066	0.001	39960	17910	29000	4.024	1.947	2.985
2 1/2	0.00132	0.00068	0.001	38280	19720	29000	4.283	2.207	3.245
3	0.00130	0.00070	0.001	37700	20300	29000	5.062	2.726	3.894
3 1/2	0.00126	0.00074	0.001	36460	21540	29000	5.711	3.375	4.543
4	0.00125	0.00075	0.001	36250	21750	29000	6.490	3.894	5.192
4 1/2	0.00122	0.00078	0.001	35440	22560	29000	7.139	4.543	5.841
5	0.00120	0.00080	0.001	34800	23200	29000	7.788	5.192	6.490
6	0.00118	0.00082	0.001	34317	23680	29000	9.216	6.360	7.788
7	0.00116	0.00084	0.001	33570	24440	29000	10.514	7.658	9.086
8	0.00115	0.00085	0.001	33350	24650	29000	11.942	8.826	10.384
9	0.00113	0.00087	0.001	32879	25130	29000	13.240	10.124	11.682
10	0.00113	0.00087	0.001	32770	25233	29000	14.667	11.293	12.980
12	0.00112	0.00088	0.001	32390	25610	29000	17.393	13.759	15.576
14	0.00110	0.00090	0.001	31900	26100	29000	19.989	16.355	18.172
16	0.00109	0.00091	0.001	31720	26290	29000	22.715	18.821	20.768
18	0.00108	0.00091	0.001	31580	26420	29000	25.441	21.287	23.364
20	0.00108	0.00092	0.001	31370	26680	29000	28.037	23.883	25.960
24	0.00107	0.00093	0.001	31050	26940	29000	33.359	28.945	31.152
28	0.00106	0.00094	0.001	30860	27130	29000	38.680	34.008	36.344
32	0.00106	0.00094	0.001	30710	27280	29000	44.002	39.070	41.536
36	0.00106	0.00094	0.001	30600	27390	29000	49.324	44.132	46.728
40	0.00105	0.00095	0.001	30510	27480	29000	54.646	49.194	51.920
48	0.00105	0.00095	0.001	30330	27680	29000	65.160	59.448	62.304
56	0.00104	0.00096	0.001	30190	27810	29000	75.673	69.703	72.688
64	0.00104	0.00096	0.001	30080	27900	29000	86.187	79.957	83.072
72	0.00103	0.00097	0.001	30000	27990	29000	96.701	90.211	93.456
80	0.00103	0.00097	0.001	29940	28060	29000	107.113	100.465	103.840
96	0.00103	0.00097	0.001	29810	28180	29000	128.115	121.103	124.608
112	0.00103	0.00097	0.001	29750	28250	29000	149.140	141.612	145.376
128	0.00102	0.00098	0.001	29690	28320	29000	170.038	162.250	166.144
				Stress = 29000000 A/d			Force = 1298 A		

A = allowance, d = mean size.

Values for stress and force are true only when hub diameter equals twice the hole diameter.

The force values are for a fit one inch long. For other lengths multiply by the length of fit in inches.

* No values, due to smallest shaft member being smaller than the largest hole member.

The values of greatest hole stress are for mean diameter sizes only. For other sizes in a step, the interference per inch of diameter and hence the greatest hole stress will vary from the values given in the table by less than 10 per cent. Where greater accuracy for the intermediate steps is required, the interference per inch of diameter should be obtained by use of the formula: Average interference of metal = $0.001d$, in which d is the diameter of the hole for which the size of the shaft is being computed.

TABLE 7–14

American Standard Basic Dimensions of Self-Holding Machine Tapers

ASA B5.10–1953

<table>
<tr><th>No. of Taper</th><th>Taper per Foot</th><th>Diameter at Gage Line[1] A</th><th colspan="4">Means of Driving and Holding</th><th>*Origin of Series</th></tr>
<tr><td>.239</td><td>0.50200</td><td>0.23922</td><td rowspan="22">Tang Drive With Shank Held in by Friction (See Table 7-15)</td><td rowspan="22">Tang Drive With Shank Held in by Key (See Table 7–16)</td><td rowspan="22">Key Drive With Shank Held in by Key (See Table 7-17)</td><td rowspan="22">Key Drive With Shank Held in by Draw-bolt (See Table 7-18)</td><td rowspan="3">Brown and Sharpe Taper Series</td></tr>
<tr><td>.299</td><td>0.50200</td><td>0.29968</td></tr>
<tr><td>.375</td><td>0.50200</td><td>0.37525</td></tr>
<tr><td>1</td><td>0.59858</td><td>0.47500</td><td rowspan="8">Morse Taper Series</td></tr>
<tr><td>2</td><td>0.59941</td><td>0.70000</td></tr>
<tr><td>3</td><td>0.60235</td><td>0.93800</td></tr>
<tr><td>4</td><td>0.62326</td><td>1.23100</td></tr>
<tr><td>4½</td><td>0.62400</td><td>1.50000</td></tr>
<tr><td>5</td><td>0.63151</td><td>1.74800</td></tr>
<tr><td>6</td><td>0.62565</td><td>2.49400</td></tr>
<tr><td>7[2]</td><td>0.62400</td><td>3.2700</td></tr>
<tr><td>200</td><td>0.750</td><td>2.000</td><td rowspan="11">3/4 Inch per Foot Taper Series</td></tr>
<tr><td>250</td><td>0.750</td><td>2.500</td></tr>
<tr><td>300</td><td>0.750</td><td>3.000</td></tr>
<tr><td>350</td><td>0.750</td><td>3.500</td></tr>
<tr><td>400</td><td>0.750</td><td>4.000</td></tr>
<tr><td>450</td><td>0.750</td><td>4.500</td></tr>
<tr><td>500</td><td>0.750</td><td>5,000</td></tr>
<tr><td>600</td><td>0.750</td><td>6.000</td></tr>
<tr><td>800</td><td>0.750</td><td>8,000</td></tr>
<tr><td>1000</td><td>0.750</td><td>10.000</td></tr>
<tr><td>1200</td><td>0.750</td><td>12.000</td></tr>
</table>

All dimensions given in inches.

[1]See illustration above Tables 7–15, 7–16, 7–17 and 7–18.

[2]This size is continued in the Tang Drive series for the present to meet special needs.

*This standard is a good illustration of the consolidation of several well-recognized systems into a single standard.

TABLE 7–15

American Standard Machine Tapers Having Tang Drive and Shank Retained by Friction

ASA B5.10–1953

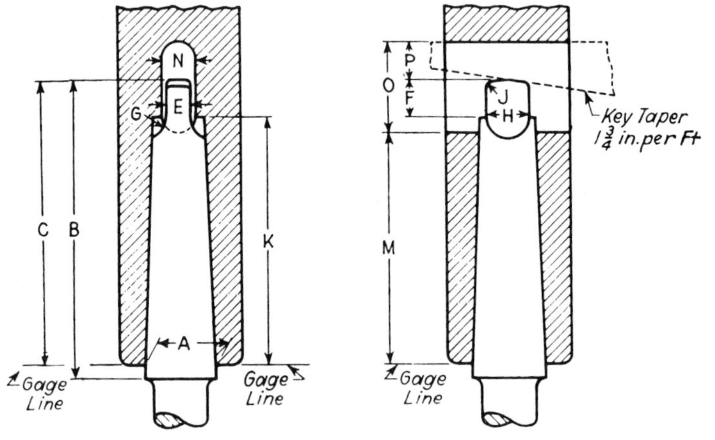

	Diameter at Gage Line	Shank		Tang					Socket		Tang Slot		
		Total Length of Shank	Gage Line to End of Shank	Thickness	Length	Radius of Mill	Diameter	Radius	Min. Depth of Tapered Hole	Gage Line to Tang Slot	Width	Length	Shank End to Back of Tang Slot
	A	B	C	E	F	G	H	I	K	M	N	O	P
.239	0.23922	1 9/32	1 3/16	0.125	3/16	3/16	11/64	1/32	1 1/16	15/16	0.141	3/8	1/8
.299	0.29968	1 19/32	1 1/2	0.156	1/4	3/16	7/32	1/32	1 5/16	1 11/64	0.172	1/2	11/64
.375	0.37525	1 31/32	1 7/8	0.187	5/16	3/16	9/32	3/64	1 5/8	1 15/32	0.203	5/8	7/32
1	0.47500	2 9/16	2 7/16	0.203	3/8	3/16	11/32	3/64	2 3/16	2 1/16	0.213	3/4	3/8
2	0.70000	3 1/8	2 15/16	0.250	7/16	1/4	17/32	1/16	2 21/32	2 1/2	0.260	7/8	7/16
3	0.93800	3 7/8	3 11/16	0.312	9/16	9/32	23/32	5/64	3 5/16	3 1/16	0.322	1 3/16	9/16
4	1.23100	4 7/8	4 5/8	0.469	5/8	5/16	31/32	3/32	4 3/16	3 7/8	0.479	1 1/4	1/2
4 1/2	1.50000	5 3/8	5 1/8	0.562	11/16	3/8	1 13/64	1/8	4 5/8	4 5/16	0.573	1 3/8	9/16
5	1.74800	6 1/8	5 7/8	0.625	3/4	3/8	1 13/32	1/8	5 5/16	4 15/16	0.635	1 1/2	9/16
6	2.49400	8 9/16	8 1/4	0.750	1 1/8	1/2	2	5/32	7 13/32	7	0.760	1 3/4	1/2
7[1]	3.27000	11 5/8	11 1/4	1.125	1 3/8	3/4	2 5/8	3/16	10 5/32	9 1/2	1.135	2 5/8	7/8

All dimensions given in inches.

[1] This size is continued in the Tang Drive series for the present to meet special needs.

Tolerances

For Diameter of Shank at Gage Line (A)
All sizes, +0.002 − 0.000

For Diameter of Hole at Gage Line (A)
All sizes, +0.000 − 0.002

For Thickness of Tang (E)
Up to and including No. 5, + 0.000 − 0.006
Larger than No. 5, + 0.000 − 0.008

For Width of Tang Slot (N)
Up to and including No. 5, + 0.006 − 0.000
Larger than No. 5, + 0.008 − 0.000

For Concentricity of Tang (E) with center line of taper
Up to and including No. 5, 0.0035 (Indicator reading)
Larger than No. 5, 0.005 (Indicator reading)

For Concentricity of Tang Slot (N) with center line of taper
Up to and including No. 5, 0.0035 (Indicator reading)
Larger than No. 5, 0.005 (Indicator reading)

Tolerances on Fractional Dimensions, ±0.010 unless otherwise specified

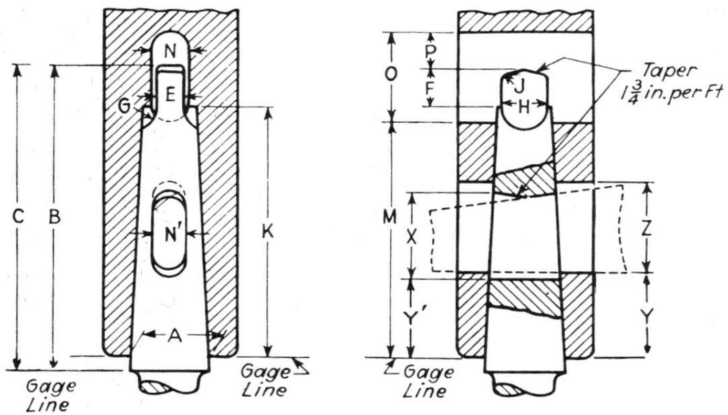

TABLE 7-16

American Standard Machine Tapers Having Tang Drive and Shank Retained by Key

ASA B5.10-1953

No. of Taper	Diameter at Gage Line	Shank		Tang					Holdback Key Slot Shank[1]			Socket		Tang Slot[1]			Holdback Key Slot Socket[1]		
		Total Length of Shank	Gage Line to End of Shank	Thickness	Length	Radius of Mill	Diameter	Radius	Gage Line to Bottom of Key Slot	Length	Width	Min Depth of Tapered Hole	Gage Line To Tang Slot	Width	Length	Shank End to Back of Tang Slot	Gage Line to Front of Key Slot	Length	Width
	A	B	C	E	F	G	H	J	Y'	X	N'	K	M	N	O	P	Y	Z	N'
3	0.93800	3 7/8	3 11/16	5/16	9/16	9/32	23/32	5/64	1 1/32	1 1/8	0.260	3 1/4	3 1/16	0.322	1 3/16	9/16	1 1/8	1 3/16	0.260
4	1.23100	4 7/8	4 5/8	15/32	5/8	5/16	31/32	3/32	1 13/32	1 3/16	0.385	4 1/8	3 7/8	0.479	1 1/4	1/2	1 1/2	1 1/4	0.385
4 1/2	1.50000	5 3/8	5 1/8	9/16	11/16	3/8	1 13/64	1/8	1 23/32	1 1/4	0.447	4 9/16	4 5/16	0.573	1 3/8	9/16	1 13/16	1 3/8	0.447
5	1.74800	6 1/8	5 7/8	5/8	3/4	3/8	1 13/32	1/8	2	1 3/8	0.510	5 1/4	4 15/16	0.635	1 1/2	9/16	2 1/8	1 1/2	0.510
6	2.49400	8 9/16	8 1/4	3/4	1 1/8	1/2	2	5/32	2 1/8	1 5/8	0.635	7 3/8	7	0.760	1 3/4	1/2	2 1/4	1 3/4	0.635
7[2]	3.27000	11 5/8	11 1/4	1 1/8	1 3/8	3/4	2 5/8	3/16	2 1/2	1 11/16	0.760	10 1/8	9 1/2	1.135	2 5/8	7/8	2 5/8	1 13/16	0.760

All dimensions given in inches.

[1] Edges at entrance side of slots N and N' shall be chamfered at 45 deg as follows: No. 3, 3/64 in. and all other sizes 1/16 in. deep.

[2] This size is continued in the Tang Drive series for the present to meet special needs.

Tolerances

For Thickness of Tang (E)
Up to and including No. 5, +0.000 − 0.006
Larger than No. 5, +0.000 − 0.008

For Width of Slots (N) and (N')
Up to and including No. 5, +0.006 − 0.000
Larger than No. 5, +0.008 − 0.000

For Diameter of Shank at Gage Line (A)
All sizes, +0.002 − 0.000

For Diameter of Hole at Gage Line (A)
All sizes, + 0.000 − 0.002

For Concentricity of Tang (E) with center line of taper
Up to and including No. 5, 0.0035 (Indicator reading)
Larger than No. 5, 0.005 (Indicator reading)

For Concentricity of Slots (N) and (N') with center line of taper
Up to and including No. 5, 0.0035 (Indicator reading)
Larger than No. 5, 0.005 (Indicator reading)

Tolerances on Fractional Dimensions, ±0.010 unless otherwise specified

TABLE 7-17

American Standard Machine Tapers Having Key Drive and Shank Retained by Key

ASA B5.10-1953

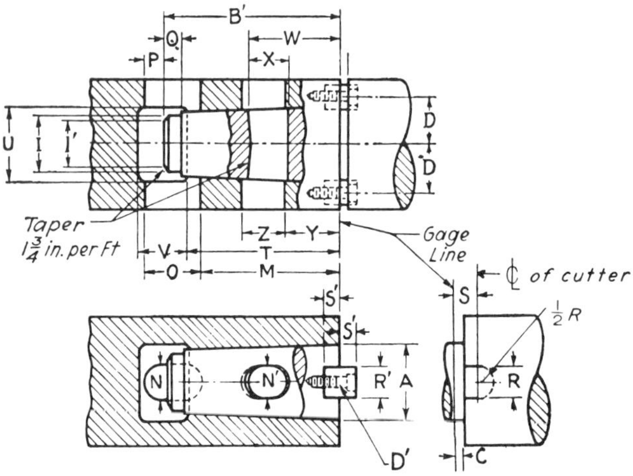

Tolerances

For Diameter of Shank at Gage Line (A)
All sizes, +0.002 − 0.000
For Diameter of Hole at Gage Line (A)
All sizes, + 0.000 − 0.002
For Width of Slots (N) and (N')
+0.008 − 0.000
For Width of Drive Keyway (R')
In socket, +0.000 − 0.001

For Width of Drive Keyway (R)
In shank, +0.010 −0.000
For Concentricity of Slots (N) and (N')
With center line of spindle, 0.007
For Concentricity of Keyway (R')
With center line of spindle, 0.002
Tolerances on Fractional Dimensions, ±0.010 unless otherwise specified

No. of Taper	Diameter at Gage Line	Shank							Drive Key Screw Holes		Holdback Key Slot in Shank[1]		
		Length From Gage Line	Exposed Length	Length of Relief	Diameter of Flat	Diameter of Relief	Drive Keyway		Center Line to Center of Screw	UNF-2B Hole UNF-2A Screw	Gage Line to Back of Key Slot	Length	Width
							Width	Depth					
	A	B'	C	Q	I'	I	R	S	D	D'	W	X	N'
200	2.000	5 1/8		1/4	1 1/8	1 5/8	1.005	9/16	1 13/32	3/8	3 7/16	1 9/16	0.656
250	2.500	5 7/8		1/4	1 1/8	2 1/16	1.005	9/16	1 21/32	3/8	3 11/16	1 9/16	0.781
300	3.000	6 5/8	Min	1/4	1 3/8	2 1/2	2.005	9/16	2 1/4	3/8	4 1/16	1 9/16	1.031
350	3.500	7 7/16	0.003	5/16	2	2 15/16	2.005	9/16	2 1/2	3/8	4 7/8	2	1.031
400	4.000	8 3/16	Max	5/16	2 3/8	3 5/16	2.005	9/16	2 3/4	3/8	5 5/16	2 1/4	1.031
450	4.500	9	0.067	3/8	2 3/8	3 13/16	3.005	13/16	3	1/2	5 7/8	2 7/16	1.031
500	5.000	9 3/4	for	3/8	2 1/2	4 1/4	3.005	13/16	3 1/4	1/2	6 7/16	2 5/8	1.031
600	6.000	11 5/16	all	7/16	2 5/8	5 3/16	3.005	13/16	3 3/4	1/2	7 7/16	3	1.281
800	8.000	14 3/8	Sizes	1/2	3 1/2	7	4.010	1 1/16	4 3/4	1/2	9 9/16	4	1.781
1000	10.000	17 7/16		5/8	4 1/2	8 3/4	4.010	1 1/16	...	...	11 1/2	4 3/4	2.031
1200	12.000	20 1/2		3/4	5 3/8	10 1/2	4.010	1 1/16	...	...	13 3/4	5 3/4	2.531

No. of Taper	Socket						Tang Slot[1]			Holdback Key Slot[1] in Socket		
	Drive Keyway		Gage Line to Front of Relief	Diameter Relief	Depth Relief	Gage Line to Keyway	Width	Length	Shank End to Back of Tang Slot	Gage Line to Front of Key Slot	Length	Width
	Width	Depth										
	R'	S'	T	U	V	M	N	O	P	Y	Z	N'
200	1.000	1/2	4 3/4	1 13/16	1	4 1/2	0.656	1 9/16	15/16	2	1 11/16	0.656
250	1.000	1/2	5 1/2	2 1/4	1	5 3/16	0.781	1 15/16	1 1/4	2 1/4	1 11/16	0.781
300	2.000	1/2	6 1/4	2 3/4	1	5 15/16	1.031	2 3/16	1 1/2	2 5/8	1 11/16	1.031
350	2.000	1/2	6 15/16	3 3/16	1 1/4	6 3/4	1.031	2 3/16	1 1/2	3	2 1/8	1.031
400	2.000	1/2	7 11/16	3 5/8	1 1/4	7 1/2	1.031	2 3/16	1 1/2	3 1/4	2 3/8	1.031
450	3.000	3/4	8 3/8	4 3/16	1 1/2	8	1.031	2 3/4	1 3/4	3 5/8	2 9/16	1.031
500	3.000	3/4	9 1/8	4 5/8	1 1/2	8 3/4	1.031	2 3/4	1 3/4	4	2 3/4	1.031
600	3.000	3/4	10 9/16	5 1/2	1 3/4	10 1/8	1.281	3 1/4	2 1/16	4 5/8	3 1/4	1.281
800	4.000	1	13 1/2	7 3/8	2	12 7/8	1.781	4 1/4	2 3/4	5 3/4	4 1/4	1.781
1000	4.000	1	16 5/16	9 3/16	2 1/2	15 3/4	2.031	5	3 5/16	7	5	2.031
1200	4.000	1	19	11	3	18 1/2	2.531	6	4	8 1/4	6	2.531

All dimensions given in inches.

[1] Edges at entrance side of slots N and N' shall be chamfered at 45 deg as follows: Nos. 200 to 350, inclusive, 1/16 in. deep; Nos. 400 to 600, inclusive, 3/32 in. deep; Nos. 800 to 1200, inclusive, 1/8 in. deep.

TABLE 7-18

American Standard Machine Tapers Having Key Drive and Shank Retained by Draw Bolts

ASA B5.10-1953

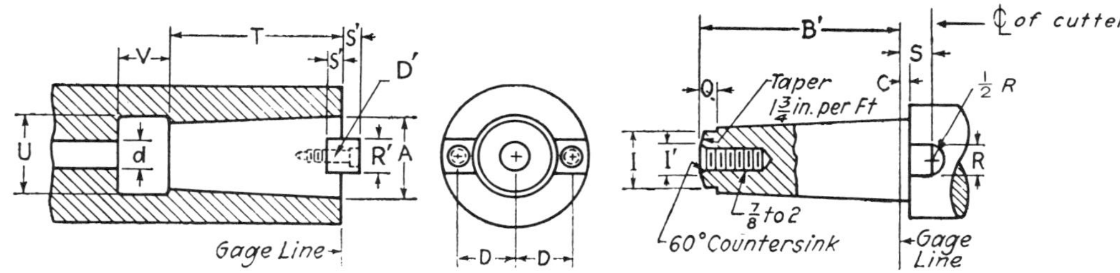

No. of Taper	Diameter at Gage Line	Shank											Drive Key Screw Holes		Socket					
		Length From Gage Line	Exposed Length	Screw Thread Holdback Rod		Drive Keyway		Diameter of Flat	Depth of 60 Deg Center	Length of Relief	Diameter of Relief	Center Line to Center of Screw	UNF-2B Hole UNF-2A Screw	Gage Line to Front of Relief	Diameter Relief	Depth Relief	Diameter Draw Bolt Hole	Drive Keyway	
				Diameter UNC-2B	Depth of Thread	Width	Depth											Width	Depth
	A	B'	C			R	S	I'		Q	I	D	D'	T	U	V	d	R'	S'
200	2.000	5 1/8		7/8	1 3/4	1.005	9/16	1 1/8	1/8	1/4	1 5/8	1 13/32	3/8	4 3/4	1 13/16	1	1	1.000	1/2
250	2.500	5 7/8		7/8	1 3/4	1.005	9/16	1 1/8	1/8	1/4	2 1/16	1 21/32	3/8	5 1/2	2 1/4	1	1	1.000	1/2
300	3.000	6 5/8	Min	1	2	2.005	9/16	1 3/8	3/16	1/4	2 1/2	2 1/4	3/8	6 1/4	2 3/4	1	1 1/8	2.000	1/2
350	3.500	7 7/16	0.003	1	2	2.005	9/16	2	3/16	5/16	2 15/16	2 1/2	3/8	6 15/16	3 3/16	1 1/4	1 1/8	2.000	1/2
400	4.000	8 3/16	Max	1 1/4	2 1/2	2.005	9/16	2 3/8	1/4	5/16	3 5/16	2 3/4	3/8	7 11/16	3 5/8	1 1/4	1 3/8	2.000	1/2
450	4.500	9	0.067	1 1/4	2 1/2	3.005	13/16	2 3/8	1/4	3/8	3 13/16	3	1/2	8 3/8	4 3/16	1 1/2	1 3/8	3.000	3/4
500	5.000	9 3/4	for	1 1/4	2 1/2	3.005	13/16	2 1/2	3/8	3/8	4 1/4	3 1/4	1/2	9 1/8	4 5/8	1 1/2	1 3/8	3.000	3/4
600	6.000	11 5/16	All	1 1/2	3	3.005	13/16	2 5/8	3/8	7/16	5 3/16	3 3/4	1/2	10 9/16	5 1/2	1 3/4	1 5/8	3.000	3/4
800	8.000	14 3/8	Sizes	1 1/2	3	4.010	1 1/16	3 1/2	1/2	1/2	7	4 3/4	1/2	13 1/2	7 3/8	2	1 5/8	4.000	1
1000	10.000	17 7/16		2	4	4.010	1 1/16	4 1/2	1/2	5/8	8 3/4	...	..	16 5/16	9 3/16	2 1/2	2 1/4	4.000	1
1200	12.000	20 1/2		2	4	4.010	1 1/16	5 3/8	1/2	3/4	10 1/2	...	..	19	11	3	2 1/4	4.000	1

All dimensions given in inches.

Tolerances

For Diameter of Hole at Gage Line (A)
All sizes, +0.000 − 0.002

For Diameter of Shank at Gage Line (A)
All sizes, + 0.002 − 0.000

For Width of Drive Keyway (R')
In socket, +0.000 − 0.001

For Width of Drive Keyway (R)
In shank, +0.010 − 0.000

For Concentricity of Drive Keyway (R)
With center line of spindle, 0.003

For Concentricity of Drive Keyway (R')
With center line of spindle, 0.001

Tolerances of Fractional Dimensions,
±0.010 unless otherwise specified

TABLE 7–19

American Standard Dimensions of Steep Machine Tapers

ASA B5.10-1953

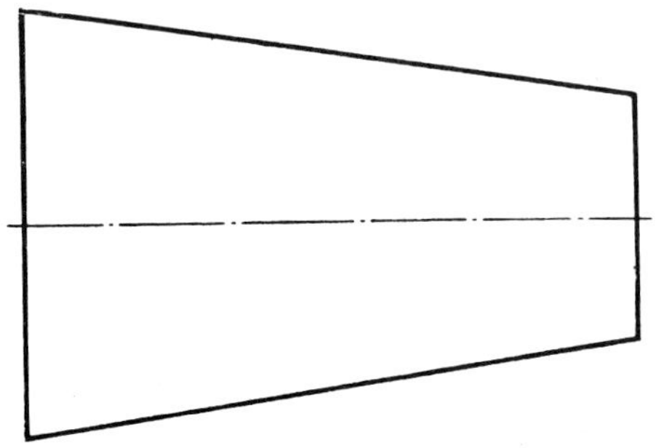

No. of Taper	Taper per Foot[1]	Diameter at Gage Line		Length Along Axis	
5	3.500	1/2	0.500	11/16	0.6875
10	**3.500**	**5/8**	**0.625**	**7/8**	**0.8750**
15	3.500	3/4	0.750	1 1/16	1.0625
20	**3.500**	**7/8**	**0.875**	**1 5/16**	**1.3125**
25	3.500	1	1.000	1 9/16	1.5625
30	**3.500**	**1 1/4**	**1.250**	**1 7/8**	**1.8750**
35	3.500	1 1/2	1.500	2 1/4	2.2500
40	**3.500**	**1 3/4**	**1.750**	**2 11/16**	**2.6875**
45	3.500	2 1/4	2.250	3 5/16	3.3125
50	**3.500**	**2 3/4**	**2.750**	**4**	**4.0000**
55	3.500	3 1/2	3.500	5 3/16	5.1875
60	**3.500**	**4 1/4**	4.250	**6 3/8**	**6.3750**

All dimensions given in inches.

[1] This taper corresponds to an included angle of 16°, 35′, 33.4″.

The tapers numbered 10, 20, 30, 40, 50, and 60 that are printed in heavy-faced type are designated as the "Preferred Series." The tapers numbered 5, 15, 25, 35, 45, and 55 that are printed in light-faced type are designated as the "Intermediate Series."

TABLE 7–20

SAE Standard Taper Shaft End

SAE Handbook 1954

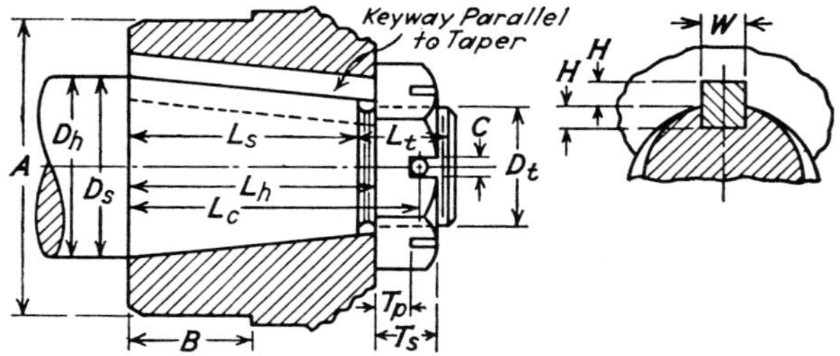

Taper per foot = 1.500 ± 0.002 in. Dimension H is measured normal to the key and at the large end of the taper.

C = cotter-pin hole or slot. The centerline of the cotterpin hole shall be 90 deg from the position of the keyway, as shown in figure.

NOM SHAFT DIAM	D_s, SHAFT DIAM Max	D_s, SHAFT DIAM Min	D_h, HOLE DIAM Max	D_h, HOLE DIAM Min	L_c	L_s	L_h	L_t	D_t	tpi	T_s	T_p	NUT WIDTH (FLATS)	C	W Max	W Min	H +0.004 −0.000	SQUARE KEY Max	SQUARE KEY Min	A[a]	B
1/4	0.250	0.249	0.248	0.247	9/16	5/16	3/8	5/16	#10	40	7/32	9/64	5/16	5/64	0.0625	0.0615	0.033	0.0635	0.0625	1/2	3/16
3/8	0.375	0.374	0.373	0.372	47/64	7/16	1/2	23/64	5/16	32	17/64	3/16	1/2	5/64	0.0937	0.0927	0.049	0.0947	0.0937	11/16	1/4
1/2	0.500	0.499	0.498	0.497	63/64	11/16	3/4	23/64	5/16	32	17/64	3/16	1/2	5/64	0.1250	0.1240	0.065	0.1260	0.1250	7/8	3/8
5/8	0.625	0.624	0.623	0.622	1 3/32	11/16	3/4	17/32	1/2	28	7/16	1/4	3/4	1/8	0.1562	0.1552	0.080	0.1572	0.1562	1 1/16	3/8
3/4	0.750	0.749	0.748	0.747	1 11/32	15/16	1	17/32	1/2	28	7/16	1/4	3/4	1/8	0.1875	0.1865	0.096	0.1885	0.1875	1 1/4	5/8
7/8	0.875	0.874	0.873	0.872	1 11/16	1 1/8	1 1/4	11/16	5/8	24	1/2	5/16	15/16	5/32	0.2500	0.2490	0.127	0.2510	0.2500	1 1/2	3/4
1	1.001	0.999	0.997	0.995	1 15/16	1 3/8	1 1/2	11/16	3/4	20	1/2	5/16	1 1/16	5/32	0.2500	0.2490	0.127	0.2510	0.2500	1 3/4	7/8
1 1/8	1.126	1.124	1.122	1.120	1 15/16	1 3/8	1 1/2	11/16	7/8	20	1/2	5/16	1 1/4	5/32	0.3125	0.3115	0.158	0.3135	0.3125	2	7/8
1 1/4	1.251	1.249	1.247	1.245	1 15/16	1 3/8	1 1/2	11/16	1	20	1/2	5/16	1 7/16	5/32	0.3125	0.3115	0.158	0.3135	0.3125	2 1/8	7/8
1 3/8	1.376	1.374	1.372	1.370	2 7/16	1 7/8	2	11/16	1	20	1/2	5/16	1 7/16	5/32	0.3750	0.3740	0.190	0.3760	0.3750	2 1/4	1
1 1/2	1.501	1.499	1.497	1.495	2 7/16	1 7/8	2	11/16	1	20	1/2	5/16	1 7/16	5/32	0.3750	0.3740	0.190	0.3760	0.3750	2 1/2	1
1 5/8	1.626	1.624	1.622	1.620	2 13/16	2 1/8	2 1/4	13/16	1 1/4	18	5/8	7/16	2 3/16	5/32	0.4375	0.4365	0.221	0.4385	0.4375	2 3/4	1 1/4
1 3/4	1.751	1.749	1.747	1.745	2 13/16	2 1/8	2 1/4	13/16	1 1/4	18	5/8	7/16	2 3/16	5/32	0.4375	0.4365	0.221	0.4385	0.4375	3	1 1/4
1 7/8	1.876	1.874	1.872	1.870	3 1/16	2 3/8	2 1/2	13/16	1 1/4	18	5/8	7/16	2 3/16	5/32	0.4375	0.4365	0.221	0.4385	0.4375	3 1/8	1 1/4
2	2.001	1.999	1.997	1.995	3 9/16	2 7/8	3	13/16	1 1/4	18	5/8	7/16	2 3/16	5/32	0.5000	0.4990	0.252	0.5010	0.5000	3 1/4	1 1/2
2 1/4	2.252	2.248	2.245	2.242	3 9/16	2 7/8	3	13/16	1 1/2	18	5/8	7/16	2 3/8	5/32	0.5625	0.5610	0.283	0.5640	0.5625	3 1/2	1 1/2
2 1/2	2.502	2.498	2.495	2.492	4 9/32	3 3/8	3 1/2	1 1/4	2	16	1	5/8	3 1/8	7/32	0.6250	0.6235	0.315	0.6265	0.6250	4	1 3/4
2 3/4	2.752	2.748	2.745	2.742	4 9/32	3 3/8	3 1/2	1 1/4	2	16	1	5/8	3 1/8	7/32	0.6875	0.6860	0.346	0.6890	0.6875	4 3/8	1 3/4
3	3.002	2.998	2.995	2.992	4 25/32	3 7/8	4	1 1/4	2	16	1	5/8	3 1/8	7/32	0.7500	0.7485	0.377	0.7515	0.7500	4 3/4	2
3 1/4	3.252	3.248	3.245	3.242	5 1/32	4 1/8	4 1/4	1 1/4	2	16	1	5/8	3 1/8	7/32	0.7500	0.7485	0.377	0.7515	0.7500	5	2 1/8
3 1/2	3.502	3.498	3.495	3.492	5 7/16	4 3/8	4 1/2	1 3/8	2 1/2	16	1 1/8	3/4	3 7/8	9/32	0.8750	0.8735	0.440	0.8765	0.8750	5 1/2	2 1/4
4	4.002	3.998	3.995	3.992	6 7/16	5 3/8	5 1/2	1 3/8	2 1/2	16	1 1/8	3/4	3 7/8	9/32	1.0000	0.9985	0.502	1.0015	1.0000	6 1/4	2 3/4

[a] Tolerance for ground finish, nominal +0.003, −0.002; and when specified, the maximum eccentricity with respect to the hole shall be 0.002 (indicator reading 0.004). Tolerance for lathe finish, nominal +1/32, −0.

TABLE 7-21

Brown and Sharpe Tapers

B & S Small Tools Catalog No. 35

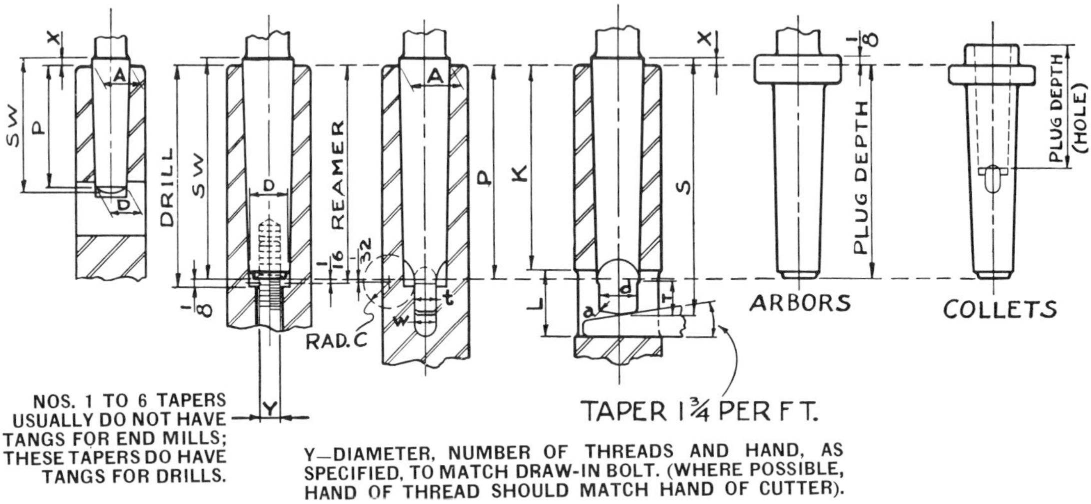

No. of Taper	Taper per Foot	Diam. of Plug at Small End	Diam. at End of Socket	Plug Depth P			Key-way from End of Spindle	Shank Length with Tang	Shank Length with-out Tang	Shank Pro-jects from End of Socket	Length of Key-way†	Width of Key-way	Length of Arbor Tongue	Diam-eter of Arbor Tongue	Thick-ness of Arbor Tongue	Radius of Tongue Circle	Radius of Tongue at a
				B & S** Stand-ard	For Mill. Mach.	Miscell.											
		D	A				K	S	SW	X	L	W	T	d	t	c	a
*1	.50200	.20000	.2392	15/16‡			15/16	1 9/32	1 1/16	3/32	3/8	.135	3/16	.170	1/8	3/16	.030
*2	.50200	.25000	.2997	1 3/16‡			1 11/64	1 19/32	1 11/32	3/32	1/2	.166	1/4	.220	5/32	3/16	.030
			.3752	1 1/2‡			1 15/32	1 31/32	1 41/64	3/32	5/8	.197	5/16	.282	3/16	3/16	.040
*3	.50200	.31250				1 3/4	1 23/32				5/8	.197	5/16	.282	3/16	3/16	.040
						2	1 31/32				5/8	.197	5/16	.282	3/16	3/16	.040
4	.50240	.35000	.4023		1 1/4‡		1 13/16	1 3/4	1 25/64	3/32	11/16	.228	11/32	.320	7/32	5/16	.050
				1 11/16			1 41/64				11/16	.228	11/32	.320	7/32	5/16	.050
			.5231		1 3/4‡		1 11/16	2 9/32	1 29/32	3/32	3/4	.260	3/8	.420	1/4	5/16	.060
5	.50160	.45000				2	1 15/16				3/4	.260	3/8	.420	1/4	5/16	.060
				2 1/8			2 1/16				3/4	.260	3/8	.420	1/4	5/16	.060
6	.50329	.50000	.5996	2 3/8‡			2 19/64	2 31/32	2 17/32	3/32	7/8	.291	7/16	.460	9/32	5/16	.060
						2 1/2	2 13/32				15/16	.322	15/32	.560	5/16	3/8	.070
7	.50147	.60000		2 7/8			2 25/32				15/16	.322	15/32	.560	5/16	3/8	.070
			.7254		3‡		2 29/32	3 5/8	3 3/32	3/32	15/16	.322	15/32	.560	5/16	3/8	.070
8	.50100	.75000	.8987	3 9/16‡			3 29/64	4 1/4	3 11/16	1/8	1	.353	1/2	.710	11/32	3/8	.080
9	.50085	.90010	1.0670		4‡		3 7/8	4 3/4	4 1/8	1/8	1 1/8	.385	9/16	.860	3/8	7/16	.100
				4 1/4			4 1/8				1 1/8	.385	9/16	.860	3/8	7/16	.100
				5			4 27/32				1 5/16	.447	21/32	1.010	7/16	7/16	.110
10	.51612	1.04465	1.2892		5 11/16‡		5 17/32	6 17/32	5 13/16	1/8	1 5/16	.447	21/32	1.010	7/16	7/16	.110
						6 7/32	6 1/16				1 5/16	.447	21/32	1.010	7/16	7/16	.110
11	.50100	1.24995		5 15/16			5 25/32				1 5/16	.447	21/32	1.210	7/16	1/2	.130
			1.5318		6 3/4‡		6 19/32	7 19/32	6 7/8	1/8	1 5/16	.447	21/32	1.210	7/16	1/2	.130
12	.49973	1.50010	1.7968	7 1/8	7 1/8‡		6 15/16	8 1/16	7 1/4	1/8	1 1/2	.510	3/4	1.460	1/2	1/2	.150
						6 1/4											
13	.50020	1.75005	2.0730	7 3/4‡			7 9/16	8 11/16	7 7/8	1/8	1 1/2	.510	3/4	1.710	1/2	5/8	.170
14	.50000	2.00000	2.3437	8 1/4‡	8 1/4		8 1/32	9 9/32	8 3/8	1/8	1 11/16	.572	27/32	1.960	9/16	3/4	.190
15	.50000	2.25000	2.6146	8 3/4‡			8 17/32	9 25/32	8 7/8	1/8	1 11/16	.572	27/32	2.210	9/16	7/8	.210
16	.50000	2.50000	2.8854	9 1/4‡			9	10 3/8	9 3/8	1/8	1 7/8	.635	15/16	2.450	5/8	1	.230
17	.50000	2.75000	3.1562	9 3/4‡					9 7/8	1/8							
18	.50000	3.00000	3.4271	10 1/4‡					10 3/8	1/8							

All dimensions in inches.

* Adopted by American Standards Association.

** "B & S Standard" Plug Depths are not used in all cases.

† Special lengths of keyway are used instead of standard lengths in some places. Standard lengths need not be used when keyway is for driving only and not for admitting key to force out tool.

‡ These lengths are standard for shank cutters.

TABLE 7-22

Amount of Taper in Terms of Length

Brown and Sharpe Small Tools Catalog No. 35

TAPERS

Tapers from 1/16 to 1 1/4 Inch per Foot — Amount of Taper for Lengths Up to 24 Inches

Length Tapered Inches	Taper per Foot									
	1/16	3/32	1/8	1/4	3/8	1/2	5/8	3/4	1	1 1/4
1/32	.0002	.0002	.0003	.0007	.0010	.0013	.0016	.0020	.0026	.0033
1/16	.0003	.0005	.0007	.0013	.0020	.0026	.0033	.0039	.0052	.0065
1/8	.0007	.0010	.0013	.0026	.0039	.0052	.0065	.0078	.0104	.0130
3/16	.0010	.0015	.0020	.0039	.0059	.0078	.0098	.0117	.0156	.0195
1/4	.0013	.0020	.0026	.0052	.0078	.0104	.0130	.0156	.0208	.0260
5/16	.0016	.0024	.0033	.0065	.0098	.0130	.0163	.0195	.0260	.0326
3/8	.0020	.0029	.0039	.0078	.0117	.0156	.0195	.0234	.0312	.0391
7/16	.0023	.0034	.0046	.0091	.0137	.0182	.0228	.0273	.0365	.0456
1/2	.0026	.0039	.0052	.0104	.0156	.0208	.0260	.0312	.0417	.0521
9/16	.0029	.0044	.0059	.0117	.0176	.0234	.0293	.0352	.0469	.0586
5/8	.0033	.0049	.0065	.0130	.0195	.0260	.0326	.0391	.0521	.0651
11/16	.0036	.0054	.0072	.0143	.0215	.0286	.0358	.0430	.0573	.0716
3/4	.0039	.0059	.0078	.0156	.0234	.0312	.0391	.0469	.0625	.0781
13/16	.0042	.0063	.0085	.0169	.0254	.0339	.0423	.0508	.0677	.0846
7/8	.0046	.0068	.0091	.0182	.0273	.0365	.0456	.0547	.0729	.0911
15/16	.0049	.0073	.0098	.0195	.0293	.0391	.0488	.0586	.0781	.0977
1	.0052	.0078	.0104	.0208	.0312	.0417	.0521	.0625	.0833	.1042
2	.0104	.0156	.0208	.0417	.0625	.0833	.1042	.125	.1667	.2083
3	.0156	.0234	.0312	.0625	.0937	.1250	.1562	.1875	.250	.3125
4	.0208	.0312	.0417	.0833	.125	.1667	.2083	.250	.3333	.4167
5	.0260	.0391	.0521	.1042	.1562	.2083	.2604	.3125	.4167	.5208
6	.0312	.0469	.0625	.125	.1875	.250	.3125	.375	.500	.625
7	.0365	.0547	.0729	.1458	.2187	.2917	.3646	.4375	.5833	.7292
8	.0417	.0625	.0833	.1667	.250	.3333	.4167	.500	.6667	.8333
9	.0469	.0703	.0937	.1875	.2812	.375	.4687	.5625	.750	.9375
10	.0521	.0781	.1042	.2083	.3125	.4167	.5208	.625	.8333	1.0417
11	.0573	.0859	.1146	.2292	.3437	.4583	.5729	.6875	.9167	1.1458
12	.0625	.0937	.125	.250	.375	.500	.625	.750	1.000	1.250
13	.0677	.1016	.1354	.2708	.4062	.5417	.6771	.8125	1.0833	1.3542
14	.0729	.1094	.1458	.2917	.4375	.5833	.7292	.875	1.1667	1.4583
15	.0781	.1172	.1562	.3125	.4687	.625	.7812	.9375	1.250	1.5625
16	.0833	.125	.1667	.3333	.500	.6667	.8333	1.000	1.3333	1.6667
17	.0885	.1328	.1771	.3542	.5312	.7083	.8854	1.0625	1.4167	1.7708
18	.0937	.1406	.1875	.3750	.5625	.750	.9375	1.125	1.500	1.875
19	.0990	.1484	.1979	.3958	.5937	.7917	.9896	1.1875	1.5833	1.9792
20	.1042	.1562	.2083	.4167	.625	.8333	1.0417	1.250	1.6667	2.0833
21	.1094	.1641	.2187	.4375	.6562	.875	1.0937	1.3125	1.750	2.1875
22	.1146	.1719	.2292	.4583	.6875	.9167	1.1458	1.375	1.8333	2.2917
23	.1198	.1797	.2396	.4792	.7187	.9583	1.1970	1.4375	1.9167	2.3958
24	.125	.1875	.250	.500	.750	1.000	1.250	1.500	2.000	2.500

TABLE 7–23*

ABEC 1 Tolerances on Metric Ball Bearings, Except Type BM (Magneto) RBEC 1 Tolerances on Roller Bearings

AFBMA Standards, Section 3 – June 1951

Bore in Millimeters		Inner Ring Tolerance, Inch		Outer Diameter in Millimeters		Outer Ring Tolerance, Inch	
Over	Inclusive	Bore Diameter +0.0000	Radial Runout	Over	Inclusive	Outer Diameter +0.0000	Radial Runout
0	9	-0.0003	0.0003	0	18	-0.0004	0.0006
9	18	-0.0003	0.0004	18	30	-0.0004	0.0006
18	30	-0.0004	0.0005	30	50	-0.0005	0.0008
30	50	-0.0005	0.0006	50	80	-0.0005	0.0010
50	80	-0.0006	0.0008	80	120	-0.0006	0.0014
80	120	-0.0008	0.0010	120	150	-0.0008	0.0016
120	180	-0.0010	0.0012	150	180	-0.0010	0.0018
180	250	-0.0012	0.0016	180	250	-0.0012	0.0020
250	315	-0.0014	0.0020	250	315	-0.0014	0.0024
315	400	-0.0016	0.0024	315	400	-0.0016	0.0028
400	500	-0.0018	0.0026	400	500	-0.0018	0.0032
				500	630	-0.0020	0.0040

*This table is a condensation of the corresponding table in Section 3 of the AFBMA Standards. Other recognized ABEC specifications for ball bearings can be found by subtracting column 3 from 2, Tables 7–30 to 7–34, inclusive, in case tolerances are needed for design purposes. The reader is referred to the original sources for matter pertinent to the specification and inspection of bearings, because the tabular matter from Section 3 is inadequate without the descriptive matter of Section 4 on Standard Gaging Practices.

TABLE 7–24

Tolerances on Width of Individual Inner or Outer Ring, Metric Ball and Roller Bearings
AFBMA Standards, Section 3 – June 1951
ABEC 1, 3, 5 and 7; RBEC 1 and 5

Nominal Bore, mm Over	Inclusive	Tolerance on Width for Bearings Other Than Duplex, + 0.0000, Inch
0	180	– 0.005
180	315	– 0.010
315	400	– 0.016
400	500	– 0.018
500	630	– 0.022

TABLE 7–25

Tolerances on Diameters of Tapered Roller Bearings

AFBMA Standards, Section 3 – June 1951

Cone Bore, In.		Tolerance on Cone Bore by Class				
		4	2	3	0	4B
Over	Inclusive	– 0.000	– 0.000	– 0.000	– 0.000	– 0.000
0	2-1/2	+ 0.0005	+ 0.0005	+ 0.0005	+ 0.0005	+ 0.0005
2-1/2	12	+ 0.001	+ 0.001	+ 0.0005	+ 0.0005	+ 0.001
12	24	+ 0.002	--	+ 0.001	--	+ 0.002
Cup Outer Diameter, In.		**Tolerance on Outer Diameter by Class**				
		4	2	3	0	4B
Over	Inclusive	– 0.000	– 0.000	– 0.000	– 0.000	– 0.000
0	12	+ 0.001	+ 0.001	+ 0.0005	+ 0.0005	+ 0.001
12	24	+ .002	--	+ .001	--	+ .002
24	36	+ .003	--	+ .0015	--	+ .003

All dimensions in inches

TABLE 7-26

Tolerances for Industrial Needle Bearings

AFBMA Standards, Section 3 – June 1951

Bore or Outside Diameter, In.		Tolerance, In.
over	inclusive	+ 0.0000
0	.7500	- 0.0004
.7500	2.0000	– 0.0005
2.0000	3.2500	– 0.0006
3.2500	4.7500	– 0.0008
4.7500	7.2500	– 0.0010
7.2500	10.2500	– 0.0012
10.2500	12.5000	– 0.0014
12.5000	15.7500	– 0.0016
15.7500	19.7500	– 0.0018

Bearing widths

Outer rings	+ 0.000	– 0.005
Inner rings whose outer diameter is 5 in. or less	+ 0.005	+ 0.010
Inner rings whose outer diameter is greater than 5 in.	+ 0.010	+ 0.015

Shaft diameters for series NAA and NBA

Size of shaft		Tolerance
from	to	+ 0.0000
0	4	– 0.0005
4	6	– 0.0007
6 and over		– 0.0010

All dimensions in inches

TABLE 7-27

Industrial Equipment, Cup and Cone, Fitting Practices for Tapered Roller Bearings

AFBMA Standards, Section 7 – March 1951

MOUNTING CONDITIONS		CLASS: No. 4 and No. 2 See Table 7-25 — CUP OUTSIDE DIAMETER								
		UP TO 3 IN. INCLUSIVE			OVER 3 TO 5 IN. INCLUSIVE			OVER 5 TO 12 IN. INCLUSIVE †		
		Variation in Cup OD	Fit	Cup Seat Tolerance	Variation in Cup OD	Fit	Cup Seat Tolerance	Variation in Cup OD	Fit	Cup Seat Tolerance
STATIONARY CUP	Adjustable or Moveable	+.0010 −.0000	.0010 loose .0010 tight	+.0000 +.0010	+.0010 −.0000	.0010 loose .0010 tight	+.0000 +.0010	+.0010 −.0000	.0020 loose .0010 tight	+.0000 +.0020
STATIONARY CUP	Floating Type TDO-TNA* Etc.	+.0010 −.0000	.0020 loose .0000 loose	+.0010 +.0020	+.0010 −.0000	.0020 loose .0000 loose	+.0010 +.0020	+.0010 −.0000	.0020 loose .0000 loose	+.0010 +.0020
ROTATING OR STATIONARY CUP	Non-Adjustable or Non-Moveable	+.0010 −.0000	.0005 tight .0025 tight	−.0015 −.0005	+.0010 −.0000	.0010 tight .0030 tight	−.0020 −.0010	+.0010 −.0000	.0010 tight .0030 tight	−.0020 −.0010
ROTATING CUP	Sheaves-Unclamped Type TDO-TNA*	+.0010 −.0000	.0020 tight .0040 tight	−.0030 −.0020	+.0010 −.0000	.0020 tight .0040 tight	−.0030 −.0020	+.0010 −.0000	.0020 tight .0040 tight	−.0030 −.0020

SPINDLE MOUNTING CONDITIONS		CLASS: No. 3 and No. 0						CLASS: No. 3 only		
		UP TO 6 IN. INCLUSIVE			OVER 6 TO 12 IN. INCLUSIVE			OVER 12 TO 24 IN. INCLUSIVE		
STATIONARY CUP	Adjustable or Moveable	+.0005 −.0000	.0005 loose .0005 tight	+.0000 +.0005	+.0005 −.0000	.0010 loose .0005 tight	+.0000 +.0005	+.0010 −.0000	.0010 loose .0010 tight	+.0000 +.0010
STATIONARY CUP	Non-Adjustable (Fixed)	+.0005 −.0000	.0000 tight .0010 tight	−.0005 −.0000	+.0005 −.0000	.0000 tight .0015 tight	−.0010 −.0000	+.0010 −.0000	.0000 tight .0020 tight	−.0010 −.0000
ROTATING OR STATIONARY CUP	Floating-Type TDO–TNA* Etc.	+.0005 −.0000	.0010 loose .0000 loose	+.0005 +.0010	+.0005 −.0000	.0010 loose .0000 loose	+.0005 +.0010	+.0010 −.0000	.0020 loose .0000 loose	+.0010 +.0020
ROTATING CUP	Non-Adjustable (Fixed)	+.0005 −.0000	.0005 tight .0015 tight	−.0010 −.0005	+.0005 −.0000	.0005 tight .0020 tight	−.0015 −.0005	+.0010 −.0000	.0005 tight .0025 tight	−.0015 −.0005

*See Table 2–25 for description of types.

†See AFBMA Standards for larger sizes or see The Timken Engineering Journal

continued on next page

TABLE 7-27, continued

MOUNTING CONDITIONS	SHAFT FINISH	SERVICE	CLASS: No. 4 and No. 2 See Table 7-25						CLASS: No. 4 only		
			CONE BORE								
			UP TO 2-1/2 IN. INCLUSIVE			OVER 2-1/2 TO 12 IN. INCLUSIVE			OVER 12 TO 24 IN. INCLUSIVE		
			Cone Bore Variation	Fit	Cone Seat Tolerance	Cone Bore Variation	Fit	Cone Seat Tolerance	Cone Bore Variation	Fit	Cone Seat Tolerance
ROTATING CONE	GROUND	STEADY LOADS WITH MODERATE SHOCK	+.0005 -.0000	.0005 tight .0015 tight	+.0015 +.0010	+.0010 -.0000	.0005 tight .0025 tight	+.0025 +.0015	+.0020 -.0000	.0010 tight .0050 tight	+.0050 +.0030
ROTATING OR STATIONARY CONE	TURNED	HEAVY LOADS HIGH SPEEDS or SHOCK	+.0005 -.0000	.0010 tight .0025 tight	+.0025 +.0015	+.0010 -.0000	Footnote*		+.0020 -.0000	Footnote*	
STATIONARY CONE	TURNED	MEDIUM LOADS NO SHOCK	+.0005 -.0000	.0005 loose .0005 tight	+.0005 +.0000	+.0010 -.0000	.0010 loose .0010 tight	+.0010 +.0000	+.0020 -.0000	.0020 loose .0020 tight	+.0020 +.0000
	TURNED GROUND	SPECIAL Sheaves Wheels, Etc. MEDIUM LOADS NO SHOCK	+.0005 -.0000	.0010 loose .0000 loose	-.0000 -.0005	+.0010 -.0000	.0020 loose .0000 loose	-.0000 -.0010	+.0020 -.0000	.0040 loose .0000 loose	-.0000 -.0020
	HARDENED and GROUND	WHEEL SPINDLES	+.0005 -.0000	.0012 loose .0002 loose	-.0002 -.0007	+.0010 -.0000	.0022 loose .0002 loose	-.0002 -.0012	--	--	--
			CLASS: No. 3 and No. 0						CLASS: No. 3 only		
ROTATING OR STATIONARY CONE	GROUND	Machine Tool Spindles	+.0005 -.0000	.0002 tight .0012 tight	+.0012 +.0007	+.0005 -.0000	.0002 tight .0012 tight	+.0012 +.0007	+.0010 -.0000	.0005 tight .0025 tight	+.0025 +.0015

*It is recommended that all cone seats be ground. In those cases where grinding is impossible a minimum cone seat should be provided equal to the nominal cone bore plus 0.0005 inch per inch of cone bore. To this value add the cone bore tolerance.

TABLE 7–28

Automotive, Cup and Cone, Fitting Practice for Tapered Roller Bearings

AFBMA Standards, Section 7 – March 1951

Industry	Type of Application		Cone Bore					
			5/8 to 2½ In. Inclusive			Over 2½ In.		
			Tolerance	Recommended Fit	Cone Seat Equals Nominal Cone Bore	Tolerance	Recommended Fit	Cone Seat Equals Nominal Cone Bore
Automotive Rotating Shafts	Pinion, Transmission, Rear Wheels, Cross Shaft, Transfer Case	Adjustable Cones	+.0005 –.0000	.0005 tight .0005 loose	+.0005 max +.0000 min	+.0010 –.0000	.0010 tight .0010 loose	+.0010 max +.0000 min
		Non-Adjustable Cones	+.0005 –.0000	.0015 tight .0005 tight	+.0015 max +.0010 min	+.0010 –.0000	.0025 tight .0005 tight	+.0025 max +.0015 min
	Differential	Non-Adjustable Cones	+.0005 –.0000	.0025 tight .0010 tight	+.0025 max +.0015 min	+.0010 –.0000	.0035 tight .0015 tight	+.0035 max +.0025 min
Automotive Stationary Shafts	Front Wheels, Full Floating Rear Wheels, etc., Trailer Wheels	Adjustable Cones	+.0005 –.0000	.0002 loose .0012 loose	–.0002 max –.0007 min	+.0010 –.0000	.0002 loose .0022 loose	–.0002 max –.0012 min

Industry	Type of Application		Cup Outside Diameter					
			Less Than 3 In.			3 to 5 In. Inclusive		
			Tolerance	Recommended Fit	Cup Seat Equals Nominal Cup O.D.	Tolerance	Recommended Fit	Cup Seat Equals Nominal Cup O.D.
Automotive	Front Wheels, Full Floating Rear Wheels, Pinion, Differential	Non-Adjustable Cups	+.0010 –.0000	.0005 tight .0025 tight	–.0005 max –.0015 min	+.0010 –.0000	.0010 tight .0030 tight	–.0010 max –.0020 min
	Differential	Adjustable Cups	+.0010 –.0000	.0020 loose .0000	+.0020 max +.0010 min	+.0010 –.0000	.0020 loose .0000	+.0020 max +.0010 min
	Rear Wheels, Transmission, Cross Shaft, Other Applications	Adjustable Cups	+.0010 –.0000	.0010 loose .0010 tight	+.0010 max +.0000 min	+.0010 –.0000	.0010 loose .0010 tight	+.0010 max +.0000 min

continued on next page

TABLE 7-28, continued

Industry	Type of Application		Cup Outside Diameter		
			Over 5 In.		
			Toler- ance	Recommended Fit	Cup Seat Equals Nominal Cup O D
Automotive	Front Wheels, Full Floating Rear Wheels, Pinion, Differential	Non-Adjustable Cups	+.0010 -.0000	.0010 tight .0040 tight	-.0010 max -.0030 min
	Differential	Adjustable Cups	+.0010 -.0000	.0020 loose .0010 tight	+.0020 max -.0000 min
	Rear Wheels, Transmission, Cross Shaft, Other Applications	Adjustable Cups	+.0010 -.0000	.0020 loose .0010 tight	+.0020 max +.0000 min

TABLE 7–29

ABEC-1 Shaft Diameters for Bearing Seats of Metric Annular Ball Bearings

AFBMA Standards, Section 7 – March 1951

	Light Loads					Normal Loads						Heavy Loads			
	For Light Loads Where the Shaft Rotates in Relation to the Direction of the Load or Where the Direction of the Load Is Indeterminate					Shaft Rotation with Relation to the Direction of the Load			Shaft Stationary With Relation to the Direction of the Load			Shaft Rotating With Relation to the Direction of the Load			
	BEARING BORE Inches		SHAFT DIAMETER Inches			SHAFT DIAMETER Inches			SHAFT DIAMETER Inches			SHAFT DIAMETER Inches			
mm	Max	Min	Max	Min	Mean Fit	Max	Min	Mean Fit	Max	Min	Mean Fit	Max	Min	Mean Fit	mm
4	.1575	.1572	.1575	.1573	T.00000	.1577	.1575	T.0002	.1573	.1570	L.0002				4
5	.1969	.1966	.1969	.1967		.1971	.1969		.1967	.1964					5
6	.2362	.2359	.2362	.2360		.2364	.2362		.2360	.2357					6
7	.2756	.2753	.2756	.2754		.2758	.2755		.2754	.2750	.0003				7
8	.3150	.3147	.3150	.3148		.3152	.3149		.3148	.3144					8
9	.3543	.3540	.3543	.3541		.3545	.3542		.3541	.3537			same		9
10	.3937	.3934	.3937	.3935		.3939	.3936		.3935	.3931					10
12	.4724	.4721	.4724	.4721		.4726	.4723		.4722	.4717			as		12
15	.5906	.5903	.5906	.5903		.5908	.5905		.5904	.5899					15
17	.6693	.6690	.6693	.6690		.6695	.6692		.6691	.6686			normal load		17
20	.7874	.7870	.7878	.7872	.0003	.7878	.7875	.0005	.7871	.7866	.0004				20
25	.9843	.9839	.9847	.9841		.9847	.9844		.9840	.9835					25
30	1.1811	1.1807	1.1815	1.1809		1.1815	1.1812		1.1808	1.1803					30
35	1.3780	1.3775	1.3784	1.3778	.0004	1.3785	1.3781	.0006	1.3776	1.3770	.0005				35
40	1.5748	1.5743	1.5752	1.5746		1.5753	1.5749		1.5744	1.5738					40
45	1.7717	1.7712	1.7721	1.7715		1.7722	1.7718		1.7713	1.7707					45
50	1.9685	1.9680	1.9689	1.9683		1.9690	1.9686		1.9681	1.9675					50
55	2.1654	2.1648	2.1659	2.1651		2.1660	2.1655	.0007	2.1650	2.1643					55
60	2.3622	2.3616	2.3627	2.3619		2.3628	2.3623		2.3618	2.3611					60
65	2.5591	2.5585	2.5596	2.5588		2.5597	2.5592		2.5587	2.5580					65
70	2.7559	2.7553	2.7564	2.7556		2.7565	2.7560		2.7555	2.7548					70
75	2.9528	2.9522	2.9533	2.9525		2.9534	2.9529		2.9524	2.9517					75
80	3.1496	3.1490	3.1501	3.1493		3.1502	3.1497		3.1492	3.1485					80

continued on next page

TABLE 7–29, continued

	Light Loads					Normal Loads						Heavy Loads			
	For Light Loads Where the Shaft Rotates in Relation to the Direction of the Load or Where the Direction of the Load Is Indeterminate					Shaft Rotation with Relation to the Direction of the Load			Shaft Stationary With Relation to the Direction of the Load			Shaft Rotating With Relation to the Direction of the Load			
	BEARING BORE		SHAFT DIAMETER			SHAFT DIAMETER			SHAFT DIAMETER			SHAFT DIAMETER			
	Inches		Inches			Inches			Inches			Inches			
mm	Max	Min	Max	Min	Mean Fit	Max	Min	Mean Fit	Max	Min	Mean Fit	Max	Min	Mean Fit	mm
85	3.3465	3.3457	3.3470	3.3461	T.0005	3.3472	3.3466	T.0008	3.3460	3.3452	L. 0005	Same as normal load			85
90	3.5433	3.5425	3.5438	3.5429		3.5440	3.5434		3.5428	3.5420					90
95	3.7402	3.7394	3.7407	3.7398		3.7409	3.7403		3.7397	3.7389					95
100	3.9370	3.9362	3.9375	3.9366		3.9377	3.9371		3.9365	3.9357					100
105	4.1339	4.1331	4.1344	4.1335		4.1346	4.1340		4.1334	4.1326		4.1350	4.1344	T.0012	105
110	4.3307	4.3299	4.3312	4.3303		4.3314	4.3308		4.3302	4.3294		4.3318	4.3312		110
120	4.7244	4.7236	4.7249	4.7240		4.7251	4.7245		4.7239	4.7231		4.7255	4.7249		120
130	5.1181	5.1171	5.1187	5.1177	.0006	5.1189	5.1182	.0010	5.1175	5.1166	.0006	5.1194	5.1187	.0015	130
140	5.5118	5.5108	5.5124	5.5114		5.5126	5.5119		5.5112	5.5103		5.5131	5.5124		140
150	5.9055	5.9045	5.9061	5.9051		5.9063	5.9056		5.9049	5.9040		5.9068	5.9061		150
160	6.2992	6.2982	6.2998	6.2988		6.3000	6.2993		6.2986	6.2977		6.3005	6.2998		160
170	6.6929	6.6919	6.6935	6.6925		6.6937	6.6930		6.6923	6.6914		6.6942	6.6935		170
180	7.0866	7.0856	7.0872	7.0862		7.0874	7.0867		7.0860	7.0851		7.0879	7.0872		180
190	7.4803	7.4791	7.4809	7.4798	.0007	7.4813	7.4805	.0012	7.4797	7.4786		7.4818	7.4810	.0017	190
200	7.8740	7.8728	7.8746	7.8735		7.8750	7.8742		7.8734	7.8723		7.8755	7.8747		200
210	8.2677	8.2665	8.2683	8.2672		8.2687	8.2679		8.2671	8.2660		8.2692	8.2684		210
220	8.6614	8.6602	8.6620	8.6609		8.6624	8.6616		8.6608	8.6597		8.6629	8.6621		220
230	9.0551	9.0539	9.0557	9.0546		9.0561	9.0553		9.0545	9.0534		9.0566	9.0558		230
240	9.4488	9.4476	9.4494	9.4483		9.4498	9.4490		9.4482	9.4471		9.4503	9.4495		240
260	10.2362	10.2348	10.2368	10.2356		10.2373	10.2364	.0014	10.2355	10.2343		10.2379	10.2370	.0020	260
280	11.0236	11.0222	11.0242	11.0230		11.0247	11.0238		11.0229	11.0217		11.0253	11.0244		280
300	11.8110	11.8096	11.8116	11.8104		11.8121	11.8112		11.8103	11.8091		11.8127	11.8118		300
320	12.5984	12.5968	12.5991	12.5977	.0008	12.5996	12.5986	.0015	12.5977	12.5963	.0006	12.6002	12.5993	.0021	320

TABLE 7–30

ABEC-1 Housing Seat Diameters for Metric Annular Ball Bearings

AFBMA Standards, Section 7 – March 1951

For normal loads where housing is stationary with relation to direction of load.

	Bearing Outside Diameter Inches		Solid Housing Bore Inches		Solid Housing	Split Housing Bore Inches		Split Housing
mm	max	min	min	max	Mean Fit Loose	min	max	Mean Fit Loose
16	.6299	.6295	.6299	.6303	.0004	.6299	.6306	.0006
19	.7480	.7476	.7480	.7485	.0004	.7480	.7488	.0006
22	.8661	.8657	.8661	.8666	.0004	.8661	.8669	.0006
24	.9449	.9445	.9449	.9454	.0004	.9449	.9457	.0006
26	1.0236	1.0232	1.0236	1.0241	.0004	1.0236	1.0244	.0006
30	1.1811	1.1807	1.1811	1.1816	.0004	1.1811	1.1819	.0006
32	1.2598	1.2593	1.2598	1.2604	.0005	1.2598	1.2608	.0007
35	1.3780	1.3775	1.3780	1.3786	.0005	1.3780	1.3790	.0007
37	1.4567	1.4562	1.4567	1.4573	.0005	1.4567	1.4577	.0007
40	1.5748	1.5743	1.5748	1.5754	.0005	1.5748	1.5758	.0007
42	1.6535	1.6530	1.6535	1.6541	.0005	1.6535	1.6545	.0007
47	1.8504	1.8499	1.8504	1.8510	.0005	1.8504	1.8514	.0007
52	2.0472	2.0467	2.0472	2.0479	.0006	2.0472	2.0484	.0009
55	2.1654	2.1649	2.1654	2.1661	.0006	2.1654	2.1666	.0009
62	2.4409	2.4404	2.4409	2.4416	.0006	2.4409	2.4421	.0009
68	2.6772	2.6767	2.6772	2.6779	.0006	2.6772	2.6784	.0009
72	2.8346	2.8341	2.8346	2.8353	.0006	2.8346	2.8358	.0009
75	2.9528	2.9523	2.9528	2.9535	.0006	2.9528	2.9540	.0009
80	3.1496	3.1491	3.1496	3.1503	.0006	3.1496	3.1508	.0009
85	3.3465	3.3459	3.3465	3.3474	.0008	3.3465	3.3479	.0010
90	3.5433	3.5427	3.5433	3.5442	.0008	3.5433	3.5447	.0010
95	3.7402	3.7396	3.7402	3.7411	.0008	3.7402	3.7416	.0010
100	3.9370	3.9364	3.9370	3.9379	.0008	3.9370	3.9384	.0010
110	4.3307	4.3301	4.3307	4.3316	.0008	4.3307	4.3321	.0010
115	4.5276	4.5270	4.5276	4.5285	.0008	4.5276	4.5290	.0010
120	4.7244	4.7238	4.7244	4.7253	.0008	4.7244	4.7258	.0010
125	4.9213	4.9205	4.9213	4.9223	.0009	4.9213	4.9229	.0012
130	5.1181	5.1173	5.1181	5.1191	.0009	5.1181	5.1197	.0012
140	5.5118	5.5110	5.5118	5.5128	.0009	5.5118	5.5134	.0012
145	5.7087	5.7079	5.7087	5.7097	.0009	5.7087	5.7103	.0012
150	5.9055	5.9047	5.9055	5.9065	.0009	5.9055	5.9071	.0012
160	6.2992	6.2982	6.2992	6.3002	.0009	6.2992	6.3008	.0013
170	6.6929	6.6919	6.6929	6.6939	.0009	6.6929	6.6945	.0013
180	7.0866	7.0856	7.0866	7.0876	.0009	7.0866	7.0882	.0013
190	7.4803	7.4791	7.4803	7.4814	.0012	7.4803	7.4821	.0015
200	7.8740	7.8728	7.8740	7.8751	.0012	7.8740	7.8758	.0015
210	8.2677	8.2665	8.2677	8.2688	.0012	8.2677	8.2695	.0015
215	8.4646	8.4634	8.4646	8.4657	.0012	8.4646	8.4664	.0015

continued on next page

TABLE 7–30, continued

For loads where the housing rotates with relation to the direction of load.			Where the direction of the load is indeterminate						
Housing			Light loads, solid housing or split housing			Normal loads, solid housing only			
Bore Inches		Mean Fit	Bore Inches		Mean Fit	Bore Inches		Mean Fit	
min	max	Tight	min	max	Loose	min	max	Loose	mm
.6292	.6299	.0002	.6296	.6303	.0003	.6294	.6301	.0000	16
.7472	.7480		.7476	.7485	.0003	.7474	.7482	.0000	19
.8653	.8661		.8657	.8666	.0003	.8655	.8663	.0000	22
.9441	.9449		.9445	.9454	.0003	.9443	.9451	.0000	24
1.0228	1.0236		1.0232	1.0241	.0003	1.0230	1.0238	.0000	26
1.1803	1.1811		1.1807	1.1816	.0003	1.1805	1.1813	.0000	30
1.2588	1.2598	.0003	1.2594	1.2604	.0003	1.2591	1.2601	.0000	32
1.3770	1.3780		1.3776	1.3786	.0003	1.3773	1.3783	.0000	35
1.4557	1.4567		1.4563	1.4573	.0003	1.4560	1.4570	.0000	37
1.5738	1.5748		1.5744	1.5754	.0003	1.5741	1.5751	.0000	40
1.6525	1.6535		1.6531	1.6541	.0003	1.6528	1.6538	.0000	42
1.8494	1.8504		1.8500	1.8510	.0003	1.8497	1.8507	.0000	47
2.0460	2.0472		2.0467	2.0479	.0004	2.0464	2.0476	.0000	52
2.1642	2.1654		2.1649	2.1661	.0004	2.1646	2.1658	.0000	55
2.4397	2.4409		2.4404	2.4416	.0004	2.4401	2.4413	.0000	62
2.6760	2.6772		2.6767	2.6779	.0004	2.6764	2.6776	.0000	68
2.8334	2.8346		2.8341	2.8353	.0004	2.8338	2.8350	.0000	72
2.9516	2.9528		2.9523	2.9535	.0004	2.9520	2.9532	.0000	75
3.1484	3.1496		3.1491	3.1503	.0004	3.1488	3.1500	.0000	80
3.3451	3.3465	.0004	3.3460	3.3474	.0005	3.3455	3.3469	.0000	85
3.5419	3.5433		3.5428	3.5442	.0005	3.5423	3.5437	.0000	90
3.7388	3.7402		3.7397	3.7411	.0005	3.7392	3.7406	.0000	95
3.9356	3.9370		3.9365	3.9379	.0005	3.9360	3.9374	.0000	100
4.3293	4.3307		4.3302	4.3316	.0005	4.3297	4.3311	.0000	110
4.5262	4.5276		4.5271	4.5285	.0005	4.5266	4.5280	.0000	115
4.7230	4.7244		4.7239	4.7253	.0005	4.7234	4.7248	.0000	120
4.9197	4.9213		4.9207	4.9223	.0006	4.9202	4.9218	.0001	125
5.1165	5.1181		5.1175	5.1191	.0006	5.1170	5.1186	.0001	130
5.5102	5.5118		5.5112	5.5128	.0006	5.5107	5.5123	.0001	140
5.7071	5.7087		5.7081	5.7097	.0006	5.7076	5.7092	.0001	145
5.9039	5.9055		5.9049	5.9065	.0006	5.9044	5.9060	.0001	150
6.2976	6.2992	.0003	6.2986	6.3002	.0007	6.2981	6.2997	.0002	160
6.6913	6.6929		6.6923	6.6939	.0007	6.6918	6.6934	.0002	170
7.0850	7.0866		7.0860	7.0876	.0007	7.0855	7.0871	.0002	180
7.4785	7.4803		7.4797	7.4815	.0009	7.4790	7.4808	.0002	190
7.8722	7.8740		7.8734	7.8752	.0009	7.8727	7.8745	.0002	200
8.2659	8.2677		8.2671	8.2689	.0009	8.2664	8.2682	.0002	210
8.4628	8.4646		8.4640	8.4658	.0009	8.4633	8.4651	.0002	215

continued on next page

TABLE 7–30, continued

For normal loads where housing is stationary with relation to direction of load.

Bearing			Solid Housing			Split Housing		
Outside Diameter Inches			Bore Inches		Mean Fit	Bore Inches		Mean Fit
mm	max	min	min	max	Loose	min	max	Loose
220	8.6614	8.6602	8.6614	8.6625	.0012	8.6614	8.6632	.0015
225	8.8583	8.8571	8.8583	8.8594	.0012	8.8583	8.8601	.0015
230	9.0551	9.0539	9.0551	9.0562	.0012	9.0551	9.0569	.0015
240	9.4488	9.4476	9.4488	9.4499	.0012	9.4488	9.4506	.0015
250	9.8425	9.8413	9.8425	9.8436	.0012	9.8425	9.8443	.0015
260	10.2362	10.2348	10.2362	10.2375	.0014	10.2362	10.2382	.0017
265	10.4331	10.4317	10.4331	10.4344	.0014	10.4331	10.4351	.0017
270	10.6299	10.6285	10.6299	10.6312	.0014	10.6299	10.6319	.0017
280	11.0236	11.0222	11.0236	11.0249	.0014	11.0236	11.0256	.0017
290	11.4173	11.4159	11.4173	11.4186	.0014	11.4173	11.4193	.0017
300	11.8110	11.8096	11.8110	11.8123	.0014	11.8110	11.8130	.0017
310	12.2047	12.2033	12.2047	12.2060	.0014	12.2047	12.2067	.0017
320	12.5984	12.5968	12.5984	12.5998	.0015	12.5984	12.6006	.0019
330	12.9921	12.9905	12.9921	12.9935	.0015	12.9921	12.9943	.0019
340	13.3858	13.3842	13.3858	13.3872	.0015	13.3858	13.3880	.0019
350	13.7795	13.7779	13.7795	13.7809	.0015	13.7795	13.7817	.0019
360	14.1732	14.1716	14.1732	14.1746	.0015	14.1732	14.1754	.0019
370	14.5669	14.5653	14.5669	14.5683	.0015	14.5669	14.5691	.0019
380	14.9606	14.9590	14.9606	14.9620	.0015	14.9606	14.9628	.0019
390	15.3543	15.3527	15.3543	15.3557	.0015	15.3543	15.3565	.0019
400	15.7480	15.7464	15.7480	15.7494	.0015	15.7480	15.7502	.0019
410	16.1417	16.1399	16.1417	16.1433	.0017	16.1417	16.1442	.0022
420	16.5354	16.5336	16.5354	16.5370	.0017	16.5354	16.5379	.0022
430	16.9291	16.9273	16.9291	16.9307	.0017	16.9291	16.9316	.0022
440	17.3228	17.3210	17.3228	17.3244	.0017	17.3228	17.3253	.0022
450	17.7165	17.7147	17.7165	17.7181	.0017	17.7165	17.7190	.0022
460	18.1102	18.1084	18.1102	18.1118	.0017	18.1102	18.1127	.0022
465	18.3071	18.3053	18.3071	18.3087	.0017	18.3071	18.3096	.0022
480	18.8976	18.8958	18.8976	18.8992	.0017	18.8976	18.9001	.0022
490	19.2913	19.2895	19.2913	19.2929	.0017	19.2913	19.2938	.0022
500	19.6850	19.6832	19.6850	19.6866	.0017	19.6850	19.6875	.0022
520	20.4724	20.4704	20.4724	20.4742	.0019	20.4724	20.4751	.0024
530	20.8661	20.8641	20.8661	20.8679	.0019	20.8661	20.8688	.0024
540	21.2598	21.2578	21.2598	21.2616	.0019	21.2598	21.2625	.0024
560	22.0472	22.0452	22.0472	22.0490	.0019	22.0472	22.0499	.0024
580	22.8346	22.8326	22.8346	22.8364	.0019	22.8346	22.8373	.0024
590	23.2283	23.2263	23.2283	23.2301	.0019	23.2283	23.2310	.0024
600	23.6220	23.6200	23.6220	23.6238	.0019	23.6220	23.6247	.0024
620	24.4094	24.4074	24.4094	24.4112	.0019	24.4094	24.4121	.0024

continued on next page

TABLE 7–30, continued

For loads where the housing rotates with relation to the direction of the load.			Where the direction of the load is indeteminate						
Housing			Light loads, solid housing or split housing			Normal loads, solid housing only			
Bore Inches min	Bore Inches max	Mean Fit Tight	Bore Inches min	Bore Inches max	Mean Fit Loose	Bore Inches min	Bore Inches max	Mean Fit Loose	mm
8.6596	8.6614		8.6608	8.6626	.0009	8.6601	8.6619	.0002	220
8.8565	8.8583		8.8577	8.8595	.0009	8.8570	8.8588	.0002	225
9.0533	9.0551		9.0545	9.0563	.0009	9.0538	9.0556	.0002	230
9.4470	9.4488		9.4482	9.4500	.0009	9.4475	9.4493	.0002	240
9.8407	9.8425		9.8419	9.8437	.0009	9.8412	9.8430	.0002	250
10.2342	10.2362		10.2356	10.2376	.0011	10.2348	10.2368	.0003	260
10.4311	10.4331		10.4325	10.4345	.0011	10.4317	10.4337	.0003	265
10.6279	10.6299		10.6293	10.6313	.0011	10.6285	10.6305	.0003	270
11.0216	11.0236		11.0230	11.0250	.0011	11.0222	11.0242	.0003	280
11.4153	11.4173		11.4167	11.4187	.0011	11.4159	11.4179	.0003	290
11.8090	11.8110		11.8104	11.8124	.0011	11.8096	11.8116	.0003	300
12.2027	12.2047		12.2041	12.2061	.0011	12.2033	12.2053	.0003	310
12.5962	12.5984		12.5977	12.5999	.0012	12.5968	12.5991	.0004	320
12.9899	12.9921		12.9914	12.9936	.0012	12.9905	12.9928	.0004	330
13.3836	13.3858		13.3851	13.3873	.0012	13.3842	13.3865	.0004	340
13.7773	13.7795		13.7788	13.7810	.0012	13.7779	13.7802	.0004	350
14.1710	14.1732		14.1725	14.1747	.0012	14.1716	14.1739	.0004	360
14.5647	14.5669		14.5662	14.5684	.0012	14.5653	14.5676	.0004	370
14.9584	14.9606		14.9599	14.9621	.0012	14.9590	14.9613	.0004	380
15.3521	15.3543		15.3536	15.3558	.0012	15.3527	15.3550	.0004	390
15.7458	15.7480		15.7473	15.7495	.0012	15.7464	15.7487	.0004	400
16.1392	16.1417		16.1409	16.1434	.0014	16.1399	16.1424	.0004	410
16.5329	16.5354		16.5346	16.5371	.0014	16.5336	16.5361	.0004	420
16.9266	16.9291		16.9283	16.9308	.0014	16.9273	16.9298	.0004	430
17.3203	17.3228		17.3220	17.3245	.0014	17.3210	17.3235	.0004	440
17.7140	17.7165		17.7157	17.7182	.0014	17.7147	17.7172	.0004	450
18.1077	18.1102		18.1094	18.1119	.0014	18.1084	18.1109	.0004	460
18.3046	18.3071		18.3063	18.3088	.0014	18.3053	18.3078	.0004	465
18.8951	18.8976		18.8968	18.8993	.0014	18.8958	18.8983	.0004	480
19.2888	19.2913		19.2905	19.2930	.0014	19.2895	19.2920	.0004	490
19.6825	19.6850		19.6842	19.6867	.0014	19.6832	19.6857	.0004	500
20.4697	20.4724	.0004	20.4715	20.4742	.0015	20.4705	20.4732	.0005	520
20.8634	20.8661		20.8652	20.8679	.0015	20.8642	20.8669	.0005	530
21.2571	21.2598		21.2589	21.2616	.0015	21.2579	21.2606	.0005	540
22.0445	22.0472		22.0463	22.0490	.0015	22.0453	22.0480	.0005	560
22.8319	22.8346		22.8337	22.8364	.0015	22.8327	22.8354	.0005	580
23.2256	23.2283		23.2274	23.2301	.0015	23.2264	23.2291	.0005	590
23.6193	23.6220		23.6211	23.6238	.0015	23.6201	23.6228	.0005	600
24.4067	24.4094		24.4085	24.4112	.0015	24.4075	24.4102	.0005	620

TABLE 7–31

ABEC–3 Shaft Diameters for Bearing Seats of Metric Annular Ball Bearings AFBMA Standards, Section 7 – March 1951.

	LIGHT LOADS For light loads where the shaft rotates in Relation to the direction of the load or where the direction of the load is indeterminate.					NORMAL LOADS Shaft rotating with relation to the direction of the load.		
	Bearing Bore Inches		Shaft Diameter Inches			Shaft Diameter Inches		
mm	max	min	max	min	Mean fit	max	min	Mean fit
4	.1575	.1573	.1575	.1573	T.0000	.1577	.1575	T.0002
5	.1969	.1967	.1969	.1967		.1971	.1969	
6	.2362	.2360	.2362	.2360		.2364	.2362	
7	.2756	.2754	.2756	.2754		.2758	.2755	.00015
8	.3150	.3148	.3150	.3148		.3152	.3149	
9	.3543	.3541	.3543	.3541		.3545	.3542	
10	.3937	.3935	.3937	.3935		.3939	.3936	
12	.4724	.4722	.4724	.4721	.00005	.4726	.4723	
15	.5906	.5904	.5906	.5903		.5908	.5905	
17	.6693	.6691	.6693	.6690		.6695	.6692	
20	.7874	.7872	.7878	.7872	.0002	.7878	.7875	.00035
25	.9843	.9841	.9847	.9841		.9847	.9844	
30	1.1811	1.1809	1.1815	1.1809		1.1815	1.1812	
35	1.3780	1.3777	1.3784	1.3778	.00025	1.3785	1.3781	.00045
40	1.5748	1.5745	1.5752	1.5746		1.5753	1.5749	
45	1.7717	1.7714	1.7721	1.7715		1.7722	1.7718	
50	1.9685	1.9682	1.9689	1.9683		1.9690	1.9686	
55	2.1654	2.1650	2.1659	2.1651	.0003	2.1660	2.1655	.00055
60	2.3622	2.3618	2.3627	2.3619		2.3628	2.3623	
65	2.5591	2.5587	2.5596	2.5588		2.5597	2.5592	
70	2.7559	2.7555	2.7564	2.7556		2.7565	2.7560	
75	2.9528	2.9524	2.9533	2.9525		2.9534	2.9529	
80	3.1496	3.1492	3.1501	3.1493		3.1502	3.1497	
85	3.3465	3.3460	3.3470	3.3461		3.3472	3.3466	.00065
90	3.5433	3.5428	3.5438	3.5429		3.5440	3.5434	
95	3.7402	3.7397	3.7407	3.7398		3.7409	3.7403	
100	3.9370	3.9365	3.9375	3.9366		3.9377	3.9371	
105	4.1339	4.1334	4.1344	4.1335		4.1346	4.1340	
110	4.3307	4.3302	4.3312	4.3303		4.3314	4.3308	
120	4.7244	4.7239	4.7249	4.7240		4.7251	4.7245	
130	5.1181	5.1175	5.1187	5.1177	.0004	5.1189	5.1182	.00075
140	5.5118	5.5112	5.5124	5.5114		5.5126	5.5119	
150	5.9055	5.9049	5.9061	5.9051		5.9063	5.9056	
160	6.2992	6.2986	6.2998	6.2988		6.3000	6.2993	
170	6.6929	6.6923	6.6935	6.6925		6.6937	6.6930	
180	7.0866	7.0860	7.0872	7.0862		7.0874	7.0867	
190	7.4803	7.4796	7.4809	7.4798		7.4813	7.4805	.0009
200	7.8740	7.8733	7.8746	7.8735		7.8750	7.8742	
210	8.2677	8.2670	8.2683	8.2672		8.2687	8.2679	
220	8.6614	8.6607	8.6620	8.6609		8.6624	8.6616	
230	9.0551	9.0544	9.0557	9.0546		9.0561	9.0553	
240	9.4488	9.4481	9.4494	9.4483		9.4498	9.4490	
260	10.2362	10.2354	10.2368	10.2356		10.2373	10.2364	.00105
280	11.0236	11.0228	11.0242	11.0230		11.0247	11.0238	
300	11.8110	11.8102	11.8116	11.8104		11.8121	11.8112	

continued on next page

TABLE 7–31, continued

NORMAL LOADS Shaft stationary with relation to the direction of the load.			Heavy Loads Shaft rotating with relation to the direction of the load.			
Shaft Diameter Inches			Shaft Diameter Inches			
max	min	Mean fit	max	min	Mean fit	mm
.1573	.1570	L.00025			T	4
.1967	.1964					5
.2360	.2357					6
.2754	.2750	.0003				7
.3148	.3144					8
.3541	.3537					9
.3935	.3931			SAME AS NORMAL LOAD		10
.4722	.4717	.00035				12
.5904	.5899					15
.6691	.6686					17
.7871	.7866	.00045				20
.9840	.9835					25
1.1808	1.1803					30
1.3776	1.3770	.00055				35
1.5744	1.5738					40
1.7713	1.7707					45
1.9681	1.9675					50
2.1650	2.1643					55
2.3618	2.3611					60
2.5587	2.5580					65
2.7555	2.7548					70
2.9524	2.9517					75
3.1492	3.1485					80
3.3460	3.3452	.00065				85
3.5428	3.5420					90
3.7397	3.7389					95
3.9365	3.9357					100
4.1334	4.1326		4.1350	4.1344	.00105	105
4.3302	4.3294		4.3318	4.3312		110
4.7239	4.7231		4.7255	4.7249		120
5.1175	5.1166	.00075	5.1194	5.1187	.00125	130
5.5112	5.5103		5.5131	5.5124		140
5.9049	5.9040		5.9068	5.9061		150
6.2986	6.2977		6.3005	6.2998		160
6.6923	6.6914		6.6942	6.6935		170
7.0860	7.0851		7.0879	7.0872		180
7.4797	7.4786	.0008	7.4818	7.4810	.00145	190
7.8734	7.8723		7.8755	7.8747		200
8.2671	8.2660		8.2692	8.2684		210
8.6608	8.6597		8.6629	8.6621		220
9.0545	9.0534		9.0566	9.0558		230
9.4482	9.4471		9.4503	9.4495		240
10.2355	10.2343	.0009	10.2379	10.2370	.00165	260
11.0229	11.0217		11.0253	11.0244		280
11.8103	11.8091		11.8127	11.8118		300

TABLE 7–32

ABEC–3 Housing Seat Diameters for Metric Annular Ball Bearings

AFBMA Standards, Section 7 – March 1951

For normal load where housing is stationary with relation to the direction of load.

Bearing Outside Diameter Inches			Solid Housing Bore Inches			Split Housing Bore Inches		
mm	max	min	min	max	Mean Fit Loose	min	max	Mean Fit Loose
16	.6299	.6296	.6299	.6303	.00035	.6299	.6306	.00055
19	.7480	.7477	.7480	.7485	.0004	.7480	.7488	.00055
22	.8661	.8658	.8661	.8666	.0004	.8661	.8669	.00055
24	.9449	.9446	.9449	.9454	.0004	.9449	.9457	.00055
26	1.0236	1.0233	1.0236	1.0241	.0004	1.0236	1.0244	.00055
30	1.1811	1.1808	1.1811	1.1816	.0004	1.1811	1.1819	.00055
32	1.2598	1.2595	1.2598	1.2604	.00045	1.2598	1.2608	.00065
35	1.3780	1.3777	1.3780	1.3786	.00045	1.3780	1.3790	.00065
37	1.4567	1.4564	1.4567	1.4573	.00045	1.4567	1.4577	.00065
40	1.5748	1.5745	1.5748	1.5754	.00045	1.5748	1.5758	.00065
42	1.6535	1.6532	1.6535	1.6541	.00045	1.6535	1.6545	.00065
47	1.8504	1.8501	1.8504	1.8510	.00045	1.8504	1.8514	.00065
52	2.0472	2.0468	2.0472	2.0479	.00055	2.0472	2.0484	.00080
55	2.1654	2.1650	2.1654	2.1661	.00055	2.1654	2.1666	.00080
62	2.4409	2.4405	2.4409	2.4416	.00055	2.4409	2.4421	.00080
68	2.6772	2.6768	2.6772	2.6779	.00055	2.6772	2.6784	.00080
72	2.8346	2.8342	2.8346	2.8353	.00055	2.8346	2.8358	.00080
75	2.9528	2.9524	2.9528	2.9535	.00055	2.9528	2.9540	.00080
80	3.1496	3.1492	3.1496	3.1503	.00055	3.1496	3.1508	.00080
85	3.3465	3.3461	3.3465	3.3474	.00065	3.3465	3.3479	.00090
90	3.5433	3.5429	3.5433	3.5442	.00065	3.5433	3.5447	.00090
95	3.7402	3.7398	3.7402	3.7411	.00065	3.7402	3.7416	.00090
100	3.9370	3.9366	3.9370	3.9379	.00065	3.9370	3.9384	.00090
110	4.3307	4.3303	4.3307	4.3316	.00065	4.3307	4.3321	.00090
115	4.5276	4.5272	4.5276	4.5285	.00065	4.5276	4.5290	.00090
120	4.7244	4.7240	4.7244	4.7253	.00065	4.7244	4.7258	.00090
125	4.9213	4.9208	4.9213	4.9223	.00075	4.9213	4.9229	.00105
130	5.1181	5.1176	5.1181	5.1191	.00075	5.1181	5.1197	.00105
140	5.5118	5.5113	5.5118	5.5128	.00075	5.5118	5.5134	.00105
145	5.7087	5.7082	5.7087	5.7097	.00075	5.7087	5.7103	.00105
150	5.9055	5.9050	5.9055	5.9065	.00075	5.9055	5.9071	.00105
160	6.2992	6.2986	6.2992	6.3002	.00080	6.2992	6.3008	.00110
170	6.6929	6.6923	6.6929	6.6939	.00080	6.6929	6.6945	.00110
180	7.0866	7.0860	7.0866	7.0876	.00080	7.0866	7.0882	.00110
190	7.4803	7.4796	7.4803	7.4814	.00090	7.4803	7.4821	.00125
200	7.8740	7.8733	7.8740	7.8751	.00090	7.8740	7.8758	.00125
210	8.2677	8.2670	8.2677	8.2688	.00090	8.2677	8.2695	.00125
215	8.4646	8.4639	8.4646	8.4657	.00090	8.4646	8.4664	.00125

continued on next page

TABLE 7–32, continued

For normal load where the housing rotates with relation to the direction of load.			Where the direction of the load is indeterminate.						
Housing Bore Inches			Housing, Solid or Split Bore Inches			Housing, Solid Only Bore Inches			
min	max	Mean Fit Tight	min	max	Mean Fit Tight	min	max	Mean Fit Loose	mm
.6292	.6299	.0002	.6296	.6303	.0002	.6294	.6301	.000	16
.7472	.7480	.00025	.7476	.7485	.0002	.7474	.7482	.00005	19
.8653	.8661	.00025	.8657	.8666	.0002	.8655	.8663	.00005	22
.9441	.9449	.00025	.9445	.9454	.0002	.9443	.9451	.00005	24
1.0228	1.0236	.00025	1.0232	1.0241	.0002	1.0230	1.0238	.00005	26
1.1803	1.1811	.00025	1.1807	1.1816	.0002	1.1805	1.1813	.00005	30
1.2588	1.2598	.00035	1.2594	1.2604	.00025	1.2591	1.2601	.00005	32
1.3770	1.3780	.00035	1.3776	1.3786	.00025	1.3773	1.3783	.00005	35
1.4557	1.4567	.00035	1.4563	1.4573	.00025	1.4560	1.4570	.00005	37
1.5738	1.5748	.00035	1.5744	1.5754	.00025	1.5741	1.5751	.00005	40
1.6525	1.6535	.00035	1.6531	1.6541	.00025	1.6528	1.6538	.00005	42
1.8494	1.8504	.00035	1.8500	1.8510	.00025	1.8497	1.8507	.00005	47
2.0460	2.0472	.00040	2.0467	2.0479	.00030	2.0464	2.0476	.0000	52
2.1642	2.1654	.00040	2.1649	2.1661	.00030	2.1646	2.1658	.0000	55
2.4397	2.4409	.00040	2.4404	2.4416	.00030	2.4401	2.4413	.0000	62
2.6760	2.6772	.00040	2.6767	2.6779	.00030	2.6764	2.6776	.0000	68
2.8334	2.8346	.00040	2.8341	2.8353	.00030	2.8338	2.8350	.0000	72
2.9516	2.9528	.00040	2.9523	2.9535	.00030	2.9520	2.9532	.0000	75
3.1484	3.1496	.00040	3.1491	3.1503	.00030	3.1488	3.1500	.0000	80
3.3451	3.3465	.00050	3.3460	3.3474	.00040	3.3455	3.3469	.0001	85
3.5419	3.5433	.00050	3.5428	3.5442	.00040	3.5423	3.5437	.0001	90
3.7388	3.7402	.00050	3.7397	3.7411	.00040	3.7392	3.7406	.0001	95
3.9356	3.9370	.00050	3.9365	3.9379	.00040	3.9360	3.9374	.0001	100
4.3293	4.3307	.00050	4.3302	4.3316	.00040	4.3297	4.3311	.0001	110
4.5262	4.5276	.00050	4.5271	4.5285	.00040	4.5266	4.5280	.0001	115
4.7230	4.7244	.00050	4.7239	4.7253	.00040	4.7234	4.7248	.0001	120
4.9197	4.9213	.00055	4.9207	4.9223	.00045	4.9202	4.9218	.00005	125
5.1165	5.1181	.00055	5.1175	5.1191	.00045	5.1170	5.1186	.00005	130
5.5102	5.5118	.00055	5.5112	5.5128	.00045	5.5107	5.5123	.00005	140
5.7071	5.7087	.00055	5.7081	5.7097	.00045	5.7076	5.7092	.00005	145
5.9039	5.9055	.00055	5.9049	5.9065	.00045	5.9044	5.9060	.00005	150
6.2976	6.2992	.00050	6.2986	6.3002	.00050	6.2981	6.2997	.0000	160
6.6913	6.6929	.00050	6.6923	6.6939	.00050	6.6918	6.6934	.0000	170
7.0850	7.0866	.00050	7.0860	7.0876	.00050	7.0855	7.0871	.0000	180
7.4785	7.4803	.00055	7.4797	7.4815	.00065	7.4790	7.4808	.00005	190
7.8722	7.8740	.00055	7.8734	7.8752	.00065	7.8727	7.8745	.00005	200
8.2659	8.2677	.00055	8.2671	8.2689	.00065	8.2664	8.2682	.00005	210
8.4628	8.4646	.00055	8.4640	8.4658	.00065	8.4633	8.4651	.00005	215

continued on next page

TABLE 7–32, continued

For normal load where housing is stationary with relation to the direction of load.

Bearing Outside Diameter	Inches		Solid Housing Bore Inches			Split Housing Bore Inches		
mm	max	min	min	max	Mean Fit Loose	min	max	Mean Fit Loose
220	8.6614	8.6607	8.6614	8.6625	.00090	8.6614	8.6632	.00125
225	8.8583	8.8576	8.8583	8.8594	.00090	8.8583	8.8601	.00125
230	9.0551	9.0544	9.0551	9.0562	.00090	9.0551	9.0569	.00125
240	9.4488	9.4481	9.4488	9.4499	.00090	9.4488	9.4506	.00125
250	9.8425	9.8418	9.8425	9.8436	.00090	9.8425	9.8443	.00125
260	10.2362	10.2354	10.2362	10.2375	.00105	10.2362	10.2382	.00140
265	10.4331	10.4323	10.4331	10.4344	.00105	10.4331	10.4351	.00140
270	10.6299	10.6291	10.6299	10.6312	.00105	10.6299	10.6319	.00140
280	11.0236	11.0228	11.0236	11.0249	.00105	11.0236	11.0256	.00140
290	11.4173	11.4165	11.4173	11.4186	.00105	11.4173	11.4193	.00140
300	11.8110	11.8102	11.8110	11.8123	.00105	11.8110	11.8130	.00140
310	12.2047	12.2039	12.2047	12.2060	.00105	12.2047	12.2067	.00140
320	12.5984	12.5975	12.5984	12.5998	.00115	12.5984	12.6007	.00165
330	12.9921	12.9912	12.9921	12.9935	.00115	12.9921	12.9944	.00165
340	13.3858	13.3849	13.3858	13.3872	.00115	13.3858	13.3881	.00165
350	13.7795	13.7786	13.7795	13.7809	.00115	13.7795	13.7818	.00165
360	14.1732	14.1723	14.1732	14.1744	.00115	14.1732	14.1755	.00165
370	14.5669	14.5660	14.5669	14.5683	.00115	14.5669	14.5692	.00165
380	14.9606	14.9597	14.9606	14.9620	.00115	14.9606	14.9629	.00165
390	15.3543	15.3534	15.3543	15.3557	.00115	15.3543	15.3566	.00165
400	15.7480	15.7471	15.7480	15.7494	.00115	15.7480	15.7503	.00165
410	16.1417	16.1407	16.1417	16.1433	.00130	16.1417	16.1442	.00175
420	16.5354	16.5344	16.5354	16.5370	.00130	16.5354	16.5379	.00175
430	16.9291	16.9281	16.9291	16.9307	.00130	16.9291	16.9316	.00175
440	17.3228	17.3218	17.3228	17.3244	.00130	17.3228	17.3253	.00175
450	17.7165	17.7155	17.7165	17.7181	.00130	17.7165	17.7190	.00175
460	18.1102	18.1092	18.1102	18.1118	.00130	18.1102	18.1127	.00175
465	18.3071	18.3061	18.3071	18.3087	.00130	18.3071	18.3096	.00175
480	18.8976	18.8966	18.8976	18.8992	.00130	18.8976	18.9001	.00175
490	19.2913	19.2903	19.2913	19.2929	.00130	19.2913	19.2938	.00175
500	19.6850	19.6840	19.6850	19.6866	.00130	19.6850	19.6875	.00175

continued on next page

TABLE 7–32, continued

For normal load where the housing rotates with relation to the direction of load.			Where the direction of the load is indeteminate.						
Housing Bore Inches			Housing, Solid or Split Bore Inches			Housing, Solid Only Bore Inches			
min	max	Mean Fit Tight	min	max	Mean Fit Loose	min	max	Mean Fit Tight	mm
8.6596	8.6614	.00055	8.6608	8.6626	.00065	8.6601	8.6619	.00005	220
8.8565	8.8583	.00055	8.8577	8.8595	.00065	8.8570	8.8588	.00005	225
9.0533	9.0551	.00055	9.0545	9.0563	.00065	9.0538	9.0556	.00005	230
9.4470	9.4488	.00055	9.4482	9.4500	.00065	9.4475	9.4493	.00005	240
9.8407	9.8425	.00055	9.8419	9.8437	.00065	9.8412	9.8430	.00005	250
10.2342	10.2362	.00060	10.2356	10.2376	.00080	10.2348	10.2368	.0000	260
10.4311	10.4331	.00060	10.4325	10.4345	.00080	10.4317	10.4337	.0000	265
10.6279	10.6299	.00060	10.6293	10.6313	.00080	10.6285	10.6305	.0000	270
11.0216	11.0236	.00060	11.0230	11.0250	.00080	11.0222	11.0242	.0000	280
11.4153	11.4173	.00060	11.4167	11.4187	.00080	11.4159	11.4179	.0000	290
11.8090	11.8110	.00060	11.8104	11.8124	.00080	11.8096	11.8116	.0000	300
12.2027	12.2047	.00060	12.2041	12.2061	.00080	12.2033	12.2053	.0000	310
12.5961	12.5984	.00070	12.5977	12.6000	.00090	12.5968	12.5991	.0000	320
12.9898	12.9921	.00070	12.9914	12.9937	.00090	12.9872	12.9928	.0000	330
13.3835	13.3858	.00070	13.3851	13.3874	.00090	13.3842	13.3865	.0000	340
13.7772	13.7795	.00070	13.7788	13.7811	.00090	13.7779	13.7802	.0000	350
14.1709	14.1732	.00070	14.1725	14.1748	.00090	14.1716	14.1739	.0000	360
14.5646	14.5669	.00070	14.5662	14.5685	.00090	14.5653	14.5676	.0000	370
14.9583	14.9606	.00070	14.9599	14.9622	.00090	14.9596	14.9613	.0000	380
15.3520	15.3543	.00070	15.3536	15.3559	.00090	15.3527	15.3550	.0000	390
15.7457	15.7480	.00070	15.7473	15.7496	.00090	15.7464	15.7487	.0000	400
16.1392	16.1417	.00075	16.1409	16.1434	.00095	16.1399	16.1424	.00005	410
16.5329	16.5354	.00075	16.5346	16.5371	.00095	16.5336	16.5361	.00005	420
16.9266	16.9291	.00075	16.9283	16.9308	.00095	16.9273	16.9298	.00005	430
17.3203	17.3228	.00075	17.3220	17.3245	.00095	17.3210	17.3235	.00005	440
17.7140	17.7165	.00075	17.7157	17.7182	.00095	17.7147	17.7172	.00005	450
18.1077	18.1102	.00075	18.1094	18.1119	.00095	18.1084	18.1109	.00005	460
18.3046	18.3071	.00075	18.3062	18.3087	.00095	18.3053	18.3078	.00005	465
18.8951	18.8976	.00075	18.8968	18.8993	.00095	18.8958	18.8983	.00005	480
19.2888	19.2913	.00075	19.2905	19.2930	.00095	19.2895	19.2920	.00005	490
19.6825	19.6850	.00075	19.6842	19.6867	.00095	19.6832	19.6857	.00005	500

TABLE 7–33

ABEC–5 and ABEC–7 Housing Bearing Seat Diameters for Metric Annular Ball Bearings

AFBMA Standards, Section 7 – March 1951

HOUSING DIMENSIONS FOR ABEC–5 TOLERANCE BEARINGS

Housing is stationary with relation to the direction of the load.

mm	Bearing O D Inches max	Bearing O D Inches min	Housing Bore Inches max	Housing Bore Inches min	Mean Fit Loose
16	.6299	.6297	.6298	.6301	.00015
19	.7480	.7478	.7479	.7482	.00015
22	.8661	.8659	.8660	.8663	.00015
24	.9449	.9447	.9448	.9451	.00015
26	1.0236	1.0234	1.0235	1.0238	.00015
30	1.1811	1.1809	1.1810	1.1813	.00015
32	1.2598	1.2596	1.2597	1.2600	.00015
35	1.3780	1.3778	1.3779	1.3782	.00015
37	1.4567	1.4565	1.4566	1.4569	.00015
40	1.5748	1.5746	1.5747	1.5750	.00015
42	1.6535	1.6533	1.6534	1.6537	.00015
47	1.8504	1.8502	1.8503	1.8506	.00015
52	2.0472	2.0469	2.0471	2.0474	.0002
55	2.1654	2.1651	2.1653	2.1656	.0002
62	2.4409	2.4406	2.4408	2.4411	.0002
68	2.6772	2.6769	2.6771	2.6774	.0002
72	2.8346	2.8343	2.8345	2.8348	.0002
75	2.9528	2.9525	2.9527	2.9530	.0002
80	3.1496	3.1493	3.1495	3.1498	.0002
85	3.3465	3.3462	3.3464	3.3468	.00025
90	3.5433	3.5430	3.5432	3.5436	.00025
95	3.7402	3.7399	3.7401	3.7405	.00025
100	3.9370	3.9367	3.9369	3.9373	.00025
110	4.3307	4.3304	4.3306	4.3310	.00025
115	4.5276	4.5273	4.5275	4.5279	.00025
120	4.7244	4.7241	4.7243	4.7247	.00025
125	4.9213	4.9209	4.9211	4.9216	.00025
130	5.1181	5.1177	5.1179	5.1184	.00025
140	5.5118	5.5114	5.5116	5.5121	.00025
145	5.7087	5.7083	5.7085	5.7090	.00025
150	5.9055	5.9051	5.9053	5.9058	.00025
160	6.2992	6.2987	6.2990	6.2995	.0003
170	6.6929	6.6924	6.6927	6.6932	.0003
180	7.0866	7.0861	7.0864	7.0869	.0003

HOUSING DIMENSIONS FOR ABEC–7 TOLERANCE BEARINGS

Housing is stationary with relation to the direction of the load.

mm	Bearing O D Inches max	Bearing O D Inches min	Housing Bore Inches min	Housing Bore Inches max	Mean Fit Loose
16	.6299	.6297	.6298	.6301	.00015
19	.7480	.7478	.7479	.7482	.00015
22	.8661	.8659	.8660	.8663	.00015
24	.9449	.9447	.9448	.9451	.00015
26	1.0236	1.0234	1.0235	1.0238	.00015
30	1.1811	1.1809	1.1810	1.1813	.00015
32	1.2598	1.2596	1.2597	1.2600	.00015
35	1.3780	1.3778	1.3779	1.3782	.00015
37	1.4567	1.4565	1.4566	1.4569	.00015
40	1.5748	1.5746	1.5747	1.5750	.00015
42	1.6535	1.6533	1.6534	1.6537	.00015
47	1.8504	1.8502	1.8503	1.8506	.00015
52	2.0472	2.0470	2.0471	2.0474	.00015
55	2.1654	2.1652	2.1653	2.1656	.00015
62	2.4409	2.4407	2.4408	2.4411	.00015
68	2.6772	2.6770	2.6771	2.6774	.00015
72	2.8346	2.8344	2.8345	2.8348	.00015
75	2.9528	2.9526	2.9527	2.9530	.00015
80	3.1496	3.1494	3.1495	3.1498	.00015
85	3.3465	3.3462	3.3464	3.3468	.00025
90	3.5433	3.5430	3.5432	3.5436	.00025
95	3.7402	3.7399	3.7401	3.7405	.00025
100	3.9370	3.9367	3.9369	3.9373	.00025
110	4.3307	4.3304	4.3306	4.3310	.00025
115	4.5276	4.5273	4.5275	4.5279	.00025
120	4.7244	4.7241	4.7243	4.7247	.00025
125	4.9213	4.9209	4.9211	4.9216	.00025
130	5.1181	5.1177	5.1179	5.1184	.00025
140	5.5118	5.5114	5.5116	5.5121	.00025
145	5.7087	5.7083	5.7085	5.7090	.00025
150	5.9055	5.9051	5.9053	5.9058	.00025
160	6.2992	6.2988	6.2990	6.2995	.00025
170	6.6929	6.6925	6.6927	6.6932	.00025
180	7.0866	7.0862	7.0864	7.0869	.00025

continued on next page

TABLE 7-33 continued

HOUSING DIMENSIONS FOR ABEC-5 TOLERANCE BEARINGS

Housing is stationary with relation to the direction of the load.

mm	Bearing O D Inches max	Bearing O D Inches min	Housing Bore Inches min	Housing Bore Inches max	Mean Fit Loose
190	7.4803	7.4798	7.4801	7.4807	.00035
200	7.8740	7.8735	7.8738	7.8744	.00035
210	8.2677	8.2672	8.2675	8.2681	.00035
215	8.4646	8.4641	8.4644	8.4650	.00035
220	8.6614	8.6609	8.6612	8.6618	.00035
225	8.8583	8.8578	8.8581	8.8587	.00035
230	9.0551	9.0546	9.0549	9.0555	.00035
240	9.4488	9.4483	9.4486	9.4492	.00035
250	9.8425	9.8420	9.8423	9.8429	.00035
260	10.2362	10.2357	10.2360	10.2366	.00035
265	10.4331	10.4326	10.4329	10.4335	.00035
270	10.6299	10.6294	10.6297	10.6303	.00035
280	11.0236	11.0231	11.0234	11.0240	.00035
290	11.4173	11.4168	11.4171	11.4177	.00035
300	11.8110	11.8105	11.8108	11.8114	.00035
310	12.2047	12.2042	12.2045	12.2051	.00035
320	12.5984	12.5978	12.5981	12.5988	.00035
330	12.9921	12.9915	12.9918	12.9925	.00035
340	13.3858	13.3852	13.3855	13.3862	.00035
350	13.7795	13.7789	13.7792	13.7799	.00035
360	14.1732	14.1726	14.1729	14.1736	.00035
370	14.5669	14.5663	14.5666	14.5673	.00035
380	14.9606	14.9600	14.9603	14.9610	.00035
390	15.3543	15.3537	15.3540	15.3547	.00035
400	15.7480	15.7474	15.7477	15.7484	.00035
410	16.1417	16.1410	16.1414	16.1422	.00045
420	16.5354	16.5347	16.5351	16.5359	.00045
430	16.9291	16.9284	16.9288	16.9296	.00045
440	17.3228	17.3221	17.3225	17.3233	.00045
450	17.7165	17.7158	17.7162	17.7170	.00045
460	18.1102	18.1095	18.1099	18.1107	.00045
465	18.3071	18.3064	18.3068	18.3076	.00045
480	18.8976	18.8969	18.8973	18.8981	.00045
490	19.2913	19.2906	19.2910	19.2918	.00045
500	19.6850	19.6843	19.6847	19.6855	.00045

HOUSING DIMENSIONS FOR ABEC-7 TOLERANCE BEARINGS

Housing is stationary with relation to the direction of the load.

mm	Bearing O D Inches max	Bearing O D Inches min	Housing Bore Inches min	Housing Bore Inches max	Mean Fit Loose
190	7.4803	7.4799	7.4801	7.4807	.0003
200	7.8740	7.8736	7.8738	7.8744	.0003
210	8.2677	8.2673	8.2675	8.2681	.0003
215	8.4646	8.4642	8.4644	8.4650	.0003
220	8.6614	8.6610	8.6612	8.6618	.0003
225	8.8583	8.8579	8.8581	8.8587	.0003
230	9.0551	9.0547	9.0549	9.0555	.0003
240	9.4488	9.4484	9.4486	9.4492	.0003
250	9.8425	9.8421	9.8423	9.8429	.0003
260	10.2362	10.2357	10.2360	10.2366	.00035
265	10.4331	10.4326	10.4329	10.4335	.00035
270	10.6299	10.6294	10.6297	10.6303	.00035
280	11.0236	11.0231	11.0234	11.0240	.00035
290	11.4173	11.4168	11.4171	11.4177	.00035
300	11.8110	11.8105	11.8108	11.8114	.00035
310	12.2047	12.2042	12.2045	12.2051	.00035
320	12.5984	12.5979	12.5981	12.5988	.0003
330	12.9921	12.9916	12.9918	12.9925	.0003
340	13.3858	13.3853	13.3855	13.3862	.0003
350	13.7795	13.7790	13.7792	13.7799	.0003
360	14.1732	14.1727	14.1729	14.1736	.0003
370	14.5669	14.5664	14.5666	14.5673	.0003
380	14.9606	14.9601	14.9603	14.9610	.0003
390	15.3543	15.3538	15.3540	15.3547	.0003
400	15.7480	15.7475	15.7477	15.7484	.0003

TABLE 7–34

ABEC–7 and ABEC–5 Shaft Diameters for Bearing Seats of Metric Annular Ball Bearings

AFBMA Standards, Section 7 – March 1951

SHAFT DIMENSIONS FOR ABEC–7 TOLERANCE BEARINGS

For normal loads, shaft rotating with relation to direction of load

	Bearing Bore Inches		Shaft Diameter Inches		Mean Fit
mm	max	min	max	min	
4	.1575	.15735	.1575	.1573	L.000025
5	.1969	.19675	.1969	.1967	L.000025
6	.2362	.23605	.2362	.2360	L.000025
7	.2756	.27545	.2756	.2754	L.000025
8	.3150	.31485	.3150	.3148	L.000025
9	.3543	.35415	.3543	.3541	L.000025
10	.3937	.39355	.3937	.3935	L.000025
12	.4724	.47225	.4724	.4722	L.000025
15	.5906	.59045	.5906	.5904	L.000025
17	.6693	.66915	.6693	.6691	L.000025
20	.7874	.78725	.7875	.7873	T.000075
25	.9843	.98415	.9844	.9842	T.000075
30	1.1811	1.18095	1.1812	1.1810	T.000075
35	1.3780	1.37780	1.3782	1.3779	T.00015
40	1.5748	1.57460	1.5750	1.5747	T.00015
45	1.7717	1.77150	1.7719	1.7716	T.00015
50	1.9685	1.96830	1.9687	1.9684	T.00015
55	2.1654	2.16520	2.1656	2.1652	T.00010
60	2.3622	2.36200	2.3624	2.3620	T.00010
65	2.5591	2.55890	2.5593	2.5589	T.00010
70	2.7559	2.75570	2.7561	2.7557	T.00010
75	2.9528	2.95260	2.9530	2.9526	T.00010
80	3.1496	3.14940	3.1498	3.1494	T.00010
85	3.3465	3.34625	3.3467	3.3463	T.000125
90	3.5433	3.54305	3.5435	3.5431	T.000125
95	3.7402	3.73995	3.7404	3.7400	T.000125
100	3.9370	3.93675	3.9372	3.9368	T.000125
105	4.1339	4.13365	4.1341	4.1337	T.000125
110	4.3307	4.33045	4.3309	4.3305	T.000125
120	4.7244	4.72415	4.7246	4.7242	T.000125
130	5.1181	5.11780	5.1183	5.1179	T.00015
140	5.5118	5.51150	5.5120	5.5116	T.00015
150	5.9055	5.90520	5.9057	5.9053	T.00015
160	6.2992	6.29890	6.2994	6.2990	T.00015
170	6.6929	6.69260	6.6931	6.6927	T.00015
180	7.0866	7.08630	7.0868	7.0864	T.00015
190	7.4803	7.47990	7.4806	7.4800	T.00020
200	7.8740	7.87360	7.8743	7.8737	T.00020
210	8.2677	8.26730	8.2680	8.2674	T.00020
220	8.6614	8.66100	8.6617	8.6611	T.00020
230	9.0551	9.05470	9.0554	9.0548	T.00020
240	9.4488	9.44840	9.4491	9.4485	T.00020

SHAFT DIMENSIONS FOR ABEC–5 TOLERANCE BEARINGS

For normal loads, shaft rotating with relation to direction of load

	Bearing Bore Inches		Shaft Diameter Inches		Mean Fit
mm	max	min	max	min	Tight
4	.1575	.1573	.1575	.1573	.00000
5	.1969	.1967	.1969	.1967	.00000
6	.2362	.2360	.2362	.2360	.00000
7	.2756	.2754	.2756	.2754	.00000
8	.3150	.3148	.3150	.3148	.00000
9	.3543	.3541	.3543	.3541	.00000
10	.3937	.3935	.3937	.3935	.00000
12	.4724	.4722	.4724	.4722	.00000
15	.5906	.5904	.5906	.5904	.00000
17	.6693	.6691	.6693	.6691	.00000
20	.7874	.7872	.7875	.7873	.00010
25	.9843	.9841	.9844	.9842	.00010
30	1.1811	1.1809	1.1812	1.1810	.00010
35	1.3780	1.3778	1.3782	1.3779	.00015
40	1.5748	1.5746	1.5750	1.5747	.00015
45	1.7717	1.7715	1.7719	1.7716	.00015
50	1.9685	1.9683	1.9687	1.9684	.00015
55	2.1654	2.1651	2.1656	2.1652	.00015
60	2.3622	2.3619	2.3624	2.3620	.00015
65	2.5591	2.5588	2.5593	2.5589	.00015
70	2.7559	2.7556	2.7561	2.7557	.00015
75	2.9528	2.9525	2.9530	2.9526	.00015
80	3.1496	3.1493	3.1498	3.1494	.00015
85	3.3465	3.3462	3.3467	3.3463	.00015
90	3.5433	3.5430	3.5435	3.5431	.00015
95	3.7402	3.7399	3.7404	3.7400	.00015
100	3.9370	3.9367	3.9372	3.9368	.00015
105	4.1339	4.1336	4.1341	4.1337	.00015
110	4.3307	4.3304	4.3309	4.3305	.00015
120	4.7244	4.7241	4.7246	4.7242	.00015
130	5.1181	5.1177	5.1183	5.1179	.00020
140	5.5118	5.5114	5.5120	5.5116	.00020
150	5.9055	5.9051	5.9057	5.9053	.00020
160	6.2992	6.2988	6.2994	6.2990	.00020
170	6.6929	6.6925	6.6931	6.6927	.00020
180	7.0866	7.0862	7.0868	7.0864	.00020
190	7.4803	7.4798	7.4806	7.4800	.00025
200	7.8740	7.8735	7.8743	7.8737	.00025
210	8.2677	8.2672	8.2680	8.2674	.00025
220	8.6614	8.6609	8.6617	8.6611	.00025
230	9.0551	9.0546	9.0554	9.0548	.00025
240	9.4488	9.4483	9.4491	9.4485	.00025
260	10.2362	10.2357	10.2365	10.2359	.00025
280	11.0236	11.0231	11.0239	11.0233	.00025
300	11.8110	11.8105	11.8113	11.8107	.00025

TABLE 7–35

ABEC–1 Shaft Diameters for Bearing Seats of Inch Dimension Ball Bearings

AFBMA Standards, Section 7 – March 1951

LIGHT LOADS						NORMAL LOADS						HEAVY LOADS
For light loads where the shaft rotates in relation to the direction of the load or where the direction of the load is indeterminate.						Shaft rotating with relation to the direction of the load.			Shaft stationary with relation to the direction of the load.			Shaft rotating with relation to the direction of the load.
Bearing Bore Inches			Shaft Size Inches			Shaft Size Inches			Shaft Size Inches			
Fract.	max	min	max	min	Mean Fit	max	min	Mean Fit	max	min	Mean Fit	
1/8	.1250	.1247	.1250	.1248	T.0000	.1252	.1250	T.0002	.1248	.1245	L.0002	
3/16	.1875	.1872	.1875	.1873		.1877	.1875		.1873	.1870		
1/4	.2500	.2497	.2500	.2498		.2502	.2499		.2498	.2494	L.0003	SAME AS NORMAL LOAD
3/8	.3750	.3747	.3750	.3748		.3752	.3749		.3748	.3744		
1/2	.5000	.4997	.5000	.4997		.5002	.4999		.4998	.4993		
5/8	.6250	.6247	.6250	.6247		.6252	.6249		.6248	.6243		
3/4	.7500	.7496	.7504	.7498	T.0003	.7504	.7501	T.0005	.7497	.7492	L.0004	
7/8	.8750	.8746	.8754	.8748		.8754	.8751		.8747	.8742		
1	1.0000	.9996	1.0004	.9998		1.0004	1.0001		.9997	.9992		
1-1/8	1.1250	1.1246	1.1254	1.1248		1.1254	1.1251		1.1247	1.1242		
1-1/4	1.2500	1.2495	1.2504	1.2498	T.0004	1.2505	1.2501	T.0006	1.2496	1.2490	L.0005	
1-3/8	1.3750	1.3745	1.3754	1.3748		1.3755	1.3751		1.3746	1.3740		
1-1/2	1.5000	1.4995	1.5004	1.4998		1.5005	1.5001		1.4996	1.4990		

TABLE 7–36

ABEC–1 Housing Seat Diameters for Inch Dimension Ball Bearings

AFBMA Standards, Section 7 – March 1951

			For normal loads where housing is stationary with relation to direction of load.						For loads where the housing rotates with relation to the direction of load.			Where the direction of the load is indeterminate					
Bearing OD Inches			Solid Housing Bore Inches			Split Housing Bore Inches			Solid or Split Housing Bore Inches			LIGHT LOADS Solid or Split Housing Bore Inches			NORMAL LOADS Solid Housing Bore Inches		
Fraction	max	min	min	max	Mean Fit Loose	max	min	Mean Fit Loose	min	max	Mean Fit Tight	min	max	Mean Fit Loose	min	max	Mean Fit Loose
3/8	.3750	.3746	.3750	.3754	.0004	.3750	.3757	.0006	.3743	.3750	.0002	.3747	.3754	.0003	.3745	.3752	.0000
1/2	.5000	.4996	.5000	.5004		.5000	.5007		.4993	.5000		.4997	.5004		.4995	.5002	
5/8	.6250	.6246	.6250	.6254		.6250	.6257		.6243	.6250		.6247	.6254		.6245	.6252	
3/4	.7500	.7496	.7500	.7505		.7500	.7508		.7492	.7500		.7496	.7505		.7494	.7502	.0000
7/8	.8750	.8746	.8750	.8755		.8750	.8758		.8742	.8750		.8746	.8755		.8744	.8752	
1-1/8	1.1250	1.1246	1.1250	1.1255		1.1250	1.1258		1.1242	1.1250		1.1246	1.1255		1.1244	1.1252	
1-3/8	1.3750	1.3745	1.3750	1.3756	.0005	1.3750	1.3760	.0007	1.3740	1.3750	.0003	1.3746	1.3756		1.3743	1.3753	.0000
1-5/8	1.6250	1.6245	1.6250	1.6256		1.6250	1.6260		1.6240	1.6250		1.6246	1.6256		1.6243	1.6253	
1-7/8	1.8750	1.8745	1.8750	1.8756		1.8750	1.8760		1.8740	1.8750		1.8746	1.8756		1.8743	1.8753	
2	2.0000	1.9995	2.0000	2.0007	.0006	2.0000	2.0012	.0009	1.9988	2.0000		1.9995	2.0007	.0004	1.9992	2.0004	.0000
2-1/8	2.1250	2.1245	2.1250	2.1257		2.1250	2.1262		2.1238	2.1250		2.1245	2.1257		2.1242	2.1254	
2-1/4	2.2500	2.2495	2.2500	2.2507		2.2500	2.2512		2.2488	2.2500		2.2495	2.2507		2.2492	2.2504	
2-1/2	2.5000	2.4995	2.5000	2.5007		2.5000	2.5012		2.4988	2.5000		2.4995	2.5007		2.4992	2.5004	
2-5/8	2.6250	2.6245	2.6250	2.6257		2.6250	2.6262		2.6238	2.6250		2.6245	2.6257		2.6242	2.6254	

INDEX TO

SECTION 8

Keys and Keyseating

TABLE 8–1

American Standard, Square and Flat, Plain Parallel Stock Keys According to Shaft Sizes

ASA B17.1–1943

Shaft Diameter	Square Key *W* x *H**	Flat Key *W* x *H**	Shaft Diameter	Square Key *W* x *H**	Flat Key *W* x *H**
1/2	1/8 x 1/8	1/8 x 3/32	2-3/8	5/8 x 5/8	5/8 x 7/16
9/16	1/8 x 1/8	1/8 x 3/32	2-7/16	5/8 x 5/8	5/8 x 7/16
5/8	3/16 x 3/16	3/16 x 1/8	2-1/2	5/8 x 5/8	5/8 x 7/16
11/16	3/16 x 3/16	3/16 x 1/8	2-5/8	5/8 x 5/8	5/8 x 7/16
3/4	3/16 x 3/16	3/16 x 1/8	2-3/4	5/8 x 5/8	5/8 x 7/16
13/16	3/16 x 3/16	3/16 x 1/8	2-7/8	3/4 x 3/4	3/4 x 1/2
7/8	3/16 x 3/16	3/16 x 1/8	2-15/16	3/4 x 3/4	3/4 x 1/2
15/16	1/4 x 1/4	1/4 x 3/16	3	3/4 x 3/4	3/4 x 1/2
1	1/4 x 1/4	1/4 x 3/16	3-1/8	3/4 x 3/4	3/4 x 1/2
1-1/16	1/4 x 1/4	1/4 x 3/16	3-1/4	3/4 x 3/4	3/4 x 1/2
1-1/8	1/4 x 1/4	1/4 x 3/16	3-3/8	7/8 x 7/8	7/8 x 5/8
1-3/16	1/4 x 1/4	1/4 x 3/16	3-7/16	7/8 x 7/8	7/8 x 5/8
1-1/4	1/4 x 1/4	1/4 x 3/16	3-1/2	7/8 x 7/8	7/8 x 5/8
1-5/16	5/16 x 5/16	5/16 x 1/4	3-5/8	7/8 x 7/8	7/8 x 5/8
1-3/8	5/16 x 5/16	5/16 x 1/4	3-3/4	7/8 x 7/8	7/8 x 5/8
1-7/16	3/8 x 3/8	3/8 x 1/4	3-7/8	1 x 1	1 x 3/4
1-1/2	3/8 x 3/8	3/8 x 1/4	3-15/16	1 x 1	1 x 3/4
1-9/16	3/8 x 3/8	3/8 x 1/4	4	1 x 1	1 x 3/4
1-5/8	3/8 x 3/8	3/8 x 1/4	4-1/4	1 x 1	1 x 3/4
1-11/16	3/8 x 3/8	3/8 x 1/4	4-7/16	1 x 1	1 x 3/4
1-3/4	3/8 x 3/8	3/8 x 1/4	4-1/2	1 x 1	1 x 3/4
1-13/16	1/2 x 1/2	1/2 x 3/8	4-3/4	1-1/4 x 1-1/4	1-1/4 x 7/8
1-7/8	1/2 x 1/2	1/2 x 3/8	4-15/16	1-1/4 x 1-1/4	1-1/4 x 7/8
1-15/16	1/2 x 1/2	1/2 x 3/8	5	1-1/4 x 1-1/4	1-1/4 x 7/8
2	1/2 x 1/2	1/2 x 3/8	5-1/4	1-1/4 x 1-1/4	1-1/4 x 7/8
2-1/16	1/2 x 1/2	1/2 x 3/8	5-7/16	1-1/4 x 1-1/4	1-1/4 x 7/8
2-1/8	1/2 x 1/2	1/2 x 3/8	5-1/2	1-1/4 x 1-1/4	1-1/4 x 7/8
2-3/16	1/2 x 1/2	1/2 x 3/8	5-3/4	1-1/2 x 1-1/2	1-1/2 x 1
2-1/4	1/2 x 1/2	1/2 x 3/8	5-15/16	1-1/2 x 1-1/2	1-1/2 x 1
2-5/16	5/8 x 5/8	5/8 x 7/16	6	1-1/2 x 1-1/2	1-1/2 x 1

All dimensions given in inches.

*W and H are width and depth of key, respectively. Tolerances on width and depth of keys cut from cold-finished stock are given in Table 8–4.

TABLE 8–2

Dimensioning* Shaft Member to Fit Square, Flat and Taper Keys

Dimension	Illustration		Tolerance for Ease of Assembly: Interchangeable	Selective	Drive or Tight
Width	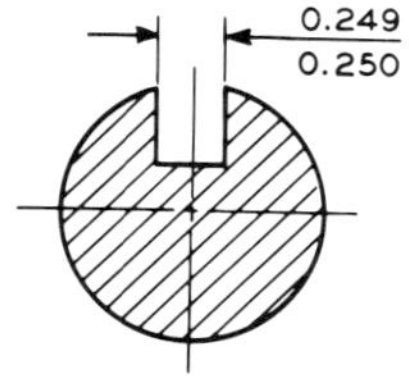		Positive tolerance† on slot equivalent to negative tolerance on key stock.	Same negative tolerance on slot as on key stock.	Interference from 0.0005 to 0.0020 or more for medium and large keys.
Depth	Case 1. When depth gage is likely to be used in checking depth, as a keyslot milled midway between the ends of a shaft, dimensioning the bottom of the slot with respect to the intersecting edges has advantages. Case 1 fulfills exactly the requirement that the key be sunk half the depth in each of the mating members.	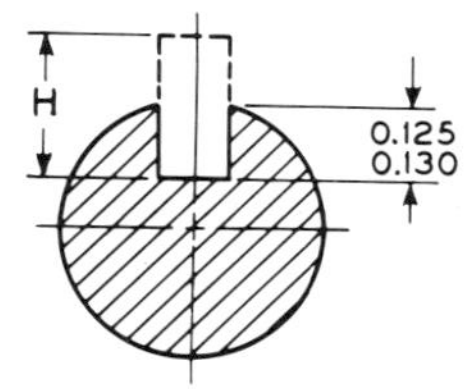 $\frac{H}{2}$ + tolerance −0.000	A tolerance of $+0.005$ -0.000 is practical for general design purposes. Square and flat keys generally are fitted snugly to the bottom and sides of the slot in the shaft member, with clearance at the top in the hub member. The taper for tapered keys is put in hub member. A well-fitted taper key has no clearance on the top, bottom or sides; hence closer tolerances on depth may be desired on slots for taper than for parallel keys.		
	Case 2. When depth of slot is likely to be determined by feeding cutter into shaft a known depth, then the dimension from bottom of slot to near side of shaft may be preferable.	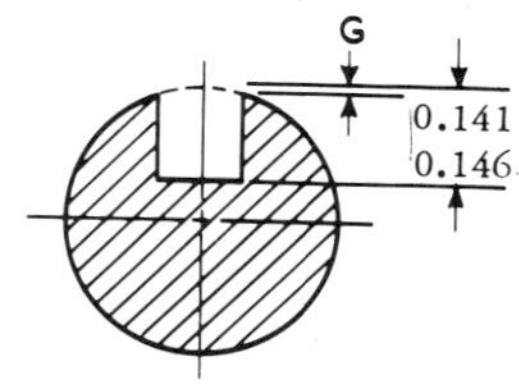 $\frac{H}{2} + G$ + tolerance −0.000 See Table 8-22 for G values.			

*To insure good fit, the width, depth and length of the keyslot need to be clearly dimensioned with tolerances. This practice facilitates inspection and absolves the designer of blame in case unsatisfactory fits happen because reasonable tolerances are not maintained during milling or slotting operations.

† If necessary, ground key stock, especially hard keys, can be held to closer tolerances than indicated in Table 8–4.

continued on next page

TABLE 8–2, continued

Depth (continued)

If keyslot is to be cut in a shaft which is afterward to be hardened and ground to finish diameter, the stock allowance for grinding influences the depth of keyslotting.

Case 3. When depth can be checked by micrometer calipers, as in an end of the shaft, then dimension S from the bottom of the slot to the far side of the shaft is a desirable dimension.

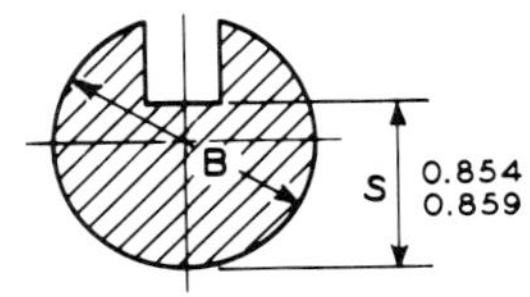

$$S = B - \frac{H}{2} - G \begin{array}{l} +0.000 \\ -\text{tolerance} \end{array}$$

or take from Table 8-5.

NOTE: Case 3 is the most generally recommended method of dimensioning by books and manuals. It definitely sidesteps any misunderstanding regarding the limits to the dimension; whereas Cases 1 and 2 can be confused unless the designer is careful in depicting the limits meant by the leader lines.

Let B = nominal finish diameter of shaft = 1.000
B' = maximum diameter of shaft with stock for grinding = 1.006

H = depth of key = 1/4
G = Table 8–22 = 0.016

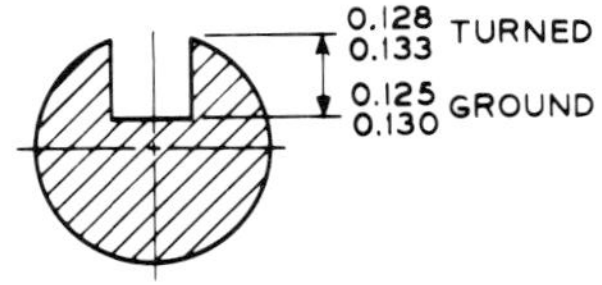

Case 1.

$$\text{min depth} = \frac{H}{2} + \frac{B'-B}{2}$$

max depth = min + tolerance

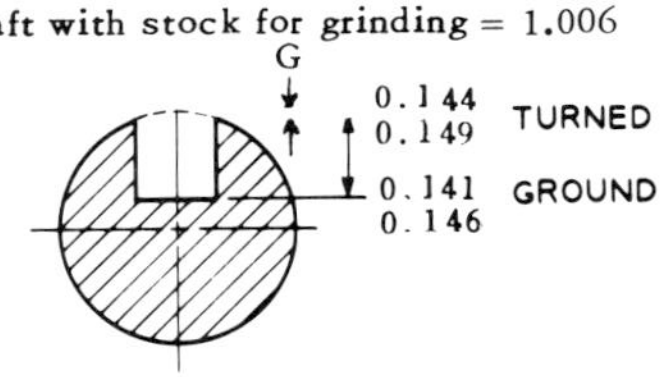

Case 2.

$$\text{min} = \frac{H}{2} + G + \frac{B'-B}{2}$$

max = min + tolerance

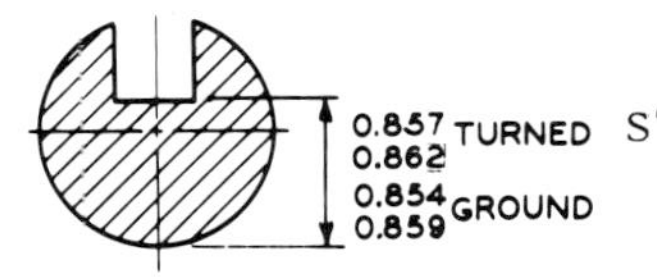

Case 3.

$$\max S' = B + \frac{B'-B}{2} - \frac{H}{2} - G$$

min S' = max − tolerance

Length

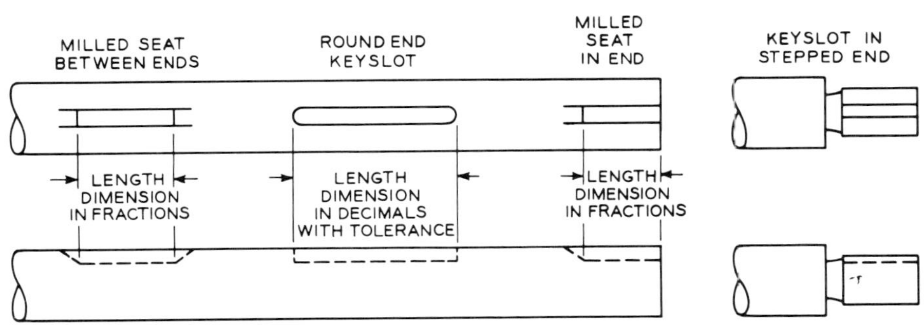

Ordinarily only the round-end keyseat, and the key to fit it, need be specified to decimal tolerances.

TABLE 8–3

Dimensioning Hub Member to Fit Square, Flat and Taper Keys

Dimension	Illustration		Tolerance for Ease of Assembly: Interchangeable	Selective	Drive or Tight
Width	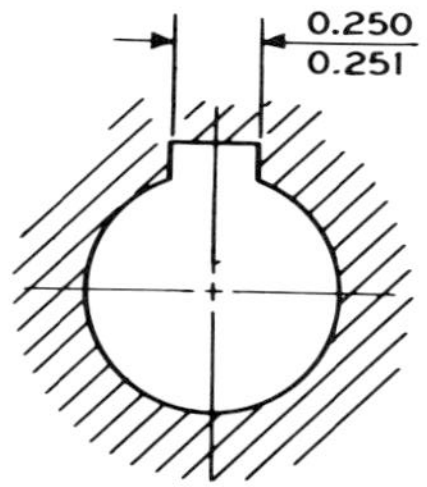		Positive tolerance* on slot equivalent to negative tolerance on key stock.	Same negative tolerance on slot as on key stock.	Interference from 0.0005 to 0.0020 or more for medium and large keys.
			Sometimes the side tolerances on the slot in the hub are specified so that the key will fit on the sides of the hub less tightly than against those of the shaft.		
Depth	PARALLEL 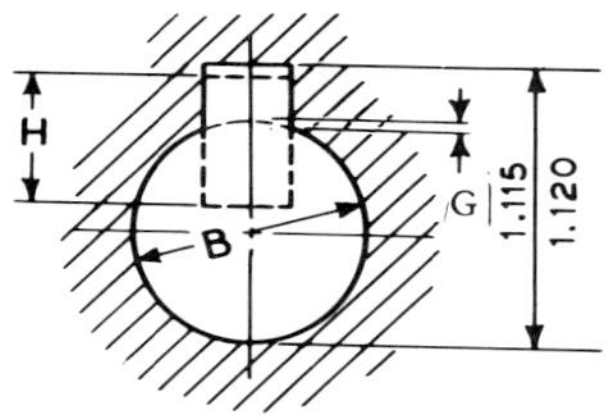	TAPER KEYSEAT	Square and flat parallel keys are generally fitted so as to have clearance (at least 0.005) between the top surface in the key and the bottom surface in the hub. Taper keys, on the other hand, should have no clearance between these surfaces. Accordingly, the depth tolerances in both members needs to be specified and held more closely on taper keyseats than on keyseats for parallel keys.		
Table 8–5	S + H + clearance + tolerances T + H + clearance + tolerances	Compute for large end of taper in same manner as for parallel key. See Table 8-5 for manner of measurement of taper on keys.			
Length	PARALLEL 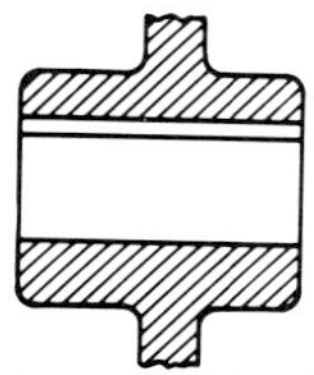 TAPER 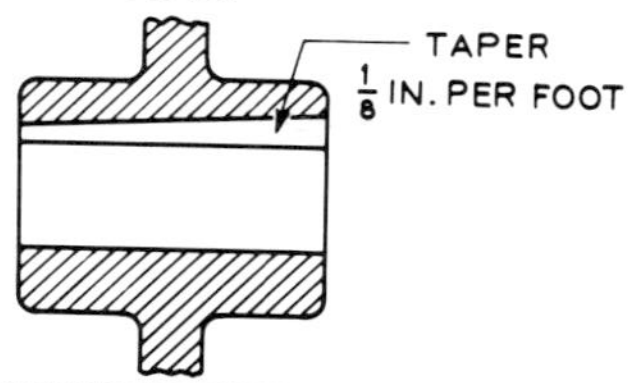		Broaching parallel internal keyslots has advantages on alignment and width tolerances. See Table 9–49. The lengths of keyslots in blind holes should terminate in recesses for keyslotting.		

*If necessary, ground key stock, especially hard keys, can be held to closer tolerances than indicated in Table 8–4.

TABLE 8-4

Tolerances* on Square and Flat Stock Keys

ASA B17.1-1943

Square Keys from	Square Keys to	Flat Keys from	Flat Keys to	Tolerance on Width and Depth, Minus
1/8	3/8	1/8 × 3/32	3/8 × 1/4	0.0020
1/2	3/4	1/2 × 3/8	3/4 × 1/2	0.0025
7/8	1-1/2	7/8 × 5/8	1-1/2 × 1	0.0030
		1-3/4 × 1-1/4	3 × 2	0.0040
		3-1/2 × 2-1/2	6 × 4	0.0050

All dimensions in inches

*These tolerances permit cutting the keys from cold-finished stock. They are suitable for general applications and where the tolerances on the keyslots are specified to provide interchangeable assembly. To get good fits on high-grade work, selective fitting can be called for or the keyslots can be dimensioned so as to require hand fitting. Since keys can be machined or ground to tolerances closer than specified by this standard, the extra machining cost on the keys may be offset by a saving in assembly cost.

TABLE 8-5

Dimensions From Bottom of Keyseat and From Top of Key To Opposite Side of Shaft, Plain Parallel Stock Keys

ASA B17.1-1943

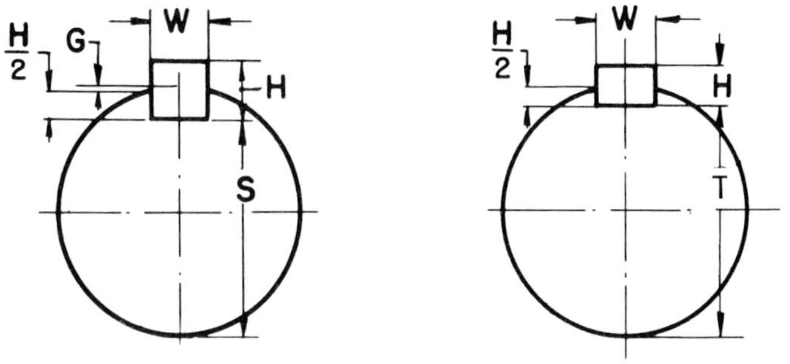

Shaft Diameter Inches	Square Key Depth H Fraction	Square Key Depth H Decimal	Square Key Bottom of Keyseat to Opposite Side of Shaft, S	Square Key Top of Key to Bottom of Shaft, S + H	Flat Key Depth H Fraction	Flat Key Depth H Decimal	Flat Key Bottom of Keyseat to Opposite Side of Shaft, T	Flat Key Top of Key to Bottom of Shaft, T + H
1/2	1/8	0.125	0.430	0.555	3/32	0.094	0.445	0.539
9/16	1/8	0.125	0.493	0.618	3/32	0.094	0.509	0.603
5/8	3/16	0.188	0.517	0.705	1/8	0.125	0.548	0.673
11/16	3/16	0.188	0.581	0.769	1/8	0.125	0.612	0.737
3/4	3/16	0.188	0.644	0.832	1/8	0.125	0.676	0.801
13/16	3/16	0.188	0.708	0.896	1/8	0.125	0.739	0.864
7/8	3/16	0.188	0.771	0.959	1/8	0.125	0.802	0.927
15/16	1/4	0.250	0.796	1.046	3/16	0.188	0.827	1.015
1	1/4	0.250	0.859	1.109	3/16	0.188	0.890	1.078
1-1/16	1/4	0.250	0.923	1.173	3/16	0.188	0.954	1.142
1-1/8	1/4	0.250	0.986	1.236	3/16	0.188	1.017	1.205
1-3/16	1/4	0.250	1.049	1.299	3/16	0.188	1.081	1.269
1-1/4	1/4	0.250	1.112	1.362	3/16	0.188	1.144	1.332
1-5/16	5/16	0.313	1.137	1.450	1/4	0.250	1.169	1.419
1-3/8	5/16	0.313	1.201	1.514	1/4	0.250	1.232	1.482
1-7/16	3/8	0.375	1.225	1.600	1/4	0.250	1.288	1.538
1-1/2	3/8	0.375	1.289	1.664	1/4	0.250	1.351	1.601
1-9/16	3/8	0.375	1.352	1.727	1/4	0.250	1.415	1.665
1-5/8	3/8	0.375	1.416	1.791	1/4	0.250	1.478	1.728
1-11/16	3/8	0.375	1.479	1.854	1/4	0.250	1.542	1.792
1-3/4	3/8	0.375	1.542	1.917	1/4	0.250	1.605	1.855
1-13/16	1/2	0.500	1.527	2.027	3/8	0.375	1.590	1.965
1-7/8	1/2	0.500	1.591	2.091	3/8	0.375	1.654	2.029
1-15/16	1/2	0.500	1.655	2.155	3/8	0.375	1.717	2.092

All dimensions given in inches.

continued on next page

TABLE 8-5 continued

Shaft Diameter Inches	Square Key Depth H Fraction	Square Key Depth H Decimal	Square Key Bottom of Keyseat to Opposite Side of Shaft, S	Square Key Top of Key to Bottom of Shaft, $S+H$	Flat Key Depth H Fraction	Flat Key Depth H Decimal	Flat Key Bottom of Keyseat to Opposite Side of Shaft, T	Flat Key Top of Key to Bottom of Shaft, $T+H$
2	1/2	0.500	1.718	2.218	3/8	0.375	1.781	2.156
2-1/16	1/2	0.500	1.782	2.282	3/8	0.375	1.843	2.218
2-1/8	1/2	0.500	1.845	2.345	3/8	0.375	1.908	2.283
2-3/16	1/2	0.500	1.909	2.409	3/8	0.375	1.971	2.346
2-1/4	1/2	0.500	1.972	2.472	3/8	0.375	2.034	2.409
2-5/16	5/8	0.625	1.957	2.582	7/16	0.438	2.051	2.489
2-3/8	5/8	0.625	2.021	2.646	7/16	0.438	2.114	2.552
2-7/16	5/8	0.625	2.084	2.709	7/16	0.438	2.178	2.616
2-1/2	5/8	0.625	2.148	2.773	7/16	0.438	2.242	2.680
2-5/8	5/8	0.625	2.275	2.900	7/16	0.438	2.368	2.806
2-3/4	5/8	0.625	2.402	3.027	7/16	0.438	2.495	2.933
2-7/8	3/4	0.750	2.450	3.200	1/2	0.500	2.575	3.075
2-15/16	3/4	0.750	2.514	3.264	1/2	0.500	2.639	3.139
3	3/4	0.750	2.577	3.327	1/2	0.500	2.702	3.202
3-1/8	3/4	0.750	2.704	3.454	1/2	0.500	2.829	3.329
3-1/4	3/4	0.750	2.831	3.581	1/2	0.500	2.956	3.456
3-3/8	7/8	0.875	2.880	3.755	5/8	0.625	3.005	3.630
3-7/16	7/8	0.875	2.944	3.819	5/8	0.625	3.069	3.694
3-1/2	7/8	0.875	3.007	3.882	5/8	0.625	3.132	3.757
3-5/8	7/8	0.875	3.140	4.015	5/8	0.625	3.259	3.884
3-3/4	7/8	0.875	3.261	4.136	5/8	0.625	3.386	4.011
3-7/8	1	1.000	3.309	4.309	3/4	0.750	3.434	4.184
3-15/16	1	1.000	3.373	4.373	3/4	0.750	3.498	4.248
4	1	1.000	3.437	4.437	3/4	0.750	3.562	4.312
4-1/4	1	1.000	3.690	4.690	3/4	0.750	3.815	4.565
4-7/16	1	1.000	3.881	4.881	3/4	0.750	4.006	4.756
4-1/2	1	1.000	3.944	4.944	3/4	0.750	4.069	4.819
4-3/4	1-1/4	1.250	4.042	5.292	7/8	0.875	4.229	5.104
4-15/16	1-1/4	1.250	4.232	5.482	7/8	0.875	4.420	5.295
5	1-1/4	1.250	4.296	5.546	7/8	0.875	4.483	5.358
5-1/4	1-1/4	1.250	4.550	5.800	7/8	0.875	4.733	5.608
5-7/16	1-1/4	1.250	4.740	5.990	7/8	0.875	4.927	5.802
5-1/2	1-1/4	1.250	4.803	6.053	7/8	0.875	4.991	5.866
5-3/4	1-1/2	1.500	4.900	6.400	1	1.000	5.150	6.150
5-15/16	1-1/2	1.500	5.091	6.591	1	1.000	5.341	6.341
6	1-1/2	1.500	5.155	6.655	1	1.000	5.405	6.405

All dimensions given in inches.

TABLE 8–6

Dimensions of Square and Flat Plain Taper Stock Keys*

ASA B17.1 1943

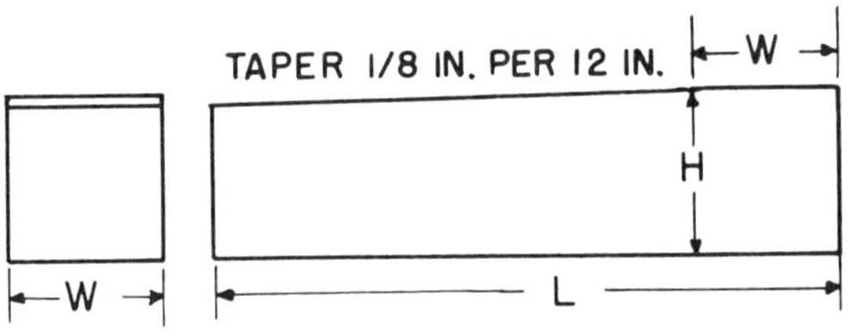

Shaft Diameter (incl)	Square Type Maximum Width W	Square Type Height at Large End** H	Flat Type Maximum Width W	Flat Type Height at Large End** H	Tolerance* On Width (−)	Tolerance* On Height (+)
1/2 - 9/16	1/8	1/8	1/8	3/32	0.0020	0.0020
5/8 - 7/8	3/16	3/16	3/16	1/8	0.0020	0.0020
15/16 - 1-1/4	1/4	1/4	1/4	3/16	0.0020	0.0020
1-5/16 - 1-3/8	5/16	5/16	5/16	1/4	0.0020	0.0020
1-7/16 - 1-3/4	3/8	3/8	3/8	1/4	0.0020	0.0020
1-13/16 - 2-1/4	1/2	1/2	1/2	3/8	0.0025	0.0025
2-5/16 - 2-3/4	5/8	5/8	5/8	7/16	0.0025	0.0025
2-7/8 - 3-1/4	3/4	3/4	3/4	1/2	0.0025	0.0025
3-3/8 - 3-3/4	7/8	7/8	7/8	5/8	0.0030	0.0030
3-7/8 - 4-1/2	1	1	1	3/4	0.0030	0.0030
4-3/4 - 5-1/2	1-1/4	1-1/4	1-1/4	7/8	0.0030	0.0030
5-3/4 - 6	1-1/2	1-1/2	1-1/2	1	0.0030	0.0030

All dimensions given in inches.

*Stock keys are applicable to the general run of work and the tolerances have been set accordingly. They are not intended to cover the finer applications where a closer fit may be required.

**This height of the key is measured at the distance W, equal to the width of the key, from the large end.

TABLE 8–7

Stock Lengths of Plain Taper Stock Keys

ASA B17.1–1943

Shaft Diameter (Incl.)	Length of Key,* L						
1/2 – 9/16	1/2	3/4	1	1-1/4	1-1/2	1-3/4	2
5/8 – 7/8	3/4	1-1/8	1-1/2	1-7/8	2-1/4	2-5/8	3
15/16 – 1-1/4	1	1-1/2	2	2-1/2	3	3-1/2	4
1-5/16 – 1-3/8	1-1/4	1-7/8	2-1/2	3-1/8	3-3/4	4-1/2	5-1/4
1-7/16 – 1-3/4	1-1/2	2-1/4	3	3-3/4	4-1/2	5-1/4	6
1-13/16 – 2-1/4	2	3	4	5	6	7	8
2-5/16 – 2-3/4	2-1/2	3-3/4	5	6-1/4	7-1/2	8-3/4	10
2-7/8 – 3-1/4	3	4-1/2	6	7-1/2	9	10-1/2	12
3-3/8 – 3-3/4	3-1/2	5-1/4	7	8-3/4	10-1/2	12-1/4	14
3-7/8 – 4-1/2	4	6	8	10	12	14	16
4-3/4 – 5-1/2	5	7-1/2	10	12-1/2	15	17-1/2	20
5-3/4 – 6	6	9	12	15	18	21	24

All dimensions given in inches.

*The minimum stock length of keys is equal to four times the key width, and the maximum stock length is equal to sixteen times the key width. The increments of increase in length are equal to twice the width.

TABLE 8–8
Dimensions of Square and Flat Gib-Head Taper Stock Keys*
ASA B17.1 1943

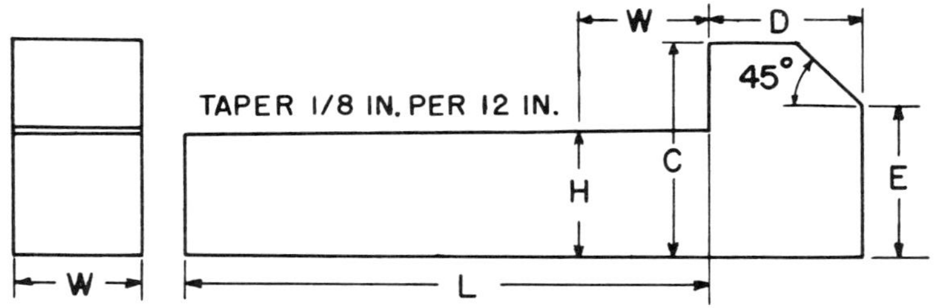

Shaft Diameter (incl)	Square Type Key Maximum Width W	Square Type Key Height at Large End** H	Square Type Gib Head Height C	Square Type Gib Head Length D	Square Type Gib Head Height Edge of Chamfer E	Flat Type Key Maximum Width W	Flat Type Key Height at Large End** H	Flat Type Gib Head Height C	Flat Type Gib Head Length D	Flat Type Gib Head Height Edge of Chamfer E	Tolerance* On Width (−)	Tolerance* On Height (+)
1/2 - 9/16	1/8	1/8	1/4	7/32	5/32	1/8	3/32	3/16	1/8	1/8	0.0020	0.0020
5/8 - 7/8	3/16	3/16	5/16	9/32	7/32	3/16	1/8	1/4	3/16	5/32	0.0020	0.0020
15/16 - 1-1/4	1/4	1/4	7/16	11/32	11/32	1/4	3/16	5/16	1/4	3/16	0.0020	0.0020
1-5/16 - 1-3/8	5/16	5/16	9/16	13/32	13/32	5/16	1/4	3/8	5/16	1/4	0.0020	0.0020
1-7/16 - 1-3/4	3/8	3/8	11/16	15/32	15/32	3/8	1/4	7/16	3/8	5/16	0.0020	0.0020
1-13/16 - 2-1/4	1/2	1/2	7/8	19/32	5/8	1/2	3/8	5/8	1/2	7/16	0.0025	0.0025
2-5/16 - 2-3/4	5/8	5/8	1-1/16	23/32	3/4	5/8	7/16	3/4	5/8	1/2	0.0025	0.0025
2-7/8 - 3-1/4	3/4	3/4	1-1/4	7/8	7/8	3/4	1/2	7/8	3/4	5/8	0.0025	0.0025
3-3/8 - 3-3/4	7/8	7/8	1-1/2	1	1	7/8	5/8	1-1/16	7/8	3/4	0.0030	0.0030
3-7/8 - 4-1/2	1	1	1-3/4	1-3/16	1-3/16	1	3/4	1-1/4	1	13/16	0.0030	0.0030
4-3/4 - 5-1/2	1-1/4	1-1/4	2	1-7/16	1-7/16	1-1/4	7/8	1-1/2	1-1/4	1	0.0030	0.0030
5-3/4 - 6	1-1/2	1-1/2	2-1/2	1-3/4	1-3/4	1-1/2	1	1-3/4	1-1/2	1-1/4	0.0030	0.0030

All dimensions given in inches.

* Stock keys are applicable to the general run of work and the tolerances have been set accordingly. They are not intended to cover the finer applications where a closer fit may be required.

** This height of the key is measured at the distance W, equal to the width of the key, from the gib head.

TABLE 8-9

Stock Lengths of Gib-Head Taper Stock Keys

ASA B17.1 1943

Shaft Diameter (Incl)	Length of Key,* L					
1/2 - 9/16	1/2	3/4	1	1-1/4	1-1/2	1-3/4	2
5/8 - 7/8	3/4	1-1/8	1-1/2	1-7/8	2-1/4	2-5/8	3
15/16 - 1-1/4	1	1-1/2	2	2-1/2	3	3-1/2	4
1-5/16 - 1-3/8	1-1/4	1-7/8	2-1/2	3-1/8	3-3/4	4-3/8	5-1/2
1-7/16 - 1-3/4	1-1/2	2-1/4	3	3-3/4	4-1/2	5-1/4	6
1-13/16 - 2-1/4	2	3	4	5	6	7	8
2-5/16 - 2-3/4	2-1/2	3-3/4	5	6-1/4	7-1/2	8-3/4	10
2-7/8 - 3-1/4	3	4-1/2	6	7-1/2	9	10-1/2	12
3-3/8 - 3-3/4	3-1/2	5-1/4	7	8-3/4	10-1/2	12-1/4	14
3-7/8 - 4-1/2	4	6	8	10	12	14	16
4-3/4 - 5-1/2	5	7-1/2	10	12-1/2	15	17-1/2	20
5-3/4 - 6	6	9	12	15	18	21	24

All dimensions given in inches.

*The minimum stock length of keys is equal to four times the key width, and the maximum stock length is equal to sixteen times the key width. The increments of increase in length are equal to twice the width.

TABLE 8–10

Keys for NEMA Foot-Mounted Motors and Generators

NEMA Standard Dimensions, Part 3 – July 1953

Frame Number	Shaft Diameter, Inches	Length of Key Inches	Square Key
56	5/8	1-3/8*	3/16
66	3/4	1-7/8*	3/16
203	3/4	1-3/8	3/16
204	3/4	1-3/8	3/16
224	1	2	1/4
225	1	2	1/4
254	1-1/8	2-3/8	1/4
284	1-1/4	2-3/4	1/4
324	1-5/8	3-3/4	3/8
326	1-5/8	3-3/4	3/8
364	1-7/8	4-1/4	1/2
364S	1-5/8	1-7/8	3/8
365	1-7/8	4-1/4	1/2
365S	1-5/8	1-7/8	3/8
404	2-1/8	5	1/2
404S	1-7/8	2	1/2
405	2-1/8	5	1/2
405S	1-7/8	2	1/2
444	2-3/8	5-1/2	5/8
444S	2-1/8	2-3/4	1/2
445	2-3/8	5-1/2	5/8
445S	2-1/8	2-3/4	1/2
504U	2-7/8	7-1/4	3/4
504S	2-1/8	2-3/4	1/2
505	2-7/8	7-1/4	3/4
505S	2-1/8	2-3/4	1/2

*Effective length of keyway.

TABLE 8-11

Keyways for Holes in Gears for General Industrial Practice

AGMA Standard Dimensions 261.01 – 1946

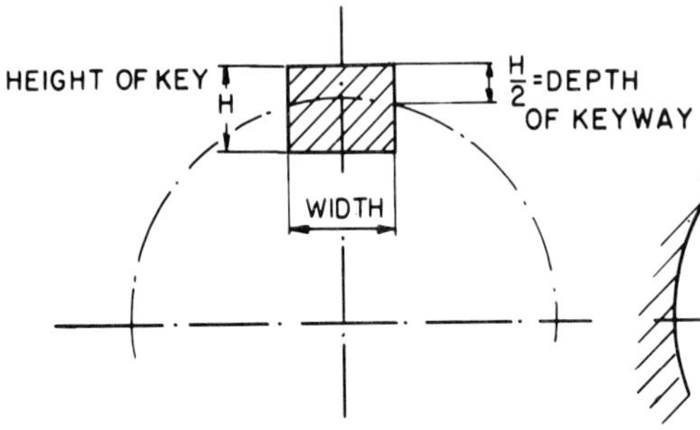

Recommended depth of Keyway shall be one-half Key height

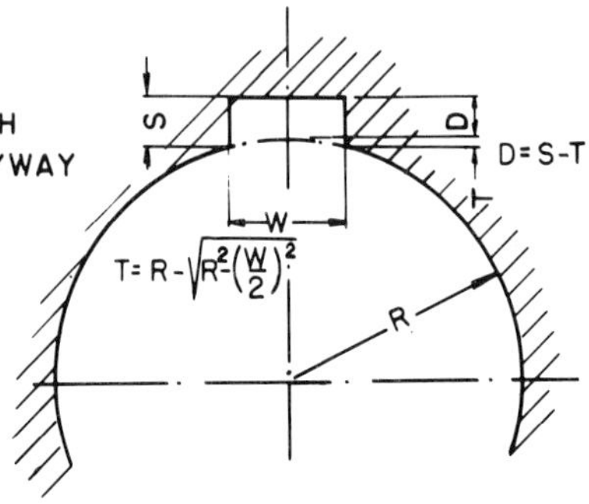

Alternate method

Recommended Keyways and Key Stock for Holes in Gears

Diameter of Holes Inclusive Inches	Standard Keyways and Keys: Keyways* Width	Keyways* Depth	Key Stock Cold Rolled Steel 0.10 to 0.20 Carbon	Tolerance On Key Stock
5/16 to 7/16	3/32	3/64	3/32 x 3/32	
1/2 to 9/16	1/8	1/16	1/8 x 1/8	
5/8 to 7/8	3/16	3/32	3/16 x 3/16	+0.000
15/16 to 1-1/4	1/4	1/8	1/4 x 1/4	-0.002
1-5/16 to 1-3/8	5/16	5/32	5/16 x 5/16	
1-7/16 to 1-3/4	3/8	3/16	3/8 x 3/8	
1-13/16 to 2-1/4	1/2	1/4	1/2 x 1/2	
2-5/16 to 2-3/4	5/8	5/16	5/8 x 5/8	+0.0000
2-13/16 to 3-1/4	3/4	3/8	3/4 x 3/4	-0.0025
3-5/16 to 3-3/4	7/8	7/16	7/8 x 7/8	
3-13/16 to 4-1/2	1	1/2	1 x 1	+0.000
4-9/16 to 5-1/2	1-1/4	7/16	1-1/4 x 7/8	-0.003
5-9/16 to 6-1/2	1-1/2	1/2	1-1/2 x 1	
6-9/16 to 7-1/2	1-3/4	5/8	1-3/4 x 1-1/4	
7-9/16 to 8-15/16	2	3/4	2 x 1-1/2	+0.000
9 to 10-15/16	2-1/2	7/8	2-1/2 x 1-3/4	-0.004
11 to 12-15/16	3	1	3 x 2	
13 to 14-15/16	3-1/2	1-1/4	3-1/2 x 2-1/2	
15 to 17-15/16	4	1-1/2	4 x 3	+0.000
18 to 21	5	1-3/4	5 x 3-1/2	-0.005

*Width tolerance –0.000, +0.002; depth tolerance, nominal to +1/64 for straight keys, nominal to –1/64 for taper keys. On heat treated pinions keyway depth shall be 1/32 to 3/64 over nominal with minimum radius in keyway corners of 1/32. Keyway corners and key stock to be rounded from 1/32 to not more than 1/5 keyway depth for impact or alternating loads.

TABLE 8–12

Formulas for Calculating Shear* and Compressive Strength of Square and Flat Keys

To Find	From	Formula
Shearing unit stress, S_S, psi	Width of key, W, in. Length of key, L, in. Shaft diameter, B, in. Torque, M, in.-lb	$S_S = \frac{2M}{WBL^*}$
Compressive unit stress, S_c, psi	Height of key, H, in. Length of key, L, in. Shaft diameter, B, in. Torque, M, in.-lb	$S_c = \frac{4M}{HLB}$

*The key area at the shear line is WL, values of which for Woodruff keys are listed in Table 8–18; hence S_s can be found for the Woodruff-type key as well as for square keys.

TABLE 8–13

Allowable Compressive Stresses in Square and Flat Keys

Practice of Caterpillar Tractor Co.

Designation	Material	Recommended Use
Soft keys	Low-carbon steel SAE 1018	When compressive stress is less than 18,000 psi, one direction, with clamped hub 15,000 psi, one direction, without clamped hub 13,000 psi, both directions, with clamped hub 10,000 psi, both directions, without clamped hub
Hard keys	SAE 4140 Rockwell C 42–50 Ground on all four sides from 1/64 oversize stock to insure removal of all decarburization	When compressive stress equals or exceeds psi listed above

TABLE 8-14

Tolerance on Alignment of Keyslots

Engineering Standards – 1954 The Caterpillar Tractor Company

Class	Tolerance on Alignment	Comments
1	within 1 degree	These tolerances are used to specify the alignment of a keyslot with a drilled hole, or any other machine operation, as well as for the alignment of two keyslots.
2	within 2 degrees	
3	within 10 degrees	
Timing gear keyways	± 5 minutes	

TABLE 8-15

Fatigue Stress Concentration Factors in Solid Steel Shafts Caused by Keyways (or Splines)

Lipson, Noll and Clock "Stress and Strength of Manufactured Parts", McGraw-Hill

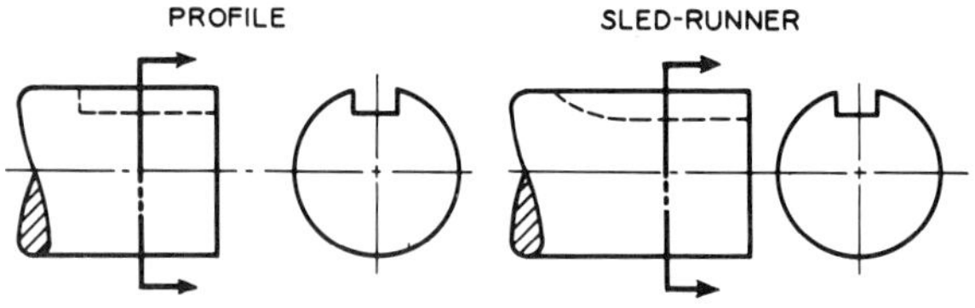

Material and State	Factor	Profile Bending	Profile Torsion	Sled-runner Bending	Sled-runner Torsion
Annealed Steels	K_f	1.6	1.3	1.3	1.3
Quenched and Drawn Steels	K_f	2.0	1.6	1.6	1.6

TABLE 8-16

Geometric Stress Concentration Factors in Hollow Circular Shaft Caused by Keyway

Lipson, Noll and Clock "Stress and Strength of Manufactured Parts," McGraw-Hill

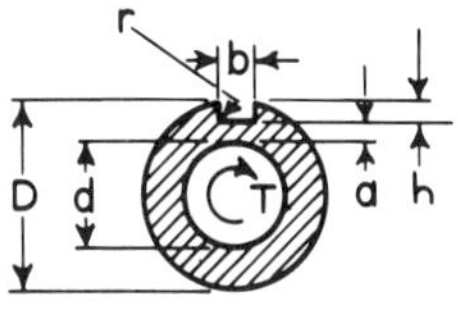

TORSION

$$S_{NOM} = \frac{2T}{\pi a d^2}$$

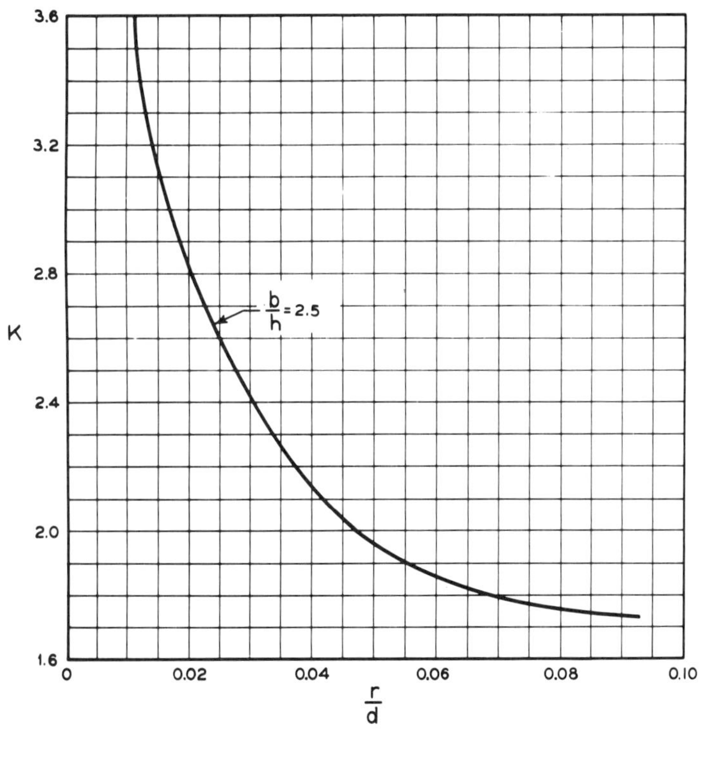

TABLE 8–1,7

Fatigue Stress Concentration Factors in Hollow Steel Shaft Caused by Keyway

Lipson, Noll and Clock "Stress and Strength of Manufactured Parts," McGraw-Hill

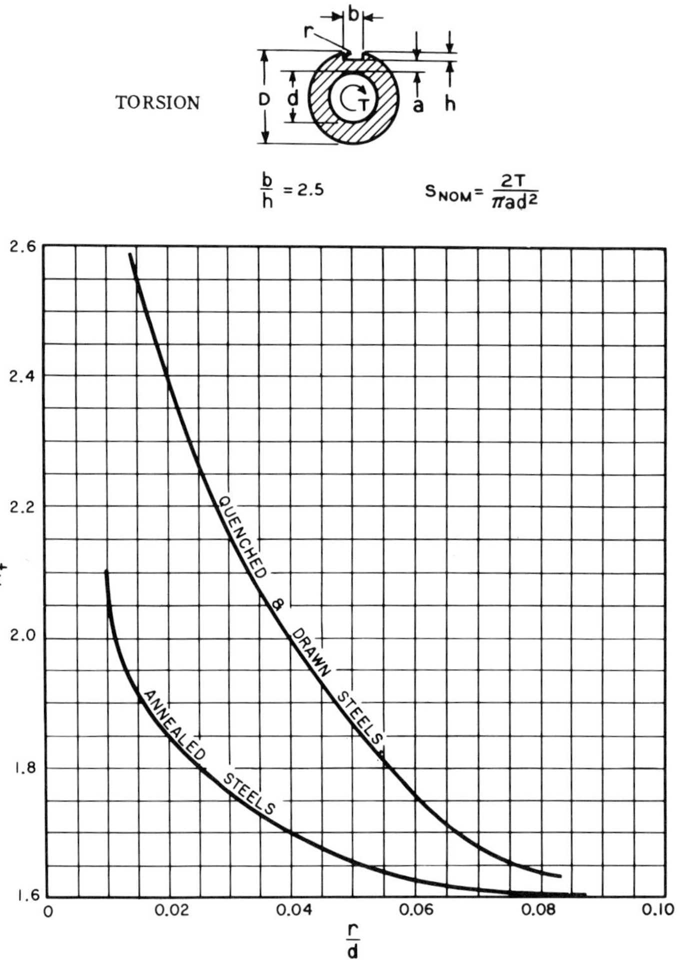

TABLE 8–18

Modified Radii on Woodruff Keys and Keyslots to Reduce Notch Effect

Engineering Standards – 1953 The Caterpillar Tractor Company

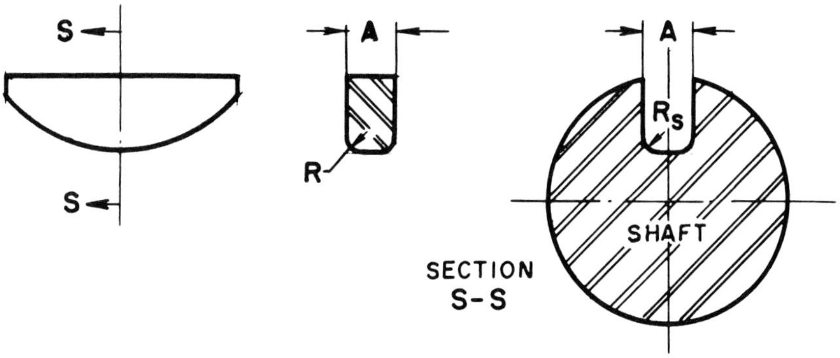

Width of Key	Radius of Key $R = \frac{A}{3}$ +0.010 −0.000		Radius of Keyslot $R_s = \frac{A}{3}$ +0.000 −0.010	
A	Min	Max	Min	Max
3/32	.032	.042	.022	.032
1/8	.042	.052	.032	.042
5/32	.047	.057	.037	.047
3/16	.062	.072	.052	.062
1/4	.083	.093	.073	.083
5/16	.104	.114	.094	.104
3/8	.125	.135	.115	.125
7/16	.146	.156	.136	.146
1/2	.167	.177	.157	.167
9/16	.188	.198	.178	.188
5/8	.208	.218	.198	.208
11/16	.229	.239	.219	.229
3/4	.250	.260	.240	.250

TABLE 8–19

Identification and Sizes of Woodruff Keys Regularly Available

Key No.	PROFILE DIMENSIONS
201	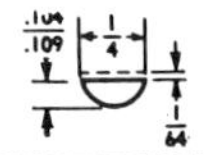
206 207	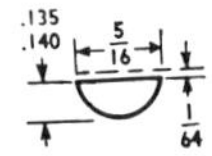
211 212 213	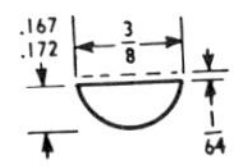
1 2 3	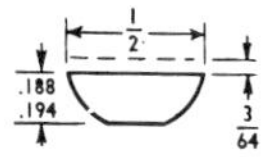
4 5 6 61	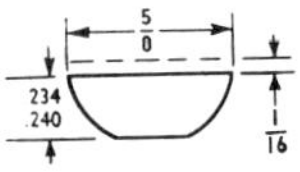
7 8 9 91	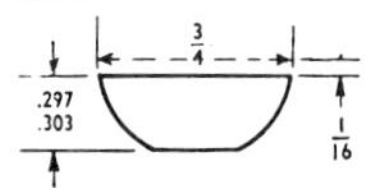
10 11 12 A	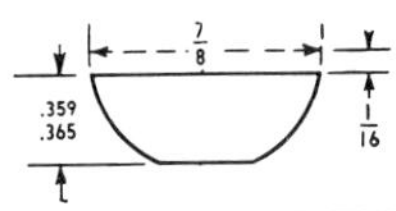
13 14 15 B	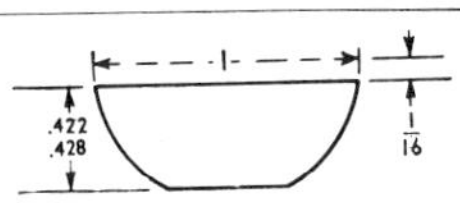
16 17 18 C	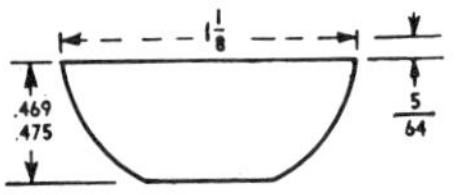
19 20 21 D E	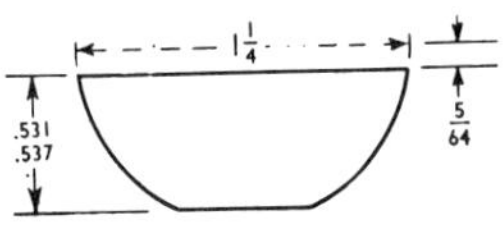
22 23 F	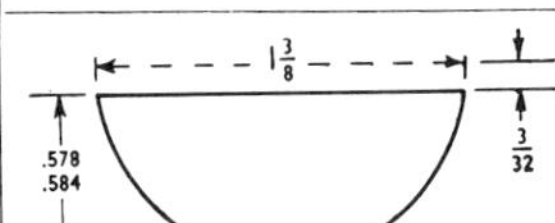
24 25 G	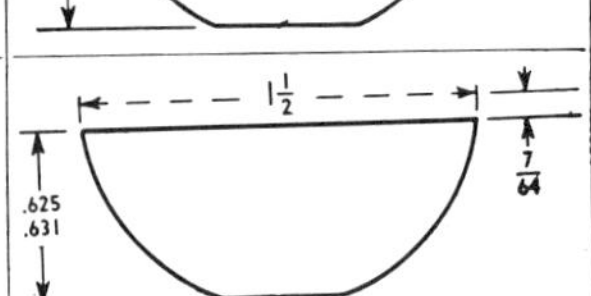

Key No.	PROFILE DIMENSIONS
126 127 128 129	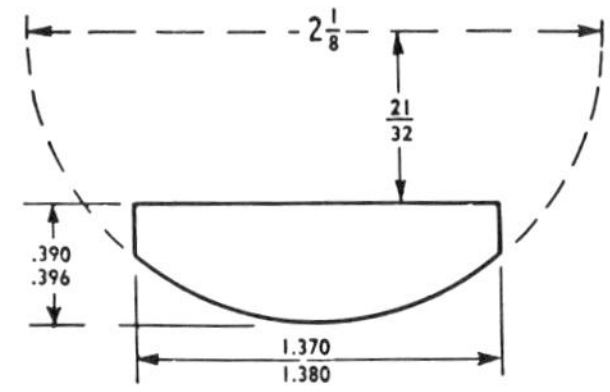
26 27 28 29	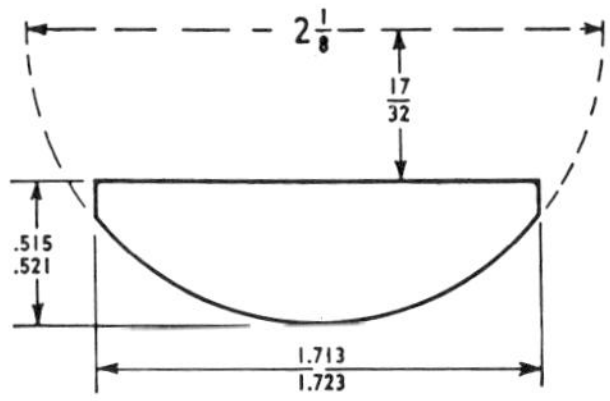
Rx Sx Tx Ux Vx	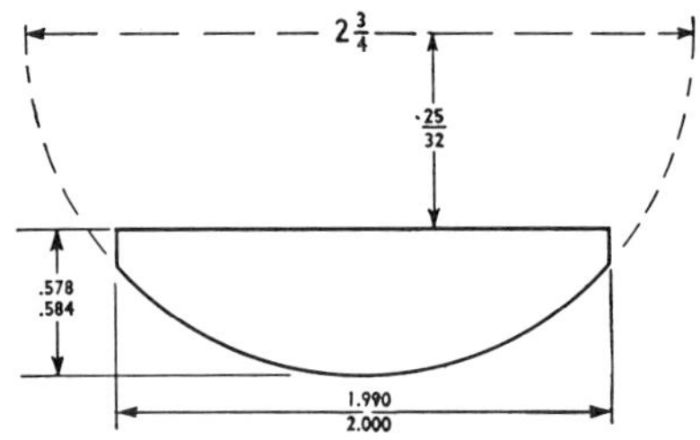
R S T U V	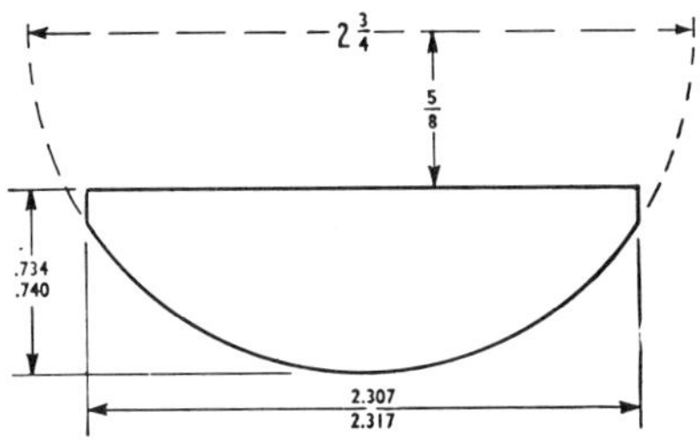
30 31 32 33 34 35 36	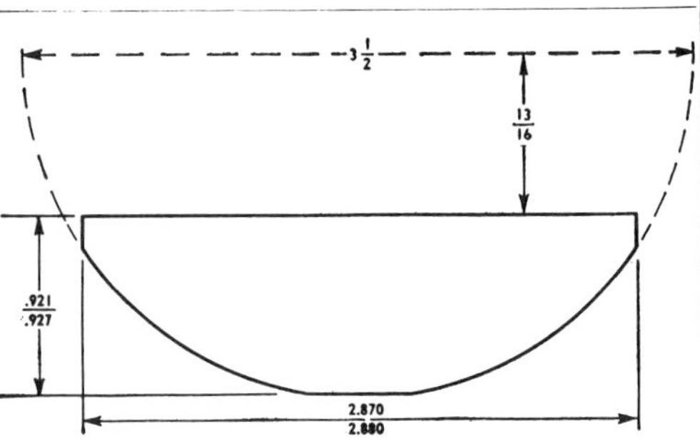

TABLE 8-20

Woodruff Key Dimensions

SAE Handbook 1954

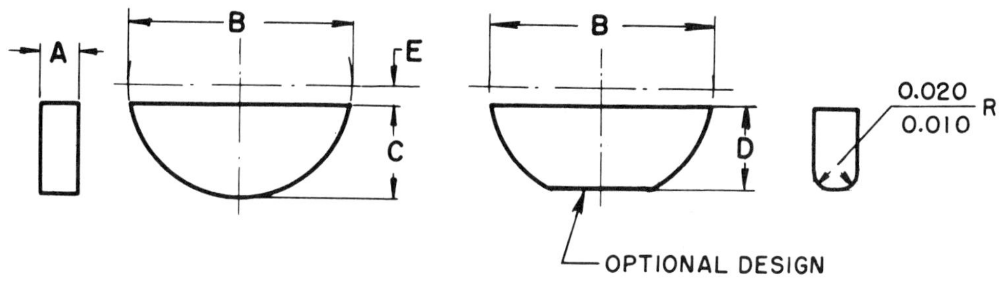

Key Numbers 201 to *G* inclusive, Table 8–19

ASA* B5.3 1950	Key No.	Nominal Key Size A x B	Width A		Diameter B		Height C		Height D		Height E	Key Area at Shearing Line	App. Weight Per 1000 Key in Lbs
			Min	Max	Min	Max	Min	Max	Min	Max	Nom		
202	201	1/16 x 1/4	.0625	.0635	.249	.250	.104	.109			1/64	.01456	.6
202½	206	1/16 x 5/16	.0625	.0635	.302	.312	.135	.140			1/64	.01845	.7
302½	207	3/32 x 5/16	.0938	.0948	.302	.312	.135	.140			1/64	.02646	.9
203	211	1/16 x 3/8	.0625	.0635	.365	.375	.167	.172			1/64	.02255	.9
303	212	3/32 x 3/8	.0938	.0948	.365	.375	.167	.172			1/64	.03280	1.3
403	213	1/8 x 3/8	.1250	.1260	.365	.375	.167	.172			1/64	.04209	1.5
204	1	1/16 x 1/2	.0625	.0635	.490	.500	.198	.203	.188	.194	3/64	.02968	1.3
304	2	3/32 x 1/2	.0938	.0948	.490	.500	.198	.203	.188	.194	3/64	.04344	1.9
404	3	1/8 x 1/2	.1250	.1260	.490	.500	.198	.203	.188	.194	3/64	.05128	2.5
305	4	3/32 x 5/8	.0938	.0948	.615	.625		.250	.234	.240	1/16	.05233	3.0
405	5	1/8 x 5/8	.1250	.1260	.615	.625		.250	.234	.240	1/16	.07160	3.9
505	6	5/32 x 5/8	.1563	.1573	.615	.625		.250	.234	.240	1/16	.08719	4.9
605	61	3/16 x 5/8	.1875	.1885	.615	.625		.250	.234	.240	1/16	.1005	5.8
406	7	1/8 x 3/4	.1250	.1260	.740	.750	.308	.313	.297	.303	1/16	.08843	6.1
506	8	5/32 x 3/4	.1563	.1573	.740	.750	.308	.313	.297	.303	1/16	.10869	7.5
606	9	3/16 x 3/4	.1875	.1885	.740	.750	.308	.313	.297	.303	1/16	.12791	9.0
806	91	1/4 x 3/4	.2500	.2510	.740	.750	.308	.313	.297	.303	1/16	.16235	12.0
507	10	5/32 x 7/8	.1563	.1573	.865	.875	.370	.375	.359	.365	1/16	.12944	11.0
607	11	3/16 x 7/8	.1875	.1885	.865	.875	.370	.375	.359	.365	1/16	.15310	13.0
707	12	7/32 x 7/8	.2188	.2198	.865	.875	.370	.375	.359	.365	1/16	.18137	14.9
807	A	1/4 x 7/8	.2500	.2510	.865	.875	.370	.375	.359	.365	1/16	.19760	17.0
608	13	3/16 x 1	.1875	.1885	.990	1.000	.433	.438	.422	.428	1/16	.17816	17.0
708	14	7/32 x 1	.2188	.2198	.990	1.000	.433	.438	.422	.428	1/16	.21002	20.1
808	15	1/4 x 1	.2500	.2510	.990	1.000	.433	.438	.422	.428	1/16	.23200	23.0
1008	B	5/16 x 1	.3125	.3135	.990	1.000	.433	.438	.422	.428	1/16	.28113	29.0
609	16	3/16 x 1 1/8	.1875	.1885	1.115	1.125	.479	.484	.469	.475	5/64	.20078	22.0
709	17	7/32 x 1 1/8	.2188	.2198	1.115	1.125	.479	.484	.469	.475	5/64	.23200	25.0
809	18	1/4 x 1 1/8	.2500	.2510	1.115	1.125	.479	.484	.469	.475	5/64	.26220	29.0
1009	C	5/16 x 1 1/8	.3125	.3135	1.115	1.125	.479	.484	.469	.475	5/64	.31938	36.0
610	19	3/16 x 1 1/4	.1875	.1885	1.240	1.250	.542	.547	.531	.537	5/64	.22844	27.1
710	20	7/32 x 1 1/4	.2188	.2198	1.240	1.250	.542	.547	.531	.537	5/64	.26084	31.8
810	21	1/4 x 1 1/4	.2500	.2510	1.240	1.250	.542	.547	.531	.537	5/64	.29556	36.0
1010	D	5/16 x 1 1/4	.3125	.3135	1.240	1.250	.542	.547	.531	.537	5/64	.36213	45.0
1210	E	3/8 x 1 1/4	.3750	.3760	1.240	1.250	.542	.547	.531	.537	5/64	.42435	54.0
811	22	1/4 x 1 3/8	.2500	.2510	1.365	1.375	.589	.594	.578	.584	3/32	.32590	43.0
1011	23	5/16 x 1 3/8	.3125	.3135	1.365	1.375	.589	.594	.578	.584	3/32	.40031	54.0
1211	F	3/8 x 1 3/8	.3750	.3760	1.365	1.375	.589	.594	.578	.584	3/32	.47055	65.0
812	24	1/4 x 1 1/2	.2500	.2510	1.490	1.500	.636	.641	.625	.631	7/64	.35625	50.0
1012	25	5/16 x 1 1/2	.3125	.3135	1.490	1.500	.636	.641	.625	.631	7/64	.43844	63.0
1212	G	3/8 x 1 1/2	.3750	.3760	1.490	1.500	.636	.641	.625	.631	7/64	.51668	75.0

*Table 9-40

continued on next page

TABLE 8–20, continued

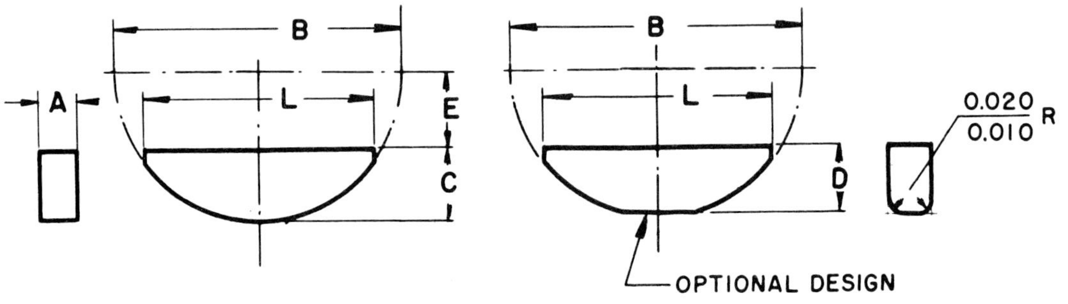

Key Numbers 126 to 36 inclusive, Table 8–19

ASA* B5.3 1950	Key No.	Nominal Key Size A x B	Width A		Diameter B		Height C		Height D		Height E	Key Area at Shearing Line	App. Weight Per 1000 Key in Lbs
			Min	Max	Min	Max	Min	Max	Min	Max	Nom		
617	126	$\frac{3}{16} \times 2\frac{1}{8}$	.1875	.1885	2.115	2.125	.401	.406	.390	.396	$\frac{21}{32}$	.25781	23.4
817	127	$\frac{1}{4} \times 2\frac{1}{8}$	.2500	.2510	2.115	2.125	.401	.406	.390	.396	$\frac{21}{32}$	.34375	31.2
1017	128	$\frac{5}{16} \times 2\frac{1}{8}$	.3125	.3135	2.115	2.125	.401	.406	.390	.396	$\frac{21}{32}$	.42969	39.3
1217	129	$\frac{3}{8} \times 2\frac{1}{8}$	.3750	.3760	2.115	2.125	.401	.406	.390	.396	$\frac{21}{32}$	.48331	47.2
617	26	$\frac{3}{16} \times 2\frac{1}{8}$	.1875	.1885	2.115	2.125	.526	.531	.515	.521	$\frac{17}{32}$	.32227	36.3
817	27	$\frac{1}{4} \times 2\frac{1}{8}$	.2500	.2510	2.115	2.125	.526	.531	.515	.521	$\frac{17}{32}$	.41780	48.2
1017	28	$\frac{5}{16} \times 2\frac{1}{8}$	.3125	.3135	2.115	2.125	.526	.531	.515	.521	$\frac{17}{32}$	.50625	60.1
1217	29	$\frac{3}{8} \times 2\frac{1}{8}$	.3750	.3760	2.115	2.125	.526	.531	.515	.521	$\frac{17}{32}$	.58687	72.3
822	Rx	$\frac{1}{4} \times 2\frac{3}{4}$	.2500	.2510	2.740	2.750	.589	.594	.578	.584	$\frac{25}{32}$	.50000	64.8
1022	Sx	$\frac{5}{16} \times 2\frac{3}{4}$	.3125	.3135	2.740	2.750	.589	.594	.578	.584	$\frac{25}{32}$	.62863	80.8
1222	Tx	$\frac{3}{8} \times 2\frac{3}{4}$	.3750	.3760	2.740	2.750	.589	.594	.578	.584	$\frac{25}{32}$	.69433	96.6
1422	Ux	$\frac{7}{16} \times 2\frac{3}{4}$	.4375	.4385	2.740	2.750	.589	.594	.578	.584	$\frac{25}{32}$	.82530	112.9
1622	Vx	$\frac{1}{2} \times 2\frac{3}{4}$	.5000	.5010	2.740	2.750	.589	.594	.578	.584	$\frac{25}{32}$	.90947	129.3
822	R	$\frac{1}{4} \times 2\frac{3}{4}$	.2500	.2510	2.740	2.750	.745	.750	.734	.740	$\frac{5}{8}$	.5718	91.6
1022	S	$\frac{5}{16} \times 2\frac{3}{4}$	.3125	.3135	2.740	2.750	.745	.750	.734	.740	$\frac{5}{8}$	.70718	114.2
1222	T	$\frac{3}{8} \times 2\frac{3}{4}$	.3750	.3760	2.740	2.750	.745	.750	.734	.740	$\frac{5}{8}$	.83190	136.6
1422	U	$\frac{7}{16} \times 2\frac{3}{4}$	.4375	.4385	2.740	2.750	.745	.750	.734	.740	$\frac{5}{8}$	.94997	159.2
1622	V	$\frac{1}{2} \times 2\frac{3}{4}$	.5000	.5010	2.740	2.750	.745	.750	.734	.740	$\frac{5}{8}$	1.06060	191.8
1228	30	$\frac{3}{8} \times 3\frac{1}{2}$	.3750	.3760	3.490	3.500	.933	.938	.921	.927	$\frac{13}{16}$	1.07813	216.0
1428	31	$\frac{7}{16} \times 3\frac{1}{2}$	.4375	.4385	3.490	3.500	.933	.938	.921	.927	$\frac{13}{16}$	1.23713	252.0
1628	32	$\frac{1}{2} \times 3\frac{1}{2}$	.5000	.5010	3.490	3.500	.933	.938	.921	.927	$\frac{13}{16}$	1.39050	288.0
1828	33	$\frac{9}{16} \times 3\frac{1}{2}$	.5625	.5635	3.490	3.500	.933	.938	.921	.927	$\frac{13}{16}$	1.53685	325.0
2028	34	$\frac{5}{8} \times 3\frac{1}{2}$	.6250	.6260	3.490	3.500	.933	.938	.921	.927	$\frac{13}{16}$	1.67550	359.0
2228	35	$\frac{11}{16} \times 3\frac{1}{2}$	.6875	.6885	3.490	3.500	.933	.938	.921	.927	$\frac{13}{16}$	1.80621	399.0
2428	36	$\frac{3}{4} \times 3\frac{1}{2}$	.7500	.7510	3.490	3.500	.933	.938	.921	.927	$\frac{13}{16}$	1.92810	435.0

*Table 9–40

TABLE 8–21

Woodruff Keyseat, Key Above Shaft, and Keyway Dimensions

SAE Handbook 1954

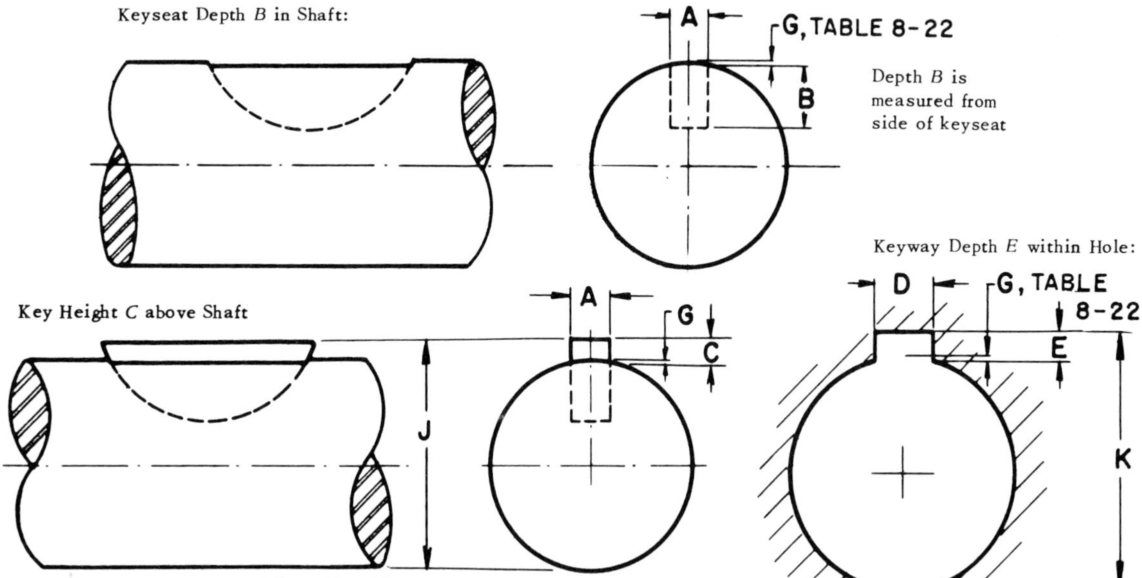

Height *C*, which is measured from side of keyseat is one-half the basic width, A.

J = shaft diameter + *C* − *G*

K = Hole diameter + *E* − *G*

ASA* B5.3 1950	Key No.	Nominal Key Size A x B	Keyseat Width A Basic	Keyseat Width A Min	Keyseat Width A Max	Keyseat Depth B Min	Keyseat Depth B Max	Key Above Shaft Height C Basic	Key Above Shaft Height C Min	Key Above Shaft Height C Max	Keyway Width D Class 2 Min	Keyway Width D Class 2 Max	Keyway Depth E Min	Keyway Depth E Max
202	201	1/16 x 1/4	.0625	.0615	.0630	.0728	.0778	.0312	.0262	.0362	.0635	.0655	.0372	.0422
202½	206	1/16 x 5/16	.0625	.0615	.0630	.1038	.1088	.0312	.0262	.0362	.0635	.0655	.0372	.0422
302½	207	3/32 x 5/16	.0938	.0928	.0943	.0882	.0932	.0469	.0419	.0519	.0948	.0968	.0529	.0579
203	211	1/16 x 3/8	.0625	.0615	.0630	.1358	.1408	.0312	.0262	.0362	.0635	.0655	.0372	.0422
303	212	3/32 x 3/8	.0938	.0928	.0943	.1202	.1252	.0469	.0419	.0519	.0948	.0968	.0529	.0579
403	213	1/8 x 3/8	.1250	.1240	.1255	.1045	.1095	.0625	.0575	.0675	.1260	.1280	.0685	.0735
204	1	1/16 x 1/2	.0625	.0615	.0630	.1668	.1718	.0312	.0262	.0362	.0635	.0655	.0372	.0422
304	2	3/32 x 1/2	.0938	.0928	.0943	.1511	.1561	.0469	.0419	.0519	.0948	.0968	.0529	.0579
404	3	1/8 x 1/2	.1250	.1240	.1255	.1355	.1405	.0625	.0575	.0675	.1260	.1280	.0685	.0735
305	4	3/32 x 5/8	.0938	.0928	.0943	.1981	.2031	.0469	.0419	.0519	.0948	.0968	.0529	.0579
405	5	1/8 x 5/8	.1250	.1240	.1255	.1825	.1875	.0625	.0575	.0675	.1260	.1280	.0685	.0735
505	6	5/32 x 5/8	.1563	.1553	.1568	.1669	.1719	.0781	.0731	.0831	.1573	.1593	.0841	.0891
605	61	3/16 x 5/8	.1875	.1863	.1880	.1513	.1563	.0937	.0887	.0987	.1885	.1905	.0997	.1047
406	7	1/8 x 3/4	.1250	.1240	.1255	.2455	.2505	.0625	.0575	.0675	.1260	.1280	.0685	.0735
506	8	5/32 x 3/4	.1563	.1553	.1568	.2299	.2349	.0781	.0731	.0831	.1573	.1593	.0841	.0891
606	9	3/16 x 3/4	.1875	.1863	.1880	.2143	.2193	.0937	.0887	.0987	.1885	.1905	.0997	.1047
806	91	1/4 x 3/4	.2500	.2487	.2505	.1830	.1880	.1250	.1200	.1300	.2510	.2530	.1310	.1360
507	10	5/32 x 7/8	.1563	.1553	.1568	.2919	.2969	.0781	.0731	.0831	.1573	.1593	.0841	.0891
607	11	3/16 x 7/8	.1875	.1863	.1880	.2763	.2813	.0937	.0887	.0987	.1885	.1905	.0997	.1047
707	12	7/32 x 7/8	.2188	.2175	.2193	.2607	.2657	.1093	.1043	.1143	.2198	.2218	.1153	.1203
807	A	1/4 x 7/8	.2500	.2487	.2505	.2450	.2500	.1250	.1200	.1300	.2510	.2530	.1310	.1360
608	13	3/16 x 1	.1875	.1863	.1880	.3393	.3443	.0937	.0887	.0987	.1885	.1905	.0997	.1047
708	14	7/32 x 1	.2188	.2175	.2193	.3237	.3287	.1093	.1043	.1143	.2198	.2218	.1153	.1203
808	15	1/4 x 1	.2500	.2487	.2505	.3080	.3130	.1250	.1200	.1300	.2510	.2530	.1310	.1360
1008	B	5/16 x 1	.3125	.3111	.3130	.2768	.2818	.1562	.1512	.1612	.3135	.3155	.1622	.1672
609	16	3/16 x 1 1/8	.1875	.1863	.1880	.3853	.3903	.0937	.0887	.0987	.1885	.1905	.0997	.1047
709	17	7/32 x 1 1/8	.2188	.2175	.2193	.3697	.3747	.1093	.1043	.1143	.2198	.2218	.1153	.1203
809	18	1/4 x 1 1/8	.2500	.2487	.2505	.3540	.3590	.1250	.1200	.1300	.2510	.2530	.1310	.1360
1009	C	5/16 x 1 1/8	.3125	.3111	.3130	.3228	.3278	.1562	.1512	.1612	.3135	.3155	.1622	.1672
610	19	3/16 x 1 1/4	.1875	.1863	.1880	.4483	.4533	.0937	.0887	.0987	.1885	.1905	.0997	.1047
710	20	7/32 x 1 1/4	.2188	.2175	.2193	.4327	.4377	.1093	.1043	.1143	.2198	.2218	.1153	.1203
810	21	1/4 x 1 1/4	.2500	.2487	.2505	.4170	.4220	.1250	.1200	.1300	.2510	.2530	.1310	.1360
1010	D	5/16 x 1 1/4	.3125	.3111	.3130	.3858	.3908	.1562	.1512	.1612	.3135	.3155	.1622	.1672
1210	E	3/8 x 1 1/4	.3750	.3735	.3755	.3545	.3595	.1875	.1825	.1925	.3760	.3780	.1935	.1985
811	22	1/4 x 1 3/8	.2500	.2487	.2505	.4640	.4690	.1250	.1200	.1300	.2510	.2530	.1310	.1360
1011	23	5/16 x 1 3/8	.3125	.3111	.3130	.4328	.4378	.1562	.1512	.1612	.3135	.3155	.1622	.1672
1211	F	3/8 x 1 3/8	.3750	.3735	.3755	.4015	.4065	.1875	.1825	.1925	.3760	.3780	.1935	.1985
812	24	1/4 x 1 1/2	.2500	.2487	.2505	.5110	.5160	.1250	.1200	.1300	.2510	.2530	.1310	.1360
1012	25	5/16 x 1 1/2	.3125	.3111	.3130	.4798	.4848	.1562	.1512	.1612	.3135	.3155	.1622	.1672
1212	G	3/8 x 1 1/2	.3750	.3735	.3755	.4485	.4535	.1875	.1825	.1925	.3760	.3780	.1935	.1985

*Table 9-40

continued on next page

ASA* B5.3 1950	Key No.	Nominal Key Size A x B	Keyseat					Key Above Shaft			Keyway			
			Width A			Depth B*		Height C*			Width D Class 2		Depth E	
			Basic	Min	Max	Min	Max	Basic	Min	Max	Min	Max	Min	Max
617	126	3/16 x 2 1/8	.1875	.1863	.1880	.3073	.3123	.0937	.0887	.0987	.1885	.1905	.0997	.1047
817	127	1/4 x 2 1/8	.2500	.2487	.2505	.2760	.2810	.1250	.1200	.1300	.2510	.2530	.1310	.1360
1017	128	5/16 x 2 1/8	.3125	.3111	.3130	.2448	.2498	.1562	.1512	.1612	.3135	.3155	.1622	.1672
1217	129	3/8 x 2 1/8	.3750	.3735	.3755	.2135	.2185	.1875	.1825	.1925	.3760	.3780	.1935	.1985
617	26	3/16 x 2 1/8	.1875	.1863	.1880	.4323	.4373	.0937	.0887	.0987	.1885	.1905	.0997	.1047
817	27	1/4 x 2 1/8	.2500	.2487	.2505	.4010	.4060	.1250	.1200	.1300	.2510	.2530	.1310	.1360
1017	28	5/16 x 2 1/8	.3125	.3111	.3130	.3698	.3748	.1562	.1512	.1612	.3135	.3155	.1622	.1672
1217	29	3/8 x 2 1/8	.3750	.3735	.3755	.3385	.3435	.1875	.1825	.1925	.3760	.3780	.1935	.1985
822	Rx	1/4 x 2 3/4	.2500	.2487	.2505	.4640	.4690	.1250	.1200	.1300	.2510	.2530	.1310	.1360
1022	Sx	5/16 x 2 3/4	.3125	.3111	.3130	.4328	.4378	.1562	.1512	.1612	.3135	.3155	.1622	.1672
1222	Tx	3/8 x 2 3/4	.3750	.3735	.3755	.4015	.4065	.1875	.1825	.1925	.3760	.3780	.1935	.1985
1422	Ux	7/16 x 2 3/4	.4375	.4360	.4380	.3703	.3753	.2187	.2137	.2237	.4385	.4405	.2247	.2297
1622	Vx	1/2 x 2 3/4	.5000	.4985	.5005	.3390	.3440	.2500	.2450	.2550	.5010	.5030	.2560	.2610
822	R	1/4 x 2 3/4	.2500	.2487	.2505	.6200	.6250	.1250	.1200	.1300	.2510	.2530	.1310	.1360
1022	S	5/16 x 2 3/4	.3125	.3111	.3130	.5888	.5938	.1562	.1512	.1612	.3135	.3155	.1622	.1672
1222	T	3/8 x 2 3/4	.3750	.3735	.3755	.5575	.5625	.1875	.1825	.1925	.3760	.3780	.1935	.1985
1422	U	7/16 x 2 3/4	.4375	.4360	.4380	.5263	.5313	.2187	.2137	.2237	.4385	.4405	.2247	.2297
1622	V	1/2 x 2 3/4	.5000	.4985	.5005	.4950	.5000	.2500	.2450	.2550	.5010	.5030	.2560	.2610
1228	30	3/8 x 3 1/2	.3750	.3735	.3755	.7455	.7505	.1875	.1825	.1925	.3760	.3780	.1935	.1985
1428	31	7/16 x 3 1/2	.4375	.4360	.4380	.7143	.7193	.2187	.2137	.2237	.4385	.4405	.2247	.2297
1628	32	1/2 x 3 1/2	.5000	.4985	.5005	.6830	.6880	.2500	.2450	.2550	.5010	.5030	.2560	.2610
1828	33	9/16 x 3 1/2	.5625	.5610	.5630	.6518	.6568	.2812	.2762	.2862	.5635	.5655	.2872	.2922
2028	34	5/8 x 3 1/2	.6250	.6235	.6255	.6205	.6255	.3125	.3075	.3175	.6260	.6280	.3185	.3235
2228	35	11/16 x 3 1/2	.6875	.6860	.6880	.5893	.5943	.3437	.3387	.3487	.6885	.6905	.3497	.3547
2428	36	3/4 x 3 1/2	.7500	.7485	.7505	.5580	.5630	.3750	.3700	.3800	.7510	.7530	.3810	.3860

*Table 9–40

TABLE 8–22

Versed Sine Dimension *G*
SAE Handbook – 1954

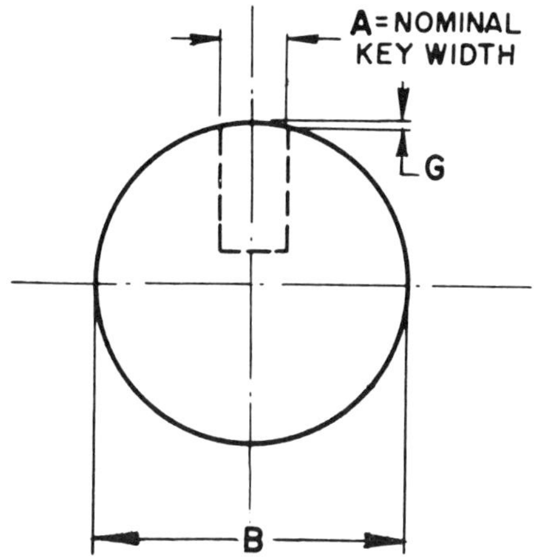

Shaft Diam *B*	Keyway Width, *A* 1/16	3/32	1/8	5/32	3/16	7/32	1/4	5/16	3/8	7/16	1/2	9/16	5/8	11/16	3/4
.3125	.0032														
.3437	.0029	.0065													
.3750	.0026	.0060	.0107												
.4060	.0024	.0055	.0099												
.4375	.0022	.0051	.0091												
.4687	.0021	.0047	.0085	.0134											
.5000	.0020	.0044	.0079	.0125											
.5625		.0039	.0070	.0111	.0161										
.6250		.0035	.0063	.0099	.0144	.0198									
.6875		.0032	.0057	.0090	.0130	.0179	.0235								
.7500		.0029	.0052	.0082	.0119	.0163	.0214	.0341							
.8125		.0027	.0048	.0076	.0110	.0150	.0197	.0312							
.8750		.0025	.0045	.0070	.0102	.0139	.0182	.0288							
.9375			.0042	.0066	.0095	.0129	.0170	.0268	.0391						
1.0000			.0039	.0061	.0089	.0121	.0159	.0250	.0365						
1.0625			.0037	.0058	.0083	.0114	.0149	.0235	.0342						
1.1250			.0035	.0055	.0079	.0107	.0141	.0221	.0322	.0443					
1.1875			.0033	.0052	.0074	.0102	.0133	.0209	.0304	.0418					
1.2500			.0031	.0049	.0071	.0097	.0126	.0198	.0288	.0395					
1.3750				.0045	.0064	.0088	.0115	.0180	.0261	.0357	.0471				
1.5000				.0041	.0059	.0080	.0105	.0165	.0238	.0326	.0429				
1.6250				.0038	.0054	.0074	.0097	.0152	.0219	.0300	.0394	.0502			
1.7500					.0050	.0069	.0090	.0141	.0203	.0278	.0365	.0464			
1.8750					.0047	.0064	.0084	.0131	.0189	.0259	.0340	.0432	.0536		
2.0000					.0044	.0060	.0078	.0123	.0177	.0242	.0318	.0404	.0501		
2.1250						.0056	.0074	.0116	.0167	.0228	.0298	.0379	.0470	.0572	.0684
2.2500							.0070	.0109	.0157	.0215	.0281	.0357	.0443	.0538	.0643
2.3750								.0103	.0149	.0203	.0266	.0338	.0419	.0509	.0608
2.5000									.0141	.0193	.0253	.0321	.0397	.0482	.0576
2.6250									.0135	.0184	.0240	.0305	.0377	.0457	.0547
2.7500										.0175	.0229	.0291	.0360	.0437	.0521
2.8750										.0168	.0219	.0278	.0344	.0417	.0498
3.0000											.0210	.0266	.0329	.0399	.0476

$$G = \frac{B}{2} - \sqrt{\frac{B^2 - A^2}{4}}$$

INDEX TO
SECTION 9

Bolts — Counterbores — Broached, Drilled, Reamed, and Tapped Holes — Slots — Screw Threads

TABLE 9–1

Decimal Equivalents of Regular Sizes of Drills

Metal Cutting Tool Handbook   Metal Cutting Tool Institute

Decimal	Inch	Wire	mm
.0135		80	
.0145		79	
.0156	1/64		
.0160		78	
.0180		77	
.0197			.5
.0200		76	
.0210		75	
.0225		74	
.0236			.6
.0240		73	
.0250		72	
.0260		71	
.0276			.7
.0280		70	
.0292		69	
.0295			.75
.0310		68	
.0312	1/32		
.0315			.8
.0320		67	
.0330		66	
.0350		65	
.0354			.9
.0360		64	
.0370		63	
.0380		62	
.0390		61	
.0394			1.
.0400		60	
.0410		59	
.0420		58	
.0430		57	
.0433			1.1
.0465		56	
.0469	3/64		
.0472			1.2
.0492			1.25
.0512			1.3
.0520		55	

Decimal	Inch	Wire	mm
.0550		54	
.0551			1.4
.0591			1.5
.0595		53	
.0625	1/16		
.0630			1.6
.0635		52	
.0669			1.7
.0670		51	
.0689			1.75
.0700		50	
.0709			1.8
.0730		49	
.0748			1.9
.0760		48	
.0781	5/64		
.0785		47	
.0787			2.
.0810		46	
.0820		45	
.0827			2.1
.0860		44	
.0866			2.2
.0886			2.25
.0890		43	
.0906			2.3
.0935		42	
.0938	3/32		
.0945			2.4
.0960		41	
.0980		40	
.0984			2.5
.0995		39	
.1015		38	
.1024			2.6
.1040		37	
.1063			2.7
.1065		36	
.1083			2.75
.1094	7/64		

Decimal	Inch	Wire	mm
.1100		35	
.1102			2.8
.1110		34	
.1130		33	
.1142			2.9
.1160		32	
.1181			3.
.1200		31	
.1220			3.1
.1250	1/8		
.1260			3.2
.1280			3.25
.1285		30	
.1299			3.3
.1339			3.4
.1360		29	
.1378			3.5
.1405		28	
.1406	9/64		
.1417			3.6
.1440		27	
.1457			3.7
.1470		26	
.1476			3.75
.1495		25	
.1496			3.8
.1520		24	
.1535			3.9
.1540		23	
.1562	5/32		
.1570		22	
.1575			4.
.1590		21	
.1610		20	
.1614			4.1
.1654			4.2
.1660		19	
.1673			4.25
.1693			4.3

Decimal	Inch	Wire	mm
.1695		18	
.1719	11/64		
.1730		17	
.1732			4.4
.1770		16	
.1772			4.5
.1800		15	
.1811			4.6
.1820		14	
.1850		13	4.7
.1870			4.75
.1875	3/16		
.1890		12	4.8
.1910		11	
.1929			4.9
.1935		10	
.1960		9	
.1968			5.
.1990		8	
.2008			5.1
.2010		7	
.2031	13/64		
.2040		6	
.2047			5.2
.2055		5	
.2067			5.25
.2087			5.3
.2090		4	
.2126			5.4
.2130		3	
.2165			5.5
.2188	7/32		
.2205			5.6
.2210		2	
.2244			5.7
.2264			5.75
.2280		1	
.2283			5.8
.2323			5.9

Decimal	Inch	Letter	mm
.2340		A	
.2344	15/64		
.2362			6.
.2380		B	
.2402			6.1
.2420		C	
.2441			6.2
.2460		D	
.2461			6.25
.2480			6.3
.2500	1/4	E	
.2520			6.4
.2559			6.5
.2570		F	
.2598			6.6
.2610		G	
.2638			6.7
.2656	17/64		
.2657			6.75
.2660		H	
.2677			6.8
.2717			6.9
.2720		I	
.2756			7.
.2770		J	
.2795			7.1
.2810		K	
.2812	9/32		
.2835			7.2
.2854			7.25
.2874			7.3
.2900		L	
.2913			7.4
.2950		M	
.2953			7.5

Decimal	Inch	Letter	mm
.2969	19/64		
.2992			7.6
.3020		N	
.3031			7.7
.3051			7.75
.3071			7.8
.3110			7.9
.3125	5/16		
.3150			8.
.3160		O	
.3189			8.1
.3228			8.2
.3230		P	
.3248			8.25
.3268			8.3
.3281	21/64		
.3307			8.4
.3320		Q	
.3346			8.5
.3386			8.6
.3390		R	
.3425			8.7
.3438	11/32		
.3445			8.75
.3465			8.8
.3480		S	
.3504			8.9
.3543			9.
.3580		T	
.3583			9.1
.3594	23/64		
.3622			9.2
.3642			9.25
.3661			9.3

Decimal	Inch	Letter	mm
.3680		U	
.3701			9.4
.3740			9.5
.3750	3/8		
.3770		V	
.3780			9.6
.3819			9.7
.3839			9.75
.3858			9.8
.3860		W	
.3898			9.9
.3906	25/64		
.3937			10.
.3970		X	
.4040		Y	
.4062	13/32		
.4130		Z	
.4134			10.5
.4219	27/64		
.4331			11.
.4375	7/16		
.4528			11.5
.4531	29/64		
.4688	15/32		
.4724			12.
.4844	31/64		
.4921			12.5
.5000	1/2		
.5118			13.
.5156	33/64		
.5312	17/32		
.5315			13.5
.5469	35/64		
.5512			14.

continued on next page

TABLE 9-1, continued

Decimal	Inch	mm
.5625	9/16	
.5709		14.5
.5781	37/64	
.5906		15.
.5938	19/32	
.6094	39/64	
.6102		15.5
.6250	5/8	
.6299		16.
.6406	41/64	
.6496		16.5
.6562	21/32	
.6693		17.
.6719	43/64	
.6875	11/16	
.6890		17.5
.7031	45/64	
.7087		18.
.7188	23/32	
.7283		18.5
.7344	47/64	
.7480		19.
.7500	3/4	
.7656	49/64	
.7677		19.5
.7812	25/32	
.7874		20.
.7969	51/64	
.8071		20.5
.8125	13/16	
.8268		21.
.8281	53/64	
.8438	27/32	
.8465		21.5
.8594	55/64	
.8661		22.
.8750	7/8	
.8858		22.5
.8906	57/64	
.9055		23.
.9062	29/32	
.9219	59/64	
.9252		23.5

Decimal	Inch	mm
.9375	15/16	
.9449		24.
.9531	61/64	
.9646		24.5
.9688	31/32	
.9843		25.
.9844	63/64	
1.0000	1	
1.0039		25.5
1.0156	1 1/64	
1.0236		26.
1.0312	1 1/32	
1.0433		26.5
1.0469	1 3/64	
1.0625	1 1/16	
1.0630		27.
1.0781	1 5/64	
1.0827		27.5
1.0938	1 3/32	
1.1024		28.
1.1094	1 7/64	
1.1220		28.5
1.1250	1 1/8	
1.1406	1 9/64	
1.1417		29.
1.1562	1 5/32	
1.1614		29.5
1.1719	1 11/64	
1.1811		30.
1.1875	1 3/16	
1.2008		30.5
1.2031	1 13/64	
1.2188	1 7/32	
1.2205		31.
1.2344	1 15/64	
1.2402		31.5
1.2500	1 1/4	
1.2598		32.
1.2656	1 17/64	
1.2795		32.5
1.2812	1 9/32	
1.2969	1 19/64	

Decimal	Inch	mm
1.2992		33.
1.3125	1 5/16	
1.3189		33.5
1.3281	1 21/64	
1.3386		34.
1.3438	1 11/32	
1.3583		34.5
1.3594	1 23/64	
1.3750	1 3/8	
1.3780		35.
1.3906	1 25/64	
1.3976		35.5
1.4062	1 13/32	
1.4173		36.
1.4219	1 27/64	
1.4370		36.5
1.4375	1 7/16	
1.4531	1 29/64	
1.4567		37.
1.4688	1 15/32	
1.4764		37.5
1.4844	1 31/64	
1.4961		38.
1.5000	1 1/2	
1.5156	1 33/64	
1.5157		38.5
1.5312	1 17/32	
1.5354		39.
1.5469	1 35/64	
1.5551		39.5
1.5625	1 9/16	
1.5748		40.
1.5781	1 37/64	
1.5938	1 19/32	
1.5945		40.5
1.6094	1 39/64	
1.6142		41.
1.6250	1 5/8	
1.6339		41.5
1.6406	1 41/64	
1.6535		42.
1.6562	1 21/32	

Decimal	Inch	mm
1.6719	1 43/64	
1.6732		42.5
1.6875	1 11/16	
1.6929		43.
1.7031	1 45/64	
1.7126		43.5
1.7188	1 23/32	
1.7323		44.
1.7344	1 47/64	
1.7500	1 3/4	
1.7520		44.5
1.7656	1 49/64	
1.7717		45.
1.7812	1 25/32	
1.7913		45.5
1.7969	1 51/64	
1.8110		46.
1.8125	1 13/16	
1.8281	1 53/64	
1.8307		46.5
1.8438	1 27/32	
1.8504		47.
1.8594	1 55/64	
1.8701		47.5
1.8750	1 7/8	
1.8898		48.
1.8906	1 57/64	
1.9062	1 29/32	
1.9094		48.5
1.9219	1 59/64	
1.9291		49.
1.9375	1 15/16	
1.9488		49.5
1.9531	1 61/64	
1.9685		50.
1.9688	1 31/32	
1.9844	1 63/64	
1.9882		50.5
2.0000	2	
2.0079		51.
2.0156	2 1/64	
2.0276		51.5

Decimal	Inch	mm
2.0312	2 1/32	
2.0469	2 3/64	
2.0472		52.
2.0625	2 1/16	
2.0669		52.5
2.0781	2 5/64	
2.0866		53.
2.0938	2 3/32	
2.1063		53.5
2.1094	2 7/64	
2.1250	2 1/8	
2.1260		54.
2.1406	2 9/64	
2.1457		54.5
2.1562	2 5/32	
2.1654		55.
2.1719	2 11/64	
2.1850		55.5
2.1875	2 3/16	
2.2031	2 13/64	
2.2047		56.
2.2188	2 7/32	
2.2244		56.5
2.2344	2 15/64	
2.2441		57.
2.2500	2 1/4	
2.2638		57.5
2.2656	2 17/64	
2.2812	2 9/32	
2.2835		58.
2.2969	2 19/64	
2.3031		58.5
2.3125	2 5/16	
2.3228		59.
2.3281	2 21/64	
2.3425		59.5
2.3438	2 11/32	
2.3594	2 23/64	
2.3622		60.
2.3750	2 3/8	
2.3819		60.5
2.3906	2 25/64	

Decimal	Inch	mm
2.4016		61.
2.4062	2 13/32	
2.4213		61.5
2.4219	2 27/64	
2.4375	2 7/16	
2.4409		62.
2.4531	2 29/64	
2.4606		62.5
2.4688	2 15/32	
2.4803		63.
2.4844	2 31/64	
2.5000	2 1/2	
2.5156	2 33/64	
2.5197		64.
2.5312	2 17/32	
2.5394		64.5
2.5469	2 35/64	
2.5591		65.
2.5625	2 9/16	
2.5781	2 37/64	
2.5787		65.5
2.5938	2 19/32	
2.5984		66.
2.6094	2 19/64	
2.6181		66.5
2.6250	2 5/8	
2.6378		67.
2.6406	2 41/64	
2.6562	2 21/32	
2.6575		67.5
2.6719	2 43/64	
2.6772		68.
2.6875	2 11/16	
2.6968		68.5
2.7031	2 45/64	
2.7165		69.
2.7188	2 23/32	
2.7344	2 47/64	
2.7362		69.5
2.7500	2 3/4	
2.7559		70.
2.7656	2 49/64	

Decimal	Inch	mm
2.7756		70.5
2.7812	2 25/32	
2.7953		71.
2.7969	2 51/64	
2.8125	2 13/16	
2.8150		71.5
2.8281	2 53/64	
2.8346		72.
2.8438	2 27/32	
2.8543		72.5
2.8594	2 55/64	
2.8740		73.
2.8750	2 7/8	
2.8906	2 57/64	
2.8937		73.5
2.9062	2 29/32	
2.9134		74.
2.9219	2 59/64	
2.9331		74.5
2.9375	2 15/16	
2.9528		75.
2.9531	2 61/64	
2.9688	2 31/32	
2.9724		75.5
2.9844	2 63/64	
2.9921		76.
3.0000	3	
3.0156	3 1/64	
3.0312	3 1/32	
3.0469	3 3/64	
3.0625	3 1/16	
3.0781	3 5/64	
3.0938	3 3/32	
3.1094	3 7/64	
3.1250	3 1/8	
3.1406	3 9/64	
3.1562	3 5/32	
3.1719	3 11/64	
3.1875	3 3/16	
3.2031	3 13/64	
3.2188	3 7/32	
3.2344	3 15/64	

Decimal	Inch
3.2500	3 1/4
3.2656	3 17/64
3.2812	3 9/32
3.2969	3 19/64
3.3125	3 5/16
3.3281	3 21/64
3.3438	3 11/32
3.3594	3 23/64
3.3750	3 3/8
3.3906	3 25/64
3.4062	3 13/32
3.4219	3 27/64
3.4375	3 7/16
3.4531	3 29/64
3.4688	3 15/32
3.4844	3 31/64
3.5000	3 1/2

TABLE 9–2

Drill Sizes Having Tang Dimensions to Fit Drill Drivers

ASA B5.12–1950   Twist Drills

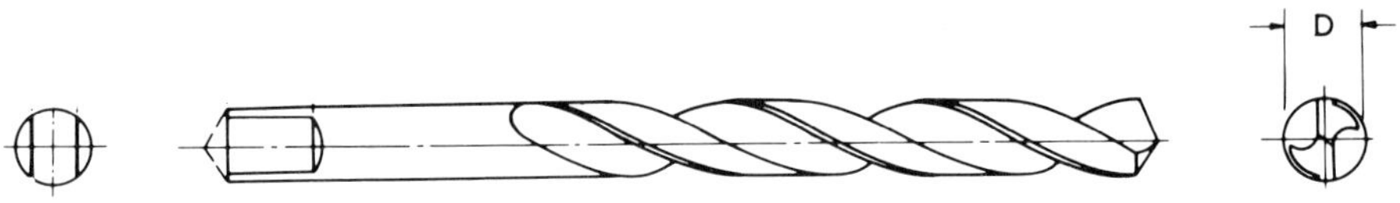

Diameter of Drill Inches D	Comparable Letter, Fraction, No., or MM Drill
0.1250	1/8
0.1285	30
0.1299	3.30 MM
0.1339	3.40 MM
0.1360	29
0.1378	3.50 MM
0.1406	9/64
0.1440	27
0.1470	26
0.1520	24
0.1562	5/32
0.1610	20
0.1660	19
0.1695	18
0.1719	11/64
0.1730	17
0.1770	16
0.1800	15
0.1850	13
0.1875	3/16
0.1910	11
0.1935	10
0.1960	9
0.1990	8
0.2031	13/64
0.2090	4
0.2130	3
0.2188	7/32
0.2244	5.70 MM
0.2280	1
0.2344	15/64
0.2402	6.10 MM
0.2460	D
0.2500	1/4
0.2520	6.40 MM
0.2570	F
0.2610	G
0.2656	17/64
0.2720	I
0.2770	J
0.2812	9/32
0.2854	7.25 MM
0.2913	7.40 MM
0.2969	19/64
0.3020	N
0.3071	7.80 MM
0.3125	5/16
0.3160	O
0.3230	P
0.3281	21/64
0.3320	Q
0.3390	R
0.3438	11/32
0.3480	S
0.3543	9 MM
0.3594	23/64
0.3680	U
0.3750	3/8
0.3860	W
0.3906	25/64
0.3970	X
0.4062	13/32
0.4219	27/64
0.4375	7/16
0.4531	29/64
0.4688	15/32
0.4844	31/64
0.5000	1/2
0.5156	33/64
0.5312	17/32
0.5469	35/64
0.5625	9/16
0.5781	37/64
0.5938	19/32
0.6094	39/64
0.6250	5/8
0.6406	41/64
0.6562	21/32
0.6719	43/64
0.6875	11/16
0.7031	45/64
0.7188	23/32
0.7344	47/64
0.7500	3/4
0.7656	49/64
0.7812	25/32
0.7969	51/64
0.8125	13/16
0.8281	53/64
0.8438	27/32
0.8594	55/64
0.8750	7/8
0.8906	57/64
0.9062	29/32
0.9219	59/64
0.9375	15/16
0.9531	61/64
0.9688	31/32
0.9844	63/64
1.0000	1
1.0156	1 1/64
1.0312	1 1/32
1.0469	1 3/64
1.0625	1 1/16
1.0781	1 5/64
1.0938	1 3/32
1.1094	1 7/64
1.1250	1 1/8
1.1406	1 9/64
1.1562	1 5/32
1.1719	1 11/64
1.1875	1 3/16
1.2031	1 13/64
1.2188	1 7/32
1.2344	1 15/64
1.2500	1 1/4
1.2812	1 9/32
1.3125	1 5/16
1.3438	1 11/32
1.3750	1 3/8

TABLE 9–3

Diameters of Center Drills

		1/32			3/16		
			3/64				13/64
	1/16					7/32	
			5/64				15/64
		3/32		1/4			
			7/64				17/64
1/8						9/32	
			9/64				19/64
		5/32			5/16		
			11/64			11/32	

TABLE 9–4

Sizes of Center Reamers

ASA B5.14–1949 Reamers

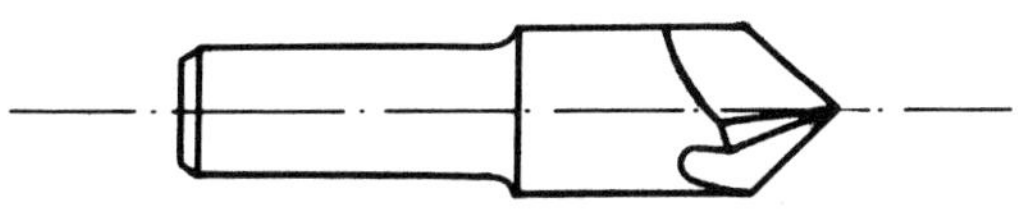

Size of Cut: 1/4, 3/8, 1/2, 5/8, 3/4 inch

Included angle is 60° for centers in shafts or tools and 82° for countersinking heads of flat-head screws.

TABLE 9–5

Sizes of Machine Countersinks

ASA B5.14–1949 Reamers

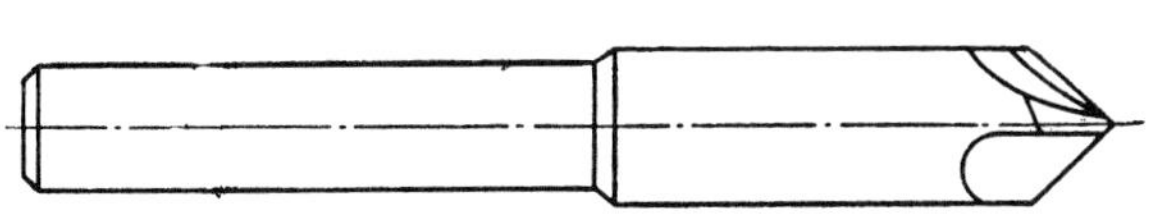

Size of Cut: 1/2, 5/8, 3/4, 7/8, and 1 inch

Included angle is 60° for centers in shafts or tools and 82° for countersinking heads of flat-head screws.

TABLE 9-6

Sizes of Combined Drills and Countersinks

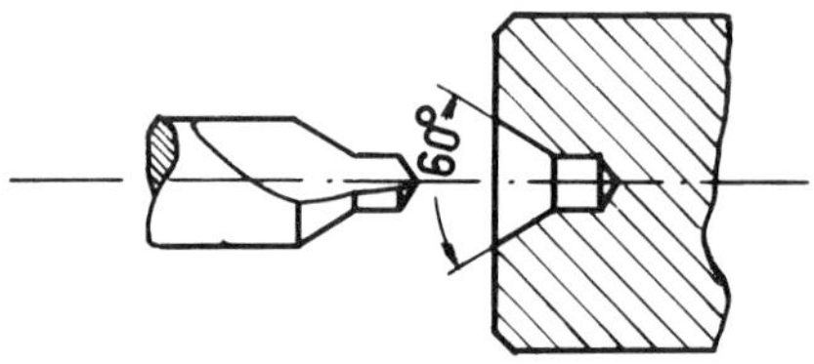

Regular or Plain Type Center

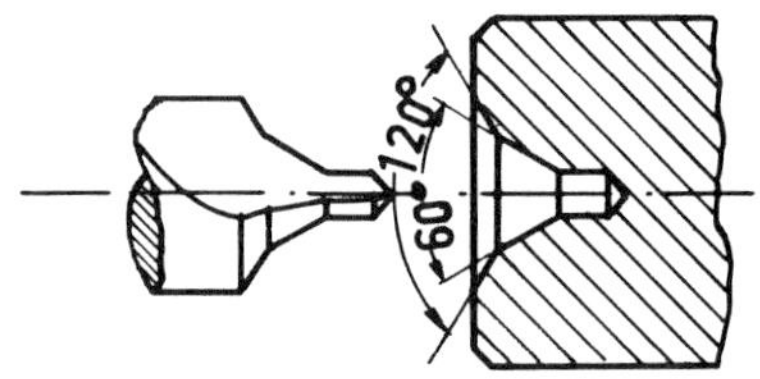

Bell Type Center

Size of Plain Type	Dimensions in Inches			Size of Bell Type	Dimensions in Inches		
	Diameter of Body	Diameter of Drill	Length of Drill		Diameter of Body	Diameter of Drill	Length of Drill
1	1/8	3/64	3/64	11	1/8	3/64	3/64
2	3/16	5/64	5/64	12	3/16	1/16	1/16
3	1/4	7/64	7/64	13	1/4	3/32	3/32
4	5/16	1/8	1/8	14	5/16	7/64	7/32
5	7/16	3/16	3/16	15	7/16	5/32	5/32
6	1/2	7/32	7/32	16	1/2	3/16	3/16
7	5/8	1/4	1/4	17	5/8	7/32	7/32
8	3/4	5/16	5/16	18	3/4	1/4	1/4

Regular centers are used on parts such as shafts that are to be turned or ground in ordinary production. The bell type center protects the outer edge of bearing area against damage during a series of operations such as grinding after turning and heat treatment. A third type of center (not shown here) is available when maximum protection to the center is required. It is specified for tools like arbors where the centers are used repeatedly, and sometimes on production parts that must be faced to length as a final operation.

TABLE 9-7

Machine Centers for Shaft Sizes*

Practice of Caterpillar Tractor Company – 1954

Shaft Diameter,* Inches	Sizes as indicated in Table 9-6	
	Regular	Bell type
3/8 to 1/2	2	
1/2 to 1-1/2	4	14
1-1/2 to 2-1/4	5	15
2-1/4 to 3	7	17
3 and over	8	18

*Length, weight of piece and amount of machining influence selection of machine center as well as diameter.

TABLE 9–8

Recommended Hole Size Limits before Threading for Different Lengths of Engagement, UNC, UNF, UNEF, UN, NC, NF, NEF and N Series, Classes 1B, 2B and 3B

ASA B1.1–1949 Unified and American Screw Threads, Third Edition

Designation		Minor Diameter Internal Threads				Recommended Hole Size Limits for Different Lengths of Engagement							
Thread Size	Threads per Inch	Minimum	Percent Basic Thread Height	Maximum†	Percent Basic Thread Height	To and Including 1/3 *D*		Above 1/3 *D* to 2/3 *D*		Above 2/3 *D* to 1 1/2 *D*		Above 1 1/2 *D* to 3 *D*	
						Min	Max	Min	Max	Min	Max	Min	Max
0 (*.060*)	80	0.0465	83.1	0.0514	52.9	0.0465	0.0500	0.0479	0.0514	0.0479	0.0514	0.0479	0.0514
1 (*.073*)	64	0.0561	83.3	0.0623	52.7	0.0561	0.0599	0.0585	0.0623	0.0585	0.0623	0.0585	0.0623
1 (*.073*)	72	0.0580	83.1	0.0635	52.7	0.0580	0.0613	0.0596	0.0629	0.0602	0.0635	0.0602	0.0635
2 (*.086*)	56	0.0667	83.2	0.0737	53.0	0.0667	0.0705	0.0686	0.0724	0.0699	0.0737	0.0699	0.0737
2 (*.086*)	64	0.0691	83.3	0.0753	52.7	0.0691	0.0724	0.0707	0.0740	0.0720	0.0753	0.0720	0.0753
3 (*.099*)	48	0.0764	83.5	0.0845	53.6	0.0764	0.0804	0.0785	0.0825	0.0805	0.0845	0.0806	0.0846
3 (*.099*)	56	0.0797	83.2	0.0865	53.9	0.0797	0.0831	0.0814	0.0848	0.0831	0.0865	0.0833	0.0867
4 (*.112*)	40	0.0849	83.4	0.0939	55.7	0.0849	0.0894	0.0871	0.0916	0.0894	0.0939	0.0902	0.0947
4 (*.112*)	48	0.0894	83.5	0.0968	56.2	0.0894	0.0931	0.0912	0.0949	0.0931	0.0968	0.0939	0.0976
5 (*.125*)	40	0.0979	83.4	0.1062	57.9	0.0979	0.1020	0.1000	0.1041	0.1021	0.1062	0.1036	0.1077
5 (*.125*)	44	0.1004	83.3	0.1079	57.9	0.1004	0.1041	0.1023	0.1060	0.1042	0.1079	0.1060	0.1097
6 (*.138*)	32	0.1042	83.3	0.1140	59.1	0.1042	0.1091	0.1066	0.1115	0.1091	0.1140	0.1115	0.1164
6 (*.138*)	40	0.1109	83.4	0.1186	59.7	0.1109	0.1148	0.1128	0.1167	0.1147	0.1186	0.1166	0.1205
8 (*.164*)	32	0.1302	83.3	0.1389	61.8	0.1302	0.1345	0.1324	0.1367	0.1346	0.1389	0.1367	0.1410
8 (*.164*)	36	0.1339	83.4	0.1416	62.1	0.1339	0.1377	0.1359	0.1397	0.1378	0.1416	0.1397	0.1435
10 (*.190*)	24	0.1449	83.3	0.1555	63.7	0.1449	0.1502	0.1475	0.1528	0.1502	0.1555	0.1528	0.1581
10 (*.190*)	32	0.1562	83.3	0.1641	63.8	0.1562	0.1601	0.1581	0.1621	0.1601	0.1641	0.1621	0.1661
12 (*.216*)	24	0.1709	83.3	0.1807	65.2	0.1709	0.1758	0.1733	0.1782	0.1758	0.1807	0.1782	0.1831
12 (*.216*)	28	0.1773	83.4	0.1857	65.3	0.1773	0.1815	0.1794	0.1836	0.1815	0.1857	0.1836	0.1878
12 (*.216*)	32	0.1822	83.3	0.1895	65.3	0.1822	0.1858	0.1837	0.1877	0.1855	0.1895	0.1873	0.1913
1/4	20	0.1959	83.3	0.2067	66.7	0.1959	0.2013	0.1986	0.2040	0.2013	0.2067	0.2040	0.2094
1/4	28	0.2113	83.4	0.2190	66.8	0.2113	0.2152	0.2131	0.2171	0.2150	0.2190	0.2169	0.2209
1/4	32	0.2162	83.3	0.2229	66.8	0.2162	0.2196	0.2172	0.2212	0.2189	0.2229	0.2206	0.2246
1/4	36	0.2199	83.4	0.2258	67.1	0.2199	0.2243	0.2199	0.2243	0.2214	0.2258	0.2229	0.2273
5/16	18	0.2524	83.3	0.2630	68.6	0.2524	0.2577	0.2551	0.2604	0.2577	0.2630	0.2604	0.2657
5/16	24	0.2674	83.3	0.2754	68.5	0.2674	0.2714	0.2694	0.2734	0.2714	0.2754	0.2734	0.2774
5/16	32	0.2787	83.3	0.2847	68.5	0.2787	0.2817	0.2792	0.2832	0.2807	0.2847	0.2822	0.2862
5/16	36	0.2824	83.4	0.2877	68.7	0.2824	0.2863	0.2824	0.2863	0.2837	0.2877	0.2850	0.2890
3/8	16	0.3073	83.4	0.3182	70.0	0.3073	0.3127	0.3101	0.3155	0.3128	0.3182	0.3155	0.3209
3/8	24	0.3299	83.3	0.3372	69.8	0.3299	0.3336	0.3314	0.3354	0.3332	0.3372	0.3351	0.3391
3/8	32	0.3412	83.3	0.3469	69.2	0.3412	0.3441	0.3415	0.3455	0.3429	0.3469	0.3444	0.3484
3/8	36	0.3449	83.4	0.3501	69.0	0.3449	0.3488	0.3449	0.3488	0.3461	0.3501	0.3474	0.3514

† Based on a length of engagement equal to the nominal diameter.

continued on next page

TABLE 9–8, continued

Designation		Minor Diameter Internal Threads				Recommended Hole Size Limits for Different Lengths of Engagement							
Thread Size	Threads per Inch	Minimum	Percent Basic Thread Height	Maximum†	Percent Basic Thread Height	To and Including 1/3 *D*		Above 1/3 *D* to 2/3 *D*		Above 2/3 *D* to 1 1/2 *D*		Above 1 1/2 *D* to 3 *D*	
						Min	Max	Min	Max	Min	Max	Min	Max
7/16	14	0.3602	83.3	0.3717	70.9	0.3602	0.3660	0.3630	0.3688	0.3659	0.3717	0.3688	0.3746
7/16	20	0.3834	83.3	0.3916	70.7	0.3834	0.3875	0.3855	0.3896	0.3875	0.3916	0.3896	0.3937
7/16	28	0.3988	83.4	0.4051	70.0	0.3988	0.4020	0.3995	0.4035	0.4011	0.4051	0.4017	0.4067
1/2	13	0.4167	83.4	0.4284	71.7	0.4167	0.4225	0.4196	0.4254	0.4226	0.4284	0.4255	0.4313
1/2	12	0.4098	83.3	0.4223	71.8	0.4098	0.4161	0.4129	0.4192	0.4160	0.4223	0.4192	0.4255
1/2	20	0.4459	83.3	0.4537	71.3	0.4459	0.4498	0.4477	0.4517	0.4497	0.4537	0.4516	0.4556
1/2	28	0.4613	83.4	0.4676	70.0	0.4613	0.4645	0.4620	0.4660	0.4636	0.4676	0.4652	0.4692
9/16	12	0.4723	83.3	0.4843	72.2	0.4723	0.4783	0.4753	0.4813	0.4783	0.4843	0.4813	0.4873
9/16	18	0.5024	83.3	0.5106	71.9	0.5024	0.5065	0.5045	0.5086	0.5065	0.5106	0.5086	0.5127
9/16	24	0.5174	83.3	0.5244	70.4	0.5174	0.5209	0.5186	0.5226	0.5204	0.5244	0.5221	0.5261
9/16	28	0.5238	83.4	0.5301	69.8	0.5238	0.5270	0.5245	0.5285	0.5261	0.5301	0.5277	0.5317
5/8	11	0.5266	83.3	0.5391	72.7	0.5266	0.5328	0.5298	0.5360	0.5329	0.5391	0.5360	0.5422
5/8	12	0.5348	83.3	0.5463	72.7	0.5348	0.5406	0.5377	0.5435	0.5405	0.5463	0.5434	0.5492
5/8	18	0.5649	83.3	0.5730	72.1	0.5649	0.5690	0.5670	0.5711	0.5690	0.5730	0.5711	0.5752
5/8	24	0.5799	83.3	0.5869	70.4	0.5799	0.5834	0.5811	0.5851	0.5829	0.5869	0.5846	0.5886
5/8	28	0.5863	83.4	0.5926	69.8	0.5863	0.5895	0.5870	0.5910	0.5886	0.5926	0.5902	0.5942
11/16	12	0.5973	83.3	0.6085	73.0	0.5973	0.6029	0.6001	0.6057	0.6029	0.6085	0.6057	0.6113
11/16	24	0.6424	83.3	0.6494	70.4	0.6424	0.6459	0.6436	0.6476	0.6454	0.6494	0.6471	0.6511
3/4	10	0.6417	83.4	0.6545	73.5	0.6417	0.6481	0.6449	0.6513	0.6481	0.6545	0.6513	0.6577
3/4	12	0.6598	83.3	0.6707	73.3	0.6598	0.6652	0.6626	0.6680	0.6653	0.6707	0.6680	0.6734
3/4*	16*	0.6823	83.4	0.6908	72.9	0.6823	0.6866	0.6844	0.6887	0.6865	0.6908	0.6886	0.6929
3/4	20	0.6959	83.3	0.7037	71.3	0.6959	0.6998	0.6977	0.7017	0.6997	0.7037	0.7016	0.7056
3/4	28	0.7113	83.4	0.7176	69.8	0.7113	0.7145	0.7120	0.7160	0.7136	0.7176	0.7152	0.7192
13/16	12	0.7223	83.3	0.7329	73.5	0.7223	0.7276	0.7250	0.7303	0.7276	0.7329	0.7303	0.7356
13/16	16	0.7448	83.4	0.7533	72.9	0.7448	0.7491	0.7469	0.7512	0.7490	0.7533	0.7511	0.7554
13/16	20	0.7584	83.3	0.7662	71.3	0.7584	0.7623	0.7602	0.7642	0.7622	0.7662	0.7641	0.7681
7/8	9	0.7547	83.3	0.7681	74.1	0.7547	0.7614	0.7580	0.7647	0.7614	0.7681	0.7647	0.7714
7/8	12	0.7848	83.3	0.7952	73.7	0.7848	0.7900	0.7874	0.7926	0.7900	0.7952	0.7926	0.7978
7/8	14	0.7977	83.3	0.8068	73.5	0.7977	0.8022	0.8000	0.8045	0.8023	0.8068	0.8045	0.8090
7/8	16	0.8073	83.4	0.8158	72.9	0.8073	0.8116	0.8094	0.8137	0.8115	0.8158	0.8136	0.8179
7/8	20	0.8209	83.3	0.8287	71.3	0.8209	0.8248	0.8227	0.8267	0.8247	0.8287	0.8266	0.8306
7/8	28	0.8363	83.4	0.8426	69.8	0.8363	0.8395	0.8370	0.8410	0.8386	0.8426	0.8402	0.8442

*To find a suitable tap drill size for a specified diameter and thread series, as for example, 3/4–16UNF–2B (1¼ in. length of engagement) enter the table and find the hole limits 0.6886 to 0.6929. Next enter Table 9–1 to find a drill size, in this case 17.5 mm. An inch dimension drill of 11/16 is close to the lower limit and might be preferable. Since the ratio of length of engagement to the nominal diameter exceeds 1½ D, as well as the standard of 1D, the question might be raised whether the thread series should be special or UNF, as shown. The answer is negative so long as the symbol is to be used on drawings to specify the hole for production purposes.

†Based on a length of engagement equal to the nominal diameter.

continued on next page

TABLE 9–8, continued

Designation		Minor Diameter Internal Threads				Recommended Hole Size Limits for Different Lengths of Engagement							
Thread Size	Threads per Inch	Minimum	Percent Basic Thread Height	Maximum†	Percent Basic Thread Height	To and Including 1/3 *D*		Above 1/3 *D* to 2/3 *D*		Above 2/3 *D* to 1 1/2 *D*		Above 1 1/2 *D* to 3 *D*	
						Min	Max	Min	Max	Min	Max	Min	Max
15/16	12	0.8473	83.3	0.8575	73.9	0.8473	0.8524	0.8499	0.8550	0.8524	0.8575	0.8550	0.8601
15/16	16	0.8698	83.4	0.8783	72.9	0.8698	0.8741	0.8719	0.8762	0.8740	0.8783	0.8761	0.8804
15/16	20	0.8834	83.3	0.8912	71.3	0.8834	0.8873	0.8852	0.8892	0.8872	0.8912	0.8891	0.8931
1	8	0.8647	83.3	0.8797	74.1	0.8647	0.8722	0.8684	0.8759	0.8722	0.8797	0.8760	0.8835
1	12	0.9098	83.3	0.9198	74.1	0.9098	0.9148	0.9123	0.9173	0.9148	0.9198	0.9173	0.9223
1	14	0.9227	83.3	0.9315	73.8	0.9227	0.9271	0.9249	0.9293	0.9271	0.9315	0.9293	0.9337
1	16	0.9323	83.4	0.9408	72.9	0.9323	0.9366	0.9344	0.9387	0.9365	0.9408	0.9386	0.9429
1	20	0.9459	83.3	0.9537	71.3	0.9459	0.9498	0.9477	0.9517	0.9497	0.9537	0.9516	0.9556
1	28	0.9613	83.4	0.9676	69.8	0.9613	0.9645	0.9620	0.9660	0.9636	0.9676	0.9652	0.9692
1 1/16	12	0.9723	83.3	0.9823	74.1	0.9723	0.9773	0.9748	0.9798	0.9773	0.9823	0.9798	0.9848
1 1/16	16	0.9948	83.4	1.0033	72.9	0.9948	0.9991	0.9969	1.0012	0.9990	1.0033	1.0011	1.0054
1 1/16	18	1.0024	83.3	1.0105	72.1	1.0024	1.0065	1.0044	1.0085	1.0064	1.0105	1.0085	1.0126
1 1/8	7	0.9704	83.3	0.9875	74.1	0.9704	0.9790	0.9747	0.9833	0.9789	0.9875	0.9832	0.9918
1 1/8	8	0.9897	83.3	1.0047	74.1	0.9897	0.9972	0.9934	1.0009	0.9972	1.0047	1.0010	1.0085
1 1/8	12	1.0348	83.3	1.0448	74.1	1.0348	1.0398	1.0373	1.0423	1.0398	1.0448	1.0423	1.0473
1 1/8	16	1.0573	83.4	1.0658	72.9	1.0573	1.0616	1.0594	1.0637	1.0615	1.0658	1.0636	1.0679
1 1/8	18	1.0649	83.3	1.0730	72.1	1.0649	1.0690	1.0669	1.0710	1.0689	1.0730	1.0710	1.0751
1 1/8	20	1.0709	83.3	1.0787	71.3	1.0709	1.0748	1.0727	1.0767	1.0747	1.0787	1.0766	1.0806
1 1/8	28	1.0863	83.4	1.0926	69.8	1.0863	1.0895	1.0870	1.0910	1.0886	1.0926	1.0902	1.0942
1 3/16	12	1.0973	83.3	1.1073	74.1	1.0973	1.1023	1.0998	1.1048	1.1023	1.1073	1.1048	1.1098
1 3/16	16	1.1198	83.4	1.1283	72.9	1.1198	1.1241	1.1219	1.1262	1.1240	1.1283	1.1261	1.1304
1 3/16	18	1.1274	83.3	1.1355	72.1	1.1274	1.1315	1.1294	1.1335	1.1314	1.1355	1.1335	1.1376
1 1/4	7	1.0954	83.3	1.1125	74.1	1.0954	1.1040	1.0997	1.1083	1.1039	1.1125	1.1082	1.1168
1 1/4	8	1.1147	83.3	1.1297	74.1	1.1147	1.1222	1.1184	1.1259	1.1222	1.1297	1.1260	1.1335
1 1/4	12	1.1598	83.3	1.1698	74.1	1.1598	1.1648	1.1623	1.1673	1.1648	1.1698	1.1673	1.1723
1 1/4	16	1.1823	83.4	1.1908	72.9	1.1823	1.1866	1.1844	1.1887	1.1865	1.1908	1.1886	1.1929
1 1/4	18	1.1899	83.3	1.1980	72.1	1.1899	1.1940	1.1919	1.1960	1.1939	1.1980	1.1960	1.2001
1 1/4	20	1.1959	83.3	1.2037	71.3	1.1959	1.1998	1.1977	1.2017	1.1997	1.2037	1.2016	1.2056
1 5/16	12	1.2223	83.3	1.2323	74.1	1.2223	1.2273	1.2248	1.2298	1.2273	1.2323	1.2298	1.2348
1 5/16	16	1.2448	83.4	1.2533	72.9	1.2448	1.2491	1.2469	1.2512	1.2490	1.2533	1.2511	1.2554
1 5/16	18	1.2524	83.3	1.2605	72.1	1.2524	1.2565	1.2544	1.2585	1.2564	1.2605	1.2585	1.2626

†Based on a length of engagement equal to the nominal diameter.

continued on next page

TABLE 9–8, continued

Designation		Minor Diameter Internal Threads				Recommended Hole Size Limits for Different Lengths of Engagement							
Thread Size	Threads per Inch	Minimum	Percent Basic Thread Height	Maximum†	Percent Basic Thread Height	To and Including 1/3 *D*		Above 1/3 *D* to 2/3 *D*		Above 2/3 *D* to 1 1/2 *D*		Above 1 1/2 *D* to 3 *D*	
						Min	Max	Min	Max	Min	Max	Min	Max
1 3/8	6	1.1946	83.3	1.2146	74.1	1.1946	1.2046	1.1996	1.2096	1.2046	1.2146	1.2096	1.2196
1 3/8	8	1.2397	83.3	1.2547	74.1	1.2397	1.2472	1.2434	1.2509	1.2472	1.2547	1.2510	1.2585
1 3/8	12	1.2848	83.3	1.2948	74.1	1.2848	1.2898	1.2873	1.2923	1.2898	1.2948	1.2923	1.2973
1 3/8	16	1.3073	83.4	1.3158	72.9	1.3073	1.3116	1.3094	1.3137	1.3115	1.3158	1.3136	1.3179
1 3/8	18	1.3149	83.3	1.3230	72.1	1.3149	1.3190	1.3169	1.3210	1.3189	1.3230	1.3210	1.3251
1 7/16	12	1.3473	83.3	1.3573	74.1	1.3473	1.3523	1.3498	1.3548	1.3523	1.3573	1.3548	1.3598
1 7/16	16	1.3698	83.4	1.3783	72.9	1.3698	1.3741	1.3719	1.3762	1.3740	1.3783	1.3761	1.3804
1 7/16	18	1.3774	83.3	1.3855	72.1	1.3774	1.3815	1.3794	1.3835	1.3814	1.3855	1.3835	1.3876
1 1/2	6	1.3196	83.3	1.3396	74.1	1.3196	1.3296	1.3246	1.3346	1.3296	1.3396	1.3346	1.3446
1 1/2	8	1.3647	83.3	1.3797	74.1	1.3647	1.3722	1.3684	1.3759	1.3722	1.3797	1.3760	1.3835
1 1/2	12	1.4098	83.3	1.4198	74.1	1.4098	1.4148	1.4123	1.4173	1.4148	1.4198	1.4173	1.4223
1 1/2	16	1.4323	83.4	1.4408	72.9	1.4323	1.4366	1.4344	1.4387	1.4365	1.4408	1.4386	1.4429
1 1/2	18	1.4399	83.3	1.4480	72.1	1.4399	1.4440	1.4419	1.4460	1.4439	1.4480	1.4460	1.4501
1 1/2	20	1.4459	83.3	1.4537	71.3	1.4459	1.4498	1.4477	1.4517	1.4497	1.4537	1.4516	1.4556
1 9/16	16	1.4948	83.4	1.5033	72.9	1.4948	1.4991	1.4969	1.5012	1.4990	1.5033	1.5011	1.5054
1 9/16	18	1.5024	83.3	1.5105	72.1	1.5024	1.5065	1.5044	1.5085	1.5064	1.5105	1.5085	1.5126
1 5/8	8	1.4897	83.3	1.5047	74.1	1.4897	1.4972	1.4934	1.5009	1.4972	1.5047	1.5010	1.5085
1 5/8	12	1.5348	83.3	1.5448	74.1	1.5348	1.5398	1.5373	1.5423	1.5398	1.5448	1.5423	1.5473
1 5/8	16	1.5573	83.4	1.5658	72.9	1.5573	1.5616	1.5594	1.5637	1.5615	1.5658	1.5636	1.5679
1 5/8	18	1.5649	83.3	1.5730	72.1	1.5649	1.5690	1.5669	1.5710	1.5689	1.5730	1.5710	1.5751
1 11/16	16	1.6198	83.4	1.6283	72.9	1.6198	1.6241	1.6219	1.6262	1.6240	1.6283	1.6261	1.6304
1 11/16	18	1.6274	83.3	1.6355	72.1	1.6274	1.6315	1.6294	1.6335	1.6314	1.6355	1.6335	1.6376
1 3/4	5	1.5335	83.3	1.5575	74.1	1.5335	1.5455	1.5395	1.5515	1.5455	1.5575	1.5515	1.5635
1 3/4	8	1.6147	83.3	1.6297	74.1	1.6147	1.6222	1.6184	1.6259	1.6222	1.6297	1.6260	1.6335
1 3/4	12	1.6598	83.3	1.6698	74.1	1.6598	1.6648	1.6623	1.6673	1.6648	1.6698	1.6673	1.6723
1 3/4	16	1.6823	83.4	1.6908	72.9	1.6823	1.6866	1.6844	1.6887	1.6865	1.6908	1.6886	1.6929
1 3/4	20	1.6959	83.3	1.7037	71.3	1.6959	1.6998	1.6977	1.7017	1.6997	1.7037	1.7016	1.7056
1 13/16	16	1.7448	83.4	1.7533	72.9	1.7448	1.7491	1.7469	1.7512	1.7490	1.7533	1.7511	1.7554
1 7/8	8	1.7397	83.3	1.7547	74.1	1.7397	1.7472	1.7434	1.7509	1.7472	1.7547	1.7510	1.7585
1 7/8	12	1.7848	83.3	1.7948	74.1	1.7848	1.7898	1.7873	1.7923	1.7898	1.7948	1.7923	1.7973
1 7/8	16	1.8073	83.4	1.8158	72.9	1.8073	1.8116	1.8094	1.8137	1.8115	1.8158	1.8136	1.8179

†Based on a length of engagement equal to the nominal diameter.

continued on next page

TABLE 9–8, continued

Designation		Minor Diameter Internal Threads				Recommended Hole Size Limits for Different Lengths of Engagement							
Thread Size	Threads per Inch	Minimum	Percent Basic Thread Height	Maximum†	Percent Basic Thread Height	To and Including 1/3 *D*		Above 1/3 *D* to 2/3 *D*		Above 2/3 *D* to 1 1/2 *D*		Above 1 1/2 *D* to 3 *D*	
						Min	Max	Min	Max	Min	Max	Min	Max
1 15/16	16	1.8698	83.4	1.8783	72.9	1.8698	1.8741	1.8719	1.8762	1.8740	1.8783	1.8761	1.8804
2*	4 ½*	1.7594	83.3	1.7861	74.1	1.7594	1.7727	1.7661	1.7794	1.7728	1.7861	1.7794	1.7927
2	8	1.8647	83.3	1.8797	74.1	1.8647	1.8722	1.8684	1.8759	1.8722	1.8797	1.8760	1.8835
2	12	1.9098	83.3	1.9198	74.1	1.9098	1.9148	1.9123	1.9173	1.9148	1.9198	1.9173	1.9223
2	16	1.9323	83.4	1.9408	72.9	1.9323	1.9366	1.9344	1.9387	1.9365	1.9408	1.9386	1.9429
2	20	1.9459	83.3	1.9537	71.3	1.9459	1.9498	1.9477	1.9517	1.9497	1.9537	1.9516	1.9556
2 1/16	16	1.9948	83.4	2.0033	72.9	1.9948	1.9991	1.9969	2.0012	1.9990	2.0033	2.0011	2.0054
2 1/8	8	1.9897	83.3	2.0047	74.1	1.9897	1.9972	1.9934	2.0009	1.9972	2.0047	2.0010	2.0085
2 1/8	12	2.0348	83.3	2.0448	74.1	2.0348	2.0398	2.0373	2.0423	2.0398	2.0448	2.0423	2.0473
2 1/8	16	2.0573	83.4	2.0658	72.9	2.0573	2.0616	2.0594	2.0637	2.0615	2.0658	2.0636	2.0679
2 3/16	16	2.1198	83.4	2.1283	72.9	2.1198	2.1241	2.1219	2.1262	2.1240	2.1283	2.1261	2.1304
2 1/4	4 ½	2.0094	83.3	2.0361	74.1	2.0094	2.0227	2.0161	2.0294	2.0228	2.0361	2.0294	2.0427
2 1/4	8	2.1147	83.3	2.1297	74.1	2.1147	2.1222	2.1184	2.1259	2.1222	2.1297	2.1260	2.1335
2 1/4	12	2.1598	83.3	2.1698	74.1	2.1598	2.1648	2.1623	2.1673	2.1648	2.1698	2.1673	2.1723
2 1/4	16	2.1823	83.4	2.1908	72.9	2.1823	2.1866	2.1844	2.1887	2.1865	2.1908	2.1886	2.1929
2 1/4	20	2.1959	83.3	2.2037	71.3	2.1959	2.1998	2.1977	2.2017	2.1997	2.2037	2.2016	2.2056
2 5/16	16	2.2448	83.4	2.2533	72.9	2.2448	2.2491	2.2469	2.2512	2.2490	2.2533	2.2511	2.2554
2 3/8	12	2.2848	83.3	2.2948	74.1	2.2848	2.2898	2.2873	2.2923	2.2898	2.2948	2.2923	2.2973
2 3/8	16	2.3073	83.4	2.3158	72.9	2.3073	2.3116	2.3094	2.3137	2.3115	2.3158	2.3136	2.3179
2 7/16	16	2.3698	83.4	2.3783	72.9	2.3698	2.3741	2.3719	2.3762	2.3740	2.3783	2.3761	2.3804
2 1/2	4	2.2294	83.3	2.2594	74.1	2.2294	2.2444	2.2369	2.2519	2.2444	2.2594	2.2519	2.2669
2 1/2	8	2.3647	83.3	2.3797	74.1	2.3647	2.3722	2.3684	2.3759	2.3722	2.3797	2.3760	2.3835
2 1/2	12	2.4098	83.3	2.4198	74.1	2.4098	2.4148	2.4123	2.4173	2.4148	2.4198	2.4173	2.4223
2 1/2	16	2.4323	83.4	2.4408	72.9	2.4323	2.4366	2.4344	2.4387	2.4365	2.4408	2.4386	2.4429
2 1/2	20	2.4459	83.3	2.4537	71.3	2.4459	2.4498	2.4478	2.4517	2.4497	2.4537	2.4516	2.4556
2 5/8	12	2.5348	83.3	2.5448	74.1	2.5348	2.5398	2.5373	2.5423	2.5398	2.5448	2.5423	2.5473
2 5/8	16	2.5573	83.4	2.5658	72.9	2.5573	2.5616	2.5594	2.5637	2.5615	2.5658	2.5636	2.5679

*To find a suitable tap drill size for a specified diameter and thread series, as for example, 2—4½ UNC—2B, enter the table and find the hole limits over the range from 1.7594 to 1.7927. Next enter Table 9–1 to find drill sizes 1-49/64 in., 45 mm, 1-25/32 in. and 45.5 mm. Only two diameters remain if the choice is restricted to inch dimensions and a choice between them can readily be made to suit length of engagement.

†Based on a length of engagement equal to the nominal diameter.

continued on next page

TABLE 9–8, continued

Designation		Minor Diameter Internal Threads				Recommended Hole Size Limits for Different Lengths of Engagement							
Thread Size	Threads per Inch	Minimum	Percent Basic Thread Height	Maximum†	Percent Basic Thread Height	To and Including 1/3 *D*		Above 1/3 *D* to 2/3 *D*		Above 2/3 *D* to 1 1/2 *D*		Above 1 1/2 *D* to 3 *D*	
						Min	Max	Min	Max	Min	Max	Min	Max
2 3/4	4	2.4794	83.3	2.5094	74.1	2.4794	2.4944	2.4869	2.5019	2.4944	2.5094	2.5019	2.5169
2 3/4	8	2.6147	83.3	2.6297	74.1	2.6147	2.6222	2.6184	2.6259	2.6222	2.6297	2.6260	2.6335
2 3/4	12	2.6598	83.3	2.6698	74.1	2.6598	2.6648	2.6623	2.6673	2.6648	2.6698	2.6673	2.6723
2 3/4	16	2.6823	83.4	2.6908	72.9	2.6823	2.6866	2.6844	2.6887	2.6865	2.6908	2.6886	2.6929
2 7/8	12	2.7848	83.3	2.7948	74.1	2.7848	2.7898	2.7873	2.7923	2.7898	2.7948	2.7923	2.7973
2 7/8	16	2.8073	83.4	2.8158	72.9	2.8073	2.8116	2.8094	2.8137	2.8115	2.8158	2.8136	2.8179
3	4	2.7294	83.3	2.7594	74.1	2.7294	2.7444	2.7369	2.7519	2.7444	2.7594	2.7519	2.7669
3	8	2.8647	83.3	2.8797	74.1	2.8647	2.8722	2.8684	2.8759	2.8722	2.8797	2.8760	2.8835
3	12	2.9098	83.3	2.9198	74.1	2.9098	2.9148	2.9123	2.9173	2.9148	2.9198	2.9173	2.9223
3	16	2.9323	83.4	2.9408	72.9	2.9323	2.9366	2.9344	2.9387	2.9365	2.9408	2.9386	2.9429
3 1/8	12	3.0348	83.3	3.0448	74.1	3.0348	3.0398	3.0373	3.0423	3.0398	3.0448	3.0423	3.0473
3 1/8	16	3.0573	83.4	3.0658	72.9	3.0573	3.0616	3.0594	3.0637	3.0615	3.0658	3.0636	3.0679
3 1/4	4	2.9794	83.3	3.0094	74.1	2.9794	2.9944	2.9869	3.0019	2.9944	3.0094	3.0019	3.0169
3 1/4	8	3.1147	83.3	3.1297	74.1	3.1147	3.1222	3.1184	3.1259	3.1222	3.1297	3.1260	3.1335
3 1/4	12	3.1598	83.3	3.1698	74.1	3.1598	3.1648	3.1623	3.1673	3.1648	3.1698	3.1673	3.1723
3 1/4	16	3.1823	83.4	3.1908	72.9	3.1823	3.1866	3.1844	3.1887	3.1865	3.1908	3.1886	3.1929
3 3/8	12	3.2848	83.3	3.2948	74.1	3.2848	3.2898	3.2873	3.2923	3.2898	3.2948	3.2923	3.2973
3 3/8	16	3.3073	83.4	3.3158	72.9	3.3073	3.3116	3.3094	3.3137	3.3115	3.3158	3.3136	3.3179
3 1/2	4	3.2294	83.3	3.2594	74.1	3.2294	3.2444	3.2369	3.2519	3.2444	3.2594	3.2519	3.2669
3 1/2	8	3.3647	83.3	3.3797	74.1	3.3647	3.3722	3.3684	3.3759	3.3722	3.3797	3.3760	3.3835
3 1/2	12	3.4098	83.3	3.4198	74.1	3.4098	3.4148	3.4123	3.4173	3.4148	3.4198	3.4173	3.4223
3 1/2	16	3.4323	83.4	3.4408	72.9	3.4323	3.4366	3.4344	3.4387	3.4365	3.4408	3.4386	3.4429
3 5/8	12	3.5348	83.3	3.5448	74.1	3.5348	3.5398	3.5373	3.5423	3.5398	3.5448	3.5423	3.5473
3 5/8	16	3.5573	83.4	3.5658	72.9	3.5573	3.5616	3.5594	3.5637	3.5615	3.5658	3.5636	3.5679
3 3/4	4	3.4794	83.3	3.5094	74.1	3.4794	3.4944	3.4869	3.5019	3.4944	3.5094	3.5019	3.5169
3 3/4	8	3.6147	83.3	3.6297	74.1	3.6147	3.6222	3.6184	3.6259	3.6222	3.6297	3.6260	3.6335
3 3/4	12	3.6598	83.3	3.6698	74.1	3.6598	3.6648	3.6623	3.6673	3.6648	3.6698	3.6673	3.6723
3 3/4	16	3.6823	83.4	3.6908	72.9	3.6823	3.6866	3.6844	3.6887	3.6865	3.6908	3.6886	3.6929
3 7/8	12	3.7848	83.3	3.7948	74.1	3.7848	3.7898	3.7873	3.7923	3.7898	3.7948	3.7923	3.7973
3 7/8	16	3.8073	83.4	3.8158	72.9	3.8073	3.8116	3.8094	3.8137	3.8115	3.8158	3.8136	3.8179

†Based on a length of engagement equal to the nominal diameter.

continued on next page

TABLE 9–8, continued

Designation		Minor Diameter Internal Threads				Recommended Hole Size Limits for Different Lengths of Engagement							
Thread Size	Threads per Inch	Minimum	Percent Basic Thread Height	Maximum†	Percent Basic Thread Height	To and Including 1/3 *D*		Above 1/3 *D* to 2/3 *D*		Above 2/3 *D* to 1 1/2 *D*		Above 1 1/2 *D* to 3 *D*	
						Min	Max	Min	Max	Min	Max	Min	Max
4	4	3.7294	83.3	3.7594	74.1	3.7294	3.7444	3.7369	3.7519	3.7444	3.7594	3.7519	3.7669
4	8	3.8647	83.3	3.8797	74.1	3.8647	3.8722	3.8684	3.8759	3.8722	3.8797	3.8760	3.8835
4	12	3.9098	83.3	3.9198	74.1	3.9098	3.9148	3.9123	3.9173	3.9148	3.9198	3.9173	3.9223
4	16	3.9323	83.4	3.9408	72.9	3.9323	3.9366	3.9344	3.9387	3.9365	3.9408	3.9386	3.9429
4 1/4	4	3.9794	83.3	4.0094	74.1	3.9794	3.9944	3.9869	4.0019	3.9944	4.0094	4.0019	4.0169
4 1/4	8	4.1147	83.3	4.1297	74.1	4.1147	4.1222	4.1184	4.1259	4.1222	4.1297	4.1260	4.1335
4 1/4	12	4.1598	83.3	4.1698	74.1	4.1598	4.1648	4.1623	4.1673	4.1648	4.1698	4.1673	4.1723
4 1/4	16	4.1823	83.4	4.1908	72.9	4.1823	4.1866	4.1844	4.1887	4.1865	4.1908	4.1886	4.1929
4 1/2	4	4.2294	83.3	4.2594	74.1	4.2294	4.2444	4.2369	4.2519	4.2444	4.2594	4.2519	4.2669
4 1/2	8	4.3647	83.3	4.3797	74.1	4.3647	4.3722	4.3684	4.3759	4.3722	4.3797	4.3760	4.3835
4 1/2	12	4.4098	83.3	4.4198	74.1	4.4098	4.4148	4.4123	4.4173	4.4148	4.4198	4.4173	4.4223
4 1/2	16	4.4323	83.4	4.4408	72.9	4.4323	4.4366	4.4344	4.4387	4.4365	4.4408	4.4386	4.4429
4 3/4	8	4.6147	83.3	4.6297	74.1	4.6147	4.6222	4.6184	4.6259	4.6222	4.6297	4.6260	4.6335
4 3/4	12	4.6598	83.3	4.6698	74.1	4.6598	4.6648	4.6623	4.6673	4.6648	4.6698	4.6673	4.6723
4 3/4	16	4.6823	83.4	4.6908	72.9	4.6823	4.6866	4.6844	4.6887	4.6865	4.6908	4.6886	4.6929
5	8	4.8647	83.3	4.8797	74.1	4.8647	4.8722	4.8684	4.8759	4.8722	4.8797	4.8760	4.8835
5	12	4.9098	83.3	4.9198	74.1	4.9098	4.9148	4.9123	4.9173	4.9148	4.9198	4.9173	4.9223
5	16	4.9323	83.4	4.9408	72.9	4.9323	4.9366	4.9344	4.9387	4.9365	4.9408	4.9386	4.9429
5 1/4	8	5.1147	83.3	5.1297	74.1	5.1147	5.1222	5.1184	5.1259	5.1222	5.1297	5.1260	5.1335
5 1/4	12	5.1598	83.3	5.1698	74.1	5.1598	5.1648	5.1623	5.1673	5.1648	5.1698	5.1673	5.1723
5 1/4	16	5.1823	83.4	5.1908	72.9	5.1823	5.1866	5.1844	5.1887	5.1865	5.1908	5.1886	5.1929
5 1/2	8	5.3647	83.3	5.3797	74.1	5.3647	5.3722	5.3684	5.3759	5.3722	5.3797	5.3760	5.3835
5 1/2	12	5.4098	83.3	5.4198	74.1	5.4098	5.4148	5.4123	5.4173	5.4148	5.4198	5.4173	5.4223
5 1/2	16	5.4323	83.4	5.4408	72.9	5.4323	5.4366	5.4344	5.4387	5.4365	5.4408	5.4386	5.4429
5 3/4	8	5.6147	83.3	5.6297	74.1	5.6147	5.6222	5.6184	5.6259	5.6222	5.6297	5.6260	5.6335
5 3/4	12	5.6598	83.3	5.6698	74.1	5.6598	5.6648	5.6623	5.6673	5.6648	5.6698	5.6673	5.6723
5 3/4	16	5.6823	83.4	5.6908	72.9	5.6823	5.6866	5.6844	5.6887	5.6865	5.6908	5.6886	5.6929
6	8	5.8647	83.3	5.8797	74.1	5.8647	5.8722	5.8684	5.8759	5.8722	5.8797	5.8760	5.8835
6	12	5.9098	83.3	5.9198	74.1	5.9098	5.9148	5.9123	5.9173	5.9148	5.9198	5.9173	5.9223
6	16	5.9323	83.4	5.9408	72.9	5.9323	5.9366	5.9344	5.9387	5.9365	5.9408	5.9386	5.9429

†Based on a length of engagement equal to the nominal diameter.

TABLE 9–9

Formula for Computing Tap Drill Sizes Corresponding to 75 per cent Thread Depth

Metal Cutting Tool Handbook — Metal Cutting Tool Institute

$$\text{Drill size for 75 per cent thread depth} = \text{Outside diameter of thread} - \frac{0.974}{\text{Number threads per inch}}$$

$$\text{Percentage of thread depth} = \text{Number threads per inch} \times \left(\frac{\begin{matrix}\text{Outside diameter} \\ \text{of thread}\end{matrix} - \begin{matrix}\text{Selected drill} \\ \text{diameter}\end{matrix}}{0.01299} \right)$$

"Except where it is desirable to have all the bearing surface possible for screws that are to be adjusted frequently, it is a costly practice to tap a greater thread depth than necessary. Much tap breakage and many cases of production difficulties may be traced to selection of tap drills that are too small.

A common nut drilled out so that it contains 50 per cent of a full depth thread will break the bolt before it will strip.

A full depth of thread in a common nut is only about 5 per cent stronger than a 75 per cent depth of thread, yet it requires three times the power to tap.

On an average, 75 per cent thread depth in the nut is stronger than the tensile strength of the screw and is recommended for most applications. For small screws and in deep holes, less thread depth will give ample strength. It should be noted that a drill will cut a hole somewhat larger than its nominal size. This should be taken into account when the percentage of thread depth is a critical factor."

TABLE 9–10

Unified and American Screw Thread Series

SAE Handbook 1954

SIZE	BASIC MAJOR DIAMETER	THREADS PER INCH						SIZE
		Coarse (UNC or NC)[c]	Fine[a] (UNF or NF)[c]	Extra fine[b] (UNEF or NEF)[c]	8-Thread series (UN, N or NS)	12-Thread series (UN or N)	16-Thread series (UN or N)	
0	0.0600	—	80	—	—	—	—	0
1	0.0730	64	72	—	—	—	—	1
2	0.0860	56	64	—	—	—	—	2
3	0.0990	48	56	—	—	—	—	3
4	0.1120	40	48	—	—	—	—	4
5	0.1250	40	44	—	—	—	—	5
6	0.1380	32	40	—	—	—	—	6
8	0.1640	32	36	—	—	—	—	8
10	0.1900	24	32	—	—	—	—	10
12	0.2160	24	28	32	—	—	—	12
1/4	0.2500	20	28	32	—	—	—	1/4
5/16	0.3125	18	24	32	—	—	—	5/16
3/8	0.3750	16	24	32	—	—	—	3/8
7/16	0.4375	14	20	28	—	—	—	7/16
1/2	0.5000	13	20	28	—	12	—	1/2
9/16	0.5625	12	18	24	—	12	—	9/16
5/8	0.6250	11	18	24	—	12	—	5/8
11/16	0.6875	—	—	24	—	12	—	11/16
3/4	0.7500	10	16	20	—	12	16	3/4
13/16	0.8125	—	—	20	—	12	16	13/16
7/8	0.8750	9	14	20	—	12	16	7/8
15/16	0.9375	—	—	20	—	12	16	15/16
1	1.0000	—	14	—	—	—	—	1
1	1.0000	8	12	20	8	12	16	1
1 1/16	1.0625	—	—	18	—	12	16	1 1/16
1 1/8	1.1250	7	12	18	8	12	16	1 1/8
1 3/16	1.1875	—	—	18	—	12	16	1 3/16
1 1/4	1.2500	7	12	18	8	12	16	1 1/4
1 5/16	1.3125	—	—	18	—	12	16	1 5/16
1 3/8	1.3750	6	12	18	8	12	16	1 3/8
1 7/16	1.4375	—	—	18	—	12	16	1 7/16
1 1/2	1.5000	6	12	18	8	12	16	1 1/2
1 9/16	1.5625	—	—	18	—	—	16	1 9/16
1 5/8	1.6250	—	—	18	8	12	16	1 5/8
1 11/16	1.6875	—	—	18	—		16	1 11/16
1 3/4	1.7500	5	—	16	8	12	16	1 3/4
1 13/16	1.8125	—	—	—	—	—	16	1 13/16
1 7/8	1.8750	—	—	—	8	12	16	1 7/8
1 15/16	1.9375	—	—	—	—	—	16	1 15/16
2	2.0000	4 1/2	—	16	8	12	16	2
2 1/16	2.0625	—	—	—	—	—	16	2 1/16
2 1/8	2.1250	—	—	—	8	12	16	2 1/8
2 3/16	2.1875	—	—	—	—	—	16	2 3/16
2 1/4	2.2500	4 1/2	—	—	8	12	16	2 1/4
2 5/16	2.3125	—	—	—	—	—	16	2 5/16
2 3/8	2.3750	—	—	—	—	12	16	2 3/8
2 7/16	2.4375	—	—	—	—	—	16	2 7/16
2 1/2	2.5000	4	—	—	8	12	16	2 1/2
2 5/8	2.6250	—	—	—	—	12	16	2 5/8
2 3/4	2.7500	4	—	—	8	12	16	2 3/4
2 7/8	2.8750	—	—	—	—	12	16	2 7/8
3	3.0000	4	—	—	8	12	16	3
3 1/8	3.1250	—	—	—	—	12	16	3 1/8
3 1/4	3.2500	4	—	—	8	12	16	3 1/4
3 3/8	3.3750	—	—	—	—	12	16	3 3/8
3 1/2	3.5000	4	—	—	8	12	16	3 1/2
3 5/8	3.6250	—	—	—	—	12	16	3 5/8
3 3/4	3.7500	4	—	—	8	12	16	3 3/4
3 7/8	3.8750	—	—	—	—	12	16	3 7/8
4	4.0000	4	—	—	8	12	16	4
4 1/4	4.2500	—	—	—	8	12	16	4 1/4
4 1/2	4.5000	—	—	—	8	12	16	4 1/2
4 3/4	4.7500	—	—	—	8	12	16	4 3/4
5	5.0000	—	—	—	8	12	16	5
5 1/4	5.2500	—	—	—	8	12	16	5 1/4
5 1/2	5.5000	—	—	—	8	12	16	5 1/2
5 3/4	5.7500	—	—	—	8	12	16	5 3/4
6	6.0000	—	—	—	8	12	16	6

[a] For diameters over 1 1/2 in., use 12-Thread Series.
[b] For diameters over 2 in., use 16-Thread Series.

[c] Unified and American Thread Series except as indicated

TABLE 9-11

Fractional Size Taps – Cut Thread

ASA B5.4-1948 Taps Cut and *Ground Threads

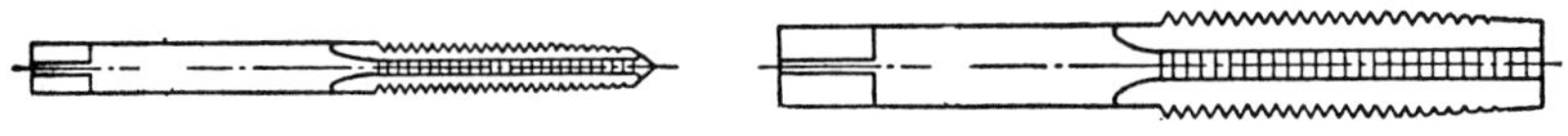

TAPER

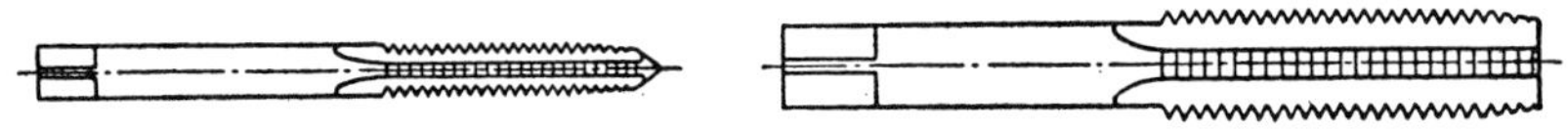

PLUG

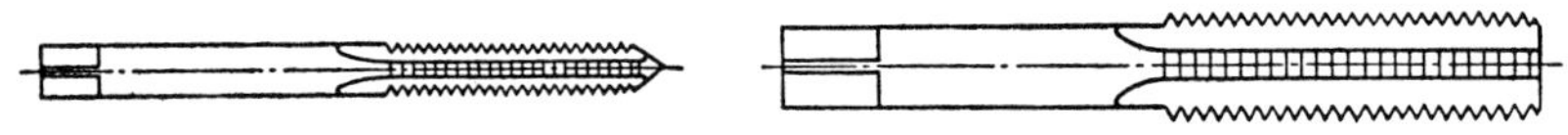

BOTTOMING

†Size	Threads per Inch		
	NC	NF	NS
1/16	..	..	64
3/32	..	..	48
1/8	..	..	40
5/32	..	..	32
5/32	..	..	36
3/16	..	..	24
3/16	..	..	32
7/32	..	..	24
7/32	..	..	32
1/4	20	..	..
1/4	..	..	24
1/4	..	..	27
1/4	..	28	..
1/4	..	..	32
5/16	18	..	..
5/16	..	..	20
5/16	..	24	..
5/16	..	..	27
5/16	..	..	32
3/8	16	..	..
3/8	..	..	20
3/8	..	24	..
3/8	..	..	27
7/16	14	..	..
7/16	..	20	..
7/16	..	..	24
7/16	..	..	27
1/2	..	..	12
1/2	13	..	..
1/2	..	20	..
1/2	..	..	24
1/2	..	..	27

†Size	Threads per Inch		
	NC	NF	NS
9/16	12	..	..
9/16	..	18	..
9/16	..	..	27
5/8	11	..	..
5/8	..	..	12
5/8	..	18	..
5/8	..	..	27
11/16	..	..	11
11/16	..	..	16
3/4	10	..	..
3/4	..	..	12
3/4	..	16	..
3/4	..	..	27
7/8	9	..	..
7/8	..	..	12
7/8	..	14	..
7/8	..	..	27
1	8	..	..
1	..	..	12
1	..	14	..
1	..	..	27
1 1/8	7	..	..
1 1/8	..	12	..
1 1/4	7	..	..
1 1/4	..	12	..
1 3/8	6	..	..
1 3/8	..	12	..
1 1/2	6	..	..
1 1/2	..	12	..
1 5/8	..	..	5 1/2
1 3/4	5	..	..
1 7/8	..	..	5
2	4 1/2	..	..

*Sizes of commercial ground threads range from 1/4 to 1½ inch; sizes of precision ground threads range from 1/4 to 1 inch, inclusive.

†The 1948 ASA Standard on taps antedates the adoption, in 1949, of the Standard on Unified and American Screw Threads. The Unified and the preceding American Standard are alike in thread form, the chief differences between them being the application of allowances, the variation of tolerances with size, and the differences in the amounts of pitch diameter tolerances on external and internal threads. For ordinary design and production purposes, these differences are unimportant because Unified threads are mechanically interchangeable with American National threads of the same diameter and number of threads per inch. A significant difference for drawing purpose is the manner of designating class, for example, 2A for external and 2B for internal threads in the new standard rather than 2 alone as in the previous American Standard.

TABLE 9–12

Sizes of Regular (Standard) Machine Screw Taps

ASA B5.4–1948 Taps Cut and Ground Threads

TAPER

PLUG

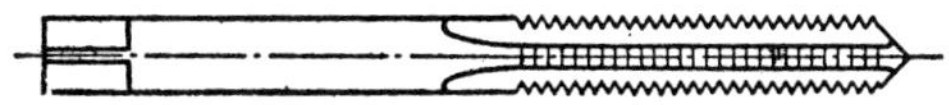

BOTTOMING

Screw Gage Number	Basic Major Diameter	Cut Thread Threads per Inch			Commercial Ground Thread Threads per Inch			Precision Ground Thread Threads per Inch			Screw Gage Number
		NC†	NF	NS	NC†	NF	NS	NC†	NF	NS	
0	0.0600		80						80		0
1	0.0730	64	72	56				64	72	56	1
2	0.0860	56	64					56	64		2
3	0.0990	48	56		48	56		48	56		3
4	0.1120	40	48	32,36	40	48	36	40	48	36	4
5	0.1250	40	44		40	44		40	44		5
6	0.1380	32	40	36	32	40		32	40		6
8	0.1640	32	36	40	32	36		32	36		8
10	0.1900	24	32	30	24	32		24	32		10
12	0.2160	24	28	32	24	28		24	28		12
14	0.2420			20,24			20,24				14

†The 1948 ASA Standard on taps antedates the adoption, in 1949, of the Standard on Unified and American Screw Threads. The Unified and the preceding American Standard are alike in thread form, the chief differences between them being the application of allowances, the variation of tolerances with size, and the differences in the amounts of pitch diameter tolerances on external and internal threads. For ordinary design and production purposes, these differences are unimportant because Unified threads are mechanically interchangeable with American National threads of the same diameter and number of threads per inch. A significant difference for drawing purpose is the manner of designating class, for example, 2A for external and 2B for internal threads in the new standard rather than 2 alone as in the previous American Standard.

TABLE 9–13

Sizes of Boiler and Staybolt Taps — Cut Thread

ASA B5.4–1948 Taps Cut and Ground Threads

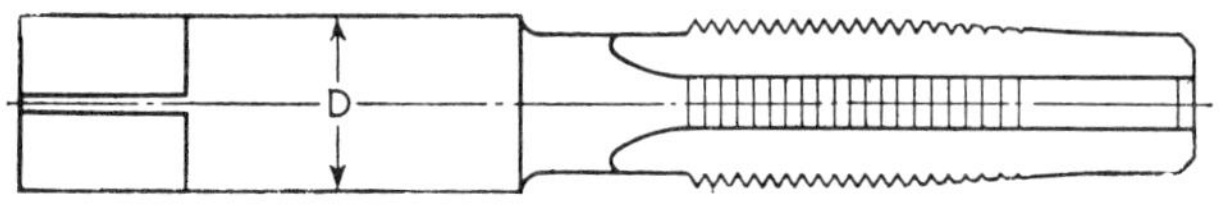

STRAIGHT BOILER TAP

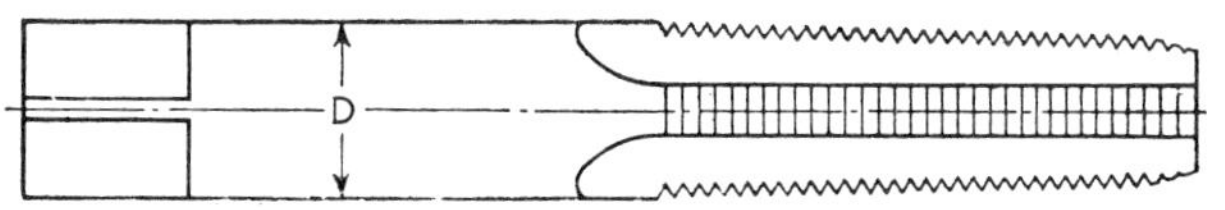

TAPER BOILER TAP

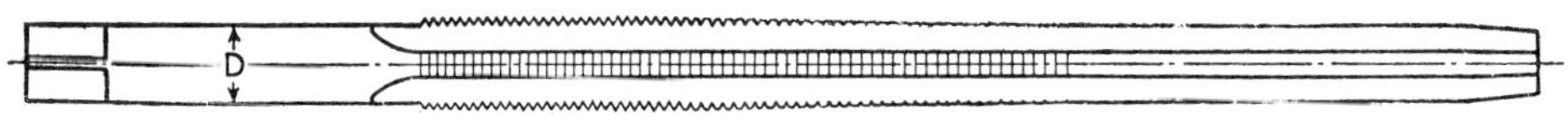

STAYBOLT TAP

NOTE: These taps are furnished with American National Form or "V" Form of thread and all sizes have twelve threads per inch.

Diameter of Tap	Diameter of Shank D Straight Boiler Tap	Taper* Boiler Tap	Staybolt Tap
1/2	0.5000	0.5000	
9/16	0.5625	0.5625	
5/8	0.6250	0.6250	
11/16	0.6875	0.6875	
3/4	0.7500	0.7500	
13/16	0.8125	0.8125	
7/8	0.8750	0.8750	0.750
15/16	0.9375	0.9375	0.812
1	1.0000	1.0000	0.875
1-1/16	1.0625	1.0625	0.937
1-1/8	1.1250	1.1250	1.000
1-3/16	1.1875	1.1875	1.062
1-1/4	1.2500	1.2500	1.125
1-5/16	1.3125	1.3125	1.187
1-3/8	1.3750	1.3750	1.250
1-7/16	1.4375	1.4375	1.312
1-1/2	1.5000	1.5000	1.375

*Taper boiler taps have a taper of 3/4 inch to the foot and the diameter is measured 5/8 inch from the large end of the thread.

*TABLE 9-14

Sizes of Taper Pipe Taps with Thread Tolerances — Cut and Ground Threads

ASA B5.4-1948 Taps Cut and Ground Threads

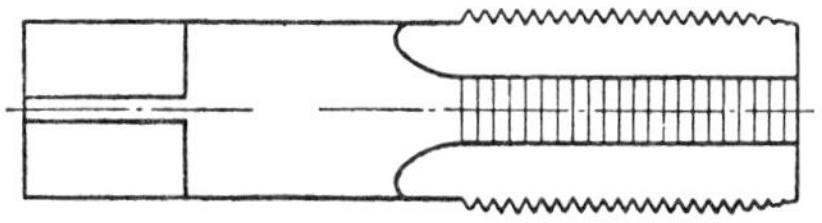

Nominal Size	Threads per Inch, NPT	Gage Measurement[1]			Lead Tolerance per Inch of Thread, Plus or Minus		Angle Tolerance			Taper per Foot, Inches			
		Projection, Inches	Tolerance, Plus or Minus				Half Angle, Plus or Minus		Full Angle	Cut Thread		Ground Thread	
			Cut Thread	Ground Thread	Cut Thread	Ground Thread	Cut Thread	Ground Thread	Cut Thread	Min	Max	Min	Max
1/16	27	0.312	1/16	1/16	0.003	0.0005	45′	30′	68′	23/32	27/32	23/32	25/32
1/8	27	0.312	1/16	1/16	0.003	0.0005	45′	30′	68′	23/32	27/32	23/32	25/32
1/4	18	0.459	1/16	1/16	0.003	0.0005	45′	30′	68′	23/32	27/32	23/32	25/32
3/8	18	0.454	1/16	1/16	0.003	0.0005	45′	30′	68′	23/32	27/32	23/32	25/32
1/2	14	0.579	1/16	1/16	0.003	0.0005	45′	30′	68′	23/32	13/16	23/32	25/32
3/4	14	0.565	1/16	1/16	0.003	0.0005	45′	30′	68′	23/32	13/16	23/32	25/32
1	11 1/2	0.678	3/32	3/32	0.003	0.0005	45′	30′	68′	23/32	13/16	23/32	25/32
1 1/4	11 1/2	0.686	3/32	3/32	0.003	0.0005	45′	30′	68′	23/32	13/16	23/32	25/32
1 1/2	11 1/2	0.699	3/32	3/32	0.003	0.0005	45′	30′	68′	23/32	13/16	23/32	25/32
2	11 1/2	0.667	3/32	3/32	0.003	0.0005	45′	30′	68′	23/32	13/16	23/32	25/32
2 1/2	8	0.925	3/32	3/32	0.003	0.0005	40′	25′	60′	47/64	51/64	47/64	25/32
3	8	0.925	3/32	3/32	0.003	0.0005	40′	25′	60′	47/64	51/64	47/64	25/32
3 1/2	8	0.938	1/8	1/8	0.003	0.0005	40′	25′	60′	47/64	51/64	47/64	25/32
4	8	0.950	1/8	1/8	0.003	0.0005	40′	25′	60′	47/64	51/64	47/64	25/32

All dimensions are given in inches.

LEAD TOLERANCE. For cut thread taps a maximum lead error of plus or minus 0.003 in. in one inch of thread is permitted, and ground thread taps a maximum lead error of plus or minus 0.0005 in. in one inch of thread is permitted.

[1]Distance that the small end of tap projects through American Standard Pipe Thread Ring Gage.

*Compare with Table 13-70

TABLE 9–15

Sizes of Straight Pipe Taps with Thread Limits — Cut Thread

ASA B5.4–1948 Taps Cut and Ground Threads

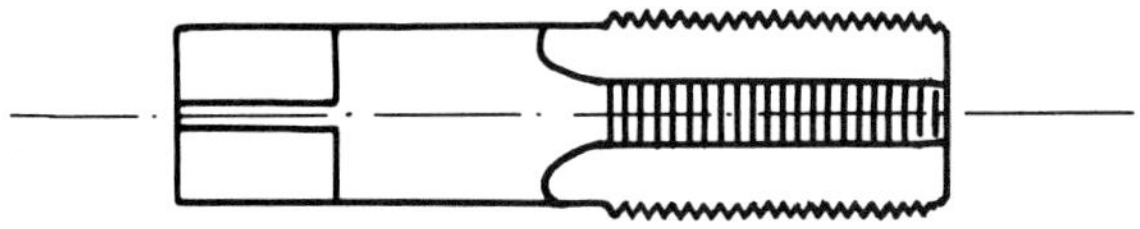

Nominal Size	Threads per Inch, NPS	Size at Gaging Notch	Pitch Diameter	
			Min	Max
1/16	27	0.2812	0.2797	0.2827
1/8	27	0.3748	0.3733	0.3763
1/4	18	0.4899	0.4884	0.4914
3/8	18	0.6270	0.6253	0.6288
1/2	14	0.7784	0.7767	0.7802
3/4	14	0.9889	0.9869	0.9909
1	11 1/2	1.2386	1.2366	1.2406
1 1/4	11 1/2	1.5834	1.5811	1.5856
1 1/2	11 1/2	1.8223	1.8201	1.8246
2	11 1/2	2.2963	2.2938	2.2988

All dimensions are given in inches.

NOTE: As the American Standard Pipe Thread Form is to be maintained the major and minor diameters vary with the pitch diameter. See formulas below. Either a flat or a rounded form is allowable at both the crest and root.

FORMULAS FOR AMERICAN STANDARD PIPE FORM
(Approximate)

Major Diameter, Min = Measured pitch diameter plus A
Major Diameter, Max = Measured pitch diameter plus B
Minor Diameter, Min = Measured pitch diameter minus B
Minor Diameter, Max = Measured pitch diameter minus C

Pitch Diameter, Min = Size at gaging notch minus one-half tolerance.
Pitch Diameter, Max = Minimum plus tolerance.

FORMULA VALUES

Threads per Inch NPS	A	B	C
27	0.0267	0.0296	0.0257
18	0.0408	0.0444	0.0401
14	0.0535	0.0571	0.0525
11 1/2	0.0658	0.0696	0.0647

LEAD TOLERANCE. A maximum lead error of plus or minus 0.003 in. in one inch of thread is permitted.

ANGLE TOLERANCE

Threads per Inch	Error in Half Angle	Error in Full Angle
11 1/2 to 27, incl.	45 min plus or minus	68 min

TABLE 9-16

Sizes of Straight Pipe Taps with Thread Limits — Ground Thread

ASA B5.4-1948 Taps Cut and Ground Threads

Nominal Size	Threads per Inch, NPS	Major Diameter			Pitch Diameter		
		Plug at Gaging Notch	Min G	Max H	Plug at Gaging Notch E	Min K	Max L
1/16	27	0.3059	0.3098	0.3108	0.2812	0.2817	0.2827
1/8	27	0.3994	0.4034	0.4044	0.3748	0.3753	0.3763
1/4	18	0.5269	0.5323	0.5333	0.4899	0.4904	0.4914
3/8	18	0.6640	0.6694	0.6704	0.6270	0.6275	0.6285
1/2	14	0.8260	0.8335	0.8345	0.7784	0.7789	0.7799
3/4	14	1.0364	1.0440	1.0450	0.9889	0.9894	0.9904
1	11 1/2	1.2966	1.3057	1.3072	1.2386	1.2396	1.2407

All dimensions are given in inches.

FORMULAS FOR AMERICAN STANDARD PIPE FORM

Nominal Size, Inches	Major Diameter Min G	Major Diameter Max H	Minor Diameter Min	Minor Diameter Max	Pitch Diameter Min K	Pitch Diameter Max L
1/8	H −0.0010	(K +A) −0.0005	M−A	M−B	E +0.0005	K + D
1/4 to 3/4 incl.	H −0.0010	(K +A) −0.0015	M−A	M−B	E +0.0005	K + D
1	H −0.0015	(K +A) −0.0020	M−A	M−B	E +0.0010	K + D

All dimensions are given in inches.

FORMULA VALUES

Threads per Inch NPS	A	B	D	E	M
27	0.0296	0.0257	0.0010	Pitch diam	Actual
18	0.0444	0.0401	0.0010	of plug	measured
14	0.0571	0.0525	0.0010	at gaging	pitch
11 1/2	0.0696	0.0647	0.0011	notch	diameter

LEAD TOLERANCE. A maximum lead error of plus or minus 0.0005 in. in one inch of thread is permitted.

ANGLE TOLERANCE.

Threads per Inch	Error in Half Angle
11 1/2 to 27, incl.	30 min plus or minus

TABLE 9-17

Twist Drill Diameters for Tapped Holes for Pipe Threads

ASA B2.1-1945 Pipe Threads

Nominal Pipe Size	Taper Thread: With Use of Reamer*	Taper Thread: Without Use of Reamer	Straight Pipe Thread
1/16	0.240	0.246	1/4 0.250
1/8	21/64 0.328	0.332	11/32 0.344
1/4	27/64 0.422	7/16 0.438	7/16 0.438
3/8	9/16 0.562	9/16 0.562	37/64 0.578
1/2	11/16 0.688	45/64 0.703	23/32 0.719
3/4	57/64 0.891	29/32 0.906	59/64 0.922
1	1 1/8 1.125	1 9/64 1.141	1 5/32 1.156
1 1/4	1 15/32 1.469	1 31/64 1.484	1 1/2 1.500
1 1/2	1 23/32 1.719	1 47/64 1.734	1 3/4 1.750
2	2 3/16 2.188	2 13/64 2.203	2 7/32 2.219
2 1/2	2 19/32 2.594	2 5/8 2.625	2 21/32 2.656

All dimensions are given in inches.

*These reamers are tapered 3/4 inch to the foot and are intended for reaming holes to be tapped with American Standard Taper Pipe Threads. See Tables 9-14 and 13-70.

TABLE 9–18

Taper *Pipe Reamers

ASA B5.14–1949 Reamers

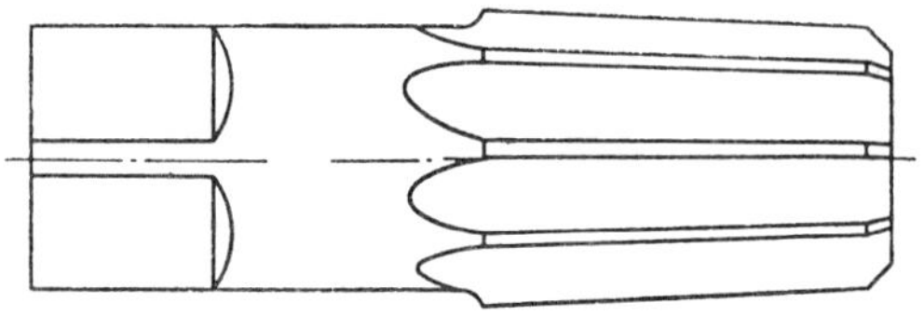

Nominal Size	Diameter Large End	Diameter Small End	Nominal Size	Diameter Large End	Diameter Small End	Nominal Size	Diameter Large End	Diameter Small End
1/8	0.362	0.316	1/2	0.751	0.665	1-1/4	1.553	1.103
1/4	0.472	0.406	3/4	0.962	0.876	1-1/2	1.793	1.684
3/8	0.606	0.540	1	1.212	1.103	2	2.268	2.159

All dimensions are given in inches.

*These reamers are tapered 3/4 inch to the foot and are intended for reaming holes to be tapped with American Standard Taper Pipe Threads. See Tables 9–14 and 13–70.

TABLE 9–19

Sizes of Pipe Burring Reamers

Catalog No. 46 The Cleveland Twist Drill Company

Capacity Pipe Inches	Reamer Diameter, Inches At Point	Reamer Diameter, Inches At Large End
1/8 to 1/2	3/16	47/64
1/8 to 1	3/16	1-1/4
1/4 to 1-1/4	1/4	1-15/32
1/4 to 2	1/4	2-17/64

TABLE 9-20

Sizes of Hand Reamers — Straight* Flutes

ASA B5.14-1949 Reamers

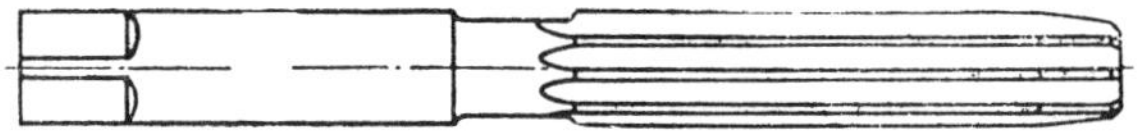

Diameter of Reamer, Inches	Tolerance on Diameter, Inch	Diameter of Reamer, Inches	Tolerance on Diameter, Inch	Diameter of Reamer, Inches	Tolerance on Diameter, Inch
1/8	+0.0001 to +0.0004	3/8	+0.0001 to +0.0005	13/16	0.0001 to 0.0005
9/64		25/64		27/32	
5/32		13/32		7/8	
11/64		27/64		29/32	
3/16		7/16		15/16	
13/64		29/64		31/32	
7/32		15/32		1	
15/64		31/64		$1\frac{1}{16}$	0.0002 to 0.0006
1/4		1/2		$1\frac{1}{8}$	
17/64	+0.0001 to +0.0005	17/32		$1\frac{3}{16}$	
9/32		9/16		$1\frac{1}{4}$	
19/64		19/32		$1\frac{5}{16}$	
5/16		5/8		$1\frac{3}{8}$	
21/64		21/32		$1\frac{7}{16}$	
11/32		11/16		$1\frac{1}{2}$	
23/64		23/32			
		3/4			
		25/32			

*Spiral Flute reamers are also available over same range of sizes, i.e., from 1/8 to 1-1/2 inches, inclusive, but exclusive of the intermediary 64th sizes.

TABLE 9–21

Sizes of Expansion Hand Reamers* — Straight† Flutes

ASA B5.14–1949 Reamers

Diameter of Reamer Inches	Maximum Expansion of Straight Flute Hand Reamer, Inch	Diameter of Reamer Inches	Maximum Expansion of Straight Flute Hand Reamer, Inch
1/4		23/32	
9/32		3/4	
5/16		13/16	
11/32		27/32	
3/8	0.006	7/8	0.010
13/32		29/32	
7/16		15/16	
15/32		31/32	
1/2		1	
17/32		1-1/16	
9/16		1-1/8	
19/32		1-3/16	
5/8	0.010	1-1/4	
21/32		1-5/16	0.012
11/16		1-3/8	
23/32		1-7/16	
		1-1/2	

*Expansion hand reamers are primarily designed for work where it is necessary to enlarge reamed holes by a few thousandths.

†Spiral flute expansion hand reamers are also available over same range of sizes, i.e., from 1/4 to 1-1/2 inches, inclusive, but exclusive of intermediary 32nd sizes.

TABLE 9-22

Sizes of Expansion Chucking* Reamers — Straight Flutes and Shank

ASA B5.14-1949 Reamers

Diameter of Reamer Inches			Diameter of Reamer Inches			Diameter of Reamer Inches	
3/8		13/32	1		1-1/32	2	2-1/16
	7/16	15/32		1-1/16	1-3/32	2-1/8	2-3/16
1/2		17/32	1-1/8		1-5/32	2-1/4	2-5/16
	9/16	19/32		1-3/16	1-7/32	2-3/8	2-7/16
5/8		21/32	1-1/4	1-5/16		2-1/2	2-9/16
	11/16	23/32	1-3/8	1-7/16		2-5/8	2-11/16
3/4		25/32	1-1/2	1-9/16		2-3/4	2-13/16
	13/16	27/32	1-5/8	1-11/16		2-7/8	2-15/16
7/8		29/32	1-3/4	1-13/16		3	
	15/16	31/32	1-7/8	1-15/16			

*Chucking reamers are available in a variety of styles, including fractional sizes as small as 1/8 inch. See Table 9—23 for decimal sizes for particular jobs. In general, reamers are tools for enlarging holes to size by the removal of small amounts of metal. Not only the thickness of stock to be removed but also the diameter of the hole influence greatly the size of reamer that can be driven by hand power. See Table 9—53.

TABLE 9-23

Stub* Screw Machine Reamers — Spiral Flutes

ASA B5.14-1949 Reamers

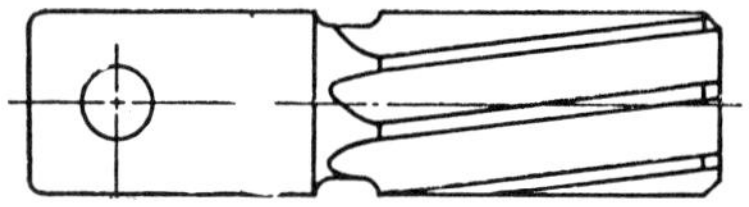

Series Number	Diameter Range Inches	Series Number	Diameter Range Inches	Series Number	Diameter Range Inches
00	0.0600 to 0.066 incl.	7	0.2191 to 0.251 incl.	15	0.4701 to 0.505 incl.
0	0.0661 to 0.074 incl.	8	0.2511 to 0.282 incl.	16	0.5051 to 0.567 incl.
1	0.0741 to 0.084 incl.	9	0.2821 to 0.313 incl.	17	0.5671 to 0.630 incl.
2	0.0841 to 0.096 incl.	10	0.3131 to 0.344 incl.	18	0.6301 to 0.692 incl.
3	0.0961 to 0.126 incl.	11	0.3441 to 0.376 incl.	19	0.6921 to 0.755 incl.
4	0.1261 to 0.158 incl.	12	0.3761 to 0.407 incl.	20	0.7551 to 0.817 incl.
5	0.1581 to 0.188 incl.	13	0.4071 to 0.439 incl.	21	0.8171 to 0.880 incl.
6	0.1881 to 0.219 incl.	14	0.4391 to 0.470 incl.	22	0.8801 to 0.942 incl.
				23	0.9421 to 1.010 incl.

*As the name implies, these short length reamers are used widely in screw machines. They can be purchased either finish ground to size or unfinished, the grinding to size, relieving and chamfering being left to the user.

TABLE 9–24

Millimeter Sizes of Hand Reamers — Straight Flutes

Catalog No. 46 The Cleveland Twist Drill Company

Diameter, millimeters	Diameter, millimeters	Diameter, millimeters	Diameter, millimeters
3	7	11	18
3.5	7.5	11.5	19
4	8	12	20
4.5	8.5	13	21
5	9	14	22
5.5	9.5	15	23
6	10	16	24
6.5	10.5	17	25

TABLE 9–25

Sizes of Taper* Pin Reamers — Straight Flutes, Spiral Flutes, Helical

ASA B5.14–1949 Reamers

Size Number of Reamer	Diameter, In. Small End	Diameter, In. Large End	Length of Flute B, Inches	Size Number of Reamer	Diameter, In. Small End	Diameter, In. Large End	Length of Flute B, Inches
7/0	0.0497	0.0666	13/16	3	0.1813	0.2294	2-5/16
6/0	0.0611	0.0806	15/16	4	0.2071	0.2604	2-9/16
5/0	0.0719	0.0966	1-3/16	5	0.2409	0.2994	2-13/16
4/0	0.0869	0.1142	1-5/16	6	0.2773	0.354	3-11/16
3/0	0.1029	0.1302	1-5/16	7	0.3297	0.422	4-7/16
2/0	0.1137	0.1462	1-9/16	8	0.3971	0.505	5-3/16
0	0.1287	0.1638	1-11/16	9	0.4805	0.6066	6-1/16
1	0.1447	0.1798	1-11/16	10	0.5799	0.7216	6-13/16
2	0.1605	0.2008	1-15/16				

*Taper of 1/4 inch per foot.

TABLE 9–26

Chart for Finding More Than One Drill Size for Holes to be Finished by Straight* Flute Taper Reamers

ASA B5.20–1947 Machine Pins

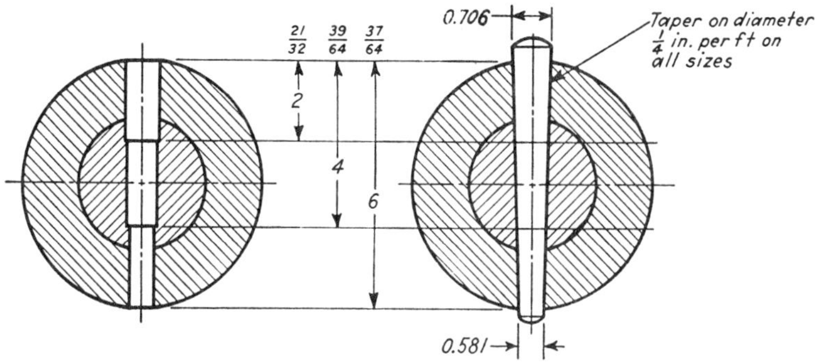

EXAMPLE — NO. 10 x 6" TAPER PIN USING 3 DRILLS FOR STEP DRILLING AND STRAIGHT FLUTED REAMERS

NOTE—— SEE DRILL CHART AT BOTTOM OF PAGE FOR SIZE OF DRILL AND NUMBER REQUIRED.

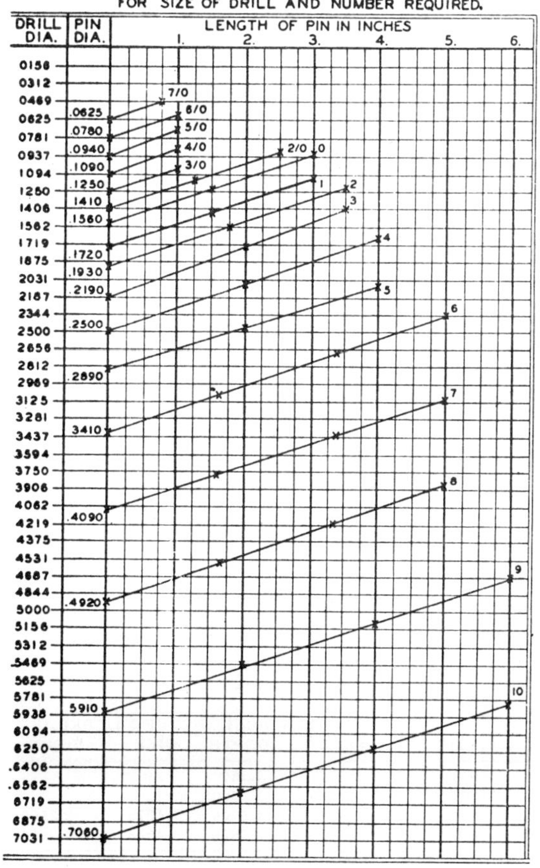

TO OBTAIN DRILL SIZES

Determine depth of hole (pin length).

Determine intersection of depth line with taper line.

Drill diameter will be the next smaller diameter on the horizontal line above the intersection.

Note the number of drills recommended for maximum length.

If chart calls for 3 drills, divide the drilling depth into 3 equal spaces (nearest 1/4 inch). If pin indicates 2 drills, divide the drilling depth into 2 spaces.

EXAMPLE A	#10 pin — 6 in. long Use 3 drills 5781 drill through 6094 drill 4 in. deep 6562 drill 2 in. deep
EXAMPLE B	#10 pin — 4 in. long Use 2 drills 6094 drill through 6562 drill 2 in. deep

All detail drawings using taper pins should carry the above type of note as indicated in (A) or (B).

To obtain the diameter at the small end, multiply the length by 0.02083 and subtract from the large diameter.

DRILL CHART

SIZE PIN	1st DRILL THROUGH SIZE	SECOND DRILL SIZE	SECOND DRILL DEPTH	THIRD DRILL SIZE	THIRD DRILL DEPTH
7/0	.0469				
6/0	.0469				
5/0	.0825				
4/0	.0781				
3/0	.0938				
2/0	.0938	.1094	1¼		
0	.0938	.1250	1½		
1	.1094	.1406	1½		
2	.1094	.1406	1¾		
3	.1406	.1719	1¼		
4	.1719	.2031	1¾		
5	.1875	.2344	2		
6	.2344	.2656	3¼	.2969	1¾
7	.2989	.3281	3¼	.3750	1-5/8
8	.3908	.4219	3¼	.4531	1-5/8
9	.4688	.5000	4	.5469	2
10	.5781	.6094	4	.6562	2

*If helically fluted taper reamers are used instead of step drilling and straight fluted reamers, the diameter at the small end of the pin is the size for the through drill.

continued on next page

NOTE — SEE DRILL CHART AT BOTTOM OF PAGE FOR SIZE OF DRILL AND NUMBER REQUIRED

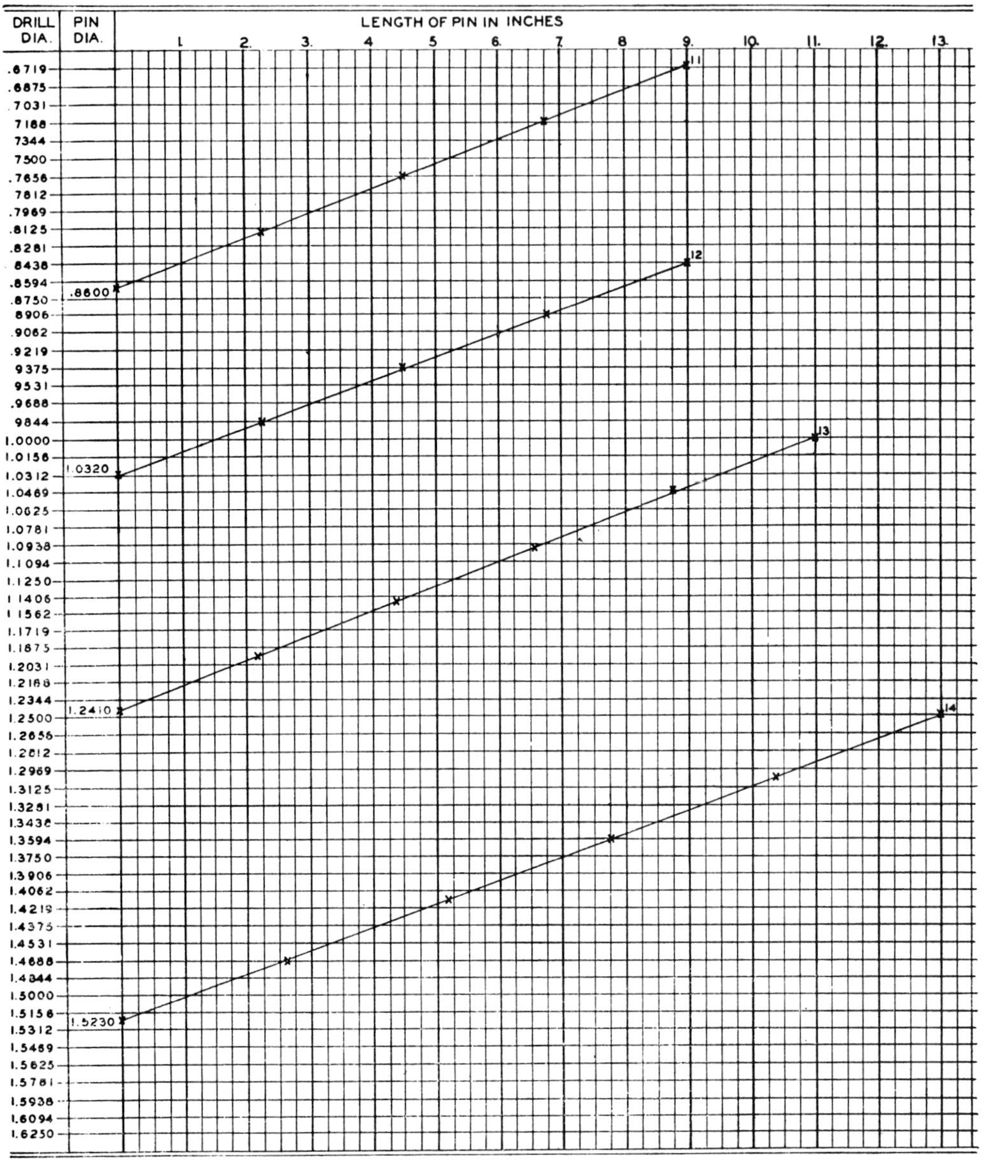

DRILL CHART

PIN SIZE	FIRST DRILL THROUGH SIZE	SECOND DRILL SIZE	SECOND DRILL DEPTH	THIRD DRILL SIZE	THIRD DRILL DEPTH	FOURTH DRILL SIZE	FOURTH DRILL DEPTH	FIFTH DRILL SIZE	FIFTH DRILL DEPTH
11	.6719	.7188	6.7500	.7658	4.5000	.8125	2.2500		
12	.8438	.8906	6.7500	.9375	4.5000	.9844	2.2500		
13	1.0000	1.0469	8.7500	1.0938	6.7500	1.1406	4.5000	1.1875	2.2500
14	1.2500	1.2969	10.5000	1.3594	7.7500	1.4062	5.2500	1.4688	2.7500

TABLE 9–27

Morse Taper Reamers
With Taper Shank

ASA B5.14–1949 Reamers

Roughing, Straight Flute, High Speed Steel

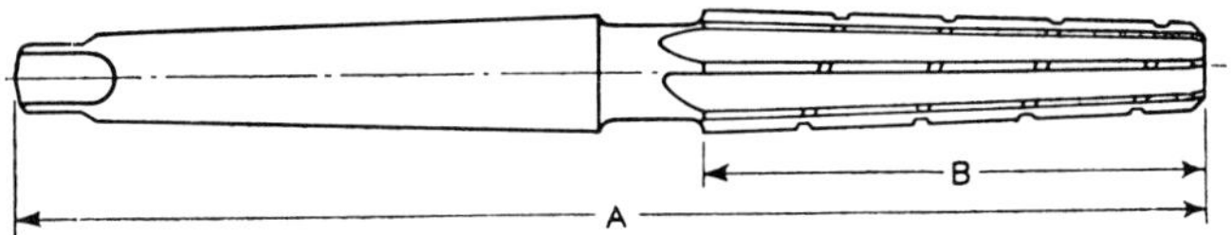

ROUGHING, STRAIGHT FLUTE

Finishing, Straight Flute, High Speed Steel

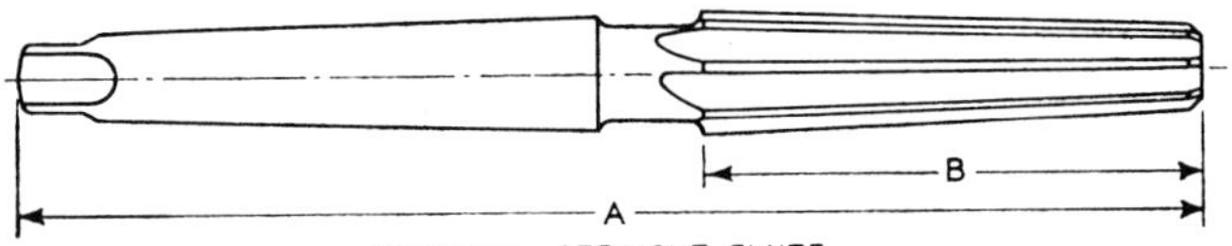

FINISHING, STRAIGHT FLUTE

Finishing Spiral Flute, High Speed Steel

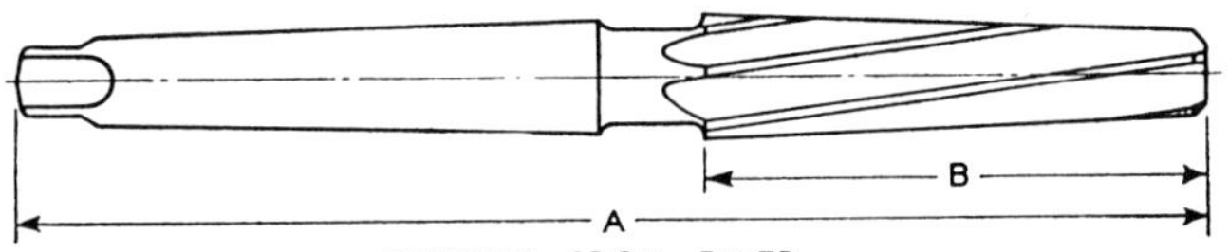

FINISHING, SPIRAL FLUTE

Size Number of Taper	Dimensions: Finishing Reamer Diameter, Small End	Dimensions: Finishing Reamer Diameter, Large End	Dimensions: Length Overall A	Dimensions: Length of Flute B	Number of Taper Shank, Am. Stand.*	Number of Flutes: Roughing	Number of Flutes: Finishing
0	0.2503	0.3674	5 11/32	2 1/4	0	4 to 6 incl.	4 to 6 incl.
1	0.3674	0.5170	6 5/16	3	1	4 to 6 incl.	6 to 8 incl.
2	0.5696	0.7444	7 3/8	3 1/2	2	4 to 6 incl.	6 to 8 incl.
3	0.7748	0.9881	8 7/8	4 1/4	3	4 to 6 incl.	8 to 10 incl.
4	1.0167	1.2893	10 7/8	5 1/4	4	4 to 8 incl.	8 to 10 incl.
5	1.4717	1.8005	13 1/8	6 1/4	5	6 to 10 incl.	10 to 12 incl.
6	2.1119	2.5550	17 13/16	8 1/2	6	6 to 12 incl.	12 to 14 incl.

All dimensions are given in inches.

These reamers are designed for use in reaming out Morse standard taper sockets. Number of flutes may vary in accordance with manufacturer's standard practice but must fall within the range specified in the table.

Sizes No. 1 to 5 incl. have ASA standard taper.

*See Tables 7-14 and 7-15.

Tolerances

Element	Range Size Number	Direction	Tolerance
Length Overall (A)	0 to 3 incl.	Plus or Minus	1/16
	4 to 5 incl.	Plus or Minus	3/32
	6	Plus or Minus	1/8
Length of Flute (B)	0 to 3 incl.	Plus or Minus	1/16
	4 to 5 incl.	Plus or Minus	3/32
	6	Plus or Minus	1/8

All dimensions are given in inches.

TABLE 9–28

Brown and Sharpe Taper Reamers With Squared Shank

ASA B5.14–1949 Reamers

Finishing, Straight Flute, Carbon Steel
Finishing, Straight Flute, High Speed Steel
Finishing, Spiral Flute, High Speed Steel

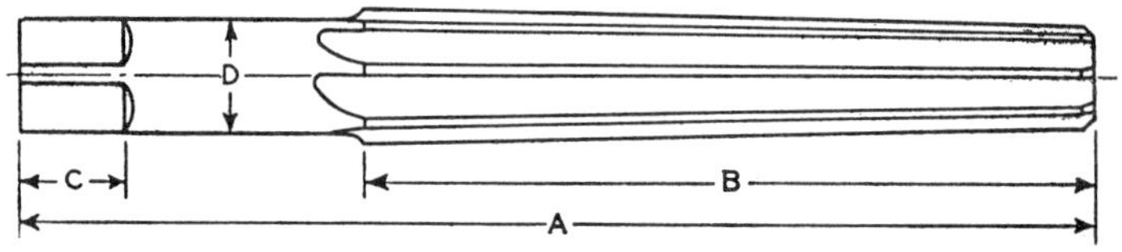

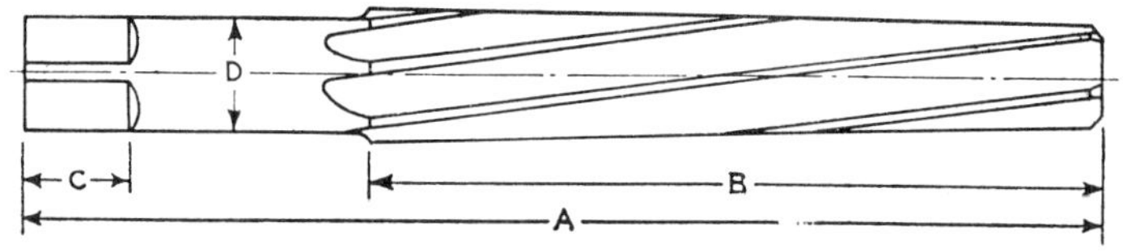

Size Number of Taper	Dimensions							Number of Flutes
	Diameter		Length Overall	Length of Flute	Length of Square	Diameter of Shank	Size of Square	
	Small End	Large End	A	B	C	D		
1	0.1974	0.3176	4 3/4	2 7/8	1/4	9/32	7/32	4 to 6
2	0.2474	0.3781	5 1/8	3 1/8	5/16	11/32	1/4	4 to 6
3	0.3099	0.4510	5 1/2	3 3/8	3/8	13/32	5/16	4 to 6
4	0.3474	0.5017	5 7/8	3 11/16	7/16	7/16	11/32	4 to 6
5	0.4474	0.6145	6 3/8	4	1/2	9/16	7/16	4 to 6
6	0.4974	0.6808	6 7/8	4 3/8	5/8	5/8	15/32	4 to 6
7	0.5974	0.8011	7 1/2	4 7/8	3/4	3/4	9/16	6 to 8
8	0.7474	0.9770	8 1/8	5 1/2	13/16	13/16	5/8	6 to 8
9	0.8974	1.1530	8 7/8	6 1/8	7/8	1	3/4	6 to 8
10	1.0420	1.3376	9 3/4	6 7/8	1	1 1/8	27/32	6 to 8
11	1.2474	1.5657	10 5/8	7 5/8	1 1/16	1 1/4	15/16	6 to 8
12	1.4974	1.8409	11 3/8	8 1/4	1 1/8	1 1/2	1 1/8	8 to 10

All dimensions are given in inches.

These reamers are designed for use in reaming out Brown & Sharpe standard taper sockets. Sizes No. 1, 2, and 3 have ASA Standard Taper. See Table 7–21.

Tolerances

Element	Range Size Number	Direction	Tolerance
Length Overall (A)	1 to 7 incl.	Plus or Minus	1/16
	8 to 10 incl.	Plus or Minus	3/32
	11 to 12 incl.	Plus or Minus	1/8
Length of Flute (B)	1 to 7 incl.	Plus or Minus	1/16
	8 to 10 incl.	Plus or Minus	3/32
	11 to 12 incl.	Plus or Minus	1/8
Length of Square (C)	1 to 9 incl.	Plus or Minus	1/32
	10 to 12 incl.	Plus or Minus	1/16
Diameter of Shank (D)	1 to 12 incl.	Minus	0.0005 to 0.002
Size of Square	1 to 3 incl.	Minus	0.004
	4 to 9 incl.	Minus	0.006
	10 to 12 incl.	Minus	0.008

TABLE 9–29

Sizes of Taper Bridge Reamers — Straight and Spiral Flutes

ASA B5.14–1949 Reamers

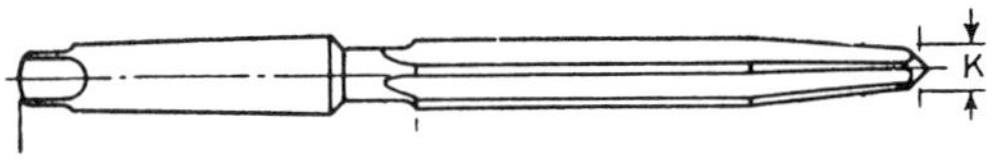

Diameter* of Reamer Inches			Diameter Small End K	Diameter* of Reamer Inches		Diameter Small End K
		13/32	7/32		15/16	5/8
	7/16	15/32	1/4	1		11/16
1/2			9/32		1-1/16	3/4
		17/32	5/16	1-1/8		13/16
	9/16		11/32		1-3/16	7/8
5/8			3/8	1-1/4		15/16
	11/16		25/64		1-5/16	1
3/4			7/16	1-3/8		1-1/16
	13/16		1/2		1-7/16	1-1/8
7/8			9/16	1-1/2		1-3/16

Taper bridge reamers are particularly adapted for reaming rivet and bolt holes in structural iron and steel, boiler plate, etc.

*Tolerance of +0.010 on all sizes.

TABLE 9–30

Short Length Taper Car Reamers

ASA B5.14–1949 Reamers

Diameter* of Reamer Inches			Diameter Small End K	Diameter* of Reamer Inches			Diameter Small End K
1/4			1/8		5/8		5/16
		9/32	5/32			11/16	3/8
	5/16		11/64		3/4		13/32
		11/32	13/64			13/16	15/32
3/8			15/64		7/8		17/32
		13/32	17/64			15/16	19/32
	7/16		1/4	1			21/32
		15/32	9/32			1-1/16	23/32
1/2			19/64	1-1/8			25/32
		17/32	1/4			1-3/16	27/32
	9/16		9/32	1-1/4			29/32

Taper car reamers are similar to taper bridge reamers, for use in tighter places. They are especially adapted for reaming rivet and bolt holes in thin structural sections.

*Tolerance of 0.010 on all sizes.

TABLE 9–31

Sizes of Counterbores and Spot* Facers with Pilots

Catalog No. 46 The Cleveland Twist Drill Company

Diameter Inches	Range of Pilot Sizes Inches	Diameter Inches	Range of Pilot Sizes Inches	Diameter Inches	Range of Pilot Sizes Inches
1/4	1/8 to 3/16	3/4	5/16 to 11/16	1-5/8	1/2 to 1-9/16
9/32	1/8 to 7/32	25/32	5/16 to 23/32	1-11/16	1/2 to 1-5/8
5/16	1/8 to 1/4	13/16	5/16 to 3/4	1-3/4	1/2 to 1-11/16
11/32	1/8 to 9/32	7/8	5/16 to 13/16	1-13/16	1/2 to 1-3/4
3/8	3/16 to 5/16	15/16	5/16 to 7/8	1-7/8	1/2 to 1-13/16
13/32	3/16 to 11/32	1	3/8 to 15/16	1-15/16	1/2 to 1-7/8
7/16	3/16 to 3/8	1-1/16	3/8 to 1	2	9/16 to 1-15/16
15/32	1/4 to 13/32	1-1/8	3/8 to 1-1/16	2-1/8	9/16 to 2-1/16
1/2	1/4 to 7/16	1-3/16	3/8 to 1-1/8	2-1/4	9/16 to 2-3/16
17/32	1/4 to 15/32	1-1/4	7/16 to 1-3/16	2-3/8	9/16 to 2-5/16
9/16	1/4 to 1/2	1-5/16	7/16 to 1-1/4	2-1/2	9/16 to 2-7/16
19/32	1/4 to 17/32	1-3/8	7/16 to 1-5/16	2-5/8	9/16 to 2-9/16
5/8	1/4 to 9/16	1-7/16	7/16 to 1-3/8	2-3/4	9/16 to 2-11/16
21/32	1/4 to 19/32	1-1/2	7/16 to 1-7/16	2-7/8	9/16 to 2-13/16
11/16	1/4 to 5/8	1-9/16	1/2 to 1-1/2	3	9/16 to 2-15/16
23/32	5/16 to 21/32				

*Whether a hole is enlarged by counterboring or by spot-facing is judged by the depth of cut. If the depth of the enlarged hole is shallow, 1/8 inch or less, the operation is termed spot-facing. Flat seats on cast surfaces for bolt heads and nuts are often spot-faced. Enlarged holes to accommodate fillister-head cap screws, on the contrary, are counterbored. Tools for spot-facing alone need no peripheral relief; those for counterboring do.

TABLE 9–32

Counterbore Sizes for Cap Screws and Machine Screws

Catalog No. 46 The Cleveland Twist Drill Company

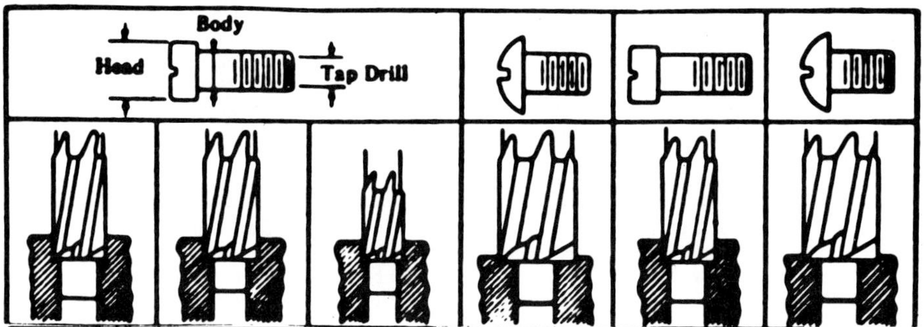

	Fillister Head Cap Screw		Fillister Head Cap Screw		Body and Tap Hole for any Screw		Round or Hexagon Head Cap Screw		Fillister Head Machine Screw		Round or Hex. Head Machine Screw		
	Head and Body		Head and Tap Hole				Head and Body		Head and Body		Head and Body		
Size Thread N.F. and N.C.	Cutter for Head	Pilot for Body	Cutter for Head	Pilot for Tap Hole	Cutter for Body	Pilot for Tap Hole	Cutter for Head	Pilot for Body	Cutter for Head	Pilot for Body	Cutter for Head	Pilot for Body	Tap Drills
1/4-28	.375	.250	.375	.213	.250	.213	.500	.250	.437	.250	.500	.250	3
1/4-20	.375	.250	.375	.201	.250	.201	.500	.250	.437	.250	.500	.250	7
5/16-24	.437	.312	.437	.272	.312	.272	.625	.312	.531	.312	.625	.312	I
5/16-18	.437	.312	.437	.257	.312	.257	.625	.312	.531	.312	.625	.312	F
3/8-24	.562	.375	.562	.332	.375	.332	.687	.375	.625	.375	.750	.375	Q
3/8-16	.562	.375	.562	.312	.375	.312	.687	.375	.625	.375	.750	.375	5/16
7/16-20	.625	.437	.625	.390	.437	.390	.813	.437	.718	.437	.875	.437	25/64
7/16-14	.625	.437	.625	.368	.437	.368	.813	.437	.718	.437	.875	.437	U
1/2-20	.750	.500	.750	.453	.500	.453	.875	.500	.843	.500	1.000	.500	29/64
1/2-13	.750	.500	.750	.421	.500	.421	.875	.500	.843	.500	1.000	.500	27/64
9/16-18	.812	.562	.812	.515	.562	.515	1.000	.562					33/64
9/16-12	.812	.562	.812	.484	.562	.484	1.000	.562					31/64
5/8-18	.875	.625	.875	.578	.625	.578	1.062	.625					37/64
5/8-11	.875	.625	.875	.531	.625	.531	1.062	.625					17/32
3/4-16	1.000	.750	1.000	.687	.750	.687	1.312	.750					11/16
3/4-10	1.000	.750	1.000	.656	.750	.656	1.312	.750					21/32
7/8-14	1.125	.875	1.125	.812	.875	.812	1.375	.875					13/16
7/8-9	1.125	.875	1.125	.765	.875	.765	1.375	.875					49/64
1-14	1.312	1.000	1.312	.937	1.000	.937	1.500	1.000					15/16
1-8	1.312	1.000	1.312	.875	1.000	.875	1.500	1.000					7/8

TABLE 9–33

Screw Threads and Threaded Fastenings — References Only

Important as threaded parts are in the design of machinery of all kinds, relatively few tables in this volume pertain to them. The sheer mass of information about screw threads and threaded fittings, such as bolts, nuts, setscrews, alone would fill a volume, and still be incomplete. Moreover, an abundance of published data, already carefully compiled by reliable sponsoring organizations, renders further duplication unnecessary. The few tables given herein are representative. They supply approximate dimensions that are deemed sufficient for drafting purposes. More complete and detailed data are found in the references, as follows, and in Table 11–15.

Description of Publication	Can Be Procured From
Unified and American Screw Threads ASA B1.1–1949	Published by The American Society of Mechanical Engineers, 29 West 39th St., New York 18, N. Y.
ASME Screw Thread Manual for Shop and Drafting Room — 1952	Same Source
Screw Thread Gages and Gaging ASA B1.2–1951	Same Source
Acme Screw Threads, B1.6–1952	Same Source
Stub Acme Screw Threads, B1.8–1952	Same Source
Slotted and Recessed Head Screws, B18.6–1947	Same Source
Plow Bolts, B18.9–1950	Same Source
Track Bolts and Nuts, B18.10–1952	Same Source
Round Head Bolts, B18.5–1952	Same Source
Square and Hexagon Bolts and Nuts, B18.2–1952	Same Source
High-Strength, High Temperature Internal Wrenching Bolts, B18.8–1950	Same Source
SAE Handbook	The Society of Automotive Engineers, Inc., 29 West 39th St., New York 18, N. Y.
Handbook H28, Screw-Thread Standards for Federal Services (1944) and 1950 Supplement thereto.	Superintendent of Documents, U.S. Government Printing Office, Washington 25, D. C.
Bolt, Nut and Rivet Standards	Industrial Fasteners Institute, 3648 Euclid Avenue, Cleveland 15, Ohio

TABLE 9–34

Regular Hexagon Bolts*

ASA B18.2–1952 Square and Hexagon Bolts and Nuts

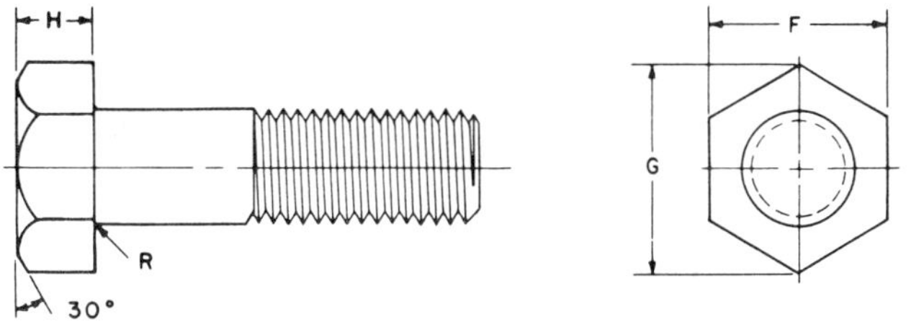

Nominal Size or Basic Major Diameter of Thread	Body Diam Max	Width Across Flats F Max (Basic)	Width Across Flats F Min	Width Across Corners G Max	Width Across Corners G Min	Height H Nom	Height H Max	Height H Min	Radius of Fillet R Max
1/4 0.2500	0.280	7/16 0.4375	0.425	0.505	0.484	11/64	0.188	0.150	0.031
5/16 0.3125	0.342	1/2 0.5000	0.484	0.577	0.552	7/32	0.235	0.195	0.031
3/8 0.3750	0.405	9/16 0.5625	0.544	0.650	0.620	1/4	0.268	0.226	0.031
7/16 0.4375	0.468	5/8 0.6250	0.603	0.722	0.687	19/64	0.316	0.272	0.031
1/2 0.5000	0.530	3/4 0.7500	0.725	0.866	0.826	11/32	0.364	0.302	0.031
5/8 0.6250	0.675	15/16 0.9375	0.906	1.083	1.033	27/64	0.444	0.378	0.062
3/4 0.7500	0.800	1 1/8 1.1250	1.088	1.299	1.240	1/2	0.524	0.455	0.062
7/8 0.8750	0.938	1 5/16 1.3125	1.269	1.516	1.447	37/64	0.604	0.531	0.062
1 1.0000	1.063	1 1/2 1.5000	1.450	1.732	1.653	43/64	0.700	0.591	0.062
1 1/8 1.1250	1.188	1 11/16 1.6875	1.631	1.949	1.859	3/4	0.780	0.658	0.125
1 1/4 1.2500	1.313	1 7/8 1.8750	1.812	2.165	2.066	27/32	0.876	0.749	0.125
1 3/8 1.3750	1.469	2 1/16 2.0625	1.994	2.382	2.273	29/32	0.940	0.810	0.125
1 1/2 1.5000	1.594	2 1/4 2.2500	2.175	2.598	2.480	1	1.036	0.902	0.125
1 5/8 1.6250	1.719	2 7/16 2.4375	2.356	2.815	2.686	1 1/16	1.100	0.962	0.125
1 3/4 1.7500	1.844	2 5/8 2.6250	2.538	3.031	2.893	1 5/32	1.196	1.054	0.125
1 7/8 1.8750	1.969	2 13/16 2.8125	2.719	3.248	3.100	1 7/32	1.260	1.114	0.125
2 2.0000	2.094	3 3.0000	2.900	3.464	3.306	1 11/32	1.388	1.175	0.125
2 1/4 2.2500	2.375	3 3/8 3.3750	3.262	3.897	3.719	1 1/2	1.548	1.327	0.188
2 1/2 2.5000	2.625	3 3/4 3.7500	3.625	4.330	4.133	1 21/32	1.708	1.479	0.188
2 3/4 2.7500	2.875	4 1/8 4.1250	3.988	4.763	4.546	1 13/16	1.869	1.632	0.188
3 3.0000	3.125	4 1/2 4.5000	4.350	5.196	4.959	2	2.060	1.815	0.188
3 1/4 3.2500	3.438	4 7/8 4.8750	4.712	5.629	5.372	2 3/16	2.251	1.936	0.188
3 1/2 3.5000	3.688	5 1/4 5.2500	5.075	6.062	5.786	2 5/16	2.380	2.057	0.188
3 3/4 3.7500	3.938	5 5/8 5.6250	5.437	6.495	6.198	2 1/2	2.572	2.241	0.188
4 4.0000	4.188	6 6.0000	5.800	6.928	6.612	2 11/16	2.764	2.424	0.188

All dimensions given in inches.

Bolt is not finished on any surface.

Taper of head (angle between one side and axis) shall not exceed 2 deg, specified width across flats being the largest dimension.

Top of bolt head shall be flat and chamfered. Diameter of top circle shall be maximum width across flats, within a tolerance of minus 15 per cent.

Bearing surface shall be at right angles to axis of body within a tolerance of 3 deg for 1-in. size or smaller and 2 deg for sizes larger than 1 in. The bearing surface shall be concentric with axis of body within a tolerance of 3 per cent of maximum width across flats.

Minimum thread length shall be twice the diameter plus 1/4 in. for lengths up to and including 6 in. and twice the diameter plus 1/2 in. for lengths over 6 in. Bolts too short for the formula thread length shall be threaded as close to the head as practical.

Thread shall be coarse-thread series, class 2A.

Tolerance on bolt length for bolts 6 in. and under in length shall be plus or minus 1/32 in. for sizes 1/4 to 3/8 in., plus or minus 1/16 in. for sizes 7/16 and 1/2 in., plus or minus 1/8 in. for sizes 5/8 to 1¼ in., and plus or minus 1/4 in. for sizes 1-3/8 to 4 in. Length tolerance for bolts over 6 in. in length shall be plus or minus 1/16 in. for sizes 1/4 to 3/8 in., plus or minus 3/32 in. for sizes 7/16 and 1/2 in., plus or minus 3/16 in. for sizes 5/8 to 1¼ in., and plus or minus 1/4 in. for sizes 1-3/8 to 4 in.

Bolts shall enter ring gage to head; ring gage shall have a maximum thickness of one diameter, a minimum fillet equal to the maximum fillet of the bolt, and an internal diameter equal to the specified maximum body diameter.

Suitable material for steel bolt is covered in Tentative Specification for Steel Machine Bolts and Nuts and Tap Bolts of the American Society for Testing Materials (ASTM A-307).

*See Tables 11–9 to 11–11 for nuts.

TABLE 9-35

Heavy Finished Hexagon Bolts*

ASA B18.2-1952 Square and Hexagon Bolts and Nuts

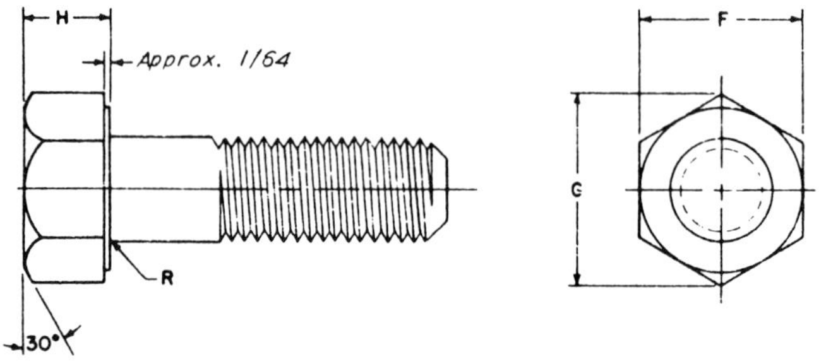

Nominal Size or Basic Major Diameter of Thread	Body Diam Min (Max Equal to Nominal Size)	Width Across Flats F		Width Across Corners G		Height H			Radius of Fillet R	
		Max (Basic)	Min	Max	Min	Nom	Max	Min	Max	Min
1/2 0.5000	**0.4940**	**7/8 0.8750**	**0.850**	**0.010**	**0.969**	**13/32**	**0.426**	**0.386**	**0.031**	**0.016**
5/8 0.6250	**0.6190**	**1 1/16 1.0625**	**1.031**	**1.227**	**1.175**	**1/2**	**0.522**	**0.478**	**0.031**	**0.016**
3/4 0.7500	**0.7440**	**1 1/4 1.2500**	**1.212**	**1.443**	**1.383**	**19/32**	**0.618**	**0.570**	**0.047**	**0.031**
7/8 0.8750	**0.8690**	**1 7/16 1.4375**	**1.394**	**1.660**	**1.589**	**11/16**	**0.714**	**0.662**	**0.047**	**0.031**
1 1.0000	**0.9940**	**1 5/8 1.6250**	**1.575**	**1.876**	**1.796**	**3/4**	**0.778**	**0.722**	**0.047**	**0.031**
1 1/8 1.1250	**1.1170**	**1 13/16 1.8125**	**1.756**	**2.093**	**2.002**	**27/32**	**0.874**	**0.814**	**0.062**	**0.047**
1 1/4 1.2500	**1.2420**	**2 2.0000**	**1.938**	**2.309**	**2.209**	**15/16**	**0.970**	**0.906**	**0.062**	**0.047**
1 3/8 1.3750	**1.3670**	**2 3/16 2.1875**	**2.119**	**2.526**	**2.416**	**1 1/32**	**1.065**	**0.997**	**0.062**	**0.047**
1 1/2 1.5000	**1.4920**	**2 3/8 2.3750**	**2.300**	**2.742**	**2.622**	**1 1/8**	**1.161**	**1.089**	**0.062**	**0.047**
1 5/8 1.6250	1.6170	2 9/16 2.5625	2.481	2.959	2.828	1 7/32	1.257	1.181	0.062	0.047
1 3/4 1.7500	**1.7420**	**2 3/4 2.7500**	**2.662**	**3.175**	**3.035**	**1 5/16**	**1.352**	**1.272**	**0.062**	**0.047**
1 7/8 1.8750	1.8670	2 15/16 2.9375	2.844	3.392	3.242	1 13/32	1.448	1.364	0.062	0.047
2 2.0000	**1.9900**	**3 1/8 3.1250**	**3.025**	**3.608**	**3.449**	**1 7/16**	**1.482**	**1.394**	**0.062**	**0.047**
2 1/4 2.2500	2.2400	3 1/2 3.5000	3.388	4.041	3.862	1 5/8	1.673	1.577	0.062	0.047
2 1/2 2.5000	2.4900	3 7/8 3.8750	3.750	4.474	4.275	1 13/16	1.864	1.760	0.062	0.047
2 3/4 2.7500	2.7400	4 1/4 4.2500	4.112	4.907	4.688	2	2.056	1.944	0.062	0.047
3 3.0000	2.9900	4 5/8 4.6250	4.475	5.340	5.102	2 3/16	2.248	2.128	0.062	0.047

All dimensions given in inches.

BOLD TYPE INDICATES PRODUCTS UNIFIED DIMENSIONALLY WITH BRITISH AND CANADIAN STANDARDS.

"Finished" in the title refers to the quality of manufacture and the closeness of tolerance and does not indicate that surfaces are completely machined.

Taper of head (angle between one side and axis) shall not exceed 2 deg, specified width across flats being the largest dimension.

Top of head shall be flat and chamfered. Diameter of top circle shall be maximum width across flats within a tolerance of minus 15 per cent.

Bearing surface shall be flat and washer faced. Diameter of washer face shall be 95 per cent of maximum width across flats within a tolerance of plus or minus 5 per cent.

Bearing surface shall be at right angles to axis of body within a tolerance of 2 deg for sizes up to and including 1 in.; and within a tolerance of 1 deg for sizes larger than 1 in. The bearing surface shall be concentric with axis of body within a tolerance of 3 per cent of the maximum width across flats.

Minimum thread length shall be twice the diameter plus 1/4 in. for lengths up to and including 6 in; twice the diameter plus 1/2 in. for lengths over 6 in. The tolerance shall be plus 3/16 in. or 2½ threads, whichever is greater. On products that are too short for minimum thread lengths, the distance from the bearing surface of the head to the first complete thread shall not exceed the length of 2½ threads, as measured with a ring thread gage, for sizes to to and including 1 in. and 3½ thread for sizes larger than 1 in.

Threads shall be coarse-, fine-, or 8-thread series, class 2A for plain (unplated) bolts. For plated bolts, the diameters may be increased by the amount of class 2A allowance. Thickness or quality of plating shall be measured or tested on the side of the bolt head.

Point shall be flat and chamfered or rounded at manufacturer's option, length of point to first full thread not to exceed 1½ threads.

Tolerance on bolt length for bolts 6 in. and under in length shall be plus or minus 1/16 in. for 1/2 in. size, plus or minus 1/8 in. for sizes 5/8 to 1¼ in., and plus or minus 1/4 in. for sizes 1-3/8 to 3 in. Length tolerance for bolts over 6 in. in length shall be plus or minus 3/32 in. for 1/2 in. size, plus or minus 3/16 in. for sizes 5/8 to 1¼ in., and plus or minus 1/4 in. for sizes 1-3/8 to 3 in.

Maximum deviation of shank from surface plate on which it is rolled shall be 0.0020 in. per inch of length.

Suitable material for steel bolt is covered by Tentative Specification for Steel Machine Bolts and Nuts and Tap Bolts of the American Society for Testing Materials (ASTM A-307); suitable material for high-strength steel bolt is covered by ASTM Tentative Spec. for Quenched and Tempered Steel Bolts and Studs with Suitable Nuts and Plain Washers (ASTM A-325).

*See Tables 11-13 and 11-14 for heavy nuts.

TABLE 9–36

Square Head Set Screws

ASA B18.2–1952 Square and Hexagon Bolts and Nuts

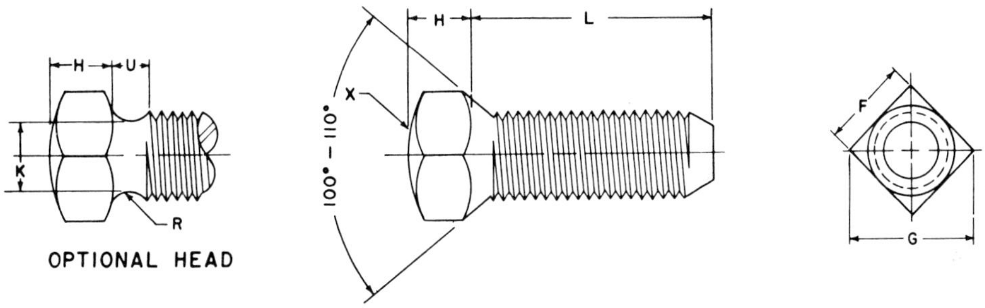

Nominal Size		Width Across Flats F		Width Across Corners G	Height of Head H			Diameter of Neck Relief K		Radius of Head X	Rad of Neck Relief R	Width of Neck Relief U
		Max	Min	Min	Nom	Max	Min	Max	Min	Nom	Max	Max
#10	0.190	0.1875	0.180	0.247	9/64	0.148	0.134	0.145	0.140	15/32	0.027	0.083
#12	0.216	0.216	0.208	0.292	5/32	0.163	0.147	0.162	0.156	35/64	0.029	0.091
1/4	0.250	0.250	0.241	0.331	3/16	0.196	0.178	0.185	0.170	5/8	0.032	0.100
5/16	0.3125	0.3125	0.302	0.415	15/64	0.245	0.224	0.240	0.225	25/32	0.036	0.111
3/8	0.3750	0.375	0.362	0.497	9/32	0.293	0.270	0.294	0.279	15/16	0.041	0.125
7/16	0.4375	0.4375	0.423	0.581	21/64	0.341	0.315	0.345	0.330	1 3/32	0.046	0.143
1/2	0.500	0.500	0.484	0.665	3/8	0.389	0.361	0.400	0.385	1 1/4	0.050	0.154
9/16	0.5625	0.5625	0.545	0.748	27/64	0.437	0.407	0.454	0.439	1 13/32	0.054	0.167
5/8	0.6250	0.625	0.606	0.833	15/32	0.485	0.452	0.507	0.492	1 9/16	0.059	0.182
3/4	0.750	0.750	0.729	1.001	9/16	0.582	0.544	0.620	0.605	1 7/8	0.065	0.200
7/8	0.875	0.875	0.852	1.170	21/32	0.678	0.635	0.731	0.716	2 3/16	0.072	0.222
1	1.000	1.000	0.974	1.337	3/4	0.774	0.726	0.838	0.823	2 1/2	0.081	0.250
1 1/8	1.125	1.125	1.096	1.505	27/32	0.870	0.817	0.939	0.914	2 13/16	0.092	0.283
1 1/4	1.250	1.250	1.219	1.674	15/16	0.966	0.908	1.064	1.039	3 1/8	0.092	0.283
1 3/8	1.375	1.375	1.342	1.843	1 1/32	1.063	1.000	1.159	1.134	3 7/16	0.109	0.333
1 1/2	1.500	1.500	1.464	2.010	1 1/8	1.159	1.091	1.284	1.259	3 3/4	0.109	0.333

All dimensions given in inches.

Threads shall be coarse-, fine-, or 8-thread series, class 2A. Square head set screws 1/4 in. size and larger are normally stocked in coarse thread series only.

Tolerance on screw length for sizes up to and including 5/8 in. shall be: minus 1/32 in. for lengths up to and including 1 in; minus 1/16 in. for lengths over 1 in. to and including 2 in.; and minus 3/32 in. for lengths over 2 in. The tolerance shall be doubled for larger size screws of comparable length.

Square head set screws shall be made from alloy or carbon steel suitably hardened. Screws made from nonferrous material or corrosion-resisting steel shall be made from a material mutually agreed upon by manufacturer and user.

For various types of points, see Table 9–37.

TABLE 9–37

Square Head Set Screw Points

ASA B18.2–1952 Square and Hexagon Bolts and Nuts

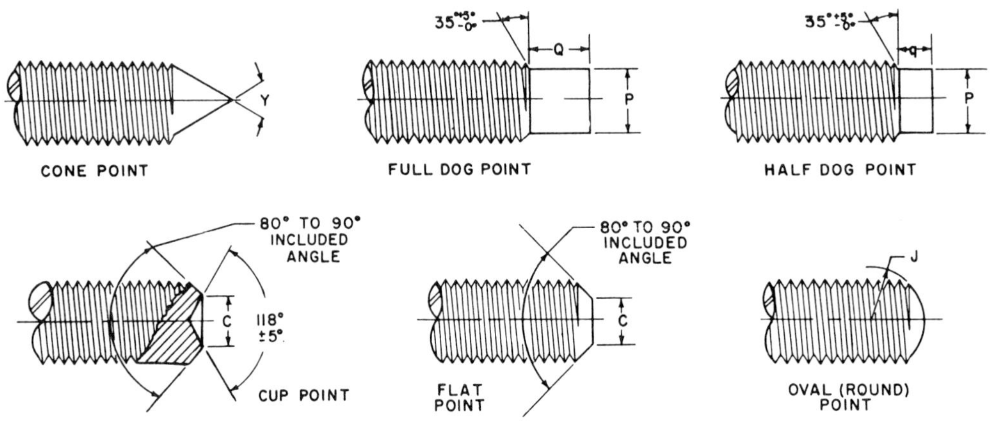

Nominal Size	Cup and Flat Point Diameter C			Oval (Round) Point Radius J	Full Dog, Half Dog Pivot Point			
					Diameter P		Full Dog Pvt. Q	Half Dog Pvt. q
	Nom	Max	Min	Nom	Max	Min		
#10	3/32	0.102	0.088	0.141	0.127	0.120	0.090	0.045
#12	7/64	0.115	0.101	0.156	0.144	0.137	0.110	0.055
1/4	1/8	0.132	0.118	0.188	0.156	0.149	0.125	0.063
5/16	11/64	0.172	0.156	0.234	0.203	0.195	0.156	0.078
3/8	13/64	0.212	0.194	0.281	0.250	0.241	0.188	0.094
7/16	15/64	0.252	0.232	0.328	0.297	0.287	0.219	0.109
1/2	9/32	0.291	0.270	0.375	0.344	0.334	0.250	0.125
9/16	5/16	0.332	0.309	0.422	0.391	0.379	0.281	0.140
5/8	23/64	0.371	0.347	0.469	0.469	0.456	0.313	0.156
3/4	7/16	0.450	0.425	0.563	0.563	0.549	0.375	0.188
7/8	33/64	0.530	0.502	0.656	0.656	0.642	0.438	0.219
1	19/32	0.609	0.579	0.750	0.750	0.734	0.500	0.250
1 1/8	43/64	0.689	0.655	0.844	0.844	0.826	0.562	0.281
1 1/4	3/4	0.767	0.733	0.938	0.938	0.920	0.625	0.312
1 3/8	53/64	0.848	0.808	1.031	1.031	1.011	0.688	0.344
1 1/2	29/32	0.926	0.886	1.125	1.125	1.105	0.750	0.375

All dimensions given in inches.

Pivot points are similar to full dog point except that the point is rounded by a radius equal to J.

Where usable length of thread is less than the nominal diameter, half-dog point shall be used.

When length equals nominal diameter or less, Y = 118 deg ± 2 deg; when length exceeds nominal diameter, Y = 90 deg ± 2 deg.

TABLE 9–38

Sizes of End Mills

ASA B5.3–1950 Milling Cutters

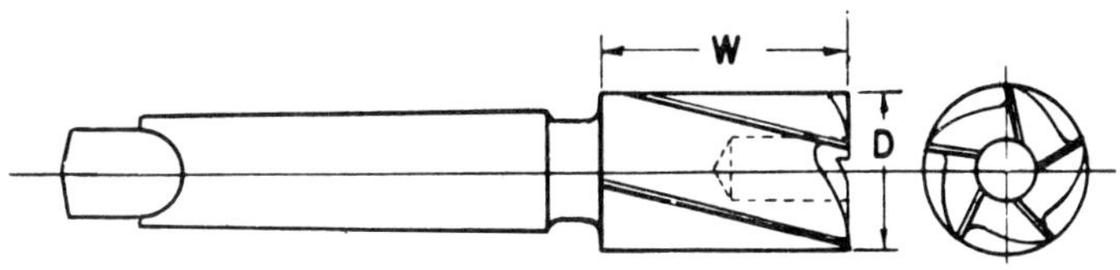

Multiple-Flute American Standard

Diameter* of Cutter D Inches	Length of Cut W Inches	Diameter* of Cutter D Inches	Length of Cut W Inches	Diameter* of Cutter D Inches	Length of Cut W Inches
1/4	5/8	9/16	1	1-1/8	1-3/4
5/16	11/16	5/8	1-1/8	1-1/4	2
3/8	3/4	3/4	1-1/4	1-1/2	2-1/4
7/16	7/8	7/8	1-7/16	1-3/4	2-1/2
1/2	15/16	1	1-5/8	2	2-3/4

*Tolerance of +0.005 on all sizes.

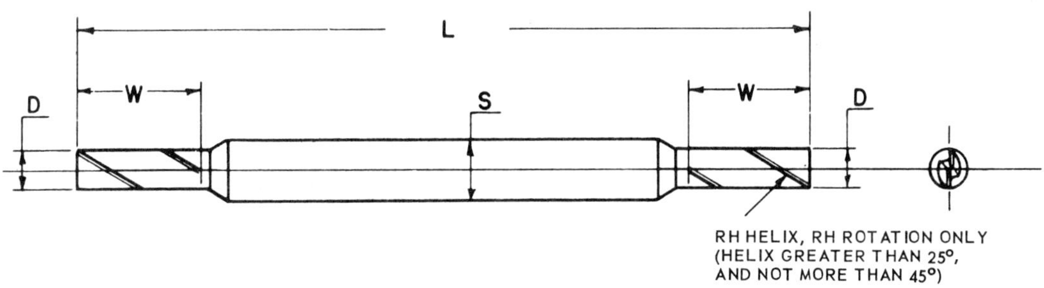

Two-Flute, Fast Helix, Double-End

Diameter* of Cutter D Inches	Length of Cut W Inches	Diameter of Shank S Inches	Length Over-all L, Inches
1/16	7/32	3/16	2-1/4
3/32	5/16	all sizes	2-1/4
1/8	3/8	Tolerance	2-1/4
5/32	7/16	of -0.0005	2-1/4
3/16	1/2		2-1/4

*Tolerance of −0.0015 on all sizes.

TABLE 9–39

Widths of Side Milling Cutters

ASA B5.3–1950 Milling Cutters

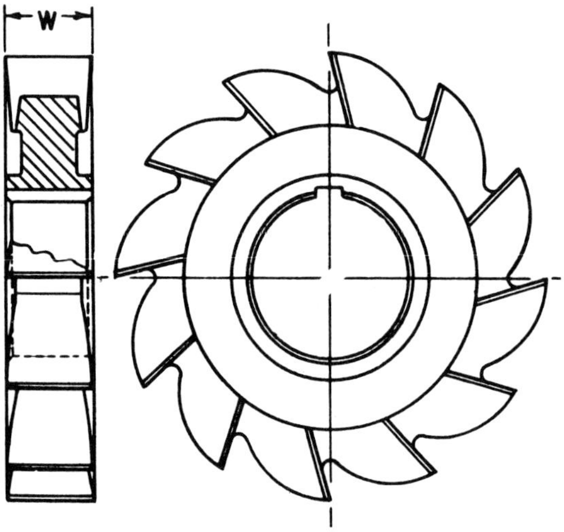

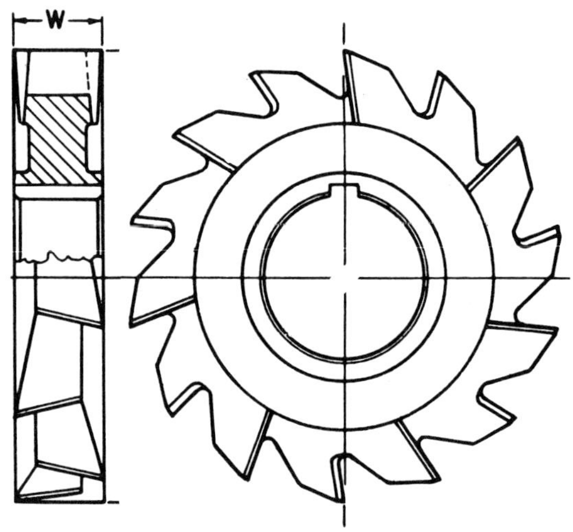

Plain Side Milling Cutter

Width of Face, W Inches	Tolerance on Width of Face
3/16	
1/4	
5/16	
3/8	-0.001
7/16	+0.002
1/2	on all
5/8	sizes
3/4	
7/8	
1	

Staggered-Tooth* Type

Width of Face, W Inches	Tolerance on Width of Face
1/4	-0.0005
5/16	-0.0005
3/8	-0.0005
7/16	-0.0005
1/2	-0.0005
5/8	-0.0005
3/4	-0.0005
7/8	-0.001
1	-0.001

*The staggered-tooth side milling cutter is preferred on deep, heavy cuts. This type is frequently used in keyslotting to close tolerances.

TABLE 9–40

Cutter Numbers of Woodruff Keyslot Milling Cutters

ASA B5.3–1950 Milling Cutters

Number of Cutters American* Standard	Old Standard	Cutter Face Width† Inches	Number of Cutters American* Standard	Old Standard	Cutter Face Width† Inches
202	201	1/16	810	21	1/4
202-1/2	206	1/16	811	22	1/4
302-1/2	207	3/32	812	24	1/4
203	211	1/16	1008	B	5/16
303	212	3/32	1009	C	5/16
403	213	1/8	1010	D	5/16
204	1	1/16	1011	23	5/16
304	2	3/32	1012	25	5/16
305	4	3/32	1210	E	3/8
404	3	1/8	1211	F	3/8
405	5	1/8	1212	G	3/8
406	7	1/8			
505	6	5/32			
605	61	3/16			
506	8	5/32	617	26	3/16
806	91	1/4	817	27	1/4
507	10	5/32	1017	28	5/16
606	9	3/16	1217	29	3/8
607	11	3/16	822	R	1/4
707	12	7/32	1022	S	5/16
608	13	3/16	1222	T	3/8
708	14	7/32	1422	U	7/16
1208	152	3/8	1622	V	1/2
609	16	3/16	1228	30	3/8
807	A	1/4	1428	31	7/16
808	15	1/4	1628	32	1/2
709	17	7/32	1828	33	9/16
809	18	1/4	2028	34	5/8
610	19	3/16	2228	35	11/16
710	20	7/32	2428	36	3/4

*The cutter numbers shown in the column headed "American Standard" indicates the nominal key dimension or size cutter, i.e., the last two digits give the nominal diameter in 8ths of an inch and the digits preceding the last two give the nominal width in 32nds of an inch. Thus, cutter No. 204 indicates a size 2/32 x 4/8 in. or 1/16 in. thick x 1/2 in. diameter.

†Tolerance on face width of all sizes of –0.0005.

For Key No. 121, use Cutter No. 807
For Key No. 141, use Cutter No. 808
For Key No. 131, use Cutter No. 1008
For Key No. 161, use Cutter No. 1009
For Key No. 126, use Cutter No. 617
For Key No. 127, use Cutter No. 817
For Key No. 128, use Cutter No. 1017
For Key No. 129, use Cutter No. 1217
For Key No. RX, use Cutter No. 822
For Key No. S X, use Cutter No. 1022
For Key No. TX, use Cutter No. 1222
For Key No. UX, use Cutter No. 1422
For Key No. VX, use Cutter No. 1622

TABLE 9–41

Dimensions of T-Slots

ASA B5.1–1949 T-Slots—Their Bolts, Nuts, Tongues and Cutters

BASIC DIMENSIONS

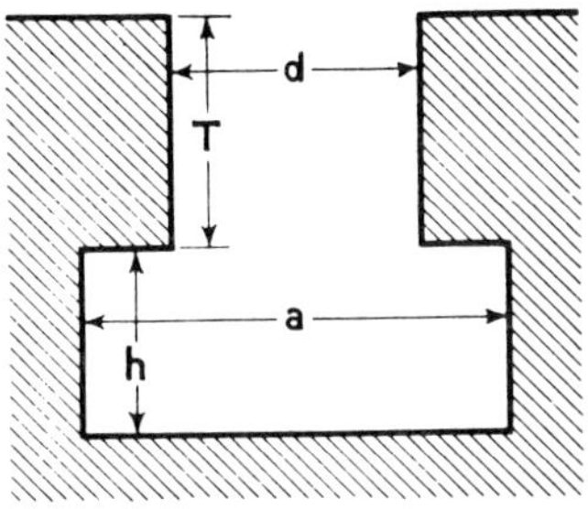

SUGGESTED DIMENSIONS FOR OPTIONAL ROUNDING AND BREAKING OF CORNERS

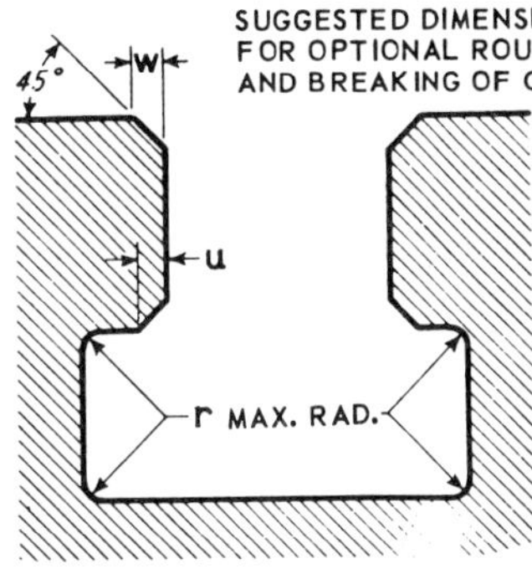

Diameter of T-bolt [2]	Width of Throat [1, 2] d	Depth of Throat T		Head Space Dimensions and Tolerances						Rounding or Breaking of Corners		
				Width a			Depth h			r	w	u
		Maximum	Minimum	Maximum Basic	Tolerance Minus	Minimum	Maximum Basic	Tolerance Minus	Minimum	Maximum	Maximum	Maximum
1/4	9/32	3/8	1/8	9/16	0.063	1/2	15/64	0.031	13/64	1/64	1/64	1/32
5/16	11/32	7/16	5/32	21/32	0.063	19/32	17/64	0.031	15/64	1/64	1/32	1/32
3/8	7/16	9/16	7/32	25/32	0.063	23/32	21/64	0.031	19/64	1/64	1/32	1/32
1/2	9/16	11/16	5/16	31/32	0.063	29/32	25/64	0.031	23/64	1/64	1/32	1/32
5/8	11/16	7/8	7/16	1 1/4	0.063	1 3/16	31/64	0.031	29/64	1/32	1/32	3/64
3/4	13/16	1 1/16	9/16	1 15/32	0.094	1 3/8	5/8	0.031	19/32	1/32	1/32	3/64
1	1 1/16	1 1/4	3/4	1 27/32	0.094	1 3/4	53/64	0.047	25/32	1/32	1/16	3/64
1 1/4	1 5/16	1 9/16	1	2 7/32	0.094	2 1/8	1 3/32	0.063	1 1/32	1/32	1/16	3/64
1 1/2	1 9/16	1 15/16	1 1/4	2 21/32	0.094	2 9/16	1 11/32	0.063	1 9/32	1/32	1/16	3/64

All dimensions in inches.

[1]A tolerance of plus 0.001 is allowed for "width of throat" when tongues or other parts must fit.

[2]In addition to the "width of throat" given above, a secondary standard is recognized, having the "width of throat" the same as the nominal diameter of the T-bolt. This is to provide for the use during the transition period of this standard on many machine tools where it is already established.

TABLE 9-44

Sizes of Metal Slitting Saws

ASA B5.3-1950 Milling Cutters

Width* of Face, Inch	Diameter of Cutter, Inches					
	2½	3	4	5	6	8
1/32	x	x	x			
3/64	x	x	x			
1/16	x	x	x	x	x	
3/32	x	x	x	x	x	
1/8	x	x	x	x	x	x
5/32		x	x	x		
3/16			x	x	x	x
1/4				x	x	x

*Tolerance on all sizes of ±0.001.

TABLE 9-45

Dimensions of Double-Angle Milling Cutters

ASA B5.3-1950 Milling Cutters

Symmetrical double-angle cutters are designated by included angle. Unsymmetrical double-angle cutters are designated by specifying the angle of each side with plane of intersection at right angles to the axis. Angle tolerance is plus or minus 10 minutes each side of center.

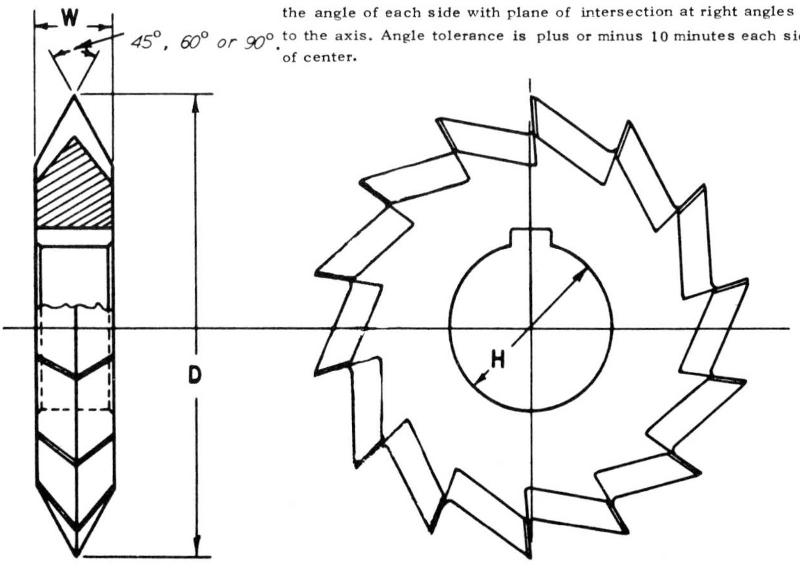

Diameter of Cutter D	Width of Face W	Diameter of Hole H
2 3/4	1/2	1

All dimensions are given in inches.

TABLE 9-46

Dimensions of Convex Milling Cutters

ASA B5.3-1950 Milling Cutters

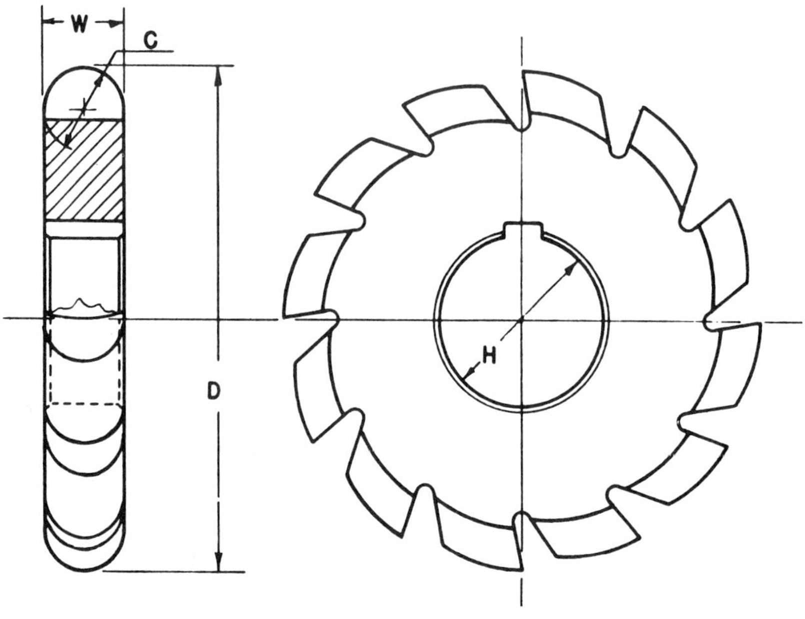

DIAMETER OF CIRCLE C	DIAMETER OF CUTTER D	WIDTH OF FACE W	DIAMETER OF HOLE H
1/8	2 1/4	1/8	1
3/16	2 1/4	3/16	1
1/4	2 1/2	1/4	1
5/16	2 3/4	5/16	1
3/8	2 3/4	3/8	1
7/16	3	7/16	1
1/2	3	1/2	1
5/8	3 1/2	5/8	1 1/4
3/4	3 3/4	3/4	1 1/4
7/8	4	7/8	1 1/4
1	4 1/4	1	1 1/4

TOLERANCES

ELEMENT	RANGE	DIRECTION	TOLERANCE
Diameter of Cutter	All Sizes	Plus or Minus	1/16
Diameter of Circle	All Sizes	Plus or Minus	0.002
Diameter of Hole	Up to 1 Inclusive	Plus	0.00075
	Over 1	Plus	0.001

All dimensions are given in inches.
Size of cutter is designated by specifying diameter (C) of circular form.

TABLE 9–47

Dimensions of Concave Milling Cutters

ASA B5.3–1950 Milling Cutters

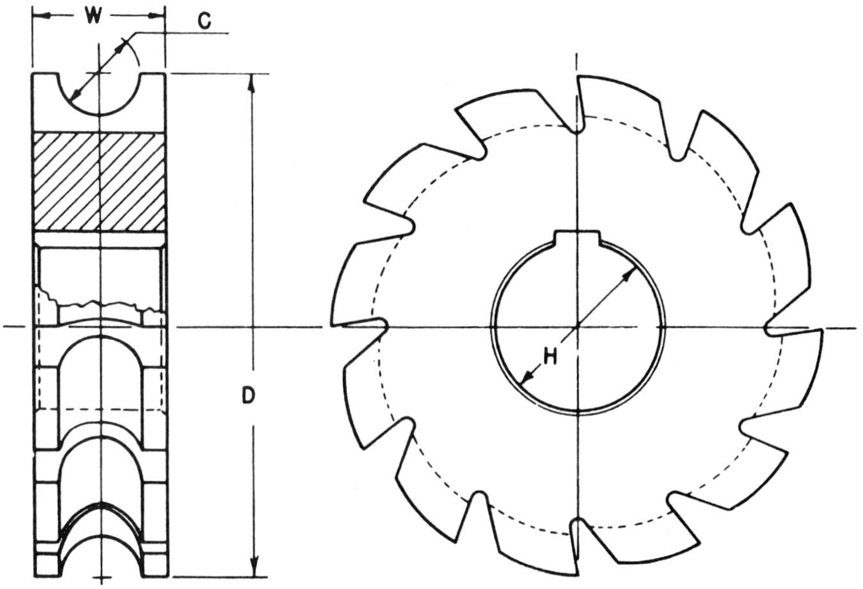

DIAMETER OF CIRCLE C	DIAMETER OF CUTTER D	WIDTH OF FACE W	DIAMETER OF HOLE H
1/8	2 1/4	1/4	1
3/16	2 1/4	3/8	1
1/4	2 1/2	7/16	1
5/16	2 3/4	9/16	1
3/8	2 3/4	5/8	1
7/16	3	3/4	1
1/2	3	13/16	1
5/8	3 1/2	1	1 1/4
3/4	3 3/4	1 3/16	1 1/4
7/8	4	1 3/8	1 1/4
1	4 1/4	1 9/16	1 1/4

TOLERANCES

ELEMENT	RANGE	DIRECTION	TOLERANCE
Diameter of Cutter	All Sizes	Plus or Minus	1/16
Diameter of Circle	Up to 7/16 Inclusive	Minus	0.001
		Plus	0.002
	Over 7/16	Minus	0.002
		Plus	0.004
Width of Face	All Sizes	Plus or Minus	0.010
Diameter of Hole	Up to 1 Inclusive	Plus	0.00075
	Over 1	Plus	0.001

All dimensions are given in inches.
Size of cutter is designated by specifying diameter (C) of circular form.

TABLE 9–48

Dimensions of Corner Rounding Cutters

ASA B5.3–1950 Milling Cutters

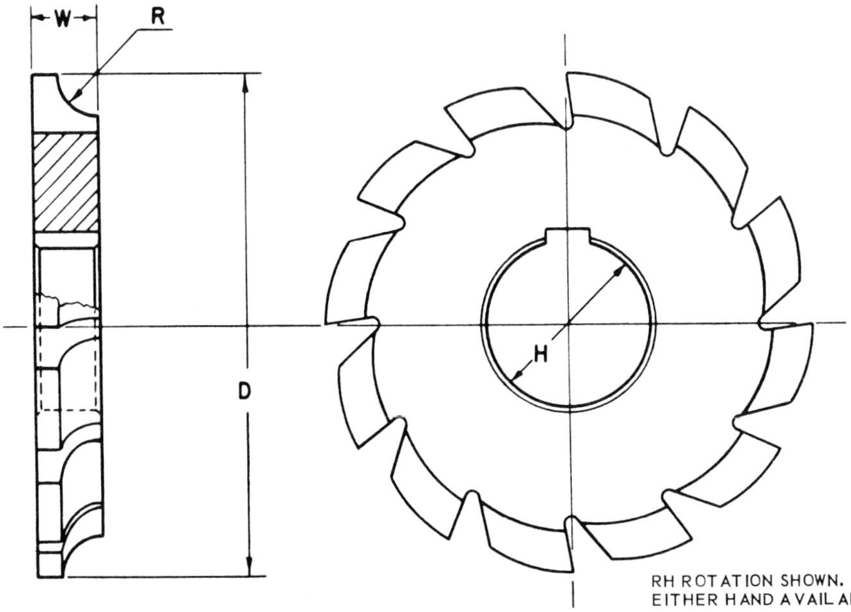

RADIUS OF CIRCLE R	DIAMETER OF CUTTER D	WIDTH OF FACE W	DIAMETER OF HOLE H
1/8	2 1/2	1/4	1
1/4	3	13/32	1
3/8	3 3/4	9/16	1 1/4
1/2	4 1/4	3/4	1 1/4
5/8	4 1/4	15/16	1 1/4

All dimensions are given in inches.
Size of cutter is designated by specifying radius (R) of circular form.

TOLERANCES

ELEMENT	RANGE	DIRECTION	TOLERANCE
Diameter of Cutter	All Sizes	Plus or Minus	1/16
Radius of Circle	1/8	Plus or Minus	0.001
	Over 1/8	Minus Plus	0.001 0.002
Width of Face	All Sizes	Plus or Minus	0.010
Diameter of Hole	Up to 1 Inclusive	Plus	0.00075
	Over 1	Plus	0.001

TABLE 9–49

Standard Keyway Broaches

Metal Cutting Tool Handbook Metal Cutting Tool Institute

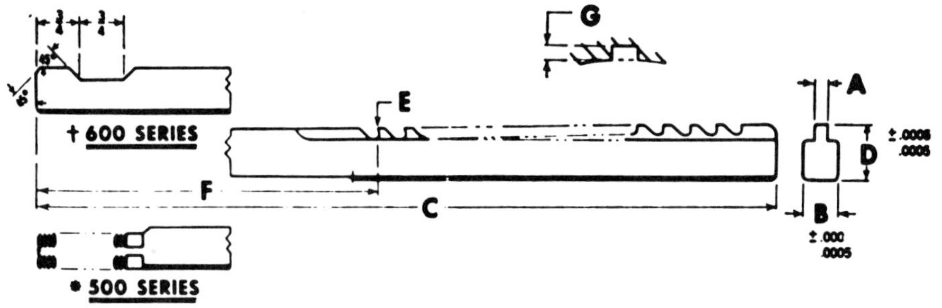

500 Series	600 Series	Nom Dim	A Decimal Dim	Toler-ance	Min Hole Cut	‡Min Length Cut	Max Length Cut	B	C	D	E	F	G	No. of Cuts	Thread Size
501	601	1/16	.0635	±.0002	3/8	3/8	1¼	.1552	20	.313	.271	7 13/16	.042	1	¼-20
502	602	3/32	.0948	±.0002	7/16	½	1½	.1865	24	.367	.309	8¼	.058	1	5/16-18
503	603	3/32	.0948	±.0002	5/8	5/8	2½	.249	33	.491	.433	10	.058	1	3/8-16
504	604	1/8	.126	±.0002	½	½	1½	.249	30	.438	.364	9	.074	1	3/8-16
505	605	1/8	.126	±.0002	7/8	5/8	2½	.3115	36	.594	.520	10	.074	1	½-13
506	606	5/32	.1572	±.0002	19/32	½	1½	.249	30	.525	.436	9	.089	1	3/8-16
507	607	5/32	.1572	±.0002	23/32	5/8	2½	.3115	33	.625	.536	10	.089	1	½-13
508	608	3/16	.1885	±.0002	11/16	5/8	2½	.374	36	.581	.476	10	.105	1	½-13
509	609	3/16	.1885	±.0002	15/16	11/16	3½	.374	36	.796	.691	10 11/16	.105	1	½-13
510	610	7/32	.2198	±.0002	11/16	5/8	2½	.374	33	.557	.437	10	.120	1	½-13
511	611	7/32	.2198	±.0002	15/16	11/16	3½	.374	42	.813	.693	11 1/16	.120	1	½-13
512	612	¼	.251	±.0002	11/16	5/8	2½	.374	36	.612	.476	10	.136	1	½-13
513	613	¼	.251	±.0002	1	11/16	4	.499	45	.877	.741	11 13/16	.136	1	5/8-11
514	614	¼	.251	±.0002	1 7/16	7/8	6	.624	51	1.250	1.114	13½	.136	1	¾-10
515	615	9/32	.2828	±.0002	7/8	11/16	4	.499	42	.716	.564	11 5/8	.152	1	5/8-11
516	616	9/32	.2828	±.0002	1¼	7/8	6	.499	51	1.093	.941	13½	.152	1	5/8-11
517	617	5/16	.314	±.0002	1	11/16	4	.499	45	.908	.741	11 13/16	.167	1	5/8-11
518	618	5/16	.314	±.0002	1 5/16	7/8	6	.499	51	1.158	.991	13½	.167	1	5/8-11
519	619	3/8	.3765	±.0002	1 1/16	11/16	4	.499	45	.938	.739	11 13/16	.199	1	5/8-11
520	620	3/8	.3765	±.0002	1 5/16	7/8	6	.499	54	1.189	.990	13½	.199	1	5/8-11
521	621	7/16	.439	±.0002	1 9/16	11/16	4	.624	48	1.390	1.160	12	.230	1	¾-10
522	622	7/16	.439	±.0002	2	1	8	.624	48	1.611	1.496	15 5/8	.230	2	¾-10
523	623	½	.5015	±.0002	1½	11/16	4	.624	48	1.312	1.051	12	.261	1	¾-10
524	624	½	.5015	±.0002	1½	1	8	.624	48	1.377	1.246	16½	.261	2	¾-10
525	625	9/16	.5645	±.0003	1¾	11/16	4	.6865	54	1.438	1.146	11 13/16	.292	1	1 - 8
526	626	9/16	.5645	±.0003	1 5/8	1	8	.6865	51	1.391	1.245	16	.292	2	1 - 8
527	627	9/16	.5645	±.0003	2¼	1 1/8	12	.874	60	1.641	1.495	20	.292	2	1 - 8
528	628	5/8	.627	±.0003	1 7/8	11/16	4	.749	60	1.625	1.301	12 3/16	.324	1	1 - 8
529	629	5/8	.627	±.0003	2½	1	8	.874	54	1.657	1.495	16 3/8	.324	2	1 - 8
530	630	5/8	.627	±.0003	2¼	1 1/8	12	.874	57	1.657	1.495	20	.324	2	1 - 8
531	631	¾	.752	±.0003	1 7/8	11/16	4	.874	60	1.625	1.239	12 3/16	.386	1	1 - 8
532	632	¾	.752	±.0003	2	1	8	.999	60	1.688	1.495	16¼	.386	2	1¼- 7
533	633	¾	.752	±.0003	2¼	1 1/8	12	.999	57	1.688	1.560	20	.386	3	1¼- 7
534	634	7/8	.877	±.0003	2¼	11/16	4	1.124	63	1.875	1.426	12 3/8	.449	1	1¼- 7
535	635	7/8	.877	±.0003	2¼	1	8	1.124	63	1.719	1.494	15¾	.449	2	1¼- 7
536	636	7/8	.877	±.0003	2¼	1 1/8	12	1.124	63.	1.719	1.569	20	.449	3	1¼- 7
537	637	1	1.002	±.0003	2¼	5/8	2½	1.249	63	1.750	1.239	10½	.511	1	1½- 6
538	638	1	1.002	±.0003	2¼	7/8	6	1.249	63	1.750	1.494	14½	.511	2	1½- 6
539	639	1	1.002	±.0003	2¼	1 1/8	12	1.249	60	1.750	1.580	20	.511	3	1½- 6

*Designates Thread Type Shank.

†Designates Notched Type Shank.

‡Minimum length of part recommended to prevent part from dropping in between teeth of broach.

TABLE 9–50

Square Shaft Ends — Broaches*

Catalog and Manual CB–46 Colonial Broach Company

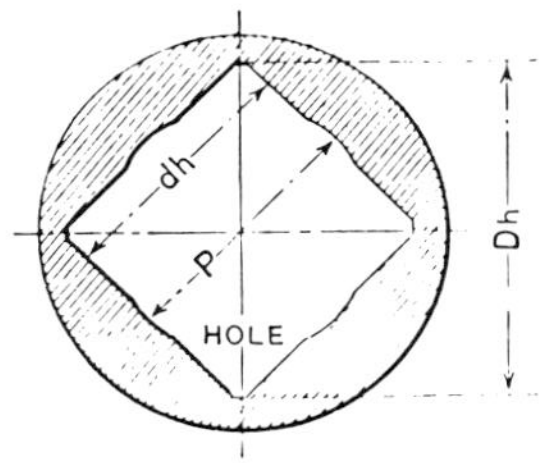

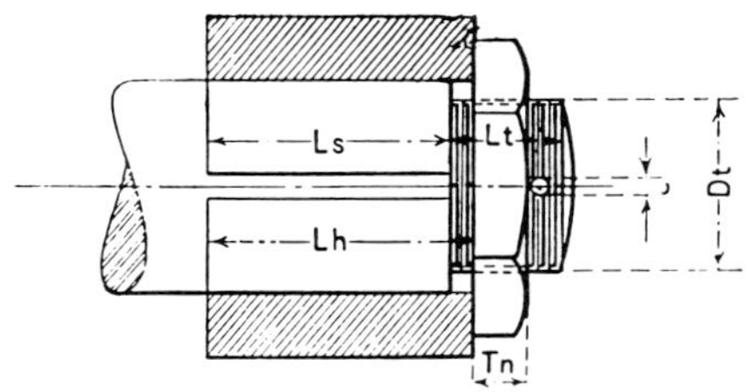

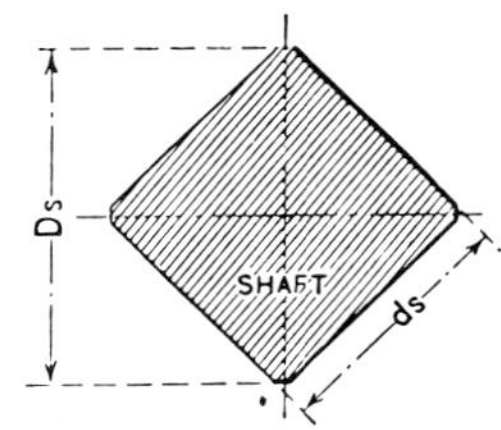

ALL FITS		PERMANENT FITS							SLIP FITS						
NOM DIAM	Length of Part	P and Drill Size	ds	dh	Ds	Dh	Approx Broach Length	Approx Pull Required	P and Drill Size	ds	dh	Ds	Dh	Approx Broach Length	Approx Pull Required
¼	⅜	.193	.1895 .1885	.1875 .1865	.250 .245	.260 .252	15	450	.257	.248 .247	.250 .249	.3537 .3337	.3537 .3457	18	500
⅜	½	.290	.2832 .2822	.2812 .2802	.375 .370	.385 .377	19	900	.386	.373 .372	.375 .374	.5156 .5106	.5256 .5176	24	1,000
½	¾	.386	.3770 .3760	.3750 .3740	.500 .495	.516 .502	25	1,500	33/64	.498 .497	.500 .499	.6875 .6825	.6975 .6895	31	1,500
⅝	¾	33/64	.5020 .5010	.5000 .4990	.625 .620	.635 .627	25	2,000	41/64	.623 .622	.625 .624	.8437 .8387	.8537 .8457	32	2,500
¾	1	37/64	.5645 .5635	.5625 .5615	.750 .745	.760 .752	26	3,500	49/64	.748 .747	.750 .749	1.031 1.026	1.051 1.036	36	5,000
⅞	1 ¼	45/64	.6895 .6885	.6875 .6865	.875 .870	.885 .877	31	4,500	29/32	.873 .872	.875 .874	1.187 1.182	1.207 1.192	44	5,500
1	1 ½	27/32	.8155 .8145	.8125 .8115	1.000 .995	1.020 1.005	35	5,500	1 1/32	.998 .997	1.000 .999	1.375 1.370	1.395 1.380	45	6,500
1 ⅛	1 ½	29/32	.8780 .8770	.8750 .8740	1.125 1.120	1.145 1.130	41	7,000	1 5/32	1.123 1.122	1.125 1.124	1.562 1.557	1.582 1.567	39-(2)	8,500
1 ¼	1 ½	1 1/32	1.003 1.002	1.000 .999	1.250 1.245	1.270 1.255	41	8,000	1 9/32	1.248 1.247	1.250 1.249	1.687 1.682	1.707 1.692	39-(2)	9,500
1 ⅜	2	1 5/32	1.128 1.127	1.125 1.124	1.375 1.370	1.395 1.380	47	11,000	1 27/64	1.373 1.372	1.375 1.374	1.875 1.870	1.895 1.880	42-(2)	17,000

*Although holes and splines and slots of practically any size or shape can be cut by broaching, keyway broaches, Table 9–49, are the nearest to being recognized as standards.

continued on next page

TABLE 9–50, continued

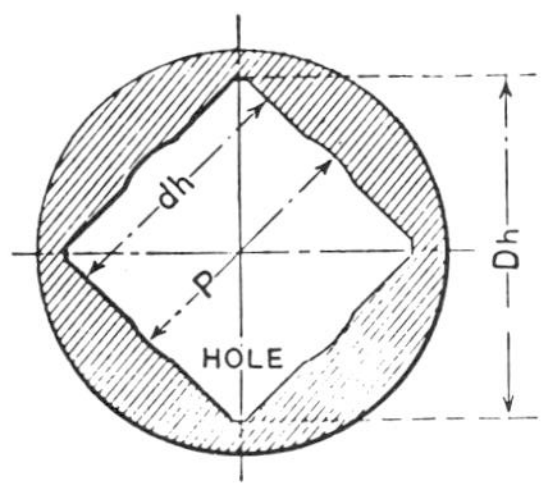

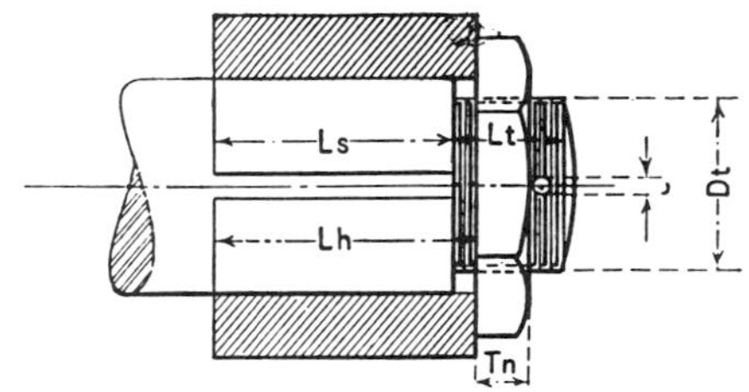

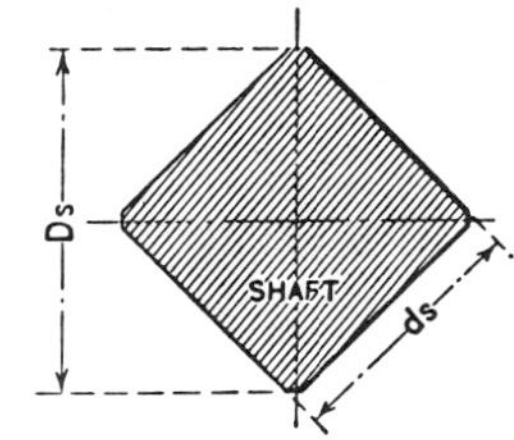

ALL FITS		PERMANENT FITS							SLIP FITS						
NOM DIAM	Length of Part	P and Drill Size	ds	dh	Ds	Dh	Approx Broach Length	Approx Pull Required	P and Drill Size	ds	dh	Ds	Dh	Approx Broach Length	Approx Pull Required
1½	2	1 5/32	1.128 1.127	1.125 1.124	1.500 1.495	1.520 1.505	55	13,000	1 35/64	1.498 1.497	1.500 1.499	2.062 2.057	2.082 2.067	45-(2)	18,000
1¾	2¼	1 27/64	1.378 1.377	1.375 1.374	1.750 1.745	1.770 1.755	60	16,000	1 13/16	1.748 1.747	1.750 1.749	2.375 2.370	2.395 2.380	52-(2)	20,000
2	3	1 35/64	1.504 1.503	1.5000 1.4985	2.000 1.995	2.020 2.005	43-(2)	17,000	2 1/16	1.9975 1.9965	2.0000 1.9985	2.750 2.745	2.770 2.755	54-(3)	20,000
2¼	3	1 13/16	1.754 1.753	1.7500 1.7485	2.250 2.245	2.270 2.255	43-(2)	18,000	2 5/16	2.2475 2.2465	2.2500 2.2485	3.062 3.057	3.082 3.067	55-(3)	27,000
2½	3½	2 1/16	2.004 2.003	2.0000 1.9985	2.500 2.495	2.520 2.505	52-(2)	27,000	2 37/64	2.4975 2.4965	2.5000 2.4985	3.437 3.432	3.457 3.442	59-(3)	35,000
2¾	3½	2 5/16	2.254 2.253	2.2500 2.2485	2.750 2.745	2.770 2.755	52-(2)	30,000	2 55/64	2.7475 2.7465	2.7500 2.7485	3.750 3.745	3.770 3.755	61-(3)	35,000
3	4	2 37/64	2.504 2.503	2.500 2.498	3.000 2.995	3.020 3.005	57-(2)	31,000	3 3/32	2.997 2.996	3.000 2.998	4.125 4.120	4.145 4.130	67-(3)	35,000
3½	4½	2 55/64	2.754 2.753	2.750 2.748	3.500 3.495	3.520 3.505	65-(3)	30,000	3 39/64	3.497 3.496	3.500 3.498	4.750 4.745	4.770 4.755		
4	5½	3 23/64	3.254 3.253	3.250 3.248	4.000 3.995	4.020 4.005	60-(3)	30,000	4 1/8	3.997 3.996	4.000 3.998	5.500 5.495	5.520 5.505		

TABLE 9–51
Serrated Shafts Ends — Broaches*

Catalog and Manual CB-46 Colonial Broach Company

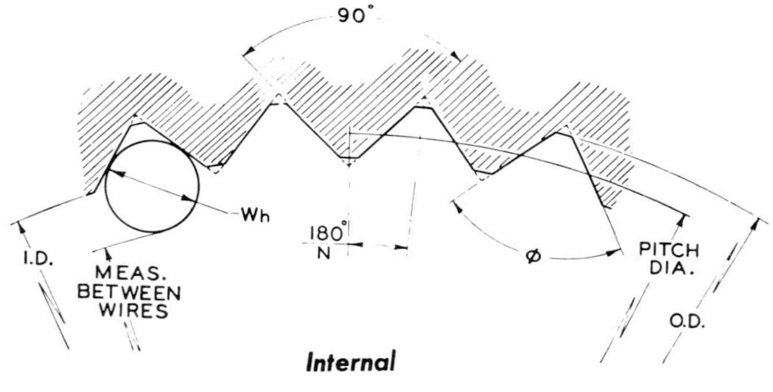

Internal

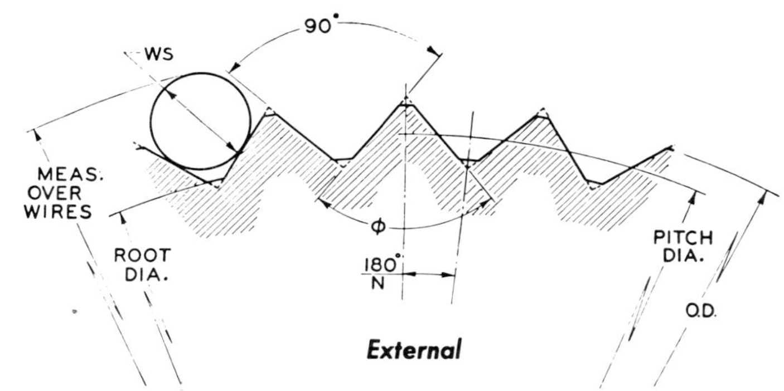

External

HOLE SIZES FOR ALL FITS											MAXIMUM CUT LENGTH (In.)	BROACH LENGTH (In.)	BASIC DIMENSIONS			SHAFT SIZES														
																FOR ALL FITS					LOOSE FITS			LIGHT DRIVE FITS			FORCE FITS			
O D		PITCH DIA		I D		THEO DIAM OF POINTS		MEAS BTW'N WIRES		Wh			NOM DIA	N	ϕ	O D		ROOT		Ws	PITCH DIA		Meas. Over Wires	PITCH DIA		Meas. Over Wires	PITCH DIA		Meas Over Wires
Max	Min	Max	Min	Max	Min	O D	R D	Max	Min							Max	Min	Max	Min		Max	Min	Max	Max	Min	Max	Max	Min	Max
.....	.125	.122	.120	.118	.117	1283	.1166	.1028	.1006	.010	¼	10	⅛	36	80°	.124	.123	.116	.....	.010	.1195	.1185	.1383	.1207	.1202	.1395	.1223	.1218	.1410
.....	.1875	.182	.180	.176	.175	.1913	.1739	.1657	.1636	.010	⅜	12	3/16	36	80°	.186	.185	.174	.....	.010	.1795	.1785	.1956	.1807	.1802	.1968	.1823	.1818	.1983
.....	.250	.243	.241	.235	.234	.2554	.2322	.2171	.2150	.015	½	15	¼	36	80°	.249	.248	.233	.....	.015	.2405	.2395	.2660	.2417	.2412	.2671	.2433	.2428	.2687
.....	.3125	.303	.301	.293	.292	.3185	.2895	.2802	.2780	.015	⅝	16	5/16	36	80°	.311	.310	.291	.....	.020	.3000	.2990	.3349	.3017	.3012	.3366	.3033	.3028	.3381
.....	.375	.363	.361	.352	.351	3815	.3468	.3304	.3283	.020	¾	17	⅜	36	80°	.374	.373	.350	.....	.020	.3605	.3595	.3927	.3617	.3612	.3939	.3633	.3628	.3954
.....	.500	.485	.483	.469	.468	.5098	.4634	.4459	.4438	.025	1	18	½	36	80°	.499	.498	467	.....	.0312	.4823	.4808	.5362	.4836	.4831	.5374	.4856	.4851	.5394
.....	.625	.605	.603	.584	.583	.6359	.5780	.5465	.5443	.035	1¼	22	⅝	36	80°	.624	.623	.582	.....	.035	.6023	.6008	.6600	.6036	.6031	.6612	.6056	.6051	.6631
.760	.755	.733	.731	.716	.714	.7600	.7086	.6815	.6793	.0312	1½	22	¾	48	82½°	.749	.747	.713	711	.035	.7303	.7288	.7905	.7316	.7311	.7918	.7336	.7331	.7937
.885	.880	.855	.853	.835	.833	.8865	.8266	.7984	.7963	.035	1¾	23	⅞	48	82½°	.874	.872	.832	.830	.040	.8523	.8508	.9206	.8536	.8531	.9218	.8556	.8551	.9237
1.010	1.005	.977	.975	.954	.952	1.0130	.9445	.8998	.8977	.045	2	25	1	48	82½°	.999	.997	.951	.949	.045	.974	.9720	1.0503	.9757	.9747	1.0519	.9783	.9773	1.0544
1.135	1.130	1.098	1.096	1.071	1.069	1.1384	1.0615	1.0126	1.0105	.050	2¼	30	1⅛	48	82½°	1.124	1.122	1.068	1.066	.050	1.095	1.0930	1.1793	1.0967	1.0957	1.1809	1:0993	1.0983	1.1834
1.260	1.255	1.220	1.218	1.190	1.188	1.2649	1.1794	1.1265	1.1244	.055	2½	32	1¼	48	82½°	1.249	1.247	1.187	1.185	.055	1.217	1.2150	1.3094	1.2187	1.2177	1.3111	1.2213	1.2203	1.3136
1.385	1.380	1.342	1.340	1.309	1.307	1.3914	1.2974	1.2404	1.2384	.060	2¾	37	1⅜	48	82½°	1.374	1.372	1.306	1.304	.0625	1.339	1.3370	1.4454	1.3407	1.3397	1.4470	1.3433	1.3423	1.4495
1.510	1.505	1.464	1.462	1.428	1.426	1.5178	1.4153	1.3605	1.3585	.0625	3	41	1½	48	82½°	1.499	1.497	1.425	1.423	.070	1.461	1.4580	1.5814	1.4629	1.4614	1.5833	1.4658	1.4643	1.5861
1.760	1.755	1.708	1.706	1.666	1.664	1.7708	1.6521	1.5821	1.5800	.075	3½	49	1¾	48	82½°	1.749	1.747	1.663	1.661	.080	1.705	1.7020	1.8414	1.7069	1.7054	1.8433	1.7098	1.7083	1.8461
2.010	2.005	1.952	1.949	1.904	1.902	2.0237	1.8871	1.8098	1.8068	.085	4	52	2	48	82½°	1.999	1.997	1.901	1.899	090	1.948	1.9450	2.1005	1.9504	1.9489	2.1028	1.9533	1.9518	2.1056
2.260	2.255	2.196	2.193	2.142	2.140	2.2768	2.1230	2.0377	2.0346	.095	4½	61	2¼	48	82½°	2.249	2.247	2.139	2.137	.100	2.192	2.1890	2.3605	2.1942	2.1927	2.3627	2.1977	2.1962	2.3660
2.510	2.505	2.440	2.437	2.380	2.378	2.5298	2.3589	2.253	2.2498	.110	5	70	2½	48	82½°	2.499	2.497	2.377	2.375	.110	2.436	2.4330	2.6206	2.4385	2.4365	2.6230	2.4420	2.4400	2.6264
2.760	2.755	2.684	2.681	2.618	2.616	2.7828	2.5947	2.4808	2.4776	.120	5½	71	2¾	48	82½°	2.749	2.747	2.615	2.613	.125	2.680	2.6770	2.8927	2.6825	2.6805	2.8951	2.6860	2.6840	2.8985
3.010	3.005	2.928	2.925	2.856	2.854	3.0357	2.8306	2.7211	2.7180	.125	6	74	3	48	82½°	2.999	2.997	2.853	2.851	.140	2.924	2.9210	3.1648	2.9265	2.9245	3.1672	2.9300	2.9280	3.1706

*Although holes and splines and slots of practically any size or shape can be cut by broaching, keyway broaches, Table 9–49, are the nearest to being recognized as standards.

TABLE 9–52
Round Broaches, Stock Allowance for Broaching, Broaching Loads

Catalog and Manual CB-46 Colonial Broach Company

The data in this table are more informative than defining in character: They indicate the manner in which stock allowance ought to increase with size of work piece, the huge forces needed to actuate broaches, how these forces increase with increases in the amounts of metal to be removed, and so on.

Nom Diam (in.)	Length of Part (in.)	Drill Size (in.)	Length of Broach (in.)	Pull Required in Pounds
3/8	1	11/32	25	1,000
	2	11/32	30	2,000
	3	11/32	33	2,000
	4	11/32	39	2,500
1/2	1	15/32	25	1,500
	2	15/32	30	2,500
	3	15/32	33	3,000
	4	15/32	39	3,000
5/8	1	19/32	25	2,000
	2	19/32	30	3,000
	3	19/32	33	3,500
	4	19/32	39	4,000
	5	19/32	45	4,500
3/4	1	23/32	25	2,000
	2	23/32	30	3,500
	3	23/32	33	4,500
	4	23/32	39	4,500
	5	23/32	45	5,500
7/8	1	27/32	25	2,500
	2	27/32	30	4,000
	3	27/32	33	5,000
	4	27/32	39	5,500
	5	27/32	45	6,500
1	1	61/64	28	3,000
	2	61/64	33	6,000
	3	61/64	38	7,000
	4	61/64	45	7,500
	5	61/64	51	8,500
	6	61/64	54	9,000
1 1/8	1	1 5/64	28	3,000
	2	1 5/64	33	7,000
	3	1 5/64	38	8,000
	4	1 5/64	45	8,000
	5	1 5/64	51	9,500
	6	1 5/64	54	10,000
1 1/4	1	1 13/64	28	3,500
	2	1 13/64	33	7,500
	3	1 13/64	38	9,000
	4	1 13/64	45	9,000
	5	1 13/64	51	10,000
	6	1 13/64	54	12,000
1 3/8	1	1 21/64	28	4,000
	2	1 21/64	33	8,500
	3	1 21/64	38	10,000
	4	1 21/64	45	10,000
	5	1 21/64	51	12,000
	6	1 21/64	54	13,000
1 1/2	1	1 29/64	28	4,000
	2	1 29/64	33	9,500
	3	1 29/64	38	10,000
	4	1 29/64	45	11,000
	5	1 29/64	51	13,000
	6	1 29/64	54	14,000
1 5/8	1	1 9/16	31	5,500
	2	1 9/16	37	11,000
	3	1 9/16	41	13,000
	4	1 9/16	48	14,000
	5	1 9/16	55	16,000
	6	1 9/16	58	17,000
1 3/4	1	1 11/16	31	6,000
	2	1 11/16	37	11,000
	3	1 11/16	41	14,000
	4	1 11/16	48	15,000
	5	1 11/16	55	17,000
	6	1 11/16	58	18,000
1 7/8	1	1 13/16	31	6,500
	2	1 13/16	37	12,000
	3	1 13/16	41	15,000
	4	1 13/16	48	16,000
	5	1 13/16	55	19,000
	6	1 13/16	58	20,000
2	1	1 15/16	31	7,000
	2	1 15/16	37	13,000
	3	1 15/16	41	16,000
	4	1 15/16	48	17,000
	5	1 15/16	55	20,000
	6	1 15/16	58	21,000
2 1/8	1	2 1/16	31	7,000
	2	2 1/16	37	14,000
	3	2 1/16	41	17,000
	4	2 1/16	48	18,000
	5	2 1/16	55	21,000
	6	2 1/16	58	22,000
2 1/4	1	2 3/16	31	7,500
	2	2 3/16	37	15,000
	3	2 3/16	41	18,000
	4	2 3/16	48	19,000
	5	2 3/16	55	22,000
	6	2 3/16	58	24,000
2 3/8	1	2 5/16	31	8,000
	2	2 5/16	37	16,000
	3	2 5/16	41	19,000
	4	2 5/16	48	20,000
	5	2 5/16	55	23,000
	6	2 5/16	58	25,000
2 1/2	1	2 7/16	31	8,500
	2	2 7/16	37	16,000
	3	2 7/16	41	20,000
	4	2 7/16	48	21,000
	5	2 7/16	55	25,000
	6	2 7/16	58	26,000
2 3/4	1	2 11/16	31	9,500
	2	2 11/16	37	18,000
	3	2 11/16	41	22,000
	4	2 11/16	48	23,000
	5	2 11/16	55	27,000
	6	2 11/16	58	29,000
3	1	2 15/16	31	10,000
	2	2 15/16	37	20,000
	3	2 15/16	41	24,000
	4	2 15/16	48	25,000
	5	2 15/16	55	30,000
	6	2 15/16	58	30,000

TABLE 9-53

Stock Allowance for Broaching, Milling, Reaming

Operation	Amount and Remarks	References
Broaching	1/32 inch minimum. The maximum depends much on machine capacity and broach design — possibly 1/2 inch in single-pass surface broaching and 5/8 or 3/4 inch for slots and internal splines. Broaches are designed for the job so that each tooth removes a prescribed thickness of stock. Too heavy a cut overloads the tooth; too light a cut may mean a burnishing or glazing action with no stock removal at all. A cut of 0.0003 inch is perhaps a minimum and 0.006 or 0.008 inch a reasonable maximum. Naturally, finishing teeth are designed to remove less stock in proportion than roughing or semi-finishing teeth.	Table 9-52 *Production Processes* by Bolz, The Penton Publishing Co.
Milling	Roughing: 1/8 inch or more Finishing: a few thousandths of an inch to as much as 1/16 inch. As for broaching, satisfactory milling operations require substantial feeds per tooth. For example, Brown and Sharpe suggest the following: Type of Cut — Starting feed per tooth, inch Face Milling — 0.008 Straddle Milling — 0.008 Channel or Slot Milling — 0.008 Slab Milling — 0.007 End Milling (1/2 inch diameter and larger) or Profiling — 0.004 Sawing — 0.003 Thread Milling — 0.002	*A Treatise on Milling and Milling Machines*, Third Edition—1951, The Cincinnati Milling Machine Co. *Small Tools Catalog* No. 35, 1951, Brown and Sharpe Manufacturing Co.
Reaming	Machine: 0.010 in a 1/4 inch hole 0.015 in a 1/2 inch hole 0.025 in a 1-1/2 inch hole Hand: 0.001 to 0.003 inch	*Metal Cutting Tool Handbook*, Second Edition – 1950, Metal Cutting Tool Institute

TABLE 9–54

Tolerances for Broaching, Drilling, Milling, Reaming

Operation	Tolerances and Remarks	Sources or References
Broaching	Holes: ±0.0005 to 0.001 Splines: ±0.001 to 0.002 Surfaces: ±0.0001 to 0.001 Narrow Slots: ±0.0005 Straddle-broached Lugs: ±0.0003	*Production Processes by* Bolz, The Penton Publishing Co.
Drilling	No. 80 to No. 71 +0.002 (−0.001) or (−0.000) No. 70 to No. 52 +0.003 All sizes No. 51 to No. 31 +0.004 1/8 inch to No. 3 +0.005 7/32 inch to size R +0.006 11/32 to 1/2 inch +0.007 33/64 to 23/32 inch +0.008 47/64 to 63/64 inch +0.009 1 inch to 2 inch +0.010 Standard drills are likely to be slightly undersize because tolerances vary from zero to −0.0006 on diameters less than, and including, 3/64 inch, to as much as zero to −0.0025 on diameters over 1-1/2 and including 3-1/2 in.	*Production Processes* by Bolz, The Penton Publishing Co.
Milling	Keyslots: +0.0005 to −0.0015 Two-Flute, Fast helix end mills: +0.002 to −0.001 Straddle Milling: +0.003 Narrow Slots: +0.005 to −0.0005 Small Flat Surfaces: 0.002 to 0.005	Compare Tolerances in Table 8–19 Table 9–38 Table 9–39
Reaming	Under 1/2 inch: +0.0008 to −0.0015 1/2 to 1 inch: +0.001 to −0.002 over one inch: +0.0015 to −0.003 Diameter tolerances on new reamers are positive, Table 9–20. After wear a reamer may be as much as 0.003 undersize. An advantage of the expansion reamer is that it can be expanded to exact size, thereby making closer tolerances economical in tool use.	Compare *Production Processes by* Bolz, The Penton Publishing Co.
	Machine Drawing by Lent recommends the following tolerances for reamed holes: under 3/4 inch: +0.0005, − zero 3/4 to 1-1/4: +0.001, − zero 1-1/4 to 3: +0.0015, − zero	*Machine Drawing* by Deane Lent, Prentice-Hall, Inc.

TABLE 9–55

Chamfer on Cotter-Pin Hole in Bolts and Capscrews

Practice of Caterpillar Tractor Co. – 1954

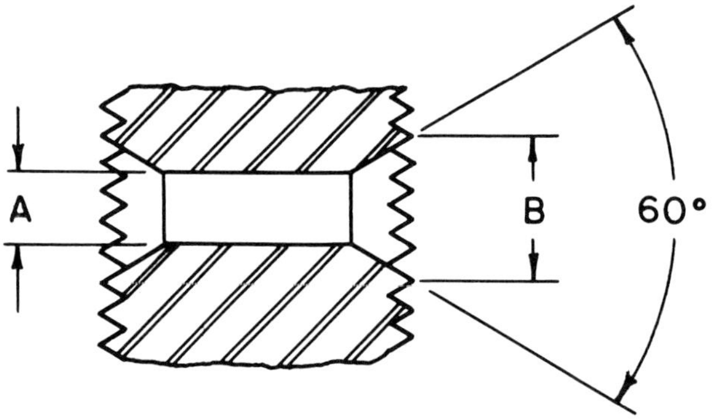

Nominal Size	Threads per Inch	Drill Size A	Chamfer Diameter B
5/16	24	5/64	1/8
5/16	18	5/64	9/64
3/8	24	7/64	5/32
3/8	16	7/64	11/64
7/16	20	7/64	5/32
7/16	14	7/64	11/64
1/2	20	7/64	5/32
1/2	13	7/64	3/16
9/16	18	9/64	13/64
9/16	12	9/64	7/32
5/8	18	9/64	13/64
5/8	11	9/64	7/32
11/16	16	9/64	13/64
3/4	16	9/64	13/64
3/4	10	9/64	15/64

TABLE 9-56

Methods of Showing Drilled and Reamed Holes on Drawings

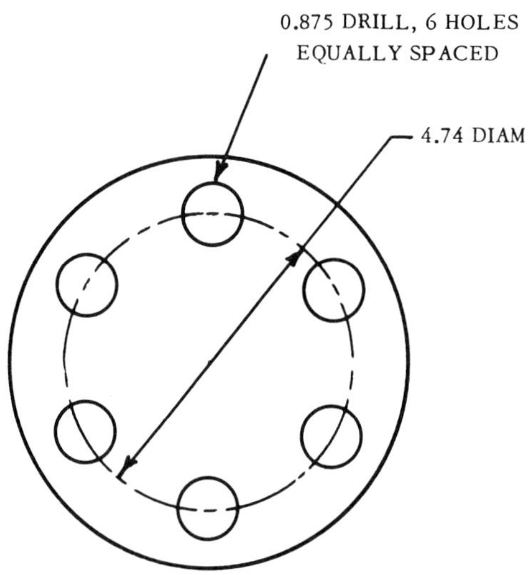

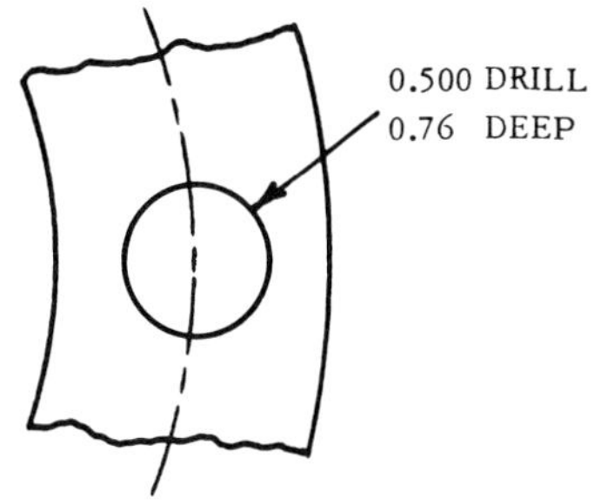

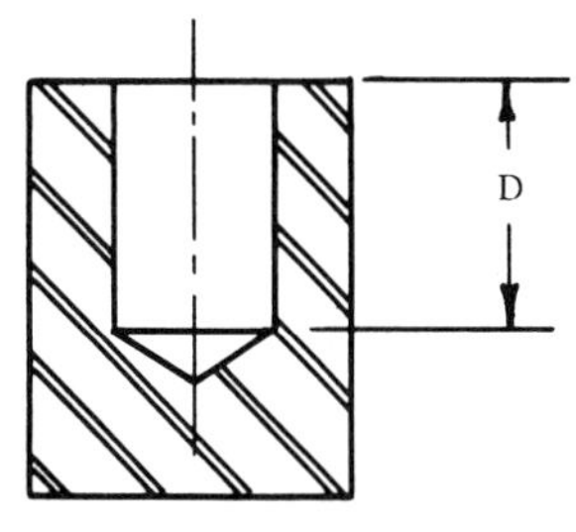

NOTE: DEPTH D OF BLIND HOLE
IS SOMETIMES SHOWN
ON ANOTHER VIEW

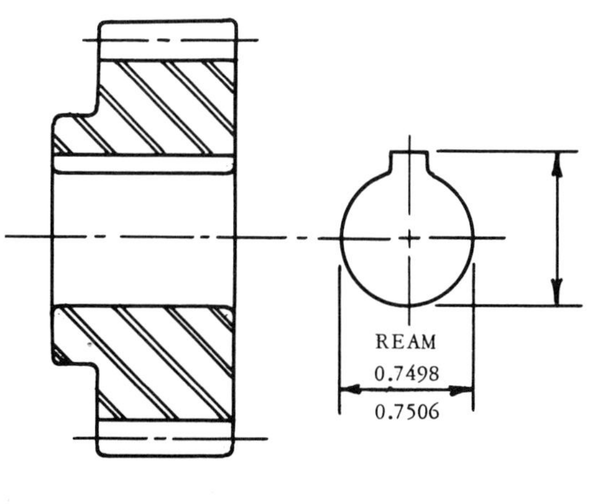

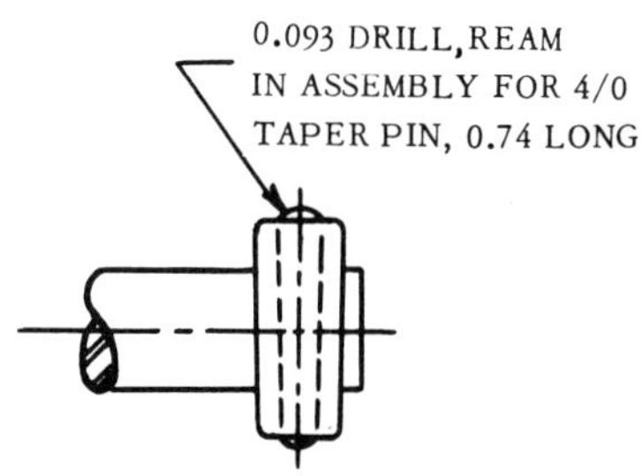

TABLE 9–57

Methods of Showing Drilled and Counterbored Holes on Drawings

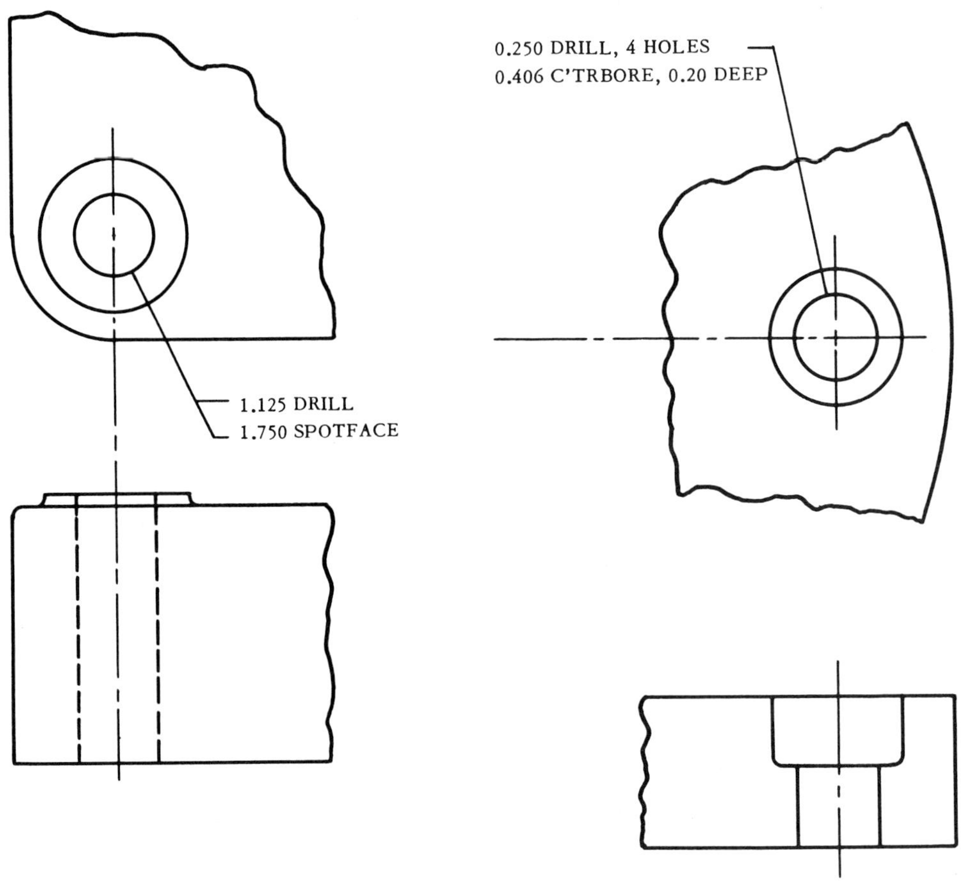

TABLE 9–58

Method of Specifying Screw Threads on Drawings

ASA B1.1–1949 Third Edition Unified and American Screw Threads

"Screw threads are designated by the initial letters of the thread series, preceded by the nominal size (diameter in inches, or the screw number) and number of threads per inch, all in Arabic characters, and followed by the class designation, with or without the pitch diameter tolerances or limits of size."
Examples

$\frac{1}{2}$ – 13 UNC – 2B
- class, internal
- thread series
- number threads per inch
- nominal size

$\frac{1}{4}$ – 20 UNC – 2A – LH
- left hand
- class, external
- thread series
- number threads per inch
- nominal size

2 – 8N – 2B (1.9188 to 1.9289 pitch diam.)

$\frac{3}{4}$ – 16 UNF – 2A (1.50 length of engagement)

"If a standard thread is modified by the inclusion of some non-standard feature, such as a smaller major diameter, etc., the word 'modified' should be added with an asterisk and the non-standard feature or dimension of the thread should be enclosed in brackets and likewise marked with an asterisk. If a standard thread has a long length of engagement for which standard allowances and tolerances are not applicable, such special length should be noted on the drawing or included in the designation, as shown above."

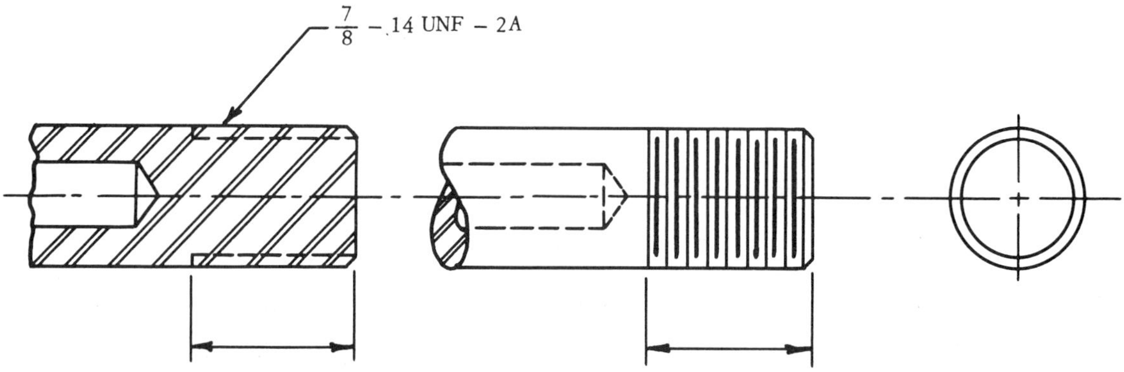

No one of the conventional ways of representing screw threads on drawings possesses all the advantages. Nor are all the ways illustrated here.

continued on next page

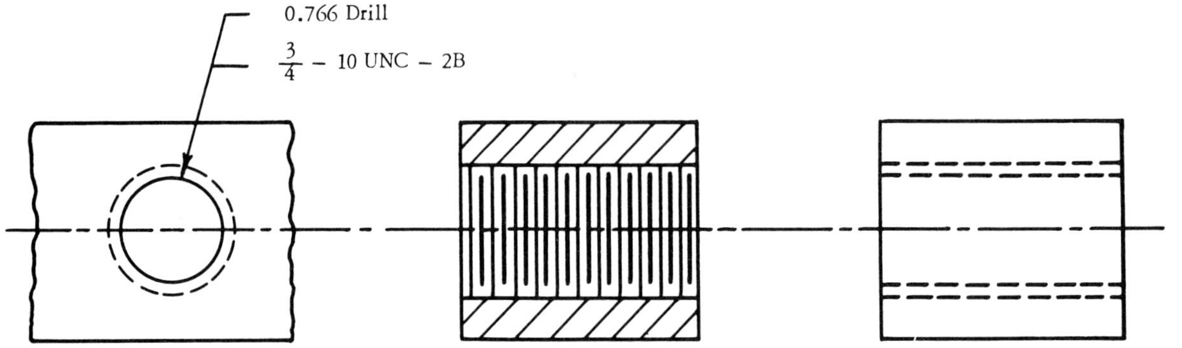

Through holes are preferable for tapping. In the case of blind holes, the next best practice is to allow plenty of extra drill depth over tap depth. This practice provides space for chips, and if great enough obviates the use of a bottoming tap.

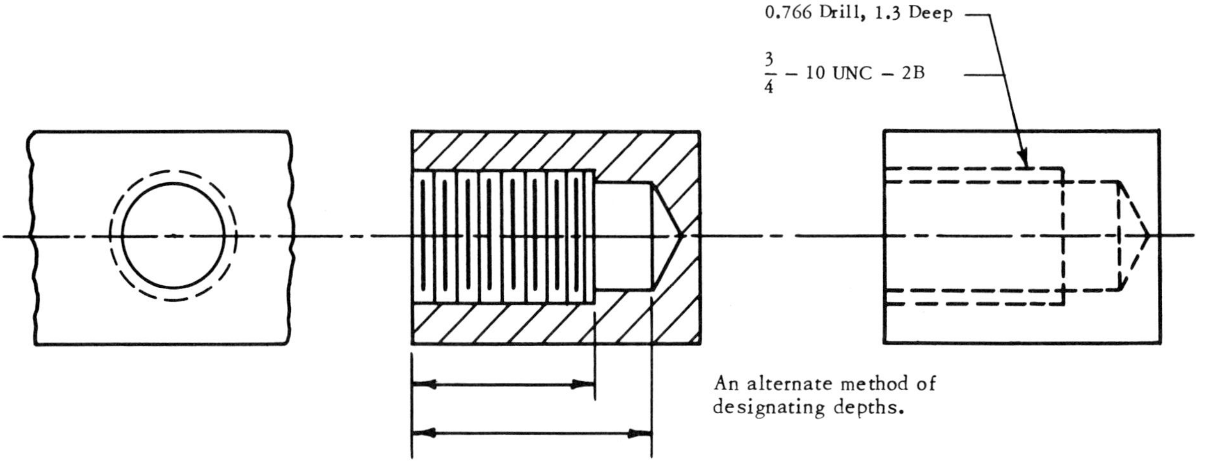

An alternate method of designating depths.

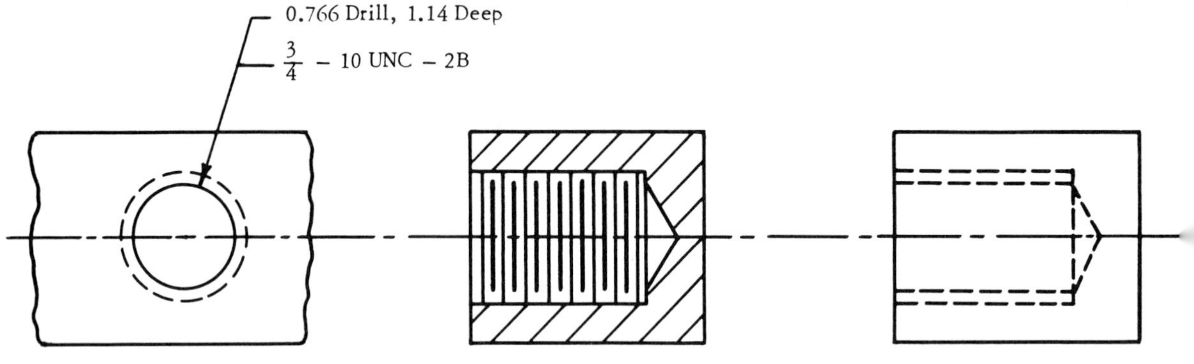

Bottoming taps have a thread and a half of chamfer; hence the thread at the very bottom is not fully cut.

INDEX TO
SECTION 10

Compiled and Edited by

EDWARD FITZGERALD

Mechanical Engineer, Battelle Memorial Institute

Serrations and Splines

TABLE 10-1

Range of Involute Splines, SAE Standard

1954 SAE Handbook

Range of Diametral Pitches*	1/2 to 48/96
Range of Number of Teeth	6 to 50
Range of Nominal Diameters	.14 inch diameter (6 teeth 48/96 D.P.) to 51 inches diameter (50 teeth 1/2 D.P.)

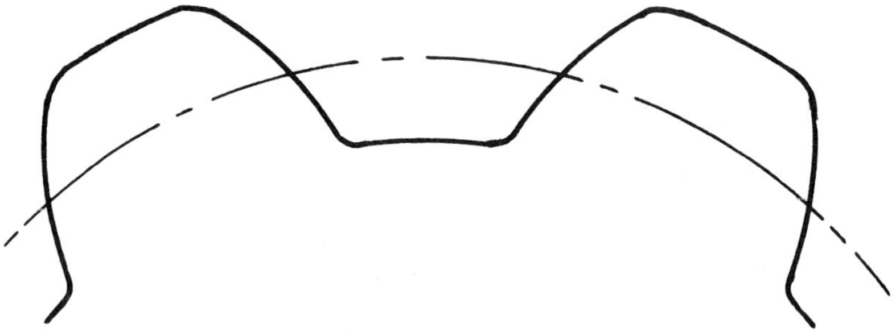

1/2 Pitch (actual size)

5/10 Pitch (actual size)

10/20 Pitch (actual size)

20/40 Pitch (actual size)

*Diametral pitch is designated by two numbers written in the form of a fraction. The numerator is the number of teeth divided by the pitch diameter in inches and the denominator is the reciprocal of the addendum. (The addendum is half the difference between the major diameter of the external spline and the pitch diameter.)

TABLE 10–2

Data To Be Placed on Drawing of External Involute* Spline

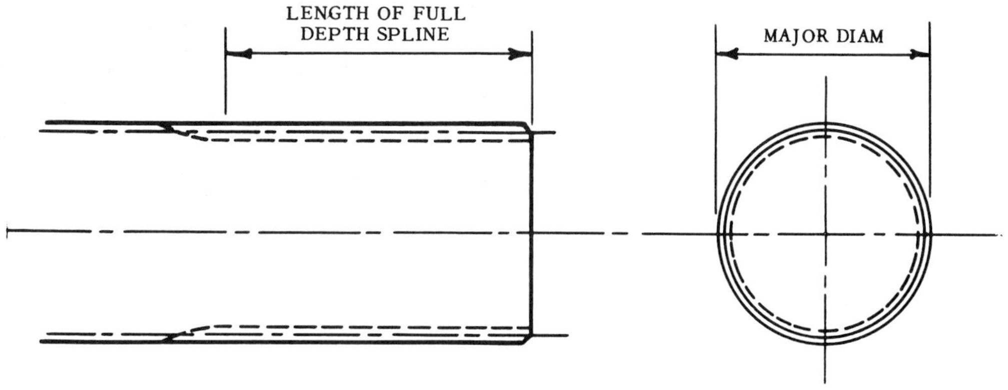

Involute Spline Data	Example	Table No.	Column
Number of Teeth	20		
Type of Spline (fillet root or flat root)	fillet root		
Diametral Pitch	1/2		
Type and Class of Fit	side of tooth, Class A	10-8	
Pitch Diameter	20,0000	10-5	2
Base Circle Diameter	17.3205	10-5	3
Major Diameter	21.0000 20.9900	10-5	24
Minor Diameter	18.0950 18.2000	10-5	25
Tooth Thickness	1.5703 1.5649	10-5	31
Measuring Pin Diameter	1.9200	10-5	Sub-heading
Measurement Across Pins	22.9559 (min.)†	10-5	29
‡TIF Diameter	19.0000	10-5	27
Chamfer Dimension	.237	10-11	2
Chamfer Height	.104	10-11	3
Minor Diameter Fillet (for fillet root splines only)	.338 (min.)	10-5	26
Hob Number			
Drawing Number of Mating Part			

*Boundary dimensions plus tabular data, as indicated here, have had limited usage in the specification of involute splines and involute serrations on drawings. Splines have become so commonplace in today's machinery that there is a real need for a simplified method of designating them on drawings, analogous perhaps to the methods of indicating screw threads or spur gears. Noteworthy references like American Standard ASA B5.15–1950, the SAE Handbook, Metal Cutting Tool Handbook and others depict and recommend spline drawings of minute detail. It is indispensable, of course, that the specification of a spline be unmistakably clear in every detail, but both an elaborate diagram and tabular data seem an unnecessary duplication of statement. Moreover, when spline bores are broached and the designer calls for the spline by broach number, even less information seems adequate; namely, the boundary dimensions with tolerances, the broach number, and such data as may be necessary for inspection purposes after broaching.

†For maximum measurement across pins, see Table 10–13.

‡ The TIF diameter or "true involute form" diameter is the diameter beyond which the tooth profiles are true involutes.

TABLE 10–3

Data To Be Placed on Drawing of Internal Involute* Spline

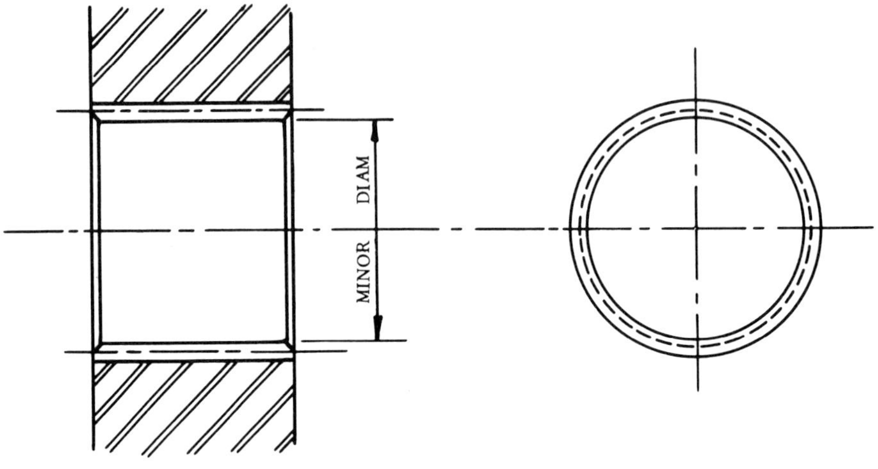

Involute Spline Data	Example	Table No.	Column
Number of Teeth	20		
Type of Spline (flat root or fillet root)	fillet root		
Diametral Pitch	1/2		
Type and Class of Fit	side of tooth, Class A	10-8	
Pitch Diameter	20.0000	10-5	2
Base Circle Diameter	17.3205	10-5	3
Major Diameter	21.9050 21.8000	10-5	12
Minor Diameter	19.0000 19.0050	10-5	11
Space Width	1.5708 1.5761	10-5	15
Measuring Pin Diameter	1.4400	10-5	Sub-heading
Measurement Between Pins	18.4081 (max.)†	10-5	14
‡ TIF Diameter	21.0000	10-5	10
Chamfer Angle	53°24'	10-12	2
Chamfer Height	.102 (min.)	10-12	3
Major Diameter Fillet (fillet root splines only)	.269	10-5	13
Broach Number			
Drawing Number of Mating Part			

*Boundary dimensions plus tabular data, as indicated here, have had limited usage in the specification of involute splines and involute serrations on drawings. Splines have become so commonplace in today's machinery that there is a real need for a simplified method of designating them on drawings, analogous perhaps to the methods of indicating screw threads or spur gears. Noteworthy references like American Standard ASA B5.15–1950, the SAE Handbook, Metal Cutting Tool Handbook and others depict and recommend spline drawings of minute detail. It is indispensable, of course, that the specification of a spline be unmistakeably clear in every detail, but both an elaborate diagram and tabular data seem an unnecessary duplication of statement. Moreover, when spline bores are broached and the designer calls for the spline by broach number, even less information seems adequate; namely, the boundary dimensions with tolerances, the broach number, and such data as may be necessary for inspection purposes after broaching.

†For minimum dimension between pins, see Table 10–14.

‡ TIF diameter or "true involute form" diameter is the diameter within which the tooth profiles are true involutes.

TABLE 10–4
Basic Data for 1/2 Diametral Pitch [a]
1954 SAE Handbook

1/2 Diametral Pitch Pressure Angle, 30°	Addendum (Basic), 0.5000 Dedendum (See Sec 6)	Circular Pitch, 3.1416 Meas-Pin Diam, 1.4400

INTERNAL AND EXTERNAL			INTERNAL: Major-diameter fit								INTERNAL: Fillet-root side fit			INTERNAL: All fits	
					Flat-Root Side Fit										
						Major-diam fillet		Major diam							Space width
N	Pitch diam, ref	Base-circle diam, ref	Major diam, basic	*TIF* diam	Minor diam	Rad	Ht	Full deden[b]	Short deden[c]	*TIF* diam	Minor diam	Major diam	Fillet rad	Measurement between pins	Min effective 1.5708
1	2	3	4	5	6	7	8	9A	9B	10	11	12	13	14	15
Recommended tolerance ⟶			+0.0007 + L[d] −0.0000	Min	+0.0050 −0.0000	App	Max	+0.1050 −0.0000	+0.0007 + L[d] −(0.0007 + L)	Min	+0.0050 −0.0000	+0.1050 −0.0000	Min[e]	Max	Dimensional Max[f]
6	6.0000	5.1962	7.0000	6.897	5.2062	0.145	0.052	7.2040	7.0000	7.0000	5.2062	7.8000	0.166	4.4022	1.5755
7	7.0000	6.0622	8.0000	7.881	6.0722	0.160	0.060	8.2040	8.0000	8.0000	6.0722	8.8000	0.194	5.2320	1.5756
8	8.0000	6.9282	9.0000	8.870	7.0400	0.170	0.066	9.2040	9.0000	9.0000	7.0400	9.8000	0.208	6.4043	1.5756
9	9.0000	7.7942	10.0000	9.860	8.0000	0.178	0.071	10.2040	10.0000	10.0000	8.0000	10.8000	0.220	7.2707	1.5757
10	10.0000	8.6603	11.0000	10.852	9.0000	0.184	0.075	11.2040	11.0000	11.0000	9.0000	11.8000	0.230	8.4055	1.5757
11	11.0000	9.5263	12.0000	11.845	10.0000	0.189	0.079	12.2040	12.0000	12.0000	10.0000	12.8000	0.238	9.2955	1.5757
12	12.0000	10.3923	13.0000	12.839	11.0000	0.193	0.081	13.2040	13.0000	13.0000	11.0000	13.8000	0.245	10.4063	1.5758
13	13.0000	11.2583	14.0000	13.834	12.0000	0.196	0.084	14.2040	14.0000	14.0000	12.0000	14.8000	0.249	11.3129	1.5758
14	14.0000	12.1244	15.0000	14.830	13.0000	0.199	0.086	15.2040	15.0000	15.0000	13.0000	15.8000	0.253	12.4070	1.5759
15	15.0000	12.9904	16.0000	15.826	14.0000	0.201	0.088	16.2040	16.0000	16.0000	14.0000	16.8000	0.257	13.3258	1.5759
16	16.0000	13.8564	17.0000	16.823	15.0000	0.203	0.090	17.2040	17.0000	17.0000	15.0000	17.8000	0.260	14.4073	1.5759
17	17.0000	14.7224	18.0000	17.820	16.0000	0.205	0.091	18.2040	18.0000	18.0000	16.0000	18.8000	0.262	15.3357	1.5760
18	18.0000	15.5885	19.0000	18.818	17.0000	0.207	0.093	19.2040	19.0000	19.0000	17.0000	19.8000	0.265	16.4078	1.5760
19	19.0000	16.4545	20.0000	19.815	18.0000	0.208	0.094	20.2040	20.0000	20.0000	18.0000	20.8000	0.267	17.3437	1.5761
20	20.0000	17.3205	21.0000	20.813	19.0000	0.209	0.095	21.2040	21.0000	21.0000	19.0000	21.8000	0.269	18.4081	1.5761
21	21.0000	18.1865	22.0000	21.811	20.0000	0.210	0.096	22.2040	22.0000	22.0000	20.0000	22.8000	0.270	19.3499	1.5761
22	22.0000	19.0526	23.0000	22.810	21.0000	0.212	0.097	23.2040	23.0000	23.0000	21.0000	23.8000	0.272	20.4085	1.5762
23	23.0000	19.9186	24.0000	23.808	22.0000	0.212	0.097	24.2040	24.0000	24.0000	22.0000	24.8000	0.273	21.3552	1.5762
24	24.0000	20.7846	25.0000	24.807	23.0000	0.213	0.098	25.2040	25.0000	25.0000	23.0000	25.8000	0.274	22.4088	1.5763
25	25.0000	21.6506	26.0000	25.805	24.0000	0.214	0.099	26.2040	26.0000	26.0000	24.0000	26.8000	0.275	23.3598	1.5763
26	26.0000	22.5167	27.0000	26.804	25.0000	0.215	0.100	27.2040	27.0000	27.0000	25.0000	27.8000	0.276	24.4089	1.5763
27	27.0000	23.3827	28.0000	27.803	26.0000	0.215	0.100	28.2040	28.0000	28.0000	26.0000	28.8000	0.277	25.3637	1.5764
28	28.0000	24.2487	29.0000	28.802	27.0000	0.216	0.101	29.2040	29.0000	29.0000	27.0000	29.8000	0.278	26.4092	1.5764
29	29.0000	25.1147	30.0000	29.801	28.0000	0.217	0.101	30.2040	30.0000	30.0000	28.0000	30.8000	0.279	27.3669	1.5764
30	30.0000	25.9808	31.0000	30.801	29.0000	0.217	0.102	31.2040	31.0000	31.0000	29.0000	31.8000	0.280	28.4094	1.5765
31	31.0000	26.8468	32.0000	31.800	30.0000	0.218	0.102	32.2040	32.0000	32.0000	30.0000	32.8000	0.281	29.3699	1.5765
32	32.0000	27.7128	33.0000	32.799	31.0000	0.218	0.103	33.2040	33.0000	33.0000	31.0000	33.8000	0.282	30.4097	1.5766
33	33.0000	28.5788	34.0000	33.798	32.0000	0.219	0.103	34.2040	34.0000	34.0000	32.0000	34.8000	0.282	31.3725	1.5766
34	34.0000	29.4449	35.0000	34.798	33.0000	0.219	0.103	35.2040	35.0000	35.0000	33.0000	35.8000	0.283	32.4097	1.5766
35	35.0000	30.3109	36.0000	35.797	34.0000	0.219	0.104	36.2040	36.0000	36.0000	34.0000	36.8000	0.284	33.3748	1.5767
36	36.0000	31.1769	37.0000	36.796	35.0000	0.220	0.104	37.2040	37.0000	37.0000	35.0000	37.8000	0.284	34.4100	1.5767
37	37.0000	32.0429	38.0000	37.796	36.0000	0.220	0.104	38.2040	38.0000	38.0000	36.0000	38.8000	0.285	35.3770	1.5768
38	38.0000	32.9090	39.0000	38.795	37.0000	0.220	0.105	39.2040	39.0000	39.0000	37.0000	39.8000	0.286	36.4101	1.5768
39	39.0000	33.7750	40.0000	39.795	38.0000	0.221	0.105	40.2040	40.0000	40.0000	38.0000	40.8000	0.286	37.3787	1.5768
40	40.0000	34.6410	41.0000	40.794	39.0000	0.221	0.105	41.2040	41.0000	41.0000	39.0000	41.8000	0.286	38.4104	1.5769
41	41.0000	35.5070	42.0000	41.794	40.0000	0.221	0.106	42.2040	42.0000	42.0000	40.0000	42.8000	0.287	39.3805	1.5769
42	42.0000	36.3731	43.0000	42.794	41.0000	0.222	0.106	43.2040	43.0000	43.0000	41.0000	43.8000	0.287	40.4106	1.5770
43	43.0000	37.2391	44.0000	43.793	42.0000	0.222	0.106	44.2040	44.0000	44.0000	42.0000	44.8000	0.287	41.3821	1.5770
44	44.0000	38.1051	45.0000	44.793	43.0000	0.222	0.106	45.2040	45.0000	45.0000	43.0000	45.8000	0.288	42.4107	1.5770
45	45.0000	38.9711	46.0000	45.793	44.0000	0.222	0.106	46.2040	46.0000	46.0000	44.0000	46.8000	0.288	43.3835	1.5771
46	46.0000	39.8372	47.0000	46.792	45.0000	0.222	0.107	47.2040	47.0000	47.0000	45.0000	47.7000	0.288	44.4108	1.5771
47	47.0000	40.7032	48.0000	47.792	46.0000	0.223	0.107	48.2040	48.0000	48.0000	46.0000	48.8000	0.288	45.3849	1.5772
48	48.0000	41.5692	49.0000	48.792	47.0000	0.223	0.107	49.2040	49.0000	49.0000	47.0000	49.8000	0.288	46.4111	1.5772
49	49.0000	42.4352	50.0000	49.791	48.0000	0.223	0.107	50.2040	50.0000	50.0000	48.0000	50.8000	0.289	47.3860	1.5772
50	50.0000	43.3013	51.0000	50.791	49.0000	0.223	0.107	51.2040	51.0000	51.0000	49.0000	51.8000	0.289	48.4113	1.5773

[a] The fits and tolerances suggested in this table are to be considered binding on a manufacturer or seller only when specifically agreed to in writing.

[b] Intended for cutting by a generating process.

[c] If this dimension is used, the dimension in Col. 24 should be decreased by twice the amount of maximum dimensional tooth clearance and the chamfer applied. The TIF diameter in Col. 5 should be used.

[d] $L = 0.0001 \times$ diameter (Col. 4).

[e] Represents minimum allowable radius of curvature, and is based on 75% of the full tangent radius for maximum depth.

[f] Allowable errors, except lead, have been added to the machining tolerance in computing the maximum space width. When allowances for lead errors must be made, add 60% of the lead error to this dimension.

continued on next page

1/2 Diametral Pitch Pressure Angle, 30°	Addendum (Basic), 0.5000 Dedendum (See Sec 6)	Circular Pitch, 3.1416 Meas-Pin Diam, 1.9200

EXTERNAL

N	Major-diameter fit[h]									Fillet root[i]		Dimensions for all fits						
	Major diameter			Major-diam chamfer		Flat-root side fit							Measurement over pins			Tooth thickness		
						Minor diam	Minor-diam fillet		Major diam[j]	Minor diam	Fil rad	TIF diam				Cl A	Cl B	Cl C
																Max effective		
	Cl I	Cl II	Cl III	Dim	Ht		Rad	Ht					Cl A	Cl B	Cl C	1.5703	1.5723	1.5753
1	16	17	18	19	20	21	22	23	24	25	26	27	28	29	30	31	32	33
Rec tol	+0.0000 −(0.0019 + Jk)	+0.0000 −(0.0006 + Jk)	+0.0009 + Jk −0.0000	App	Min	+0.0000 −0.1050	App	Max	+0.0000 −0.0100	+0.0000 −0.1050	Min[l]	Max	Min	Min	Min	Min dimensional[m]		
6	6.9985	6.9999	7.0015	0.228	0.091	4.7960	0.170	0.168	7.0000	4.2000	—	5.2062	8.8590	8.8622	8.8661	1.5654	1.5679	1.5709
7	7.9985	7.9999	8.0016	0.230	0.094	5.7960	0.150	0.150	8.0000	5.2000	—	6.0722	9.6747	9.6780	9.6819	1.5653	1.5678	1.5708
8	8.9985	8.9999	9.0017	0.230	0.095	6.7960	0.140	0.137	9.0000	6.2000	0.379	7.0400	10.8869	10.8903	10.8914	1.5653	1.5678	1.5708
9	9.9985	9.9999	10.0018	0.232	0.097	7.7960	0.130	0.127	10.0000	7.2000	0.373	8.0000	11.7460	11.7494	11.7535	1.5653	1.5678	1.5708
10	10.9985	10.9999	11.0019	0.233	0.098	8.7960	0.120	0.119	11.0000	8.2000	0.367	9.0000	12.9065	12.9100	12.9143	1.5652	1.5677	1.5707
11	11.9985	11.9999	12.0020	0.234	0.099	9.7960	0.116	0.113	12.0000	9.2000	0.362	10.0000	13.7924	13.7959	13.8002	1.5652	1.5677	1.5707
12	12.9985	12.9999	13.0021	0.234	0.100	10.7960	0.110	0.107	13.0000	10.2000	0.357	11.0000	14.9215	14.9250	14.9294	1.5652	1.5677	1.5707
13	13.9985	13.9999	14.0022	0.235	0.100	11.7960	0.108	0.103	14.0000	11.2000	0.353	12.0000	15.8253	15.8289	15.8332	1.5651	1.5676	1.5706
14	14.9985	14.9999	15.0023	0.235	0.101	12.7960	0.105	0.099	15.0000	12.2000	0.350	13.0000	16.9328	16.9365	16.9409	1.5651	1.5676	1.5706
15	15.9985	15.9999	16.0024	0.236	0.102	13.7960	0.102	0.096	16.0000	13.2000	0.348	14.0000	17.8500	17.8537	17.8581	1.5651	1.5676	1.5706
16	16.9985	16.9999	17.0025	0.236	0.102	14.7960	0.100	0.093	17.0000	14.2000	0.345	15.0000	18.9421	18.9458	18.9503	1.5650	1.5675	1.5705
17	17.9985	17.9999	18.0026	0.236	0.103	15.7960	0.100	0.091	18.0000	15.2000	0.343	16.0000	19.8691	19.8729	19.8774	1.5650	1.5675	1.5705
18	18.9985	18.9999	19.0027	0.237	0.103	16.7960	0.100	0.089	19.0000	16.2000	0.341	17.0000	20.9497	20.9534	20.9579	1.5650	1.5675	1.5705
19	19.9985	19.9999	20.0028	0.237	0.103	17.7960	0.100	0.087	20.0000	17.2000	0.339	18.0000	21.8845	21.8883	21.8928	1.5649	1.5674	1.5704
20	20.9985	20.9999	21.0029	0.237	0.104	18.7960	0.100	0.085	21.0000	18.2000	0.338	19.0000	22.9559	22.9597	22.9643	1.5649	1.5674	1.5704
21	21.9985	21.9999	22.0030	0.237	0.104	19.7960	0.100	0.083	22.0000	19.2000	0.336	20.0000	23.8970	23.9009	23.9054	1.5648	1.5673	1.5703
22	22.9985	22.9999	23.0031	0.237	0.104	20.7960	0.100	0.082	23.0000	20.2000	0.334	21.0000	24.9612	24.9650	24.9696	1.5648	1.5673	1.5703
23	23.9985	23.9999	24.0032	0.238	0.104	21.7960	0.100	0.081	24.0000	21.2000	0.333	22.0000	25.9075	25.9114	25.9160	1.5648	1.5673	1.5703
24	24.9985	24.9999	25.0033	0.238	0.105	22.7960	0.100	0.080	25.0000	22.2000	0.332	23.0000	26.9657	26.9696	26.9743	1.5647	1.5672	1.5702
25	25.9985	25.9999	26.0034	0.238	0.105	23.7960	0.100	0.079	26.0000	23.2000	0.331	24.0000	27.9164	27.9203	27.9250	1.5647	1.5672	1.5702
26	26.9985	26.9999	27.0035	0.238	0.105	24.7960	0.100	0.078	27.0000	24.2000	0.330	25.0000	28.9698	28.9737	28.9784	1.5647	1.5672	1.5702
27	27.9985	27.9999	28.0036	0.239	0.105	25.7960	0.100	0.077	28.0000	25.2000	0.330	26.0000	29.9240	29.9279	29.9326	1.5646	1.5671	1.5701
28	28.9985	28.9999	29.0037	0.239	0.105	26.7960	0.100	0.076	29.0000	26.2000	0.329	27.0000	30.9732	30.9771	30.9818	1.5646	1.5671	1.5701
29	29.9985	29.9999	30.0038	0.239	0.106	27.7960	0.100	0.075	30.0000	27.2000	0.328	28.0000	31.9307	31.9347	31.9394	1.5646	1.5671	1.5701
30	30.9985	30.9999	31.0039	0.239	0.106	28.7960	0.100	0.074	31.0000	28.2000	0.327	29.0000	32.9763	32.9802	32.9849	1.5645	1.5670	1.5700
31	31.9985	31.9999	32.0040	0.239	0.106	29.7960	0.100	0.074	32.0000	29.2000	0.327	30.0000	33.9365	33.9405	33.9452	1.5645	1.5670	1.5700
32	32.9985	32.9999	33.0041	0.239	0.106	30.7960	0.100	0.073	33.0000	30.2000	0.326	31.0000	34.9789	34.9828	34.9876	1.5644	1.5669	1.5699
33	33.9985	33.9999	34.0042	0.239	0.106	31.7960	0.100	0.072	34.0000	31.2000	0.326	32.0000	35.9415	35.9455	35.9503	1.5644	1.5669	1.5699
34	34.9985	34.9999	35.0043	0.239	0.106	32.7960	0.100	0.072	35.0000	32.2000	0.325	33.0000	36.9814	36.9854	36.9902	1.5644	1.5669	1.5699
35	35.9985	35.9999	36.0044	0.239	0.106	33.7960	0.100	0.071	36.0000	33.2000	0.325	34.0000	37.9461	37.9501	37.9549	1.5643	1.5668	1.5698
36	36.9985	36.9999	37.0045	0.240	0.106	34.7960	0.100	0.071	37.0000	34.2000	0.324	35.0000	38.9835	38.9875	38.9923	1.5643	1.5668	1.5698
37	37.9985	37.9999	38.0046	0.240	0.107	35.7960	0.100	0.070	38.0000	35.2000	0.324	36.0000	39.9503	39.9543	39.9591	1.5643	1.5668	1.5698
38	38.9985	38.9999	39.0047	0.240	0.107	36.7960	0.100	0.070	39.0000	36.2000	0.323	37.0000	40.9854	40.9894	40.9942	1.5642	1.5667	1.5697
39	39.9985	39.9999	40.0048	0.240	0.107	37.7960	0.100	0.069	40.0000	37.2000	0.323	38.0000	41.9539	41.9579	41.9627	1.5642	1.5667	1.5697
40	40.9985	40.9999	41.0049	0.240	0.107	38.7960	0.100	0.069	41.0000	38.2000	0.323	39.0000	42.9874	42.9914	42.9962	1.5642	1.5667	1.5697
41	41.9985	41.9999	42.0050	0.240	0.107	39.7960	0.100	0.069	42.0000	39.2000	0.322	40.0000	43.9572	43.9612	43.9661	1.5641	1.5666	1.5696
42	42.9985	42.9999	43.0051	0.240	0.107	40.7960	0.100	0.068	43.0000	40.2000	0.322	41.0000	44.9888	44.9928	44.9927	1.5641	1.5666	1.5696
43	43.9985	43.9999	44.0052	0.240	0.107	41.7960	0.100	0.068	44.0000	41.2000	0.322	42.0000	45.9603	45.9643	45.9692	1.5641	1.5666	1.5696
44	44.9985	44.9999	45.0053	0.240	0.107	42.7960	0.100	0.068	45.0000	42.2000	0.322	43.0000	46.9903	46.9943	46.9992	1.5640	1.5665	1.5695
45	45.9985	45.9999	46.0054	0.240	0.107	43.7960	0.100	0.067	46.0000	43.2000	0.321	44.0000	47.9630	47.9670	47.9719	1.5640	1.5665	1.5695
46	46.9985	46.9999	47.0055	0.240	0.107	44.7960	0.100	0.067	47.0000	44.2000	0.321	45.0000	48.9916	48.9957	49.0005	1.5639	1.5664	1.5694
47	47.9985	47.9999	48.0056	0.240	0.107	45.7960	0.100	0.067	48.0000	45.2000	0.320	46.0000	49.9654	49.9694	49.9743	1.5639	1.5664	1.5694
48	48.9985	48.9999	49.0057	0.240	0.107	46.7960	0.100	0.066	49.0000	46.2000	0.320	47.0000	50.9929	50.9969	51.0018	1.5639	1.5664	1.5694
49	49.9985	49.9999	50.0058	0.240	0.107	47.7960	0.100	0.066	50.0000	47.2000	0.320	48.0000	51.9676	51.9717	51.9766	1.5638	1.5663	1.5693
50	50.9985	50.9999	51.0059	0.240	0.107	48.7960	0.100	0.066	51.0000	48.2000	0.319	49.0000	52.9940	52.9981	53.0030	1.5638	1.5663	1.5693

[g]The fits and tolerances suggested in this table are to be considered binding on a manufacturer or seller only when specifically agreed to in writing.

[h]Measurement over pins for Class A is recommended (Col. 28), but if tighter fits are required, Class B (Col. 29) may be used.

[i]This may be used for a major-diameter fit by using dimension in Cols. 16, 17, or 18 instead of that in Col. 24.

[j]When Col. 9B is used for the internal spline, reduce this dimension as covered in footnote c.

[k]J = 0.0002 X diameter (Col. 24).

[l]Represents minimum allowable radius of curvature, and is based on 75% of the full tangent radius for maximum depth.

[m]Allowable errors, except lead, have been added to the machining tolerance in computing the minimum tooth thickness. When allowance for lead errors must be made, subtract 60% of the lead error from this dimension.

TABLE 10–5

Basic Data for 5/10 Diametral Pitch — Minor-Diameter Fits[a]

1954 SAE Handbook

5/10 Diametral Pitch Pressure Angle, 30°	Addendum (Basic), 0.1000 Meas-Pin Diam (Int), 0.2880	Circular Pitch, 0.6283 Meas-Pin Diam (Ext), 0.3840

Minor–Diameter Fits

Internal and External			Internal										External					
N	Pitch diam, ref	Base-circle diam	*TIF* diam	Major-diam fillet		Major diam	Minor-diam chamfer		Minor diameter			Measure-ment between pins	Major diam	Minor diam, basic	Minor-diam fillet		*TIF* diam	Measure-ment over pins
				Rad	Ht		Ang	Ht	Cl X	Cl Y	Cl Z				Rad	Ht		
1	2	3	4	5	6	7	8	9	10	11	12	13	14	15	16	17	18	19
Recommended tolerance→			Min	App	Max	+0.0250 −0.0000	Deg	Min	+0.0019 + L^b −0.0000	+0.0006 + L^b −0.0000	+0.0000 −(0.0009 + L^b)	Max	+0.0000 −0.0070	+0.0000 −(0.0006 + J^c)	App	Max	Max	Min
6	1.2000	1.0392	1.4000	0.029	0.010	1.4440	53.5	0.040	1.0015	1.0001	0.9991	0.8859	1.4000	1.0000	0.034	0.034	1.067	1.7676
7	1.4000	1.2124	1.6000	0.032	0.012	1.6440	53.5	0.036	1.2015	1.2001	1.1991	1.0505	1.6000	1.2000	0.030	0.030	1.259	1.9307
8	1.6000	1.3856	1.8000	0.034	0.013	1.8440	53.5	0.033	1.4015	1.4001	1.3990	1.2861	1.8000	1.4000	0.028	0.027	1.453	2.1729
9	1.8000	1.5588	2.0000	0.036	0.014	2.0440	53.5	0.030	1.6015	1.6001	1.5990	1.4593	2.0000	1.6000	0.026	0.025	1.649	2.3447
10	2.0000	1.7321	2.2000	0.037	0.015	2.2440	53.5	0.029	1.8015	1.8001	1.7989	1.6862	2.2000	1.8000	0.024	0.024	1.847	2.5767
11	2.2000	1.9053	2.4000	0.038	0.016	2.4440	53.5	0.027	2.0015	2.0001	1.9989	1.8641	2.4000	2.0000	0.023	0.023	2.045	2.7539
12	2.4000	2.0785	2.6000	0.039	0.016	2.6440	53.5	0.026	2.2015	2.2001	2.1989	2.0864	2.6000	2.2000	0.022	0.021	2.241	2.9796
13	2.6000	2.2517	2.8000	0.039	0.017	2.8440	53.5	0.025	2.4015	2.4001	2.3988	2.2676	2.8000	2.4000	0.022	0.021	2.441	3.1604
14	2.8000	2.4249	3.0000	0.040	0.017	3.0440	53.5	0.024	2.6015	2.6001	2.5988	2.4864	3.0000	2.6000	0.021	0.020	2.639	3.3817
15	3.0000	2.5981	3.2000	0.040	0.018	3.2440	53.5	0.023	2.8015	2.8001	2.7987	2.6701	3.2000	2.8000	0.020	0.019	2.837	3.5652
16	3.2000	2.7713	3.4000	0.041	0.018	3.4440	53.5	0.022	3.0015	3.0001	2.9987	2.8866	3.4000	3.0000	0.020	0.019	3.037	3.7835
17	3.4000	2.9445	3.6000	0.041	0.018	3.6440	53.5	0.022	3.2015	3.2001	3.1987	3.0722	3.6000	3.2000	0.020	0.018	3.235	3.9690
18	3.6000	3.1177	3.8000	0.041	0.019	3.8440	53.5	0.021	3.4015	3.4001	3.3986	3.2866	3.8000	3.4000	0.020	0.018	3.435	4.1851
19	3.8000	3.2909	4.0000	0.042	0.019	4.0440	53.5	0.021	3.6015	3.6001	3.5986	3.4738	4.0000	3.6000	0.020	0.017	3.633	4.3720
20	4.0000	3.4641	4.2000	0.042	0.019	4.2440	53.5	0.020	3.8015	3.8001	3.7985	3.6866	4.2000	3.8000	0.020	0.017	3.833	4.5863
21	4.2000	3.6373	4.4000	0.042	0.019	4.4440	53.5	0.020	4.0015	4.0001	3.9985	3.8750	4.4000	4.0000	0.020	0.017	4.033	4.7742
22	4.4000	3.8105	4.6000	0.042	0.019	4.6440	53.5	0.020	4.2015	4.2001	4.1985	4.0867	4.6000	4.2000	0.020	0.016	4.231	4.9872
23	4.6000	3.9837	4.8000	0.042	0.019	4.8440	52.5	0.019	4.4015	4.4001	4.3984	4.2760	4.8000	4.4000	0.020	0.016	4.431	5.1765
24	4.8000	4.1569	5.0000	0.043	0.020	5.0440	53.5	0.019	4.6015	4.6001	4.5984	4.4867	5.0000	4.6000	0.020	0.016	4.631	5.3880
25	5.0000	4.3301	5.2000	0.043	0.020	5.2440	53.5	0.019	4.8015	4.8001	4.7983	4.6769	5.2000	4.8000	0.020	0.016	4.830	5.5782
26	5.2000	4.5033	5.4000	0.043	0.020	5.4440	53.5	0.019	5.0015	5.0001	4.9983	4.8867	5.4000	5.0000	0.020	0.016	5.030	5.7889
27	5.4000	4.6765	5.6000	0.043	0.020	5.6440	53.5	0.018	5.2015	5.2001	5.1983	5.0778	5.6000	5.2000	0.020	0.015	5.228	5.9797
28	5.6000	4.8497	5.8000	0.043	0.020	5.8440	53.5	0.018	5.4015	5.4001	5.3982	5.2870	5.8000	5.4000	0.020	0.015	5.428	6.1895
29	5.8000	5.0229	6.0000	0.043	0.020	6.0440	53.5	0.018	5.6015	5.6001	5.5982	5.4784	6.0000	5.6000	0.020	0.015	5.628	6.3810
30	6.0000	5.1962	6.2000	0.043	0.020	6.2440	53.5	0.018	5.8015	5.8001	5.7981	5.6869	6.2000	5.8000	0.020	0.015	5.828	6.5901
31	6.2000	5.3694	6.4000	0.044	0.020	6.4440	53.5	0.018	6.0015	6.0001	5.9981	5.8790	6.4000	6.0000	0.020	0.015	6.028	6.7822
32	6.4000	5.5426	6.6000	0.044	0.021	6.6440	53.5	0.018	6.2015	6.2001	6.1981	6.0869	6.6000	6.2000	0.020	0.015	6.228	6.9907
33	6.6000	5.7158	6.8000	0.044	0.021	6.8440	53.5	0.017	6.4015	6.4001	6.3980	6.2795	6.8000	6.4000	0.020	0.014	6.426	7.1832
34	6.8000	5.8890	7.0000	0.044	0.021	7.0440	53.5	0.017	6.6015	6.6001	6.5980	6.4869	7.0000	6.6000	0.020	0.014	6.626	7.3911
35	7.0000	6.0622	7.2000	0.044	0.021	7.2440	53.5	0.017	6.8015	6.8001	6.7979	6.6799	7.2000	6.8000	0.020	0.014	6.826	7.5840
36	7.2000	6.2354	7.4000	0.044	0.021	7.4440	53.5	0.017	7.0015	7.0001	6.9979	6.8869	7.4000	7.0000	0.020	0.014	7.026	7.7914
37	7.4000	6.4086	7.6000	0.044	0.021	7.6440	53.5	0.017	7.2015	7.2001	7.1979	7.0805	7.6000	7.2000	0.020	0.014	7.226	7.9847
38	7.6000	6.5818	7.8000	0.044	0.021	7.8440	54.0	0.017	7.4015	7.4001	7.3978	7.2871	7.8000	7.4000	0.020	0.014	7.426	8.1918
39	7.8000	6.7550	8.0000	0.044	0.021	8.0440	54.0	0.017	7.6015	7.6001	7.5978	7.4808	8.0000	7.6000	0.020	0.014	7.626	8.3855
40	8.0000	6.9282	8.2000	0.044	0.021	8.2440	54.0	0.017	7.8015	7.8001	7.7977	7.6871	8.2000	7.8000	0.020	0.014	7.826	8.5922
41	8.2000	7.1014	8.4000	0.044	0.021	8.4440	54.0	0.017	8.0015	8.0001	7.9977	7.8812	8.4000	8.0000	0.020	0.014	8.026	8.7862
42	8.4000	7.2746	8.6000	0.044	0.021	8.6440	54.0	0.016	8.2015	8.2001	8.1977	8.0871	8.6000	8.2000	0.020	0.014	8.226	8.9925
43	8.6000	7.4478	8.8000	0.044	0.021	8.8440	54.0	0.016	8.4015	8.4001	8.3976	8.2814	8.8000	8.4000	0.020	0.014	8.426	9.1868
44	8.8000	7.6210	9.0000	0.044	0.021	9.0440	54.0	0.016	8.6015	8.6001	8.5976	8.4872	9.0000	8.6000	0.020	0.014	8.626	9.3928
45	9.0000	7.7942	9.2000	0.044	0.021	9.2440	54.0	0.016	8.8015	8.8001	8.7975	8.6817	9.2000	8.8000	0.020	0.013	8.824	9.5873
46	9.2000	7.9674	9.4000	0.044	0.021	9.4440	54.0	0.016	9.0015	9.0001	8.9975	8.8872	9.4000	9.0000	0.020	0.013	9.024	9.7931
47	9.4000	8.1406	9.6000	0.045	0.021	9.6440	54.0	0.016	9.2015	9.2001	9.1975	9.0820	9.6000	9.2000	0.020	0.013	9.224	9.9878
48	9.6000	8.3138	9.8000	0.045	0.021	9.8440	54.0	0.016	9.4015	9.4001	9.3974	9.2873	9.8000	9.4000	0.020	0.013	9.424	10.1933
49	9.8000	8.4870	10.0000	0.045	0.021	10.0440	54.0	0.016	9.6015	9.6001	9.5974	9.4823	10.0000	9.6000	0.020	0.013	9.624	10.3881
50	10.0000	8.6603	10.2000	0.045	0.021	10.2440	54.0	0.016	9.8015	9.8001	9.7973	9.6873	10.2000	9.8000	0.020	0.013	9.823	10.5934

[a] The fits and tolerances suggested in this table are to be considered binding on a manufacturer or seller only when specifically agreed to in writing.

[b] L = 0.0001 X diameter (Col. 15).

[c] J = 0.0002 X diameter (Col. 15).

TABLE 10-6

Dimensional and Effective Clearances[a]

1954 SAE Handbook

DIAMETRAL PITCH	CLASS A				CLASS B				CLASS C			
	Dimensional clearance[b,c]		Effective clearance[d]		Dimensional clearance[b,c]		Effective clearance[d]		Dimensional clearance[b,c]		Effective clearance[d]	
	Min	Max	Min	Max	Min	Max	Min	Max	Min	Max	Min	Max
1/2	+0.0076	+0.0116	+0.0005	+0.0045	+0.0056	+0.0091	−0.0015	+0.0020	+0.0026	+0.0061	−0.0045	−0.0010
2.5/5	+0.0057	+0.0097	+0.0005	+0.0045	+0.0037	+0.0072	−0.0015	+0.0020	+0.0010	+0.0045	−0.0042	−0.0007
3/6	+0.0052	+0.0092	+0.0005	+0.0045	+0.0032	+0.0067	−0.0015	+0.0020	+0.0005	+0.0040	−0.0042	−0.0007
4/8 and 5/10	+0.0044	+0.0084	+0.0005	+0.0045	+0.0024	+0.0059	−0.0015	+0.0020	+0.0003	+0.0038	−0.0036	−0.0001
6/12 and 8/16	+0.0039	+0.0069	+0.0005	+0.0035	+0.0021	+0.0051	−0.0013	+0.0017	+0.0004	+0.0034	−0.0030	−0.0000
10/20 thru 20/40	+0.0037	+0.0062	+0.0005	+0.0030	+0.0021	+0.0046	−0.0011	+0.0014	+0.0002	+0.0027	−0.0030	−0.0005
24/48 thru 48/96	+0.0037	+0.0057	+0.0005	+0.0025	+0.0023	+0.0043	−0.0009	+0.0011	+0.0002	+0.0022	−0.0030	−0.0010

[a] Based on using machining tolerances plus 60% of all allowable errors cumulatively, not including lead.

[b] For 25 teeth.

[c] The maximum dimensional clearance is obtained by subtracting dimension in Col. 31, 32, or 33 from that in Col. 15 in Table 10–5. The minimum dimensional clearance is obtained by subtracting the sum of the machining tolerances for Col. 15 and that for either Col. 31, 32, or 33, as listed in Table 10–5, from the maximum dimensional clearance.

[d] The minimum effective clearance is obtained by subtracting the dimension (maximum effective) in the heading over Col. 31, 32 or 33 from the dimension (minimum effective) in the heading over Col. 15 (Table 10–5). The maximum effective clearance is obtained by adding to the minimum effective clearance the machining tolerances for Col. 15 and that for either Col. 31, 32, or 33.

TABLE 10-7

Types of Fits for Involute Splines

1954 SAE Handbook

Type of Fit	Major Diameter.			Sides of Teeth			Minor Diameter		
Dimension Varied to Control Fit	Major diameter of external spline			Tooth thickness			Minor diameter of internal spline		
Classes of Fit	Sliding Fit	Close Fit	Press Fit	Sliding Fit	Close Fit	Press Fit	Sliding Fit	Close Fit	Press Fit
Designation used in Tables 10-5 and 10-6.	I	II	III	A	B	C	X	Y	Z

TABLE 10–8

Basic Data – All Involute Splines

1954 SAE Handbook

N	1/2 DIAMETRAL PITCH			ALL PITCHES						
	Internal and external		External	Internal			Int and ext	External		
	D_b	$D_b \cos \frac{90°}{N}$	Measurement over pins	$\frac{t_s}{D}$	$\frac{d_i}{D_b}$	F	$\cos \frac{90°}{N}$	$\frac{d_e}{D_b}$	$\frac{\pi}{N}$	E
1	2	3	4	5	6	7	8	9	10	11
6	5.196152		8.8660	0.261799	0.277128	1.91		0.369504	0.523599	1.305
7	6.062178	5.910186	9.6819	0.224399	0.237538	1.83	0.974928	0.316718	0.448799	1.302
8	6.928203		10.8944	0.196350	0.207846	1.86		0.277128	0.392699	1.362
9	7.794229	7.675817	11.7535	0.174533	0.184752	1.81	0.984808	0.246336	0.349066	1.364
10	8.660254		12.9144	0.157080	0.166277	1.83		0.221703	0.314159	1.406
11	9.526279	9.429316	13.8003	0.142800	0.151161	1.80	0.989821	0.201548	0.285599	1.409
12	10.392305		14.9295	0.130900	0.138564	1.81		0.184752	0.261799	1.440
13	11.258330	11.176244	15.8335	0.120830	0.127905	1.79	0.992709	0.170540	0.241661	1.443
14	12.124356		16.9412	0.112200	0.118769	1.80		0.158359	0.224399	1.467
15	12.990381	12.919218	17.8584	0.104720	0.110851	1.78	0.994522	0.147802	0.209440	1.471
16	13.856406		18.9507	0.098175	0.103923	1.79		0.138564	0.196350	1.490
17	14.722432	14.659629	19.8778	0.092400	0.097810	1.78	0.995734	0.130413	0.184800	1.493
18	15.588457		20.9584	0.087266	0.092376	1.78		0.123168	0.174533	1.509
19	16.454483	16.398282	21.8934	0.082673	0.087514	1.78	0.996584	0.116686	0.165347	1.512
20	17.320508		22.9649	0.078540	0.083138	1.78		0.110851	0.157080	1.525
21	18.186533	18.135680	23.9062	0.074800	0.079179	1.77	0.997204	0.105573	0.149600	1.528
22	19.052559		24.9704	0.071400	0.075580	1.77		0.100774	0.142800	1.539
23	19.918548	19.872149	25.9168	0.068295	0.072294	1.77	0.997669	0.096392	0.136591	1.541
24	20.784610		26.9752	0.065450	0.069282	1.77		0.092376	0.130900	1.551
25	21.650635	21.607912	27.9259	0.062832	0.066511	1.77	0.998027	0.088681	0.125664	1.553
26	22.516660		28.9793	0.060415	0.063953	1.76		0.085270	0.120830	1.562
27	23.382686	23.343126	29.9337	0.058178	0.061584	1.76	0.998308	0.082112	0.116355	1.564
28	24.248711		30.9829	0.056100	0.059385	1.76		0.079179	0.112200	1.571
29	25.114737	25.077904	31.9405	0.054165	0.057337	1.76	0.998533	0.076449	0.108331	1.573
30	25.980762		32.9862	0.052360	0.055426	1.76		0.073901	0.104720	1.580
31	26.846787	26.812337	33.9465	0.050671	0.053638	1.76	0.998717	0.071517	0.101342	1.581
32	27.712813		34.9890	0.049087	0.051962	1.76		0.069282	0.098175	1.587
33	28.578838	28.546468	35.9517	0.047600	0.050387	1.76	0.998867	0.067183	0.095200	1.589
34	29.444864		36.9916	0.046200	0.048905	1.76		0.065207	0.092400	1.594
35	30.310889	30.280368	37.9565	0.044880	0.047508	1.76	0.998993	0.063344	0.089760	1.596
36	31.176914		38.9939	0.043633	0.046188	1.76		0.061584	0.087266	1.600
37	32.042940	32.014068	39.9607	0.042454	0.044940	1.76	0.999100	0.059920	0.084908	1.602
38	32.908965		40.9960	0.041337	0.043757	1.75		0.058343	0.082673	1.606
39	33.774991	33.747599	41.9645	0.040277	0.042635	1.75	0.999189	0.056847	0.080554	1.608
40	34.641016		42.9980	0.039270	0.041569	1.75		0.055426	0.078540	1.611
41	35.507041	35.480986	43.9680	0.038312	0.040555	1.75	0.999266	0.054074	0.076624	1.613
42	36.373067		44.9996	0.037400	0.039590	1.75		0.052786	0.074800	1.616
43	37.239092	37.214248	45.9711	0.036530	0.038669	1.75	0.999333	0.051559	0.073060	1.617
44	38.105118		47.0013	0.035700	0.037790	1.75		0.050387	0.071400	1.621
45	38.971143	38.947403	47.9740	0.034907	0.036950	1.75	0.999391	0.049267	0.069813	1.622
46	39.837168		49.0028	0.034148	0.036147	1.75		0.048196	0.068295	1.625
47	40.703194	40.680464	49.9766	0.033421	0.035378	1.75	0.999442	0.047171	0.066842	1.626
48	41.569219		51.0041	0.032725	0.034641	1.75		0.046188	0.065450	1.629
49	42.435245	42.413442	51.9790	0.032057	0.033934	1.75	0.999486	0.045245	0.064114	1.630
50	43.301270		53.0055	0.031416	0.033255	1.75		0.044341	0.062832	1.632

Constants, $\pi = 3.141593$

$\cos 30° = 0.866025$

$\text{Inv } 30° = 0.053751$

$$E = \frac{\cos 30°}{\sin \phi_e} = \frac{\text{Rate of change of } M_e}{\text{Rate of change of } t} \left(\text{For } N \text{ odd, } E = \frac{\cos 30°}{\sin \phi_e} \cos \frac{90°}{N}\right)$$

$$F = \frac{\cos 30°}{\sin \phi_i} = \frac{\text{Rate of change of } M_i}{\text{Rate of change of } t_e} \left(\text{For } N \text{ odd, } F \approx \frac{\cos 30°}{\sin \phi_i} \cos \frac{90°}{N}\right)$$

TABLE 10–9

Allowable Errors[a] on Involute Splines

1954 SAE Handbook

DIAMETRAL PITCH	INVOLUTE PROFILE ERRORS	ACCUMULATED PITCH ERROR BETWEEN ANY TWO TEETH — Use as given or interpolate as required — Number of teeth													LEAD ERRORS — Length of spline					
		6	8	10	12	14	16	20	25	30	35	40	45	50	0 to 0.49	0.50 to 1.24	1.25 to 2.49	2 50 to 3.99	4.00 to 5.24	5.25 to 6.50
1/2	15	22	22	22	22	23	24	27	30	33	36	39	42	45	0	4	5	6	7	8
2.5/5	11	20	20	20	20	20	20	20	20	20	21	22	23	24	0	4	5	6	7	8
3/6	10	18	18	18	18	18	18	18	18	18	18	19	20	20	0	4	5	6	7	8
4/8 and 5/10	8	15	15	15	15	15	15	15	15	15	15	15	15	15	0	4	5	6	7	8
6/12 and 8/16	6	15	15	15	15	15	15	15	15	15	15	15	15	15	0	3	3	4	5	6
10/20 up	5	15	15	15	15	15	15	15	15	15	15	15	15	15	0	3	3	4	5	6

DIAMETRAL PITCH	OUT OF ROUNDNESS — Internal spline, number of teeth													External spline, number of teeth												
	6	8	10	12	14	16	20	25	30	35	40	45	50	6	8	10	12	14	16	20	25	30	35	40	45	50
1/2	17	19	21	24	23	23	24	24	24	25	25	26	26	16	18	20	23	23	24	24	24	25	25	25	25	25
2.5/5	13	13	14	15	16	17	18	20	23	23	23	23	24	12	13	14	15	16	17	18	20	22	22	22	22	22
3/6	12	13	13	14	14	15	17	19	20	22	23	24	24	12	12	14	14	15	15	17	18	20	21	21	21	21
4/8 and 5/10	11	12	12	13	13	13	15	16	17	18	20	21	22	10	11	11	12	13	13	14	15	16	18	19	20	21
6/12 thru 48/96	9	9	10	10	10	11	11	12	13	14	15	16	17	8	9	10	10	10	11	12	12	14	14	15	16	17

[a] All table figures are ten thousandths of an inch.

TABLE 10–10

Basic Formulas for Involute Splines

1954 SAE Handbook

$$\text{Circular Pitch} = \frac{3.141593}{\text{Diametral Pitch}}$$

$$\text{Pitch Diameter} = \frac{\text{Number of Teeth}}{\text{Diametral Pitch}}$$

$$\text{Addendum} = \text{Dedendum} = \frac{.500}{\text{Diametral Pitch}}$$

$$\text{Major Diameter} = \frac{\text{No. of Teeth} + 1}{\text{Diametral Pitch}}$$

$$\text{Circular Tooth Thickness} = \frac{1.570796}{\text{Diametral Pitch}}$$

$$\text{Minor Diameter} = \frac{\text{No. of Teeth} - 1}{\text{Diametral Pitch}}$$

As used in the above formulae, diametral pitch means the numerator of the expression used to specify the size of the spline. In the case of a 1/2 diametral pitch spline, 1 should be substituted for diametral pitch in the above formulae.

TABLE 10-11

Chamfer Data – External Involute Splines

1954 SAE Handbook

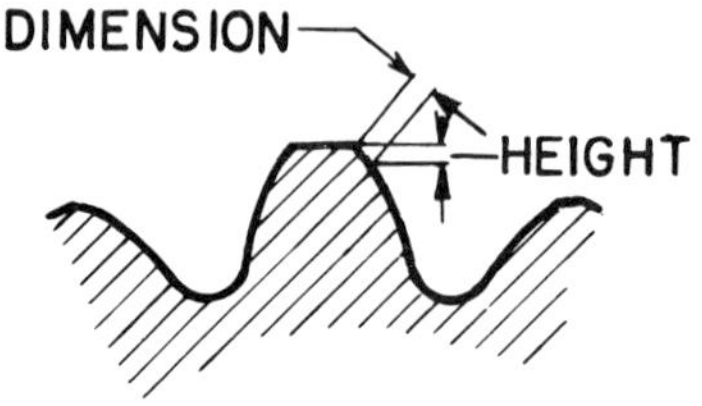

Number of Teeth	Diametral Pitch															
	1/2		2.5/5		3/6		4/8		5/10		6/12		8/16		10/20	
	Major Diameter Chamfer															
	Dim	Hgt	Dim	Hgt	Dim	Hgt	Dim	Hgt	Dim	Hgt	Dim	Hgt	Dim	Hgt	Dim	Hgt
1	2	3	4	5	6	7	8	9	10	11	12	13	14	15	16	17
	Approx	Min	Approx	Min	Approx	Min	Approx	Min	Approx	Min	Approx	Min	Approx	Min	Approx	Min
6	.228	.091	.091	.036	.076	.030	.057	.023	.046	.018	.038	.015	.029	.011	.023	.009
7	.230	.094	.092	.037	.077	.031	.057	.023	.046	.019	.038	.016	.029	.012	.023	.009
8	.230	.095	.092	.038	.077	.032	.057	.024	.046	.019	.038	.016	.029	.012	.023	.009
9	.232	.097	.093	.039	.077	.032	.058	.024	.046	.019	.039	.016	.029	.012	.023	.010
10	.233	.098	.093	.039	.078	.033	.058	.024	.047	.020	.039	.016	.029	.012	.023	.010
11	.234	.099	.094	.039	.078	.033	.058	.025	.047	.020	.039	.017	.029	.012	.023	.010
12	.234	.100	.094	.040	.078	.033	.058	.025	.047	.020	.039	.017	.029	.013	.023	.010
13	.235	.100	.094	.040	.078	.033	.059	.025	.047	.020	.039	.017	.029	.013	.023	.010
14	.235	.101	.094	.040	.078	.034	.059	.025	.047	.020	.039	.017	.029	.013	.023	.010
15	.236	.102	.094	.040	.079	.034	.059	.025	.047	.020	.039	.017	.030	.013	.024	.010
16	.236	.102	.094	.040	.079	.034	.059	.025	.047	.020	.039	.017	.030	.013	.024	.010
17	.236	.103	.094	.041	.079	.034	.059	.026	.047	.021	.039	.017	.030	.013	.024	.010
18	.237	.103	.095	.041	.079	.034	.059	.026	.047	.021	.040	.017	.030	.013	.024	.010
19	.237	.103	.095	.041	.079	.034	.059	.026	.047	.021	.040	.017	.030	.013	.024	.010
20	.237	.104	.095	.041	.079	.035	.059	.026	.047	.021	.040	.017	.030	.013	.024	.010
21	.237	.104	.095	.042	.079	.035	.059	.026	.047	.021	.040	.017	.030	.013	.024	.010
22	.237	.104	.095	.042	.079	.035	.059	.026	.047	.021	.040	.017	.030	.013	.024	.010
23	.238	.104	.095	.042	.079	.035	.060	.026	.048	.021	.040	.017	.030	.013	.024	.010
24	.238	.105	.095	.042	.079	.035	.060	.026	.048	.021	.040	.018	.030	.013	.024	.011
25	.238	.105	.095	.042	.079	.035	.060	.026	.048	.021	.040	.018	.030	.013	.024	.011
26	.238	.105	.095	.042	.079	.035	.060	.026	.048	.021	.040	.018	.030	.013	.024	.011
27	.239	.105	.096	.042	.080	.035	.060	.026	.048	.021	.040	.018	.030	.013	.024	.011
28	.239	.105	.096	.042	.080	.035	.060	.026	.048	.021	.040	.018	.030	.013	.024	.011
29	.239	.106	.096	.042	.080	.035	.060	.026	.048	.021	.040	.018	.030	.013	.024	.011
30	.239	.106	.096	.042	.080	.035	.060	.026	.048	.021	.040	.018	.030	.013	.024	.011
31	.239	.106	.096	.042	.080	.035	.060	.026	.048	.021	.040	.018	.030	.013	.024	.011
32	.239	.106	.096	.042	.080	.035	.060	.027	.048	.021	.040	.018	.030	.013	.024	.011
33	.239	.106	.096	.042	.080	.035	.060	.027	.048	.021	.040	.018	.030	.013	.024	.011
34	.239	.106	.096	.042	.080	.035	.060	.027	.048	.021	.040	.018	.030	.013	.024	.011
35	.239	.106	.096	.043	.080	.035	.060	.027	.048	.021	.040	.018	.030	.013	.024	.011

continued on next page

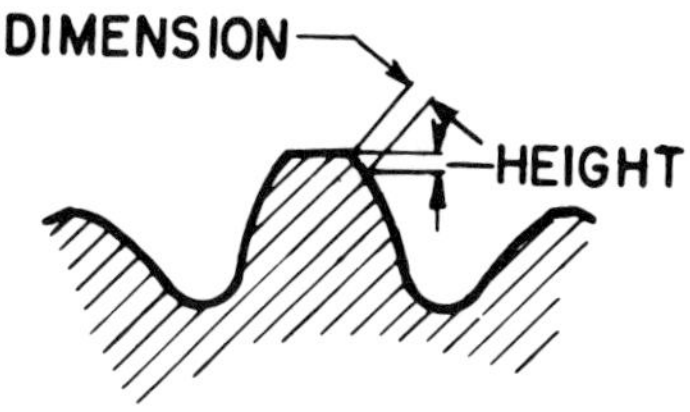

Diametral Pitch 12/24		16/32		20/40		24/48		32/64		40/80		48/96		Number of Teeth
Major Diameter Chamfer														
Dim 18 Approx	Hgt 19 Min	Dim 20 Approx	Hgt 21 Min	Dim 22 Approx	Hgt 23 Min	Dim 24 Approx	Hgt 25 Min	Dim 26 Approx	Hgt 27 Min	Dim 28 Approx	Hgt 29 Min	Dim 30 Approx	Hgt 31 Min	1
.019	.008	.014	.006	.011	.005	.009	.004	.007	.003	.006	.002	.005	.002	6
.019	.008	.014	.006	.012	.005	.010	.004	.007	.003	.006	.002	.005	.002	7
.019	.008	.014	.006	.012	.005	.010	.004	.007	.003	.006	.002	.005	.002	8
.019	.008	.015	.006	.012	.005	.010	.004	.007	.003	.006	.002	.005	.002	9
.019	.008	.015	.066	.012	.005	.010	.004	.007	.003	.006	.002	.005	.002	10
.020	.008	.015	.006	.012	.005	.010	.004	.007	.003	.006	.002	.005	.002	11
.020	.008	.015	.006	.012	.005	.010	.004	.007	.003	.006	.003	.005	.002	12
.020	.008	.015	.006	.012	.005	.010	.004	.007	.003	.006	.003	.005	.002	13
.020	.008	.015	.006	.012	.005	.010	.004	.007	.003	.006	.003	.005	.002	14
.020	.009	.015	.006	.012	.005	.010	.004	.007	.003	.006	.003	.005	.002	15
.020	.009	.015	.006	.012	.005	.010.	.004	.007	.003	.006	.003	.005	.002	16
.020	.009	.015	.006	.012	.005	.010	.004	.007	.003	.006	.003	.005	.002	17
.020	.009	.015	.006	.012	.005	.010	.004	.007	.003	.006	.003	.005	.002	18
.020	.009	.015	.006	.012	.005	.010	.004	.007	.003	.006	.003	.005	.002	19
.020	.009	.015	.007	.012	.005	.010	.004	.007	.003	.006	.003	.005	.002	20
.020	.009	.015	.007	.012	.005	.010	.004	.007	.003	.006	.003	.005	.002	21
.020	.009	.015	.007	.012	.005	.010	.004	.007	.003	.006	.003	.005	.002	22
.020	.009	.015	.007	.012	.005	.010	.004	.007	.003	.006	.003	.005	.002	23
.020	.009	.015	.007	.012	.005	.010	.004	.007	.003	.006	.003	.005	.002	24
.020	.009	.015	.007	.012	.005	.010	.004	.007	.003	.006	.003	.005	.002	25
.020	.009	.015	.007	.012	.005	.010	.004	.007	.003	.006	.003	.005	.002	26
.020	.009	.015	.007	.012	.005	.010	.004	.007	.003	.006	.003	.005	.002	27
.020	.009	.015	.007	.012	.005	.010	.004	.007	.003	.006	.003	.005	.002	28
.020	.009	.015	.007	.012	.005	.010	.004	.007	.003	.006	.003	.005	.002	29
.020	.009	.015	.007	.012	.005	.010	.004	.007	.003	.006	.003	.005	.002	30
.020	.009	.015	.007	.012	.005	.010	.004	.007	.003	.006	.003	.005	.002	31
.020	.009	.015	.007	.012	.005	.010	.004	.007	.003	.006	.003	.005	.002	32
.020	.009	.015	.007	.012	.005	.010	.004	.007	.003	.006	.003	.005	.002	33
.020	.009	.015	.007	.012	.005	.010	.004	.007	.003	.006	.003	.005	.002	34
.020	.009	.015	.007	.012	.005	.010	.004	.007	.003	.006	.003	.005	.002	35

continued on next page

Number of Teeth	Diametral Pitch 1/2		2.5/5		3/6		4/8		5/10		6/12		8/16		10/20	
	Major Diameter Chamfer															
	Dim	Hgt	Dim	Hgt	Dim	Hgt	Dim	Hgt	Dim	Hgt	Dim	Hgt	Dim	Hgt	Dim	Hgt
1	2	3	4	5	6	7	8	9	10	11	12	13	14	15	16	17
	Approx	Min	Approx	Min	Approx	Min	Approx	Min	Approx	Min	Approx	Min	Approx	Min	Approx	Min
36	.240	.106	.096	.043	.080	.035	.060	.027	.048	.021	.040	.018	.030	.013	.024	.011
37	.240	.107	.096	.043	.080	.036	.060	.027	.048	.021	.040	.018	.030	.013	.024	.011
38	.240	.107	.096	.043	.080	.036	.060	.027	.048	.021	.040	.018	.030	.013	.024	.011
39	.240	.107	.096	.043	.080	.036	.060	.027	.048	.021	.040	.018	.030	.013	.024	.011
40	.240	.107	.096	.043	.080	.036	.060	.027	.048	.021	.040	.018	.030	.013	.024	.011
41	.240	.107	.096	.043	.080	.036	.060	.027	.048	.021	.040	.018	.030	.013	.024	.011
42	.240	.107	.096	.043	.080	.036	.060	.027	.048	.021	.040	.018	.030	.013	.024	.011
43	.240	.107	.096	.043	.080	.036	.060	.027	.048	.021	.040	.018	.030	.013	.024	.011
44	.240	.107	.096	.043	.080	.036	.060	.027	.048	.021	.040	.018	.030	.013	.024	.011
45	.240	.107	.096	.043	.080	.036	.060	.027	.048	.021	.040	.018	.030	.013	.024	.011
46	.240	.107	.096	.043	.080	.036	.060	.027	.048	.021	.040	.018	.030	.013	.024	.011
47	.240	.107	.096	.043	.080	.036	.060	.027	.048	.021	.040	.018	.030	.013	.024	.011
48	.240	.107	.096	.043	.080	.036	.060	.027	.048	.021	.040	.018	.030	.013	.024	.011
49	.240	.107	.096	.043	.080	.036	.060	.027	.048	.021	.040	.018	.030	.013	.024	.011
50	.240	.107	.096	.043	.080	.036	.060	.027	.048	.021	.040	.018	.030	.013	.024	.011

TABLE 10-12

Chamfer Data – Internal Involute Splines 1954 SAE Handbook

Number of Teeth	Diametral Pitch 1/2		2.5/5		3/6		4/8		5/10		6/12		8/16		10/20	
	Minor Diameter Chamfer															
	Ang	Hgt	Ang	Hgt	Ang	Hgt	Ang	Hgt	Ang	Hgt	Ang	Hgt	Ang	Hgt	Ang	Hgt
1	2	3	4	5	6	7	8	9	10	11	12	13	14	15	16	17
	D°-M′	Min	D°-M′	Min	D°-M′	Min	D°-M′	Min	D°-M′	Min	D°-M′	Min	D°-M′	Min	D°-M′	Min
6	53-37	.202	53-37	.081	53-37	.067	53-37	.051	53-37	.040	53-37	.034	53-37	.025	53-37	.020
7	53-34	.180	53-34	.072	53-34	.060	53-34	.045	53-34	.036	53-34	.030	53-34	.022	53-34	.018
8	53-31	.164	53-31	.066	53-31	.055	53-31	.041	53-31	.033	53-31	.027	53-31	.020	53-31	.016
9	53-29	.152	53-29	.061	53-29	.051	53-29	.038	53-29	.030	53-29	.025	53-29	.019	53-29	.015
10	53-29	.143	53-29	.057	53-29	.048	53-29	.036	53-29	.029	53-29	.024	53-29	.018	53-29	.014
11	53-28	.135	53-28	.054	53-28	.045	53-28	.034	53-28	.027	53-28	.023	53-28	.017	53-28	.014
12	53-29	.129	53-29	.052	53-29	.043	53-29	.032	53-29	.026	53-29	.022	53-29	.016	53-29	.013
13	53-29	.124	53-29	.049	53-29	.041	53-29	.031	53-29	.025	53-29	.021	53-29	.016	53-29	.012
14	53-30	.119	53-30	.048	53-30	.040	53-30	.030	53-30	.024	53-30	.020	53-30	.015	53-30	.012
15	53-31	.115	53-31	.046	53-31	.038	53-31	.029	53-31	.023	53-31	.019	53-31	.014	53-31	.012

continued on next page

TABLE 10–11, continued

Diametral Pitch 12/24 Dim 18 Approx	12/24 Hgt 19 Min	16/32 Dim 20 Approx	16/32 Hgt 21 Min	20/40 Dim 22 Approx	20/40 Hgt 23 Min	24/48 Dim 24 Approx	24/48 Hgt 25 Min	32/64 Dim 26 Approx	32/64 Hgt 27 Min	40/80 Dim 28 Approx	40/80 Hgt 29 Min	48/96 Dim 30 Approx	48/96 Hgt 31 Min	Number of Teeth 1
						Major Diameter Chamfer								
.020	.009	.015	.007	.012	.005	.010	.004	.008	.003	.006	.003	.005	.002	36
.020	.009	.015	.007	.012	.005	.010	.004	.008	.003	.006	.003	.005	.002	37
.020	.009	.015	.007	.012	.005	.010	.004	.008	.003	.006	.003	.005	.002	38
.020	.009	.015	.007	.012	.005	.010	.004	.008	.003	.006	.003	.005	.002	39
.020	.009	.015	.007	.012	.005	.010	.004	.008	.003	.006	.003	.005	.002	40
.020	.009	.015	.007	.012	.005	.010	.004	.008	.003	.006	.003	.005	.002	41
.020	.009	.015	.007	.012	.005	.010	.004	.008	.003	.006	.003	.005	.002	42
.020	.009	.015	.007	.012	.005	.010	.004	.008	.003	.006	.003	.005	.002	43
.020	.009	.015	.007	.012	.005	.010	.004	.008	.003	.006	.003	.005	.002	44
.020	.009	.015	.007	.012	.005	.010	.004	.008	.003	.006	.003	.005	.002	45
.020	.009	.015	.007	.012	.005	.010	.004	.008	.003	.006	.003	.005	.002	46
.020	.009	.015	.007	.012	.005	.010	.004	.008	.003	.006	.003	.005	.002	47
.020	.009	.015	.007	.012	.005	.010	.004	.008	.003	.006	.003	.005	.002	48
.020	.009	.015	.007	.012	.005	.010	.004	.008	.003	.006	.003	.005	.002	49
.020	.009	.015	.007	.012	.005	.010	.004	.008	.003	.006	.003	.005	.002	50

TABLE 10–12, continued

Diametral Pitch 12/24 Ang 18 D°-M′	12/24 Hgt 19 Min	16/32 Ang 20 D°-M′	16/32 Hgt 21 Min	20/40 Ang 22 D°-M′	20/40 Hgt 23 Min	24/48 Ang 24 D°-M′	24/48 Hgt 25 Min	32/64 Ang 26 D°-M′	32/64 Hgt 27 Min	40/80 Ang 28 D°-M′	40/80 Hgt 29 Min	48/96 Ang 30 D°-M′	48/96 Hgt 31 Min	Number of Teeth 1
						Minor Diameter Chamfer								
53-37	.017	53-37	.013	53-37	.010	53-37	.008	53-37	.006	53-37	.005	53-37	.004	6
53-34	.015	53-34	.011	53-34	.009	53-34	.008	53-34	.006	53-34	.005	53-34	.004	7
53-31	.014	53-31	.010	53-31	.008	53-31	.007	53-31	.005	53-31	.004	53-31	.003	8
53-29	.013	53-29	.010	53-29	.008	53-29	.006	53-29	.005	53-29	.004	53-29	.003	9
53-29	.012	53-29	.009	53-29	.007	53-29	.006	53-29	.004	53-29	.004	53-29	.003	10
53-28	.011	53-28	.008	53-28	.007	53-28	.006	53-28	.004	53-28	.003	53-28	.003	11
53-29	.011	53-29	.008	53-29	.006	53-29	.005	53-29	.004	53-29	.003	53-29	.003	12
53-29	.010	53-29	.008	53-29	.006	53-29	.005	53-29	.004	53-29	.003	53-29	.003	13
53-30	.010	53-30	.007	53-30	.006	53-30	.005	53-30	.004	53-30	.003	53-30	.002	14
53-31	.010	53-31	.007	53-31	.006	53-31	.005	53-31	.004	53-31	.003	53-31	.002	15

continued on next page

TABLE 10–12, continued

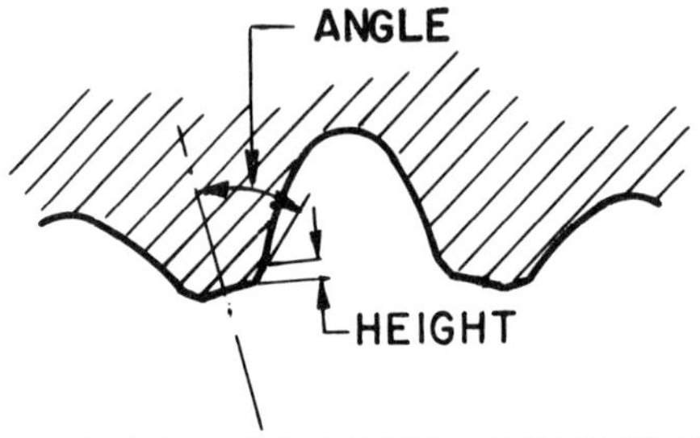

Number of Teeth	Diametral Pitch 1/2		2.5/5		3/6		4/8		5/10		6/12		8/16		10/20	
	Minor Diameter Chamfer															
	Ang	Hgt	Ang	Hgt	Ang	Hgt	Ang	Hgt	Ang	Hgt	Ang	Hgt	Ang	Hgt	Ang	Hgt
1	2	3	4	5	6	7	8	9	10	11	12	13	14	15	16	17
	D°-M′	Min	D°-M′	Min	D°-M′	Min	D°-M′	Min	D°-M′	Min	D°-M′	Min	D°-M′	Min	D°-M′	Min
16	53-31	.112	53-31	.045	53-31	.037	53-31	.028	53-31	.022	53-31	.019	53-31	.014	53-31	.011
17	53-32	.109	53-32	.044	53-32	.036	53-32	.027	53-32	.022	53-32	.018	53-32	.014	53-32	.011
18	53-32	.106	53-32	.043	53-32	.035	53-32	.027	53-32	.021	53-32	.018	53-32	.013	53-32	.011
19	53-33	.104	53-33	.042	53-33	.035	53-33	.026	53-33	.021	53-33	.017	53-33	.013	53-33	.010
20	53-34	.102	53-34	.041	53-34	.034	53-34	.026	53-34	.020	53-34	.017	53-34	.013	53-34	.010
21	53-35	.100	53-35	.040	53-35	.033	53-35	.025	53-35	.020	53-35	.017	53-35	.013	53-35	.010
22	53-36	.098	53-36	.039	53-36	.033	53-36	.025	53-36	.020	53-36	.016	53-36	.012	53-36	.010
23	53-37	.097	53-37	.039	53-37	.032	53-37	.024	53-37	.019	53-37	.016	53-37	.012	53-37	.010
24	53-38	.096	53-38	.038	53-38	.032	53-38	.024	53-38	.019	53-38	.016	53-38	.012	53-38	.010
25	53-38	.094	53-38	.038	53-38	.031	53-38	.024	53-38	.019	53-38	.016	53-38	.012	53-38	.009
26	53-39	.093	53-39	.037	53-39	.031	53-39	.023	53-39	.019	53-39	.016	53-39	.012	53-39	.009
27	53-40	.092	53-40	.037	53-40	.031	53-39	.023	53-40	.018	53-40	.015	53-40	.011	53-40	.009
28	53-40	.091	53-40	.036	53-40	.030	53-40	.023	53-40	.018	53-40	.015	53-40	.011	53-40	.009
29	53-41	.090	53-41	.036	53-41	.030	53-41	.023	53-41	.018	53-41	.015	53-41	.011	53-41	.009
30	53-41	.089	53-41	.036	53-41	.030	53-41	.022	53-41	.018	53-41	.015	53-41	.011	53-41	.009
31	53-41	.088	53-41	.035	53-41	.029	53-41	.022	53-41	.018	53-41	.015	53-41	.011	53-41	.009
32	53-42	.088	53-42	.035	53-42	.029	53-42	.022	53-42	.018	53-42	.015	53-42	.011	53-42	.009
33	53-43	.087	53-43	.035	53-43	.029	53-43	.022	53-43	.017	53-43	.015	53-43	.011	53-43	.009
34	53-43	.086	53-43	.035	53-43	.029	53-43	.022	53-43	.017	53-43	.014	53-43	.011	53-43	.009
35	53-44	.086	53-44	.034	53-44	.029	53-44	.021	53-44	.017	53-44	.014	53-44	.011	53-44	.009
36	53-44	.085	53-44	.034	53-44	.028	53-44	.021	53-44	.017	53-44	.014	53-44	.011	53-44	.009
37	53-44	.085	53-44	.034	53-44	.028	53-44	.021	53-44	.017	53-44	.014	53-44	.011	53-44	.009
38	53-45	.084	53-45	.034	53-45	.028	53-45	.021	53-45	.017	53-45	.014	53-45	.011	53-45	.008
39	53-46	.084	53-46	.034	53-46	.028	53-46	.021	53-46	.017	53-46	.014	53-46	.010	53-46	.008
40	53-46	.083	53-46	.033	53-46	.028	53-46	.021	53-46	.017	53-46	.014	53-46	.010	53-46	.008
41	53-46	.083	53-46	.033	53-46	.028	53-46	.021	53-46	.017	53-46	.014	53-46	.010	53-46	.008
42	53-47	.082	53-47	.033	53-47	.027	53-47	.021	53-47	.016	53-47	.014	53-47	.010	53-47	.008
43	53-47	.082	53-47	.033	53-47	.027	53-47	.020	53-47	.016	53-47	.014	53-47	.010	53-47	.008
44	53-47	.081	53-47	.032	53-47	.027	53-47	.020	53-47	.016	53-47	.014	53-47	.010	53-47	.008
45	53-48	.081	53-48	.032	53-48	.027	53-48	.020	53-48	.016	53-48	.014	53-48	.010	53-48	.008
46	53-49	.080	53-49	.032	53-49	.027	53-49	.020	53-49	.016	53-49	.013	53-49	.010	53-49	.008
47	53-49	.080	53-49	.032	53-49	.027	53-49	.020	53-49	.016	53-49	.013	53-49	.010	53-49	.008
48	53-49	.080	53-49	.032	53-49	.027	53-49	.020	53-49	.016	53-49	.013	53-49	.010	53-49	.008
49	53-50	.079	53-50	.032	53-50	.026	53-50	.020	53-50	.016	53-50	.013	53-50	.010	53-50	.008
50	53-50	.079	53-50	.032	53-50	.026	53-50	.020	53-50	.016	53-50	.013	53-50	.010	53-50	.008

continued on next page

TABLE 10–12, continued

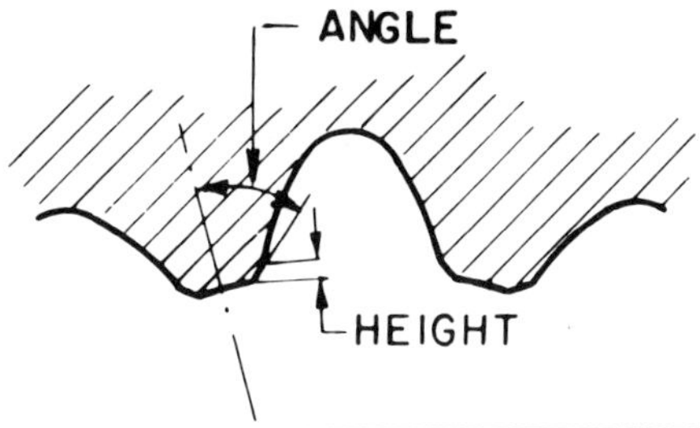

Diametral Pitch														
12/24		16/32		20/40		24/48		32/64		40/80		48/96		Number of Teeth
Minor Diameter Chamfer														
Ang 18 D°-M′	Hgt 19 Min	Ang 20 D°-M′	Hgt 21 Min	Ang 22 D°-M′	Hgt 23 Min	Ang 24 D°-M′	Hgt 25 Min	Ang 26 D°-M′	Hgt 27 Min	Ang 28 D°-M′	Hgt 29 Min	Ang 30 D°-M′	Hgt 31 Min	1
53-31	.009	53-31	.007	53-31	.006	53-31	.005	53-31	.003	53-31	.003	53-31	.002	16
53-32	.009	53-32	.007	53-32	.005	53-32	.005	53-32	.003	53-32	.003	53-32	.002	17
53-32	.009	53-32	.007	53-32	.005	53-32	.004	53-32	.003	53-32	.003	53-32	.002	18
53-33	.009	53-33	.007	53-33	.005	53-33	.004	53-33	.003	53-33	.003	53-33	.002	19
53-34	.009	53-34	.006	53-34	.005	53-34	.004	53-34	.003	53-34	.003	53-34	.002	20
53-35	.008	53-35	.006	53-35	.005	53-35	.004	53-35	.003	53-35	.003	53-35	.002	21
53-36	.008	53-36	.006	53-36	.005	53-36	.004	53-36	.003	53-36	.002	53-36	.002	22
53-37	.008	53-37	.006	53-37	.005	53-37	.004	53-37	.003	53-37	.002	53-37	.002	23
53-38	.008	53-38	.006	53-38	.005	53-38	.004	53-38	.003	53-38	.002	53-38	.002	24
53-38	.008	53-38	.006	53-38	.005	53-38	.004	53-38	.003	53-38	.002	53-38	.002	25
53-39	.008	53-39	.006	53-39	.005	53-39	.004	53-39	.003	53.39	.002	53-39	.002	26
53-40	.008	53-40	.006	53-40	.005	53-40	.004	53-40	.003	53-40	.002	53-40	.002	27
53-40	.008	53-40	.006	53-40	.005	53-40	.004	53-40	.003	53-40	.002	53-40	.002	28
53-41	.008	53-41	.006	53-41	.004	53-41	.004	53-41	.003	53-41	.002	53-41	.002	29
53-41	.007	53-41	.006	53-41	.004	53-41	.004	53-41	.003	53-41	.002	53-41	.002	30
53-41	.007	53-41	.006	53-41	.004	53-41	.004	53-41	.003	53-41	.002	53-41	.002	31
53-42	.007	53-42	.006	53-42	.004	53-42	.004	53-42	.003	53-42	.002	53-42	.002	32
53-43	.007	53-43	.005	53-43	.004	53-43	.004	53-43	.003	53-43	.002	53-43	.002	33
53-43	.007	53-43	.005	53-43	.004	53-43	.004	53-43	.003	53-43	.002	53-43	.002	34
53-44	.007	53-44	.005	53-44	.004	53-44	.004	53-44	.003	53-44	.002	53-44	.002	35
53-44	.007	53-44	.005	53-44	.004	53-44	.004	53-44	.003	53-44	.002	53-44	.002	36
53-44	.007	53-44	.005	53-44	.004	53-44	.004	53-44	.003	53-44	.002	53-44	.002	37
53-45	.007	53-45	.005	53-45	.004	53-45	.003	53-45	.003	53-45	.002	53-45	.002	38
53-46	.007	53-46	.005	53-46	.004	53-46	.003	53-46	.003	53-46	.002	53-46	.002	39
53-46	.007	53-46	.005	53-46	.004	53-46	.003	53-46	.003	53-46	.002	53-46	.002	40
53-46	.007	53-46	.005	53-46	.004	53-46	.003	53-46	.003	53-46	.002	53-46	.002	41
53-47	.007	53-47	.005	53-47	.004	53-47	.003	53-47	.003	53-47	.002	53-47	.002	42
53-47	.007	53-47	.005	53-47	.004	53-47	.003	53-47	.003	53-47	.002	53-47	.002	43
53-47	.007	53-47	.005	53-47	.004	53-47	.003	53-47	.003	53-47	.002	53-47	.002	44
53-48	.007	53-48	.005	53-48	.004	53-48	.003	53-48	.003	53-48	.002	53-48	.002	45
53-49	.007	53-49	.005	53-49	.004	53-49	.003	53-49	.003	53-49	.002	53-49	.002	46
53-49	.007	53-49	.005	53-49	.004	53-49	.003	53-49	.003	53-49	.002	53-49	.002	47
53-49	.007	53-49	.005	53-49	.004	53-49	.003	53-49	.003	53-49	.002	53-49	.002	48
53-50	.007	53-50	.005	53-50	.004	53-50	.003	53-50	.002	53-50	.002	53-50	.002	49
53-50	.007	53-50	.005	53-50	.004	53-50	.003	53-50	.002	53-50	.002	53-50	.002	50

TABLE 10–13

Pin Diameters & Measurements over Pins for External Involute Splines

1954 SAE Handbook

N	DIAMETRAL PITCH														
	1/2	2.5/5	3/6	4/8	5/10	6/12	8/16	10/20	12/24	16/32	20/40	24/48	32/64	40/80	48/96
6	8.8660	3.5464	2.9553	2.2165	1.7732	1.4777	1.1083	0.8866	0.7388	0.5541	0.4433	0.3694	0.2771	0.2217	0.1847
7	9.6819	3.8728	3.2273	2.4205	1.9364	1.6137	1.2102	0.9682	0.8068	0.6051	0.4841	0.4034	0.3026	0.2421	0.2017
8	10.8944	4.3578	3.6315	2.7236	2.1789	1.8157	1.3618	1.0894	0.9079	0.6809	0.5447	0.4539	0.3405	0.2724	0.2270
9	11.7535	4.7014	3.9178	2.9384	2.3507	1.9589	1.4692	1.1754	0.9795	0.7346	0.5877	0.4897	0.3673	0.2938	0.2449
10	12.9144	5.1658	4.3048	3.2286	2.5829	2.1524	1.6143	1.2914	1.0762	0.8072	0.6457	0.5381	0.4036	0.3229	0.2690
11	13.8003	5.5201	4.6001	3.4501	2.7601	2.3001	1.7250	1.3800	1.1500	0.8625	0.6900	0.5750	0.4313	0.3450	0.2875
12	14.9295	5.9718	4.9765	3.7324	2.9859	2.4883	1.8662	1.4930	1.2441	0.9331	0.7465	0.6221	0.4665	0.3732	0.3110
13	15.8335	6.3334	5.2778	3.9584	3.1667	2.6389	1.9792	1.5834	1.3195	0.9896	0.7917	0.6597	0.4948	0.3958	0.3299
14	16.9412	6.7765	5.6471	4.2353	3.3882	2.8235	2.1177	1.6941	1.4118	1.0588	0.8471	0.7059	0.5294	0.4235	0.3529
15	17.8584	7.1434	5.9528	4.4646	3.5717	2.9764	2.2323	1.7858	1.4882	1.1162	0.8929	0.7441	0.5581	0.4465	0.3721
16	18.9507	7.5803	6.3169	4.7377	3.7901	3.1585	2.3688	1.8951	1.5792	1.1844	0.9475	0.7896	0.5922	0.4738	0.3948
17	19.8778	7.9511	6.6259	4.9695	3.9756	3.3130	2.4847	1.9878	1.6565	1.2424	0.9939	0.8282	0.6212	0.4970	0.4141
18	20.9584	8.3834	6.9861	5.2396	4.1917	3.4931	2.6198	2.0958	1.7465	1.3099	1.0479	0.8733	0.6550	0.5240	0.4366
19	21.8934	8.7574	7.2978	5.4734	4.3787	3.6489	2.7367	2.1893	1.8245	1.3683	1.0947	0.9122	0.6842	0.5473	0.4561
20	22.9649	9.1860	7.6550	5.7412	4.5930	3.8275	2.8706	2.2965	1.9137	1.4353	1.1482	0.9569	0.7177	0.5741	0.4784
21	23.9062	9.5625	7.9687	5.9766	4.7812	3.9844	2.9883	2.3906	1.9922	1.4941	1.1953	0.9961	0.7471	0.5977	0.4980
22	24.9704	9.9882	8.3235	6.2426	4.9941	4.1617	3.1213	2.4970	2.0809	1.5607	1.2485	1.0404	0.7803	0.6243	0.5202
23	25.9168	10.3667	8.6389	6.4792	5.1834	4.3195	3.2396	2.5917	2.1597	1.6198	1.2958	1.0799	0.8099	0.6479	0.5399
24	26.9752	10.7901	8.9917	6.7438	5.3950	4.4959	3.3719	2.6975	2.2479	1.6860	1.3488	1.1240	0.8430	0.6744	0.5620
25	27.9259	11.1704	9.3086	6.9815	5.5852	4.6543	3.4907	2.7926	2.3272	1.7454	1.3963	1.1636	0.8727	0.6982	0.5818
26	28.9793	11.5917	9.6598	7.2448	5.7959	4.8299	3.6224	2.8979	2.4149	1.8112	1.4490	1.2075	0.9056	0.7245	0.6037
27	29.9337	11.9735	9.9779	7.4834	5.9867	4.9890	3.7417	2.9934	2.4945	1.8709	1.4967	1.2472	0.9354	0.7483	0.6236
28	30.9829	12.3932	10.3276	7.7457	6.1966	5.1638	3.8729	3.0983	2.5819	1.9364	1.5491	1.2910	0.9682	0.7746	0.6455
29	31.9405	12.7762	10.6468	7.9851	6.3881	5.3234	3.9926	3.1941	2.6617	1.9963	1.5970	1.3309	0.9981	0.7985	0.6654
30	32.9862	13.1945	10.9954	8.2466	6.5972	5.4977	4.1233	3.2986	2.7489	2.0616	1.6493	1.3744	1.0308	0.8247	0.6872
31	33.9465	13.5786	11.3155	8.4866	6.7893	5.6578	4.2433	3.3947	2.8289	2.1217	1.6973	1.4144	1.0608	0.8487	0.7072
32	34.9890	13.9956	11.6630	8.7473	6.9978	5.8315	4.3736	3.4989	2.9158	2.1868	1.7495	1.4579	1.0934	0.8747	0.7289
33	35.9517	14.3807	11.9839	8.9879	7.1903	5.9920	4.4940	3.5952	2.9960	2.2470	1.7976	1.4980	1.1235	0.8988	0.7490
34	36.9916	14.7966	12.3305	9.2479	7.3983	6.1653	4.6240	3.6992	3.0826	2.3120	1.8496	1.5413	1.1560	0.9248	0.7707
35	37.9565	15.1826	12.6522	9.4891	7.5913	6.3261	4.7446	3.7957	3.1630	2.3723	1.8978	1.5815	1.1861	0.9489	0.7908
36	38.9939	15.5976	12.9980	9.7485	7.7988	6.4990	4.8742	3.8994	3.2495	2.4371	1.9497	1.6247	1.2186	0.9749	0.8124
37	39.9607	15.9843	13.3202	9.9902	7.9921	6.6601	4.9951	3.9961	3.3301	2.4975	1.9980	1.6650	1.2488	0.9990	0.8325
38	40.9960	16.3984	13.6653	10.2490	8.1992	6.8327	5.1245	4.0996	3.4163	2.5623	2.0498	1.7082	1.2811	1.0249	0.8541
39	41.9645	16.7858	13.9882	10.4911	8.3929	6.9941	5.2456	4.1965	3.4970	2.6228	2.0982	1.7485	1.3114	1.0491	0.8743
40	42.9980	17.1992	14.3327	10.7495	8.5996	7.1663	5.3748	4.2998	3.5832	2.6874	2.1499	1.7916	1.3437	1.0750	0.8958
41	43.9680	17.5872	14.6560	10.9920	8.7936	7.3280	5.4960	4.3968	3.6640	2.7480	2.1984	1.8320	1.3740	1.0992	0.9160
42	44.9996	17.9998	14.9999	11.2499	8.9999	7.4999	5.6245	4.5000	3.7500	2.8125	2.2500	1.8750	1.4062	1.1250	0.9375
43	45.9711	18.3884	15.3237	11.4928	9.1942	7.6619	5.7464	4.5971	3.8309	2.8732	2.2986	1.9155	1.4366	1.1493	0.9577
44	47.0013	18.8005	15.6671	11.7503	9.4003	7.8336	5.8752	4.7001	3.9168	2.9376	2.3501	1.9584	1.4688	1.1750	0.9792
45	47.9740	19.1896	15.9913	11.9935	9.5948	7.9957	5.9968	4.7974	3.9978	2.9984	2.3987	1.9989	1.4992	1.1994	0.9995
46	49.0028	19.6011	16.3343	12.2507	9.8006	8.1671	6.1253	4.9003	4.0836	3.0627	2.4501	2.0418	1.5313	1.2251	1.0209
47	49.9766	19.9906	16.6589	12.4942	9.9953	8.3294	6.2471	4.9977	4.1647	3.1235	2.4988	2.0824	1.5618	1.2494	1.0412
48	51.0041	20.4016	17.0014	12.7510	10.2008	8.5007	6.3755	5.1004	4.2503	3.1878	2.5502	2.1252	1.5939	1.2751	1.0626
49	51.9790	20.7916	17.3263	12.9948	10.3958	8.6632	6.4974	5.1979	4.3316	3.2487	2.5990	2.1658	1.6245	1.2995	1.0829
50	53.0054	21.2022	17.6685	13.2514	10.6011	8.8343	6.6257	5.3006	4.4171	3.3128	2.6503	2.2086	1.6564	1.3251	1.1043

Pin diameter, $d_e = \frac{1.9200}{P}$

TABLE 10–14

Pin Diameters and Measurements between Pins for Internal Involute Splines

1954 SAE Handbook

N	DIAMETRAL PITCH														
	1/2	2.5/5	3/6	4/8	5/10	6/12	8/16	10/20	12/24	16/32	20/40	24/48	32/64	40/80	.48/96
6	4.3933	1.7573	1.4644	1.0983	0.8787	0.7322	0.5492	0.4393	0.3661	0.2746	0.2197	0.1831	0.1373	0.1098	0.0915
7	5.2232	2.0892	1.7410	1.3058	1.0446	0.8705	0.6529	0.5223	0.4353	0.3264	0.2612	0.2176	0.1632	0.1306	0.1088
8	6.3954	2.5582	2.1318	1.5989	1.2791	1.0659	0.7994	0.6395	0.5329	0.3997	0.3198	0.2665	0.1999	0.1599	0.1332
9	7.2618	2.9047	2.4206	1.8155	1.4524	1.2103	0.9077	0.7262	0.6052	0.4539	0.3631	0.3026	0.2269	0.1815	0.1513
10	8.3966	3.3586	2.7989	2.0992	1.6793	1.3994	1.0496	0.8397	0.6997	0.5248	0.4198	0.3499	0.2624	0.2099	0.1749
11	9.2867	3.7147	3.0956	2.3217	1.8573	1.5478	1.1608	0.9287	0.7739	0.5804	0.4643	0.3869	0.2902	0.2322	0.1935
12	10.3973	4.1589	3.4658	2.5993	2.0795	1.7329	1.2997	1.0397	0.8664	0.6498	0.5199	0.4332	0.3249	0.2599	0.2166
13	11.3040	4.5216	3.7680	2.8260	2.2608	1.8840	1.4130	1.1304	0.9420	0.7065	0.5652	0.4710	0.3533	0.2826	0.2355
14	12.3978	4.9591	4.1326	3.0995	2.4796	2.0663	1.5497	1.2398	1.0332	0.7749	0.6199	0.5166	0.3874	0.3099	0.2583
15	13.3167	5.3267	4.4389	3.3292	2.6633	2.2195	1.6646	1.3317	1.1097	0.8323	0.6658	0.5549	0.4161	0.3329	0.2774
16	14.3982	5.7593	4.7994	3.5996	2.8796	2.3997	1.7998	1.4398	1.1999	0.8999	0.7199	0.5999	0.4499	0.3600	0.3000
17	15.3265	6.1306	5.1089	3.8317	3.0653	2.5544	1.9158	1.5327	1.2772	0.9579	0.7663	0.6386	0.4790	0.3832	0.3193
18	16.3985	6.5594	5.4662	4.0996	3.2797	2.7331	2.0498	1.6399	1.3665	1.0249	0.8199	0.6833	0.5125	0.4100	0.3416
19	17.3343	6.9337	5.7781	4.3336	3.4669	2.8891	2.1668	1.7335	1.4445	1.0834	0.8667	0.7223	0.5417	0.4334	0.3611
20	18.3987	7.3595	6.1329	4.5997	3.6797	3.0665	2.2998	1.8399	1.5332	1.1499	0.9199	0.7666	0.5750	0.4600	0.3833
21	19.3405	7.7362	6.4468	4.8351	3.8681	3.2234	2.4176	1.9341	1.6118	1.2088	0.9670	0.8059	0.6044	0.4835	0.4029
22	20.3989	8.1596	6.7996	5.0997	4.0798	3.3998	2.5499	2.0399	1.6999	1.2749	1.0199	0.8500	0.6375	0.5100	0.4250
23	21.3457	8.5383	7.1152	5.3364	4.2691	3.5576	2.6682	2.1346	1.7788	1.3341	1.0673	0.8894	0.6671	0.5337	0.4447
24	22.3991	8.9596	7.4664	5.5998	4.4798	3.7332	2.7999	2.2399	1.8666	1.3999	1.1199	0.9333	0.7000	0.5600	0.4666
25	23.3501	9.3400	7.7834	5.8375	4.6700	3.8917	2.9188	2.3350	1.9459	1.4595	1.1676	0.9730	0.7297	0.5838	0.4865
26	24.3992	9.7597	8.1331	6.0998	4.8798	4.0665	3.0499	2.4399	2.0333	1.5250	1.2200	1.0166	0.7625	0.6100	0.5083
27	25.3538	10.1415	8.4513	6.3385	5.0708	4.2256	3.1692	2.5354	2.1128	1.5846	1.2677	1.0564	0.7923	0.6339	0.5282
28	26.3993	10.5597	8.7998	6.5998	5.2799	4.3999	3.2999	2.6399	2.1999	1.6500	1.3200	1.1000	0.8250	0.6600	0.5500
29	27.3571	10.9428	9.1190	6.8393	5.4714	4.5595	3.4196	2.7357	2.2797	1.7098	1.3679	1.1399	0.8549	0.6840	0.5700
30	28.3994	11.3598	9.4665	7.0999	5.6799	4.7332	3.5499	2.8399	2.3666	1.7750	1.4200	1.1833	0.8875	0.7100	0.5917
31	29.3599	11.7440	9.7866	7.3400	5.8720	4.8933	3.6700	2.9360	2.4467	1.8350	1.4680	1.2234	0.9175	0.7340	0.6117
32	30.3995	12.1598	10.1332	7.5999	6.0799	5.0666	3.7999	3.0400	2.5333	1.9000	1.5200	1.2666	0.9500	0.7600	0.6333
33	31.3623	12.5449	10.4541	7.8406	6.2725	5.2271	3.9203	3.1363	2.6136	1.9602	1.5681	1.3068	0.9801	0.7841	0.6534
34	32.3995	12.9598	10.7998	8.0999	6.4799	5.3999	4.0499	3.2400	2.7000	2.0250	1.6200	1.3500	1.0125	0.8100	0.6750
35	33.3645	13.3458	11.1215	8.3411	6.6729	5.5608	4.1706	3.3365	2.7804	2.0853	1.6682	1.3902	1.0426	0.8341	0.6951
36	34.3996	13.7598	11.4665	8.5999	6.8799	5.7333	4.3000	3.4400	2.8666	2.1500	1.7200	1.4333	1.0750	0.8600	0.7167
37	35.3665	14.1466	11.7888	8.8416	7.0733	5.8944	4.4208	3.5366	2.9472	2.2104	1.7683	1.4736	1.1052	0.8842	0.7368
38	36.3996	14.5598	12.1332	9.0999	7.2799	6.0666	4.5500	3.6400	3.0333	2.2750	1.8200	1.5166	1.1375	0.9100	0.7583
39	37.3682	14.9473	12.4561	9.3421	7.4736	6.2280	4.6710	3.7368	3.1140	2.3355	1.8684	1.5570	1.1677	0.9342	0.7785
40	38.3997	15.3599	12.7999	9.5999	7.6799	6.4000	4.8000	3.8400	3.2000	2.4000	1.9200	1.6000	1.2000	0.9600	0.8000
41	39.3698	15.7479	13.1233	9.8425	7.8740	6.5616	4.9212	3.9370	3.2808	2.4606	1.9685	1.6404	1.2303	0.9842	0.8202
42	40.3997	16.1599	13.4666	10.0999	8.0799	6.7333	5.0500	4.0400	3.3666	2.5250	2.0200	1.6833	1.2625	1.0100	0.8417
43	41.3712	16.5485	13.7904	10.3428	8.2742	6.8952	5.1714	4.1371	3.4476	2.5857	2.0685	1.7238	1.2928	1.0343	0.8619
44	42.3998	16.9599	14.1333	10.6000	8.4800	7.0666	5.3000	4.2400	3.5333	2.6500	2.1200	1.7667	1.3250	1.0600	0.8833
45	43.3725	17.3490	14.4575	10.8431	8.6745	7.2288	5.4216	4.3373	3.6144	2.7108	2.1686	1.8072	1.3554	1.0843	0.9036
46	44.3998	17.7599	14.7999	11.1000	8.8800	7.4000	5.5500	4.4400	3.7000	2,7750	2.2200	1.8500	1.3875	1.1100	0.9250
47	45.3737	18.1495	15.1246	11.3434	9.0747	7.5623	5.6717	4.5373	3.7811	2.8338	2.2687	1.8905	1.4179	1.1343	0.9453
48	46.3999	18.5600	15.4666	11.6000	9.2800	7.7333	5.8000	4.6400	3.8667	2.9000	2.3200	1.9333	1.4500	1.1600	0.9667
49	47.3748	18.9499	15.7916	11.8437	9.4750	7.8958	5.9218	4.7375	3.9479	2.9609	2.3687	1.9739	1.5805	1.1844	0.9870
50	48.3999	19.3600	16.1333	12.1000	9.6800	8.0667	6.0500	4.8400	4.0333	3.0250	2.4200	2.0167	1.5125	1.2100	1.0083

Pin diameter, $d_s = \frac{1.9200}{P}$

TABLE 10-15
Involute Spline Selections Based on Standard Ball Bearing Bores
1954 SAE Handbook

BEARING BORE		DIAMETRAL PITCH																											
		2.5/5		3/6		4/8		5/10		6/12		8/16		10/20		12/24		16/32		20/40		24/48		32/64		40/80		48/96	
Mm	Inches	*N*	Diam	*N*	Diam	*N*	Diam	*N*	Diam	*N*	Diam	*N*	Diam	*N*	Diam	*N*	Diam	*N*	Diam	*N*	Diam	*N*	Diam	*N*	Diam	*N*	Diam	*N*	Diam
—	0.1900	—	—	—	—	—	—	—	—	—	—	—	—	—	—	—	—	—	—	—	—	—	—	—	—	6	0.175	8	0.188
—	0.2500	—	—	—	—	—	—	—	—	—	—	—	—	—	—	—	—	—	—	—	—	—	—	6	0.219	8	0.225	10	0.229
—	0.3125	—	—	—	—	—	—	—	—	—	—	—	—	—	—	—	—	—	—	—	—	6	0.292	8	0.281	11	0.300	13	0.292
—	0.3750	—	—	—	—	—	—	—	—	—	—	—	—	—	—	—	—	—	—	6	0.350	7	0.333	10	0.344	13	0.350	16	0.354
10	0.3937	—	—	—	—	—	—	—	—	—	—	—	—	—	—	—	—	—	—	6	0.350	8	0.375	11	0.375	14	0.375	17	0.375
12	0.4724	—	—	—	—	—	—	—	—	—	—	—	—	—	—	—	—	6	0.438	8	0.450	10	0.458	14	0.469	17	0.450	21	0.458
—	0.5000	—	—	—	—	—	—	—	—	—	—	—	—	—	—	—	—	6	0.438	8	0.450	10	0.458	14	0.469	18	0.475	22	0.479
15	0.5906	—	—	—	—	—	—	—	—	—	—	—	—	—	—	6	0.583	8	0.563	10	0.550	13	0.583	17	0.563	22	0.575	27	0.583
—	0.6250	—	—	—	—	—	—	—	—	—	—	—	—	—	—	6	0.583	8	0.563	11	0.600	13	0.583	18	0.594	23	0.600	28	0.604
17	0.6693	—	—	—	—	—	—	—	—	—	—	—	—	—	—	7	0.667	9	0.625	12	0.650	15	0.667	20	0.656	25	0.650	31	0.667
—	0.7500	—	—	—	—	—	—	—	—	—	—	—	—	6	0.700	7	0.667	10	0.688	13	0.700	16	0.708	22	0.719	28	0.725	34	0.729
20	0.7874	—	—	—	—	—	—	—	—	—	—	—	—	6	0.700	8	0.750	11	0.750	14	0.750	17	0.750	24	0.781	30	0.775	36	0.771
25	0.9843	—	—	—	—	—	—	—	—	—	—	6	0.875	8	0.900	10	0.917	14	0.938	18	0.950	22	0.958	30	0.969	38	0.975	46	0.979
—	1.0000	—	—	—	—	—	—	—	—	—	—	6	0.875	8	0.900	10	0.917	14	0.938	18	0.950	22	0.958	30	0.969	38	0.975	46	0.979
30	1.1811	—	—	—	—	—	—	—	—	6	1.167	8	1.125	10	1.100	13	1.167	17	1.125	22	1.150	27	1.167	36	1.156	46	1.175	55	1.167
—	0.2500	—	—	—	—	—	—	—	—	6	1.167	8	1.125	11	1.200	13	1.167	18	1.188	23	1.200	28	1.208	38	1.219	48	1.225	58	1.229
—	0.3130	—	—	—	—	—	—	—	—	6	1.167	9	1.250	12	1.300	14	1.250	19	1.250	25	1.300	30	1.292	40	1.281	51	1.300	61	1.292
35	1.3780	—	—	—	—	—	—	—	—	7	1.333	10	1.375	12	1.300	15	1.333	21	1.375	26	1.350	32	1.375	43	1.375	54	1.375	65	1.375
—	1.4380	—	—	—	—	—	—	—	—	7	1.333	10	1.375	13	1.400	16	1.417	21	1.375	27	1.400	33	1.417	44	1.406	56	1.425	67	1.417
—	1.5630	—	—	—	—	—	—	—	—	8	1.500	11	1.500	14	1.500	17	1.500	23	1.500	30	1.550	36	1.542	48	1.541	61	1.550	73	1.542
40	1.5748	—	—	—	—	—	—	6	1.400	8	1.500	11	1.500	14	1.500	17	1.500	24	1.563	30	1.550	36	1.542	49	1.563	61	1.550	74	1.563
45	1.7717	—	—	—	—	6	1.750	7	1.600	9	1.667	13	1.750	16	1.700	20	1.750	27	1.750	34	1.750	41	1.750	55	1.750	69	1.750	83	1.750
—	1.8130	—	—	—	—	6	1.750	8	1.800	9	1.667	13	1.750	17	1.800	20	1.750	27	1.750	35	1.800	42	1.792	56	1.781	71	1.800	85	1.792
50	1.9685	—	—	—	—	6	1.750	8	1.800	10	1.833	14	1.875	18	1.900	22	1.917	30	1.938	38	1.950	46	1.958	62	1.968	77	1.950	93	1.958
—	2.0630	—	—	—	—	7	2.000	9	2.000	11	2.000	15	2.000	19	2.000	23	2.000	31	2.000	40	2.050	48	2.042	64	2.031	81	2.050	97	2.042
55	2.1654	—	—	—	—	7	2.000	9	2.000	11	2.000	16	2.125	20	2.100	24	2.083	33	2.125	42	2.150	50	2.125	68	2.156	85	2.150	—	—
—	2.3130	—	—	—	—	8	2.250	10	2.200	12	2.167	17	2.250	22	2.300	26	2.250	35	2.250	45	2.300	54	2.292	72	2.281	91	2.300	—	—
60	2.3622	—	—	6	2.333	8	2.250	10	2.200	13	2.333	17	2.250	22	2.300	27	2.333	36	2.313	46	2.350	55	2.333	74	2.343	93	2.350	—	—
65	2.5591	—	—	6	2.333	9	2.500	11	2.400	14	2.500	19	2.500	24	2.500	29	2.500	39	2.500	50	2.550	60	2.542	80	2.531	—	—	—	—
70	2.7559	—	—	7	2.667	10	2.750	12	2.600	15	2.667	21	2.750	26	2.700	32	2.750	43	2.750	54	2.750	65	2.750	87	2.750	—	—	—	—
—	2.9380	6	2.800	7	2.667	10	2.750	13	2.800	16	2.833	22	2.900	28	2.900	34	2.917	45	2.875	57	2.900	69	2.917	92	2.906	—	—	—	—
75	2.9528	6	2.800	7	2.667	10	2.750	13	2.800	16	2.833	22	2.875	28	2.900	34	2.917	46	2.938	58	2.950	69	2.917	93	2.938	—	—	—	—
—	3.0630	6	2.800	8	3.000	11	3.000	14	3.000	17	3.000	23	3.000	29	3.000	35	3.000	47	3.000	60	3.050	72	3.042	96	3.031	—	—	—	—
80	3.1496	6	2.800	8	3.000	11	3.000	14	3.000	17	3.000	24	3.125	30	3.100	36	3.083	49	3.125	61	3.100	74	3.125	99	3.125	—	—	—	—
85	3.3465	7	3.200	9	3.333	12	3.250	15	3.200	19	3.333	25	3.250	32	3.300	39	3.333	52	3.313	65	3.300	79	3.333	—	—	—	—	—	—
90	3.5433	7	3.200	9	3.333	13	3.500	16	3.400	20	3.500	27	3.500	34	3.500	41	3.500	55	3.500	69	3.500	84	3.542	—	—	—	—	—	—
95	3.7402	8	3.600	10	3.667	13	3.500	17	3.600	21	3.667	28	3.625	36	3.700	43	3.667	58	3.688	73	3.700	88	3.708	—	—	—	—	—	—
100	3.9370	8	3.600	10	3.667	14	3.750	18	3.800	22	3.833	30	3.875	38	3.900	46	3.917	61	3.875	77	3.900	93	3.917	—	—	—	—	—	—
105	4.1339	9	4.000	11	4.000	15	4.000	19	4.000	23	4.000	32	4.125	40	4.100	48	4.083	65	4.125	81	4.100	98	4.125	—	—	—	—	—	—
110	4.3307	9	4.000	11	4.000	16	4.250	20	4.200	24	4.167	33	4.250	42	4.300	50	4.250	68	4.312	85	4.300	—	—	—	—	—	—	—	—
120	4.7244	10	4.400	13	4.667	17	4.500	22	4.600	27	4.667	36	4.625	46	4.700	55	4.667	74	4.688	93	4.700	—	—	—	—	—	—	—	—
130	5.1181	11	4.800	14	5.000	19	5.000	24	5.000	29	5.000	39	5.000	50	5.100	60	5.083	80	5.062	—	—	—	—	—	—	—	—	—	—
140	5.5118	12	5.200	15	5.333	21	5.500	26	5.400	32	5.500	43	5.500	54	5.500	65	5.500	87	5.500	—	—	—	—	—	—	—	—	—	—
150	5.9055	13	5.600	16	5.667	22	5.750	28	5.800	34	5.833	46	5.875	58	5.900	69	5.833	93	5.875	—	—	—	—	—	—	—	—	—	—
160	6.2992	14	6.000	17	6.000	24	6.250	30	6.200	36	6.167	49	6.250	61	6.200	74	6.250	99	6.250	—	—	—	—	—	—	—	—	—	—
170	6.6929	15	6.400	19	6.667	25	6.500	32	6.600	39	6.667	52	6.625	65	6.600	79	6.667	—	—	—	—	—	—	—	—	—	—	—	—
180	7.0866	16	6.800	20	7.000	27	7.000	34	7.000	41	7.000	55	7.000	69	7.000	84	7.083	—	—	—	—	—	—	—	—	—	—	—	—
190	7.4803	17	7.200	21	7.333	28	7.250	36	7.400	43	7.333	58	7.375	73	7.400	88	7.417	—	—	—	—	—	—	—	—	—	—	—	—
200	7.8740	18	7.600	22	7.667	30	7.750	38	7.800	46	7.833	61	7.750	77	7.800	93	7.833	—	—	—	—	—	—	—	—	—	—	—	—
210	8.2677	19	8.000	23	8.000	32	8.250	40	8.200	48	8.167	65	8.250	81	8.200	98	8.250	—	—	—	—	—	—	—	—	—	—	—	—
220	8.6614	20	8.400	24	8.333	33	8.500	42	8.600	50	8.500	68	8.625	85	8.600	—	—	—	—	—	—	—	—	—	—	—	—	—	—
230	9.0551	21	8.800	26	9.000	35	9.000	44	9.000	53	9.000	71	9.000	89	9.000	—	—	—	—	—	—	—	—	—	—	—	—	—	—
240	9.4488	22	9.200	27	9.333	36	9.250	46	9.400	55	9.333	74	9.375	93	9.400	—	—	—	—	—	—	—	—	—	—	—	—	—	—
250	9.8425	23	9.600	28	9.667	38	9.750	48	9.800	58	9.833	77	9.750	97	9.800	—	—	—	—	—	—	—	—	—	—	—	—	—	—
260	10.2362	24	10.000	29	10.000	39	10.000	50	10.200	60	10.167	80	10.125	—	—	—	—	—	—	—	—	—	—	—	—	—	—	—	—
280	11.0236	26	10.800	32	11.000	43	11.000	54	11.000	65	11.000	87	11.000	—	—	—	—	—	—	—	—	—	—	—	—	—	—	—	—
320	12.5984	30	12.400	36	12.333	49	12.500	61	12.400	74	12.500	99	12.500	—	—	—	—	—	—	—	—	—	—	—	—	—	—	—	—

[a]Dimensions given are number of teeth and major diameters nearest under bearing bore.

TABLE 10–16

Range of Involute Serrations, SAE Standard

1954 SAE Handbook

Range of Diametral Pitches*	10/20 to 128/256
Range of Number of Teeth	6 to 100
Range of Nominal Diameters	.10 inch diameter (11 teeth 128/256 D.P.) to 10.14 inch diameter (100 teeth 10/20 D.P.)

10/20 Pitch (actual size)

24/48 Pitch (actual size)

*Diametral pitch is designated by two numbers written in the form of a fraction. The numerator is the number of teeth divided by the pitch diameter in inches and the denominator is the reciprocal of the addendum. (The addendum is half the difference between the major diameter of the external serration and the pitch diameter.)

TABLE 10–17

Data To Be Placed on Drawing of External Involute* Serrations

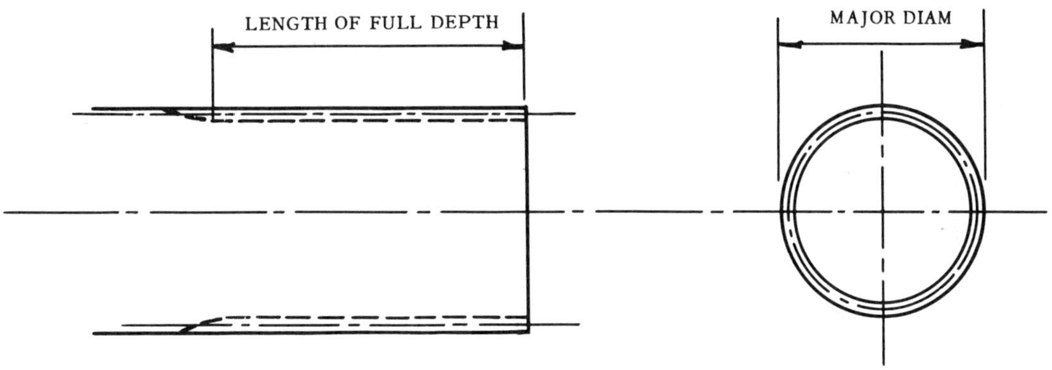

Involute Serration Data	Example	Table No.	Column
Number of Teeth	20		
Diametral Pitch	10/20		
Pitch Diameter	2.0000	10-20	2
Base Circle Diameter	1.4142	10-20	3
Major Diameter	2.1000 2.0900	10-20	8
Minor Diameter	1.8900 1.9000	10-20	9
Tooth Thickness	.1771	10-20	Sub-heading
Measuring Pin Diameter	.1920	10-20	Sub-heading
Measurement Across Pins	2.3189 (min)	10-20	11
†TIF Diameter	1.9300 (max)	10-20	10
Class of Fit	Class A	10-21	
Drawing Number of Mating Part			

*See footnote to Table 10–2.

†The TIF diameter or "true involute form" diameter is the diameter beyond which the tooth profiles are true involutes.

TABLE 10–18

Data To Be Placed on Drawing of Internal Involute* Serrations

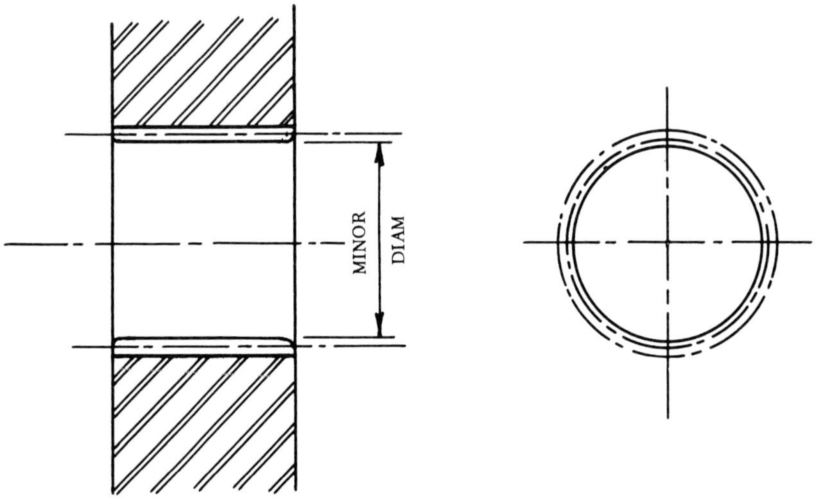

Involute Serration Data	Example	Table No.	Column
Number of Teeth	20		
Diametral Pitch	10/20		
Pitch Diameter	2.0000	10–20	2
Base Circle Diameter	1.4142	10–20	3
Major Diameter	2.1500 2.1400	10-20	4
Minor Diameter	1.9400 1.9500	10-20	5
Measuring Pin Diameter	.1920	10–20	Sub-heading
Measurement Between Pins	1.7139 (max)	10–20	7
†TIF Diameter	2.1100 (min)	10–20	6
Class of Fit	Class A	10–21	
Drawing Number of Mating Part			

*See footnote to Table 10–2.

† TIF diameter or "true involute form" diameter is the diameter within which the tooth profiles are true involutes.

TABLE 10-19

Basic Formulas for Involute Serrations

1954 SAE Handbook

$$\text{Circular Pitch} = \frac{3.141593}{\text{diametral pitch}}$$

$$\text{Addendum (external)} = \text{Dedendum (external)} = \frac{.500}{\text{diametral pitch}}$$

$$\text{Addendum (internal)} = \frac{.300}{\text{diametral pitch}}$$

$$\text{Dedendum (internal)} = \frac{.700}{\text{diametral pitch}}$$

Circular Tooth Thickness (external) = Circular Tooth Space Width

$$\text{(Internal)} = \frac{1.770796}{\text{diametral pitch}}$$

$$\text{Pitch Diameter} = \frac{\text{Number of Teeth}}{\text{Diametral Pitch}}$$

As used in the above formulas, diametral pitch means the numerator of the expression used to specify the size of the serration. In the case of a 1/2 diametral pitch serration, 1 should be substituted for diametral pitch in the above formulas.

TABLE 10–20

Basic Data for 10/20 Diametral Pitch Involute Serrations

1954 SAE Handbook

10/20 Diametral Pitch Pressure Angle, 45°	Circular Pitch, 0.3142 Circular Tooth Thickness, 0.1771	Addendum, External, 0.0500 Addendum, Internal, 0.0300
Cutting-tool radius, 0.045	Measuring-pin diameter, internal and external, 0.1920	

INT AND EXT			INTERNAL				EXTERNAL						
N	Pitch diam	Base-circle diam	Major diam	Minor diam	*TIF* diam	Meas between pins	Major diam	Minor diam	*TIF* diam	Measurement over pins			Factor *E*
	Ref	Ref	Min	Min	Min		Basic	Max	Max	Class A	Class B	Class C	
1	2	3	4	5	6	7	8	9	10	11	12	13	14
Recommended tolerance⟶			+0.0100 −0.0000	+0.0100 −0.0000		Max	+0.0000 −0.0100	+0.0000 −0.0100		Min	Min	Min	
6	0.6000	0.4243	0.7400	0.5400	0.7100	0.3044	0.7000	0.5000	0.5300	0.9131	0.9144	0.9161	0.872
7	0.7000	0.4950	0.8400	0.6400	0.8100	0.3923	0.8000	0.6000	0.6300	0.9934	0.9948	0.9965	0.884
8	0.8000	0.5657	0.9400	0.7400	0.9100	0.5088	0.9000	0.7000	0.7300	1.1149	1.1162	1.1180	0.893
9	0.9000	0.6364	1.0400	0.8400	1.0100	0.5978	1.0000	0.8000	0.8300	1.2000	1.2013	1.2031	0.901
10	1.0000	0.7071	1.1400	0.9400	1.1100	0.7108	1.1000	0.9000	0.9300	1.3161	1.3174	1.3192	0.908
11	1.1000	0.7778	1.2400	1.0400	1.2100	0.8012	1.2000	1.0000	1.0300	1.4040	1.4054	1.4072	0.914
12	1.2000	0.8485	1.3400	1.1400	1.3100	0.9120	1.3000	1.1000	1.1300	1.5169	1.5183	1.5201	0.919
13	1.3000	0.9192	1.4400	1.2400	1.4100	1.0035	1.4000	1.2000	1.2300	1.6069	1.6083	1.6101	0.924
14	1.4000	0.9900	1.5400	1.3400	1.5100	1.1126	1.5000	1.3000	1.3300	1.7176	1.7190	1.7208	0.928
15	1.5000	1.0607	1.6400	1.4400	1.6100	1.2053	1.6000	1.4000	1.4300	1.8090	1.8104	1.8122	0.931
16	1.6000	1.1314	1.7400	1.5400	1.7100	1.3132	1.7000	1.5000	1.5300	1.9182	1.9196	1.9214	0.935
17	1.7000	1.2021	1.8400	1.6400	1.8100	1.4066	1.8000	1.6000	1.6300	2.0105	2.0119	2.0138	0.938
18	1.8000	1.2728	1.9400	1.7400	1.9100	1.5136	1.9000	1.7000	1.7300	2.1186	2.1200	2.1219	0.941
19	1.9000	1.3435	2.0400	1.8400	2.0100	1.6076	2.0000	1.8000	1.8300	2.2118	2.2132	2.2151	0.943
20	2.0000	1.4142	2.1400	1.9400	2.1100	1.7139	2.1000	1.9000	1.9300	2.3189	2.3203	2.3222	0.945
21	2.1000	1.4849	2.2400	2.0400	2.2100	1.8086	2.2000	2.0000	2.0300	2.4129	2.4143	2.4162	0.948
22	2.2000	1.5556	2.3400	2.1400	2.3100	1.9143	2.3000	2.1000	2.1300	2.5192	2.5206	2.5225	0.950
23	2.3000	1.6263	2.4400	2.2400	2.4100	2.0092	2.4000	2.2000	2.2300	2.6137	2.6151	2.6170	0.951
24	2.4000	1.6971	2.5400	2.3400	2.5100	2.1144	2.5000	2.3000	2.3300	2.7195	2.7209	2.7228	0.953
25	2.5000	1.7678	2.6400	2.4400	2.6100	2.2098	2.6000	2.4000	2.4300	2.8143	2.8157	2.8176	0.955
26	2.6000	1.8385	2.7400	2.5400	2.7100	2.3146	2.7000	2.5000	2.5300	2.9195	2.9210	2.9229	0.956
27	2.7000	1.9092	2.8400	2.6400	2.8100	2.4102	2.8000	2.6000	2.6300	3.0150	3.0164	3.0183	0.958
28	2.8000	1.9799	2.9400	2.7400	2.9100	2.5147	2.9000	2.7000	2.7300	3.1198	3.1212	3.1231	0.959
29	2.9000	2.0506	3.0400	2.8400	3.0100	2.6107	3.0000	2.8000	2.8300	3.2155	3.2169	3.2188	0.960
30	3.0000	2.1213	3.1400	2.9400	3.1100	2.7148	3.1000	2.9000	2.9300	3.3200	3.3214	3.3233	0.961
31	3.1000	2.1920	3.2400	3.0400	3.2100	2.8110	3.2000	3.0000	3.0300	3.4159	3.4174	3.4193	0.962
32	3.2000	2.2627	3.3400	3.1400	3.3100	2.9150	3.3000	3.1000	3.1300	3.5201	3.5216	3.5235	0.963
33	3.3000	2.3335	3.4400	3.2400	3.4100	3.0114	3.4000	3.2000	3.2300	3.6163	3.6178	3.6197	0.964
34	3.4000	2.4042	3.5400	3.3400	3.5100	3.1151	3.5000	3.3000	3.3300	3.7202	3.7217	3.7236	0.965
35	3.5000	2.4749	3.6400	3.4400	3.6100	3.2118	3.6000	3.4000	3.4300	3.8166	3.8181	3.8200	0.966
36	3.6000	2.5456	3.7400	3.5400	3.7100	3.3152	3.7000	3.5000	3.5300	3.9203	3.9218	3.9237	0.967
37	3.7000	2.6163	3.8400	3.6400	3.8100	3.4120	3.8000	3.6000	3.6300	4.0170	4.0185	4.0204	0.968
38	3.8000	2.6870	3.9400	3.7400	3.9100	3.5153	3.9000	3.7000	3.7300	4.1205	4.1220	4.1239	0.968
39	3.9000	2.7577	4.0400	3.8400	4.0100	3.6123	4.0000	3.8000	3.8300	4.2172	4.2187	4.2206	0.969
40	4.0000	2.8284	4.1400	3.9400	4.1100	3.7154	4.1000	3.9000	3.9300	4.3206	4.3221	4.3240	0.970
41	4.1000	2.8991	4.2400	4.0400	4.2100	3.8125	4.2000	4.0000	4.0300	4.4175	4.4190	4.4209	0.971
42	4.2000	2.9698	4.3400	4.1400	4.3100	3.9154	4.3000	4.1000	4.1300	4.5205	4.5220	4.5239	0.971
43	4.3000	3.0406	4.4400	4.2400	4.4100	4.0127	4.4000	4.2000	4.2300	4.6176	4.6191	4.6210	0.972
44	4.4000	3.1113	4.5400	4.3400	4.5100	4.1155	4.5000	4.3000	4.3300	4.7206	4.7221	4.7240	0.972
45	4.5000	3.1820	4.6400	4.4400	4.6100	4.2128	4.6000	4.4000	4.4300	4.8179	4.8194	4.8213	0.973
46	4.6000	3.2527	4.7400	4.5400	4.7100	4.3155	4.7000	4.5000	4.5300	4.9207	4.9222	4.9241	0.973
47	4.7000	3.3234	4.8400	4.6400	4.8100	4.4130	4.8000	4.6000	4.6300	5.0181	5.0196	5.0215	0.974
48	4.8000	3.3941	4.9400	4.7400	4.9100	4.5156	4.9000	4.7000	4.7300	5.1208	5.1223	5.1242	0.974
49	4.9000	3.4648	5.0400	4.8400	5.0100	4.6132	5.0000	4.8000	4.8300	5.2182	5.2197	5.2216	0.975
50	5.0000	3.5355	5.1400	4.9400	5.1100	4.7157	5.1000	4.9000	4.9300	5.3209	5.3224	5.3243	0.975

[a]The fits and tolerances suggested in this table are to be considered binding on a manufacturer or seller only when specifically agreed to in writing.

continued on next page

10/20 Diametral Pitch Pressure Angle, 45°	Circular Pitch, 0.3142 Circular Tooth Thickness, 0.1771	Addendum, External, 0.0500 Addendum, Internal, 0.0300

Cutting-tool radius, 0.045	Measuring-pin diameter, internal and external, 0.1920

INT AND EXT			INTERNAL				EXTERNAL						
N	Pitch diam	Base-circle diam	Major diam	Minor diam	*TIF* diam	Meas between pins	Major diam	Minor diam	*TIF* diam	Measurement over pins			Factor *E*
	Ref	Ref	Min	Min	Min		Basic	Max	Max	Class A	Class B	Class C	
1	2	3	4	5	6	7	8	9	10	11	12	13	14
Recommended tolerance⟶			+0.0100 −0.0000	+0.0100 −0.0000		Max	+0.0000 −0.0100	+0.0000 −0.0100		Min	Min	Min	
51	5.1000	3.6062	5.2400	5.0400	5.2100	4.8134	5.2000	5.0000	5.0300	5.4184	5.4199	5.4218	0.976
52	5.2000	3.6770	5.3400	5.1400	5.3100	4.9158	5.3000	5.1000	5.1300	5.5209	5.5224	5.5243	0.976
53	5.3000	3.7477	5.4400	5.2400	5.4100	5.0135	5.4000	5.2000	5.2300	5.6186	5.6201	5.6220	0.977
54	5.4000	3.8184	5.5400	5.3400	5.5100	5.1158	5.5000	5.3000	5.3300	5.7210	5.7225	5.7244	0.977
55	5.5000	3.8891	5.6400	5.4400	5.6100	5.2136	5.6000	5.4000	5.4300	5.8187	5.8202	5.8221	0.977
56	5.6000	3.9598	5.7400	5.5400	5.7100	5.3158	5.7000	5.5000	5.5300	5.9210	5.9225	5.9244	0.978
57	5.7000	4.0305	5.8400	5.6400	5.8100	5.4137	5.8000	5.6000	5.6300	6.0188	6.0203	6.0222	0.978
58	5.8000	4.1012	5.9400	5.7400	5.9100	5.5159	5.9000	5.7000	5.7300	6.1211	6.1226	6.1245	0.979
59	5.9000	4.1719	6.0400	5.8400	6.0100	5.6138	6.0000	5.8000	5.8300	6.2189	6.2204	6.2223	0.979
60	6.0000	4.2426	6.1400	5.9400	6.1100	5.7159	6.1000	5.9000	5.9300	6.3210	6.3225	6.3244	0.979
61	6.1000	4.3134	6.2400	6.0400	6.2100	5.8139	6.2000	6.0000	6.0300	6.4190	6.4205	6.4224	0.980
62	6.2000	4.3841	6.3400	6.1400	6.3100	5.9159	6.3000	6.1000	6.1300	6.5211	6.5226	6.5245	0.980
63	6.3000	4.4548	6.4400	6.2400	6.4100	6.0140	6.4000	6.2000	6.2300	6.6191	6.6206	6.6225	0.980
64	6.4000	4.5255	6.5400	6.3400	6.5100	6.1159	6.5000	6.3000	6.3300	6.7211	6.7226	6.7245	0.980
65	6.5000	4.5962	6.6400	6.4400	6.6100	6.2141	6.6000	6.4000	6.4300	6.8192	6.8207	6.8226	0.981
66	6.6000	4.6669	6.7400	6.5400	6.7100	6.3160	6.7000	6.5000	6.5300	6.9211	6.9226	6.9245	0.981
67	6.7000	4.7376	6.8400	6.6400	6.8100	6.4142	6.8000	6.6000	6.6300	7.0193	7.0208	7.0227	0.981
68	6.8000	4.8083	6.9400	6.7400	6.9100	6.5160	6.9000	6.7000	6.7300	7.1212	7.1227	7.1246	0.982
69	6.9000	4.8790	7.0400	6.8400	7.0100	6.6142	7.0000	6.8000	6.8300	7.2194	7.2209	7.2228	0.982
70	7.0000	4.9497	7.1400	6.9400	7.1100	6.7160	7.1000	6.9000	6.9300	7.3212	7.3226	7.3246	0.982
71	7.1000	5.0205	7.2400	7.0400	7.2100	6.8143	7.2000	7.0000	7.0300	7.4195	7.4209	7.4229	0.982
72	7.2000	5.0912	7.3400	7.1400	7.3100	6.9160	7.3000	7.1000	7.1300	7.5213	7.5227	7.5247	0.982
73	7.3000	5.1619	7.4400	7.2400	7.4100	7.0144	7.4000	7.2000	7.2300	7.6195	7.6209	7.6229	0.983
74	7.4000	5.2326	7.5400	7.3400	7.5100	7.1160	7.5000	7.3000	7.3300	7.7213	7.7227	7.7247	0.983
75	7.5000	5.3033	7.6400	7.4400	7.6100	7.2144	7.6000	7.4000	7.4300	7.8195	7.8209	7.8229	0.983
76	7.6000	5.3740	7.7400	7.5400	7.7100	7.3161	7.7000	7.5000	7.5300	7.9212	7.9226	7.9246	0.983
77	7.7000	5.4447	7.8400	7.6400	7.8100	7.4146	7.8000	7.6000	7.6300	8.0196	8.0210	8.0230	0.984
78	7.8000	5.5154	7.9400	7.7400	7.9100	7.5162	7.9000	7.7000	7.7300	8.1212	8.1226	8.1246	0.984
79	7.9000	5.5861	8.0400	7.8400	8.0100	7.6146	8.0000	7.8000	7.8300	8.2197	8.2211	8.2231	0.984
80	8.0000	5.6569	8.1400	7.9400	8.1100	7.7162	8.1000	7.9000	7.9300	8.3213	8.3227	8.3247	0.984
81	8.1000	5.7276	8.2400	8.0400	8.2100	7.8147	8.2000	8.0000	8.0300	8.4197	8.4211	8.4231	0.984
82	8.2000	5.7983	8.3400	8.1400	8.3100	7.9162	8.3000	8.1000	8.1300	8.5213	8.5227	8.5247	0.985
83	8.3000	5.8690	8.4400	8.2400	8.4100	8.0147	8.4000	8.2000	8.2300	8.6198	8.6212	8.6232	0.985
84	8.4000	5.9397	8.5400	8.3400	8.5100	8.1162	8.5000	8.3000	8.3300	8.7213	8.7227	8.7247	0.985
85	8.5000	6.0104	8.6400	8.4400	8.6100	8.2148	8.6000	8.4000	8.4300	8.8199	8.8213	8.8233	0.985
86	8.6000	6.0811	8.7400	8.5400	8.7100	8.3162	8.7000	8.5000	8.5300	8.9213	8.9227	8.9247	0.985
87	8.7000	6.1518	8.8400	8.6400	8.8100	8.4148	8.8000	8.6000	8.6300	9.0199	9.0213	9.0233	0.985
88	8.8000	6.2225	8.9400	8.7400	8.9100	8.5162	8.9000	8.7000	8.7300	9.1214	9.1228	9.1248	0.985
89	8.9000	6.2933	9.0400	8.8400	9.0100	8.6149	9.0000	8.8000	8.8300	9.2200	9.2214	9.2234	0.986
90	9.0000	6.3640	9.1400	8.9400	9.1100	8.7162	9.1000	8.9000	8.9300	9.3214	9.3228	9.3248	0.986
91	9.1000	6.4347	9.2400	9.0400	9.2100	8.8150	9.2000	9.0000	9.0300	9.4200	9.4214	9.4234	0.986
92	9.2000	6.5054	9.3400	9.1400	9.3100	8.9164	9.3000	9.1000	9.1300	9.5213	9.5227	9.5247	0.986
93	9.3000	6.5761	9.4400	9.2400	9.4100	9.0151	9.4000	9.2000	9.2300	9.6200	9.6214	9.6234	0.986
94	9.4000	6.6468	9.5400	9.3400	9.5100	9.1164	9.5000	9.3000	9.3300	9.7213	9.7227	9.7247	0.986
95	9.5000	6.7175	9.6400	9.4400	9.6100	9.2151	9.6000	9.4000	9.4300	9.8200	9.8214	9.8234	0.986
96	9.6000	6.7882	9.7400	9.5400	9.7100	9.3164	9.7000	9.5000	9.5300	9.9213	9.9227	9.9247	0.987
97	9.7000	6.8589	9.8400	9.6400	9.8100	9.4151	9.8000	9.6000	9.6300	10.0201	10.0215	10.0235	0.987
98	9.8000	6.9296	9.9400	9.7400	9.9100	9.5164	9.9000	9.7000	9.7300	10.1214	10.1228	10.1248	0.987
99	9.9000	7.0004	10.0400	9.8400	10.0100	9.6152	10.0000	9.8000	9.8300	10.2201	10.2215	10.2235	0.987
100	10.0000	7.0711	10.1400	9.9400	10.1100	9.7164	10.1000	9.9000	9.9300	10.3214	10.3228	10.3248	0.987

TABLE 10–21
Class of Fits for Involute Serrations
1954 SAE Handbook

Classes of Fit	Sliding Fit	Close Fit	Press Fit
Designation used in Table 10–22	A	B	C

TABLE 10–22
Basic Tooth Dimensions for Involute Serrations
1954 SAE Handbook

DIAMETRAL PITCH	EXTERNAL ADDENDUM, DEDENDUM	INTERNAL SERRATION		CIRCULAR PITCH	MINIMUM EFFECTIVE SPACE	MAXIMUM EFFECTIVE TOOTH THICKNESS			MEASURING PIN DIAMETER
		Addendum	Dedendum			Class of Fit			
P	$a = b$	a_1	b_1	p	t_s	A	B	C	d
1/2	0.5000	0.3000	0.7000	3.1416	1.7703	1.7703	1.7718	1.7738	1.9200
10/20	0.0500	0.0300	0.0700	0.3142	0.1771	0.1766	0.1781	0.1801	0.1920
16/32	0.0313	0.0188	0.0438	0.1963	0.1107	0.1102	0.1117	0.1137	0.1200
24/48	0.0208	0.0125	0.0292	0.1309	0.0738	0.0733	0.0748	0.0768	0.0800
32/64	0.0156	0.0094	0.0218	0.0982	0.0553	0.0548	0.0563	0.0583	0.0600
40/80	0.0125	0.0075	0.0175	0.0785	0.0443	0.0438	0.0453	0.0473	0.0480
48/96	0.0104	0.0063	0.0145	0.0654	0.0369	0.0364	0.0379	0.0399	0.0400
64/128	0.0078	0.0047	0.0109	0.0491	0.0277	0.0272	0.0287	0.0307	0.0300
80/160	0.0063	0.0038	0.0088	0.0393	0.0221	0.0216	0.0231	0.0251	0.0240
128/256	0.0039	0.0023	0.0055	0.0245	0.0138	0.0133	0.0148	0.0168	0.0150

TABLE 10–23
Allowable Errors[a] for Involute Serrations
1954 SAE Handbook

DIAMETRAL PITCH	INVOLUTE PROFILE	ACCUMULATED PITCH ERROR BETWEEN ANY TWO TEETH	LEAD ERRORS					
			Length of serration					
			0 to 0.49	0.50 to 1.24	1.25 to 2.49	2.50 to 3.99	4.00 to 5.24	5.25 to 6.50
10/20	5	15	0	3	3	4	5	6
16/32	5	15	0	3	3	4	5	6
24/48	5	15	0	3	3	4	5	6
32/64 up	5	15	0	3	3	4	5	6

OUT OF ROUNDNESS—INTERNAL (Number of Teeth) / **MACHINING TOLERANCE** (Internal)

DIAMETRAL PITCH	6	8	10	12	14	16	20	25	30	35	40	45	50	60	70	80	90	100	Internal
10/20	11	10	9	9	9	10	10	10	11	12	12	13	13	14	16	17	18	19	10
16/32	10	9	9	9	9	9	9	9	10	10	10	11	11	12	13	13	14	15	10
24/48	10	9	9	9	9	9	9	9	9	9	10	10	10	11	11	12	13	13	10
32/64 up	10	9	8	8	8	8	8	8	8	9	9	9	9	9	10	10	11	11	10

OUT OF ROUNDNESS—EXTERNAL (Number of Teeth) / **MACHINING TOLERANCE** (External)

DIAMETRAL PITCH	6	8	10	12	14	16	20	25	30	35	40	45	50	60	70	80	90	100	External
10/20	8	9	9	9	10	10	10	11	12	12	13	13	14	15	16	17	18	19	15
16/32	8	8	9	9	9	9	10	10	10	11	11	11	12	12	13	14	14	15	15
24/48	8	8	8	9	9	9	9	9	10	10	10	10	10	11	11	11	12	12	10
32/64 up	8	8	8	8	8	8	9	9	9	9	9	10	10	10	10	10	11	11	10

[a] All table figures are ten thousandths of inch. Add the profile error, accumulated pitch error, and the out of roundness divided by F or E. If serration is longer than its pitch diameter, a lead error may be added.

TABLE 10–24
Basic Measurements Between Pins — Internal Involute Serrations
1954 SAE Handbook

	Diametral Pitch									
	½	10/20	16/32	24/48	32/64	40/80	48/96	64/128	80/160	128/256
	Pin Diam									
N	1.920	.1920	.1200	.0800	.0600	.0480	.0400	.0300	.0240	.0150
6	3.0079	0.3008	0.1880	0.1253	0.0940	0.0752	0.0626	—	—	—
7	3.8885	0.3889	0.2430	0.1620	0.1215	0.0972	0.0810	—	—	—
8	5.0560	0.5056	0.3160	0.2107	0.1580	0.1264	0.1054	—	—	—
9	5.9470	0.5947	0.3717	0.2478	0.1859	0.1487	0.1239	—	—	—
10	7.0773	0.7077	0.4423	0.2949	0.2212	0.1769	0.1475	—	—	—
11	7.9823	0.7982	0.4989	0.3326	0.2495	0.1996	0.1663	0.1247	0.0998	0.0624
12	9.0895	0.9090	0.5680	0.3788	0.2840	0.2272	0.1894	0.1420	0.1136	0.0710
13	10.0062	1.0006	0.6254	0.4169	0.3127	0.2501	0.2085	0.1563	0.1251	0.0782
14	11.0974	1.1097	0.6936	0.4624	0.3468	0.2774	0.2312	0.1734	0.1387	0.0867
15	12.0236	1.2024	0.7515	0.5010	0.3758	0.3006	0.2505	0.1879	0.1503	0.0939
16	13.1030	1.3103	0.8189	0.5460	0.4095	0.3276	0.2780	0.2047	0.1638	0.1024
17	14.0369	1.4037	0.8773	0.5848	0.4387	0.3509	0.2924	0.2193	0.1755	0.1097
18	15.1072	1.5107	0.9442	0.6295	0.4721	0.3777	0.3148	0.2361	0.1888	0.1180
19	16.0473	1.6047	1.0030	0.6686	0.5015	0.4012	0.3343	0.2508	0.2006	0.1254
20	17.1104	1.7110	1.0694	0.7129	0.5347	0.4277	0.3565	0.2674	0.2139	0.1337
21	18.0557	1.8056	1.1285	0.7523	0.5643	0.4514	0.3762	0.2822	0.2257	0.1411
22	19.1129	1.9113	1.1946	0.7964	0.5973	0.4778	0.3982	0.2986	0.2389	0.1493
23	20.0626	2.0063	1.2539	0.8360	0.6270	0.5015	0.4180	0.3135	0.2508	0.1567
24	21.1150	2.1115	1.3197	0.8798	0.6598	0.5279	0.4399	0.3299	0.2639	0.1650
25	22.0685	2.2069	1.3793	0.9195	0.6897	0.5517	0.4598	0.3449	0.2758	0.1725
26	23.1167	2.3117	1.4448	0.9632	0.7224	0.5779	0.4816	0.3612	0.2890	0.1806
27	24.0734	2.4073	1.5046	1.0030	0.7523	0.6018	0.5015	0.3762	0.3009	0.1881
28	25.1182	2.5118	1.5699	1.0466	0.7849	0.6280	0.5233	0.3925	0.3140	0.1962
29	26.0777	2.6078	1.6298	1.0866	0.8149	0.6519	0.5433	0.4075	0.3255	0.2037
30	27.1194	2.7119	1.6950	1.1300	0.8475	0.6780	0.5650	0.4237	0.3390	0.2119
31	28.0814	2.8081	1.7551	1.1700	0.8775	0.7020	0.5850	0.4388	0.3510	—
32	29.1205	2.9121	1.8200	1.2134	0.9100	0.7280	0.6067	0.4550	0.3640	—
33	30.0847	3.0085	1.8803	1.2535	0.9402	0.7521	0.6268	0.4701	0.3761	—
34	31.1215	3.1122	1.9451	1.2968	0.9726	0.7781	0.6484	0.4863	0.3890	—
35	32.0876	3.2088	2.0055	1.3370	1.0028	0.8022	0.6685	0.5014	0.4011	—
36	33.1223	3.3122	2.0701	1.3800	1.0351	0.8281	0.6900	0.5175	—	—
37	34.0902	3.4090	2.1306	1.4204	1.0653	0.8522	0.7102	0.5327	—	—
38	35.1230	3.5123	2.1952	1.4635	1.0976	0.8781	0.7317	0.5488	—	—
39	36.0925	3.6093	2.2558	1.5039	1.1279	0.9023	0.7519	0.5639	—	—
40	37.1237	3.7124	2.3202	1.5468	1.1601	0.9281	0.7734	0.5801	—	—
41	38.0946	3.8095	2.3809	1.5873	1.1905	0.9524	0.7937	—	—	—
42	39.1243	3.9124	2.4453	1.6302	1.2226	0.9781	0.8151	—	—	—
43	40.0965	4.0097	2.5060	1.6707	1.2530	1.0024	0.8354	—	—	—
44	41.1248	4.1125	2.5703	1.7135	1.2852	1.0281	0.8568	—	—	—
45	42.0983	4.2098	2.6311	1.7541	1.3156	1.0524	0.8770	—	—	—
46	43.1253	4.3125	2.6953	1.7969	1.3477	1.0781	0.8984	—	—	—
47	44.0998	4.4100	2.7562	1.8375	1.3781	1.1025	0.9188	—	—	—
48	45.1258	4.5126	2.8204	1.8803	1.4102	1.1282	0.9401	—	—	—
49	46.1013	4.6101	2.8813	1.9209	1.4407	1.1525	0.9605	—	—	—
50	47.1262	4.7126	2.9454	1.9636	1.4727	1.1782	0.9818	—	—	—
51	48.1026	4.8103	3.0064	2.0043	1.5032	1.2026	1.0022	—	—	—
52	49.1265	4.9127	3.0704	2.0470	1.5352	1.2282	1.0235	—	—	—
53	50.1039	5.0104	3.1315	2.0877	1.5658	1.2526	1.0439	—	—	—
54	51.1269	5.1127	3.1954	2.1303	1.5977	1.2782	1.0652	—	—	—
55	52.1050	5.2105	3.2566	2.1710	1.6283	1.3026	1.0855	—	—	—
56	53.1272	5.3127	3.3205	2.2136	1.6602	1.3282	1.1068	—	—	—
57	54.1061	5.4106	3.3816	2.2544	1.6908	1.3526	1.1272	—	—	—
58	55.1275	5.5128	3.4455	2.2970	1.7228	1.3782	1.1485	—	—	—
59	56.1071	5.6107	3.5067	2.3378	1.7533	1.4027	1.1689	—	—	—
60	57.1278	5.7128	3.5705	2.3803	1.7853	1.4282	1.1902	—	—	—
61	58.1080	5.8108	3.6317	2.4212	1.8159	1.4527	1.2106	—	—	—
62	59.1280	5.9128	3.6955	2.4637	1.8478	1.4782	1.2319	—	—	—
63	60.1089	6.0109	3.7568	2.5045	1.8784	1.5027	1.2523	—	—	—
64	61.1283	6.1128	3.8205	2.5470	1.9103	1.5282	1.2735	—	—	—
65	62.1097	6.2110	3.8819	2.5879	1.9409	1.5528	1.2940	—	—	—
66	63.1285	6.3129	3.9455	2.6304						
67	64.1105	6.4111	4.0069	2.6713						
68	65.1287	6.5129	4.0705	2.7137						
69	66.1112	6.6111	4.1320	2.7546						
70	67.1289	6.7129	4.1956	2.7970						
71	68.1119	6.8112	4.2570	2.8380						
72	69.1291	6.9129	4.3206	2.8804						
73	70.1125	7.0113	4.3820	2.9214						
74	71.1293	7.1129	4.4456	2.9637						
75	72.1131	7.2113	4.5071	3.0047						
76	73.1295	7.3130	4.5706	3.0471						
77	74.1137	7.4114	4.6321	3.0881						
78	75.1296	7.5130	4.6956	3.1304						
79	76.1143	7.6114	4.7571	3.1714						
80	77.1298	7.7130	4.8206	3.2138						
81	78.1148	7.8115	4.8822	3.2548						
82	79.1299	7.9130	4.9456	3.2971						
83	80.1153	8.0115	5.0072	3.3381						
84	81.1301	8.1130	5.0706	3.3804						
85	82.1158	8.2116	5.1322	3.4215						
86	83.1302	8.3130	5.1956	3.4638						
87	84.1162	8.4116	5.2573	3.5048						
88	85.1303	8.5130	5.3206	3.5471						
89	86.1167	8.6117	5.3823	3.5882						
90	87.1304	8.7130	5.4457	3.6304						
91	88.1171	8.8117	5.5073	3.6715						
92	89.1305	8.9131	5.5707	3.7138						
93	90.1175	9.0118	5.6323	3.7549						
94	91.1306	9.1131	5.6957	3.7971						
95	92.1178	9.2118	5.7574	3.8383						
96	93.1308	9.3131	5.8207	3.8805						
97	94.1182	9.4118	5.8824	3.9216						
98	95.1309	9.5131	5.9457	3.9638						
99	96.1186	9.6119	6.0074	4.0050						
100	97.1309	9.7131	6.0707	4.0471						

TABLE 10–25
Basic Measurements Over Pins — External Involute Serrations
1954 SAE Handbook

	Diametral Pitch									
	½	10/20	16/32	24/48	32/64	40/80	48/96	64/128	80/160	128/256
	Pin Diam									
N	1.920	.1920	.1200	.0800	.0600	.0480	.0400	.0300	.0240	.0150
6	9.1631	0.9163	0.5727	0.3818	0.2864	0.2291	0.1909	—	—	—
7	9.9666	0.9967	0.6229	0.4152	0.3115	0.2492	0.2076	—	—	—
8	11.1820	1.1182	0.6989	0.4659	0.3494	0.2795	0.2329	—	—	—
9	12.0330	1.2033	0.7521	0.5014	0.3760	0.3008	0.2507	—	—	—
10	13.1949	1.3195	0.8247	0.5498	0.4123	0.3299	0.2749	—	—	—
11	14.0749	1.4075	0.8797	0.5865	0.4399	0.3519	0.2933	0.2200	0.1759	0.1100
12	15.2042	1.5204	0.9503	0.6335	0.4751	0.3801	0.3168	0.2376	0.1901	0.1188
13	16.1039	1.6104	1.0065	0.6710	0.5033	0.4026	0.3355	0.2517	0.2013	0.1259
14	17.2114	1.7211	1.0757	0.7171	0.5379	0.4303	0.3586	0.2690	0.2151	0.1345
15	18.1251	1.8125	1.1328	0.7552	0.5664	0.4531	0.3776	0.2832	0.2266	0.1416
16	19.2171	1.9217	1.2011	0.8007	0.6006	0.4804	0.4004	0.3003	0.2402	0.1501
17	20.1413	2.0141	1.2588	0.8392	0.6294	0.5035	0.4196	0.3147	0.2518	0.1574
18	21.2215	2.1222	1.3263	0.8842	0.6632	0.5305	0.4421	0.3316	0.2653	0.1658
19	22.1541	2.2154	1.3846	0.9231	0.6923	0.5538	0.4616	0.3462	0.2769	0.1731
20	23.2253	2.3225	1.4516	0.9677	0.7258	0.5806	0.4839	0.3629	0.2903	0.1814
21	24.1645	2.4165	1.5103	1.0069	0.7552	0.6041	0.5035	0.3776	0.3021	0.1888
22	25.2284	2.5228	1.5768	1.0512	0.7884	0.6307	0.5256	0.3942	0.3154	0.1971
23	26.1731	2.6173	1.6358	1.0905	0.8179	0.6543	0.5453	0.4090	0.3272	0.2045
24	27.2311	2.7231	1.7019	1.1346	0.8510	0.6803	0.5673	0.4255	0.3404	0.2128
25	28.1804	2.8180	1.7613	1.1742	0.8807	0.7045	0.5871	0.4403	0.3523	0.2202
26	29.2334	2.9233	1.8271	1.2181	0.9136	0.7303	0.6090	0.4568	0.3654	0.2284
27	30.1865	3.0187	1.8866	1.2578	0.9433	0.7547	0.6289	0.4717	0.3773	0.2359
28	31.2354	3.1235	1.9522	1.3015	0.9761	0.7809	0.6507	0.4881	0.3904	0.2440
29	32.1919	3.2192	2.0119	1.3413	1.0060	0.8048	0.6706	0.5030	0.4024	0.2515
30	33.2372	3.3237	2.0773	1.3849	1.0387	0.8309	0.6924	0.5193	0.4155	0.2597
31	34.1965	3.4197	2.1373	1.4249	1.0687	0.8549	0.7124	0.5343	0.4275	0.2672
32	35.2387	3.5239	2.2024	1.4683	1.1012	0.8810	0.7341	0.5506	0.4405	0.2753
33	36.2006	3.6201	2.2625	1.5084	1.1313	0.9050	0.7542	0.5657	0.4525	0.2829
34	37.2401	3.7240	2.3275	1.5517	1.1638	0.9310	0.7758	0.5819	0.4655	0.2909
35	38.2042	3.8204	2.3878	1.5918	1.1939	0.9551	0.7959	0.5970	0.4776	0.2985
36	39.2414	3.9241	2.4526	1.6351	1.2263	0.9810	0.8175	0.6131	0.4905	0.3066
37	40.2075	4.0208	2.5129	1.6753	1.2565	1.0052	0.8376	0.6283	0.5026	0.3141
38	41.2425	4.1243	2.5776	1.7184	1.2888	1.0311	0.8592	0.6444	0.5155	0.3222
39	42.2103	4.2210	2.6381	1.7588	1.3191	1.0552	0.8794	0.6596	0.5276	0.3298
40	43.2435	4.3244	2.7027	1.8018	1.3514	1.0811	0.9009	0.6757	0.5405	0.3379
41	44.2129	4.4213	2.7633	1.8422	1.3817	1.1053	0.9211	0.6909	0.5527	—
42	45.2444	4.5244	2.8277	1.8852	1.4139	1.1311	0.9426	0.7069	0.5656	—
43	46.2153	4.6215	2.8884	1.9256	1.4442	1.1554	0.9628	0.7221	0.5777	—
44	47.2453	4.7245	2.9528	1.9686	1.4764	1.1811	0.9843	0.7382	0.5906	—
45	48.2175	4.8218	3.0136	2.0091	1.5068	1.2054	1.0045	0.7534	0.6027	—
46	49.2461	4.9246	3.0779	2.0519	1.5389	1.2311	1.0260	0.7695	—	—
47	50.2195	5.0220	3.1387	2.0925	1.5694	1.2555	1.0461	0.7847	—	—
48	51.2468	5.1247	3.2029	2.1353	1.6015	1.2811	1.0676	0.8007	—	—
49	52.2213	5.2221	3.2638	2.1759	1.6319	1.3055	1.0879	0.8159	—	—
50	53.2475	5.3248	3.3279	2.2186	1.6640	1.3312	1.1093	0.8320	—	—
51	54.2230	5.4223	3.3889	2.2593	1.6945	1.3556	1.1297	—	—	—
52	55.2481	5.5248	3.4530	2.3020	1.7265	1.3812	1.1510	—	—	—
53	56.2245	5.6225	3.5140	2.3427	1.7570	1.4056	1.1714	—	—	—
54	57.2487	5.7249	3.5780	2.3854	1.7890	1.4312	1.1927	—	—	—
55	58.2260	5.8226	3.6391	2.4261	1.8196	1.4556	1.2131	—	—	—
56	59.2492	5.9249	3.7031	2.4687	1.8516	1.4812	1.2344	—	—	—
57	60.2273	6.0227	3.7642	2.5095	1.8821	1.5057	1.2548	—	—	—
58	61.2497	6.1250	3.8281	2.5521	1.9141	1.5312	1.2761	—	—	—
59	62.2286	6.2229	3.8893	2.5929	1.9447	1.5557	1.2965	—	—	—
60	63.2502	6.3250	3.9531	2.6354	1.9766	1.5813	1.3177	—	—	—
61	64.2297	6.4230	4.0144	2.6762	2.0072	1.6057	1.3381	—	—	—
62	65.2506	6.5251	4.0782	2.7188	2.0391	1.6313	1.3594	—	—	—
63	66.2308	6.6231	4.1394	2.7596	2.0697	1.6558	1.3798	—	—	—
64	67.2510	6.7251	4.2032	2.8021	2.1016	1.6813	1.4011	—	—	—
65	68.2319	6.8232	4.2645	2.8430	2.1323	1.7058	1.4215	—	—	—
66	69.2514	6.9251	4.3282	2.8855						
67	70.2328	7.0233	4.3896	2.9264						
68	71.2518	7.1252	4.4532	2.9688						
69	72.2337	7.2234	4.5146	3.0097						
70	73.2521	7.3252	4.5783	3.0522						
71	74.2346	7.4235	4.6397	3.0931						
72	75.2525	7.5253	4.7033	3.1355						
73	76.2354	7.6235	4.7647	3.1765						
74	77.2528	7.7253	4.8283	3.2189						
75	78.2362	7.8236	4.8898	3.2598						
76	79.2531	7.9253	4.9533	3.3022						
77	80.2369	8.0237	5.0148	3.3432						
78	81.2534	8.1253	5.0783	3.3856						
79	82.2376	8.2238	5.1399	3.4266						
80	83.2536	8.3254	5.2034	3.4689						
81	84.2383	8.4238	5.2649	3.5099						
82	85.2539	8.5254	5.3284	3.5522						
83	86.2389	8.6239	5.3899	3.5933						
84	87.2541	8.7254	5.4534	3.6356						
85	88.2395	8.8240	5.5150	3.6766						

	Diametral Pitch			
	½	10/20	16/32	24/48
	Pin Diam			
N	1.920	.1920	.1200	.0800
86	89.2544	8.9254	5.5784	3.7189
87	90.2401	9.0240	5.6400	3.7600
88	91.2546	9.1255	5.7034	3.8023
89	92.2406	9.2241	5.7650	3.8436
90	93.2548	9.3255	5.8284	3.8856
91	94.2411	9.4241	5.8901	3.9267
92	95.2550	9.5255	5.9534	3.9690
93	96.2416	9.6242	6.0151	4.0101
94	97.2552	9.7255	6.0785	4.0523
95	98.2421	9.8242	6.1401	4.0934
96	99.2554	9.9255	6.2035	4.1356
97	100.2426	10.0243	6.2652	4.1768
98	101.2556	10.1256	6.3285	4.2190
99	102.2430	10.2243	6.3902	4.2601
100	103.2557	10.3256	6.4535	4.3023

TABLE 10–26

SAE Standard 4-Spline Fittings

1954 SAE Handbook

Nom Diam	For All Fits D Min	D Max	W Min	W Max	Permanent Fit d Min	d Max	h Min	h Max	T	To Slide When Not Under Load d Min	d Max	h Min	h Max	T
3/4	0.749	0.750	0.179	0.181	0.636	0.637	0.055	0.056	78	0.561	0.562	0.093	0.094	123
7/8	0.874	0.875	0.209	0.211	0.743	0.744	0.065	0.066	107	0.655	0.656	0.108	0.109	167
1	0.999	1.000	0.239	0.241	0.849	0.850	0.074	0.075	139	0.749	0.750	0.124	0.125	219
1-1/8	1.124	1.125	0.269	0.271	0.955	0.956	0.083	0.084	175	0.843	0.844	0.140	0.141	277
1-1/4	1.249	1.250	0.299	0.301	1.061	1.062	0.093	0.094	217	0.936	0.937	0.155	0.156	341
1-3/8	1.374	1.375	0.329	0.331	1.168	1.169	0.102	0.103	262	1.030	1.031	0.171	0.172	414
1-1/2	1.499	1.500	0.359	0.361	1.274	1.275	0.111	0.112	311	1.124	1.125	0.186	0.187	491
1-5/8	1.624	1.625	0.389	0.391	1.380	1.381	0.121	0.122	367	1.218	1.219	0.202	0.203	577
1-3/4	1.749	1.750	0.420	0.422	1.486	1.487	0.130	0.131	424	1.311	1.312	0.218	0.219	670
2	1.998	2.000	0.479	0.482	1.698	1.700	0.148	0.150	555	1.498	1.500	0.248	0.250	875
2-1/4	2.248	2.250	0.539	0.542	1.910	1.912	0.167	0.169	703	1.685	1.687	0.279	0.281	1106
2-1/2	2.498	2.500	0.599	0.602	2.123	2.125	0.185	0.187	865	1.873	1.875	0.310	0.312	1365
3	2.998	3.000	0.720	0.723	2.548	2.550	0.223	0.225	1249	2.248	2.250	0.373	0.375	1969

TABLE 10–27

SAE Standard 6-Spline Fittings

1954 SAE Handbook

Nom Dia	For All Fits D Min	D Max	W Min	W Max	Permanent Fit d Min	d Max	T	To Slide When Not Under Load d Min	d Max	T	To Slide When Under Load d Min	d Max	T
3/4	0.749	0.750	0.186	0.188	0.674	0.675	80	0.637	0.638	117	0.599	0.600	152
7/8	0.874	0.875	0.217	0.219	0.787	0.788	109	0.743	0.744	159	0.699	0.700	207
1	0.999	1.000	0.248	0.250	0.899	0.900	143	0.849	0.850	208	0.799	0.800	270
1-1/8	1.124	1.125	0.279	0.281	1.012	1.013	180	0.955	0.956	263	0.899	0.900	342
1-1/4	1.249	1.250	0.311	0.313	1.124	1.125	223	1.062	1.063	325	0.999	1.000	421
1-3/8	1.374	1.375	0.342	0.344	1.237	1.238	269	1.168	1.169	393	1.099	1.100	510
1-1/2	1.499	1.500	0.373	0.375	1.349	1.350	321	1.274	1.275	468	1.199	1.200	608
1-5/8	1.624	1.625	0.404	0.406	1.462	1.463	376	1.380	1.381	550	1.299	1.300	713
1-3/4	1.749	1.750	0.436	0.438	1.574	1.575	436	1.487	1.488	637	1.399	1.400	827
2	1.998	2.000	0.497	0.500	1.798	1.800	570	1.698	1.700	833	1.598	1.600	1080
2-1/4	2.248	2.250	0.560	0.563	2.023	2.025	721	1.911	1.913	1052	1.798	1.800	1367
2-1/2	2.498	2.500	0.622	0.625	2.248	2.250	891	2.123	2.125	1300	1.998	2.000	1688
3	2.998	3.000	0.747	0.750	2.698	2.700	1283	2.548	2.550	1873	2.398	2.400	2430

TABLE 10–28

SAE Standard 10-Spline Fittings

1954 SAE Handbook

NOMINAL DIAMETER	FOR ALL FITS				PERMANENT FIT			TO SLIDE WHEN NOT UNDER LOAD			TO SLIDE WHEN UNDER LOAD		
	D		*W*		*d*		*T*	*d*		*T*	*d*		*T*
	Min	Max	Min	Max	Min	Max		Min	Max		Min	Max	
¾	0.749	0.750	0.115	0.117	0.682	0.683	120	0.644	0.645	183	0.607	0.608	241
⅞	0.874	0.875	0.135	0.137	0.795	0.796	165	0.752	0.753	248	0.708	0.709	329
1	0.999	1.000	0.154	0.156	0.909	0.910	215	0.859	0.860	326	0.809	0.810	430
1⅛	1.124	1.125	0.174	0.176	1.023	1.024	271	0.967	0.968	412	0.910	0.911	545
1¼	1.249	1.250	0.193	0.195	1.137	1.138	336	1.074	1.075	508	1.012	1.013	672
1⅜	1.374	1.375	0.213	0.215	1.250	1.251	406	1.182	1.183	614	1.113	1.114	813
1½	1.499	1.500	0.232	0.234	1.364	1.365	483	1.289	1.290	732	1.214	1.215	967
1⅝	1.624	1.625	0.252	0.254	1.478	1.479	566	1.397	1.398	860	1.315	1.316	1135
1¾	1.749	1.750	0.271	0.273	1.592	1.593	658	1.504	1.505	997	1.417	1.418	1316
2	1.998	2.000	0.309	0.312	1.818	1.820	860	1.718	1.720	1302	1.618	1.620	1720
2¼	2.248	2.250	0.348	0.351	2.046	2.048	1088	1.933	1.935	1647	1.821	1.823	2176
2½	2.498	2.500	0.387	0.390	2.273	2.275	1343	2.148	2.150	2034	2.023	2.025	2688
3	2.998	3.000	0.465	0.468	2.728	2.730	1934	2.578	2.580	2929	2.428	2.430	3869
3½	3.497	3.500	0.543	0.546	3.182	3.185	2632	3.007	3.010	3987	2.832	2.835	5266
4	3.997	4.000	0.621	0.624	3.637	3.640	3438	3.437	3.440	5208	3.237	3.240	6878
4½	4.497	4.500	0.699	0.702	4.092	4.095	4351	3.867	3.870	6591	3.642	3.645	8705
5	4.997	5.000	0.777	0.780	4.547	4.550	5371	4.297	4.300	8137	4.047	4.050	10746
5½	5.497	5.500	0.855	0.858	5.002	5.005	6500	4.727	4.730	9846	4.452	4.455	13003
6	5.997	6.000	0.933	0.936	5.457	5.460	7735	5.157	5.160	11718	4.857	4.860	15475

TABLE 10–29

SAE Standard 16-Spline Fittings

1954 SAE Handbook

NOMINAL DIAMETER	FOR ALL FITS				PERMANENT FIT			TO SLIDE WHEN NOT UNDER LOAD			TO SLIDE WHEN UNDER LOAD		
	D		*W*		*d*		*T*	*d*		*T*	*d*		*T*
	Min	Max	Min	Max	Min	Max		Min	Max		Min	Max	
2	1.997	2.000	0.193	0.196	1.817	1.820	1375	1.717	1.720	2083	1.617	1.620	2751
2½	2.497	2.500	0.242	0.245	2.273	2.275	2149	2.147	2.150	3255	2.022	2.025	4299
3	2.997	3.000	0.291	0.294	2.727	2.730	3094	2.577	2.580	4687	2.427	2.430	6190
3½	3.497	3.500	0.340	0.343	3.182	3.185	4212	3.007	3.010	6378	2.832	2.835	8426
4	3.997	4.000	0.389	0.392	3.637	3.640	5501	3.437	3.440	8333	3.237	3.240	11005
4½	4.497	4.500	0.438	0.441	4.092	4.095	6962	3.867	3.870	10546	3.642	3.645	13928
5	4.997	5.000	0.487	0.490	4.547	4.550	8595	4.297	4.300	13020	4.047	4.050	17195
5½	5.497	5.500	0.536	0.539	5.002	5.005	10395	4.727	4.730	15754	4.452	4.455	20806
6	5.997	6.000	0.585	0.588	5.457	5.460	12377	5.157	5.160	18749	4.857	4.860	24760

TABLE 10-30

Table of Dimensions for 6-Spline Minor Diameter Fit

ASME Paper No. 48-SA-21 J. B. Armitage, Straight Sided Splines

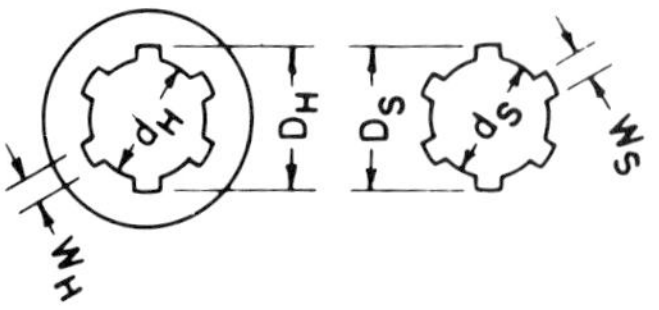

1. Free
2. Sliding Fit
3. Push Fit
4. Light Drive Fit
5. Press Fit

Nom Dia	D_h	d_h 1,2	d_h 3	d_h 4	d_h 5	W_h	D_s	d_s 1	d_s 2,3,4,5	W_s 1	W_s 2,3,4,5
.750	.755	.6380	.6378	.6369	.6364	.186	.750	.6362	.6377	.183	.184
	.765	.6383	.6381	.6372	.6367	.188	.745	.6354	.6373	.181	.183
.875	.880	.7443	.7441	.7432	.7427	.217	.875	.7425	.7440	.213	.214
	.890	.7446	.7444	.7435	.7430	.219	.870	.7417	.7436	.211	.213
1.000	1.005	.8505	.8503	.8494	.9489	.248	1.000	.8487	.8502	.244	.245
	1.015	.8508	.8506	.8497	.9492	.250	.995	.8479	.8498	.242	.244
1.125	1.130	.9569	.9566	.9555	.9549	.279	1.125	.9550	.9565	.275	.276
	1.140	.9573	.9570	.9559	.9553	.281	1.120	.9541	.9560	.273	.275
1.250	1.255	1.0631	1.0628	1.0617	1.0611	.311	1.250	1.0612	1.0627	.307	.308
	1.265	1.0635	1.0632	1.0621	1.0615	.313	1.245	1.0603	1.0622	.305	.307
1.375	1.380	1.1694	1.1691	1.1680	1.1674	.342	1.375	1.1675	1.1690	.338	.339
	1.390	1.1698	1.1695	1.1684	1.1678	.344	1.370	1.1666	1.1685	.336	.338
1.500	1.505	1.2758	1.2754	1.2742	1.2736	.373	1.500	1.2738	1.2753	.368	.369
	1.515	1.2762	1.2758	1.2746	1.2740	.375	1.495	1.2728	1.2747	.366	.368
1.625	1.630	1.3821	1.3817	1.3805	1.3799	.404	1.625	1.3801	1.3816	.399	.400
	1.640	1.3825	1.3821	1.3809	1.3803	.406	1.620	1.3791	1.3810	.397	.399
1.750	1.755	1.4883	1.4879	1.4867	1.4861	.436	1.750	1.4863	1.4878	.431	.432
	1.765	1.4887	1.4883	1.4871	1.4865	.438	1.745	1.4853	1.4872	.429	.431
1.875	1.880	1.5946	1.5942	1.5930	1.5924	.467	1.875	1.5926	1.5941	.462	.463
	1.890	1.5950	1.5946	1.5934	1.5928	.469	1.870	1.5916	1.5935	.460	.462
2.000	2.005	1.7010	1.7005	1.6991	1.6984	.498	2.000	1.6989	1.7004	.493	.494
	2.015	1.7015	1.7010	1.6996	1.6989	.500	1.995	1.6978	1.6997	.491	.493
2.250	2.255	1.9135	1.9130	1.9116	1.9109	.560	2.250	1.9114	1.9129	.555	.556
	2.265	1.9140	1.9135	1.9121	1.9114	.562	2.245	1.9105	1.9122	.553	.555
2.500	2.505	2.1256	2.1251	2.1237	2.1230	.623	2.500	2.1235	2.1250	.617	.619
	2.515	2.1261	2.1256	2.1242	2.1235	.625	2.495	2.1224	2.1243	.615	.617
2.750	2.755	2.3406	2.3401	2.3387	2.3380	.686	2.750	2.3385	2.3400	.680	.682
	2.765	2.3411	2.3406	2.3392	2.3385	.688	2.745	2.3374	2.3393	.678	.680
3.000	3.005	2.5512	2.5507	2.5492	2.5484	.748	3.000	2.5491	2.5506	.742	.744
	3.015	2.5517	2.5512	2.5497	2.5489	.750	2.995	2.5479	2.5498	.740	.742
3.250	3.255	2.7656	2.7651	2.7636	2.7628	.810	3.250	2.7635	2.7650	.804	.806
	3.265	2.7661	2.7656	2.7641	2.7633	.812	3.245	2.7623	2.7642	.802	.804
3.500	3.505	2.9786	2.9781	2.9766	2.9758	.873	3.500	2.9765	2.9780	.867	.869
	3.515	2.9791	2.9786	2.9771	2.9763	.875	3.495	2.9753	2.9772	.865	.867
3.750	3.755	3.1906	3.1901	3.1886	3.1878	.936	3.750	3.1885	3.1900	.930	.932
	3.765	3.1911	3.1906	3.1891	3.1883	.938	3.745	3.1873	3.1892	.928	.930
4.000	4.005	3.4015	3.4009	3.3992	3.3983	.988	4.000	3.3993	3.4008	.981	.983
	4.015	3.4021	3.4015	3.3998	3.3989	.990	3.990	3.3980	3.3999	.979	.981
4.500	4.505	3.8307	3.8301	3.8284	3.8275	1.123	4.500	3.8285	3.8300	1.116	1.118
	4.515	3.8313	3.8307	3.8290	3.8281	1.125	4.490	3.8272	3.8291	1.114	1.116

continued on next page

TABLE 10–30, continued

Nom Dia	D_h	d_h 1,2	d_h 3	d_h 4	d_h 5	W_h	D_s	d_s 1	d_s 2,3,4,5	W_s 1	W_s 2,3,4,5
5.000	5.005	4.2517	4.2511	4.2494	4.2485	1.248	5.000	4.2495	4.2510	1.241	1.243
	5.015	4.2523	4.2517	4.2500	4.2491	1.250	4.990	4.2482	4.2501	1.239	1.241
5.500	5.505	4.6807	4.6801	4.6783	4.6774	1.373	5.500	4.6785	4.6800	1.366	1.368
	5.515	4.6813	4.6807	4.6789	4.6780	1.375	5.490	4.6871	4.6790	1.364	1.366
6.000	6.005	5.1019	5.1013	5.0995	5.0986	1.498	6.000	5.0997	5.1012	1.491	1.493
	6.015	5.1025	5.1019	5.1001	5.0992	1.500	5.990	5.0983	5.1002	1.489	1.491

TABLE 10–31

Table of Dimensions for 6-Spline Key Fit

ASME Paper No. 48–SA–21 J. B. Armitage, Straight Sided Splines

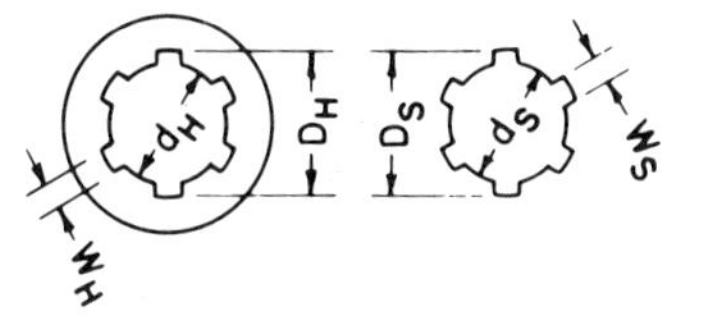

1. Free
2. Sliding Fit
3. Push Fit
4. Light Drive Fit
5. Press Fit

Nom Dia	D_h	d_h	W_h (all fits)	D_s	d_s	W_s 1	W_s 2	W_s 3	W_s 4	W_s 5
.750	.755	.638	.1880	.750	.633	.1862	.1877	.1879	.1892	.1897
	.765	.643	.1888	.745	.628	.1859	.1874	.1876	.1889	.1894
.875	.880	.744	.2188	.875	.739	.2170	.2185	.2187	.2200	.2205
	.890	.749	.2196	.870	.734	.2167	.2182	.2184	.2197	.2202
1.000	1.005	.850	.2500	1.000	.845	.2482	.2497	.2499	.2512	.2517
	1.015	.855	.2508	.995	.840	.2479	.2494	.2496	.2509	.2514
1.125	1.130	.956	.2812	1.125	.951	.2793	.2808	.2811	.2826	.2832
	1.140	.961	.2821	1.120	.946	.2789	.2804	.2807	.2822	.2828
1.250	1.255	1.063	.3125	1.250	1.058	.3106	.3121	.3124	.3139	.3145
	1.265	1.068	.3134	1.245	1.053	.3102	.3117	.3120	.3135	.3141
1.375	1.380	1.169	.3438	1.375	1.164	.3419	.3434	.3437	.3452	.3458
	1.390	1.174	.3447	1.370	1.159	.3415	.3430	.3433	.3448	.3454
1.500	1.505	1.275	.3750	1.500	1.270	.3730	.3745	.3749	.3765	.3771
	1.515	1.280	.3760	1.495	1.265	.3726	.3741	.3745	.3761	.3767
1.625	1.630	1.382	.4062	1.625	1.377	.4042	.4057	.4061	.4077	.4083
	1.640	1.387	.4072	1.620	1.372	.4038	.4053	.4057	.4073	.4079
1.750	1.755	1.488	.4375	1.750	1.483	.4355	.4370	.4374	.4390	.4396
	1.765	1.493	.4385	1.745	1.478	.4351	.4366	.4370	.4386	.4392
1.875	1.880	1.594	.4688	1.875	1.589	.4668	.4683	.4687	.4703	.4709
	1.890	1.599	.4698	1.870	1.584	.4664	.4679	.4683	.4699	.4705

continued on next page

TABLE 10-31, continued

Nom Dia	D_h	d_h	W_h (all fits)	D_s	d_s	W_s 1	W_s 2	W_s 3	W_s 4	W_s 5
2.000	2.005	1.700	.5000	2.000	1.695	.4979	.4994	.4999	.5017	.5024
	2.015	1.705	.5011	1.995	1.690	.4974	.4989	.4994	.5012	.5019
2.250	2.255	1.913	.5625	2.250	1.908	.5604	.5619	.5624	.5642	.5649
	2.265	1.918	.5636	2.245	1.903	.5599	.5614	.5619	.5637	.5644
2.500	2.505	2.125	.6250	2.500	2.120	.6229	.6244	.6249	.6267	.6274
	2.515	2.130	.6261	2.495	2.115	.6224	.6239	.6244	.6262	.6269
2.750	2.755	2.340	.6875	2.750	2.335	.6854	.6869	.6874	.6892	.6899
	2.765	2.345	.6886	2.745	2.330	.6849	.6864	.6869	.6887	.6894
3.000	3.005	2.551	.7500	3.000	2.546	.7479	.7494	.7499	.7518	.7525
	3.015	2.556	.7512	2.995	2.541	.7474	.7489	.7494	.7513	.7520
3.250	3.255	2.765	.8125	3.250	2.760	.8104	.8119	.8124	.8143	.8150
	3.265	2.770	.8137	3.245	2.755	.8099	.8114	.8119	.8138	.8145
3.500	3.505	2.978	.8750	3.500	2.973	.8729	.8744	.8749	.8768	.8775
	3.515	2.983	.8762	3.495	2.968	.8724	.8739	.8744	.8763	.8770
3.750	3.755	3.190	.9375	3.750	3.185	.9354	.9369	.9374	.9393	.9400
	3.765	3.195	.9387	3.745	3.180	.9349	.9364	.9369	.9388	.9395
4.000	4.005	3.401	1.0000	4.000	3.396	.9978	.9993	.9999	1.0020	1.0028
	4.015	3.411	1.0013	3.990	3.386	.9972	.9987	.9993	1.0014	1.0022
4.500	4.505	3.830	1.1250	4.500	3.825	1.1228	1.1243	1.1249	1.1270	1.1278
	4.515	3.840	1.1263	4.490	3.815	1.1222	1.1237	1.1243	1.1264	1.1272
5.000	5.005	4.251	1.2500	5.000	4.246	1.2478	1.2493	1.2499	1.2520	1.2528
	5.015	4.261	1.2513	4.990	4.236	1.2472	1.2487	1.2493	1.2514	1.2522
5.500	5.505	4.680	1.3750	5.500	4.670	1.3728	1.3743	1.3749	1.3771	1.3779
	5.515	4.690	1.3764	5.490	4.660	1.3722	1.3737	1.3743	1.3765	1.3773
6.000	6.005	5.101	1.5000	6.000	5.091	1.4978	1.4993	1.4999	1.5021	1.5029
	6.015	5.111	1.5014	5.990	5.081	1.4972	1.4987	1.4993	1.5015	1.5023

TABLE 10–32

Table of Dimensions for 6-Spline Major Diameter Fit

ASME Paper No. 48–SA–21 J. B. Armitage, Straight Sided Splines

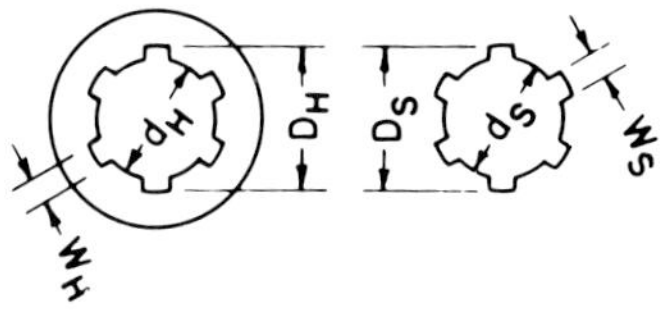

1. Free
2. Sliding Fit
3. Push Fit
4. Light Drive Fit
5. Press Fit

Nom Dia	D_h (all fits)	d_h	W_h	D_s 1	D_s 2	D_s 3	D_s 4	D_s 5	d_s (all fits)	W_s 1	W_s 2,3,4,5
.750	.7500	.638	.186	.7482	.7497	.7499	.7512	.7517	.633	.183	.184
	.7508	.643	.188	.7479	.7494	.7496	.7509	.7514	.628	.181	.183
.875	.8750	.744	.217	.8732	.8747	.8749	.8762	.8767	.739	.213	.214
	.8758	.749	.219	.8729	.8744	.8746	.8759	.8764	.734	.211	.213
1.000	1.0000	.850	.248	.9982	.9997	.9999	1.0012	1.0017	.845	.244	.245
	1.0008	.855	.250	.9979	.9994	.9996	1.0009	1.0014	.840	.242	.244
1.125	1.1250	.956	.279	1.1231	1.1246	1.1249	1.1264	1.1270	.951	.275	.276
	1.1259	.961	.281	1.1227	1.1242	1.1245	1.1260	1.1266	.946	.273	.275
1.250	1.2500	1.063	.311	1.2481	1.2496	1.2499	1.2514	1.2520	1.058	.307	.308
	1.2509	1.068	.313	1.2477	1.2492	1.2495	1.2510	1.2516	1.053	.305	.307
1.375	1.3750	1.169	.342	1.3731	1.3746	1.3749	1.3764	1.3770	1.164	.338	.339
	1.3759	1.174	.344	1.3727	1.3742	1.3745	1.3760	1.3766	1.159	.336	.338
1.500	1.5000	1.275	.373	1.4980	1.4995	1.4999	1.5015	1.5021	1.270	.368	.369
	1.5010	1.280	.375	1.4976	1.4991	1.4995	1.5011	1.5017	1.265	.366	.368
1.625	1.6250	1.382	.404	1.6230	1.6245	1.6249	1.6265	1.6271	1.377	.399	.400
	1.6260	1.387	.406	1.6226	1.6241	1.6245	1.6261	1.6267	1.372	.397	.399
1.750	1.7500	1.488	.436	1.7480	1.7495	1.7499	1.7515	1.7521	1.483	.431	.432
	1.7510	1.493	.438	1.7476	1.7491	1.7495	1.7511	1.7517	1.478	.429	.431
1.875	1.8750	1.594	.467	1.8730	1.8745	1.8749	1.8765	1.8771	1.589	.462	.463
	1.8760	1.599	.469	1.8726	1.8741	1.8745	1.8761	1.8767	1.584	.460	.462
2.000	2.0000	1.700	.498	1.9979	1.9994	1.9999	2.0017	2.0024	1.695	.493	.494
	2.0011	1.705	.500	1.9974	1.9989	1.9994	2.0012	2.0019	1.690	.491	.493
2.250	2.2500	1.913	.560	2.2479	2.2494	2.2499	2.2517	2.2524	1.908	.555	.556
	2.2511	1.918	.562	2.2474	2.2489	2.2494	2.2512	2.2519	1.903	.553	.555
2.500	2.5000	2.125	.623	2.4979	2.4994	2.4999	2.5017	2.5024	2.120	.617	.619
	2.5011	2.130	.625	2.4974	2.4789	2.4994	2.5012	2.5019	2.115	.615	.617
2.750	2.7500	2.340	.686	2.7479	2.7494	2.7499	2.7517	2.7524	2.335	.680	.682
	2.7511	2.345	.688	2.7474	2.7489	2.7494	2.7512	2.7519	2.330	.678	.680
3.000	3.0000	2.551	.748	2.9979	2.9994	2.9999	3.0018	3.0025	2.546	.742	.744
	3.0012	2.556	.750	2.9974	2.9989	2.9994	3.0013	3.0020	2.541	.740	.742
3.250	3.2500	2.765	.810	3.2479	3.2494	3.2499	3.2518	3.2525	2.760	.804	.806
	3.2512	2.770	.812	3.2474	3.2489	3.2494	3.2513	3.2520	2.755	.802	.804
3.500	3.5000	2.978	.873	3.4979	3.4994	3.4999	3.5018	3.5025	2.973	.867	.869
	3.5012	2.983	.875	3.4974	3.4989	3.4994	3.5013	3.5020	2.968	.865	.867
3.750	3.7500	3.190	.936	3.7479	3.7494	3.7499	3.7518	3.7525	3.185	.930	.932
	3.7512	3.195	.938	3.7474	3.7489	3.7494	3.7513	3.7520	3.180	.928	.930
4.000	4.0000	3.401	.988	3.9978	3.9993	3.9999	4.0020	4.0028	3.396	.981	.983
	4.0013	3.411	.990	3.9972	3.9987	3.9993	4.0014	4.0022	3.386	.979	.981
4.500	4.5000	3.830	1.123	4.4978	4.4993	4.4999	4.5020	4.5028	3.825	1.116	1.118
	4.5013	3.840	1.125	4.4972	4.4987	4.4993	4.5014	4.5022	3.815	1.114	1.116

continued on next page

TABLE 10–32, continued

Nom Dia	D_h (all fits)	d_h	W_h	D_s 1	D_s 2	D_s 3	D_s 4	D_s 5	d_s (all fits)	W_s 1	W_s 2,3,4,5
5.000	5.0000	4.251	1.248	4.9978	4.9993	4.9999	5.0020	5.0028	4.246	1.241	1.243
	5.0013	4.261	1.250	4.9972	4.9987	4.9993	5.0014	5.0022	4.236	1.239	1.241
5.500	5.5000	4.680	1.373	5.4978	5.4993	5.4999	5.5021	5.5029	4.670	1.366	1.368
	5.5014	4.690	1.375	5.4972	5.4987	5.4993	5.5015	5.5023	4.660	1.364	1.366
6.000	6.0000	5.101	1.498	5.9978	5.9993	5.9999	6.0021	6.0029	5.091	1.491	1.493
	6.0014	5.111	1.500	5.9972	5.9987	5.9993	6.0015	6.0023	5.081	1.489	1.491

TABLE 10–33

Table of Dimensions for 10-Spline Minor Diameter Fit

ASME Paper No. 48–SA–21 J. B. Armitage, Straight Sided Splines

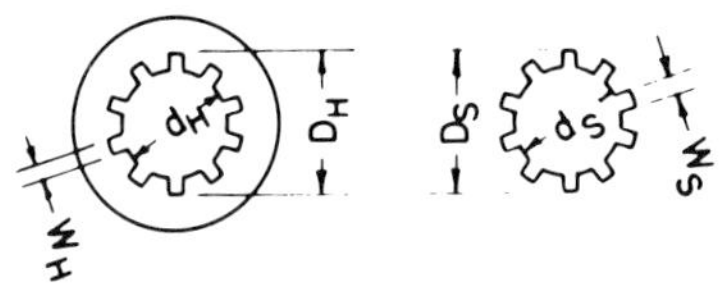

1. Free
2. Sliding Fit
3. Push Fit
4. Light Drive Fit
5. Press Fit

Nom Dia	D_h	d_h 1,2	d_h 3	d_h 4	d_h 5	W_h	D_s	d_s 1	d_s 2,3,4,5	W_s 1	W_s 2,3,4,5
.750	.755	.6453	.6451	.6442	.6437	.115	.750	.6435	.6450	.112	.113
	.765	.6456	.6454	.6445	.6440	.117	.745	.6427	.6446	.110	.112
.875	.880	.7528	.7526	.7517	.7512	.135	.875	.7510	.7525	.131	.132
	.890	.7531	.7529	.7520	.7515	.137	.870	.7502	.7521	.129	.131
1.000	1.005	.8603	.8601	.8592	.8587	.154	1.000	.8585	.8600	.150	.151
	1.015	.8606	.8604	.8595	.8590	.156	.995	.8577	.8596	.148	.150
1.125	1.130	.9679	.9676	.9665	.9659	.174	1.125	.9660	.9675	.170	.171
	1.140	.9683	.9680	.9669	.9663	.176	1.120	.9651	.9670	.168	.170
1.250	1.255	1.0754	1.0751	1.0740	1.0734	.193	1.250	1.0735	1.0750	.189	.190
	1.265	1.0758	1.0755	1.0744	1.0738	.195	1.245	1.0726	1.0745	.187	.189
1.375	1.380	1.1829	1.1826	1.1815	1.1809	.213	1.375	1.1810	1.1825	.209	.210
	1.390	1.1833	1.1830	1.1819	1.1813	.215	1.370	1.1801	1.1820	.207	.209
1.500	1.505	1.2905	1.2901	1.2889	1.2883	.232	1.500	1.2885	1.2900	.228	.228
	1.515	1.2909	1.2905	1.2893	1.2887	.234	1.495	1.2875	1.2894	.226	.227
1.625	1.630	1.3980	1.3976	1.3964	1.3958	.252	1.625	1.3960	1.3975	.247	.248
	1.640	1.3984	1.3980	1.3968	1.3962	.254	1.620	1.3950	1.3969	.245	.247
1.750	1.755	1.5055	1.5051	1.5039	1.5033	.271	1.750	1.5035	1.5050	.266	.267
	1.765	1.5059	1.5055	1.5043	1.5037	.273	1.745	1.5025	1.5044	.264	.266
1.875	1.880	1.6130	1.6126	1.6114	1.6108	.291	1.875	1.6110	1.6125	.286	.287
	1.890	1.6134	1.6130	1.6118	1.6112	.293	1.870	1.6100	1.6119	.284	.286

continued on next page

TABLE 10–33, continued

Nom Dia	D_h	d_h				W_h	D_s	d_s		W_s	
		1,2	3	4	5			1	2,3,4,5	1	2,3,4,5
2.000	2.005	1.7206	1.7201	1.7187	1.7180	.310	2.000	1.7185	1.7200	.305	.306
	2.015	1.7211	1.7206	1.7192	1.7185	.312	1.995	1.7174	1.7193	.303	.305
2.250	2.255	1.9356	1.9351	1.9337	1.9330	.349	2.250	1.9335	1.9350	.344	.345
	2.265	1.9361	1.9356	1.9342	1.9335	.351	2.245	1.9324	1.9343	.342	.344
2.500	2.505	2.1506	2.1501	2.1487	2.1480	.388	2.500	2.1485	2.1500	.482	.384
	2.515	2.1511	2.1506	2.1492	2.1485	.390	2.495	2.1474	2.1493	.480	.383
2.750	2.755	2.3656	2.3651	2.3637	2.3630	.427	2.750	2.3635	2.3650	.521	.423
	2.765	2.3661	2.3656	2.3642	2.3635	.429	2.745	2.3624	2.3643	.519	.421
3.000	3.005	2.5806	2.5801	2.5786	2.5778	.466	3.000	2.5785	2.5800	.460	.462
	3.015	2.5811	2.5806	2.5791	2.5783	.468	2.995	2.5773	2.5792	.458	.460
3.250	3.255	2.7956	2.7951	2.7936	2.7928	.505	3.250	2.7935	2.7950	.499	.501
	3.265	2.7961	2.7956	2.7941	2.7933	.507	3.245	2.7923	2.7942	.497	.499
3.500	3.505	3.0106	3.0101	3.0086	3.0078	.544	3.500	3.0085	3.0100	.498	.540
	3.515	3.0111	3.0106	3.0091	3.0083	.546	3.495	3.0073	3.0092	.496	.538
3.750	3.755	3.2256	3.2251	3.2236	3.2228	.583	3.750	3.2235	3.2250	.497	.579
	3.765	3.2261	3.2256	3.2241	3.2233	.585	3.745	3.2223	3.2242	.495	.577
4.000	4.005	3.4407	3.4401	3.4384	3.4375	.622	4.000	3.4385	3.4400	.615	.617
	4.015	3.4413	3.4407	3.4390	3.4381	.624	3.990	3.4372	3.4391	.613	.615
4.500	4.505	3.8707	3.8701	3.8684	3.8675	.700	4.500	3.8685	3.8700	.693	.695
	4.515	3.8713	3.8707	3.8690	3.8681	.702	4.490	3.8672	3.8691	.691	.693
5.000	5.005	4.3007	4.3001	4.2984	4.2975	.778	5.000	4.2985	4.3000	.771	.773
	5.015	4.3013	4.3007	4.2990	4.2981	.780	4.990	4.2972	4.2991	.769	.771
5.500	5.505	4.7307	4.7301	4.7283	4.7274	.856	5.500	4.7285	4.7300	.849	.851
	5.515	4.7313	4.7307	4.7289	4.7280	.858	5.490	4.7271	4.7290	.847	.849
6.000	6.005	5.1607	5.1601	5.1583	5.1574	.934	6.000	5.1585	5.1600	.927	.929
	6.015	5.1613	5.1607	5.1589	5.1580	.936	5.990	5.1571	5.1590	.925	.927

TABLE 10–34
Table of Dimensions for 16-Spline Minor Diameter Fit
ASME Paper No. 48–SA–21 J. B. Armitage, Straight Sided Splines

Nom Dia	D_h	d_h				W_h	D_s	d_s		W_s	
		1,2	3	4	5			1	2,3,4,5	1	2,3,4,5
.750	.755	.6453	.6451	.6442	.6437	.072	.750	.6435	.6450	.069	.070
	.765	.6456	.6454	.6445	.6440	.074	.745	.6427	.6446	.067	.069
.875	.880	.7528	.7526	.7517	.7512	.084	.875	.7510	.7525	.080	.081
	.890	.7531	.7529	.7520	.7515	.086	.870	.7502	.7521	.078	.080
1.000	1.005	.8603	.8601	.8592	.8587	.096	1.000	.8585	.8600	.092	.093
	1.015	.8606	.8604	.8595	.8590	.098	.995	.8577	.8596	.090	.092
1.125	1.130	.9679	.9676	.9665	.9659	.108	1.125	.9660	.9675	.104	.105
	1.140	.9683	.9680	.9669	.9663	.110	1.120	.9651	.9670	.102	.104
1.250	1.255	1.0754	1.0751	1.0740	1.0734	.120	1.250	1.0735	1.0750	.116	.117
	1.265	1.0758	1.0755	1.0744	1.0738	.122	1.245	1.0726	1.0745	.114	.116

continued on next page

TABLE 10–34, continued

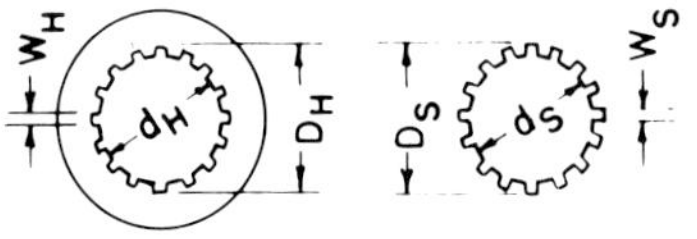

1. Free
2. Sliding Fit
3. Push Fit
4. Light Drive Fit
5. Press Fit

Nom Dia	D_h	d_h 1,2	d_h 3	d_h 4	d_h 5	W_h	D_s	d_s 1	d_s 2,3,4,5	W_s 1	W_s 2,3,4,5
1.375	1.380	1.1829	1.1826	1.1815	1.1809	.132	1.375	1.1810	1.1825	.128	.129
	1.390	1.1833	1.1830	1.1819	1.1813	.134	1.370	1.1801	1.1820	.126	.128
1.500	1.505	1.2905	1.2901	1.2889	1.2883	.145	1.500	1.2885	1.2900	.140	.141
	1.515	1.2909	1.2905	1.2893	1.2887	.147	1.495	1.2875	1.2894	.138	.140
1.625	1.630	1.3980	1.3976	1.3964	1.3958	.157	1.625	1.3960	1.3975	.152	.153
	1.640	1.3984	1.3980	1.3968	1.3962	.159	1.620	1.3950	1.3969	.150	.152
1.750	1.755	1.5055	1.5051	1.5039	1.5033	.169	1.750	1.5035	1.5050	.164	.165
	1.765	1.5059	1.5055	1.5043	1.5037	.171	1.745	1.5025	1.5044	.162	.164
1.875	1.880	1.6130	1.6126	1.6114	1.6108	.181	1.875	1.6110	1.6125	.176	.177
	1.890	1.6134	1.6130	1.6118	1.6112	.183	1.870	1.6100	1.6119	.174	.176
2.000	2.005	1.7206	1.7201	1.7187	1.7180	.194	2.000	1.7185	1.7200	.189	.190
	2.015	1.7211	1.7206	1.7192	1.7185	.196	1.995	1.7174	1.7193	.187	.189
2.250	2.255	1.9356	1.9351	1.9337	1.9330	.218	2.250	1.9335	1.9350	.213	.214
	2.265	1.9361	1.9355	1.9342	1.9335	.220	2.245	1.9324	1.9343	.211	.213
2.500	2.505	2.1506	2.1501	2.1487	2.1480	.243	2.500	2.1485	2.1500	.237	.239
	2.515	2.1511	2.1506	2.1492	2.1485	.245	2.495	2.1474	2.1493	.235	.237
2.750	2.755	2.3656	2.3651	2.3637	2.3630	.267	2.750	2.3635	2.3650	.261	.263
	2.765	2.3661	2.3656	2.3642	2.3635	.269	2.745	2.3624	2.3643	.259	.261
3.000	3.005	2.5806	2.5801	2.5786	2.5778	.292	3.000	2.5785	2.5800	.286	.288
	3.015	2.5811	2.5806	2.5791	2.5783	.294	2.995	2.5773	2.5792	.284	.286
3.250	3.255	2.7956	2.7951	2.7936	2.7928	.316	3.250	2.7935	2.7950	.310	.312
	3.265	2.7961	2.7956	2.7941	2.7933	.318	3.245	2.7923	2.7942	.308	.310
3.500	3.505	3.0106	3.0101	3.0086	3,0078	.341	3.500	3.0085	3.0100	.335	.337
	3.515	3.0111	3.0106	3.0091	3.0083	.343	3.495	3.0073	3.0092	.333	.335
3.750	3.755	3.2256	3.2251	3.2236	3.2228	.365	3.750	3.2235	3.2250	.359	.361
	3.765	3.2261	3.2256	3.2241	3.2233	.367	3.745	3.2223	3.2242	.357	.359
4.000	4.005	3.4407	3.4401	3.4384	3.4375	.390	4.000	3.4385	3.4400	.383	.385
	4.015	3.4413	3.4407	3.4390	3.4381	.392	3.990	3.4372	3.4391	.381	.383
4.500	4.505	3.8707	3.8701	3.8684	3.8675	.439	4.500	3.8685	3.8700	.432	.434
	4.515	3.8713	3.8707	3.8690	3.8681	.441	4.490	3.8672	3.8691	.430	.432
5.000	5.005	4.3007	4.3001	4.2984	4.2975	.488	5.000	4.2985	4.3000	.481	.483
	5.015	4.3013	4.3007	4.2990	4.2981	.490	4.990	4.2972	4.2991	.479	.481
5.500	5.505	4.7307	4.7301	4.7283	4.7274	.537	5.500	4.7285	4.7300	.530	.532
	5.515	4.7313	4.7307	4.7289	4.7280	.539	5.490	4.7271	4.7290	.528	.530
6.000	6.005	5.1607	5.1601	5.1583	5.1574	.586	6.000	5.1585	5.1600	.579	.581
	6.015	5.1613	5.1607	5.1589	5.1580	.588	5.990	5.1571	5.1590	.577	.579

TABLE 10-35
Basic Dimensions for Major Diameter Fits

ASME Paper No. 48-SA-21 J. B. Armitage, Straight Sided Splines

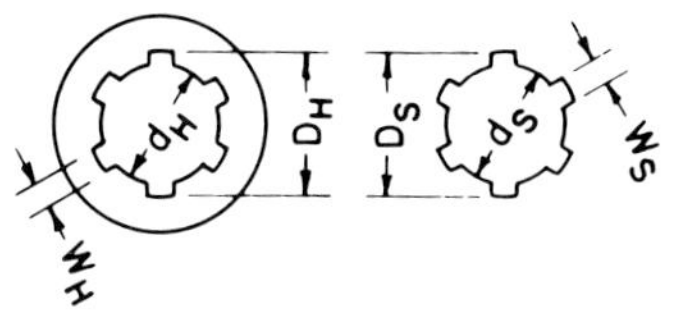

1. Free
2. Sliding Fit
3. Push Fit
4. Light Drive Fit
5. Press Fit

D_h (minimum) — Nominal Diameter

Nom Dia	Minimum Clearance or Maximum Interference Clearance is Plus, Interference is Minus				
	1	2	3	4	5
3/4	+.0018	+.0003	+.0001	-.0012	-.0017
7/8	+.0018	+.0003	+.0001	-.0012	-.0017
1	+.0018	+.0003	+.0001	-.0012	-.0017
1-1/8	+.0019	+.0004	+.0001	-.0014	-.0020
1-1/4	+.0019	+.0004	+.0001	-.0014	-.0020
1-3/8	+.0019	+.0004	+.0001	-.0014	-.0020
1-1/2	+.0020	+.0005	+.0001	-.0015	-.0021
1-5/8	+.0020	+.0005	+.0001	-.0015	-.0021
1-3/4	+.0020	+.0005	+.0001	-.0015	-.0021
1-7/8	+.0020	+.0005	+.0001	-.0015	-.0021
2	+.0021	+.0006	+.0001	-.0017	-.0024
2-1/4	+.0021	+.0006	+.0001	-.0017	-.0024
2-1/2	+.0021	+.0006	+.0001	-.0017	-.0024
2-3/4	+.0021	+.0006	+.0001	-.0017	-.0024
3	+.0021	+.0006	+.0001	-.0018	-.0025
3-1/4	+.0021	+.0006	+.0001	-.0018	-.0025
3-1/2	+.0021	+.0006	+.0001	-.0018	-.0025
3-3/4	+.0021	+.0006	+.0001	-.0018	-.0025
4	+.0022	+.0007	+.0001	-.0020	-.0028
4-1/2	+.0022	+.0007	+.0001	-.0020	-.0028
5	+.0022	+.0007	+.0001	-.0020	-.0028
5-1/2	+.0022	+.0007	+.0001	-.0021	-.0029
6	+.0022	+.0007	+.0001	-.0021	-.0029

TABLE 10–36
Tolerances for Major Diameter Fits

ASME Paper No. 48–SA–21 J. B. Armitage, Straight Sided Splines

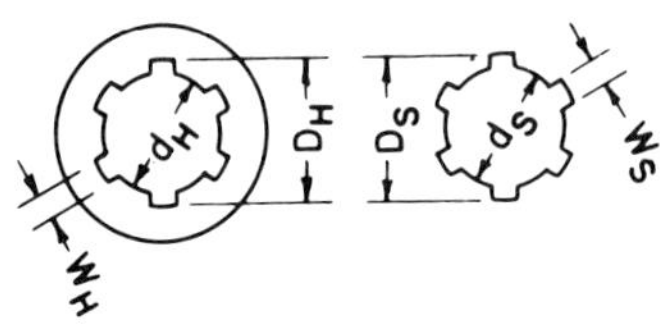

1. Free
2. Sliding Fit
3. Push Fit
4. Light Drive Fit
5. Press Fit

	Tolerance	
Nom Dia	D_h 1,2,3,4,5	D_s 1,2,3,4,5
3/4	.0008	.0003
7/8	.0008	.0003
1	.0008	.0003
1-1/8	.0009	.0004
1-1/4	.0009	.0004
1-3/8	.0009	.0004
1-1/2	.0010	.0004
1-5/8	.0010	.0004
1-3/4	.0010	.0004
1-7/8	.0010	.0004
2	.0011	.0005
2-1/4	.0011	.0005
2-1/2	.0011	.0005
2-3/4	.0011	.0005
3	.0012	.0005
3-1/4	.0012	.0005
3-1/2	.0012	.0005
3-3/4	.0012	.0005
4	.0013	.0006
4-1/2	.0013	.0006
5	.0013	.0006
5-1/2	.0014	.0006
6	.0014	.0006

TABLE 10–37
Tolerances on Key Fits

ASME Paper No. 48–SA–21 J. B. Armitage, Straight Sided Splines

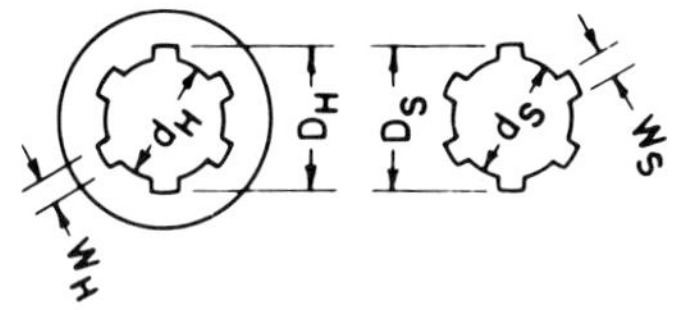

1. Free
2. Sliding Fit
3. Push Fit
4. Light Drive Fit
5. Press Fit

Nom Dia	Tolerance D_h 1,2,3,4,5	Tolerance D_s 1,2,3,4,5
3/4	.0008	.0003
7/8	.0008	.0003
1	.0008	.0003
1-1/8	.0009	.0004
1-1/4	.0009	.0004
1-3/8	.0009	.0004
1-1/2	.0010	.0004
1-5/8	.0010	.0004
1-3/4	.0010	.0004
1-7/8	.0010	.0004
2	.0011	.0005
2-1/4	.0011	.0005
2-1/2	.0011	.0005
2-3/4	.0011	.0005
3	.0012	.0005
3-1/4	.0012	.0005
3-1/2	.0012	.0005
3-3/4	.0012	.0005
4	.0013	.0006
4-1/2	.0013	.0006
5	.0013	.0006
5-1/2	.0014	.0006
6	.0014	.0006

TABLE 10-38
Basic Dimensions for Key Fits

ASME Paper No. 48-SA-21 J. B. Armitage, Straight Sided Splines

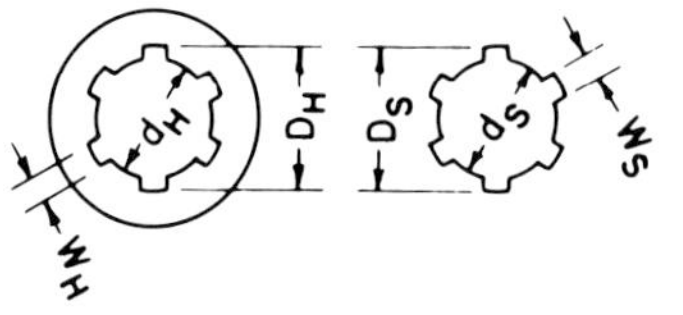

1. Free
2. Sliding Fit
3. Push Fit
4. Light Drive Fit
5. Press Fit

W_h (Minimum) for 6 splines — 1/4 Nom Dia
W_h (Minimum) for 10 splines — .156 Nom Dia
W_h (Minimum) for 16 splines — .098 Nom Dia

Nom Dia	Minimum Clearance or Maximum Interference Clearance is Plus, Interference is Minus				
	1	2	3	4	5
3/4	+.0018	+.0003	+.0001	-.0012	-.0017
7/8	+.0018	+.0003	+.0001	-.0012	-.0017
1	+.0018	+.0003	+.0001	-.0012	-.0017
1-1/8	+.0019	+.0004	+.0001	-.0014	-.0020
1-1/4	+.0019	+.0004	+.0001	-.0014	-.0020
1-3/8	+.0019	+.0004	+.0001	-.0014	-.0020
1-1/2	+.0020	+.0005	+.0001	-.0015	-.0021
1-5/8	+.0020	+.0005	+.0001	-.0015	-.0021
1-3/4	+.0020	+.0005	+.0001	-.0015	-.0021
1-7/8	+.0020	+.0005	+.0001	-.0015	-.0021
2	+.0021	+.0006	+.0001	-.0017	-.0024
2-1/4	+.0021	+.0006	+.0001	-.0017	-.0024
2-1/2	+.0021	+.0006	+.0001	-.0017	-.0024
2-3/4	+.0021	+.0006	+.0001	-.0017	-.0024
3	+.0021	+.0006	+.0001	-.0018	-.0025
3-1/4	+.0021	+.0006	+.0001	-.0018	-.0025
3-1/2	+.0021	+.0006	+.0001	-.0018	-.0025
3-3/4	+.0021	+.0006	+.0001	-.0018	-.0025
4	+.0022	+.0007	+.0001	-.0020	-.0028
4-1/2	+.0022	+.0007	+.0001	-.0020	-.0028
5	+.0022	+.0007	+.0001	-.0020	-.0028
5-1/2	+.0022	+.0007	+.0001	-.0021	-.0029
6	+.0022	+.0007	+.0001	-.0021	-.0029

TABLE 10–39

Stress Analysis of Splined and Serrated Fits

T = Maximum allowable torque (inch pounds)

S = Allowable shear stress in pounds per square inch of either shaft or hub, whichever is lower.

D = Pitch diameter of spline or serration (in inches). In the case of straight sided splines, D is the numerical average of major and minor diameters (see example below).

L = Length of contact of splined or serrated fit (in inches).

For all spline fits

$$T = (0.7854)\ D^2 LS$$

Example: A bronze worm gear on a C-1045 shaft.
2-inch nominal diameter straight-sided six spline.
Allowable shear stress in shaft: 10,000 psi.
Allowable shear stress in worm gear: 5,000 psi.
Length of spline: 2¼ inches.

To determine D:

For a 2-inch six-sided spline, the minor diameter is approximately 1.700 (Table 10–31).

D = 1/2(2 + 1.700)

= 1/2(3.700)

= 1.850

S = 5,000 psi, the allowable shear stress of the weaker of the two materials

$T = 0.1965\ D^2LS$

$= (0.7854)\ (1.850)^2 (2\text{-}1/4)\ (5000)$

T = 24,000 lb. in.

INDEX TO

SECTION 11

Snaprings
Pins

Nuts

Wrench Openings
Washers

TABLE 11-1

Dimensions of Standard and Oversize, Hardened and Ground, Dowel Pins

ASA B5.20-1947 Machine Pins

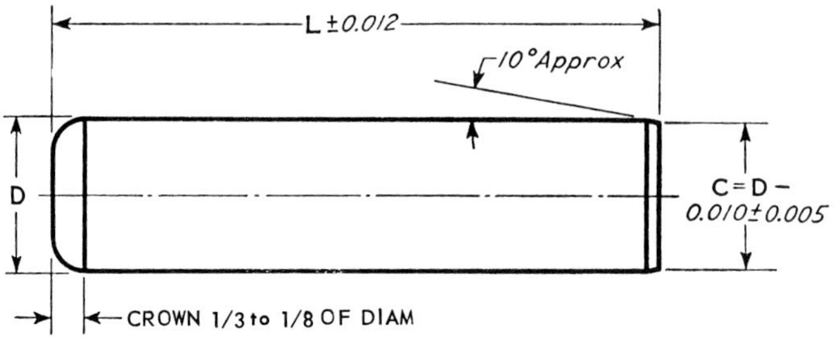

Length, L	Nominal Diameter D									
	1/8	3/16	1/4	5/16	3/8	7/16	1/2	5/8	3/4	7/8
	Diameter Standard Pins ±0.0001									
	0.1252	0.1877	0.2502	0.3127	0.3752	0.4377	0.5002	0.6252	0.7502	0.8752
	Diameter Oversize Pins ±0.0001									
	0.1260	0.1885	0.2510	0.3135	0.3760	0.4385	0.5010	0.6260	0.7510	0.8760
1/2	X	X	X	X						
5/8	X	X	X	X						
3/4	X	X	X	X	X					
7/8	X	X	X	X	X	X				
1	X	X	X	X	X	X				
1 1/4		X	X	X	X	X	X	X		
1 1/2		X	X	X	X	X	X	X	X	
1 3/4		X	X	X	X	X	X	X	X	
2		X	X	X	X	X	X	X	X	X
2 1/4				X	X	X	X			
2 1/2				X	X	X	X	X	X	X
3							X	X	X	X
3 1/2							X	X		
4							X	X	X	X
4 1/2								X	X	X
5									X	X
5 1/2									X	X

All dimensions are given in inches.

These pins are extensively used in the tool and machine industry and a machine reamer of nominal size may be used to produce the holes into which these pins tap or press fit. They must be straight and free from any defects that will affect their serviceability.

TABLE 11-2

Dimensions of Straight Pins — Chamfered and Square End

ASA B5.20-1947 Machine Pins

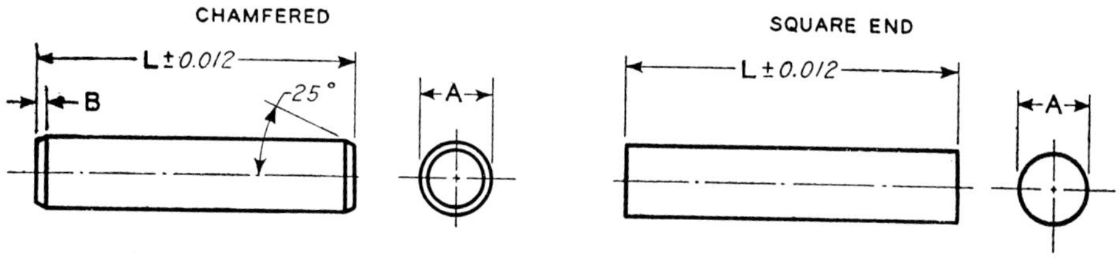

Nominal Diameter	Diameter A		Chamfer B
	Max	Min	
0.062	0.0625	0.0605	0.010
0.094	0.0937	0.0917	0.010
0.109	0.1094	0.1074	0.010
0.125	0.1250	0.1230	0.010
0.156	0.1562	0.1542	1/64
0.188	0.1875	0.1855	1/64
0.219	0.2187	0.2167	1/64
0.250	0.2500	0.2480	1/64
0.312	0.3125	0.3095	1/32
0.375	0.3750	0.3720	1/32
0.438	0.4375	0.4345	1/32
0.500	0.500	0.4970	1/32

All dimensions are given in inches.

These pins must be straight and free from burrs or any other defects that will affect their serviceability.

TABLE 11–3

Dimensions of Ground Dowel Pins (Not Hardened)

ASA B5.20–1947 Machine Pins

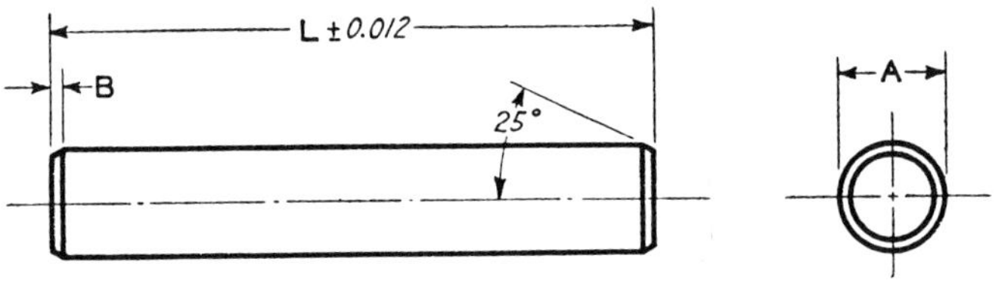

Nominal Diameter	Diameter A Max	Diameter A Min	Chamfer B
0.062	0.0600	0.0595	0.010
0.094	0.0912	0.0907	0.010
0.109	0.1068	0.1063	0.010
0.125	0.1223	0.1218	0.010
0.156	0.1535	0.1530	1/64
0.188	0.1847	0.1842	1/64
0.219	0.2159	0.2154	1/64
0.250	0.2470	0.2465	1/64
0.312	0.3094	0.3089	1/32
0.375	0.3717	0.3712	1/32
0.438	0.4341	0.4336	1/32
0.500	0.4964	0.4959	1/32
0.625	0.6211	0.6206	3/64
0.750	0.7458	0.7453	3/64
0.875	0.8705	0.8700	1/16
1.000	0.9952	0.9947	1/16

All dimensions are given in inches.

Maximum diameters are graduated from 0.0005 on 1/16 in. pins to 0.0028 on 1-in. pins under the minimum commercial bar stock sizes.

TABLE 11-4

Sizes and Dimensions of Taper* Pins

ASA B5.20-1947 Machine Pins

Number	7/0	6/0	5/0	4/0	3/0	2/0	0	1	2	3	4	5	6	7	8	9	10
Size (Large End)	0.0625	0.0780	0.0940	0.1090	0.1250	0.1410	0.1560	0.1720	0.1930	0.2190	0.2500	0.2890	0.3410	0.4090	0.4920	0.5910	0.7060
Length, L																	
0.375	X	X															
0.500	X	X	X	X	X	X	X										
0.625	X	X	X	X	X	X	X										
0.750		X	X	X	X	X	X	X	X	X							
0.875					X	X	X	X	X	X							
1.000			X	X	X	X	X	X	X	X	X	X					
1.250						X	X	X	X	X	X	X	X				
1.500							X	X	X	X	X	X	X				
1.750								X	X	X	X	X	X				
2.000								X	X	X	X	X	X	X	X		
2.250									X	X	X	X	X	X	X		
2.500									X	X	X	X	X	X	X		
2.750										X	X	X	X	X	X	X	
3.000										X	X	X	X	X	X	X	
3.250													X	X	X	X	
3.500													X	X	X	X	X
3.750													X	X	X	X	X
4.000													X	X	X	X	X
4.250															X	X	X
4.500															X	X	X
4.750															X	X	X
5.000															X	X	X
5.250																X	X
5.500																X	X
5.750																X	X
6.000																X	X

*Standard reamers are available for pins given above the line, Table 9-25. All dimensions are given in inches.

Pins Nos. 11 (size 0.8600), 12 (size 1.032), 13 (size 1.241), and 14 (1.523) are special sizes — hence their lengths are special.

To find small diameter of pin, multiply the length by 0.02083 and subtract the result from the large diameter.

TYPES	COMMERCIAL TYPE	PRECISION TYPE
Sizes	7/0 to 14	7/0 to 10
Tolerance on Diameter	(+0.0013, −0.0007)	(+0.0013, −0.0007)
Taper	1/4 In. per Ft.	1/4 In. per Ft.
Length Tolerance	(±0.030)	(±0.030)
Concavity Tolerance	None	0.0005 up to 1 in. long 0.001 1-1/16 to 2 in. long 0.002 2-1/16 and longer

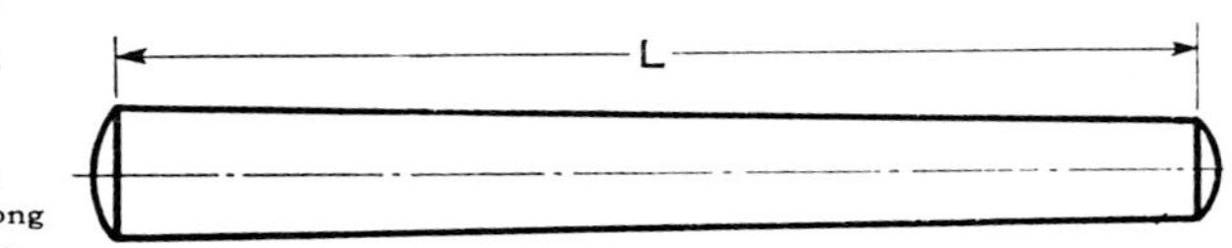

TABLE 11-5

Rollpin Dimensions, Hole Sizes, and Shear Strengths

Catalog No. 825 Elastic Stop Nut Corporation of America

Tolerance on specified length "L":

0.187 to 1.000 ±.015
1.001 to 2.000 ±.020
2.001 to 3.000 ±.025
3.001 to 4.000 ±.030
4.001 & above ±.035

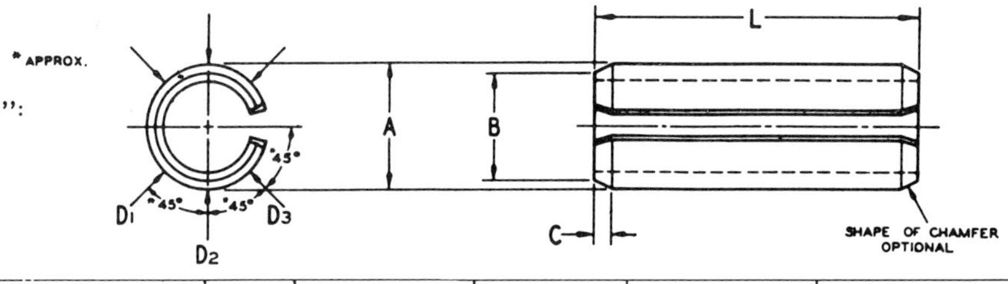

Nominal	A Maximum (Go Ring Gage)	A Minimum $^{1}/_{3}(D_1 + D_2 + D_3)$	B Max	C Min	C Max	Stock Thickness	Recommended Hole Size Min	Recommended Hole Size Max	Minimum Double Shear Strength, Pounds Carbon Steel and Stainless Steel
.062	.069	.066	.059	.007	.028	.012	.062	.065	425
.078	.086	.083	.075	.008	.032	.018	.078	.081	650
.094	.103	.099	.091	.008	.038	.022	.094	.097	1,000
.125	.135	.131	.122	.008	.044	.028	.125	.129	2,100
.140	.149	.145	.137	.008	.044	.028	.140	.144	2,200
.156	.167	.162	.151	.010	.048	.032	.156	.160	3,000
.187	.199	.194	.182	.011	.055	.040	.187	.192	4,400
.219	.232	.226	.214	.011	.065	.048	.219	.224	5,700
.250	.264	.258	.245	.012	.065	.048	.250	.256	7,700
.312	.328	.321	.306	.014	.080	.062	.312	.318	11,500
.375	.392	.385	.368	.016	.095	.077	.375	.382	17,600
.437	.456	.448	.430	.017	.095	.077	.437	.445	20,000
.500	.521	.513	.485	.025	.110	.094	.500	.510	25,800

All dimensions are given in inches.

Materials

Standard Rollpins are manufactured from carbon steel and Type 420 corrosion resistant steel.

Temperatures at which Rollpins can be operated satisfactorily usually exceed the permissible operating temperatures of the members in which they are inserted. However, for special applications the following values may be taken as safe limits before the physical properties of the Rollpin are substantially altered:

Carbon steel . . . . 500°F.
Type 420 Corrosion Resistant Steel . . . . . 700°F.

Finishes

Carbon steel Rollpins are customarily supplied plain, with the black oiled finish characteristic of heat treating. Zinc and cadmium plated finishes are also available on special order. Corrosion resistant steel pins are furnished only with passivated finish. In cases where Government limitations apply to raw material or plating, certifications under these limitations must accompany purchase orders. Where a mild degree of corrosion resistance is necessary, a phosphate coating is available.

continued on next page

TABLE 11-6

Rollpin Availability and Numbering System

Catalog No. 825 Elastic Stop Nut Corporation of America

PLAIN CARBON STEEL ROLLPIN AVAILABILITY

NOMINAL DIAMETER

HARDNESS: Carbon-Rockwell "C" 46-53 or equivalent
Corrosion resistant-Rockwell "C" 43-52 or equivalent

LENGTH	.062	.078	.094	.125	.140	.156	.187	.219	.250	.312	.375	.437	.500
0.125"	†												
0.187	*	*	†										
0.250	*	*	*	†	Δ x	Δ x							
0.312	*	*	*	†	Δ	Δ	Δ x						
0.375	*	*	*	*	Δ	†	†	Δ x	Δ x				
0.437	*	*	*	*	Δ	†	†	Δ	Δ x	Δ x			
0.500	*	*	*	*	*	*	*	Δ	Δ	Δ x	Δ x		
0.562	*	*	*	*	Δ	*	*	Δ	Δ	Δ x	Δ x		
0.625	*	*	*	*	*	*	*	†	*	Δ	Δ x		
0.687	*	*	*	*	Δ	*	*	Δ	†	Δ	Δ x		
0.750	*	*	*	*	*	*	*	*	*	†	*		
0.812	Δ	*	*	*	Δ	*	*	Δ	†	Δ	Δ		
0.875	Δ	*	*	*	*	*	*	*	*	†	†		
0.937	Δ	*	*	*	Δ	*	*	Δ	†	Δ	Δ		
1.000	Δ	*	*	*	*	*	*	*	*	*	*	†	*
1.125	Δ x	*	*	*	Δ	*	*	*	*	†	†	Δ	†
1.250		*	*	*	*	*	*	*	*	*	*	†	*
1.375		*	*	*	Δ	*	*	*	*	†	†	Δ	†
1.500		*	*	*	Δ	*	*	*	*	*	*	†	*
1.625		Δ	Δ	*	Δ	*	*	*	*	†	†	Δ	†
1.750		Δ	Δ	*	Δ	*	*	*	*	*	*	†	*
1.875		Δ	Δ	Δ	Δ	*	*	*	*	†	†	Δ	†
2.000		Δ	Δ	Δ	Δ	*	*	*	*	*	*	†	*
2.250				Δ	Δ	Δ	†	*	*	*	*	†	*
2.500				Δ	Δ	Δ	†	*	*	*	*	†	*
2.750						Δ	Δ	Δ	*	*	*	†	*
3.000						Δ	Δ	Δ	*	*	*	†	*
3.250						Δ	Δ	Δ	*	*	*	†	*
3.500						Δ	Δ	Δ	*	*	*	†	*
3.750						Δ	Δ	Δ	Δ	*	*	†	*
4.000						Δ	Δ	Δ	Δ	*	*	†	*
4.250							Δ	Δ	Δ	Δ	Δ	Δ	Δ
4.500							Δ	Δ	Δ	Δ	Δ	Δ	Δ
4.750							Δ	Δ	Δ	Δ	Δ	Δ	Δ
5.000							Δ	Δ	Δ	Δ	Δ	Δ	Δ
5.250											Δ	Δ	Δ
5.500											Δ	Δ	Δ

* STOCK SIZES
† STANDARD PARTS IN STOCK OR AVAILABLE WITHIN NORMAL DELIVERY SCHEDULES
Δ STANDARD PARTS AVAILABLE WITHIN NORMAL DELIVERY SCHEDULES BUT FOR WHICH MINIMUM PRODUCTION RUN REQUIREMENTS ARE NECESSARY

*In addition to the stock lengths appearing in this table, Rollpins less than 1 in. long can be made available in length increments of 1/32 in. Rollpins longer than 1 in. can be made available in length increments of 1/16 in, but in both cases such parts are non-stock items and must be made to order. Rollpins of special decimal lengths are subject to special order requirements.

x Indicates Sizes Stocked Only in Carbon Steel.

Diameters .140, .312, .375, .437 and .500 for Corrosion Resistant Steel Rollpins are available within normal delivery schedules but minimum production run requirements are necessary.

TABLE 11–6, continued

ROLLPINS are identified by a series of dash numbers in accordance with the example shown below.

EXAMPLE

5 9 - 028 - 125 - 0750 = ROLLPIN, .125 Diam., .028 wall, 3/4 long, Plain Carbon Steel.

THIRD DASH NUMBER indicates length in thousandths of an inch.

SECOND DASH NUMBER indicates nominal diameter in thousandths of an inch and is determined from "A-Nominal" column in tabulation on page 14.

FIRST DASH NUMBER indicates wall thickness in thousandths of an inch and is determined from "stock thickness" column in tabulation on page 14.

SECOND DIGIT indicates finish:
9—Plain
3—Phosphate coating, Army Spec 57-0-2C, Type 2, Class B.
2—Cadmium plate, Federal Spec QQ-P-416, Type 1, Class C.
1—Zinc plate, Federal Spec QQ-Z-325, Type 1, Class 3.

FIRST DIGIT indicates material:
7—Corrosion Resistant Steel, AISI 420. (Magnetic)
5—High Carbon Steel.
9—Beryllium Copper

TABLE 11-7

Rollpin Diameters for Different Shaft Diameters

Catalog No. 825 Elastic Stop Nut Corporation of America

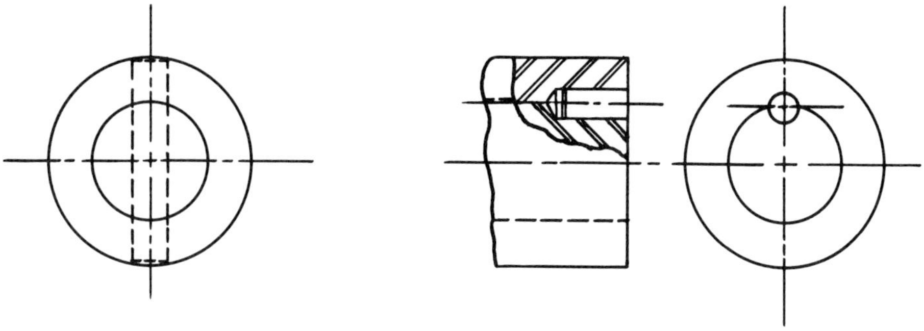

Shaft Diameter Inches	Nominal Diameter of Transverse* Rollpin	Nominal Diameter of Rollpin Used as †Key
3/16	1/16	
7/32	5/64	
1/4	3/32	1/16
5/16	1/8	5/64
3/8	1/8	3/32
7/16	5/32	1/8
1/2	5/32	1/8
9/16	3/16	5/32
5/8	3/16	5/32
11/16	7/32	3/16
3/4	7/32	3/16
13/16	1/4	7/32
7/8	5/16	7/32
15/16	5/16	1/4
1	3/8	1/4
1-1/4	7/16	5/16
1-3/8	7/16	3/8
1-1/2	1/2	7/16
1-5/8		7/16
2		1/2

*Double shear strength of rollpins in transverse shear is given in Table 11-5, last column.

†To find the shear strength, pounds, of a rollpin in longitudinal shear in the manner of a key, multiply the double shear strength listed in the last column of Table 11–5, corresponding to the proper diameter, by two-fifths the number of diameters in the length of the pin. Thus a pin of 0.125 nominal diameter and 3/4 inch long, and which has a double sheer strength of 2,1000 pounds, has a longitudinal shear strength of

$$2100 \times \frac{2}{5} \times \frac{3/4}{1/8} = 5040 \text{ pounds, minimum.}$$

TABLE 11-8

Dimensions of Cotter Pins

ASA B5.20-1947 Machine Pins

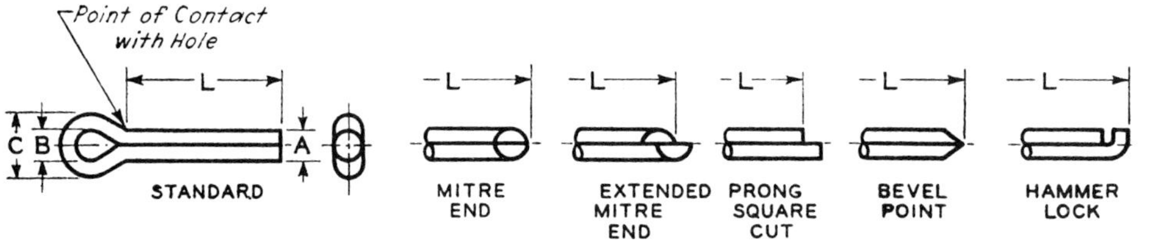

Diameter Nominal	Diameter A Max	Diameter A Min	Inside Eye Diameter B Min	Outside Eye Diameter C Min	Hole Sizes Recom-mended	Lengths of Carried-in-Stock Sizes The Atlas Bolt and Screw Co
0.031	0.032	0.028	1/32	1/16	3/64	
0.047	0.048	0.044	3/64	3/32	1/16	
0.062	0.060	0.056	1/16	1/8	5/64	1/2 to 2 inch by quarter inch
0.078	0.076	0.072	5/64	5/32	3/32	
0.094	0.090	0.086	3/32	3/16	7/64	1/2 to 2 by quarters; 2½
0.109	0.104	0.100	7/64	7/32	1/8	
0.125	0.120	0.116	1/8	1/4	9/64	{ 1/2 to 2 by quarters; to 3 by half inches
0.141	0.134	0.130	9/64	9/32	5/32	
0.156	0.150	0.146	5/32	5/16	11/64	3/4 to 2½ by quarters; 3
0.188	0.176	0.172	3/16	3/8	13/64	3/4 to 2½ by quarters; 3, 3½, 4
0.219	0.207	0.202	7/32	7/16	15/64	
0.250	0.225	0.220	1/4	1/2	17/64	1 to 2½ by quarters; 3, 3½, 4
0.312	0.280	0.275	5/16	5/8	5/16	1 to 2½ by quarters; 3, 3½, 4, 5
0.375	0.335	0.329	3/8	3/4	3/8	1½ to 4 by half inches; 5
0.438	0.406	0.400	7/16	7/8	7/16	
0.500	0.473	0.467	1/2	1	1/2	2 to 4 by half inches; 5
0.625	0.598	0.590	5/8	1 1/4	5/8	3, 3½, 4, 5
0.750	0.723	0.715	3/4	1 1/2	3/4	

All dimensions are given in inches.

A certain amount of leeway is permitted in the design of the head; however, the inside diameters and outside diameters given should be adhered to.

Prongs are to be parallel, ends shall not be open.
Points may be blunt, bevel, extended-prong, mitre, etc., and purchaser may specify type required.
Lengths shall be measured as shown on the above illustration (dimension).
Cotter pins shall be free from burrs or any defects that will affect their serviceability.

TABLE 11-9

Regular Square Nuts

ASA B18.2-1952 Square and Hexagon Bolts and Nuts

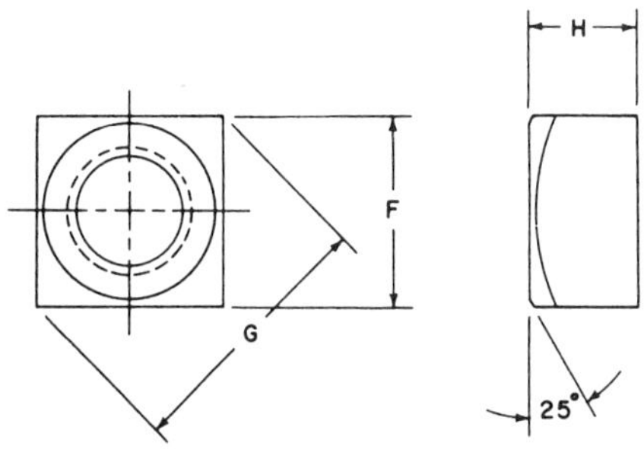

Nominal *Size or Basic Major Diameter of Thread		Width Across Flats F			Width Across Corners G		Thickness H		
		Max (Basic)		Min	Max	Min	Nom	Max	Min
1/4	0.2500	7/16	0.4375	0.425	0.619	0.584	7/32	0.235	0.203
5/16	0.3125	9/16	0.5625	0.547	0.795	0.751	17/64	0.283	0.249
3/8	0.3750	5/8	0.6250	0.606	0.884	0.832	21/64	0.346	0.310
7/16	0.4375	3/4	0.7500	0.728	1.061	1.000	3/8	0.394	0.356
1/2	0.5000	13/16	0.8125	0.788	1.149	1.082	7/16	0.458	0.418
5/8	0.6250	1	1.0000	0.969	1.414	1.330	35/64	0.569	0.525
3/4	0.7500	1 1/8	1.1250	1.088	1.591	1.494	21/32	0.680	0.632
7/8	0.8750	1 5/16	1.3125	1.269	1.856	1.742	49/64	0.792	0.740
1	1.0000	1 1/2	1.5000	1.450	2.121	1.991	7/8	0.903	0.847
1 1/8	1.1250	1 11/16	1.6875	1.631	2.386	2.239	1	1.030	0.970
1 1/4	1.2500	1 7/8	1.8750	1.812	2.652	2.489	1 3/32	1.126	1.062
1 3/8	1.3750	2 1/16	2.0625	1.994	2.917	2.738	1 13/64	1.237	1.169
1 1/2	1.5000	2 1/4	2.2500	2.175	3.182	2.986	1 5/16	1.348	.276
1 5/8	1.6250	2 7/16	2.4375	2.356	3.447	3.235	1 27/64	1.460	1.384

*Bolts are given in Table 9-34.

All dimensions given in inches.

Regular square nuts are not finished on any surface but are threaded.

Taper of the sides of nuts (angle between one side and the axis) shall not exceed 2 deg, the specified width across flats being the largest dimension.

Tops of nuts shall be flat and chamfered or washer crowned. Diameter of the top circle shall be the maximum width across the flats within a tolerance of minus 15 per cent.

Bearing surface shall be at right angles to the axis of the threaded hole within a tolerance of 3 deg for 1-in. nuts or smaller and 2 deg for nuts larger than 1 in.

Threads shall be coarse-thread series, class 2B.

Suitable material for steel nuts is covered by ASTM A-307; other materials will be as agreed upon by manufacturer and user.

TABLE 11-10

Finished Hexagon and Hexagon-Jam Nuts

ASA B18.2-1952 Square and Hexagon Bolts and Nuts

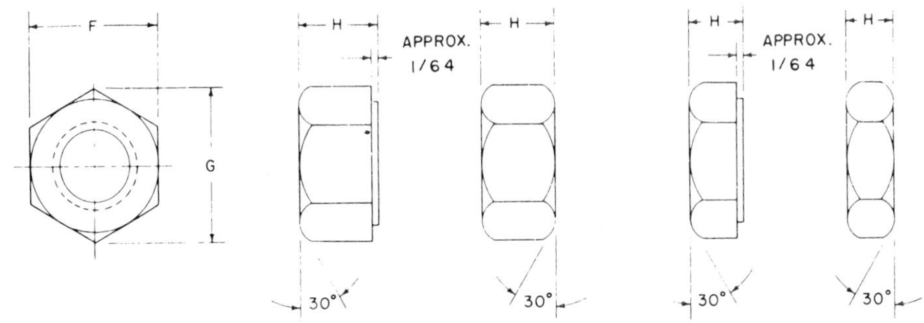

Nominal *Size or Basic Major Diameter of Thread	Width Across Flats F			Width Across Corners G		Thickness Nuts H			Thickness Jam Nuts H		
	Max	(Basic)	Min	Max	Min	Nom	Max	Min	Nom	Max	Min
1/4 0.2500	**7/16**	**0.4375**	**0.428**	**0.505**	**0.488**	**7/32**	**0.226**	**0.212**	**5/32**	**0.163**	**0.150**
5/16 0.3125	**1/2**	**0.5000**	**0.489**	**0.577**	**0.557**	**17/64**	**0.273**	**0.258**	**3/16**	**0.195**	**0.180**
3/8 0.3750	**9/16**	**0.5625**	**0.551**	**0.650**	**0.628**	**21/64**	**0.337**	**0.320**	**7/32**	**0.227**	**0.210**
7/16 0.4375	**11/16**	**0.6875**	**0.675**	**0.794**	**0.768**	**3/8**	**0.385**	**0.365**	**1/4**	**0.260**	**0.240**
1/2 0.5000	**3/4**	**0.7500**	**0.736**	**0.866**	**0,840**	**7/16**	**0.448**	**0.427**	**5/16**	**0.323**	**0.302**
9/16 0.5625	**7/8**	**0.8750**	**0.861**	**1.010**	**0.982**	**31/64**	**0.496**	**0.473**	**6/16**	**0.324**	**0.301**
5/8 0.6250	**15/16**	**0.9375**	**0.922**	**1.083**	**1.051**	**35/64**	**0.559**	**0.535**	**3/8**	**0.387**	**0.363**
3/4 0.7500	**1 1/8**	**1.1250**	**1.088**	**1.299**	**1.240**	**41/64**	**0.665**	**0.617**	**27/64**	**0.446**	**0.398**
7/8 0.8750	**1 5/16**	**1.3125**	**1.269**	**1.516**	**1.447**	**3/4**	**0.776**	**0.724**	**31/64**	**0.510**	**0.458**
1 1.0000	**1 1/2**	**1.5000**	**1.450**	**1.732**	**1.653**	**55/64**	**0.887**	**0.831**	**35/64**	**0.575**	**0.519**
1 1/8 1.1250	1 11/16	1.6875	1.631	1.949	1.859	31/32	0.999	0.939	39/64	0.639	0.579
1 1/4 1.2500	1 7/8	1.8750	1.812	2.165	2.066	1 1/16	1.094	1.030	23/32	0.751	0.687
1 3/8 1.3750	2 1/16	2.0625	1.994	2.382	2.273	1 11/64	1.206	1.138	25/32	0.815	0.747
1 1/2 1.5000	2 1/4	2.2500	2.175	2.598	2.480	1 9/32	1.317	1.245	27/32	0.880	0.808
1 5/8 1.6250	2 7/16	2.4375	2.356	2.815	2.686	1 25/64	1.429	1.353	29/32	0.944	0.868
1 3/4 1.7500	2 5/8	2.6250	2.538	3.031	2.893	1 1/2	1.540	1.460	31/32	1.009	0.929
1 7/8 1.8750	2 13/16	2.8125	2.719	3.248	3.100	1 39/64	1.651	1.567	1 1/32	1.073	0.989
2 2.0000	3	3.0000	2.900	3.464	3.306	1 23/32	1.763	1.675	1 3/32	1.138	1.050
2 1/4 2.2500	3 3/8	3.3750	3.262	3.897	3.719	1 59/64	1.970	1.874	1 13/64	1.251	1.155
2 1/2 2.5000	3 3/4	3.7500	3.625	4.330	4.133	2 9/64	2.193	2.089	1 29/64	1.505	1.401
2 3/4 2.7500	4 1/8	4.1250	3.988	4.763	4.546	2 23/64	2.415	2.303	1 37/64	1.634	1.522
3 3.0000	4 1/2	4.5000	4.350	5.196	4.959	2 37/64	2.638	2.518	1 45/64	1.763	1.643

*Bolts are given in Table 9-34.

All dimensions given in inches.

BOLD TYPE INDICATES PRODUCTS UNIFIED DIMENSIONALLY WITH BRITISH AND CANADIAN STANDARDS.

"Finished" in the title refers to the quality of manufacture and the closeness of tolerance and does not indicate that surfaces are completely machined.

Taper of the sides of nuts (angle between one side and the axis) shall not exceed 2 deg, the specified width across flats being the largest dimension.

Tops of nuts shall be flat and chamfered. Diameter of top circle shall be the maximum width across flats within a tolerance of minus 15 per cent for washer faced nuts and within a tolerance of minus 5 per cent for double chamfered nuts.

Bearing surface shall be washer faced or with chamfered corners. Diameter of circle bearing surface shall be the maximum width across flats within a tolerance of minus 5 per cent. Tapped hole shall be counter-sunk 1/64 in. over the major diameter of thread for nuts up to and including 1/2 in., and 1/32 in. over tne major diameter of thread for nuts over 1/2 in. size.

Bearing surface shall be at right angles to the axis of the threaded hole within a tolerance of 2 deg for 5/8 in. nuts or smaller and 1 deg for nuts larger than 5/8 in.; therefore, the maximum total runout of bearing face would equal the tangent of specified angle times the distance across flats.

Thread shall be coarse-, fine-, or 8-thread series; class 2B.

Suitable material for steel nuts is covered by ASTM A-307; other materials will be as agreed upon by manufacturer and user.

Tolerance on width across flats may be increased 0.015 in. for hot formed nuts 5/8 in. and smaller.

TABLE 11-11

Finished Hexagon Slotted Nuts

ASA B18.2-1952 Square and Hexagon Bolts and Nuts

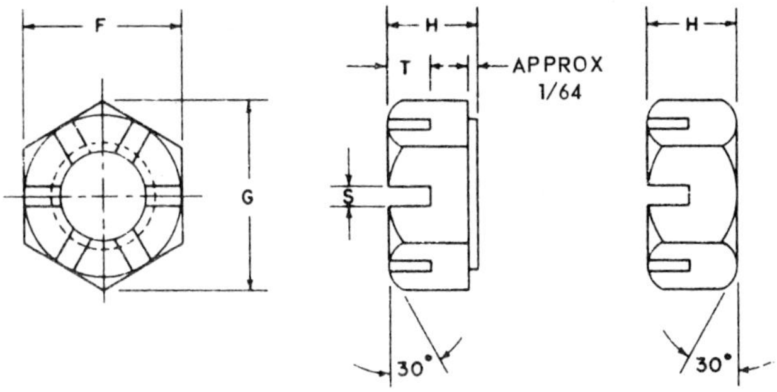

Nominal Size or Basic Major Diameter of Thread		Width Across Flats F			Width Across Corners G		Thickness H			Slot	
		Max (Basic)		Min	Max	Min	Nom	Max	Min	Width S	Depth T
1/4	**0.2500**	**7/16**	**0.4375**	**0.428**	**0.505**	**0.488**	**7/32**	**0.226**	**0.212**	0.078	0.094
5/16	**0.3125**	**1/2**	**0.5000**	**0.489**	**0.577**	**0.557**	**17/64**	**0.273**	**0.258**	0.094	0.094
3/8	**0.3750**	**9/16**	**0.5625**	**0.551**	**0.650**	**0.628**	**21/64**	**0.337**	**0.320**	0.125	0.125
7/16	**0.4375**	**11/16**	**0.6875**	**0.675**	**0.794**	**0.768**	**3/8**	**0.385**	**0.365**	0.125	0.156
1/2	**0.5000**	**3/4**	**0.7500**	**0.736**	**0.866**	**0.840**	**7/16**	**0.448**	**0.427**	0.156	0.156
9/16	**0.5625**	**7/8**	**0.8750**	**0.861**	**1.010**	**0.982**	**31/64**	**0.496**	**0.473**	0.156	0.188
5/8	**0.6250**	**15/16**	**0.9375**	**0.922**	**1.083**	**1.051**	**35/64**	**0.559**	**0.535**	0.188	0.219
3/4	**0.7500**	**1 1/8**	**1.1250**	**1.088**	**1.299**	**1.240**	**41/64**	**0.665**	**0.617**	0.188	0.250
7/8	**0.8750**	**1 5/16**	**1.3125**	**1.269**	**1.516**	**1.447**	**3/4**	**0.776**	**0.724**	0.188	0.250
1	**1.0000**	**1 1/2**	**1.5000**	**1.450**	**1.732**	**1.653**	**55/64**	**0.887**	**0.831**	0.250	0.281
1 1/8	1.1250	1 11/16	1.6875	1.631	1.949	1.859	31/32	0.999	0.939	0.250	0.344
1 1/4	1.2500	1 7/8	1.8750	1.812	2.165	2.066	1 1/16	1.094	1.030	0.312	0.375
1 3/8	1.3750	2 1/16	2.0625	1.994	2.382	2.273	1 11/64	1.206	1.138	0.312	0.375
1 1/2	1.5000	2 1/4	2.2500	2.175	2.598	2.480	1 9/32	1.317	1.245	0.375	0.438
1 5/8	1.6250	2 7/16	2.4375	2.356	2.815	2.686	1 25/64	1.429	1.353	0.375	0.438
1 3/4	1.7500	2 5/8	2.6250	2.538	3.031	2.893	1 1/2	1.540	1.460	0.438	0.500
1 7/8	1.8750	2 13/16	2.8125	2.719	3.248	3.100	1 39/64	1.651	1.567	0.438	0.562
2	2.0000	3	3.0000	2.900	3.464	3.306	1 23/32	1.763	1.675	0.438	0.562
2 1/4	2.2500	3 3/8	3.3750	3.262	3.897	3.719	1 59/64	1.970	1.874	0.438	0.562
2 1/2	2.5000	3 3/4	3.7500	3.625	4.330	4.133	2 9/64	2.193	2.089	0.562	0.688
2 3/4	2.7500	4 1/8	4.1250	3.988	4.763	4.546	2 23/64	2.415	2.303	0.562	0.688
3	3.0000	4 1/2	4.5000	4.350	5.196	4.959	2 37/64	2.638	2.518	0.625	0.750

All dimensions given in inches.

BOLD TYPE INDICATES PRODUCTS UNIFIED DIMENSIONALLY WITH BRITISH AND CANADIAN STANDARDS.

"Finished in the title refers to the quality of manufacturer and the closeness of tolerance and does not indicate that surfaces are completely machined.

Taper of the sides of nuts (angle between one side and the axis) shall not exceed 2 deg, the specified width across flats being the largest dimension.

Tops of nuts shall be flat and chamfered. Diameter of top circle shall be the maximum width across flats within a tolerance of minus 15 per cent for washer faced nuts and within a tolerance of minus 5 per cent for double chamfered nuts.

Bearing surface shall be washer faced or with chamfered corners. Diameter of washer face and the diameter of circle of bearing surface of double chamfered nuts shall be the maximum width across flats within a tolerance of minus 5 per cent. Tapped hole shall be countersunk 1/64 in. over the major diameter of thread for nuts up to and including 1/2 in. and 1/32 in. over the major diameter of thread for nuts over 1/2 in. size.

Bearing surface shall be at right angles to the axis of the threaded hole within a tolerance of 2 deg for 5/8 in. nuts or smaller and 1 deg for nuts larger than 5/8 in.; therefore, the maximum total runout of bearing face would equal the tangent of specified angle times the distance across flats.

Slots may have square or round bottoms at option of manufacturer.

Thread shall be coarse-, fine-, or 8-thread series; class 2B.

Maximum width across corners equals 1.1547 times maximum width across flats.

Suitable material for steel nuts is covered by ASTM A-307; other materials will be agreed upon by manufacturer and user.

Tolerance on width across flats may be increased 0.015 in. for hot formed nuts 5/8 in. and smaller.

TABLE 11-12

Machine Screw and Stove Bolt Nuts

ASA B18.2-1952 Square and Hexagon Bolts and Nuts

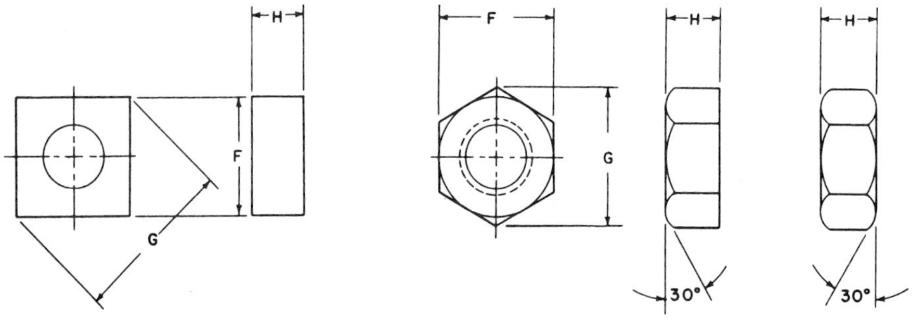

Nominal Size or Basic Major Diameter of Thread		Width Across Flats F			Width Across Corners G				Thickness H		
		Maximum (Basic)		Min	Square Max	Square Min	Hex Max	Hex Min	Nominal	Max	Min
No. 0	0.0600	5/32	0.1562	0.150	0.221	0.206	0.180	0.171	3/64	0.050	0.043
No. 1	0.0730	5/32	0.1562	0.150	0.221	0.206	0.180	0.171	3/64	0.050	0.043
No. 2	0.0860	3/16	0.1875	0.180	0.265	0.247	0.217	0.205	1/16	0.066	0.057
No. 3	0.0990	3/16	0.1875	0.180	0.265	0.247	0.217	0.205	1/16	0.066	0.057
No. 4	0.1120	1/4	0.2500	0.241	0.354	0.331	0.289	0.275	3/32	0.098	0.087
No. 5	0.1250	5/16	0.3125	0.302	0.442	0.415	0.361	0.344	7/64	0.114	0.102
No. 6	0.1380	5/16	0.3125	0.302	0.442	0.415	0.361	0.344	7/64	0.114	0.102
No. 8	0.1640	11/32	0.3438	0.332	0.486	0.456	0.397	0.378	1/8	0.130	0.117
No. 10	0.1900	3/8	0.3750	0.362	0.530	0.497	0.433	0.413	1/8	0.130	0.117
No. 12	0.2160	7/16	0.4375	0.423	0.619	0.581	0.505	0.482	5/32	0.161	0.148
1/4	0.2500	7/16	0.4375	0.423	0.619	0.581	0.505	0.482	3/16	0.193	0.178
5/16	0.3125	9/16	0.5625	0.545	0.795	0.748	0.650	0.621	7/32	0.225	0.208
3/8	0.3750	5/8	0.6250	0.607	0.884	0.833	0.722	0.692	1/4	0.257	0.239

All dimensions given in inches.

Hexagon machine screw nuts shall have tops flat and chamfered. Diameter of top circle shall be the maximum width across flats within a tolerance of minus 15 per cent. Bottoms are flat but for special purposes may be chamfered or washer faced if so specified.

Square machine screw nuts and stove bolt nuts shall have tops and bottoms flat without chamfer.

Thread shall be coarse-thread series for square machine-screw or stove-bolt nuts, coarse- or fine-thread series for hexagon machine screw nut; class 2B.

Suitable material for steel nuts is covered by ASTM A-307; other materials will be as agreed upon by manufacturer and user.

TABLE 11-13

Heavy Square, Hexagon and Hexagon-Jam Nuts

ASA B18.2–1952 Square and Hexagon Bolts and Nuts

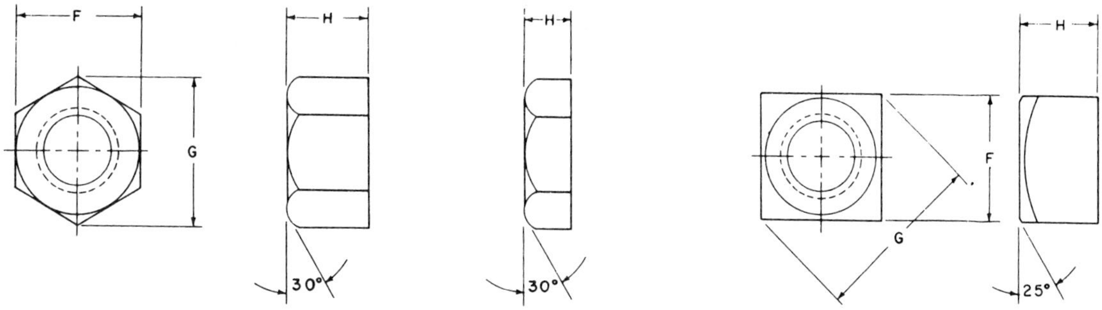

Nominal *Size or Basic Major Diameter of Thread	Width Across Flats F Max (Basic)	Width Across Flats F Min	Width Across Corners G Square Max	Square Min	Hex Max	Hex Min	Thickness Heavy Nuts H Nom	Max	Min	Thickness Heavy Jam Nuts H Nom	Max	Min
1/4 0.2500	1/2 0.5000	0.488	0.707	0.670	0.577	0.556	1/4	0.266	0.218	3/16	0.204	0.156
5/16 0.3125	9/16 0.5625	0.546	0.795	0.750	0.650	0.622	5/16	0.330	0.280	7/32	0.236	0.186
3/8 0.3750	11/16 0.6875	0.669	0.973	0.919	0.794	0.763	3/8	0.393	0.341	1/4	0.268	0.216
7/16 0.4375	3/4 0.7500	0.728	1.060	1.000	0.866	0.830	7/16	0.456	0.403	9/32	0.300	0.247
1/2 0.5000	**7/8 0.8750**	**0.850**	1.237	1.167	**1.010**	**0.969**	**1/2**	**0.520**	**0.464**	**5/16**	**0.332**	**0.277**
5/8 0.6250	**1 1/16 1.0625**	**1.031**	1.503	1.416	**1.227**	**1.175**	**5/8**	**0.647**	**0.587**	**3/8**	**0.397**	**0.337**
3/4 0.7500	**1 1/4 1.2500**	**1.212**	1.768	1.665	**1.443**	**1.382**	**3/4**	**0.774**	**0.710**	**7/16**	**0.462**	**0.398**
7/8 0.8750	**1 7/16 1.4375**	**1.394**	2.033	1.914	**1.660**	**1.589**	**7/8**	**0.901**	**0.833**	**1/2**	**0.526**	**0.458**
1 1.0000	**1 5/8 1.6250**	**1.575**	2.298	2.162	**1.876**	**1.796**	**1**	**1.028**	**0.956**	**9/16**	**0.590**	**0.519**
1 1/8 1.1250	**1 13/16 1.8125**	**1.756**	2.563	2.411	**2.093**	**2.002**	**1 1/8**	**1.155**	**1.079**	**5/8**	**0.655**	**0.579**
1 1/4 1.2500	**2 2.0000**	**1.938**	2.828	2.661	**2.309**	**2.209**	**1 1/4**	**1.282**	**1.187**	**3/4**	**0.782**	**0.687**
1 3/8 1.3750	**2 3/16 2.1875**	**2.119**	3.094	2.909	**2.526**	**2.416**	**1 3/8**	**1.409**	**1.310**	**13/16**	**0.846**	**0.747**
1 1/2 1.5000	**2 3/8 2.3750**	**2.300**	3.359	3.158	**2.742**	**2.622**	**1 1/2**	**1.536**	**1.433**	**7/8**	**0.911**	**0.808**
1 5/8 1.6250	2 9/16 2.5625	2.481	↑	↑	2.959	2.828	1 5/8	1.663	1.556	15/16	0.976	0.868
1 3/4 1.7500	**2 3/4 2.7500**	**2.662**			**3.175**	**3.035**	**1 3/4**	**1.790**	**1.679**	**1**	**1.040**	**0.929**
1 7/8 1.8750	2 15/16 2.9375	2.844			3.392	3.242	1 7/8	1.917	1.802	1 1/16	1.104	0.989
2 2.0000	**3 1/8 3.1250**	**3.025**	Sizes Not Standard	Sizes Not Standard	**3.608**	**3.449**	**2**	**2.044**	**1.925**	**1 1/8**	**1.169**	**1.050**
2 1/4 2.2500	3 1/2 3.5000	3.388			4.041	3.862	2 1/4	2.298	2.155	1 1/4	1.298	1.155
2 1/2 2.5000	3 7/8 3.8750	3.750			4.474	4.275	2 1/2	2.552	2.401	1 1/2	1.552	1.401
2 3/4 2.7500	4 1/4 4.2500	4.112			4.907	4.688	2 3/4	2.806	2.647	1 5/8	1.681	1.522
3 3.0000	4 5/8 4.6250	4.475			5.340	5.102	3	3.060	2.893	1 3/4	1.810	1.643
3 1/4 3.2500	5 5.0000	4.838			5.774	5.515	3 1/4	3.314	3.124	1 7/8	1.939	1.748
3 1/2 3.5000	5 3/8 5.3750	5.200			6.207	5.928	3 1/2	3.568	3.370	2	2.068	1.870
3 3/4 3.7500	5 3/4 5.7500	5.562			6.640	6.341	3 3/4	3.822	3.616	2 1/8	2.197	1.990
4 4.0000	6 1/8 6.1250	5.925	↓	↓	7.073	6.755	4	4.076	3.862	2 1/4	2.326	2.112

*Bolts are given in Table 9-35.

All dimensions given in inches.

BOLD TYPE INDICATES PRODUCTS UNIFIED DIMENSIONALLY WITH BRITISH AND CANADIAN STANDARDS.

Nuts are not finished on any surface but are threaded.

Taper of the sides of nuts (angle between one side and the axis) shall not exceed 2 deg, the specified width across flats being the largest dimension.

Tops of nuts shall be flat and chamfered or (except jam nuts) washer crowned. Diameter of top circle shall be the maximum width across flats within a tolerance of minus 15 per cent.

Bearing surface shall be at right angles to the axis of the threaded hole within a tolerance of 3 deg for 1-in. nuts or smaller and 2 deg for nuts larger than 1 in.; therefore, the maximum total runout of bearing face would equal the tangent of specified angle times the distance across flats.

Thread shall be coarse-thread series, class 2B.

Suitable material for steel nuts is covered by ASTM A-307; other materials will be as agreed upon by manufacturer and user.

TABLE 11-14

Heavy Semifinished Hexagon and Hexagon-Jam Nuts
ASA B18.2-1952 Square and Hexagon Bolts and Nuts

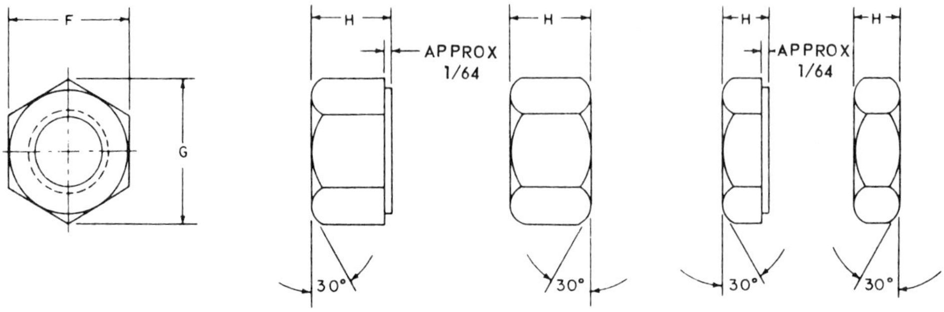

Nominal *Size or Basic Major Diameter of Thread	Width Across Flats F		Width Across Corners G		Thickness Heavy Nuts H			Thickness Heavy Jam Nuts H		
	Max (Basic)	Min	Max	Min	Nom	Max	Min	Nom	Max	Min
1/4 0.2500	1/2 0.5000	0.488	0.577	0.556	15/64	0.250	0.218	11/64	0.188	0.156
5/16 0.3125	9/16 0.5625	0.546	0.650	0.622	19/64	0.314	0.280	13/64	0.220	0.186
3/8 0.3750	11/16 0.6875	0.669	0.794	0.763	23/64	0.377	0.341	15/64	0.252	0.216
7/16 0.4375	3/4 0.7500	0.728	0.866	0.830	27/64	0.441	0.403	17/64	0.285	0.247
1/2 0.5000	**7/8 0.8750**	**0.850**	**1.010**	**0.969**	**31/64**	**0.504**	**0.464**	**19/64**	**0.317**	**0.277**
9/16 0.5625	15/16 0.9375	0.909	1.083	1.037	35/64	0.568	0.526	21/64	0.349	0.307
5/8 0.6250	**1 1/16 1.0625**	**1.031**	**1.227**	**1.175**	**39/64**	**0.631**	**0.587**	**23/64**	**0.381**	**0.337**
3/4 0.7500	**1 1/4 1.2500**	**1.212**	**1.443**	**1.382**	**47/64**	**0.758**	**0.710**	**27/64**	**0.446**	**0.398**
7/8 0.8750	**1 7/16 1.4375**	**1.394**	**1.660**	**1.589**	**55/64**	**0.885**	**0.833**	**31/64**	**0.510**	**0.458**
1 1.0000	**1 5/8 1.6250**	**1.575**	**1.876**	**1.796**	**63/64**	**1.012**	**0.956**	**35/64**	**0.575**	**0.519**
1 1/8 1.1250	**1 13/16 1.8125**	**1.756**	**2.093**	**2.002**	**1 7/64**	**1.139**	**1.079**	**39/64**	**0.639**	**0.579**
1 1/4 1.2500	**2 2.0000**	**1.938**	**2.309**	**2.209**	**1 7/32**	**1.251**	**1.187**	**23/32**	**0.751**	**0.687**
1 3/8 1.3750	**2 3/16 2.1875**	**2.119**	**2.526**	**2.416**	**1 11/32**	**1.378**	**1.310**	**25/32**	**0.815**	**0.747**
1 1/2 1.5000	**2 3/8 2.3750**	**2.300**	**2.742**	**2.622**	**1 15/32**	**1.505**	**1.433**	**27/32**	**0.830**	**0.808**
1 5/8 1.6250	2 9/16 2.5625	2.481	2.959	2.828	1 19/32	1.632	1.556	29/32	0.944	0.868
1 3/4 1.7500	**2 3/4 2.7500**	**2.662**	**3.175**	**3.035**	**1 23/32**	**1.759**	**1.679**	**31/32**	**1.009**	**0.929**
1 7/8 1.8750	2 15/16 2.9375	2.844	3.392	3.242	1 27/32	1.886	1.802	1 1/32	1.073	0.989
2 2.0000	**3 1/8 3.1250**	**3.025**	**3.608**	**3.449**	**1 31/32**	**2.013**	**1.925**	**1 3/32**	**1.138**	**1.050**
2 1/4 2.2500	3 1/2 3.5000	3.388	4.041	3.862	2 13/64	2.251	2.155	1 13/64	1.251	1.155
2 1/2 2.5000	3 7/8 3.8750	3.750	4.474	4.275	2 29/64	2.505	2.401	1 29/64	1.505	1.401
2 3/4 2.7500	4 1/4 4.2500	4.112	4.907	4.688	2 45/64	2.759	2.647	1 37/64	1.634	1.522
3 3.0000	4 5/8 4.6250	4.475	5.340	5.102	2 61/64	3.013	2.893	1 45/64	1.763	1.643
3 1/4 3.2500	5 5.0000	4.838	5.774	5.515	3 3/16	3.252	3.124	1 13/16	1.876	1.748
3 1/2 3.5000	5 3/8 5.3750	5.200	6.207	5.928	3 7/16	3.506	3.370	1 15/16	2.006	1.870
3 3/4 3.7500	5 3/4 5.7500	5.562	6.640	6.341	3 11/16	3.760	3.616	2 1/16	2.134	1.990
4 4.0000	6 1/8 6.1250	5.925	7.073	6.755	3 15/16	4.014	3.862	2 3/16	2.264	2.112

All dimensions given in inches.

BOLD TYPE INDICATES PRODUCTS UNIFIED DIMENSIONALLY WITH BRITISH AND CANADIAN STANDARDS.

Semifinished nuts are finished on bearing surface and threaded.

Taper of the sides of nuts (angle between one side and the axis) shall not exceed 2 deg, the specified width across flats being the largest dimension.

Tops of nuts shall be flat and chamfered. Diameter of top circle shall be the maximum width across flats within a tolerance of minus 15 per cent for washer faced nuts and within a tolerance of minus 5 per cent for double chamfered nuts.

Bearing surface shall be washer faced or with chamfered corners. Diameter of washer face and the diameter of circle of bearing surface of double chamfered nuts shall be the maximum width across flats within a tolerance of minus 5 per cent. Tapped hole shall be countersunk 1/64 in. over the major diameter of thread for nuts up to and including 1/2 in. and 1/32 in. over the major diameter of thread for nuts over 1/2 in. size.

Bearing surface shall be at right angles to the axis of the threaded hole within a tolerance of 2 deg for 5/8 in. nuts or smaller and 1 deg for nuts larger than 5/8 in.; therefore, the maximum total runout of bearing face would equal the tangent of specified angle times the distance across flats.

Thread may be coarse-, fine-, or 8-thread series; class 2B tolerance; unless otherwise specified, coarse-thread series will be furnished.

Suitable material for steel nuts is covered by ASTM A-307; other materials will be as agreed upon by manufacturer and user.

*Bolts are given in Table 9–35.

TABLE 11–15

Manufacturers of Locknuts

1953 ASME Mechanical Catalog and Directory

Company and Address	Some Trade-Names
Allen Manufacturing Co, 133 Sheldon St, Hartford, Conn.	
Boots Aircraft Nut Corp, Stamford, Conn.	Bellows, Roltop Tri-Lok
Boss Bolt and Nut Co, 3403 W. 47th St, Chicago, Ill.	
Columbia Nut and Bolt Co, 945 Main St, Bridgeport, Conn.	
Elastic Stop Nut Corp of America, 2330 Vauxhall Road, Union, N.J.	Esna
Grip Nut Co, 300 S Michigan Ave, Chicago, Ill.	Gripco, Grip
Laminated Shim Co, Inc, 81 Union St, Glenbrook, Conn.	An-cor-lox
Lamson and Sessions Co, 1979 W. 85th St, Cleveland, Ohio.	Dardelet, Lamson
MacLean-Fogg Lock Nut Co, 5533 N. Wolcott St, Chicago 39, Ill.	"M-F" Lock Nuts "M-F" Unitary
National Screw and Mfg Co, 2440 E. 75th St, Cleveland, O.	Drake, Dynamic Huglock, Marsden
Palnut Co, 64 Cordier St, Irvington, N. J.	Pal
Russell, Burdsall and Ward Bolt and Nut Co, Port Chester, N. Y.	Huglock, Marsden, Nylok
Security Locknut Corp, 1500 North Ave, Melrose Park, Ill.	Security Caploc
Selflock Screw Products Co, East Syracuse, N. Y.	
Shakeproof, Inc, St. Charles Road, Elgin, Ill.	Shakeproof Lokut
SKF Industries, Inc, Philadelphia 32, Pa.	
Standard Locknut and Lockwasher Co, 510 N. Capital Ave, Indianapolis, Ind.	
Standard Pressed Steel Co, Box 806, Jenkintown, Pa.	Flexloc, Unbrako
Stover Lock Nut and Machinery Corp, Hilton and Holly Sts, Easton, Pa.	Stover
Stronghold Screw Products, Inc, 216-222 W. Hubbard St, Chicago 10, Ill.	
Thompson-Bremer and Co, 1640 W. Hubbard St, Chicago, Ill.	
Tinnerman Products, Inc, 2124 Fulton Rd, Cleveland 13, Ohio	Hi-Stress, Speed Grip Speed Nut
Townsend Co, New Brighton, Pa.	Nylok, Tufflok
United Screw and Bolt Corp, 2513 W. Cullerton St, Chicago, Ill.	
Wood and Spencer Co, 1930 E. 61st St, Cleveland, Ohio	

TABLE 11-16

Wrench Openings

Appendix 1, ASA B18.2-1952 Square and Hexagon Bolts and Nuts

For Finished, Finished Thick, Regular and Heavy Series Nuts; Regular, Heavy, Finished and Finished Heavy Bolts; Cap Screws; Set Screws; Lag Bolts; Machine Screw and Stove Bolt Nuts

Nominal Size of Wrench, also Basic or Maximum Width Across Flats of Bolt and Screw Heads and Nuts		Allowance between Bolt Head or Nuts and Jaws of Wrench	Wrench Openings			NOMINAL BOLT, SCREW AND NUT DIAMETERS							
						NUTS			Regular Series Bolts / Finished Bolt / Hexagon Head Cap Screw	Heavy Series Bolts / Finished Heavy Bolt	Lag Bolts	Set Screws	Machine Screw Nuts and Stove Bolt Nuts
			Min	Tol	Max	Finished and Finished Thick Series Nuts	Regular Series Nuts	Heavy Series Nuts					
5/32	0.1562	0.002	0.158	0.005	0.163								No. 0 and No. 1
3/16	0.1875	0.002	0.190	0.005	0.195							No. 10	No. 2 and No. 3
1/4	0.2500	0.002	0.252	0.005	0.257							1/4	No. 4
5/16	0.3125	0.003	0.316	0.006	0.322						No. 10	5/16	No. 5 and No. 6
11/32	0.3438	0.003	0.347	0.006	0.353								No. 8
3/8	0.3750	0.003	0.378	0.006	0.384				1/4*		1/4	3/8	No. 10
7/16	0.4375	0.003	0.440	0.006	0.446	1/4	1/4		1/4			7/16	No. 12 and 1/4
1/2	0.5000	0.004	0.504	0.006	0.510	5/16		1/4	5/16		5/16	1/2	
9/16	0.5625	0.004	0.566	0.007	0.573	3/8	5/16	5/16	3/8		3/8	9/16	5/16
19/32	0.5938	0.004	0.598	0.007	0.605								
5/8	0.6250	0.004	0.629	0.007	0.636		3/8		7/16		7/16	5/8	3/8
11/16	0.6875	0.004	0.692	0.007	0.699	7/16		3/8					
3/4	0.7500	0.005	0.755	0.008	0.763	1/2	7/16	7/16	1/2		1/2	3/4	
25/32	0.7812	0.005	0.786	0.008	0.794								
13/16	0.8125	0.005	0.818	0.008	0.826		1/2		9/16				
7/8	0.8750	0.005	0.880	0.008	0.888	9/16	9/16	1/2		1/2		7/8	
15/16	0.9375	0.006	0.944	0.009	0.953	5/8		9/16	5/8		5/8		
1	1.0000	0.006	1.006	0.009	1.015		5/8					1	
1 1/16	1.0625	0.006	1.068	0.009	1.077			5/8		5/8			
1 1/8	1.1250	0.007	1.132	0.010	1.142	3/4	3/4		3/4		3/4	1 1/8	
1 1/4	1.2500	0.007	1.257	0.010	1.267			3/4		3/4		1 1/4	
1 5/16	1.3125	0.008	1.320	0.011	1.331	7/8	7/8		7/8		7/8		
1 3/8	1.3750	0.008	1.383	0.011	1.394							1 3/8	
1 7/16	1.4375	0.008	1.446	0.011	1.457			7/8		7/8			
1 1/2	1.5000	0.008	1.508	0.012	1.520	1	1		1		1	1 1/2	
1 5/8	1.6250	0.009	1.634	0.012	1.646			1		1			
1 11/16	1.6875	0.009	1.696	0.012	1.708	1 1/8	1 1/8		1 1/8		1 1/8		
1 13/16	1.8125	0.010	1.822	0.013	1.835			1 1/8		1 1/8			
1 7/8	1.8750	0.010	1.885	0.013	1.898	1 1/4	1 1/4		1 1/4		1 1/4		
2	2.0000	0.011	2.011	0.014	2.025			1 1/4		1 1/4			
2 1/16	2.0625	0.011	2.074	0.014	2.088	1 3/8	1 3/8		1 3/8				
2 3/16	2.1875	0.012	2.200	0.015	2.215			1 3/8		1 3/8			
2 1/4	2.2500	0.012	2.262	0.015	2.277	1 1/2	1 1/2		1 1/2				
2 3/8	2.3750	0.013	2.388	0.016	2.404			1 1/2		1 1/2			
2 7/16	2.4375	0.013	2.450	0.016	2.466	1 5/8	1 5/8		1 5/8				
2 9/16	2.5625	0.014	2.576	0.017	2.593			1 5/8		1 5/8			
2 5/8	2.6250	0.014	2.639	0.017	2.656	1 3/4			1 3/4				
2 3/4	2.7500	0.014	2.766	0.017	2.783			1 3/4		1 3/4			
2 13/16	2.8125	0.015	2.827	0.018	2.845	1 7/8			1 7/8				
2 15/16	2.9375	0.016	2.954	0.019	2.973			1 7/8		1 7/8			
3	3.0000	0.016	3.016	0.019	3.035	2			2				
3 1/8	3.1250	0.017	3.142	0.020	3.162			2		2			
3 3/8	3.3750	0.018	3.393	0.021	3.414	2 1/4			2 1/4				
3 1/2	3.5000	0.019	3.518	0.022	3.540			2 1/4		2 1/4			
3 3/4	3.7500	0.020	3.770	0.023	3.793	2 1/2			2 1/2				
3 7/8	3.8750	0.020	3.895	0.023	3.918			2 1/2		2 1/2			
4 1/8	4.1250	0.022	4.147	0.025	4.172	2 3/4			2 3/4				
4 1/4	4.2500	0.022	4.272	0.025	4.297			2 3/4		2 3/4			
4 1/2	4.5000	0.024	4.524	0.026	4.550	3			3				
4 5/8	4.6250	0.024	4.649	0.027	4.676			3		3			
5	5.0000	0.026	5.026	0.029	5.055			3 1/4	3 1/4				
5 3/8	5.3750	0.028	5.403	0.031	5.434			3 1/2	3 1/2				
5 3/4	5.7500	0.030	5.780	0.033	5.813			3 3/4	3 3/4				
6 1/8	6.1250	0.032	6.157	0.035	6.192			4	4				

All dimensions given in inches.

*Regular square only.

Wrenches shall be marked with the "Nominal Size of Wrench" which is equal to the basic or maximum width across flats of the corresponding bolt head or nut.

Allowance (minimum clearance) between maximum width across flats of nut or bolt head and jaws of wrench equals (1.005 W + .004 from minimum). (W equals nominal size of wrench.)

TABLE 11-17
Dimensions of AFBMA Standard Ball and Roller Bearing Lock Nuts

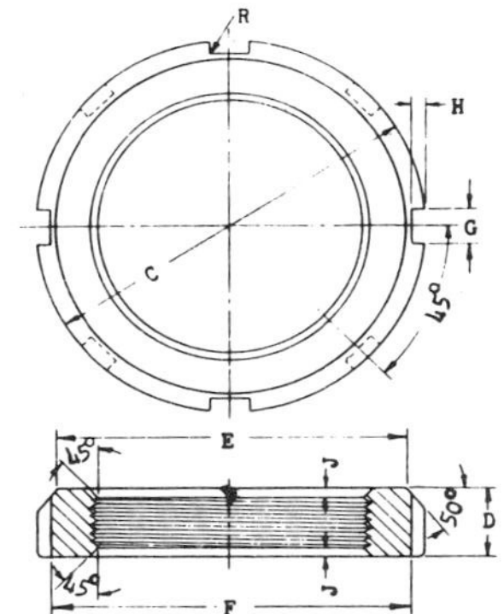

AFBMA Standards Section No. 8

BALL AND ROLLER BEARING LOCK NUTS
Runout and parallelism of faces, measured mounted on a tight fitting threaded arbor

N-00 to N-06 = .002 in. Max.
N-07 to AN-22 = .004 in. Max.
AN-24 to AN-40 = .006 in. Max.

8-N-00 to 8-AN-140 = .002 in. Max.

Material Specification
AISI 1112, 1113, 1115, 1015, 1020, 1035, 1315.

Thds. per Inch	Thread Minor Diam			Thread Pitch Diam			Thread Major Diam	Outside Diam C	Face Diam E		Ctsk. Depth	Slot Dimensions					Ball Bearings, Cylindrical and Spherical Roller Bearings 4 Slots			Tapered Roller Bearings 8 Slots		
												Diam F	Width G		Height	Fillet	Thickness D			Thickness D		
	Min	Tol	Max	Min	Tol	Max	Min	+.005 -.015	Min	Max	J	+.010 -.020	Min	Max	H ±.010	R	Min	Max	Nut No.	Min	Max	Nut No.
32	0.3572	.0034	0.3606	0.3707	.0026	0.3733	0.391	3/4	.605	.625	3/64	5/8	.120	.130	1/16	.010	.209	.229	N-00	.209	.229	8-N-00
32	0.4352	.0034	0.4386	0.4487	.0026	0.4513	0.469	7/8	.699	.719	3/64	3/4	.120	.130	1/16	.010	.303	.323	N-01	.303	.323	8-N-01
32	0.5522	.0034	0.5556	0.5657	.0030	0.5687	0.586	1	.793	.813	3/64	13/16	.120	.130	3/32	.010	.303	.323	N-02	.303	.323	8-N-02
32	0.6302	.0034	0.6336	0.6437	.0030	0.6467	0.664	1- 1/8	.918	.938	3/64	15/16	.120	.130	3/32	.010	.334	.354	N-03	.334	.354	8-N-03
32	0.7472	.0034	0.7506	0.7607	.0034	0.7641	0.781	1- 3/8	1.105	1.125	3/64	1- 3/16	.178	.198	3/32	.015	.365	.385	N-04	.365	.385	8-N-04
32	0.9352	.0034	0.9386	0.9487	.0034	0.9521	0.969	1- 9/16	1.261	1.281	3/64	1- 3/8	.178	.198	3/32	.015	.396	.416	N-05	.396	.416	8-N-05
18	1.1129	.0060	1.1189	1.1369	.0040	1.1409	1.173	1- 3/4	1.480	1.500	5/64	1- 9/16	.178	.198	3/32	.015	.396	.416	N-06	.396	.416	8-N-06
18	1.2524	.0060	1.2584	1.2764	.0040	1.2804	1.3125	2- 1/16	1.793	1.813	5/64	1- 7/8	.178	.198	3/32	.015				.428	.448	8-N-065
18	1.3159	.0060	1.3219	1.3399	.0040	1.3439	1.376	2- 1/16	1.793	1.813	5/64	1- 7/8	.178	.198	3/32	.015	.428	.448	N-07	.428	.448	8-N-07
18	1.5029	.0060	1.5089	1.5269	.0045	1.5314	1.563	2- 1/4	1.980	2.000	5/64	2- 1/16	.240	.260	3/32	.020	.428	.448	N-08	.428	.448	8-N-08
18	1.7069	.0060	1.7129	1.7309	.0045	1.7354	1.767	2-17/32	2.261	2.281	5/64	2-11/32	.240	.260	3/32	.020	.428	.448	N-09	.428	.448	8-N-09
18	1.9069	.0060	1.9129	1.9309	.0045	1.9354	1.967	2-11/16	2.418	2.438	5/64	2- 1/2	.240	.260	3/32	.020	.490	.510	N-10	.490	.510	8-N-10
18	2.0969	.0060	2.1029	2.1209	.0051	2.1260	2.157	2-31/32	2.636	2.656	5/64	2-23/32	.240	.260	1/8	.020	.490	.510	N-11	.490	.510	8-N-11
18	2.2999	.0060	2.3059	2.3239	.0051	2.3290	2.360	3- 5/32	2.824	2.844	5/64	2-29/32	.240	.260	1/8	.020	.521	.541	N-12	.521	.541	8-N-12
18	2.4879	.0060	2.4939	2.5119	.0051	2.5170	2.548	3- 3/8	3.043	3.063	5/64	3- 1/8	.240	.260	1/8	.020	.553	.573	N-13	.553	.573	8-N-13
18	2.6909	.0060	2.6969	2.7149	.0051	2.7200	2.751	3- 5/8	3.283	3.313	5/64	3- 3/8	.240	.260	1/8	.020	.553	.573	N-14	.553	.573	8-N-14
12	2.8428	.0090	2.8518	2.8789	.0054	2.8843	2.933	3- 7/8	3.533	3.563	1/8	3- 5/8	.360	.385	1/8	.025	.584	.604	AN-15	.584	.604	8-AN-15
12	3.0468	.0090	3.0558	3.0829	.0059	3.0888	3.137	4- 5/32	3.814	3.844	1/8	3-29/32	.360	.385	1/8	.025	.584	.604	AN-16	.584	.604	8-AN-16
12	3.2498	.0090	3.2588	3.2859	.0074	3.2933	3.340	4-13/32	4.001	4.031	1/8	4- 3/32	.360	.385	5/32	.025	.615	.635	AN-17	.615	.635	8-AN-17
12	3.4368	.0090	3.4458	3.4729	.0074	3.4803	3.527	4-21/32	4.251	4.281	1/8	4-11/32	.360	.385	5/32	.025	.678	.698	AN-18	.678	.698	8-AN-18
12	3.6398	.0090	3.6488	3.6759	.0074	3.6833	3.730	4-15/16	4.533	4.563	1/8	4- 5/8	.360	.385	5/32	.025	.709	.729	AN-19	.709	.729	8-AN-19
12	3.8278	.0090	3.8368	3.8639	.0074	3.8713	3.918	5- 3/16	4.783	4.813	1/8	4- 7/8	.360	.385	5/32	.025	.735	.760	AN-20	.735	.760	8-AN-20
12	4.0318	.0090	4.0408	4.0679	.0083	4.0762	4.122	5- 7/16	4.970	5.000	1/8	5- 1/16	.485	.510	3/16	.030	.735	.760	AN-21	.735	.760	8-AN-21
12	4.2348	.0090	4.2438	4.2709	.0083	4.2793	4.325	5-23/32	5.251	5.281	1/8	5-11/32	.485	.510	3/16	.030	.766	.791	AN-22	.766	.791	8-AN-22
12	4.6258	.0090	4.6348	4.6619	.0083	4.6702	4.716	6- 1/8	5.658	5.688	1/8	5- 3/4	.485	.510	3/16	.030	.798	.823	AN-24	.798	.823	8-AN-24
12	5.0158	.0090	5.0248	5.0519	.0083	5.0602	5.106	6- 3/4	6.158	6.188	1/8	6- 1/4	.610	.635	1/4	.030	.860	.885	AN-26	.860	.885	8-AN-26
12	5.4068	.0090	5.4158	5.4429	.0083	5.4512	5.497	7- 3/32	6.501	6.531	1/8	6-19/32	.610	.635	1/4	.030	.923	.948	AN-28	1.173	1.198	8-AN-128
12	5.7978	.0090	5.8068	5.8339	.0083	5.8422	5.888	7-11/16	7.033	7.063	1/8	7- 1/8	.610	.635	9/32	.030	.954	.979	AN-30	1.235	1.260	8-AN-130
8	6.1487	.0135	6.1622	6.2028	.0091	6.2119	6.284	8- 1/16	7.398	7.438	5/32	7- 1/2	.610	.635	9/32	.030	1.016	1.041	AN-32	1.266	1.291	8-AN-132
8	6.5237	.0135	6.5372	6.5778	.0091	6.5869	6.659	8-21/32	7.991	8.031	5/32	8- 3/32	.610	.635	9/32	.030	1.048	1.073	AN-34	1.329	1.354	8-AN-134
8	6.9307	.0135	6.9442	6.9848	.0091	6.9939	7.066	9- 1/16	8.335	8.375	5/32	8- 7/16	.735	.760	5/16	.030	1.079	1.104	AN-36	1.391	1.416	8-AN-136
8	7.3367	.0135	7.3502	7.3908	.0091	7.3999	7.472	9-15/32	8.741	8.781	5/32	8-27/32	.735	.760	5/16	.030	1.110	1.135	AN-38	1.391	1.416	8-AN-138
8	7.7117	.0135	7.7252	7.7658	.0114	7.7772	7.847	9-27/32	9.116	9.156	5/32	9- 7/32	.735	.760	5/16	.030	1.173	1.198	AN-40	1.485	1.510	8-AN-140

Threads are American National Form NS, Class 3.

TABLE 11–18

Dimensions of AFBMA Standard Ball and Roller Bearing Lock Washers

AFBMA Standards   Section No. 8

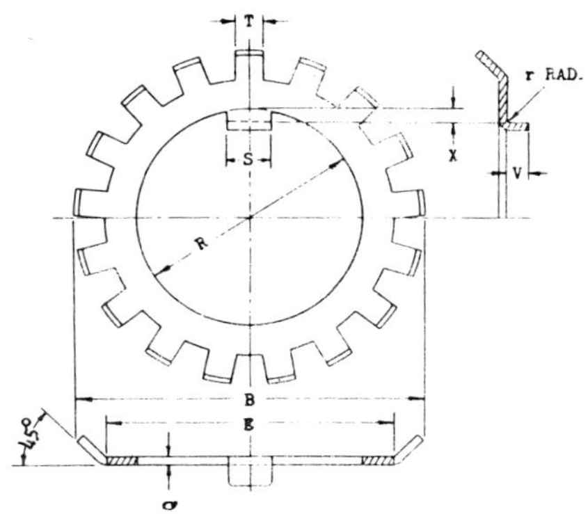

BALL AND ROLLER BEARING LOCK WASHERS

Material shall be a low carbon steel of such analysis and temper as to permit its being bent back flat, without fracture, either way of the grain, over a radius equal to 1/2 the stock thickness.

continued on next page

TABLE 11–18, continued

Stock Thickness Q				Tangs			Key							Bore R		Diameter		Diam B over Tangs
					Width		Projection		Width S		Bend X		Rad					
Washer Number	Q	Washer Number	Q	No.	T	Tol	V	Tol	Min	Max	Min	Max	*r*	Min	Max	*E*	Tol	Max
W-00	.042			9	.115	±.007	1/16	+1/32	.110	.120	.020	.030	1/64	.406	.421	5/8	+.015	7/8
W-01	.042			9	.115	±.007	1/16	+1/32	.110	.120	.020	.030	1/64	.484	.499	23/32	+.015	1- 1/64
W-02	.042	W-102	.058	11	.115	±.007	1/16	+1/32	.110	.120	.020	.030	1/64	.601	.616	13/16	+.015	1- 5/32
W-03	.042	W-103	.058	11	.115	±.007	1/16	+1/32	.110	.120	.020	.030	1/64	.679	.694	15/16	+.015	1-21/64
W-04	.042	W-104	.058	11	.156	±.010	1/16	+1/32	.156	.176	.020	.030	1/64	.801	.816	1- 1/8	+.015	1-17/32
W-05	.050	W-105	.062	13	.156	±.010	3/32	+1/32	.156	.176	.020	.030	1/64	.989	1.009	1- 9/32	+.015	1-23/32
W-06	.050	W-106	.062	13	.156	±.010	3/32	+1/32	.156	.176	.035	.050	1/32	1.193	1.213	1- 1/2	+.015	1-59/64
		W-065	.062	15	.156	±.010	3/32	+1/32	.156	.176	.035	.050	1/32	1.333	1.353	1-13/16	+.015	2- 1/4
W-07	.050	W-107	.062	15	.156	±.010	3/32	+1/32	.156	.176	.035	.050	1/32	1.396	1.416	1-13/16	+.015	2- 1/4
W-08	.058	W-108	.072	15	.219	±.015	3/32	+1/32	.250	.290	.035	.050	1/32	1.583	1.603	2	+.030	2-15/32
W-09	.058	W-109	.072	17	.219	±.015	1/8	+1/32	.250	.290	.035	.050	1/32	1.792	1.817	2- 9/32	+.030	2-47/64
W-10	.058	W-110	.072	17	.219	±.015	1/8	+1/32	.250	.290	.035	.050	1/32	1.992	2.017	2- 7/16	+.030	2-59/64
W-11	.063	W-111	.072	17	.219	±.015	1/8	+1/32	.250	.290	.035	.050	1/32	2.182	2.207	2-21/32	+.030	3- 7/64
W-12	.063	W-112	.082	17	.219	±.015	1/8	+1/32	.250	.290	.055	.070	3/64	2.400	2.425	2-27/32	+.030	3-11/32
W-13	.063	W-113	.082	19	.219	±.015	1/8	+1/32	.250	.290	.055	.070	3/64	2.588	2.613	3- 1/16	+.030	3-37/64
W-14	.063	W-114	.082	19	.219	±.015	3/16	+1/16	.250	.290	.055	.070	3/64	2.791	2.816	3- 5/16	+.030	3-53/64
W-15	.072	W-115	.095	19	.313	±.015	3/16	+1/16	.250	.290	.055	.070	3/64	2.973	3.003	3- 9/16	+.030	4- 7/64
W-16	.072	W-116	.095	19	.313	±.015	3/16	+1/16	.313	.353	.055	.070	3/64	3.177	3.207	3-27/32	+.030	4- 3/8
W-17	.072	W-117	.095	19	.313	±.015	3/16	+1/16	.313	.353	.055	.070	3/64	3.395	3.425	4- 1/32	+.030	4- 5/8
W-18	.094	W-118	.125	19	.313	±.015	3/16	+1/16	.313	.353	.055	.070	3/64	3.582	3.612	4- 9/32	+.045	4-15/16
W-19	.094	W-119	.125	19	.313	±.015	3/16	+1/16	.313	.353	.055	.070	3/64	3.800	3.830	4- 9/16	+.045	5- 7/32
W-20	.094	W-120	.125	19	.313	±.015	1/4	+1/16	.313	.353	.065	.085	3/64	3.988	4.018	4-13/16	+.045	5- 1/2
W-21	.094	W-121	.125	19	.375	±.015	1/4	+1/16	.313	.353	.065	.085	3/64	4.192	4.222	5	+.045	5-45/64
W-22	.125	W-122	.140	19	.375	±.015	1/4	+1/16	.313	.353	.065	.085	3/64	4.395	4.425	5- 9/32	+.045	6- 1/16
W-24	.125	W-124	.165	19	.375	±.015	1/4	+1/16	.313	.353	.065	.085	3/64	4.801	4.831	5-11/16	+.045	6-15/32
W-26	.125	W-126	.165	19	.500	±.020	1/4	+1/16	.375	.435	.085	.105	1/16	5.191	5.226	6- 3/16	+.045	7- 1/32
W-28	.125	W-128	.165	19	.500	±.020	1/4	+1/16	.500	.590	.085	.105	1/16	5.582	5.617	6-17/32	+.045	7- 7/16
W-30	.156	W-130	.203	19	.500	±.020	5/16	+1/16	.500	.590	.085	.105	1/16	5.983	6.018	7- 1/16	+.060	8- 1/16
W-32	.156	W-132	.203	19	.500	±.020	5/16	+1/16	.500	.590	.085	.105	1/16	6.389	6.424	7- 7/16	+.060	8- 7/16
W-34	.156	W-134	.203	19	.500	±.020	5/16	+1/16	.625	.715	.085	.105	1/16	6.764	6.799	8- 1/32	+.060	9- 1/16
W-36	.156	W-136	.203	19	.625	±.020	5/16	+1/16	.625	.715	.085	.105	1/16	7.171	7.206	8- 3/8	+.060	9- 7/16
W-38	.156	W-138	.203	19	.625	±.020	5/16	+1/16	.625	.715	.085	.105	1/16	7.577	7.612	8-25/32	+.060	9- 7/8
W-40	.156	W-140	.203	19	.625	±.020	5/16	+1/16	.750	.840	.085	.105	1/16	7.982	8.017	9- 5/32	+.060	10- 5/16

TABLE 11 – 19

Shafts for Ball and Roller Bearing Lock Nuts

AFBMA Standards   Section No. 8

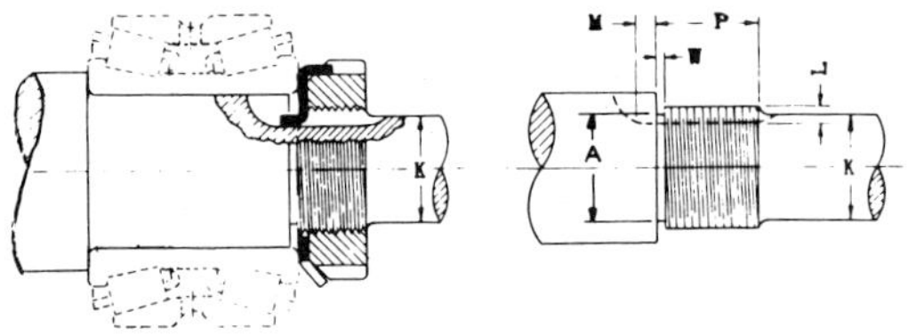

SHAFTS FOR TAPERED ROLLER BEARING LOCK NUTS
CLAMPED MOUNTING

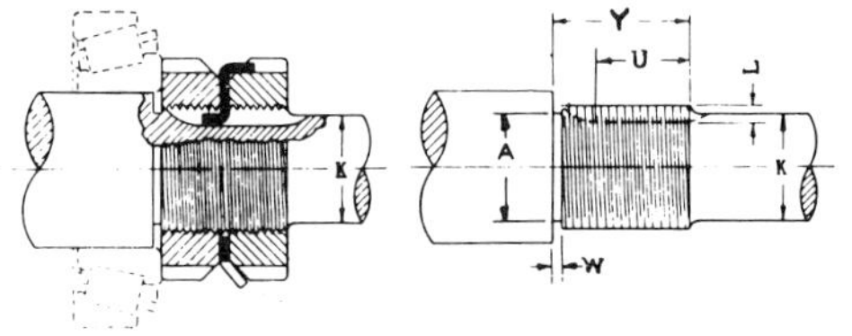

SHAFTS FOR TAPERED ROLLER BEARING LOCK NUTS
ADJUSTABLE MOUNTING

Thds. per Inch	Thread Major Diam			Thread Pitch Diam			Thread Minor Diam Max	Thread Relief W +1/64 -0	Ball Bearings, Cylindrical and Double Row Self-aligning Roller Bearings						
	Max	Tol	Min	Max	Tol	Min			Brg. Bore	Diam K Max	Thread Length P +1/64 -0	Keyway Depth L +1/64 -0	Keyway Width N +1/64 -0	Keyway Length M +1/64 -0	Nut No.
32	0.391	.0054	0.3856	0.3707	.0026	0.3681	0.3527	1/16	0.3937	5/16	9/32	1/16	1/8	3/32	N-00
32	0.469	.0054	0.4636	0.4487	.0026	0.4461	0.4307	1/16	0.4724	13/32	3/8	1/16	1/8	3/32	N-01
32	0.586	.0054	0.5806	0.5657	.0030	0.5627	0.5477	1/16	0.5906	1/2	3/8	5/64	1/8	3/32	N-02
32	0.664	.0054	0.6586	0.6437	.0030	0.6407	0.6257	1/16	0.6693	9/16	13/32	5/64	1/8	3/32	N-03
32	0.781	.0054	0.7756	0.7607	.0034	0.7573	0.7427	1/16	0.7874	23/32	7/16	5/64	3/16	3/32	N-04
32	0.969	.0054	0.9636	0.9487	.0034	0.9453	0.9307	1/16	0.9843	7/8	15/32	3/32	3/16	1/8	N-05
18	1.173	.0082	1.1648	1.1369	.0040	1.1329	1.1048	3/32	1.1811	1- 1/16	15/32	3/32	3/16	1/8	N-06
18	1.3125	.0082	1.3043	1.2764	.0040	1.2724	1.2443	3/32							
18	1.376	.0082	1.3678	1.3399	.0040	1.3359	1.3078	3/32	1.3780	1- 1/4	1/2	3/32	3/16	1/8	N-07
18	1.563	.0082	1.5548	1.5269	.0045	1.5224	1.4948	3/32	1.5748	1-15/32	17/32	3/32	5/16	1/8	N-08
18	1.767	.0082	1.7588	1.7309	.0045	1.7264	1.6988	1/8	1.7717	1-11/16	17/32	3/32	5/16	5/32	N-09
18	1.967	.0082	1.9588	1.9309	.0045	1.9264	1.8988	1/8	1.9685	1- 7/8	19/32	3/32	5/16	5/32	N-10
18	2.157	.0082	2.1488	2.1209	.0051	2.1158	2.0888	1/8	2.1654	2- 1/16	19/32	1/8	5/16	5/32	N-11
18	2.360	.0082	2.3518	2.3239	.0051	2.3188	2.2918	1/8	2.3622	2- 1/4	5/8	1/8	5/16	5/32	N-12
18	2.548	.0082	2.5398	2.5119	.0051	2.5068	2.4798	1/8	2.5591	2- 7/16	21/32	1/8	5/16	5/32	N-13
18	2.751	.0082	2.7528	2.7149	.0051	2.7098	2.6828	1/8	2.7559	2- 5/8	21/32	1/8	5/16	1/4	N-14
12	2.933	.0112	2.9218	2.8789	.0054	2.8735	2.8308	5/32	2.9528	2-25/32	11/16	1/8	5/16	1/4	AN-15
12	3.137	.0112	3.1258	3.0829	.0059	3.0770	3.0348	5/32	3.1496	3	11/16	1/8	3/8	1/4	AN-16
12	3.340	.0112	3.3288	3.2859	.0074	3.2785	3.2378	5/32	3.3465	3- 3/16	23/32	1/8	3/8	1/4	AN-17
12	3.527	.0112	3.5158	3.4729	.0074	3.4655	3.4248	5/32	3.5433	3- 3/8	13/16	5/32	3/8	1/4	AN-18
12	3.730	.0112	3.7188	3.6759	.0074	3.6685	3.6278	5/32	3.7402	3- 9/16	27/32	5/32	3/8	1/4	AN-19
12	3.918	.0112	3.9068	3.8639	.0074	3.8565	3.8158	5/32	3.9370	3-25/32	7/8	5/32	3/8	5/16	AN-20
12	4.122	.0112	4.1108	4.0679	.0083	4.0596	4.0198	5/32	4.1339	3-15/16	7/8	5/32	3/8	5/16	AN-21
12	4.325	.0112	4.3138	4.2709	.0083	4.2626	4.2228	5/32	4.3307	4- 3/16	29/32	3/16	3/8	5/16	AN-22
12	4.716	.0112	4.7048	4.6619	.0083	4.6536	4.6138	5/32	4.7244	4- 9/16	15/16	3/16	3/8	5/16	AN-24
12	5.106	.0112	5.0948	5.0519	.0083	5.0436	5.0038	5/32	5.1181	4-15/16	1	3/16	1/2	5/16	AN-26
12	5.497	.0112	5.4858	5.4429	.0083	5.4346	5.3948	5/32	5.5118	5- 5/16	1- 1/16	3/16	5/8	5/16	AN-28
12	5.888	.0112	5.8768	5.8339	.0083	5.8256	5.7858	5/32	5.9055	5-23/32	1- 1/8	7/32	5/8	3/8	AN-30
8	6.284	.0152	6.2688	6.2028	.0091	6.1937	6.1306	1/4	6.2992	6- 1/8	1- 3/16	15/64	5/8	3/8	AN-32
8	6.659	.0152	6.6438	6.5778	.0091	6.5687	6.5056	1/4	6.6929	6- 1/2	1- 7/32	15/64	3/4	3/8	AN-34
8	7.066	.0152	7.0508	6.9848	.0091	6.9757	6.9126	1/4	7.0866	6-29/32	1- 1/4	15/64	3/4	3/8	AN-36
8	7.472	.0152	7.4568	7.3908	.0091	7.3817	7.3186	1/4	7.4803	6- 5/16	1- 9/32	15/64	3/4	3/8	AN-38
8	7.847	.0152	7.8318	7.7658	.0114	7.7544	7.6936	1/4	7.8740	7-11/16	1-11/32	15/64	7/8	3/8	AN-40

Thread Relief Diameter "A" = Max Minor Thread Diameter – 1/64 in. ±.005 in. for 32 and 18 Pitch Threads, and Max Minor Thread diameter - 1/32 in. ±.010 in. for 12 and 8 Pitch Threads. Length of Bearing Seat "Z" = Minimum Bearing Width — 1/64 in. ±.010 in. Threads are American National Form NS, Class 3.

continued on next page

TABLE 11–19 continued

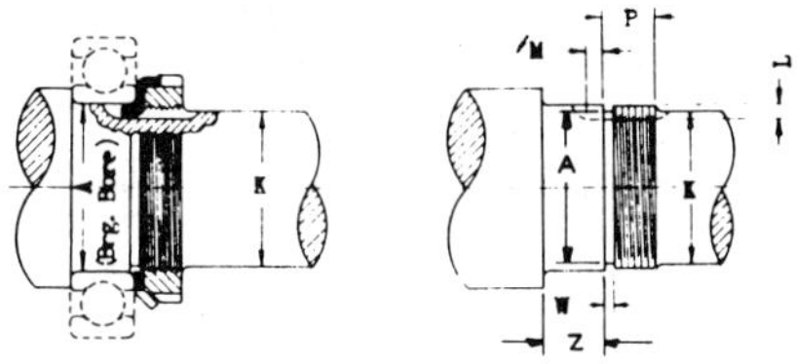

SHAFTS FOR BALL BEARING LOCK NUTS

Tapered Roller Bearings								Thds.
			Keyway					
Diam K Max	Thread Length Y +1/64 −0	Thread Length P +1/64 −0	Depth L +1/64 −0	Width N +1/64 −0	Length M +1/64 −0	Length U +1/64 −0	Nut No.	per Inch
5/16	19/32	3/8	3/32	1/8	3/32	15/32	8-N-00	32
13/32	25/32	15/32	3/32	1/8	3/32	9/16	8-N-01	32
1/2	13/16	1/2	3/32	1/8	3/32	19/32	8-N-02	32
9/16	7/8	17/32	3/32	1/8	3/32	5/8	8-N-03	32
45/64	29/32	17/32	3/32	3/16	3/32	5/8	8-N-04	32
7/8	1-	19/32	1/8	3/16	1/8	23/32	8-N-05	32
1- 1/16	1-	19/32	1/8	3/16	1/8	23/32	8-N-06	18
1- 3/16	1- 1/16	5/8	1/8	3/16	1/8	3/4	8-N-065	18
1- 1/4	1- 1/16	5/8	1/8	3/16	1/8	3/4	8-N-07	18
1- 7/16	1- 1/16	5/8	1/8	5/16	1/8	3/4	8-N-08	18
1-21/32	1- 1/16	5/8	1/8	5/16	5/32	25/32	8-N-09	18
1-55/64	1- 3/16	11/16	1/8	5/16	5/32	27/32	8-N-10	18
2- 3/64	1- 3/16	11/16	1/8	5/16	5/32	27/32	8-N-11	18
2- 1/4	1- 9/32	3/4	5/32	5/16	5/32	29/32	8-N-12	18
2-27/64	1-11/32	25/32	5/32	5/16	5/32	15/16	8-N-13	18
2- 5/8	1-11/32	25/32	5/32	5/16	1/4	1-	8-N-14	18
2-25/32	1-13/32	13/16	3/16	5/16	1/4	1- 1/32	8-AN-15	12
3-	1-13/32	13/16	3/16	3/8	1/4	1- 1/32	8-AN-16	12
3- 3/16	1-15/32	27/32	3/16	3/8	1/4	1- 1/16	8-AN-17	12
3- 3/8	1- 5/8	15/16	7/32	3/8	1/4	1- 5/32	8-AN-18	12
3- 9/16	1-11/16	31/32	7/32	3/8	1/4	1- 3/16	8-AN-19	12
3-49/64	1- 3/4	1-	7/32	3/8	5/16	1- 9/32	8-AN-20	12
3-16/16	1- 3/4	1-	7/32	3/8	5/16	1- 9/32	8-AN-21	12
4- 5/32	1-13/16	1- 1/32	7/32	3/8	5/16	1- 5/16	8-AN-22	12
4-17/32	1-29/32	1- 3/32	1/4	3/8	5/16	1- 3/8	8-AN-24	12
4-29/32	2- 1/32	1- 5/32	1/4	1/2	5/16	1- 7/16	8-AN-26	12
5-19/64	2-21/32	1-15/32	1/4	5/8	5/16	1- 3/4	8-AN-128	12
5-21/32	2-13/16	1- 9/16	9/32	5/8	3/8	1-29/32	8-AN-130	12
6- 1/16	2- 7/8	1-19/32	5/16	5/8	3/8	1-15/16	8-AN-132	8
6- 7/16	3-	1-21/32	5/16	3/4	3/8	2-	8-AN-134	8
6-27/32	3- 1/8	1-23/32	5/16	3/4	3/8	2- 1/16	8-AN-136	8
7- 1/4	3- 1/8	1-23/32	5/16	3/4	3/8	2- 1/16	8-AN-138	8
7- 5/8	3- 5/16	1-13/16	5/16	7/8	3/8	2- 1/8	8-AN-140	8

Thread Relief Diameter "A" = Max Minor Thread Diameter — 1/64 in. ± .005 in. for 32 and 18 Pitch Threads, and Max Minor Thread diameter - 1/32 in. ± .010 in. for 12 and 8 Pitch Threads. Length of Bearing Seat "Z" = Minimum Bearing Width — 1/64 in. ± .010 in. Threads are American National Form NS, Class 3.

TABLE 11-20

Dimensions of Spring Lock Washers (Carbon Steel)

ASA B27.1-1950 Lock Washers

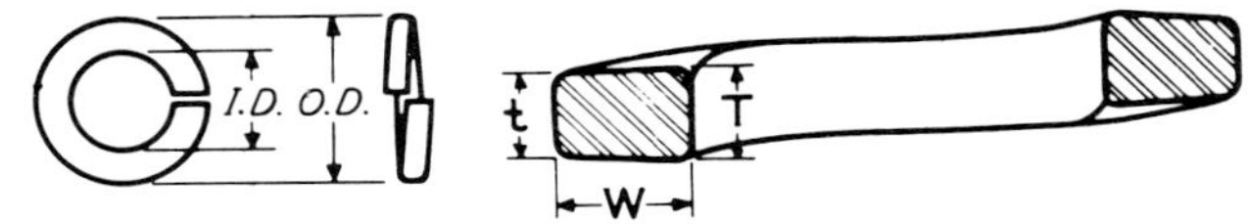

Nominal Size	Inside Diameter Min	Clearance of Nominal Bolt Size Min	Clearance of Nominal Bolt Size Max	Light Washer Sections (Min) Width w	Light Thickness (T + t)/2	Light Outside Diam Max*	Medium Washer Sections (Min) Width w	Medium Thickness (T + t)/2	Medium Outside Diam Max*	Heavy Washer Sections (Min) Width w	Heavy Thickness (T + t)/2	Heavy Outside Diam Max*	Extra Heavy Washer Sections (Min) Width w	Extra Heavy Thickness (T + t)/2	Extra Heavy Outside Diam Max*
0.086 (No. 2)	0.088	0.002	0.011	0.030	0.015	0.165	0.035	0.020	0.175	0.040	0.025	0.185	0.053	0.027	0.211
0.099 (No. 3)	0.102	0.002	0.011	0.035	0.020	0.188	0.040	0.025	0.198	0.047	0.031	0.212	0.062	0.034	0.242
0.112 (No. 4)	0.115	0.003	0.012	0.035	0.020	0.202	0.040	0.025	0.212	0.047	0.031	0.226	0.062	0.034	0.256
0.125 (No. 5)	0.128	0.003	0.012	0.040	0.025	0.225	0.047	0.031	0.239	0.055	0.040	0.255	0.079	0.045	0.303
0.138 (No. 6)	0.141	0.003	0.013	0.040	0.025	0.239	0.047	0.031	0.253	0.055	0.040	0.269	0.079	0.045	0.317
0.164 (No. 8)	0.168	0.004	0.014	0.047	0.031	0.280	0.055	0.040	0.296	0.062	0.047	0.310	0.096	0.057	0.378
0.190 (No. 10)	0.194	0.004	0.015	0.055	0.040	0.323	0.062	0.047	0.337	0.070	0.056	0.353	0.112	0.068	0.437
0.216 (No. 12)	0.221	0.005	0.016	0.062	0.047	0.364	0.070	0.056	0.380	0.077	0.063	0.394	0.130	0.080	0.500
1/4	0.255	0.005	0.017	0.107	0.047	0.489	0.109	0.062	0.493	0.110	0.077	0.495	0.132	0.084	0.539
5/16	0.319	0.006	0.020	0.117	0.056	0.575	0.125	0.078	0.591	0.130	0.097	0.601	0.143	0.108	0.627
3/8	0.382	0.007	0.023	0.136	0.070	0.678	0.141	0.094	0.688	0.145	0.115	0.696	0.170	0.123	0.746
7/16	0.446	0.008	0.026	0.154	0.085	0.780	0.156	0.109	0.784	0.160	0.133	0.792	0.186	0.143	0.844
1/2	0.509	0.009	0.029	0.170	0.099	0.877	0.171	0.125	0.879	0.176	0.151	0.889	0.204	0.162	0.945
9/16	0.573	0.010	0.032	0.186	0.113	0.975	0.188	0.141	0.979	0.193	0.170	0.989	0.223	0.182	1.049
5/8	0.636	0.011	0.035	0.201	0.126	1.082	0.203	0.156	1.086	0.210	0.189	1.100	0.242	0.202	1.164
11/16	0.700	0.012	0.038	0.216	0.138	1.178	0.219	0.172	1.184	0.227	0.207	1.200	0.260	0.221	1.266
3/4	0.763	0.013	0.041	0.233	0.153	1.277	0.234	0.188	1.279	0.244	0.226	1.299	0.279	0.241	1.369
13/16	0.827	0.014	0.044	0.249	0.168	1.375	0.250	0.203	1.377	0.262	0.246	1.401	0.298	0.261	1.473
7/8	0.890	0.015	0.047	0.264	0.179	1.470	0.266	0.219	1.474	0.281	0.266	1.504	0.322	0.285	1.586
15/16	0.954	0.016	0.050	0.277	0.191	1.562	0.281	0.234	1.570	0.298	0.284	1.604	0.345	0.308	1.698
1	1.017	0.017	0.053	0.289	0.202	1.656	0.297	0.250	1.672	0.319	0.306	1.716	0.366	0.330	1.810
1 1/16	1.081	0.018	0.056	0.301	0.213	1.746	0.312	0.266	1.768	0.338	0.326	1.820	0.389	0.352	1.922
1 1/8	1.144	0.019	0.059	0.314	0.224	1.837	0.328	0.281	1.865	0.356	0.345	1.921	0.411	0.375	2.031
1 3/16	1.208	0.020	0.062	0.324	0.234	1.923	0.344	0.297	1.963	0.373	0.364	2.021	0.431	0.396	2.137
1 1/4	1.271	0.021	0.065	0.336	0.244	2.012	0.359	0.312	2.058	0.393	0.384	2.126	0.452	0.417	2.244
1 5/16	1.335	0.022	0.068	0.346	0.254	2.098	0.375	0.328	2.156	0.410	0.403	2.226	0.472	0.438	2.350
1 3/8	1.398	0.023	0.071	0.356	0.264	2.183	0.391	0.344	2.253	0.427	0.422	2.325	0.491	0.458	2.453
1 7/16	1.462	0.024	0.074	0.366	0.273	2.269	0.406	0.359	2.349	0.442	0.440	2.421	0.509	0.478	2.555
1 1/2	1.525	0.025	0.077	0.375	0.282	2.352	0.422	0.375	2.446	0.458	0.458	2.518	0.526	0.496	2.654

All dimensions are given in inches.

*The maximum outside diameters specified allow for the commercial tolerances on cold drawn wire and hot rolled rod.

**Besides carbon steel, these series are available in stainless steel, types 302 and 420, Phosphor-Bronze, Silicon-Bronze and K-Monel.

When carbon steel spring lock washers are to be hot-dipped galvanized for use with hot-dipped galvanized bolts or screws, they shall be coiled to limits 0.020 in excess of those specified above for minimum inside diameter, minimum and maximum clearance, and maximum outside diameter. Galvanized lock washers under 1/4 inch nominal size are impractical.

TABLE 11-21

Internal and Heavy Internal Tooth Lock Washers

ASA B27.1-1950 Lock Washers

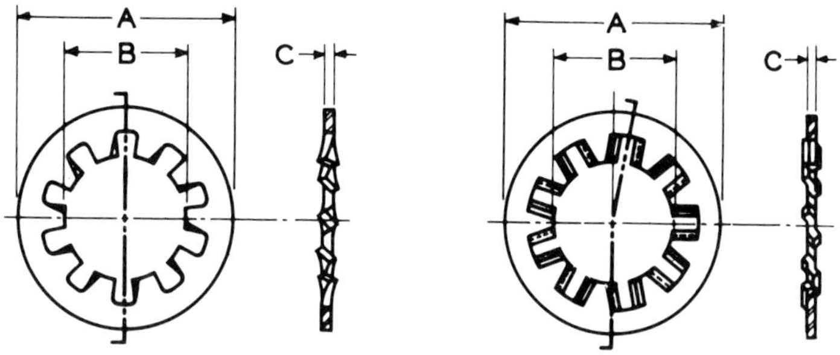

TYPE A TYPE B

Internal Tooth Lock Washers

	Size	#2	#3	#4	#5	#6	#8	#10	#12	1/4	5/16	3/8
A	Max	0.200	0.232	0.270	0.280	0.295	0.340	0.381	0.410	0.478	0.610	0.692
	Min	0.175	0.215	0.255	0.245	0.275	0.325	0.365	0.394	0.460	0.594	0.670
B	Max	0.095	0.109	0.123	0.136	0.150	0.176	0.204	0.231	0.267	0.332	0.398
	Min	0.089	0.102	0.115	0.129	0.141	0.168	0.195	0.221	0.256	0.320	0.384
C	Max	0.015	0.019	0.019	0.021	0.021	0.023	0.025	0.025	0.028	0.034	0.040
	Min	0.010	0.012	0.015	0.017	0.017	0.018	0.020	0.020	0.023	0.028	0.032

	Size	7/16	1/2	9/16	5/8	11/16	3/4	13/16	7/8	1	1 1/8	1 1/4
A	Max	0.789	0.900	0.985	1.071	1.166	1.245	1.315	1.410	1.637	1.830	1.975
	Min	0.740	0.867	0.957	1.045	1.130	1.220	1.290	1.364	1.590	1.799	1.921
B	Max	0.464	0.530	0.596	0.663	0.728	0.795	0.861	0.927	1.060	1.192	1.325
	Min	0.448	0.512	0.576	0.640	0.704	0.769	0.832	0.894	1.019	1.144	1.275
C	Max	0.040	0.045	0.045	0.050	0.050	0.055	0.055	0.060	0.067	0.067	0.067
	Min	0.032	0.037	0.037	0.042	0.042	0.047	0.047	0.052	0.059	0.059	0.059

All dimensions are given in inches.

Heavy Internal Tooth Lock Washers

	Size	1/4	5/16	3/8	7/16	1/2	9/16	5/8	3/4	7/8
A	Max	0.536	0.607	0.748	0.858	0.924	1.034	1.135	1.265	1.447
	Min	0.500	0.590	0.700	0.800	0.880	0.990	1.100	1.240	1.400
B	Max	0.267	0.332	0.398	0.464	0.530	0.596	0.663	0.795	0.927
	Min	0.256	0.320	0.384	0.448	0.512	0.576	0.640	0.768	0.894
C	Max	0.045	0.050	0.050	0.067	0.067	0.067	0.067	0.084	0.084
	Min	0.035	0.040	0.042	0.050	0.055	0.055	0.059	0.070	0.075

All dimensions are given in inches.

TABLE 11-22

External Tooth Lock Washers

ASA B27.1-1950 Lock Washers

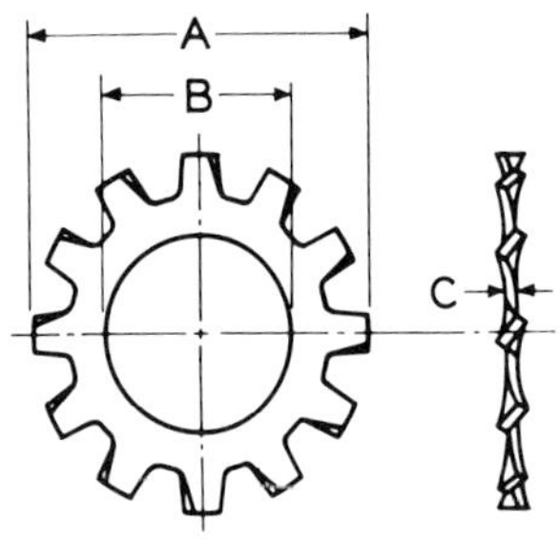

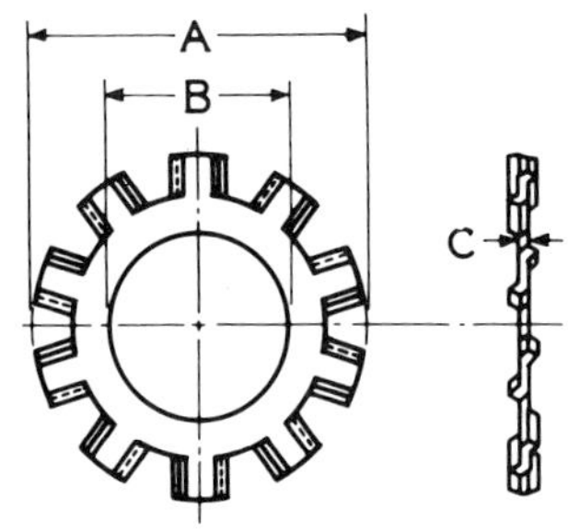

	Size	#2	#3	#4	#5	#6	#8	#10	#12	1/4	5/16	3/8
A	Max	.....	.....	0.290	.....	0.320	0.381	0.410	0.475	0.510	0.610	0.694
	Min	.....	.....	0.275	.....	0.305	0.365	0.395	0.460	0.494	0.588	0.670
B	Max	.....	.....	0.123	.....	0.150	0.176	0.204	0.231	0.267	0.332	0.398
	Min	.....	.....	0.115	.....	0.141	0.168	0.195	0.221	0.256	0.320	0.384
C	Max	.....	.....	0.019	.....	0.022	0.023	0.025	0.028	0.028	0.034	0.040
	Min	.....	.....	0.015	.....	0.016	0.018	0.020	0.023	0.023	0.028	0.032

	Size	7/16	1/2	9/16	5/8	11/16	3/4	13/16	7/8	1	1 1/8	1 1/4
A	Max	0.760	0.900	0.985	1.070	1.155	1.260	1.315	1.410	1.620	.....	.....
	Min	0.740	0.880	0.960	1.045	1.130	1.220	1.290	1.380	1.590	.....	.....
B	Max	0.464	0.530	0.596	0.663	0.728	0.795	0.861	0.927	1.060	.....	.....
	Min	0.448	0.513	0.576	0.641	0.704	0.768	0.833	0.897	1.025	.....	.....
C	Max	0.040	0.045	0.045	0.050	0.050	0.055	0.055	0.060	0.067	.....	.....
	Min	0.032	0.037	0.037	0.042	0.042	0.047	0.047	0.052	0.059	.....	.....

All dimensions are given in inches.

TABLE 11-23

American Standard for Plain Washers

ASA B27.2-1953 Plain Washers

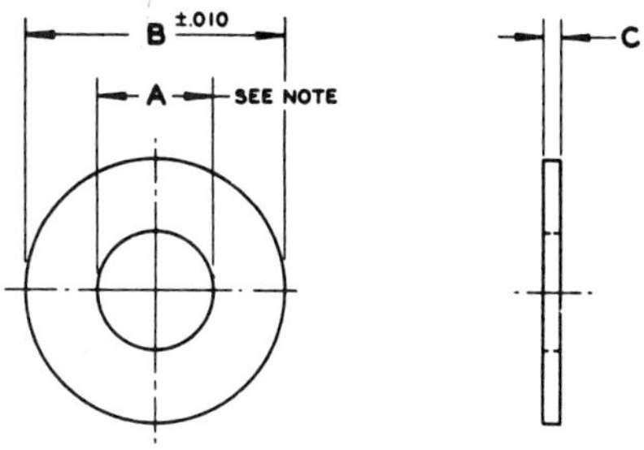

Inside Diameter	Outside Diameter	Thickness, C			
A	B	Gage	Nom	Max	Min
5/64	3/16	25	0.020	0.025	0.016
3/32	7/32	25	0.020	0.025	0.016
3/32	1/4	25	0.020	0.025	0.016
1/8	1/4	24	0.022	0.028	0.017
1/8	5/16	21	0.032	0.040	0.025
5/32	5/16	20	0.035	0.048	0.027
5/32	3/8	18	0.049	0.065	0.036
11/64	13/32	18	0.049	0.065	0.036
3/16	3/8	18	0.049	0.065	0.036
3/16	7/16	18	0.049	0.065	0.036
13/64	15/32	18	0.049	0.065	0.036
7/32	7/16	18	0.049	0.065	0.036
7/32	1/2	18	0.049	0.065	0.036
15/64	17/32	18	0.049	0.065	0.036
1/4	1/2	18	0.049	0.065	0.036
1/4*	9/16	18	0.049	0.065	0.036
1/4*	9/16	16	0.065	0.080	0.051
17/64	5/8	18	0.049	0.065	0.036
9/32	5/8	16	0.065	0.080	0.051
5/16	3/4	16	0.065	0.080	0.051
5/16	7/8	16	0.065	0.080	0.051
11/32	11/16	16	0.065	0.080	0.051
3/8	3/4	16	0.065	0.080	0.051
3/8	7/8	14	0.083	0.104	0.064
3/8	1 1/8	16	0.065	0.080	0.051
13/32	13/16	16	0.065	0.080	0.051
7/16	7/8	14	0.083	0.104	0.064
7/16	1	14	0.083	0.104	0.064
7/16	1 3/8	14	0.083	0.104	0.064
15/32	59/64	16	0.065	0.080	0.051
1/2	1 1/8	14	0.083	0.104	0.064
1/2	1 1/4	14	0.083	0.104	0.064
1/2	1 5/8	14	0.083	0.104	0.064
17/32	1 1/16	13	0.095	0.121	0.074
9/16	1 1/4	12	0.109	0.132	0.086
9/16	1 3/8	12	0.109	0.132	0.086
9/16	1 7/8	12	0.109	0.132	0.086
19/32	1 3/16	13	0.095	0.121	0.074
5/8	1 3/8	12	0.109	0.132	0.086
5/8	1 1/2	12	0.109	0.132	0.086
5/8	2 1/8	10	0.134	0.160	0.108
21/32	1 5/16	13	0.095	0.121	0.074
11/16	1 1/2	10	0.134	0.160	0.108
11/16	1 3/4	10	0.134	0.160	0.108
11/16	2 3/8	8	0.165	0.192	0.136
13/16	1 1/2	10	0.134	0.160	0.108
13/16	1 3/4	9	0.148	0.177	0.122
13/16	2	9	0.148	0.177	0.122
13/16	2 7/8	8	0.165	0.192	0.136
15/16	1 3/4	10	0.134	0.160	0.108
15/16	2	8	0.165	0.192	0.136
15/16	2 1/4	8	0.165	0.192	0.136
15/16	3 3/8	7	0.180	0.213	0.153
1 1/16	2	10	0.134	0.160	0.108
1 1/16	2 1/4	8	0.165	0.192	0.136
1 1/16	2 1/2	8	0.165	0.192	0.136
1 1/16	3 7/8	4	0.238	0.280	0.210
1 3/16	2 1/2	8	0.165	0.192	0.136
1 1/4	2 3/4	8	0.165	0.192	0.136
1 5/16	2 3/4	8	0.165	0.192	0.136
1 3/8	3	8	0.165	0.192	0.136
1 7/16	3	7	0.180	0.213	0.153
1 1/2	3 1/4	7	0.180	0.213	0.153
1 9/16	3 1/4	7	0.180	0.213	0.153
1 5/8	3 1/2	7	0.180	0.213	0.153
1 11/16	3 1/2	7	0.180	0.213	0.153
1 3/4	3 3/4	7	0.180	0.213	0.153
1 13/16	3 3/4	7	0.180	0.213	0.153
1 7/8	4	7	0.180	0.213	0.153
1 15/16	4	7	0.180	0.213	0.153
2	4 1/4	7	0.180	0.213	0.153
2 1/16	4 1/4	7	0.180	0.213	0.153
2 1/8	4 1/2	7	0.180	0.213	0.153
2 3/8	4 3/4	5	0.220	0.248	0.193
2 5/8	5	4	0.238	0.280	0.210
2 7/8	5 1/4	3	0.259	0.310	0.228
3 1/8	5 1/2	2	0.284	0.327	0.249

NOTE: Tolerance ±0.005 on inside diameter to and including 7/32 inside diameter; ±0.010 on inside diameter greater than 7/32 with exception of two 1/4x9/16 sizes marked with an asterisk on which two sizes the tolerance is ±0.005.

GENERAL: Plain Washers covered by this specification are intended for general industrial applications.

MATERIAL: Plain Washers shall be made of ferrous or nonferrous metal, plastic, etc., as specified.

THICKNESS: Reflecting long established trade practice, the nominal thicknesses of washers are Birmingham gage sizes. The limits specified represent a tolerance of plus or minus one gage or the spread from the minimum of one gage minus to the maximum of one gage plus.

DEFECTS: Washers must be free from burrs, loose scale, sharp edges and all other defects that might affect their serviceability.

TABLE 11-24

Lug, Sections and Hole Dimensions of Series 5000 INTERNAL Waldes Truarc Retaining Rings

Engineering Data and Specifications, Oct, 1952 Waldes Kohinoor, Inc.

Ring No. 5000	Lug Dim B	sect J	sect E	Hole Diam P
5000- 25	.063	.015	.025	.031
5000- 31	.066	.018	.033	.031
5000- 37	.082	.028	.040	.041
5000- 43	.098	.029	.049	.047
5000- 45	.098	.030	.050	.047
5000- 50	.114	.035	.053	.047
5000- 51	.114	.035	.053	.047
5000- 56	.132	.035	.053	.047
5000- 62	.132	.035	.060	.062
5000- 68	.132	.036	.063	.062
5000- 75	.142	.040	.070	.062
5000- 77	.146	.044	.074	.062
5000- 81	.155	.044	.077	.062
5000- 86	.155	.045	.081	.062
5000- 87	.155	.045	.084	.062
5000- 90	.155	.047	.087	.062
5000- 93	.155	.050	.091	.062
5000- 100	.155	.052	.096	.062
5000- 102	.155	.052	.098	.062
5000- 106	.180	.055	.101	.078
5000- 112	.180	.056	.105	.078
5000- 118	.180	.059	.112	.078
5000- 125	.180	.059	.114	.078
5000- 131	.180	.061	.120	.078
5000- 137	.180	.063	.124	.078
5000- 143	.180	.065	.126	.078
5000- 145	.180	.064	.128	.078
5000- 150	.180	.065	.130	.078
5000- 156	.202	.070	.138	.078
5000- 162	.227	.076	.146	.078
5000- 165	.227	.072	.144	.078
5000- 168	.227	.071	.145	.078
5000- 175	.234	.073	.144	.078
5000- 181	.234	.073	.147	.093
5000- 185	.234	.075	.149	.093
5000- 187	.234	.074	.150	.093
5000- 193	.234	.076	.155	.093
5000- 200	.240	.082	.165	.093
5000- 206	.250	.085	.171	.093
5000- 212	.260	.085	.173	.093
5000- 218	.264	.089	.179	.093
5000- 225	.270	.090	.180	.093
5000- 231	.270	.095	.192	.093
5000- 237	.270	.096	.195	.093
5000- 244	.280	.098	.201	.110
5000- 250	.280	.106	.204	.110
5000- 253	.280	.101	.205	.110

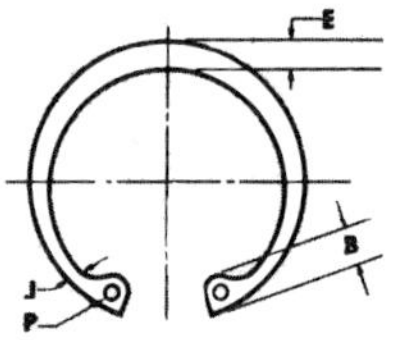

Since tolerance dimensions are not included, it is recommended that this chart be used only as a reference chart of working dimensions *and not for purposes of inspection.*

Ring No. 5000	Lug Dim B	sect J	sect E	Hole Diam P
5000- 256	.290	.104	.207	.110
5000- 262	.290	.105	.210	.110
5000- 268	.300	.106	.218	.110
5000- 275	.300	.108	.225	.110
5000- 281	.300	.131	.266	.110
5000- 283	.300	.111	.230	.110
5000- 287	.310	.117	.234	.110
5000- 300	.310	.121	.249	.110
5000- 306	.320	.121	.252	.110
5000- 312	.320	.122	.255	.110
5000- 315	.320	.142	.258	.125
5000- 325	.344	.128	.264	.125
5000- 334	.344	.132	.272	.125
5000- 347	.340	.137	.282	.125
5000- 350	.340	.132	.285	.125
5000- 354	.350	.133	.288	.125
5000- 356	.350	.134	.291	.125
5000- 362	.350	.135	.296	.125
5000- 375	.360	.142	.309	.125
5000- 387	.370	.145	.318	.125
5000- 393	.370	.147	.323	.125
5000- 400	.370	.150	.330	.125
5000- 412	.370	.144	.321	.125
5000- 425	.370	.148	.333	.125
5000- 433	.400	.153	.339	.125
5000- 450	.400	.157	.351	.125
5000- 462	.400	.159	.360	.125
5000- 475	.410	.163	.369	.125
5000- 500	.446	.172	.390	.156
5000- 525	.450	.179	.408	.156
5000- 537	.468	.181	.408	.156
5000- 550	.468	.181	.408	.156
5000- 575	.468	.180	.408	.156
5000- 600	.468	.177	.408	.156
5000- 625	.478	.184	.423	.156
5000- 650	.488	.189	.438	.156
5000- 662	.524	.192	.447	.187
5000- 675	.530	.199	.456	.187
5000- 700	.544	.203	.474	.187
5000- 725	.561	.211	.489	.187
5000- 750	.564	.218	.507	.187
5000- 800	.586	.230	.540	.187
5000- 825	.598	.239	.558	.187
5000- 850	.608	.246	.573	.187
5000- 875	.620	.251	.591	.187
5000- 900	.632	.257	.609	.187
5000- 950	.654	.271	.642	.187
5000-1000	.676	.284	.675	.187

TABLE 11–25

INTERNAL Snaprings — Series 5000 Waldes Truarc Retaining Rings

Engineering Data and Specifications, Oct. 1952 Waldes Kohinoor, Inc.

SEE ENLARGED DETAIL OF GROOVE

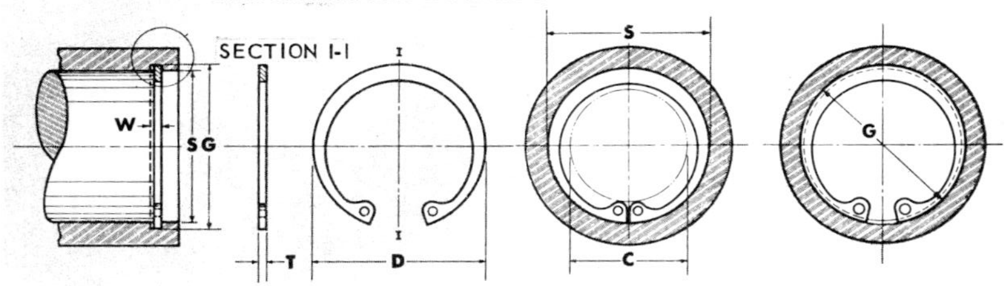

See Table 11–24 for additional ring dimensions.

Maximum bottom radii:

rings no. 25 to 100	.005
rings no. 102 and over	.010

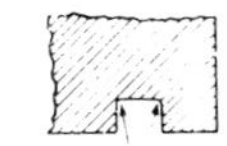

enlarged detail of groove

C = actual clearance diameter of ring when sprung into housing, prior to installation into groove.

Ring No. 5000	HOUSING Diam Inches S Dec Equiv	S Fract Equiv	mm S Approx Equiv	RING DIMENSIONS Free Diam D	Tol	Thickness T	Tol	Approx wt lb per 1000 pcs	GROOVE DIMENSIONS Diam G	Tol	Width W	Tol	Depth	Ring Clearance C	Allowable Thrust Load (lbs) Safety Factor = 4	Min Gap Width (Ring in Groove)
25	.250	1/4	6.4	.272	+.005	.015	±.0013	.08	.268	±.001	.017	+.002	.009	.115	350	.047
31	.312	5/16	7.9	.336	−.002	.015		.11	.330		.017	−.000	.009	.173	440	.055
37	.375	3/8	9.5	.403	+.005	.025		.25	.397		.028		.011	.208	870	.063
43	.438	7/16	11.1	.468	−.002	.025	±.0015	.37	.461	±.002	.028		.012	.23	1020	.063
45	.453	29/64	11.5	.484		.025		.43	.477		.028		.012	.25	1070	
50	.500	1/2	12.7	.533		.035		.70	.524		.039	+.003	.012	.27	1650	.071
51	.512	—	13.0	.544		.035		.77	.536	±.003	.039	−.000	.012	.27	1690	.069
56	.562	9/16	14.3	.603	+.010	.035		.86	.592		.039		.015	.275	1850	.064
62	.625	5/8	15.9	.675	−.005	.035	±.002	1.0	.659		.039		.017	.34	2060	.083
68	.688	11/16	17.5	.742		.035		1.2	.724	.002	.039		.0185	.40	2250	.091
75	.750	3/4	19.0	.808		.035		1.3	.790	T.I.R.	.039		.020	.45	2500	.122

Material is carbon spring steel, but Waldes Truarc Retaining Rings can also be made from aluminum, beryllium copper, phosphor bronze, SAE 1060-1090, and stainless steel No. 420.

T.I.R. (total indicator reading) is the maximum allowable deviation of concentricity between the groove and the housing.

continued on next page

Ring No. 5000	HOUSING Diam Inches S Dec Equiv	S Fract Equiv	mm S Approx Equiv	RING DIMENSIONS Free Diam D	Tol	Thickness T	Tol	Approx wt lb per 1000 pcs	GROOVE DIMENSIONS Diam G	Tol	Width W	Tol	Depth	Ring Clearance C	Allowable Thrust Load (lbs) Safety Factor = 4	Min Gap Width (Ring in Groove)
77	.777	—	19.7	.836		.042		1.7	.819		.046		.021	.475	3800	.127
81	.812	13/16	20.6	.877		.042		1.9	.857		.046		.022	.49	4000	.138
86	.866	—	22.0	.934		.042		2.0	.912	±.003	.046		.023	.55	4250	.148
87	.875	7/8	22.2	.944	+.010	.042	±.002	2.1	.922		.046	+.003	.0235	.545	4300	.152
90	.901	—	22.9	.970	−.005	.042		2.2	.950		.046	−.000	.0245	.565	4450	.161
93	.938	15/16	23.8	1.015		.042		2.4	.989	.002	.046		.0255	.62	4650	.167
100	1.000	1	25.4	1.081		.042		2.7	1.055	T.I.R.	.046		.0275	.665	4950	.169
102	1.023	—	26.0	1.106		.042		2.8	1.079		.046		.028	.69	5050	.182
106	1.062	1 1/16	27.0	1.150		.050		3.7	1.120		.056		.0285	.685	6200	.177
112	1.125	1 1/8	28.6	1.217		.050		4.0	1.185		.056		.030	.745	6600	.192
118	1.188	1 3/16	30.2	1.283		.050		4.3	1.250	±.004	.056		.0315	.80	7000	.213
125	1.250	1 1/4	31.7	1.351		.050		4.5	1.320		.056		.035	.875	7350	.242
131	1.312	1 5/16	33.3	1.418	+.015	.050	±.002	4.6	1.385		.056	+.004	.0365	.93	7750	.249
137	1.375	1 3/8	34.9	1.486	−.010	.050		5.3	1.450		.056	−.000	.0375	.99	8100	.263
143	1.438	1 7/16	36.5	1.552		.050		5.7	1.515	.003	.056		.039	1.06	8500	.273
145	1.456	—	37.0	1.572		.050		5.9	1.535	T.I.R.	.056		.0395	1.08	8600	.286
150	1.500	1 1/2	38.1	1.622		.050		6.6	1.580		.056		.040	1.13	8800	.287
156	1.562	1 9/16	39.6	1.688		.062		8.8	1.647		.068		.0425	1.15	11400	.305
162	1.625	1 5/8	41.2	1.756		.062		9.7	1.715		.068		.045	1.16	11850	.332
165	1.653	—	42.0	1.786		.062		9.8	1.745		.068		.046	1.17	12100	.335
168	1.688	1 11/16	42.8	1.823		.062		10.0	1.780	±.005	.068		.0465	1.22	12350	.337
175	1.750	1 3/4	44.4	1.891	+.020	.062		10.1	1.845		.068	+.004	.0475	1.26	12800	.349
181	1.812	1 13/16	46.0	1.958	−.013	.062	±.003	10.5	1.910		.068	−.000	.049	1.33	13250	.351
185	1.850	—	47.0	1.998		.062		10.8	1.949		.068		.0495	1.36	13500	.360
187	1.875	1 7/8	47.6	2.025		.062		11.1	1.975	.003	.068		.050	1.39	13700	.364
193	1.938	1 15/16	49.2	2.095		.062		11.4	2.040	T.I.R.	.068		.0515	1.46	14150	.376
200	2.000	2	50.8	2.160		.062		13.2	2.110		.068		.055	1.50	14600	.419

Material is carbon spring steel, but Waldes Truarc Retaining Rings can also be made from aluminum, beryllium copper, phosphor bronze, SAE 1060-1090, and stainless steel No. 420.

T.I.R. (total indicator reading) is the maximum allowable deviation of concentricity between the groove and the housing.

continued on next page

TABLE 11–25, continued

Ring No. 5000	HOUSING Diam S Dec Equiv	S Fract Equiv	S Approx Equiv	RING DIMENSIONS Free Diam D	Tol	Thickness T	Tol	Approx wt lb per 1000 pcs	GROOVE DIMENSIONS Diam G	Tol	Width W	Tol	Depth	Ring Clear-ance C	Allow-able Thrust Load (lbs) Safety Factor = 4	Min Gap Width (Ring in Groove)
206	2.062	2 1/16	52.3	2.225		.078		17.0	2.175		.086		.0565	1.54	18950	.416
212	2.125	2 1/8	53.9	2.295		.078		18.0	2.240		.086		.057	1.59	19500	.418
218	2.188	2 3/16	55.5	2.365		.078		20.0	2.305		.086		.059	1.65	20100	.420
225	2.250	2 1/4	57.1	2.435		.078		20.5	2.370		.086		.060	1.69	20700	.434
231	2.312	2 5/16	58.7	2.500	+.025	.078		21.5	2.440		.086		.0635	1.75	21200	.470
237	2.375	2 3/8	60.3	2.567	−.015	.078		22.0	2.505		.086		.065	1.81	21800	.487
244	2.440	2 7/16	61.9	2.634		.078		24.0	2.570		.086		.066	1.86	22400	.492
250	2.500	2 1/2	63.5	2.700		.078		26.0	2.635		.086		.0675	1.90	23000	.496
253	2.531	2 17/32	64.2	2.733		.078		26.0	2.668		.086		.0685	1.94	23300	.514
256	2.562	2 9/16	65.0	2.760		.093		30.0	2.700		.103		.0685	1.95	28100	.534
262	2.625	2 5/8	66.6	2.840		.093		33.5	2.765		.103		.070	2.02	28800	.524
268	2.688	2 11/16	68.2	2.907		.093		34.5	2.834	±.006	.103		.072	2.05	29500	.540
275	2.750	2 3/4	69.8	2.975	+.030	.093		35.0	2.900		.103	+.005	.075	2.12	30100	.550
281	2.813	2 13/16	71.4	3.040	−.020	.093	±.003	43.0	2.965		.103	−.000	.076	2.18	30800	.570
283	2.834	—	71.9	3.063		.093		38.0	2.987	.004	.103		.0765	2.21	31100	.570
287	2.875	2 7/8	73.0	3.105		.093		41.0	3.030	T.I.R.	.103		.0775	2.22	31500	.590
300	3.000	3	76.1	3.245		.093		45.0	3.165		.103		.0825	2.34	32900	.670
306	3.062	3 1/16	77.7	3.310		.109		51.0	3.230		.120		.0835	2.41	39300	.690
312	3.125	3 1/8	79.3	3.377		.109		53.0	3.295		.120		.085	2.47	40100	.704
315	3.156	3 5/32	80.1	3.408		.109		54.0	3.328		.120		.086	2.50	40500	.724
325	3.250	3 1/4	82.5	3.509		.109		57.0	3.426		.120		.088	2.56	41700	.734
334	3.346	3 11/32	85.0	3.611		.109		62.0	3.525		.120		.090	2.62	43000	.754
347	3.469	3 15/32	88.0	3.746		.109		63.0	3.657		.120		.094	2.77	44500	.800
350	3.500	3 1/2	88.8	3.780		.109		66.0	3.690		.120		.095	2.81	44900	.807
354	3.543	—	89.9	3.826		.109		68.0	3.735		.120		.096	2.82	45500	.810
356	3.562	3 9/16	90.4	3.850		.109		69.0	3.756		.120		.097	2.84	45700	.810
362	3.625	3 5/8	92.0	3.920		.109		70.0	3.822		.120		.0985	2.90	46600	.814

Material is carbon spring steel, but Waldes Truarc Retaining Rings can also be made from aluminum, beryllium copper, phosphor bronze, SAE 1060-1090, and stainless steel No. 420.

T.I.R. (total indicator reading) is the maximum allowable deviation of concentricity between the groove and the housing.

continued on next page

TABLE 11–25, continued

Ring No. 5000	HOUSING Diam Inches S Dec Equiv	S Fract Equiv	mm S Approx Equiv	RING DIMENSIONS Free Diam D	Tol	Thickness T	Tol	Approx wt lb per 1000 pcs	GROOVE DIMENSIONS Diam G	Tol	Width W	Tol	Depth	Ring Clearance C	Allowable Thrust Load (lbs) Safety Factor = 4	Min Gap Width (Ring in Groove)
375	3.750	3 3/4	95.2	4.060		.109		77.0	3.955		.120		.1025	3.01	48100	.834
387	3.875	3 7/8	98.3	4.205		.109		81.0	4.087		.120		.106	3.12	49700	.840
393	3.938	3 15/16	99.9	4.283		.109		84.0	4.150	±.006	.120		.106	3.18	50600	.814
400	4.000	4	101.5	4.350		.109		87.0	4.220		.120		.110	3.23	51400	.844
412	4.125	4 1/8	104.7	4.496		.109		88.0	4.339		.120		.107	3.36	53000	
425	4.250	4 1/4	107.9	4.632	+.030	.109	±.003	93.0	4.470		.120	+.005	.110	3.48	54600	
433	4.330	—	109.9	4.719	−.020	.109		95.0	4.556		.120	−.000	.113	3.51	55500	
450	4.500	4 1/2	114.2	4.905		.109		112.0	4.735	.004	.120		.117	3.66	57800	
462	4.625	4 5/8	117.4	5.041		.109		111.0	4.865	T.I.R.	.120		.120	3.79	59400	
475	4.750	4 3/4	120.6	5.177		.109		116.0	4.995		.120		.122	3.90	61000	
500	5.000	5	126.9	5.450		.109		130.0	5.260		.120		.130	4.08	64200	
525	5.250	5 1/4	133.2	5.723		.125		162.0	5.520	±.007	.139		.135	4.33	77300	
537	5.375	5 3/8	136.4	5.763		.125		162.0	5.650		.139		.135	4.40	79100	
550	5.500	5 1/2	139.6	5.885	+.040	.125	±.004	168.0	5.770		.139	+.006	.135	4.52	81000	
575	5.750	5 3/4	145.9	6.140	−.020	.125		176.0	6.020	.005	.139	−.000	.135	4.70	84700	
600	6.000	6	152.3	6.395		.125		182.0	6.270	T.I.R.	.139		.135	5.02	88300	
625	6.250	6 1/4	158.6	6.662		.156		245.0	6.530	±.008	.174		.140	5.26	114800	
650	6.500	6 1/2	165.0	6.928	+.050	.156		265.0	6.790		.174		.145	5.49	119400	
662	6.625	6 5/8	168.1	7.062		.156	±.005	275.0	6.925		.174	+.008	.150	5.54	121700	
675	6.750	6 3/4	171.3	7.194	−.020	.156		283.0	7.055	.006	.174	−.000	.152	5.65	124000	
700	7.000	7	177.7	7.461		.156		304.0	7.315	T.I.R.	.174		.157	5.88	128600	
725	7.250	7 1/4	184.0	7.728		.187		392.0	7.575		.209		.162	6.08	159600	
750	7.500	7 1/2	190.4	7.994		.187		427.0	7.840	±.008	.209		.170	6.33	165200	
800	8.000	8	203.0	8.527		.187		489.0	8.360		.209		.180	6.78	176200	
825	8.250	8 1/4	209.4	8.793	+.060	.187		531.0	8.620		.209		.185	7.00	181700	
850	8.500	8 1/2	215.7	9.060		.187		552.0	8.880		.209		.190	7.24	187200	
875	8.750	8 3/4	222.1	9.327	−.020	.187		576.0	9.145		.209		.197	7.44	192000	
900	9.000	9	228.4	9.593		.187		645.0	9.405	.006	.209		.202	7.68	198200	
950	9.500	9 1/2	241.1	10.126		.187		707.0	9.930	T.I.R.	.209		.215	8.13	209200	
1000	10.000	10	253.8	10.659		.187		754.0	10.450		.209		.225	8.59	220200	

Material is carbon spring steel, but Waldes Truarc Retaining Rings can also be made from aluminum, beryllium copper, phosphor bronze, SAE 1060-1090, and stainless steel No. 420.

T.I.R. (total indicator reading) is the maximum allowable deviation of concentricity between the groove and the housing.

TABLE 11–26

BEVELED, INTERNAL Retaining Rings — Series 5002

Engineering Data and Specifications, Oct, 1952 Waldes Kohinoor, Inc.

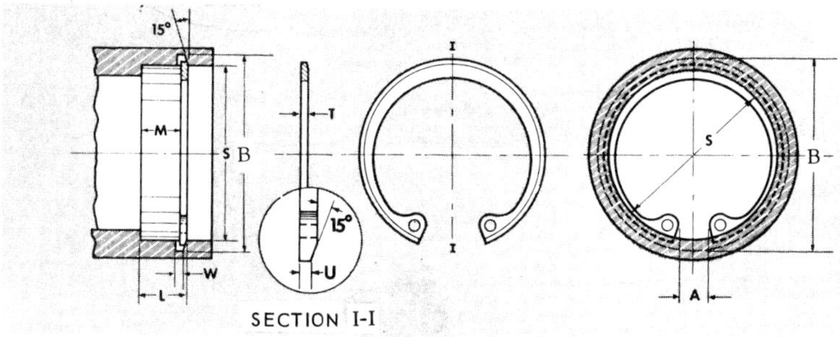

Tolerances on dimensions *L* and *M* should be so chosen that any combination of them cannot exceed the end-play take-up.

Clearance dimension is same as *C* in Table 11–25.

Ring No. 5002	HOUSING Diam Inches S Dec Equiv	S Fract Equiv	mm S Approx Equiv	DIMENSIONS OF BEVELED RING U	Tol	Thickness T	Tol	DIMENSIONS OF BEVELED GROOVE Diam B	Tol −.000	Width W	Tol −.000	Min Width of Gap A (Ring in Groove)	Max End-Play Take-Up	Allowable Thrust Load (lbs) Safety Factor = 4
100	1.000	1	25.4	.033		.042		1.081	+.003	.036		.150	.005	4950
102	1.023	--	26.0	.033		.042		1.106		.036		.155	.005	5050
106	1.062	1 1/16	27.0	.041		.050		1.150		.044		.160	.006	6200
112	1.125	1 1/8	28.6	.040		.050		1.217		.043		.170	.006	6600
118	1.188	1 3/16	30.2	.040		.050	±.002	1.283	+.004	.043		.185	.006	7000
125	1.250	1 1/4	31.7	.039		.050		1.351		.042		.200	.006	7350
131	1.312	1 5/16	33.3	.039		.050		1.418		.042		.215	.006	7750
137	1.375	1 3/8	34.9	.038		.050		1.486		.041		.230	.007	8100
143	1.438	1 7/16	36.5	.037		.050		1.552		.040		.245	.007	8500
145	1.456	--	37.0	.037		.050		1.572		.040		.245	.007	8600
150	1.500	1 1/2	38.1	.037		.050		1.622		.040		.255	.008	8800
156	1.562	1 9/16	39.6	.048		.062		1.688		.052		.265	.008	11400
162	1.625	1 5/8	41.2	.047		.062		1.756		.051		.270	.008	11850
165	1.653	--	42.0	.047		.062		1.786		.051		.280	.008	12100
168	1.688	1 11/16	42.8	.046		.062		1.823		.050		.295	.008	12350
175	1.750	1 3/4	44.4	.046		.062		1.891		.050		.295	.009	12800
181	1.812	1 13/16	46.0	.046	±.001	.062	±.003	1.958	+.005	.050	+.001	.310	.009	13250
185	1.850	--	47.0	.046		.062		1.998		.050		.300	.010	13500
187	1.875	1 7/8	47.6	.046		.062		2.025		.050		.310	.010	13700
193	1.938	1 15/16	49.2	.045		.062		2.095		.049		.325	.010	14150
200	2.000	2	50.8	.044		.062		2.160		.048		.340	.010	14600
206	2.062	2 1/16	52.3	.060		.078		2.225		.065		.355	.010	18950
212	2.125	2 1/8	53.9	.060		.078		2.295		.065		.365	.011	19500
218	2.188	2 3/16	55.5	.059		.078		2.365		.064		.365	.012	20100
225	2.250	2 1/4	57.1	.059		.078		2.435		.064		.380	.012	20700
231	2.312	2 5/16	58.7	.058		.078		2.500		.063		.400	.012	21200
237	2.375	2 3/8	60.3	.058	±.0015	.078		2.567	+.006	.063	+.0015	.420	.012	21800
244	2.440	2 7/16	61.9	.057		.078		2.634		.062		.415	.013	22400
250	2.500	2 1/2	63.5	.057		.078		2.700		.062		.430	.013	23000
253	2.531	2 17/32	64.2	.057		.078		2.733		.062		.440	.013	23300

Material is carbon spring steel, but Waldes Truarc Retaining Rings can also be made from aluminum, beryllium copper, phosphor bronze, SAE 1060- 1090, and stainless steel No. 420.

Series 5002 not available in phosphor bronze and aluminum.

continued on next page

TABLE 11-26, continued

Ring No. 5002	HOUSING Diam Inches S Dec Equiv	S Fract Equiv	mm S Approx Equiv	DIMENSIONS OF BEVELED RING U	Tol	Thickness T	Tol	DIMENSIONS OF BEVELED GROOVE Diam B	Tol −.000	Width W	Tol −.000	Min Width of Gap A (Ring in Groove)	Min End-Play Take-Up	Allowable Thrust Load (lbs) Safety Factor = 4
256	2.562	2 9/16	65.0	.072		.093		2.760		.078		.445	.013	28100
262	2.625	2 5/8	66.6	.071		.093		2.840		.077		.440	.014	28800
268	2.688	2 11/16	68.2	.071		.093		2.907		.077		.440	.015	29500
275	2.750	2 3/4	69.8	.070		.093		2.975		.076		.450	.015	30100
281	2.813	2 13/16	71.4	.070	±.0015	.093	±.003	3.040	+.006	.076	+.0015	.470	.015	30800
283	2.834	--	71.9	.070		.093		3.063		.076		.485	.015	31100
287	2.875	2 7/8	73.0	.070		.093		3.105		.076		.500	.015	31500
300	3.000	3	76.1	.068		.093		3.245		.074		.570	.016	32900
306	3.062	3 1/16	77.7	.084		.109		3.310		.091		.575	.016	39300
312	3.125	3 1/8	79.3	.083		.109		3.377		.090		.600	.016	40100
315	3.156	3 5/32	80.1	.083		.109		3.408		.090		.605	.016	40500
325	3.250	3 1/4	82.5	.082		.109		3.509		.089		.610	.017	41700
334	3.346	3 11/32	87.5	.082		.109		3.611		.089		.645	.017	43000
347	3.469	3 15/32	88.0	.081		.109		3.746		.088		.665	.018	44500
350	3.500	3 1/2	88.8	.081		.109		3.780		.088		.675	.018	44900
354	3.543	--	89.9	.080		.109		3.826		.087		.685	.018	45500
356	3.562	3 9/16	90.4	.080		.109		3.850		.087		.670	.019	45700
362	3.625	3 5/8	92.0	.079		.109		3.920		.086		.690	.019	46600
375	3.750	3 3/4	95.2	.078	±.002	.109		4.060		.085	+.002	.690	.021	48100
387	3.875	3 7/8	98.3	.077		.109		4.205		.084		.695	.023	49700
393	3.938	3 15/16	99.9	.077		.109		4.283		.084		.680	.025	50600
400	4.000	4	101.5	.076		.109		4.350		.083		.695	.025	51400
412	4.125	4 1/8	104.7	.084		.109		4.496		.091		.675	.028	53000
425	4.250	4 1/4	107.9	.084		.109		4.632		.091		.690	.029	54600
433	4.330	--	109.9	.083		.109		4.719		.090		.700	.029	55500
450	4.500	4 1/2	114.2	.082		.109		4.905		.089		.730	.030	57800
462	4.625	4 5/8	117.4	.081		.109		5.041		.088		.760	.031	59400
475	4.750	4 3/4	120.6	.081		.109		5.177		.088		.770	.032	61000
500	5.000	5	126.9	.081		.109		5.450		.088		.810	.034	64200

Material is carbon spring steel, but Waldes Truarc Retaining Rings can also be made from aluminum, beryllium copper, phosphor bronze, SAE 1060-1090, and stainless steel No. 420.

TABLE 11–27

EXTERNAL Snaprings — Series 5100 Waldes Truarc Retaining Rings

Engineering Data and Specifications, Oct, 1952

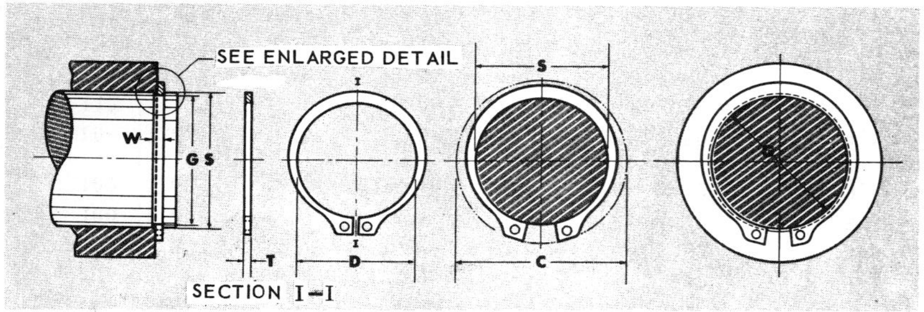

Maximum bottom radii:

rings no. 12 to 23 incl.	sharp corners
rings no. 25 to 35 incl.	.003
rings no. 37 to 100 incl.	.005
rings no. 102 and over	.010

C = actual clearance diameter of ring when sprung over shaft, prior to installation into groove.

lugs of rings no. 12 to 23 inclusive.

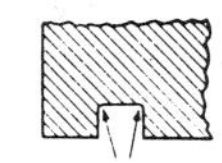

ENLARGED DETAIL OF GROOVE

Ring No. 5100	SHAFT Diam Inches S Dec Equiv	SHAFT Diam Inches S Fract Equiv	SHAFT Diam mm S Approx Equiv	RING DIMENSIONS Free Diam D	Free Diam Tol	Thickness T	Thickness Tol	Approx wt lb per 1000 pcs	GROOVE DIMENSIONS Diam G	Diam Tol	Width W	Width Tol	Depth	Ring Clearance C	Allowable Thrust Load (lbs) Safety Factor = 4
12	.125	1/8	3.2	.112		.010	±.001	.018	.117		.012		.004	.222	110
15	.156	5/32	4.0	.142		.010		.037	.146		.012		.005	.270	130
18	.188	3/16	4.8	.168		.015		.059	.175	±.0015	.017		.006	.298	240
19	.197	--	5.0	.179	+.002	.015	±.0013	.063	.185		.017	+.002	.006	.319	250
21	.219	7/32	5.6	.196	–.004	.015		.074	.205		.017	–.000	.007	.338	280
23	.236	15/64	6.0	.215		.015		.086	.222	.0015 T.I.R.	.017		.007	.355	310
25	.250	1/4	6.3	.225		.025		.21	.230		.028		.010	.45	590
27	.275	--	7.0	.250		.025		.23	.255		.028		.010	.48	650
28	.281	9/32	7.1	.256		.025		.24	.261	±.002	.028		.010	.49	660
31	.312	5/16	7.9	.281		.025		.27	.290		.028		.011	.54	740
34	.344	11/32	8.7	.309	+.002	.025	±.0015	.31	.321	.002 T.I.R.	.028	+.003	.011	.57	800
35	.354	--	9.0	.320	–.005	.025		.35	.330		.028	–.000	.012	.59	820

Material for sizes 5100–12 to 5100–23, inclusive, is beryllium copper. Material for other sizes is carbon spring steel. The other sizes can also be made from aluminum, beryllium copper, phosphor bronze, SAE 1060-1090, and stainless steel No. 420.

continued on next page

TABLE 11–27, continued

Copyright 1952

Ring No. 5100	SHAFT Diam Inches S Dec Equiv	S Fract Equiv	mm S Approx Equiv	RING DIMENSIONS Free Diam D	Tol	Thickness T	Tol	Approx wt lb per 1000 pcs	GROOVE DIMENSIONS Diam G	Tol	Width W	Tol	Depth	Ring Clearance C	Allowable Thrust Load (lbs) Safety Factor = 4
37	.375	3/8	9.5	.338		.025		.39	.352	±.002	.028		.012	.63	870
39	.393	--	10.0	.354		.025	±.0015	.42	.369		.028		.012	.62	940
40	.406	13/32	10.3	.366		.025		.43	.382		.028		.012	.61	950
43	.438	7/16	11.1	.395	+.002	.025		.50	.412	.002	.028		.013	.66	1020
46	.469	15/32	11.9	.428	-.005	.025		.54	.443	T.I.R.	.028		.013	.68	1100
50	.500	1/2	12.7	.461		.035		.91	.474		.039		.013	.77	1650
55	.551	--	14.0	.509		.035		.90	.524		.039		.0135	.81	1800
56	.562	9/16	14.3	.521		.035		1.1	.535	±.003	.039	+.003	.014	.82	1850
59	.594	19/32	15.1	.550		.035		1.2	.565		.039	-.000	.0145	.86	1950
62	.625	5/8	15.9	.579		.035		1.3	.596		.039		.0145	.90	2060
66	.669	--	17.0	.618		.035		1.4	.638	.002	.039		.0155	.93	2200
66	.672	43/64	17.1	.618	+.005	.035		1.4	.640	T.I.R.	.039		.016	.93	2200
68	.688	11/16	17.5	.635	-.010	.042		1.8	.655		.046		.016	1.01	2750
75	.750	3/4	19.0	.693		.042		2.1	.715		.046		.0175	1.09	2900
78	.781	25/32	19.8	.722		.042		2.2	.745	±.003	.046		.018	1.12	3900
81	.812	13/16	20.6	.751		.042		2.5	.776		.046		.018	1.15	4000
87	.875	7/8	22.2	.810		.042	±.002	2.8	.835		.046		.021	1.21	4300
93	.938	15/16	23.8	.867		.042		3.1	.894		.046		.021	1.34	4650
98	.984	63/64	25.0	.910		.042		3.5	.940	.002	.046		.022	1.39	4850
100	1.000	1	25.4	.925		.042		3.6	.955	T.I.R.	.046		.0225	1.41	4950
102	1.023	--	26.0	.946		.042		3.9	.977		.046		.023	1.43	5050
106	1.062	1 1/16	27.0	.982		.050		4.8	1.015		.056		.0235	1.50	6200
112	1.125	1 1/8	29.0	1.041		.050		5.1	1.075	±.004	.056		.025	1.55	6600
118	1.188	1 3/16	30.2	1.098		.050		5.6	1.135		.056		.026	1.61	7000
125	1.250	1 1/4	31.7	1.156	+.010	.050		5.9	1.195		.056	+.004	.0275	1.69	7350
131	1.312	1 5/16	33.3	1.214	-.015	.050		6.8	1.250		.056	-.000	.031	1.75	7750
137	1.375	1 3/8	34.9	1.272		.050		7.2	1.310	.003	.056		.0325	1.80	8100
143	1.438	1 7/16	36.5	1.333		.050		8.1	1.370	T.I.R.	.056		.034	1.87	8500
150	1.500	1 1/2	38.1	1.387		.050		9.0	1.430		.056		.035	1.99	8800

Material for sizes 5100-12 to 5100-23, inclusive, is beryllium copper. Material for other sizes is carbon spring steel. The other sizes can also be made from aluminum, beryllium copper, phosphor bronze, SAE 1060-1090, and stainless steel No. 420.

continued on next page

TABLE 11–27, continued

Ring No. 5100	SHAFT Diam Inches S Dec Equiv	S Fract Equiv	mm S Approx Equiv	RING DIMENSIONS Free Diam D	Tol	Thickness T	Tol	Approx wt lb per 1000 pcs	GROOVE DIMENSIONS Diam G	Tol	Width W	Tol	Depth	Ring Clearance C	Allowable Thrust Load (lbs) Safety Factor = 4
156	1.562	1 9/16	39.6	1.446		.062		12.4	1.490		.068		.0365	2.10	11400
162	1.625	1 5/8	41.7	1.503		.062		13.2	1.550		.068		.0375	2.17	11850
168	1.687	1 11/16	42.8	1.560		.062		14.8	1.610		.068		.038	2.24	12350
175	1.750	1 3/4	44.4	1.618		.062		15.3	1.670	±.005	.068		.040	2.31	12800
177	1.771	--	44.9	1.637		.062		15.4	1.689		.068		.041	2.33	12950
181	1.812	1 13/16	46.0	1.675	+.013	.062		16.2	1.730		.068	+.004	.0415	2.38	13250
187	1.875	1 7/8	47.6	1.735	-.020	.062		17.3	1.790	.003	.068	-.000	.0425	2.44	13700
196	1.969	1 31/32	50.0	1.819		.062		18.0	1.879	T.I.R.	.068		.045	2.54	14350
200	2.000	2	50.8	1.850		.062		19.0	1.910		.068		.045	2.55	14600
206	2.062	2 1/16	52.3	1.906		.078		25.0	1.970		.086		.0465	2.68	18950
212	2.125	2 1/8	53.9	1.964		.078		26.1	2.027		.086		.049	2.75	19500
215	2.156	2 5/32	54.7	1.993		.078		26.3	2.057	±.006	.086		.0495	2.78	19800
225	2.250	2 1/4	57.1	2.081	+.015	.078		27.7	2.145		.086		.0525	2.87	20700
231	2.312	2 5/16	58.7	2.139	-.025	.078		28.0	2.205		.086		.054	2.94	21200
237	2.375	2 3/8	60.3	2.197		.078		29.2	2.265		.086	+.005	.055	3.01	21800
243	2.437	2 7/16	61.9	2.255		.078	±.003	29.5	2.325		.086	-.000	.056	3.07	22400
250	2.500	2 1/2	63.5	2.313		.078		29.7	2.385	.004	.086		.0575	3.12	23000
255	2.559	--	64.9	2.377		.078		33.9	2.451	T.I.R.	.086		.054	3.18	23500
262	2.625	2 5/8	66.6	2.428		.078		35.0	2.505		.086		.060	3.25	24100
268	2.687	2 11/16	68.2	2.485		.078		36.0	2.565		.086		.061	3.32	24700
275	2.750	2 3/4	69.8	2.543		.093		47.0	2.625		.103		.0625	3.45	30100
287	2.875	2 7/8	73.0	2.659		.093		48.5	2.742		.103		.066	3.57	31500
293	2.937	2 15/16	74.5	2.717		.093		50.0	2.801		.103		.068	3.64	32200
300	3.000	3	76.1	2.775		.093		52.0	2.860		.103		.070	3.69	32900
306	3.062	3 1/16	77.7	2.832	+.020	.093		47.0	2.920		.103		.071	3.74	33500
312	3.125	3 1/8	79.3	2.892	-.030	.093		58.0	2.980		.103		.0725	3.82	34300
315	3.156	3 5/32	80.1	2.920		.093		59.0	3.010		.103		.073	3.85	34600
325	3.250	3 1/4	82.5	3.006		.093		62.0	3.100		.103		.075	3.95	35600
334	3.344	3 11/32	84.9	3.092		.093		64.0	3.190		.103		.077	4.04	36700
343	3.437	3 7/16	87.2	3.179		.093		66.0	3.280		.103		.078	4.14	37700

Material for sizes 5100–12 to 5100–23, inclusive, is beryllium copper. Material for other sizes is carbon spring steel. The other sizes can also be made from aluminum, beryllium copper, phosphor bronze, SAE 1060-1090, and stainless steel No. 420.

continued on next page

TABLE 11–27, continued

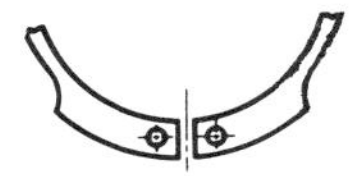

rings no. 425 and over are balanced

Ring No. 5100	SHAFT Diam Inches S Dec Equiv	S Fract Equiv	mm S Approx Equiv	RING DIMENSIONS Free Diam D	Tol	Thickness T	Tol	Approx wt lb per 1000 pcs	GROOVE DIMENSIONS Diam G	Tol	Width W	Tol	Depth	Ring Clear-ance C	Allow-able Thrust Load (lbs) Safety Factor = 4
350	3.500	3 1/2	88.8	3.237		.109		72.0	3.340		.120		.080	4.25	44900
354	3.543	--	89.9	3.277		.109		73.0	3.381		.120		.081	4.29	45500
362	3.625	3 5/8	92.0	3.352		.109		76.0	3.458	±.006	.120		.083	4.37	46600
368	3.687	3 11/16	93.6	3.410		.109		80.0	3.517		.120		.085	4.43	47300
375	3.750	3 3/4	95.2	3.468		.109		83.0	3.577		.120		.0865	4.50	48100
387	3.875	3 7/8	98.3	3.584		.109		88.0	3.696		.120		.089	4.60	49700
393	3.938	3 15/16	99.9	3.642	+.020	.109	±.003	95.0	3.756		.120	+.005	.0905	4.70	50600
400	4.000	4	101.5	3.700	-.030	.109		101.0	3.815		.120	-.000	.092	4.78	51400
425	4.250	4 1/4	107.9	3.989		.109		112.0	4.065		.120		.092	5.09	54600
437	4.375	4 3/8	111,0	4.106		.109		115.0	4.190	.004	.120		.092	5.22	56200
450	4.500	4 1/2	114.2	4.223		.109		101.0	4.310	T.I.R.	.120		.095	5.37	57800
475	4.750	4 3/4	120.6	4.458		.109		113.0	4.550		.120		.100	5.67	61000
500	5.000	5	126.9	4.692		.109		149.0	4.790		.120		.105	5.96	64200
525	5.250	5 1/4	133.2	4.927		.125		190.0	5.030		.139		.110	6.27	77300
550	5.500	5 1/2	140.0	5.162		.125		201.0	5.265	±.007	.139	+.006	.117	6.57	81000
575	5.750	5 3/4	145.9	5.396		.125	±.004	199.0	5.505	.005	.139	-.000	.122	6.86	84700
600	6.000	6	152.3	5.631		.125		210.0	5.745	T.I.R.	.139		.127	7.16	88300
625	6.250	6 1/4	158.6	5.866		.156		282.0	5.985		.174		.132	7.46	114800
650	6.500	6 1/2	165.0	6.100	+.020	.156		330.0	6.225	±.008	.174	+.008	.137	7.87	119400
675	6.750	6 3/4	171.3	6.335	-.050	.156	±.005	356.0	6.465	.006	.174	-.000	.142	8.06	124000
700	7.000	7	177.7	6.570		.156		388.0	6.705	T.I.R.	.174		.147	8.36	128600

Material for sizes 5100-12 to 5100-23, inclusive, is beryllium copper. Material for other sizes is carbon spring steel. The other sizes can also be made from aluminum, beryllium copper, phosphor bronze, SAE 1060-1090, and stainless steel No. 420.

T.I.R. (total indicator reading) is the maximum allowable deviation of concentricity between the groove and the shaft.

TABLE 11–28

BEVELED, EXTERNAL Retaining Rings — Series 5102

Engineering Data and Specifications, Oct, 1952 Waldes Kohinoor, Inc.

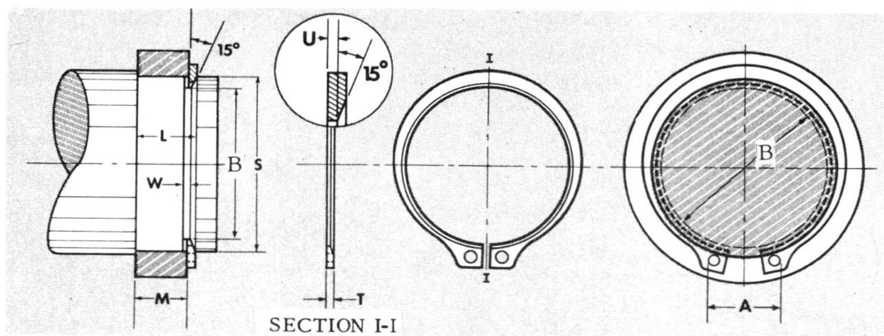

SECTION I-I

Tolerances on dimensions L and M should be so chosen that any combination of them cannot exceed the end-play take-up.

Clearance dimension is same as C in Table 11–27.

Ring No. 5102	SHAFT Diam Inches S Dec Equiv	S Fract Equiv	mm S Approx Equiv	DIMENSIONS OF BEVELED RING U	Tol	Thickness T	Tol	DIMENSIONS OF BEVELED GROOVE Diam B	Tol −.000	Width W	Tol −.000	Max Outside Distance of Holes A (Ring in Groove)	Max End-Play Take-Up	Allowable Thrust Load (lbs) Safety Factor = 4
100	1.000	1	25.4	.034		.042		.925	+.003	.037		.370	.005	4950
102	1.023	--	26.0	.033		.042		.946		.036		.370	.005	5050
106	1.062	1 1/16	27.0	.041		.050		.982		.044		.495	.006	6200
112	1.125	1 1/8	29.0	.041		.050		1.041		.044		.500	.006	6600
118	1.188	1 3/16	30.2	.041		.050		1.098		.044		.520	.006	7000
125	1.250	1 1/4	31.7	.040	±.001	.050	±.002	1.156	+.004	.043	+.001	.510	.006	7350
131	1.312	1 5/16	33.3	.039		.050		1.214		.042		.540	.006	7750
137	1.375	1 3/8	34.9	.039		.050		1.272		.042		.540	.007	8100
143	1.438	1 7/16	36.5	.039		.050		1.333		.042		.540	.007	8500
150	1.500	1 1/2	38.1	.038		.050		1.387		.041		.605	.008	8800

Material is carbon spring steel, but Waldes Truarc Retaining Rings can also be made from aluminum, beryllium copper, phosphor bronze, SAE 1060-1090, and stainless steel No. 420.

Series 5102 not available in phosphor bronze.

continued on next page

TABLE 11–28, continued

Ring No. 5102	SHAFT Diam Inches S Dec Equiv	SHAFT Diam Inches S Fract Equiv	SHAFT Diam mm S Approx Equiv	DIMENSIONS OF BEVELED RING U	DIMENSIONS OF BEVELED RING Tol	DIMENSIONS OF BEVELED RING Thick-ness T	DIMENSIONS OF BEVELED RING Tol	DIMENSIONS OF BEVELED GROOVE Diam B	DIMENSIONS OF BEVELED GROOVE Tol −.000	DIMENSIONS OF BEVELED GROOVE Width W	DIMENSIONS OF BEVELED GROOVE Tol −.000	Max Outside Distance of Holes A (Ring in Groove)	Max End-Play Take-Up	Allow-able Thrust Load (lbs) Safety Factor = 4
156	1.562	1 9/16	39.6	.049		.062		1.446		.053		.675	.008	11400
162	1.625	1 5/8	41.2	.049		.062		1.503		.053		.700	.009	11850
168	1.687	1 11/16	42.8	.048		.062		1.560		.052		.700	.009	12350
175	1.750	1 3/4	44.4	.048		.062		1.618		.052		.700	.009	12800
177	1.771	--	44.9	.048	±.001	.062		1.637	+.005	.052	+.001	.725	.010	12950
181	1.812	1 13/16	46.0	.048		.062		1.675		.052		.725	.010	13250
187	1.875	1 7/8	47.6	.048		.062		1.735		.052		.725	.010	13700
196	1.969	1 31/32	50.0	.047		.062		1.819		.051		.750	.011	14350
200	2.000	2	51.0	.047		.062		1.850		.051		.745	.011	14600
206	2.062	2 1/16	52.3	.062		.078		1.906		.067		.800	.011	18950
212	2.125	2 1/8	53.9	.062		.078		1.946		.067		.820	.012	19500
215	2.156	2 5/32	54.7	.062		.078		1.993		.067		.825	.012	19800
225	2.250	2 1/4	57.1	.061		.078		2.081		.066		.820	.012	20700
231	2.312	2 5/16	58.7	.060		.078		2.139		.065		.825	.012	21200
237	2.375	2 3/8	60.3	.060		.078		2.197		.065		.825	.012	21800
243	2.437	2 7/16	61.9	.060	±.0015	.078	±.003	2.255	+.006	.065	+.0015	.850	.013	22400
250	2.500	2 1/2	63.5	.059		.078		2.313		.064		.850	.013	23000
255	2.559	--	64.9	.059		.078		2.377		.064		.830	.013	23500
262	2.625	2 5/8	66.6	.059		.078		2.428		.064		.870	.014	24100
268	2.687	2 11/16	68.2	.059		.078		2.485		.064		.870	.014	24700
275	2.750	2 3/4	69.8	.073		.093		2.543		.079		.950	.015	30100
287	2.875	2 7/8	73.0	.072		.093		2.659		.078		.950	.015	31500
293	2.937	2 15/16	74.5	.072		.093		2.[illegible]17		.078		.950	.015	32200
300	3.000	3	76.1	.071		.093		2.775		.077		.950	.015	32900

Material is carbon spring steel, but Waldes Truarc Retaining Rings can also be made from aluminum, beryllium copper, phosphor bronze, SAE 1060-1090, and stainless steel No. 420.

TABLE 11–29

Waldes Truarc Retaining Rings for Use With Ball Bearings

Engineering Data and Specifications, Oct, 1952 Waldes Kohinoor, Inc.

EXTERNAL	WALDES TRUARC RETAINING RINGS to be used with →						BALL BEARING			WALDES TRUARC RETAINING RINGS ← to be used with						INTERNAL
GROOVE DIAM for Groove of Standard Depth	5100- (Table 11–27) Max Allowable Ball Bearing Corner Radii or Chamfer			5108- Max Allowable Ball Bearing Corner Radii or Chamfer			Bore	SAE	OD	5000- (Table 11–25) Max Allowable Ball Bearing Corner Radii or Chamfer			5008- Max Allowable Ball Bearing Corner Radii or Chamfer			GROOVE DIAM for Groove of Standard Depth
Inch	No.	Inch	mm	No.	Inch	mm	mm	No.	mm	No.	Inch	mm	No.	Inch	mm	Inch
.146	15	.012	0.3				4	34	16	62	.033	0.85	--	--	--	.659
.185	19	.016	0.4				5	--	16	62	.033	0.85	--	--	--	.659
								35	19	75	.038	0.95	75	.054	1.4	.790
.222	23	.019	0.5				6	--	19	75	.038	0.95	75	.054	1.4	.790
								--	21	81	.039	1.0	81	.058	1.5	.871
								--	22	86	.043	1.1	--	--	--	.912
								--	24	93	.048	1.2	93	.068	1.7	.996
.255	27	.021	0.55				7	37	22	86	.043	1.1	--	--	--	.912
								--	24	93	.048	1.2	93	.068	1.7	.996
								--	26	102	.050	1.25	--	--	--	1.079
.290	31	.024	0.6				8	--	22	86	.043	1.1	--	--	--	.912
								--	24	93	.048	1.2	93	.068	1.7	.996
.330	35	.022	0.55				9	--	22	86	.049	1.1	--	--	--	.912
								39	26	102	.050	1.25	--	--	--	1.079
								--	28	--	--	--	--	--	--	
								--	30	118	.058	1.45	118	.079	2.0	1.244
.369	39	.022	0.55				10	--	26	102	.050	1.25	--	--	--	1.079
								--	28	--	--	--	--	--	--	--
								200	30	118	.058	1.45	118	.079	2.0	1.244
								--	32	125	.056	1.4	125	.079	2.0	1.330
								300	35	137	.060	1.5	137	.082	2.1	1.450

continued on next page

EXTERNAL	WALDES TRUARC RETAINING RINGS to be used with →						BALL BEARING Bore SAE OD			WALDES TRUARC RETAINING RINGS ← to be used with						INTERNAL
GROOVE DIAM for Groove of Standard Depth	5100- (Table 11–27) Max Allowable Ball Bearing Corner Radii or Chamfer			5108- Max Allowable Ball Bearing Corner Radii or Chamfer						5000- (Table 11–25) Max Allowable Ball Bearing Corner Radii or Chamfer			5008- Max Allowable Ball Bearing Corner Radii or Chamfer			GROOVE DIAM for Groove of Standard Depth
Inch	No.	Inch	mm	No.	Inch	mm	mm	No.	mm	No.	Inch	mm	No.	Inch	mm	Inch
.412	43	.024	0.6				11	--	32	125	.056	1.4	125	.079	2.0	1.330
								--	43	168	.065	1.65	168	.100	2.5	1.786
.443	46	.025	0.65				12	--	28	--	--	--	--	--	--	--
								201	32	125	.056	1.4	125	.079	2.0	1.330
								301	37	145	.061	1.55	--	--	--	1.535
								--	46	181	.067	1.7	*181	.108	2.7	1.910
.486	50	.039	1.0	50	.056	1.4	13	--	28	--	--	--	--	--	--	--
								--	30	118	.C58	1.45	118	.079	2.0	1.244
								--	32	125	.056	1.4	125	.079	2.0	1.330
								--	50	193	.070	1.8	--	--	--	2.071
.524	55	.033	0.85				14	--	35	137	.060	1.5	137	.082	2.1	1.450
.565	59	.030	0.75	59	.064	1.6	15	--	32	125	.C56	1.4	125	.079	2.0	1.330
								202	35	137	.060	1.5	137	.082	2.1	1.450
								--	37	145	.061	1.55	--	--	--	1.535
								--	40	156	.066	1.65	156	.097	2.5	1.660
								302	42	165	.066	1.65	--	--	--	1.745
								--	59	231	.087	2.2	*231	.146	3.7	2.450
.596	62	.030	0.75	62	.068	1.7	16	--	32	125	.056	1.4	125	.079	2.0	1.330
								--	35	137	.060	1.5	137	.082	2.1	1.450
								--	38	150	.062	1.6	150	.090	2.3	1.580
.638	66	.036	0.9	68	.074	1.9	17	--	35	137	.060	1.5	137	.082	2.1	1.450
								203	40	156	.066	1.65	156	.097	2.5	1.660
								--	44	175	.066	1.65	175	.103	2.6	1.827
								303	47	185	.067	1.7	--	--	--	1.949
								--	52	206	.077	1.95	206	.135	3.4	2.160
								403	62	244	.090	2.3	243	.152	3.8	2.570
								--	72	283	.110	2.8	--	--	--	2.987

*Production dies not available as of date of printing.

continued on next page

TABLE 11–29, continued

EXTERNAL	WALDES TRUARC RETAINING RINGS to be used with →						BALL BEARING			WALDES TRUARC RETAINING RING ← to be used with						INTERNAL
GROOVE DIAM for Groove of standard Depth	5100- (Table 11–27) Max Allowable Ball Bearing Corner Radii or Chamfer			5108- Max Allowable Ball Bearing Corner Radii or Chamfer			Bore	SAE	OD	5000- (Table 11–25) Max Allowable Ball Bearing Corner Radii or Chamfer			5008- Max Allowable Ball Bearing Corner Radii or Chamfer			GROOVE DIAM for Groove of Standard Depth
Inch	No.	Inch	mm	No.	Inch	mm	mm	No.	mm	No.	Inch	mm	No.	Inch	mm	Inch
.676	68	.040	1.0	68	.074	1.9	18	--	40	156	.066	1.65	156	.097	2.5	1.660
.715	75	.041	1.05	75	.079	2	19	--	40	156	.066	1.65	156	.097	2.5	1.660
.745	78	.045	1.15	78	.082	2.1	20	--	40	156	.066	1.65	156	.097	2.5	1.660
								--	42	165	.066	1.65	--	--	--	1.745
								204	47	185	.067	1.7	--	--	--	1.949
								304	52	206	.077	1.95	206	.135	3.4	2.160
								--	65	256	.095	2.4	--	--	--	2.700
								404	72	283	.110	2.8	--	--	--	2.987
.824	87	.040	1.0	87	.089	2.3	22	--	42	165	.066	1.65	--	--	--	1.745
								--	72	283	.110	2.8	--	--	--	2.987
.940	98	.063	1.6	100	.101	2.6	25	--	47	185	.067	1.7	--	--	--	1.949
								205	52	206	.077	1.95	206	.135	3.4	2.160
								305	62	244	.090	2.3	243	.152	3.8	2.570
								405	80	315	.112	2.85	315	.190	4.8	3.328
.977	102	.043	1.1	100	.101	2.6	26	--	52	206	.077	1.95	206	.135	3.4	2.160
1.015	106	.058	1.5	106	.103	2.6	27	--	88	347	.118	3.0				3.657
1.135	118	.065	1.65	118	.106	2.7	30	--	52	206	.077	1.95	206	.135	3.4	2.160
								--	55	212	.079	2.0	*212	.138	3.5	2.279
								206	62	244	.090	2.3	243	.152	3.8	2.570
								306	72	283	.110	2.8	--	--	--	2.987
								406	90	354	.120	3.05	--	--	--	3.735
									95	375	.128	3.25	*375	.191	4.8	3.945
1.310	137	.068	1.7	137	.110	2.8	35	--	62	244	.090	2.3	243	.152	3.8	2.570
								--	70	275	.099	2.5	275	.169	4.3	2.900
								207	72	283	.110	2.8	--	--	--	2.987
								307	80	315	.112	2.85	315	.190	4.8	3.328
								407	100	393	.134	3.4	--	--	--	4.150
								--	103	400	.136	3.45	*400	.190	4.8	4.275

*Production dies not available as of date of printing.

continued on next page

EXTERNAL	WALDES TRUARC RETAINING RINGS to be used with →						BALL BEARING			WALDES TRUARC RETAINING RINGS ← to be used with						INTERNAL
GROOVE DIAM for Groove of Standard Depth	5100- (Table 11–27) Max Allowable Ball Bearing Corner Radii or Chamfer			5108- Max Allowable Ball Bearing Corner Radii or Chamfer			Bore	SAE	OD	5000- (Table 11–25) Max Allowable Ball Bearing Corner Radii or Chamfer			5008- Max Allowable Ball Bearing Corner Radii or Chamfer			GROOVE DIAM for Groove of Standard Depth
Inch	No.	Inch	mm	No.	Inch	mm	mm	No.	mm	No.	Inch	mm	No.	Inch	mm	Inch
1.502	156	.073	1.85	156	.118	3.0	40	--	68	268	.095	2.4	--	--	--	2.821
								--	72	283	.110	2.8	--	--	--	2.987
								208	80	315	.112	2.85	315	.190	4.8	3.328
								308	90	354	.120	3.05	--	--	--	3.735
								408	110	433	.140	3.55	--	--	--	4.556
1.689	177	.075	1.9	177	.129	3.3	45	--	75	--	--	--	300	.183	4.6	3.118
								--	80	315	.112	2.85	315	.190	4.8	3.328
								209	85	334	.120	3.05	334	.193	4.9	3.525
								309	100	393	.134	3.4	--	--	--	4.150
								409	120	475	.152	3.85	--	--	--	4.968
1.879	196	.065	1.65	196	.139	3.5	50	--	80	315	.112	2.85	315	.190	4.8	3.328
								210	90	354	.120	3.05	--	--	--	3.735
								310	110	433	.140	3.55	--	--	--	4.556
								410	130	500	.159	4.05	--	--	--	5.378
2.066	215	.071	1.8	215	.149	3.8	55	--	90	354	.120	3.05				3.735
								211	100	393	.134	3.4				4.150
								--	117	462	.148	3.75				4.846
								311	120	475	.152	3.85				4.986
								411	140	550	.168	4.25				5.782
2.252	237	.078	2.0	*237	.161	4.1	60	--	95	375	.128	3.25	*375	.191	4.8	3.945
								--	100	393	.134	3.4	--	--	--	4.150
								--	105	412	.136	3.45	--	--	--	4.348
								212	110	433	.144	3.65	--	--	--	4.556
								--	127	500	.159	4.05	--	--	--	5.260
								312	130	500	.159	4.05	--	--	--	5.378
								412	150	600	.165	4.2	--	--	--	6.175

*Production dies not available as of date of printing.

continued on next page

TABLE 11-29, continued

EXTERNAL	WALDES TRUARC RETAINING RINGS to be used with →						BALL BEARING			WALDES TRUARC RETAINING RING ← to be used with						INTERNAL
GROOVE DIAM for Groove of Standard Depth	5100- (Table 11-27) Max Allowable Ball Bearing Corner Radii or Chamfer			5108- Max Allowable Ball Bearing Corner Radii or Chamfer			Bore	SAE	OD	5000- (Table 11-25) Max Allowable Ball Bearing Corner Radii or Chamfer			5008- Max Allowable Ball Bearing Corner Radii or Chamfer			GROOVE DIAM for Groove of Standard Depth
Inch	No.	Inch	mm	No.	Inch	mm	mm	No.	mm	No.	Inch	mm	No.	Inch	mm	Inch
2.451	255	.110	2.8	*255	.176	4.5	65	--	100	393	.134	3.4				4.150
								--	105	412	.136	3.45				4.348
								--	115	450	.144	3.65				4.761
								213	120	475	.152	3.85				4.986
								--	137	537	.167	4.25				5.664
								313	140	550	.168	4.25				5.782
								413	160	625	.171	4.35				6.579
2.625	275	.090	2.3	*275	.181	4.6	70	--	110	433	.144	3.65				4.556
								--	115	450	.144	3.65				4.761
								--	120	475	.152	3.85				4.986
								214	125	475	.152	3.85				5.145
								--	147	575	.166	4.2				6.057
								314	150	600	.165	4.2				6.175
								414	180	700	.192	4.9				7.400
2.817	293	.096	2.4	*293	.190	4.8	75	--	115	450	.144	3.65				4.761
								--	120	475	.152	3.85				4.986
								215	130	500	.159	4.05				5.378
								--	157	625	.171	4.35				6.461
								315	160	625	.171	4.35				6.579
								415	190	750	.203	5.15				7.840
3.010	315	.095	2.4	315	.203	5.2	80	--	125	475	.152	3.85				5.165
								--	135	537	.167	4.25				5.585
								216	140	550	.168	4.25				5.782
								--	168	662	.167	4.25				6.925
								316	170	675	.166	4.2				6.997
								416	200	800	.216	5.5				8.234

*Production dies not available as of date of printing.

continued on next page

TABLE 11–29, continued

EXTERNAL	WALDES TRUARC RETAINING RINGS to be used with →						BALL BEARING			WALDES TRUARC RETAINING RINGS ← to be used with						INTERNAL
GROOVE DIAM for Groove of Standard Depth	5100- (Table 11–27) Max Allowable Ball Bearing Corner Radii or Chamfer			5108- Max Allowable Ball Bearing Corner Radii or Chamfer			Bore	SAE	OD	5000- (Table 11–25) Max Allowable Ball Bearing Corner Radii or Chamfer			5008- Max Allowable Ball Bearing Corner Radii or Chamfer			GROOVE DIAM for Groove of Standard Depth
Inch	No.	Inch	mm	No.	Inch	mm	mm	No.	mm	No.	Inch	mm	No.	Inch	mm	Inch
3.190	334	.100	2.5	334	.213	5.4	85	--	130	500	.159	4.05				5.387
								--	145	575	.166	4.2				5.979
								217	150	600	.165	4.2				6.175
								317	180	700	.192	4.9				7.400
								417	210	825	.224	5.7				8.620
3.381	354	.155	3.9	*354	.222	5.6	90	--	125	475	.152	3.85				5.165
								--	140	550	.168	4.25				5.782
								--	150	600	.165	4.2				6.175
								218	160	625	.171	4.35				6.579
								--	170	675	.184	4.7				6.997
								318	190	750	.203	5.15				7.840
								--	200	800	.216	5.5				8.234
								418	225	875	.235	6.0				9.252
								--	250	*1000	.267	6.8				10.292
3.577	375	.112	2.9	*375	.226	5.7	95	--	135	537	.167	4.25				5.582
								--	145	575	.166	4.2				5.979
								--	150	600	.165	4.2				6.175
								--	160	625	.171	4.35				6.579
								219	170	675	.184	4.7				6.997
								319	200	800	.216	5.5				8.234
								--	250	*1000	.267	6.8				10.292
3.756	393	.117	3.0	393	.231	5.9	100	--	150	600	.165	4.2				6.175
								--	160	625	.171	4.35				6.579
								--	165	650	.177	4.5				6.790
								220	180	700	.192	4.9				7.400
								320	215	850	.229	5.8				8.844
								--	265	...	...	...				...
								--	320	...	...	...				...

*Production dies not available as of date of printing.

continued on next page

TABLE 11–29, continued

EXTERNAL	WALDES TRUARC RETAINING RINGS to be used with →						BALL BEARING			WALDES TRUARC RETAINING RINGS ← to be used with						INTERNAL
GROOVE DIAM for Groove of Standard Depth	5100- (Table 11–27) Max Allowable Ball Bearing Corner Radii or Chamfer			5108- Max Allowable Ball Bearing Corner Radii or Chamfer			Bore	SAE	O D	5000- (Table 11–25) Max Allowable Ball Bearing Corner Radii or Chamfer			5008- Max Allowable Ball Bearing Corner Radii or Chamfer			GROOVE DIAM for Groove of Standard Depth
Inch	No.	Inch	mm	No.	Inch	mm	mm	No.	mm	No.	Inch	mm	No.	Inch	mm	Inch
4.000	425	.190	4.8				105	--	160	625	.171	4.35				6.579
								--	180	700	.192	4.9				7.400
								221	190	750	.203	5.15				7.840
								321	225	875	.235	6.0				9.252
								--	290	...	...	...				...
4.146	437	.190	4.8				110	--	170	675	.184	4.7				6.997
								--	175	700	.192	4.9				7.204
								--	185	725	.198	5.0				7.607
								222	200	800	.216	5.5				8.234
								322	240	*950	.254	6.5				9.879
								--	320	...	...	...				...
4.524	475	.150	3.8				120	--	180	700	.192	4.9				7.400
									190	750	.203	5.15				7.840
									215	850	.229	5.8				8.844
									260	*1000	.267	6.8				10.536
4.908	500	.161	4.1				130		200	800	.216	5.5				8.234
									203	800	.216	5.5				8.352
									205	800	.216	5.5				8.431
									230	*900	.244	6.2				9.459
									238	*950	.254	6.5				9.800
									280	...	...	...				...

*Production dies not available as of date of printing

TABLE 11-30

SAE Standard Ball-Bearing Snap Rings

SAE Handbook 1954

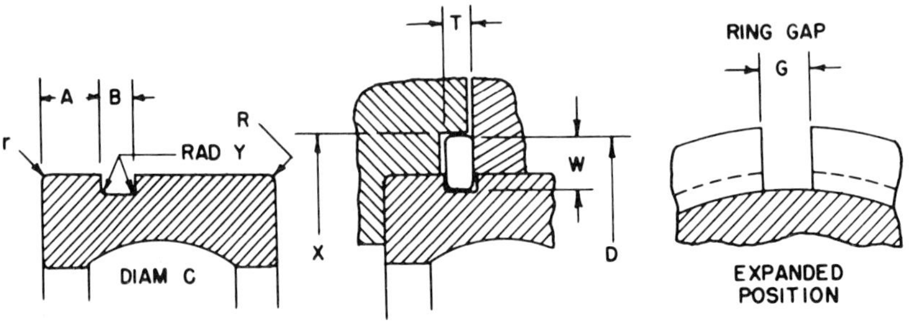

Dimension R — To clear standard fillet radius
Dimension r — To clear 0.020 inch radius
Radius Y — Maximum 4 mm (0.016 in.) up to 52 mm OD, and 6 mm (0.024 in.) above 52 mm OD.

For Metric Annular Ball Bearings, All Types (Single and Double Row)

Bearing Bores, mm			A								
Extra Light	Light	Medium	Extra Light	Light Medium	Width B	Diam C	Diam D	Thick. T	Width W	Gap G	Ctr Bore Min
--	10	--	--	.078	.056	1.109	1-23/64	.042	.125	1/8	1-25/64
15	12	--	.078	.078	.056	1.187	1-7/16	.042	.125	1/8	1-15/32
17	15	10	.078	.078	.056	1.306	1-35/64	.042	.125	1/8	1-37/64
--	--	12	--	.078	.056	1.369	1-39/64	.042	.125	1/8	1-41/64
--	17	--	--	.078	.056	1.500	1-3/4	.042	.125	1/8	1-25/32
20	--	15	.078	.078	.056	1.565	1-13/16	.042	.125	1/8	1-27/32
25	20	17	.078	.094	.056	1.756	2-1/16	.042	.156	1/8	2-3/32
--	25	20	--	.094	.056	1.958	2-17/64	.042	.156	3/16	2-19/64
30	--	--	.078	--	.056	2.071	2-3/8	.042	.156	3/16	2-13/32
35	30	25	.078	.125	.078	2.347	2-21/32	.065	.156	3/16	2-11/16
40	--	--	.094	--	.078	2.552	2-59/64	.065	.188	3/16	2-63/64
--	35	30	--	.125	.078	2.709	3-5/64	.065	.188	3/16	3-9/64
45	--	--	.094	--	.078	2.828	3-13/64	.065	.188	3/16	3-17/64
50	40	35	.094	.125	.078	3.024	3-13/32	.065	.188	3/16	3-15/32
--	45	--	--	.125	.078	3.221	3-19/32	.065	.188	3/16	3-21/32
55	50	40	.109	.125	.109	3.417	3-51/64	.095	.188	3/16	3-55/64
60	--	--	.109	--	.109	3.615	3-63/64	.095	.188	3/16	4-3/64
65	55	45	.109	.125	.109	3.811	4-3/16	.095	.188	3/16	4-1/4
70	60	50	.109	.125	.109	4.205	4-37/64	.095	.188	3/16	4-41/64

TABLE 11–31

Commercial Tolerances for Snap Rings (ID, OD and Gap)

Handbook of Mechanical Spring Design — Associated Spring Corporation

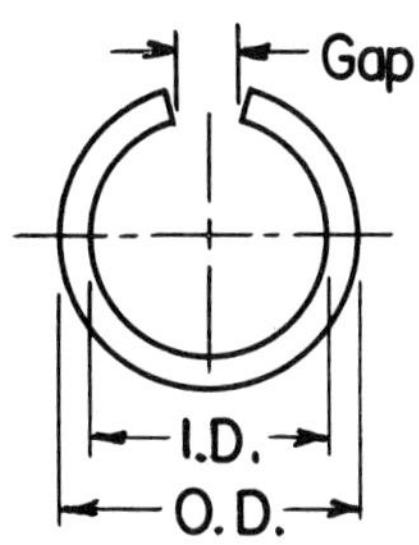

When measuring ring diameters, the scale shall be held at right angles to the center line passing through the gap.

Inside Diam of Ring (Free)	Tolerances (Free Dimensions) Inside Diam External Rings	Internal Rings	Gap
Under 1¼	+ .000 − .015	+ .015 − .000	± .031
1¼ to 3	+ .000 − .031	+ .031 − .000	± .062
3 to 5	+ .000 − .062	+ .062 − .000	± .093
5 and over	+ .000 − .125	+ .125 + .000	± .125

All dimensions are in inches.

TABLE 11-32

EXTERNAL, Round Section Reliance Snap Rings

Eaton Manufacturing Co, Reliance Division

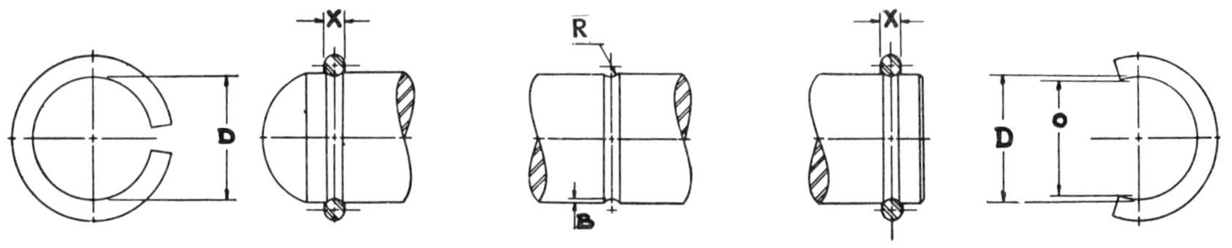

Shaft diam	Shaft series	Ring Dimensions X**	Ring Dimensions D*	Ring Dimensions O*	Groove Dimensions 2R	Groove Dimensions B
1/8"	1					
	2	.022	.110	.097	.022	.007
3/16"	1					
	2	.022	.172	.152	.022	.007
1/4"	1					
	2	.029	.230	.203	.029	.009
5/16"	1	.022	.296	.261	.022	.007
	2	.035	.288	.254	.035	.011
3/8"	1	.029	.353	.322	.029	.009
	2	.043	.345	.304	.043	.013
7/16"	1	.035	.412	.363	.035	.011
	2	.051	.402	.355	.051	.016
1/2"	1	.043	.468	.412	.043	.013
	2	.059	.458	.404	.059	.018
9/16"	1	.045	.529	.467	.045	.014
	2	.062	.518	.457	.062	.019
5/8"	1	.051	.587	.518	.051	.016
	2	.071	.575	.507	.071	.022
11/16"	1	.051	.649	.572	.051	.016
	2	.071	.637	.562	.071	.022
3/4"	1	.059	.706	.623	.059	.018
	2	.085	.690	.609	.085	.026
13/16"	1	.059	.769	.678	.059	.018
	2	.085	.753	.664	.085	.026
7/8"	1	.071	.823	.726	.071	.022
	2	.100	.804	.709	.100	.031
15/16"	1	.071	.885	.780	.071	.022
	2	.100	.867	.764	.100	.031
1"	1	.085	.938	.827	.085	.026
	2	.118	.917	.809	.118	.037
1 1/16"	1	.085	1.000	.882	.085	.026
	2	.118	.979	.863	.118	.037
1 1/8"	1	.100	1.051	.927	.100	.031
	2	.130	1.034	.912	.130	.040
1 3/16"	1	.100	1.114	.983	.100	.031
	2	.130	1.096	.967	.130	.040
1 1/4"	1	.118	1.164	1.027	.118	.037
	2	.140	1.150	1.014	.140	.044
1 5/16"	1	.118	1.226	1.081	.118	.037
	2	.140	1.212	1.069	.140	.044
1 3/8"	1	.130	1.281	1.130	.130	.040
	2	.156	1.264	1.115	.156	.049
1 7/16"	1	.130	1.344	1.185	.130	.040
	2	.156	1.326	1.170	.156	.049
1 1/2"	1	.140	1.398	1.233	.140	.044
	2	.172	1.378	1.215	.172	.054
1 5/8"	1	.140	1.522	1.342	.140	.044
	2	.172	1.502	1.325	.172	.054
1 3/4"	1	.172	1.626	1.434	.172	.054
	2	.203	1.608	1.418	.203	.063
2"	1	.203	1.855	1.636	.203	.063
	2	.232	1.837	1.620	.232	.072
2 1/4"	1	.203	2.103	1.855	.203	.063
	2	.232	2.085	1.839	.232	.072
2 1/2"	1	.232	2.332	2.057	.232	.072
	2	.250	2.321	2.047	.250	.078
3"	1	.232	2.827	2.494	.232	.072
	2	.250	2.816	2.483	.250	.078

*Tolerances on D and O dimensions are minus as follows: 1/8 in. to 3/8 in. sizes 6%, 7/16 in. to 1 in. sizes 5%, 1-1/8 in. to 3 in. sizes 4% of respective D dimension.

**Tolerance on X dimensions ±2%.

TABLE 11-33

EXTERNAL, Square Section Reliance Snap Rings

Eaton Manufacturing Co, Reliance Division

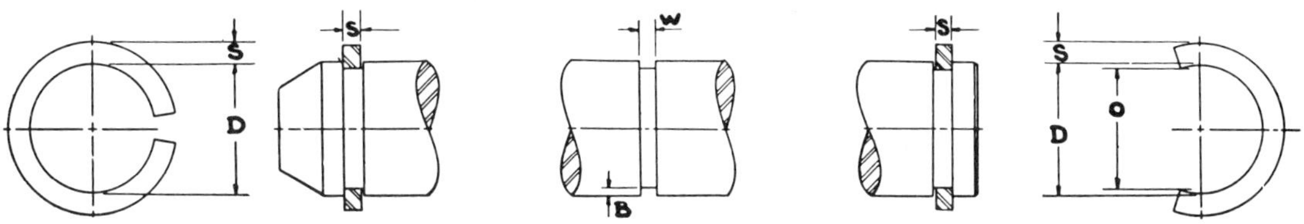

Shaft diam	Shaft series	Ring Dimensions S**	Ring Dimensions D*	Ring Dimensions O*	Groove Dimensions W	Groove Dimensions B
1/8"	1					
	2	.020	.114	.101	.024	.005
3/16"	1	.020	.175	.154	.024	.005
	2	.025	.173	.153	.029	.006
1/4"	1	.025	.236	.208	.030	.006
	2	.031	.232	.205	.036	.008
5/16"	1	.031	.293	.258	.036	.008
	2	.039	.289	.255	.045	.010
3/8"	1	.035	.353	.311	.041	.009
	2	.046	.347	.305	.052	.012
7/16"	1	.039	.413	.364	.045	.010
	2	.055	.405	.357	.062	.014
1/2"	1	.046	.471	.414	.052	.012
	2	.062	.463	.408	.069	.016
9/16"	1	.062	.525	.463	.069	.016
	2	.071	.521	.460	.078	.018
5/8"	1	.055	.591	.521	.062	.014
	2	.078	.579	.511	.085	.020
11/16"	1	.055	.653	.576	.062	.014
	2	.078	.641	.565	.085	.020
3/4"	1	.062	.712	.628	.069	.016
	2	.093	.697	.615	.100	.023
13/16"	1	.062	.773	.682	.069	.016
	2	.093	.759	.669	.100	.023
7/8"	1	.071	.831	.732	.078	.018
	2	.109	.813	.717	.117	.027
15/16"	1	.071	.893	.787	.078	.018
	2	.109	.875	.771	.117	.027
1"	1	.078	.950	.838	.085	.020
	2	.125	.929	.819	.133	.031

Shaft diam	Shaft series	Ring Dimensions S**	Ring Dimensions D*	Ring Dimensions O*	Groove Dimensions W	Groove Dimensions B
1 1/16"	1	.078	1.012	.893	.085	.020
	2	.125	.990	.873	.133	.031
1 1/8"	1	.093	1.068	.940	.100	.023
	2	.140	1.044	.921	.148	.035
1 3/16"	1	.093	1.130	.997	.100	.023
	2	.140	1.106	.975	.148	.035
1 1/4"	1	.109	1.184	1.044	.117	.027
	2	.156	1.160	1.023	.164	.039
1 5/16"	1	.109	1.246	1.099	.117	.027
	2	.156	1.222	1.078	.164	.039
1 3/8"	1	.120	1.304	1.150	.128	.030
	2	.172	1.276	1.125	.180	.043
1 7/16"	1	.120	1.364	1.203	.128	.030
	2	.172	1.338	1.180	.180	.043
1 1/2"	1	.125	1.424	1.256	.133	.031
	2	.187	1.392	1.228	.195	.047
1 5/8"	1	.125	1.547	1.364	.133	.031
	2	.187	1.516	1.337	.195	.047
1 3/4"	1	.156	1.657	1.461	.164	.039
	2	.218	1.624	1.432	.227	.055
2"	1	.187	1.887	1.664	.195	.047
	2	.250	1.855	1.636	.260	.063
2 1/4"	1	.187	2.134	1.882	.195	.047
	2	.250	2.103	1.855	.260	.063
2 1/2"	1	.250	2.350	2.073	.265	.063
	2	.3125	2.321	2.047	.327	.078
3"	1	.250	2.845	2.510	.265	.063
	2	.3125	2.816	2.483	.327	.078

*Tolerances on D and O dimensions are minus as follows: 1/8 in. to 3/8 in. sizes 6%, 7/16 in. to 1 in. sizes 5%, 1-1/8 in. to 3 in sizes 4% of respective D dimension.

**Tolerance on S dimensions ±2%.

TABLE 11–34

Allowable Stress and Deflection in Snap Rings

Handbook of Mechanical Spring Design   Associated Spring Corporation

"If a ring is to deflect without set, the maximum stress should not exceed 200,000 psi. For conditions in which some set is acceptable, stresses up to 260,000 may be used where necessary." The following problem is solved to illustrate a method of finding the allowable deflection of rings when the maximum stress is approximately 200,000 psi: Find the allowable deflections of a ring of 1/4 by 1/8 rectangular spring steel "coiled on edge", i.e., with the 1/4 dimension radial, and of $1\frac{1}{4}$ inch mean diameter, free.

Desired Deflection f, inch	Formula	Definition of Symbols
Ring compressed for placement in hole	$f = \dfrac{S\,h}{E\,K}$	S = Allowable stress, psi = 200,000
	$= \dfrac{200{,}000 \times 0.25}{30{,}000{,}000 \times 0.041}$	D = mean diameter of free ring, inches
	$= 0.0406$ inch	h = radial dimension
		E = modulus of elasticity of material, psi
	D/h = 5, giving K = 0.041	K = factor from Table 11–35
Ring expanded for placement externally in axial direction	$f = \dfrac{S\,h}{E\,K}$	K = factor from Table 11–36
	$= \dfrac{200{,}000 \times 0.25}{30{,}000{,}000 \times 0.062}$	
	$= 0.0269$ inch	
	D/h = 5, giving K = 0.062	
Ring expanded as by placement in radial direction	$F = \dfrac{3\,h\,S}{E\,K}$	K = factor from Table 11–36
	$= \dfrac{3 \times 0.25 \times 200{,}000}{30{,}000{,}000 \times 0.062}$	
	$= 0.0806$ inch	

Differences in the amounts of deflections for the same stress of 200,000 psi vary greatly with the D/h ratio. For a ratio of 4 they would be still more pronounced, while the differences would be slight for a ratio of 14.

TABLE 11-35

Snap Rings in Contraction, Round or Rectangular Sections. Maximum Tensile Stress at Outside. Hardened and Drawn After Forming

Handbook of Mechanical Spring Design   Associated Spring Corporation

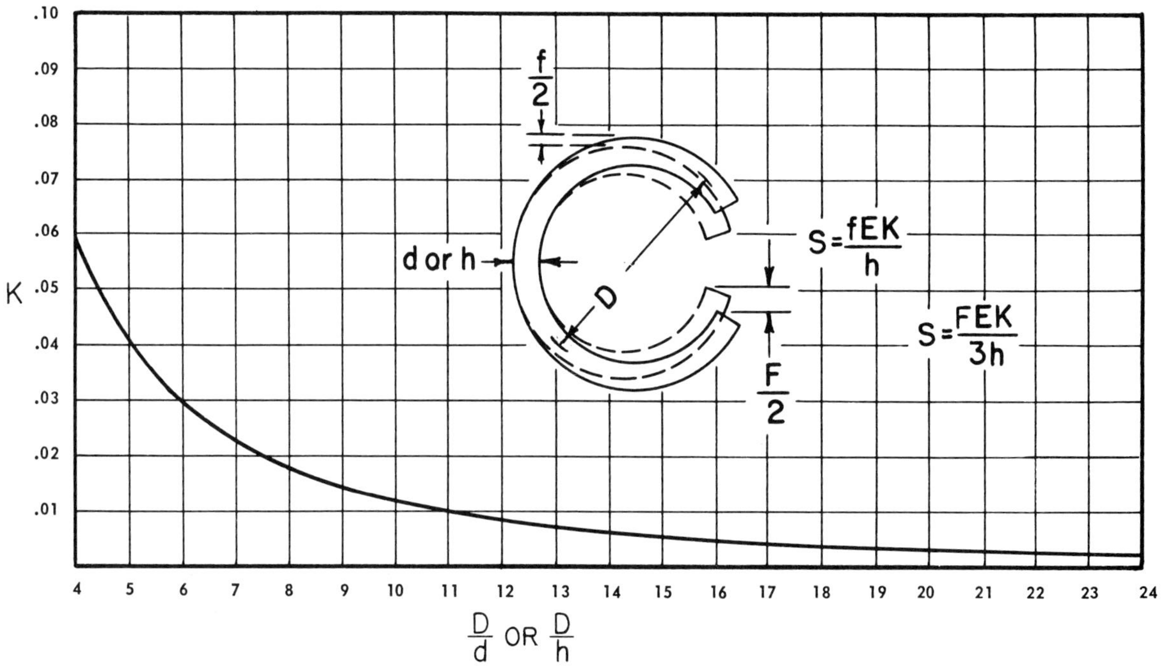

For all practical purposes, the stress calculation for ratios greater than 24 can be based on the straight beam formula:

$$S = \frac{4Efd}{\pi D^2} \quad \text{or} \quad S = \frac{4EFd}{3\pi D^2}$$

TABLE 11–36

Snap Rings in Expansion, Round or Rectangular Sections. Maximum Tensile Stress at Inside. Hardened and Drawn After Forming

Handbook of Mechanical Spring Design   Associated Spring Corporation

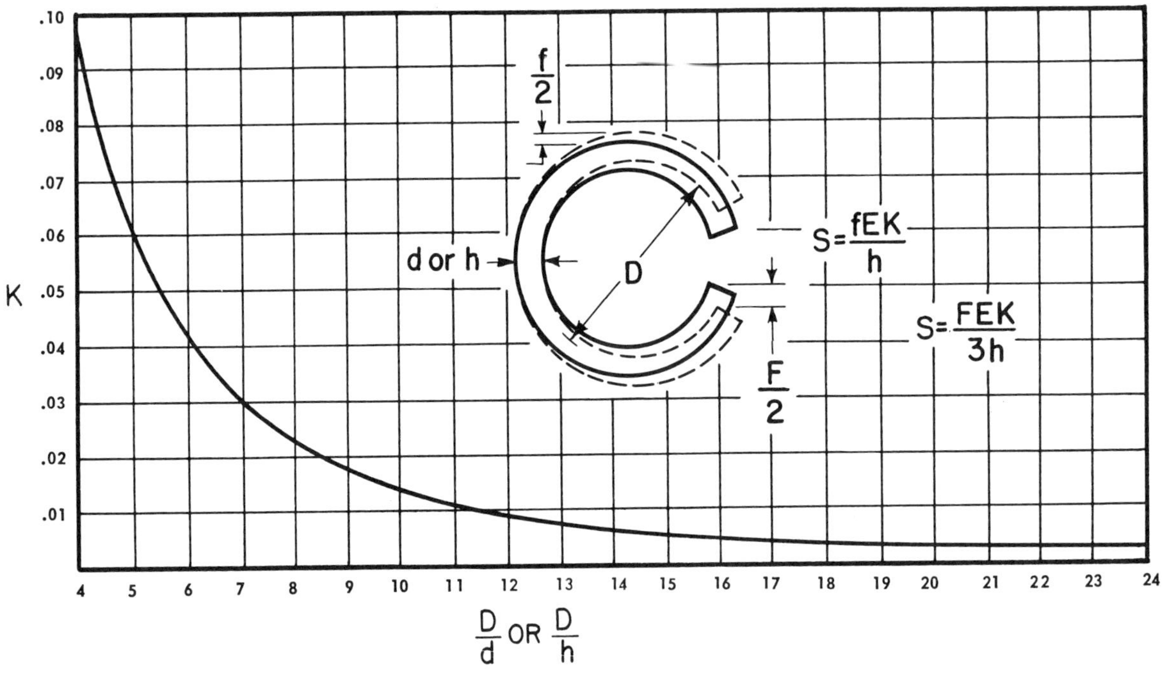

INDEX TO
SECTION 12
Mechanical Springs

by
A. M. Wahl
Advisory Engineer, Westinghouse Research Labs
East Pittsburgh, Pennsylvania

MECHANICAL SPRINGS

by

A. M. Wahl*

SPRING TYPES†

The more important types of spring are:

Helical Compression or Tension Springs. These are made of bar stock or wire coiled into a helical form, the load being applied along the helix axis. In a compression spring, the helix is compressed; in a tension spring, it is extended.

Helical Torsion Springs. Similar in form to the helical compression spring, these are loaded by a torque about the helix axis.

Leaf Springs. These consist essentially of flat bars of varying lengths clamped together to obtain greater efficiency and resilience (automobile leaf springs). Such springs may be *full elliptic, semi-elliptic,* or *cantilever.*

Spiral springs consist of flat strip wound in the form of a spiral (clock springs) and are loaded in torsion.

Belleville springs are essentially coned disks; these may be stacked up to give a variety of spring load-deflection characteristics.

HELICAL TENSION OR COMPRESSION SPRINGS

Because of inherent advantages of low cost, compactness, and efficient use of material, helical tension or compression springs are widely used in machine design and mechanical engineering.

Basic Formulas – Round Wire Springs. *Uncorrected Stress.* The usual method of calculating stress in round-wire helical springs axially loaded is to assume that the bar acts essentially as a straight bar under a torsion or twisting moment equal to load P times mean coil radius r (Fig. 1). Thus on this basis the shearing stress S_s in the spring will be equal to the moment Pr divided by the section modulus in torsion, which is $\pi d^3/16$ where d is the wire or bar diameter. This gives the commonly used formula for stress in helical springs:

$$S_S = \frac{16Pr}{\pi d^3} \tag{1}$$

This stress is frequently known as the "uncorrected" stress (to distinguish it from the corrected stress figured by taking into account direct shear and curvature effects).

Although this formula neglects a great many factors which may modify the stress distribution in actual helical springs, it is surprisingly good as a criterion of load-carrying ability without excessive set, where only static loads are involved. The reason for this is that stresses figured by Eq. (1) are approximately 133 percent of the stresses which would exist in the spring after complete yielding has taken place over the cross section (assuming yielding at constant stress). Thus the formula gives an indication of the load at which excessive set may occur.[1,2]

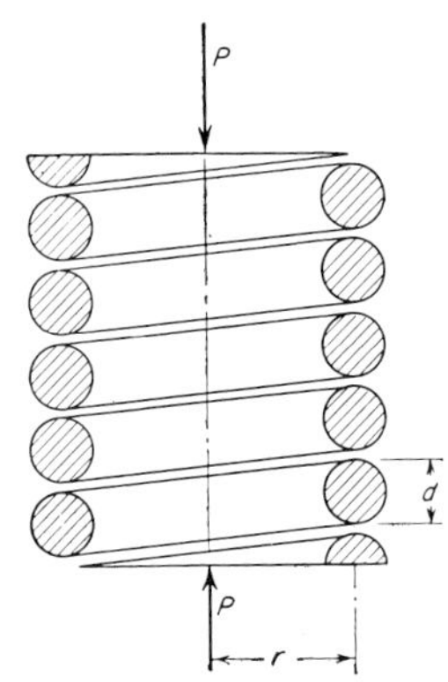

Fig. 1. Helical compression spring.

Effects of curvature and eccentricity of loading are discussed below.

Corrected Stress. Because of the curvature of the bar or wire, particularly in helical springs of small index (ratio of coil to wire diameter) the torsional shearing deformations (or strains) on the

†Much of the data in this section is based on the writer's book (ref. 1).

*Advisory Engineer, Westinghouse Research Labs., East Pittsburgh, Pa.

inside of the coil are considerably greater than those on the outside. In the plan view of the coil of a helical spring (Fig. 2), the torsional shearing deformations (or strains) at a near the inside of the coil are considerably greater than those at b. The reason for this is that the fiber length of an element

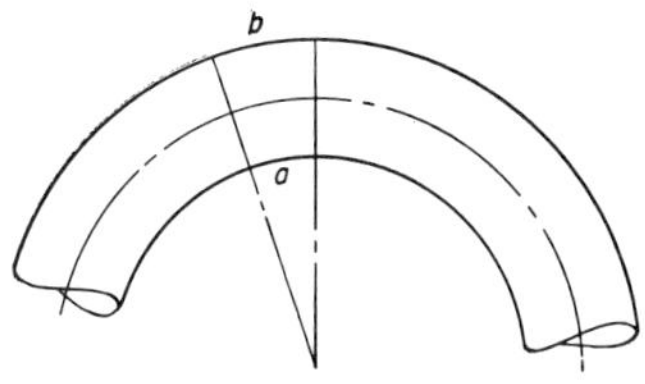

Fig. 2. Coil of helical spring.

at a (between radial cross sections as shown) is much less than that at b. This has been confirmed by strain measurements on actual springs.[3] In addition, the spring cross section must carry a direct shear load equal to the axial load; this further increases stress at the inside of the coil at a. Thus as long as the elastic limit is not exceeded and no residual or trapped stresses are present, stress distribution across a transverse diameter of the

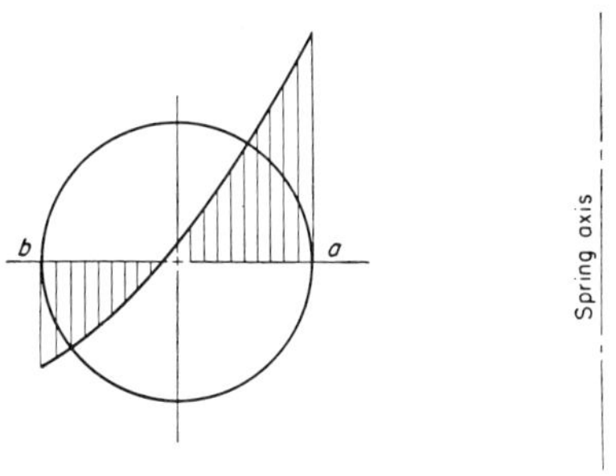

Fig. 3. Elastic shearing stress distribution across transverse diameter of helical spring of small index.

Fig. 4. Elastic shearing stress distribution across transverse diameter of helical spring of large index.

cross section of a helical spring of small index will be somewhat as shown in Fig. 3, with the stress at the inside of the coil at a much larger than that at the outside at b. For a large index spring, however, the torsional stress distribution approximates the linear distribution obtained in a straight bar under torsion (Fig. 4).

To calculate the elastic stress at point a in Fig. 3, at the inside of the coil, the stress given by the ordinary formula [Eq. (1)] should be multiplied by a factor K known as a "curvature correction factor[3]" which depends on the spring index $c = 2r/d$. This gives

$$S_s' = K \frac{16Pr}{\pi d^3} \tag{2}$$

where

$$K = \frac{4c-1}{4c-4} + \frac{0.615}{c} \tag{3}$$

The stress S_s' is frequently called the "corrected" stress to distinguish it from the uncorrected value

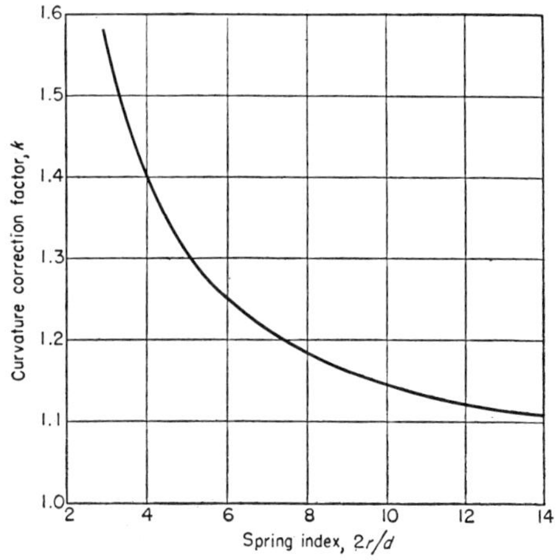

Fig. 5. Correction factor K for helical spring.

of Eq. (1). A plot of the correction factor K vs. spring index $c = 2r/d$ is given in Fig. 5.

Although the value of K given by Eq. 3 is approximate, it agrees closely with results obtained using more exact theory for practical springs with indexes greater than three[1].

Most highly stressed compression springs are cold-set by the manufacturer. This consists of coiling the spring initially to a length greater than the specified free length and compressing it beyond the elastic limit so plastic flow or set occurs, with resultant reduction of free length. This plastic flow or yielding occurs first at the inside of the coil at points of maximum stress. On unloading, trapped stresses are set up which subtract from the peak stress at *a* in Fig. 3. Thus the actual peak stress in a spring will usually be considerably lower than the corrected-stress value. However, the corrected stress formula [Eq. (2)] will, in general, still yield the stress *range* for repeated loading if load *P* is taken as load range. This stress range is important where fatigue loading is present.

Stress Augment from Direct Shear. Where static loading is involved, it appears reasonable to neglect stresses in helical springs from bar or wire curvature, as these are highly localized and may be relieved by yielding of the material. However, the stress augment from the direct shear load is more or less uniformly distributed over the cross section. For conservative design where static loading is involved, it appears reasonable to add this stress to that given by Eq. (1).[1,4] This gives

$$S_s'' = \frac{16Pr}{\pi d^3} + \frac{4P}{\pi d^2}$$

Letting $c = 2r/d$ = spring index or ratio of coil to wire diameter, this equation may be written

$$S_s'' = K_s \frac{16Pr}{\pi d^3} \tag{4}$$

$$K_s = 1 + \frac{0.5}{c} \tag{5}$$

This formula results in somewhat more conservative load ratings for low-index springs than those obtained using the uncorrected stress [Eq. (1)]. This is considered desirable by some engineers since for such low-index springs some reduction in static load-carrying ability may be expected due to the direct shear load.

Deflection. To calculate *deflection* in a round-wire helical spring, the latter is considered to be essentially a straight bar under a torsion moment *Pr*. Length *l* of this bar will be equal to $2\pi rn$ where *n* is the number of *active* coils in the spring. Calculating the angular deflection due to twisting of this bar under a moment *Pr*, and multiplying by the coil radius, the deflection δ of the spring becomes

$$\delta = \frac{64Pr^3n}{Gd^4} \tag{6}$$

Spring rate P/δ is

$$\frac{P}{\delta} = \frac{Gd^4}{64r^3n} \tag{7}$$

Although derived on the basis of simplifying assumptions, these formulas will give sufficiently accurate results in most practical cases provided that the proper values of *G* and *n* are used (ref. 1, page 48). The effects of large pitch angles are discussed in Ref. 1.

Allowance for End Turns. For most compression springs with ends squared and ground, the number of active turns *n* used in Eq. (6) or (7) to calculate deflection or rate may be taken equal to total turns (tip to tip of bar) minus 1¾ turn (ref. 1. page 157). For usual tension springs with a full coil turned up, *n* may be taken as one plus the number of coils between points where the loops begin.

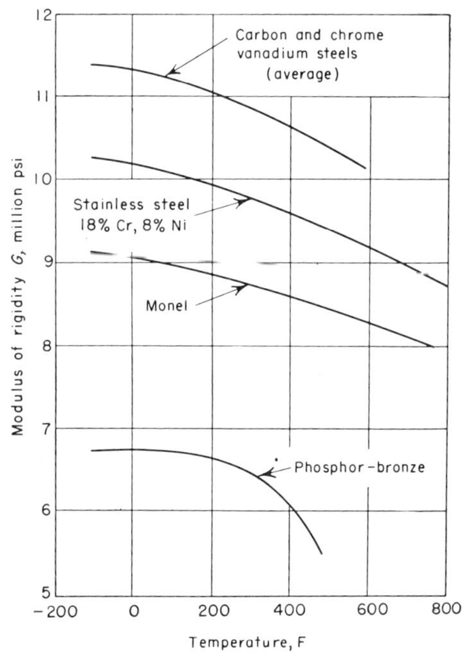

Fig. 6. Effect of temperature on modulus of rigidity of spring materials. (Based on tests by Zimmerli, Proc. ASTM 1930, Part II, page 356.)

Modulus of Rigidity. For springs made of music wire, carbon or alloy steels, the modulus of rigidity *G* may be taken as 11.5×10^6 psi and for 18-8 stainless steel 10×10^6 (for other nonferrous spring materials, see Table 16). Values of *G* will drop off at elevated temperatures (Fig. 6).

Compression Springs – Round Wire. *Working Stresses.* For purposes of choosing working stress, helical-spring applications may conveniently be divided into (1) static loading, (2) intermediate loading, and (3) fatigue loading.

Static Loading – Normal Temperature. This category includes cases in which the load is steady, or applied infrequently, say less than 1,000 times during the life of the spring. An example is a spring used to apply gasket pressure. It is also assumed that normal temperatures prevail. The choice of working stress in such cases depends to some extent on whether some set may be tolerated; if so, a higher working stress may be used than would be the case otherwise.

In calculating springs for static loading, present practice in most cases is to use the *uncorrected* stress formula [Eq. (1)], stress augments from curvature and direct shear being thus neglected. For somewhat more conservative design where the spring index is small, springs may be figured by Eq. (4), which takes into account the stress augment from direct-shear loading but not that from curvature.

Extremely high working stresses are practical for helical springs where space is at a premium

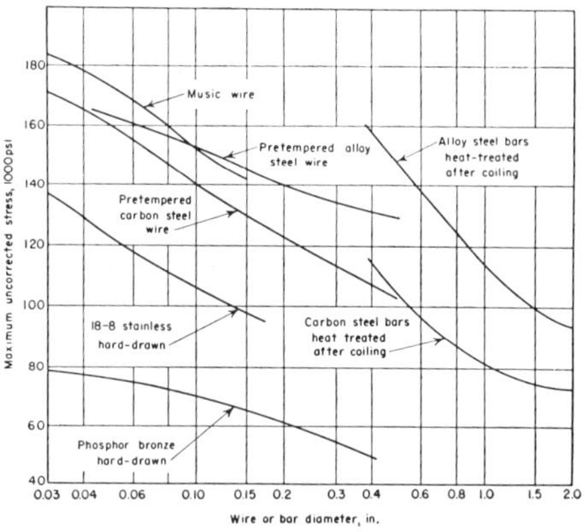

Fig. 7. Maximum design stresses for compression springs as used in ordnance. (SAE spring manual).

and only limited life is required (Fig. 7). The curves in Fig. 7, taken from data published by the SAE Spring Committee,[5] represent *uncorrected* design stress values [Eq. (1)] for various spring materials and various wire or bar sizes, for helical compression springs in ordnance applications where a relatively short spring life is ordinarily required. These curves apply only under the following conditions: springs cold-set and shot-peened for sizes over 1/16 in.; normal temperature operation; spring index between 4 and 9. Where these conditions are not fulfilled, lower stresses should be used.

Stress values of Fig. 7 are *maximum* values which may be used *when space is at a premium and only limited life is expected.* In general, stresses at solid compression should not exceed values given in this figure. Where fatigue loading is present (as in automotive valve springs), the use of such stresses would generally result in early fatigue failure. For industrial applications where more space is usually available, it is suggested that about 25 per cent lower stresses be used, even for infrequent or static loading.

As another example, suggested stress values (ref. 6, 1944 Ed.) are given in Table 1 for various spring materials. These apply to springs under occasional loading at normal temperature. Values in this table are more conservative than those of Fig. 7.

Table 1: Safe Working Stress for Helical Springs[6]
(Under occasional loading only)

Material	Wire, size, in.	Corrected stress, psi [Eq. (2)]
Music wire................	0.015	180,000
	0.060	155,000
	0.156	128,000
Chrome-vanadium wire.....	0.040	157,000
	0.100	140,000
	0.500	115,000
Oil-tempered or carbon valve-spring wire.......	0.020	164,000
	0.100	127,000
	0.500	92,000
18-8 stainless wire.......	0.025	122,000
	0.150	93,000
	0.500	70,000
Phosphor bronze wire.....	0.025	77,000
	0.150	65,000
	0.500	46,000

NOTE: If uncorrected stresses are used, values in this table should be divided by 1.2.

Static Loading – Elevated Temperature. Tests show that temperatures as low as 250 F may greatly increase the amount of relaxation, or set, which occurs in statically loaded springs as compared with that which occurs at room temperature. Thus a spring clamped between parallel plates at elevated temperature may gradually lose part of its load over a period of time. This creep, or set, must be allowed for in design, and lower working stresses should be used where low amounts of set are permissible. Unfortunately, not many data are available in the literature on the amounts of relaxation to be expected in springs when operating for long periods of time. The results of relaxation tests made by Zimmerli[7], which covered periods from 3 to 10 days are summarized in Table 2. In these the springs were compressed to a given amount and placed in a furnace for a given time, the load loss being estimated from the loss in free height. These tests indicate that, even at as low as 80,000 psi corrected stress, music-wire springs will lose 2 to 3 per cent of their initial loads at 250 F, and 7 per cent at 350 F, within 3 days. At 100,000 psi (90,000 uncorrected), load losses in 3 days of about 5 per cent at 250 F and 10 per cent at 350 F may be expected. Extrapolated relaxation curves[8] for music-wire springs indicate that, in a year or so, two or more times as much relaxation may occur as that in 3 days. Hence it appears that, for operation in the range 250 to 350 F, and uncorrected stresses around 90,000 psi, music-wire springs may lose around 20 per cent of initial load in a year due to relaxation. For this reason it is in general desirable to use springs of 18-8 stainless-steel wire above 250 F because tests show that these will have much lower load losses.

Relaxation-test data (Table 2) indicate that at 450 F stainless-steel springs will lose about 6 per cent in load after 10 days at uncorrected stresses around 90,000 psi. In a year's time, a considerably larger amount of relaxation, or load loss, would be expected. For applications at temperatures above 500 F, springs made from an alloy such as Z-nickel or Inconel may be advisable.[9]

Relaxation, or set, during operation may be reduced by loading the springs under temperature at a load somewhat above the working load for a period of time. This operation, called "heat-setting," removes a considerable part of the initial set.

Intermediate Loading (Load Applied Say 1,000 to 100,000 Cycles.) Applications intermediate between static loading and fatigue loading – a peak load say 1,000 to 100,000 cycles during spring life (such as automobile suspension springs) – are in this class.

A typical scatter band obtained by fatigue tests on heavy helical springs, hot-wound, of carbon steel[10] is shown in Fig. 8, where corrected stress is plotted against cycles to failure. These results were obtained on carbon-steel springs (¾ inch bar diameter, index 5) tested in a zero-to-maximum

Table 2: Percentage Loss in Load for Helical Springs at Elevated Temperatures (based on tests by Zimmerli)

Material	Diameter, in.	Test Temp., deg F	Bluing temp.,* deg F	Loss in load at 80,000 psi stress,† %	Loss in load at 100,000 psi Stress† %	Rockwell hardness
Music wire.................	0.148	250	700	2.5	4.7	48
0.91% C, 0.31% Mn..........	0.148	350	700	7	10	48
Music wire.................	0.062	250	700	2.5	3.5	51
0.91% C, 0.31% Mn..........	0.062	350	700	7	7.5	51
Carbon steel...............	0.148	250	800	3	4.5	45
0.66% C, 0.76% Mn..........	0.148	350	700	6	10	45
Carbon steel...............	0.062	250	800	3	3.5	43
0.59% C, 0.75% Mn..........	0.062	350	700	6	8.5	47
Cr–V steel.................	0.148	250	800	2	4	45.5
0.87% Cr, 0.18% V..........	0.148	350	800	4	7	45.5
Cr–V steel.................	0.062	250	700	1.5	2.5	49
0.97% Cr, 0.18% V..........	0.062	350	800	3.5	5	46.5
	0.062	350	800	3	3	43.5
Stainless steel............	0.148	450	700	3.5	5.5	45
18.2% Cr, 9.2% Ni..........	0.148	550	800	9.5	11.5	43.5
	0.148	250	800	1.3	2	43
Stainless steel............	0.062	350	800	2	4	45.5
19.2% Cr, 9.1% Ni..........	0.062	450	800	3.5	5.5	45.5
	0.062	550	800	9	11.5	45.5

*Values have been reported for optimum bluing (or stress-relieving) temperatures. Greater losses in load may usually be expected for bluing temperatures other than those listed. Tests on stainless steel springs run 10 days; all others run 3 days.

†Stresses calculated with curvature correction. If calculated without curvature correction, about 10% lower values would be obtained.

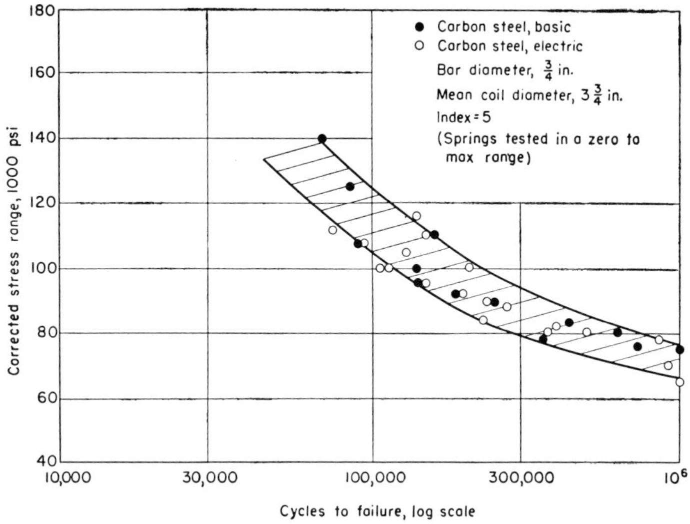

Fig. 8. Typical scatter band obtained in fatigue tests of hot-wound heavy helical springs. (Tests by Edgerton.)

stress range. In this case, the allowable stress range for a life of 100,000 cycles would be about 50 per cent higher than for infinite life.

The following method may be used, provided that sufficient fatigue test data are available for the given spring material: Stress at maximum load should be figured as uncorrected stress (*i.e.*, without curvature correction) and should not exceed the allowable value for static loading (see page 4). The stress range, obtained from the upper and lower limits of the load range, should be figured as a corrected stress range and should be equal to the fatigue-stress range for the required number of cycles, divided by a factor of safety. It is assumed that the ratio of minimum to maximum stress of the range in the fatigue test is equal to that in the application and that the spring index is the same. For higher indexes than those used in the test, somewhat lower fatigue ranges may be obtained.

Unfortunately, in most instances not enough fatigue data are at hand for application of this method; in such cases it is necessary to resort to empirical working-stress values. An example of design stresses suggested for average service conditions is given in Table 3. Average service conditions in this case are defined as noncorrosive

Table 3: Recommended Maximum Corrected Stresses for Helical Springs under Average Service*

Wire or bar diameter, in.	Kind of wire or bar (plain carbon steels)		
	Music	Tempered	Hard-drawn
0.020 – 0.030	100,000	100,000	90,000
0.031 – 0.092	90,000	100,000	80,000
0.093 – 0.176	90,000	90,000	80,000
0.177 – 0.282	....	90,000	70,000
0.283 – 0.436	....	85,000	
0.437 – 0.624	....	80,000	OD under 3 in. (cold-wound)
	....	90,000	OD over 3 in. (hot-wound)
0.625 – 0.874	....	90,000	Hot-wound
0.875 – 1.249	....	80,000	Hot-wound
1.250 – 1.5	....	80,000	Hot-wound

*Defined as noncorrosive atmosphere, temperatures not over 150 F, slowly varying or static loads. All stresses in pounds per square inch. (Data from ref. 11.)

atmosphere, temperature not over 150 F, and relatively slowly varying loads. Since corrected stresses are used, values in this table are in general more conservative than those of Fig. 7.

Example. Hot-wound Alloy-steel Spring for Flexible Drive. Springs in a flexible drive for a turbine locomotive are: OD 3.56 in., bar diameter 0.87 in., index 3.1, 8¾ active turns, free height 10.51 in. Solid load 7,650 lb. Spring made from chrome-vanadium steel bars, centerless-ground, and shot-peened. Uncorrected stress at solid load 80,000 psi, which occurs only at rare intervals. For normal starting conditions with 30 per cent maximum adhesion between wheel and rail, uncorrected stress will be 42,000 psi and corrected stress 65,000 psi. This corrected stress is less than half the expected endurance range for 100,000 cycles for shot-peened springs of index 3.

Fatigue Loading. In this category are included springs, such as valve springs, subject to say a million or more repeated load cycles in service. In the typical endurance diagram for a helical spring (Fig. 9) the limiting stress range is read vertically between the line marked "maximum stress" and that marked "minimum stress." Thus a spring operating between points *A* and *B* would operating in a range from 20,000 minimum stress to 87,000 maximum stress, the corrected stress *range* in this case being 67,000 psi.

Where fatigue is involved, a common practice is to use the corrected stress range as a basis for design because this gives the range at the inside of the coil where fatigue failures usually start. Thus for the spring of Fig. 9 operating from a minimum stress of 20,000 psi (point *A*), with 1.5 factor of safety, the working-stress range would be 67,000/1.5 = 45,000 (approx) or from 20,000 minimum to 65,000 maximum stress.

In actual springs, curvature of the bar acts to produce stress concentration[1] or a localized peak stress near the inside of the coil. Because of lack of sensitivity of some spring materials to such stress-concentration effects (low *notch sensitivity*), endurance ranges calculated by this method will in general be higher for the lower index springs (small *r/d* than for the high-index springs, the difference depending on the material. Available data indicate that this difference will be considerably greater for heavy hot-wound springs[10] than for springs of high-quality material such as valve-spring wire. Thus if the endurance range, calculated as a corrected stress, is determined from tests of large-index springs, it will, in general, be safe to apply the results to lower index springs of the same material and wire size. However, caution should be exercised in applying the results of tests on low- or medium-index springs to those of high-index of the same

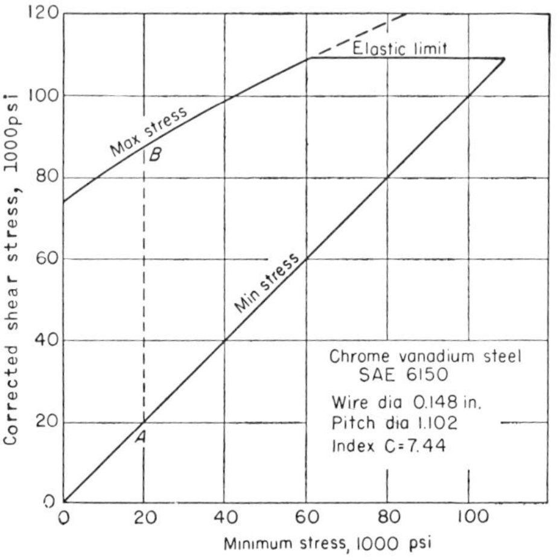

Fig. 9. Typical endurance diagram for helical compression springs. (Based on tests by Zimmerli, ref. 12.)

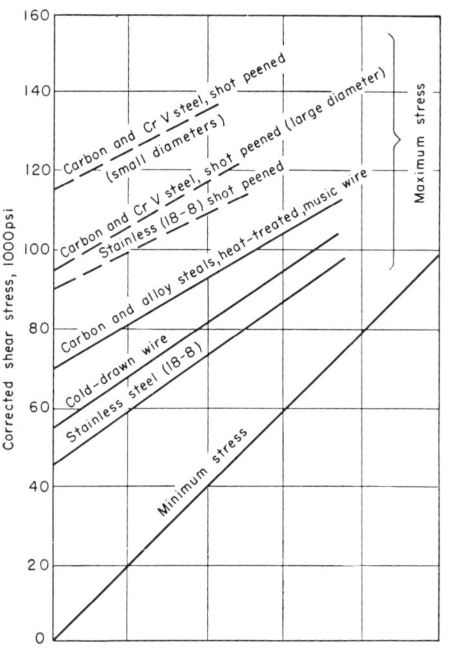

Fig. 10. Approximate endurance diagrams for good quality helical compression springs. (Indexes around 5-10.)

wire size, because the latter will frequently show lower endurance-range values. Also it should be remembered that the endurance range of the material decreases somewhat as the maximum stress of the range increases (Fig. 9), and this should be considered in design. For peak stresses below the elastic limit, this decrease in range is usually quite small, however.

Approximate endurance diagrams which may be expected from good-quality helical compression springs of different materials are shown in Fig. 10. In all cases, the limiting endurance range in terms of corrected stress is read vertically between the lower line representing minimum stress and the upper line corresponding to the given spring material. These curves hold roughly for springs having indexes between 5 and 10. As these curves represent rough average values, considerable deviation in individual instances may be noted. For small-index springs, higher values than those of Fig. 10 may be obtained in some cases. The effect of shot peening in raising the endurance limit is shown by the three upper dashed lines. It should be emphasized that in practical design lower stress ranges than those indicated by these endurance values should be used to provide a factor of safety. This factor of safety may range from 1.25 to 2, depending on the seriousness of failure, in cases where sufficient fatigue-test data are available.

As a further example stresses given in Table 4 are used as a basis for design for severe service at normal temperatures by Westinghouse Electric Corp. (ref. 1, page 135). These stresses, which are corrected values, apply to music or oil-tempered wire in the smaller sizes and to hot-wound carbon steel springs, heat-treated after forming in the larger sizes. In general, stress values of Table 4 are considered conservative; in many cases higher values may be used if a careful analysis of the application is carried out.

Table 4: Working Stresses for Severe Service—Helical Compression Springs*

Diameter of Wire, In.	Working Stresses, Psi
Up to 0.085	60,000
0.085 – 0.185	55,000
0.185 – 0.320	48,000
0.320 – 0.530	42,000
0.530 – 0.970	36,000
0.970 – 1.5	32,000

*For springs made of music or oil-tempered wire in smaller sizes and hot-wound springs, heat-treated after forming, in larger sizes. For phosphor bronze 50% of these values and for stainless steel 75% of these values are used. Table does not apply where corrosion effects or elevated temperature are present. Values represent *corrected* stresses.

Stress at Solid Compression. It is usually desirable to proportion compression springs so no appreciable permanent set will occur when compressed solid. In operation, springs may at times be compressed solid so any set occurring under these conditions will change the free height, hence the operating characteristics. To avoid such set it is suggested[13] that the stress at solid compression be limited to values given in Table 5. One column of this table gives the stress at solid compression for springs not cold set; the other column gives stress at solid compression which may be used after all set is removed by means of the cold setting operation.

Effect of Eccentricity of Loading. If a compression spring having the usual shape of end turns is compressed between two parallel plates as in a testing machine, the resultant load P is displaced by a small amount e from the axis of the spring (Fig. 11). The effect of this eccentricity of loading is to increase torsional stress on one side of the spring and decrease it on the other. Analysis[14] indicates that load eccentricity e depends primarily on the number of coils and may be expressed as

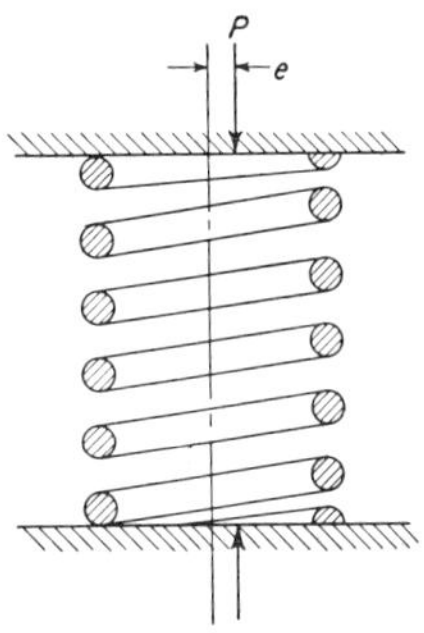

Fig. 11. Compression spring loaded eccentrically.

$$e = 1.12r\left[\frac{0.504}{N} + \frac{0.121}{N^2} + \frac{2.06}{N^3}\right] \quad (8)$$

Where r = mean coil radius

N = number of solid coils, or total number of coils minus 1/2, for the usual type of squared and ground ends

Stress in the spring will be increased approximately in the ratio $1 + e/r$ as compared with the stress in a spring under a purely axial load. Thus this effect is of greater importance for springs with a small number of active coils. For example, in the case of a spring with five total coils the correction amounts to about 15 per cent. In practical springs, as a result of variations in the shape of the end turns, eccentricity values may vary over a wide range; so Eq. (8) should be considered as giving only a rough indication.

Table 5: Suggested Corrected Stresses at Solid Compression,[13] Helical Springs

Material	Solid stress, psi (springs not cold set)	Solid stress, psi (springs cold set)
Music wire:		
Up to 0.032 diameter	130,000	180,000
0.032 – 0.062	110,000	170,000
0.062 – 0.125	100,000	160,000
0.125 and over	90,000	150,000
Hard-drawn spring wire:		
Up to 0.032 diameter	120,000	170,000
0.032 – 0.062	100,000	160,000
0.062 – 0.125	90,000	150,000
0.125 and over	80,000	140,000
Oil-tempered wire, 0.125 and over	80,000	140,000
18-8 stainless (hard-drawn):		
Up to 0.125	85,000	140,000
Over 0.125	75,000	120,000
Phosphor bronze (general sizes)	40,000	70,000

Tolerances in Spring Dimensions and Rates. As spring rate (pounds per inch deflection) is proportional to the fourth power of the wire diameter and inversely proportional to the cube of the coil diameter, small variations in these quantities will produce much larger variations in rate. Because of such variations, in general, it is not practical to hold the rate closer than about 5 to 10 per cent in actual springs except at extra cost.

Tolerances in coil diameter, free length, wire diameter, load, and rate for compression or extension springs as listed in the SAE Manual[5] are given in Tables 6 to 9.

Table 6: Coil-diameter Tolerances*

Coil diameter, In.	Tolerance, Plus or Minus, In.
1/16 or less	0.004
1/16 – 1/4	0.007
1/4 – 1/2	0.010
1/2 – 1	0.020
1 – 2	0.030
2 – 4	0.050
4 – 7	0.090
7 – 12	0.125
Over 12	0.156

*These apply to springs with indexes between 4 and 9.

Buckling of Compression Springs. If compression springs are made too long, buckling may occur from column action under load. In design it is necessary to guard against this by choosing spring proportions so working load will always be less than critical buckling load; if this is not practical, guides must be provided to prevent sideways motion.

Analysis of buckling[15] shows that *critical deflection* at which buckling just occurs depends on the ratio between free length l_o and mean coil diameter 2r, and on the method of fastening the spring ends, *i.e.*, whether fixed or hinged (in Fig. 12 *a*, a spring with fixed ends is indicated; in Fig. 12 *b*, one with hinged ends). Distance *h* in Fig. 12*b* is assumed very small compared to free length of spring. In Fig. 13, curve *a* shows the ratio of the critical buckling deflection to the free length, plotted against ratio of free length l_o to mean coil diameter, for a spring with hinged ends. Curve *b* represents the same for springs with built-in ends.

Table 7: Free-length Tolerances*

Free length, In.	Open-coiled springs, tolerances, plus or minus, In.	Close-coiled springs, tolerances, plus or minus, In.
1/2 or less	0.025	0.015
1/2 – 1	0.050	0.030
1 – 2	0.080	0.040
2 – 4	0.110	0.050
4 – 8	0.150	0.075
8 – 16	0.190	0.130
16 – 24	0.312	0.200
24 – 40	0.750	0.375
40 – 60	1.125	0.750

*When a load or loads is specified free length may be shown as an approximation and no tolerances applied. Tolerances for extension springs are for measurements between inside of hooks.

Actual springs compressed between plates which are approximately parallel may be expected to show intermediate results. For this reason, a conservative method of design is to use the hinged-end curve *a*, although this may yield somewhat low values for the buckling load. Critical buckling load for any

Table 8: Wire-diameter Tolerances*

Wire Diameter, In.	Tolerance, Plus or Minus, In.
Under 0.080	0.001
0.080 – 0.187............	0.002
0.187 – 0.406............	0.003
0.406 and larger..........	0.008

*Music and 'valve-spring wire may be obtained to to ±0.001 in. for sizes under 0.148 in.

Table 9: Load and Rate Tolerances*

No. active coils	Load tolerance, plus or minus, %	Rate tolerance, plus or minus, %
Under 3	15	10
3 – 9	10	8
9 – 15	8	6
15 and over	7	5

*Rate to be determined between 10 and 50% of total deflection when not already established by two specified loads.

spring is the critical deflection multiplied by the spring rate P/δ [Eq. (7)]. Tests[15] show good agreement with the curve for springs with hinged ends.

In practical springs, some deviation from these theoretical curves must be expected because of uncertainty as to the end condition and the effects of eccentricity of loading. From Fig. 13 it is seen that, if the ratio $l_o/2r$ between free length and mean coil diameter is less than about 2.7 for the hinged end condition, no buckling will occur before the spring is compressed solid. For the built-in end condition, buckling will not occur for a ratio $l_o/2r$ less than about 5.3.

Example. Required, critical buckling load for a steel compression spring of the following dimensions: free length $l_o = 6$ in., mean coil radius r = 0.75 in., $l_o/2r = 4$, OD = 1.75 in., wire diameter = 0.25 in., active turns $n = 12$. Calculated spring rate [Eq. (7)] is 139 lb per in. Assuming the most unfavorable condition (hinged ends), from Fig. 13 the ratio of critical deflection to free length for $l_o/2r = 4$ is 0.2. Thus buckling for such a spring would occur at a deflection $0.2\, l_o = 0.2(6) = 1.2$ in., or at a load of 1.2×139 lb per in. = 167 lb. If the built-in end condition were realized, however, no buckling would be expected in this spring at any practical deflection.

Lateral Loading of Compression Springs. When a compression spring is loaded as in Fig. 14 by an axial force P and a lateral force Q, the lateral deflection δ_1 for a given Q in general increases as P increases. For steel springs where the modulus of rigidity is 11.5×10^6 psi and the modulus of elasticity 30×10^6 psi, the following formula may be used to calculate lateral deflection δ_1.

$$\delta_1 = \frac{C\, Qnr}{10^6 d^4}(0.408 l^2 + 2.12 r^2) \qquad (9)$$

where n = number of active coils
r = mean coil radius
d = wire diameter
l = compressed length (Fig. 14)
C = a factor which depends on the ratio $l_o/2r$ between free length and mean coil diameter, and on the ratio δ/l_o between compression δ due to axial load P and free length (to find C, the chart of Fig. 15[15] may be used).

It is assumed here that the ends are constrained to move parallel during lateral deflection.

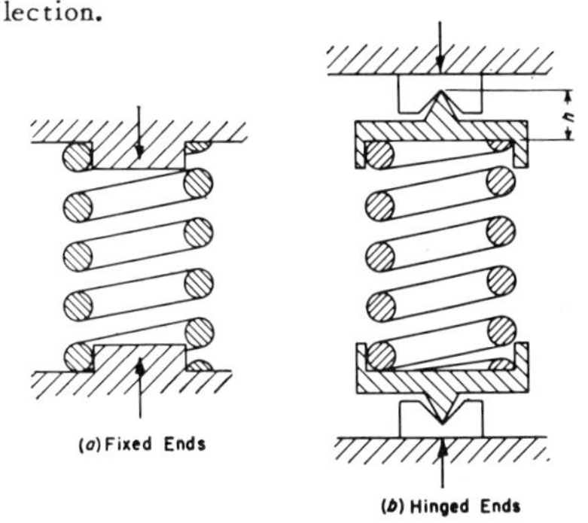

Fig. 12. Springs with fixed and hinged ends.

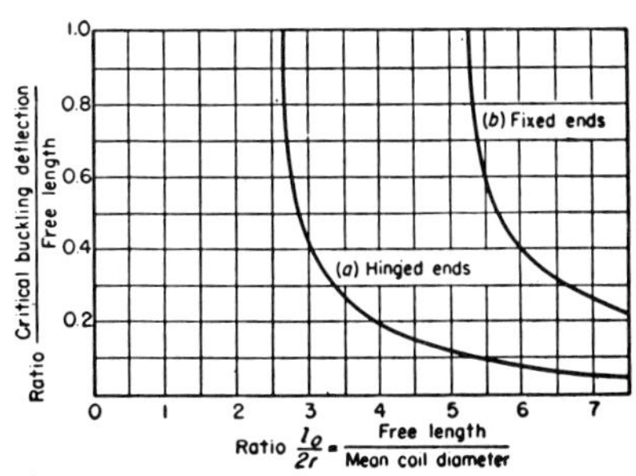

Fig. 13. Curves for finding critical buckling deflection. (Buckling load = buckling deflection times spring rate.) (Due to Haringx.)

Because of imperfect clamping of the ends in practical springs, and other deviations from the idealized case in Fig. 14, considerable differences may occur between test and theoretical values for lateral deflections. Hence Eq. (9) should be used as a rough guide only.[16] Also because of these lateral deflections, additional stresses are set up (ref. 1, page 180), which should be allowed for in design.

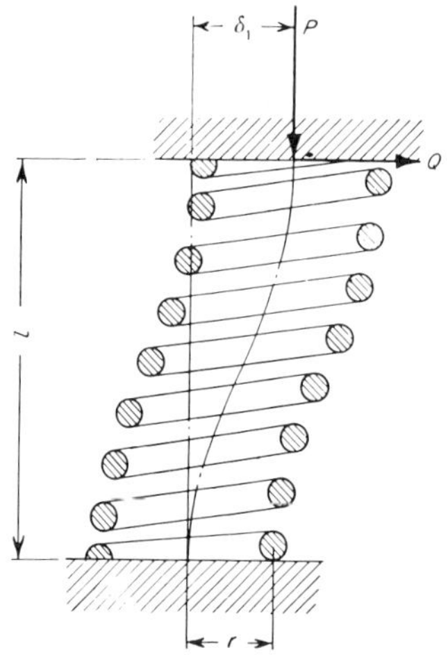

Fig. 14. Spring under combined lateral and axial loading.

Example. A laterally loaded steel spring has the following dimensions: mean coil diameter $2r = 4.25$ in., bar diameter $d = 3/4$ in., free length $l_o = 9.5$, ratio $l_o/2r = 2.24$, active coils $n = 8$. Calculated spring rate [Eq. (7)] is 732 lb per in.

Assume axial load P on this spring is 2,400 lb and lateral load $Q = 200$ lb. Then axial deflection δ at this load is $2{,}400/732 = 3.28$ in. The ratio δ/l_o between deflection and free length is $3.28/9.5 = 0.345$. From Fig. 15 for $\delta/l_o = 0.345$ and $l_o/2r = 2.24$, factor $C = 2.15$. Using this value and taking $Q = 200$ lb, from Eq. (9) the lateral deflection $\delta = 0.59$ in. In this case, the lateral deflection is about twice the value obtained for a very small axial load where C will be approximately unity.

Dynamical Effects. A spring, such as a valve spring, subject to rapid variation in load, may experience additional dynamical or surging effects which seriously increase the stress. Particularly severe dynamical effects may occur in such cases when a harmonic in the motion of the spring end (or

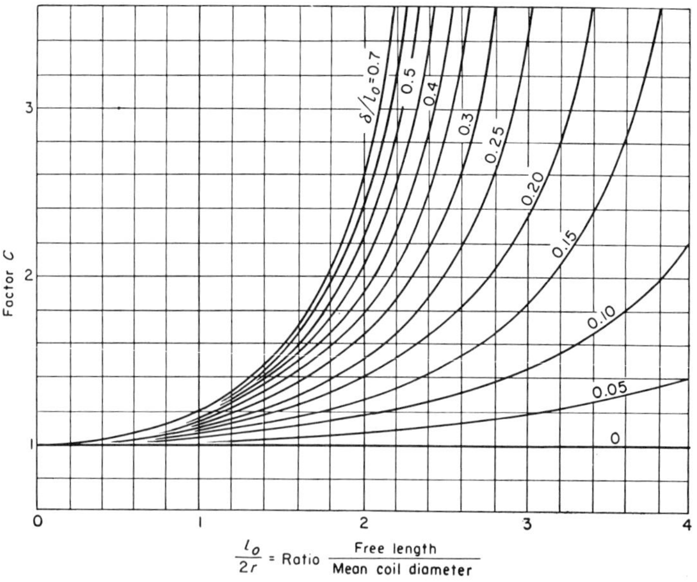

Fig. 15. Chart for finding factor C. (Due to Haringx.)

in the valve-lift curve in the case of valve springs) has a frequency which coincides with one of the natural frequencies of the spring (corresponding to one of the modes of vibration). Harmonics of very high order may have to be considered in certain cases. Usually the lowest frequency is of the most importance. For a spring compressed between parallel plates, the first mode of vibration (corresponding to the lowest natural frequency) will consist of a vibratory motion of the middle part of the spring with the ends remaining stationary. The second mode of vibration (corresponding to a higher frequency) will have a node (or point of zero motion of the coils) in the middle of the spring, while maximum motion of the coils occurs at points one-fourth and three-fourths of the length distant from a given end of the spring.

Natural Frequencies. The lowest natural frequency f in cycles per second for a helical spring clamped between two parallel plates (ref. 1, page 232) is given by

$$f = \frac{d}{2\pi r^2 n}\sqrt{\frac{Gg}{32\gamma}} \tag{10}$$

where d = wire size in.
r = mean coil radius, in.
G = modulus of rigidity, psi
g = acceleration of gravity, in. per sec^2
g = 386 in. per sec^2
γ = specific weight of material, lb per cu in.
n = number of active coils

For steel springs having both ends clamped, where $G = 11.5 \times 10^6$ psi and $\gamma = 0.285$ lb per cu in., the lowest natural frequency f (cycles per second) reduces to

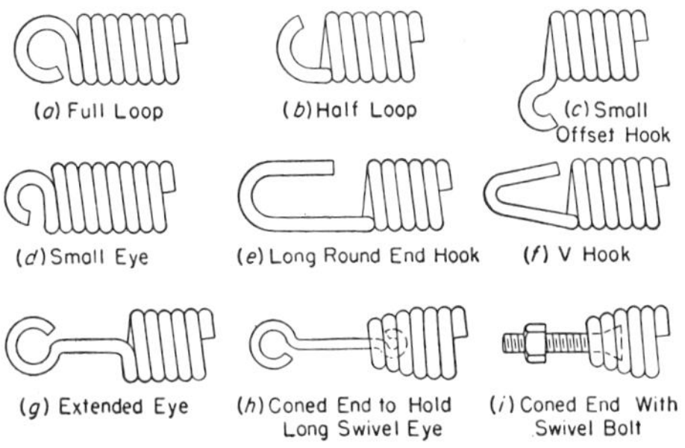

Fig. 16. Tension-spring end designs.

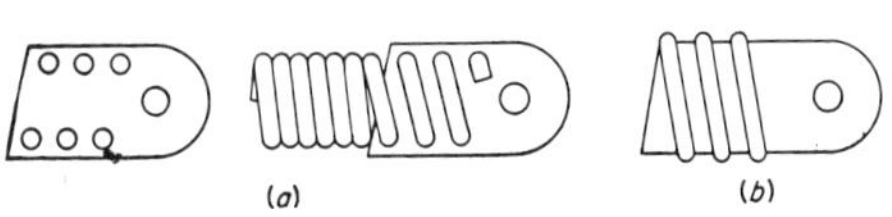

Fig. 17. Spring ends.

$$f = \frac{3{,}510\,d}{r^2 n} \tag{11}$$

Frequencies in the higher modes of vibration will be 2, 3, 4, etc., times this frequency.

For a spring with one end free and the other clamped, the lowest natural frequency would be equal to that of a similar spring twice as long but with both ends clamped. For such a spring, Eq. (11) applies if the number of turns n is taken as twice the actual number in the spring.

Example. A steel spring with $d = 0.3$ in., $r = 1$ in., $n = 6$, is clamped at both ends. Using Eq. (11), the lowest natural frequency becomes

$$f = \frac{3{,}510 \times 0.3}{(1)^2 \times 6} = 175 \text{ cycles per sec}$$

In the second mode of vibration, the spring frequency will be double this (or 350 cycles per sec). If such a spring were used as a valve spring, at an engine speed such as to give a frequency of motion of one of the harmonic components of the valve-lift curve equal to one of these frequencies, excessive vibration would be expected (ref. 1, page 222).

The following methods are useful in reducing stresses in valve springs from dynamical effects: use of a high natural frequency to avoid resonance with lower harmonics; cam contour shaped to reduce amplitudes of the harmonics of importance in the operating-speed range; reducing pitch of coils near ends of the spring to change natural frequency when oscillation occurs, closing these coils; use of friction dampers pressing against center coils.

Tension Springs – Round Wire. *Spring Ends.* Tension springs may be wound with a variety of end designs (Fig. 16). Some of these may result in considerably higher stress than calculated from Eq. (1). For example, the design of Fig. 16 *b* (a half loop turned up sharply) may have rather sharp curvature and severe stress concentration, resulting in a decrease in strength particularly under fatigue loading. Much lower stress-concentration effects are present in the full loop of Fig. 16*a*. If the loop is at the side (Fig. 16*c*), the moment arm of the load is practically twice that which would exist if a purely axial load were applied, thus resulting in an approximate doubling of the stress in the spring. The designs shown in Fig. 16*h* and *i* will have relatively low stress in the end turns.

Often tension springs are made with plain ends. Special fixtures called "spring ends" are attached to these as indicated in Fig. 17*a*. With these, the spring is close-wound and the ends are spread apart by screwing the spring into the holes. In this case an initial stress is set up corresponding to the spreading apart of the turns near the ends. The spring end shown in Fig. 17*b* is screwed into the end of a spring coil having plain square-cut ends.

Initial Tension. Tension springs are usually wound with *initial tension, i.e.,* a certain initial load must be applied to the spring before the coils start to separate. After separation, spring rate is the same as that which would be obtained for a spring with no initial tension. This effect is important in the operation of many mechanisms.

Values of initial load P_1 which can be obtained without difficulty in practice are given by the formula (ref. 6, page 48).

$$P_1 = \frac{\pi S_S d^3}{16r} \qquad (12)$$

where S_S = torsion stress due to initial tension
d = wire diameter
r = mean coil radius

Table 10: Torsional Stress Corresponding to Initial Tension (Tension Springs)

Spring Index $2r/d$	Initial tension stress, psi
3	24,000
4	22,500
5	20,000
6	18,000
7	16,200
8	14,500
9	13,000
10	11,600
11	10,600
12	9,700
13	8,800
14	7,900
15	7,000

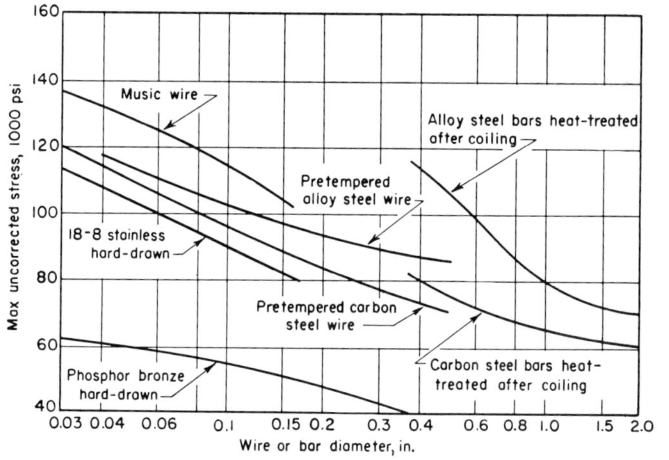

Fig. 18. Stresses for tension springs in ordnance applications where space is limited and only limited life is required. (SAE Manual.)

Values of S_S depend on the spring index $2r/d$ and may be taken from Table 10.

Example. Dimensions of a tension spring are: 2 in. OD, 1/4 in. wire diameter, and index $2r/d = 7$. From Table 10 for an index of 7, the maximum stress from initial tension will be $S_S = 16{,}200$ psi. From Eq. (12), initial tension load becomes

$$P_1 = \frac{\pi\,(16{,}200)\,(0.25)^3}{16\,(0.375)} = 57 \text{ lb}$$

By special methods of winding, much higher values of initial tension than those in Table 10 may be obtained if necessary, but usually at extra expense.

Working Stresses. Because tension springs cannot be cold-set without losing initial tension or producing excessive space between turns, lower working stresses (about 20 per cent lower) should be used than for compression springs of the same material and wire size.

An example of maximum allowable working stresses in tension springs for ordnance applications where space is limited and only a limited life is required, is shown in Fig. 18 for various materials (data from SAE Manual[5]). In general, where more space is available and longer life is desired, stresses 20 or 25 per cent lower than these values should be used.

Data on working stresses[11] for average conditions are also given in Table 3.

If fatigue is present, because of the weakening effect of the end turns, it is suggested that the design be carried out in the same manner as for compression springs (page 7), using stresses about 20 per cent lower for the case of the full-loop ends (Fig. 16*a*). For other types of ends, still lower stresses may be advisable, depending on the shape of the end.

Spring Tables – Tension or Compression Springs. *Corrected Stress Tables.* To facilitate spring selection in practical design where fatigue loading is involved, Table 11 due to Ross[4] and based on the corrected stress formula [Eq. (2)] may be used. For each OD and wire size two figures are given, an upper one representing load in pounds at 100,000 psi *corrected* stress and a lower one representing spring scale or rate in pounds per inch per single coil based on a shear modulus G of 11.5×10^6 psi. For any actual spring, the spring scale per coil given in the tables must be divided by the number of active coils. This scale is independent of the load. To obtain the spring scale per coil at any modulus G, values in the table should be multiplied by $G/11.5 \times 10^6$. This factor is approximately 0.87 for 18-8 stainless steel and 0.55 for phosphor bronze.

The figure of 100,000 psi corrected stress is used mainly for convenience and is not necessarily the working stress. For any other corrected stress S_s, loads given may be multiplied by the ratio $S_s/100{,}000$. Also the loads given in the table may be considered to represent load *ranges* at corrected stress ranges of 100,000 psi.

If it is desired to determine the load at an uncorrected stress of 100,000 psi, the loads given in Table 11 should be multiplied by the correction factor K (Fig. 5) corresponding to the spring index.

Example. A compression spring with a rate of 200 lb per in. is required for a mechanism. Maximum allowable OD is 2 in. and maximum load 160 lb. Assuming this spring to operate under severe conditions, (approximately zero to maximum stress range) the allowable stress will be taken from Table 4 and is 48,000 psi for 0.263 wire. A load of 160 lb at 48,000 psi would correspond to $160 \times 100{,}000/48{,}000 = 335$ lb at 100,000 psi. From Table 11, this could be obtained for 0.263-in. wire and 2 in. OD with spring scale per coil equal to 1,316 lb. As the scale wanted is 200 lb per in., number of active coils = 1,316/200 or approximately 6½. Total number of coils would be around 8 to 8½. This would require a spring of about 1.1 (8.5) (0.263) + 0.8 = 3.26 in. free length, allowing 10 per cent extra length for space between the turns when the spring is compressed at a load of 160 lb. The length at a load of 160 lb would be 3.26 − 0.8; or

Table 11: Helical Compression or Tension Springs*

Wire dia, d	Outside diameter of coil, inches																		
	1/8	9/64	5/32	11/64	3/16	13/64	7/32	15/64	1/4	17/64	9/32	19/64	5/16	21/64	11/32	23/64	3/8	25/64	13/32
.010	.305 9.47																		
.011	.401 14.1	.360 9.63	.325 6.90																
.012	.522 20.7	.464 13.8	.421 10.0	.383 7.32	.350 5.43														
.013	.659 29.3	.592 19.6	.534 14.0	.486 10.3	.447 7.51	.414 6.01	.385 4.72												
.014	.823 10.3	.732 25.9	.669 19.4	.603 14.0	.560 10.5	.519 8.19	.480 6.45	.450 5.20	.422 4.18										
.016	1.21 72.9	1.08 48.2	.983 34.4	.907 24.8	.824 18.4	.768 14.4	.714 11.3	.673 9.13	.626 7.37	.592 6.06	.561 5.05								
.018	1.71 124	1.53 81.0	1.40 57.4	1.27 41.2	1.17 30.7	1.08 23.7	1.01 18.6	.947 15.0	.889 12.1	.839 9.87	.793 8.30	.754 6.96							
.020	2.32 200	2.08 130	1.89 91.3	1.74 65.4	1.60 48.6	1.48 37.5	1.39 29.3	1.30 23.5	1.22 18.9	1.14 15.4	1.08 12.8	1.03 10.9	.972 9.06	.936 7.87					
.022	3.03 306	2.75 199	2.50 140	2.29 100	2.12 73.9	1.96 56.8	1.83 44.0	1.72 35.5	1.60 28.2	1.52 23.2	1.44 19.3	1.37 16.2	1.30 13.7	1.24 11.7	1.18 10.1				
.024	3.90 464	3.51 297	3.22 208	2.96 147	2.72 108	2.53 83.2	2.36 64.3	2.23 51.5	2.09 41.4	1.97 33.6	1.87 28.2	1.77 23.4	1.68 19.9	1.61 16.9	1.53 14.6	1.47 12.6	1.41 11.1	1.36 9.67	1.31 8.62
.025	4.36 566	3.95 359	3.63 250	3.32 177	3.06 130	2.85 99.3	2.66 77.1	2.51 61.5	2.35 49.2	2.22 40.3	2.11 33.7	1.99 27.7	1.91 23.7	1.81 20.2	1.74 17.6	1.65 15.0	1.59 13.1	1.53 11.5	1.48 10.2
.026	4.88 678	4.42 433	4.06 299	3.72 210	3.44 155	3.20 119	2.98 91.1	2.81 72.8	2.63 58.4	2.48 47.5	2.35 39.6	2.24 33.2	2.14 28.0	2.04 23.9	1.94 20.5	1.87 17.8	1.79 15.5	1.73 13.6	1.65 12.0
.028	5.98 965	5.45 612	4.99 419	4.61 297	4.25 216	3.98 165	3.70 127	3.49 101	3.29 80.6	3.10 65.7	2.94 54.5	2.79 45.4	2.65 38.0	2.55 32.9	2.41 28.1	2.33 24.4	2.24 21.1	2.14 18.4	2.07 16.4
.029	6.61 1160	6.03 727	5.55 496	5.08 348	4.71 253	4.41 193	4.10 149	3.86 118	3.64 94.1	3.43 76.6	3.25 63.7	3.11 53.0	2.94 44.6	2.81 38.1	2.71 32.9	2.59 28.2	2.50 24.6	2.38 21.4	2.29 18.9
.031	7.94 1588	7.24 992	6.69 683	6.15 473	5.74 344	5.36 260	4.99 200	4.69 159	4.42 126	4.15 102	3.97 85.2	3.76 70.7	3.58 59.2	3.43 50.7	3.28 43.5	3.15 37.5	3.01 32.7	2.91 28.6	2.79 25.1
.032		7.87 1157	7.30 793	6.76 545	6.25 398	5.84 301	5.47 231	5.14 184	4.84 146	4.59 118	4.35 97.8	4.12 81.1	3.95 68.0	3.75 58.1	3.60 49.9	3.46 43.3	3.33 37.5	3.18 32.7	3.07 28.7
.033		8.61 1367	7.97 916	7.37 635	6.85 460	6.39 345	5.97 265	5.62 210	5.29 166	5.01 135	4.74 112	4.53 92.4	4.31 77.5	4.13 66.5	3.96 56.9	3.77 49.3	3.61 42.4	3.51 37.5	3.35 32.8
.035		10.1 1804	9.34 1213	8.67 834	8.10 604	7.54 457	7.10 346	6.66 274	6.32 217	5.93 175	5.66 145	5.36 120	5.14 101	4.92 85.9	4.72 74.0	4.48 63.4	4.32 55.0	4.14 47.6	4.02 42.2
.037			10.9 1580	10.1 1098	9.42 785	8.88 588	8.28 448	7.84 353	7.41 280	7.01 225	6.64 186	6.33 154	6.01 128	5.75 110	5.51 94.2	5.29 80.9	5.07 69.6	4.89 60.7	4.73 53.9
.039			12.6 2066	11.7 1410	10.9 1009	10.3 752	9.66 572	9.13 448	8.63 354	8.15 284	7.76 234	7.34 193	7.03 162	6.72 138	6.45 118	6.17 101	5.93 87.9	5.71 76.2	5.52 67.3
.041				13.4 1811	12.5 1276	11.9 960	11.1 716	10.5 561	9.96 445	9.40 356	8.96 294	8.54 243	8.10 203	7.80 172	7.47 147	7.15 126	6.85 109	6.61 94.7	6.39 83.5
.043				15.3 2284	14.4 1618	13.5 1205	12.7 901	12.0 706	11.4 559	10.8 443	10.3 368	9.76 300	9.33 249	8.91 212	8.60 183	8.24 156	7.91 134	7.61 117	7.36 103
.045					16.3 2012	15.3 1485	14.5 1115	13.7 873	12.9 679	12.3 549	11.8 452	11.2 368	10.7 306	10.2 260	9.77 233	9.42 190	9.05 165	8.70 142	8.41 126
.047					18.3 2507	17.3 1860	16.4 1378	15.5 1069	14.7 840	13.9 671	13.3 547	12.6 448	12.1 373	11.6 316	11.1 270	10.6 230	10.3 199	9.88 172	9.54 151
.049						19.4 2232	18.4 1688	17.3 1301	16.4 1019	15.7 813	15.0 667	14.3 544	13.7 452	13.1 383	12.6 328	12.0 278	11.6 240	11.2 208	10.8 182
.051							20.4 2061	19.5 1585	18.5 1233	17.6 983	16.8 800	15.9 652	15.3 543	14.7 458	14.2 389	13.5 332	13.0 288	12.6 248	12.1 217
.054							23.9 2716	22.8 2092	21.6 1624	20.5 1281	19.8 1048	18.9 855	18.0 706	17.4 596	16.6 509	16.0 431	15.4 370	14.8 320	14.3 281
.055								24.1 2295	22.8 1767	21.7 1400	20.8 1143	19.9 934	18.9 768	18.3 649	17.6 553	16.8 468	16.2 401	15.6 347	15.1 304
.059									27.6 2509	26.3 1963	25.3 1591	24.2 1294	23.2 1059	22.2 895	21.4 770	20.7 647	19.9 554	19.1 475	18.5 418
.063										31.7 2709	30.3 2196	29.0 1773	27.9 1461	26.8 1213	25.8 1032	24.8 873	23.8 721	23.0 641	22.4 562
.067											35.9 2967	34.5 2379	33.1 1947	31.9 1628	30.8 1381	29.5 1161	28.6 986	27.6 849	26.8 744
.071												4.05 3189	38.7 2563	37.5 2155	36.2 1783	34.9 1531	33.4 1295	32.5 1113	31.5 972
.072												42.0 3415	40.3 2760	38.9 2302	37.6 1938	36.2 1631	35.1 1382	33.8 1190	32.8 1035
.075													45.3 3381	43.7 2819	42.1 2379	40.8 1981	39.2 1690	38.0 1445	36.9 1259
.077														46.7 3199	45.3 2696	43.9 2263	42.4 1910	40.8 1632	39.5 1421
.080														52.0 3852	50.3 3245	48.7 2706	47.1 2300	45.5 1953	44.1 1705
.085															59.3 4328	57.6 3646	55.5 3083	54.0 2621	52.2 2270
.090																	65.2 4075	63.1 3467	61.3 2990
.092																	69.3 4530	66.9 3845	64.9 3330
.095																		73.3 4497	71.4 3902
.100																			81.8 5018

*Based on 100,000 psi corrected stress Eq.(2). For any other stress S_s load should be multiplied by $S_s/100{,}000$.

Upper figure for each wire size is load in pounds at 100,000 psi corrected stress; lower figure is spring scale per single coil.

Table 11: Helical Compression or Tension Springs *(Continued)*

Wire dia., d	Outside diameter of coil, inches: 27/64	7/16	29/64	15/32	31/64	1/2	17/32	9/16	19/32	5/8	21/32	11/16	23/32	3/4	25/32	13/16	27/32	7/8	29/32
.024	1.26 7.54	1.21 6.67	1.17 6.02	1.14 5.42															
.025	1.42 8.98	1.38 8.01	1.33 7.19	1.28 6.40	1.24 5.81														
.026	1.60 10.6	1.54 9.43	1.50 8.47	1.44 7.55	1.40 6.87	1.35 6.18	1.28 5.11												
.028	1.99 14.4	1.92 12.9	1.86 11.5	1.79 10.3	1.73 9.29	1.69 8.37	1.58 6.91	1.51 5.79											
.029	2.21 16.7	2.13 14.9	2.06 13.4	2.00 12.0	1.93 10.8	1.87 9.75	1.77 8.07	1.66 6.66	1.58 5.60										
.031	2.70 22.3	2.60 19.8	2.52 17.7	2.44 15.9	2.35 14.3	2.29 12.9	2.15 10.7	2.03 8.77	1.92 7.40	1.84 6.31									
.032	2.95 25.4	2.85 22.4	2.75 20.1	2.67 17.9	2.59 16.4	2.51 14.8	2.37 12.1	2.24 10.0	2.12 8.47	2.03 7.27	1.93 6.20								
.033	3.24 28.9	3.11 25.6	3.04 23.2	2.92 20.5	2.84 18.5	2.74 16.7	2.60 13.8	2.45 11.4	2.33 9.67	2.21 8.26	2.12 7.07	2.02 6.11							
.035	3.84 37.0	3.73 33.0	3.61 29.8	3.50 26.3	3.39 24.0	3.29 21.4	3.08 17.7	2.93 14.7	2.62 11.7	2.64 10.4	2.52 9.04	2.41 7.72	2.31 6.77						
.037	4.54 47.3	4.41 42.2	4.25 37.7	4.11 32.8	3.97 30.1	3.88 27.3	3.62 22.2	3.44 18.5	3.27 15.5	3.10 13.3	2.98 11.4	2.83 9.77	2.73 8.52	2.61 7.43	2.50 6.56				
.039	5.29 59.2	5.14 52.8	4.99 47.0	4.80 41.1	4.68 37.8	4.52 34.1	4.27 28.0	4.06 23.3	3.83 19.6	3.66 16.6	3.47 14.2	3.33 12.2	3.18 10.6	3.04 9.30	2.94 8.17	2.82 7.20			
.041	6.18 73.5	5.92 64.7	5.80 58.8	5.60 51.8	5.41 46.8	5.26 41.9	4.94 34.6	4.69 28.8	4.45 24.0	4.23 20.5	4.04 17.5	3.84 14.9	3.71 13.1	3.54 11.4	3.40 10.1	3.28 8.84	3.15 7.84		
.043	7.11 90.5	6.82 80.0	6.63 71.4	6.43 63.7	6.19 56.8	6.07 51.6	5.69 42.6	5.36 34.9	5.11 29.4	4.89 25.0	4.62 21.2	4.44 18.3	4.25 16.0	4.10 14.0	3.91 12.2	3.76 10.8	3.65 9.60	3.52 8.60	3.38 7.63
.045	8.12 110	7.79 97.01	7.56 86.7	7.35 77.5	7.09 69.6	6.91 62.5	6.52 51.5	6.18 42.5	5.82 35.6	5.56 30.3	5.32 25.8	5.06 22.2	4.87 19.2	4.65 16.8	4.49 14.7	4.31 13.0	4.14 11.5	4.00 10.4	3.89 9.27
.047	9.22 133	8.90 117	8.59 105	8.33 93.3	8.12 84.1	7.83 75.6	7.39 61.9	6.99 51.0	6.68 43.2	6.35 36.2	6.02 30.9	5.81 26.6	5.52 23.1	5.31 20.1	5.11 17.8	4.93 15.6	4.73 13.8	4.57 12.4	4.39 11.0
.049	10.4 160	10.1 141	9.70 126	9.41 112	9.16 101	8.84 90.8	8.41 73.8	7.95 60.7	7.52 51.5	7.14 43.2	6.86 37.0	6.54 32.0	6.26 27.4	6.00 24.1	5.81 21.3	5.55 18.6	5.34 16.5	5.17 14.6	4.99 13.1
.051	11.7 191	11.3 168	10.9 150	10.6 134	10.3 120	10.0 107	9.45 88.3	8.97 73.0	8.53 61.4	8.01 51.1	7.67 43.6	7.37 37.5	7.18 33.3	6.78 28.6	6.50 25.1	6.29 22.1	6.05 19.7	5.79 17.3	5.63 15.5
.054	13.8 246	13.3 216	12.9 192	12.5 171	12.2 154	11.8 138	11.2 113	10.6 93.1	10.1 78.4	9.57 65.1	9.14 56.7	8.75 48.3	8.39 41.8	8.00 36.2	7.72 31.7	7.39 27.8	7.14 24.9	6.92 22.1	6.65 19.7
.055	14.6 266	14.1 234	13.7 209	13.3 187	12.8 167	12.4 149	11.7 122	11.2 101	10.65 83.8	10.0 70.4	9.62 60.5	9.23 51.9	8.77 45.0	8.50 39.5	8.11 34.4	7.83 30.1	7.56 27.0	7.32 23.9	7.04 21.3
.059	17.8 365	17.3 322	16.8 285	16.3 253	15.7 227	15.3 204	14.5 166	13.7 137	13.0 113	12.4 96.3	11.9 82.2	11.3 69.4	10.8 60.4	10.4 52.7	10.0 46.7	9.63 40.5	9.27 36.1	9.01 32.2	8.64 28.5
.063	21.5 489	20.9 429	20.3 382	19.7 338	19.1 303	18.6 272	17.5 221	16.6 181	15.8 151	15.1 128	14.4 109	13.8 92.6	13.2 80.9	12.7 70.0	12.2 61.3	11.7 53.7	11.3 47.3	10.9 42.3	10.5 37.7
.067	25.8 648	25.1 569	24.3 504	23.5 446	22.9 401	22.2 358	21.0 290	19.9 239	18.8 198	17.9 166	17.3 142	16.5 121	15.8 105	15.2 90.7	14.6 79.3	14.1 70.5	13.0 61.9	13.0 54.5	12.7 49.1
.071	30.6 845	29.4 741	28.8 656	27.8 580	27.0 518	26.1 463	24.8 375	23.6 307	22.4 256	21.3 215	20.3 183	19.5 156	18.6 139	18.0 117	17.4 102	16.6 88.9	16.1 79.1	15.5 70.6	15.0 62.4
.072	31.8 903	30.8 786	29.9 700	28.9 618	28.0 552	27.4 495	26.0 400	24.5 328	23.4 273	22.3 229	21.3 193	20.3 165	19.5 142	18.8 124	18.1 108	17.4 95.3	16.8 84.4	16.1 74.0	15.7 66.6
.075	35.6 1085	34.6 953	33.5 844	32.6 748	31.7 666	30.7 592	29.1 480	27.6 390	26.4 325	25.1 274	24.0 232	22.9 197	22.0 170	21.2 148	20.4 130	19.7 113	18.8 99.2	18.1 88.2	17.6 78.9
.077	38.5 1230	37.1 1075	36.2 953	35.2 840	34.1 748	33.2 669	31.3 541	29.8 442	28.5 368	27.0 305	25.8 260	24.9 221	23.9 192	22.8 166	22.0 145	21.2 127	20.5 112	19.7 98.9	19.1 88.1
.080	42.9 1469	41.3 1285	40.3 1138	39.2 1000	38.0 894	36.9 795	34.5 640	33.4 525	31.8 432	30.2 364	28.9 309	27.8 262	26.7 227	25.7 196	24.8 171	23.7 150	22.9 132	22.2 117	21.4 105
.085	50.9 1965	49.2 1708	47.9 1506	46.5 1325	45.2 1183	44.1 1053	41.9 845	39.8 689	37.9 569	36.1 478	34.6 403	33.1 343	31.7 294	30.5 255	29.4 223	28.4 195	27.4 172	26.6 153	25.6 136
.090	59.5 2576	57.9 2236	56.4 1979	54.8 1734	53.4 1539	51.7 1368	49.2 1101	46.9 890	41.8 738	42.9 617	40.8 520	39.0 442	37.6 379	36.2 329	34.8 286	33.6 250	32.5 220	31.2 195	30.3 174
.092	63.0 2864	61.0 2490	59.8 2199	58.4 1920	56.6 1710	55.1 1515	52.3 1220	50.0 986	47.7 816	45.6 683	43.4 574	41.7 488	40.0 418	38.4 363	37.0 315	35.7 276	34.5 242	33.4 215	32.4 191
.095	69.2 3359	67.3 2901	65.3 2551	63.8 2231	61.9 1997	60.3 1763	57.7 1414	54.5 1140	52.4 944	50.0 786	47.6 662	45.8 562	43.9 482	42.2 417	40.6 363	39.1 316	37.8 279	36.6 247	35.5 220
.100	79.8 4314	77.5 3726	75.7 3263	73.5 2860	71.6 2539	69.7 2248	66.6 1800	63.3 1449	60.7 1193	58.0 993	55.7 838	53.0 707	51.2 607	49.2 525	47.3 455	45.6 397	44.0 349	42.6 309	41.3 275
.105	90.8 5470	88.6 4720	86.1 4159	84.5 3610	82.2 3211	79.8 2830	76.2 2250	72.5 1820	69.4 1495	66.7 1240	64.0 1047	60.8 880	58.8 756	56.9 653	54.7 566	52.7 494	50.9 434	49.3 384	47.7 340
.106		91.0 4973	88.8 4353	86.7 3803	84.1 3351	81.9 2967	78.1 2367	74.8 1903	71.6 1563	68.3 1298	65.4 1090	62.8 921	60.6 789	58.1 679	55.9 590	54.3 514	52.4 452	50.8 400	48.7 354
.109			95.3 4964	93.7 4358	91.1 3844	89.1 3388	85.0 2681	81.1 2163	77.2 1775	74.2 1481	71.0 1241	68.2 1048	65.7 894	63.0 771	61.0 664	58.8 583	56.8 513	54.9 452	53.2 402
.112				100 4950	98.3 4408	95.6 3886	91.5 3081	87.4 2462	83.7 2027	79.7 1678	76.8 1404	73.8 1186	71.1 1013	68.1 871	66.1 757	63.5 657	61.3 577	59.3 509	57.5 452
.118					113 5678	110 4977	106 3970	101 3176	97.0 2594	92.9 2136	88.9 1792	85.8 1505	82.7 1286	79.1 1103	76.7 956	73.7 830	71.2 730	69.3 642	67.1 569
.120					119 6198	116 5420	110 4300	105.5 3410	101 2800	97.4 2310	93.0 1930	90.0 1633	86.5 1386	83.6 1198	80.2 1028	77.7 898	75.1 789	72.6 693	70.2 613
.125						130 6701	124 5232	119 4205	114 3393	109 2817	105 2354	101 1977	97.2 1679	93.8 1436	90.7 1248	87.2 1083	84.8 945	81.9 832	79.2 733
.135								147 6080	141 4930	135 4060	130 3394	125 2820	121 2406	116 2035	112 1761	109 1535	106 1354	102 1176	98.8 1046
.148									181 7802	175 6350	168 5250	163 4380	156 3705	151 3166	147 2717	142 2343	138 2060	133 1802	129 1583
.156										201 8238	193 6772	188 5646	181 4763	176 4074	170 3462	165 2995	159 2619	155 2293	150 2030
.162											214 8150	207 6800	202 5706	195 4887	189 4200	183 3581	178 3134	172 2734	168 2421
.177													258 8866	250 7485	241 6427	235 5504	227 4759	222 4150	215 3626
.187														289 9863	281 8388	273 7147	265 6221	258 5386	250 4735
.192															303 9558	293 8184	284 7030	277 6128	269 5359
.207																	350 10204	341 8857	332 7721
.218																		393 11491	381 9948
.225																			416 11653

Wire dia., d	Outside diameter of coil, inches: 15/16	31/32	1	1 1/32	1 1/16	1 3/32	1 1/8	1 5/32	1 3/16	1 7/32	1 1/4	1 5/16	1 3/8	1 7/16	1 1/2	1 9/16	1 5/8	1 11/16	1 3/4
.045	3.76 8.33																		
.047	4.27 9.89	4.14 8.99	4.00 8.09																
.049	4.83 11.9	4.69 10.6	4.55 9.66	4.41 8.80															
.051	5.44 14.0	5.26 12.6	5.14 11.4	4.98 10.3	4.82 9.43														
.054	6.47 17.6	6.28 16.0	6.06 14.5	5.86 13.1	5.73 11.9	5.55 10.8	5.41 10.0												
.055	6.77 19.0	6.63 17.3	6.40 15.6	6.22 14.2	6.02 12.9	5.88 11.7	5.70 10.7	5.56 9.90											
.059	8.42 25.6	8.15 23.2	7.89 21.1	7.68 18.9	7.45 17.3	7.24 15.8	7.00 14.3	6.81 13.2	6.69 12.2	6.49 11.1	6.33 10.3								
.063	10.2 33.7	9.85 30.4	9.60 27.5	9.29 24.8	9.00 22.6	8.81 20.6	8.55 18.8	8.35 17.4	8.07 15.8	7.89 14.7	7.75 13.6	7.36 11.6							
.067	12.2 43.8	11.9 39.3	11.5 35.6	11.1 32.2	10.9 29.3	10.6 26.9	10.2 24.3	9.93 22.3	9.78 20.7	9.50 19.0	9.23 17.4	8.86 15.0	8.46 13.0						
.071	14.5 56.2	14.2 51.0	13.6 45.3	13.3 41.3	12.9 37.4	12.5 34.2	12.2 31.5	11.9 28.6	11.6 26.4	11.2 24.0	11.1 22.4	10.5 19.1	9.97 16.4	9.59 14.3	9.21 12.6				
.072	15.2 59.9	14.7 53.1	14.2 48.2	13.8 44.0	13.4 39.4	13.0 36.0	12.7 33.3	12.4 30.2	12.1 27.9	11.8 25.8	11.5 23.5	11.0 20.4	10.4 17.4	10.0 15.1	9.62 13.3				
.075	17.2 71.1	16.6 63.9	16.0 57.6	15.7 52.5	15.1 47.0	14.8 43.0	14.4 39.5	14.0 36.2	13.6 33.2	13.3 30.3	12.9 27.9	12.4 24.0	11.8 20.7	11.2 17.9	10.9 15.8	10.4 13.8			
.077	18.4 79.1	17.9 70.8	17.3 64.0	16.9 58.1	16.4 52.8	15.9 48.1	15.6 44.0	15.1 40.3	14.7 37.0	14.3 34.1	14.1 31.6	13.3 26.6	12.8 23.0	12.2 20.0	11.7 17.6	11.3 15.4			
.080	20.8 93.7	20.1 84.4	19.5 75.6	18.9 68.5	18.3 62.1	17.8 56.1	17.3 51.4	16.9 46.8	16.4 42.9	16.1 40.1	15.7 37.0	15.0 31.5	14.2 27.0	13.7 23.5	13.1 20.5	12.6 18.2	12.2 16.0		
.085	24.9 122	23.9 108	23.3 97.5	22.6 89.0	22.0 80.2	21.3 72.7	20.8 67.2	20.3 61.1	19.7 55.8	19.2 51.7	18.8 47.4	17.9 40.8	17.1 34.8	16.4 30.4	15.8 26.6	15.1 23.2	14.5 20.6	14.0 18.1	
.090	29.4 155	28.3 139	27.6 125	26.6 112	26.1 103	25.2 92.7	24.7 85.1	24.1 78.6	23.3 71.3	22.9 65.9	22.2 60.3	21.3 51.6	20.2 44.3	19.5 38.5	18.6 33.5	18.0 29.5	17.2 25.9	16.6 23.1	
.092	31.2 171	30.3 153	29.6 138	28.6 124	27.8 111	27.0 102	26.3 94.1	25.6 85.1	25.0 78.6	24.1 71.2	23.8 66.4	22.5 56.4	21.8 49.1	20.7 42.4	19.9 36.8	19.1 32.4	18.4 28.4	17.8 25.5	17.1 22.6
.095	34.5 196	33.2 175	32.4 158	31.3 143	30.5 129	29.6 118	29.1 108	28.0 97.5	27.5 89.7	26.8 82.8	26.0 75.6	26.9 70.1	23.9 55.7	22.9 48.7	21.8 42.2	21.0 36.8	20.2 32.7	19.5 28.8	18.9 25.9
.100	40.1 245	38.7 220	37.6 197	36.7 178	35.5 161	34.7 147	33.5 132	32.8 121	31.9 111	31.1 103	30.5 94.5	29.0 81.0	27.7 69.0	26.4 59.8	25.5 52.4	24.5 46.2	23.6 40.4	22.7 35.9	22.1 32.1
.105	46.3 302	44.6 271	43.5 244	42.4 221	41.0 199	40.0 181	38.8 165	38.0 151	36.9 138	36.2 127	35.2 117	33.6 99.3	32.0 85.3	30.7 73.6	29.3 64.3	28.3 56.2	27.2 49.7	26.3 43.8	25.3 39.1
.106	47.3 316	46.0 283	44.8 255	43.2 229	42.2 207	41.2 188	40.0 172	39.1 157	38.0 144	36.9 132	36.2 121	34.6 103	32.9 88.6	31.6 76.5	30.1 66.5	29.3 59.1	28.1 52.1	27.2 46.0	26.1 40.9
.109	51.6 357	49.7 318	48.4 287	47.2 259	45.6 234	44.5 212	43.6 194	42.2 176	41.4 162	40.2 148	39.0 136	37.6 117	35.9 100	34.2 86.5	32.9 75.0	31.6 66.4	30.5 58.3	29.3 51.6	28.4 45.7
.112	55.7 401	54.2 360	52.2 323	50.9 291	49.6 263	48.0 239	47.0 218	46.0 199	44.6 182	43.8 167	42.4 152	40.8 131	38.6 112	37.3 97.7	35.8 84.5	34.2 73.7	33.2 65.4	31.8 57.6	30.7 51.7
.118	64.5 505	62.7 453	61.0 407	59.4 366	57.9 330	56.1 300	54.8 273	53.6 249	52.0 227	51.0 209	49.6 192	47.5 164	45.3 139	43.2 121	41.8 106	40.0 93.0	38.5 81.1	37.1 72.2	36.0 64.4
.120	68.0 546	66.1 488	64.3 438	62.7 395	60.5 355	59.1 323	57.7 293	56.0 269	54.8 245	53.3 225	52.2 207	49.9 176	47.3 150	45.5 130	43.9 114	42.1 99.7	40.8 87.9	38.9 77.0	37.8 68.6
.125	76.8 651	74.6 587	72.5 525	70.6 473	68.2 426	66.6 387	65.0 350	63.1 321	61.7 292	60.5 269	58.8 246	56.2 209	53.9 180	51.3 155	49.4 135	47.7 118	45.7 104	44.3 92.1	42.6 82.0
.135	96.3 919	93.5 822	90.8 739	88.4 666	85.5 600	83.4 543	81.4 493	79.6 449	77.2 410	75.6 376	73.5 344	70.1 291	67.2 251	64.5 216	62.2 188	59.8 163	57.6 145	55.6 128	53.3 113
.148	125 1403	122 1240	119 1119	115 998	112 896	110 815	106 741	104 678	101 611	99.1 561	96.3 514	91.9 437	88.0 375	84.4 322	81.2 279	78.3 244	75.7 214	72.6 189	70.5 168
.156	146 1787	141 1593	138 1412	134 1270	131 1146	127 1031	124 932	121 850	117 770	115 709	113 651	107 546	103 471	98.6 404	94.8 351	91.4 306	88.3 269	84.7 237	82.3 211
.162	162 2112	158 1881	153 1685	150 1512	146 1353	142 1225	139 1114	135 1002	131 918	129 844	126 769	120 651	115 554	110 478	106 414	102 361	98.5 317	95.2 279	91.5 247
.177	209 3215	204 2849	198 2526	194 2266	188 2035	183 1830	178 1664	174 1505	170 1367	166 1245	163 1145	156 964	149 819	143 703	137 609	132 529	128 468	123 409	119 361
.187	244 4136	238 3684	231 3272	226 2935	219 2620	215 2345	209 2122	204 1921	199 1758	194 1591	191 1470	183 1236	175 1054	168 895	162 778	156 676	150 591	145 520	141 463
.192	262 4712	256 4156	249 3694	244 3324	236 2950	232 2682	226 2407	220 2183	215 1982	210 1806	205 1640	197 1386	190 1184	181 1007	174 873	168 761	162 666	156 582	151 515
.207	322 6765	316 5951	307 5266	300 4710	293 4210	287 3781	279 3419	274 3086	267 2796	261 2534	255 2320	244 1955	235 1658	226 1405	217 1218	209 1058	202 930	195 817	188 716
.218	372 8712	364 7679	356 6781	348 6042	339 5372	332 4819	323 4359	317 3948	309 3560	304 3258	297 2970	284 2480	273 2097	263 1787	252 1541	244 1333	235 1166	227 1026	220 905
.225	408 10175	398 8924	388 7918	378 7052	371 6246	363 5654	353 5072	346 4583	339 4109	331 3753	323 3407	311 2851	299 2415	288 2070	276 1770	267 1531	258 1344	249 1178	241 1043
.244			484 11805	473 10419	461 9257	454 8315	444 7462	434 6718	426 6051	415 5519	406 4988	390 4176	376 3498	362 2987	349 2562	338 2219	326 1935	314 1691	306 1488
.250				508 11841	494 10466	484 9326	474 8389	465 7573	455 6842	446 6152	436 5640	419 4703	401 3953	389 3359	375 2874	363 2491	349 2169	339 1890	328 1667
.263					570 13443	559 11996	545 10728	533 9638	522 8686	513 7904	503 7175	483 5970	466 4984	450 4249	432 3664	420 3125	404 2721	392 2379	379 2102
.283								655 13907	639 12432	626 11239	614 10199	592 8469	574 7086	551 6028	534 5154	516 4437	499 3836	484 3320	467 2919
.307											799 15248	743 12530	719 10435	694 8830	671 7565	647 6425	630 5605	610 4845	588 4258
.3125												779 13715	753 11444	730 9605	706 8143	680 7047	662 6051	641 5284	623 4611
.331													882 15207	854 12765	824 10785	798 9183	776 7984	750 6906	733 6048
.343													972 18101	941 15177	915 12833	884 10914	858 9429	831 8195	806 7139
.362															1058 16794	1027 14224	997 12293	964 10664	940 9207
.375															1166 19966	1132 16921	1099 14556	1067 12538	1038 10903
.394																	1262 18641	1224 16042	1190 13967
.4375																			1587 23202

Table 11: Helical Compression or Tension Springs *(Concluded)*

Wire dia., d	Outside diameter of coil, inches																														
	1 13/16	1 7/8	1 15/16	2	2 1/16	2 1/8	2 3/16	2 1/4	2 5/16	2 3/8	2 7/16	2 1/2	2 9/16	2 5/8	2 11/16	2 3/4	2 13/16	2 7/8	2 15/16	3	3 1/8	3 1/4	3 3/8	3 1/2	3 5/8	3 3/4	3 7/8	4	4 1/8	4 1/4	4 3/8	4 1/2
.092	16.6	16.0	15.5																													
	20.2	18.1	16.4																													
.095	18.1	17.7	17.1	16.5	16.0																											
	23.0	20.8	18.7	16.9	15.4																											
.100	21.3	20.4	19.8	19.3	18.7	18.1																										
	28.7	25.5	23.1	20.9	19.0	17.2																										
.105	24.4	23.7	22.9	22.3	21.8	21.1	20.5																									
	35.0	31.3	28.2	25.8	23.4	21.3	19.5																									
.106	25.2	24.5	23.6	22.8	22.1	21.7	21.1	20.4																								
	36.6	32.8	29.4	26.6	24.1	22.2	20.2	18.4																								
.109	27.5	26.4	25.7	25.0	24.2	23.4	22.8	22.3																								
	41.3	36.8	33.0	30.2	27.3	24.8	22.5	20.8																								
.112	29.8	28.8	27.7	27.0	26.2	25.4	24.7	24.1	23.4																							
	46.0	41.5	37.1	33.5	30.5	27.7	25.4	23.1	21.1																							
.118	34.5	33.7	32.6	31.6	30.7	29.8	29.0	28.0	27.2	26.8	26.0																					
	56.9	51.4	46.4	42.1	37.8	34.5	31.6	28.7	26.3	24.3	22.3																					
.120	36.5	35.2	34.2	33.1	32.0	31.4	30.5	29.4	28.6	28.1	27.4	26.7																				
	61.6	55.3	49.4	44.7	40.5	37.1	33.9	30.7	28.1	25.9	24.0	22.2																				
.125	41.4	39.9	38.5	37.6	36.3	37.7	34.5	33.4	32.5	31.6	30.7	30.2	29.4	28.7																		
	73.3	65.6	59.0	53.3	48.3	51.3	40.1	36.6	33.6	30.8	28.4	26.2	24.2	22.5																		
.135	52.0	50.0	48.2	47.2	45.5	44.3	43.2	41.9	40.8	40.0	38.8	37.9	36.8	36.0	35.4	34.5	33.8															
	101	90.7	81.1	73.9	66.4	60.9	55.3	50.4	46.4	42.5	38.9	36.2	33.3	31.0	28.8	26.6	25.0															
.148	67.4	65.7	63.5	62.0	60.1	57.9	56.7	55.1	53.6	52.6	51.0	49.7	48.5	47.7	46.3	45.3	44.3	43.3	42.8	41.7	40.0											
	148	133	120	109	99.0	88.8	81.1	74.3	68.2	62.8	57.4	53.1	49.2	45.6	41.9	39.1	36.5	34.1	32.0	29.7	26.2											
.156	79.8	76.9	74.9	72.4	70.0	68.4	66.3	64.3	63.0	61.7	59.6	58.5	57.0	55.5	54.2	53.3	52.1	50.9	49.8	48.7	47.0	45.1	43.4									
	187	168	151	137	123	112	102	93.2	85.2	78.2	72.0	66.3	61.3	56.7	52.7	48.9	45.6	42.6	39.8	37.2	32.6	28.8	25.6									
.162	88.7	86.1	83.0	80.8	78.7	76.1	74.3	72.0	69.9	68.4	67.0	65.1	63.4	62.3	60.7	59.1	58.2	57.2	55.8	54.6	52.2	50.5	48.7	46.8								
	219	196	175	161	145	131	119	109	99.3	90.9	84.8	77.9	71.9	66.4	61.5	56.9	52.9	50.0	46.5	43.5	38.0	33.5	30.0	26.7								
.177	115	112	109	105	102	99.1	96.4	94.0	90.9	89.5	86.7	84.7	82.9	81.1	78.8	77.2	75.8	74.3	72.4	71.0	68.3	65.5	63.0	61.2	59.0	57.0						
	324	289	259	234	209	191	173	159	144	133	122	113	104	98.7	89.0	83.3	77.0	72.3	67.1	63.2	54.9	48.4	43.0	38.4	34.4	30.9						
.187	135	131	128	123	121	116	114	111	108	105	102	99.8	97.5	95.2	92.4	91.2	89.2	86.8	85.0	84.1	80.3	77.6	74.9	71.9	69.2	67.2	65.2	63.0				
	408	364	329	295	269	240	221	202	183	168	155	141	131	122	112	105	97.8	90.4	84.7	79.5	69.5	61.1	54.6	48.5	43.3	38.7	35.1	31.7				
.192	147	142	137	134	129	126	122	120	117	113	111	108	105	103	100	98.1	96.1	94.1	92.1	90.2	86.9	83.6	80.8	78.0	74.9	72.5	70.5	68.4	66.0			
	460	408	367	331	298	269	245	226	207	186	173	160	148	135	125	117	108	101	94.5	88.6	77.2	68.6	60.4	54.2	48.2	43.5	39.1	35.5	32.1			
.207	182	177	172	166	162	158	153	150	145	141	138	134	132	129	125	122	120	118	115	113	109	104	101	97.2	94.5	91.2	88.1	85.2	83.3	80.7	78.3	76.0
	637	569	509	457	413	375	340	310	281	258	237	218	202	187	172	160	149	139	130	121	107	93.6	83.1	74.1	66.2	59.6	53.7	48.5	44.1	40.1	36.6	33.5
.218	213	207	199	194	189	184	178	174	169	164	161	156	153	150	146	144	140	137	135	131	127	122	117	113	110	106	103	100	96.6	93.5	91.4	89.1
	801	714	637	577	519	468	424	386	353	324	297	271	250	234	216	201	186	173	161	149	133	117	103	91.0	82.6	73.9	66.3	60.4	54.6	49.5	45.1	41.6
.225	233	226	220	212	206	201	195	191	186	181	177	173	168	165	162	157	154	151	147	144	139	134	129	124	121	116	113	110	106	103	100	97.9
	921	822	735	659	593	536	489	445	404	371	339	314	288	266	249	229	212	198	184	173	151	134	118	104	94.0	83.6	76.2	68.6	62.1	56.5	51.8	47.2
.244	296	287	279	269	262	255	249	241	236	229	224	218	214	208	205	199	195	192	188	183	177	170	165	158	153	148	144	139	135	132	128	125
	1319	1179	1048	941	847	764	693	631	575	526	481	443	408	376	350	322	302	280	262	243	213	188	167	148	131	118	106	95.9	87.3	79.4	72.5	66.7
.250	317	307	298	290	282	273	266	260	252	247	240	235	229	225	219	215	209	205	203	198	191	183	177	170	165	159	154	150	145	141	138	134
	1472	1306	1169	1051	943	853	770	700	641	586	537	492	454	419	388	359	331	310	292	270	237	208	184	163	146	131	118	106	96.3	87.9	80.1	73.3
.263	366	355	344	335	326	317	309	300	293	287	279	273	266	261	254	250	244	240	234	229	221	213	206	197	191	184	180	174	168	164	160	155
	1845	1650	1464	1316	1180	1068	963	876	798	731	665	615	566	523	482	448	415	386	358	336	293	256	230	203	180	161	146	132	119	108	99.5	90.6
.283	455	440	427	415	404	393	383	374	363	355	348	338	331	322	316	309	304	296	292	285	276	263	256	247	238	228	223	216	210	204	197	194
	2576	2281	2030	1820	1637	1474	1331	1210	1104	1007	924	849	778	718	663	616	572	530	494	460	405	352	314	276	247	219	199	180	162	148	134	123
.307	572	558	541	525	510	501	483	472	461	450	441	429	420	409	401	394	385	379	369	361	348	336	325	312	304	293	285	276	269	261	251	246
	3739	3298	2951	2625	2364	2123	1923	1742	1589	1441	1318	1213	1113	1027	945	877	813	757	701	654	571	501	442	392	351	314	282	255	231	210	188	172
.3125	601	586	568	555	540	525	512	499	488	473	483	453	441	432	424	413	407	397	391	381	368	355	341	330	320	309	301	290	281	274	266	258
	4058	3586	3196	2842	2559	2313	2078	1892	1715	1568	1430	1311	1205	1085	1021	947	876	816	759	705	617	541	478	424	377	338	304	273	248	224	205	187
.331	712	690	673	652	638	621	604	589	575	562	545	534	524	514	500	491	478	471	460	453	436	421	403	391	380	367	356	343	334	326	315	308
	5309	4716	4152	3713	3319	2988	2701	2449	2212	2020	1841	1693	1555	1433	1315	1217	1127	1050	977	911	792	694	611	543	484	435	388	348	314	289	262	238
.343	787	761	742	725	704	684	671	655	639	620	605	593	581	570	554	544	531	522	510	501	483	466	451	433	420	408	394	382	372	362	350	341
	6271	5535	4898	4370	3907	3515	3170	2869	2606	2375	2161	1982	1818	1675	1540	1425	1319	1226	1142	1059	925	810	714	633	563	503	452	404	370	333	301	278
.362	917	886	864	844	818	801	780	765	746	729	713	692	678	665	651	634	624	613	599	589	567	547	525	508	493	478	462	452	434	426	411	403
	8023	7162	6297	5653	5028	4518	4054	3655	3327	3026	2756	2535	2322	2131	1958	1814	1681	1555	1449	1346	1171	1025	905	800	712	634	571	518	463	423	382	350
.375	1009	980	954	932	910	883	867	844	822	804	785	767	752	737	722	704	691	679	663	652	628	605	585	567	545	529	515	497	485	471	457	445
	9555	8426	7447	6652	5905	5281	4787	4302	3914	3551	3252	2960	2723	2497	2290	2124	1959	1820	1692	1575	1368	1195	1053	933	828	739	664	597	540	492	442	406
.394	1161	1127	1096	1068	1043	1022	993	974	949	926	904	884	865	849	832	816	795	781	769	750	722	701	677	651	631	612	595	575	560	542	530	515
	12106	10602	9432	8350	7418	6684	6040	5420	4912	4476	4052	3714	3375	3117	2872	2656	2448	2268	2111	1955	1704	1487	1307	1159	1029	917	821	740	667	604	550	503
.4375	1546	1504	1472	1433	1395	1363	1334	1307	1280	1249	1219	1201	1173	1148	1126	1102	1082	1063	1037	1019	980	951	919	888	860	835	804	783	763	738	721	704
	20315	17715	15560	13859	12180	11010	9845	8861	8005	7257	6582	6005	5471	5013	4647	4243	3935	3644	3377	3132	2718	2367	2082	1835	1626	1450	1296	1166	1051	952	864	785
.500				2073	2025	1976	1938	1897	1854	1819	1783	1742	1711	1675	1649	1617	1584	1555	1528	1500	1451	1396	1356	1310	1268	1229	1193	1160	1129	1092	1065	1041
				26645	23410	20999	18635	16817	15036	13718	12348	11268	10209	9358	8526	7880	7293	6694	6221	5736	4971	4325	3783	3330	2947	2619	2338	2097	1887	1705	1545	1405
.5625								2623	2563	2508	2473	2426	2376	2337	2284	2251	2205	2177	2135	2096	2020	1956	1898	1832	1786	1727	1676	1628	1584	1541	1503	1468
								29977	26894	24232	21885	19853	17946	16388	14987	13751	12643	11654	10761	9957	8541	7412	6471	5686	5022	4442	3962	3546	3188	2868	2597	2359
.625												3239	3174	3115	3061	3010	2964	2921	2861	2825	2722	2649	2565	2491	2405	2344	2271	2221	2159	2101	2047	1997
												33289	30171	27445	25008	22872	20962	19268	17748	16386	14045	12135	10551	9240	8130	7192	6394	5710	5121	4608	4162	3773

2.46 in., and the solid length about 10 per cent less, or about 2.2 in.

Tables for Static Loading. Where springs are *statically loaded*, Table 12 based on 100,000 psi stress, calculated using the factor K_S [Eq. (5)] which takes into account direct shear but not curvature effects, may be used. Loads P and deflections per turn y at a stress of 100,000 psi are given for various wire sizes and outside coil diameters. Deflection y is based on a shear modulus of 11.4×10^6 psi. Although stresses of 100,000 psi may be used in many cases, this figure is used in Table 12 merely for convenience. For any other stress S_S loads P and deflections y should be multiplied by the ratio $S_S/100{,}000$ and for any other modulus G, deflections y should be multiplied by $11.4 \times 10^6/G$. Interpolation for intermediate wire sizes or OD values may be used with sufficient accuracy for most purposes.

In general, for most springs, loads and deflections per turn as given in Table 12 are about 5 to 10 per cent lower than would be the case if calculated using the uncorrected stress formula [Eq. (1)]. To find loads and deflections at 100,000 psi uncorrected stress, values of P and y given in the table should be multiplied by the factor $K_S = 1 + 0.5/c$ where c = spring index.

For calculating stresses and deflections in helical springs, special spring slide rules, furnished by many manufacturers, are very convenient.

Table 12: Loads and Deflections per Turn for Statically Loaded Helical Springs

Wire dia, in.	Outside Diameter Of Springs, Inches																													
	1/8	5/32	3/16	1/4	5/16	3/8	7/16	1/2	5/8	3/4	7/8	1	1 1/8	1 1/4	1 3/8	1 1/2	1 5/8	1 3/4	1 7/8	2	2 1/4	2 1/2	2 3/4	3	3 1/2	4	4 1/2	5	5 1/2	6
.014 P	.912	.720	.596	.441	.352	.293																								
y	.0228	.0378	.0570	.106	.172	.253																								
.016 P	1.372	1.09	.900	.664	.529	.438	.376																							
y	.0189	.0319	.0478	.0910	.1465	.218	.302																							
.018 P	1.98	1.55	1.275	.952	.753	.625	.538	.470																						
y	.0162	.0273	.0413	.0794	.1285	.191	.266	.352																						
.020 P	2.73	2.15	1.76	1.31	1.04	.860	.735	.644																						
y	.0139	.0238	.0365	.0702	.114	.169	.234	.312																						
.022 P	3.66	2.87	2.37	1.75	1.395	1.145	.981	.855	.686																					
y	.0119	.0207	.0319	.0620	.102	.151	.208	.279	.449																					
.024 P	4.77	3.76	3.07	2.28	1.81	1.50	1.27	1.115	.888																					
y	.0104	.0183	.0284	.0557	.0918	.137	.191	.254	.410																					
.026 P	6.15	4.80	3.97	2.92	2.31	1.91	1.62	1.42	1.13																					
y	.00915	.0163	.0257	.0499	.0830	.125	.173	.231	.373																					
.028 P	7.76	6.08	4.93	3.66	2.90	2.39	2.03	1.77	1.41	1.18																				
y	.00811	.0146	.0228	.0457	.0756	.114	.160	.214	.344	.506																				
.030 P	9.60	7.5	6.12	4.52	3.59	2.93	2.52	2.19	1.74	1.445																				
y	.00715	.0130	.0206	.0416	.0685	.1045	.1475	.198	.317	.468																				
.032 P	11.8	9.12	7.52	5.47	4.31	3.56	3.07	2.66	2.12	1.77	1.51																			
y	.00638	.0116	.0188	.0383	.0638	.0960	.1365	.182	.294	.437	.603																			
.034 P	14.4	11.1	9.07	6.64	5.22	4.32	3.68	3.23	2.54	2.12	1.80	1.56																		
y	.00574	.0106	.0172	.0351	.0591	.0899	.1265	.170	.275	.407	.563	.762																		
.036 P		13.3	10.8	7.94	6.21	5.14	4.35	3.82	3.01	2.51	2.15	1.86																		
y		.00962	.0156	.0327	.0549	.0835	.118	.159	.257	.382	.529	.701																		
.038 P		15.6	12.8	9.32	7.30	6.05	5.15	4.52	3.54	2.97	2.54	2.20	1.96																	
y		.00869	.01435	.0299	.0508	.0782	.1105	.150	.241	.361	.500	.662	.842																	
.040 P		18.3	15.1	10.9	8.55	7.05	6.05	5.23	4.16	3.42	2.96	2.57	2.28																	
y		.00792	.0132	.0277	.0475	.0728	.104	.140	.228	.338	.472	.625	.800																	
.042 P		21.6	17.5	12.8	9.80	8.24	7.00	6.06	4.84	3.99	3.40	2.98	2.65	2.37																
y		.0073	.0121	.0259	.0439	.0685	.0978	.132	.217	.320	.445	.591	.760	.941																
.044 P			20.1	14.6	11.45	9.46	8.06	6.98	5.57	4.62	3.92	3.41	3.03	2.72																
y			.0111	.0239	.0415	.0641	.0910	.1235	.204	.304	.419	.558	.717	.897																
.046 P			23.2	16.8	13.2	10.85	9.21	8.01	6.35	5.27	4.48	3.92	3.48	3.14																
y			.0104	.0224	.0389	.0605	.0858	.118	.193	.289	.401	.533	.685	.856																
.048 P			26.6	19.2	15.0	12.3	10.5	9.17	7.24	6.00	5.10	4.46	3.94	3.55	3.25															
y			.00952	.0209	.0368	.0571	.0821	.112	.185	.275	.380	.509	.651	.820	.998															
.051 P			32.5	23.1	18.1	14.9	12.6	10.9	8.68	7.20	6.13	5.35	4.74	4.27	3.89	3.55														
y			.00856	.0189	.0338	.0525	.0752	.103	.170	.256	.356	.476	.611	.763	.937	1.12														
.055 P				29.3	22.9	18.8	15.9	13.8	10.9	9.05	7.72	6.70	5.95	5.35	4.85	4.45	4.11													
y				.0166	.0302	.0472	.0685	.0933	.155	.234	.326	.434	.558	.704	.860	1.03	1.22													
.059 P				36.5	28.4	23.3	19.8	17.1	13.5	11.1	9.56	8.35	7.37	6.63	6.01	5.50	5.08													
y				.0147	.0268	.0426	.0620	.0853	.143	.212	.300	.403	.518	.649	.782	.955	1.13													
.063 P				45.3	34.9	28.6	24.1	20.9	16.6	13.6	11.6	10.1	9.02	8.10	7.33	6.71	6.20	5.75												
y				.0131	.0241	.0385	.0565	.0778	.131	.198	.278	.372	.482	.603	.737	.887	1.05	1.23												
.067 P				54.8	42.1	34.7	29.3	25.2	19.9	16.4	14.1	12.2	10.8	9.70	8.82	8.06	7.45	6.90	6.45											
y				.0117	.0218	.0354	.0519	.0715	.121	.183	.258	.347	.447	.557	.686	.828	.985	1.15	1.32											
.071 P					50.6	41.1	34.5	30.2	23.7	19.7	16.7	14.6	12.9	11.6	10.5	9.58	8.87	8.22	7.68	7.20										
y					.0197	.0319	.047	.0659	.111	.170	.241	.325	.417	.524	.645	.775	.922	1.07	1.24	1.43										
.074 P					57.4	46.8	39.7	34.2	27.1	22.3	18.9	16.5	14.7	13.2	11.9	10.9	10.0	9.33	8.68	8.12										
y					.0183	.0302	.0446	.0621	.106	.161	.229	.307	.398	.503	.610	.739	.876	1.03	1.19	1.36										
.078 P					68.0	55.5	46.8	40.3	31.8	26.0	22.3	19.4	17.2	15.4	14.0	12.7	11.7	10.9	10.2	9.54										
y					.0166	.0275	.0414	.0576	.0987	.150	.214	.289	.375	.471	.579	.696	.825	.970	1.12	1.28										
.082 P					80.2	64.7	54.8	47.0	37.1	30.4	26.0	22.5	20.0	18.0	16.3	14.8	13.6	12.7	11.8	11.1										
y					.0152	.0252	.0382	.0535	.0925	.141	.202	.272	.353	.444	.548	.656	.780	.915	1.06	1.22										
.086 P						75.5	63.3	54.9	42.9	35.1	29.9	26.0	23.1	20.8	18.7	17.2	15.7	14.7	13.6	12.8	11.3									
y						.0234	.0354	.050	.086	.132	.190	.256	.332	.420	.515	.624	.740	.870	.989	1.15	1.48									
.090 P						83.5	70.3	60.8	47.5	39.2	33.2	28.8	25.7	23.2	21.9	19.1	17.6	16.2	15.0	14.7	12.6									
y						.0212	.0326	.0462	.0808	.124	.178	.241	.314	.397	.490	.592	.702	.825	.946	1.09	1.40									
.106 P							122.	104.	81.2	66.8	57.0	49.2	43.4	38.9	35.4	32.5	29.9	27.6	25.7	24.1	21.2	19.1	17.3							
y							.0246	.0356	.0630	.0987	.143	.196	.255	.324	.402	.487	.582	.681	.790	.905	1.15	1.46	1.78							
.121 P								159.	121.	101.	85.5	73.5	65.3	58.0	52.9	48.4	44.4	41.1	38.4	36.1	31.7	28.5	25.9	23.8						
y								.0283	.0516	.082	.120	.164	.217	.273	.342	.415	.494	.582	.677	.780	1.00	1.25	1.53	1.85						
.135 P								224.	174.	141.	120.	103	91.0	81.7	73.6	66.7	62.0	57.5	53.6	50.0	44.4	39.5	35.9	32.9						
y								.0230	.0431	.0695	.102	.141	.187	.239	.297	.367	.434	.510	.595	.685	.890	1.11	1.36	1.63						
.148 P									231.	188.	159.	137	121.	108.	98.	88.7	81.5	75.5	70.8	66.0	59.1	52.5	47.5	43.5						
y									.0367	.060	.0891	.124	.165	.212	.264	.320	.385	.455	.534	.612	.797	1.00	1.22	1.47						
.162 P									309.	249.	211.	183	160.	142.	128.	117.	107.	100.	92.5	87.3	77.2	69.0	62.5	57.1						
y									.0311	.0517	.0778	.110	.145	.186	.235	.285	.342	.405	.473	.550	.715	.903	1.11	1.33						
.177 P										330.	266.	240	210.	187.	168.	154.	141.	131.	121.	114.	101.	90.5	81.8	74.8						
y										.0442	.0668	.0955	.127	.165	.208	.255	.307	.364	.425	.493	.642	.811	.995	1.20						
.207 P													343.	303.	275.	250.	228.	211.	196.	183.	161.	145.	130.	121.	103.					
y													.101	.132	.167	.206	.247	.296	.347	.403	.525	.670	.829	1.01	1.41					
.225 P														394.	355.	321.	297.	271.	252.	236.	208.	187.	170.	154.	132.	114.				
y														.116	.147	.183	.223	.264	.311	.361	.475	.600	.746	.905	1.27	1.69				
.263 P															572.	520.	477.	437.	410.	381.	335.	300.	270.	249.	211.	184.	164.			
y															.115	.145	.177	.212	.252	.295	.386	.492	.610	.748	1.05	1.41	1.83			
.283 P																653.	600.	550.	513.	478.	422.	375.	339.	311.	266.	232.	204.	183.		
y																.129	.159	.190	.226	.264	.351	.447	.556	.680	.962	1.30	1.67	2.10		
.307 P																	780.	709.	660.	615.	543.	483.	434.	396.	339.	296.	262.	236.	212.	193.
y																	.140	.169	.201	.235	.313	.401	.501	.611	.870	1.17	1.52	1.93	2.36	2.82
.331 P																		897.	833.	775.	682.	608.	547.	498.	426.	370.	329.	294.	268.	244.
y																		.150	.180	.210	.282	.364	.454	.557	.793	1.07	1.39	1.75	2.16	2.60
.363 P																			1110	1030	908.	811	730	668.	563.	487.	436.	390.	356.	323
y																			.154	.182	.246	.320	.402	.494	.702	.950	1.24	1.57	1.94	2.34
.394 P																				1330	1170	1040	938	858	724	630	558	499	453	416
y																				.161	.218	.282	.358	.442	.631	.859	1.12	1.42	1.78	2.13
.430 P																					1530	1370	1230	1115	950	824	725	652	591	540
y																					.189	.249	.315	.391	.565	.768	1.00	1.28	1.58	1.92
.460 P																						1690	1510	1380	1170	1010	892	797	725	663
y																						.224	.286	.354	.515	.700	.922	1.17	1.45	1.77
.500 P																						2170	1960	1780	1510	1310	1150	1030	940	854
y																						.196	.251	.311	.456	.630	.827	1.05	1.31	1.59

NOTE: P = load in pounds, y = deflection per turn at load P. Table based on stress of 100,000 psi (not corrected for curvature, Eq. 5). For any other stress S_s, values of P and y should be multiplied by $S_s/100{,}000$. Deflections are based on a torsion modulus $G = 11.4 \times 10^6$ psi. For any other value of G, deflections given should be multiplied by the ratio $11{,}400{,}000/G$.

Springs of Square or Rectangular Bar Section. Such springs are sometimes used where it is desired to provide maximum energy storage within a limited space. One reason for this is that more material may be provided within a given OD and length of spring, particularly if the spring is coiled **flatwise,** and hence greater energy storage is possible. This advantage is partially offset by the fact that it is difficult to obtain material in the square or the rectangular section of equal quality to that in the round section.

Square-bar Springs. Neglecting correction for curvature and direct shear, the *uncorrected* stress in a helical spring of **square section** is

$$S_S = \frac{4.8\,Pr}{a^3} \tag{13}$$

where P = load
a = side of square cross section
r = mean coil radius

Deflection δ of a square-bar helical spring is given by

$$\delta = \frac{44.6 P r^3 n}{G a^4} \tag{14}$$

where n = number of active coils
G = modulus of rigidity

This formula is theoretically around 2 to 4 per cent in error for indexes between 3 and 4, but for most practical cases it is sufficiently accurate.

For square wire, the *corrected* stress is found[17] by multiplying the stress of Eq. (13) by a factor K' where

$$K' = 1 + \frac{1.2}{c} + \frac{0.56}{c^2} + \frac{0.5}{c^3} \tag{15}$$

This gives

$$S_s' = K' S_s$$

In this equation $c = 2r/a$ = spring index. The factor K' is slightly below the factor K for round wire [Eq. (3)], but the latter may be used as an approximation if desired.

In general when a square-wire spring is wound, plastic deformation during coiling makes the section trapezoidal (Fig. 19). In such cases an approximation may be had by taking an average value of a equal to

$$a = \frac{b_1 + b_2 + 2\,a_1}{4}$$

and taking the spring index equal to $2r/a_1$ for finding K'.

Table 11, which applies to round-wire helical springs, may also be used for a calculation of loads and deflections in square wire or bar springs at given uncorrected stresses. It is merely necessary to determine the load and deflection at the given *uncorrected* stress in the corresponding round-wire spring, *i.e.*, one having the same outside coil diameter, number of turns, and a wire diameter equal to the average side of the square cross section. Loads thus found are multiplied by the factor 1.06 and the deflections by 0.738 to find those for the square-wire spring at the given stress. This procedure may also be used if corrected stresses are taken as a basis, but there will be a small error for low-index springs because the correction factor K' for square wire is slightly different from the corresponding factor K for round wire.

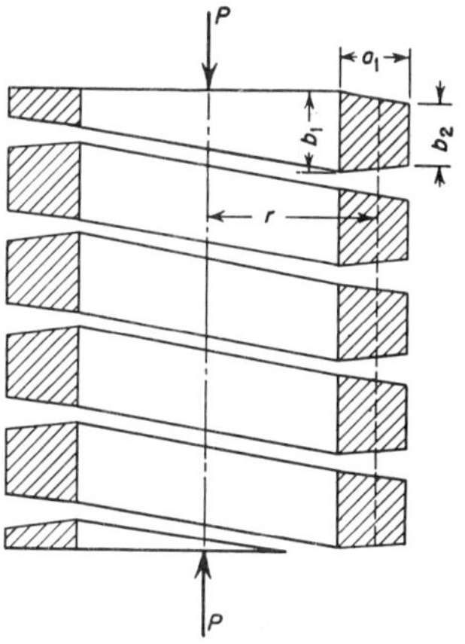

Fig. 19. Helical spring of square wire axially loaded. (Note: The wire becomes somewhat trapezoidal from the coiling operation.)

Rectangular-bar Springs. The *uncorrected* stress in a spring of rectangular wire (Fig. 20) is

$$S_S = \frac{Pr\,(3a + 1.8\,b)}{a^2 b^2} \tag{16}$$

where a = long side and b = short side of rectangle. This formula holds whether the spring is coiled flatwise as in Fig. 20 or lengthwise with the long side of the section parallel to the spring axis. However, the correction for curvature is different in the two cases.

To calculate *corrected* stress S_s' in rectangular bar springs, the chart of Fig. 21 may be used.[18] This gives

$$S_s' = \beta \frac{Pr}{ab\sqrt{ab}} \tag{17}$$

In this case a and b are sides perpendicular and parallel, respectively, to the spring axis (Fig. 21) and β is a factor depending on the ratio a/b or b/a

and on the ratio $2r/a$. (In this case b always represents the side parallel to the spring axis.)

To calculate deflections in rectangular-bar springs having large indexes (say greater than 8), the following formula, based on torsion of a straight bar of rectangular section, will yield results accurate to within a few per cent where the pitch angle is not large:

$$\delta = \frac{19.6 P r^3 n}{G b^3 (a - 0.56 b)} \tag{18}$$

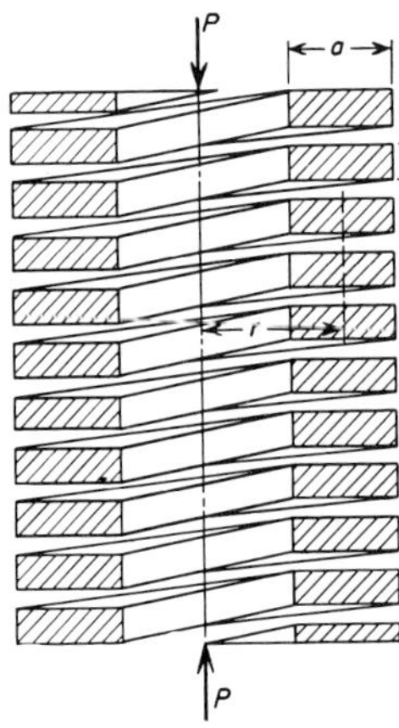

Fig. 20. Helical spring of rectangular wire coiled flatwise.

where a = long side and b = short side of cross section (Fig. 20).

For both small and large indexes, the deflection of a rectangular-bar spring may be calculated by using the chart of Fig. 22[18]. Deflection is

$$\delta = \frac{8 \gamma P r^3 n}{a^2 b^2 G} \tag{19}$$

where the constant γ depends on the ratio a/b or b/a and may be taken from Fig. 22, where in this case b is the side parallel to the spring axis and a the side perpendicular thereto. If the spring is coiled flatwise as in Fig. 20, the ratio a/b is taken, while if it is coiled with the long side parallel to the axis a ratio b/a is taken (Fig. 22).

Example. A rectangular-bar spring is coiled flatwise with $a = 1/2$ in., $b = 1/4$, $r = 1.5$ in., $P = 300$ lb, number of active turns $n = 5$, $G \doteq 11.5 \times 10^6$ psi (steel). From Fig. 22, the constant $\gamma = 6.7$ for $a/b = 2$, $2r/a = 6$. From Eq. (19) the deflection is

$$\delta = \frac{8 \gamma P r^3 n}{a^2 b^2 G} = \frac{8 \times 6.7 \times 300 \times 3.37 \times 5}{0.25 \times 0.0625 \times 11.5 \times 10^6} = 1.51 \text{ in.}$$

From Fig. 21, the factor $\beta = 5.88$, and by Eq. (17), corrected stress is

$$S_S^1 = \beta \frac{P r}{a b \sqrt{a b}} = \frac{5.88 \times 300 \times 1.5}{0.25 (0.5) \sqrt{0.125}} = 60{,}000 \text{ psi}$$

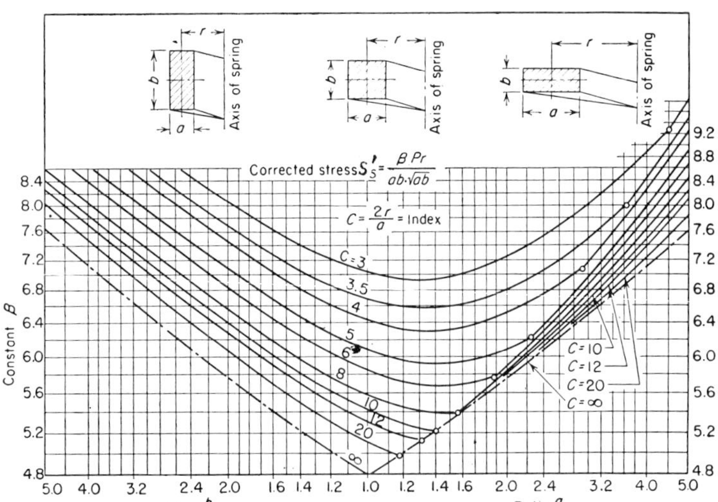

Fig. 21. Curves for finding stress factor β for rectangular bar spring. (Based on charts by Liesecke.)

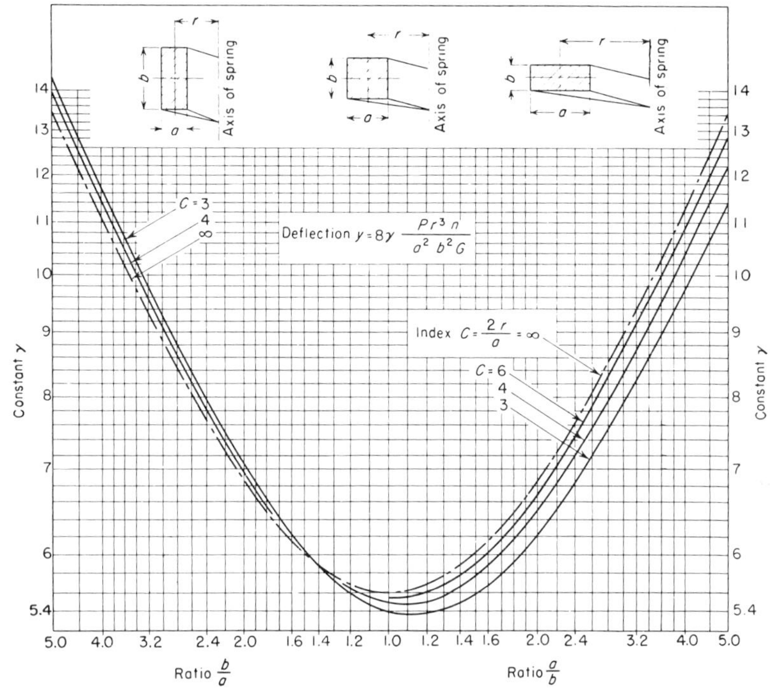

Fig. 22. Curves for finding deflection factor γ for rectangular bar springs. (Based on charts by Liesecke.)

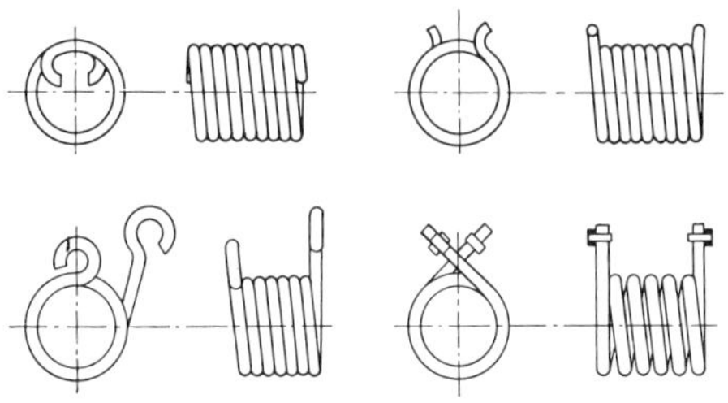

Fig. 23. Torsion springs.

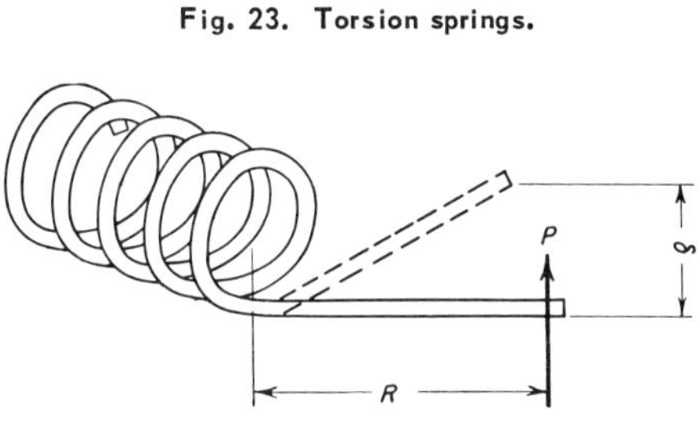

Fig. 24. Typical method of loading torsion spring.

HELICAL TORSION SPRINGS

General Notes. Springs of this type in most cases are subject to a torque about the spring axis so the primary stress is *flexural*, in contrast to the helical compression or tension spring where the primary stress is *torsional*. Torque may be transmitted to the spring by variously shaped ends (Fig. 23) or the spring may have an arm loaded as in Fig. 24. Applications of such springs include: door-hinge springs, springs for starters in automobiles, springs for brush holders in electric motors.

In most cases it is advisable to load torsion springs so the spring tends to wind up as the load is applied. If this is done, residual or trapped stresses set up by cold winding are in such a direction as to subtract from the stress as a result of loading; and a lower peak stress results. In cases where the direction of loading is such as to unwind the spring, a low-temperature heat-treatment to remove trapped coiling stresses is advisable.

Loading the spring in the manner shown in Fig. 24 also results in a lower moment arm to the point of maximum stress since the reaction is against the arbor.

Where the ends of the spring are clamped, or if special ends are used, some stress concentration may be expected and should be taken into account if the spring is subject to repeated loading.

Because a torsion spring (for usual applications) tends to wind up with load, its diameter decreases. To prevent binding and excessive stress, it is important that sufficient clearance be provided between the arbor or rod, about which the spring is wound, and the inner diameter of the spring. Clearance necessary may be estimated from the calculated deflection of the ends of the spring as given by Eq. (22) or (26) below. Thus if the spring end deflects 90 deg or 1/4 turn and the spring has 8 turns, the diameter will change in the ratio of 1/4 to 8, or by about 3 per cent. Likewise, if the spring fits inside a tube and is loaded to unwind, sufficient clearance must be allowed between the OD of the spring and the ID of the tube.

Where relatively long torsion springs are used, there is always the possibility of torsional buckling if the torque exceeds a certain value, and in such cases the use of guides may be necessary. Sometimes the application of a tension load to the spring may be sufficient to avoid buckling.

Usually for manufacturing reasons, torsion springs are made of round wire, but where maximum energy storage is required in a given space, square or rectangular wire may be used.

Design Formulas. *Uncorrected stress* for circular wire, curvature effects being neglected

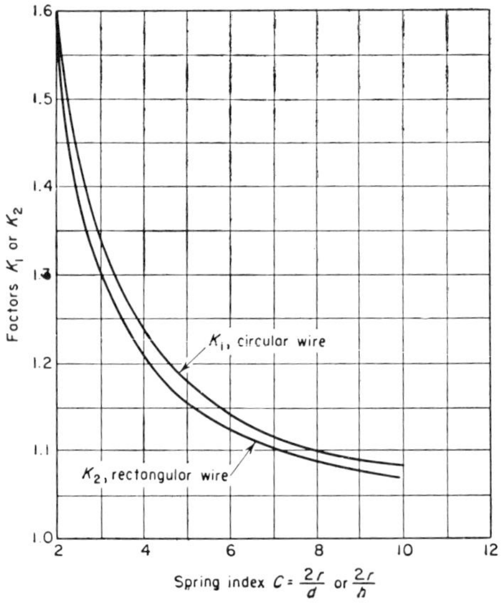

Fig. 25. Curved-bar factors for torsion springs of circular and rectangular wire.

$$S = \frac{32M}{\pi d^3} \tag{20}$$

where M = bending moment
d = wire diameter

For the spring shown in Fig. 24, $M = PR$.

Corrected stress S' (which takes into account stress augment from bar curvature is (ref. 1, page 317)

$$S' = K_1 \frac{32M}{\pi d^3} \tag{21}$$

The factor K_1 depends on the spring index $2r/d$ (Fig. 25). Angular deflection ϕ in radians as a result of a given moment M is

$$\phi = \frac{128Mrn}{Ed^4} \tag{22}$$

where r = mean coil radius
n = number of active coils
E = modulus of elasticity

Deflection δ along arc from load P applied at end of lever arm of length R (Fig. 24) is

$$\delta = \frac{128PR^2rn}{Ed^4} \tag{23}$$

For rectangular-bar torsion springs, the corresponding formulas are:

Uncorrected stress:

$$S = \frac{6M}{bh^2} \tag{24}$$

where b = width
h = radial depth of section

Corrected stress:

$$S' = K_2 \frac{6M}{bh^2} \tag{25}$$

where K_2 depends on the ratio $2r/h$ (Fig. 25).

Angular deflection ϕ in radians:

$$\phi = \frac{24\pi Mrn}{Ebh^3} \tag{26}$$

Deflection δ along arc as a result of load P applied at radius R (Fig. 24) is

$$\delta = \frac{24\pi PR^2rn}{Ebh^3} \tag{27}$$

Example. A brush-holder spring for a small motor is loaded as indicated in Fig. 24. Dimensions are: load arm $R = 7/8$ in., mean coil radius $r = 3/16$ in., wire size $d = 0.04$ in., spring index $c = 2r/d = 9.4$. Load P at the end of arm is 1 1/4 lb. From Fig. 25 factor K_1 for $c = 9.4$ is 1.08. Using Eq. (21), corrected stress is (taking $M = PR$)

$$S' = K_1 \frac{32M}{\pi d^3} = \frac{1.08 \times 32 \times 1.25 \times 0.875}{\pi \times (0.04)^3} =$$
188,000 psi

If there are 10 active turns, from Eq. (22), the deflection in radians is (for $E = 30 \times 10^6$ psi

$$\phi = \frac{128\,Mrn}{Ed^4} = \frac{128 \times 1.25 \times 0.875 \times 0.187 \times 10}{30 \times 10^6 \times (0.04)^4} =$$
3.4 radians, or 195 deg

Working Stresses. Suggested maximum corrected stresses[11] for torsion springs of different wire sizes and materials for average service conditions are given in Table 13. These conditions are defined as: noncorrosive atmosphere, temperatures not exceeding 150 F, and relatively slowly varying or static loads.

Where the spring is subject to repeated or fatigue stresses through a considerable range or to elevated temperatures, in general it will be necessary to use much lower working stresses than those suggested in Table 13.

Table 13: Recommended Maximum Design Stresses* – Helical Torsion Springs – for Average Service Conditions, Psi

	Kind of wire (plain carbon steels)		
Diameter, d	Hard-drawn	Tempered	Music
0.004–0.009	....	....	280,000
0.010–0.020	....	....	270,000
0.021–0.040	160,000	....	240,000
0.041–0.060	160,000	180,000	220,000
0.061–0.080	160,000	180,000	210,000
0.081–0.100	140,000	180,000	200,000
0.101–0.150	130,000	165,000	185,000
0.151–0.225	110,000	145,000	165,000
0.226–0.400	....	135,000	
0.401–0.625	....	125,000	

Stainless steel 18-8....... 140,000
Monel metal.............. 60,000
Brass.................... 30,000

*Computed by taking curvature effects into account using Eqs. 21 or 25.

SPIRAL SPRINGS

Spiral springs such as clock or watch springs consist essentially of flat strip wound to form a spiral shape (Fig. 26). Usually the inner end is clamped to an arbor, while the outer end may be either pinned or clamped. If individual turns of the spring do not come in contact, as in the case of the spring for the balance wheel of a watch, the analysis of such springs may be carried out with considerable accuracy.

Clamped Outer End (Fig. 26). In this case the outer end A of the spring is clamped or built in while the inner end is pivoted about O and is acted on by a moment M_o. Assuming that the number of turns is large and that adjacent coils do not come in contact during operation, angular deflection ϕ of the arbor in radians is (ref. 1, page 334)

$$\phi = \frac{M\ l}{EI} \tag{28}$$

where l = total length of spiral
E = modulus of elasticity
I = moment of inertia of section

To obtain the rotation in degrees, the value given by Eq. (28) must be multiplied by 57.3.

Analysis shows that the moment is constant along the length; hence stress is

$$S = \frac{6\,M_o}{bh^2} \tag{29}$$

where b = width of spring
h = thickness

Usually, however, there is some stress concentration at the clamped ends of the spring, and this will reduce the fatigue strength somewhat.

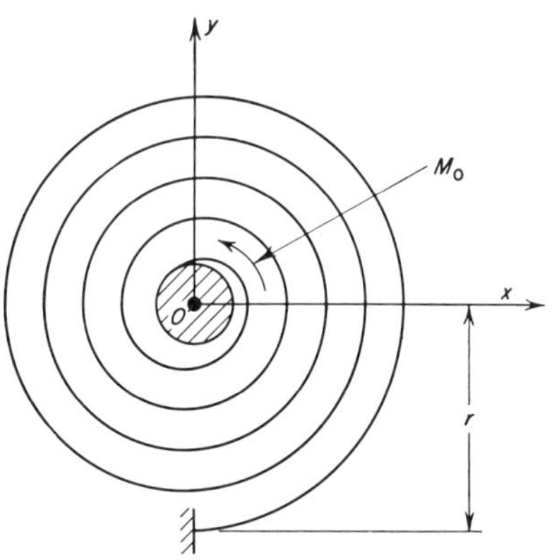

Fig. 26. Spiral spring with large number of turns (clamped outer end).

For a spring with a small number of turns, say less than two, angular deflection ϕ will in general be 3 to 15 per cent less than that given by Eq. (28). The peak stress will also be considerably greater than the value given by Eq. (29) if the number of turns is less than about three. For example, for one turn the ratio of the peak stress to the value given by Eq. (29) varies from 1.35 to about 1.6 depending on the ratio of ID to OD of the wound portion; for three turns this ratio varies from 1.12 to 1.19[19].

Pinned Outer End (Fig. 27). Frequently in practice the outer end of a spiral spring is hinged instead of clamped. In this case, external moment M_O may be taken as Pr and an analysis[1] shows a value for

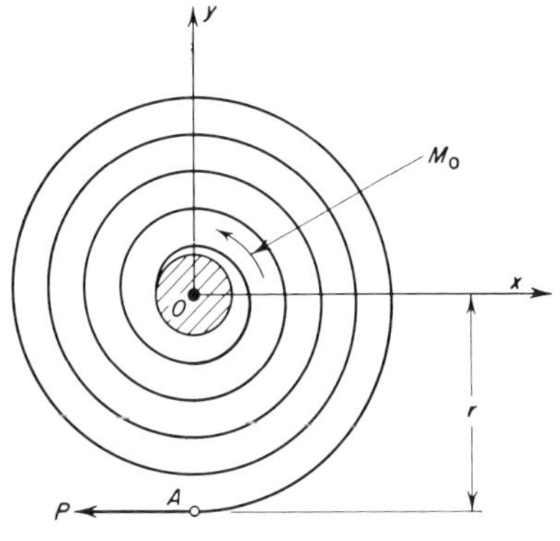

Fig. 27. Spiral spring with large number of turns. (Pinned outer end.)

angular deflection about 25 per cent greater than for the clamped-end condition (Fig. 26) for the same applied moment and assuming the coils do not touch. The maximum stress in this case will be twice that given by Eq. (29), but it occurs at a point where there is no stress concentration.

Example. Torsion spring of steel with pinned end (Fig. 27). $M_o = 25$ in.-lb, $r = 1$ in., strip section $1/2 \times 0.06$ in., $l = 15$ in., $I = 9 \times 10^{-6}$ in.^{4}, $E = 30 \times 10^6$ psi. From Eq. (28) for a clamped end:

$$\phi = \frac{Ml}{EI} = \frac{25 \times 15}{30 \times 10^6 \times 9 \times 10^{-6}} = 1.39 \text{ radians}$$

For a pinned end, the angle ϕ will be 25 per cent greater, or $1.25 \times 1.39 = 1.74$ radians = 100-deg. From Eq. (29), stress at point O is $6M_O/bh^2 =$ 83,500 psi. Maximum stress will be twice this, or 167,000 psi.

Power Springs – Coils in Contact. Where spiral springs are wound up tightly (as in the power spring of a phonograph) the previous analysis does not apply. In such cases the spring is usually placed inside a hollow case (Fig. 28).

To find the total number of turns N delivered by the spring in unwinding from the woundup position of Fig. 28 to the unwound position, the following approximate expression may be used (ref. 1, page 346).

$$N = \frac{K}{2h}\left[\sqrt{\frac{4}{\pi}lh + d^2} + \sqrt{D^2 - \frac{4}{\pi}lh} - (D + d)\right] \qquad (30)$$

where l = total length of spring
h = strip thickness
d, D are dimensions given in Fig. 28

Factor K depends on the ratio m given by

$$m = \frac{\pi}{4lh}(D^2 - d^2) \qquad (31)$$

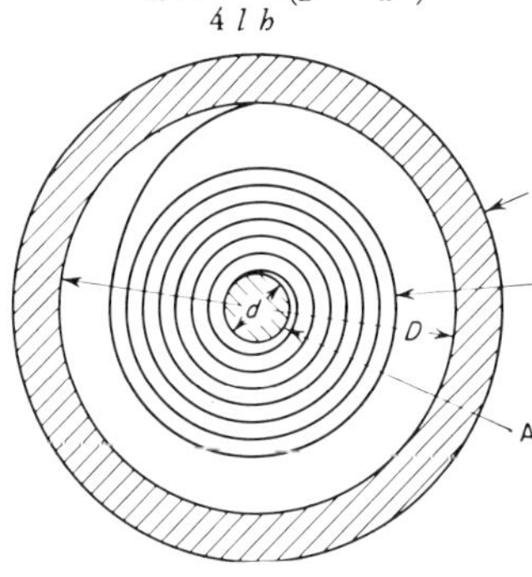

Fig. 28. Spiral spring wound on arbor.

Values of K for various values of m are given in Table 14.

Table 14

m	5	4	3	2	1.5
K	0.672	0.702	0.739	0.796	0.85

For good practice, $m = 2$ in which case $D = \sqrt{2.55lh + d^2}$ and Eq. 30 simplifies to[6]

$$N = \frac{1}{2.55h}\left[\sqrt{2(D^2 + d^2)} - (D + d)\right] \qquad (31a)$$

In practice to avoid excessive stress, the arbor diameter is usually made around fifteen to twenty-five times the strip thickness.

Example. A power spring has the following dimensions (Fig. 28): $h = 0.015$, $l = 129$, $D = 2.25$, $d = 0.375$ in. From Eq. (31)

$$m = \frac{\pi}{4(0.015)(129)}(2.25^2 - 0.375^2) = 2.00$$

Since $m = 2$, Eq. (31a) may be used.

Using the given values of D, h, and d in Eq. (31a), the number of turns N delivered in unwinding becomes 16.0

Working Stresses. Working stresses in spiral springs calculated from Eq. (29) may run as high as 175,000 psi for 1/8-in.-thick steel strip and

250,000 psi for 1/32-in. strip, in applications where fatigue conditions are not a factor. For example, an ordinary clock spring during its life may be subject to less than 5,000 cycles, hence may be stressed much higher than would be the case where millions of cycles are involved. When fatigue conditions are present (as, for example, in the spiral spring for the balance wheel of a watch), the stress range should be kept well below the endurance range of the material, stress-concentration conditions at the clamped edges being considered.

FLAT SPRINGS

Simple Cantilever Spring. In this category are included springs made from flat strip or bar stock even though they may be formed into more or less complicated shapes. Frequently, however, such springs may be considered as simple cantilevers (Fig. 29). Deflections δ and stress S for this case are given by

$$\delta = \frac{4Pl^3}{E\, b_o\, h^3} \tag{32}$$

$$\delta = \frac{6Pl}{b_o h^2} \tag{33}$$

where l = length
b_o = width
h = thickness
E = modulus of elasticity of the material, psi

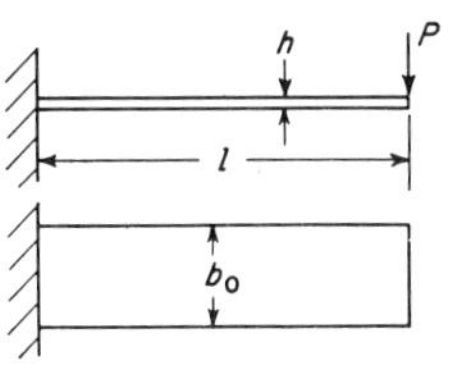

Fig. 29. Simple cantilever spring.

Where the strip is very wide compared with the thickness, slightly lower deflections than given by Eq.(32), (up to about 10% less) may result. This stiffening effect which occurs because lateral expansion or contraction of elements near the surface of the strip is prevented.

Trapezoidal Profile Spring. Somewhat more efficient use of material is obtained by the use of the cantilever spring of trapezoidal profile (Fig. 30). In this case maximum stress is given by Eq. (33), where b_o is the width at the built-in end, and deflection is obtained by multiplying the simple cantilever deflection [Eq.(32] by a factor K_1 which gives

$$\delta = K_1 \frac{4Pl^3}{E\, b_o\, h^3} \tag{34}$$

Factor K_1 may be taken from the curve for Fig. 30 for various values of the ratio b_1/b_o, where b_1 = width at loaded end.

Stress-concentration Effects. *Sharp Bends.* Frequently flat springs are formed with sharp bends as in the spring clip (Fig. 31), and these produce stress-concentration effects. To obtain corrected

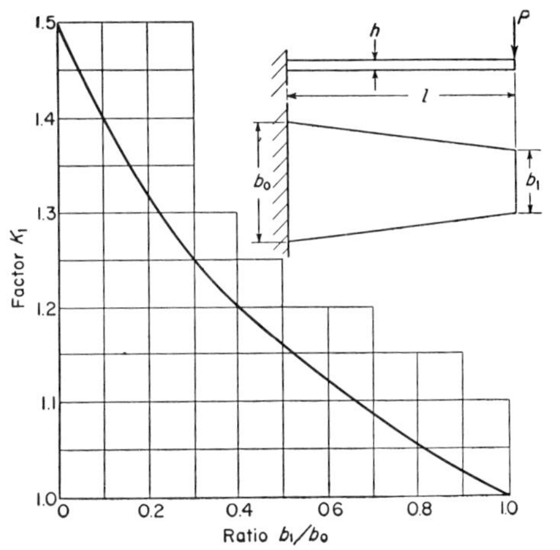

Fig. 30. Curve for finding factor K_1. (Trapezoidal profile spring.)

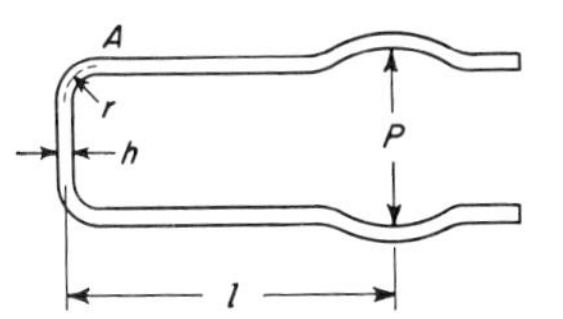

Fig. 31. Spring clip (stress concentration is present from sharp curvature of bend at "A").

stress in such cases, the nominal stress calculated from the bending moment must be multiplied by a concentration factor K_2 which may be taken from the curve marked K_2 in Fig. 25, using given r and h values. Where fatigue loading is involved, in such cases the stress *range* should be calculated using the factor K_2 while for static or infrequently repeated loading K_2 may be assumed as unity, which is equivalent to using an uncorrected stress. In general, sharp bends should be avoided to reduce stress concentration and also for manufacturing reasons.

Holes or Notches. For a hole in a flat strip subjected to bending (Fig. 32), theoretical stress-concentration factor K_t depends on the ratio d/h between hole diameter and strip thickness and on the ratio d/b between hole diameter and strip width. Values of K_t may be obtained for various values of these ratios with the curves of Sec. 6.6 (Fig. 10.)* In this case, the corrected stress is K_t $[6M/(b-d)h^2]$. For small d/h and d/b, $K_t = 3.0$; for large d/h with small d/b, $K_t = 1.85$ (Fig. 10, Sec. 6.6 [20]).

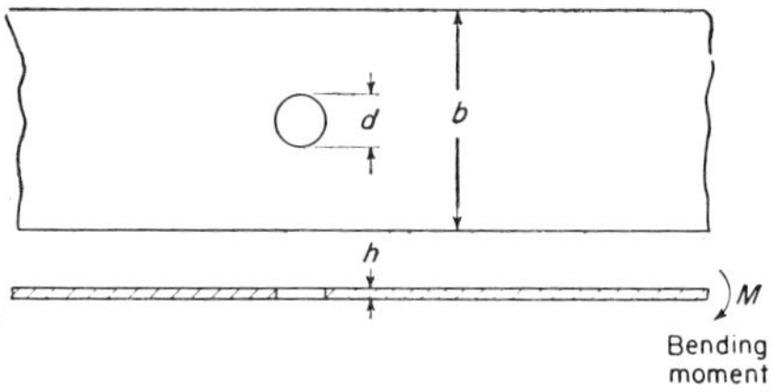

Fig. 32. Flat spring with hole in bending.

For semicircular notches of diameter d, small compared with the strip width b, values of K_t may be taken as approximately the same as those for holes of the same diameter, and assuming the same ratios d/b and d/h (Fig. 10, Sec. 6.6).

Working Stresses. For static loads or loads infrequently repeated, working stresses for flat springs made of 0.7 to 0.8 carbon steel (Rockwell C 45 to 48) may run as high as 180,000 psi for 1/8-in.-thick strip to 240,000 psi for 0.025-in.-thick strip.

Where fatigue or repeated loading is present, much lower stresses than these should be used and the stress range should be calculated taking stress-concentration effects into account.

LEAF SPRINGS

General Notes. Leaf springs are somewhat less efficient in terms of energy storage capacity per pound of metal as compared with helical springs. Balancing this disadvantage is the fact that such springs may be applied to function as structural members.[21] Sketches of a typical semielliptic leaf spring and of some typical ends and center clamps are shown in Figs. 33 to 35. Rebound and alignment clips are illustrated in Fig. 36.

Design Formulas. The following formulas for leaf springs are based on the assumption of a beam of uniform strength.[21]

Symmetric Semielliptic Leaf Spring (Fig. 37*a*). Spring rate in pounds per inch is

$$\frac{P}{\delta} = \frac{8Enbh^3}{3L^3} \tag{35}$$

where n = number of leaves
b = width
h = thickness of leaf
L = length, in.

Stress
$$S = \frac{3PL}{2nbh^2} \tag{36}$$

Unsymmetric Semielliptic Leaf Spring (Fig. 37*b*)

Rate
$$\frac{P}{\delta} = \frac{Ebnh^3 L}{6 l_1^2 l_2^2} \tag{37}$$

Stress
$$S = \frac{6P l_1 l_2}{Lnbh^2} \tag{38}$$

In these, l_1 and l_2 are dimensions shown in Fig. 37*b*.

Cantilever Leaf Spring (Fig. 37*c*) having n leaves, l being the length of the longest leaf.

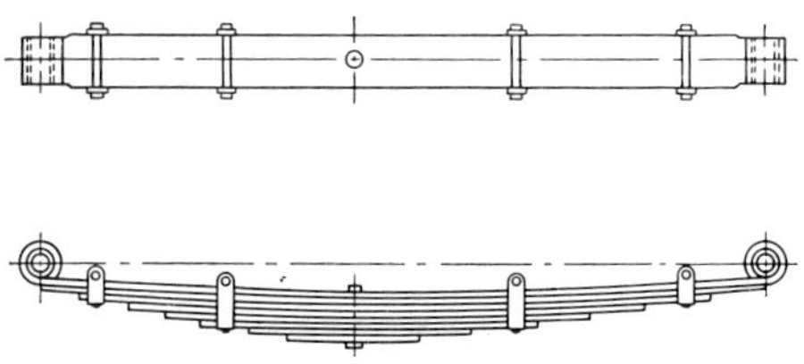

Fig. 33. Semi-elliptic spring.

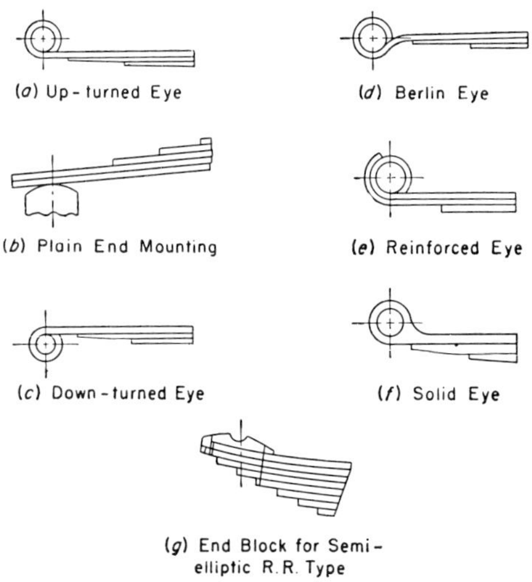

Fig. 34. Leaf spring ends.

*Metals Engineering—Design, p. 110.

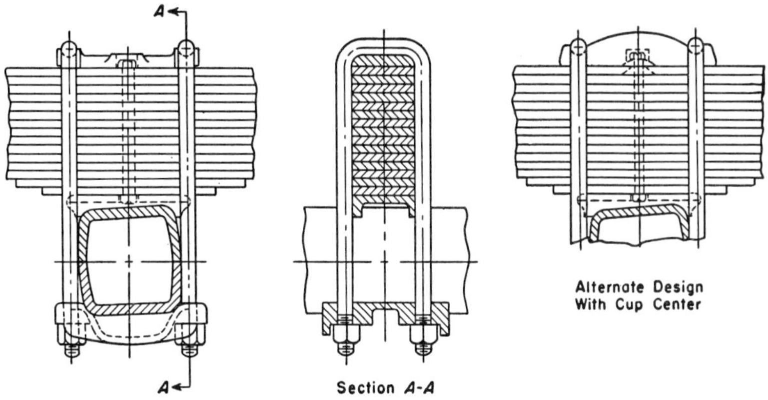

Fig. 35. Center clamps, leaf spring.

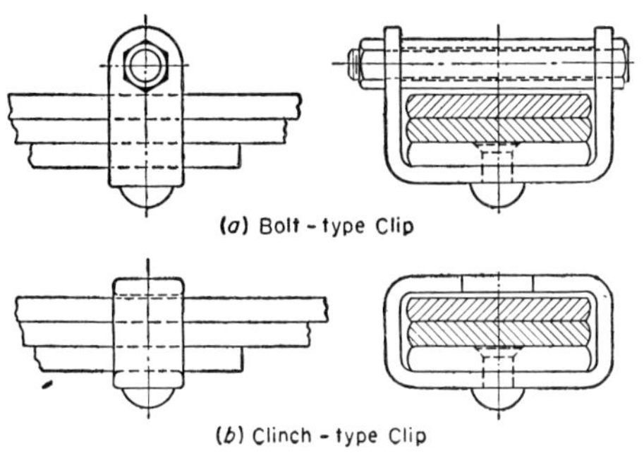

Fig. 36. Rebound and alignment clips for leaf springs.

Rate $$\frac{P}{\delta} = \frac{E\,b\,n\,h^3}{6\,l^3} \tag{39}$$

Stress $$S = \frac{6\,P\,l}{n\,b\,h^2} \tag{40}$$

Equations (35) to (40), although based on the assumption that the spring may be considered as a beam of uniform strength and composed of leaves of equal thickness, are frequently used for preliminary design even for cases where these conditions are not fulfilled. Effects which cause deviations from the ideal conditions assumed are: use of leaves of different thicknesses; use of more than one main leaf; stiffening effects of center clamp and leaf ends; interleaf friction; effects of spring shackles which result in angular loading at the ends. These effects may result in rather large deviations from values calculated from the formulas. Methods for applying Eqs. (35) to (40) and correcting the results to take some of these factors into account are discussed in the SAE Manual.[21]

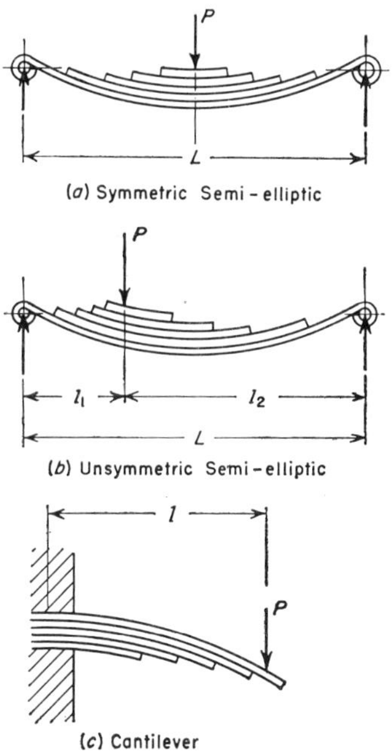

Fig. 37. Leaf springs.

Working Stresses. Design stresses based on the use of Eqs. (35) to (40) modified as suggested in the SAE Manual for automotive and railroad leaf springs, are given in Fig. 38. If the design is based on the static-loading condition, *i.e.*, on the static deflection resulting from the load carried, design stress increases as static deflection increases (lower curves of Fig. 38). Where possible, however, the design should be based on the maximum stress which occurs

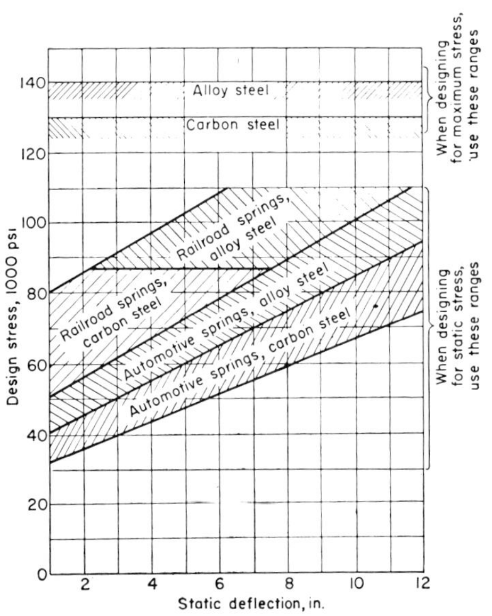

Fig. 38. Design stresses for leaf springs.

under conditions of dynamic deflection, a value of 130,000 psi being suggested for carbon steel and 140,000 psi for alloy steel.

Use of these stresses in general will result in a life of about 100,000 full-load cycles of stress (assuming that the springs have been shot-peened and cold-set). Also a set of from 3 to 5 per cent of full-load deflection at such stresses may be expected. Where the spring operates under conditions of appreciable negative loading in rebound, lower stresses than those of Fig. 38 must be used. Where other stresses than those resulting from vertical loading (such as windup torque stresses) are present, these must be considered in calculating maximum stress, or lower stresses must be used if the design is based on the static loading condition.[21]

Materials. Materials commonly used for leaf springs include SAE Nos. 1095, 4063, 5150, 6150, and 9260. In general, alloy steels are used for automotive springs, and carbon steels for railway springs. Heat-treatments giving Brinell hardnesses 415 to 460 will usually yield satisfactory results for spring steels. Also shot-peening and cold-setting operations are desirable for satisfactory life in service.

For data on endurance limits of leaf spring materials see Table 19.

CONED-DISK (BELLEVILLE) SPRINGS

General Notes. Such springs consist essentially of coned or dished disks with diametral cross sections and loading as indicated in Fig. 39. The load-deflection characteristic depends primarily on the ratio h/t between initial cone height (or dish)

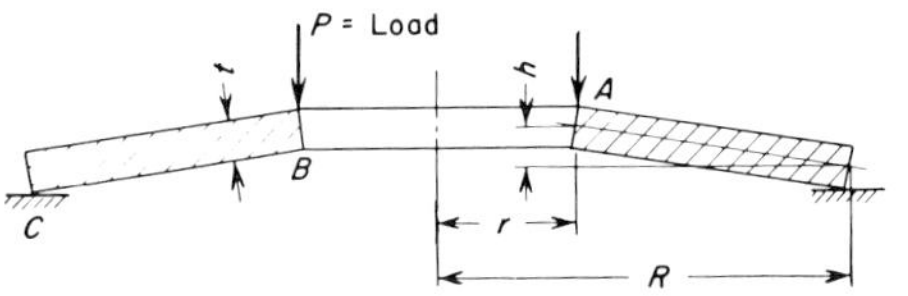

Fig. 39. Belleville spring.

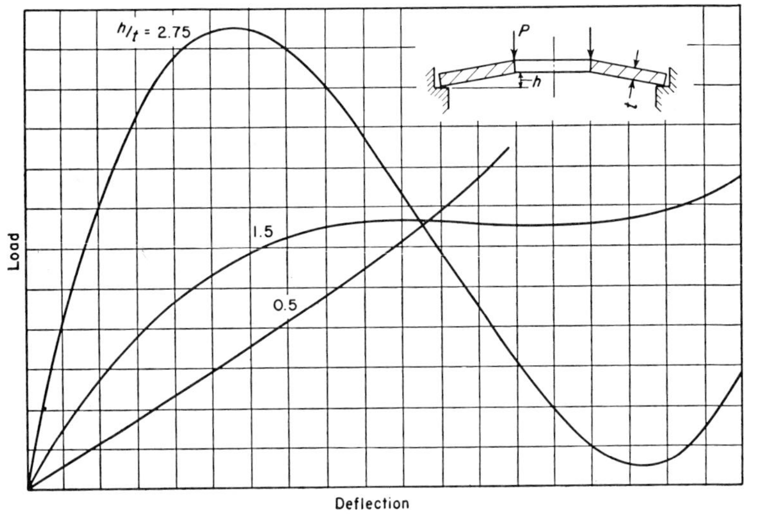

Fig. 40. Shapes of load-deflection diagrams obtained by varying ratio h/t.

and thickness. Some of these characteristics are indicated in Fig. 40 for various values of h/t. For ratios h/t around 0.5, the curve approximates a straight line up to a deflection equal to half the thickness while for h/t equal to 1.5, the load is nearly constant (within a few per cent) over a considerable range of deflection. Springs with ratios h/t approximating 1.5 are known as *constant-load* or *zero-rate* springs. By stacking Belleville springs in parallel (Fig. 41*a*) load capacity is increased, while the series arrangement (Fig. 41*b*) gives increased deflection for a given load. The latter method should not be used, however, for ratios of h/t greater than 1.3 because instability and an irregular load-deflection characteristic may result. Guides are usually advisable to prevent buckling or lateral deflection.

Advantages of Belleville springs include: small space requirement in direction of load application; ability to carry lateral loads; characteristics variable by adding or removing disks. Disadvantages include nonuniformity of stress distribution, particularly for large ratios of OD to ID.

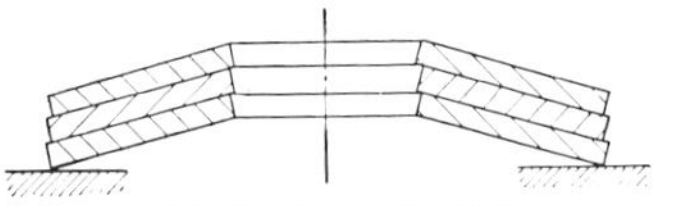

(*a*) Stacked in Parallel

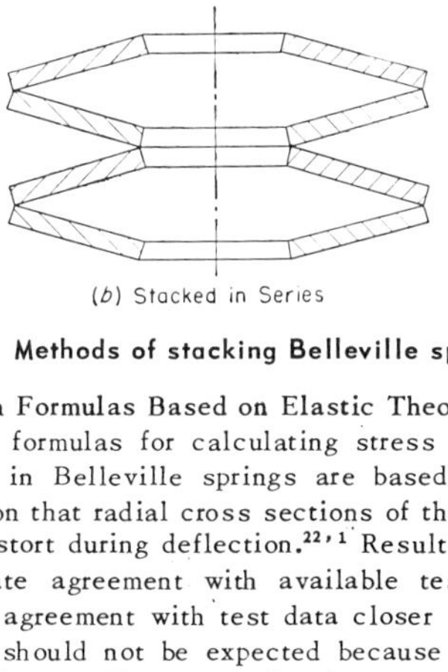

(*b*) Stacked in Series

Fig. 41. Methods of stacking Belleville springs.

Design Formulas Based on Elastic Theory. The following formulas for calculating stress and deflections in Belleville springs are based on the assumption that radial cross sections of the spring do not distort during deflection.[22,1] Results are in approximate agreement with available test data; however, agreement with test data closer than ±5 per cent should not be expected because of frictional and other effects not taken into account.

Load P at deflection δ is

$$P = \frac{1.1\,\delta E C}{R^2}\left[(h-\delta)\left(h-\frac{\delta}{2}\right)t + t^3\right] \qquad (41)$$

This formula may be written

$$P = \frac{C\,C_1 E t^4}{R^2} \qquad (42)$$

where t = thickness
R = outer radius (Fig. 39)
E = modulus of elasticity

Factor C depends on R/r (Fig. 42) and the factor C_1 depends on both h/t and δ/t where h = initial cone height. Values of C_1 may be taken from Fig. 43. Shapes of the curves for C_1 also represent the shapes of load-deflection diagrams for the various ratios h/t.

Theoretical load P_1 required to flatten the spring is

$$P_1 = \frac{1.1\,E C h t^3}{R^2} \qquad (43)$$

If the spring is tested against a flat plate, actual flattening loads will be higher than P_1 because of the tendency of edge loads to move inward as the spring approaches the flattened position.

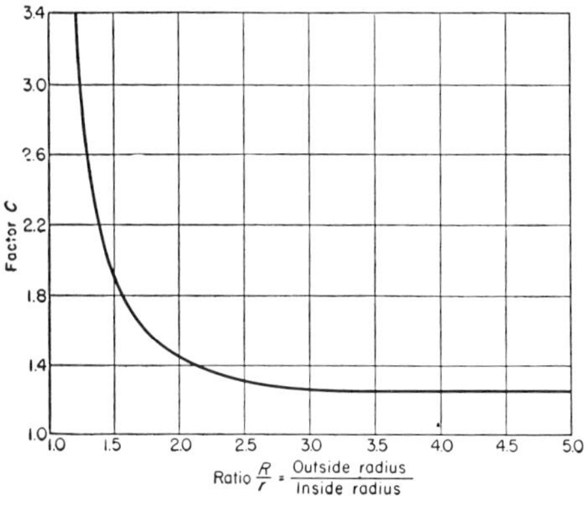

Fig. 42. Factor C.

Compression stress S_C at upper inner edge A (*Fig. 39*) at deflection δ is[22]

$$S_C = \frac{-1.1\,E\delta C}{R^2}\left[C_2\left(h-\frac{\delta}{2}\right)+C_3 t\right] \qquad (44)$$

(Normally where $\delta < 2h$ this is the maximum stress in the spring.)

Stress S_{t1} (normally tension) at lower inner edge B (Fig. 39) is

$$S_{t_1} = \frac{1.1\,E\delta C}{R^2}\left[-C_2\left(h-\frac{\delta}{2}\right)+C_3 t\right] \qquad (45)$$

(For $\delta > 2h$, S_{t_1} is greater than S_C)

Stress S_{t_2} at lower outer edge C (Fig. 39) is

$$S_{t_2} = \frac{1.1\,E\delta C}{R^2}\left[C_4\left(h-\frac{\delta}{2}\right)+C_5 t\right] \qquad (46)$$

In these equations C_2, C_3, C_4, *and* C_5 may be taken

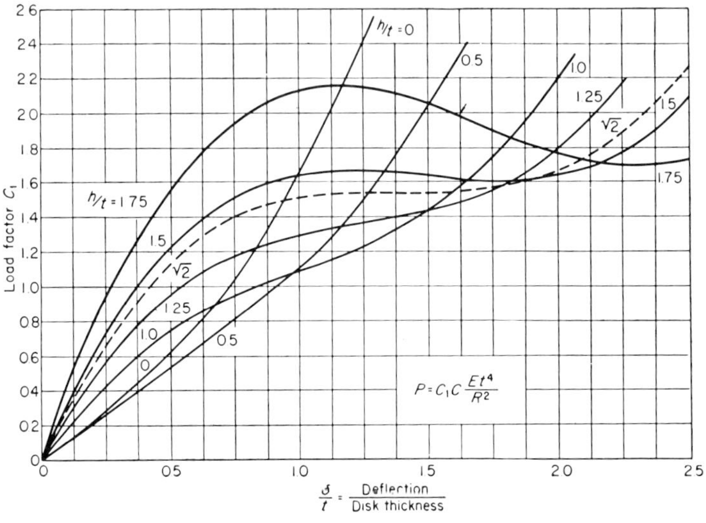

Fig. 43. Curves for determining load factor C_1.

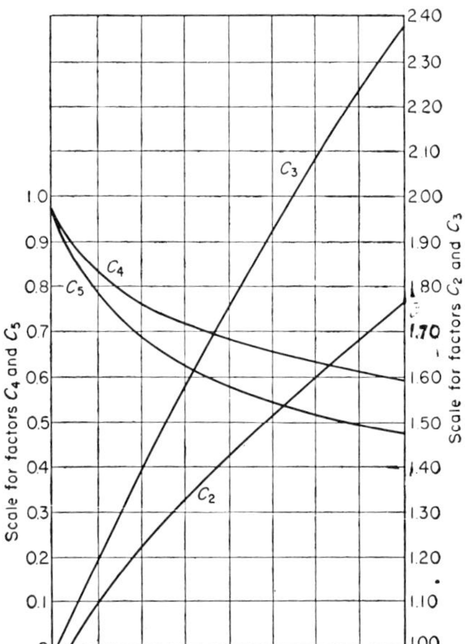

Fig. 44. Curves for finding C_2, C_3, C_4 and C.

from the curves of Fig. 44. As fatigue fractures frequently start at the lower *outer* edge of the spring at C, Fig. 39), tension stress S_{t2} may be critical even though numerically smaller than the compression stress S_c at A.

In all cases in Eqs. (44) to (46) a negative sign indicates compression, a positive sign tension.

Equations for calculating stress may be simplified to the following:

$$S = \frac{K E t^2}{R^2} \tag{47}$$

where K is a stress factor depending on ratios R/r, δ/t, and h/t. It may be taken from the chart of Fig. 45 for $R/r = 1.5$ and from Fig. 46 for R/r between 2 and 2.5. In these figures, the lower group of curves represents stress at upper inner edge for various h/t values; the upper curve represents stress at lower inner edge for $h/t = 0$ (flat plate). Additional charts for calculating stresses at points B and C (Fig. 39) are given in the SAE Manual.[22]

It should be noted that, because of the presence of trapped stresses in actual springs, the peak stresses calculated by Eqs. (44) to (47) may be considerably higher than the actual stresses present.

Formulas for Small Deflections and Cone Heights. Where both cone-height and deflection are small relative to thickness (say h/t and δ/t both less 0.5) a rough estimate of load and stress can be obtained by using the following simplified formulas based on the elastic flat plate theory.[23]

Note. As dishing action is neglected in deriving these equations, results calculated from them may

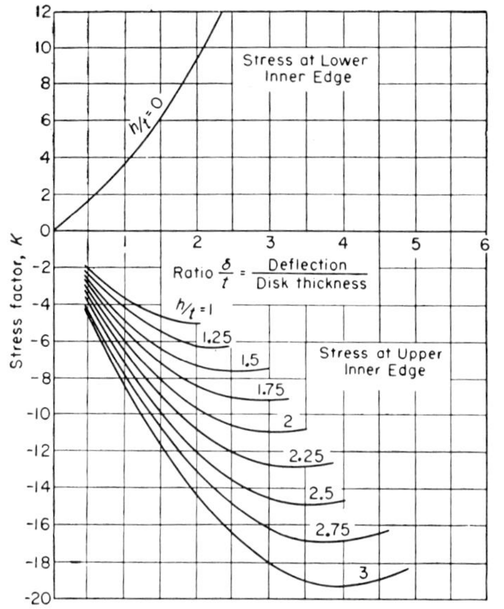

Fig. 45. Curves for determining stress factor K for Belleville springs. ($R/r = 1.5$).

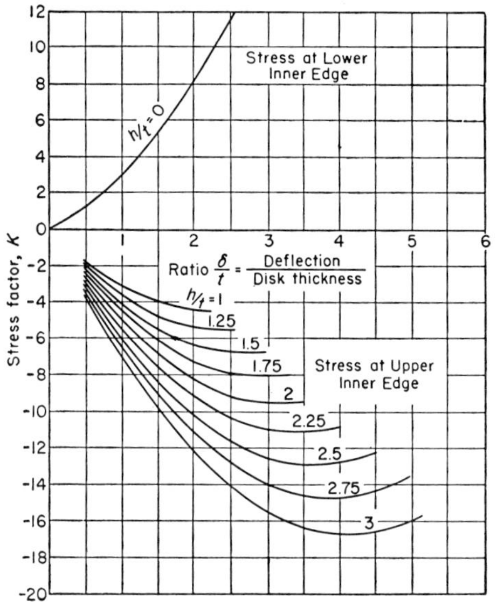

Fig. 46. Curves for determining stress factor K for Belleville springs ($R/r = 2.0$ to 2.5)

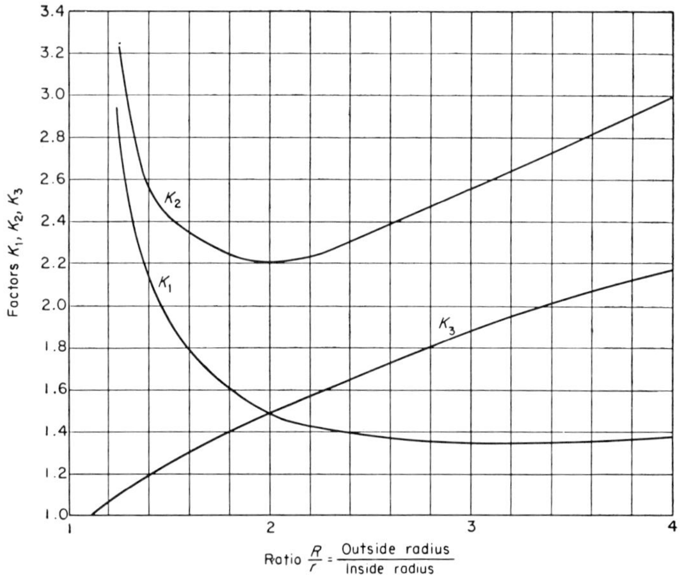

Fig. 47. Curves for finding factors K_1, K_2, and K_3. (Belleville springs with δ/t and h/t less than 0.5.)

differ somewhat from those of Eqs. (41) to (46), and it is suggested that for best accuracy the latter be used.

$$P = \frac{K_1 \delta E t^3}{R^2} \quad (48)$$

$$S = \frac{K_2 \delta E t}{R^2} \quad (49)$$

$$S = \frac{K_3 P}{t^2} \quad (50)$$

$$P = \frac{S t^2}{K_3} \quad (51)$$

Factors K_1, K_2 and K_3 in these formulas depend on ratio R/r and may be taken from the curves of Fig. 47.

Formula for Nominal Stress across Diametral Section. If the spring is approximately flat when loaded, as is true in most practical cases, a nominal stress S_n across a diametral section may be calculated simply by dividing the external bending moment by the section modulus. This gives

$$S_n = \frac{3P}{\pi t^2} = .95 \frac{P}{t^2} \quad (52)$$

Derived in this way, this formula yields a kind of average stress, the stress-concentration effect from the presence of the hole being neglected. In general it is believed to offer a somewhat better indication of the load-carrying capacity of the spring under static-loading conditions than Eqs. (44) or (50), which yield a localized stress at one corner of the spring. However, Eq. (52) should not be used where fatigue loading is involved.

Correction for Load Acting Inside Edge. In cases where the disk spring is loaded, not at the edges, but on two rims, as shown in Fig. 48, the following modifications are necessary.

Calculate load P and stresses S_c, S_{t1} and S_{t2} as functions of deflection δ using Eqs. (42), (44), (45), and (46). Then correct thus:

Deflection δ' between rims (Fig. 48)

$$\delta' = \delta \frac{a}{R - r} \quad (53)$$

Load between rims

$$P' = P \frac{R - r}{a} \quad (54)$$

Stresses remain unchanged as functions of δ and can be restated as functions of δ' if conversion according to Eq. (53) is made.

Design Example. A constant-load type of Belleville spring is required for a gasket application where a load of 5,000 to 6,000 lb is desired in a deflection range of 0.09 in. OD = 8 1/2 in., ID = 4 1/4 in., $h/t = 1.5$. Stress S_c limited to 220,000 psi. Maximum deflection $\delta = 1.3t$. From Fig. 46, for $h/t = 1.5$, $\delta/t = 1.3$, $R/r = 2$, $K = -5.1$. Taking $S_c = -220{,}000$ psi (compression), $E = 30 \times 10^6$ psi for steel, by solving Eq. (47) for t and substituting these values, required thickness t becomes, since $S = S_s$

$$t = R \sqrt{\frac{S_c}{KE}} = 4.25 \sqrt{\frac{220{,}000}{5.1 \times 30 \times 10^6}} = 0.161 \text{ in.}$$

From Fig. 43, C_1 is practically constant from $\delta = 0.75t$ to $\delta = 1.3t$, a range of about 0.09 in. This will give approximately constant load within the desired range.

From Fig. 43, for $h/t = 1.5$, $\delta/t = 1.3$, load factor $C_1 = 1.67$, and from Fig. 42, for $R/r = 2$, factor $C = 1.45$. Load per disk at $\delta - 1.3t$, or 0.21 in., is

$$P = \frac{C_1 C E t^4}{R^2} = 2{,}700 \text{ lb}$$

By using two springs in parallel, the desired load is obtained.

Working Stresses. *Static Loading.* Where Belleville springs are subject only to static or infrequently repeated loading, calculated stresses S_c based on Eq. (44), equal to 200,000 psi or higher have been used for steel springs. Although such stresses seem high, localized yielding will generally occur allowing relief of the most highly stressed portions of the spring.

An alternative method of design for static loading is to use the nominal stress formula [Eq. (52)] keeping the value of S_n well below the yield point of the material.

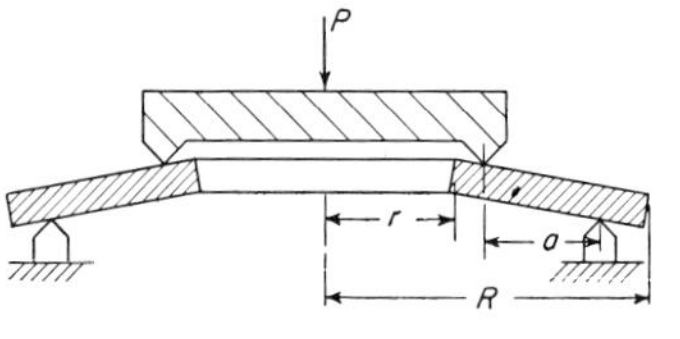

Fig. 48. Belleville spring with load inside edges.

Fatigue Loading. For springs subject to fatigue or repeated loading, lower stresses should be used than for those subject to static loading. Although not much data are available, the indications are that Belleville springs made of good-quality spring steel will withstand repeated loading up to perhaps 500,000 cycles even for peak stresses as high as

200,000 psi calculated from Eq. (44).[6] For longer spring life or higher operating temperatures, it is necessary to use lower stresses.

A more logical method of design where fatigue is involved is to calculate the stress *ranges* from Eqs. (44) to (46) at critical points and compare these values with the data available on the endurance properties of the material. The reduction in fatigue properties from the surface condition of the spring (effect of decarburization, etc.) should be taken into account. For further discussion of working stresses, see SAE Manual (ref. 22).

Materials. Materials used in the smaller sizes of Belleville springs are SAE 1074, 1085, and 1095. For larger thicknesses and for highest quality, silicomanganese steel SAE 9260 or chrome-vanadium steel SAE 6150 are preferred. Phosphor bronze (spring temper) may also be used, in which case stresses about half those allowable for steel springs should be used and the modulus E should be taken as 15×10^6 psi.

SPRING MATERIALS

Physical Properties. In Table 15 are given physical properties, including torsional and tensile ultimate strengths, modulus of elasticity, shear modulus, and elastic limits for the more widely used spring steels. Similar data for stainless steels and nonferrous materials are given in Table 16.[1,6]

Fatigue Properties and Endurance Ranges. Data on endurance ranges and physical properties for various spring materials in torsion and bending are given in Tables 17 and 18. On each table, besides the limiting endurance range, pertinent information is given including kind of material, heat-treatment, surface condition (*i.e.*, whether ground and polished, or untouched), ultimate and yield strengths in tension, modulus of rupture, and yield strength in torsion. An endurance range from 0 to 110,000 psi as listed means that the bar will withstand indefinitely a stress range between these limits while completely reversed stress is indicated by the sign ±. Similar data for leaf- and flat-spring materials in bending are given in Table 19. For further information, refer to literature references given on these tables.

Data given in these tables indicate that, where the test specimen is ground and polished, in general a considerably higher endurance limit is obtained than is the case where the surface is in the "as-received" condition. This difference in endurance limits may probably be attributed largely to surface decarburization from heat-treatment, although other factors (such as surface flaws or defects) may also be present. Such defects may account for the low values of endurance range reported in some cases.

Notes on Spring Wires and Materials (refer to Tables 15 and 16 for SAE or ASTM numbers.

Music Wire. A high-quality carbon steel, this wire is widely used for small-sized helical springs, particularly those subject to severe stress conditions. The high strength of this material is obtained by using a steel of about 0.70 to 1.00 per cent carbon, patenting and cold drawing to size. Variations of tensile strength for different wire sizes are shown in Fig. 49, maximum and minimum values being indicated. In forming helical springs of music wire, the winding is done cold over a mandrel. After winding, a low-temperature heat-treatment to relieve coiling stresses is usually given. The optimum temperature for this treatment may vary from 500 to 800F, the higher values being advisable for springs subject to elevated temperatures.[25]

Oil-tempered Spring Wire. This is a good-quality high-carbon-steel wire, made by the open-hearth or electric-furnace process. In manufacturing, the wire is cold-drawn to size, then heat-treated. Variations of tensile strength with wire size are indicated in Fig. 50. Springs made of this wire are wound cold, then usually given a thermal treatment.

Hard-drawn Spring Wire. A carbon steel of lower quality than music or oil-tempered wire, this wire

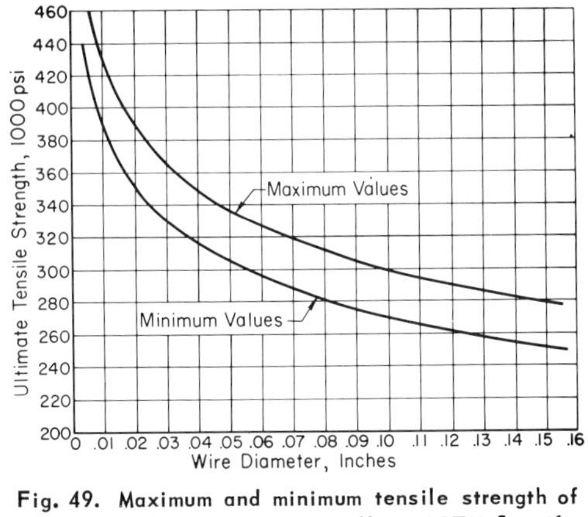

Fig. 49. Maximum and minimum tensile strength of music wire for various sizes. (from ASTM Specification A 228–51.)

is used in cases where the stresses are low or where a high degree of uniformity is not so essential. This material has tensile-strength properties slightly below those of oil-tempered wire Fig. 50. It should not, in general, be used where fatigue loading is involved.

Hot-wound Helical Springs, Heat-treated after Forming. For the larger sizes of helical springs

Table 15: Approximate Physical Properties of Typical Spring Steels[1]

Material	ASTM or SAE No.	Composition, %	Ultimate tensile strength, psi	Elastic limit in tension, psi	Ultimate strength in torsion, psi	Elastic limit in torsion, psi
Hard-drawn spring wire......	A-227-47	C 0.6–0.7 Mn 0.9–1.2	150,000– 300,000[2]	100,000– 200,000	120,000– 220,000	75,000– 130,000
Oil-tempered spring wire.....	A-229-41	C 0.6–0.7 Mn 0.6–0.9	155,000– 300,000[2]	120,000– 250,000	115,000– 200,000	80,000– 130,000
Music wire	A-228-41	C 0.7–1.0 Mn 0.3–0.6	250,000– 500,000[2]	150,000– 350,000	150,000– 300,000	90,000– 180,000
Hot-rolled bars.............	SAE 1095	C 0.9–1.05 Mn 0.25–0.5	175,000– 200,000	105,000– 140,000	110,000– 140,000	75,000– 110,000
Chrome-vanadium alloy steel.	SAE 6150	C 0.48–0.53 Mn 0.7–0.9 Cr 0.8–1.1 Va 0.15 min	200,000– 250,000	180,000– 230,000	140,000– 175,000	100,000– 130,000
Silicomanganese steel.......	SAE 9260	C 0.55–0.65 Mn 0.6–0.9 Si 1.8–2.2	200,000 to –250,000	180,000– 230,000	140,000– 175,000	100,000– 130,000
Clock-spring steel..........	SAE 1095	C 0.9–1.05 Mn 0.3–0.5	180,000– 340,000	150,000 to –310,000		

[1]Data from ref. 6 1952 Ed. Modulus of elasticity approximately 30×10^6 psi and modulus of rigidity 11.5×10^6 psi for all steels.
[2]Depending on size (see Figs. 49 and 50)
[3]Low value probably a result of surface decarburization.

Table 16: Approximate Physical Properties of Stainless Steels and Nonferrous Metals*(as Used in Springs)

Material	Composition, %	Ultimate tensile strength, psi	Elastic limit in tension, psi	Modulus of elasticity, psi	Ultimate strength in torsion, psi	Elastic limit in torsion, psi	Modulus of rigidity, psi
Stainless steel 18-8, Type 302...............	C 0.08–0.15 Cr 17–20 Ni 6–10	160,000– 330,000[1]	60,000– 260,000	28×10^6	120,000– 240,000	45,000– 140,000	10×10^6
Phosphor bronze	Cu 91–93 Tin 7–9, or Cu 94–96 Tin 4–6	100,000– 150,000	60,000– 110,000	15×10^6	80,000– 105,000	50,000– 80,000	6.2×10^6
Monel	Ni 64 Cu 26 Mn 2.5 Fe 2.25	100,000– 140,000	80,000– 120,000	26×10^6	75,000– 110,000	45,000– 70,000	9.5×10^6
Beryllium copper	Cu 98 Be 2	160,000– 200,000	100,000– 150,000	16×10^6– 18.5×10^6	100,000– 130,000	65,000– 95,000	6×10^6+, 7×10^6
Permanickel............	Ni 97†	180,000– 230,000	130,000– 170,000	30×10^6	120,000– 150,000	60,000– 90,000	11×10^6

1Depending on size
*Ref 6, 1952 Ed.
†Plus small amounts of Cu, Mn, Fe, Si, Mg, Ti.

Table 17: Endurance Ranges and Physical Properties of Spring Materials in Torsion

Material	Specimen or wire dia, in.	Heat-treatment[1]	Condition of surface	Hardness	Ultimate strength (tension), psi	Yield point (tension), psi	Modulus of rupture (torsion), psi	Yield point (torsion)[2], psi	Limiting endurance range in torsion[3] psi	Investigator
Tempered Swedish steel wire 0.60% C	0.125	....	As received	Rockwell C 45	226,000	203,000	166,000	139,000	0–110,000[4]	Weibel[7]
	0.187	....	As received	Rockwell C 43	211,000	190,000	162,000	125,000	0–107,000[4]	Weibel[7]
0.35 Mn approximate ...	0.225	....	As received	Rockwell C 42	204,000	181,000	166,000	120,000	±56,000 0–100,000[4]	Weibel[7]
0.6% C spring steel....	0.13[5]	O.Q. 950 C T. 500 C	Ground and polished	Bhn 365	177,000	161,000	150,000	......	±52,600 4,500–103,000 55,000–125,000	**Hankins**[8]
Cr-Va spring steel.....	0.13[5]	O.Q. 850 C T. 600 C	Ground and polished	Bhn 385	177,000	168,000	148,000	......	±56,000 10,000–111,000 57,000–131,000	Hankins[8]
Cold-drawn steel wire..	0.160	....	As received	Bhn 350	202,000	......	155,000	......	±36,400[6]	Lea and Dick[9]
High-carbon O.H. steel, 0.91% C, 0.38% Mn	0.25	O.Q. 1575 F D. 940 F	Ground and polished	Bhn 438–450	225,000	179,000	173,000	118,000	±52,000 0–102,000	Johnson[10]
Cr-Va electric, 0.52% C, 0.88% Cr, 0.21% Va....	0.25	O.Q. 1600 F D. 810 F	Ground and polished	Bhn 477–488	237,000	229,000	183,000	141,000	±75,000 0–128,000	Johnson[10]
High-carbon electric steel	0.25	O.Q. 1550 F D. 800 F	Ground and polished	Bhn 430–470	237,000	194,000	194,000	126,000	0–123,000	Johnson[10]
Beryllium bronze	0.25	....	....	Bhn 303	166,000	132,000	110,000	95,000	±16,000 0–30,000	Johnson[10]

[1] O.Q. = oil-quenched. T. = tempered. D. = drawn.
[2] Based on 0.2% plastic strain.
[3] Where more than one figure is listed for a given material, this means that tests were made at different stress ranges, *i.e.*, ± or completely reversed stress, 0 to maximum, or intermediate to maximum stress as indicated.
[4] These figures probably represent ideal conditions when no flaws are present.
[5] Specimen turned down to this diameter.
[6] Low value attributed by investigators to defects set up by drawing operation.
[7] *Trans. ASME.* November, 1935, p. 501.
[8] *Dept. Sci. Ind. Research (British) Eng. Res. Spec. Rep. No. 9.*
[9] *Proc. Inst. Mech. Engrs. (London)*, 1931, p. 661.
[10] *Iron Age*, Mar. 15, 1934, p. 12.

Table 18: Endurance Ranges and Physical Properties of Spring Materials in Bending (Round Specimens)

Material	Dia of specimen or wire, in.	Heat-treatment[1]	Condition of surface	Hardness	Ultimate strength (tension), psi	Yield point (tension), psi	Elongation, %	Endurance limit in reversed bending psi	Investigator
0.6% C steel..........	......	O.Q. 950 C, T. 500 C	Ground and polished	Bhn 365	177,000	161,000	7[2]	±85,000	Hankins[4]
Cr-Va spring steel......	......	O.Q. 850 C, T. 600 C	Ground and polished	Bhn 385	177,000	168,000	7[2]	±95,000	Hankins[4]
0.65% C steel..........	0.162	....	As received	....	221,000	....	.....	±76,000	Shelton, Swanger[5]
0.65% C steel..........	0.130	....	Ground and polished	....	221,000	....	.....	±126,000	Shelton, Swanger[5]
0.65% C steel..........	0.148	....	As received	....	217,000	....	.....	±65,000	Shelton, Swanger[5]
Tempered Swedish valve-spring wire	0.225	....	As received	Rockwell C 42	204,000	181,000	10.5[3]	±67,000	Weibel[6]
Tempered Swedish valve-spring wire.......	0.225	....	Rough-ground 60-grit wheel, 1 pass	Rockwell C 42	204,000	181,000	10.5[3]	±81,000	Weibel[6]
Tempered Swedish valve-spring wire.......	0.225	....	Shot-blasted	Rockwell C 42	204,000	181,000	10.5[3]	±85,000	Weibel[6]
High-carbon O.H. steel, 0.91% C, 0.38% Mn......	0.273	O.Q. 1575 F, D. 940 F	Ground and polished	Bhn 438–450	225,000	179,000	7[3]	±80,000	Johnson[7]
Cr-Va steel, 0.52% C, 0.88% Cr, 0.21% Va	......	O.Q. 1600 F, D. 810 F	Ground and polished	Bhn 477–488	237,000	229,000	11[3]	±104,000	Johnson[7]
High-carbon electric steel 1.04% C, 0.36% Mn.	......	O.Q. 1550 F, D. 800 F	Ground and polished	Bhn 430–470	237,000	194,000	5[3]	±98,000	Johnson[7]

[1]O.Q. = oil-quenched. T. = tempered. D. = drawn.
[2]In 8 in.
[3]In 2 in.
[4]*Dept. Sci. Ind. Research (British) Eng. Res. Spec. Rep. No. 9.*
[5]*J. Research Natl. Bur. Standards*, vol 14, RP 754, p. 17-32, 1935.
[6]*Trans. ASME*, November, 1935, p. 501.
[7]*Iron Age*, Mar. 15, 1934, p. 12.

Table 19: Endurance Ranges and Physical Properties of Leaf and Flat Spring Materials in Bending

Material	Thickness of specimen, in.	Heat treatment[1]	Condition of Surface	Brinell Hardness	Ultimate strength (tension), psi	Endurance limit or limiting stress range in bending[2], psi	Investigator
0.6% commercial, carbon-spring steel.....	3/8	Hardened and tempered	As received	350–370	....	0–42,000	Batson and Bradley[3]
0.6% commercial, carbon-spring steel.....	3/8	Hardened and tempered	0.062 in. machined from surface after heat-treatment	350–370	....	0–128,000	Batson and Bradley[3]
Silicomanganese steel ..	3/8	O.Q. 900 C, T. 540 C	As received	390–400	....	0–63,000	Batson and Bradley[3]
Silicomanganese steel ..	3/8	O.Q. 900 C, T. 540 C	0.062 in. machined from surface	390–400	....	0–110,000	Batson and Bradley[3]
0.6% C spring steel.....	3/8	Quenched and tempered	As rolled	....	....	±49,000	Hankins[4]
Swedish spring steel....	0.006	....	Polished	....	275,000	±130,000	Hengstenberg[5]
Carbon spring steel 1% C	1¼	Quenched and tempered	As rolled	Approx 400	....	0–86,000	Beeuwkes[5]
Carbon spring steel 0.5% C, 0.56% Mn...........	1¼	....	As rolled	....	156,000	34,000–79,000 51,000–91,000	Lehr[6]
Cr-Va spring steel......	.....	....	As rolled	....	164,000	21,000–92,000	Lehr[6]
Cr-Va spring steel......	.....	....	Polished	....	171,000	−12,000–126,000	Lehr[6]
Manganese spring steel .	.....	....	As rolled	....	188,000	30,000–84,000	Lehr[6]

[1]O.Q. = oil-quenched. T. = tempered.
[2]Where two values are given for the same material, these correspond to actual ranges used in tests.
[3]*Proc. Inst. Mech. Engrs. (London)*, vol. 120, p. 301, 1931.
[4]*Dept. Sci. Ind. Research (British) Eng. Res. Spec. Rep. No. 5.*
[5]Tests made at Westinghouse Research Laboratories.
[6]*Stahl u. Eisen*, July 7, 1932, p. 653.

(say over about 3/8 to 1/2 in. wire diameter) it is not practical to wind the springs cold. In such cases, the spring may be made from carbon- or alloy-steel bars wound hot and then heat-treated. For winding of these springs, ASTM specification A125-52 calls for heating to a temperature of 1700 F and coiling on a preheated mandrel. The springs are then allowed to cool uniformly to a black heat, after which they are heated uniformly to a temperature sufficient to refine the grain, then oil-quenched. After quenching, they are tempered by heating to a temperature below the critical for a sufficient time to yield the required hardness values.

47T). Stainless steels having a composition of about 18 per cent chromium and 8 per cent nickel are widely used for springs subject to corrosion conditions and are also of value for elevated temperature conditions.

The tensile strength of this wire is developed by cold drawing and may vary from 160,000 psi in 0.312-in.-diameter size to 320,000 psi in the 0.01-in.-diameter size. Springs may be wound cold and stress-relieved at 750 F for 15 min to 1 hr, depending on wire size.

Phosphor Bronze. This material finds its greatest use in cases where a spring with good electrical conductivity is desired; it is also used for applications where corrosion resistance is important.

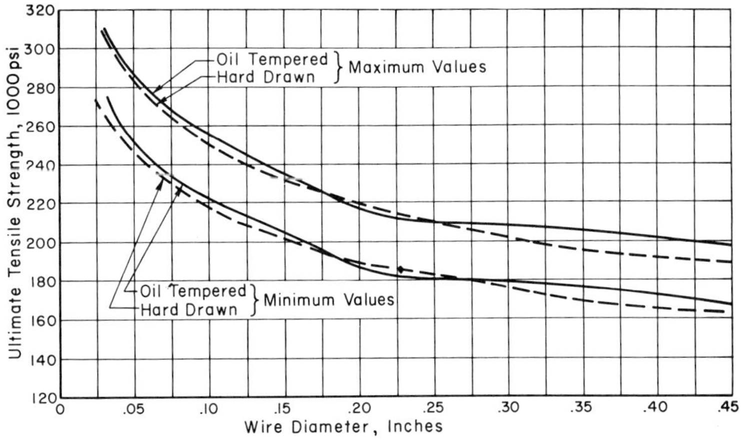

Fig. 50 Maximum and minimum values of ultimate tensile strength for hard-drawn and oil-tempered spring wire. (ASTM Specifications A227-47 and A229-41.) NOTE: ASTM tentative revision, 1951, of A229-41 shows 5000 psi higher minimum values for oil-tempered wire than those given on this curve.

Chrome-vanadium Steel Wire. In the past this alloy-steel wire has been frequently specified where a high-quality material is needed and where temperatures are somewhat higher than normal, such as is the case for automotive-valve springs. It may be obtained in either annealed or heat-treated condition. When wound from annealed wire, springs must be heat-treated after coiling. After winding from oil-tempered chrome-vanadium wire, a low-temperature heat-treatment at around 500 to 700 F should be given, the higher temperatures being preferred for applications involving elevated temperatures.

Stainless Steel (18-8) Spring Wire (ASTM: A313-

Beryllium Copper. Used frequently where high electrical conductivity is desired, this is a copper alloy containing about 2 per cent beryllium and small amounts of alloying materials. In general, wire made from it is quenched from 1475 F, then cold-drawn to increase the hardness. After coiling, it is heat-treated to increase physical properties. This heat-treatment may also be varied to change the modulus of elasticity or the amount of drift or creep.

Many other materials are also used for springs including Inconel, Inconel X, K-monel, Permanickel.[25]

REFERENCES

1. Wahl, A. M., "Mechanical Springs," Penton Publishing Company, 1944.
2. "The Mainspring," Assoc. Spring Corp., February, 1948.
3. Wahl, A. M., Stresses in Heavy Closely-coiled Helical Springs, *Trans. ASME*, 1929, paper APM 51-17, p. 191. Also ref. 1, p. 33.
4. Ross, H. F., Application of Tables for Helical Compression and Extension Spring Design, *Trans. ASME*, October, 1947, p. 727.
5. "Manual on Design and Application of Helical and Spiral Springs for Ordnance," SAE, New York, 1945.
6. "Mechanical Springs, Their Engineering and Design," Assoc. Spring Corp., 1944.
7. Zimmerli, F. P., Effects of Temperature on Coiled Steel Springs at Various Loadings, *Trans. ASME*, May, 1941, p. 363.
8. Eakin, C. T., Unpublished test data, Westinghouse Electric Corp.
9. Betty *et al.*, Relaxation Resistance of Nickel Alloy Springs, *Trans. ASME*, 1942, p. 465.
10. "Progress Report No. 3 on Heavy Helical Springs," ASME Special Research Committee on Mechanical Springs. Abstract by C. T. Edgerton, *Trans. ASME*, October, 1937, p. 609.
11. "Manual of Spring Engineering," American Steel and Wire Co., 1941.
12. Zimmerli, F. P., Permissible Stress Range for Small Helical Springs, *Univ. Mich. Eng. Research Bull.* 26, 1934.
13. "The Mainspring", *Assoc. Spring Corp.*, Feb., 1941.
14. Keysor, H. C., *Trans. ASME*, May, 1940, p. 319.
15. Haringx, J. A., On the Buckling and Lateral Rigidity of Helical Compression Springs, *Proc. Nederl, Akademie van Wetenschappen*, 1942, Nos. 6, 7. Also *Philips Research Reports* V. 3, 1948, p. 401 and V. 4, 1949, p. 49.
16. Burdick *et al., Trans. ASME*, October, 1939, p. 623.
17. Gohner, O., *Z. VDI*, vol. 76, p. 269.
18. Liesecke, *Z. VDI*, vol. 77, pp. 892, 425.
19. Kroon and Davenport, Spiral Springs with Small Number of Turns, *J. Franklin Inst.*, 1938, p. 171.
20. Discussion by D. C. Drucker, *J. Applied Mechanics*, September, 1946, p. A 251.
21. "Manual on Design and Application of Leaf Springs," *SAE*, 1944.
22. Almen and Laszlo, The Uniform Section Disk Spring, *Trans. ASME*, 1936, p. 305. Also "Manual on Design and Manufacture of Coned Disk Springs or Belleville Springs", SAE 1950.
23. Wahl, A. M., and G. Lobo, Jr. Stresses and Deflections in Flat Circular Plates with Central Holes, *Trans. ASME*, 1930, APM 52-3.
24. Gross und Lehr, "Die Federn," VDI, 1938 (Edwards Bros., Inc.).
25. F. P. Zimmerli, *Metal Progress* May and June, 1952.

INDEX TO
SECTION 13

Aircraft Tubing
Pipe

Mechanical Tubing
Pipe Fittings and Threads

continued on next page

Words or phrases, more or less descriptive of the categories of tubular products, have become associated with the variety of uses to which such products are put in design.

Pressure Tubes is the term often used to indicate that the tubes convey fluids at elevated temperatures or pressures, or both, while heat is being transferred to, or extracted from, the fluid. Boiler and superheater tubes, oil-still tubes, heat-exchanger and condenser tubes are examples.

Pressure Piping connotes tubes for conveying fluids at normal or elevated temperatures or pressures, or both, but without heat transfer.

Mechanical Tubing is used in the manufacture of machine parts where the tubular section has advantages, as for example, hollow shafts, shock absorbers, motor parts, incasements.

Pipe is named and designated in many ways, depending on materials, method of manufacture, method of installation, sponsor practice, use of, and a host of other defining characteristics. Of necessity the less frequently used materials and sizes are omitted here. So are the special sizes and shapes of tubular sections. The references grouped in Tables 13-36, 13-37, and 13-42, as well as those at the tables themselves, will serve as a source of additional matter pertaining to tubes for particular purposes, special sizes, and out-of-the-ordinary shapes.

TABLE 13–1

Cold-Finished, Round, Seamless, Carbon-Steel Mechanical Tubing (List No. 531) Sizes for Warehouse Stocks

Steel Products Manual AISI, Section 18 – 1951
Steel and Iron Wrought Products – 1953, Supply and Logistics, Standardization H-8

Wall thickness		Outside diameter in inches																												
Decimal of inch	Equivalent GA or fraction	3/8	7/16	1/2	9/16	5/8	11/16	3/4	13/16	7/8	15/16	1	1 1/16	1 1/8	1 3/16	1 1/4	1 5/16	1 3/8	1 7/16	1 1/2	1 9/16	1 5/8	1 3/4	1 7/8	2	2 1/8	2 1/4	2 3/8	2 1/2	2 5/8
.035	20 GA	.035	.035	.035	.035	.035	.035	.035		.035		.035	.035	.035	.035	.035				.035										
.049	18 GA	.049	.049	.049	.049	.049	.049	.049	.049	.049		.049		.049		.049		.049		.049		.049	.049		.049					
.058	17 GA			.058		.058		.058				.058																		
.065	16 GA	.065	.065	.065	.065	.065	.065	.065	.065	.065	.065	.065	.065	.065	.065	.065	.065	.065	.065	.065		.065	.065	.065	.065	.065	.065	.065	.065	.065
.083	14 GA		.083	.083	.083	.083	.083	.083	.083	.083		.083		.083		.083		.083		.083					.083					
.095	13 GA		.095	.095	.095	.095	.095	.095	.095	.095	.095	.095	.095	.095	.095	.095	.095	.095	.095	.095		.095	.095	.095	.095	.095	.095	.095	.095	.095
.109	12 GA			.109		.109		.109		.109		.109		.109		.109				.109										
.120	11 GA			.120	.120	.120	.120	.120	.120	.120	.120	.120	.120	.120	.120	.120	.120	.120	.120	.120	.120	.120	.120	.120	.120	.120	.120	.120	.120	.120
.134	10 GA					.134		.134				.134		.134		.134		.134		.134		.134	.134		.134					
.156	5/32					.156	.156	.156	.156	.156	.156	.156	.156	.156	.156	.156	.156	.156	.156	.156	.156	.156	.156	.156	.156	.156	.156	.156	.156	.156
.188	3/16							.188	.188	.188	.188	.188	.188	.188	.188	.188	.188	.188	.188	.188		.188	.188	.188	.188	.188	.188	.188	.188	.188
.219	7/32									.219		.219	.219	.219	.219	.219	.219	.219	.219	.219		.219	.219	.219	.219	.219	.219	.219	.219	.219
.250	1/4							.250		.250		.250	.250	.250	.250	.250	.250	.250	.250	.250	.250	.250	.250	.250	.250	.250	.250	.250	.250	.250
.281	9/32															.281	.281			.281		.281	.281	.281	.281		.281	.281	.281	.281
.313	5/16															.313	.313	.313		.313		.313	.313	.313	.313	.313	.313	.313	.313	.313
.375	3/8															.375	.375	.375		.375		.375	.375	.375	.375	.375	.375	.375	.375	.375
.438	7/16																			.438		.438	.438	.438	.438	.438	.438	.438	.438	.438
.500	1/2																			.500		.500	.500	.500	.500	.500	.500	.500	.500	.500
.625	5/8																								.625		.625	.625	.625	.625
.750	3/4																												.750	
.875	7/8																													
1.000	1																													

continued on next page

TABLE 13–1, continued

Wall thickness		Outside diameter in inches																												
Decimal of inch	Equivalent GA or fraction	2¾	2⅞	3	3⅛	3¼	3⅜	3½	3¾	4	4¼	4½	4¾	5	5¼	5½	5¾	6	6¼	6½	6¾	7	7¼	7½	8	8½	9	9½	10	10½
.035	20 GA	....	....	....	....	....	....	....	....	....	....	....	....	....	....	....	....	....	....	....	....	....	....	....	....	....	....	....	....	....
.049	18 GA	....	....	....	....	....	....	....	....	....	....	....	....	....	....	....	....	....	....	....	....	....	....	....	....	....	....	....	....	....
.058	17 GA	....	....	....	....	....	....	....	....	....	....	....	....	....	....	....	....	....	....	....	....	....	....	....	....	....	....	....	....	....
.065	16 GA	.065	.065	.065	....	.065	....	.065	....	.065	....	....	....	....	....	....	....	....	....	....	....	....	....	....	....	....	....	....	....	....
.083	14 GA	....	....	....	....	....	....	....	....	....	....	....	....	....	....	....	....	....	....	....	....	....	....	....	....	....	....	....	....	....
.095	13 GA	.095	....	.095	....	.095	....	.095	.095	.095	....	....	....	....	....	....	....	....	....	....	....	....	....	....	....	....	....	....	....	....
.109	12 GA	....	....	....	....	....	....	....	....	....	....	....	....	....	....	....	....	....	....	....	....	....	....	....	....	....	....	....	....	....
.120	11 GA	.120	.120	.120	.120	.120	.120	.120	.120	.120	.120	.120	.120	.120	.120	....	....	....	....	....	....	....	....	....	....	....	....	....	....	....
.134	10 GA	....	....	....	....	....	....	....	....	....	....	....	....	....	....	....	....	....	....	....	....	....	....	....	....	....	....	....	....	....
.156	5/32	.156	.156	.156	....	.156	....	.156	....	.156	.156	....	....	....	....	....	....	....	....	....	....	....	....	....	....	....	....	....	....	....
.188	3/16	.188	.188	.188	.188	.188	.188	.188	.188	.188	.188	.188	.188	.188	.188	.188	.188	.188	....	....	....	....	....	....	....	....	....	....	....	....
.219	7/32	.219	.219	.219	.219	....	....	.219	....	.219	....	....	....	....	....	....	....	....	....	....	....	....	....	....	....	....	....	....	....	....
.250	¼	.250	.250	.250	.250	.250	.250	.250	.250	.250	.250	.250	.250	.250	.250	.250	.250	.250	.250	.250	.250	.250	....	.250	.250	.250	.250	.250	.250	.250
.281	9/32	.281	.281	....	....	....	....	....	....	....	....	....	....	....	....	....	....	....	....	....	....	....	....	....	....	....	....	....	....	....
.313	5/16	.313	.313	.313	.313	.313	.313	.313	.313	.313	.313	.313	.313	.313	.313	.313	.313	....	....	....	....	....	....	....	....	....	....	....	....	....
.375	⅜	.375	.375	.375	.375	.375	.375	.375	.375	.375	.375	.375	.375	.375	.375	.375	.375	.375	.375	.375	.375	.375	.375	.375	.375	.375	....	....	....	....
.438	7/16	.438	.438	.438	.438	.438	.438	.438	.438	.438	.438	.438	....	....	....	....	....	....	....	....	....	....	....	....	....	....	....	....	....	....
.500	½	.500	.500	.500	.500	.500	.500	.500	.500	.500	.500	.500	.500	.500	.500	.500	.500	.500	.500	.500	.500	.500	.500	.500	.500	.500	.500	....	.500	....
.625	⅝	.625	.625	.625	.625	.625	....	.625	.625	.625	.625	.625	.625	.625	.625	.625	.625	.625	....	.625	....	....	....	....	....	....	....	....	....	....
.750	¾	.750	....	.750	....	.750	....	.750	.750	.750	.750	.750	.750	.750	.750	.750	.750	.750	.750	.750	.750	.750	....	.750	.750	.750	....	....	....	....
.875	⅞	....	....	.875	....	.875	....	.875	.875	.875	.875	.875	.875	.875	....	....	....	....	....	....	....	....	....	....	....	....	....	....	....	....
1.000	1	....	....	1.000	....	1.000	....	1.000	1.000	1.000	1.000	1.000	1.000	1.000	1.000	1.000	1.000	1.000	1.000	1.000	1.000	1.000	....	1.000	1.000	1.000	....	....	....	....

TABLE 13–2

Dimensional Tolerances for Seamless Cold Finished Mechanical Tubing, Carbon and Alloy Steel

Steel Products Manual AISI, Section 18 – 1951
Tolerances for Steel and Iron Wrought Products – 1954 Federal Standard No. 48

Size O D, in.	Variations from Diameters and Wall Thickness					
	Outside Diam., in		Inside Diam., in		Wall Thickness, per cent	
	Over	Under	Over	Under	Over	Under
3/16 to 1/2 excl.(a) (b)	0.004	0				
1/2 to 1½ excl.(a) (b) (c)	0.005	0	0	0.005	10	10
1½ to 3½ excl.(a) (b) (c)	0.010	0	0	0.010	10	10
3½ to 5½ excl.(a) (b) (c)	0.015	0	0.005	0.015	10	10
5½ to 8 excl.(c) when wall is less than 5% of O D	0.030	0.030	0.035	0.035	10	10
5½ to 8 excl. when wall is from 5% to 7.5% of O D	0.020	0.020	0.025	0.025	10	10
5½ to 8 excl.(a) when wall is over 7.5% of O D	0.030	0	0.015	0.030	10	10
8 to 10¾ incl.(c) when wall is less than 5% of O D	0.045	0.045	0.050	0.050	10	10
8 to 10¾ incl. when wall is from 5% to 7.5% of O D	0.035	0.035	0.040	0.040	10	10
8 to 10¾ incl.(a) when wall is over 7.5% of O D	0.045	0	0.015	0.040	10	10

NOTE – Requirements for sizes over 10¾ in. outside diameter are commonly negotiated between purchaser and producer.

(a) For tubes with inside diameter less than 50 per cent of outside diameter or with wall thickness more than 25 per cent of outside diameter, or with wall thickness over 1¼ in., or weighing more than 90 lb. per ft. which are difficult to draw over a stationary mandrel, the inside diameter may vary over or under by an amount equal to 10 per cent of the wall thickness. The wall thickness may vary 12½ per cent over and under that specified.

(b) For tubes with inside diameter less than ½ in. (or less than ⅝ in. when the wall thickness is more than 20 per cent of the outside diameter), which cannot be successfully drawn over a mandrel, the wall thickness may vary 15 per cent over and under that specified and the inside diameter will be governed by the outside diameter and wall thickness variations.

(c) Tubing having a wall thickness less than 3 per cent of the outside diameter cannot be straightened properly without a certain amount of distortion. Consequently such tubes, while having an average outside diameter and inside diameter within the tolerances shown in Table 13-2, will require an ovality tolerance of ½ per cent over and under the nominal outside diameter and inside diameter, this being in addition to the tolerances indicated in the table.

TABLE 13–3

Dimensional Variations in Diameter of Cold-Finished Welded, Carbon-Steel, Mechanical Tubing

Steel Products Manual AISI, Section 18 – 1951

	Wall Thickness	Flash-In Tube (1)		Standard Tube			O D Same as Standard I D Up to 55% Greater Than Standard *Nominal (2)			O D Up to 35% Closer than Standard (3) I D No Specification		O D Up to 35% Closer than Standard I D Up to 40% Closer than Standard		
Size of Tube	B.W. Gage	O D	Oval	O D	I D	Oval	O D	I D	Oval	O D	Oval	O D	I D	Oval
¼ to ⅜	16 to 22	±.003	.003	±.003	±.008	.003				±.002	.003	±.002	±.005	.003
¼ to ⅜	14	±.003	.003	±.003	±.010	.003				±.002	.003	±.002	±.007	.003
½ to ⅝	20 to 22	±.004	.005	±.004	±.005	.004				±.003	.003	±.003	±.003	.003
½ to ⅝	16 to 18	±.004	.004	±.004	±.005	.004				±.003	.003	±.002	±.002	.003
½ to ⅝	12 to 14	±.004	.004	±.004	±.009	.004				±.002	.003	±.002	±.005	.003
¾ to 1⅛	18 to 22	±.004	.007	±.004	±.005	.005				±.003	.004	±.003	±.003	.004
¾ to 1⅛	14 to 16	±.004	.007	±.004	±.005	.005				±.003	.004	±.003	±.003	.004
¾ to 1⅛	11 to 13	±.004	.005	±.004	±.008	.005				±.003	.004	±.003	±.003	.004
1¼ to 2	18 to 22	±.005	.008	±.005	±.006	.008	±.005	±.010	.008	±.004	.006	±.004	±.004	.005
1¼ to 2	14 to 16	±.005	.007	±.005	±.006	.006	±.005	±.010	.007	±.004	.005	±.003	±.003	.005
1¼ to 2	9 to 13	±.005	.006	±.005	±.008	.006	±.005	±.011	.006	±.003	.004	±.003	±.003	.004
2⅛ to 2½	18 to 20	±.006	.010	±.006	±.007	.010	±.006	±.012	.010	±.004	.007	±.004	±.004	.005
2⅛ to 2½	14 to 16	±.006	.008	±.006	±.007	.008	±.006	±.012	.008	±.004	.007	±.004	±.004	.005
2⅛ to 2½	9 to 13	±.006	.008	±.006	±.009	.008	±.006	±.014	.008	±.004	.005	±.004	±.004	.004
2⅝ to 3	18 to 20	±.010	.020	±.010	±.012	.020	±.010	±.015	.020	±.007	.010			
2⅝ to 3	14 to 16	±.008	.015	±.008	±.010	.015	±.008	±.014	.015	±.006	.008	±.006	±.006	.008
2⅝ to 3	9 to 13	±.008	.012	±.008	±.010	.012	±.008	±.016	.012	±.006	.008	±.006	±.006	.008
3⅛ to 3½	16 to 18	±.010	.018	±.010	±.012	.016	±.010	±.017	.018	±.008	.012			
3⅛ to 3½	9 to 14	±.008	.014	±.008	±.012	.014	±.008	±.016	.014	±.006	.010			
3⅝ to 4	14 to 16	±.010	.020	±.010	±.014	.018	±.010	±.016	.020	±.008	.014			
3⅝ to 4	8 to 13	±.010	.014	±.010	±.016	.014	±.010	±.023	.014	±.007	.008			
4⅛ to 5	14 to 16	±.020	.025	±.020	±.020	.025	±.020	±.028	.025					
4⅛ to 5	8 to 13	±.015	.020	±.015	±.018	.020	±.015	±.025	.020					
5⅛ to 5½	10 to 16	±.020	.025				±.020	±.032	.025					

* The actual inside diameter variations are computed by using the outside diameter and wall thickness tolerances shown in Tables 13-7 and 13-8.

HEIGHT OF I D FLASH

(1) *Flash-In Tube:* The maximum height of the inside welding flash does not customarily exceed the wall thickness or in any case ⅛ in.

(2) *O D Same as Std.; I D 55% Greater than Std.:* The maximum height of the inside welding flash is commonly 0.010.

(3) *O D to 35% Closer than Std.; I D No Specifications:* The maximum height of the inside welding flash for flash-in-grade does not customarily exceed the wall thickness, or in any case ⅛ in. If flash removed grade is specified, the maximum height of the inside welding flash is commonly 0.010.

TABLE 13–4

Wall-Thickness Variations in Electric-Welded, Cold Rolled Carbon-Steel Mechanical Tubing

Steel Products Manual AISI, Section 18 – 1951
Tolerances for Steel and Iron Wrought Products – 1954 Federal Standard No. 48

Wall Thickness B.W. Gage	Outside Diameter of Tubes				
	3/8—7/8	1—1 7/8	2—3 3/4	3 7/8—5	5 1/8—5 1/2
22 (.028)	+.000 —.003	+.000 —.003			
20 (.035)	+.000 —.004	+.000 —.004	+.000 —.004		
18 (.049)	+.000 —.004	+.000 —.005	+.000 —.005		
16 (.065)	+.000 —.004	+.000 —.005	+.000 —.005	+.002 —.006	+.002 —.007
14 (.083)	+.000 —.004	+.000 —.005	+.000 —.006	+.002 —.007	+.002 —.007
13 (.095)	+.000 —.004	+.000 —.005	+.000 —.006	+.002 —.007	+.002 —.007
12 (.109)		+.000 —.006	+.000 —.006	+.003 —.007	+.003 —.009
11 (.120)		+.000 —.006	+.000 —.006	+.003 —.007	+.003 —.009
10 (.134)		+.000 —.006	+.000 —.006	+.003 —.007	+.003 —.009
9 (.148)		+.000 —.006	+.000 —.006	+.003 —.007	+.003 —.009
8 (.165)		+.000 —.007	+.000 —.007	+.003 —.008	+.003 —.010

The following additional plus tolerances apply to the plus limits shown, due to the crown on single width strip or wider variations on multiple slit strip:

3/8 to 1 7/8 in. O D	.025 to .065 = .0015
	.065 to .187 = .002
Over 1 7/8 to 3 3/4 in. O D	.025 to .065 = .002
	.065 to .187 = .0025

TABLE 13–5

Dimensional Tolerances for Ground Seamless, Carbon-Steel, Mechanical Tubing

Steel Products Manual AISI, Section 18 – 1951

Size, O D in.	Diameter Variations for Sizes and Lengths Given, in.			
	Over	Under	Over	Under
Lengths:	Up to 16 ft. incl.		Over 16 ft.	
Up to 1¼ incl.	0.003	0.000	0.004	0.000
Over 1¼ to 2 incl.	0.005	0.000	0.006	0.000
Lengths:	Up to 12 ft. incl.		Over 12 ft. to 16 ft.	
Over 2 to 3 incl.	0.005	0.000	0.006	0.000
Over 3 to 4 incl.	0.006	0.000	0.008	0.000

Note—The wall thickness and inside diameter tolerances are the same as for cold drawn mechanical tubing shown in Table 13-2.

TABLE 13–6

Dimensional Tolerances for Hot-Finished, Seamless, Carbon-Steel, Mechanical Tubing

Steel Products Manual, Section 18 – 1951

Specified Size, Outside Diameter, in.	Ratio of Wall Thickness to Outside Diameter	Variations from Diameter and Wall Thickness									
		Outside Diameter, in.		Wall Thickness, per cent							
				.109" and under		Over .109" to .172" incl.		Over .172" to .203" incl.		Over .203"	
		Over	Under	Over	Under	Over	Under	Over	Under	Over	Under
Under 3	All wall thicknesses	0.023	0.023	16.5	16.5	15	15	14	14	12.5	12.5
3 to 5½ excl.	All wall thicknesses	.031	.031	16.5	16.5	15	15	14	14	12.5	12.5
5½ to 8 excl.	All wall thicknesses	.047	.047					14	14	12.5	12.5
8 to 10¾ incl.	5% and over	.047	.047							12.5	12.5
8 to 10¾ incl.	Under 5%	.063	.063							12.5	12.5

NOTE: The common range of sizes of hot finished tubes is 1½ in. to and including 10¾ in. outside diameter with wall thickness not less than 0.095 in. (No. 13 BWG) or 3 per cent or more of the outside diameter. For sizes under 1½ in. or over 10¾ in. outside diameter the variations are negotiated between the purchaser and producer.

TABLE 13–7

Dimensional Variations in Diameter of Hot-Rolled Welded, Carbon-Steel, Mechanical Tubing

Steel Products Manual AISI, Section 18 – 1951

	Wall Thickness	Flash-In Tube (1)		Standard Tube			O D Same as Standard I D Up to 55% Greater than Standard (2)			O D Up to 35% Closer than Standard I D No Specification (3)	
								*Nominal			
Size of Tube	B.W. Gage	O D	Oval	O D	I D	Oval	O D	I D	Oval	O D	Oval
¼ to ⅜	16 to 22	±.003	.003	±.003		.003				±.002	.003
¼ to ⅜	14	±.003	.003	±.003		.003				±.002	.003
½ to ⅝	20 to 22	±.004	.005	±.004		.004				±.003	.003
½ to ⅝	16 to 18	±.004	.004	±.004		.004				±.003	.003
½ to ⅝	12 to 14	±.004	.004	±.004		.004				±.002	.003
¾ to 1⅛	18 to 22	±.004	.007	±.004	±.007	.005				±.003	.004
¾ to 1⅛	14 to 16	±.004	.007	±.004	±.012	.005				±.003	.004
¾ to 1⅛	11 to 13	±.004	.005	±.004	±.012	.005				±.003	.004
1¼ to 2	18 to 22	±.005	.008	±.005	±.008	.008	±.005	±.019	.008	±.004	.006
1¼ to 2	14 to 16	±.005	.007	±.005	±.010	.006	±.005	±.017	.007	±.004	.005
1¼ to 2	7 to 13	±.005	.006	±.005	±.015	.006	±.005	±.023	.006	±.003	.004
2⅛ to 2½	18 to 20	±.006	.010	±.006	±.010	.010	±.006	±.014	.010	±.004	.007
2⅛ to 2½	14 to 16	±.006	.008	±.006	±.012	.008	±.006	±.019	.008	±.004	.007
2⅛ to 2½	6 to 13	±.006	.008	±.006	±.016	.008	±.006	±.024	.008	±.004	.005
2⅝ to 3	18 to 20	±.010	.020	±.010	±.014	.020	±.010	±.022	.020	±.007	.010
2⅝ to 3	14 to 16	±.008	.015	±.008	±.012	.015	±.008	±.019	.015	±.006	.008
2⅝ to 3	4 to 13	±.008	.012	±.008	±.016	.012	±.008	±.024	.012	±.006	.008
3⅛ to 3½	16 to 18	±.010	.018	±.010	±.015	.016	±.010	±.023	.018	±.008	.012
3⅛ to 3½	4 to 14	±.008	.014	±.008	±.020	.014	±.008	±.028	.014	±.006	.010
3⅝ to 4	14 to 16	±.010	.020	±.010	±.017	.018	±.010	±.026	.020	±.008	.014
3⅝ to 4	4 to 13	±.010	.014	±.010	±.021	.014	±.010	±.029	.014	±.007	.008
4⅛ to 5	14 to 16	±.020	.025	±.020	±.022	.025	±.020	±.034	.025		
4⅛ to 5	9 to 13	±.015	.020	±.015	±.026	.020	±.015	±.040	.020		
5⅛ to 5½	10 to 16	±.020	.025				±.020	±.032	.025		

*The actual inside diameter variations are computed by using the outside diameter and wall thickness tolerances shown in Tables 13-7 and 13-8.

HEIGHT OF I D FLASH

(1) *Flash-in Tube:* The maximum height of the inside welding flash does not customarily exceed the wall thickness or in any case 1/8 in.

(2) *O D Same as Std.; I D 55% Greater than Std.:* The maximum height of the inside welding flash is commonly 0.010.

(3) *O D to 35% Closer than St.; I D No Specifications:* The maximum height of the inside welding flash for flash-in grade does not customarily exceed the wall thickness, or in any case 1/8 in. If flash remove grade is specified, the maximum height of the inside welding flash is commonly 0.010.

TABLE 13–8

Wall Thickness Variations in Hot-Rolled, Electric Welded, Carbon-Steel, Mechanical Tubing

Steel Products Manual AISI, Section 18 – 1951
Tolerances for Steel and Iron Wrought Products – 1954 Federal Standard No. 48

Wall Thickness B. W. Gage	Outside Diameter of Tube					
	5/8—1	1 1/8—1 3/4	1 7/8—3 1/4	3 3/8—3 3/4	4—4 1/2	4 3/4—5 1/2
			Wall Thickness Variation			
20 (.035)	+.002 —.004					
18 (.049)	+.002 —.004	+.002 —.004				
16 (.065)	+.002 —.006	+.002 —.008	+.002 —.008	+.002 —.008	+.002 —.010	+.002 —.010
14 (.083)	+.002 —.006	+.002 —.008	+.002 —.008	+.002 —.008	+.003 —.010	+.002 —.012
13 (.095)	+.002 —.006	+.002 —.008	+.002 —.008	+.002 —.008	+.003 —.010	+.002 —.012
12 (.109)	+.002 —.006	+.002 —.008	+.002 —.008	+.002 —.008	+.005 —.010	+.004 —.012
11 (.120)	+.002 —.008	+.002 —.008	+.002 —.008	+.002 —.008	+.005 —.010	+.004 —.012
10 (.134)		+.002 —.008	+.002 —.008	+.002 —.008	+.005 —.010	+.004 —.012
9 (.148)		+.002 —.008	+.002 —.008	+.002 —.008	+.005 —.011	+.005 —.012
8 (.165)		+.002 —.008	+.002 —.008	+.002 —.008	+.005 —.011	+.005 —.012
7 (.180)		+.002 —.008	+.002 —.008	+.002 —.008	+.005 —.011	
6 (.203)			+.002 —.010	+.002 —.010	+.005 —.012	
5 (.220)			+.002 —.010	+.002 —.010	+.005 —.012	
4 (.238)			+.002 —.010	+.002 —.010	+.005 —.012	

The following additional plus tolerances apply to the plus limits shown due to the crown on single width strip or wider variations on multiple slit strip:

5/8 to 1 in. O D	.002
Over 1 to 1 3/4 in. O D	.003
Over 1 3/4 to 3 3/4 in. O D	.004

TABLE 13–9

Out-of-Straightness Tolerances for Round, Carbon-Steel, Mechanical Tubing

Steel Products Manual AISI, Section 18 – 1951
Tolerances for Steel and Iron Wrought Products – 1954 Federal Standard No. 48

Size Limits	Max. Curvature in any 3 Feet	Maximum Curvature in Total Lengths	Maximum Curvature for Lengths under 3 Feet
O D 5″ and smaller. Wall thickness, over 3% of O D, but not over 0.5″	0.030″	$0.030'' \times \dfrac{\text{Number of feet of length}}{3}$	Ratio of 0.010″ per foot
O D over 5″ to 8″, incl. Wall thickness, over 4% of O D, but not over 0.75″	0.045″	$0.045'' \times \dfrac{\text{Number of feet of length}}{3}$	Ratio of 0.015″ per foot
O D over 8″ to 10¾″, incl. Wall thickness, over 4% of O D, but not over 1″	0.060″	$0.060'' \times \dfrac{\text{Number of feet of length}}{3}$	Ratio of 0.020″ per foot

TABLE 13–10

Cleanup or Machining Allowances for Carbon and Alloy Steel, Seamless Mechanical Tubing

Steel Products Manual AISI, Section 18 – 1951

For Machined Parts Size, O D inches	Machining Allowances on Diameter, inch	
	O D	I D
Less than 3/32	0.008	0.008
3/32 to 3/16, excl.	0.012	0.012
3/16 to 1/2, excl.	0.015	0.015
1/2 to 1½, excl.	0.020	0.020
1½ to 3, excl.	0.040	0.040
3 to 5½, excl.	0.060	0.060
5½ to 8, excl.	0.080	0.080

NOTE: Machining allowances for sizes 8 inches and over are customarily negotiated between purchaser and producer.

When tubing is finished to size by grinding instead of machining, the cleanup grinding allowances are shown in Table 13-11.

TABLE 13–11

Cleanup Grinding Allowances for Carbon and Alloy Steel, Seamless Mechanical Tubing

Steel Products Manual AISI, Section 18 – 1951

Size: O D inch	Grinding Allowances on Diameter, inch	
	O D	I D
Less than 3/32	0.005	0.005
3/32 to 3/16, excl.	0.006	0.006
3/16 to 1/2 , excl.	0.008	0.008

TABLE 13–12

Stainless Steel – Wall Thickness Variations on Welded, Mechanical and Aircraft Tubing

Steel Products Manual AISI, Section 18 – 1951

Wall Thickness B. W. Gage	SIZE OF TUBE, IN INCH				
	Up to 3/16	¼—⅞	1—1⅞	2—3¾	4—6
	Wall Thickness Variation				
23 (.025)	+0.000 —0.0025	+0.000 —0.003	+0.002 —0.003		
22 (.028)	+0.000 —0.0025	+0.002 —0.003	+0.002 —0.003		
21 (.032)	+0.000 —0.0025	+0.002 —0.003	+0.003 —0.003	+0.004 —0.003	
20 (.035)	+0.001 —0.003	+0.003 —0.003	+0.004 —0.003	+0.005 —0.003	+0.006 —0.003
19 (.042)	+0.000 —0.004	+0.000 —0.006	+0.001 —0.006	+0.001 —0.007	+0.004 —0.007
18 (.049)	+0.000 —0.004	+0.000 —0.006	+0.001 —0.006	+0.001 —0.007	+0.004 —0.007
17 (.058)		+0.000 —0.006	+0.002 —0.006	+0.001 —0.007	+0.004 —0.007
16 (.065)		+0.000 —0.006	+0.002 —0.006	+0.002 —0.006	+0.004 —0.007
15 (.072)		+0.000 —0.006	+0.003 —0.006	+0.002 —0.007	+0.004 —0.007
14 (.083)		+0.000 —0.006	+0.003 —0.006	+0.001 —0.008	+0.001 —0.010
13 (.095)		+0.000 —0.006	+0.003 —0.006	+0.001 —0.008	+0.001 —0.010
12 (.095)		+0.000 —0.006	+0.003 —0.006	+0.003 —0.008	+0.003 —0.010
11 (.120)		+0.000 —0.006	+0.003 —0.006	+0.003 —0.008	+0.003 —0.010
10 (.134)				+0.003 —0.008	+0.003 —0.010

TABLE 13–13

Annealed Stainless Steel – Dimensional Tolerances on Welded, Mechanical and Aircraft Tubing

Steel Products Manual AISI, Section 18 – 1951
Tolerances for Steel and Iron Wrought Products – 1954 Federal Standard No. 48

Dimensions		Standard Tube I D Specified			O D Up to 25 Per Cent Closer than Standard I D No Specification		O D Up to 25 Per Cent Closer than Standard I D Up to 40 Per Cent Closer than Standard		
		Variation in inch			Variation in inch		Variation in inch		
Size of Tube, in.	Wall Thickness B. W. Gage	O D	I D	Ovality	O D	Ovality	O D	I D	Ovality
Up to 3/16	20 Ga. & lighter	0.002	0.003	0.003	0.0015	0.002	0.0015	0.0015	0.002
¼	18-24 incl.	0.003	0.005	0.004	0.002	0.003	0.002	0.002	0.003
Over ¼— ½ incl.	13-23 "	0.003	0.005	0.004	0.002	0.003	0.002	0.003	0.003
" ½— ⅞ "	16-23 "	0.004	0.006	0.007	0.003	0.005	0.003	0.004	0.005
" ½— ⅞ "	12-15 "	0.004	0.006	0.005	0.003	0.004	0.003	0.004	0.004
" ⅞—1½ "	17-23 "	0.005	0.007	0.009	0.004	0.006	0.004	0.004	0.006
" ⅞—1½ "	10-16 "	0.005	0.007	0.007	0.004	0.005	0.004	0.004	0.005
" 1½—2 "	17-23 "	0.006	0.008	0.010	0.004	0.007	0.004	0.004	0.007
" 1½—2 "	10-16 "	0.006	0.008	0.008	0.004	0.006	0.004	0.004	0.006
" 2 —2½ "	15-21 "	0.007	0.010	0.012	0.005	0.009	0.005	0.005	0.009
" 2 —2½ "	10-14 "	0.007	0.010	0.010	0.005	0.007	0.005	0.005	0.007
" 2½—3 "	15-20 "	0.010	0.012	0.018	0.008	0.014	0.007	0.007	0.014
" 2½—3 "	10-14 "	0.010	0.012	0.015	0.008	0.012	0.007	0.007	0.012
" 3 —4 "	14-18 "	0.012	0.014	0.024	0.010	0.018			
" 3 —4 "	10-13 "	0.012	0.014	0.020	0.010	0.015			
" 4 —5 "	10-16 "	0.016	0.020	0.030	0.012	0.020			
" 5 —6 "	10-16 "	0.020	0.030	0.035					

NOTE 1. Tubing is ordinarily specified to outside diameter and wall thickness. If inside diameter is the more important dimension, tubing should be specified to O D and I D dimensions. Tolerances in any tube are applicable only to two cross-sectional dimensions. Thus, if outside diameter and wall thickness are specified, the inside diameter may not necessarily conform to these tolerances, and if inside and outside diameters are specified, the wall thicknesses may not necessarily conform to these tolerances.

NOTE 2. All tolerances, except ovality, as shown above are plus and minus.

TABLE 13–14

Annealed Stainless Steel – Dimensional Tolerances on Seamless, Cold-Finished Mechanical Tubing

Steel Products Manual AISI, Section 18 – 1951
Tolerances for Steel and Iron Wrought Products – 1954 Federal Standard No. 48

O D Size, in.	Variations from Outside Diameter, in.		Ovality, Double O D Tolerance when wall is:	Wall Thickness, per cent See Notes a, b	
	Over	Under		Over	Under
Under 1/2	0.005	0.005	Lighter than 0.015″	15	15
1/2 to 1½ excl.	0.005	0.005	Lighter than 0.065″	10	10
1½ to 3½ excl.	0.010	0.010	Lighter than 0.095″	10	10
3½ to 5½ excl.	0.015	0.015	Lighter than 0.150″	10	10
5½ to 8 excl.	0.030	0.030	Lighter than 0.240″	10	10

(a) Tubes with wall thicknesses more than 25 per cent of their outside diameter or with wall thicknesses greater than 1¼ in. or weighing more than 90 pounds per foot may vary in wall thickness plus and minus 12½ per cent.

(b) For tubes with inside diameter less than 1/2 in. (or less than 5/8 in. when the wall thickness is more than 20 per cent of the outside diameter), which are difficult to draw over a stationary mandrel, the wall thickness may vary 15 per cent over or under that specified and the inside diameter will be governed by the outside diameter and wall thickness variations.

TABLE 13-15

Aluminum-Alloy, Drawn, Seamless Tubes Permissible Variations in Dimensions

ASTM Designation: B210-53

PERMISSIBLE VARIATIONS IN DIAMETER (APPLIES TO OUTSIDE DIAMETER UNLESS OTHERWISE SPECIFIED)

Nominal Diameter, in.	Permissible Variations in Diameter, plus or minus, in.		
	Mean Diameter[a] or Pi-Tape Measurement—Alloys M1A, G1A, GR20A, CG 42A, GS11A	Individual Measurement of Diameter (Out-of-Roundness) Except Soft or Thin Wall Tubes[b]	
		Alloys G1A, GR 20A, M1A	Alloys CG42A, GS 11A
1/8 to 1/2, incl	0.003	0.003	0.006
Over 1/2 to 1, incl	0.004	0.004	0.008
Over 1 to 2, incl	0.005	0.005	0.010
Over 2 to 3, incl	0.006	0.006	0.012
Over 3 to 5, incl	0.008	0.008	0.016
Over 5 to 6, incl	0.010	0.010	0.020
Over 6 to 8, incl	0.015	0.015	0.030
Over 8 to 10, incl	0.020	0.020	0.040
Over 10 to 12, incl	0.025	0.025	0.050

[a] Mean diameter is the average of any two measurements of diameter taken at right angles to each other at any point along the length of the tube.

[b] Thin wall tubes, that is tubes having a wall thickness less than 2.5 per cent of the diameter or less than 0.020 in., and tubes in the soft temper shall be commercially round. The deviations of individual measurements from the nominal will vary with the alloy and the ratio of wall thickness to diameter.

PERMISSIBLE VARIATIONS IN WIDTH OR DEPTH OF SQUARE, RECTANGULAR, HEXAGONAL AND OCTAGONAL TUBES

Nominal Width or Depth, in.[a]	Permissible Variations in Width or Depth, plus or minus, in.		
	At Corners	Not at Corners	
		Square, Hexagonal, and Octagonal	Rectangular
0.500 and less	0.003	0.006	The permissible variation in width is the value from column 3 which corresponds to the depth dimension, provided that this value is not less than the permissible variation in width when measured at the corners. The permissible variation in depth is the value from column 3 corresponding to the width dimension, provided that this value is not less than the permissible variation in depth when measured at the corners.
0.501 to 1.00	0.004	0.008	
1.01 to 2.00	0.005	0.010	
2.01 to 3.00	0.006	0.012	
3.01 to 5.00	0.008	0.016	
5.01 to 6.00	0.010	0.020	
6.01 to 8.00	0.015	0.030	
8.01 to 10.00	0.020	0.040	
Over 10.00	0.025	0.050	

[a] Intermediate widths or depths shall be rounded off to the third decimal place, if the dimension is less than 1.00 in. and the second, if the dimension is 1.00 in. or more, in accordance with the Recommended Practices for Designating Significant Places in Specified Limiting Values (A.S.T.M. Designation E 29).

PERMISSIBLE VARIATIONS IN WALL THICKNESS

Nominal Wall Thickness, in.[b]	Permissible Variations in Wall thickness, plus or minus, in.			
	Mean[a]	Individual Measurements		
	Alloys GS11A CG42A, Round Only	Alloys G1A, GR20A, M1A, Round Only	Alloys GS11A, CG42A Round Only	All Alloys Other than Round
0.010 to 0.035	0.002	0.002	10 per cent of nominal wall thickness, but not less than 0.003 in.	
0.036 to 0.049	0.003	0.003		
0.050 to 0.120	0.004	0.004		
0.121 to 0.203	0.005	0.005		
0.204 to 0.300	0.008	0.008		
0.301 to 0.375	0.012	0.012		
0.376 to 0.500	0.032	0.032		

[a] Mean wall thickness is the average of the two measurements of wall thickness taken at opposite ends of any diameter.

[b] Intermediate wall thicknesses shall be rounded off to the third decimal place in accordance with the Recommended Practices for Designating Significant Places in Specified Limiting Values (A.S.T.M. Designation E 29).

TABLE 13–16

Aluminum-Alloy Extruded Tubes
Permissible Variation in Dimensions

ASTM Designation: B235–53T

PERMISSIBLE VARIATIONS IN DIAMETER FOR ROUND TUBES

Specified Outside or Inside Diameter, in.	Permissible Variations in Diameter, plus or minus, in.	
	Mean Diameter[a] from Specified Diameter (Size)	Diameter at Any Point from Specified Diameter[b] (Ovalness)
1/2 to 1, excl	0.010	0.020
1 to 2, excl	0.012	0.025
2 to 4, excl	0.015	0.030
4 to 6, excl	0.025	0.050
6 to 8, excl	0.035	0.075
8 to 10, excl	0.045	0.100
10 to 12, excl	0.055	0.125
12 to 12 1/4, incl	0.065	0.150

[a] The "mean diameter" is determined by the average of two measurements taken at right angles to each other.
[b] Not applicable in the annealed (O) temper or if wall thickness is less than 2 1/2 per cent of the outside diameter.

PERMISSIBLE VARIATIONS IN WIDTH OR DEPTH FOR SQUARE RECTANGULAR, HEXAGONAL, AND OCTAGONAL TUBES

Specified Width or Depth, in.	Permissible Variations in Width or Depth, plus or minus, in.		
	Width or Depth at Corners from Specified-Width or Depth	Width or Depth not at Corners from Specified Width or Depth	
		Square, Hexagonal, and Octagonal	Rectangular
1/2 to 3/4, excl	0.012	0.020	The permissible variation for the width is the value for square, hexagonal, and octagonal tubing for a dimension equal to the depth, and conversely, but in no case is the permissible variation less than at the corners.[a]
3/4 to 1, excl	0.014	0.020	
1 to 2, excl	0.018	0.025	
2 to 4, excl	0.025	0.035	
4 to 5, incl	0.035	0.045	

[a] Example: The width permissible variation of 1 by 3 in. rectangular tubing is plus or minus 0.025 in. and the depth permissible variation is plus or minus 0.035 in.

PERMISSIBLE VARIATIONS IN WALL THICKNESS FOR ROUND TUBES

Specified Wall Thickness, in.	Permissible Variations in Wall Thickness, plus or minus, in.			
	Mean Wall Thickness[a] from Specified Wall Thickness			Wall Thickness at Any Point from Mean Wall Thickness[a] (Eccentricity)
	Outside Diameter, in.			
	Under 3	3 to 5, excl	5 and over	
Under 0.063	0.007	0.008	0.010	10 per cent of the mean wall thickness: maximum of 0.060 minimum of 0.010
0.063 to 0.125, excl	0.008	0.010	0.015	
0.125 to 0.250, excl	0.009	0.013	0.020	
0.250 to 0.375, excl	0.011	0.016	0.025	
0.375 to 0.500, excl	0.015	0.021	0.035	
0.500 to 0.750, excl	0.020	0.028	0.045	
0.750 to 1.000, excl	...	0.035	0.055	
1.000 to 1.500, excl	...	0.045	0.065	

[a] The "mean wall thickness" is determined by the average of two measurements taken opposite each other.

PERMISSIBLE VARIATIONS IN WALL THICKNESS FOR SQUARE, RECTANGULAR, HEXAGONAL, AND OCTAGONAL TUBES

Permissible variations of wall thickness at any point from specified wall thickness shall not be greater than plus or minus 10 per cent of the specified wall thickness: maximum 0.060 in., minimum 0.010 in.

TABLE 13-17

Preferred Sizes of Round, Seamless Brass and Copper Tubing

Simplified Practice Recommendations R235-48, Sup't. of Documents, Gov't. Printing Office

ASTM Designations: B135–52, B75–52, and B251–53T

PREFERRED SIZES OF ROUND, SEAMLESS COPPER TUBE,[a,b]

[X indicates the preferred sizes]

Outside Diameter In.	Wall Thicknesses, In.																		
	0.010	0.013	0.016	0.020	0.025	0.032	0.040	0.049	0.065	0.083	0.100	1/8	5/32	3/16	1/4	5/16	3/8	1/2	5/8
1/8	x	x	x	x	x	x	x	..	..	..	..	..	..	..	..	..	..	..	..
3/16	x	x	x	x	x	x	x	x	..	..	..	..	..	..	..	..	..	..	..
1/4	..	..	x	x	x	x	x	x	x	x	..	..	..	..	..	..	..	..	..
5/16	..	..	x	x	x	x	x	x	x	x	x	..	..	..	..	..	..	..	..
3/8	..	..	x	x	x	x	x	x	x	x	x	x	..	..	..	..	..	..	..
1/2	..	..	x	x	x	x	x	x	x	x	x	x	x	..	..	..	..	..	..
5/8	..	..	x	x	x	x	x	x	x	x	x	x	x	..	..	..	..	..	..
3/4	..	..	x	x	x	x	x	x	x	x	x	x	x	x	..	..	..	..	..
7/8	..	..	x	x	x	x	x	x	x	x	x	x	x	x	..	..	..	..	..
1	..	..	..	x	x	x	x	x	x	x	x	x	x	x	x	..	..	..	..
1 1/4	..	..	..	x	x	x	x	x	x	x	x	x	x	x	x	x	..	..	..
1 1/2	..	..	..	x	x	x	x	x	x	x	x	x	x	x	x	x	x	..	..
1 3/4	..	..	..	x	x	x	x	x	x	x	x	x	x	x	x	x	x	..	..
2	..	..	..	..	x	x	x	x	x	x	x	x	x	x	x	x	x	x	..
2 1/4	..	..	..	..	x	x	x	x	x	x	x	x	x	x	x	x	x	x	..
2 1/2	..	..	..	..	x	x	x	x	x	x	x	x	x	x	x	x	x	x	x
2 3/4	..	..	..	..	x	x	x	x	x	x	x	x	x	x	x	x	x	x	x
3	..	..	..	..	..	x	x	x	x	x	x	x	x	x	x	x	x	x	x
3 1/2	..	..	..	..	..	x	x	x	x	x	x	x	x	x	x	x	x	x	x
4	..	..	..	..	..	..	x	x	x	x	x	x	x	x	x	x	x	x	x
4 1/2	..	..	..	..	..	..	x	x	x	x	x	x	x	x	x	x	x	x	x
5	..	..	..	..	..	..	..	x	x	x	x	x	x	x	x	x	x	x	x
5 1/2	..	..	..	..	..	..	..	x	x	x	x	x	x	x	x	x	x	x	x
6	..	..	..	..	..	..	..	..	x	x	x	x	x	x	x	x	x	x	x
7	..	..	..	..	..	..	..	..	x	x	x	x	x	x	x	x	x	x	x
8	..	..	..	..	..	..	..	..	..	x	x	x	x	x	x	x	x	x	x
9	..	..	..	..	..	..	..	..	..	x	x	x	x	x	x	x	x	x	x
10	..	..	..	..	..	..	..	..	..	..	x	x	x	x	x	x	x	x	x
11	..	..	..	..	..	..	..	..	..	..	x	x	x	x	x	x	x	x	x
12	..	..	..	..	..	..	..	..	..	..	..	x	x	x	x	x	x	x	..

[a] In conformance with the Simplified Practice Recommendations R 235-48 for Copper and Copper-Alloy Round Seamless Tube issued by the U. S. Department of Commerce.

[b] This tube is not necessarily available in all alloys in the full range of sizes shown.

TABLE 13-18

Wall Thickness Tolerances* for Copper and Copper-Alloy Tube
(Not Applicable to Pipe,† Water Tube, and Type B Sizes.)

ASTM Designation: B251–53T (Applicable to ASTM Designations B68, B75 and B135)

Wall Thickness in.	Outside Diameter in.						
	1/32 to 1/8 Incl.	Over 1/8 to 5/8 Incl.	Over 5/8 to 1 Incl.	Over 1 to 2 Incl.	Over 2 to 4 Incl.	Over 4 to 7 Incl.	Over 7 to 10 Incl.
Under 0.018	0.002	0.001	0.0015	0.002			
0.018, incl., to 0.025	0.003	0.002	0.002	0.0025			
0.025, incl., to 0.035	0.003	0.0025	0.0025	0.003	0.004		
0.035, incl., to 0.058	0.003	0.003	0.0035	0.0035	0.005	0.007	
0.058, incl., to 0.083		0.0035	0.004	0.004	0.006	0.008	0.010
0.083, incl., to 0.120		0.004	0.005	0.005	0.007	0.009	0.011
0.120, incl., to 0.165		0.005	0.006	0.006	0.008	0.010	0.012
0.165, incl., to 0.220		0.007	0.0075	0.008	0.010	0.012	0.014
0.220, incl., to 0.284			0.009	0.010	0.012	0.014	0.016
0.284, incl., to 0.380			0.011	0.012	0.014	0.016	0.018
0.380 and over				5%	5%	6%	6%

*If tolerances of all plus or all minus are desired, double the values given.
†Tables 13–36, 13–37.

TABLE 13-19

Average Diameter Tolerances for Copper and Copper-Alloy Tube

ASTM Designation: B251–53T (Applicable to ASTM Designations B68, B75 and B135).

Specified Diameter in.	Diameter to Which Tolerance Applies	Tolerance, Plus and Minus
Up to ⅛, incl.	inside or outside	0.002
Over 1/8 to 5/8 incl.	inside or outside	0.002
Over 5/8 to 1 incl.	Inside or outside	0.0025
Over 1 to 2 incl.	inside or outside	0.003
Over 2 to 3 incl.	inside or outside	0.004
Over 3 to 4 incl.	inside or outside	0.005
Over 4 to 5 incl.	inside or outside	0.006
Over 5 to 6 incl.	inside or outside	0.007
Over 6 to 8 incl.	inside or outside	0.008
Over 8 to 10 incl.	inside or outside	0.010

TABLE 13-20

Permissible Variations on Diameter for Magnesium-Base Alloy, Extruded, Round Tubes

ASTM Designation: B217-53T

Specified Diameter, In.*	Permissible Variation, Plus or Minus: Deviation of Mean Diameter† from Specified Diameter	Permissible Variation, Plus or Minus: Deviation of Diameter at Any Point from Specified Diameter‡
0.999 and under	0.010	0.020
1.000 to 1.999	0.012	0.025
2.000 to 3.999	0.015	0.030
4.000 to 5.999	0.025	0.050
6.000 to 7.999	0.037	0.075
8.000 to 9.999	0.045	0.100

*Intermediate diameters shall be rounded off to the third decimal place.

†The "mean diameter" is determined by the average of two measurements taken at right angles to each other.

‡Not applicable if the wall thickness is less than 2.5 per cent of the outside diameter.

TABLE 13-21

Permissible Variations in Wall Thickness, Plus or Minus, of Magnesium-Base Alloy, Extruded, Round Tubes

ASTM Destination: B217-53T

Specified Thickness, In.*	Deviation of Mean Wall Thickness† from Specified Wall Thickness: Outside Diameter 2.99 and Under	Outside Diameter 3 to 4.99	Outside Diameter 5 and Over	Deviation of Wall Thickness at Any Point from Specified Wall Thickness
0.062 and under	0.007	0.008	0.010	10 per cent of the mean wall thickness† with a maximum of 0.060 and a minimum of 0.010
0.063 to 0.124	0.008	0.010	0.015	
0.125 to 0.249	0.009	0.013	0.020	
0.250 to 0.374	0.011	0.016	0.025	
0.375 to 0.499	0.015	0.021	0.035	
0.500 to 0.749	0.020	0.028	0.045	
0.750 to 0.999		0.035	0.055	
1.000 to 1.499		0.045	0.065	

*Intermediate wall thickness shall be rounded off to the third decimal place.

†The "mean wall thickness" is determined by the average of two measurements taken at 180 deg to each other.

TABLE 13–22

Sizes for Seamless Airframe Tubing
Carbon and Alloy Steel

Steel Products Manual AISI, Section 18 – 1951

Wall Thickness, inch	Outside Diameter, inches																																		
	3/16	1/4	5/16	3/8	7/16	1/2	9/16	5/8	11/16	3/4	13/16	7/8	15/16	1	1 1/16	1 1/8	1 3/16	1 1/4	1 5/16	1 3/8	1 7/16	1 1/2	1 9/16	1 5/8	1 3/4	1 7/8	2	2 1/8	2 1/4	2 3/8	2 1/2	2 3/4	3	3 1/4	3 1/2
.022	X	X	X	X																															
.028	X	X	X	X	X	X	X	X		X	X	X		X		X	X																		
.035	X	X	X	X	X	X	X	X	X	X		X		X		X		X		X		X		X	X	X	X								
.042		X		X	X	X	X			X		X										X													
.049	X	X	X	X	X	X	X	X	X	X		X		X	X	X	X	X		X	X	X		X	X	X	X	X	X		X				
.058		X	X	X	X	X	X	X		X	X	X		X		X		X		X		X		X	X	X	X	X	X	X	X	X	X		
.065		X	X	X	X	X	X	X		X	X	X	X	X	X	X	X	X	X	X		X	X	X	X	X	X	X	X	X	X	X	X		
.072					X	X		X		X		X				X						X					X								
.083			X	X	X	X	X	X	X	X	X	X	X	X		X	X	X		X		X	X	X	X	X	X	X	X	X	X	X	X		
.095			X	X	X	X	X	X	X	X	X	X		X	X	X	X	X		X	X	X	X	X	X	X	X	X	X	X	X	X	X	X	X
.109						X	X	X	X						X		X	X							X	X	X		X			X	X		
.120				X	X	X	X	X	X	X	X	X	X	X	X	X	X	X	X	X		X		X	X	X	X	X	X	X	X	X	X		X
.125						X	X	X		X		X		X		X		X		X		X		X	X	X	X		X		X	X	X		
.134						X		X		X		X	X					X									X								
.156						X	X	X		X		X	X	X	X	X		X	X	X	X	X	X	X	X	X	X	X	X	X	X	X	X		
.188						X	X	X	X	X	X	X		X		X		X	X	X		X	X	X	X	X	X	X	X	X	X	X	X		
.219										X		X		X		X		X	X	X		X		X	X	X	X	X			X		X	X	X
.250										X		X		X		X		X		X	X	X		X	X	X	X	X	X		X	X	X	X	

TABLE 13–23

Carbon Steel – Sizes for Welded Airframe Tubing

Steel Products Manual AISI, Section 18 – 1951

Wall	Outside Diameter in Inches																																			
	5/16	3/8	7/16	1/2	9/16	5/8	11/16	3/4	13/16	7/8	15/16	1	1 1/8	1 1/4	1 3/8	1 1/2	1 5/8	1 3/4	1 7/8	2	2 1/8	2 1/4	2 3/8	2 1/2	2 5/8	2 3/4	2 7/8	3	3 1/8	3 1/4	3 1/2	3 3/4	4	4 1/8	4 1/2	5
.028				X		X	X	X	X	X		X	X	X	X						X															
.032				X		X	X	X	X	X		X	X	X	X	X					X															
.035	X	X		X	X	X	X	X	X	X		X	X	X	X	X	X	X	X	X	X	X	X	X												
.042	X	X		X	X	X	X	X	X	X		X	X	X	X	X	X	X	X	X	X	X	X	X				X								
.049	X			X	X	X	X	X	X	X		X	X	X	X	X	X	X	X	X	X	X	X	X	X	X	X	X	X	X						
.058				X		X	X	X	X	X		X	X	X	X	X	X	X	X	X	X	X	X	X	X	X	X	X	X	X						
.065				X		X	X	X	X	X		X	X	X	X	X	X	X	X	X	X	X	X	X	X	X	X	X	X	X	X	X	X	X		X
.072								X		X		X	X	X	X	X	X	X	X	X	X	X	X	X	X	X	X	X	X	X	X	X	X	X		X
.083								X		X		X	X	X	X	X	X	X	X	X	X	X	X	X	X	X	X	X	X	X	X	X	X	X		X
.095												X	X	X	X	X	X	X	X	X	X	X	X	X	X	X	X	X	X	X	X	X	X	X		X
.109												X	X	X	X	X	X	X	X	X		X	X	X	X	X	X	X	X	X	X	X	X		X	X
.120														X		X	X	X	X	X		X	X	X	X	X	X	X		X	X		X		X	X
.134																X		X	X	X		X	X	X	X	X	X	X		X	X		X		X	X
.148																		X	X	X		X	X	X	X	X	X	X-		X	X		X		X	X
.165																				X		X	X	X	X	X	X	X		X	X		X		X	
.180																				X		X	X	X	X	X	X	X		X	X		X		X	
.203																				X		X	X	X	X	X	X	X		X	X		X		X	
.220																								X	X	X	X	X		X	X		X		X	
.238																												X		X	X		X		X	
.259																														X	X					

TABLE 13–24

Alloy Steel – Sizes for Welded Airframe Tubing

Steel Products Manual AISI, Section 18 – 1951

Wall	Outside Diameter in Inches																											
	5/16	3/8	7/16	1/2	9/16	5/8	11/16	3/4	13/16	7/8	15/16	1	1 1/8	1 1/4	1 3/8	1 1/2	1 5/8	1 3/4	1 7/8	2	2 1/8	2 1/4	2 3/8	2 1/2	2 5/8	2 3/4	2 7/8	3
.028			X	X	X	X		X		X		X	X															
.032			X	X	X	X		X		X		X	X	X	X	X	X											
.035		X	X	X	X	X		X		X		X	X	X	X	X	X											
.042		X	X	X	X	X		X		X		X	X	X	X	X	X											
.049		X	X	X	X	X		X		X		X	X	X	X	X	X	X	X	X								
.058				X	X	X		X		X		X	X	X	X	X	X	X	X	X								
.065				X	X	X		X		X		X	X	X	X	X	X	X	X	X		X		X	X	X	X	
.072																		X		X		X		X	X	X	X	
.083																		X		X		X		X	X	X	X	X
.095																				X		X		X	X	X	X	X
.109																										X	X	X

TABLE 13–25

Stainless Steel – Sizes of Seamless Aircraft Tubing

Steel Products Manual AISI, Section 18 – 1951

Wall Thickness	Outside Diameter in Inches																												
	1/16	3/32	1/8	3/16	1/4	5/16	3/8	7/16	1/2	9/16	5/8	11/16	3/4	7/8	1	1 1/8	1 1/4	1 3/8	1 1/2	1 5/8	1 3/4	1 7/8	2	2 1/4	2 1/2	2 3/4	3	3 1/4	3 1/2
.010	X	X																											
.012	X	X	X	X	X	X																							
.016			X	X	X	X																							
.020			X	X	X	X	X	X	X		X		X																
.028				X	X	X	X	X	X		X		X		X														
.035			X	X	X	X	X	X	X		X		X	X	X	X	X	X	X	X	X		X	X	X	X	X		
.042							X		X		X								X			X							
.049				X	X	X	X	X	X		X		X		X	X	X	X	X	X	X		X	X	X	X	X	X	X
.058						X	X	X	X		X		X	X				X	X	X									
.065						X	X	X	X		X	X	X		X	X	X		X	X	X		X	X	X		X		
.083							X	X	X				X	X	X		X	X											
.095									X		X			X				X			X		X				X		X
.109										X																			
.120									X		X		X		X	X	X												
.134																													
.156																													
.188																							X						
.219																										X	X	X	
.250																								X					

TABLE 13–26

Outside Diameters and Wall Thicknesses of Seamless, Cold-Finished, Round, Annealed and Hard-Drawn Stainless Aircraft Tubing

Steel Products Manual AISI, Section 18 – 1951

Nominal OD inches	Nominal Wall Thickness, inch	OD Tolerance inch		*Ovality Tolerance (Annealed tubing only) in addition to OD tolerance, inch		Wall Thickness, per cent	
		Over	Under	Over	Under	Over	Under
Under 0.5	Under .015	.004	.004	.004	.004	10	10
Under 0.5	.015 & over	.004	.004	—	—		
0.5 to 1.5 excl.	Under .065	.005	.005	.005	.005	10	10
0.5 to 1.5 excl.	.065 & over	.005	.005	—	—		
1.5 to 3.5 excl.	Under .095	.010	.010	.010	.010	10	10
1.5 to 3.5 excl.	.095 & over	.010	.010	—	—		
3.5 to 5.5 excl.	Under .150	.015	.015	.015	.015	10	10
3.5 to 5.5 excl.	0.150 & over	.015	.015	—	—		
5.5 to 8.0 excl.	Under .240	.030	.030	.030	.030	10	10
5.5 to 8.0 excl.	0.240 & over	.030	.030	—	—		

TABLE 13–27

Stainless Steel – Sizes of Welded Aircraft Tubing

Steel Products Manual AISI, Section 18 – 1951

Wall Thickness, inch	Outside Diameter, inches																														
	¼	⅜	½	⅝	¾	⅞	1	1⅛	1¼	1⅜	1½	1⅝	1¾	1⅞	2	2⅛	2¼	2⅜	2½	2⅝	2¾	2⅞	3	3¼	3½	3¾	4	4¼	4½	4¾	5
.028	X	X	X	X	X	X	X	X	X	X	X	X	X	X	X																
.035	X	X	X	X	X	X	X	X	X	X	X	X	X	X	X	X	X	X	X	X	X	X	X								
.042	X	X	X	X	X	X	X	X	X	X	X	X	X	X	X	X	X	X	X	X	X	X	X								
.049	X	X	X	X	X	X	X	X	X	X	X	X	X	X	X	X	X	X	X	X	X	X	X	X	X	X	X				
.065		X	X	X	X	X	X	X	X	X	X	X	X	X	X	X	X	X	X	X	X	X	X	X	X	X	X	X	X	X	X
.083		X	X	X	X	X	X	X	X	X	X	X	X	X	X	X	X	X	X	X	X	X	X	X	X	X	X	X	X	X	X
.095		X	X	X	X	X	X	X	X	X	X	X	X	X	X	X	X	X	X	X	X	X	X	X	X	X	X	X	X	X	X
.109				X	X	X	X	X	X	X	X	X	X	X	X	X	X	X	X	X	X	X	X	X	X	X	X	X	X	X	X
.120							X	X	X	X	X	X	X	X	X	X	X	X	X	X	X	X	X	X	X	X	X	X	X	X	X
.134							X	X	X	X	X	X	X	X	X	X	X	X	X	X	X	X	X	X	X	X	X	X	X	X	X

TABLE 13–28
Standard Weight, Threaded, Line Pipe
Steel Products Manual AISI. Section 18 – 1951

Size: Nom. in.	Weights per Foot				Diameters		Couplings			Test Pressures, psi.				
		Calculated								Steel				Iron
	Nom. Thd. & Cplg., lb.*	Plain Ends, lb.	Thd. & Cplg., lb.	Wall Thickness, in.	Outside, in.	Inside, in.	Length, in.	Outside Diam., in.	Cal. Weight, lb.	Butt-Welded	Lap-Welded or Grade A	Grade B	Grade C	Open Hearth and Wrought †
⅛	.25	.24	.25	.068	.405	.269	1 1/16	.563	.04	700	700	700	700	700
¼	.43	.42	.43	.088	.540	.364	1⅝	.719	.09	700	700	700	700	700
⅜	.57	.57	.57	.091	.675	.493	1⅝	.875	.13	700	700	700	700	700
½	.86	.85	.86	.109	.840	.622	2⅛	1.063	.24	700	700	700	700	700
¾	1.14	1.13	1.14	.113	1.050	.824	2⅛	1.313	.34	700	700	700	700	700
1	1.70	1.68	1.69	.133	1.315	1.049	2⅝	1.576	.54	700	700	700	700	700
1¼	2.30	2.27	2.30	.140	1.660	1.380	2¾	2.054	1.03	800	1000	1100	1300	800
1½	2.75	2.72	2.74	.145	1.900	1.610	2¾	2.200	.90	800	1000	1100	1300	800
2	3.75	3.65	3.71	.154	2.375	2.067	2⅞	2.875	1.86	800	1000	1100	1300	800
2½	5.90	5.79	5.88	.203	2.875	2.469	4⅛	3.375	3.27	800	1000	1100	1300	800
3	7.70	7.58	7.67	.216	3.500	3.068	4¼	4.000	4.09	800	1000	1100	1300	800
3½	9.25	9.11	9.27	.226	4.000	3.548	4⅜	4.625	5.92	1200	1200	1300	1600	950
4	11.00	10.79	11.01	.237	4.500	4.026	4½	5.200	7.59	1200	1200	1300	1600	950
5	15.00	14.62	14.90	.258	5.563	5.047	4⅝	6.296	9.98	—	1200	1300	1600	950
6	19.45	18.97	19.33	.280	6.625	6.065	4⅞	7.390	12.92	—	1200	1300	1600	950
8	25.55	24.70	25.44	.277	8.625	8.071	5¼	9.625	23.18	—	1200	1300	1600	950
8	29.35	28.55	29.25	.322	8.625	7.981	5¼	9.625	23.18	—	1300	1600	1600	950
10	32.75	31.20	32.20	.279	10.750	10.192	5¾	11.750	31.55	—	1000	1200	1400	800
10	35.75	34.24	35.20	.307	10.750	10.136	5¾	11.750	31.55	—	1000	1200	1400	800
10	41.85	40.48	41.35	.365	10.750	10.020	5¾	11.750	31.55	—	1200	1400	1400	800
12	45.45	43.77	45.40	.330	12.750	12.090	6⅛	14.000	49.27	—	1000	1200	1400	800
12	51.15	49.56	51.10	.375	12.750	12.000	6⅛	14.000	49.27	—	1100	1200	1400	800
14D	57.00	54.57	55.80	.375	14.000	13.250	6⅜	15.000	45.83	—	950	1100	1400	750
16D	65.30	62.58	64.08	.375	16.000	15.250	6¾	17.000	55.83	—	850	1000	1300	700
18D	73.00	70.59	72.37	.375	18.000	17.250	7⅛	19.000	66.53	—	750	900	1100	600
20D	81.00	78.60	80.70	.375	20.000	19.250	7⅝	21.000	79.37	—	700	800	1000	550

*Nominal weights, threads and couplings are shown for purposes of identification in specifying weight of pipe.

†Test pressures shown apply to seamless, electric-welded, and lap-welded open hearth iron pipe and to lap-welded wrought iron pipe. Pressures shown for sizes 4-in. nominal and smaller apply also to butt-welded open hearth iron and in sizes 2-in. nominal and smaller apply to wrought iron pipe.

(a) The customary variation in weight for any length of pipe is 10 per cent above and 3½ per cent below; and carload weight is customarily not more than 1¾ per cent under the calculated weight.

(b) Taper of threads is ¾ in. per foot on diameter for all sizes.

(c) API STD 5-L covers these data in detail.

TABLE 13–29

Wall Thicknesses and Test Pressures for Plain-End Line Pipe

Steel Products Manual AISI, Section 18 – 1951

Size: Outside Diameter in.	Wall Thick-ness in.	Test pressures, psi.			
		Steel			Iron
		Lap-Welded and Grade A	Grade B	Grade C	Open Hearth and Wrought [1]
3½	.188	1900	2200	2500	1500
	.216	2200	2500	2500	1800
	.250	2500	2500	2500	2100
	.281	2500	2500	2500	2300
4½	.188	1500	1800	2200	1200
	.219	1700	2000	2500	1400
	.237	1900	2200	2500	1500
	.250	2000	2300	2500	1600
	.281	2200	2500	2500	1800
	.312	2500	2500	2500	2000
6⅝	.188	1000	1200	1500	800
	.219	1200	1400	1800	950
	.250	1400	1600	2000	1100
	.280	1500	1800	2300	1200
	.312	1700	2000	2500	1400
	.344	1900	2200	2500	1500
	.375	2000	2400	2500	1600
8⅝	.188	800	900	1200	650
	.219	900	1100	1400	750
	.250	1000	1200	1600	850
	.277	1200	1300	1700	950
	.312	1300	1500	2000	1000
	.322	1300	1600	2000	1000
	.344	1400	1700	2200	1100
	.375	1600	1800	2300	1300
	.438	1800	2100	2500	1500
10¾	.219	750	850	1100	600
	.250	850	1000	1300	700
	.279	1000	1200	1400	800
	.307	1000	1200	1500	800
10¾	.344	1100	1300	1700	900
	.365	1200	1400	1800	1000
	.438	1500	1700	2200	1200
12¾	.250	700	800	1100	600
	.281	800	950	1200	650
	.312	900	1000	1300	700
	.330	1000	1200	1400	800
	.344	1000	1200	1500	800
	.375	1100	1200	1600	850
	.438	1200	1400	1900	1000
14	.312	800	950	1200	650
	.344	900	1000	1300	700
	.375	950	1100	1400	750
	.438	1100	1300	1700	900
	.500	1300	1500	1900	1000
16	.312	700	800	1100	550
	.344	750	900	1200	600
	.375	850	1000	1300	700
	.438	1000	1100	1500	800
	.500	1100	1300	1700	900
18	.312	600	750	950	500
	.344	700	800	1000	550
	.375	750	900	1100	600
	.438	900	1000	1300	700
	.500	1000	1200	1500	800
20	.312	550	650	850	450
	.344	600	700	950	500
	.375	700	800	1000	550
	.438	800	900	1200	650
	.500	900	1000	1400	700

(a) The customary variation in weight for any length of pipe is 10 per cent above and 3½ per cent below; and carload weight is customarily not more than 1¾ per cent under the calculated weight.

(b) API STD 5 L covers these data in detail.

[1] Test pressures shown apply to seamless, electric-welded, and lap-welded open hearth iron pipe and to lap-welded wrought iron pipe. Test pressures for sizes 4½-inch outside diameter and smaller apply also to butt-welded open hearth iron pipe.

TABLE 13-30

Extra Strong Standard Pipe
Plain Ends or Threaded and Coupled

Steel Products Manual AISI, Section 18 – 1951

Nominal Size, in.	Weight per Foot Calculated Plain Ends, lb	Wall Thickness, in.	Diameters		Test Pressures				
					Steel				Iron
			Outside, in.	Inside, in.	Butt Welded, psi	*Lap Welded or Grade A, psi	Grade B, psi	Grade C, psi	Open Hearth and Wrought†
1/8	.31	.095	.405	.215	850	850	850	850	850
1/4	.54	.119	.540	.302	850	850	850	850	850
3/8	.74	.126	.675	.423	850	850	850	850	850
1/2	1.09	.147	.840	.546	850	850	850	850	850
3/4	1.47	.154	1.050	.742	850	850	850	850	850
1	2.17	.179	1.315	.957	850	850	850	850	850
1¼	3.00	.191	1.660	1.278	1300	1800	1900	2300	1400
1½	3.63	.200	1.900	1.500	1300	1800	1900	2300	1400
2	5.02	.218	2.375	1.939	1300	1800	1900	2300	1400
2½	7.66	.276	2.875	2.323	1300	1800	1900	2300	1400
3	10.25	.300	3.500	2.900	1300	2500	2500	2500	2500
3½	12.51	.318	4.000	3.364	1700	2500	2500	2500	2300
4	14.98	.337	4.500	3.826	1700	2500	2500	2500	2200
5	20.78	.375	5.563	4.813	—	2400	2500	2500	1900
6	28.57	.432	6.625	5.761	—	2300	2500	2500	1900
8	43.39	.500	8.625	7.625	—	2100	2400	2500	1700
10	54.74	.500	10.750	9.750	—	1700	2000	2500	1300
12	65.42	.500	12.750	11.750	—	1400	1600	2100	1100

*Lap-welded pipe is not commonly produced in sizes smaller than 1¼ inch.

†Test pressures shown apply to seamless, electric-welded, and lap-welded, open hearth iron pipe and lap-welded wrought iron pipe. Test pressures for sizes 4½-in. outside diameter and smaller apply also to butt-welded open hearth iron pipe.

(a) The customary variation in weight for any length of pipe is 10 per cent above and 3½ per cent below and the carload weight is customarily not more than 1¾ per cent under the calculated weight.

(b) API Standard Specification 5-L covers these data in detail.

TABLE 13–31

Double Extra Strong Pipe
Plain Ends or Threaded and Coupled

Steel Products Manual AISI, Section 18 – 1951

Size: Nominal, in.	Weight per Foot Calculated Plain Ends, lb.	Wall Thickness, in.	Diameters		Test Pressures		
			Outside, in.	Inside, in.	Butt Welded, psi.	Lap Welded or Grade A, psi.	Grade B, psi.
1/2	1.71	.294	.840	.252	1000	1000	1000
3/4	2.44	.308	1.050	.434	1000	1000	1000
1	3.66	.358	1.315	.599	1000	1000	1000
1¼	5.21	.382	1.660	.896	1200	1800	1900
1½	6.41	.400	1.900	1.100	1200	1800	1900
2	9.03	.436	2.375	1.503	1200	1800	1900
2½	13.70	.552	2.875	1.771	1200	1800	1900
3	18.58	.600	3.500	2.300	—	1800	1900
4	27.54	.674	4.500	3.152	—	2000	2100
5	38.55	.750	5.563	4.063	—	2000	2100
6	53.16	.864	6.625	4.897	—	2000	2100
8	72.42	.875	8.625	6.875	—	2800	2800

(a) The customary variation in weight is 10 per cent above and 10 per cent below.

TABLE 13–32

Formulas for Pressures in Pipe and Pipe Fittings

ASA B31.1a–1953, Supplement No. 1, Code for Pressure Piping

To Find	Formula	In Which
Minimum pipe wall thickness, t_m, inches	$t_m = \left[\frac{P\,D}{2S + 2yP}\right] + C$	D = outside diameter of pipe, inches
Maximum internal service pressure, P, psig	$P = \frac{2S(t_m - C)}{D - 2y(t_m - C)}$	S = allowable stress, psi, in material due to internal pressure, at the operating temperature, deg F. The Sections of the Code for Pressure Piping, B31.1, provide not only tabular values of S for specific materials, temperatures and service, as illustrated in Table 13–33, but also instructions to govern special and unusual conditions C = allowance for threading, mechanical strength, and/or corrosion, in inches y = a coefficient. Table at left is suitable for interpolation.

Table of y values

Temp Deg F	Up to 900	950	1000	1050	1100	1150 and up
Ferritic Steels	0.4	0.5	0.7	0.7	0.7	0.7
Austenitic Steels	0.4	0.4	0.4	0.4	0.5	0.7

To Find	Formula	In Which
Hydrostatic test pressures	The Code, B31.1, depending upon the Section, prescribes hydrostatic test pressures for pipe fittings, and piping systems, both before and after erection. Quite generally the hydrostatic test after erection calls for a test pressure of not less than one and one-half times the maximum operating pressure, adjusted to 100°F.	
Bursting strength of fittings, P_B, psig ASA B16.9–1951 Steel Butt-Welding Fittings Table 13–42 lists pressure ratings of fittings	$P_B = \frac{2St}{D}$, and which applies strictly to calculating the bursting strength of straight pipe.	t = thickness of pipe wall, inches Strengths of fittings are determined by comparing their test strengths with the bursting pressures exhibited by straight seamless pipe of the designated wall thickness and material.

TABLE 13–33

Typical Allowable Stresses for Pipe in Power Piping Systems

ASA B31.1a–1953, Supplement No. 1 Code for Pressure Piping

Material[1]	ASTM Specification	Grade	Minimum Ultimate Tensile Strength	Values of S Psi for Temperatures in Deg F Not to Exceed[2]							
				–20 to 100	200	300	400*	450	500	600	650
Seamless steel:											
4.0-6.0 chromium	A 158	P5a	60,000	15,000	15,000	15,000	15,000		14,500	14,000	13,700
0.50 molybdenum	A 335										
	A 158	P5b	60,000	15,000	15,000	15,000	15,000		14,500	14,000	13,700
	A 335										
Chromium-nickel	A 213 A 158	TP321 P8b	75,000	18,750	18,750	17,000	15,800		15,200	14,900	14,850
	A 213 A 158	TP347 P8d	75,000	18,750	18,750	17,000	15,800		15,200	14,900	14,850
Carbon steel	A 120			10,800	10,600	10,200	9,800	9,600			
Electric-fusion-welded steel:	A 134	A 245 A	48,000	8,800	8,800	8,800	8,800		8,800	8,800	8,800
		A 245 B	52,000	9,600	9,600	9,600	9,600		9,600	9,600	9,600
		A 245 C	55,000	10,100	10,100	10,100	10,100		10,100	10,100	10,100
		A 283 A	45,000	8,300	8,300	8,300	8,300		8,300	8,300	8,300
		A 283 B	50,000	9,200	9,200	9,200	9,200		9,200	9,200	9,200
		A 283 C	55,000	10,100	10,100	10,100	10,100		10,100	10,100	10,100
		A 283 D	60,000	10,100	10,100	10,100	10,100		10,100	10,100	10,100
	A 139	A[4]	48,000	9,600	9,600	9,600	9,600		9,600	9,600	9,600
		B[4]	60,000	12,000	12,000	12,000	12,000		12,000	12,000	12,000
Electric-resistance-welded steel	A 53	A[3]	48,000	10,200	10,200	10,200	10,200		10,200	10,200	10,200
		B[3]	60,000	12,750	12,750	12,750	12,750		12,750	12,750	12,750
	A 135	A[3]	48,000	10,200	10,200	10,200	10,200		10,200	10,200	10,200
		B[3]	60,000	12,750	12,750	12,750	12,750		12,750	12,750	12,750
Lap welded											
Steel	A 53		45,000	9,000	9,000	9,000	9,000		9,000	9,000	9,000
Steel	A 120			8,800	8,600	8,200	7,800	7,600			
Wrought iron	A 72		40,000	8,000	8,000	8,000	8,000		8,000	8,000	8,000
Butt welded											
Steel	A 53		45,000	6,750	6,750	6,750	6,750		6,750	6,750	6,750
Steel	A 120			6,500	6,350	6,100	5,850	5,700			
Wrought iron	A 72		40,000	6,000	6,000	6,000	6,000		6,000	6,000	6,000
Seamless											
Red brass	B 43			8,000	8,000	7,000	3,000				
Copper – 2 in. & smaller	B 42			6,000	5,500	4,750	3,000				
Copper – over 2 in.	B 42			6,000	5,500	4,750	3,000				
Copper tubing	B 75			6,000	5,500	4,750	3,000				
Annealed	B 88		30,000	6,000	5,500	4,750	3,000				
Bright annealed	B 68		30,000	6,000	5,500	4,750	3,000				
Copper brazed steel	A 254	Class I	42,000	6,000	5,500	4,750	3,000				
		Class II	42,000	3,600	3,300	2,850	1,800				
Cast iron[5]											
Centrifugally cast	FSB WWP-421	Types I & II		6,000	6,000	6,000	6,000	6,000			
Pit cast	ASA A21.2			4,000	4,000	4,000	4,000	4,000			

[1] Pipe in accordance with API Specification 5L may be used. *For steam at 250 psi (406 F) the values given may be used.

[2] The several types and grades of pipe tabulated above shall not be used at temperatures in excess of the maximum temperatures for which the S values are indicated. (See also specific requirements for service conditions contemplated.) Allowable S values for intermediate temperatures may be obtained by interpolation.

[3] For electric-resistance-welded pipe for applications where the temperature is below 650 F, and where pipe furnished under this classification is subjected to supplemental tests and/or heat treatments as agreed to by the supplier and the purchaser, and whereby such supplemental tests and/or heat treatments demonstrate the strength characteristics of the weld to be equal to the minimum tensile strength specified for the pipe, the S values equal to the corresponding seamless grades may be used.

[4] If plate material having physical properties other than stated in Section 6 of the ASTM Specification A 139 is used in the manufacture of ordinary electric-fusion-welded steel pipe, the allowable stress shall be taken as 0.20 times the tensile strength for temperatures of 450 F and below.

[5] Cast-iron pipe shall not be used for lubricating oil lines for machinery and in any case not for oil having a temperature above 300 F.

TABLE 13–34

American Standard Dimensions of Welded and Seamless Steel Pipe

ASA B36.10–1950 Wrought-Steel and *Wrought-Iron Pipe

Nominal Pipe Size	Outside Diam	NOMINAL WALL THICKNESS (Listed by Schedule Numbers) Sched 10	Sched 20	Sched 30	Sched 40	Sched 60	Sched 80	Sched 100	Sched 120	Sched 140	Sched 160	NOMINAL WALL THICKNESS Standard Wall	Extra Strong Wall	Double Extra Strong Wall
1/8	0.405	.....	.....	.....	0.068	.....	0.095	.....	.....	.....	.....	0.068	0.095	.....
1/4	0.540	.....	.....	.....	0.088	.....	0.119	.....	.....	.....	.....	0.088	0.119	.....
3/8	0.675	.....	.....	.....	0.091	.....	0.126	.....	.....	.....	.....	0.091	0.126	.....
1/2	0.840	.....	.....	.....	0.109	.....	0.147	.....	.....	.....	0.187	0.109	0.147	0.294
3/4	1.050	.....	.....	.....	0.113	.....	0.154	.....	.....	.....	0.218	0.113	0.154	0.308
1	1.315	.....	.....	.....	0.133	.....	0.179	.....	.....	.....	0.250	0.133	0.179	0.358
1 1/4	1.660	.....	.....	.....	0.140	.....	0.191	.....	.....	.....	0.250	0.140	0.191	0.382
1 1/2	1.900	.....	.....	.....	0.145	.....	0.200	.....	.....	.....	0.281	0.145	0.200	0.400
2	2.375	.....	.....	.....	0.154	.....	0.218	.....	.....	.....	0.343	0.154	0.218	0.436
2 1/2	2.875	.....	.....	.....	0.203	.....	0.276	.....	.....	.....	0.375	0.203	0.276	0.552
3	3.500	.....	.....	.....	0.216	.....	0.300	.....	.....	.....	0.438	0.216	0.300	0.600
3 1/2	4.000	.....	.....	.....	0.226	.....	0.318	.....	.....	.....	.....	0.226	0.318	.....
4	4.500	.....	.....	.....	0.237	.....	0.337	.....	0.438	.....	0.531	0.237	0.337	0.674
5	5.563	.....	.....	.....	0.258	.....	0.375	.....	0.500	.....	0.625	0.258	0.375	0.750
6	6.625	.....	.....	.....	0.280	.....	0.432	.....	0.562	.....	0.718	0.280	0.432	0.864
8	8.625	.....	0.250	0.277	0.322	0.406	0.500	0.593	0.718	0.812	0.906	0.322	0.500	0.875
10	10.750	.....	0.250	0.307	0.365	0.500	0.593	0.718	0.843	1.000	1.125	0.365	0.500	.....
12	12.750	.....	0.250	0.330	0.406	0.562	0.687	0.843	1.000	1.125	1.312	0.375	0.500	.....
14	14.000	0.250	0.312	0.375	0.438	0.593	0.750	0.937	1.093	1.250	1.406	0.375	0.500	.....
16	16.000	0.250	0.312	0.375	0.500	0.656	0.843	1.031	1.218	1.438	1.593	0.375	0.500	.....
18	18.000	0.250	0.312	0.438	0.562	0.750	0.937	1.156	1.375	1.562	1.781	0.375	0.500	.....
20	20.000	0.250	0.375	0.500	0.593	0.812	1.031	1.281	1.500	1.750	1.968	0.375	0.500	.....
24	24.000	0.250	0.375	0.562	0.687	0.968	1.218	1.531	1.812	2.062	2.343	0.375	0.500	.....
30	30.000	0.312	0.500	0.625	.....	.....	.....	.....	.....	.....	.....	.....	.....	.....

All dimensions are given in inches.

The decimal thicknesses listed for the respective pipe sizes represent their nominal or average wall dimensions. For tolerances on wall thicknesses, see appropriate material specifications. Thickness shown in bold face type for Schedule 40 are identical with thicknesses shown in bold face type for *Standard Wall*. Those in bold face type in Schedules 60 and 80 are identical with thicknesses in bold face type for *Extra Strong Wall*. *Double Extra Strong Wall* has no corresponding schedule number.

Some of the larger, heavier wall sections are beyond the capabilities of seamless mill production and must be obtained from turned-and-bored billets or other sources.

*Wall thicknesses of standard wrought-iron pipe are slightly thicker than the corresponding sizes for steel pipe as given in the table.

TABLE 13–35

American Standard Welded and Seamless Stainless Steel Pipe

ASA B36.19–1952 Stainless Steel Pipe

NOMINAL PIPE SIZE	OUTSIDE DIAMETER	NOMINAL WALL THICKNESS			
		SCHEDULE 5S**	SCHEDULE 10S**	SCHEDULE 40S	SCHEDULE 80S
1/8	0.405	.....	0.049	0.068	0.095
1/4	0.540	.....	0.065	0.088	0.119
3/8	0.675	.....	0.065	0.091	0.126
1/2	0.840	0.065	0.083	0.109	0.147
3/4	1.050	0.065	0.083	0.113	0.154
1	1.315	0.065	0.109	0.133	0.179
1 1/4	1.660	0.065	0.109	0.140	0.191
1 1/2	1.900	0.065	0.109	0.145	0.200
2	2.375	0.065	0.109	0.154	0.218
2 1/2	2.875	0.083	0.120	0.203	0.276
3	3.500	0.083	0.120	0.216	0.300
3 1/2	4.000	0.083	0.120	0.226	0.318
4	4.500	0.083	0.120	0.237	0.337
5	5.563	0.109	0.134	0.258	0.375
6	6.625	0.109	0.134	0.280	0.432
8	8.625	0.109	0.148	0.322	0.500
10	10.750	0.134	0.165	0.365	0.500*
12	12.750	0.165	0.180	0.375*	0.500*

All dimensions are given in inches.
The decimal thicknesses listed for the respective pipe sizes represent their nominal or average wall dimensions.
*These do not conform to ASA B36.10.
**Schedule 5S and 10S wall thicknesses do not permit threading in accordance with ASA B2.1.

TABLE 13–36

Outside Diameter Variations According to Different Specifications for Various Kinds of Pipe and Pressure Tubing

Steel Products Manual AISI, Section 18 – 1951

Type	Size, in.	Reference Specifications	Variations Over	Variations Under
PIPE *				
Seamless and Welded Standard Weight Extra Strong Double Extra Strong	To 1½, incl. 2 and over	ASTM A-53, A-120, A-253	1/64 in. 1%	1/32 in. 1%
Seamless	⅛ to 1½, incl. Over 1½ to 4, incl. Over 4 to 8, incl. Over 8 to 18, incl. Over 18 to 24, incl.	ASTM A-106, A-158, A-206, A-280, A-315, A-333, A-335	1/64 in. 1/32 in. 1/16 in. 3/32 in. 1/8 in.	1/32 in. 1/32 in. 1/32 in. 1/32 in. 1/32 in.
Electric Welded	All sizes	ASTM A-135	1%	1%
LINE PIPE * Seamless and Welded	To 1½ incl. 2 and over	API STD 5-L	1/64 in. 1%	1/32 in. 1%
OIL COUNTRY PIPE Casing, Drill Pipe and Tubing Seamless and Welded	4 and under Over 4	API STD 5-A	1/32 in. 0.75%	1/32 in. 0.75%
PRESSURE TUBES*				
BOILER AND SUPERHEATER (Seamless and Lap-Welded) Hot Finished and Hot Rolled	4 and under Over 4	ASTM A-83, A-192, A-209, A-210, A-213	1/64 in. 1/64 in.	1/32 in. 3/64 in.
(Seamless and Electric Welded) Cold Finished and Cold Rolled	Under 1 1 to 1½, incl. Over 1½ to 2, excl. 2 to 2½, excl. 2½ to 3, excl. 3 to 4, incl. Over 4	ASTM A-83, A-178, A-192 A-209, A-210, A-213, A-226, A-249, A-250	0.004 in. 0.006 in. 0.008 in. 0.010 in. 0.012 in. 0.015 in. 0.015 in.	0.004 in. 0.006 in. 0.008 in. 0.010 in. 0.012 in. 0.015 in. 0.025 in.
STILL TUBES (Seamless) Hot Finished Cold Finished	4 and under Over 4 to 7½ 4 and under Over 4 to 7½	ASTM A-161, A-200, A-271	1/64 in. 1/64 in. 0.015 in. 0.015 in.	1/32 in. 3/64 in. 0.015 in. 0.025 in.
HEAT EXCHANGER AND CONDENSER TUBES* (Seamless and Electric Welded) Cold Finished and Cold Rolled	Under 1 1 to 1½, incl. Over 1½ to 2, excl.	ASTM A-179, A-199, A-214 A213, A249	0.004 in. 0.006 in. 0.008 in.	0.004 in. 0.006 in. 0.008 in.

*Identical tolerances on diameters are specified in Federal Standard No. 48–1954, *Tolerances for Steel and Iron Wrought Products*.

TABLE 13–37

Wall Thickness Variations According to Different Specifications for Various Kinds of Pipe and Pressure Tubing

Steel Products Manual AISI, Section 18 – 1951

Type	Wall Thickness or Size, in.	Reference Specifications	Variations, Per Cent	
			Over	Under
PIPE				
Seamless or Welded*				
Standard Weight Extra Strong Double Extra Strong	All sizes	ASTM A-53, A-120, A-135, A-253	——	12½
Seamless	All sizes	ASTM A-106, A-206, A-158, A-280, A-315, A-333, A-335	——	12½
LINE PIPE* Seamless and Welded	All sizes	API STD 5-L	——	12½
OIL COUNTRY PIPE CASING, DRILL PIPE, TUBING Seamless and Welded	All sizes	API STD 5-A	——	12½
PRESSURE TUBES*				
BOILER AND SUPERHEATER TUBES (Seamless and Lap Welded)				
Hot Finished	.095 in. thick and under	ASTM A-83, A-192, A-209, A-210, A-213	40	0
	Over .095 in to .150 in. incl.		35	0
	Over .150 in. to .180 in. incl.		33	0
	Over .180 in.		28	0
BOILER AND SUPERHEATER TUBES Cold Finished (Seamless)	Up to 1½ in. O.D.	ASTM A-83, A-192, A-209, A-210, A-213	20	0
	Over 1½ in. O.D.		22	0
(Electric Welded) Cold Rolled	All sizes	ASTM A-178, A-226, A-249, A-250	18	0
STILL TUBES (Seamless)				
Hot Finished	All sizes	ASTM A-161, A-200, A-271	28	0
Cold Finished	All sizes		22	0
HEAT EXCHANGER AND CONDENSER TUBES* Cold Finished				
(Seamless) Cold Finished	All sizes	ASTM A-179, A-199, A-213	22	0
(Electric Welded) Cold Rolled	All sizes	ASTM A-214, A-249	18	0

*Identical tolerances on wall thickness are specified in Federal Standard No. 48—1954, *Tolerances for Steel and Iron Wrought Products*.

TABLE 13–38

Dimensions and Weights of Copper and Red Brass Pipe – Standard Pipe Sizes

ASTM Designation: B251–52T (Applicable to ASTM Designations B42, B43 and B188)

Pipe Size, In.	Nominal Dimensions, In. Outside Diameter	Inside Diameter	Wall Thickness	Cross-Sectional Area of Bore, Sq. In.	Nominal Weight, Lb per Ft Copper	Red Brass
			Regular			
1/8	0.405	0.281	0.062	0.062	0.259	0.253
1/4	0.540	0.376	0.082	0.110	0.457	0.447
3/8	0.675	0.495	0.090	0.192	0.641	0.627
1/2	0.840	0.625	0.107	0.307	0.955	0.934
3/4	1.050	0.822	0.114	0.531	1.30	1.27
1	1.315	1.063	0.126	0.887	1.82	1.78
1-1/4	1.660	1.368	0.146	1.47	2.69	2.63
1-1/2	1.900	1.600	0.150	2.01	3.20	3.13
2	2.375	2.063	0.156	3.34	4.22	4.12
2-1/2	2.875	2.501	0.187	4.91	6.12	5.99
3	3.500	3.062	0.219	7.37	8.75	8.56
3-1/2	4.000	3.500	0.250	9.62	11.4	11.2
4	4.500	4.000	0.250	12.6	12.9	12.7
5	5.562	5.062	0.250	20.1	16.2	15.8
6	6.625	6.125	0.250	29.5	19.4	19.0
8	8.625	8.001	0.312	50.3	31.6	30.9
10	10.750	10.020	0.365	78.8	46.2	45.2
12	12.750	12.000	0.375	113	56.5	55.3
			Extra Strong			
1/8	0.405	0.205	0.100	0.033	0.371	0.363
1/4	0.540	0.294	0.123	0.068	0.625	0.611
3/8	0.675	0.421	0.127	0.139	0.847	0.829
1/2	0.840	0.542	0.149	0.231	1.25	1.23
3/4	1.050	0.736	0.157	0.425	1.71	1.67
1	1.315	0.951	0.182	0.710	2.51	2.46
1-1/4	1.660	1.272	0.194	1.27	3.46	3.39
1-1/2	1.900	1.494	0.203	1.75	4.19	4.10
2	2.375	1.933	0.221	2.94	5.80	5.67
2-1/2	2.875	2.315	0.280	4.21	8.85	8.66
3	3.500	2.892	0.304	6.57	11.8	11.6
3-1/2	4.000	3.358	0.321	8.86	14.4	14.1
4	4.500	3.818	0.341	11.5	17.3	16.9
5	5.562	4.812	0.375	18.2	23.7	23.2
6	6.625	5.751	0.437	26.0	32.9	32.2
8	8.625	7.625	0.500	45.7	49.5	48.4
10	10.750	9.750	0.500	74.7	62.4	61.1

TABLE 13–39

Standard Copper Water Tube – Dimensions and Weights, and Tolerances in Diameter and Wall Thickness

ASTM Designation: B251–53T (Applicable to ASTM Designations B-88 and B-188)

All tolerances in this table are plus and minus except as otherwise indicated.

Standard Water Tube Size, in.	Actual Outside Diameter, in.	Average Outside Diameter Tolerances, in.		Wall Thickness, In.						Theoretical Weight, lb per ft		
				Type K		Type L		Type M				
		Annealed	Drawn	Nominal	Tolerance	Nominal	Tolerance	Nominal	Tolerance	Type K	Type L	Type M
1/4	0.375	0.002	0.001	0.035	0.0035	0.030	0.0035			0.145	0.126	
3/8	0.500	0.0025	0.001	0.049	0.004	0.035	0.0035	...	...	0.269	0.198	...
1/2	0.625	0.0025	0.001	0.049	0.004	0.040	0.0035	...	...	0.344	0.285	...
5/8	0.750	0.0025	0.001	0.049	0.004	0.042	0.0035	...	...	0.418	0.362	...
3/4	0.875	0.003	0.001	0.065	0.0045	0.045	0.004	...	...	0.641	0.455	...
1	1.125	0.0035	0.0015	0.065	0.0045	0.050	0.004	...	...	0.839	0.655	...
1 1/4	1.375	0.004	0.0015	0.065	0.0045	0.055	0.0045	0.042	0.0035	1.04	0.884	0.682
1 1/2	1.625	0.0045	0.002	0.072	0.005	0.060	0.0045	0.049	0.004	1.36	1.14	0.940
2	2.125	0.005	0.002	0.083	0.007	0.070	0.006	0.058	0.006	2.06	1.75	1.46
2 1/2	2.625	0.005	0.002	0.095	0.007	0.080	0.006	0.065	0.006	2.93	2.48	2.03
3	3.125	0.005	0.002	0.109	0.007	0.090	0.007	0.072	0.006	4.00	3.33	2.68
3 1/2	3.625	0.005	0.002	0.120	0.008	0.100	0.007	0.083	0.007	5.12	4.29	3.58
4	4.125	0.005	0.002	0.134	0.010	0.110	0.009	0.095	0.009	6.51	5.38	4.66
5	5.125	0.005	0.002	0.160	0.010	0.125	0.010	0.109	0.009	9.67	7.61	6.66
6	6.125	0.005	0.002	0.192	0.012	0.140	0.010	0.122	0.010	13.9	10.2	8.92
8	8.125	0.006	+0.002 −0.004	0.271	0.016	0.200	0.014	0.170	0.014	25.9	19.3	16.5
10	10.125	0.008	+0.002 −0.006	0.338	0.018	0.250	0.016	0.212	0.015	40.3	30.1	25.6
12	12.125	0.008	+0.002 −0.006	0.405	0.020	0.280	0.018	0.254	0.016	57.8	40.4	36.7

STANDARD DIMENSIONS AND WEIGHTS OF COPPER TUBE, TYPE B[a]

(Applicable to ASTM Designation B188)

Actual Outside Diameter, in.	Thickness, min, in.	Theoretical Weight lb per ft
0.125	0.035	0.0384
0.188	0.035	0.0652
0.250	0.035	0.0916
0.313	0.035	0.118
0.375	0.035	0.145
0.438	0.035	0.172
0.500	0.035	0.198
0.500	0.065	0.344
0.540	0.065	0.376
0.675	0.065	0.483
0.840	0.065	0.613
1.050	0.065	0.780
1.315	0.065	0.989
1.660	0.065	1.262
1.900	0.065	1.452
2.375	0.065	1.828

[a]These sizes correspond with Type B sizes of Military Specification MIL-T-873.

(Applicable to ASTM Designations B42, B43 and B188)

Weight Tolerances.—The weight of the pipe shall not vary from the nominal weight per foot prescribed in Table 13–38 by more than the following:

Pipe Size, in.	Plus and Minus Tolerance, per cent
6 and under	5
Over 6 to 8, incl.	7
Over 8	8

Thickness Tolerances.—The thickness of the pipe at any point shall not be less than that prescribed in Table 13–38 by more than the following.

Pipe Size, in.	Minus Tolerance, per cent[a]
6 and under	5
Over 6 to 8, incl.	7
Over 8	8

[a] Expressed to the nearest 0.001 in.

TABLE 13–40

Aluminum-Alloy Pipe Nominal Dimensions and Weight

ASTM Designation: B241–53T

Size, in.	Outside Diameter, in.	Standard Pipe (ASA Schedule 40)			Extra Heavy Pipe (ASA Schedule 80)		
		Wall Thickness, in.	Weight, lb per ft: Alloys GS10A and GS11A	Weight, lb per ft: Alloy M1A	Wall Thickness, in.	Weight, lb per ft: Alloys GS10A and GS11A	Weight, lb per ft: Alloy M1A
1/8	0.405	0.068	0.085	0.086	0.095	0.109	0.110
1/4	0.540	0.088	0.147	0.149	0.119	0.185	0.187
3/8	0.675	0.091	0.196	0.198	0.126	0.256	0.259
1/2	0.840	0.109	0.294	0.297	0.147	0.376	0.380
3/4	1.050	0.113	0.391	0.395	0.154	0.510	0.515
1	1.315	0.133	0.581	0.587	0.179	0.751	0.759
1 1/4	1.660	0.140	0.786	0.794	0.191	1.037	1.047
1 1/2	1.900	0.145	0.940	0.949	0.200	1.256	1.269
2	2.375	0.154	1.264	1.277	0.218	1.737	1.751
2 1/2	2.875	0.203	2.004	2.024	0.276	2.650	2.677
3	3.500	0.216	2.621	2.647	0.300	3.547	3.582
3 1/2	4.000	0.226	3.151	3.183	0.318	4.326	4.369
4	4.500	0.237	3.733	3.770	0.337	5.183	5.235
5	5.563	0.258	5.057	5.108	0.375	7.188	7.260
6	6.625	0.280	6.564	6.630	0.432	9.884	9.983
8	8.625	0.277[a]	8.543	8.628	0.500	15.01	15.16
8	8.625	0.322	9.878	9.977	...	...	...
10	10.750	0.279[c]	10.79	10.90	0.500[b]	18.93	19.12
10	10.750	0.307[a]	11.84	11.96	...	...	...
10	10.750	0.365	14.00	14.14	...	...	...
12	12.750	0.330[a]	15.14	15.29	0.500[c]	22.63	22.86

[a] Schedule 30. [b] Schedule 60. [c] No ASA Schedule.

TABLE 13–41

Nickel and Nickel-Alloy Seamless Pipe and Tubing – Sizes Regularly Available

ASTM Designations: B161–49T, B165–49T, B167–49T

Nominal Pipe Size	Outside Diameter	Nominal Wall Thickness: Schedule No. 10	Schedule No. 40 or Standard	Schedule No. 80 or Extra Strong
1/8	0.405	0.049	0.068	0.095
1/4	0.540	0.065	0.088	0.119
3/8	0.674	0.065	0.091	0.126
1/2	0.840	0.083	0.109	0.147
3/4	1.050	0.083	0.113	0.154
1	1.315	0.109	0.133	0.179
1-1/4	1.660	0.109	0.140	0.191
1-1/2	1.900	0.109	0.145	0.200
2	2.375	0.109	0.154	0.218
2-1/2	2.875	0.120	0.203	0.276
3	3.500	0.120	0.216	0.300
3-1/2	4.000	0.120	0.226	0.318
4	4.500	0.120	0.237	0.337
5	5.563	0.134	0.258	0.375
6	6.625	0.134	0.280	0.432
8	8.625		0.322	0.500

TABLE 13–42

Pressure Ratings of *Some American Standard Pipe Fittings

ASA Designation of Standard	Table Numbers	Title of Standard	Titular Pressure Rating	Rating Specified by the Standard
B16.11–1946 Reaffirmed 1952	13–48 to 13–50	Steel Socket-Welding Fittings		Fittings may be used for the same pressure-temperature ratings as pipe of the same schedule number, if they are made of a material having allowable stresses which are equal to or greater than the allowable stresses for the pipe material.
B16.14–1949 Reaffirmed 1953	13–67 to 13–69	Ferrous Plugs, Bushings, and Locknuts		Plugs and bushings have no definite ratings and are used with regular 125-lb cast-iron and 150-lb malleable-iron screwed fittings. For higher pressures solid plugs (not cored) are quite commonly used, and face bushings are recommended.
B16.15–1947 Reaffirmed 1952	13–62 to 13–66	Brass or Bronze Screwed Fittings	125 lb	125 psi max steam pressure, Table 13–64 175 psi max gas or liquid service at 150°F.
B16.17–1949 Reaffirmed 1953	13–60 to 13–61	Brass or Bronze Screwed Fittings	250 lb	250 psi max steam pressure, Table 13–61 400 psi max gas or liquid service at 150°F.
B16.18–1950	None here	Cast-Brass Solder-Joint Fittings		Primarily for use with copper water tube Temperature, Degrees F / 2-inch size and smaller / 2-1/2 inch size and larger: 100 / 175 psi / 150 psi 150 / 125 / 100 200 / 100 / 75 250 / 75 / 50
B16.19–1951	13–51 to 13–56	Malleable-Iron Screwed Fittings	300 lb	Steam and oil: 300 psi at 550°F for sizes 1/4 to 3 in., incl. Liquid and gas at 150°F: (Except street elbows) 2000 psi for sizes 1/4 to 1 in., incl. 1500 psi for sizes 1-1/4 to 2 in., incl. 1000 psi for sizes 2-1/2 and 3 in.

continued on next page

TABLE 13–42, continued

ASA Designation of Standard	Table Numbers	Title of Standard	Titular Pressure Rating	Rating Specified by the Standard
B16.22–1951	None here	Wrought-Copper and Wrought-Bronze Solder-Joint Fittings		Data given for B16.18 apply when solder used in joints is 50:50 tin-lead, ASTM B 32, Alloy Grade 50A. Higher temperature-pressure ratings are listed in B16.22 for other solders.
B16.24–1953	None here	Brass or Bronze Flanges and Flanged Fittings	150 lb	150 psi max steam pressure at 400 or 500°F, depending upon grade of material in fitting. 225 psi max gas or liquid service at 150°F.
			300 lb	300 psi max steam pressure at 422 and 550°F, depending upon grade of material in fitting. 500 psi max gas or liquid service at 150°F.
B16.3 –1951	None here	Malleable-Iron Screwed Fittings	150 lb	150 psig max saturated steam pressure 300 psig max liquid and gas service pressures at 150°F.
B16.4–1949 Reaffirmed 1953	13–58 to 13–59	Cast-Iron Screwed Fittings	125 lb	125 psi gage max saturated steam pressure 175 psi gage max liquid and gas service pressure at 150°F.
	13–57		250 lb	250 psi gage max saturated steam pressure 400 psi gage max liquid and gas service pressure at 150°F.
B16.9 –1951	13–43 to 13–47	Steel Butt-Welding Fittings		Fittings shall be designed so that the pressure rating may be calculated as for straight seamless pipe of the same or equivalent material in accordance with the rules established in the various sections of the *Code for Pressure Piping.

*Representative standards are presented here and in the tables of this volume. Neither is any attempt made to cover the design of systems of piping in accordance with the various codes for fired and unfired pressure vessels. For such systems the reader is referred, in general, to the ASME Boiler and Pressure Vessel Code, and specifically to the following: (1) *Code for Pressure Piping*, B31.1-1951, Including 1953 Supplement No. 1; (2) *API-ASME Code for Unfired Pressure Vessels*-1951; and (3) *Steel Pipe Flanges and Flanged Fittings*, B16.5-1953.

TABLE 13–43

Dimensions of Long-Radius Elbows – Steel, Butt-Welding, Pipe Fittings

ASA B16.9–1951

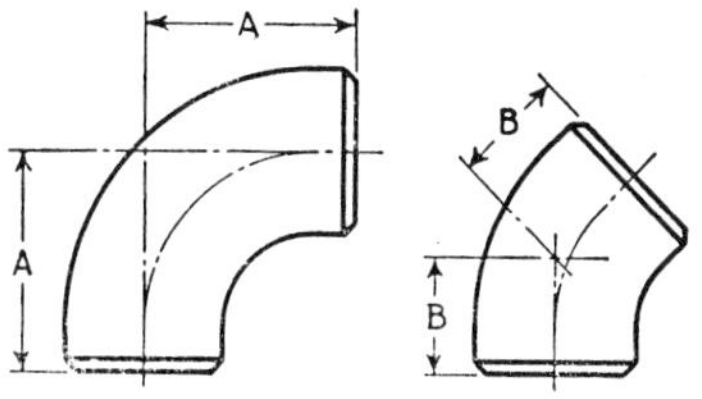

Nominal Pipe Size	Outside Diameter at Bevel	Center-to-End	
		90-Deg Elbows A	45-Deg Elbows B
1	1.315	1½	⅞
1¼	1.660	1⅞	1
1½	1.900	2¼	1⅛
2	2.375	3	1⅜
2½	2.875	3¾	1¾
3	3.500	4½	2
3½	4.000	5¼	2¼
4	4.500	6	2½
5	5.563	7½	3⅛
6	6.625	9	3¾
8	8.625	12	5
10	10.750	15	6¼
12	12.750	18	7½
14	14.000	21	8¾
16	16.000	24	10
18	18.000	27	11¼
20	20.000	30	12½
24	24.000	36	15

All dimensions are in inches.

TABLE 13–44

Dimensions of Straight Tees – Steel, Butt-Welding, Pipe Fittings

ASA B16.9–1951

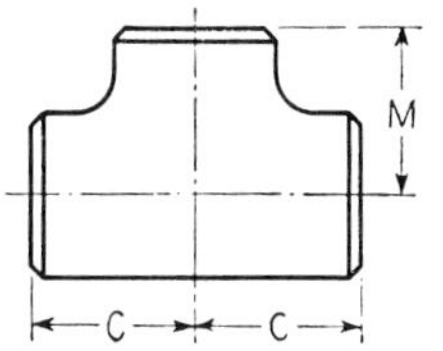

Nominal Pipe Size	Outside Diameter at Bevel	Center-to-End	
		Run C	Outlet M
1	1.315	1½	1½
1¼	1.660	1⅞	1⅞
1½	1.900	2¼	2¼
2	2.375	2½	2½
2½	2.875	3	3
3	3.500	3⅜	3⅜
3½	4.000	3¾	3¾
4	4.500	4⅛	4⅛
5	5.563	4⅞	4⅞
6	6.625	5⅝	5⅝
8	8.625	7	7
10	10.750	8½	8½
12	12.750	10	10
14	14.000	11	
16	16.000	12	Not
18	18.000	13½	stand-
20	20.000	15	ard
24	24.000	17	

All dimensions are in inches.

TOLERANCES

All Fittings				Elbows and Tees Also Table 13–45	Reducers Table 13–46
Nominal Pipe Size	Outside Diameter at Bevel	Inside Diameter at End	Wall Thickness	Center-to-End Dimension A, B, C, M	Over-all Length H
½ to 2½	+1/16 −1/32	±1/32	Not less than 87½% of nominal thickness	±1/16	±1/16
3 to 3½	±1/16	±1/16		±1/16	±1/16
4	±1/16	±1/16		±1/16	±1/16
5 to 8	+3/32 −1/16	±1/16		±1/16	±1/16
10 to 18	+5/32 −⅛	±⅛		±3/32	±3/32
20 to 24	+¼ −3/16	±3/16		±3/32	±3/32

All dimensions are in inches.

TABLE 13-45

Dimensions of Reducing Outlet Tees – Steel, Butt-Welding, Pipe Fittings

ASA B16.9-1951

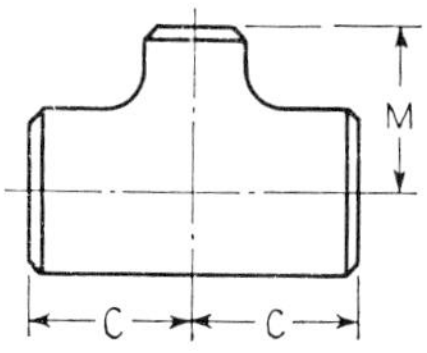

Nominal Pipe Size	Outside Diameter at Bevel		Center-to-End	
	Run	Outlet	Run C	Outlet M
1 × 1 × 3/4	1.315	1.050	1 1/2	1 1/2
1 × 1 × 1/2	1.315	0.840	1 1/2	1 1/2
1 1/4 × 1 1/4 × 1	1.660	1.315	1 7/8	1 7/8
1 1/4 × 1 1/4 × 3/4	1.660	1.050	1 7/8	1 7/8
1 1/4 × 1 1/4 × 1/2	1.660	0.840	1 7/8	1 7/8
1 1/2 × 1 1/2 × 1 1/4	1.900	1.660	2 1/4	2 1/4
1 1/2 × 1 1/2 × 1	1.900	1.315	2 1/4	2 1/4
1 1/2 × 1 1/2 × 3/4	1.900	1.050	2 1/4	2 1/4
1 1/2 × 1 1/2 × 1/2	1.900	0.840	2 1/4	2 1/4
2 × 2 × 1 1/2	2.375	1.900	2 1/2	2 3/8
2 × 2 × 1 1/4	2.375	1.660	2 1/2	2 1/4
2 × 2 × 1	2.375	1.315	2 1/2	2
2 × 2 × 3/4	2.375	1.050	2 1/2	1 3/4
2 1/2 × 2 1/2 × 2	2.875	2.375	3	2 3/4
2 1/2 × 2 1/2 × 1 1/2	2.875	1.900	3	2 5/8
2 1/2 × 2 1/2 × 1 1/4	2.875	1.660	3	2 1/2
2 1/2 × 2 1/2 × 1	2.875	1.315	3	2 1/4
3 × 3 × 2 1/2	3.500	2.875	3 3/8	3 1/4
3 × 3 × 2	3.500	2.375	3 3/8	3
3 × 3 × 1 1/2	3.500	1.900	3 3/8	2 7/8
3 × 3 × 1 1/4	3.500	1.660	3 3/8	2 3/4
3 1/2 × 3 1/2 × 3	4.000	3.500	3 3/4	3 5/8
3 1/2 × 3 1/2 × 2 1/2	4.000	2.875	3 3/4	3 1/2
3 1/2 × 3 1/2 × 2	4.000	2.375	3 3/4	3 1/4
3 1/2 × 3 1/2 × 1 1/2	4.000	1.900	3 3/4	3 1/8
4 × 4 × 3 1/2	4.500	4.000	4 1/8	4
4 × 4 × 3	4.500	3.500	4 1/8	3 7/8
4 × 4 × 2 1/2	4.500	2.875	4 1/8	3 3/4
4 × 4 × 2	4.500	2.375	4 1/8	3 1/2
4 × 4 × 1 1/2	4.500	1.900	4 1/8	3 3/8
5 × 5 × 4	5.563	4.500	4 7/8	4 5/8
5 × 5 × 3 1/2	5.563	4.000	4 7/8	4 1/2
5 × 5 × 3	5.563	3.500	4 7/8	4 3/8
5 × 5 × 2 1/2	5.563	2.875	4 7/8	4 1/4
5 × 5 × 2	5.563	2.375	4 7/8	4 1/8
6 × 6 × 5	6.625	5.563	5 5/8	5 3/8
6 × 6 × 4	6.625	4.500	5 5/8	5 1/8
6 × 6 × 3 1/2	6.625	4.000	5 5/8	5
6 × 6 × 3	6.625	3.500	5 5/8	4 7/8
6 × 6 × 2 1/2	6.625	2.875	5 5/8	4 3/4
8 × 8 × 6	8.625	6.625	7	6 5/8
8 × 8 × 5	8.625	5.563	7	6 3/8
8 × 8 × 4	8.625	4.500	7	6 1/8
8 × 8 × 3 1/2	8.625	4.000	7	6
10 × 10 × 8	10.750	8.625	8 1/2	8
10 × 10 × 6	10.750	6.625	8 1/2	7 5/8
10 × 10 × 5	10.750	5.563	8 1/2	7 1/2
10 × 10 × 4	10.750	4.500	8 1/2	7 1/4
12 × 12 × 10	12.750	10.750	10	9 1/2
12 × 12 × 8	12.750	8.625	10	9
12 × 12 × 6	12.750	6.625	10	8 5/8
12 × 12 × 5	12.750	5.563	10	8 1/2
14 × 14 × 12	14.000	12.750	11	Not Standard
14 × 14 × 10	14.000	10.750	11	Not Standard
14 × 14 × 8	14.000	8.625	11	Not Standard
14 × 14 × 6	14.000	6.625	11	Not Standard
16 × 16 × 14	16.000	14.000	12	Not Standard
16 × 16 × 12	16.000	12.750	12	Not Standard
16 × 16 × 10	16.000	10.750	12	Not Standard
16 × 16 × 8	16.000	8.625	12	Not Standard
16 × 16 × 6	16.000	6.625	12	Not Standard
18 × 18 × 16	18.000	16.000	13 1/2	Not Standard
18 × 18 × 14	18.000	14.000	13 1/2	Not Standard
18 × 18 × 12	18.000	12.750	13 1/2	Not Standard
18 × 18 × 10	18.000	10.750	13 1/2	Not Standard
18 × 18 × 8	18.000	8.625	13 1/2	Not Standard
20 × 20 × 18	20.000	18.000	15	Not Standard
20 × 20 × 16	20.000	16.000	15	Not Standard
20 × 20 × 14	20.000	14.000	15	Not Standard
20 × 20 × 12	20.000	12.750	15	Not Standard
20 × 20 × 10	20.000	10.750	15	Not Standard
20 × 20 × 8	20.000	8.625	15	Not Standard
24 × 24 × 20	24.000	20.000	17	Not Standard
24 × 24 × 18	24.000	18.000	17	Not Standard
24 × 24 × 16	24.000	16.000	17	Not Standard
24 × 24 × 14	24.000	14.000	17	Not Standard
24 × 24 × 12	24.000	12.750	17	Not Standard
24 × 24 × 10	24.000	10.750	17	Not Standard

All dimensions are in inches.

* See Table 13-43 for tolerances

TABLE 13-46

Dimensions* of Reducers – Steel, Butt-Welding, Pipe Fittings

ASA B16.9–1951

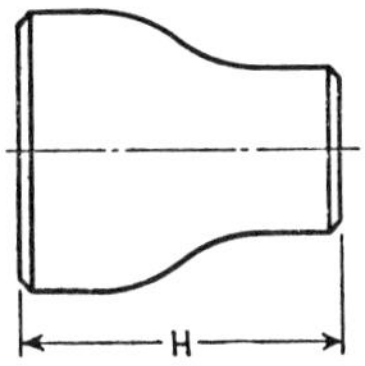

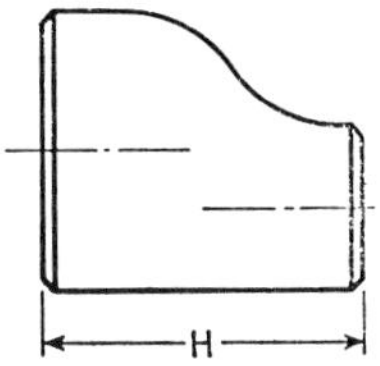

Nominal Pipe Size	Outside Diameter at Bevel		End-to-End
	Large End	Small End	H
1 × ¾	1.315	1.050	2
1 × ½	1.315	0.840	2
1¼ × 1	1.660	1.315	2
1¼ × ¾	1.660	1.050	2
1¼ × ½	1.660	0.840	2
1½ × 1¼	1.900	1.660	2½
1½ × 1	1.900	1.315	2½
1½ × ¾	1.900	1.050	2½
1½ × ½	1.900	0.840	2½
2 × 1½	2.375	1.900	3
2 × 1¼	2.375	1.660	3
2 × 1	2.375	1.315	3
2 × ¾	2.375	1.050	3
2½ × 2	2.875	2.375	3½
2½ × 1½	2.875	1.900	3½
2½ × 1¼	2.875	1.660	3½
2½ × 1	2.875	1.315	3½
3 × 2½	3.500	2.875	3½
3 × 2	3.500	2.375	3½
3 × 1½	3.500	1.900	3½
3 × 1¼	3.500	1.660	3½
3½ × 3	4.000	3.500	4
3½ × 2½	4.000	2.875	4
3½ × 2	4.000	2.375	4
3½ × 1½	4.000	1.900	4
3½ × 1¼	4.000	1.660	4
4 × 3½	4.500	4.000	4
4 × 3	4.500	3.500	4
4 × 2½	4.500	2.875	4
4 × 2	4.500	2.375	4
4 × 1½	4.500	1.900	4
5 × 4	5.563	4.500	5
5 × 3½	5.563	4.000	5
5 × 3	5.563	3.500	5
5 × 2½	5.563	2.875	5
5 × 2	5.563	2.375	5
6 × 5	6.625	5.563	5½
6 × 4	6.625	4.500	5½
6 × 3½	6.625	4.000	5½
6 × 3	6.625	3.500	5½
6 × 2½	6.625	2.875	5½
8 × 6	8.625	6.625	6
8 × 5	8.625	5.563	6
8 × 4	8.625	4.500	6
8 × 3½	8.625	4.000	6
10 × 8	10.750	8.625	7
10 × 6	10.750	6.625	7
10 × 5	10.750	5.563	7
10 × 4	10.750	4.500	7
12 × 10	12.750	10,750	8
12 × 8	12.750	8.625	8
12 × 6	12.750	6.625	8
12 × 5	12.750	5.563	8
14 × 12	14.000	12.750	13
14 × 10	14.000	10.750	13
14 × 8	14.000	8.625	13
14 × 6	14.000	6.625	13
16 × 14	16.000	14.000	14
16 × 12	16.000	12.750	14
16 × 10	16.000	10.750	14
16 × 8	16.000	8.625	14
18 × 16	18.000	16.000	15
18 × 14	18.000	14.000	15
18 × 12	18.000	12.750	15
18 × 10	18.000	10.750	15
20 × 18	20.000	18.000	20
20 × 16	20.000	16.000	20
20 × 14	20.000	14.000	20
20 × 12	20.000	12.750	20
24 × 20	24.000	20.000	20
24 × 18	24.000	18.000	20
24 × 16	24.000	16.000	20

All dimensions are in inches.

* See Table 13–43 for tolerances

TABLE 13–47

Recommended Bevel When Wall Thickness* of Fitting Equals That of Pipe – Steel, Butt-Welding, Pipe Fittings

ASA B16.9–1951

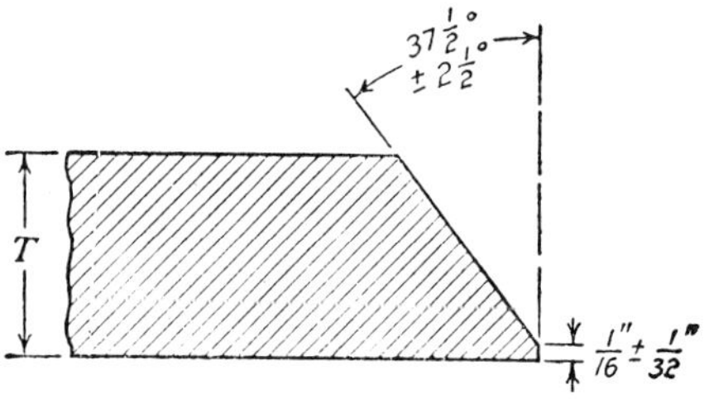

Fig. 1 Recommended Bevel for Wall Thicknesses (T) at End of Fitting, 3/4 In. or Less

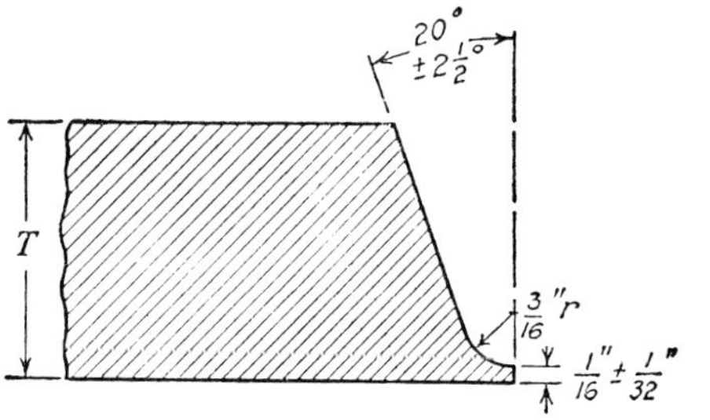

Fig. 2 Recommended Bevel for Wall Thicknesses (T) at End of Fitting, Greater Than 3/4 In. but not Greater Than 1 3/4 In.

Ends when wall thickness is less than 3/16 in. may be cut square.

*Wall thickness of fitting is usually chosen equal to that of pipe, not only for appearance but also to insure fitting of same bursting strength as pipe. Sometimes fitting of greater wall thickness than pipe is used.

TABLE 13-48

Tolerances on Steel, Socket-Welding, Pipe Fittings

ASA B16.11 – 1946 Reaffirmed 1952

(a) Center to Bottom of Socket

Tolerance for "A" Table 13–49

For sizes	
1/8, 1/4 in.	±0.03 in.
3/8, 1/2, 3/4 in	±0.06 in.
1, 1 1/4, 1 1/2, 2 in	±0.08 in.
2 1/2 in. and larger	±0.10 in.

(b) Bottom to Bottom of Sockets – Couplings

Tolerance for "E" Table 13–50

For sizes	
1/8, 1/4 in.	±0.06 in.
3/8, 1/2, 3/4 in	±0.12 in.
1, 1 1/4, 1 1/2, 2 in	±0.16 in.
2 1/2 in. and larger	±0.20 in.

(c) Bottom of Socket to Opposite Face – Half Couplings

Tolerance for "F" Table 13–50

For sizes	
1/8, 1/4 in.	±0.03 in.
3/8, 1/2, 3/4 in	±0.06 in.
1, 1 1/4, 1 1/2, 2 in	±0.08 in.
2 1/2 in. and larger	±0.10 in.

(d) Bore Diameter of Socket

Tolerance for "B" Table 13–49 and 13–50

For sizes	
2 in. and smaller	– 0 in. + 0.010 in.
2 1/2 in. and larger	– 0 in. + 0.015 in.

(e) Bore Diameter of Fitting

Tolerancefor "D" Table 13–49 and 13–50

For sizes	
2 in. and smaller	± 0.015 in.
2 1/2 in. and larger	± 0.030 in.

(f) Concentricity of Bores

The socket and fitting bores shall be concentric within a tolerance of plus or minus 0.030 in. for all sizes.

(g) Coincidence of Axes

The maximum allowable variation in the alignment of the fitting bore and socket bore axes shall be 1/16 in. in 1 ft.

TABLE 13–49

Dimensions† of Socket-Welding Elbows, Tees, and Crosses – Steel Pipe Fittings

ASA B16.11 – 1946 Reaffirmed 1952

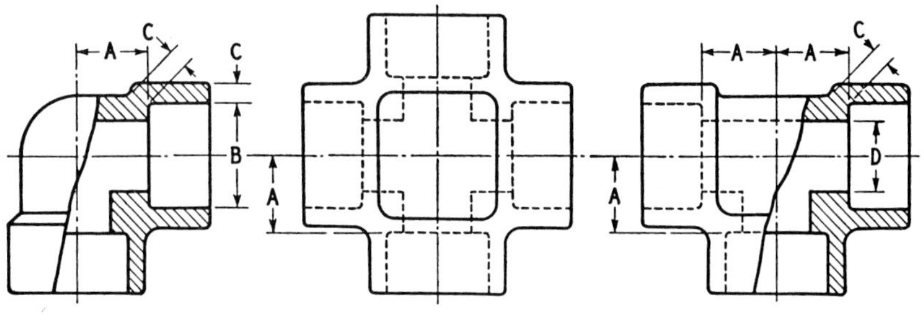

Nominal Pipe Size	Depth of Socket, Min	Center to Bottom of Socket[3] Sched* 40 and 80	Center to Bottom of Socket[3] Sched 160	Bore Diameter of Socket, Min	Socket Wall Thickness[2] Min Sched 40	Socket Wall Thickness[2] Min Sched 80	Socket Wall Thickness[2] Min Sched 160	Bore Diameter of Fitting Sched 40	Bore Diameter of Fitting Sched 80	Bore Diameter of Fitting Sched 160
		A**		B	C			D		
1/8	3/8	7/16	....	0.420	0.125	0.125	.....	0.269	0.215	.....
1/4	3/8	7/16	....	0.555	0.125	0.149	.....	0.364	0.302	.....
3/8	3/8	17/32	....	0.690	0.125	0.158	.....	0.493	0.423	.....
1/2	3/8	5/8	3/4	0.855	0.136	0.184	0.234	0.622	0.546	0.466
3/4	1/2	3/4	7/8	1.065	0.141	0.193	0.273	0.824	0.742	0.614
1	1/2	7/8	1 1/16	1.330	0.166	0.224	0.313	1.049	0.957	0.815
1 1/4	1/2	1 1/16	1 1/4	1.675	0.175	0.239	0.313	1.380	1.278	1.160
1 1/2	1/2	1 1/4	1 1/2	1.915	0.181	0.250	0.351	1.610	1.500	1.338
2	5/8	1 1/2	1 5/8	2.406	0.193	0.273	0.429	2.067	1.939	1.689
2 1/2	5/8	1 5/8	2 1/4	2.906	0.254	0.345	0.469	2.469	2.323	2.125
3	5/8	2 1/4	2 1/2	3.535	0.270	0.375	0.546	3.068	2.900	2.626

All dimensions are given in inches.

[2] Minimum dimension "C" is 1¼ times the nominal pipe thickness, but not less than ⅛ in.

[3] Reducing sizes have the same center to bottom of socket dimension as the largest size of the reducing fitting.

†See Table 13–48 for tolerances.

*Schedule 40 corresponds to standard pipe; schedule 80 to extra strong.

**A cardinal principle of Standard B16.11 is the maintenance of a fixed position for the bottom of the socket with reference to the center line of the fitting.

TABLE 13–50

Dimensions† of Socket-Welding 45-Degree Elbows, Couplings, and Half Couplings – Steel Pipe Fittings

ASA B16.11 – 1946 Reaffirmed 1952

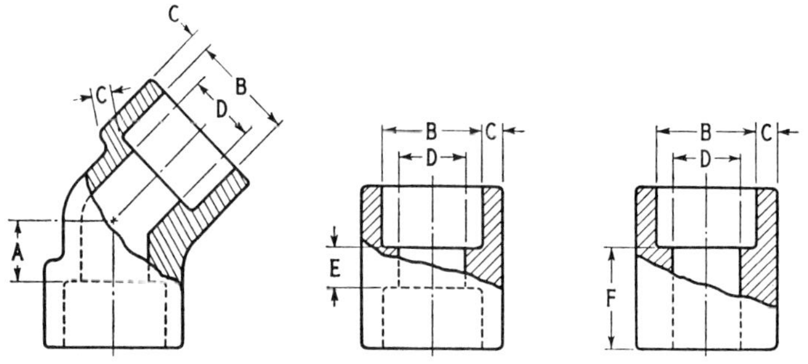

Nominal Pipe Size	Depth of Socket, Min	Center to Bottom of Socket for 45 Deg Ells[3]: Sched* 40 and 80	Center to Bottom of Socket for 45 Deg Ells[3]: Sched 160	Couplings Distance Between Bottoms of Sockets[3]	Half Couplings, Bottom of Socket to Opposite Face	Bore Diameter of Socket, Min	Socket Wall Thickness[2] Min: Sched 40	Socket Wall Thickness[2] Min: Sched 80	Socket Wall Thickness[2] Min: Sched 160	Bore Diameter of Fitting: Sched 40	Bore Diameter of Fitting: Sched 80	Bore Diameter of Fitting: Sched 160
		A**		E	F	B	C			D		
1/8	3/8	5/16	.....	1/4	5/8	0.420	0.125	0.125	.....	0.269	0.215	.....
1/4	3/8	5/16	.....	1/4	5/8	0.555	0.125	0.149	.....	0.364	0.302	.....
3/8	3/8	5/16	.....	1/4	11/16	0.690	0.125	0.158	.....	0.493	0.423	.....
1/2	3/8	7/16	1/2	3/8	7/8	0.855	0.136	0.184	0.234	0.622	0.546	0.466
3/4	1/2	1/2	9/16	3/8	15/16	1.065	0.141	0.193	0.273	0.824	0.742	0.614
1	1/2	9/16	11/16	1/2	1 1/8	1.330	0.166	0.224	0.313	1.049	0.957	0.815
1 1/4	1/2	11/16	13/16	1/2	1 3/16	1.675	0.175	0.239	0.313	1.380	1.278	1.160
1 1/2	1/2	13/16	1	1/2	1 1/4	1.915	0.181	0.250	0.351	1.610	1.500	1.338
2	5/8	1	1 1/8	3/4	1 5/8	2.406	0.193	0.273	0.429	2.067	1.939	1.689
2 1/2	5/8	1 1/8	1 1/4	3/4	1 11/16	2.906	0.254	0.345	0.469	2.469	2.323	2.125
3	5/8	1 1/4	1 3/8	3/4	1 3/4	3.535	0.270	0.375	0.546	3.068	2.900	2.626

All dimensions given in inches.

[2] Minimum dimension "C" is 1¼ times the nominal pipe thickness, but not less than ⅛ in.

[3] Reducing sizes have the same center to bottom of socket dimension as the largest size of the reducing fitting.

†See Table 13–48 for tolerances.

*Schedule 40 corresponds to Standard pipe; Schedule 80 to extra strong.

**A cardinal principle of Standard B16.11 is the maintenance of a fixed position for the bottom of the socket with reference to the line of the fitting.

TABLE 13–51

Dimensions of Elbows, Tees and Crosses – Malleable-Iron Screwed Fittings, 300 lb *

ASA B16.19–1951

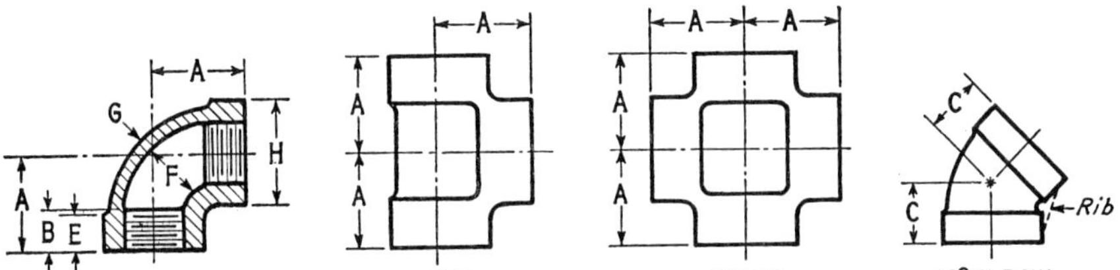

Nominal Pipe Size	Center-to-End, Elbows, Tees, and Crosses A	Center-to-End, 45-Deg Elbows C	Length of Thread, Min B	Width of Band, Min E	Inside Diameter of Fitting F Min	Inside Diameter of Fitting F Max	Metal Thickness G	Outside Diameter of Band, Min H
1/4	0.94	0.81	0.43	0.38	0.540	0.584	0.14	0.93
3/8	1.06	0.88	0.47	0.44	0.675	0.719	0.15	1.12
1/2	1.25	1.00	0.57	0.50	0.840	0.897	0.16	1.34
3/4	1.44	1.13	0.64	0.56	1.050	1.107	0.18	1.63
1	1.63	1.31	0.75	0.62	1.315	1.385	0.20	1.95
1 1/4	1.94	1.50	0.84	0.69	1.660	1.730	0.22	2.39
1 1/2	2.13	1.69	0.87	0.75	1.900	1.970	0.24	2.68
2	2.50	2.00	1.00	0.84	2.375	2.445	0.26	3.28
2 1/2	2.94	2.25	1.17	0.94	2.875	2.975	0.31	3.86
3	3.38	2.50	1.23	1.00	3.500	3.600	0.35	4.62

All dimensions are given in inches.
*See Table 13–42 for pressure rating.

TABLE 13–52

Dimensions of Couplings – Malleable-Iron Screwed Fittings, 300 lb*

ASA B16.19–1951

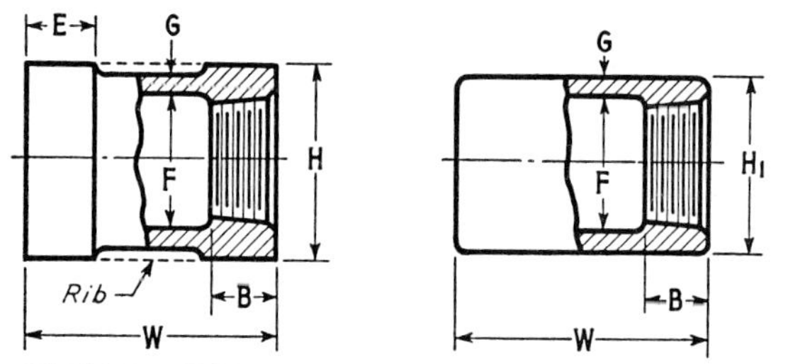

Nominal Pipe Size	Length of Thread, Min	Width of Band, Min	Inside Diameter of Fitting F		Metal Thick-ness	Outside Diameter of Coupling Min*	Outside Diameter of Band Min**	Length of Straight Coupling
	B	E	Min	Max	G	H_1	H	W
1/4	0.43	0.38	0.540	0.584	0.14	0.820	0.93	1.375
3/8	0.47	0.44	0.675	0.719	0.15	0.975	1.12	1.625
1/2	0.57	0.50	0.840	0.897	0.16	1.160	1.34	1.875
3/4	0.64	0.56	1.050	1.107	0.18	1.410	1.63	2.125
1	0.75	0.62	1.315	1.385	0.20	1.715	1.95	2.376
1 1/4	0.84	0.69	1.660	1.730	0.22	2.100	2.39	2.875
1 1/2	0.87	0.75	1.900	1.970	0.24	2.380	2.68	2.875
2	1.00	0.84	2.375	2.445	0.26	2.895	3.28	3.625
2 1/2	1.17	0.94	2.875	2.975	0.31	3.495	3.86	4.125
3	1.23	1.00	3.500	3.600	0.35	4.200	4.62	4.125

All dimensions are given in inches.

*H_1 diameter is standard for couplings without bands. H_1 equals F minimum plus 2G.
**H minimum is for couplings with bands and is optional with the manufacturer.
★See Table 13–42 for pressure rating.

TABLE 13–53

Center-to-End Dimensions of Reducing Tees – Malleable-Iron Screwed Fittings, 300 lb*

ASA B16.19–1951

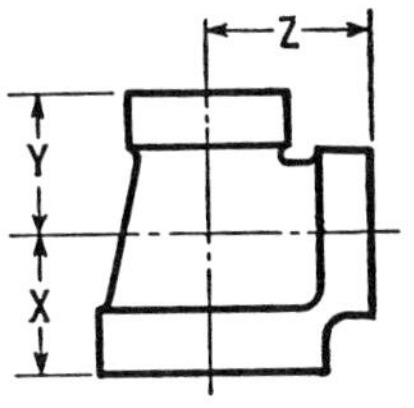

Nominal Pipe Size	Center-to-End X	Center-to-End Y	Center-to-End Z
½ × ½ × ⅜	1.19	1.19	1.19
½ × ⅜ × ½	1.25	1.19	1.25
¾ × ¾ × ½	1.31	1.31	1.38
¾ × ½ × ¾	1.44	1.38	1.44
1 × 1 × ¾	1.50	1.50	1.56
1 × 1 × ½	1.44	1.44	1.50
1 × 1 × ⅜	1.31	1.31	1.44
1 × ¾ × 1	1.63	1.56	1.63
1 ¼ × 1 ¼ × 1	1.75	1.75	1.81
1 ¼ × 1 ¼ × ¾	1.63	1.63	1.75
1 ¼ × 1 ¼ × ½	1.50	1.50	1.69
1 ¼ × 1 × 1 ¼	1.94	1.81	1.94
1 ½ × 1 ½ × 1 ¼	2.00	2.00	2.06
1 ½ × 1 ½ × 1	1.81	1.81	2.00
1 ½ × 1 ½ × ¾	1.69	1.69	1.88
1 ½ × 1 ½ × ½	1.63	1.63	1.81
1 ½ × 1 ¼ × 1 ½	2.13	2.06	2.13
2 × 2 × 1 ½	2.25	2.25	2.38
2 × 2 × 1 ¼	2.13	2.13	2.31
2 × 2 × 1	2.00	2.00	2.25
2 × 2 × ¾	1.81	1.81	2.13
2 × 2 × ½	1.75	1.75	2.06
2 × 1 ½ × 2	2.50	2.38	2.50
2 ½ × 2 ½ × 2	2.69	2.69	2.75
2 ½ × 2 ½ × 1 ½	2.44	2.44	2.63
2 ½ × 2 × 2 ½	2.94	2.75	2.94
3 × 3 × 2 ½	3.06	3.06	3.31
3 × 3 × 2	2.81	2.81	3.13
3 × 2 ½ × 3	3.38	3.31	3.38

All dimensions are given in inches.

For dimensions not given see Table 13–51

Reducing sizes of fittings for which dimensions are not given in tables may be produced from regular patterns for listed sizes by sand bushing.

*See Table 13–42 for pressure rating.

TABLE 13–54

Dimensions of Reducing Couplings – Malleable-Iron Screwed Fittings, 300 lb*

ASA B16.19–1951

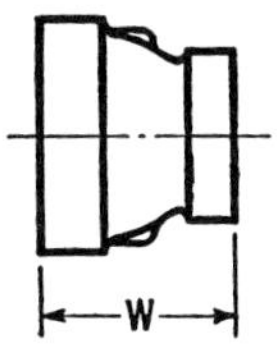

Nominal Pipe Size	Length W
⅜ × ¼	1.44
½ × ⅜	1.69
½ × ¼	1.69
¾ × ½	1.75
¾ × ⅜	1.75
¾ × ¼	1.75
1 × ¾	2.00
1 × ½	2.00
1 × ⅜	2.00
1 × ¼	2.00
1 ¼ × 1	2.38
1 ¼ × ¾	2.38
1 ¼ × ½	2.38
1 ½ × 1 ¼	2.69
1 ½ × 1	2.69
1 ½ × ¾	2.69
1 ½ × ½	2.69
2 × 1 ½	3.19
2 × 1 ¼	3.19
2 × 1	3.19
2 × ¾	3.19
2 × ½	3.19
2 ½ × 2	3.69
2 ½ × 1 ½	3.69
3 × 2 ½	4.06
3 × 2	4.06
3 × 1 ½	4.06

All dimensions are given in inches.

For dimensions not given see Table 13–51.

Reducing sizes of fittings for which dimensions are not given in the tables may be produced from regular patterns for listed sizes by sand bushing.

*See Table 13–42 for pressure rating.

TABLE 13-55

Center-to-End Dimensions of 90-Deg Reducing Elbows – Malleable-Iron Screwed Fittings, 300 lb*

ASA B16.19-1951

Nominal Pipe Size	Center-to-End X	Center-to-End Z
½ × ⅜	1.19	1.19
¾ × ½	1.31	1.38
1 × ¾	1.50	1.56
1 ¼ × 1	1.75	1.81
1 ½ × 1 ¼	2.00	2.06
2 × 1 ½	2.25	2.38
2 ½ × 2	2.69	2.75
3 × 2 ½	3.06	3.31

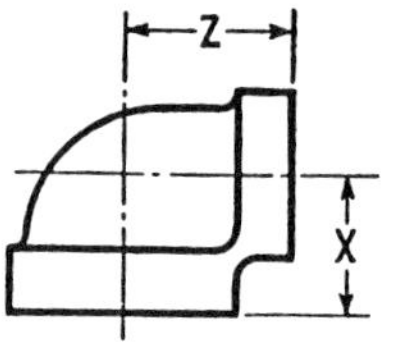

All dimensions are given in inches.

For dimensions not given see Table 13–51.

Reducing sizes of fittings for which dimensions are not given in tables may be produced from regular patterns for listed sizes by sand bushing.

*See Table 13–42 for pressure rating.

TABLE 13-56

Dimensions of Street Elbows – Malleable-Iron Screwed Fittings, 300 lb*

ASA B16.19-1951

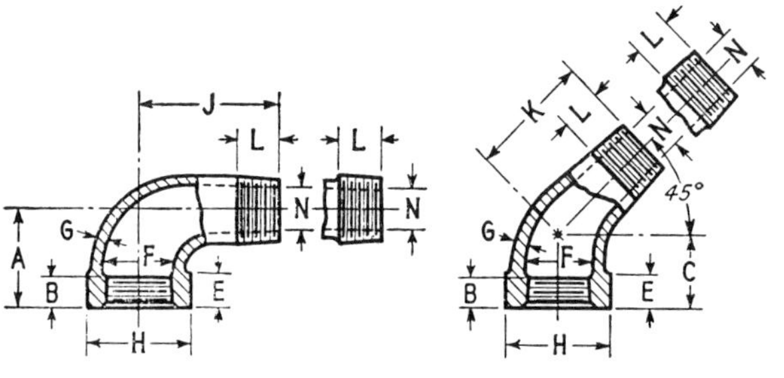

Nominal Pipe Size	90-Deg Elbows Center-to-End	90-Deg Elbows Center to Male End	45-Deg Elbows Center-to-End	45-Deg Elbows Center to Male End	Length of Thread, Min	Width of Band, Min	Inside Diameter of Fitting F Min	Inside Diameter of Fitting F Max	Metal Thick-ness	Outside Diam-eter of Band, Min	Length of External Thread, Min	Max Port Diam-eter Male End
	A	J	C	K	B	E	Min	Max	G	H	L	N
¼	0.94	1.44	..	..	0.43	0.38	0.540	0.584	0.14	0.93	0.40	0.26
⅜	1.06	1.63	..	..	0.47	0.44	0.675	0.719	0.15	1.12	0.41	0.36
½	1.25	2.00	1.00	1.38	0.57	0.50	0.840	0.897	0.16	1.34	0.53	0.49
¾	1.44	2.19	1.13	1.56	0.64	0.56	1.050	1.107	0.18	1.63	0.55	0.67
1	1.63	2.56	1.31	1.81	0.75	0.62	1.315	1.385	0.20	1.95	0.68	0.88
1 ¼	1.94	2.88	1.50	2.13	0.84	0.69	1.660	1.730	0.22	2.39	0.71	1.16
1 ½	2.13	3.13	1.69	2.31	0.87	0.75	1.900	1.970	0.24	2.68	0.72	1.35
2	2.50	3.69	2.00	2.69	1.00	0.84	2.375	2.445	0.26	3.28	0.76	1.75
2 ½	2.94	4.50	..	..	1.17	0.94	2.875	2.975	0.31	3.86	1.14	2.16
3	3.38	5.13	..	..	1.23	1.00	3.500	3.600	0.35	4.62	1.20	2.67

All dimensions are given in inches.

*Street elbows are not recommended for pressures above 600 lb.

TABLE 13-57

Dimensions of Elbows, Tees and Crosses – 250 lb* Cast Iron Screwed Fittings.

ASA B16.4–1949 Reaffirmed 1953

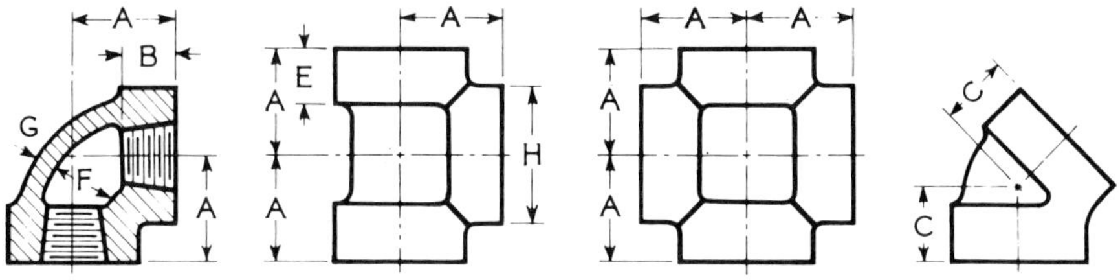

ELBOW TEE CROSS 45° ELBOW

Nominal Pipe Size	Center to End, Elbows, Tees, and Crosses	Center to End, 45-Deg Elbows	Length of Thread, Min	Width of Band, Min	Inside Diameter of Fitting F		Metal Thick-ness[1]	Outside Diameter of Band, Min
	A	C	B	E	Max	Min	G	H
1/4	0.94	0.81	0.43	0.49	0.584	0.540	0.18	1.17
3/8	1.06	0.88	0.47	0.55	0.719	0.675	0.18	1.36
1/2	1.25	1.00	0.57	0.60	0.897	0.840	0.20	1.59
3/4	1.44	1.13	0.64	0.68	1.107	1.050	0.23	1.88
1	1.63	1.31	0.75	0.76	1.385	1.315	0.28	2.24
1 1/4	1.94	1.50	0.84	0.88	1.730	1.660	0.33	2.73
1 1/2	2.13	1.69	0.87	0.97	1.970	1.900	0.35	3.07
2	2.50	2.00	1.00	1.12	2.445	2.375	0.39	3.74
2 1/2	2.94	2.25	1.17	1.30	2.975	2.875	0.43	4.60
3	3.38	2.50	1.23	1.40	3.600	3.500	0.48	5.36
3 1/2	3.75	2.63	1.28	1.49	4.100	4.000	0.52	5.98
4	4.13	2.81	1.33	1.57	4.600	4.500	0.56	6.61
5	4.88	3.19	1.43	1.74	5.663	5.563	0.66	7.92
6	5.63	3.50	1.53	1.91	6.725	6.625	0.74	9.24
8	7.00	4.31	1.72	2.24	8.725	8.625	0.90	11.73
10	8.63	5.19	1.93	2.58	10.850	10.750	1.08	14.37
12	10.00	6.00	2.13	2.91	12.850	12.750	1.24	16.84

All dimensions are given in inches.

The 250-lb standard for screwed fittings covers only the straight sizes of 90- and 45-deg elbows, tees, and crosses.

Fittings having right- and left-hand threads shall have four or more ribs or the letter "L" cast on the band at end with left-hand thread.

[1]Patterns shall be designed to produce castings of the metal thicknesses given in the tables. Metal thickness at no point shall be less than 90 per cent of thickness given in the table.

RIBS. The use of ribs is optional with the manufacturer, except that couplings having right- and left-hand threads shall have four or more ribs or the letter "L" cast on the band at end with left-hand thread.

*Maximum saturated steam pressure or 400 psi gage maximum liquid and gas pressure at 150 F.

TABLE 13–58

Dimensions of Elbows, Tees and Crosses – 125 lb** Cast Iron Screwed Fittings

ASA B16.4–1949 Reaffirmed 1953

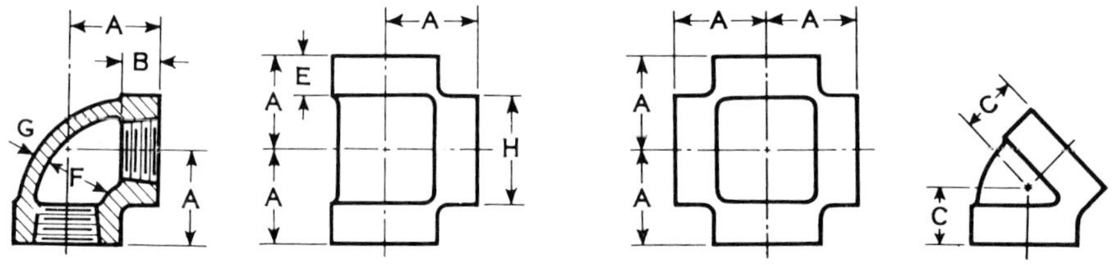

90° ELBOW TEE CROSS 45° ELBOW

**Maximum saturated steam pressure or 175 psi gage maximum liquid and gas pressure at 150 F.

Nominal Pipe Size	Center to End, Elbows, Tees, and Crosses[1] A	Center to End, 45-Deg Elbows C	Length of Thread, Min B	Width of Band, Min E	Inside Diameter of Fitting F Max	Inside Diameter of Fitting F Min	Metal Thickness[2] G	Outside Diameter of Band, Min H
1/4	0.81	0.73	0.32	0.38	0.584	0.540	0.110	0.93
3/8	0.95	0.80	0.36	0.44	0.719	0.675	0.120	1.12
1/2	1.12	0.88	0.43	0.50	0.897	0.840	0.130	1.34
3/4	1.31	0.98	0.50	0.56	1.107	1.050	0.155	1.63
1	1.50	1.12	0.58	0.62	1.385	1.315	0.170	1.95
1 1/4	1.75	1.29	0.67	0.69	1.730	1.660	0.185	2.39
1 1/2	1.94	1.43	0.70	0.75	1.970	1.900	0.200	2.68
2	2.25	1.68	0.75	0.84	2.445	2.375	0.220	3.28
2 1/2	2.70	1.95	0.92	0.94	2.975	2.875	0.240	3.86
3	3.08	2.17	0.98	1.00	3.600	3.500	0.260	4.62
3 1/2	3.42	2.39	1.03	1.06	4.100	4.000	0.280	5.20
4	3.79	2.61	1.08	1.12	4.600	4.500	0.310	5.79
5	4.50	3.05	1.18	1.18	5.663	5.563	0.380	7.05
6	5.13	3.46	1.28	1.28	6.725	6.625	0.430	8.28
8	6.56	4.28	1.47	1.47	8.725	8.625	0.550	10.63
10	*8.08	5.16	1.68	1.68	10.850	10.750	0.690	13.12
12	*9.50	5.97	1.88	1.88	12.850	12.750	0.800	15.47

All dimensions given in inches.
Fittings having right- and left-hand threads shall have four or more ribs or the letter "L" cast on the band at end with left-hand thread.
*This applies to elbows and tees only.
[1]Dimensions for reducing tees are given in Table 13–59.
[2]Patterns shall be designed to produce castings of metal thicknesses given in the table. Metal thickness at no point shall be less than 90 per cent of the thickness given in the table.
RIBS. The use of ribs is optional with the manufacturer, except that couplings having right- and left-hand threads shall have four or more ribs or the letter "L" cast on the band at end with left-hand thread.

TABLE 13–59

Dimensions of † Reducing Tees – 125 lb* Cast Iron Screwed Fittings

ASA B16.4–1949 Reaffirmed 1953

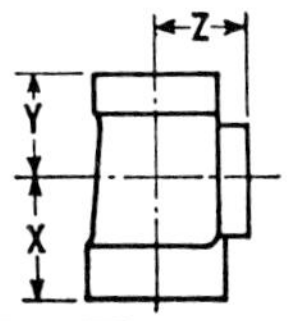

Nominal Pipe Sizes	Center to End		
	X	Y	Z
1/2 x 1/2 x 3/8	1.04	1.04	1.03
1/2 x 1/2 x 1/4	0.97	0.97	0.98
3/4 x 3/4 x 1/2	1.20	1.20	1.22
3/4 x 3/4 x 3/8	1.12	1.12	1.13
3/4 x 3/4 x 1/4	1.05	1.05	1.08
3/4 x 1/2 x 3/4	1.31	1.22	1.31
3/4 x 1/2 x 1/2	1.20	1.12	1.22
1/2 x 1/2 x 3/4	1.22	1.22	1.20
1 x 1 x 3/4	1.37	1.37	1.45
1 x 1 x 1/2	1.26	1.26	1.36
1 x 1 x 3/8	1.18	1.18	1.27
1 x 3/4 x 1	1.50	1.45	1.50
1 x 3/4 x 3/4	1.37	1.31	1.45
1 x 3/4 x 1/2	1.26	1.20	1.36
1 x 1/2 x 1	1.50	1.36	1.50
1 x 1/2 x 3/4	1.37	1.22	1.45
1 x 1/2 x 1/2	1.26	1.12	1.36
1 x 3/8 x 1	1.50	1.27	1.50
3/4 x 3/4 x 1	1.45	1.45	1.37
1 1/4 x 1 1/4 x 2	2.10	2.10	1.90
1 1/4 x 1 1/4 x 1	1.58	1.58	1.67
1 1/4 x 1 1/4 x 3/4	1.45	1.45	1.62
1 1/4 x 1 1/4 x 1/2	1.34	1.34	1.53
1 1/4 x 1 x 1 1/4	1.75	1.67	1.75
1 1/4 x 1 x 1	1.58	1.50	1.67
1 1/4 x 1 x 3/4	1.45	1.37	1.62
1 1/4 x 1 x 1/2	1.34	1.26	1.53
1 1/4 x 3/4 x 1 1/4	1.75	1.62	1.75
1 1/4 x 3/4 x 1	1.58	1.45	1.67
1 1/4 x 3/4 x 3/4	1.45	1.31	1.62
1 1/4 x 1/2 x 1 1/4	1.75	1.53	1.75
1 1/4 x 1/2 x 1	1.58	1.36	1.67
1 x 1 x 1 1/4	1.67	1.67	1.58
1 1/2 x 1 1/2 x 1 1/4	1.82	1.82	1.88
1 1/2 x 1 1/2 x 1	1.65	1.65	1.80
1 1/2 x 1 1/2 x 3/4	1.52	1.52	1.75
1 1/2 x 1 1/2 x 1/2	1.41	1.41	1.66
1 1/2 x 1 1/4 x 1 1/2	1.94	1.88	1.94
1 1/2 x 1 1/4 x 1 1/4	1.82	1.75	1.88
1 1/2 x 1 1/4 x 1	1.65	1.58	1.80
1 1/2 x 1 1/4 x 3/4	1.52	1.45	1.75
1 1/2 x 1 1/4 x 1/2	1.41	1.34	1.66
1 1/2 x 1 x 1 1/2	1.94	1.80	1.94
1 1/2 x 1 x 1 1/4	1.82	1.67	1.88
1 1/2 x 1 x 1	1.65	1.50	1.80
1 1/2 x 3/4 x 1 1/2	1.94	1.75	1.94
1 1/2 x 3/4 x 1 1/4	1.82	1.62	1.88
1 1/2 x 1/2 x 1 1/2	1.94	1.66	1.94
1 1/4 x 1 x 1 1/2	1.88	1.80	1.82
1 1/4 x 1 1/4 x 1 1/2	1.88	1.88	1.82
1 x 1 x 1 1/2	1.80	1.80	1.65
2 x 2 x 1 1/2	2.02	2.02	2.16
2 x 2 x 1 1/4	1.90	1.90	2.10
2 x 2 x 1	1.73	1.73	2.02

Nominal Pipe Sizes	Center to End		
	X	Y	Z
2 x 2 x 3/4	1.60	1.60	1.97
2 x 2 x 1/2	1.49	1.49	1.88
2 x 1 1/2 x 2	2.25	2.16	2.25
2 x 1 1/2 x 1 1/2	2.02	1.94	2.16
2 x 1 1/2 x 1 1/4	1.90	1.82	2.10
2 x 1 1/2 x 1	1.73	1.65	2.02
2 x 1 1/2 x 3/4	1.60	1.52	1.97
2 x 1 1/2 x 1/2	1.49	1.41	1.88
2 x 1 1/4 x 2	2.25	2.10	2.25
2 x 1 1/4 x 1 1/2	2.02	1.88	2.16
2 x 1 1/4 x 1 1/4	1.90	1.75	2.10
2 x 1 1/4 x 1	1.73	1.58	2.02
2 x 1 x 2	2.25	2.02	2.25
2 x 1 x 1 1/2	2.02	1.80	2.16
2 x 1 x 1 1/4	1.90	1.67	2.10
2 x 3/4 x 2	2.25	1.97	2.25
1 1/2 x 1 x 2	2.16	2.02	2.02
2 x 1/2 x 2	2.25	1.88	2.25
1 1/2 x 1 1/2 x 2	2.16	2.16	2.02
1 1/2 x 1 1/4 x 2	2.16	2.10	2.02
2 1/2 x 2 1/2 x 2	2.39	2.39	2.60
2 1/2 x 2 1/2 x 1 1/2	2.16	2.16	2.51
2 1/2 x 2 1/2 x 1 1/4	2.04	2.04	2.45
2 1/2 x 2 1/2 x 1	1.87	1.87	2.37
2 1/2 x 2 1/2 x 3/4	1.74	1.74	2.32
2 1/2 x 2 1/2 x 1/2	1.63	1.63	2.23
2 1/2 x 2 x 2 1/2	2.70	2.60	2.70
2 1/2 x 2 x 2	2.39	2.25	2.60
2 1/2 x 2 x 1 1/2	2.16	2.02	2.51
2 1/2 x 2 x 1 1/4	2.04	1.90	2.45
2 1/2 x 2 x 1	1.87	1.73	2.37
2 1/2 x 2 x 3/4	1.74	1.60	2.32
2 1/2 x 2 x 1/2	1.63	1.49	2.23
2 1/2 x 1 1/2 x 2 1/2	2.70	2.51	2.70
2 1/2 x 1 1/2 x 2	2.39	2.16	2.60
2 1/2 x 1 1/2 x 1 1/2	2.16	1.94	2.51
2 1/2 x 1 1/4 x 2 1/2	2.70	2.45	2.70
2 1/2 x 1 1/4 x 2	2.39	2.10	2.60
2 1/2 x 1 x 2 1/2	2.70	2.37	2.70
2 1/2 x 1 x 2	2.39	2.02	2.60
2 1/2 x 3/4 x 2 1/2	2.70	2.32	2.70
2 1/2 x 1/2 x 2 1/2	2.70	2.23	2.70
2 x 2 x 2 1/2	2.60	2.60	2.39
2 x 1 1/2 x 2 1/2	2.60	2.51	2.39
2 x 1 1/4 x 2 1/2	2.60	2.45	2.39
1 1/2 x 1 1/2 x 2 1/2	2.51	2.51	2.16
3 x 3 x 2 1/2	2.83	2.83	2.99
3 x 3 x 2	2.52	2.52	2.89
3 x 3 x 1 1/2	2.29	2.29	2.80
3 x 3 x 1 1/4	2.17	2.17	2.74
3 x 3 x 1	2.00	2.00	2.66
3 x 3 x 3/4	1.87	1.87	2.61
3 x 3 x 1/2	1.76	1.76	2.52
3 x 2 1/2 x 3	3.08	2.99	3.08
3 x 2 1/2 x 2 1/2	2.83	2.70	2.99
3 x 2 1/2 x 2	2.52	2.39	2.89

All dimensions given in inches.

†For dimensions not given see Table 13–58.

*Maximum saturated steam pressure or 175 psi gage maximum liquid and gas pressure at 150 F.

continued on next page

Nominal Pipe Sizes	Center to End		
	X	Y	Z
3 x 2 1/2 x 1 1/2	2.29	2.16	2.80
3 x 2 1/2 x 1 1/4	2.17	2.04	2.74
3 x 2 1/2 x 1	2.00	1.87	2.66
3 x 2 x 3	3.08	2.89	3.08
3 x 2 x 2 1/2	2.83	2.60	2.99
3 x 2 x 2	2.52	2.25	2.89
3 x 2 x 1 1/2	2.29	2.02	2.80
3 x 1 1/2 x 3	3.08	2.80	3.08
3 x 1 1/4 x 3	3.08	2.74	3.08
3 x 1 x 3	3.08	2.66	3.08
3 x 3/4 x 3	3.08	2.61	3.08
2 1/2 x 2 x 3	2.99	2.89	2.83
2 x 2 x 3	2.89	2.89	2.52
2 1/2 x 2 1/2 x 3	2.99	2.99	2.83
3 1/2 x 3 1/2 x 3	3.18	3.18	3.33
3 1/2 x 3 1/2 x 2 1/2	2.93	2.93	3.24
3 1/2 x 3 1/2 x 2	2.62	2.62	3.14
3 1/2 x 3 1/2 x 1 1/2	2.39	2.39	3.05
3 1/2 x 3 1/2 x 1 1/4	2.27	2.27	2.99
3 1/2 x 3 1/2 x 1	2.10	2.10	2.91
3 1/2 x 3 x 3	3.18	3.08	3.33
3 1/2 x 3 x 2 1/2	2.93	2.83	3.24
3 1/2 x 3 x 2	2.62	2.52	3.14
3 1/2 x 3 x 1 1/2	2.39	2.29	3.05
3 1/2 x 2 1/2 x 3 1/2	3.42	3.24	3.42
3 1/2 x 2 1/2 x 3	3.18	2.99	3.33
3 1/2 x 2 1/2 x 2 1/2	2.93	2.70	3.24
3 1/2 x 2 x 3 1/2	3.42	3.14	3.42
3 1/2 x 1 1/2 x 3 1/2	3.42	3.05	3.42
3 1/2 x 1 1/4 x 3 1/2	3.42	2.99	3.42
3 1/2 x 1 x 3 1/2	3.42	2.91	3.42
3 x 3 x 3 1/2	3.33	3.33	3.18
4 x 4 x 3 1/2	3.54	3.54	3.69
4 x 4 x 3	3.30	3.30	3.60
4 x 4 x 2 1/2	3.05	3.05	3.51
4 x 4 x 2	2.74	2.74	3.41
4 x 4 x 1 1/2	2.51	2.51	3.32
4 x 4 x 1 1/4	2.39	2.39	3.26
4 x 4 x 1	2.22	2.22	3.18
4 x 4 x 3/4	2.09	2.09	3.13
4 x 3 1/2 x 4	3.79	3.69	3.79
4 x 3 1/2 x 3 1/2	3.54	3.42	3.69
4 x 3 1/2 x 3	3.30	3.18	3.60
4 x 3 1/2 x 2 1/2	3.05	2.93	3.51
4 x 3 1/2 x 2	2.74	2.62	3.41
4 x 3 1/2 x 1 1/2	2.51	2.39	3.32
4 x 3 1/2 x 1 1/4	2.39	2.27	3.26
4 x 3 x 4	3.79	3.60	3.79
4 x 3 x 3	3.30	3.08	3.60
4 x 3 x 2 1/2	3.05	2.83	3.51
4 x 3 x 2	2.74	2.52	3.41
4 x 2 1/2 x 4	3.79	3.51	3.79
4 x 2 1/2 x 3	3.30	2.99	3.60
4 x 2 1/2 x 2 1/2	3.05	2.70	3.51
4 x 2 x 4	3.79	3.41	3.79
4 x 2 x 3	3.30	2.89	3.60
4 x 2 x 2	2.74	2.25	3.41
4 x 1 1/2 x 4	3.79	3.32	3.79
4 x 1 1/4 x 4	3.79	3.26	3.79
4 x 1 x 4	3.79	3.18	3.79
3 1/2 x 3 1/2 x 4	3.69	3.69	3.54

Nominal Pipe Sizes	Center to End		
	X	Y	Z
3 x 3 x 4	3.60	3.60	3.30
2 1/2 x 2 1/2 x 4	3.51	3.51	3.05
2 x 2 x 4	3.41	3.41	2.74
5 x 5 x 4	4.00	4.00	4.41
5 x 5 x 3 1/2	3.75	3.75	4.31
5 x 5 x 3	3.51	3.51	4.22
5 x 5 x 2 1/2	3.26	3.26	4.13
5 x 5 x 2	2.95	2.95	4.03
5 x 5 x 1 1/2	2.72	2.72	3.94
5 x 5 x 1 1/4	2.60	2.60	3.88
5 x 5 x 1	2.43	2.43	3.80
5 x 4 x 5	4.50	4.41	4.50
5 x 4 x 4	4.00	3.79	4.41
5 x 4 x 3 1/2	3.75	3.54	4.31
5 x 4 x 3	3.51	3.30	4.22
5 x 4 x 2 1/2	3.26	3.05	4.13
5 x 4 x 2	2.95	2.74	4.03
5 x 4 x 1 1/2	2.72	2.51	3.94
5 x 3 x 5	4.50	4.22	4.50
5 x 3 x 4	4.00	3.60	4.41
5 x 3 x 3	3.51	3.08	4.22
5 x 2 1/2 x 5	4.50	4.13	4.50
5 x 2 x 5	4.50	4.03	4.50
4 x 4 x 5	4.41	4.41	4.00
6 x 6 x 5	4.63	4.63	5.03
6 x 6 x 4	4.13	4.13	4.94
6 x 6 x 3	3.64	3.64	4.75
6 x 6 x 2 1/2	3.39	3.39	4.66
6 x 6 x 2	3.08	3.08	4.56
6 x 6 x 1 1/2	2.85	2.85	4.47
6 x 6 x 1 1/4	2.73	2.73	4.41
6 x 6 x 1	2.56	2.56	4.33
6 x 5 x 6	5.13	5.03	5.13
6 x 5 x 5	4.63	4.50	5.03
6 x 5 x 4	4.13	4.00	4.94
6 x 5 x 3	3.64	3.51	4.75
6 x 5 x 2 1/2	3.39	3.26	4.66
6 x 5 x 2	3.08	2.95	4.56
6 x 4 x 6	5.13	4.94	5.13
6 x 4 x 5	4.63	4.41	5.03
6 x 4 x 4	4.13	3.79	4.94
6 x 3 x 6	5.13	4.75	5.13
6 x 3 x 3	3.64	3.08	4.75
6 x 2 1/2 x 6	5.13	4.66	5.13
6 x 2 x 6	5.13	4.56	5.13
4 x 4 x 6	4.94	4.94	4.13
5 x 3 x 6	5.03	4.75	4.63
5 x 5 x 6	5.03	5.03	4.63
8 x 8 x 6	5.56	5.56	6.37
8 x 8 x 5	5.03	5.03	6.27
8 x 8 x 4	4.50	4.50	6.17
8 x 8 x 3	4.00	4.00	6.07
8 x 8 x 2 1/2	3.69	3.69	6.01
8 x 8 x 2	3.44	3.44	5.84
8 x 6 x 8	6.56	6.37	6.56
8 x 6 x 6	5.56	5.13	6.37
8 x 4 x 8	6.56	6.17	6.56
6 x 6 x 8	6.37	6.37	5.56

All dimensions given in inches.

†For dimensions not given see Table 13–58.

*Maximum saturated steam pressure or 175 psi gage maximum liquid and gas pressure at 150F.

TABLE 13–60

Dimensions of Elbows, Tees, Crosses and Couplings – 250 lb* Brass or Bronze Screwed Fittings

ASA B16.17–1949 Reaffirmed 1953

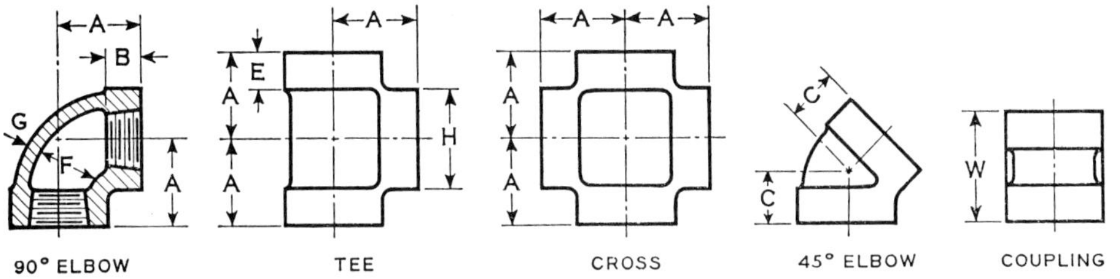

Nominal Pipe Size	Center to End, Elbows, Tees, and Crosses	Length of Thread, Min	Center to End 45-Deg Elbows	Width of Band, Min	Inside Diameter of Fitting		Metal Thickness, Min	Outside Diameter of Band, Min	End to End Coupling
					Min	Max			
	A	B	C	E	F		G	H	W
1/4	0.81	0.32	0.73	0.38	0.54	0.58	0.11	0.93	1.06
3/8	0.95	0.36	0.80	0.44	0.68	0.72	0.12	1.12	1.16
1/2	1.12	0.43	0.88	0.50	0.84	0.90	0.13	1.34	1.34
3/4	1.31	0.50	0.98	0.56	1.05	1.11	0.16	1.63	1.52
1	1.50	0.58	1.12	0.62	1.32	1.38	0.17	1.95	1.67
1 1/4	1.75	0.67	1.29	0.69	1.66	1.73	0.19	2.39	1.93
1 1/2	1.94	0.70	1.43	0.75	1.90	1.97	0.20	2.68	2.15
2	2.25	0.75	1.68	0.84	2.38	2.45	0.22	3.28	2.53
2 1/2	2.70	0.92	1.95	0.94	2.88	2.98	0.24	3.86	2.88
3	3.08	0.98	2.17	1.00	3.50	3.60	0.26	4.62	3.18
4	3.79	1.08	2.61	1.12	4.50	4.60	0.31	5.79	3.69

All dimensions are given in inches.

*See Table 13–61 for pressure-temperature ratings.

TABLE 13-61

Dimensions of Reducing Tees – 250 lb* Brass or Bronze Screwed Fittings

ASA B16.17–1949 Reaffirmed 1953

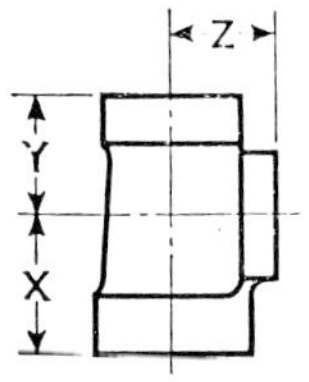

TEE, REDUCING

Nominal Pipe Size	Center to End		
	X	Y	Z
½ × ½ × ⅜	1.04	1.04	1.03
¾ × ¾ × ½	1.20	1.20	1.22
¾ × ¾ × ⅜	1.12	1.12	1.13
¾ × ½ × ¾	1.31	1.22	1.31
¾ × ½ × ½	1.20	1.12	1.22
½ × ½ × ¾	1.22	1.22	1.20
1 × 1 × ¾	1.37	1.37	1.45
1 × 1 × ½	1.26	1.26	1.36
1 × ¾ × 1	1.50	1.45	1.50
1 × ¾ × ¾	1.37	1.31	1.45
¾ × ¾ × 1	1.45	1.45	1.37
1 ¼ × 1 ¼ × 1	1.58	1.58	1.67
1 ¼ × 1 ¼ × ¾	1.45	1.45	1.62
1 ¼ × 1 ¼ × ½	1.34	1.34	1.53
1 ¼ × 1 × 1 ¼	1.75	1.67	1.75
1 ¼ × 1 × 1	1.58	1.50	1.67
1 ¼ × ¾ × 1 ¼	1.75	1.62	1.75
1 × 1 × 1 ¼	1.67	1.67	1.58
1 ½ × 1 ½ × 1 ¼	1.82	1.82	1.88
1 ½ × 1 ½ × 1	1.65	1.65	1.80
1 ½ × 1 ½ × ¾	1.52	1.52	1.75
1 ½ × 1 ½ × ½	1.41	1.41	1.66
1 ½ × 1 ¼ × 1 ¼	1.82	1.75	1.88
1 ½ × 1 ¼ × 1	1.65	1.58	1.80
1 ½ × 1 × 1 ½	1.94	1.80	1.94
1 ¼ × 1 ¼ × 1 ½	1.88	1.88	1.82
2 × 2 × 1 ½	2.02	2.02	2.16
2 × 2 × 1 ¼	1.90	1.90	2.10
2 × 2 × 1	1.73	1.73	2.02
2 × 2 × ¾	1.60	1.60	1.97
2 × 2 × ½	1.49	1.49	1.88
2 ½ × 2 ½ × 2	2.39	2.39	2.60
3 × 3 × 2	2.52	2.52	2.89
3 × 2 ½ × 3	3.08	2.99	3.08
3 × 2 × 3	3.08	2.89	3.08
4 × 4 × 3	3.30	3.30	3.60
4 × 4 × 2	2.74	2.74	3.41
4 × 3 × 4	3.79	3.60	3.79

All dimensions are given in inches.

For dimensions not given see Table 13–60

Note that dimensions are in accord with the American Standard, Table 13–59, for 125-lb Cast-Iron Screwed Fittings.

*Pressure-temperature rating and grade of material

Designated Rating	Service	Pressure psi Gage Maximum	Temperature, Deg F, Maximum: MSS SP-20, Grade C or ASTM B62	Temperature, Deg F, Maximum: MSS SP-20, Grade B or ASTM B61
250 lb	Steam	250	406	550
	Gas or liquid	400	150	150

TABLE 13–62

Dimensions of Elbows, Tees, Crosses and Couplings – 125 lb* Brass or Bronze Screwed Fittings

ASA B16.15–1947 Reaffirmed 1952

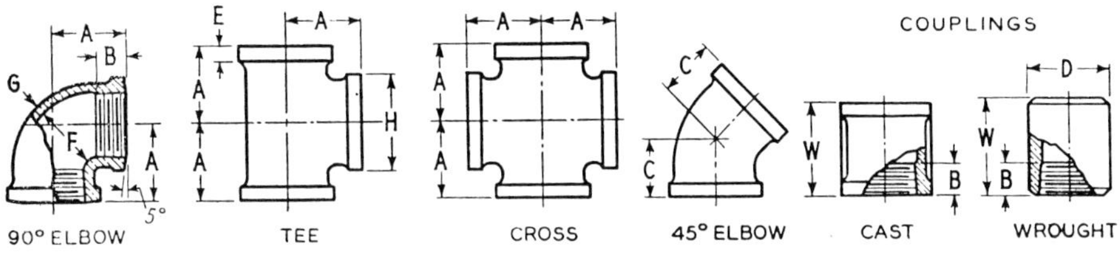

Nominal Pipe Size	Center to End Elbows, Tees, Crosses	Length of Thread,[1] Min	Center to End 45 Deg Elbows	Diameter Wrought Coupling[2]	Band Length, Min	Inside Diameter of Cast Fitting		Metal Thickness, Min	Band Diameter, Min	End to End Straight Coupling	
						Min	Max			Cast	Wrought
	A	B	C	D	E	F		G	H	W	
1/8	0.54	0.25	0.42	9/16	0.14	0.41	0.44	0.08	0.67	0.80	0.83
1/4	0.71	0.32	0.56	11/16	0.16	0.54	0.58	0.08	0.81	0.97	1.03
3/8	0.82	0.36	0.63	27/32	0.17	0.68	0.72	0.09	1.00	1.05	1.11
1/2	1.01	0.43	0.78	1 1/16	0.19	0.84	0.90	0.09	1.17	1.29	1.36
3/4	1.18	0.50	0.89	1 5/16	0.23	1.05	1.11	0.10	1.42	1.43	1.50
1	1.43	0.58	1.06	....	0.27	1.32	1.39	0.11	1.72	1.68	....
1 1/4	1.69	0.67	1.22	....	0.31	1.66	1.73	0.12	2.10	1.86	....
1 1/2	1.84	0.70	1.30	....	0.34	1.90	1.97	0.13	2.38	1.92	....
2	2.12	0.75	1.45	....	0.41	2.38	2.45	0.15	2.92	2.20	....
2 1/2	2.70	0.92	1.95	....	0.48	2.88	2.98	0.17	3.49	2.88	....
3	3.08	0.98	2.17	....	0.55	3.50	3.60	0.19	4.20	3.18	....
4	3.79	1.08	2.61	....	0.66	4.50	4.60	0.22	5.31	3.69	....

All dimensions are given in inches.

On right-hand couplings, the use of ribs is optional with the manufacturer, but no right-hand couplings shall have more than two ribs. Right-and left-hand couplings shall have four or more ribs unless the left-hand opening is clearly marked L, in which case the use of ribs is optional with the manufacturer.

[1]Dimension, B, for wrought couplings includes minimum length of effective thread only.

[2]Couplings, sizes 3/4 in. and smaller, may be cast or made from rod at the option of the manufacturer. Diameters, D, are commercial rod sizes.

*See Table 13–64 for pressure–temperature ratings.

TABLE 13–63

Dimensions of Reducing Tees – 125 lb* Brass or Bronze Screwed Fittings

ASA B16.15 – 1947 Reaffirmed 1952

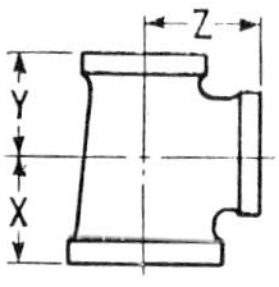

TEE, REDUCING

Nominal Pipe Size	Center to End X	Center to End Y	Center to End Z
1/4 × 1/4 × 1/8	0.65	0.65	0.60
3/8 × 3/8 × 1/4	0.75	0.75	0.78
3/8 × 1/4 × 3/8	0.82	0.78	0.82
3/8 × 1/4 × 1/4	0.75	0.71	0.78
1/2 × 1/2 × 3/8	0.93	0.93	0.90
1/2 × 1/2 × 1/4	0.87	0.87	0.87
1/2 × 3/8 × 1/2	1.01	0.90	1.01
1/2 × 3/8 × 3/8	0.93	0.82	0.90
3/8 × 3/8 × 1/2	0.90	0.90	0.93
3/4 × 3/4 × 1/2	1.08	1.08	1.11
3/4 × 3/4 × 3/8	1.00	1.00	1.00
3/4 × 1/2 × 3/4	1.18	1.11	1.18
3/4 × 1/2 × 1/2	1.08	1.01	1.11
1/2 × 1/2 × 3/4	1.11	1.11	1.08
1 × 1 × 3/4	1.30	1.30	1.31
1 × 1 × 1/2	1.20	1.20	1.24
1 × 1 × 3/8	1.12	1.12	1.13
1 × 3/4 × 1	1.43	1.31	1.43
1 × 3/4 × 3/4	1.30	1.18	1.31
1 × 3/4 × 1/2	1.20	1.08	1.24
1 × 1/2 × 1	1.43	1.24	1.43
1 × 1/2 × 3/4	1.30	1.11	1.31
3/4 × 3/4 × 1	1.31	1.31	1.30
1 1/4 × 1 1/4 × 1	1.52	1.52	1.60
1 1/4 × 1 1/4 × 3/4	1.39	1.39	1.48
1 1/4 × 1 1/4 × 1/2	1.29	1.29	1.41
1 1/4 × 1 × 1 1/4	1.69	1.60	1.69
1 1/4 × 1 × 1	1.52	1.43	1.60

Nominal Pipe Size	Center to End X	Center to End Y	Center to End Z
1 1/4 × 1 × 3/4	1.39	1.30	1.48
1 1/4 × 3/4 × 1 1/4	1.69	1.48	1.69
1 1/4 × 1/2 × 1 1/4	1.69	1.40	1.69
1 × 1 × 1 1/4	1.60	1.60	1.52
1 1/2 × 1 1/2 × 1 1/4	1.72	1.72	1.81
1 1/2 × 1 1/2 × 1	1.55	1.55	1.72
1 1/2 × 1 1/2 × 3/4	1.42	1.42	1.60
1 1/2 × 1 1/2 × 1/2	1.32	1.32	1.53
1 1/2 × 1 1/4 × 1 1/2	1.84	1.81	1.84
1 1/2 × 1 1/4 × 1 1/4	1.72	1.69	1.81
1 1/2 × 1 1/4 × 1	1.55	1.52	1.72
1 1/2 × 3/4 × 1 1/2	1.84	1.60	1.84
1 1/4 × 1 1/4 × 1 1/2	1.81	1.81	1.72
1 × 1 × 1 1/2	1.72	1.72	1.55
2 × 2 × 1 1/2	1.89	1.89	2.07
2 × 2 × 1 1/4	1.77	1.77	2.04
2 × 2 × 1	1.59	1.59	1.95
2 × 2 × 3/4	1.47	1.47	1.84
2 × 1 1/2 × 2	2.12	2.07	2.12
2 × 1 1/2 × 1 1/2	1.89	1.84	2.07
1 1/2 × 1 1/2 × 2	2.07	2.07	1.89
2 1/2 × 2 1/2 × 2	2.39	2.39	2.60
2 1/2 × 2 × 2	2.39	2.25	2.60
2 × 2 × 2 1/2	2.60	2.60	2.39
3 × 3 × 2 1/2	2.83	2.83	2.99
3 × 3 × 2	2.52	2.52	2.89
4 × 4 × 3	3.30	3.30	3.60
4 × 4 × 2	2.74	2.74	3.41

All dimensions given in inches.

For dimensions not given see Table 13–62.

Sizes 1 x 1/2 x 1/2, 1 1/2 x 1 x 1 1/2, 1 1/2 x 1 x 1 1/4, 1 1/2 x 1 x 1 and 2 x 2 x 1/2 are also available but not to standard center to end dimensions. The demand for these sizes is insufficient to warrant unification.

*See Table 13–64 for pressure-temperature ratings.

TABLE 13–64

Dimensions of Street Elbows – 125 lb* Brass or Bronze Screwed Fittings

ASA B16.15 – 1947 Reaffirmed 1952

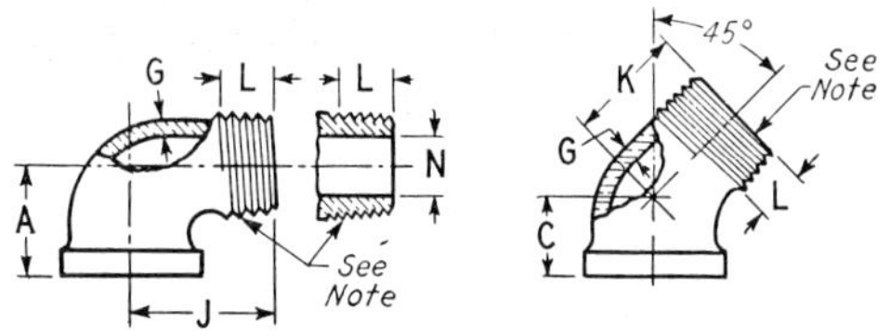

90° STREET ELBOW   45° STREET ELBOW

Nominal Pipe Size	Center to Female End 90 Deg Ell A	Center to Female End 45 Deg Ell C	Metal Thickness, Min G	Center to Male End 90 Deg Ell J	Center to Male End 45 Deg Ell K	Length of Thread Male End, Min L	Port Diameter Male End, Max N
1/8	0.54	0.42	0.08	0.92	0.78	0.27	0.22
1/4	0.71	0.56	0.08	1.11	0.88	0.41	0.28
3/8	0.82	0.63	0.09	1.24	0.92	0.41	0.40
1/2	1.01	0.78	0.09	1.48	1.06	0.54	0.53
3/4	1.18	0.89	0.10	1.65	1.23	0.55	0.72
1	1.43	1.06	0.11	1.98	1.40	0.69	0.93
1 1/4	1.69	1.22	0.12	2.24	1.64	0.71	1.25
1 1/2	1.84	1.30	0.13	2.46	1.80	0.73	1.47
2	2.12	1.45	0.15	2.88	2.14	0.76	1.91

All dimensions are given in inches.
For dimensions not given see Table 13–62.
Both designs illustrated for external thread ends of street fittings are approved.

*Pressure-Temperature rating and grade of material

Service	Pressure Psi, Max	Temperature, Deg F, Max		
		Cast MSS SP-20, Grade C ASTM B62	Wrought ASTM B16, B140	Cast MSS SP-20, Grade B ASTM B61
Steam	125	400	400	500
Gas or Liquid	175	150	150	150

TABLE 13–65

Dimensions of Square Head and Square Socket Plugs 125-lb Brass or Bronze Screwed Fittings

ASA B16.15 – 1947 Reaffirmed 1952

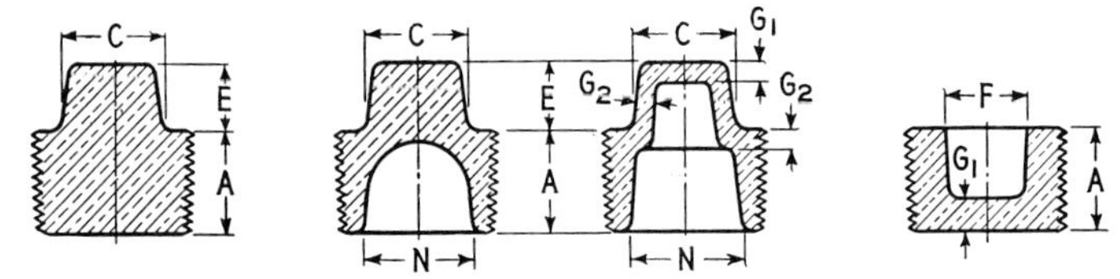

SOLID PATTERN SIZES $\frac{1}{8}$"–4" INCLUSIVE

CORED PATTERNS BOTH DESIGNS APPROVED SIZES $\frac{1}{2}$"–4" INCLUSIVE

COUNTER-SUNK PATTERN SIZES $\frac{1}{2}$"–4" INCLUSIVE

Nominal Pipe Size	Thread Length, Min	Width Across Flats[1]		Height of Plug Square, Min	Metal Thickness, Min		Inside Diameter of Plug Max	Size of Square Socket,[2] Min
		Nom	Max					
	A	C		E	G_1	G_2	N	F
1/8	0.27	9/32	0.281	0.24	....	....	....	....
1/4	0.41	3/8	0.375	0.28	....	....	....	....
3/8	0.41	7/16	0.438	0.31	....	....	....	....
1/2	0.54	9/16	0.563	0.38	0.09	0.12	0.53	3/8
3/4	0.55	5/8	0.625	0.44	0.10	0.13	0.72	1/2
1	0.69	13/16	0.813	0.50	0.11	0.14	0.93	1/2
1 1/4	0.71	15/16	0.938	0.56	0.12	0.15	1.25	3/4
1 1/2	0.73	1 1/8	1.125	0.62	0.13	0.16	1.47	3/4
2	0.76	1 5/16	1.313	0.68	0.15	0.17	1.91	7/8
2 1/2	1.07	1 1/2	1.500	0.74	0.17	0.18	2.32	1 1/8
3	1.13	1 11/16	1.688	0.80	0.19	0.19	2.90	1 3/8
4	1.22	2 1/4	2.250	0.92	0.22	0.22	3.83	2

All dimensions are given in inches.

[1] These dimensions, C, are the nominal size of wrench as given in Table 19, American Standard Wrench-Head Bolts and Nuts and Wrench Openings (ASA B18.2–1941). Square head plugs are designed to fit these wrenches. Plug squares may have opposite sides tapered a maximum of 4 deg total. Table 11–16

[2] Square socket of countersunk plugs to have dimensions, F, to fit commercial square bars of sizes indicated. Countersunk square sockets may have opposite sides tapered a maximum of 4 deg total.

TABLE 13-66

Dimensions of Outside Head, Inside Head, and Face Bushings – 125 lb Brass or Bronze Screwed Fittings

ASA B16.15 – 1947 Reaffirmed 1952

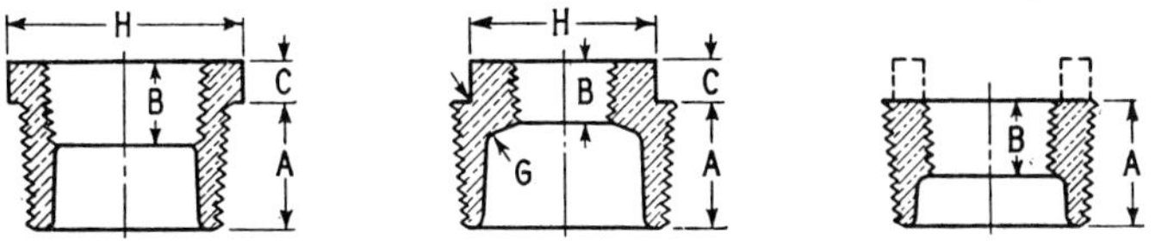

OUTSIDE HEAD INSIDE HEAD FACE

Nominal Pipe Size	Length of External Thread, Min	Length of Internal Thread, Min	Height of Head, Min	Width of Head[1] Min H		Metal Thickness,[2] Min
	A	B	C	Outside	Inside	G
¼ × ⅛	0.44	0.26*	0.14	0.64†	....	.....
⅜ × ¼	0.48	0.40*	0.16	0.68†	....	.....
⅜ × ⅛	0.48	0.25	0.16	0.68†	....	.....
½ × ⅜	0.56	0.41*	0.19	0.87†	....	.....
½ × ¼	0.56	0.32	0.19	0.87†	....	.....
½ × ⅛	0.56	0.25	0.19	0.87†	....	.....
¾ × ½	0.63	0.53*	0.22	1.15†	....	.....
¾ × ⅜	0.63	0.36	0.22	1.15†	....	.....
¾ × ¼	0.63	0.32	0.22	1.15†	....	.....
1 × ¾	0.75	0.50	0.25	1.42†	....	.....
1 × ½	0.75	0.43	0.25	1.42†	....	.....
1 × ⅜	0.75	0.36	0.30	....	1.12	.....
1 × ¼	0.75	0.32	0.30	....	1.12	.....
1¼ × 1	0.80	0.58	0.28	1.76	....	.....
1¼ × ¾	0.80	0.50	0.28	1.76	....	.....
1¼ × ½	0.80	0.43	0.34	....	1.34	0.185
1¼ × ⅜	0.80	0.36	0.34	....	1.12	0.185
1½ × 1¼	0.83	0.71*	0.31	2.00	....	.....
1½ × 1	0.83	0.58	0.31	2.00	....	.....
1½ × ¾	0.83	0.50	0.37	....	1.63	0.200
1½ × ½	0.83	0.43	0.37	....	1.34	0.200
2 × 1½	0.88	0.70	0.34	2.48	....	.....
2 × 1¼	0.88	0.67	0.34	2.48	....	.....
2 × 1	0.88	0.58	0.41	....	1.95	0.220
2 × ¾	0.88	0.50	0.41	....	1.63	0.220
2 × ½	0.88	0.43	0.41	....	1.34	0.220
2½ × 2	1.07	0.75	0.37	2.98	....	.....
2½ × 1½	1.07	0.70	0.44	....	2.68	.....
2½ × 1¼	1.07	0.67	0.44	....	2.39	0.240
2½ × 1	1.07	0.58	0.44	....	1.95	0.240
3 × 2½	1.13	0.92	0.40	3.86	....	.....
3 × 2	1.13	0.75	0.48	....	3.28	.....
3 × 1½	1.13	0.70	0.48	....	2.68	0.260
3 × 1¼	1.13	0.67	0.48	....	2.39	0.260
4 × 3	1.22	0.98	0.50	4.62	....	.....
4 × 2½	1.22	0.92	0.60	....	3.86	0.310
4 × 2	1.22	0.75	0.60	....	3.28	0.310
4 × 1½	1.22	0.70	0.60	....	2.68	0.310

All dimensions are given in inches.

The addition of lugs on face bushings shall not be prohibited.

Bushings reducing to pipe sizes smaller than given are bushed from the smallest reduction appearing in the table.

* To provide proper metal thickness these sizes shall not be cored out to diameters greater than the root diameter of the internal thread. The length of the internal thread may be equal to the minimum dimension, B, or greater up to the full length of bushing.

† Bushings in these sizes may be made from bar stock. When made from bar stock the dimension H may be $^5/_8$, $^{11}/_{16}$, $^7/_8$, $1^1/_8$, and $1^7/_{16}$ in., respectively, in order to conform with regular hexagon or octagon bar stock sizes.

[1] Heads of bushings shall be hexagonal or octagonal.

[2] Metal thickness, G, is the same as 125 Lb Cast-Iron Screwed Fittings (ASA B16d–1941).

TABLE 13–67

Dimensions of Square Head Ferrous Plugs

ASA B16.14 – 1949 Reaffirmed 1953

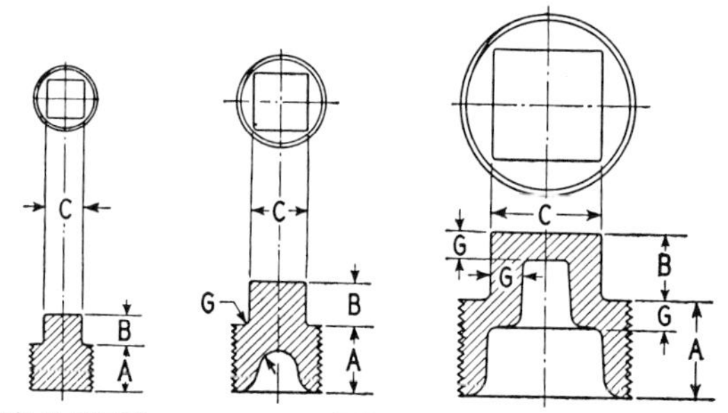

SOLID PLUGS CORED PLUGS

Nominal[1,2] Pipe Size	Thread Length, Min	Height of Square, Min	Width Across Flats C		Metal Thickness,[4] Min
	A	B	Nom[3]	Max	G
1/8	0.37	0.24	9/32	0.281	....
1/4	0.44	0.28	3/8	0.375	....
3/8	0.48	0.31	7/16	0.438	....
1/2	0.56	0.38	9/16	0.563	0.16
3/4	0.63	0.44	5/8	0.625	0.18
1	0.75	0.50	13/16	0.813	0.20
1 1/4	0.80	0.56	15/16	0.938	0.22
1 1/2	0.83	0.62	1 1/8	1.125	0.24
2	0.88	0.68	1 5/16	1.313	0.26
2 1/2	1.07	0.74	1 1/2	1.500	0.29
3	1.13	0.80	1 11/16	1.688	0.31
3 1/2	1.18	0.86	1 7/8	1.875	0.34

All dimensions given in inches.

Material to be cast iron, malleable iron, or steel.

These plugs are threaded with American Standard tapered pipe threads. (ASA B2.1.)

[1] Solid plugs are provided in sizes 1/8 to 3 1/2 in., incl.; cored plugs, 1/2 to 3 1/2 in., incl.

[2] For sizes 4-in. and larger slotted or bar pattern plugs are provided.

[3] These dimensions are the nominal size of wrench as given in Table 19, American Standard Wrench-Head Bolts and Nuts and Wrench Openings (ASA B18.2). Square head plugs are designed to fit these wrenches. Table 11–16

[4] Cored plugs have minimum metal thickness at all points, equal to dimension G, except at the end of the thread.

TABLE 13–68

Dimensions of Countersunk Ferrous Plugs

ASA B16.14 – 1949 Reaffirmed 1953

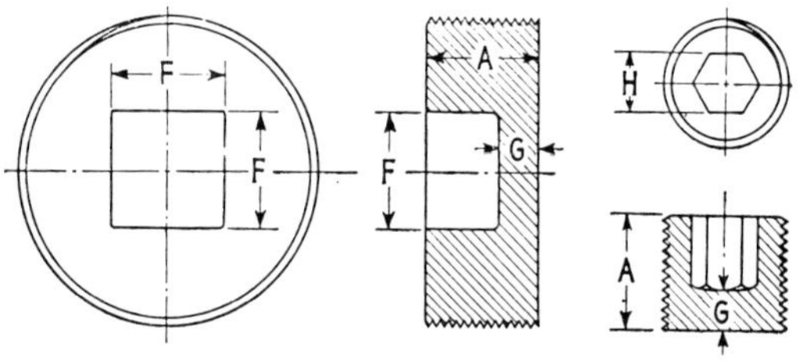

Nominal Pipe Size	Thread Length, Min	Size of Square Socket[1] F		Size of Hexagon[2]	Metal Thickness, Min
	A	Nom	Min	H	G
1/8	0.37	...	.....	3/16	0.06
1/4	0.44	...	.....	1/4	0.09
3/8	0.48	...	.....	5/16	0.13
1/2	0.56	3/8	0.382	3/8	0.16
3/4	0.63	1/2	0.508	9/16	0.18
1	0.75	1/2	0.508	5/8	0.20
1 1/4	0.80	3/4	0.759	...	0.22
1 1/2	0.83	3/4	0.759	...	0.24
2	0.88	7/8	0.884	...	0.26
2 1/2	1.07	1 1/8	1.137	...	0.29
3	1.13	1 3/8	1.391	...	0.31
3 1/2	1.18	1 1/2	1.518	...	0.34
4	1.22	2	2.022	...	0.37

All dimensions given in inches.

Material to be cast iron, malleable iron, or steel.

These plugs are threaded with American Standard tapered pipe threads. (ASA B2.1.)

[1] Square socket of countersunk pattern to have dimensions to fit commercial square bars of sizes indicated.

[2] Hexagon socket of countersunk pattern shall have dimensions to fit regular wrenches used with hexagon socket set screws.

TABLE 13-69

Dimensions of Ferrous Outside Head, Inside Head, and Face Bushings

ASA B16.14-1949 Reaffirmed 1953

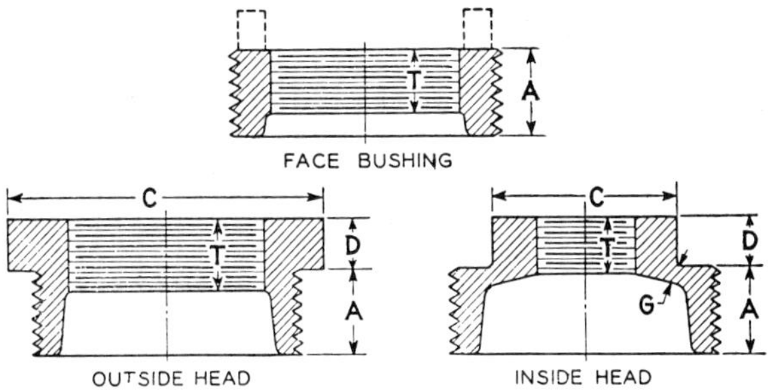

Size[1]	Length of External Thread,[2] Min A	Length of Internal Thread, Min T	Height of Head, Min D	Width of Head,[3] Min C Outside	Width of Head,[3] Min C Inside	Metal Thickness,[4] Min G
¼ × ⅛	0.44	§0.26	0.14	*0.64	..	...
⅜ × ¼	0.48	§0.40	0.16	*0.68	..	...
⅜ × ⅛	0.48	0.25	0.16	*0.68	..	...
½ × ⅜	0.56	§0.41	0.19	*0.87	..	...
½ × ¼	0.56	0.32	0.19	*0.87	..	...
½ × ⅛	0.56	0.25	0.19	*0.87	..	...
¾ × ½	0.63	§0.53	0.22	*1.15	..	...
¾ × ⅜	0.63	0.36	0.22	*1.15	..	...
¾ × ¼	0.63	0.32	0.22	*1.15	..	...
¾ × ⅛	0.63	0.25	0.22	*1.15	..	...
1 × ¾	0.75	0.50	0.25	*1.42	..	...
1 × ½	0.75	0.43	0.25	*1.42	..	...
1 × ⅜	0.75	0.36	0.30	..	1.12	...
1 × ¼	0.75	0.32	0.30	..	1.12	...
1 × ⅛	0.75	0.25	0.30	..	1.12	...
1¼ × 1	0.80	0.58	0.28	1.76	..	...
1¼ × ¾	0.80	0.50	0.28	1.76	..	...
1¼ × ½	0.80	0.43	0.34	..	1.34	0.185
1¼ × ⅜	0.80	0.36	0.34	..	1.12	0.185
1¼ × ¼	0.80	0.32	0.34	..	1.12	0.185
1½ × 1¼	0.83	§0.71	0.31	2.00	..	...
1½ × 1	0.83	0.58	0.31	2.00	..	...
1½ × ¾	0.83	0.50	0.37	..	1.63	0.200
1½ × ½	0.83	0.43	0.37	..	1.34	0.200
1½ × ⅜	0.83	0.36	0.37	..	1.12	0.200
1½ × ¼	0.83	0.32	0.37	..	1.12	0.200
2 × 1½	0.88	0.70	0.34	2.48	..	...
2 × 1¼	0.88	0.67	0.34	2.48	..	...
2 × 1	0.88	0.58	0.41	..	1.95	0.220
2 × ¾	0.88	0.50	0.41	..	1.63	0.220
2 × ½	0.88	0.43	0.41	..	1.34	0.220
2 × ⅜	0.88	0.36	0.41	..	1.12	0.220
2 × ¼	0.88	0.32	0.41	..	1.12	0.220
2½ × 2	1.07	0.75	0.37	2.98	..	...
2½ × 1½	1.07	0.70	0.44	2.68	..	...
2½ × 1¼	1.07	0.67	0.44	..	2.39	0.240
2½ × 1	1.07	0.58	0.44	..	1.95	0.240
2½ × ¾	1.07	0.50	0.44	..	1.63	0.240
2½ × ½	1.07	0.43	0.44	..	1.34	0.240

continued on next page

TABLE 13–69, *continued*

Size[1]	Length of External Thread,[2] Min	Length of Internal Thread, Min	Height of Head, Min	Width of Head,[3] Min C		Metal Thickness,[4] Min
	A	T	D	Outside	Inside	G
3 × 2 1/2	1.13	0.92	0.40	3.86	..	...
3 × 2	1.13	0.75	0.48	3.28	..	...
3 × 1 1/2	1.13	0.70	0.48	..	2.68	0.260
3 × 1 1/4	1.13	0.67	0.48	..	2.39	0.260
3 × 1	1.13	0.58	0.48	..	1.95	0.260
3 × 3/4	1.13	0.50	0.48	..	1.63	0.260
3 × 1/2	1.13	0.43	0.48	..	1.34	0.260
3 1/2 × 3	1.18	0.98	0.43	4.62	..	...
3 1/2 × 2 1/2	1.18	0.92	0.52	3.86	..	...
3 1/2 × 2	1.18	0.75	0.52	..	3.28	0.280
3 1/2 × 1 1/2	1.18	0.70	0.52	..	2.68	0.280
3 1/2 × 1 1/4	1.18	0.67	0.52	..	2.39	0.280
3 1/2 × 1	1.18	0.58	0.52	..	1.95	0.280
4 × 3 1/2	1.22	1.03	0.50	5.20	..	...
4 × 3	1.22	0.98	0.50	4.62	..	...
4 × 2 1/2	1.22	0.92	0.60	..	3.86	0.310
4 × 2	1.22	0.75	0.60	..	3.28	0.310
4 × 1 1/2	1.22	0.70	0.60	..	2.68	0.310
4 × 1 1/4	1.22	0.67	0.60	..	2.39	0.310
4 × 1	1.22	0.58	0.60	..	1.95	0.310
5 × 4	1.31	1.08	0.50	5.79	..	...
5 × 3 1/2	1.31	1.03	0.60	5.20	..	...
5 × 3	1.31	0.98	0.60	..	4.62	0.380
5 × 2 1/2	1.31	0.92	0.60	..	3.86	0.380
5 × 2	1.31	0.75	0.60	..	3.28	0.380
6 × 5	1.40	1.18	0.63	7.05	..	...
6 × 4	1.40	1.08	0.75	..	5.79	0.430
6 × 3 1/2	1.40	1.03	0.75	..	5.20	0.430
6 × 3	1.40	0.98	0.75	..	4.62	0.430
6 × 2 1/2	1.40	0.92	0.75	..	3.86	0.430
6 × 2	1.40	0.75	0.75	..	3.28	0.430
8 × 6	1.57	1.28	0.83	8.28	..	...
8 × 5	1.57	1.18	0.83	..	7.05	0.550
8 × 4	1.57	1.08	0.83	..	5.79	0.550
8 × 3 1/2	1.57	1.03	0.83	..	5.20	0.550
8 × 3	1.57	0.98	0.83	..	4.62	0.550

All dimensions given in inches.

Material to be cast iron, malleable iron, or steel. (See footnote 1.)
These bushings are threaded with American Standard tapered pipe threads. (ASA B2.1.)
The addition of lugs on face bushings shall not be prohibited.
Cored bushings have minimum metal thickness at all points, equal to dimension G, except at the end of the thread.

* When made of bar stock, the dimensions may be 5/8, 11/16, 7/8, 1 1/8, and 1 7/16 in., respectively, in order to use regular bar stock sizes.

§ To provide proper metal thickness these sizes shall not be cored out to diameters greater than the root diameter of the internal thread. The length of the internal thread may be equal to the minimum dimension, T, or greater up to the full length of bushing.

[1] Hexagon head or octagon head bushings, sizes 2 1/2 in. and smaller reducing one size may be made either of malleable iron or steel. Other sizes may be made either of cast iron or malleable iron or steel. Face bushings, sizes 2 1/2 in. and smaller may be made either of malleable iron or steel. Face bushings, 3 in. and larger reducing one size may be made either of malleable iron or steel. Face bushings, 3 in. and larger reducing two sizes or more, may be made either of cast or malleable iron or steel.

[2] In the case of Outside Head Bushings, Length A includes provisions for imperfect threads.

[3] Heads of bushings shall be hexagonal or octagonal, except that on the larger sizes of outside head bushings the heads may be made round with lugs instead of hexagonal or octagonal.

[4] G same as metal thickness for 125 Lb Cast-Iron Screwed Fittings (ASA B16d).

*TABLE 13–70

Basic Dimensions, American Standard, Taper Pipe Thread[1]

ASA B2.1 – 1945 Pipe Threads

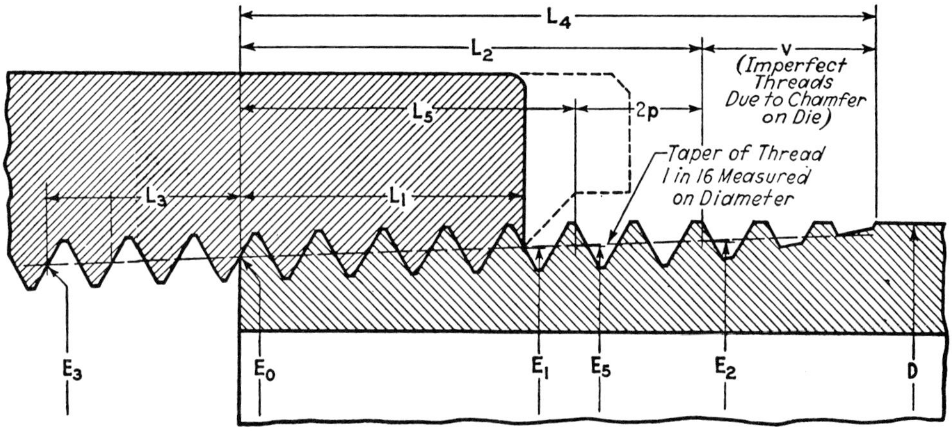

1	2	3	4	5	6	7	8	9	10	11
Nominal Pipe Size	Outside Diameter of Pipe	Threads per Inch	Pitch of Thread	Pitch Diameter at Beginning of External Thread	Hand-Tight Engagement			Effective Thread, External		
					Length[2] L_1		Diam[3]	Length[4] L_2		Diam E_2
	D	n	p	E_0	In.	Thds	E_1	In.	Thds	In.
1/16	0.3125	27	0.03704	0.27118	0.160	4.32	0.28118	0.2611	7.05	0.28750
1/8	0.405	27	0.03704	0.36351	0.180	4.86	0.37476	0.2639	7.12	0.38000
1/4	0.540	18	0.05556	0.47739	0.200	3.60	0.48989	0.4018	7.23	0.50250
3/8	0.675	18	0.05556	0.61201	0.240	4.32	0.62701	0.4078	7.34	0.63750
1/2	0.840	14	0.07143	0.75843	0.320	4.48	0.77843	0.5337	7.47	0.79179
3/4	1.050	14	0.07143	0.96768	0.339	4.75	0.98887	0.5457	7.64	1.00179
1	1.315	11 1/2	0.08696	1.21363	0.400	4.60	1.23863	0.6828	7.85	1.25630
1 1/4	1.660	11 1/2	0.08696	1.55713	0.420	4.83	1.58338	0.7068	8.13	1.60130
1 1/2	1.900	11 1/2	0.08696	1.79609	0.420	4.83	1.82234	0.7235	8.32	1.84130
2	2.375	11 1/2	0.08696	2.26902	0.436	5.01	2.29627	0.7565	8.70	2.31630
2 1/2	2.875	8	0.12500	2.71953	0.682	5.46	2.76216	1.1375	9.10	2.79062
3	3.500	8	0.12500	3.34062	0.766	6.13	3.38850	1.2000	9.60	3.41562
3 1/2	4.000	8	0.12500	3.83750	0.821	6.57	3.88881	1.2500	10.00	3.91562
4	4.500	8	0.12500	4.33438	0.844	6.75	4.38712	1.3000	10.40	4.41562
5	5.563	8	0.12500	5.39073	0.937	7.50	5.44929	1.4063	11.25	5.47862
6	6.625	8	0.12500	6.44609	0.958	7.66	6.50597	1.5125	12.10	6.54062
8	8.625	8	0.12500	8.43359	1.063	8.50	8.50003	1.7125	13.70	8.54062
10	10.750	8	0.12500	10.54531	1.210	9.68	10.62094	1.9250	15.40	10.66562
12	12.750	8	0.12500	12.53281	1.360	10.88	12.61781	2.1250	17.00	12.66562
14 OD	14.000	8	0.12500	13.77500	1.562	12.50	13.87262	2.2500	18.90	13.91562
16 OD	16.000	8	0.12500	15.76250	1.812	14.50	15.87575	2.4500	19.60	15.91562
18 OD	18.000	8	0.12500	17.75000	2.000	16.00	17.87500	2.6500	21.20	17.91562
20 OD	20.000	8	0.12500	19.73750	2.125	17.00	19.87031	2.8500	22.80	19.91562
24 OD	24.000	8	0.12500	23.71250	2.375	19.00	23.86094	3.2500	26.00	23.91562

*Compare with Table 9–14.

[1] The basic dimensions of the American Standard Taper Pipe Thread are given in inches to four or five decimal places. While this implies a greater degree of precision than is ordinarily attained, these dimensions are the basis of gage dimensions and are so expressed for the purpose of eliminating errors in computations.

[2] Also length of thin ring gage and length from gaging notch to small end of plug gage.

continued on next page

TABLE 13–70, continued

1	12	13	14	15	16	17	18	19	20	21	22
Nominal Pipe Size	Wrench Make-up Length for Internal Thread			Vanish Threads		Over-all Length External Thread	Nominal Perfect External Threads[5]		Depth of Thread	Increase in Diam per Thread 0.0625	Basic[6] Minor Diam at Small End of Pipe
	Length L_3		Diam E_3	V		L_4	Length L_5	Diam E_5	h	$\frac{0.0625}{n}$	K_0
	In.	Thds	In.	In.	Thds	In.	In.	In.	In.	In.	In.
1/16	0.1111	3	0.26424	0.1285	3.47	0.3896	0.1870	0.28287	0.02963	0.00231	0.2416
1/8	0.1111	3	0.35656	0.1285	3.47	0.3924	0.1898	0.37537	0.02963	0.00231	0.3339
1/4	0.1667	3	0.46697	0.1928	3.47	0.5946	0.2907	0.49556	0.04444	0.00347	0.4329
3/8	0.1667	3	0.60160	0.1928	3.47	0.6006	0.2967	0.63056	0.04444	0.00347	0.5676
1/2	0.2143	3	0.74504	0.2478	3.47	0.7815	0.3909	0.78286	0.05714	0.00446	0.7013
3/4	0.2143	3	0.95429	0.2478	3.47	0.7935	0.4029	0.99286	0.05714	0.00446	0.9105
1	0.2609	3	1.19733	0.3017	3.47	0.9845	0.5088	1.24543	0.06957	0.00543	1.1441
1 1/4	0.2609	3	1.54083	0.3017	3.47	1.0085	0.5329	1.59043	0.06957	0.00543	1.4876
1 1/2	0.2609	3	1.77978	0.3017	3.47	1.0252	0.5496	1.83043	0.06957	0.00543	1.7265
2	0.2609	3	2.25272	0.3017	3.47	1.0582	0.5826	2.30543	0.06957	0.00543	2.1995
2 1/2	0.2500†	2	2.70391	0.4337	3.47	1.5712	0.8875	2.77500	0.100000	0.00781	2.6195
3	0.2500†	2	3.32500	0.4337	3.47	1.6337	0.9500	3.40000	0.100000	0.00781	3.2406
3 1/2	0.2500	2	3.82188	0.4337	3.47	1.6837	1.0000	3.90000	0.100000	0.00781	3.7375
4	0.2500	2	4.31875	0.4337	3.47	1.7337	1.0500	4.40000	0.100000	0.00781	4.2344
5	0.2500	2	5.37511	0.4337	3.47	1.8400	1.1563	5.46300	0.100000	0.00781	5.2907
6	0.2500	2	6.43047	0.4337	3.47	1.9462	1.2625	6.52500	0.100000	0.00781	6.3461
8	0.2500	2	8.41797	0.4337	3.47	2.1462	1.4625	8.52500	0.100000	0.00781	8.3336
10	0.2500	2	10.52969	0.4337	3.47	2.3587	1.6750	10.65000	0.100000	0.00781	10.4453
12	0.2500	2	12.51719	0.4337	3.47	2.5587	1.8750	12.65000	0.100000	0.00781	12.4328
14 OD	0.2500	2	13.75938	0.4337	3.47	2.6837	2.0000	13.90000	0.100000	0.00781	13.6750
16 OD	0.2500	2	15.74688	0.4337	3.47	2.8837	2.2000	15.90000	0.100000	0.00781	15.6625
18 OD	0.2500	2	17.73438	0.4337	3.47	3.0837	2.4000	17.90000	0.100000	0.00781	17.6500
20 OD	0.2500	2	19.72188	0.4337	3.47	3.2837	2.6000	19.90000	0.100000	0.00781	19.6375
24 OD	0.2500	2	23.69688	0.4337	3.47	3.6837	3.0000	23.90000	0.100000	0.00781	23.6125

[3] Also pitch diameter at gaging notch (hand-tight plane.) [4] Also length of plug gage.

[5] The length L_5 from the end of the pipe determines the plane beyond which the thread form is imperfect at the crest. **The next** two threads are perfect at the root. At this plane the cone formed by the crests of the thread intersects the cylinder forming **the** external surface of the pipe. $L_5 = L_2 - 2p$.

[6] Given as information for use in selecting tap drills. Tables 9–1 and 9–17

† The Army-Navy Aeronautical Specifications are based on a wrench make-up of three threads 3 in. and smaller. **The E_3 dimensions are as follows:** Size 2 1/2 in. 2.69609 and size 3 in. 3.31719.

All dimensions are given in inches.

TABLE 13–71

Limits on Crest and Root of American Standard External and Internal Taper Pipe Threads

ASA B2.1 – 1945 Pipe Threads

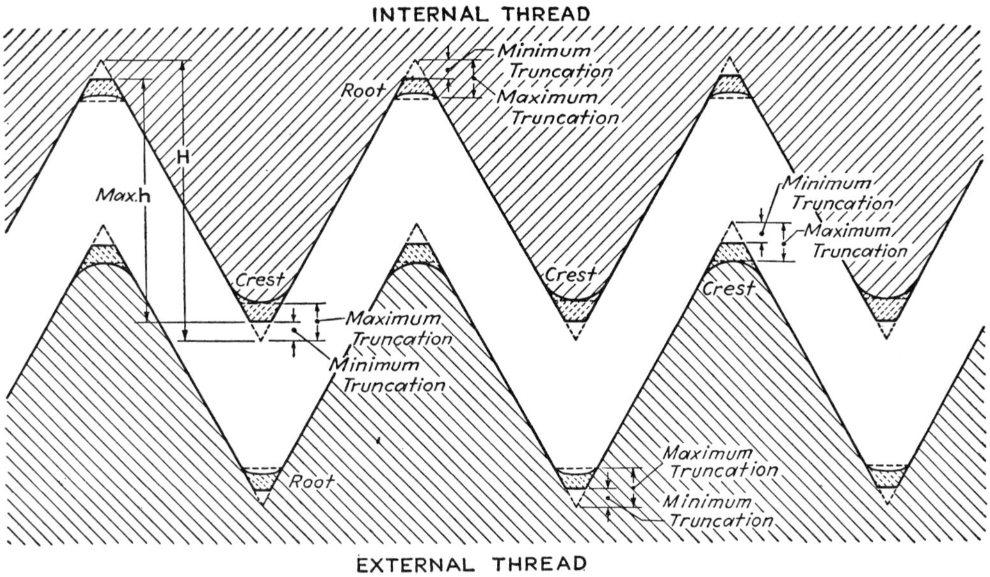

1	2	3	4	5	6	7	8	9	10	11	12	13	14
Threads per Inch	Depth Sharp V Thread H	Depth Pipe Thread[1] h		Truncation				Tolerance on Truncation, Inches	Equivalent Width of Flat				Tolerance on Equiv. Width of Flat, Inches
		Max	Min	Minimum		Maximum			Minimum		Maximum		
	Inches	Inches	Inches	Formula	Inches	Formula	Inches		Formula	Inches	Formula	Inches	
27*	0.03208	0.02963	0.02496	0.033p	0.0012	0.096p	0.0036	0.0024	0.038p	0.0014	0.111p	0.0041	0.0027
18	0.04811	0.04444	0.03833	0.033p	0.0018	0.088p	0.0049	0.0031	0.038p	0.0021	0.102p	0.0057	0.0036
14	0.06186	0.05714	0.05071	0.033p	0.0024	0.078p	0.0056	0.0032	0.038p	0.0027	0.090p	0.0064	0.0037
11 ½	0.07531	0.06957	0.06261	0.033p	0.0029	0.073p	0.0063	0.0034	0.038p	0.0033	0.084p	0.0073	0.0040
8	0.10825	0.10000	0.09275	0.033p	0.0041	0.062p	0.0078	0.0037	0.038p	0.0048	0.072p	0.0090	0.0042

The basic dimensions of the American Standard taper pipe thread are given in inches to four and five decimal places. While this implies a greater degree of precision than is ordinarily attained, these dimensions are so expressed for the purpose of eliminating errors in computations.

The limits specified in the table are intended to serve as a guide for establishing limits of the thread elements of taps, dies and thread chasers. These limits may be required on product.

*The Army-Navy Aeronautical Specification AN-GGG-P-363 agrees with all values given in this table, except those for the maximum truncation and maximum width of flat for the ⅛ in. size, 27 threads. These values are, respectively, 0.0027 in. and 0.0031 in.

[1] Dimensions of gages, such as plain taper plug and ring gages, which depend on maximum and minimum truncations, Cols. 5 to 8, inclusive, shall be determined by applying the thread depths in Cols. 3 and 4 to the basic pitch diameter, E_0 or E_1. Step values of tolerance notches are 16 times (Col. 3 – Col. 4), rather than 32 times Col. 9.

TABLE 13–72

Limits on Crest and Root of American Standard [1] Dryseal Pipe Threads

ASA B2.1 – 1945 Pipe Threads

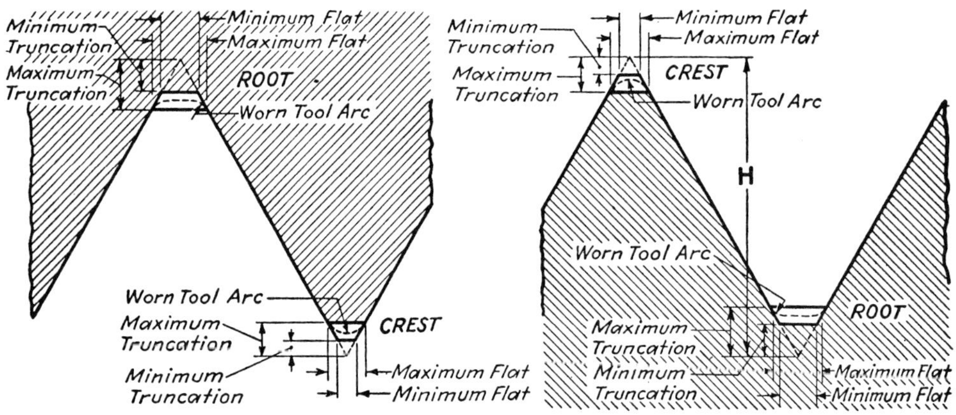

INTERNAL THREAD   EXTERNAL THREAD

1	2	3	4	5	6	7	8	9	10	11	12
Number of Threads per Inch	Depth of Sharp V Thread	Truncation[2]				Tolerance on Truncation	Equivalent Width of Flat				Tolerance on Equivalent Width of Flat
	H	Minimum		Maximum			Minimum		Maximum		
	Inches	Formula	Inches	Formula	Inches	Inches	Formula	Inches	Formula	Inches	Inches
27 Crest	0.03208	0.047*p*	0.0017	0.094*p*	0.0035	0.0018	0.054*p*	0.0020	0.108*p*	0.0040	0.0020
Root		0.094*p*	0.0035	0.140*p*	0.0052	0.0017	0.108*p*	0.0040	0.162*p*	0.0060	0.0020
18 Crest	0.04811	0.047*p*	0.0026	0.078*p*	0.0043	0.0017	0.054*p*	0.0030	0.090*p*	0.0050	0.0020
Root		0.078*p*	0.0043	0.109*p*	0.0061	0.0018	0.090*p*	0.0050	0.126*p*	0.0070	0.0020
14 Crest	0.06186	0.036*p*	0.0026	0.060*p*	0.0043	0.0017	0.042*p*	0.0030	0.070*p*	0.0050	0.0020
Root		0.060*p*	0.0043	0.085*p*	0.0061	0.0018	0.070*p*	0.0050	0.098*p*	0.0070	0.0020
11 ½ Crest	0.07531	0.040*p*	0.0035	0.060*p*	0.0052	0.0017	0.046*p*	0.0040	0.069*p*	0.0060	0.0020
Root		0.060*p*	0.0052	0.090*p*	0.0078	0.0026	0.069*p*	0.0060	0.103*p*	0.0090	0.0030
8 Crest	0.10825	0.042*p*	0.0052	0.055*p*	0.0069	0.0017	0.048*p*	0.0060	0.064*p*	0.0080	0.0020
Root		0.055*p*	0.0069	0.076*p*	0.0095	0.0026	0.064*p*	0.0080	0.088*p*	0.0110	0.0030

NOTE: Dimensions are specified to four and five decimal places. While this implies a greater degree of precision than is ordinarily attained, these dimensions are so expressed for the purpose of eliminating errors in computations.

[1] Although these threads are designed for use without a lubricant or sealer, a lubricant may be used when desired in making up these joints.

[2] The major diameter of plug gages and the minor diameter of ring gages used for gaging dryseal threads shall be truncated 0.20*p* minimum or 0.25*p* maximum for all pitches.

TABLE 13–73

Dimensions of External and Internal, Straight, Pipe Threads for Mechanical Joints

ASA B2.1 – 1945 Pipe Threads

1	2	3	4	5	6	7	8	9	10
		External Thread[1]				Internal Thread			
Nominal Pipe Size	Threads per Inch	Major Diameter		Pitch Diameter		Minor Diameter		Pitch Diameter	
		Max	Min	Max[2]	Min	Max	Min	Max	Min[2]
1/8	27	0.399	0.393	0.3748	0.3713	0.357	0.350	0.3783	0.3748
1/4	18	0.527	0.516	0.4899	0.4847	0.463	0.453	0.4951	0.4899
3/8	18	0.664	0.654	0.6270	0.6218	0.600	0.590	0.6322	0.6270
1/2	14	0.826	0.813	0.7784	0.7717	0.744	0.731	0.7851	0.7784
3/4	14	1.036	1.023	0.9889	0.9822	0.955	0.941	0.9956	0.9889
1	11 1/2	1.296	1.280	1.2386	1.2305	1.197	1.181	1.2468	1.2386
1 1/4	11 1/2	1.641	1.625	1.5834	1.5753	1.542	1.526	1.5916	1.5834
1 1/2	11 1/2	1.880	1.864	1.8223	1.8142	1.781	1.764	1.8305	1.8223
2	11 1/2	2.354	2.338	2.2963	2.2882	2.254	2.238	2.3044	2.2963
2 1/2	8	2.846	2.822	2.7622	2.7505	2.702	2.679	2.7739	2.7622
3	8	3.472	3.448	3.3885	3.3768	3.329	3.305	3.4002	3.3885
3 1/2	8	3.972	3.949	3.8888	3.8771	3.829	3.806	3.9005	3.8888
4	8	4.470	4.447	4.3871	4.3754	4.327	4.304	4.3988	4.3871
5	8	5.533	5.509	5.4493	5.4376	5.389	5.366	5.4610	5.4493
6	8	6.589	6.566	6.5060	6.4943	6.446	6.423	6.5177	6.5060

All dimensions are given in inches.

[1] For the convenience of those who might desire to use this type of straight pipe thread with an allowance, the following schedule is suggested:

for 27 threads per inch, 0.0025 in.
18 " " 0.0030 in.
14 " " 0.0040 in.
11½ " " 0.0050 in.
8 " " 0.0070 in.

It is recommended also that these allowances be subtracted from the diameter of the external screw thread given above.

[2] Columns 5 and 10 are the same as the pitch diameter at the end of internal thread, E_1, *Basic*. (See Table 13–70, Col. 8.)

The minor diameter of external threads and major diameter of internal threads are those as produced by commercial straight pipe dies and commercial ground straight pipe taps.

The *maximum minor diameter* of the external thread may be the same as the minimum minor diameter of the internal thread, as shown in Col. 8, and the *minimum major diameter* of the internal thread may be the same as the maximum major diameter of the external thread, as shown in Col. 3.

The major and minor diameters have been calculated on the basis of a truncation of $p/10$ to avoid interference at crest and root when product is gaged with gages made in accordance with Par. 28d of the Standard.

TABLE 13–74

Pipe Thread Details on Drawings of *Machine Parts

Threaded holes for pipe connections to cylinders, machine bases and similar machine parts are called for on drawings by symbols in the conventional manner by specifying pipe size and threaded connection. Compare Table 9–58. Because pipe threads must have pressure tightness, as well as fasten parts together, this requirement is important in the symbol. Proper sealing also limits the depth the pipe can enter the hole, and consequently the depth of tapped hole is often specifically dimensioned.

Since the number of threads is governed by pipe size, this item is of less importance in the designation of pipe threads than it is in the designation of ordinary screw threads. Thus 3 NPT unmistakably, though indirectly, specifies 8 threads per inch on a pipe having an outside diameter of 3-1/2 inches. The SAE Handbook, 1953, in the Section on Dryseal Pipe Threads gives the following examples: 1/8-27 Dryseal NPTF and 1/8-27 Dryseal NPSF. The letters are abbreviations as follows: N = National or American Standard; P = Pipe; T = Taper; F = Fuel or Dryseal and S = Straight. Other abbreviations are defined in the SAE Handbook and in the American Standard on Pipe Threads, ASA B2.1–1945.

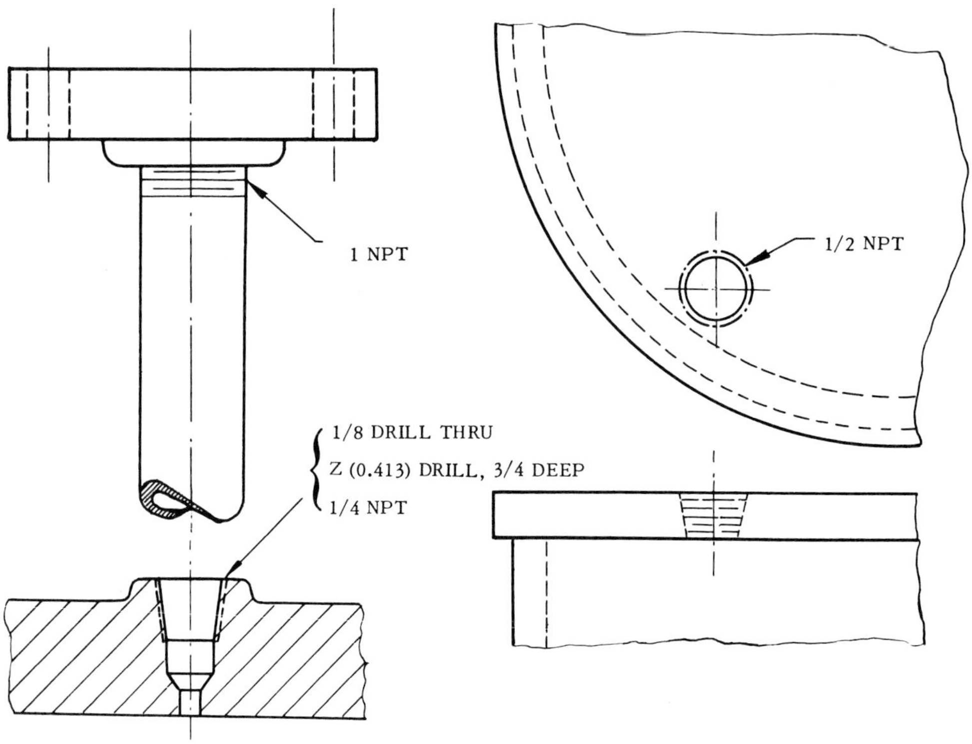

*For line diagrams of piping systems, see Tables 15–22 and 15–23.

INDEX TO

SECTION 14

Electric Motors — Graphical Symbols — Welding

TABLE 14–1

Designation of Common Welds by Symbols on Drawings

ASA Z32.2.1 – 1949, Reaffirmed 1953 Graphical Symbols for Welding

The joint is the basis of reference. Any joint, the welding of which is indicated by a symbol, always has an ARROW SIDE and an OTHER SIDE. Accordingly, the arrow plus the weld symbol can always be placed with respect to the joint so as to comply clearly and positively, with the phrasing ARROW SIDE, OTHER SIDE, and BOTH SIDES.

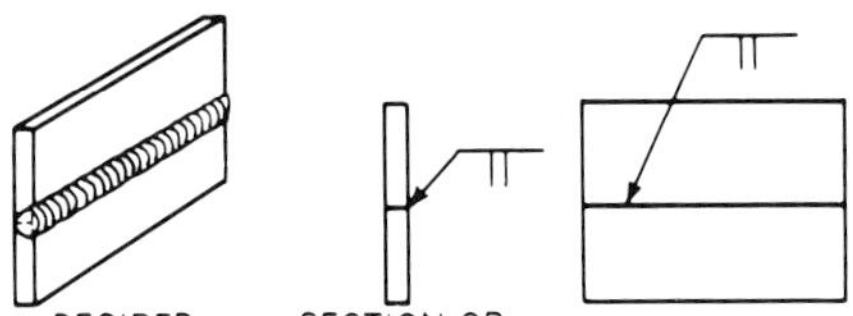

(A) ARROW-SIDE SQUARE-GROOVE WELDING SYMBOL

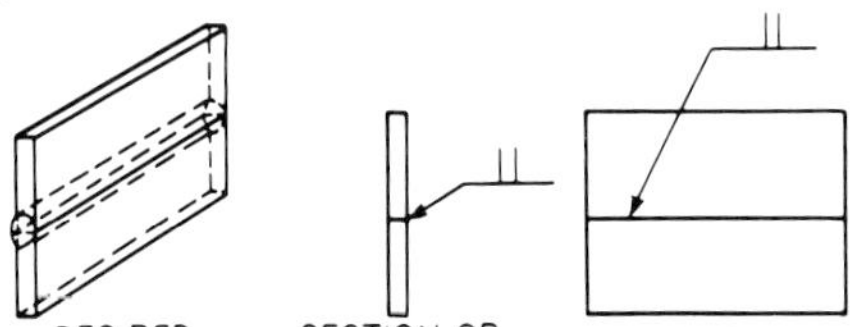

(B) OTHER-SIDE SQUARE-GROOVE WELDING SYMBOL

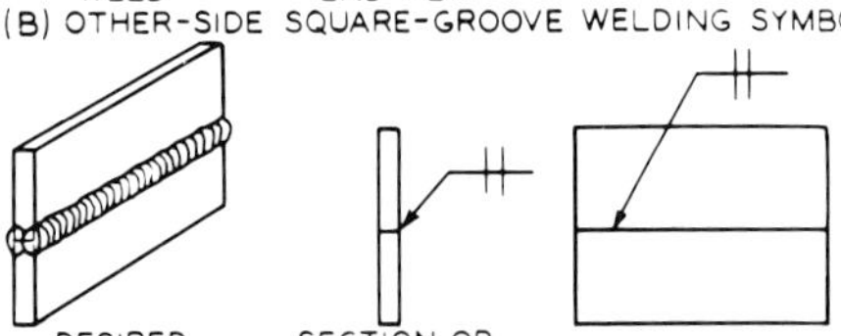

(C) BOTH-SIDES SQUARE-GROOVE WELDING SYMBOL

APPLICATION OF SQUARE-GROOVE WELDING SYMBOLS

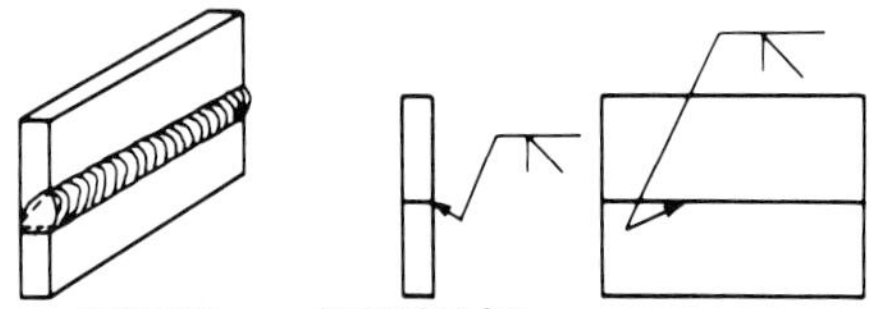

(A) ARROW-SIDE BEVEL-GROOVE WELDING SYMBOL

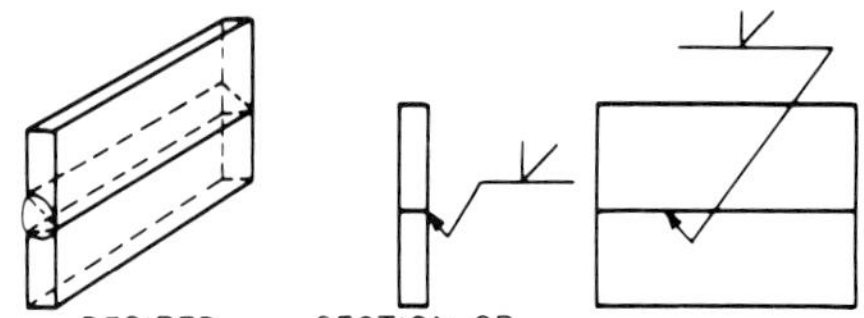

(B) OTHER-SIDE BEVEL-GROOVE WELDING SYMBOL

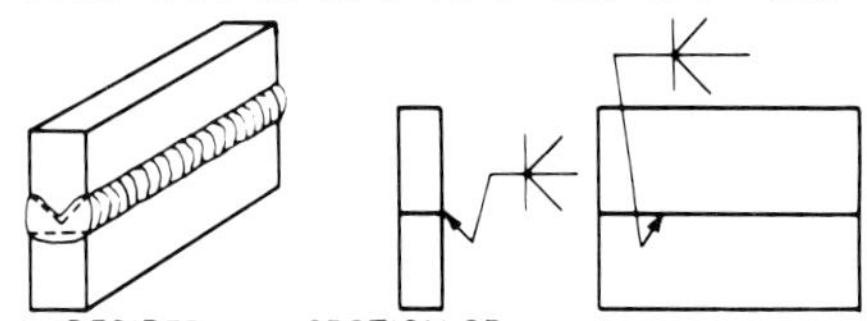

(C) BOTH-SIDES BEVEL-GROOVE WELDING SYMBOL

APPLICATION OF BEVEL-GROOVE WELDING SYMBOLS

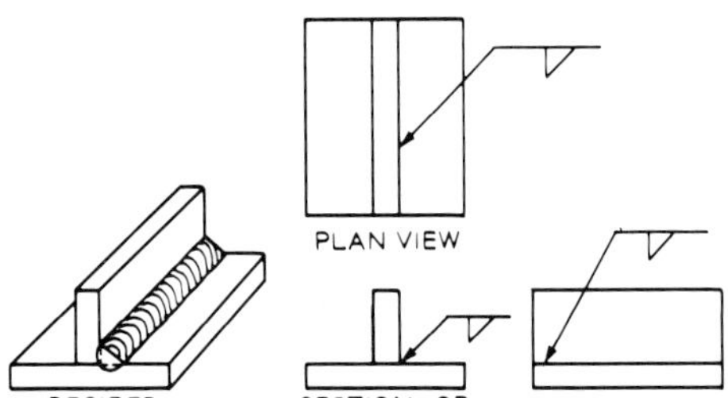

(A) ARROW-SIDE FILLET WELDING SYMBOL

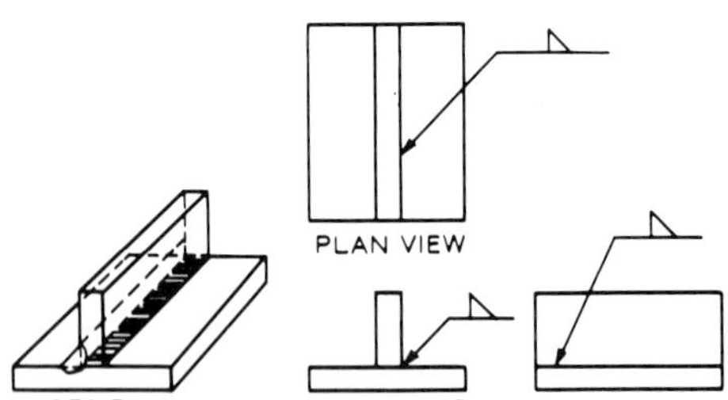

(B) OTHER-SIDE FILLET WELDING SYMBOL

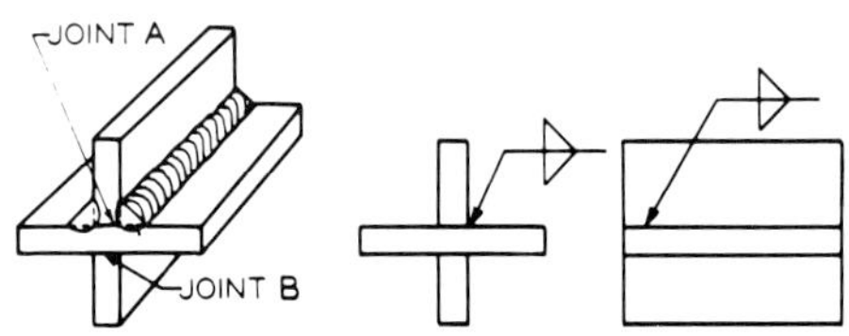

(A) BOTH-SIDES FILLET WELDING SYMBOL FOR ONE JOINT

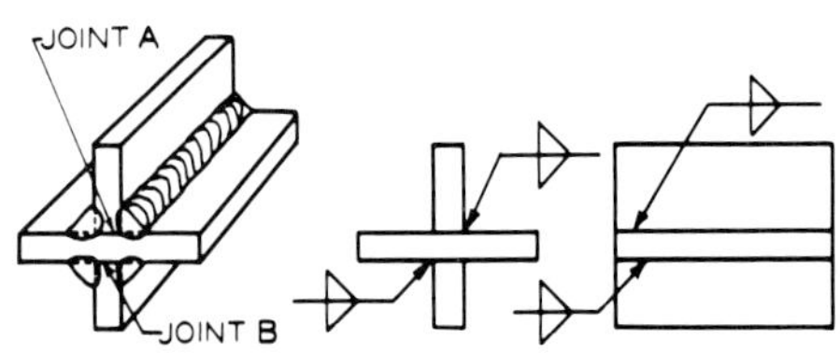

(B) BOTH-SIDES FILLET WELDING SYMBOL FOR TWO JOINTS

APPLICATION OF FILLET WELDING SYMBOLS

continued on next page

TABLE 14-1, continued

Butt welds, fillet welds and groove welds, as illustrated on the preceding page, comprise the bulk of today's welded design. The Lincoln Electric Company says, "basically, around 80 percent of all welded machinery structure connections are made by fillet welds, 15 percent are butt welds, and 5 percent require special welds. In pressure vessel welding, approximately 90 percent of all connections are butt welds."

Among the other types of welding that are often used in engineering design are spot welding, seam welding, upset welding and projection welding.

LOCATION OF WELD WITH RESPECT TO JOINT:

(a) Welds on the arrow side of the joint shall be shown by placing the weld symbol on the side of the reference line toward the reader, thus:

(b) Welds on the other side of the joint shall be shown by placing the weld symbol on the side of the reference line away from the reader, thus:

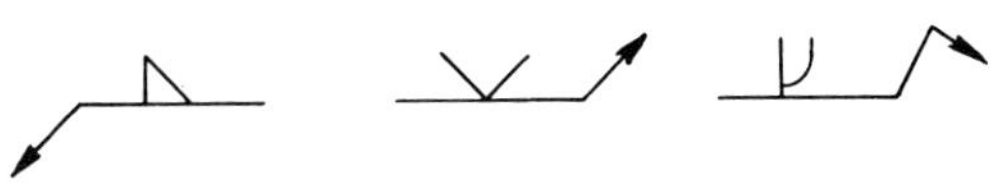

(c) Welds on both sides of the joint shall be shown by placing weld symbols on both sides of the reference line, toward and away from the reader, thus:

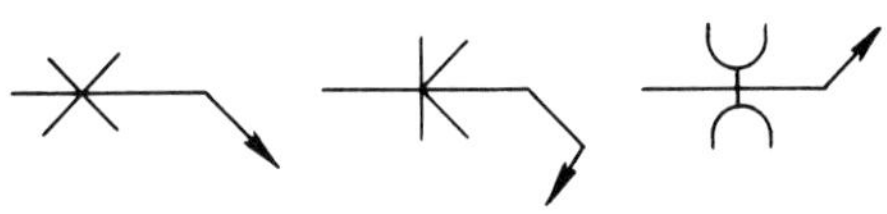

(d) Spot, seam, flash and upset weld symbols have <u>no</u> arrow-side or other-side significance in themselves, although supplementary symbols used in conjunction therewith may have such significance. Spot, seam, flash and upset weld symbols shall be centered on the reference line, thus:

LOCATION SIGNIFICANCE OF ARROW:

(a) In the case of groove, fillet, and flash or upset welding symbols, the arrow shall connect the welding symbol reference line to one side of the joint, and this side shall be considered the arrow side of the joint. The side opposite the arrow side of the joint shall be considered the other side of the joint.

(b) In the case of plug, slot, spot, seam and projection welding symbols, the arrow shall connect the welding symbol reference line to the outer surface of one of the members of the joint at the center line of the desired weld. The member

Continued on next page

to which the arrow points shall be considered the arrow-side member. The other member of the joint shall be considered the other-side member.

(c) When a joint is depicted by a single line on the drawing and the arrow of a welding symbol is directed to this line, the arrow side of the joint shall be considered as the near side of the joint in accordance with the usual conventions of drafting.

(d) When a joint is depicted as an area parallel to the plane of projection in a drawing and the arrow of a welding symbol is directed to that area, the arrow-side member of the joint shall be considered as the near member of the joint in accordance with the usual conventions of drafting.

SPOT WELDING SYMBOL:

Symbols have no arrow-side nor other-side significance.

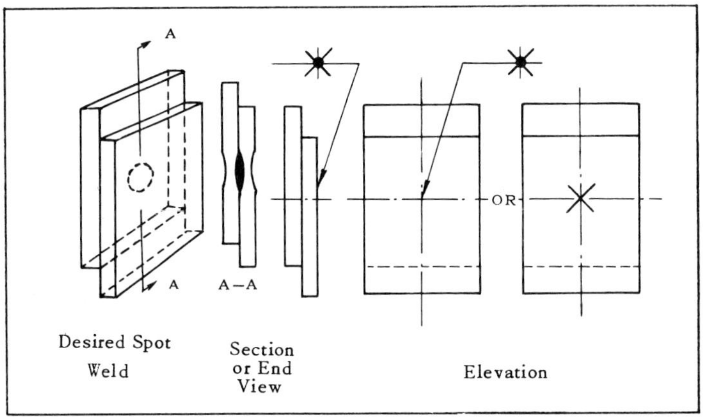

SEAM WELDING SYMBOL:

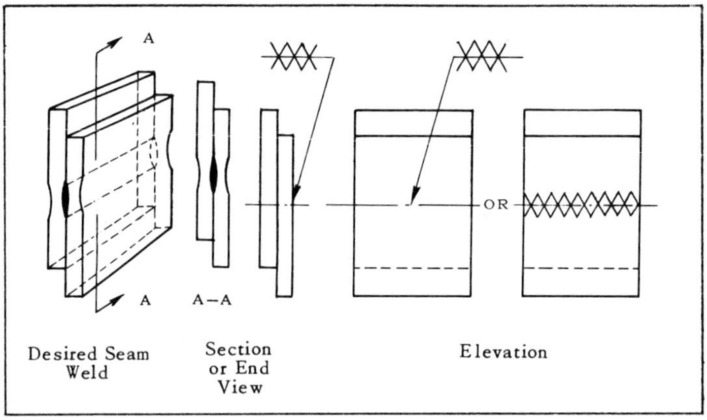

continued on next page

FLASH AND UPSET WELDING SYMBOL:

Symbol has no arrow-side nor other-side significance

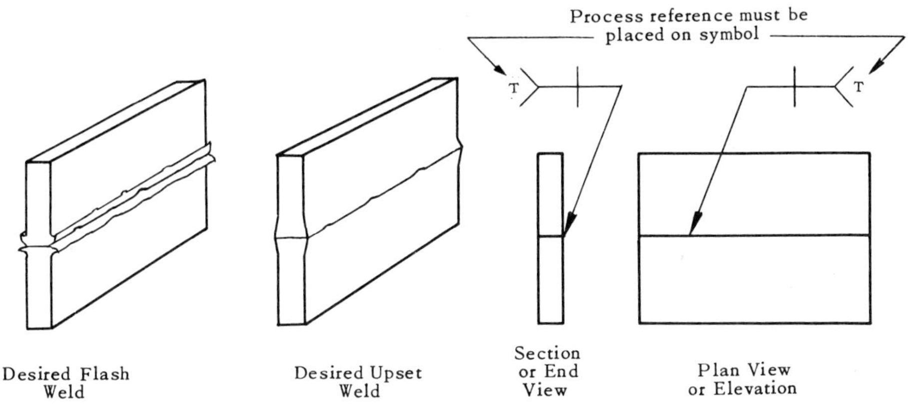

PROJECTION WELDING SYMBOLS:

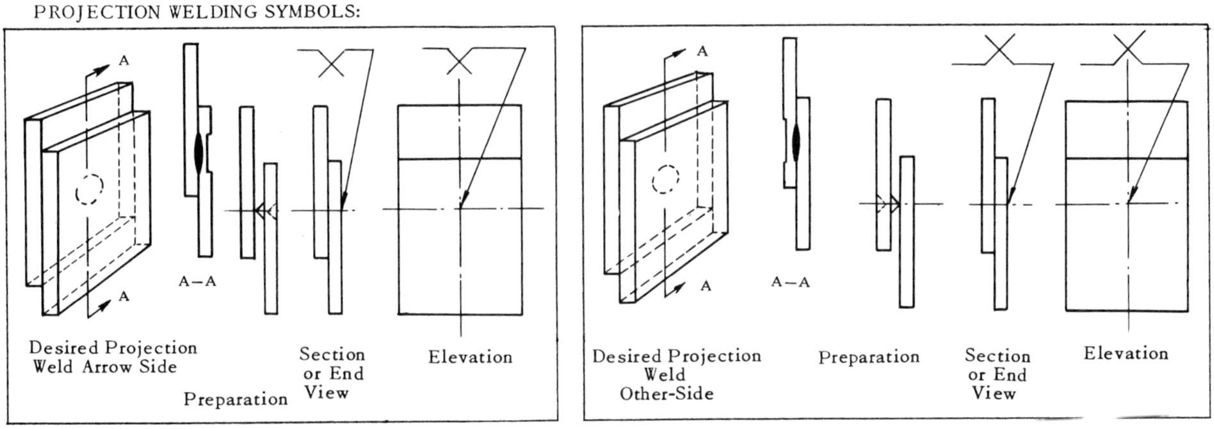

TABLE 14–2

Allowable Loads on Fillet Welds in Shear*

Size of Fillet Weld Inches	Pounds per Linear Inch	Remarks
1/8	1200	These values are recommended in the Code for Fusion Welding (Standard of the American Welding Society. They are based on a permissible unit stress of 13,600 psi, being suitable for weldments and conventional machine design. For dynamic, inbrational and lifting loads, these values should be reduced, depending upon the severity of the loading.
3/16	1800	
1/4	2400	
5/16	3000	
3/8	3600	
1/2	4800	
5/8	6000	
3/4	7200	

*In fusion welding for structural purposes the stress in a fillet weld is considered as shear for any direction of applied stress.

TABLE 14–3

Weld Stress Formulas

Welding Handbook, Third Edition   American Welding Society

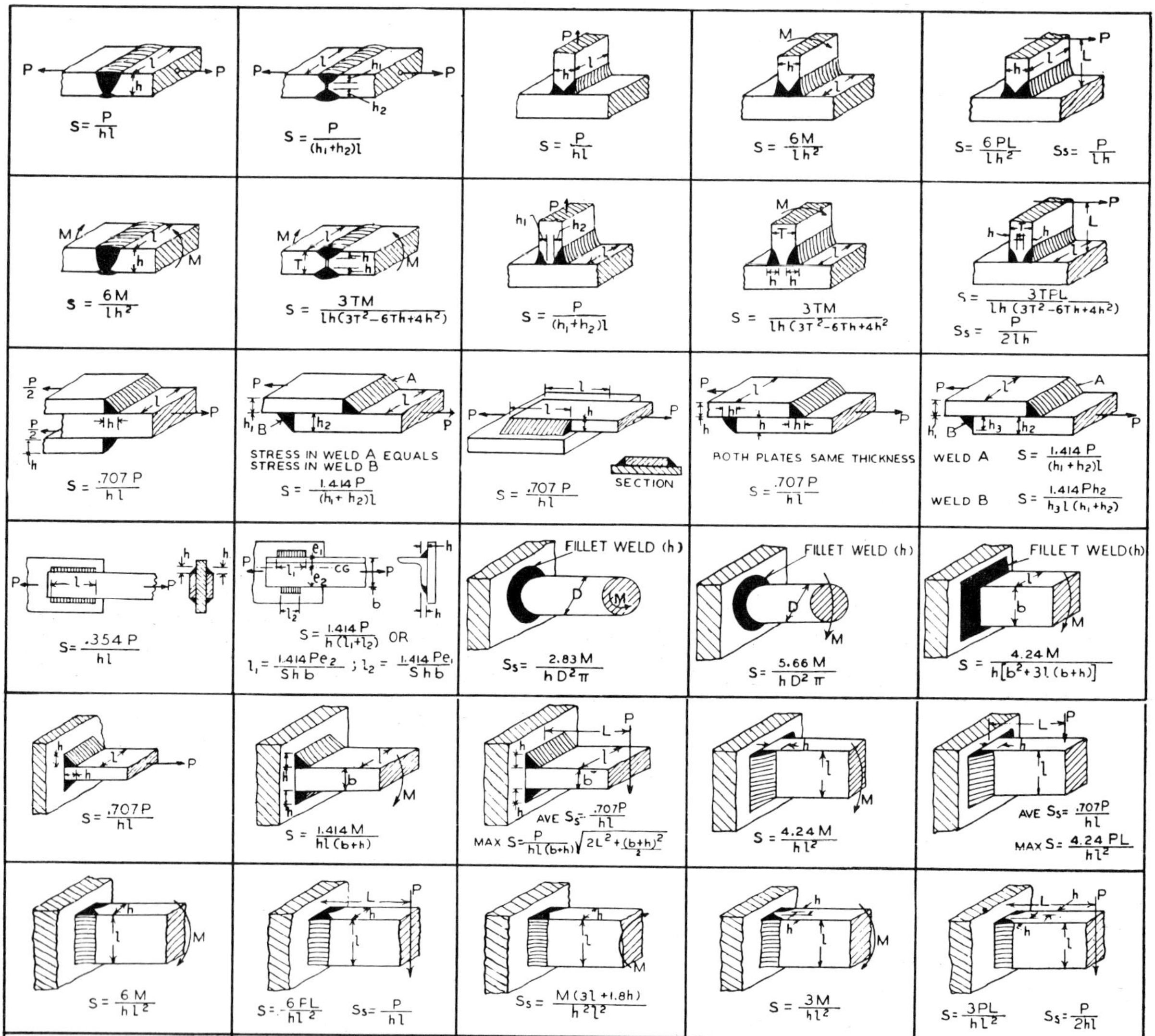

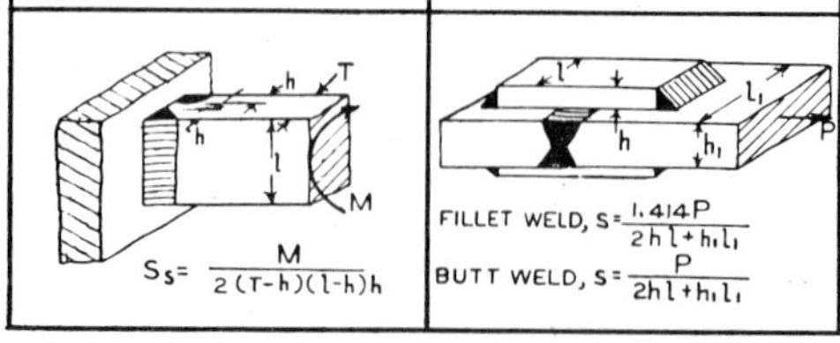

S = Normal stress, psi.		D = External load, lb.	
Ss = Unit stress, psi.		L = Linear distance	
I = Moment of inertia, in. units		H = Size of weld	
M = Bending moment, in lb.		l = Length of weld, in.	

TABLE 14–4

Recommended Proportions of Grooves for Metal-Arc Welding and * Gas Welding – Steel

Welding Handbook, Third Edition American Welding Society

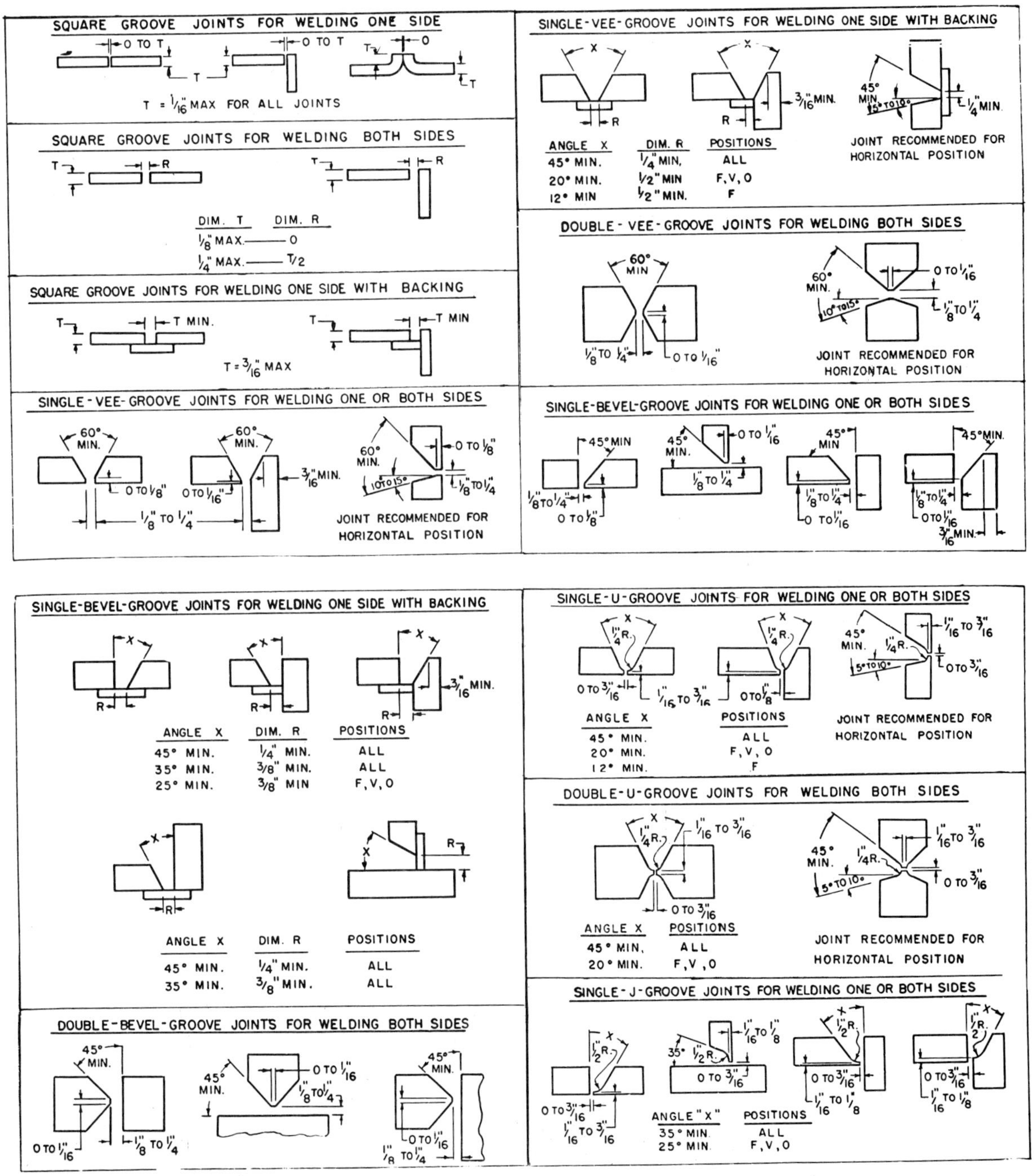

* Except Pressure Gas Welding

TABLE 14–5

Typical Edge Preparations for Welding Aluminum

Welding Handbook, Third Edition   American Welding Society

OXYHYDROGEN, OXYACETYLENE
ATOMIC HYDROGEN, CARBON ARC

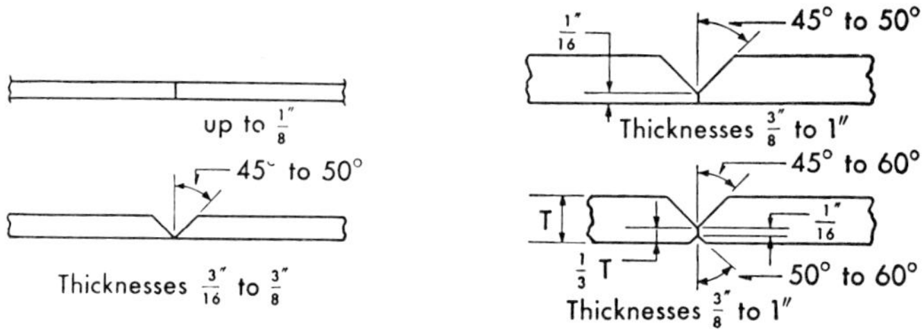

METAL ARC

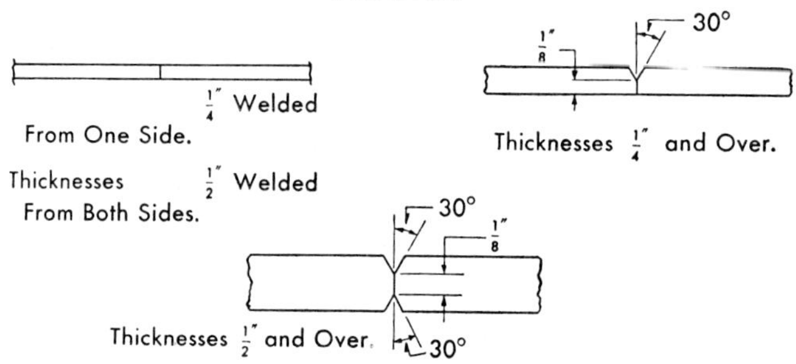

INERT GAS

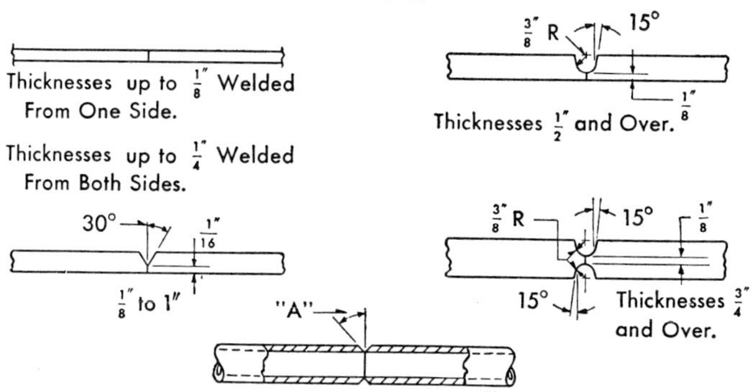

Pipe Welds up to 12" I.P.S.
A = 55° Position Weld.
A = 30° Roll Weld.

TABLE 14-6

Edge Preparation for Arc Welding Stainless Steel

Welding Handbook, Third Edition American Welding Society

(a) For plate 3/16 in. and thinner

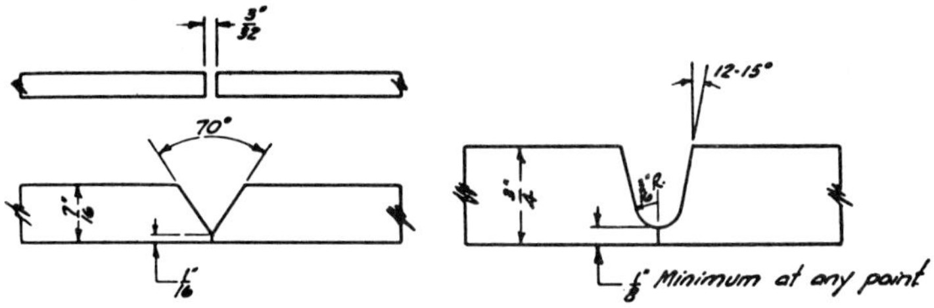

(b) For plate 1/4- to 7/16-in. thick (c) For plate 1/2 in. and thicker

TABLE 14-7

Clearances Required for Stud Welding Gun

Welding Handbook, Third Edition American Welding Society

Stud Diam	Minimum Stud Clearances A	B	C	D	E
3/16	5/16	Bend Radius + 5/16	5/16	1/2	Weld Width + 1/4
1/4	5/16	Bend Radius + 5/16	5/16	9/16	Weld Width + 1/4
5/16	3/8	Bend Radius + 3/8	3/8	11/16	Weld Width + 5/16
3/8	3/8	Bend Radius + 3/8	3/8	3/4	Weld Width + 5/16
7/16	13/32	Bend Radius + 13/32	13/32	13/16	Weld Width + 3/8
1/2	15/32	Bend Radius + 15/32	15/32	15/16	Weld Width + 7/16
5/8	9/16	Bend Radius + 9/16	9/16	1 3/32	Weld Width + 1/2
3/4	5/8	Bend Radius + 5/8	5/8	1 7/32	Weld Width + 9/16

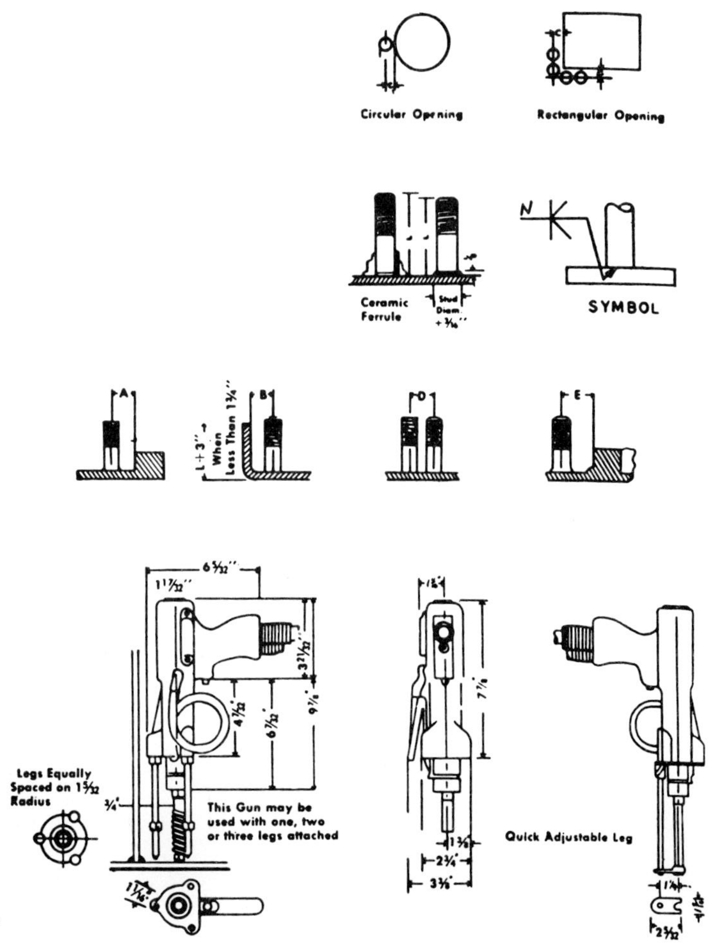

TABLE 14–8

Stud Welding Specifications

Welding Handbook, Third Edition   American Welding Society

(a) Specifications for Typical Studs

Stud Size	Stud Length, L, In. Min.	Stud Length, L, In. Max.	Threads Per Inch NC	Threads Per Inch NF	Thread Length B — Length of Stud, L: 1 In. or Less	Thread Length B — Length of Stud, L: $1\frac{1}{8}$ to $1\frac{7}{16}$ In.	Thread Length B — Length of Stud, L: $1\frac{1}{2}$ In. or Over	Burn Off, C ± 1/32 In.	Fillet Diam. D+ 0.020 In.	Approx. Size Ferrule
No. 6	13/16	8	32	40	L minus 3/16	1	$1\frac{1}{4}$	3/32	0.190	6
No. 8	13/16	8	32	36	L minus 3/16	1	$1\frac{1}{4}$	3/32	0.214	8
No. 10	13/16	8	24	32	L minus 3/16	1	$1\frac{1}{4}$	3/32	0.262	10
1/4 in.	13/16	8	20	28	L minus 3/16	1	$1\frac{1}{4}$	1/8	0.330	1/4
5/16 in.	13/16	8	18	24	L minus 3/16	1	$1\frac{1}{4}$	1/8	0.406	5/16
3/8 in.	13/16	8	16	24	L minus 3/16	1	$1\frac{1}{4}$	1/8	0.468	3/8
7/16 in.	13/16	8	14	20	L minus 3/16	1	$1\frac{1}{4}$	1/8	0.546	7/16
1/2 in.	7/8	8	13	20	L minus 3/16	1	$1\frac{1}{4}$	1/8	0.613	1/2
5/8 in.	15/16	8	11	18	For $1\frac{11}{16}$ or less L minus 1/4	For $1\frac{3}{4}$ or over $1\frac{1}{2}$		1/8	0.765	5/8
3/4 in.	$1\frac{1}{4}$	8	10	16	For $1\frac{15}{16}$ or less L minus 1/4	For 2 or over $1\frac{3}{4}$		1/8	•0.890	3/4

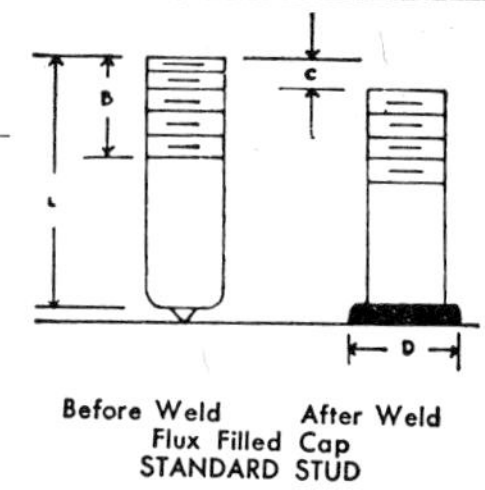

(b) Specifications for Reduced Base Studs

Stud Size, A	Stud Length, L, In. Min.	Stud Length, L, In. Max.	Red. Base Length, K, In.	Weld Diam., E, In.	Threads Per Inch NC	Threads Per Inch NF	Thread Length B, — Length of Stud — $1\frac{3}{8}$ In. or Less	Thread Length B, — Length of Stud — $1\frac{1}{2}$ In. or Over	Burn Off, C ± 1/32 In.	Approx. Fillet Diam., D+0.020 In.	Ferrule No.
No. 8	13/16	8	9/64	1/8	32	36	L minus K	$1\frac{1}{4}$	1/16	0.169	8 RRB
No. 10	13/16	8	9/64	9/64	24	32	L minus K	$1\frac{1}{4}$	1/16	0.192	10 RRB
1/4 in.	13/16	8	5/32	3/16	20	28	L minus K	$1\frac{1}{4}$	1/16	0.255	1/4 in. RRB
5/16 in.	13/16	8	1/4	1/4	18	24	L minus K	$1\frac{1}{4}$	3/32	0.343	5/16 in. RB
3/8 in.	13/16	8	1/4	5/16	16	24	L minus K	$1\frac{1}{4}$	3/32	0.406	3/8 in. RB
7/16 in.	13/16	8	9/32	3/8	14	20	L minus K	$1\frac{1}{4}$	1/8	0.468	7/16 in. RB
1/2 in.	7/8	8	9/32	7/16	13	20	L minus K	$1\frac{1}{4}$	1/8	0.560	1/2 in. RB

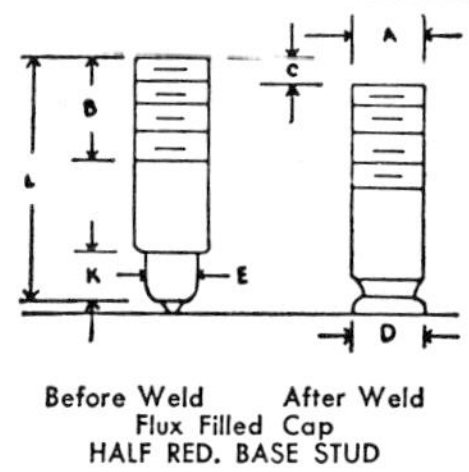

Standard stud material is AISI C-1020 or C-1015

TABLE 14–9

Relative Speeds and Costs for 3/8-In. Fillet Welds in Different Positions

Welding Handbook, Third Edition   American Welding Society

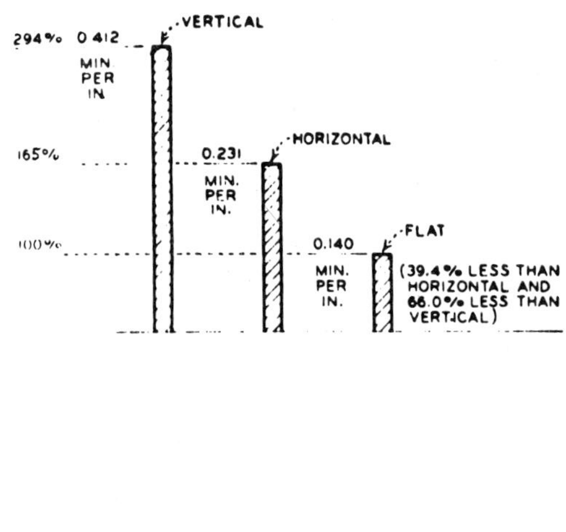

TABLE 14–10

Illustration of Cost of Poor Fit-Up in Fillet Welds

Welding Handbook, Third Edition American Welding Society

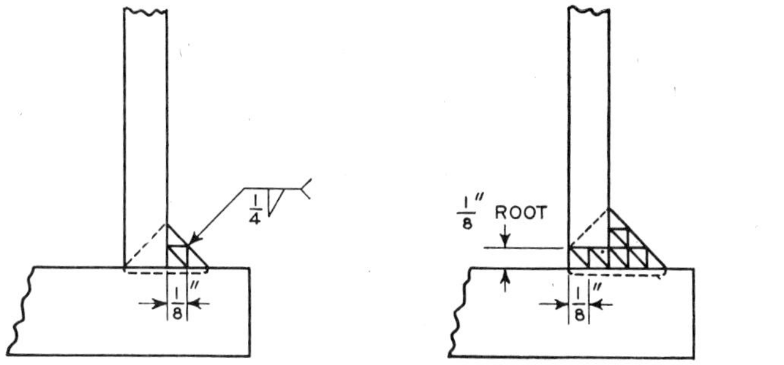

GOOD FIT-UP (4-1/8 TRIANGLE VOLUMES USED) –VS.– ROOT OPENING 1/2 THE SIZE OF WELD (9 TRIANGLE VOLUMES REQUIRED AND 13 PROBABLY USED)

TABLE 14–11

Effect of Overwelding

Welding Handbook, Third Edition American Welding Society

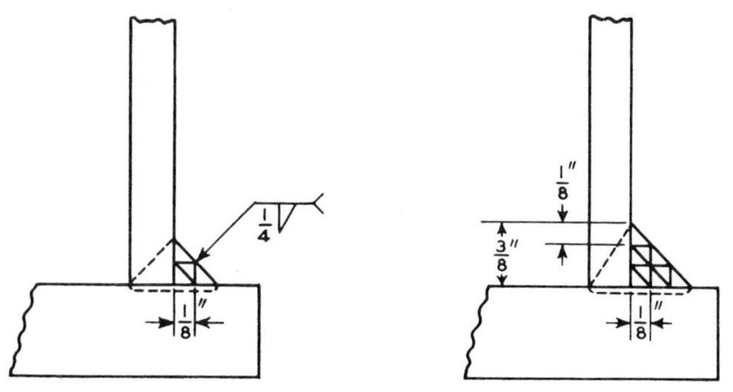

CORRECT WELD (4 TRIANGLE VOLUMES OF METAL REQUIRED) –VS.– OVERWELDED–JOINT 1/2 LARGER THAN SPECIFIED SIZE (9 TRIANGLE VOLUMES OF METAL USED)

TABLE 14–12

*Selected Electrical Standards about Industrial Equipment

Identification of Standard	Title	Remarks
ASA C1–1953	National Electrical Code	National Board of Fire Underwriters National Fire Protection Association
ASA C2	National Electrical Safety Code	National Bureau of Standards Handbook H30
ASA C6.1–1944	Terminal Markings for Electrical Apparatus	
ASA C50–1943	Rotating Electrical Machinery	
ASA Y32.1.1–1951	Graphical Symbols for Single-Line Electrical Engineering Diagrams	
ASA Z32.10–1948	Graphical Symbols for Electron Devices	Under revision – out of print
ASA Z32.12–1947	Basic Graphical Symbols for Electric Apparatus	Under revision – out of print
ASA Z32.3–1946	Graphical Symbols for Electric Power and Control	Under revision – out of print
NEMA	National Electrical Manufacturers Association	Tables 14–15 to 14–19 are illustrative.
JIC	Joint Industry Conference	Tables 14–20 and 14–21
	Electrical Standards for Industrial Equipment	Table 15–19 states purpose and source of JIC Standards

* References rather than abstracted matter are given here not only because the quantity of matter is abundant and space in this volume is limited but also because essential standards are undergoing revision. It is recognized that to power industrial equipment and safely to control it, by electrical apparatus, is often as perplexing and absorbing for the designer as is the design of the mechanical equipment itself — and no less important. Although an electrical co-designer may carry the burden of the electrical design, the space for motors, the convenience and arrangement of controls, the safety of operating personnel, operating instructions and diagrams, as well as acceptable and satisfactory over-all performance, are nevertheless responsibilities of the mechanical designer. Both designers can use the latest revisions of these references to advantage.

TABLE 14–13

Horsepower and Speed Ratings for D-C Fractional Horsepower Constant-Speed Motors

Standards for Fractional Horsepower Motors, NEMA Publ. No. MG 2–1951

Hp	Approximate Full Load, Rpm				
1/20	3450	1725	1140	850	Rated 32, 115 and 230 volts
1/12	3450	1725	1140	850	
1/8	3450	1725	1140	850	
1/6	3450	1725	1140	850	
1/4	3450	1725	1140	850	
1/3	3450	1725	1140	850	
1/2	3450	1725	1140	...	
3/4	3450	1725	...	...	
1	3450	...	...	...	

TABLE 14–14

Horsepower and Speed Ratings for Fractional Horsepower Induction Motors

Standards for Fractional Horsepower Motors, NEMA Publ. No. MG 2–1951

Rated 115 and 230 Volts Single-Phase and 110, 208* and 220 V Polyphase

Hp	60-Cycle Synchronous Rpm	All Motors Except Shaded Pole and Permanent-split Capacitor	Shaded-pole Motors	Permanent-split Capacitor Motors	50-Cycle Synchronous Rpm	All Motors Except Shaded Pole and Permanent-split Capacitor	Shaded-pole Motors	Permanent-split Capacitor Motors	25-Cycle Synchronous Rpm	All Motors Except Shaded Pole and Permanent-split Capacitor	Shaded-pole Motors	Permanent-split Capacitor Motors
		Approximate Full-load Rpm				Approximate Full-load Rpm				Approximate Full load Rpm		
1; 1.5; 2; 3; 5; 7.5; 10; 15; 25 and 35 millihorsepower	3600 1800 1200 900	3450 1725 1140 ..	3000 1550 1050 800	3000 1550 1050 800	3000 1500 1000	2850 1425 950	2500 1300 875	2500 1300 875	1500	1425	1300	1300
1/20; 1/12 and 1/8 horsepower	3600 1800 1200 900	3450 1725 1140 850	3000 1550 1050 800	3250 1625 1075 825	3000 1500 1000	2850 1425 950	2500 1300 875	2700 1350 900	1500	1425	1300	1350
1/6; 1/4 and 1/3 horsepower	3600 1800 1200 900	3450 1725 1140 850	.. ..	3250 1625 1075 825	3000 1500 1000	2850 1425 950	.. .. ..	2700 1350 900	1500	1425	..	1350
1/2 horsepower	3600 1800 1200	3450 1725 1140	.. .. ..	3250 1625 1075	3000 1500 1000	2850 1425 950	.. .. ..	2700 1350 900	1500	1425	..	1350
3/4 horsepower	3600 1800	3450 1725	.. ..	3250 1625	3000 1500	2850 1425	.. ..	2700 1350	1500	1425	..	1350
1 horsepower	3600	3450	..	3250	3000	2850	..	2700	..	..	..	..

* Applies to 60-cycle circuits only.

TABLE 14-15

Horsepower and Synchronous Speed Ratings of Single-Phase Integral-Horsepower Motors

Standards for Motors and Generators, NEMA, 1949

Rated 115 and 230 V

Hp	60 Cycles Rpm				50 Cycles Rpm				25 Cycles Rpm	
½	...	...	...	900	...	...	1000	750	...	...
¾	...	...	1200	900	...	1500	1000	750	...	...
1	...	1800	1200	900	3000	1500	1000	750	1500	...
1½	3600	1800	1200	900	3000	1500	1000	750	1500	...
2	3600	1800	1200	900	3000	1500	1000	750	1500	750
3	3600	1800	1200	900	3000	1500	1000	750	1500	750
5	3600	1800	1200	900	3000	1500	1000	750	1500	750
7½	3600	1800	1200	900	3000	1500	1000	750	1500	750
10	3600	1800	1200	900	3000	1500	1000	750	1500	750
15	3600	1800	1200	900	3000	1500	1000	750	1500	750
20	3600	1800	1200	900	3000	1500	1000	750	1500	750
25	3600	1800	1200	900	...	...	...	...	...	...

TABLE 14-16

Horsepower and Synchronous Speed Ratings of Polyphase Integral-Horsepower Induction Motors

Standards for Motors and Generators, NEMA, 1949

Rated at 110, 208† and 220, 440, 550 and 2300 V

Hp	60 Cycles Rpm								50 Cycles Rpm				25 Cycles Rpm		
½	...	...	...	900	720	600	514	450	...	...	...	750	...	750	500
¾	...	...	1200	900	720	600	514	450	...	...	1000	750	...	750	500
1	...	1800	1200	900	720	600	514	450	...	1500	1000	750	1500*	750	500
1½	3600*	1800	1200	900	720	600	514	450	3000*	1500	1000	750	1500*	750	500
2	3600*	1800	1200	900	720	600	514	450	3000*	1500	1000	750	1500*	750	500
3	3600*	1800	1200	900	720	600	514	450	3000*	1500	1000	750	1500*	750	500
5	3600*	1800	1200	900	720	600	514	450	3000*	1500	1000	750	1500*	750	500
7½	3600*	1800	1200	900	720	600	514	450	3000*	1500	1000	750	1500*	750	500
10	3600*	1800	1200	900	720	600	514	450	3000*	1500	1000	750	1500*	750	500
15	3600*	1800	1200	900	720	600	514	450	3000*	1500	1000	750	1500*	750	500
20	3600*	1800	1200	900	720	600	514	450	3000*	1500	1000	750	1500*	750	500
25	3600*	1800	1200	900	720	600	514	450	3000*	1500	1000	750	1500*	750	500
30	3600*	1800	1200	900	720	600	514	450	3000*	1500	1000	750	1500*	750	500
40	3600*	1800	1200	900	720	600	514	450	3000*	1500	1000	750	1500*	750	500
50	3600*	1800	1200	900	720	600	514	450	3000*	1500	1000	750	1500*	750	500
60	3600*	1800	1200	900	720	600	514	450	3000*	1500	1000	750	1500*	750	500
75	3600*	1800	1200	900	720	600	514	450	3000*	1500	1000	750	1500*	750	500
100	3600*	1800	1200	900	720	600	514	450	3000*	1500	1000	750	1500*	750	500
125	3600*	1800	1200	900	720	600	514	450	3000*	1500	1000	750	1500*	750	500
150	...	1800	1200	900	720	600	514	450	...	1500	1000	750	1500*	750	500
200	...	1800	1200	900	720	600	514	...	...	1500	1000	750	1500*	750	500

* Applies to squirrel-cage motors only.

† 60-cycle only.

TABLE 14–17

Standard Dimensions for Foot-Mounted Motors and Generators with Single Straight Shaft Extension

NEMA Standards for Motors and Generators, Part 3–1953

Frame Number*	Key Width	Key Thickness	Key Length†	A Max	B Max	D	E	F	BA	H	N-W ‡‡	U	V Min	AA¶ Minimum Size of Conduit	AL	AM	AO	AR	AU	AW	AX	AY Bases	AY Max Rails	BT
42	3/64 Flat			..	..	2 5/8**	1 3/4	27/32	2 1/16	9/32 slot	1 1/8	3/8	..	..	..	..	..	..	..	..	..	..	..	..
48		3/64 *Flat*		..	..	3**	2 1/8	1 3/8	2 1/2	11/32 *slot*◆	1 1/2	1/2	..	..	..	..	..	..	..	..	..	..	..	..
56	3/16	3/16	1 3/8‡	..	..	3 1/2**	2 7/16	1 1/2	2 3/4	11/32 slot	1 7/8	5/8	...	..	..	..	..	..	..	..	..	..	..	..
66	3/16	3/16	1 7/8‡	..	..	4 1/8**	2 15/16	2 1/2	3 1/8	13/32 slot	2 1/4	3/4	..	..	..	..	..	..	..	..	..	..	..	..
203	3/16	3/16	1 3/8	10	7 1/2	5††	4	2 3/4	3 1/8	13/32	2 1/4	3/4	2	3/4	14	11	5	4 3/4	1/2	3/4	1 3/4	1/2	4	3
204	3/16	3/16	1 3/8	10	8 1/2	5††	4	3 1/4	3 1/8	13/32	2 1/4	3/4	2	3/4	14	12	5	5 1/4	1/2	1 1/4	1 3/4	1/2	4	3
224	1/4	1/4	2	11	8 3/4	5 1/2††	4 1/2	3 3/8	3 1/2	13/32	3	1	2 3/4	3/4	15 1/2	12 1/4	5 1/2	5 3/8	1/2	1 3/8	1 3/4	1/2	4 1/4	3 1/2
225	1/4	1/4	2	11	9 1/2	5 1/2††	4 1/2	3 3/4	3 1/2	13/32	3	1	2 3/4	3/4	15 1/2	13	5 1/2	5 3/4	1/2	1 3/4	1 3/4	1/2	4 1/4	3 1/2
254	1/4	1/4	2 3/8	12 1/2	10 3/4	6 1/4††	5	4 1/8	4 1/4	17/32	3 3/8	1 1/8	3 1/8	1	17 3/4	15 1/8	6 1/4	6 5/8	5/8	1 5/8	2	5/8	5	4
284	1/4	1/4	2 3/4	14	12 1/2	7††	5 1/2	4 3/4	4 3/4	17/32	3 3/4	1 1/4	3 1/2	1	19 3/4	16 7/8	7	7 1/2	5/8	2	2	5/8	5 3/4	4 1/2
324	3/8	3/8	3 3/4	16	14	8††	6 1/4	5 1/4	5 1/4	21/32	4 7/8	1 5/8	4 5/8	1 1/4	22 3/4	19 1/4	8	8 1/2	3/4	2	2 1/2	3/4	6 1/2	5 1/4
326	3/8	3/8	3 3/4	16	15 1/2	8††	6 1/4	6	5 1/4	21/32	4 7/8	1 5/8	4 5/8	1 1/4	22 3/4	20 3/4	8	9 1/4	3/4	2 3/4	2 1/2	3/4	6 1/2	5 1/4
364	1/2	1/2	4 1/4	18	15 1/4	9††	7	5 5/8	5 7/8	21/32	5 5/8	1 7/8	5 3/8	1 1/2	25 1/2	20 1/2	9	9 1/8	3/4	2 1/8	2 1/2	3/4	7 1/2	6
364S	3/8	3/8	1 7/8	18	15 1/4	9††	7	5 5/8	5 7/8	21/32	3 1/4	1 5/8	3	1 1/2	..	..	..	..	..	..	..	..	..	..
365	1/2	1/2	4 1/4	18	16 1/4	9††	7	6 1/8	5 7/8	21/32	5 5/8	1 7/8	5 3/8	1 1/2	25 1/2	21 1/2	9	9 5/8	3/4	2 5/8	2 1/2	3/4	7 1/2	6
365S	3/8	3/8	1 7/8	18	16 1/4	9††	7	6 1/8	5 7/8	21/32	3 1/4	1 5/8	3	1 1/2	..	..	..	..	..	..	..	..	..	..
404	1/2	1/2	5	20	16 1/4	10††	8	6 1/8	6 5/8	13/16	6 3/8	2 1/8	6 1/8	2	28 3/4	22 3/8	10	9 7/8	7/8	2 3/8	3	7/8	7 7/8	7
404S	1/2	1/2	2	20	16 1/4	10††	8	6 1/8	6 5/8	13/16	3 3/4	1 7/8	3 1/2	2	..	..	..	..	..	..	..	..	..	..
405	1/2	1/2	5	20	17 3/4	10††	8	6 7/8	6 5/8	13/16	6 3/8	2 1/8	6 1/8	2	28 3/4	23 7/8	10	10 5/8	7/8	3 1/8	3	7/8	7 7/8	7
405S	1/2	1/2	2	20	17 3/4	10††	8	6 7/8	6 5/8	13/16	3 3/4	1 7/8	3 1/2	2	..	..	..	..	..	..	..	..	..	..
444	5/8	5/8	5 1/2	22	18 1/2	11††	9	7 1/4	7 1/2	13/16	7 1/8	2 3/8	6 7/8	2	31 1/4	24 5/8	11	11	7/8	3 1/2	3	7/8	8 3/8	7 1/2
444S	1/2	1/2	2 3/4	22	18 1/2	11††	9	7 1/4	7 1/2	13/16	4 1/4	2 1/8	4	2	..	..	..	..	..	..	..	..	..	..
445	5/8	5/8	5 1/2	22	20 1/2	11††	9	8 1/4	7 1/2	13/16	7 1/8	2 3/8	6 7/8	2	31 1/4	26 5/8	11	12	7/8	4 1/2	3	7/8	8 3/8	7 1/2
445S	1/2	1/2	2 3/4	22	20 1/2	11††	9	8 1/4	7 1/2	13/16	4 1/4	2 1/8	4	2	..	..	..	..	..	..	..	..	..	..
504U	3/4	3/4	7 1/4	25	21	12 1/2††	10	8	8 1/2	15/16	8 5/8	2 7/8	8 3/8	2 1/2	35	28	12 1/2	12 1/2	1	3 1/2	3 1/2	1	9	8
504S	1/2	1/2	2 3/4	25	21	12 1/2††	10	8	8 1/2	15/16	4 1/4	2 1/8	4	2 1/2	..	..	..	..	..	..	..	..	..	..
505	3/4	3/4	7 1/4	25	23	12 1/2††	10	9	8 1/2	15/16	8 5/8	2 7/8	8 3/8	2 1/2	35	30	12 1/2	13 1/2	1	4 1/2	3 1/2	1	9	8
505S	1/2	1/2	2 3/4	25	23	12 1/2††	10	9	8 1/2	15/16	4 1/4	2 1/8	4	2 1/2	..	..	..	..	..	..	..	..	..	..

Letter dimensions are defined on diagrams in Table 14–18

All dimensions in inches.

*The system used for obtaining frame numbers is given in a note to table 14–19.

† Tolerance on length of key—plus or minus 1/32 inch.

‡ Effective length of keyway.

** Dimension D will never be greater than the above values for rigid-base motors, but it may be less so that shims are usually required for coupled or geared machines. When the exact dimension is required, shims up to 1/32 inch may be necessary.

†† Dimension D will never be greater than the above values, but it may be less so that shims are usually required for coupled or geared machines. When the exact dimension is required, shims up to 1/32 inch may be necessary on frame sizes whose dimension D is 8 inches and less; on larger frames, shims up to 1/16 inch may be necessary.

‡‡ N-W is the length of the shaft extension from the shoulder to the end of the shaft.

¶ Size of conduit was approved as a Recommended Standard.

NOTE It is recommended that all machines with keyways cut in the shaft extension for pulley, coupling, pinion, etc., should be furnished with a key, unless otherwise specified by the purchaser.

TABLE 14–18

Standardized Letter Dimensions for Electric Motors

NEMA Standards for Motors and Generators, Part 3 – 1953

Lettering of Dimension Sheets for Horizontal Machines – Side View

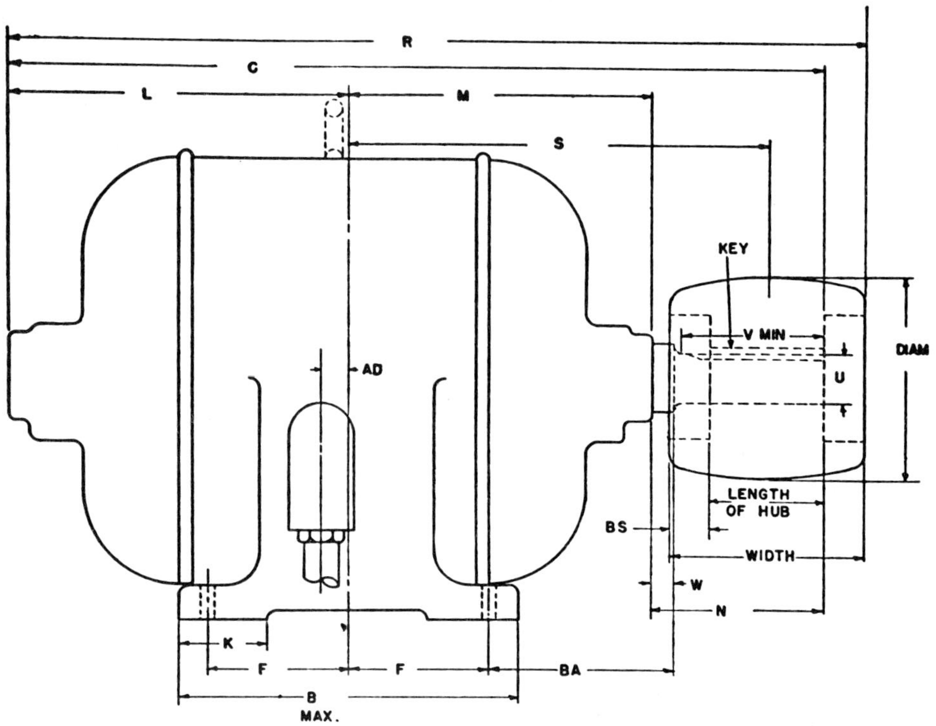

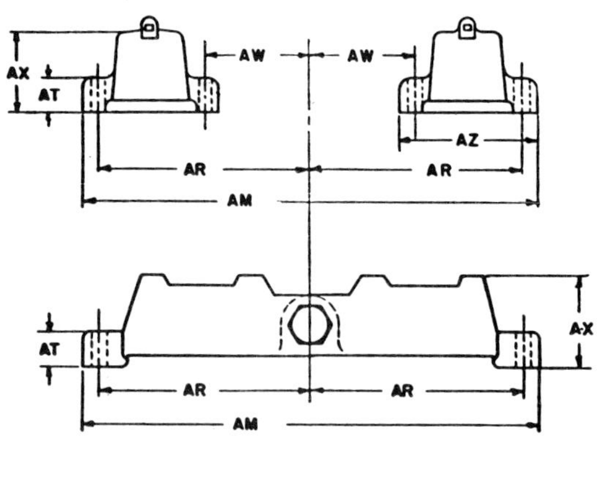

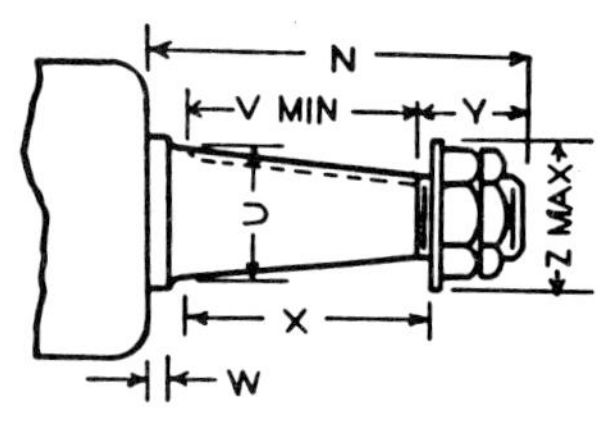

Lettering of Dimension Sheets for Horizontal Machines – Front View

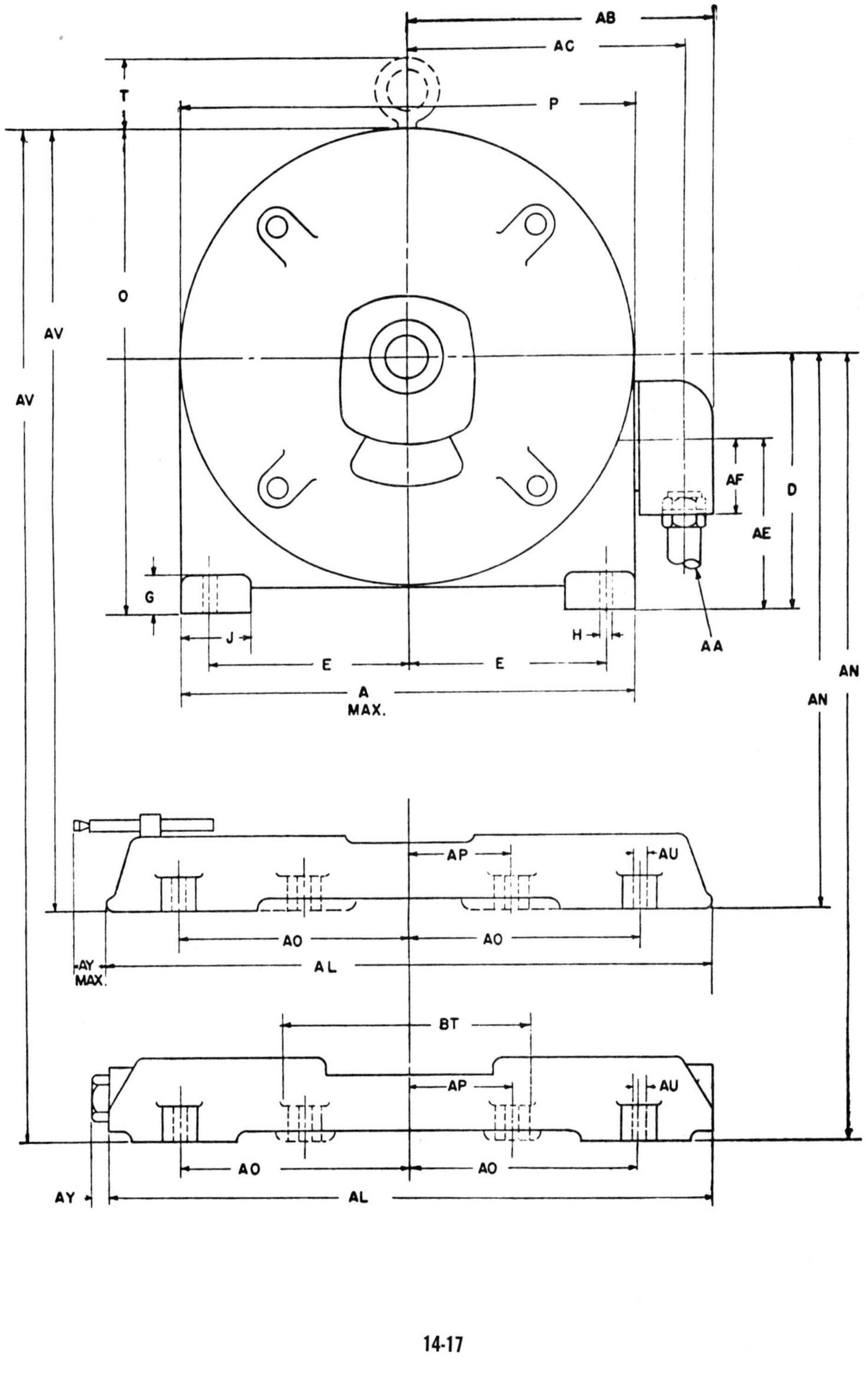

TABLE 14–19

Shaft Extension and Key Dimensions for Foot-mounted Motors and Generators with Tapered or Double Shaft Extension

NEMA Standards for Motors and Generators, Part 3 – 1953

DRIVE END TAPERED SHAFT EXTENSION

Letter dimensions are defined on diagrams in Table 14–18

Frame Number	Drive End Tapered Shaft Extension*										
									Key**		
	BA	N-W†	U	V Min	X	Y	Z Max	Shaft Thread‡	Width	Thickness	Length††
203	3 1/8	2 1/4	3/4	1 1/2	1 5/8	5/8	1 1/16	1/2–20	3/16	3/16	1 1/4
204	3 1/8	2 1/4	3/4	1 1/2	1 5/8	5/8	1 1/16	1/2–20	3/16	3/16	1 1/4
224	3 1/2	3	1	2	2 1/8	3/4	1 1/2	3/4–16	1/4	1/4	1 3/4
225	3 1/2	3	1	2	2 1/8	3/4	1 1/2	3/4–16	1/4	1/4	1 3/4
254	4 1/4	3 3/8	1 1/8	2 1/4	2 3/8	7/8	1 1/2	3/4–16	1/4	1/4	2
284	4 3/4	3 3/4	1 1/4	2 1/2	2 5/8	1	1 3/4	7/8–14	1/4	1/4	2 1/4
324	5 1/4	4 1/2	1 5/8	2 7/8	3	1 1/4	2	1 –14	3/8	3/8	2 1/2
326	5 1/4	4 1/2	1 5/8	2 7/8	3	1 1/4	2	1 –14	3/8	3/8	2 1/2
364	5 7/8	4 3/4	1 7/8	3 1/8	3 1/4	1 1/4	2 3/8	1 1/4– 8	1/2	1/2	2 7/8
364S	. . .	. . .	. . .	. . .	. . .	. . .	. . .	. . .	. . .	. . .	. . .
365	5 7/8	4 3/4	1 7/8	3 1/8	3 1/4	1 1/4	2 3/8	1 1/4– 8	1/2	1/2	2 7/8
365S	. . .	. . .	. . .	. . .	. . .	. . .	. . .	. . .	. . .	. . .	. . .
404	6 5/8	5 1/4	2 1/8	3 1/2	3 5/8	1 3/8	2 3/4	1 1/2– 8	1/2	1/2	3 1/4
404S	. . .	. . .	. . .	. . .	. . .	. . .	. . .	. . .	. . .	. . .	. . .
405	6 5/8	5 1/4	2 1/8	3 1/2	3 5/8	1 3/8	2 3/4	1 1/2– 8	1/2	1/2	3 1/4
405S	. . .	. . .	. . .	. . .	. . .	. . .	. . .	. . .	. . .	. . .	. . .
444	7 1/2	5 3/4	2 3/8	3 3/4	3 7/8	1 1/2	3 1/4	1 3/4– 8	5/8	5/8	3 1/2
444S	. . .	. . .	. . .	. . .	. . .	. . .	. . .	. . .	. . .	. . .	. . .
445	7 1/2	5 3/4	2 3/8	3 3/4	3 7/8	1 1/2	3 1/4	1 3/4– 8	5/8	5/8	3 1/2
445S	. . .	. . .	. . .	. . .	. . .	. . .	. . .	. . .	. . .	. . .	. . .
504U	8 1/2	6 5/8	2 7/8	4 3/8	4 1/2	1 3/4	3 5/8	2 – 8	3/4	3/4	4 1/8
504S	. . .	. . .	. . .	. . .	. . .	. . .	. . .	. . .	. . .	. . .	. . .
505	8 1/2	6 5/8	2 7/8	4 3/8	4 1/2	1 3/4	3 5/8	2 – 8	3/4	3/4	4 1/8
505S	. . .	. . .	. . .	. . .	. . .	. . .	. . .	. . .	. . .	. . .	. . .

All dimensions in inches.

continued on next page

* The standard taper of shaft shall be at the rate of 1 1/4 inches in diameter per foot of length.

† N-W is the length of the shaft extension from the shoulder to the end of the shaft, if shoulder is used.

‡ The threaded end of the tapered shaft shall be provided with two nuts or a nut and suitable locking device.

** It is recommended that all machines with keyways cut in the shaft extension for pulley, coupling, pinion, etc., shall be furnished with a key.

†† Tolerance on length of key is plus or minus 1/n inch.

NOTE: The frame number for a fractional horsepower motor is the D dimension in inches multiplied by 16. In the case of an integral-horsepower motor, the first two digits of the frame number are equal to 4 times the dimension D in inches. The third digit applies indirectly to the F dimension through a code.

OPPOSITE DRIVE END STRAIGHT AND TAPERED SHAFT EXTENSION

Letter dimensions are defined on diagrams in Table 14–18

Frame	Opposite Drive End Straight Shaft Extension						Opposite Drive End Tapered Shaft Extension*									
			FV	Key**					FV			FZ	Shaft	Key**		
Number	FN-FW†	FU	Min	Width	Thickness	Length††	FN-FW†	FU	Min	FX	FY	Max	Thread‡	Width	Thickness	Length††
203	2¼	¾	2	3/16	3/16	1⅜	2¼	¾	1½	1⅝	⅝	1 1/16	½–20	3/16	3/16	1¼
204	2¼	¾	2	3/16	3/16	1⅜	2¼	¾	1½	1⅝	⅝	1 1/16	½–20	3/16	3/16	1¼
224	2¼	¾	2	3/16	3/16	1⅜	2¼	¾	1½	1⅝	⅝	1 1/16	½–20	3/16	3/16	1¼
225	2¼	¾	2	3/16	3/16	1⅜	2¼	¾	1½	1⅝	⅝	1 1/16	½–20	3/16	3/16	1¼
254	3	1	2¾	¼	¼	2	3	1	2	2⅛	¾	1½	¾–16	¼	¼	1¾
284	3⅜	1⅛	3⅛	¼	¼	2⅜	3⅜	1⅛	2¼	2⅜	⅞	1½	¾–16	¼	¼	2
324	3¾	1¼	3½	¼	¼	2¾	3¾	1¼	2½	2⅝	1	1¾	⅞–14	¼	¼	2¼
326	3¾	1¼	3½	¼	¼	2¾	3¾	1¼	2½	2⅝	1	1¾	⅞–14	¼	¼	2¼
364	3¼	1⅝	3	⅜	⅜	1⅞	4½	1⅝	2⅞	3	1¼	2	1 –14	⅜	⅜	2½
364S	3¼	1⅝	3	⅜	⅜	1⅞	...	...	...	...	...	...	...	...	...	...
365	3¼	1⅝	3	⅜	⅜	1⅞	4½	1⅝	2⅞	3	1¼	2	1 –14	⅜	⅜	2½
365S	3¼	1⅝	3	⅜	⅜	1⅞	...	...	...	...	...	...	...	...	...	...
404	3¾	1⅞	3½	½	½	2	4¾	1⅞	3⅛	3¼	1¼	2⅜	1¼– 8	½	½	2⅞
404S	3¾	1⅞	3½	½	½	2	...	...	...	...	...	...	...	...	...	...
405	3¾	1⅞	3½	½	½	2	4¾	1⅞	3⅛	3¼	1¼	2⅜	1¼– 8	½	½	2⅞
405S	3¾	1⅞	3½	½	½	2	...	...	...	...	...	...	...	...	...	...
444	4¼	2⅛	4	½	½	2¾	5¼	2⅛	3½	3⅝	1⅜	2¾	1½– 8	½	½	3¼
444S	4¼	2⅛	4	½	½	2¾	...	...	...	...	...	...	...	...	...	...
445	4¼	2⅛	4	½	½	2¾	5¼	2⅛	3½	3⅝	1⅜	2¾	1½– 8	½	½	3¼
445S	4¼	2⅛	4	½	½	2¾	...	...	...	...	...	...	...	...	...	...
504U	4¼	2⅛	4	½	½	2¾	5¾	2⅜	3¾	3⅞	1½	3¼	1¾– 8	⅝	⅝	3½
504S	4¼	2⅛	4	½	½	2¾	...	...	...	...	...	...	...	...	...	...
505	4¼	2⅛	4	½	½	2¾	5¾	2⅜	3¾	3⅞	1½	3¼	1¾– 8	⅝	⅝	3½
505S	4¼	2⅛	4	½	½	2¾	...	...	...	...	...	...	...	...	...	...

All dimensions in inches.

* The standard taper of shaft shall be at the rate of 1¼ inches in diameter per foot of length.

† FN-FW is the length of the shaft extension from the shoulder to the end of the shaft, if shoulder is used.

‡ The threaded end of the tapered shaft shall be provided with two nuts or a nut and suitable locking device.

** It is recommended that all machines with keyways cut in the shaft extension for pulley, coupling, pinion, etc., shall be furnished with a key.

†† Tolerance on length of key is plus or minus 1/n inch.

NOTE: The frame number for a fractional horsepower motor is D dimension in inches multiplied by 16. In the case of an integral horsepower motor, the first two digits of the frame number are equal to four times the dimension D in inches. The third digit applies indirectly to the F dimension through a code.

TABLE 14–20

JIC Typical Graphical Symbols for Electrical Diagrams
General Motors Standards, Equipment and Operations,
Section S General Motors Corp.

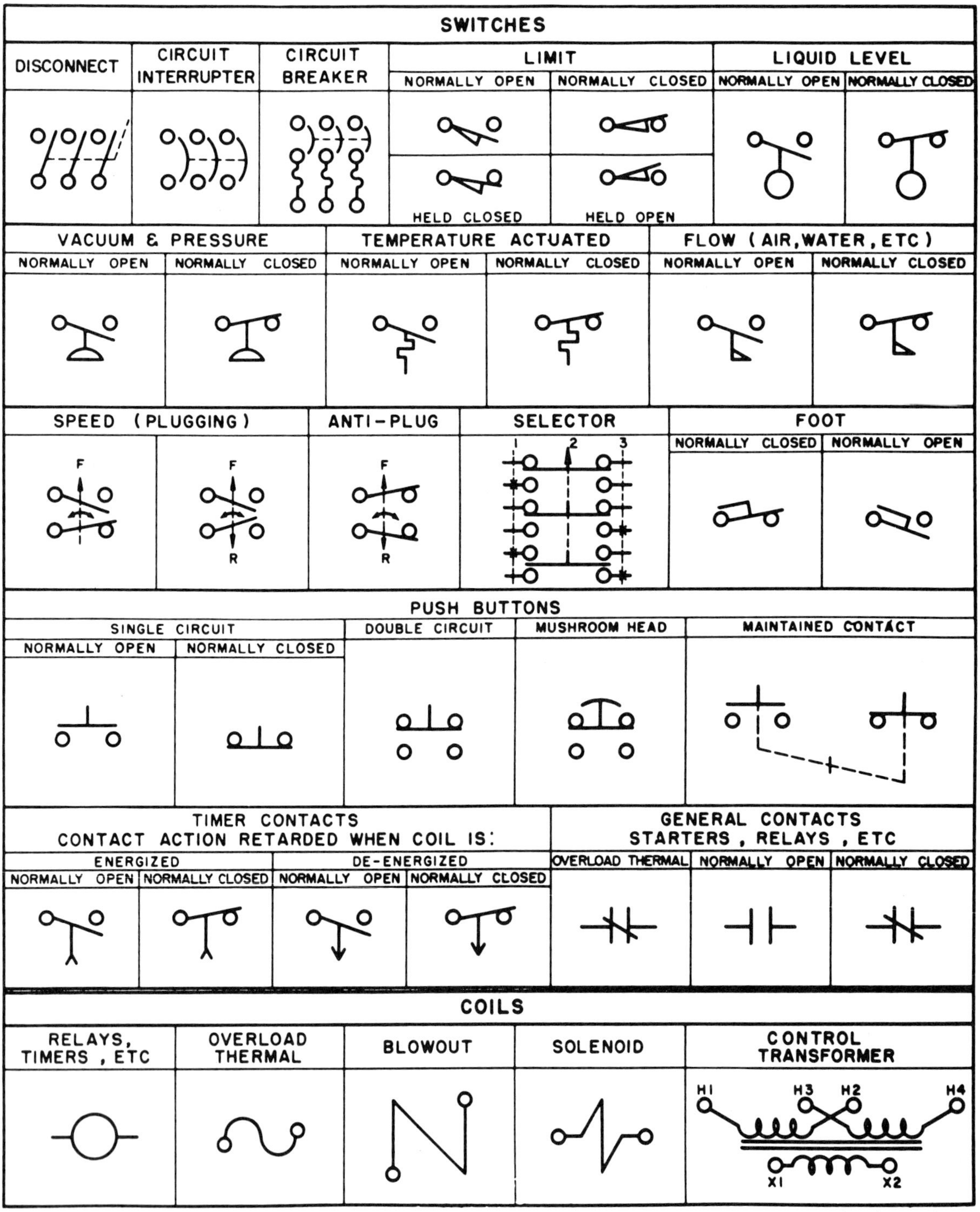

continued on next page

TABLE 14-20, continued

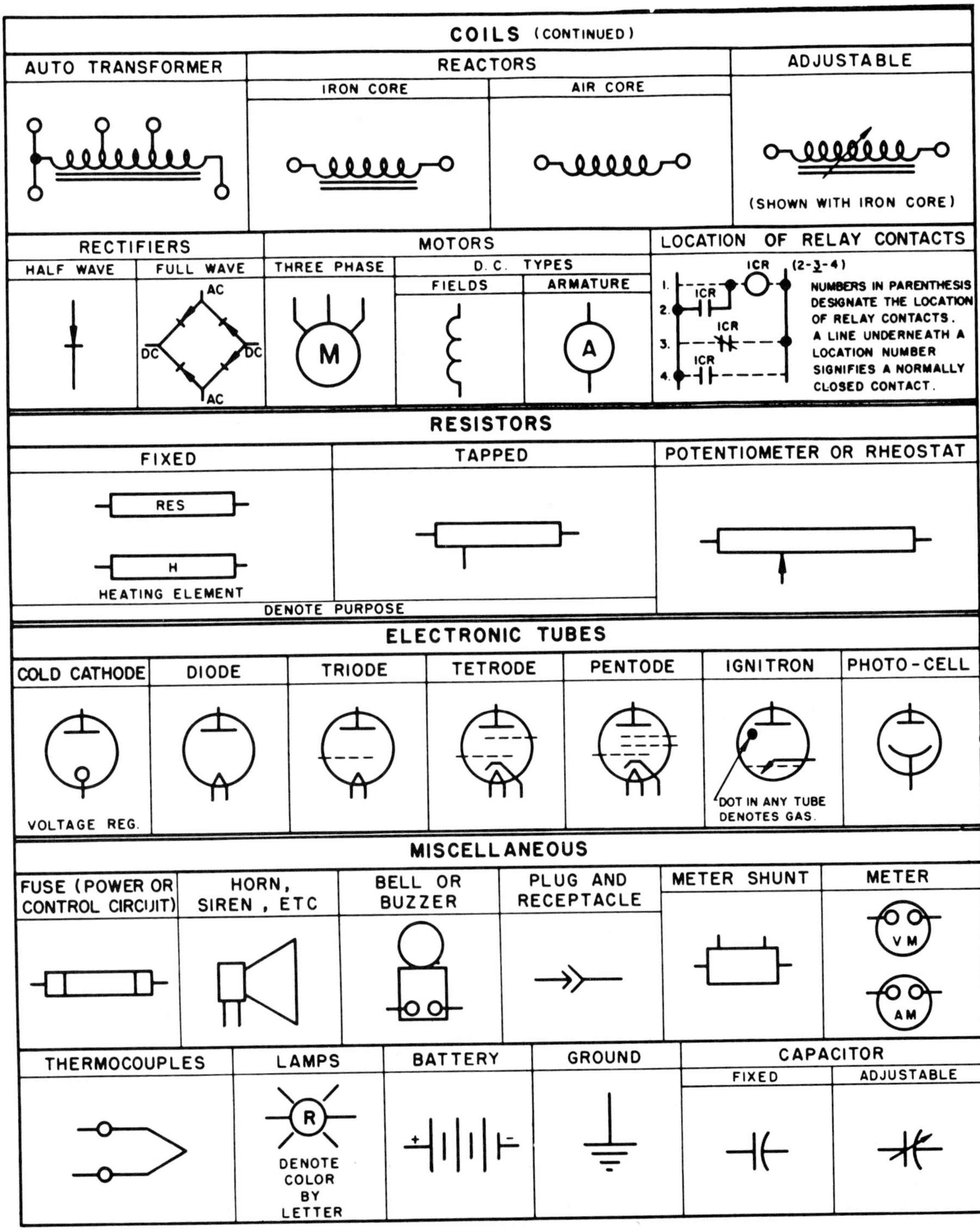

TABLE 14–21

JIC Typical Sample Electrical Diagrams

General Motors Standards, Equipment and Operation, Section S General Motors Corp.

WIRING AND INTERCONNECTION DIAGRAM

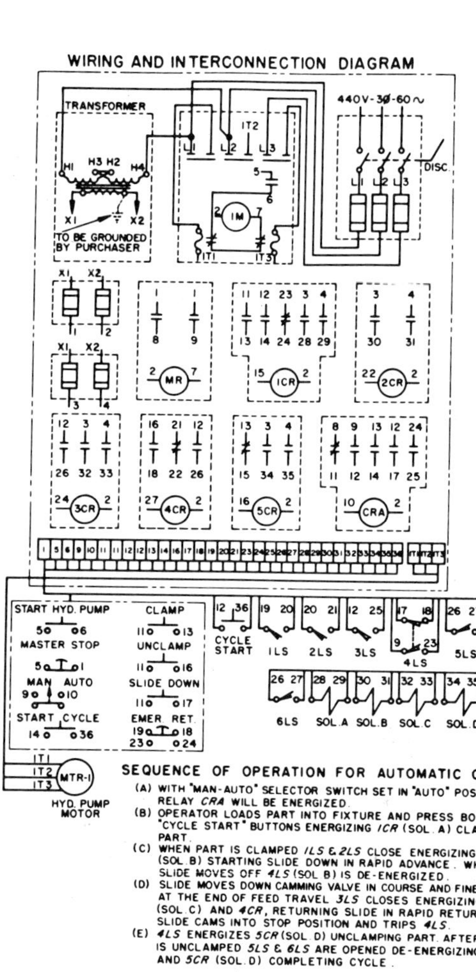

ELEMENTARY DIAGRAM

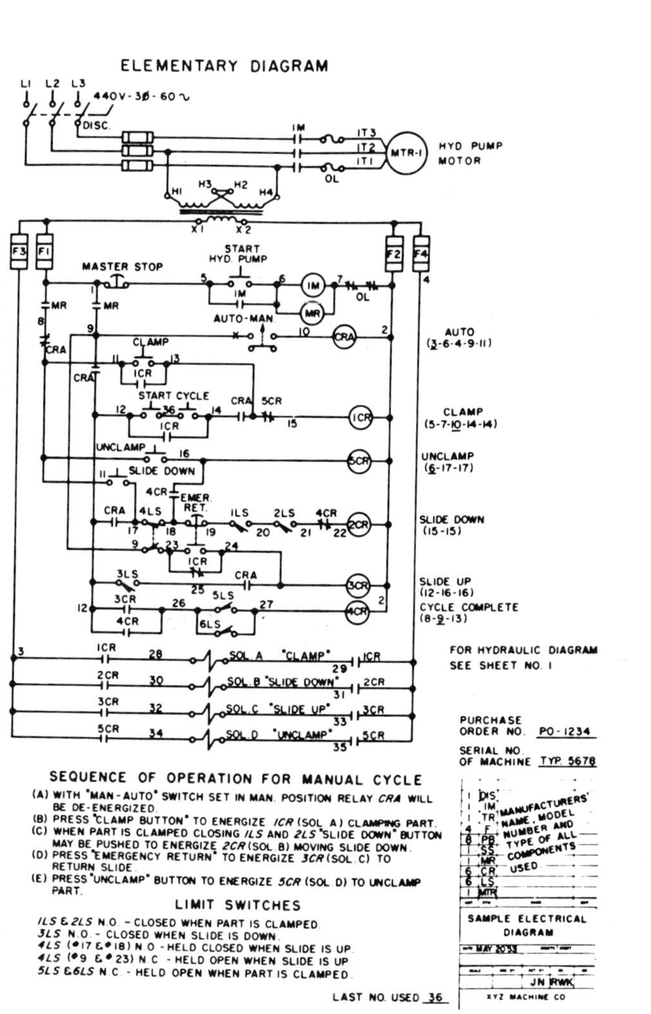

FOR HYDRAULIC DIAGRAM SEE SHEET NO. 1

SEQUENCE OF OPERATION FOR AUTOMATIC CYCLE

(A) WITH "MAN-AUTO" SELECTOR SWITCH SET IN "AUTO" POSITION RELAY *CRA* WILL BE ENERGIZED.

(B) OPERATOR LOADS PART INTO FIXTURE AND PRESS BOTH "CYCLE START" BUTTONS ENERGIZING *1CR* (SOL. A) CLAMPING PART.

(C) WHEN PART IS CLAMPED *1LS* & *2LS* CLOSE ENERGIZING *2CR* (SOL. B) STARTING SLIDE DOWN IN RAPID ADVANCE. WHEN SLIDE MOVES OFF *4LS* (SOL. B) IS DE-ENERGIZED.

(D) SLIDE MOVES DOWN CAMMING VALVE IN COURSE AND FINE FEED. AT THE END OF FEED TRAVEL *3LS* CLOSES ENERGIZING *3CR* (SOL. C) AND *4CR*, RETURNING SLIDE IN RAPID RETURN. SLIDE CAMS INTO STOP POSITION AND TRIPS *4LS*.

(E) *4LS* ENERGIZES *5CR* (SOL. D) UNCLAMPING PART. AFTER PART IS UNCLAMPED *5LS* & *6LS* ARE OPENED DE-ENERGIZING *4CR* AND *5CR* (SOL. D) COMPLETING CYCLE.

SEQUENCE OF OPERATION FOR MANUAL CYCLE

(A) WITH "MAN-AUTO" SWITCH SET IN MAN. POSITION RELAY *CRA* WILL BE DE-ENERGIZED.

(B) PRESS "CLAMP BUTTON" TO ENERGIZE *1CR* (SOL. A) CLAMPING PART.

(C) WHEN PART IS CLAMPED CLOSING *1LS* AND *2LS* "SLIDE DOWN" BUTTON MAY BE PUSHED TO ENERGIZE *2CR* (SOL. B) MOVING SLIDE DOWN.

(D) PRESS "EMERGENCY RETURN" TO ENERGIZE *3CR* (SOL. C) TO RETURN SLIDE.

(E) PRESS "UNCLAMP" BUTTON TO ENERGIZE *5CR* (SOL. D) TO UNCLAMP PART.

LIMIT SWITCHES

1LS & *2LS* N.O. – CLOSED WHEN PART IS CLAMPED.
3LS N.O. – CLOSED WHEN SLIDE IS DOWN.
4LS (•17 & •18) N.O. – HELD CLOSED WHEN SLIDE IS UP.
4LS (•9 & •23) N.C. – HELD OPEN WHEN SLIDE IS UP.
5LS & *6LS* N.C. – HELD OPEN WHEN PART IS CLAMPED.

LAST NO. USED 36

PURCHASE ORDER NO. PO-1234

SERIAL NO. OF MACHINE TYP. 5678

1	DIS	MANUFACTURERS' NAME, MODEL NUMBER AND TYPE OF ALL COMPONENTS USED
1	IM	
1	TR	
4	F	
8	PB	
1	SS	
1	MR	
6	CR	
6	LS	
1	MTR	

SAMPLE ELECTRICAL DIAGRAM

MAY 2033

JN RWK

XYZ MACHINE CO

TABLE 14–22

American Standard Graphical Symbols for Heat-Power Apparatus
ASA Z32.2.6 – 1950

1 COMPRESSOR

1.1 ROTARY

1.2 RECIPROCATING

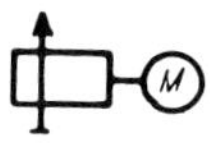

1.3 CENTRIFUGAL

M-Motor
T-Turbine

2 CONDENSER

2.1 BAROMETRIC

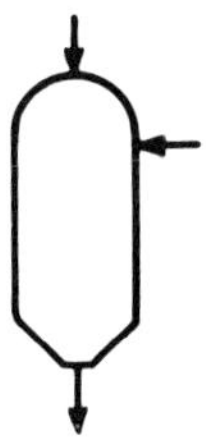

2.2 JET

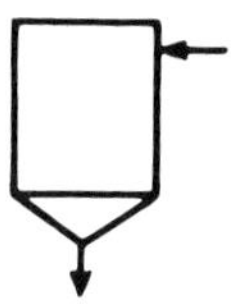

2.3 SURFACE

3 COOLER OR HEAT EXCHANGER

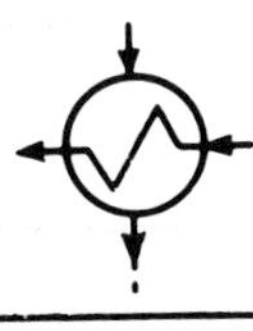

4 COOLING TOWER

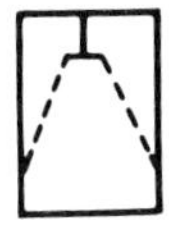

5 DEAERATOR

5.1

5.2 WITH SURGE TANK

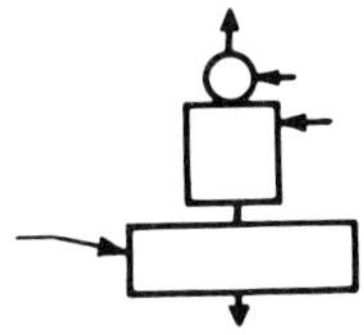

6 DRAINER OR LIQUID LEVEL CONTROLLER

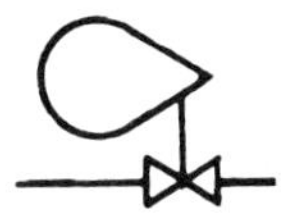

7 ENGINE

7.1 STEAM

7.2

S-Supercharger
D-Diesel

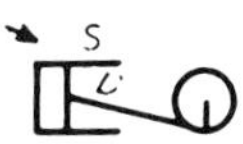

7.3

G-Gas

continued on next page

8 EVAPORATOR

8.1 SINGLE EFFECT

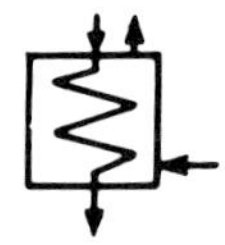

8.2 DOUBLE EFFECT

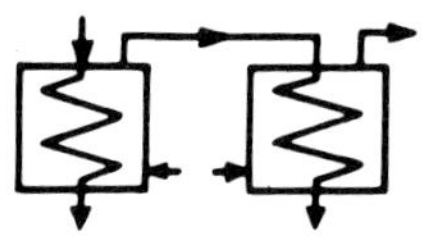

9 EXTRACTOR

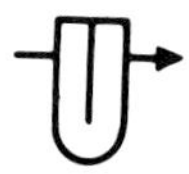

10 FAN-BLOWER

11 FILTER

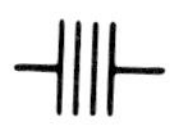

12 FLOW NOZZLE

13 FLUID DRIVE

14 HEATER

14.1 AIR (Plate or Tubular)

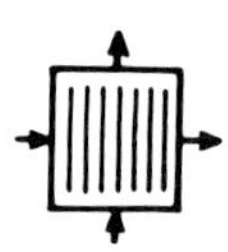

14.2 AIR (Rotating Type)

14.3 DESUPERHEATER

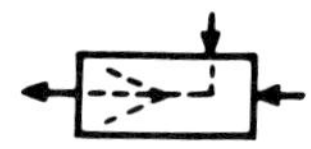

14.4 DIRECT CONTACT FEED-WATER

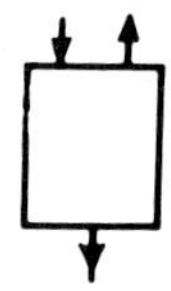

14.5 FEED WITH AIR OUTLET

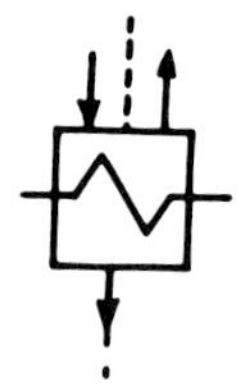

14.6 FLUE GAS REHEATER
(Intermediate Superheater)

14.7 LIVE STEAM SUPERHEATER
OR REHEATER

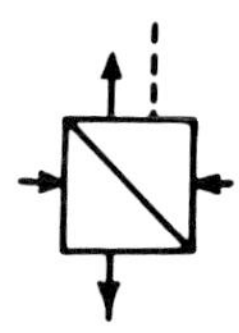

15 LIQUID LEVEL CONTROLLER

(See 6)

continued on next page

TABLE 14–22, continued

16 ORIFICE

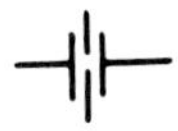

17 PRECIPITATOR

F-Electrostatic
M-Mechanical
W-Wet

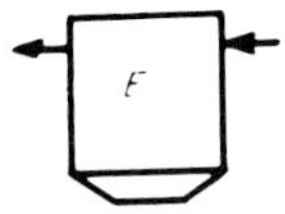

18 PUMP

18.1 CENTRIFUGAL AND ROTARY

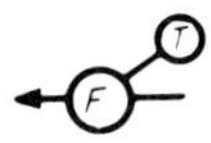

Letters Denote Service

F- Boiler Feed
S- Service
D- Condensate
C- Circ. Water
V- Air
O- Oil

M- Motor
T- Turbine
E- Steam Engine
G- Gas Engine
D- Diesel Engine

18.2 RECIPROCATING

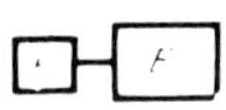

18.3 DYNAMIC (Air Ejector or Eductor)

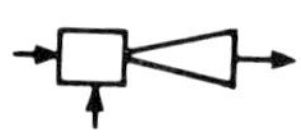

19 SEPARATOR

20 STEAM GENERATOR

(Boiler with Economizer)

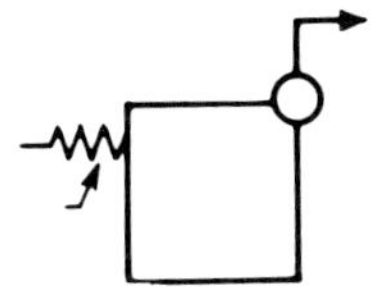

21 STEAM TRAP

22 STRAINER

22.1 SINGLE

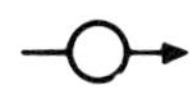

22.2 DOUBLE

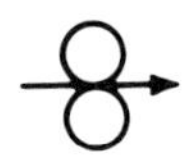

23 TANK

23.1 CLOSED

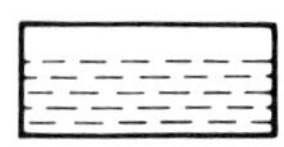

23.2 OPEN

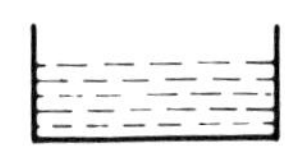

23.3 FLASH OR PRESSURE

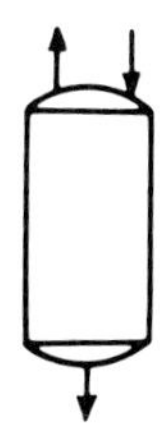

continued on next page

TABLE 14–22, continued

24 TURBINE 24.1 CONDENSING 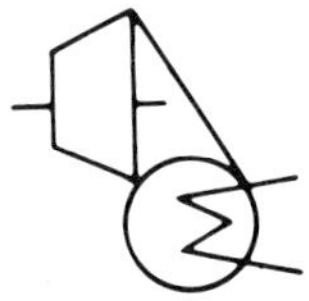	24.2 STEAM TURBINE OR AXIAL COMPRESSOR 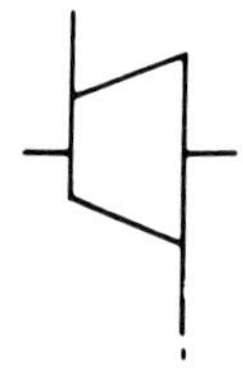 **25 VENTURI TUBE** 

TABLE 14–23

References to Standards on or Containing Graphical Symbols

Designation of Standard	Title	Remarks
ASA Z32.2.2–1949	Graphical Symbols for Plumbing	In general these stand ards provide designers, engineers, architects, contractors with recognized nomenclature for use on drawings.
ASA Z32.2.4–1949 Reaffirmed 1953	Graphical Symbols for Heating, Ventilating, and Air Conditioning	
ASA Z32.2.5–1950	Graphical Symbols for Railroad Use	
ASA Z32.13–1950	Abbreviations for Use on Drawings	
JIC Standards	Joint Industry Conference Pneumatic Standards for Industrial Equipment	Table 15-24 states purpose and source of JIC Standards.
ASA Z32.2.3–1949	Pipe and Pipe Fittings	Tables 15-22 and 15-23
ASA and others	Electrical Apparatus	Table 14-12

TABLE 14–24

American Standard Symbol to Designate Surface Irregularities
ASA B46.1–1947 Surface Roughness, Waviness, and Lay

SURFACE SYMBOL

The symbol used to designate surface irregularities is the check mark and an horizontal extension as shown.

The point of the symbol may be on the line indicating the surface, on the witness line, or on an arrow pointing to the surface. The long leg and extension shall preferably be to the right, as the drawing is read.

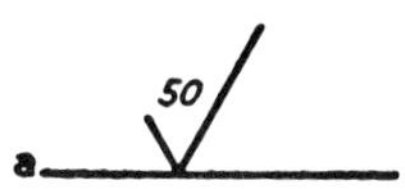

Roughness height value is placed adjacent to and on the inside of the long leg as shown.

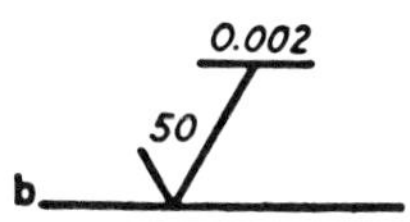

Waviness height value, when required, is placed above the horizontal extension line as shown.

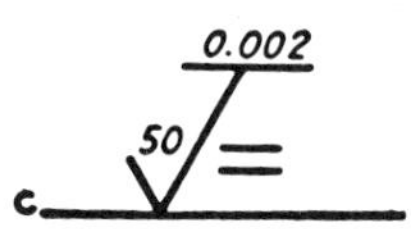

Lay designation, when required, is indicated by the lay symbol, placed under the extension to the right of the long leg line as shown.

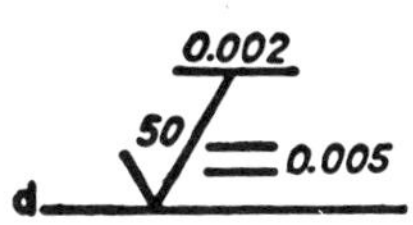

Roughness width value, when required, is placed to the right of the lay symbol as shown.

SYMBOLS INDICATING DIRECTION OF LAY

Symbol	Meaning
=	Parallel to the boundary line of the surface indicated by the symbol.
⊥	Perpendicular to the boundary line of the surface indicated by the symbol.
X	Angular in both directions to the boundary line of the surface indicated by the symbol.
M	Multi-directional.
C	Approximately circular relative to the center of the surface indicated by the symbol.
R	Approximately radial relative to the center of the surface indicated by the symbol.
= .005	The numerical value in inches of the width of spacing of roughness is added to the right of the directional indication of the lay symbol as shown.

A typical example would be the use of the symbol to express the following specification:

Roughness height	=	16 microinches
Waviness height	=	0.0005 in.
Lay	=	Circumferential
Roughness width	=	0.030 in.

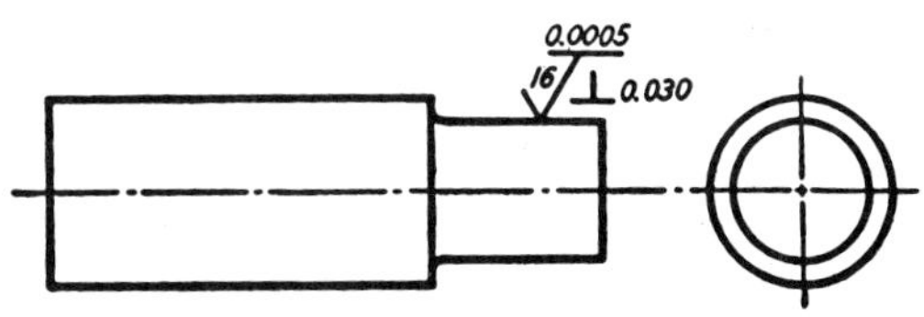

SPECIMEN APPLICATION

INDEX TO

SECTION 15

Gaskets | O-Rings | Packings | Hydraulic Standards and Symbols | Seals

TABLE 15–1

Joint Industry Conference

JIC Materials and Type-of-Packing Codes

Material	Type of Packing
F–FABRIC REINFORCED	V–VEE RINGS ONLY
	VT–VEES & FEMALE ADAPTER
H–HOMOGENEOUS (NAT. OR SYNTH.)	VB–VEES & MALE ADAPTER
	VA–VEES & BOTH ADAPTERS
	C–CUP PACKING
L–LEATHER	U–U-PACKING
M–METALLIC	O–O-RING PACKING
	O–RING GASKET (STATIC SEAL)
X–SPECIAL	F–FLANGE PACKING
	X–SPECIAL
	W–WASHER

TABLE 15–2

Typical Examples of JIC Packing – Code Identification

Joint Industry Conference Standards

(1) JIC Standard Vee Packing – No Adapters

LV – 40 – .62

Material (Leather)	Type of Pkg. (Multiple Lip No Adapters	JIC Standard Part No. (Equiv. in size to AN6225-40)	Nom Contact Ht of Pkg. Measured Along the Rod

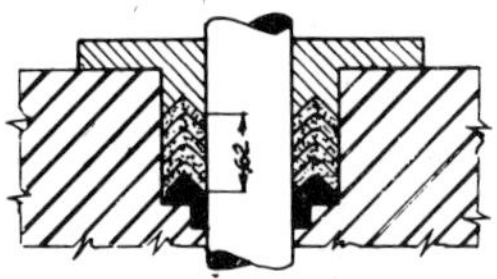

(6) U-Packing, No Adapters (Non Standard Size)

LU – 3.12 x 3.87 x .12 – .37

Material (Leather)	Type of Pkg. (U-Packing)	Nom ID of Pkg.	Nom OD of Pkg.	Nom. Height of Packing set Measured along the Rod

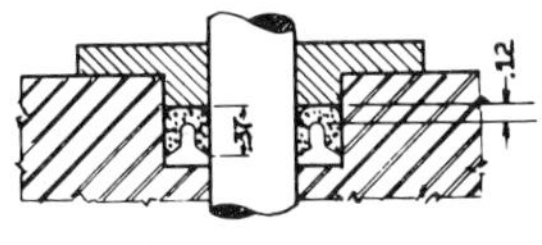

(2) JIC Standard Vees & Female Adapter

FVT – 40 – .93

Material (Fabric Reinforced)	Type of Pkg. (Multiple Lip Female Adapter only	JIC Standard Part No. (Equiv. in size to AN6225-40)	Nom Contact Ht of Pkg. Measured Along the Rod

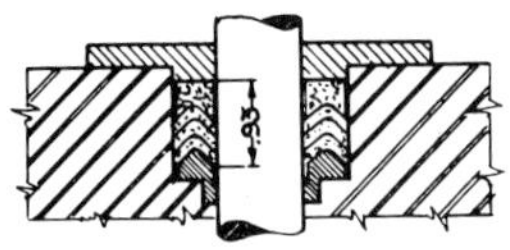

(7) JIC Standard O-Ring Packing

HO – 40

Material (Homogeneous Rubber)	Type of Pkg. O-Ring	JIC Standard Part No. (Equiv. in Size to AN6225-40)

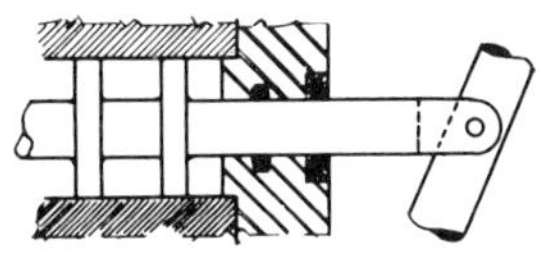

(3) Vees & Male Adapter Only (Non-Standard Size).

HVB – 3.12 x 3.87 – .68

Material (Homogeneous Rubber)	Type of Packing (Multiple Lip Male Adapter only)	Nom ID of Pkg	Nom OD of Pkg	Nom Height of Packing Set Measured along the Rod

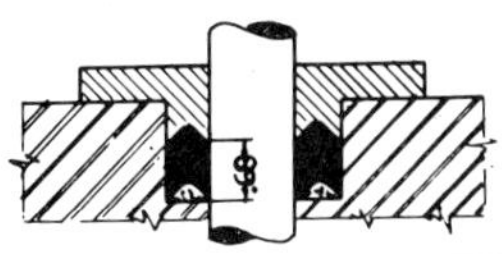

(8) JIC Standard O-Ring Gasket

HS – 25

Material (Homogeneous Rubber	Type of Pkg. (O-Ring Gasket)	JIC Standard Part No. (Equiv. in Size to AN 6230-25)

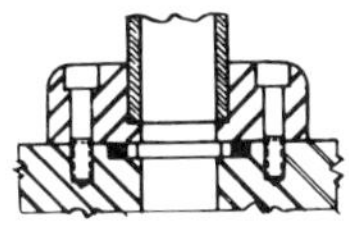

(4) JIC Standard Vees & Both Adapters

LVA – 40 – 1.25

Material (Leather)	Type of Pkg. (Multiple Lip Both Adapt	JIC Standard Part No. (Equiv. in Size to AN6225-40)	Nom Contact Ht of Pkg. Measured along the Rod

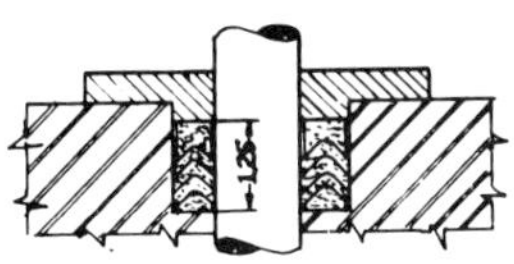

(9) Flange Packing – (Non Standard Size)

FF – .75 x 1.75 x .12 – .50

Material (Fabric Reinforced Rubber	Type of Pkg. Flange	Nom Rod Diam	Nom OD	Flange Thickness	Nom Contact Ht Measured along the Rod.

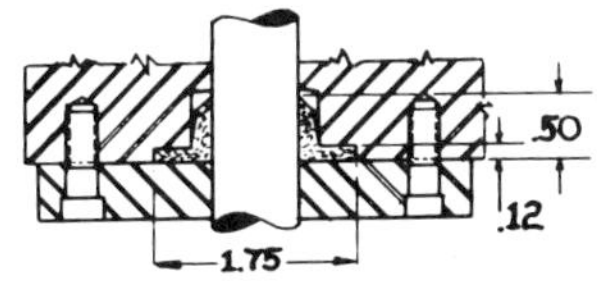

(5) Cup Packing (Non-Standard Size)

HC – 3.06 x 1.00 x 12 – .50

Material (Homogeneous Rubber	Type of Pkg. (Piston Cup)	Nom Cylinder Bore Diam.	Hole Diam.	Thickness of Base	Nom Contact Ht of Pkg Measured along the Cylinder

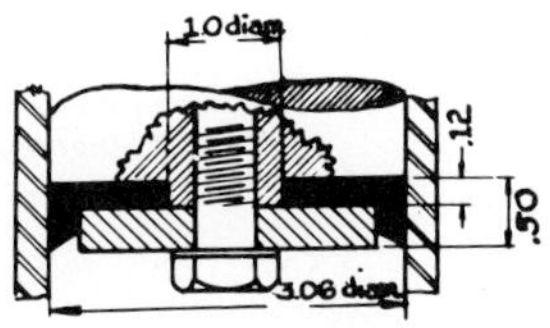

TABLE 15–3

Nominal Sizes for Leather Cup Packings

Joint Industry Conference Standards

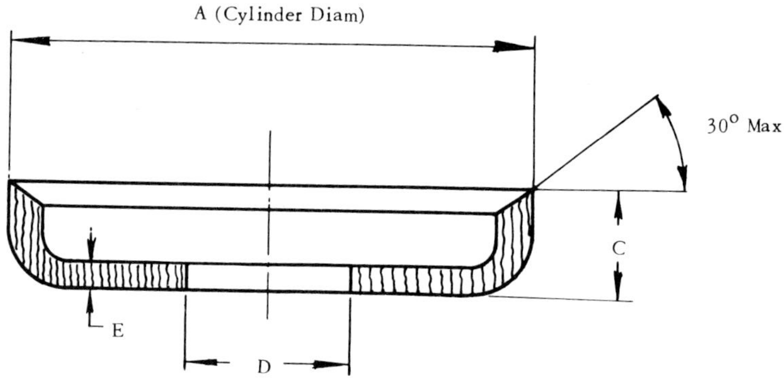

Dash numbers shown in this table are for reference only and apply specifically to this drawing.

Dash No.	A	C Max	D	E Min	E Max
1	7/16	1/4	TO SUIT	1/32	1/16
2	1/2	1/4	TO SUIT	1/32	1/16
3	9/16	1/4	TO SUIT	1/32	1/16
4	5/8	9/32	TO SUIT	1/32	3/32
5	11/16	9/32	TO SUIT	1/32	3/32
6	3/4	5/16	TO SUIT	1/32	3/32
7	13/16	5/16	TO SUIT	1/32	3/32
8	7/8	3/8	TO SUIT	1/32	3/32
9	15/16	3/8	TO SUIT	1/32	3/32
10	1	1/2	TO SUIT	1/16	1/8
11	1-1/8	1/2	TO SUIT	1/16	1/8
12	1-1/4	1/2	TO SUIT	1/16	1/8
13	1-3/8	1/2	TO SUIT	1/16	1/8
14	1-1/2	1/2	TO SUIT	3/32	1/8
15	1-5/8	1/2	TO SUIT	3/32	1/8
16	1-3/4	1/2	TO SUIT	3/32	5/32
17	1-7/8	1/2	TO SUIT	3/32	5/32
18	2	1/2	TO SUIT	3/32	5/32
19	2-1/8	1/2	TO SUIT	3/32	5/32
20	2-1/4	1/2	TO SUIT	3/32	5/32
21	2-3/8	1/2	TO SUIT	3/32	5/32
22	2-1/2	1/2	TO SUIT	3/32	5/32
23	2-5/8	1/2	TO SUIT	3/32	5/32
24	2-3/4	1/2	TO SUIT	3/32	5/32
25	2-7/8	1/2	TO SUIT	3/32	5/32
26	3	5/8	TO SUIT	1/8	3/16
27	3-1/8	5/8	TO SUIT	1/8	3/16
28	3-1/4	5/8	TO SUIT	1/8	3/16
29	3-3/8	5/8	TO SUIT	1/8	3/16
30	3-1/2	5/8	TO SUIT	1/8	3/16
31	3-5/8	5/8	TO SUIT	1/8	3/16
32	3-3/4	5/8	TO SUIT	1/8	3/16
33	3-7/8	5/8	TO SUIT	1/8	3/16
34	4	5/8	TO SUIT	1/8	3/16
35	4-1/4	5/8	TO SUIT	1/8	3/16
36	4-1/2	5/8	TO SUIT	1/8	3/16
37	4-3/4	5/8	TO SUIT	1/8	3/16
38	5	3/4	TO SUIT	1/8	3/16
39	5-1/4	3/4	TO SUIT	1/8	3/16
40	5-1/2	3/4	TO SUIT	1/8	3/16
41	5-3/4	3/4	TO SUIT	1/8	3/16
42	6	3/4	TO SUIT	1/8	3/16
43	6-1/4	3/4	TO SUIT	1/8	3/16
44	6-1/2	3/4	TO SUIT	1/8	3/16
45	6-3/4	3/4	TO SUIT	1/8	3/16
46	7	3/4	TO SUIT	1/8	3/16
47	7-1/4	3/4	TO SUIT	1/8	3/16
48	7-1/2	3/4	TO SUIT	1/8	3/16
49	7-3/4	3/4	TO SUIT	1/8	3/16
50	8	1	TO SUIT	1/8	3/16
51	8-1/2	1	TO SUIT	1/8	3/16
52	9	1	TO SUIT	1/8	3/16
53	9-1/2	1	TO SUIT	1/8	3/16
54	10	1	TO SUIT	1/8	3/16
55	10-1/2	1-1/4	TO SUIT	1/8	3/16
56	11	1-1/4	TO SUIT	1/8	3/16
57	11-1/2	1-1/4	TO SUIT	1/8	3/16
58	12	1-1/4	TO SUIT	1/8	3/16

Note 1: The above are nominal commercial sizes recommended for new designs. Other sizes for equipment already in use are available.

Note 2: Materials and sizes shown are for guidance only when these types are used. It should not be construed that other types of packing are not acceptable.

TABLE 15–4

Nominal Sizes for Leather or Homogeneous V Packings

Joint Industry Conference Standards

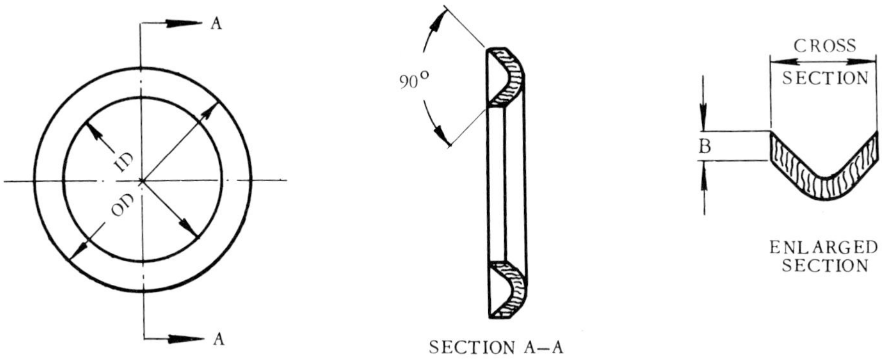

Dash numbers shown in this table are for reference only and apply specifically to this drawing.

Dash No.	Cross Section	Nominal Inside Diameter	Nominal Outside Diameter	*B* ±.010
8	1/4	1/4	3/4	.083
10	1/4	3/8	7/8	.083
12	1/4	1/2	1	.083
14	1/4	5/8	1-1/8	.083
16	1/4	3/4	1-1/4	.083
18	1/4	7/8	1-3/8	.083
20	1/4	1	1-1/2	.083
22	1/4	1-1/8	1-5/8	.083
24	1/4	1-1/4	1-3/4	.083
25	5/16	1-1/4	1-7/8	.140
26	5/16	1-3/8	2	.140
27	5/16	1-1/2	2-1/8	.140
28	5/16	1-5/8	2-1/4	.140
29	5/16	1-3/4	2-3/8	.140
30	5/16	1-7/8	2-1/2	.140
31	5/16	2	2-5/8	.140
32	5/16	2-1/8	2-3/4	.140
33	5/16	2-1/4	2-7/8	.140
34	5/16	2-3/8	3	.140
35	5/16	2-1/2	3-1/8	.140
36	3/8	2-1/2	3-1/4	.156
38	3/8	2-3/4	3-1/2	.156
40	3/8	3	3-3/4	.156
42	3/8	3-1/4	4	.156
44	3/8	3-1/2	4-1/4	.156
46	3/8	3-3/4	4-1/2	.156
49	7/16	4	4-7/8	.197
50	7/16	4-1/4	5-1/8	.197
51	7/16	4-1/2	5-3/8	.197
52	7/16	4-3/4	5-5/8	.197
53	7/16	5	5-7/8	.197
54	7/16	5-1/4	6-1/8	.197
55	7/16	5-1/2	6-3/8	.197
56	1/2	5-1/2	6-1/2	.197
58	1/2	6	7	.197
60	1/2	6-1/2	7-1/2	.197
62	1/2	7	8	.197
64	1/2	7-1/2	8-1/2	.197
66	1/2	8	9	.197
67	1/2	8-1/2	9-1/2	.197
68	1/2	9	10	.197
69	1/2	9-1/2	10-1/2	.197
70	1/2	10	11	.197
71	1/2	10-1/2	11-1/2	.197
72	1/2	11	12	.197
74	1/2	12	13	.197
76	1/2	13	14	.197
78	1/2	14	15	.197
80	1/2	15	16	.197

Note 1: The above are nominal commercial sizes recommended for new designs. Other sizes for equipment already in use are available.

Note 2: Materials and sizes shown are for guidance only when these types are used. It should not be construed that other types of packing are not acceptable.

TABLE 15-5

Nominal Sizes for Leather or Homogeneous U Packings

Joint Industry Conference Standards

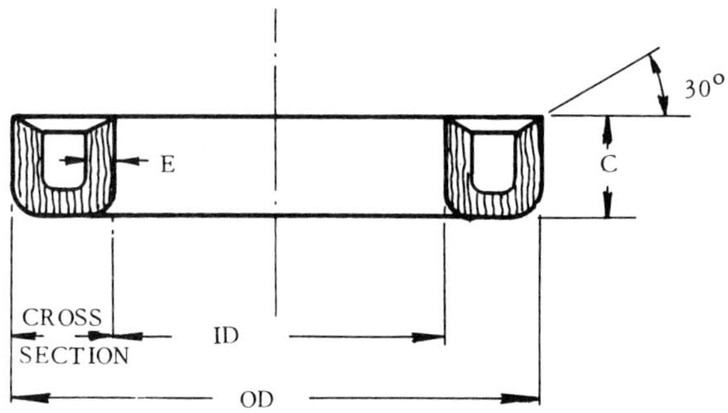

Inside Diameter			Increment	Cross Section	C
1/2	THRU	7/8	1/8	1/4	5/16
1	THRU	1-3/4	1/8	3/8	3/8
1-7/8	THRU	2-1/2	1/8	1/2	7/16
2-3/4	THRU	3-3/4	1/4	1/2	1/2
4	THRU	5-1/2	1/4	5/8	5/8
5-1/2	THRU	11	1/2	3/4	3/4
12	THRU	15	1	3/4	1
16 AND OVER			1	3/4	*
*16 in. TO 36 in.- 1-1/4					
*37 in. & UP - 1-1/2					

Dash numbers shown in this table are for reference only and apply specifically to this drawing.

Dash No.	Cross Section	Nominal Inside Diameter	Nominal Outside Diameter	C	E
12	1/4	1/2	1	5/16	1/16
14	1/4	5/8	1-1/8	5/16	1/16
16	1/4	3/4	1-1/4	5/16	1/16
18	1/4	7/8	1-3/8	5/16	1/16
20	3/8	1	1-3/4	3/8	3/32
22	3/8	1-1/8	1-7/8	3/8	3/32
24	3/8	1-1/4	2	3/8	3/32
26	3/8	1-3/8	2-1/8	3/8	3/32
27	3/8	1-1/2	2-1/4	3/8	3/32
28	3/8	1-5/8	2-3/8	3/8	3/32
29	3/8	1-3/4	2-1/2	3/8	3/32
30	1/2	1-7/8	2-7/8	7/16	1/8
31	1/2	2	3	7/16	1/8
32	1/2	2-1/8	3-1/8	7/16	1/8
33	1/2	2-1/4	3-1/4	7/16	1/8
34	1/2	2-3/8	3-3/8	7/16	1/8
36	1/2	2-1/2	3-1/2	1/2	1/8
38	1/2	2-3/4	3-3/4	1/2	1/8
40	1/2	3	4	1/2	1/8
42	1/2	3-1/4	4-1/4	1/2	1/8
44	1/2	3-1/2	4-1/2	1/2	1/8
46	1/2	3-3/4	4-3/4	1/2	1/8

Note 1: The above are nominal commercial sizes recommended for new designs. Other sizes for equipment already in use are available.

Dash No.	Cross Section	Nominal Inside Diameter	Nominal Outside Diameter	C	E
49	5/8	4	5-1/4	5/8	5/32
50	5/8	4-1/4	5-1/2	5/8	5/32
51	5/8	4-1/2	5-3/4	5/8	5/32
52	5/8	4-3/4	6	5/8	5/32
53	5/8	5	6-1/4	5/8	5/32
54	5/8	5-1/4	6-1/2	5/8	5/32
55	5/8	5-1/2	6-3/4	5/8	5/32
56	3/4	5-1/2	7	3/4	5/32
58	3/4	6	7-1/2	3/4	3/16
60	3/4	6-1/2	8	3/4	3/16
62	3/4	7	8-1/2	3/4	3/16
64	3/4	7-1/2	9	3/4	3/16
66	3/4	8	9-1/2	3/4	3/16
67	3/4	8-1/2	10	3/4	3/16
68	3/4	9	10-1/2	3/4	3/16
69	3/4	9-1/2	11	3/4	3/16
70	3/4	10	11-1/2	3/4	3/16
71	3/4	10-1/2	12	3/4	3/16
72	3/4	11	12-1/2	3/4	3/16
74	3/4	12	13-1/2	1	3/16
76	3/4	13	14-1/2	1	3/16
78	3/4	14	15-1/2	1	3/16
80	3/4	15	16-1/2	1	3/16

Note 2: Materials and sizes shown are for guidance only when these types are used. It should not be construed that other types of packing are not acceptable.

TABLE 15-6

Nominal Sizes for Homogeneous U-Cup Packings

Joint Industry Conference Standards

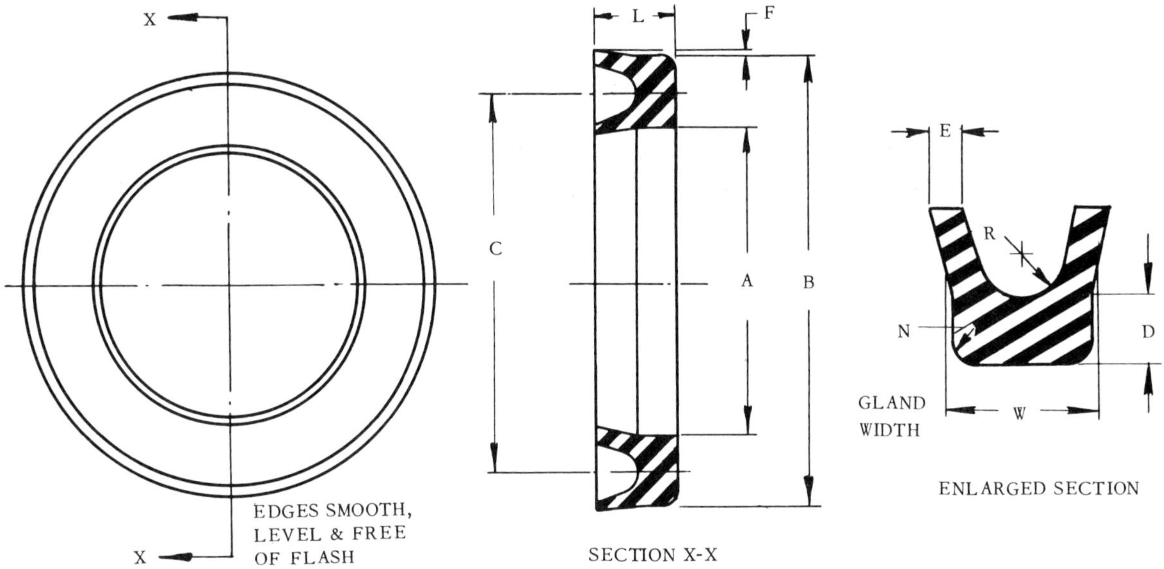

Dash No.	*W & L*	Nominal Size I.D.	Nominal Size O.D.	Diameter ±.005 *A*	Diameter ±.005 *B*	Diameter ±.005 *C*	*D*	*E*	*F*	*R*	*N*
8	1/4	1/4	3/4	.265	.735	1/2	3/32	.045	.030	.070	1/32
10	1/4	3/8	7/8	.390	.860	5/8	3/32	.045	.030	.070	1/32
12	1/4	1/2	1	.515	.985	3/4	3/32	.045	.030	.070	1/32
14	1/4	5/8	1-1/8	.640	1.110	7/8	3/32	.045	.030	.070	1/32
16	1/4	3/4	1-1/4	.765	1.235	1	3/32	.045	.030	.070	1/32
18	1/4	7/8	1-3/8	.890	1.360	1-1/8	3/32	.045	.030	.070	1/32
20	1/4	1	1-1/2	1.015	1.485	1-1/4	3/32	.045	.030	.070	1/32
22	1/4	1-1/8	1-5/8	1.140	1.610	1-3/8	3/32	.045	.030	.070	1/32
24	1/4	1-1/4	1-3/4	1.265	1.735	1-1/2	3/32	.045	.030	.070	1/32
25	5/16	1-1/4	1-7/8	1.265	1.860	1-9/16	1/8	.050	.032	.093	1/32
26	5/16	1-3/8	2	1.390	1.985	1-11/16	1/8	.050	.032	.093	1/32
27	5/16	1-1/2	2-1/8	1.515	2.110	1-13/16	1/8	.050	.032	.093	1/32
28	5/16	1-5/8	2-1/4	1.640	2.235	1-15/16	1/8	.050	.032	.093	1/32
29	5/16	1-3/4	2-3/8	1.765	2.360	2-1/16	1/8	.050	.032	.093	1/32
30	5/16	1-7/8	2-1/2	1.890	2.485	2-3/16	1/8	.050	.032	.093	1/32
31	5/16	2	2-5/8	2.015	2.610	2-5/16	1/8	.050	.032	.093	1/32
32	5/16	2-1/8	2-3/4	2.140	2.735	2-7/16	1/8	.050	.032	.093	1/32
33	5/16	2-1/4	2-7/8	2.265	2.860	2-9/16	1/8	.050	.032	.093	1/32
34	5/16	2-3/8	3	2.390	2.985	2-11/16	1/8	.050	.032	.093	1/32
35	5/16	2-1/2	3-1/8	2.515	3.110	2-13/16	1/8	.050	.032	.093	1/32
36	3/8	2-1/2	3-1/4	2.515	3.235	2-7/8	1/8	.054	.035	.125	3/64
38	3/8	2-3/4	3-1/2	2.765	3.485	3	1/8	.054	.035	.125	3/64
40	3/8	3	3-3/4	3.015	3.735	3-3/8	1/8	.054	.035	.125	3/64

Dash numbers shown in this table are for reference only and apply specifically to this drawing.

Materials and sizes shown are for guidance only when these types are used. It shoud not be construed that other types of packing are not acceptable.

TABLE 15–10

Female Adapter Rings for V Packings
Sizes Dimensionally the Same as AN 6228

Catalog and Manual 201 G and K-International

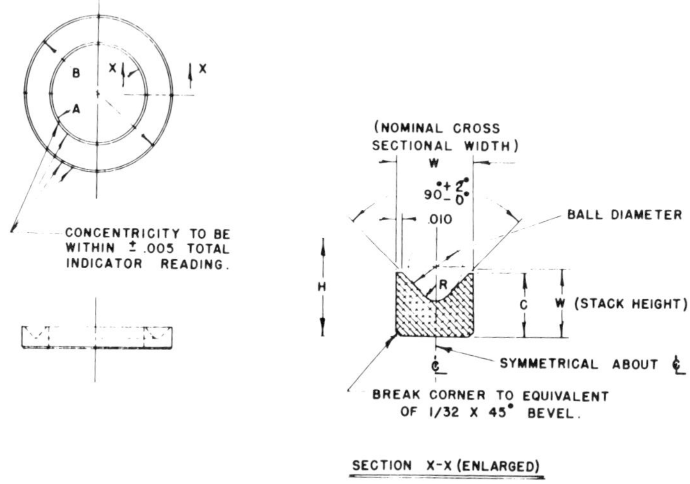

Dash No.	Nominal W	Nominal ID	Nominal OD	A +.002 −.000	B +.000 −.002	C Ref.	Ball Diam	H ±.005	R +0 −1/64
1	3/16	1/8	1/2	.127	.498	.177	.1562	.282	1/16
2	3/16	3/16	9/16	.190	.560	.177	.1562	.282	1/16
3	3/16	1/4	5/8	.252	.623	.177	.1562	.282	1/16
4	3/16	5/16	11/16	.315	.685	.177	.1562	.282	1/16
5	3/16	3/8	3/4	.377	.748	.177	.1562	.282	1/16
5	3/16	7/16	13/16	.440	.810	.177	.1562	.282	1/16
7	3/16	1/2	7/8	.502	.873	.177	.1562	.282	1/16
8	1/4	1/4	3/4	.252	.748	.240	.1875	.351	1/16
9	1/4	5/16	13/16	.315	.810	.240	.1875	.351	1/16
10	1/4	3/8	7/8	.377	.873	.240	.1875	.351	1/16
11	1/4	7/16	15/16	.440	.935	.240	.1875	.351	1/16
12	1/4	1/2	1	.502	.998	.240	.1875	.351	1/16
13	1/4	9/16	1- 1/16	.565	1.060	.240	.1875	.351	1/16
14	1/4	5/8	1- 1/8	.627	1.123	.240	.1875	.351	1/16
15	1/4	11/16	1- 3/16	.690	1.185	.240	.1875	.351	1/16
16	1/4	3/4	1- 1/4	.752	1.248	.240	.1875	.351	1/16
17	1/4	13/16	1- 5/16	.815	1.310	.240	.1875	.351	1/16
18	1/4	7/8	1- 3/8	.877	1.373	.240	.1875	.351	1/16
19	1/4	15/16	1- 7/16	.940	1.435	.240	.1875	.351	1/16
20	1/4	1	1- 1/2	1.002	1.498	.240	.1875	.351	1/16
21	1/4	1- 1/16	1- 9/16	1.065	1.560	.240	.1875	.351	1/16
22	1/4	1- 1/8	1- 5/8	1.127	1.623	.240	.1875	.351	1/16
23	1/4	1- 3/16	1-11/16	1.190	1.685	.240	.1875	.351	1/16
24	1/4	1- 1/4	1- 3/4	1.252	1.748	.240	.1875	.351	1/16
25	5/16	1- 1/4	1- 7/8	1.252	1.873	.302	.2500	.458	7/64
26	5/16	1- 3/8	2	1.377	1.998	.302	.2500	.458	7/64
27	5/16	1- 1/2	2- 1/8	1.502	2.123	.302	.2500	.458	7/64
28	5/16	1- 5/8	2- 1/4	1.627	2.248	.302	.2500	.458	7/64
29	5/16	1- 3/4	2- 3/8	1.752	2.373	.302	.2500	.458	7/64
30	5/16	1- 7/8	2- 1/2	1.877	2.498	.302	.2500	.458	7/64

continued on next page

TABLE 15–10, continued

Dash No.	Nominal W	Nominal ID	Nominal OD	A +.004 –.000	B +.000 –.004	C Ref.	Ball Diam	H ±.005	R +0 –1/64
31	5/16	2	2- 5/8	2.002	2.623	.302	.2500	.458	7/64
32	5/16	2- 1/8	2- 3/4	2.127	2.748	.302	.2500	.458	7/64
33	5/16	2- 1/4	2- 7/8	2.252	2.873	.302	.2500	.458	7/64
34	5/16	2- 3/8	3	2.377	2.998	.302	.2500	.458	7/64
35	5/16	2- 1/2	3- 1/8	2.502	3.123	.302	.2500	.458	7/64
36	3/8	2- 1/2	3- 1/4	2.502	3.248	.365	.3125	.565	1/8
37	3/8	2- 5/8	3- 3/8	2.628	3.372	.365	.3125	.565	1/8
38	3/8	2- 3/4	3- 1/2	2.753	3.497	.365	.3125	.565	1/8
39	3/8	2- 7/8	3- 5/8	2.878	3.622	.365	.3125	.565	1/8
40	3/8	3	3- 3/4	3.003	3.747	.365	.3125	.565	1/8
41	3/8	3- 1/8	3- 7/8	3.128	3.872	.365	.3125	.565	1/8
42	3/8	3- 1/4	4	3.253	3.997	.365	.3125	.565	1/8
43	3/8	3- 3/8	4- 1/8	3.378	4.122	.365	.3125	.565	1/8
44	3/8	3- 1/2	4- 1/4	3.503	4.247	.365	.3125	.565	1/8
45	3/8	3- 5/8	4- 3/8	3.628	4.372	.365	.3125	.565	1/8
46	3/8	3- 3/4	4- 1/2	3.753	4.497	.365	.3125	.565	1/8
47	3/8	3- 7/8	4- 5/8	3.878	4.622	.365	.3125	.565	1/8
48	7/16	3- 7/8	4- 3/4	3.878	4.747	.427	.3750	.671	5/32
49	7/16	4	4- 7/8	4.003	4.872	.427	.3750	.671	5/32
50	7/16	4- 1/4	5- 1/8	4.253	5.122	.427	.3750	.671	5/32
51	7/16	4- 1/2	5- 3/8	4.503	5.372	.427	.3750	.671	5/32
52	7/16	4- 3/4	5- 5/8	4.753	5.622	.427	.3750	.671	5/32
53	7/16	5	5- 7/8	5.003	5.872	.427	.3750	.671	5/32
54	7/16	5- 1/4	6- 1/8	5.253	6.122	.427	.3750	.671	5/32
55	7/16	5- 1/2	6- 3/8	5.503	6.372	.427	.3750	.671	5/32
56	1/2	5- 1/2	6- 1/2	5.503	6.497	.490	.4375	.778	5/32
57	1/2	5- 3/4	6- 3/4	5.753	6.747	.490	.4375	.778	5/32
58	1/2	6	7	6.003	6.997	.490	.4375	.778	5/32
59	1/2	6- 1/4	7- 1/4	6.253	7.247	.490	.4375	.778	5/32
60	1/2	6- 1/2	7- 1/2	6.503	7.497	.490	.4375	.778	5/32
61	1/2	6- 3/4	7- 3/4	6.753	7.747	.490	.4375	.778	5/32
62	1/2	7	8	7.003	7.997	.490	.4375	.778	5/32
63	1/2	7- 1/4	8- 1/4	7.253	8.247	.490	.4375	.778	5/32
64	1/2	7- 1/2	8- 1/2	7.503	8.497	.490	.4375	.778	5/32
65	1/2	7- 3/4	8- 3/4	7.753	8.747	.490	.4375	.778	5/32
66	1/2	8	9	8.003	8.997	.490	.4375	.778	5/32
67	1/2	8- 1/2	9- 1/2	8.503	9.497	.490	.4375	.778	5/32
68	1/2	9	10	9.003	9.997	.490	.4375	.778	5/32
69	1/2	9- 1/2	10- 1/2	9.503	10.497	.490	.4375	.778	5/32
70	1/2	10	11	10.003	10.997	.490	.4375	.778	5/32
71	1/2	10- 1/2	11- 1/2	10.503	11.497	.490	.4375	.778	5/32
72	1/2	11	12	11.003	11.997	.490	.4375	.778	5/32
73	1/2	11- 1/2	12- 1/2	11.503	12.497	.490	.4375	.778	5/32
74	1/2	12	13	12.003	12.997	.490	.4375	.778	5/32
75	1/2	12- 1/2	13- 1/2	12.503	13.497	.490	.4375	.778	5/32
76	1/2	13	14	13.003	13.997	.490	.4375	.778	5/32
77	1/2	13- 1/2	14- 1/2	13.503	14.497	.490	.4375	.778	5/32
78	1/2	14	15	14.003	14.997	.490	.4375	.778	5/32
79	1/2	14- 1/2	15- 1/2	14.503	15.497	.490	.4375	.778	5/32
80	1/2	15	16	15.003	15.997	.490	.4375	.778	5/32

TABLE 15–11

Male Adapter Rings for V Packings
Sizes Dimensionally the Same as AN 6229

Catalog and Manual 201 G and K-International

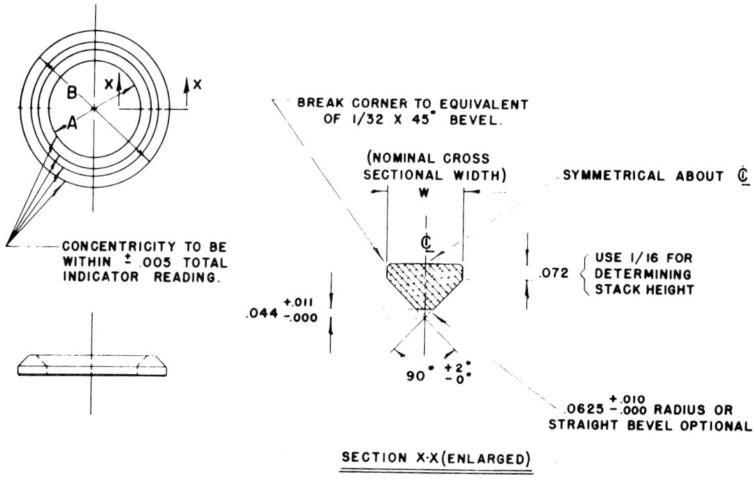

Dash No.	W	Nominal ID	Nominal OD	A +.005 −.000	B +.000 −.005
1	3/16	1/8	1/2	.145	.480
2	3/16	3/16	9/16	.200	.542
3	3/16	1/4	5/8	.270	.605
4	3/16	5/16	11/16	.333	.667
5	3/16	3/8	3/4	.395	.730
6	3/16	7/16	13/16	.458	.792
7	3/16	1/2	7/8	.520	.855
8	1/4	1/4	3/4	.270	.730
9	1/4	5/16	13/16	.333	.792
10	1/4	3/8	7/8	.395	.855
11	1/4	7/16	15/16	.458	.917
12	1/4	1/2	1	.520	.980
13	1/4	9/16	1- 1/16	.583	1.042
14	1/4	5/8	1- 1/8	.645	1.105
15	1/4	11/16	1- 3/16	.708	1.167
16	1/4	3/4	1- 1/4	.770	1.230
17	1/4	13/16	1- 5/16	.833	1.292
18	1/4	7/8	1- 3/8	.895	1.355
19	1/4	15/16	1- 7/16	.958	1.417
20	1/4	1	1- 1/2	1.020	1.480
21	1/4	1- 1/16	1- 9/16	1.083	1.542
22	1/4	1- 1/8	1- 5/8	1.145	1.605
23	1/4	1- 3/16	1-11/16	1.208	1.667
24	1/4	1- 1/4	1- 3/4	1.270	1.730
25	5/16	1- 1/4	1- 7/8	1.270	1.855

continued on next page

TABLE 15–11, continued

Dash No.	W	Nominal ID	Nominal OD	A +.005 −.000	B +.000 −.005
26	5/16	1- 3/8	2	1.395	1.980
27	5/16	1- 1/2	2- 1/8	1.520	2.105
28	5/16	1- 5/8	2- 1/4	1.645	2.230
29	5/16	1- 3/4	2- 3/8	1.770	2.355
30	5/16	1- 7/8	2- 1/2	1.895	2.480
31	5/16	2	2- 5/8	2.020	2.605
32	5/16	2- 1/8	2- 3/4	2.145	2.730
33	5/16	2- 1/4	2- 7/8	2.270	2.855
34	5/16	2- 3/8	3	2.395	2.980
35	5/16	2- 1/2	3- 1/8	2.520	3.105
36	3/8	2- 1/2	3- 1/4	2.520	3.230
37	3/8	2- 5/8	3- 3/8	2.645	3.355
38	3/8	2- 3/4	3- 1/2	2.770	3.480
39	3/8	2- 7/8	3- 5/8	2.895	3.605
40	3/8	3	3- 3/4	3.020	3.730
41	3/8	3- 1/8	3- 7/8	3.145	3.855
42	3/8	3- 1/4	4	3.270	3.980
43	3/8	3- 3/8	4- 1/8	3.395	4.105
44	3/8	3- 1/2	4- 1/4	3.520	4.230
45	3/8	3- 5/8	4- 3/8	3.645	4.355
46	3/8	3- 3/4	4- 1/2	3.770	4.480
47	3/8	3- 7/8	4- 5/8	3.895	4.605
48	7/16	3- 7/8	4- 3/4	3.895	4.730
49	7/16	4	4- 7/8	4.020	4.855
50	7/16	4- 1/4	5- 1/8	4.270	5.105
51	7/16	4- 1/2	5- 3/8	4.520	5.355
52	7/16	4- 3/4	5- 5/8	4.770	5.605
53	7/16	5	5- 7/8	5.020	5.855
54	7/16	5- 1/4	6- 1/8	5.270	6.105
55	7/16	5- 1/2	6- 3/8	5.520	6.355
56	1/2	5- 1/2	6- 1/2	5.520	6.480
57	1/2	5- 3/4	6- 3/4	5.770	6.730
58	1/2	6	7	6.020	6.980
59	1/2	6- 1/4	7- 1/4	6.270	7.230
60	1/2	6- 1/2	7- 1/2	6.520	7.480
61	1/2	6- 3/4	7- 3/4	6.770	7.730
62	1/2	7	8	7.020	7.980
63	1/2	7- 1/4	8- 1/4	7.270	8.230
64	1/2	7- 1/2	8- 1/2	7.520	8.480
65	1/2	7- 3/4	8- 3/4	7.770	8.730
66	1/2	8	9	8.020	8.980
67	1/2	8- 1/2	9- 1/2	8.520	9.480
68	1/2	9	10	9.020	9.980
69	1/2	9- 1/2	10- 1/2	9.520	10.480
70	1/2	10	11	10.020	10.980
71	1/2	10- 1/2	11- 1/2	10.520	11.480
72	1/2	11	12	11.020	11.980
73	1/2	11- 1/2	12- 1/2	11.520	12.480
74	1/2	12	13	12.020	12.980
75	1/2	12- 1/2	13- 1/2	12.520	13.480
76	1/2	13	14	13.020	13.980
77	1/2	13- 1/2	14- 1/2	13.520	14.480
78	1/2	14	15	14.020	14.980
79	1/2	14- 1/2	15- 1/2	14.520	15.480
80	1/2	15	16	15.020	15.980

TABLE 15–12

Dimensions and Groove Data for PRP 6227 O-Rings

Parco Engineering Handbook   Plastic and Rubber Products Company

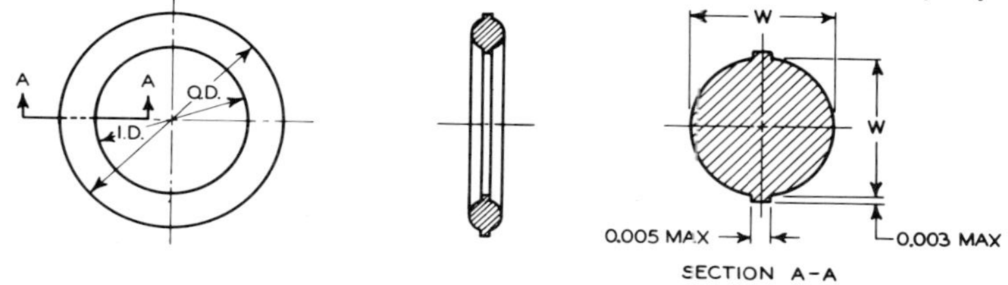

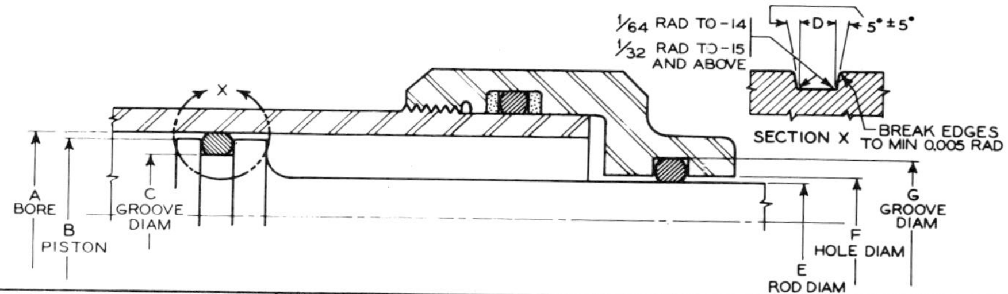

Dash No.	W	Nominal ID	Nominal OD	Actual Size W	Actual Size ID	External Groove A Bore ±.001	External Groove B Piston +.000 −.001	External Groove B′ Before Plate +.000 −.001	External Groove C Groove Diam +.000 −.002	Gland D With No Back-up ±.010	Gland D With One Back-up +.005 −.000	Gland D With Two Back-ups +.005 −.000	Internal Groove E Rod Diam +.000 −.001	Internal Groove E′ Rod Diam Before Plate +.000 −.001	Internal Groove F Bearing Diam +.001 −.000	Internal Groove G Groove Diam +.002 −.000	ϵ Max. Eccentricity T.I.R.
1	1/16	1/8	1/4	.070±.003	.114±.005	.250	.247	.244	.139	.093	.138	.205	.123	.120	.126	.234	.002
2	1/16	5/32	9/32	.070±.003	.145±.005	.281	.278	.275	.170				.154	.151	.157	.265	
3	1/16	3/16	5/16	.070±.003	.176±.005	.312	.309	.306	.201				.185	.182	.188	.296	
4	1/16	7/32	11/32	.070±.003	.208±.005	.344	.341	.338	.233				.217	.214	.220	.328	
5	1/16	1/4	3/8	.070±.003	.239±.005	.375	.372	.369	.264				.248	.245	.251	.359	
6	1/16	5/16	7/16	.070±.003	.301±.005	.438	.435	.432	.326				.310	.307	.313	.421	
7	1/16	3/8	1/2	.070±.003	.364±.005	.500	.497	.494	.389				.373	.370	.376	.484	
8	3/32	3/8	9/16	.103±.003	.363±.005	.563	.560	.557	.386	.141	.171	.238	.373	.370	.376	.550	.002
9	3/32	7/16	5/8	.103±.003	.424±.005	.625	.622	.619	.448				.435	.432	.438	.612	
10	3/32	1/2	11/16	.103±.003	.487±.005	.688	.685	.682	.511				.498	.495	.501	.675	
11	3/32	9/16	3/4	.103±.003	.549±.005	.750	.747	.744	.573				.560	.557	.563	.737	
12	3/32	5/8	13/16	.103±.003	.612±.005	.813	.810	.807	.636				.623	.620	.626	.800	
13	3/32	11/16	7/8	.103±.003	.674±.005	.875	.872	.869	.698				.685	.682	.688	.862	
14	3/32	3/4	15/16	.103±.003	.737±.005	.938	.935	.932	.761				.748	.745	.751	.925	
15	1/8	3/4	1	.139±.004	.734±.006	1.001	.997	.994	.758	.188	.208	.275	.747	.744	.751	.990	.003

continued on next page

TABLE 15–12, continued

Dash No.	W	Nominal ID	Nominal OD	Actual Size W	Actual Size ID	External Groove A Bore ±.001	External Groove B Piston +.000 −.001	External Groove B′ Before Plate +.000 −.001	External Groove C Groove Diam +.000 −.002	Gland D With No Back-up +.010	Gland D With One Back-up +.005 −.000	Gland D With Two Back-ups +.005 −.000	E Rod Diam +.000 −.001	E′ Rod Diam Before Plate +.000 −.001	Internal Groove F Bearing Diam +.001 −.000	Internal Groove G Groove Diam +.002 −.000	ε Max. Eccentricity T.I.R.
16	1/8	13/16	1 1/16	.139±.004	.796±.006	1.063	1.059	1.056	.820				.809	.806	.813	1.052	
17	1/8	7/8	1 1/8	.139±.004	.859±.006	1.126	1.122	1.119	.883				.872	.869	.876	1.115	
18	1/8	15/16	1 3/16	.139±.004	.921±.006	1.188	1.184	1.181	.945				.934	.931	.938	1.177	
19	1/8	1	1 1/4	.139±.004	.984±.006	1.251	1.247	1.244	1.008				.997	.994	1.001	1.240	
20	1/8	1 1/16	1 5/16	.139±.004	1.046±.006	1.313	1.309	1.306	1.070				1.059	1.056	1.063	1.302	
21	1/8	1 1/8	1 3/8	.139±.004	1.109±.006	1.376	1.372	1.369	1.133				1.122	1.119	1.126	1.365	
22	1/8	1 3/16	1 7/16	.139±.004	1.171±.006	1.438	1.434	1.431	1.195				1.184	1.181	1.188	1.427	
23	1/8	1 1/4	1 1/2	.139±.004	1.234±.006	1.501	1.497	1.494	1.258				1.247	1.244	1.251	1.490	
24	1/8	1 5/16	1 9/16	.139±.004	1.296±.006	1.563	1.559	1.556	1.320				1.309	1.306	1.313	1.552	
25	1/8	1 3/8	1 5/8	.139±.004	1.359±.006	1.626	1.622	1.619	1.383				1.372	1.369	1.376	1.615	
26	1/8	1 7/16	1 11/16	.139±.004	1.421±.006	1.688	1.684	1.681	1.445				1.434	1.431	1.438	1.677	
27	1/8	1 1/2	1 3/4	.139±.004	1.484±.006	1.751	1.747	1.744	1.508				1.497	1.494	1.501	1.740	
28	3/16	1 1/2	1 7/8	.210±.005	1.475±.010	1.876	1.871	1.868	1.504	.281	.311	.410	1.497	1.494	1.502	1.869	.004
29	3/16	1 5/8	2	.210±.005	1.600±.010	2.001	1.996	1.993	1.629				1.622	1.619	1.627	1.994	
30	3/16	1 3/4	2 1/8	.210±.005	1.725±.010	2.126	2.121	2.118	1.754				1.747	1.744	1.752	2.119	
31	3/16	1 7/8	2 1/4	.210±.005	1.850±.010	2.251	2.246	2.243	1.879				1.872	1.869	1.877	2.244	
32	3/16	2	2 3/8	.210±.005	1.975±.010	2.376	2.371	2.368	2.004				1.997	1.994	2.002	2.369	
33	3/16	2 1/8	2 1/2	.210±.005	2.100±.010	2.501	2.496	2.493	2.129				2.122	2.119	2.127	2.494	
34	3/16	2 1/4	2 5/8	.210±.005	2.225±.010	2.626	2.621	2.618	2.254				2.247	2.244	2.252	2.619	
35	3/16	2 3/8	2 3/4	.210±.005	2.350±.010	2.751	2.746	2.743	2.379				2.372	2.369	2.377	2.744	
36	3/16	2 1/2	2 7/8	.210±.005	2.475±.010	2.876	2.871	2.868	2.504				2.497	2.494	2.502	2.869	
37	3/16	2 5/8	3	.210±.005	2.600±.010	3.001	2.996	2.993	2.629				2.622	2.619	2.627	2.994	
38	3/16	2 3/4	3 1/8	.210±.005	2.725±.015	3.126	3.121	3.118	2.754				2.747	2.744	2.752	3.119	
39	3/16	2 7/8	3 1/4	.210±.005	2.850±.015	3.251	3.246	3.243	2.879				2.872	2.869	2.877	3.244	
40	3/16	3	3 3/8	.210±.005	2.975±.015	3.377	3.372	3.369	3.005				2.996	2.993	3.001	3.368	
41	3/16	3 1/8	3 1/2	.210±.005	3.100±.015	3.502	3.497	3.494	3.130				3.121	3.118	3.126	3.493	
42	3/16	3 1/4	3 5/8	.210±.005	3.225±.015	3.627	3.622	3.619	3.255				3.246	3.243	3.251	3.618	
43	3/16	3 3/8	3 3/4	.210±.005	3.350±.015	3.752	3.747	3.744	3.380				3.371	3.368	3.376	3.743	
44	3/16	3 1/2	3 7/8	.210±.005	3.475±.015	3.877	3.872	3.869	3.505				3.496	3.493	3.501	3.868	
45	3/16	3 5/8	4	.210±.005	3.600±.015	4.002	3.997	3.994	3.630				3.621	3.618	3.626	3.993	
46	3/16	3 3/4	4 1/8	.210±.005	3.725±.015	4.127	4.122	4.119	3.755				3.746	3.743	3.751	4.118	
47	3/16	3 7/8	4 1/4	.210±.005	3.850±.015	4.252	4.247	4.244	3.880				3.871	3.868	3.876	4.243	
48	3/16	4	4 3/8	.210±.005	3.975±.015	4.377	4.372	4.369	4.005				3.996	3.993	4.001	4.368	
49	3/16	4 1/8	4 1/2	.210±.005	4.100±.015	4.502	4.497	4.494	4.130				4.121	4.118	4.126	4.493	
50	3/16	4 1/4	4 5/8	.210±.005	4.225±.015	4.627	4.622	4.619	4.255				4.246	4.243	4.251	4.618	

continued on next page

TABLE 15–12, continued

Dash No.	W	Nominal ID	Nominal OD	Actual Size W	Actual Size ID	External Groove A Bore ±.001	External Groove B Piston +.000 −.001	External Groove B′ Before Plate +.000 −.001	External Groove C Groove Diam +.000 −.002	Gland D With No Back-up +.010	Gland D With One Back-up +.005 −.000	Gland D With Two Back-ups +.005 −.000	E Rod Diam +.000 −.001	E′ Rod Diam Before Plate +.000 −.001	Internal Groove F Bearing Diam +.001 −.000	Internal Groove G Groove Diam +.002 −.000	ε Max. Eccentricity T.I.R.
51	3/16	4 3/8	4 3/4	.210±.005	4.350±.015	4.752	4.747	4.744	4.380				4.371	4.368	4.376	4.743	
52	3/16	4 1/2	4 7/8	.210±.005	4.475±.015	4.877	4.872	4.869	4.505				4.496	4.493	4.501	4.868	
53	1/4	4 5/8	5 1/8	.275±.006	4.600±.015	5.128	5.122	5.119	4.651				4.621	4.618	4.827	5.098	
54	1/4	4 3/4	5 1/4	.275±.006	4.725±.015	5.253	5.247	5.244	4.776	.375	.413	.543	4.746	4.743	4.752	5.223	
55	1/4	4 7/8	5 3/8	.275±.006	4.850±.015	5.378	5.372	5.369	4.901				4.871	4.868	4.877	5.348	
56	1/4	5	5 1/2	.275±.006	4.975±.015	5.503	5.497	5.494	5.026				4.996	4.993	5.002	5.473	
57	1/4	5 1/8	5 5/8	.275±.006	5.100±.023	5.628	5.622	5.619	5.151				5.121	5.118	5.127	5.598	
58	1/4	5 1/4	5 3/4	.275±.006	5.225±.023	5.753	5.747	5.744	5.276				5.246	5.243	5.252	5.723	
59	1/4	5 3/8	5 7/8	.275±.006	5.350±.023	5.878	5.872	5.869	5.401				5.371	5.368	5.377	5.848	
60	1/4	5 1/2	6	.275±.006	5.475±.023	6.003	5.997	5.994	5.526				5.496	5.493	5.502	5.973	
61	1/4	5 5/8	6 1/8	.275±.006	5.600±.023	6.128	6.122	6.119	5.651				5.621	5.618	5.627	6.098	
62	1/4	5 3/4	6 1/4	.275±.006	5.725±.023	6.253	6.247	6.244	5.776				5.746	5.743	5.752	6.223	
63	1/4	5 7/8	6 3/8	.275±.006	5.850±.023	6.378	6.372	6.369	5.901				5.871	5.868	5.877	6.348	
64	1/4	6	6 1/2	.275±.006	5.975±.023	6.503	6.497	6.494	6.026				5.996	5.993	6.002	6.473	
65	1/4	6 1/4	6 3/4	.275±.006	6.225±.023	6.753	6.747	6.744	6.276				6.246	6.243	6.252	6.723	
66	1/4	6 1/2	7	.275±.006	6.475±.023	7.003	6.997	6.994	6.526				6.496	6.493	6.502	6.973	
67	1/4	6 3/4	7 1/4	.275±.006	6.725±.023	7.253	7.247	7.244	6.776				6.746	6.743	6.752	7.223	
68	1/4	7	7 1/2	.275±.006	6.975±.023	7.503	7.497	7.494	7.026				6.996	6.993	7.002	7.473	
69	1/4	7 1/4	7 3/4	.275±.006	7.225±.030	7.753	7.747	7.744	7.276				7.246	7.243	7.652	7.723	
70	1/4	7 1/2	8	.275±.006	7.475±.030	8.003	7.997	7.994	7.526				7.496	7.493	7.502	7.973	
71	1/4	7 3/4	8 1/4	.275±.006	7.725±.030	8.253	8.247	8.244	7.776				7.746	7.743	7.752	8.223	
72	1/4	8	8 1/2	.275±.006	7.975±.030	8.503	8.497	8.494	8.026				7.996	7.993	8.002	8.473	
73	1/4	8 1/2	9	.275±.006	8.475±.030	9.003	8.997	8.994	8.526				8.496	8.493	8.502	8.973	
74	1/4	9	9 1/2	.275±.006	8.975±.030	9.503	9.497	9.494	9.026				8.996	8.993	9.002	9.473	
75	1/4	9 1/2	10	.275±.006	9.475±.030	10.003	9.997	9.994	9.526				9.496	9.493	9.502	9.973	
76	1/4	10	10 1/2	.275±.006	9.975±.030	10.503	10.497	10.494	10.026				9.996	9.993	10.002	10.473	
77	1/4	10 1/2	11	.275±.006	10.475±.030	11.003	10.997	10.994	10.526				10.496	10.493	10.502	10.973	
78	1/4	11	11 1/2	.275±.006	10.975±.030	11.503	11.497	11.494	11.026				10.996	10.993	11.002	11.473	
79	1/4	11 1/2	12	.275±.006	11.475±.030	12.003	11.997	11.994	11.526				11.496	11.493	11.502	11.973	
80	1/4	12	12 1/2	.275±.006	11.975±.030	12.503	12.497	12.494	12.026				11.996	11.993	12.002	12.473	
81	1/4	12 1/2	13	.275±.006	12.475±.030	13.003	12.997	12.994	12.526				12.496	12.493	12.502	12.973	
82	1/4	13	13 1/2	.275±.006	12.975±.030	13.503	13.497	13.494	13.026				12.996	12.993	13.002	13.473	
83	1/4	13 1/2	14	.275±.006	13.475±.030	14.003	13.997	13.994	13.526				13.496	13.493	13.502	13.973	
84	1/4	14	14 1/2	.275±.006	13.975±.030	14.503	14.497	14.494	14.026				13.996	13.993	14.002	14.473	
85	1/4	14 1/2	15	.275±.006	14.475±.030	15.003	14.997	14.994	14.526				14.496	14.493	14.502	14.973	
86	1/4	15	15 1/2	.275±.006	14.975±.030	15.503	15.497	15.494	15.026				14.996	14.993	15.002	15.473	
87	1/4	15 1/2	16	.275±.006	15.475±.030	16.003	15.997	15.994	15.526				15.496	15.493	15.502	15.973	
88	1/4	4 1/2	5	.275±.006	4.475±.015	5.003	4.997	4.994	4.526				4.496	4.493	4.502	4.973	

TABLE 15–13

Dimensions and Groove Data for PRP 6230 O-Rings

Parco Engineering Handbook Plastics and Rubber Products Company

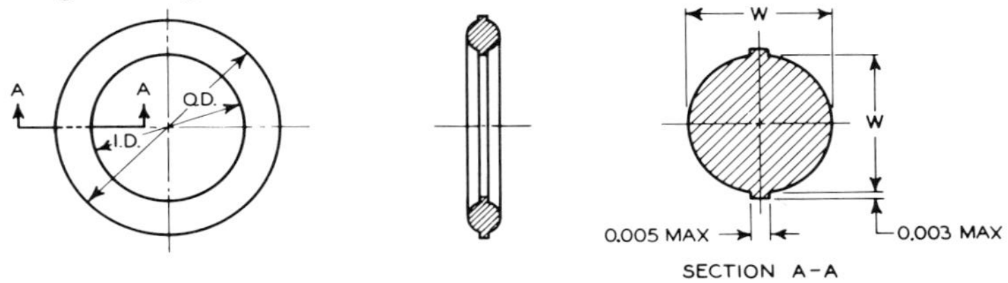

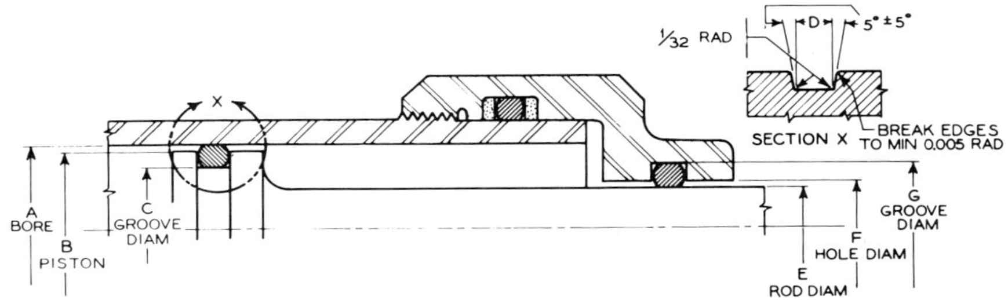

Dash No.	W	Nominal ID	Nominal OD	Actual Size W	Actual Size ID	External Groove A Bore ±.001	External Groove B Piston +.000 −.001	External Groove B′ Before Plate +.000 −.001	External Groove C Groove Diam +.000 −.002	Gland D With No Back-up ±.010	Gland D With One Back-up +.005 −.000	Gland D With Two Back-Ups +.005 −.000	Internal Groove E Rod Diam +.000 −.001	Internal Groove E′ Rod Diam Before Plate +.000 −.001	Internal Groove F Bearing Diam +.001 −.000	Internal Groove G Groove Diam +.002 −.000	ε Max. Eccentricity T.I.R.
1	1/8	1 5/8	1 7/8	.139±.004	1.609±.010	1.876	1.871	1.868	1.629	.188	.208	.275	1.622	1.619	1.627	1.869	.004
2		1 3/4	2		1.734±.010	2.001	1.996	1.993	1.754				1.747	1.744	1.752	1.994	
3		1 7/8	2 1/8		1.859±.010	2.126	2.121	2.118	1.879				1.872	1.869	1.877	2.119	
4		2	2 1/4		1.984±.010	2.251	2.246	2.243	2.004				1.997	1.994	2.002	2.244	
5		2 1/8	2 3/8		2.109±.010	2.376	2.371	2.368	2.129				2.122	2.119	2.127	2.369	
6		2 1/4	2 1/2		2.234±.010	2.501	2.496	2.493	2.254				2.247	2.244	2.252	2.494	
7		2 3/8	2 5/8		2.359±.010	2.626	2.621	2.618	2.379				2.372	2.369	2.377	2.619	
8		2 1/2	2 3/4		2.484±.010	2.751	2.746	2.743	2.504				2.497	2.494	2.502	2.744	
9		2 5/8	2 7/8		2.609±.010	2.876	2.871	2.868	2.629				2.622	2.619	2.627	2.869	
10		2 3/4	3		2.734±.015	3.001	2.996	2.993	2.754				2.747	2.744	2.752	2.994	
11		2 7/8	3 1/8		2.859±.015	3.126	3.121	3.118	2.879				2.872	2.869	2.877	3.119	
12		3	3 1/4		2.984±.015	3.251	3.246	3.243	3.005				2.996	2.993	3.001	3.243	
13		3 1/8	3 3/8		3.109±.015	3.377	3.372	3.369	3.130				3.121	3.118	3.126	3.368	
14		3 1/4	3 1/2		3.234±.015	3.502	3.497	3.494	3.255				3.246	3.243	3.251	3.493	
15		3 3/8	3 5/8		3.359±.015	3.627	3.622	3.619	3.380				3.371	3.368	3.376	3.618	

continued on next page

TABLE 15–13, continued

Dash No.	W	Nominal ID	Nominal OD	Actual Size W	Actual Size ID	External Groove A Bore ±.001	External Groove B Piston +.000 −.001	External Groove B′ Before Plate +.000 −.001	External Groove C Groove Diam +.000 −.002	Gland D With No Back-up ±.010	Gland D With One Back-up +.005 −.000	Gland D With Two Back-Ups +.005 −.000	E Rod Diam +.000 −.001	E′ Rod Diam Before Plate +.000 −.001	Internal Groove F Bearing Diam +.001 −.000	Internal Groove G Groove Diam +.002 −.000	ε Max. Eccentricity T.I.R.
16		3 1/2	3 3/4		3.484±.015	3.752	3.747	3.744	3.505				3.496	3.493	3.501	3.743	
17		3 5/8	3 7/8		3.609±.015	3.877	3.872	3.869	3.630				3.621	3.618	3.626	3.868	
18		3 3/4	4		3.734±.015	4.002	3.997	3.994	3.755				3.746	3.743	3.751	3.993	
19		3 7/8	4 1/8		3.859±.015	4.127	4.122	4.119	3.880				3.871	3.868	3.876	4.118	
20		4	4 1/4		3.984±.015	4.252	4.247	4.244	4.005				3.996	3.993	4.001	4.243	
21		4 1/8	4 3/8		4.109±.015	4.377	4.372	4.369	4.130				4.121	4.118	4.126	4.368	
22		4 1/4	4 1/2		4.234±.015	4.502	4.497	4.494	4.255				4.246	4.243	4.251	4.493	
23		4 3/8	4 5/8		4.359±.015	4.627	4.622	4.619	4.380				4.371	4.368	4.376	4.618	
24		4 1/2	4 3/4		4.484±.015	4.752	4.747	4.744	4.505				4.496	4.493	4.501	4.743	
25		4 5/8	4 7/8		4.609±.015	4.877	4.872	4.869	4.651				4.621	4.618	4.627	4.868	
26		4 3/4	5		4.734±.015	5.002	4.997	4.994	4.776				4.746	4.743	4.752	4.993	
27		4 7/8	5 1/8		4.859±.015	5.128	5.122	5.119	4.901				4.871	4.868	4.877	5.098	
28		5	5 1/4		4.984±.015	5.253	5.247	5.244	5.026				4.996	4.993	5.002	5.223	
29		5 1/8	5 3/8		5.109±.023	5.378	5.372	5.369	5.151				5.121	5.118	5.127	5.348	
30		5 1/4	5 1/2		5.234±.023	5.503	5.497	5.493	5.276				5.246	5.243	5.252	5.473	
31		5 3/8	5 5/8		5.359±.023	5.628	5.622	5.619	5.401				5.371	5.368	5.377	5.598	
32		5 1/2	5 3/4		5.484±.023	5.753	5.747	5.744	5.526				5.496	5.493	5.502	5.723	
33		5 5/8	5 7/8		5.609±.023	5.878	5.872	5.869	5.651				5.621	5.618	5.627	5.848	
34		5 3/4	6		5.734±.023	6.003	5.997	5.994	5.776				5.746	5.743	5.752	5.973	
35		5 7/8	6 1/8		5.859±.023	6.128	6.122	6.119	5.901				5.871	5.868	5.877	6.098	
36		6	6 1/4		5.984±.023	6.253	6.247	6.244	6.026				5.996	5.993	6.002	6.223	
37		6 1/4	6 1/2		6.234±.023	6.503	6.497	6.494	6.276				6.246	6.243	6.252	6.473	
38		6 1/2	6 3/4		6.484±.023	6.753	6.747	6.744	6.526				6.496	6.493	6.502	6.723	
39		6 3/4	7		6.734±.023	7.003	6.997	6.994	6.776				6.746	6.743	6.752	6.973	
40		7	7 1/4		6.984±.023	7.253	7.247	7.244	7.026				6.996	6.993	7.002	7.223	
41		7 1/4	7 1/2		7.234±.030	7.503	7.497	7.494	7.276				7.246	7.243	7.252	7.473	
42		7 1/2	7 3/4		7.484±.030	7.753	7.747	7.744	7.526				7.496	7.493	7.502	7.723	
43		7 3/4	8		7.734±.030	8.003	7.997	7.994	7.776				7.746	7.743	7.752	7.973	
44		8	8 1/4		7.984±.030	8.253	8.247	8.244	8.026				7.996	7.993	8.002	8.223	
45		8 1/4	8 1/2		8.234±.030	8.503	8.497	8.494	8.276				8.246	8.243	8.252	8.473	
46		8 1/2	8 3/4		8.484±.030	8.753	8.747	8.744	8.526				8.496	8.493	8.502	8.723	
47		8 3/4	9		8.734±.030	9.003	8.997	8.994	8.776				8.746	8.743	8.752	8.973	
48		9	9 1/4		8.984±.030	9.253	9.247	9.244	9.026				8.996	8.993	9.002	9.223	
49		9 1/4	9 1/2		9.234±.030	9.503	9.497	9.494	9.276				9.246	9.243	9.252	9.473	
50		9 1/2	9 3/4		9.484±.030	9.753	9.747	9.744	9.526				9.496	9.493	9.502	9.723	
51		9 3/4	10		9.734±.030	10.003	9.997	9.994	9.776				9.746	9.743	9.752	9.973	
52		10	10 1/4		9.984±.030	10.253	10.247	10.244	10.026				9.996	9.993	10.002	10.223	

TABLE 15–14

Leather Back-Up Rings for Use with 27 Series O-Rings (Table 15-12)
Sizes Dimensionally the Same as AN 6246 Back-Up Rings
Catalog and Manual 201 G and K-International

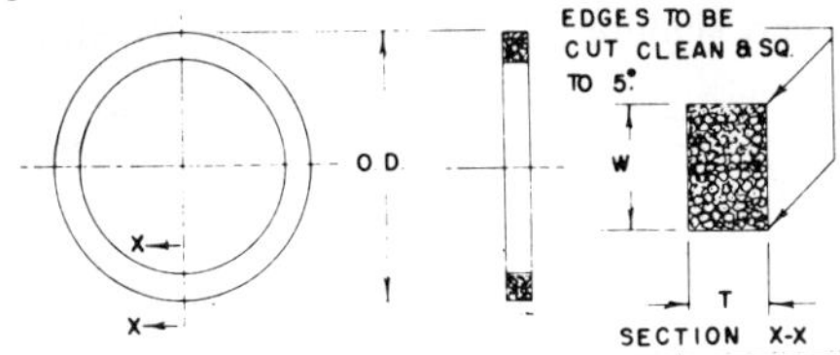

Dash No.	Actual Size OD ±.007	W ±.007	T ±.010
1	.250	.057	.062
2	.281	.057	.062
3	.312	.057	.062
4	.344	.057	.062
5	.375	.057	.062
6	.438	.057	.062
7	.500	.057	.062
8	.562	.090	.062
9	.625	.090	.062
10	.688	.090	.062
11	.750	.090	.062
12	.812	.090	.062
13	.875	.090	.062
14	.938	.090	.062

Dash No.	Actual Size OD ±1/64	W +.010 −.005	T ±.010
15	1	.123	.062
16	1-1/16	.123	.062
17	1-1/8	.123	.062
18	1-3/16	.123	.062
19	1-1/4	.123	.062
20	1-5/16	.123	.062
21	1-3/8	.123	.062
22	1-7/16	.123	.062
23	1-1/2	.123	.062
24	1-9/16	.123	.062
25	1-5/8	.123	.062
26	1-11/16	.123	.062
27	1-3/4	.123	.062
28	1-7/8	.188	.094
29	2	.188	.094
30	2-1/8	.188	.094
31	2-1/4	.188	.094
32	2-3/8	.188	.094
33	2-1/2	.188	.094
34	2-5/8	.188	.094
35	2-3/4	.188	.094
36	2-7/8	.188	.094
37	3	.188	.094
38	3-1/8	.188	.094
39	3-1/4	.188	.094
40	3-3/8	.188	.094

Dash No.	Actual Size OD ±1/64	W +.010 −.005	T ±.010
41	3-1/2	.188	.094
42	3-5/8	.188	.094
43	3-3/4	.188	.094
44	3-7/8	.188	.094
45	4	.188	.094
46	4-1/8	.188	.094
47	4-1/4	.188	.094
48	4-3/8	.188	.094
49	4-1/2	.188	.094
50	4-5/8	.188	.094
51	4-3/4	.188	.094
52	4-7/8	.188	.094
53	5-1/8	.240	.125
54	5-1/4	.240	.125
55	5-3/8	.240	.125
56	5-1/2	.240	.125
57	5-5/8	.240	.125
58	5-3/4	.240	.125
59	5-7/8	.240	.125
60	6	.240	.125
61	6-1/8	.240	.125
62	6-1/4	.240	.125
63	6-3/8	.240	.125
64	6-1/2	.240	.125
65	6-3/4	.240	.125
66	7	.240	.125
67	7-1/4	.240	.125
68	7-1/2	.240	.125
69	7-3/4	.240	.125
70	8	.240	.125
71	8-1/4	.240	.125
72	8-1/2	.240	.125
73	9	.240	.125
74	9-1/2	.240	.125
75	10	.240	.125
76	10-1/2	.240	.125
77	11	.240	.125
78	11-1/2	.240	.125
79	12	.240	.125
80	12-1/2	.240	.125
81	13	.240	.125
82	13-1/2	.240	.125
83	14	.240	.125
84	14-1/2	.240	.125
85	15	.240	.125
86	15-1/2	.240	.125
87	16	.240	.125
88	5	.240	.125

TABLE 15–15

Leather Back-Up Rings for Use with 30 Series O-Rings (Table 15-13) Sizes Dimensionally the Same as AN 6244 Back-Up Rings

Catalog and Manual 201 G and K-International

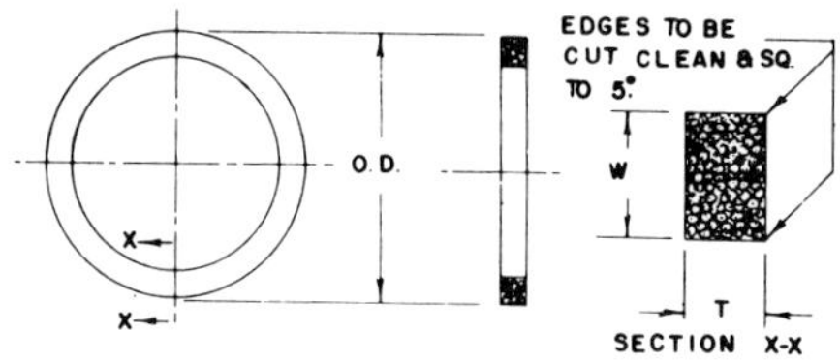

DETAIL FOR LEATHER BACK-UP RINGS – FOR PRESSURES OVER 1500 PSI

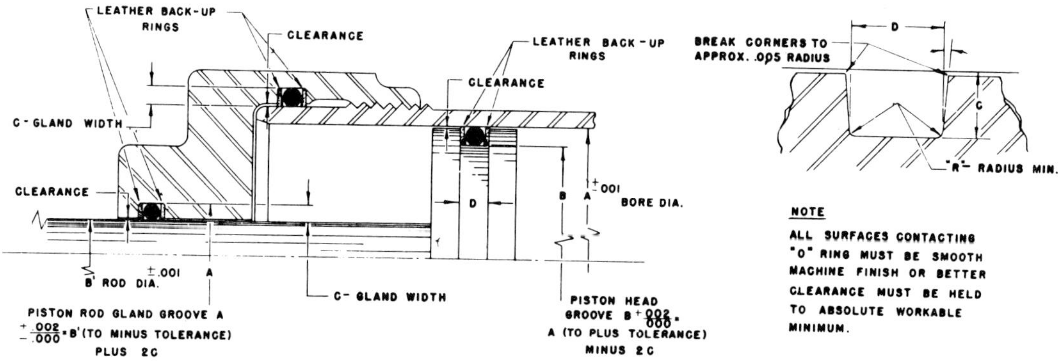

Dash No.	OD ±1/64	ID	*W* +.010 −.005	*T* +.010
1	1-7/8	1-5/8	1/8	1/16
2	2	1-3/4	1/8	1/16
3	2-1/8	1-7/8	1/8	1/16
4	2-1/4	2	1/8	1/16
5	2-3/8	2-1/8	1/8	1/16
6	2-1/2	2-1/4	1/8	1/16
7	2-5/8	2-3/8	1/8	1/16
8	2-3/4	2-1/2	1/8	1/16
9	2-7/8	2-5/8	1/8	1/16
10	3	2-3/4	1/8	1/16
11	3-1/8	2-7/8	1/8	1/16
12	3-1/4	3	1/8	1/16
13	3-3/8	3-1/8	1/8	1/16
14	3-1/2	3-1/4	1/8	1/16
15	3-5/8	3-3/8	1/8	1/16
16	3-3/4	3-1/2	1/8	1/16
17	3-7/8	3-5/8	1/8	1/16
18	4	3-3/4	1/8	1/16
19	4-1/8	3-7/8	1/8	1/16
20	4-1/4	4	1/8	1/16
21	4-3/8	4-1/8	1/8	1/16
22	4-1/2	4-1/4	1/8	1/16
23	4-5/8	4-3/8	1/8	1/16
24	4-3/4	4-1/2	1/8	1/16
25	4-7/8	4-5/8	1/8	1/16
26	5	4-3/4	1/8	1/16
27	5-1/8	4-7/8	1/8	1/16
28	5-1/4	5	1/8	1/16
29	5-3/8	5-1/8	1/8	1/16
30	5-1/2	5-1/4	1/8	1/16
31	5-5/8	5-3/8	1/8	1/16
32	5-3/4	5-1/2	1/8	1/16
33	5-7/8	5-5/8	1/8	1/16
34	6	5-3/4	1/8	1/16
35	6-1/8	5-7/8	1/8	1/16
36	6-1/4	6	1/8	1/16
37	6-1/2	6-1/4	1/8	1/16
38	6-3/4	6-1/2	1/8	1/16
39	7	6-3/4	1/8	1/16
40	7-1/4	7	1/8	1/16
41	7-1/2	7-1/4	1/8	1/16
42	7-3/4	7-1/2	1/8	1/16
43	8	7-3/4	1/8	1/16
44	8-1/4	8	1/8	1/16
45	8-1/2	8-1/4	1/8	1/16
46	8-3/4	8-1/2	1/8	1/16
47	9	8-3/4	1/8	1/16
48	9-1/4	9	1/8	1/16
49	9-1/2	9-1/4	1/8	1/16
50	9-3/4	9-1/2	1/8	1/16
51	10	9-3/4	1/8	1/16
52	10-1/4	10	1/8	1/16

NOTE: Thickness shall not vary more than .010" between any two points on a ring.

TABLE 15–16

Gasoline and Kerosene Resistant O-Rings (IPC 34) Sizes Dimensionally the Same as USAF 934 Series

Catalog and Manual 201 G and K-International

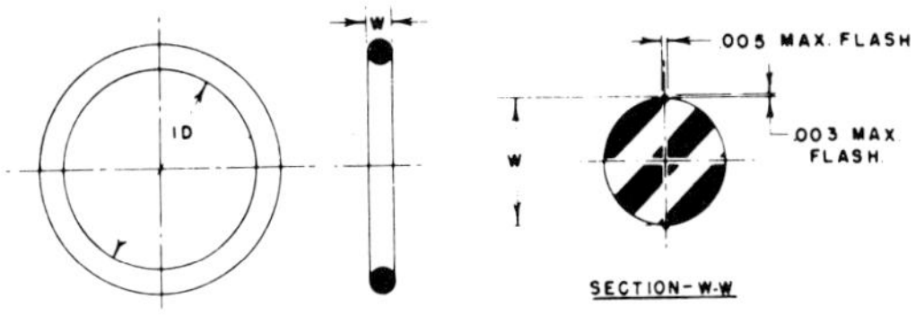

Dash No.	Nominal W in.	Nominal ID in.	Nominal OD in.	Actual Size W in.	Actual Size ID in.
1	.062	.125	.250	.070±.003	.114±.005
2	.062	.156	.281	.070±.003	.145±.005
3	.062	.188	.312	.070±.003	.176±.005
4	.062	.219	.344	.070±.003	.208±.005
5	.062	.250	.375	.070±.003	.239±.005
6	.062	.312	.438	.070±.003	.301±.005
7	.062	.375	.500	.070±.003	.364±.005
8	.094	.375	.562	.103±.003	.362±.005
9	.094	.438	.625	.103±.003	.424±.005
10	.094	.500	.688	.103±.003	.487±.005
11	.094	.562	.750	.103±.003	.549±.005
12	.094	.625	.812	.103±.003	.612±.005
13	.094	.688	.875	.103±.003	.674±.005
14	.094	.750	.938	.103±.003	.737±.005
15	.125	.750	1.000	.139±.004	.734±.006
16	.125	.812	1.062	.139±.004	.796±.006
17	.125	.875	1.125	.139±.004	.859±.006
18	.125	.938	1.188	.139±.004	.921±.006
19	.125	1.000	1.250	.139±.004	.984±.006
20	.125	1.062	1.312	.139±.004	1.046±.006
21	.125	1.125	1.375	.139±.004	1.109±.006
22	.125	1.188	1.438	.139±.004	1.171±.006
23	.125	1.250	1.500	.139±.004	1.234±.006
24	.125	1.312	1.562	.139±.004	1.296±.006
25	.125	1.375	1.625	.139±.004	1.359±.006
26	.125	1.438	1.688	.139±.004	1.421±.006
27	.125	1.500	1.750	.139±.004	1.484±.006
28	.188	1.500	1.875	.210±.005	1.475±.010
29	.188	1.625	2.000	.210±.005	1.600±.010
30	.188	1.750	2.125	.210±.005	1.725±.010
31	.188	1.875	2.250	.210±.005	1.850±.010
32	.188	2.000	2.375	.210±.005	1.975±.010
33	.188	2.125	2.500	.210±.005	2.100±.010
34	.188	2.250	2.625	.210±.005	2.225±.010
35	.188	2.375	2.750	.210±.005	2.350±.010
36	.188	2.500	2.875	.210±.005	2.475±.010
37	.188	2.625	3.000	.210±.005	2.600±.010
38	.188	2.750	3.125	.210±.005	2.725±.015
39	.188	2.875	3.250	.210±.005	2.850±.015
40	.188	3.000	3.375	.210±.005	2.975±.015
41	.188	3.125	3.500	.210±.005	3.100±.015
42	.188	3.250	3.625	.210±.005	3.225±.015
43	.188	3.375	3.750	.210±.005	3.350±.015
44	.188	3.500	3.875	.210±.005	3.475±.015
45	.188	3.625	4.000	.210±.005	3.600±.015

continued on next page

Dash No.	Nominal W in.	Nominal ID in.	Nominal OD in.	Actual Size W in.	Actual Size ID in.
46	.188	3.750	4.125	.210±.005	3.725±.015
47	.188	3.875	4.250	.210±.005	3.850±.015
48	.188	4.000	4.375	.210±.005	3.975±.015
49	.188	4.125	4.500	.210±.005	4.100±.015
50	.188	4.250	4.625	.210±.005	4.225±.015
51	.188	4.375	4.750	.210±.005	4.305±.015
52	.188	4.500	4.875	.210±.005	4.475±.015
53	.250	4.500	5.000	.275±.006	4.475±.015
54	.250	4.625	5.125	.275±.006	4.600±.015
55	.250	4.750	5.250	.275±.006	4.725±.015
56	.250	4.875	5.375	.275±.006	4.850±.015
57	.250	5.000	5.500	.275±.006	4.975±.015
58	.250	5.125	5.625	.275±.006	5.100±.023
59	.250	5.250	5.750	.275±.006	5.225±.023
60	.250	5.375	5.875	.275±.006	5.350±.023
61	.250	5.500	6.000	.275±.006	5.475±.023
62	.250	5.625	6.125	.275±.006	5.600±.023
63	.250	5.750	6.250	.275±.006	5.725±.023
64	.250	5.875	6.375	.275±.006	5.850±.023
65	.250	6.000	6.500	.275±.006	5.975±.023
66	.250	6.250	6.750	.275±.006	6.225±.023
67	.250	6.500	7.000	.275±.006	6.475±.023
68	.250	6.750	7.250	.275±.006	6.725±.023
69	.250	7.000	7.500	.275±.006	6.975±.023
70	.250	7.250	7.750	.275±.006	7.225±.030
71	.250	7.500	8.000	.275±.006	7.475±.030
72	.250	7.750	8.250	.275±.006	7.725±.030
73	.250	8.000	8.500	.275±.006	7.975±.030
74	.250	8.500	9.000	.275±.006	8.475±.030
75	.250	9.000	9.500	.275±.006	8.975±.030
76	.250	9.500	10.000	.275±.006	9.475±.030
77	.250	10.000	10.500	.275±.006	9.975±.030
78	.250	10.500	11.000	.275±.006	10.475±.030
79	.250	11.000	11.500	.275±.006	10.975±.030
80	.250	11.500	12.000	.275±.006	11.475±.030
81	.250	12.000	12.500	.275±.006	11.975±.030
82	.250	12.500	13.000	.275±.006	12.475±.030
83	.250	13.000	13.500	.275±.006	12.975±.030
84	.250	13.500	14.000	.275±.006	13.475±.030
85	.250	14.000	14.500	.275±.006	13.975±.030
86	.250	14.500	15.000	.275±.006	14.475±.030
87	.250	15.000	15.500	.275±.006	14.975±.030
88	.250	15.500	16.000	.275±.006	15.475±.030
89	.062	.438	.562	.070±.003	.426±.005
90	.062	.500	.625	.070±.003	.489±.005
91	.062	.562	.688	.070±.003	.551±.005
92	.062	.625	.750	.070±.003	.614±.005
93	.062	.688	.812	.070±.003	.676±.005
94	.062	.750	.875	.070±.003	.739±.005
95	.062	.812	.938	.070±.003	.801±.006
96	.062	.875	1.000	.070±.003	.864±.006
97	.062	.938	1.062	.070±.003	.926±.006
98	.062	1.000	1.125	.070±.003	.989±.006
99	.062	1.062	1.188	.070±.003	1.051±.006
100	.062	1.125	1.250	.070±.003	1.114±.006

continued on next page

TABLE 15–17

O-Rings as Gaskets in Flanges, Cylinder Head Covers, Static Seals

Handbook of O-Ring Hydraulic Packings   Plastic and Rubber Products, Inc.

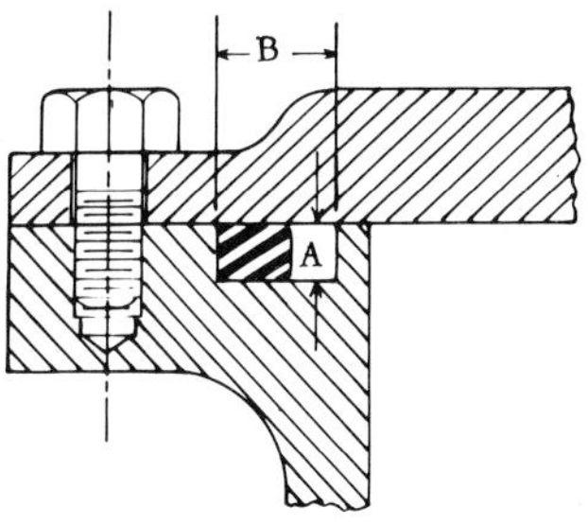

Nominal O-Ring Cross-Section	Actual O-Ring Cross-Section	Minimum Squeeze	Depth of Recess *A*	Width of Recess *B*
1/16	.070 ± .003	.017	.045 ± .005	3/32
3/32	.103 ± .003	.020	.075 ± .005	9/64
1/8	.139 ± .004	.025	.105 ± .005	3/16
3/16	.210 ± .005	.030	.170 ± .005	9/32
1/4	.275 ± .006	.040	.225 ± .005	3/8

TABLE 15–18

Service Suitability of O-Rings

Parco Engineering Handbook   Plastic and Rubber Products Company

	Buna N 209-70	Buna N 228-70	Neoprene 318-70	Thiokol 501-70	Butyl 805-70	Natural 606-70	Silicone Rubber
Acetone	P	P	F	F	E	E	F
Acids (dilute)	F	F	G	P	E	G	G
Alkalis	G	G	G	F	G	G	G
Ammonia	G	G	G				
Aniline	P	P	F		E	P	
Automotive Hydraulic Brake Fluid	F	P	F	F		E	
Butane	F	G	P	F	P	P	F
Carbon Tetrachloride	P	P	P	F	P	P	P
Dibutyl Phthalate	P	P	P		G	G	
Ethyl Acetate	P	F	P	F	E	G	
Ethyl Alcohol	E	G	G	G	G	G	
Ethylene Glycol	E	G	G		G	G	
Fuel Oil	F	E	P	G	P	P	P
Freon 12	G	G	G	G	F	F	P
Freon 22	P	P	F	G	G	F	
Gasoline (aromatic)	G	E	P	G	P	P	P
Gasoline (non-aromatic)	E	G	F	G	P	P	P
Hydraulic Fluid (mineral oil base)	E	G	F	F	P	P	P
Kerosene	E	E	P	G	P	P	P
Lubricating Oil	E	P	E	G	P	P	G
Phenol	F	F	P	F	G	G	
Propane	F	G	P	F	P	P	
Skydrol	P	P	P	P	E	P	E
Turpentine	P	F	P	G	P	P	
Vegetable Oils	G	F	F	G	G	P	

E — Excellent.
G — Good — serviceable under most conditions.
F — Fair — serviceable under some conditions.
P — Poor — not suitable.

NOTE: There are a number of types of silicone rubber. Either the particular type desired or the use to which it will be put should be specified. Because of its relatively poor physical properties, silicone rubber is recommended only for static seals.

TABLE 15-19

American Standard Nonmetallic Gaskets for Pipe Flanges

ASA B 16.21 – 1951

PIPE FLANGE SIZES

1	2			3	4	5	6	7	8	9	10	11	12	13	14	15
Reference				Full Face		Flat Ring							Male and Female		Tongue and Groove	
													Small	Large	Small	Large
MSS	Brass	SP-2	150-lb	1/2-12	...	...	.....	....	....	...	....	...	...	...	...	...
		SP-2	300-lb		1/2-8	...	.....	....	....	...	....	...	...	...	...	...
	18-8	SP-42	150-lb	1/4-12	...	...	.....	....	....	...	....	...	...	...	...	...
	Cast Iron	B16b2	25-lb	4-96[1]	...	...	.....	4	5-96	...	....	...	...	...	...	...
		B16-1	125-lb	1-48[1,2]	...	...	.....	1-48[2]	....	...	....	...	...	...	...	...
		B16g1	175-lb	2-8	...	...	.....	2-8	....	...	....	...	...	...	...	...
Example ↓		B16b	250-lb	....	...	1-3	3 1/2-48	....	....	...	....	...	...	1-24	...	1-24
ASA		B16.16	300-lb	....	...	...	.....		....	...	....	...	...	...	1/4-4	5-12
		B16b1	800-lb	....	...	2-12	.....	3 1/2	....	...	....	...	...	2-12	...	2-12
	Steel	B16e	150-lb	1-24[2]	...	...	.....	1/2-24[2]	....	...	....	...	1/2-24	1/2-24	...	1/2-24
		B16e	300-lb	....	...	1/2-3	3 1/2-24		....	...	....	...	1/2-24	1/2-24	...	1/2-24
		B16e	400-lb	....	...	1/2-3	.....	3 1/2	....	4-24	....	...	1/2-24	1/2-24	...	1/2-24
		B16e	600-lb	....	...	1/2-24	.....	3 1/2	....	...	....	...	1/2-24	1/2-24	...	1/2-24
		B16e	900-lb	....	...	...	.....		....	...	1/2-2 1/2	3-24	1/2-24	1/2-24	...	1/2-24
		B16e	1500-lb	....	...	...	.....	....	....	...	1/2-24	10	1/2-24	1/2-24	...	1/2-24

GASKET DIAMETERS

Nominal Pipe Size	Full Face and Flat Ring Gasket ID	OD	OD	OD	OD	OD	OD	OD	OD	OD	ID X OD	ID X OD	ID X OD	ID X OD
1/4	9/16	2 1/2	...	...	...	...	...	...	...	...	....	....	1 X 1 3/8	....
3/8	11/16	2 1/2	...	...	...	...	...	...	...	...	....	....	1 X 1 3/8	....
1/2	27/32	3 1/2	3 3/4	2 1/8	...	1 7/8	...	...	2 1/2	...	X 23/32	27/32 X 1 3/8	1 X 1 3/8	1 X 1 3/8
3/4	1 1/16	3 7/8	4 5/8	2 5/8	...	2 1/4	...	...	2 3/4	...	X 15/16	1 1/16 X 1 11/16	1 5/16 X 1 11/16	1 5/16 X 1 11/16
1	1 5/16	4 1/4	4 7/8	2 7/8	...	2 5/8	...	...	3 1/8	...	X 1 3/16	1 5/16 X2	1 1/2 X 1 7/8	1 1/2 X 2
1 1/4	1 21/32	4 5/8	5 1/4	3 1/4	...	3	...	...	3 1/2	...	X 1 1/2	1 21/32 X 2 1/2	1 7/8 X 2 1/4	1 7/8 X 2 1/2
1 1/2	1 29/32	5	6 1/8	3 3/4	...	3 3/8	...	...	3 7/8	...	X 1 3/4	1 29/32 X 2 7/8	2 1/8 X 2 1/2	2 1/8 X 2 7/8
2	2 3/8	6	6 1/2	4 3/8	...	4 1/8	...	...	5 5/8	...	X 2 1/4	2 3/8 X 3 5/8	2 7/8 X 3 1/4	2 7/8 X 3 5/8
2 1/2	2 7/8	7	7 1/2	5 1/8	...	4 7/8	...	...	6 1/2	...	X 2 11/16	2 7/8 X 4 1/8	3 3/8 X 3 3/4	3 3/8 X 4 1/8

continued on next page

TABLE 15-19, continued

1	2	3	4	5	6	7	8	9	10	11	12	13	14	15
		GASKET DIAMETERS												
Nominal Pipe Size	Full Face and Flat Ring Gasket ID	OD	OD	OD	OD	OD	OD	OD	OD	OD	ID X OD	ID X OD	ID X OD	ID X OD
3	3½	7½	8¼	5⅞	...	5⅜	...	...	6⅞	6⅝	Purchaser to specify ID of pipe See Par. 5.3 X 3 5/16	3½ X 5	4¼ X 4⅝	4¼ X 5
3½	4	8½	9	See Col. 7	6½	6⅜	...	...	...	...	X 3 13/16	4 X 5½	4¾ X 5⅛	4¾ X 5½
4	4½	9	10	7⅝	7⅛	6⅞	...	7	8¼	8⅛	X 4 5/16	4½ X 6 3/16	5 3/16 X 5 11/16	5 3/16 X 6 3/16
5	5 9/16	10	11	9½	8½	7¾	7⅞	8⅜	10	9¾	X 5⅜	5 9/16 X 7 5/16	....	6 5/16 X 7 5/16
6	6⅝	11	12½	10½	9⅞	8¾	8⅞	9¾	11⅛	11⅜	X 6⅜	6⅝ X 8½	....	7½ X 8⅝
8	8⅝	13½	15	12⅝	12⅛	11	11⅛	12	13⅞	14⅛	X 8⅜	8⅝ X 10⅝	....	9⅜ X 10⅝
10	10¾	16	...	15¾	14¼	13⅜	13⅝	14⅛	See Col. 11	17⅛	X 10½	10¾ X 12¾	....	11¼ X 12¾
12	12¾	19	...	18	16⅝	16⅛	16⅜	16½	20½	19⅝	X 12½	12¾ X 15	....	13½ X 15
14 OD	14	21	...	19⅜	19⅛	17¾	18	19	22¾	20½	X 13¾	14 X 16¼	....	14¾ X 16¼
16 OD	16	23½	...	22¼	21¼	20¼	20½	21⅛	25¼	22⅝	X 15¾	16 X 18½	....	16¾ X 18½
18 OD	18	25	...	24⅛	23½	21⅝	22	23⅜	27¾	25⅛	X 17¾	18 X 21	....	19¼ X 21
20 OD	20	27½	...	26⅞	25¾	23⅞	24¼	25½	29¾	27½	X 19¾	20 X 23	....	21 X 23
24 OD	24	32	...	31⅛	30½	28¼	28¾	30¼	35½	33	X 23¾	24 X 27¼	....	25¼ X 27¼
30 OD	30	38¾	...	...	37½	34¾	35⅛	...	...	...	....	....	....	....
36 OD	36	46	...	...	44	41¼	41⅞	...	...	...	....	....	....	....
42 OD	42	53	...	...	50¾	48	48½	...	...	...	....	....	....	....
48 OD	48	59½	...	...	58¾	54½	55	...	...	...	....	....	....	....
54 OD	54	66¼	...	...	...	...	61¾	...	...	...	....	....	....	....
60 OD	60	73	...	...	...	...	68⅛	...	...	...	....	....	....	....
72 OD	72	86½	...	...	...	...	81⅜	...	...	...	....	....	....	....
84 OD	84	99¾	...	...	...	...	94¼	...	...	...	....	....	....	....
96 OD	96	113¼	...	...	...	...	107¼	...	...	...	....	....	....	....

[1]25-lb and 125-lb cast-iron flanges use the same size full-face gaskets, except for diameter of bolt hole punchings.

[2]In all cases where 125-lb cast-iron flanges are bolted to a 150-lb steel flange with raised face removed from the latter, either a full-face or flat-ring gasket may be used.

USE OF TABLE. The table has been arranged to facilitate the selection of a proper gasket for the flange material, rating, and type of facing used.

Example: To select a flat-ring gasket for 2-in. 250-lb cast-iron pipe flange, consult the upper portion of the table "PIPE FLANGE SIZES," and in col. 2 "Reference" locate the material "Cast Iron" and the pressure rating "250-lb," then follow across to locate the type of gasket "Flat Ring" and the proper size group "1-3" in col. 5. Reading down col. 5 to the lower portion of the table "GASKET DIAMETERS" and opposite the 2-in. nominal pipe size (col. 1) is the dimension of the outside diameter of the gasket, 4⅜ in. The inside diameter is given in col. 2 as 2⅜ in. The **correct** size gasket is therefore 2⅜ X 4⅜.
This example is illustrated by the heavy broken line in the table.

TABLE 15–20

Dimensions of Ring-Joint Gaskets for Steel Pipe Flanges

ASA B16.20-1952 Ring-Joint Gaskets and Grooves for Steel Pipe Flanges

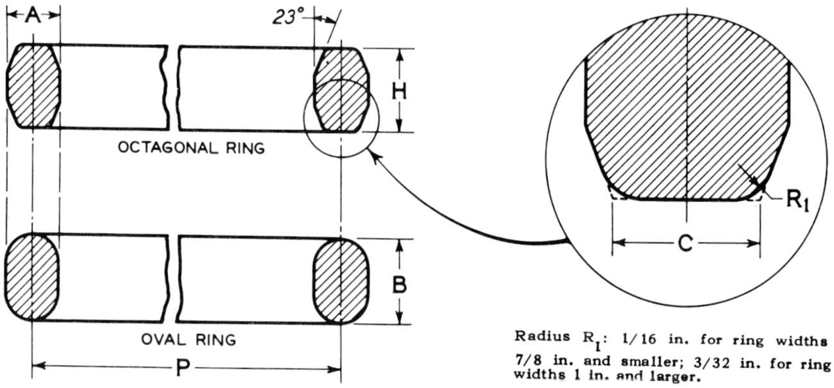

Radius R_1: 1/16 in. for ring widths 7/8 in. and smaller; 3/32 in. for ring widths 1 in. and larger.

Tolerances

P	(average pitch diameter of ring) . .	±0.007
A	(width of ring)	±0.008
*B and H	(height of ring).	±1/64
C	(width on flat of octagonal ring) . .	±0.008
23°	(angle)	±1/2 deg
R_1	(radius of ring)	±1/64

A small bead near the center of oval or octagonal shaped rings, located so that it will not enter the groove, is not objectionable.

* A plus tolerance of 3/64 in. for heights B and H is permitted providing the variation in the height of any given ring does not exceed 1/64 in. throughout its entire circumference.

1	2	3	4	5	6
Ring Number	Pitch Diameter of Ring P	Width of Ring A	Height of Ring		Width on Flat of Octagonal Ring C
			Oval B	Octagonal H	
R11	1 11/32	1/4	7/16	3/8	0.170
R12	1 9/16	5/16	9/16	1/2	0.206
R13	1 11/16	5/16	9/16	1/2	0.206
R14	1 3/4	5/16	9/16	1/2	0.206
R15	1 7/8	5/16	9/16	1/2	0.206
R16	2	5/16	9/16	1/2	0.206
R17	2 1/4	5/16	9/16	1/2	0.206
R18	2 3/8	5/16	9/16	1/2	0.206
R19	2 9/16	5/16	9/16	1/2	0.206
R20	2 11/16	5/16	9/16	1/2	0.206
R21	2 27/32	7/16	11/16	5/8	0.305
R22	3 1/4	5/16	9/16	1/2	0.206
R23	3 1/4	7/16	11/16	5/8	0.305
R24	3 3/4	7/16	11/16	5/8	0.305
R25	4	5/16	9/16	1/2	0.206

continued on next page

TABLE 15–20, continued

1	2	3	4	5	6
Ring Number	Pitch Diameter of Ring	Width of Ring	Height of Ring		Width on Flat of Octagonal Ring
			Oval	Octagonal	
	P	A	B	H	C
R26	4	7/16	11/16	5/8	0.305
R27	4 1/4	7/16	11/16	5/8	0.305
R28	4 3/8	1/2	3/4	11/16	0.341
R29	4 1/2	5/16	9/16	1/2	0.206
R30	4 5/8	7/16	11/16	5/8	0.305
R31	4 7/8	7/16	11/16	5/8	0.305
R32	5	1/2	3/4	11/16	0.341
R33	5 3/16	5/16	9/16	1/2	0.206
R34	5 3/16	7/16	11/16	5/8	0.305
R35	5 3/8	7/16	11/16	5/8	0.305
R36	5 7/8	5/16	9/16	1/2	0.206
R37	5 7/8	7/16	11/16	5/8	0.305
R38	6 3/16	5/8	7/8	13/16	0.413
R39	6 3/8	7/16	11/16	5/8	0.305
R40	6 3/4	5/16	9/16	1/2	0.206
R41	7 1/8	7/16	11/16	5/8	0.305
R42	7 1/2	3/4	1	15/16	0.485
R43	7 5/8	5/16	9/16	1/2	0.206
R44	7 5/8	7/16	11/16	5/8	0.305
R45	8 5/16	7/16	11/16	5/8	0.305
R46	8 5/16	1/2	3/4	11/16	0.341
R47	9	3/4	1	15/16	0.485
R48	9 3/4	5/16	9/16	1/2	0.206
R49	10 5/8	7/16	11/16	5/8	0.305
R50	10 5/8	5/8	7/8	13/16	0.413
R51	11	7/8	1 1/8	1 1/16	0.583
R52	12	5/16	9/16	1/2	0.206
R53	12 3/4	7/16	11/16	5/8	0.305
R54	12 3/4	5/8	7/8	13/16	0.413
R55	13 1/2	1 1/8	1 7/16	1 3/8	0.780
R56	15	5/16	9/16	1/2	0.206
R57	15	7/16	11/16	5/8	0.305
R58	15	7/8	1 1/8	1 1/16	0.583
R59	15 5/8	5/16	9/16	1/2	0.206
R60	16	1 1/4	1 9/16	1 1/2	0.879
R61	16 1/2	7/16	11/16	5/8	0.305
R62	16 1/2	5/8	7/8	13/16	0.413
R63	16 1/2	1	1 5/16	1 1/4	0.681
R64	17 7/8	5/16	9/16	1/2	0.206
R65	18 1/2	7/16	11/16	5/8	0.305
R66	18 1/2	5/8	7/8	13/16	0.413
R67	18 1/2	1 1/8	1 7/16	1 3/8	0.780
R68	20 3/8	5/16	9/16	1/2	0.206
R69	21	7/16	11/16	5/8	0.305
R70	21	3/4	1	15/16	0.485
R71	21	1 1/8	1 7/16	1 3/8	0.780
R72	22	5/16	9/16	1/2	0.206
R73	23	1/2	3/4	11/16	0.341
R74	23	3/4	1	15/16	0.485
R75	23	1 1/4	1 9/16	1 1/2	0.879
R76	26 1/2	5/16	9/16	1/2	0.206
R77	27 1/4	5/8	7/8	13/16	0.413
R78	27 1/4	1	1 5/16	1 1/4	0.681
R79	27 1/4	1 3/8	1 3/4	1 5/8	0.977
R80	24 1/4	5/16	....	1/2	0.206
R81	25	9/16	....	3/4	0.377
R82	2 1/4	7/16	....	5/8	0.305
R84	2 1/2	7/16	....	5/8	0.305
R85	3 1/8	1/2	....	11/16	0.341
R86	3 9/16	5/8	....	13/16	0.413
R87	3 15/16	5/8	....	13/16	0.413
R88	4 7/8	3/4	....	15/16	0.485
R89	4 1/2	3/4	....	15/16	0.485
R90	6 1/8	7/8	....	1 1/16	0.583
R91	10 1/4	1 1/4	....	1 1/2	0.879
R92	9	7/16	11/16	5/8	0.305
R93	29 1/2	3/4	....	15/16	0.485
R94	31 1/2	3/4	....	15/16	0.485
R95	33 3/4	3/4	....	15/16	0.485
R96	36	7/8	....	1 1/16	0.583
R97	38	7/8	....	1 1/16	0.583
R98	40 1/4	7/8	....	1 1/16	0.583

All dimensions given in inches.

TABLE 15–21

Dimensions of Ring-Joint Grooves in Steel Pipe Flanges

ASA B16.20–1952 Ring Joint Gaskets and Grooves for Steel Pipe Flanges

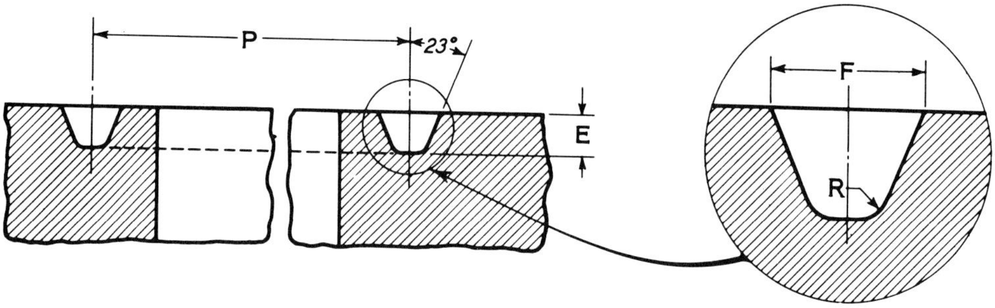

TOLERANCES

E (depth) $^{+1/64}_{-0}$
F (width) ±0.008
P (pitch diameter) ±0.005
R (radius at bottom) max
23° (angle) ±1/2°

1	2	3	4	5
Groove Number	Pitch Diameter P	Depth E	Width F	Radius at Bottom R
R11	1 11/32	7/32	9/32	1/32
R12	1 9/16	1/4	11/32	1/32
R13	1 11/16	1/4	11/32	1/32
R14	1 3/4	1/4	11/32	1/32
R15	1 7/8	1/4	11/32	1/32
R16	2	1/4	11/32	1/32
R17	2 1/4	1/4	11/32	1/32
R18	2 3/8	1/4	11/32	1/32
R19	2 9/16	1/4	11/32	1/32
R20	2 11/16	1/4	11/32	1/32
R21	2 27/32	5/16	15/32	1/32
R22	3 1/4	1/4	11/32	1/32
R23	3 1/4	5/16	15/32	1/32
R24	3 3/4	5/16	15/32	1/32
R25	4	1/4	11/32	1/32
R26	4	5/16	15/32	1/32
R27	4 1/4	5/16	15/32	1/32
R28	4 3/8	3/8	17/32	1/16
R29	4 1/2	1/4	11/32	1/32
R30	4 5/8	5/16	15/32	1/32
R31	4 7/8	5/16	15/32	1/32
R32	5	3/8	17/32	1/16
R33	5 3/16	1/4	11/32	1/32
R34	5 3/16	5/16	15/32	1/32
R35	5 3/8	5/16	15/32	1/32
R36	5 7/8	1/4	11/32	1/32
R37	5 7/8	5/16	15/32	1/32
R38	6 3/16	7/16	21/32	1/16
R39	6 3/8	5/16	15/32	1/32
R40	6 3/4	1/4	11/32	1/32
R41	7 1/8	5/16	15/32	1/32
R42	7 1/2	1/2	25/32	1/16
R43	7 5/8	1/4	11/32	1/32
R44	7 5/8	5/16	15/32	1/32
R45	8 5/16	5/16	15/32	1/32

continued on next page

TABLE 15–21, continued

1	2	3	4	5
Groove Number	Pitch Diameter P	Depth E	Width F	Radius at Bottom R
R46	8 5/16	3/8	17/32	1/16
R47	9	1/2	25/32	1/16
R48	9 3/4	1/4	11/32	1/32
R49	10 5/8	5/16	15/32	1/32
R50	10 5/8	7/16	21/32	1/16
R51	11	9/16	29/32	1/16
R52	12	1/4	11/32	1/32
R53	12 3/4	5/16	15/32	1/32
R54	12 3/4	7/16	21/32	1/16
R55	13 1/2	11/16	1 3/16	3/32
R56	15	1/4	11/32	1/32
R57	15	5/16	15/32	1/32
R58	15	9/16	29/32	1/16
R59	15 5/8	1/4	11/32	1/32
R60	16	11/16	1 5/16	3/32
R61	16 1/2	5/16	15/32	1/32
R62	16 1/2	7/16	21/32	1/16
R63	16 1/2	5/8	1 1/16	3/32
R64	17 7/8	1/4	11/32	1/32
R65	18 1/2	5/16	15/32	1/32
R66	18 1/2	7/16	21/32	1/16
R67	18 1/2	11/16	1 3/16	3/32
R68	20 3/8	1/4	11/32	1/32
R69	21	5/16	15/32	1/32
R70	21	1/2	25/32	1/16
R71	21	11/16	1 3/16	3/32
R72	22	1/4	11/32	1/32
R73	23	3/8	17/32	1/16
R74	23	1/2	25/32	1/16
R75	23	11/16	1 5/16	3/32
R76	26 1/2	1/4	11/32	1/32
R77	27 1/4	7/16	21/32	1/16
R78	27 1/4	5/8	1 1/16	3/32
R79	27 1/4	13/16	1 7/16	3/32
R80	24 1/4	1/4	11/32	1/32
R81	25	7/16	19/32	1/16
R82	2 1/4	5/16	15/32	1/32
R84	2 1/2	5/16	15/32	1/32
R85	3 1/8	3/8	17/32	1/16
R86	3 9/16	7/16	21/32	1/16
R87	3 15/16	7/16	21/32	1/16
R88	4 7/8	1/2	25/32	1/16
R89	4 1/2	1/2	25/32	1/16
R90	6 1/8	9/16	29/32	1/16
R91	10 1/4	11/16	1 15/16	3/32
R92	9	5/16	15/32	1/32
R93	29 1/2	1/2	25/32	1/16
R94	31 1/2	1/2	25/32	1/16
R95	33 3/4	1/2	25/32	1/16
R96	36	9/16	29/32	1/16
R97	38	9/16	29/32	1/16
R98	40 1/4	9/16	29/32	1/16

All dimensions given in inches.

TABLE 15–22

American Standard Graphical Symbols for Piping

ASA Z32.2.3–1949

Air Conditioning	
28 Brine Return	— — —BR— — —
29 Brine Supply	———— B ————
30 Circulation Chilled or Hot-Water Flow	———— CH ————
31 Circulation Chilled or Hot-Water Return	— — —CHR— — —
32 Condenser Water Flow	———— C ————
33 Condenser Water Return	— — —CR— — —
34 Drain	———— D ————
35 Humidification Line	——- ——H——- ——
36 Make-Up Water	——- ——- ——- ——
37 Refrigerant Discharge	———— RD ————
38 Refrigerant Liquid	———— RL ————
39 Refrigerant Suction	— — —RS— — —
Heating	
40 Air-Relief Line	— —— — —— —
41 Boiler Blow Off	—— —— —— ——
42 Compressed Air	———— A ————
43 Condensate or Vacuum Pump Discharge	—0— —0— —0—
44 Feedwater Pump Discharge	—00— —00— —00—
45 Fuel-Oil Flow	———— FOF ————
46 Fuel-Oil Return	— — —FOR— — —
47 Fuel-Oil Tank Vent	— — —FOV— — —
48 High-Pressure Return	— -//- — -//- — -//- —
49 High-Pressure Steam	——//——//——//——
50 Hot-Water Heating Return	— — — — — —
51 Hot-Water Heating Supply	————————

continued on next page

TABLE 15–22, continued

52 Low-Pressure Return	— — — — — — —
53 Low-Pressure Steam	————————————
54 Make-Up Water	——- ——— - ——— - ——
55 Medium Pressure Return	— +— — +— — +— —
56 Medium Pressure Steam	——/———/———/——
Plumbing	
57 Acid Waste	ACID
58 Cold Water	——- ——- ——- ———
59 Compressed Air	——————— A ———————
60 Drinking-Water Flow	——— - ————— - ———
61 Drinking-Water Return	——— - - ———— - - ———
62 Fire Line	—— F ————— F ——
63 Gas	—— G ————— G ——
64 Hot Water	——- - ——- - ——- - ——
65 Hot-Water Return	——- - - ——- - - ——- - - ——
66 Soil, Waste or Leader (Above Grade)	————————————
67 Soil, Waste or Leader (Below Grade)	— — — — — — —
68 Vacuum Cleaning	—— V ————— V ——
69 Vent	- - - - - - - - - - - - - -
Pneumatic Tubes	
70 Tube Runs	════════════
Sprinklers	
71 Branch and Head	——o—————o——
72 Drain	—— S — — — — S ——
73 Main Supplies	——————— S ———————

TABLE 15–23

American Standard Graphical Symbols for Pipe Fittings and Valves

ASA 232.2.3–1949

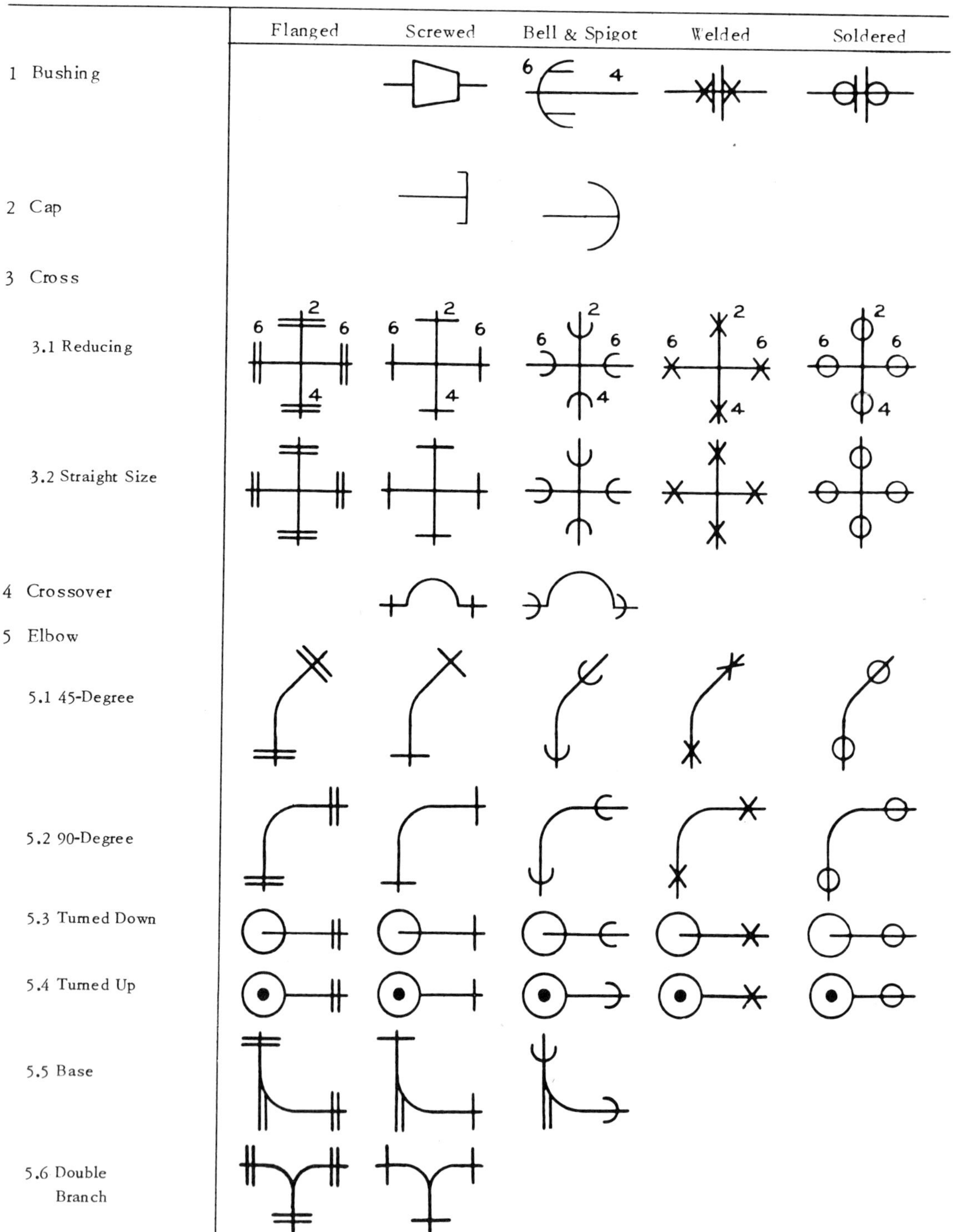

continued on next page

TABLE 15–23, continued

	Flanged / Screwed / Bell & Spigot / Welded / Soldered
5.7 Long Radius	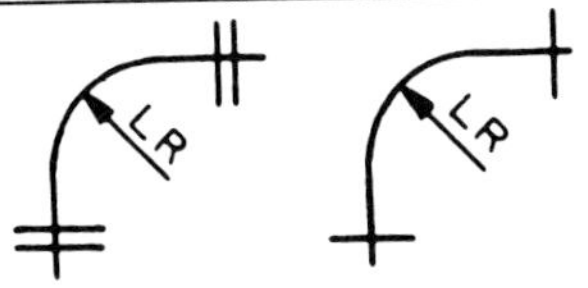
5.8 Reducing	
5.9 Side Outlet (Outlet Down)	
5.10 Side Outlet (Outlet Up)	
5.11 Street	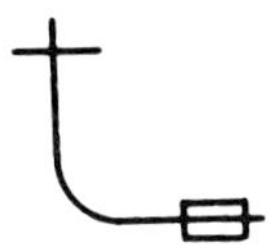
6 Joint	
6.1 Connecting Pipe	
6.2 Expansion	
7 Lateral	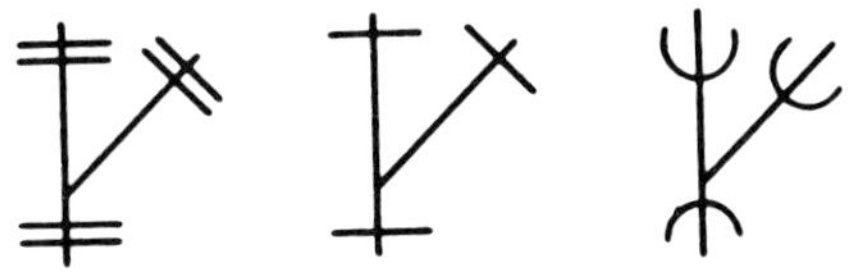
8 Orifice Flange	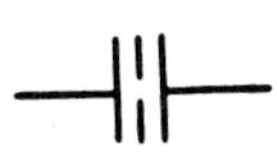
9 Reducing Flange	

continued on next page

TABLE 15–23, continued

	Flanged	Screwed	Bell & Spigot	Welded	Soldered
10 Plugs					
10.1 Bull Plug					
10.2 Pipe Plug					
11 Reducer					
11.1 Concentric					
11.2 Eccentric					
12 Sleeve					
13 Tee					
13.1 (Straight Size)					
13.2 (Outlet Up)					
13.3 (Outlet Down)					
13.4 (Double Sweep)					
13.5 Reducing					
13.6 (Single Sweep)					

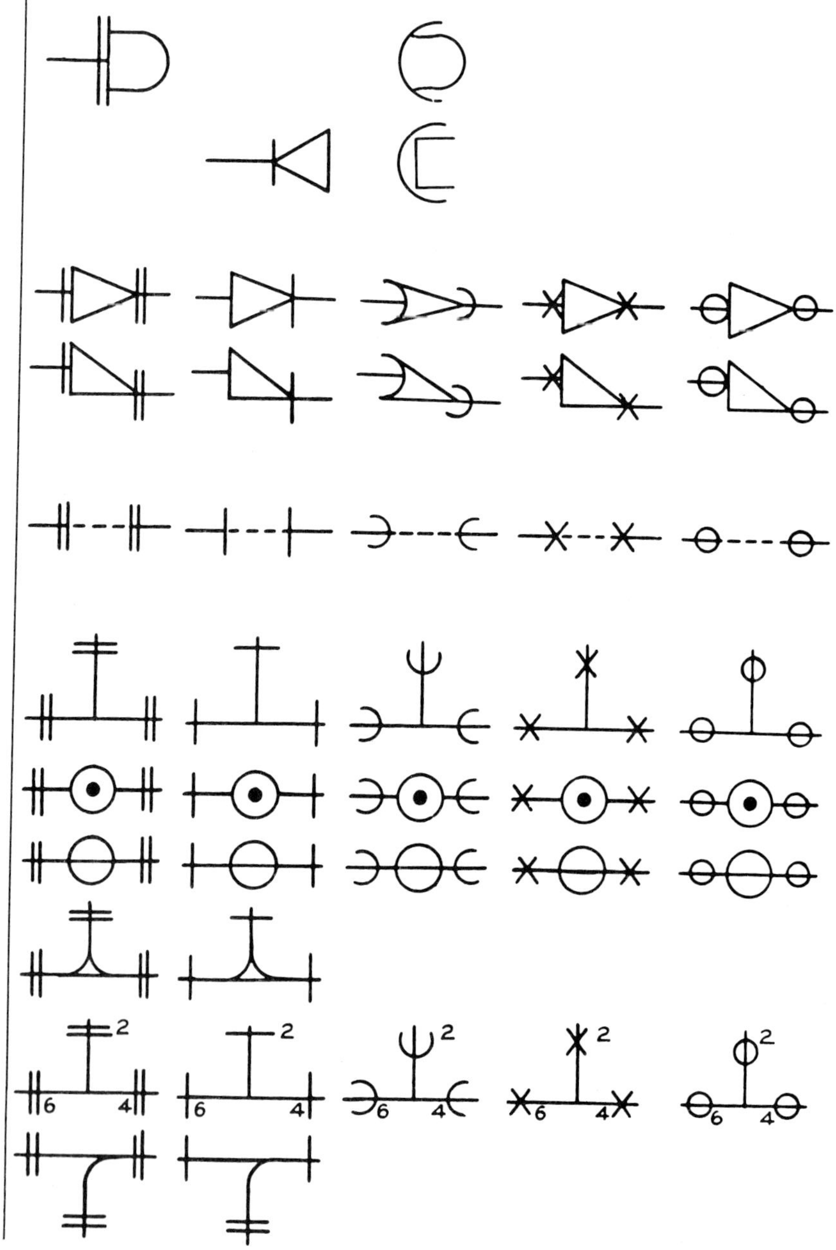

continued on next page

TABLE 15–23, continued

	Flanged	Screwed	Bell & Spigot	Welded	Soldered
13.7 Side Outlet (Outlet Down)	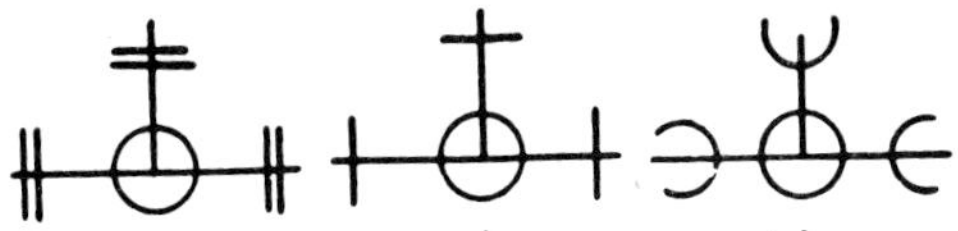				
13.8 Side Outlet (Outlet Up)	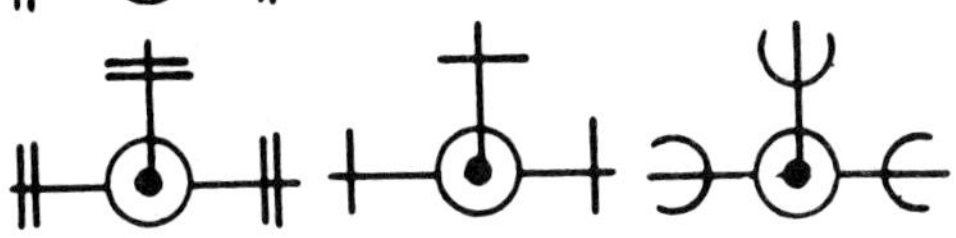				
14 Union					
15 Angle Valve					
15.1 Check					
15.2 Gate (Elevation)	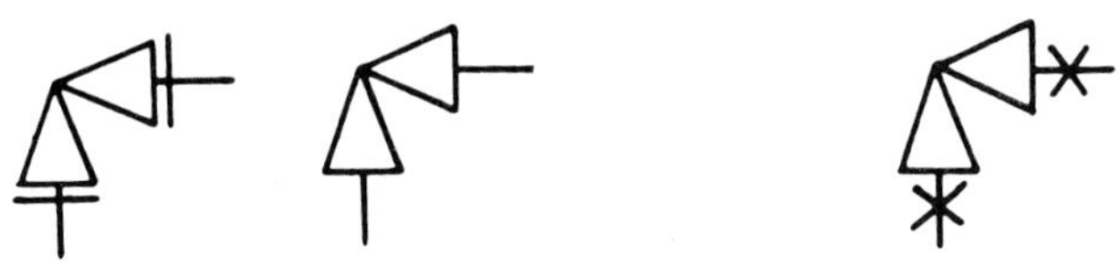				
15.3 Gate (Plan)					
15.4 Globe (Elevation)					
15.5 Globe (Plan)					
15.6 Hose Angle	Same As	Symbol	23.1		
16 Automatic Valve					
16.1 By-Pass					

continued on next page

TABLE 15-23, continued

	Flanged	Screwed	Bell & Spigot	Welded	Soldered
16.2 Governor-Operated					
16.3 Reducing					
17 Check Valve					
17.1 Angle Check	Same As	Symbol	15.1		
17.2 (Straight Way)					
18 Cock					
19 Diaphragm Valve	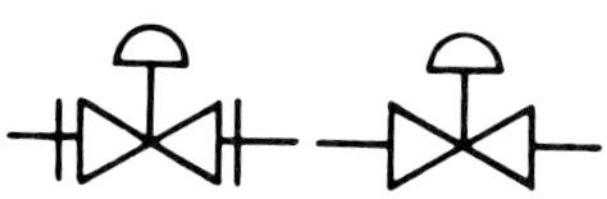				
20 Float Valve					
21 Gate Valve					
*21.1	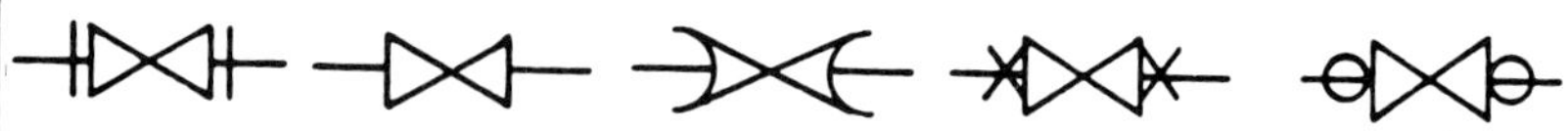				
21.2 Angle Gate	Same As	Symbol	15.2	15.3	
21.3 Hose Gate	Same As	Symbol	23.2		

*Also used for general STOP VALVE symbol when amplified by specification.

continued on next page

TABLE 15–23, continued

	Flanged	Screwed	Bell & Spigot	Welded	Soldered
21.4 Motor-Operated	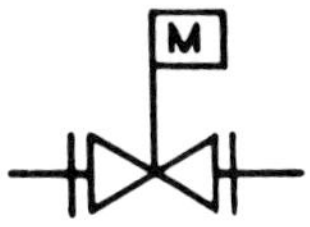	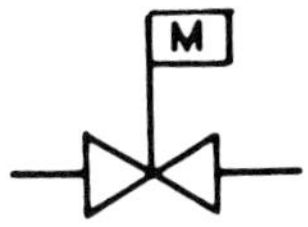		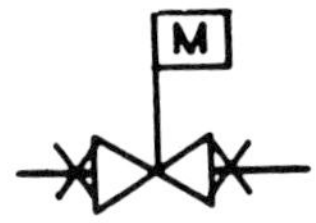	
22 Globe Valve					
22.1					
22.2 Angle Globe	Same As	Symbols	15.4 & 15.5		
22.3 Hose Globe	Same As	Symbol	23.3		
22.4 Motor-Operated	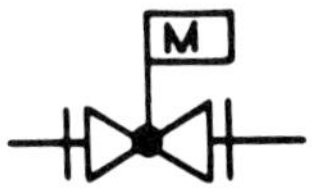	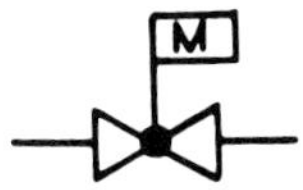		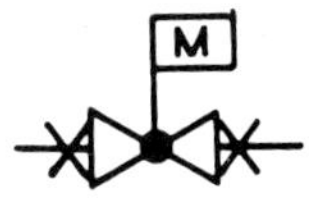	
23 Hose Valve					
23.1 Angle	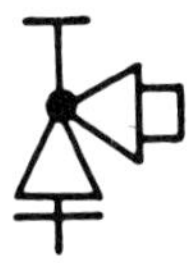				
23.2 Gate					
23.3 Globe					
24 Lockshield Valve		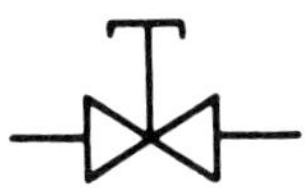			
25 Quick Opening Valve		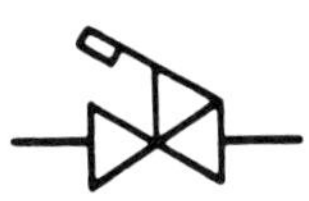			
26 Safety Valve					
27 Stop Valve	Same As	Symbol	21.1		

TABLE 15–24

Joint Industry Conference (JIC) Hydraulic Standards for Industrial Equipment

Joint Industry Conference Standards

JIC Standards are published by General Motors Corporation for distribution to Divisions of the Corporation and to manufacturers who are suppliers to those Divisions. Credit for the preparation of the JIC Hydraulic Standards for Industrial Equipment, one of several JIC Standards, is attributed to

Hydraulic Equipment Manufacturers
Hydraulic Press Manufacturers
Industrial Equipment Users
National Machine Tool Builders' Association
Resistance Welder Manufacturers Association
Tubing and Fitting Manufacturers
Packing and Seals Manufacturers

Copies of these Standards can be procured from the General Motors Production Engineering Section, General Motors Building, Detroit 2, Michigan.

Nine of the tables of this Section have been taken from the JIC Hydraulic Standards, including the next two tables and the remainder of this one, which apply to description and diagrams as follows:

"1. All hydraulic equipment shall be identified. When possible, name of component, catalog number and manufacturer's name shall be shown.
2. Size of piping (outside diameter and wall thickness).
3. Diameters of pistons and rods, length of stroke, and estimated required force of cylinders when other than maximum pressure is applied.
4. Time of cycle, when pertinent (for example, time range of cycle exclusive of loading).
5. Operating pressures.
6. Horsepower, rpm and direction of rotation of each pump drive.
7. Pump speed and delivery in gpm.
8. Reservoir capacity.
9. Recommended oil viscosity range.
10. Displacement, speed range, and torque rating of each hydraulic motor.
11. Data or text, or both, shall show operations performed with related electrical and mechanical control and actuating equipment."

TABLE 15–25

JIC Graphical Symbols for Hydraulic Devices

Joint Industry Conference Standards

LINES	
LINE, WORKING	
LINE, PILOT (L > 20W)	
LINE, DRAIN (L < 5W)	
LINE, FLEXIBLE	
CONNECTOR (Dot to be 3 × Width of Associated Line)	
DIRECTION OF FLOW	
LINE, PASSING	
LINE, JOINING (Tee Cross Etc) (Dot to be 3 × W)	
RESERVOIR (Fluid Tank)	
LINE, TO RESERVOIR ABOVE FLUID LEVEL	
BELOW FLUID LEVEL	
MANIFOLD, VENTED	
PLUG OR PLUGGED CONNECTION	X
TESTING STATION (Gage Connection)	T S
POWER TAKE-OFF	P T O
RESTRICTION, CHOKE, FIXED VISCOUS	
RESTRICTION, ORIFICE, FIXED NON-VISCOUS	

PUMPS	
PUMP, SINGLE FIXED DISPLACEMENT	PF
PUMP, SINGLE VARIABLE DISPLACEMENT	PV

MOTORS AND CYLINDERS	
MOTOR, ROTARY FIXED DISPLACEMENT	MF
MOTOR, ROTARY VARIABLE DISPLACEMENT	MV
MOTOR, OSCILLATING	MO
CYLINDER, SINGLE ACTING PLUNGER TYPE	
PISTON TYPE	
CYLINDER, DOUBLE ACTING SINGLE END ROD	
DOUBLE END ROD	

MISCELLANEOUS UNITS	
MOTOR, DRIVE, ELECTRIC	ELEC MTR
HEAT EXCHANGER	HE
INTENSIFIER	INT
ACCUMULATOR	ACC
FILTER	FILT
STRAINER	STRN
PRESSURE SWITCH	PR SW
PRESSURE GAGE	
SPRING	
SHAFT ROTATING (Arrow in Front of Shaft.)	
COMPONENT ENCLOSURE	

continued on next page

VALVE	
VALVE, CHECK	
VALVE, RESTRICTION, CHOKE VARIABLE VISCOUS	
VALVE, RESTRICTION, ORIFICE VARIABLE NON-VISCOUS	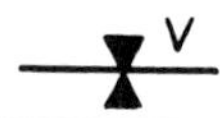
VALVE, BASIC SYMBOL (Insert Model No. for Special Valves	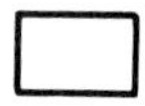
Method of indicating internal flow.	
VALVE EXAMPLES	
VALVE, MANUAL SHUT OFF	
VALVE, MAXIMUM PRESSURE	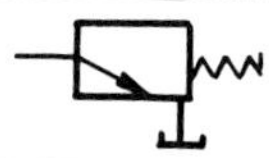
VALVE, RELIEF REMOTELY OPERATED	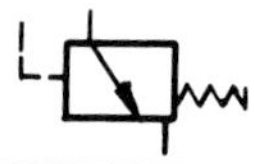
VALVE, SEQUENCE DIRECTLY OPERATED	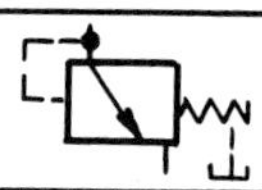
VALVE, PRESSURE REDUCING	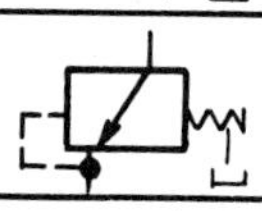
VALVE, FLOW CONTROL PRESSURE COMPENSATED VISCOUS	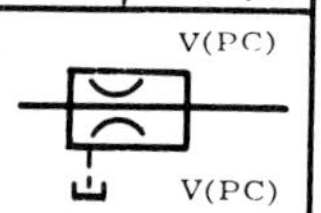
NON-VISCOUS	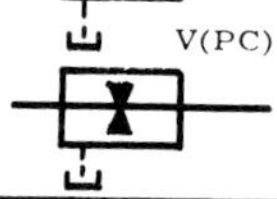
VALVE, SHUT OFF 2 POSITION – 2 CONNECTION	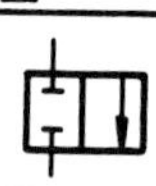
VALVE, DIRECTIONAL 2 POSITION – 4 CONNECTION	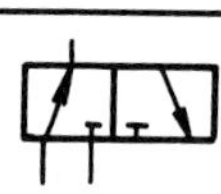
VALVE, DIRECTIONAL 2 POSITION – 3 CONNECTION	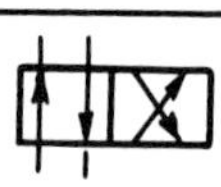
VALVE, DIRECTIONAL 3 POSITION – 4 CONNECTION OPEN CENTER	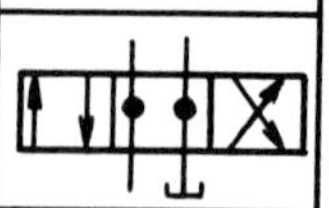
VALVE, DIRECTIONAL 3 POSITION – 4 CONNECTION CLOSED CENTER	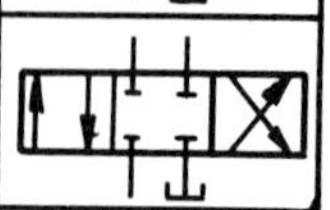

METHODS OF OPERATION	
CONTROL, BASIC SYMBOL	
CONTROL, CENTRIFUGAL	CENT
CONTROL, COMPENSATOR	COMP
CONTROL, COMPENSATOR PRESSURE	PR COMP
CONTROL, COMPENSATOR TEMPERATURE	TEMP COMP
CONTROL, CYLINDER	CYL
CONTROL, DETENT	DET
CONTROL, MANUAL	MAN
CONTROL, MECHANICAL	MECH
CONTROL, MOTOR ELECTRIC	ELEC MTR
CONTROL, MOTOR HYDRAULIC	HYD MTR
CONTROL, PILOT HYDRAULIC	HYD
CONTROL, PILOT AIR	AIR
CONTROL, SERVO	SERVO
CONTROL, SOLENOID	SOL
CONTROL, SOLENOID HYD. PILOT OPERATED	SOL HYD
CONTROL, THERMAL	THERM
CONTROL, PILOT HYD. DIFFERENTIAL AREA	HYD DIFF S L

TABLE 15–26

Diagrams Illustrating Combinations of JIC Graphical Symbols

Joint Industry Conference Standards

EXAMPLES OF COMBINATIONS

PUMP, DOUBLE-WITH ELECTRIC MOTOR
ONE FIXED DISPLACEMENT
ONE VARIABLE DISPLACEMENT
WITH PRESSURE COMPENSATOR

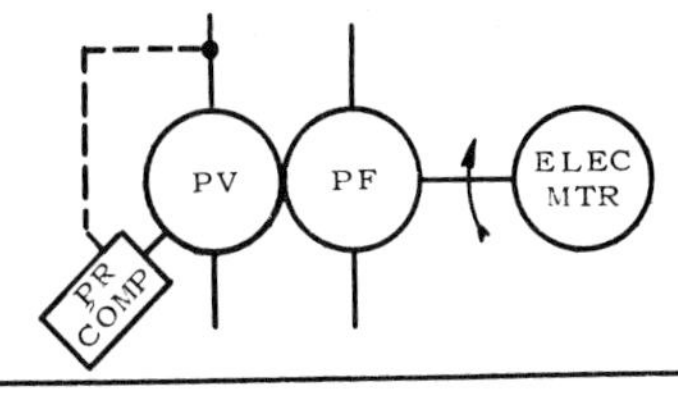

PUMP, SINGLE-WITH ELECTRIC MOTOR
VARIABLE DISPLACEMENT
HAND WHEEL & CYLINDER CONTROL

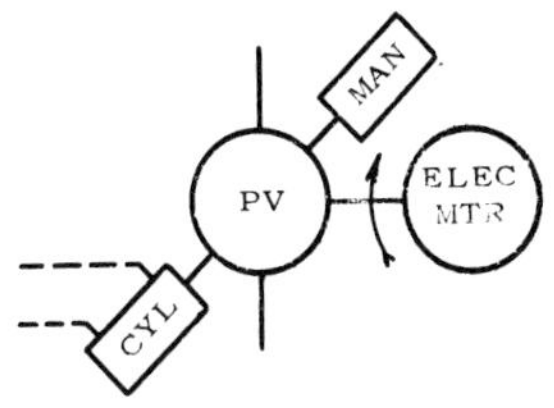

VALVE, FLOW CONTROL& MAXIMUM
PRESSURE WITH COMPENSATOR

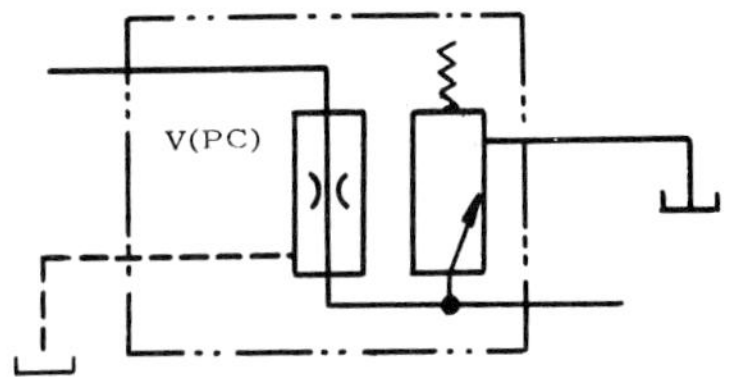

VALVE, REPLENISHING UNIT

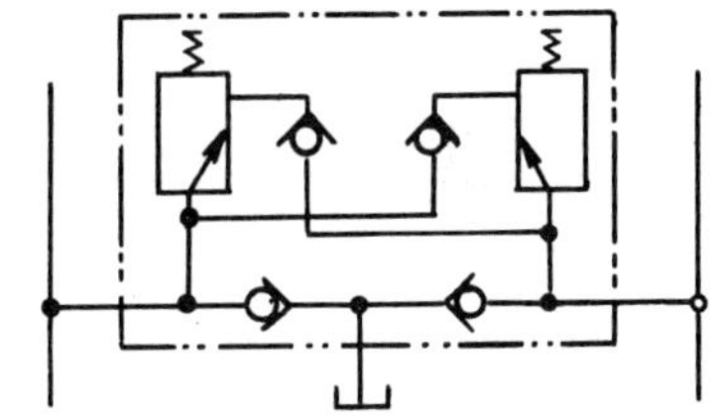

VALVE, 4 WAY
THREE POSITION-SPRING CENTERED
MANUAL CONTROL
P → T, CYL. PORT BLOCKED IN
CENTER POSITION
(Note: Symbol Shown in Center Position)

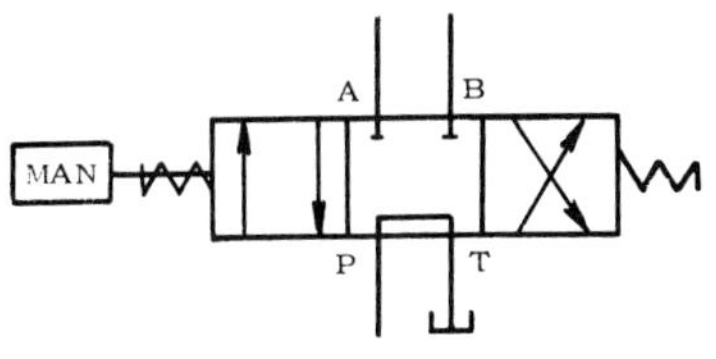

VALVE, 4 WAY
2 POSITION-SPRING OFFSET
SOLENOID CONTROL
PILOT OPERATED

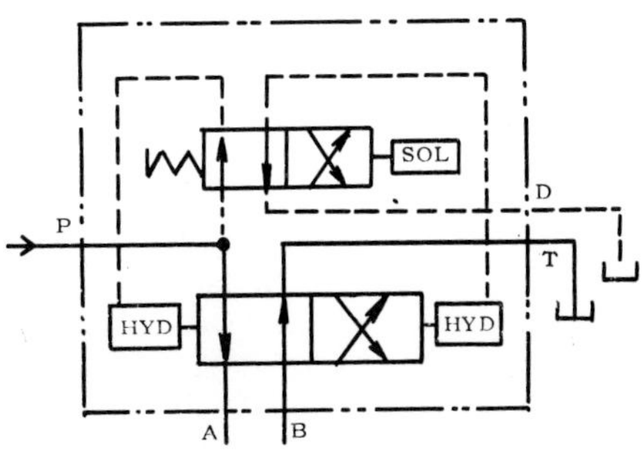

TABLE 15–27

Packing Standards Prove Popular

"Last September (1953) in these columns (The Houghton Line) we covered the progress made towards standardization of packing sizes.

"To most plant men who have grown up in their jobs in an age of standards for almost every conceivable commodity, it may seem strange that the hydraulic-pneumatic packing industry is just awakening to the advantages to be obtained from such a move.

"Probably the Joint Industry Conference thought so, too, when it spurred manufacturers on to concerted action. We thought it was going to be more of a job to get our thousands of packing customers, small and large, to change over to standard depths and thicknesses, than it was to sell our fellow manufacturers and our trade association on the idea. But it turned out that we were wrong; the customers caught on immediately and agreed to changes as called to their attention.

"During the third quarter of this year over 50% of all cup packings we made conformed to the standards as set up. We have passed the half-way mark, and soon expect that 75% of our cup packing business will be according to standards. To the customer it means quicker deliveries from stock rapidly being built up on all popular sizes. You can see the delay occasioned by making each order a la carte, so to speak. Also, as pointed out in the September Line, plant inventories of packings can be reduced; dash numbers indicate the size, regardless of the installation.

"Cup packings are made with a small bolt hole, and can be punched by the user to obtain the required hole size.

"So our packing department is quite happy about the cooperation we're getting from buyers, and wants us to tell you so."

Editors Note:

The foregoing is quoted as indicative of the kind of progress that can be made toward standardization in a relatively short time (six or seven years). Isn't it logical to expect that similar progress might be made toward a standardization of oil seals, packing expanders and other hydraulic devices. Oil seals, in particular, come in a great multitude of sizes and styles. In fact, the sizes and variety are so great that specific sizes are omitted altogether from these tables.

TABLE 15–28

Chart Showing the Resistance of the Standard Synthetic Rubber Members of Garlock Oil Seals

Catalog 10, Garlock Klozures — The Garlock Packing Company

		Fluid	Resistance
WATER (Neutral, alkaline, or weak acid solutions)	(a)	Salt solutions and brines (example: soda, sodium chloride)	Good
	(b)	Organic acid (example: acetic acid)	Good
	(c)	Alkalies (examples: ammonia, sodium hydroxide, lye, potash)	Good
	(d)	Slurries	Good
	(e)	Miscellaneous (example: soap)	Good
FOOD PRODUCTS	(a)	Aqueous (example: vinegar)	Good
	(b)	Oils and emulsions (example: salad dressing)	Good
MINERAL ACIDS (with or without dissolved salts or suspended matter)	(a)	Hydrochloric	Good
	(b)	Hydrofluoric (aqueous)	Good
	(c)	Hydrofluoric (anhydrous)	*Destroyed
	(d)	Phosphoric	Good
	(e)	Sulfuric	*Destroyed
	(f)	Nitric (including mixtures with sulfuric)	*Destroyed
OILS AND SOLVENTS	(a)	Alcohol, acetone, and glycerine	Good
	(b)	Aliphatic aldehydes (example: formaldehyde)	Good
	(c)	Aliphatic amines	Good
	(d)	Aliphatic nitro compounds	Unsatisfactory
	(e)	Silicone oils	Good
	(f)	Aliphatic ethers	Good
	(g)	Animal or vegetable oils	Good
	(h)	Aliphatic hydrocarbons (example: butane)	Good
	(i)	Freon (F-11, F-12, and F-122)	Good
	(j)	Phenols	*Unsatisfactory
	(k)	Aromatic aldehydes	*Unsatisfactory
	(l)	Aromatic ethers	*Unsatisfactory
	(m)	Esters (example: methylchloride)	*Unsatisfactory
	(n)	Aromatic or terpene hydrocarbons (examples: benzol, turpentine, toluol, xylol)	*Unsatisfactory
	(o)	All Ketones	*Unsatisfactory
	(p)	Aromatic amines	*Unsatisfactory
	(q)	Aromatic nitro compounds	Unsatisfactory
	(r)	Chlorinated hydrocarbons (example: trichlorethylene)	*Unsatisfactory
	(s)	Freon (F-21 and F-22)	*Unsatisfactory
	(t)	Carbon bisulfide	*Unsatisfactory
DRY GASES		(Examples: acetylene, carbon monoxide, steam, and sulfur dioxide)	Good

*Oil seals with Teflon sealing members will withstand these acids and solvents.

NOTE: The metal parts of oil seals are available in a choice of different metals. The kind of metal chosen for the manufacture of such parts, therefore, determines the degree to which they resist corrosion.

TABLE 15–29

Notes on Closures for Use with Timken Roller Bearings

Timken Engineering Journal — Timken Roller Bearing Co.

"*General* In selecting the proper closure design for any Timken bearing application it should be kept in mind that the consistency of the lubricant, foreign material to be excluded, speed of the application and general operating conditions are of unusual importance. Foreign material such as dust, mill scale or any hard, gritty substance will act as a lapping agent and quickly destroy the bearing. Likewise, water, acid or inferior lubricant will etch the highly finished surfaces and will also bring about early failure.

"*Felt Closures* Felt can be used with some success with grease lubrication and at low surface rubbing velocities. The effectiveness of felt closures depends upon the quality of the felt used and the degree of surface smoothness. All closure rubbing surfaces should have a high polish finish. An endless or carefully sewed wool type felt will give best results.

"When high speeds are encountered a harder felt should be used than at lower speeds.

"Under dirty conditions felt closures should be protected from the dirt as much as possible with a good machined dirt shield.

"*Metal Stampings* Metal stamping closures may be used effectively with low speed, clean to dirty conditions, in combination with other closure elements. These parts may be steel stampings manufactured within a tolerance of 0.005 in. All stampings should be designed so as to provide a clearance of 0.020 in. to 0.025 in. on diameter between rotating and stationary parts. A minimum endwise clearance of 1/16 in. should be provided between adjacent rotating and stationary parts.

"*Machined Parts* Machined parts used with other closure elements are recommended in place of stampings where it is desirable to maintain more accurate closure clearances. This results in a greater closure efficiency, either in retaining oil in the inside of the bearing housing or by keeping dirt and foreign matter out.

"*Annular Grooves* Annular groove closures are used in place of felt closures where considerable grit and dust are common. These grooved closures are very effective when used with either oil or grease at all speeds. They are especially effective when used in connection with internal or external flingers, depending upon the requirements of the application. The closer the running clearance the greater the effectiveness of the closure. On shafts up to 2 in. diameter a running clearance of 0.010 in. to 0.015 in. on diameter is recommended. On larger diameter shafts these clearances may be increased to as much as 0.040 in. to 0.050 in. due to operating speeds and temperatures.

"The closure usually has a number of grooves cut in the bore or on the outside diameter depending on the design. When used with oil these grooves tend to break the capillary action. In the case of grease the grooves pack hard with the grease and make a tight closure. The grooves are usually cut with a round nose tool and with the sides of the tool ground to an included angle of 30 degrees. The width of the groove at the widest part should be about 1/8 in. to 3/16 in. and the land between the grooves should be about one half the width of the groove. The depth should be about 5/32 in. to 3/16 in. Not less than three grooves are recommended. A greater number will make the closure more effective.

"*Commercial Seals* Various types of commercial seals are available which have been developed for a variety of uses. Most of these seals are of the rubbing type and have their rubbing elements made of felts, leather or various compositions.

"These are usually supplied as complete assemblies retained by metal stampings. The rubbing elements are invariably backed by some type of spring or resilient material to provide an automatic take-up for axial or radial movements by the shaft.

"The rubbing elements are usually carefully selected materials properly processed to provide long seal life with maximum efficiency.

"Under dirty conditions these seals should be protected from foreign matter by means of external flingers, shrouds or other types of dirt seals to avoid rapid seal and shaft wear.

"Being a rubbing type seal, its use becomes somewhat restricted at high shaft speeds where high rubbing velocities are produced. Under such conditions proper recommendations for use should be obtained from the seal manufacturer.

"Under dirty conditions the seal should be mounted so that the lips of the rubbing member point outward to permit fresh lubricant to pass through the seal thus keeping the dirt from lodging under the sealing element.

"Under such conditions where the major problem is lubricant retention the seal should be mounted with the sealing element lip pointed in toward the bearing chamber.

"Under conditions where both lubricant retention and foreign matter exclusion are involved, the use of two seals with lips turned away from each other is recommended. In such cases it is advisable to supply lubricant to the space between the closures."

BIBLIOGRAPHY

Albert, C. D., *Machine Design Drawing Room Problems*, 4th ed, John Wiley & Sons, New York

Albert and Rogers, *Kinematics of Machinery*, John Wiley & Sons, 1931

American Gear Manufacturers Association (AGMA),

112.02 – 1950	*Gear Nomenclature*
206.03 – 1950	*Fine-Pitch Straight Bevel Gears*
207.03 – 1950	*Fine-Pitch System for Spur and Helical Gears* (20 Degree Involute)
210.01 – 1946	*Surface Durability of Spur Gears*
211.01 – 1944	*Surface Durability of Helical and Herringbone Gears*
213.02 – 1952	*Surface Durability of Cylindrical Worm Gearing*
220.01 – 1946	*Strength of Spur Gear Teeth*
221.01 – 1948	*Strength of Helical and Herringbone Gear Teeth*
236.03 – 1951	*Inspection of Fine-Pitch Gears*
243.01 – 1954	*Cast Bronze Gear Blanks*
261.01 – 1946	*Keyways for Holes in Gears for General Industrial Practice*
321.03 – 1951	*Helical and Herringbone Mill Gears*
374.02 – 1950	*Design for Fine-Pitch Worm Gearing*
420.02 – 1947	*Helical and Herringbone Gear Speed Reducers*
421.02 – 1947	*High Speed Helical and Herringbone Gear Units*
440.02 – 1952	*Standard Practice for Cylindrical Worm Gear Speed Reducers*

American Iron & Steel Institute (AISI), *Steel Products Manual*,

Section 8 – August, 1952	*Hot Rolled Carbon Steel Bars*
9 – September, 1952	*Cold Finished Carbon Steel Bars*
10 – May, 1949	*Hot Rolled Alloy Steels*
18 – December, 1951	*Steel Tubular Products*
25 – April, 1949	*Tool Steel Tolerances*
29 – May, 1949	*Cold Finished Alloy Steel Bars*

American Machinist, July 4 and 11, 1929, Allen H. Candee, "Gear Geometry" McGraw-Hill Publishing Co , Inc , New York

American Society of Mechanical Engineers (ASME),
Mechanical Catalog & Directory, 1953, New York
Paper No. 48-SA-21. J. B. Armitage, "Straight and Sided Splines"
Transactions, Vol. 56, 1934, H. A. S. Howarth, "Current Practice in Pressures, Speeds, Clearances, and Lubrication of Oil-Film Bearings"
Vol. 68, 1946, Robertson & Yorgiadis, "Internal Friction in Engineering Materials"

American Society for Testing Materials (ASTM)

A 107–52aT	*Specifications for Hot-Rolled Carbon-Steel Bars*
A 108–52 T	*Specifications for Cold-Finished Carbon-Steel Bars and Shafting*
A 331–50 T	*Specifications for Cold-Finished Alloy-Steel Bars*
B 16–52	*Specifications for Free-Cutting Brass Rod, Bar and Shapes for Use in Screw Machines*
B 21–52	*Specifications for Naval Brass Rod, Bar, and Shapes*
B 42–52	*Specifications for Seamless Copper Pipe, Standard Sizes*
B 43–52	*Specifications for Seamless Red Brass Pipe, Standard Sizes*
B 68–54	*Specifications for Copper Tube, Seamless, Bright Annealed*
B 75–52	*Specifications for Seamless Copper Tube*
B 88–51	*Specifications for Seamless Copper Water Tube*
B 135–52	*Specifications for Seamless Brass Tube*
B 161–49 T	*Specifications for Nickel Seamless Pipe and Tubing*

American Society for Testing Materials (ASTM)

B 163–49 T *Specifications for Seamless Nickel and High Nickel Alloy Condenser, Evaporator and Heat Exchanger Tubes*
B 165–49 T *Specifications for Nickel-Copper Alloy Seamless Pipe and Tubing*
B 167–49 T *Specifications for Nickel-Chromium-Iron Alloy Seamless Pipe and Tubing*
B 188–52 *Specifications for Seamless Copper Bus Pipe and Tube*
B 210–53 T *Specifications for Aluminum-Alloy Drawn Seamless Tubes*
B 217–53 T *Specifications for Magnesium-Base Alloy Extruded Round Tubes*
B 235–53 T *Specifications for Aluminum-Alloy Extruded Tubes*
B 241–53 T *Specifications for Aluminum-Alloy Pipe*
B 249–53 T *Specifications for General Requirements for Wrought Copper and Copper Alloy Rod, Bar and Shapes*
B 251–54 T *Specifications for General Requirements for Wrought Seamless Copper and Copper-Alloy Pipe and Tube*

American Standards Association (ASA)

B 1.1–1949 *Unified and American Screw Threads*, 3rd edition
B 2.1–1945 *Pipe Threads*
B 4 a–1925 *Tolerances, Allowances and Gages for Metal Fits*
B 4.1–1947 *Limits and Fits for Engineering and Manufacturing* (Part I)
B 5.1–1949 *T-Slots—Their Bolts, Nuts, Tongues and Cutters*
B 5.3–1950 *Milling Cutters*
B 5.4–1948 *Taps, Cut and Ground Threads*
B 5.10–1953 *Machine Tapers, Self-Holding and Steep Taper Series*
B 5.12–1950 *Twist Drills, Straight Shank and Taper Shank*
B 5.14–1949 *Reamers*
B 5.20–1947 *Machine Pins*
B 6.1–1932 *Spur Gear Tooth Form*
B 6.6–1946 *Gear Tolerances and Inspection*
B 6.7–1950 *20-Deg Involute Fine-Pitch System for Spur and Helical Gears*
B 6.8–1950 *Fine Pitch Straight Bevel Gears*
B 6.9–1950 *Design for Fine-Pitch Worm Gearings*
B 6.10–1950 *Gear Nomenclature, Terms, Definitions and Illustrations*
B 6.11–1951 *Inspection of Fine-Pitch Gears*
B 16.4–1949 *Cast-Iron Screwed Fittings, 125 and 250 lb*
B 16.9–1951 *Steel Butt-Welding Fittings*
B 16.11–1946 *Steel Socket-Welding Fittings*
B 16.14–1949 *Ferrous Plugs, Bushings and Locknuts with Pipe Threads*
B 16.15–1947 *Brass or Bronze Screwed Fittings, 125 lb*
B 16.17–1949 *Brass or Bronze Screwed Fittings, 250 lb*
B 16.19–1951 *Malleable-Iron Screwed Fittings, 300 lb*
B 16.20–1952 *Ring-Joint Gaskets and Grooves for Steel Pipe Flanges*
B 16.21–1951 *Nonmetallic Gaskets for Pipe Flanges*
B 17.1–1943 *Shafting and Stock Keys*
B 18.2–1952 *Square and Hexagon Bolts and Nuts*
B 27.1–1950 *Lock Washers*
B 27.2–1953 *Plain Washers*
B 31.1a–1953 *Code for Pressure Piping*
B 36.10–1950 *Wrought-Steel and Wrought-Iron Pipe*
B 36.19–1952 *Stainless Steel Pipe*
B 46.1–1947 *Surface Roughness, Waviness and Lay*
Z 32.2.1–1949 *Graphical Symbols for Welding*
Z 32.2.3–1949 *American Standard Graphical Symbols for Piping*
Z 32.2.6–1950 *Graphical Symbols for Heat-Power Apparatus*
2 32.2.3–1949 *American Standard Graphical Symbols for Pipe Fittings and Valves*

BIBLIOGRAPHY *(Continued)*

American Welding Society, *Welding Handbook,* 3rd ed, New York, 1950

Anti-Friction Bearings Manufacturer's Association (AFBMA)

Standards

Section 2 – November, 1952	*Boundary Dimensions for Ball and Roller Bearings*
Section 3 – June, 1951	*Bearing Tolerances*
Section 5 – April, 1953	*Bearing Identification Code for Anti-Friction Ball and Roller Bearings*
Section 7 – March, 1951	*Bearing Mounting for Ball and Roller Bearings*
Section 8 – June, 1953	*Bearing Mounting Accessories*

Standard Cam Followers

Associated Spring Corp, *Handbook of Mechanical Spring Design*

Barber-Colman Co, Practice of

Boston Gear Works, Practice of

Brown & Sharpe Manufacturing Co, *Formulas in Gearing,* 17th ed, 1950
Small Tools Catalog No. 35

Buckingham, Earle, *Manual of Gear Design,* 1935, Industrial Press
Analytical Mechanics of Gears, 1st ed, 1949, McGraw-Hill Book Co

Caterpillar Tractor Co, Practice of, 1954

Cleveland Graphite Bronze Co, (Div. of Clevite Corp) *General Catalog*

Cleveland Twist Drill Co, *Catalog No. 46*

Cleveland Worm and Gear Co, *Catalog 400* 1st ed, 1952
Catalog 200

Colonial Broach Co, *Catalog and Manual CB-46*

Den Hartog, J. P., *Mechanical Vibrations,* 3rd ed, 1947, McGraw-Hill Book Co

Dodge Manufacturing Corp, *Catalog D55*

Eaton Manufacturing Co, Reliance Division, *Catalog 4k/3*

Elastic Stop Nut Corp of America, *Catalog No. 800,* 1952

Falk Corp, Data from the

Farrel-Birmingham Co, Inc, Practice of

Fellows Gear Shaper Co, *The Internal Gear*
The Involute Curve and Involute Gearing, 1950

G and K International, *Catalog and Manual 201*

Garlock Packing Co, *Catalog 10, Garlock Klozures*

General Motors Sales Corp, New Departure Division, *New Departures Handbook,* Volume II, Bristol, Connecticut, 1947

General Motors Corp, *General Motors Standards, Equipment and Operations, Section R*

Gleason Works

20° Straight Bevel Gear System – 1954
Bevel and Hypoid Gear Design
Spiral Bevel Gear System, 1952
Zerol Bevel Gear System, 1954

E. F. Houghton and Co, *Packing Standards*

D. O. James Manufacturing Co, *Catalog No. 1000*

Johnson Bronze Co, *Bearing Design Data*

Link-Belt Co, *General Catalog 900*

Lipson, Noll and Clark, *Stress and Strength of Manufactured Parts,* 1st ed, 1950, McGraw-Hill Book Co

Lord Manufacturing Co, *Bulletin No. 103*

Metal Cutting Tool Institute, *Metal Cutting Tool Handbook,* 1949, New York

National Electrical Manufacturer's Association, (NEMA), *Standard Dimensions,* Part 3, July 1953
Publication MG2–1951

Plastics and Rubber Products Co, *Handbook of O-Ring Hydraulic Packings*
Parco Engineering Handbook

Product Engineering, March, 1948 "Motor Driven Machines," Special Section, McGraw-Hill Publishing Co

Roark, R. J., *Formulas for Stress and Strain,* 3rd ed, 1954 McGraw-Hill Book Co

Society of Automotive Engineers, (SAE), *Handbook,* 1954
Journal, March, 1945

Timken Roller Bearing Co, *The Timken Engineering Journal,* 1935, Canton, Ohio

Torrington Co, *The Bearing Engineer,* Vol. II, No. 4

U. S. Govt Printing Office, Supt. of Documents, *Bulletin R222-46*
Simplified Practice Recommendation *R235-48, Copper and Copper Alloy Round Seamless Tube*

Vogel, W. F., *Involutometry and Trigonometry,* 1945, Michigan Tool Co, Detroit

Waldes Kohinoor, Inc, *Engineering Data and Specifications,* Oct. 1952